Handbook of Bolts an

Presenting time-tested standards as well as validated emerging knowledge on threaded fasteners and bolted joints, this updated edition covers how to design, select parts and materials, control assembly processes, predict behavior, and solve on-the-job problems.

This handbook examines key issues affecting bolting in the automotive, pressure vessel, petrochemical, aerospace, energy, and structural steel industries. The editors have successfully created a useful rather than scholarly handbook with chapters written in a straightforward, how-to manner. Theory is discussed only when necessary and the handbook's logical organization and thorough index enhance its usefulness. *Handbook of Bolts and Bolted Joints, Second Edition* includes updated chapters, solved numerical examples, and case studies.

This new edition is an essential handbook for professionals, researchers, and students in all fields in which threaded joints are used, including automotive, aerospace, structural, chemical, and naval and ocean engineering, as well as agricultural equipment, wind turbines, and medical devices.

Handbook of Bolts and Bolted Joints

Second Edition

Edited by
Sayed A. Nassar and John H. Bickford

CRC Press
Taylor & Francis Group
Boca Raton London New York

CRC Press is an imprint of the
Taylor & Francis Group, an **informa** business

Designed cover image: Sayed A. Nassar

Second edition published 2025
by CRC Press
2385 NW Executive Center Drive, Suite 320, Boca Raton FL 33431

and by CRC Press
4 Park Square, Milton Park, Abingdon, Oxon, OX14 4RN

CRC Press is an imprint of Taylor & Francis Group, LLC

[First edition published by CRC Press 1998]

ISBN: 9781032804392 (hbk)
ISBN: 9781032817521 (pbk)
ISBN: 9781003501183 (ebk)

DOI: 10.1201/9781003501183

Typeset in Times
by codeMantra

For Product Safety Concerns and Information, please contact our EU representative:
GPSR@taylorandfrancis.com
Taylor & Francis Verlag GmbH, Kaufingerstraße 24, 80331 München, Germany.

Contents

PART V Assembly of Bolted Joints

PART VI The Joint in Service

PART VII Testing and Inspection of Bolts and Joints

Preface

The second edition of the handbook of Bolts and Bolted Joints includes 14 new chapters written by many subject matter experts from global academia, industry, and research agencies. New Chapters 3, 7, 10, 14, 16, 19, 22, 33, 35–38, 41, and 45 contain up-to-date useful data and information related to various aspects of the design, analysis, testing, and validation of threaded fastener systems. Most of their data, observations, and conclusions have been peer-reviewed prior to their publication in respective archival journals. Chapter 45 includes a case study from the renewable energy sector that is ongoing at the time of preparing the manuscript of this edition of the handbook. Revised Chapter 28 on the torque-to-yield control method for bolt tightening includes useful procedures with formulas and solved numerical examples that show the actual bolt preload upon its torque-to-yield tightening; a set of practice problems is added.

A systems approach to threaded fasteners and bolted joints continues to remain a main theme in this new edition as it was in the earlier first edition of in 1998. Readers and practicing engineers are reminded of the importance of that holistic approach to performance, quality, reliability, and safety of critical threaded systems, where failure could result in costly shutdown, product recall, injury, or loss of human life. Needless to reemphasize the fact that a threaded system may have hundreds of variables and variable combinations, and that each will have to be thoroughly engineered and validated for the relative significance to system performance. These variables are typically divided into six groups: (1) threaded fastener variables, (2) joint variables, (3) assembly tool variables, (4) process control variables, (5) post-assembly loads, and (6) environmental factors. Any and all changes to system variables must be fully vetted by, and coordinated with, a collective body of system stakeholders that includes design, analysis, materials, test and validation, plant, quality, and audit engineers.

Open literature and first-hand experience show evidence of unintended consequences leading to costly failures and product recall that could be attributed to insufficient or lack of coordination will the stakeholders of threaded systems. A costly several-day shutdown of an automotive plant was caused by a seemingly simple task of insulating a robot end effector at a chassis weld station, where a robot picks up vehicle floor pan to place on the assembly line to get other chassis components welded to it. Without coordination with system stakeholders, an electrician was tasked to disassemble the bolted flange, insert insulator [sheet] between the two steel flanges, and to retighten the bolts that clamped the insulating flanges. The unintended consequence [of that seemingly simple task] was not that simple, however. The insulation sheet between the two steel flanges had actually converted a *hard join*t design to a *[soft]gasketed joint* design that has a completely different post-assembly pattern of behavior and response to the operating loads. Subsequent to initial bolt, the soft gasket [insulation sheet] continued to compress and thus reducing clamp load, followed by bolt loosening that eventually led to fatigue failure of one bolt and quickly spreading to other bolts (popping effect) until the complete rupture of robot end effector joint. That assembly line shutdown was a bottleneck that led to a complete shutdown of the entire automotive plant for several days (costing millions of dollars in lost productivity of an otherwise very profitable plant.)

Another case was that of a leaky threaded fitting installed on the exhaust system of passenger cars for detecting O_2 leakage. The operating temperature of exhaust pipe is 1,000°C. Although all stakeholders have [rightly so] agreed that loosening of the threaded fitting was the root cause of the O_2 leakage. It turned out, however, that an elevated temperature of 1,000°C had dramatically reduced the strength of fitting material to less than one-third of its room-temperature strength. Subsequent thermo-mechanical modeling and analysis using FEA with the temperature-dependent strength data on fitting material had clearly shown that thermal cycling of the pre-tightened threaded fitting has resulted in plastic deformation of the threads, leading to partial or total loosening and O_2 leakage from the loose fitting. Fortunately, the situation was remedied by replacing the existing fitting material with another alloy steel that retains most of its room-temperature strength at the elevated exhaust temperature of 1,000°C.

Credit largely goes to my co-editor, colleague, and dear friend John H. Bickford for his pioneering vision and advocacy of a holistic system's approach to bolted Joints with their hundreds of system variables and many stakeholders. His voluminous contributions to the field have enlightened an entire generation of practicing engineers and academicians around the world.

Although John Bickford is one of the most read authors on the subject of bolted joints, his readers find his writing style nearly free of the typical intimidation of associated with complex mathematical formulas and technical jargon: a rare talent to have. It would also be remiss mentioning John's sense of humor. As we assembled a few chapter co-authors of the first edition of this handbook to teach a short course in Michigan in 1999, I called his hotel room to be surprised by a musical melody playing on phone for a while before someone John burst into laughter on the other end announcing "it is me on my harmonica- Sayed; when are we having dinner?" A year or so later, I also recall him, sending him me a copy of a small sketch booklet he called "Insects Geographic Area" with a handwritten note bragging about selling four copies of it as he manned a stand on a first road in the Woods near his summer cottage in the state of Maine-USA. while manning a stand on a dirt road in the woods near his summer cottage in the state of Maine. When it came to the letter C, he wrote "C: Carpenter Ant, Geographic area: My House!"; that is my co-editor and dear friend John H. Bickford. Thank you John for all your contributions to the Bolted Joint community and to the threaded systems' archive.

Sayed A. Nassar

Editors

Sayed A. Nassar is a fellow of ASME, the John F. Dodge Chair of Engineering, Distinguished Professor of Mechanical Engineering, and founding director of the Fastening and Joining Research Institute (FAJRI), Oakland University (OU), Rochester, Michigan. He is also the founding director of the OU site of funded NSF IUCRC Center for Composite and Hybrid Material Interfacing (CHMI) which has two more sites at Georgia Institute of Technology and the University of Tennessee-Knoxville. At OU, Dr. Nassar teaches graduate and undergraduate courses in design, solid mechanics, and material joining; his current research is predominantly in the area of material joining, where he supervised more than 35 PhD and MS theses and published more than 300 peer-reviewed papers on the subject. Dr. Nassar holds seven US and International patents and is the founding president of University Research and Engineering Consultants (UREC LLC), Northville, Michigan. He is currently the chair of Fastening and Joining Committee, Design Division of ASME. Dr. Nassar received his MS and Ph.D. in Aerospace Engineering (Structural Mechanics) from the University of Cincinnati, Cincinnati, Ohio.

John H. Bickford is a private consultant, Middletown, Connecticut. He was formerly vice-president of Raymond Engineering, Inc., and manager of its Bolting Products Division in Middletown, Connecticut.

Contributors

Antoine G. Abboud
Magna International
Auburn Hills, Michigan

Ali Alkelani
BMW Manufacturing
Greer, South Carolina

Richard T. Barrett*
Barrett Engineering Consulting
Olmsted Falls, Ohio

Joseph Barron
Newport News Shipbuilding
Newport News, Virginia

William L. Biach
Biach Industries, Inc.
Cranford, New Jersey

George Bibel
Mech. Engineering
University of North Dakota
Grand Forks, North Dakota

John H. Bickford*
Consultant
Middletown, Connecticut

Chelsea Blub
One Power Company
Findlay, Ohio

Bernard Clément
École Polytechnique de Montréal
Montréal, Quebec, Canada

Dario Croccolo
Mech. Engineering
University of Bologna
Bologna, Italy

Rod Corbett
Rotabolt
West Midlands, England

Massimiliano DeAgostinis
Mech. Engineering
University of Bologna
Bologna, Italy

Michael J. Dube
NASA Engineering and Safety Center
Langley, Virginia

Joseph R. Dudley
Nylok Fastener Corporation
Macomb, Michigan

Stefano Fini
Mech. Engineering
University of Bologna
Bologna, Italy

Marco Gerini-Romagnoli
Mech. Engineering
Oakland University
Rochester, Michigan

Sherif Gindy
SensoData Technologies
Clinton Township, Michigan

Daniel P. Hess
Mech. Engineering
University of South Florida
Tampa, Florida

Alex W. Heston
SPS Technologies
Cleveland, Ohio

A. Craig Hood*
ACH Technologies
Wayne, Pennsylvania

Basil Housari
BMW Manufacturing
Greer, South Carolina

Peter F. Kasper
Vermont Fasteners Manufacturing
Swanton, Vermont

Amir Kazemi
Mech. Engineering
Oakland University
Rochester, Michigan

Jereme Kent
One Power Company
Findlay, Ohio

John Laurilliard
Consultant
Philadelphia, Pennsylvania

Marc Levinson
ITW Shakeproof
Elgin, Illinois

Tianwu Li
Mech. Engineering
Oakland University
Rochester, Michigan

Russell E. Mack*
Aluminum Company of America
Lititz, Pennsylvania

Rania Mansour
Mech. Engineering
Oakland University
Rochester, Michigan

Mattia Mele
Mechanical Engineering
University of Bologna
Bologna, Italy

William A. Milek
Consultant
Glen Ellyn, Illinois

Giorgio Olmi
Mech. Engineering
University of Bologna
Bologna, Italy

Sayed A. Nassar
Mech. Engineering
Oakland University
Rochester, Michigan

Bernard S. Nau
Consultant, BHR Group Ltd.
Cranfield, Bedford, England

Gary J. Novak
Fel-Pro Chemical Products
Skokie, Illinois

Tom Patel
Fel-Pro Chemical Products
Skokie, Illinois

Jesse A. Phebus
A. Phebus and Associates, Inc.
Safety Harbor, Florida

Louis Raymond
L.Raymond & Associates
Newport Beach, California

Shahriar M. Sadri
Huck International, Inc.
Carson, California

David Sharp
J&M Turner, Inc.
Southampton, Pennsylvania

Ralph S. Shoberg**
RS Technologies
Farmington Hills, Michigan

Andrew Q. Smail
US Army Ground Vehicle Systems Center
Warren, Michigan

Nicolò Vincenzi
Bucci Industries
Faenza, Italy

Paul W. Wallace*
Ingersoll-Rand Company
Flemington, New Jersey

John W. Wilhelm, Jr.
Skidmore-Wilhelm
Cleveland, Ohio

NOTES

* Retired.
** Deceased.

Part I

Material Properties and Selection

1 Fastener Materials

A. Craig Hood

1.1 INTRODUCTION

The threaded mechanical fastener is an engineered structural product. It is controlled in selection, manufacturing, and processing by a series of specifications under the jurisdiction of organizations such as American Society of Mechanical Engineers (ASME), American Society for Testing and Materials (ASTM), Society of Automotive Engineers (SAE), and the Department of Defense.

The use of specifications is voluntary in the United States. However, a purchase order and contract for procurement make them subject to the laws of business contracts, and they must meet these requirements. Fastener materials are selected either by the fastener manufacturer to conform to product requirements or by a customer to meet the needs of the environment to which the fastener will be subjected. The purpose of a bolt and nut is to provide clamping force, and when you purchase a bolt, you are, in effect, buying clamping force. The nut, by its design and application, enables you to achieve that clamping force. The selection of materials for bolts and nuts is based on maximum strength, high ductility, and high notch strength. Other choices are based on the environmental aspects of bolt and nut applications, and these are discussed here in separate categories.

Fastener materials are initially selected for engineering products. However, the fabrication and performance requirements of the fastener govern the ultimate choice. One objective of fastener design is to achieve a threaded product with properties that match those of the base material.

Material selection is further based on the capability of being fabricated and the material's response to heat treating and cold work strengthening mechanisms.

In this chapter, consideration is given to the above in discussing fastener selection. The reasons for selecting fastener materials for environmental conditions, such as corrosion, elevated temperatures, cryogenic temperatures, hydrogen embrittlement, and stress corrosion, are also included. Coatings selection in conjunction with the base material is included as an adjunct to the basic fastener design. Fabrication capability is covered in the appropriate section on material classes that respond best to different environments.

1.2 CARBON AND ALLOY STEELS

No other structural material offers the high strength at low cost that can be achieved by steel. For this reason, fasteners made of steel constitute the majority of fasteners used worldwide. The automotive industry, which uses about 25% of all fasteners, uses nuts and bolts of carbon or alloy steel almost exclusively in automobile and truck manufacturing.

Carbon steels, containing essentially iron and carbon, are the least expensive of the steels and are used whenever possible. Carbon steel, in addition to the carbon, contains less than 1.60% manganese (Mn), 0.60% silicon (Si), and 0.60% copper (Cu). The balance, except for residual elements such as sulfur (S) and phosphorus (P), is iron. Hot-rolled carbon steels with low carbon content are used for machine screws and non-heat-treated bolts and nuts. They are at the bottom of the strength scale.

Steel is categorized according to grades established by the American Iron and Steel Institute (AISI) and by the SAE.

Carbon steels range from SAE 1006 (0.08% max C) to 1095 (0.90% C). The low carbon range (SAE 1006–1025) is used in the hot-rolled condition. Higher carbon grades (SAE 1030–1040) are used in the heat-treated condition. Because of the loss of ductility and susceptibility to hydrogen

DOI: 10.1201/9781003501183-2

embrittlement and stress corrosion cracking, it is recommended that fasteners with carbon levels higher than 0.45% not be manufactured. Grades containing more than 0.45% carbon have been made, but they have required very special fabrication techniques.

The heat-treated grades are selected on the basis of hardenability. This is a measure of the depth to which a steel can be hardened in different quenching media during the hardening phase of heat treatment. All fasteners (with the exception of self-drilling and self-tapping screws) must be through-hardened. This provides for uniform properties in what is in many cases a small cross section.

The strength of the cross section is not only important in tension. A large number of screws are used in shear, and the shear strength must be in a consistent ratio to the tensile strength. This is not possible with a partially hardened cross section.

Screw material is selected on the basis of hardenability and cross section. A typical selection of carbon steels for frequently specified fastener grades is shown in the specifications listed in Table 1.1.

One group of carbon steels are known as the high manganese steels. They are the 1,500 series of steels. These provide higher hardenability at low cost. One permitted in the Grade 8 fastener up to 7/16 in. diameter is the SAE 1541 steel with 0.36%–0.45% C and 1.30%–1.65% Mn.

Because of their limited hardenability, carbon steels are limited in diameter as the strength increases. When the diameter becomes too large for the carbon steel to provide through-hardening during heat treatment, a choice of alloy steels must be made. The selection of alloy steels is based on their ability to provide higher hardenability. A great many alloy steels are not used because their greater strength or higher cost is not justified. Most of the alloy steels contain varying percentages of nickel (Ni), chromium (Cr), and molybdenum (Mo).

Strength levels and grades for bolts, screws, and steels are selected to conform to Table 1.1 in general usage. While the current SAE J429 specification for Grade 8 permits SAE 1541, a manganese carbon steel, to be used for diameters of 7/16 in. and under, ASTM A354BD permits only alloy steel for all diameters. As a rule, the user does not, in this category, select the carbon or alloy grade but leaves it to the discretion of the fastener manufacturer.

There is a family of special steels containing boron (B) that provide a high degree of hardenability to both low carbon (0.20% C) and medium carbon (0.35% C) steels. Although these steels have been permitted in special J429 grades, namely 5.2 and 8.2, they have not been allowed in the standard grades. A newly amended J429 (still in committee) is expected to permit medium carbon boron steels to be permitted for Grade 8 fasteners.

In fastener design considerations, the nut must be strong enough to break a fastener designed for use in tension. The nut has an advantage the bolt does not have. When a bolt fails, the failure is usually concentrated in the area of highest stress, which is the area of joint separation. The nut will fail, if insufficient threads are provided, by the phenomenon of stripping when the shear strength of the total threads cannot carry the load. Therefore, a nut designed with a greater number of threads does not need as much material strength. As an example, in the fastener research

TABLE 1.1
Strength Grades for SAE J429 Bolt Materials

SAE J429 [1]	ASTM Equivalent	Diameter (in.)	Ultimate Tensile Strength (UTS) (psi)
Grade 2	A307 [2]	¼–¾	74,000
		Over ¾–1½	60,000
Grade 5	A325 [3]	¼–1	120,000
Grade 8	A490 or A354 BD [4,5]	¼–1½	150,000

Source: As noted.

TABLE 1.2
Strength Grades for SAE J995 Nut Materials

SAE J995-79 [6]	ASTM Equivalent,[a] A563-94 [7][a]	Diameter (in.)	Proof Load Stress (psi) UNC	Proof Load Stress (psi) UNF	Carbon (SAE) (%) Min.	Carbon (SAE) (%) Max.
Grade 2	A	¼–1½	90,000	90,000		.47
Grade 5	B	¼–1	120,000	109,000	.30	.55
		>1–1½	105,000	94,000		
Grade 8	D	¼–1½	150,000	150,000	.30	.55

[a] ASTM A563, for the A and B grades of nuts, has only a 0.55% C maximum requirement, whereas Grade D has 0.30% Mn in addition.

laboratory where I worked at the time, we broke a 300,000 psi bolt with an aluminum nut by merely increasing the number of threads engaged.

There are limitations to the height of nuts that can be used, and therefore some nuts must be heat treated to strength levels higher than those provided by low carbon steel.

Nut material requirements in the specifications for commonly used grades are listed in Table 1.2. Carbon steel is permitted even for 150,000 psi nuts. This is because of the strength latitude permitted by nut dimensions and thread engagement. ASTM A563-94 lists a number of other nut strength requirements based on nut dimensions and some grades that have mild corrosion resistance to the weather.

In summary, steel bolt and nut material selection is based primarily on bolt and nut mass created by dimensions as well as steel hardenability. Material should be selected on the basis of the lowest alloy content that will provide the strength required.

1.3 DISCONTINUITIES

From a fabrication standpoint, except for those with large diameters, bolts are cold forged. Because of the severe surface deformation created during cold heading, bolting materials must have relatively seam-free surfaces. Surface discontinuities in the final product can come from seams, forging cracks, and quench cracks that occur during heat treating. Various levels of discontinuities are permitted in the final product based on depth and location. Proprietary raw material specifications issued by bolt and nut producers contain allowable discontinuity depth limits. Bolt and nut product discontinuity depth limits are found in SAE specifications J123c and J122a, respectively [8].

1.4 ELEVATED TEMPERATURES (STEEL)

A question is often raised about the suitability of carbon and alloy steel fasteners at temperatures above room temperature. SAE J429 does not provide for heat treatment of Grade 2 bolts. However, Grades 5 and 8 must be tempered at 800°F min.

A fastener must provide reliable clamping force at the in-service temperature. Clamping force depends, in part, on the modulus of elasticity, which is a measure of the elasticity or springback. The modulus drops in value as the temperature rises [9, pp. 115–117] (see Table 1.3). As a result, loss of clamping force can occur at the service temperature and must also be taken into consideration.

The strength of the fastener is reduced at elevated temperatures, but this does not occur below the tempering temperature of the steel. However, there is a phenomenon known as stress relaxation which is a form of elevated temperature creep that reduces the clamping force as well. Stress relaxation occurs when elastic deformation becomes plastic or permanent, and the loss of elasticity lowers the clamping force. Bolts and screws provide clamping force because the bolt stretches during

TABLE 1.3
Modulus of Elasticity at Various Temperatures (×10⁶ psi)

Specification	Grade	Temperature °F (°C) –325 (–198)	–200 (–129)	70 (20)	400 (204)	600 (316)	800 (427)	1,000 (538)	1,200 (649)	1,400 (760)	1,600 (871)	1,800 (982)
ASTM A193	B5	32.9	32.3	30.9	29.0	28.0	26.1					
	B6	31.2	30.7	29.2	27.3	26.1	24.7					
	B7	31.6	31.0	29.7	27.9	26.9	25.5					
	B8-CL 1	30.3	29.7	28.7	25.5	25.3	24.1					
	B16	31.6	31.0	29.7	27.9	26.9	25.5					
ASTM A307		31.4	30.8	29.5	27.7	26.7	24.2					
ASTM A320	L7	31.6	31.0	29.7	27.9	26.9	25.5					
	L43	31.6	31.0	29.7	27.9	26.9	25.5					
	B8	30.3	29.7	28.3	26.5	25.3	24.1					
ASTM A325	Type 1, 2, 3	31.4	30.8	29.5	27.7	26.7	24.2					
ASTM A354		31.2	30.6	29.3	27.5	26.5	24.0					
ASTM A449		31.2	30.6	29.3	27.5	26.5	24.0					
ASTM A453		30.3	29.7	28.3	26.5	25.3	24.1					
ASTM A490		31.2	30.6	29.3	27.5	26.5	24.0					
ASTM A540	B21, B22	31.6	31.0	29.7	27.9	26.9	25.5					
	B23, B24	29.6	29.1	27.8	26.[illegible]	25.2	23.0					
SAE J429	GR 1, 2, 4	31.4	30.8	29.5	27.7	26.7	24.2					
	GR 5, 7, 8	31.2	30.6	29.3	27.5	26.5	24.0					
High performance materials	A286			29.1				23.5	21.1	18.7	18.9	
	Hasteloy-X			28.6								18.5
	H-11			30.6				23.0	16.0			
	Inconel 600			31.4							23.1	
	Inconel 718			29.0								
	Inconel X750			31.0								20.0
	MP35N			33.6		30.8						
	Nimonic 80A			31.2							22.7	
	Ti 6AI-4V			16.5								
	Rene 41								25.9		23.6	
	Waspaloy			30.6				26.7			22.7	

(Continued)

TABLE 1.3 (*Continued*)
Modulus of Elasticity at Various Temperatures (×10⁶ psi)

		Temperature °F (°C)										
Specification	**Grade**	**–325 (–198)**	**–200 (–129)**	**70 (20)**	**400 (204)**	**600 (316)**	**800 (427)**	**1,000 (538)**	**1,200 (649)**	**1,400 (760)**	**1,600 (871)**	**1,800 (982)**
Metric	4.6/4.8/5.8	31.4	30.8	29.5	27.7	26.7	24.2					
SAE J1199	8.8/9.8/10.9	31.2	30.6	29.3	27.5	26.5	24.0					
Joint materials	Low carbon	31.4	30.8	29.5	27.7	26.7	24.2	20.4				
Steels												
	Med. Carbon	31.2	30.6	29.3	27.5	26.5	24.0	20.2				
	Carbon-moly.	31.1	30.5	29.2	27.4	26.4	23.9	20.1				
	Chrome-moly.	31.6	31.0	29.7	27.9	26.9	25.5	23.9				
	Austenitic	30.3	29.7	28.3	26.5	25.3	24.1	22.8				
	2024 Aluminum	11.7	11.4	10.6	9.2							
SAE J158	Malleable iron castings			25–26								
SAE J434	Ductile iron castings			22								
	Cast aluminum			8–10								
	Brass, cold-rolled			13.1								
	Cast iron			12–14								
	Wrought iron			26–29								
	Magnesium			6.1								
	Concrete			2.0								
	Wood (with grain)			1.2								

Source: Ref. [9].

tightening. Its desire to go back to its original length provides the compression load that holds the joint together. Lose this, and a loose joint connection occurs.

Unfortunately, very little data on stress relaxation exist for carbon and steel alloys. The most extensive data can be found in Ref. [9] (Table 4.7 and Fig. 4.3, pp. 118, 120, 121). These are reproduced here as Table 1.4 and Figure 1.1, respectively. Note that the relaxation strength is between 200°F to 400°F below the tempering temperature.

TABLE 1.4
Stress Relaxation

	Test Temp.		Initial	Residual Stress (ksi)			Final	
Bolt Information	**°F**	**°C**	**Stress (ksi)**	**100 h**	**1,000 h**	**10,000 h**	**Stress (ksi)**	**Time (h)**
ASTM A193 B7 1 in. diam.	850	454	30.0		13.2	(10.0)		1,920
ASTM A453 GR 651 1¼ in. diam.; as recd plus 1,200°F (650°C) for 48 h	1,200	649	32		5	(2.85)		1,012
ASTM A453 GR 655: 2 in. diam.	1,100	593	19.9	14.3	15.9		16.0	3,769
Type 431 stainless steel (AN6C)	700	371	95.8	68			67.8	150
	900	482	95.8	23			20	150
	900	482	63.25	20			18.8	150
Austenitic SS DIN 17006:	842	450	50.6	46.5	44.7			
X5CrNiMo 1810 M12 bolt. Nut: DIN 931 or 934: rolled threads	1,022	550	50.6	34.1	22.9			
A286 bolt: 5/16 in. diam.	1,200	649						
MS 9035-19 configuration			70	32.9				100
HiR thread form per AMS 7478			70	34				100
Nimonic 80A: 3.5 in. diam.	1,202	650	39.5	30.4	24.2	17.8	14.2	21,167
	1,292	700	37.1	28.5	22.1		15.7	2,900
	1,499	815	29.6	9.8			5.7	290
Nimonic 80A: 1⅛ in. diam.	1,157	625	33.1	28	24.2	18.6	15.7	27,200
Bolt: AISI 8740 steel (M12)	662	350	45	39	31.5			
Nut: German aircraft spec. LN 1.4944			85.2	67.5	52.5			
Vascojet 1000 (modified H-11 steel) ½–20 studs	900	482	117	78				
MP35N: ¼-28 bolt; nut of same alloy	700	371	132				113	150
Rene 41 bolt	1,400	760						
Waspaloy nut								
10–32 threads			99.7				25.5	50
¼ threads			73				32.9	50
½–20 threads			73				35.1	50
½–20 threads			46.4				28.3	50
Udimet 500 bolt, 5/16 threads	1,100	593	50				36.5	12
Waspaloy nut and bolt, MIL-S-8879 thread form:	1,400	760	100				7.7	50
¼–28 threads								

Source: Ref. [9].

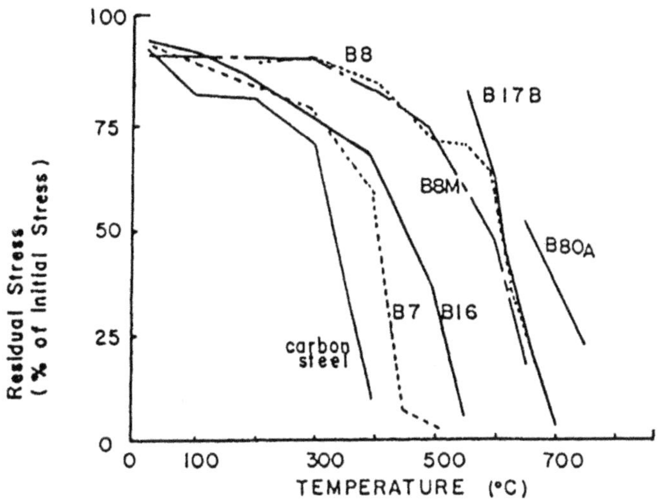

FIGURE 1.1 Stress relaxation of petrochemical bolting materials as a function of service temperature. Exposure in each case was for 1,000h at the temperature shown (from Ref. [9].)

Some comparative data exist that show the relationship between carbon and alloy steel and boron steel fasteners. A study [10] shows that fasteners made of 1,340 steel (0.38%–0.43% C; 1.60%–1.90% Mn) relax less than 10B22 fasteners. The same study with nuts shows that nuts contribute little to relaxation.

1.5 CORROSION (STEEL)

Carbon and alloy steel fasteners have little or no corrosion resistance and must be coated to be used even in low moisture environments. Coatings are used to protect the fasteners, and these present a set of problems not experienced by uncoated fasteners. This is discussed in Section 1.7 on coatings.

1.6 HIGH AND ULTRAHIGH STRENGTH STEEL BOLTS

In general industrial usage, Grade 8 and A490 are designed as "high strength steel" bolts. There are, however, a number of steel bolts and screws with higher strength. The earliest of these was the socket head cap screw (SHCS) with a minimum tensile strength of 180,000 psi. Other socket screw designs have 160,000 psi tensile strength.

The steel aircraft fastener category includes the MS20004 series of internal wrenching head bolts at 160,000 psi. At 180,000 psi are NAS624 and the military version, MS21250. These have 12-point wrenching heads, as do all fasteners above 160,000 psi.

In the late 1950s, a series of high-strength aircraft bolts were developed by one manufacturer. The material used was H-11, a hot work die steel that can withstand temperatures up to 900°F due to its excellent heat treatment response and a tempering temperature of approximately 1,000°F. The composition of H-11 is 0.40C, 0.30 Mn, 1.00 Si, 5.00 Cr, 0.04V, and 1.30 Mo. From this material, a series of fasteners were developed at 220,000 psi and 260,000 psi tensile strengths.

All aircraft fasteners are available in a shear bolt version with minimum shear strength at 60% of minimum ultimate strength (See also Chapter 21).

The highest strength commercially available bolt has a minimum tensile strength of 300,000 psi and is made from T15, whose composition is 1.50–1.60C, 0.10–0.40 Mn, 0.10 –0.40 Si, 4.50–4.75 Cr, 4.755–5.00V, 12.5–13.50 W, 0.50 Mo, and 4.75–5.25 Co.

Further research on steel and fastener fabrication resulted in fasteners with tensile strengths in the vicinity of 400,000 psi, hardnesses of Rc 65–67, and sufficient ductility to produce acceptable properties. Fatigue was measured at an endurance limit of 180,000 psi (R=0.1). When threads were rolled after heat treatment, the fatigue strength continued to increase, with no perceived upper limit. Unfortunately, at a hardness of Rc 66, the thread roll die life was only one bolt per die, and the project was abandoned with the assessment that it had no commercial potential.

These ultrahigh strength steels were in use for perhaps 4–5 years, when a field failure was observed that was determined to be stress corrosion cracking. This prompted an extensive study by one manufacturer over the next several years [11], which encompassed a coatings development program and an alternative materials program. The outcome of the research is discussed in Section 1.7. In summary, carbon and alloy steel fasteners offer a series of strength levels that will cover a wide range of application requirements. A number of design problems arise when they are used at temperatures higher than room temperature, and it is usually preferable to utilize fastener materials designed for elevated temperatures. If carbon or alloy steels are selected for economic reasons at elevated temperatures, these potential problems should be discussed with experts familiar with high temperature joint requirements.

1.7 CORROSION-RESISTANT MATERIALS AND COATINGS

The carbon and alloy steels do an excellent job of providing low to high-strength fasteners at low to moderate prices. For this reason, they dominate the choices in fastener materials. However, they will rust in normal atmospheres and will corrode substantially in more corrosive environments. For this reason, corrosion-resistant alloys and coatings have been developed. Many of these are applicable to fasteners. Some, because of their high nickel and/or chromium content, are also used at elevated and cryogenic temperatures. They are discussed in this section and in Section 1.10.

The primary selection of fastener materials for corrosion resistance is based on the properties of the basic material. Since the fastener must have corrosion compatibility with the joint material, it has been a challenge for the fastener manufacturer to provide fasteners of the same material as that required by the design metallurgist.

Initially, corrosion-resistant fasteners were made of pure nonferrous metals like copper. Later, the addition of zinc produced the brasses, and the further addition of tin created the bronzes. Nickel and nickel alloys were to follow; the most popular of these are the cupro-nickel alloys and Monel; the latter is used extensively in marine applications. Aluminum has been used to some extent, primarily in nuts, where its light weight and lower strength requirement make it an excellent choice (see Chapter 14).

Titanium came along in the 1950s after a process to produce it was developed. Although titanium is often used where high strength and low density (i.e., a high strength/density ratio) are required, it happens to be one of the most corrosion-resistant materials in the world. It rates just below platinum and gold in this regard as long as its oxide surface is intact. Titanium has been widely used in aircraft, first for military applications and later in commercial aircraft. It was not available in fastener form until the late 1950s. Titanium became widely used in aircraft after the development of the C-5A transport for the military. Today, based on C-5A technology, it is used extensively in alloy form (6A1-4V) for the skin fasteners of all the major commercial wide-bodied transports. Because of its light weight and because it develops an anodized surface to minimize corrosion, an aluminum alloy is used for the companion nuts. Many non-aircraft uses for titanium are currently being developed as its properties become more widely appreciated.

Nonferrous structural fasteners are the subject of ASTM F467-93 [12] and F468-93 [13]. Tables 1.5 and 1.6 (F467) and Tables 1.7 and 1.8 (F468) show the chemical compositions and mechanical requirements of these specifications.

Since their development in the early part of this century, stainless steels have become the most popular substitute for carbon and alloy steels where long-term corrosion resistance is required.

TABLE 1.5
ASTM F 467 Chemical Requirements, Percent Composition

UNS Designation Number	Alloy	General Name	Al	Cu, min	Fe, max	Mn, max	Ni, max	P	Si	Zn, max[a]	Pb, max	Sb	As, max
		Copper and Copper-Base Alloys											
C11000	110	ETP copper		99.9									
C27000	270	Brass		63.0–68.5	0.07					Balance	0.10		
C46200	462	Naval brass		62.0–65.0	0.10					Balance	0.20	0.5–1.0	
C46400	464	Naval brass		59.0–62.0	0.10					Balance	0.20	0.5–1.0	
C51000	510	Phosphor bronze		Balance[a]	0.10			0.03–0.35		0.30	0.05	4.2–5.8	
C61300	613	Aluminum bronze	6.0–7.5	—[b]	2.0–3.0	0.10	0.15[e]	0.015	0.10	0.05	0.01	0.20–0.50	
C61400	614	Aluminum bronze	6.0–8.0	88.0[c]	1.5–3.5	1.0							
C63000	630	Aluminum bronze	9.0–11.0	78.0[c]	2.0–4.0	1.5	4.0–5.5		0.25 max			0.20 max	
C64200	642	Aluminum silicon bronze	6.3–7.6	88.65[c]	0.30	0.10	0.25		1.5–2.2[d]	0.50	0.05	0.20 max	0.15
C65100	651	Silicon bronze		96.0[c]	0.8	0.7			0.8–2.0	1.5	0.05		
C65500	655	Silicon bronze		94.8[c]	0.3	1.5	0.6		2.8–3.8	1.5	0.05		
C66100	661	Silicon bronze		94.0[c]	0.25	1.5			2.8–3.5	1.5	0.20–0.8		
C67500	675	Manganese bronze	0.25 max	57.0–60.0	0.8–2.0	0.05–0.5				Balance	0.20	0.5–1.5	
C71000	710	Cupro-nickel		74.0[c]	0.60	1.00	19.0–23.0[e]			1.00	0.05		
C71500	715	Cupro-nickel		65.0[c]	0.40–0.7	1.00	29.0–33.0[e]			1.00	0.05		

Nickel and Nickel-Base Alloys

UNS designation number	Alloy	General name	Al	c, max	Cr	Cu[a]	Fe, max	Mn, max	Ni[a]	P, max	Si, max	Ti	Co, max	Mo	S, max	Va	W
N10001	335	Ni-Mo		0.05	1.0 max		4.0–6.0	1.0	Balance	0.025	1.00		2.50	26.0–30.0	0.030	0.2–0.4	
N10276	276	Ni-Mo-Cr		0.02	14.5–16.5		4.0–7.0	1.00	Balance	0.040	0.08		2.50	15.0–17.0	0.030	0.35 max	3.0–4.5
N04400	400	Ni-Cu Class A		0.3		Balance	2.5	2.0	63.0–70.0		0.5		[d]		0.024		
N04405	405	Ni-Cu Class B		0.3		Balance	2.5	2.0	63.0–70.0		0.5		[d]		0.025–0.060		
N05500	500	Ni-Cu-Al	2.30–3.15	0.25		Balance	2.0	1.5	63.0–70.0		0.5	0.35–0.85	[d]		0.01		

(Continued)

TABLE 1.5 (*Continued*)
ASTM F 467 Chemical Requirements, Percent Composition

Aluminum-Base Alloys[f]

UNS designation number	Alloy	General name	Al[a]	Cr	Cu	Fe, max	Mn, max	Si, max	Ti, max	Zn, max	Mg	Other elements, max	
												Each	Total
A92024	2024	Aluminum 2024	Balance	0.10 max	3.8–4.9	0.50	0.30–0.9	0.50	0.15[g]	0.25	1.2–1.8	0.05	0.15
A96061	6061	Aluminum 6061	Balance	0.04–0.35	0.15–0.40	0.7	0.15	0.40–0.8	0.15	0.25	0.8–1.2	0.05	0.15
A96262	6262	Aluminum 6262	Balance	0.14–0.14	0.15–0.40	0.7	0.15	0.40–0.8	0.15	0.25	0.8–1.2	[h]	

Titanium and Titanium-Base Alloys[i]

UNS designation number	Alloy	General name	Al	C	Fe	Ti	H	N	O	Pd	V	Cr	Mo	Zr	Ru	Residuals[j]	
																Each, max	Total, max
R50250	1	Titanium Gr 1		0.10	0.20	Balance	0.0125	0.05	0.18							0.1	0.4
R50400	2	Titanium Gr 2		0.10	0.30	Balance	0.0125	0.05	0.25							0.1	0.4
R50700	4	Titanium Gr 4		0.10	0.50	Balance	0.0125	0.07	0.40							0.1	0.4
R56400	5	Titanium Gr 5	5.5–6.75	0.10	0.40	Balance	0.0125	0.05	0.20		3.5–4.5					0.1	0.4
R56401	5 ELI	Titanium Gr 5 ELI	5.5–6.5	0.08	0.25	Balance	0.0120	0.05	0.13		3.5–4.5					0.1	0.4
R52400	7	Titanium Gr 7		0.10	0.30	Balance	0.0125	0.05	0.25	0.12–0.25						0.1	0.4
R58640	19	Titanium Ti-38-6-44	3.0–4.0	0.05	0.30	Balance	0.020	0.03	0.12	0.10[k]	7.5–8.5	5.5–6.5	3.5–4.5	3.5–4.5	0.10[k]	0.15	0.4

Source: Ref. [12]. Copyright ASTM; reprinted with permission.

[a] Elements shown as balance shall be arithmetically computed by deducting the sum of the other named elements from 100.

[b] Copper plus specified elements = 99.8 min; copper plus silver = 88.5–91.5

[c] Cobalt is to be counted as nickel.

[d] Minimum content of copper plus all other elements with specified limits shall be 99.5%.

[e] An alloy containing as high as 2.6% silicon is acceptable provided the sum of all the elements other than copper, silicon, and iron does not exceed 0.30%.

[f] Analysis shall regularly be made only for the elements in this table. If, however, the presence of other elements is suspected or indicated in amounts greater than the specified limits, further analysis shall be made to determine that these elements are not present in excess of the specified limits.

[g] Titanium + zirconium 0.20% max.

[h] Lead 0.4%–0.7%; bismuth 0.4%–0.7%.

[i] All reported values are maximums, unless a range is specified.

[j] A residual is an element present in a metal or an alloy in small quantities inherent to the manufacturing process but not added intentionally. Residual elements need not be reported unless a report is specifically required by the purchaser.

[k] Ruthenium and palladium, or both, may be added to Grade 19 for enhanced corrosion resistance as negotiated between the purchaser and vendor. Chemical analysis is not required unless specifically negotiated.

TABLE 1.6
Mechanical Property Requirements, ASTM F 467

Alloy	Mechanical Property Marking	Hardness, min[a]	Proof Stress, min (ksi)
Cu 110	F 467A	65 HRF	30
Cu 270	F 467B	55 HRF	60
Cu 462	F 467C	65 HRB	50
Cu 464	F 467D	55 HRB	50
Cu 510	F 467E	60 HRB	60
Cu 613	F 467F	70 HRB	80
Cu 614	F 467G	70 HRB	75
Cu 630	F 467H	85 HRB	100
Cu 642	F 467J	75 HRB	75
Cu 651	F 467K	75 HRB	70
Cu 655	F 467L	60 HRB	50
Cu 661	F 467M	75 HRB	70
Cu 675	F 467N	60 HRB	55
Cu 710	F 467P	50 HRB	45
Cu 715	F 467R	60 HRB	55
Ni 335	F 467S	20 HRC	115
N 276	F 467T	20 HRC	110
Ni 400	F 467U	75 HRB	80
Ni 405	F 467V	60 HRB	70
Ni 500	F 467W	24 HRC	130
Al 2024-T4[b]	F 467X	70 HRB	55
Al 6061-T6	F 467Y	40 HRB	40
Al 6262-T9	F 467Z	60 HRB	52
Ti 1	F 467AT	140 HV	40
Ti 2	F 467BT	150 HV	55
Ti 4	F 467CT	200 HV	85
Ti 5	F 467DT	30 HRC	135
Ti 7	F 467ET	160 HV	55
Ti-38-6-44	F 467FT	24 HRC	120
Ti 5 ELI	F 467GT	35 HRC	125

Source: Ref. [12]. Copyright ASTM; reprinted with permission.

[a] For aluminum and titanium alloys, hardness values are for information only.

[b] Aluminum and titanium alloy 2024-T4 shall be supplied in naturally aged condition. This material is not recommended for nuts in sizes greater than ¼ (0.250) in.

Modern fastener manufacturing techniques have made them more economical. The structural stainless steel fasteners are covered in ASTM F593-95 [14] and F594-95 [15]. Tables 1.9 and 1.10 (F593) and Tables 1.11 and 1.12 (F594) show the chemical composition and mechanical property requirements for these materials.

In spite of good corrosion resistance, the nonferrous and stainless steel fasteners could not match the strengths of the existing alloy steel bolts. Therefore, the fastener manufacturer was challenged to provide high-strength corrosion-resistant bolts and nuts. (It should be noted that bolts and nuts are usually made of the same material in order to provide corrosion compatibility.)

TABLE 1.7
ASTM F 468 Chemical Requirements, Percent Composition

UNS Designation Number	Alloy	General Name	Al	Cu, min	Fe, max	Mn, max	Ni, max	P	Si	Zn, max[a]	Pb, max	Sb	As, max
						Copper and Copper-Base Alloys							
C11000	110	ETP copper		99.9									
C27000	270	Brass		63.0–68.5	0.07					Balance	0.10		
C46200	462	Naval brass		62.0–65.0	0.10					Balance	0.20	0.5–1.0	
C46400	464	Naval brass		59.0–62.0	0.10					Balance	0.20	0.5–1.0	
C51000	510	Phosphor bronze		Balance[a]	0.10			0.03–0.35		0.30	0.05	4.2–5.8	
C61300	613	Aluminum bronze	6.0–7.5	—[b]	2.0–3.0	0.10	0.15[c]	0.015	0.10	0.05	0.01	0.20–0.50	
C61400	614	Aluminum bronze	6.0–8.0	88.0[d]	1.5–3.5	1.0							
C63000	630	Aluminum bronze	9.0–11.0	78.0[d]	2.0–4.0	1.5	4.0–5.5		0.25 max			0.20 max	
C64200	642	Aluminum silicon bronze	6.3–7.6	88.65[d]	0.30	0.10	0.25		1.5–2.2[e]	0.50	0.05	0.20 max	0.15
C65100	651	Silicon bronze		96.0[d]	0.8	0.7			0.8–2.0	1.5	0.05		
C65500	655	Silicon bronze		94.8[d]	0.8	1.5	0.6		2.8–3.8	1.5	0.05		
C66100	661	Silicon bronze	0.25 max	94.0[d]	0.25	1.5			2.8–3.5	1.5	0.20–0.8		
C67500	675	Manganese bronze		57.0–60.0	0.8–2.0	0.05–0.5				Balance	0.20	0.5–1.5	
C71000	710	Cupro-nickel		74.0[d]	0.60	1.00	19.0–23.0[e]			1.00	0.05		
C71500	745	Cupro-nickel		65.0[d]	0.40–0.7	1.00	29.0–33.0[e]			1.00	0.05		

Nickel and Nickel-Base Alloys

UNS designation number	Alloy	General name	Al	c, max	Cr	Cu[a]	Fe, max	Mn, max	Ni[a]	P, max	Si, max	Ti	Co, max	Mo	S, max	Va	W
N10001	335	Ni-Mo		0.05	1.0 max		4.0–6.0	1.0	Balance	0.025	1.00		2.50	26.0–30.0	0.030	0.2–0.4	
N10276	276	Ni-Mo-Cr		0.02	14.5–16.5		4.0–7.0	1.00	Balance	0.040	0.08		2.50	15.0–17.0	0.030	0.35 max	3.0–4.5
N04400	400	Ni-Cu Class A		0.3		Balance	2.5	2.0	63.0–70.0		0.5		c		0.024		
N04405	405	Ni-Cu Class B		0.3		Balance	2.5	2.0	63.0–70.0		0.5		c		0.025–0.060		
N05500	500	Ni-Cu-Al	2.30–3.15	0.25		Balance	2.0	1.5	63.0–70.0		0.5	0.35–0.85	c		0.01		

(Continued)

TABLE 1.7 (*Continued*)
ASTM F 468 Chemical Requirements, Percent Composition

Aluminum-Base Alloys

UNS designation number	Alloy	General name	Al[a]	Cr	Cu	Fe, max	Mn, max	Si, max	Ti, max	Zn, max	Mg	Other elements, max	
												Each	Total
A92024	2024	Aluminum 2024	Balance	0.10 max	3.8–4.9	0.50	0.30–0.9	0.50	0.15[g]	0.25	1.2–1.8	0.05	0.15
A96061	6061	Aluminum 6061	Balance	0.04–0.35	0.15–0.40	0.7	0.15	0.40–0.8	0.15	0.25	0.8–1.2	0.05	0.15
A97075	7075	Aluminum 7075	Balance	0.18–0.35	1.2–2.0	0.50	0.30	0.40	0.20[h]	5.1–6.1	2.1–2.9	0.05	0.15

Titanium and Titanium-Base Alloys

UNS designation number	Alloy	General name	Al	C	Fe	Ti	H	N	O	Pd	V	Cr	Mo	Zr	Ru	Residuals[j]	
																Each max	Total, max
R50250	1	Titanium Gr 1		0.10	0.20	Balance	0.0125	0.05	0.18							0.1	0.4
R50400	2	Titanium Gr 2		0.10	0.30	Balance	0.0125	0.05	0.25							0.1	0.4
R50700	4	Titanium Gr 4		0.10	0.50	Balance	0.0125	0.07	0.40							0.1	0.4
R56400	5	Titanium Gr 5	5.5–6.75	0.10	0.40	Balance	0.0125	0.05	0.20		3.5–4.5					0.1	0.4
R56401	5 ELI	Titanium Gr 5 ELI	5.5–6.5	0.08	0.25	Balance	0.0120	0.05	0.13		3.5–4.5					0.1	0.4
R52400	7	Titanium Gr 7		0.10	0.30	Balance	0.0125	0.05	0.25	0.12–0.25						0.1	0.4
R58640	19	Titanium Ti-38-6-44	3.0–4.0	0.05	0.30	Balance	0.020	0.03	0.12	0.10[k]	7.5–8.5	5.5–6.5	3.5–4.5	3.5–4.5	0.10[k]	0.15	0.4

Source: Ref. [3]. Copyright ASTM; reprinted with permission.

[a] Elements shown as balance shall be arithmetically computed by deducting the sum of the other named elements from 100.

[b] Copper plus specified elements=99.8 min; copper plus silver=88.5–91.5

[c] Cobalt is to be counted as nickel.

[d] Minimum content of copper plus all other elements with specified limits shall be 99.5%.

[e] An alloy containing as high as 2.6% silicon is acceptable provided the sum of all the elements other than copper, silicon, and iron does not exceed 0.30%.

[f] Analysis shall regularly be made only for the elements in this table. If, however, the presence of other elements is suspected or indicated in amounts greater than the specified limits, further analysis shall be made to determine that these elements are not present in excess of the specified limits.

[g] Titanium+zirconium 0.20% max.

[h] Titanium+zirconium 0.25% max.

[i] All reported values are maximums, unless a range is specified.

[j] A residual is an element present in a metal or an alloy in small quantities inherent to the manufacturing process but not added intentionally. Residual elements need not be reported unless a report is specifically required by the purchaser.

[k] Ruthenium and palladium, or both, may be added to Grade 19 for enhanced corrosion resistance as negotiated between the purchaser and vendor. Chemical analysis is not required unless specifically negotiated.

TABLE 1.8
Mechanical Property Requirements, ASTM F 468

Alloy	Mechanical Property Marking	Nominal Diameter, in.	Hardness[a]	Full-Size Tests[b]		Machined Specimen Tests		
				Tensile Strength (ksi)	Yield Strength, min (ksi)[c]	Tensile Strength, min (ksi)	Yield Strength, min (ksi)[c]	Elongation in 4D, min (%)[d]
				Copper				
Cu 110	F 468A	All	65–90 HRF	30–50	10	30	10	15
Cu 270	F 468B	All	55–80 HRF	60–90	50	55	50	35
Cu 462	F 468C	All	65–90 HRB	50–80	25	50	25	20
Cu 464	F 468D	All	55–75 HRB	50–80	15	50	15	25
Cu 510	F 468E	All	60–95 HRB	60–90	35	55	30	15
Cu 618	F 468F	0.250–0.500	70–95 HRB	80–110	50	80	50	30
		0.625–1.500	70–95 HRB	75–105	45	75	45	30
Cu 614	F 468G	All	70–95 HRB	75–110	35	75	35	30
Cu 630	F 468H	All	85–100 HRB	100–130	50	100	50	5
Cu 642	F 468J	All	75–95 HRB	75–110	35	75	35	10
Cu 651	F 468K	0.250–0.750	75–95 HRB	70–100	55	70	53	8
		0.875–1.500	70–95 HRB	55–90	40	54	38	8
Cu 655	F 468L	All	60–80 HRB	50–80	20	50	15	20
Cu 661	F 468M	All	75–95 HRB	70–100	35	70	35	15
Cu 675	F 468N	All	60–90 HRB	55–85	25	55	25	20
Cu 710	F 468P	All	50–85 HRB	45–75	15	45	15	40
Cu 715	F 468R	All	60–95 HRB	55–85	20	55	20	45
				Nickel				
Ni 335	F 468S	All	20–32 HRC	115–145	45	115	45	35
Ni 276	F 468T	All	20–32 HRC	110–140	45	110	45	25
Ni 400	F 468U	0.250–0.750	75 HRB–25 HRC	80–130	40	80	40	20
		0.875–1.500	60 HRB–25 HRC	70–130	30	70	30	20
Ni 400 HF[e]	F 468HF	All	60 HRB–95 HRB	70–120	30	70	30	20
Ni 405	F 468V	All	60 HRB–20 HRC	70–125	30	70	30	20

(Continued)

TABLE 1.8 (*Continued*)
Mechanical Property Requirements, ASTM F 468

Alloy	Mechanical Property Marking	Nominal Diameter, in.	Hardness[a]	Full-Size Tests[b]: Tensile Strength (ksi)	Full-Size Tests[b]: Yield Strength, min (ksi)[c]	Machined Specimen Tests: Tensile Strength, min (ksi)	Machined Specimen Tests: Yield Strength, min (ksi)[c]	Machined Specimen Tests: Elongation in 4D, min (%)[d]
Ni 500	F 468W	0.250–0.875	24–37 HRC	130–180	90	130	90	20
		1.000–1.500	24–37 HRC	130–180	85	130	85	20
				Aluminum				
Al 2024-T4[f]	F 468X	All	70–85 HRB	55–70	36	62	40	10
Al 6061-T6[f]	F 468Y	All	40–50 HRB	37–52	31	42	35	10
Al 7075-T73[f]	F 468Z	All	80–90 HRB	61–76	50	68	56	10
				Titanium[g]				
Ti 1	F 468AT	All	140–160 HV	40–70	30	35	25	24
Ti 2	F 468BT	All	160–180 HV	55–85	45	50	40	20
Ti 4	F 468CT	All	200–220 HV	85–115	75	80	70	15
Ti 5	F 468DT	All	30–39 HRC	135–165	125	130	120	10
Ti 7	F 468ET	All	160–180 HV	55–85	45	50	40	20
Ti-38-6-44	F 468FT	All	24–38 HRC	120–150	115	120	115	15
Ti 5 ELI	F 468GT	All	29–36 HRC	125–165	115	125	115	10

Source: Ref. [13]. Copyright ASTM; reprinted with permission.

[a] Where both tension and hardness tests are performed, the tension tests shall take precedence for acceptance purposes. For aluminum alloys hardness tests are for information only.

[b] The yield and tensile strength values for full-size products shall be computed by dividing the yield and maximum tensile load by the stress area for the product size and threaded series as given in the table on tensile stress areas.

[c] Yield strength is the stress at which an offset of 0.2% gage length occurs.

[d] Elongation is determined using a gage length of 4 diameters of test specimen with Test Methods E 8.

[e] "HF" denotes a hot-formed product.

[f] Aluminum alloy temper designations are in accordance with ANSI H35.1.

[g] Full-size test mechanical properties apply to fasteners with a maximum diameter of 3 in. (76 mm). Mechanical properties of larger sections shall be negotiated between the material manufacturer and the fastener producer.

TABLE 1.9
ASTM F 593 Chemical Requirements

Alloy Group	UNS Designation	Alloy	Composition (% Maximum Except as Shown)									
			C	Mn	P	S	Si	Cr	Ni	Cu	Mo	Others
Austenitic Alloys												
1	S30300	303	0.15	2.00	0.20	0.15 min	1.00	17.0–19.0	8.0–10.0		0.60 max[a]	
1	S30323	303Se	0.15	2.00	0.20	0.060	1.00	17.0–19.0	8.0–10.0			Se 0.15 min
1	S30400	304	0.08	2.00	0.045	0.030	1.00	18.0–20.0	8.0–10.5	1.00		
	S30403	304L	0.03	2.00	0.045	0.030	1.00	18.0–20.0	8.0–12.0	1.00		
1	S30500	305	0.12	2.00	0.045	0.030	1.00	17.0–19.0	10.5–13.0	1.00		
1	S38400	384	0.08	2.00	0.045	0.030	1.00	15.0–17.0	17.0–19.0	1.75–2.25	0.50 max[a]	
1	S20300	XM1	0.08	5.0–6.5	0.040	0.18–0.35	1.00	16.0–18.0	5.0–6.5	3.0–4.0		
1	S30430	XM7	0.10	2.00	0.045	0.030	1.00	17.0–19.0	8.0–10.0			
2	S31600	316	0.08	2.00	0.045	0.030	1.00	16.0–18.0	10.0–14.0		200–3.00	
	S31603	316L	0.03	2.00	0.045	0.030	1.00	16.0–18.0	10.0–14.0			
3	S32100	321	0.08	2.00	0.045	0.030	1.00	17.0–19.0	9.0–12.0			Ti 5×C min
3	S34700	347	0.08	2.00	0.045	0.030	1.00	17.0–19.0	9.0–13.0			Cb+Ta 10×C min
Ferritic Alloys												
4	S43000	430	0.12	1.00	0.040	0.030	1.00	16.0–18.0				
4	S43020	430F	0.12	1.25	0.060	0.15 min	1.00	16.0–18.0			0.60 max[8]	
Martensitic Alloys												
5	S41000	410	0.15	1.00	0.040	0.030	1.00	11.5–13.5				
5	S41600	416	0.15	1.25	0.060	0.15 min	1.00	12.0–14.0			0.60 max[a]	
5	S41623	416Se	0.15	1.25	0.060	0.060	1.00	12.0–14.0				Se 0.15 min
6	S43100	431	0.20	1.00	0.040	0.030	1.00	15.0–17.0	1.25–2.50			
Precipitation Hardening Alloy												
7	S17400	630	0.07	1.00	0.040	0.030	1.00	15.0–17.5	3.0–5.0	3.0–5.0		Cb+Ta 0.15–0.45

Source: Ref. [14]. Copyright ASTM; reprinted with permission.

[a] At the manufacturer's option, determined only when intentionally added.

TABLE 1.10
ASTM F 593 Mechanical Property Requirements[a]

				Full-Size Tests			Machined Specimen Tests		
Stainless Alloy Group	Condition[b]	Alloy Mechanical Property Marking	Nominal Diameter (in.)	Tensile Strength[c] (ksi)	Yield Strength[c,d] (ksi)	Rockwell hardness	Tensile Strength[c] (ksi)	Yield Strength[c,d] (ksi)	Elongation in 4D (%)
1 (303, 304, 304L, 305, 384, XM1, XM7, 303Se)	AF	F593A	¼–1½, incl	85 max	—	B85 max	80	50 max	40
	A	F593B	¼–1½, incl	75–100	30	B65–95	max	30	30
	CW1	F593C	¼–⅝, incl	100–150	65	B95–C32	70	60	20
	CW2	F593D	¾–1½, incl	85–140	45	B80–C32	95	40	25
	SH1	F593A	¼–⅝, incl	120–160	95	C24–C36	80	90	12
	SH2	F593B	¾–1, incl	110–150	75	C20–C32	115	70	15
	SH3	F593C	1⅛–1¼, incl	100–140	60	B95–C30	105	55	20
	SH4	F593D	1⅜–1½, incl	95–130	45	B90–C28	95	50	28
							90		
2 (316, 316L)	AF	F593E	¼–1½, incl	85 max	—	B85 max	80	50 max	40
	A	F593F	¼–1½, incl	75–100	30	B65–95	max	30	30
	CW1	F593G	¼–⅝, incl	100–150	65	B95–C32	70	60	20
	CW2	F593H	¾–1½, incl	85–140	45	B80–C32	95	40	25
	SH1	F593E	¼–⅝, incl	120–160	95	C24–C36	80	90	12
	SH2	F593F	¾–1, incl	110–150	75	C20–C32	115	70	15
	SH3	F593G	1⅛–1¼, incl	100–140	60	B95–C30	105	55	20
	SH4	F593H	1⅜–1½, incl	95–130	45	B90–C28	95	40	28
							90		
3 (321, 347)	AF	F593J	¼–1½, incl	85 max	—	B85 max	80	50 max	40
	A	F593K	¼–1½, incl	75–100	30	B65–95	max	30	30
	CW1	F593L	¼–⅝, incl	100–150	65	B95–C32	70	60	20
	CW2	F593M	¾–1½, incl	85–140	45	B80–C32	95	40	25
	SH1	F593J	¼–⅝, incl	120–160	95	C24–C36	80	90	12
	HS2	F593K	¾–1, incl	110–150	75	C20–C32	115	70	15
	SH3	F593L	1⅛–1¼, incl	100–140	60	B95–C30	105	55	20
	SH4	F593M	1⅜–1½, incl	95–130	45	B90–C28	95	40	28
							90		

(Continued)

TABLE 1.10 (*Continued*)
ASTM F 593 Mechanical Property Requirements[a]

Stainless Alloy Group	Condition[b]	Alloy Mechanical Property Marking	Nominal Diameter (in.)	Full-Size Tests			Machined Specimen Tests		
				Tensile Strength[c] (ksi)	Yield Strength[c,d] (ksi)	Rockwell hardness	Tensile Strength[c] (ksi)	Yield Strength[c,d] (ksi)	Elongation in 4D (%)
				Ferritic alloys					
4 (430, 430F)	A	F593N	¼–1½, incl	70–100	35	B65–95	70	35	25
	CW1	F593V	¼–⅝, incl	75–105	40	B75–98	70	30	20
	CW2	F593W	¾–1½, incl	70–100	35	B65–95	70	30	20
				Martensitic alloys					
5 (410, 416, 416Se)	H	F593P	¼–1½, incl	110–140	90	C20–30	110	90	18
	HT	F593R	¼–1½, incl	160–190	120	C34–45	160	120	12
6 (431)	H	F593S	¼–1½, incl	125–150	100	C25–32	125	100	15
	HT	F593T	¼–1½, incl	180–220	140	C40–48	180	140	10
				Precipitation hardening alloys					
7 (630)	AH	F593U	¼–1½, incl	135–170	105	C28–38	135	105	16

Source: Ref. [14]. Copyright ASTM; reprinted with permission.

[a] Minimum values except where shown as maximum or as a range.

[b] Legend of conditions:

A Machined from annealed or solution-annealed stock thus retaining the properties of the original material, or hot-formed and solution-annealed.

AF Headed and rolled from annealed stock and then reannealed.

AH Solution annealed and age-hardened after forming.

CW Headed and rolled from annealed stock thus acquiring a degree of cold work; sizes 0.75 in. and larger may be hot worked and solution-annealed.

H Hardened and tempered at 1,050°F (565°C) minimum.

HT Hardened and tempered at 525°F (274°C) minimum.

SH Machined from strain hardened stock or cold worked to develop the specified properties.

[c] The yield and tensile strength values for full-size products shall be computed by dividing the yield and maximum tensile load values by the stress area for the product size and thread series determined in accordance with Test Methods F 606 (see Table 1.4).

[d] Yield strength is the stress at which an offset of 0.2% gage length occurs.

TABLE 1.11
ASTM F 594 Chemical Requirements

Alloy Group	UNS Designation	Alloy	Composition (% Maximum Except as Shown)									
			C	Mn	P	S	Si	Cr	Ni	Cu	Mo	Others
Austenitic Alloys												
1	UNS30300	303	0.15	2.00	0.20	0.15 min	1.00	17.0–19.0	8.0–10.0		0.60 max[a]	
1	UNS30323	303Se	0.15	2.00	0.20	0.600	1.00	17.0–19.0	8.0–10.0			Se 0.15 min
1	UNS30400	304	0.80	2.00	0.045	0.030	1.00	18.0–20.0	8.0–10.5			
1	UNS30500	305	0.12	2.00	0.045	0.030	1.00	17.0–19.0	10.5–13.0			
1	UNS38400	384	0.08	2.00	0.045	0.030	1.00	15.0–17.0	17.0–19.0			
1	UNS20300	XM1	0.08	5.0–6.5	0.040	0.18–0.35	1.00	16.0–18.0	5.0–6.5	1.75–2.25	0.50 max[a]	
1	UNS30430	XM7	0.10	2.00	0.045	0.030	1.00	17.0–19.0	8.0–10.0	3.00–4.00		
2	UNS31600	316	0.08	2.00	0.045	0.030	1.00	16.0–18.0	10.0–14.0		2.00–3.00	
3	UNS32100	321	0.08	2.00	0.045	0.030	1.00	17.0–19.0	9.0–12.0			Ti 5×C min; Cb+Ta 10×C min
3	UNS34700	347	0.08	2.00	0.045	0.030	1.00	17.0–19.0	9.0–13.0			
Ferritic Alloys												
4	UNS43000	430	0.12	1.00	0.040	0.030	1.00	16.0–18.0				
4	UNS43020	430F	0.12	1.25	0.060	0.15 min	1.00	16.0–18.0			0.60 max[a]	
Martensitic Alloys												
5	UNS41000	410	0.15	1.00	0.040	0.030	1.00	11.5–13.5				
5	UNS41600	416	0.15	1.25	0.060	0.15 min	1.00	12.0–14.0			0.60 max[a]	
5	UNS41623	416Se	0.15	1.25	0.060	0.060	1.00	12.0–14.0				Se 0.15 min
6	UNS43100	431	0.20	1.00	0.040	0.030	1.00	15.0–17.0	1.25–2.50			
Precipitation Hardening Alloy												
7	UNS17400	630	0.07	1.00	0.040	0.030	1.00	15.0–17.5	3.00–5.00	3.00–5.00		Cb+Ta 0.15–0.45

Source: Ref. [15]. Copyright ASTM; reprinted with permission.

[a] At the manufacturer's option, determined only when intentionally added.

TABLE 1.12
ASTM F 594 Mechanical Property Requirements[a]

Stainless Alloy Group	Condition[b]	Alloy Mechanical Property Marking	Nominal Diameter, in.	Proof Stress, ksi, (min)	Rockwell Hardness
1 (303, 304, 305,	AF	F594A	¼–1½, incl	70	B85 max
384, XM1, XM7,	A	F594B	¼–1½, incl	75	B65–95, incl
303Se)	CW1	F594C	¼–⅝, incl	100	B95–C32, incl
	CW2	F594D	¾–1½, incl	85	B80–C32, incl
	SH1	F594A	¼–⅝, incl	120	C24–C36, incl
	SH2	F594B	¾–1, incl	110	C20–C32, incl
	SH3	F594C	1⅛–1¼, incl	100	B95–C30, incl
	SH4	F594D	1⅜–1½, incl	85	B90–C28, incl
2 (316)	AF	F594E	¼–1½, incl	70	B85 max
	A	F594F	¼–1½, incl	75	B65–95, incl
	CW1	F594G	¼–⅝, incl	100	B95–C32, incl
	CW2	F594H	¾–1½, incl	85	B80–C32, incl
	SH1	F594E	¼–⅝, incl	120	C24–C36, incl
	SH2	F594F	¾–1, incl	110	C20–C32, incl
	SH3	F594G	1⅛–1¼, incl	100	B95–C30, incl
	SH4	F594H	1⅜–1¼, incl	85	B90–C28, incl
3 (321, 347)	AF	F594J	¼–1½, incl	70	B85 max
	A	F594K	¼–1½, incl	75	B65–95, incl
	CW1	F594L	¼–⅝, incl	100	B95–C32, incl
	CW2	F594M	¾–1½, incl	85	B80–C32, incl
	SH1	F594J	¼–⅝, incl	120	C24–C36, incl
	SH2	F594K	¾–1 incl	110	C20–C32, incl
	SH3	F594L	1⅛–1¼, incl	100	B95–C30, incl
	SH4	F594M	1⅜–1½, incl	85	B90–C28, incl
Ferritic alloys					
4 (430, 430F)	A	F594N	¼–1½, incl	70	B65–95, incl
Martensitic alloys					
5 (410, 416,	H	F594P	¼–1½, incl	100	C20–30, incl
416Se)	HT	F594R	¼–1½, incl	160	C34–45, incl
Precipitation hardening alloys					
7 (630)	AH	F594U	¼–1½, incl	135	C28–38, incl

Source: Ref. [15]. Copyright ASTM; reprinted with permission.

[a] Minimum values except where shown as maximum or as a range.

[b] Legend of conditions:

A Machined from annealed or solution annealed stock, thus retaining the properties of the original stock, or hot formed and solution annealed.

AF Annealed after all threading is completed.

AH Solution annealed and age hardened after forming.

CW Annealed and cold worked. Sizes 0.75 in. and larger may be hot worked.

H Hardened and tempered at 1,050°F (566°C) min.

HT Hardened and tempered at 525°F (274°C) min.

SH Machined from strain-hardened stock.

First, the higher strength 400 series stainless steels were used. However, they lacked the corrosion resistance of the 300 series and presented fastener ductility problems in the higher strength levels. The materials that provided good corrosion resistance and higher strength were, in many cases, also high-temperature alloys. The earliest of these was A-286, a high Cr-Ni stainless steel with good properties to 1,000°F. A-286 can be heat treated to 125,000 psi. However, one manufacturer, using cold working prior to aging, developed a 200,000 psi version.

Another corrosion-resistant material with good strength and high-temperature capability is Inconel 718. This high nickel alloy can be heat treated to 180,000 psi. Again, this same manufacturer developed, with cold working and aging, a 220,000 psi variety.

1.8 STRESS CORROSION AND HYDROGEN EMBRITTLEMENT

In the discussion of ultrahigh strength steels in Section 1.6, it was pointed out that these steels are subject to stress corrosion cracking. Stress corrosion is a phenomenon whereby a corrosive material attacks the surface of the metal (which is under high stress), creating a series of fine intergranular and often transgranular cracks that initiate failure. As mentioned earlier, H-11, the workhorse of the 220,000 and 260,000 psi fastener steels, was susceptible to stress corrosion, and a major effort was undertaken to develop alternative materials as well as coatings to combat this problem.

Before discussing this effort, it should be understood that stress corrosion cracking is not hydrogen embrittlement (see Chapter 39.) It is possible to have hydrogen released during exposure to corrosive elements if a galvanic cell or differential aeration cell is created. This hydrogen can cause brittle failure, even in the absence of any significant corrosion. Hydrogen embrittlement will be encouraged if the fastener is exposed to hydrogen through acid cleaning and/or electroplating during manufacture. It can occur at hardness levels above Rc 35 in steel after the bolt is tightened to clamp the joint. It is interesting to note that H-11 steel has high resistance to hydrogen embrittlement but not to stress corrosion cracking (SCC).

Early research developed a test intended to detect SCC in the laboratory. The results of this effort are described in Ref. [11]. The solution to the problem did not involve reducing the clamping force, as failure was observed with loads as low as approximately 30% of the tensile strength.

Somewhat higher resistance to failure occurred when threads were rolled after heat treatment when the failure occurred in the shank instead of the threads. This was not a solution to the problem either. Some of the alloys that were examined for resistance to SCC included PH 13-8 Mo, Custom 455, PH 12-9 Mo, and Inconel 718 [16]. The alloy steels 8,740, 4,037, and 4,340, excellent for their strength capability, could not withstand corrosion or stress corrosion.

Research continued in an effort to find a material that would replace H-11 for 260,000 psi bolts. An alloy containing 35% Ni, 20% Cr, 10% Mo, and 35% Co was offered to the fastener manufacturer by the inventor. After almost 5 years of research, it was produced as a material and developed into a fastener grade material, the alloy MP35N. This was capable of strengths up to 260,000 psi with good ductility as well as excellent resistance to corrosion and stress corrosion [17]. In addition, this material offered the highest resistance to fatigue failure of any fastener material yet developed (except for the 400,000 psi steels) and is even used in applications where SCC is not a problem, such as automotive connecting rod bolts for racing cars. It is used in landing gear bolts for some commercial aircraft where high strength and high fatigue resistance as well as resistance to SCC are needed.

An extensive study of the MP35N resistance to various environments was conducted [18]. Some of these data are shown in Tables 1.13 and 1.14.

Fasteners may also be subject to crevice corrosion in which the corrosive material will invade the cracks of the joint or under the nut and head of the bolt as well as along the shank. Materials such as PH 13-8 Mo could not resist this attack and were rejected.

TABLE 1.13
Corrosion of Alloys in Nitric Acid Solutions[3] at 50°C

	Corrosion Rate (mils/yr) at Concentration (M)						
Alloy	**0.5**	**1**	**2**	**4**	**6**	**8**	**10**
MP35N	—	—	0.01	0.07	0.07	0.09	0.1
Hastelloy C	<0.01	0.03	0.08	—	—	—	—
Incoloy 825	<0.02	<0.02	<0.02	—	—	—	—
EN58J[b]	<0.02	<0.02	<0.02	—	—	—	—

Source: Data from Ref. [18].
[a] Test time 28–44 days.
[b] EN58J similar to 316 stainless steel.

TABLE 1.14
Corrosion of Alloys in Huey Test[a] (65% Nitric Acid, Boiling)

	Corrosion Rate (mils/yr)					
Alloy	**48 h**	**96 h**	**114 h**	**192 h**	**240 h**	**Avg.**
MP35N						
Annealed	13	21	36	52	63	37
Cold rolled	14	23	39	50	85	42
Cold rolled and aged	12	19	35	53	70	38
As-welded	20	14	17	21	19	18
Inconel 625, annealed	21	21	23	25	25	23
Hastelloy C276	440	770	—	—	—	605

Source: Data from Ref. [18]. Copyright SPS Technologies.
[a] Data obtained in successive 48-h periods on original specimens.

The materials that showed the most resistance to SCC, which had good fatigue life, and could be fabricated were Inconel 718 and MP35N. A study including these alloys [19] presents an evaluation of these materials.

A major aircraft manufacturer, after experiencing SCC with H-11 standard bolts, switched to Inconel 718 (220,000 psi) and MP 35N (260,000 psi). Other aircraft manufacturers soon followed.

1.9 COATINGS

An alternative to using corrosion-resistant bolt materials is to use coatings. Sacrificial coatings such as zinc and cadmium will corrode in preference to the base metal and provide limited protection in industrial and marine atmospheres. One of the reasons the resistance is limited is the very thin layer that must be controlled to avoid thread interference. Hot dip galvanizing is used in structural bolting. It gives a heavier coating, but the nut must be retapped after coating to prevent thread interference.

Another problem with sacrificial coatings is that hydrogen is liberated during plating, and steps must be taken to minimizc hydrogen embrittlement. Field exposure may also provide enough hydrogen through the corrosion mechanism to occasionally cause hydrogen embrittlement.

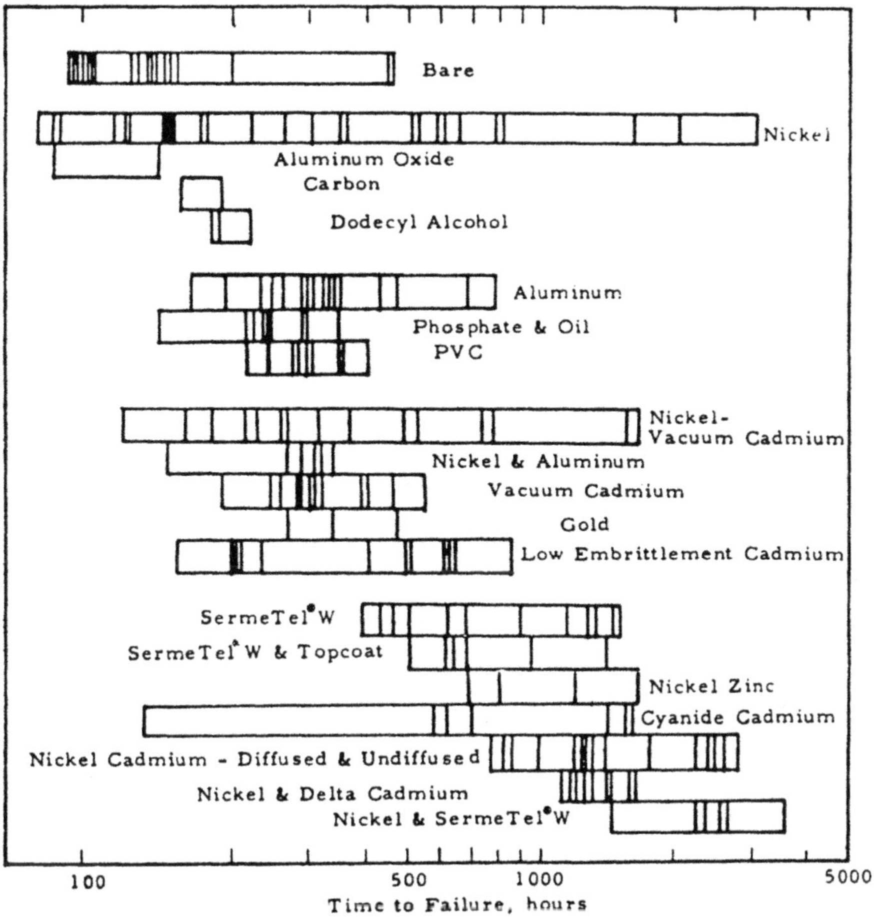

FIGURE 1.2 Effect of coating on stress corrosion cracking resistance of H-11 bolts (from Ref. [20].)

Coatings are sometimes applied for lubrication, as some corrosion-resistance materials such as titanium, stainless steel, and nickel-based alloys will gall and seize during the tightening operation. Metals such as graphite and petroleum jelly (MIL-T-5544), molybdenum disulfide (MoS_2), and silver are used.

Part of the research effort to prevent SCC included a study of single- and multiple-layer coatings [20]. Results are shown in Figure 1.2. The best of the coatings, nickel plus SermeTel W, was patented [21]. The SermeTel W is a proprietary coating but is covered by military specification MIL-C-81751B [22].

Although these coatings do not match the performance of the corrosion-resistant base metals that have been developed, they could be considered for lower strength applications, as SCC resistance is proportional to strength.

The selection of fastener materials and coatings is not the only consideration when designing joints to resist corrosion. Not only must the materials be suitable to the environment, they must also be compatible with each other. One example of this possible problem is the use of titanium fasteners with aluminum joints in wide-body aircraft. The superior corrosion resistance of the titanium causes rapid attack by galvanic and exfoliation (a form of crevice) corrosion of the aluminum. Solutions have included the use of metallic ceramic and aluminum coatings for the titanium fasteners.

Reviews of corrosion and stress corrosion are given in Refs. [23] and [24]. They cover the application of fasteners to joints and the precautionary measures that must be taken to prevent premature failures.

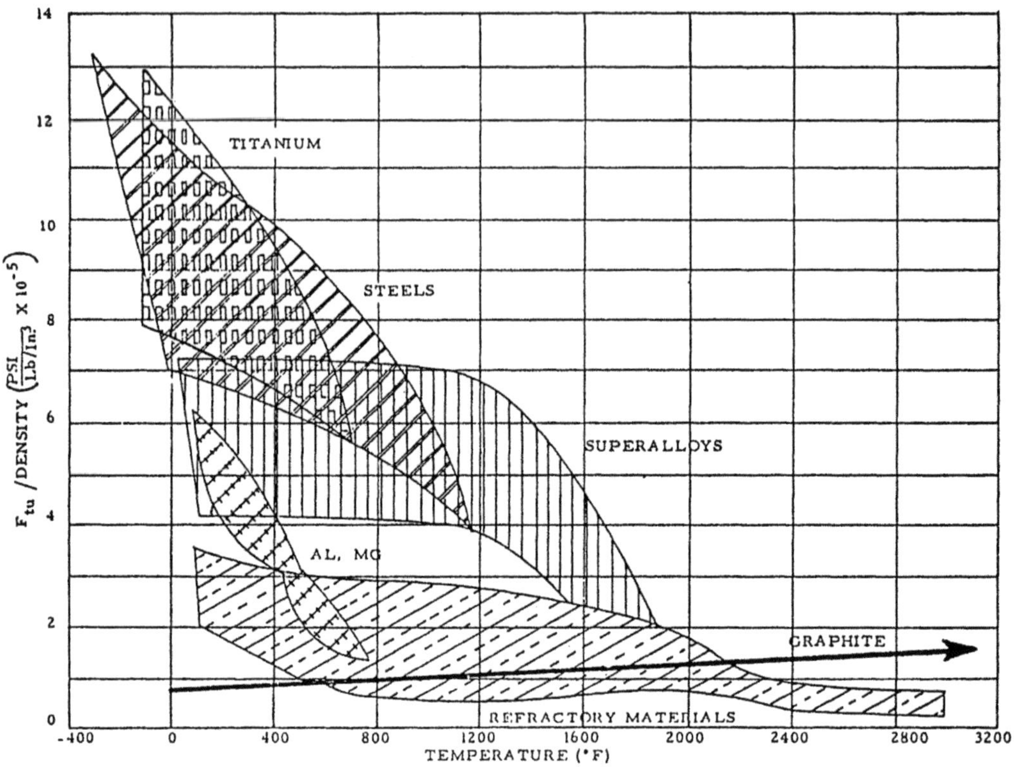

FIGURE 1.3 The useful temperature ranges of metallic structural materials (from Ref. [20].)

1.10 TEMPERATURE-RESISTANT MATERIALS

Structural materials are exposed to a wide range of temperatures both above and below the usual summer–winter exposure in buildings and bridges. Moderate temperatures are considered to be those between room temperature and 900°F. Elevated temperatures range to 1,600°F. Extreme elevated temperatures are those in the range of 1,600°F–3,000°F. Moderately low temperatures range from room temperature to –50°F. Cryogenic (extremely low) temperatures range from –50 to –423°F.

Fasteners have been developed to cover the range from –423°F to 3,000°F. The temperature ranges for various materials in terms of density versus temperature are shown in Figure 1.3 [25].

High-temperature applications range from pressure vessels and ovens to heat-treating furnaces and jet engines to space vehicles with both extremely low and extremely high-temperature requirements. Moderately high and elevated temperature fastener materials are included in ASTM specifications A193, A194, A437, and A453 [26–29]. These include special steels and stainless steels. The presence of elevated to high elevated temperatures in aircraft jet engines mandates the use of materials designed for that purpose. Materials such as A-286 and Inconel 718 were introduced as corrosion-resistant materials in Section 1.7, and they are also used in jet engines because of their good elevated temperature resistance. Other high nickel and cobalt base alloys that are used in jet engines up to 1,800°F include Waspaloy, Rene 41, and Astroloy. Waspaloy can be cold worked and aged to 260,000 psi, but its ductility is so low that it has been rejected for stress corrosion applications.

Table 1.15 lists materials that are available as fasteners from aircraft fastener manufacturers in the range from room temperature to 900°F. Table 1.16 covers fastener materials to 1,600°F, and Table 1.17 lists fastener materials for use to 3,000°F.

The refractory materials such as tantalum, columbium, and molybdenum alloys require elaborate multilayer coatings to prevent failure by oxidation. Molybdenum alloys were used in only a few applications because a pinhole break in the coating would cause the metal to vaporize.

TABLE 1.15
Ultimate Tensile Strength (Ftu) of Selected Alloys at Room to Moderately Elevated Temperatures

Material Type	Room Temp. Ftu (ksi)	Max. Temp. (°F)
	Iron-Base Alloys	
Vasco MA	300	900[a]
H-11	260	900[a]
Marage 300	260	900[a]
PH 13-8 Mo	220	650
Custom 455	220	650
C. R. A-286	200	900
B5F5	200	900
8740	180	450[a]
4340	180	450[a]
PH 17-4	160	900
15-5 PH	150	900
	Nickel-Base Alloys	
MP35N	260	700
C. R. Inco 718	220	900[b]
Inco 718	180	900[b]
	Titanium-Base Alloys	
Ti 1-8-5	200	300
Ti 6-6-2	180	500
Ti 6-4	160	500

Source: Ref. [25].

[a] Protective plating is required.

[b] Rolled threads after heat treatment.

TABLE 1.16
Ultimate Tensile Strength (Ftu) of Selected Alloys at Elevated Temperatures (to 1,600°F)

Material Type	Room Temp. Ftu (ksi)	Max. Temp. (°F)
	Iron Base	
A-286	200	1,200
A-286	140	1,200
	Nickel Base	
Rene 95	230	1,200
Astroloy	190	1,600
Inco 718	180	1,200
Waspaloy	150	1,400
Rene 41	150	1,400

Source: Ref. [25].

TABLE 1.17
Ultimate Tensile Strength (Ftu) of Selected Alloys at Extremely Elevated Temperatures

Material Type	Ftu at Max. Temp, (ksi)	Max. Temp. (°F)
	Tantalum Base	
T-222	20	3,000
90 Ta-10W	18	3,000
	Columbium Base	
Cb 752	35	2,500
C 1294	35	2,500
	Nickel Base	
TD-Ni	12	2,000
TD-Ni-Cr	18	2,000
	Cobalt Base	
HA 188	45	1,800
L 605	35	1,800

Source: Ref. [25].

Recent developments in jet engine alloys have included materials such as MP159, useful to 1,200°F, and a new alloy, Aerex 350 [30], which is useful to 1,400°F. MP159 with room temperature strength of 260,000 psi and Aerex 350 with a room temperature strength of 220,000 psi meet the need for high-temperature alloys that also have high room temperature strength.

A very extensive fastener material study was done for NASA in 1965 that was useful in selecting materials for Project Apollo and later space shuttles [31]. Materials in this study cover the range from −423°F (the boiling point of liquid hydrogen) to 1,600°F. Other properties beside the standard ones were evaluated, such as strength/density ratio, tension impact, and stress rupture. The complete report is over 400 pages long and contains extensive data on the effect of temperature on both fastener materials and fasteners. Some examples of the data are included here to provide some sense of the type of information that was developed.

Chart 1.I shows notch properties of materials in the range of −423°F to 1,600°F.

Chart 1.II shows stress relaxation of a René 41 bolt and a silver-plated Waspaloy nut under different preloads.

Table 1.18 shows tensile and angle block tensile strength at temperatures −423°F, −320°F, and 70°F for a Ti6A1-4V bolt with an A-286 (160 ksi) nut.

As noted in Section 1.5, the selection of the fastener made from a particular material involves a great deal more than the bolt and nut. Elevated temperature applications are even more complex than those at room temperature and must take into consideration the things such as lubrication, fastener material compatibility with joint materials, coefficient of thermal expansion, modulus of elasticity, and stress relaxation, not to mention the drop in yield and tensile strength of the fastener material with a change in temperature.

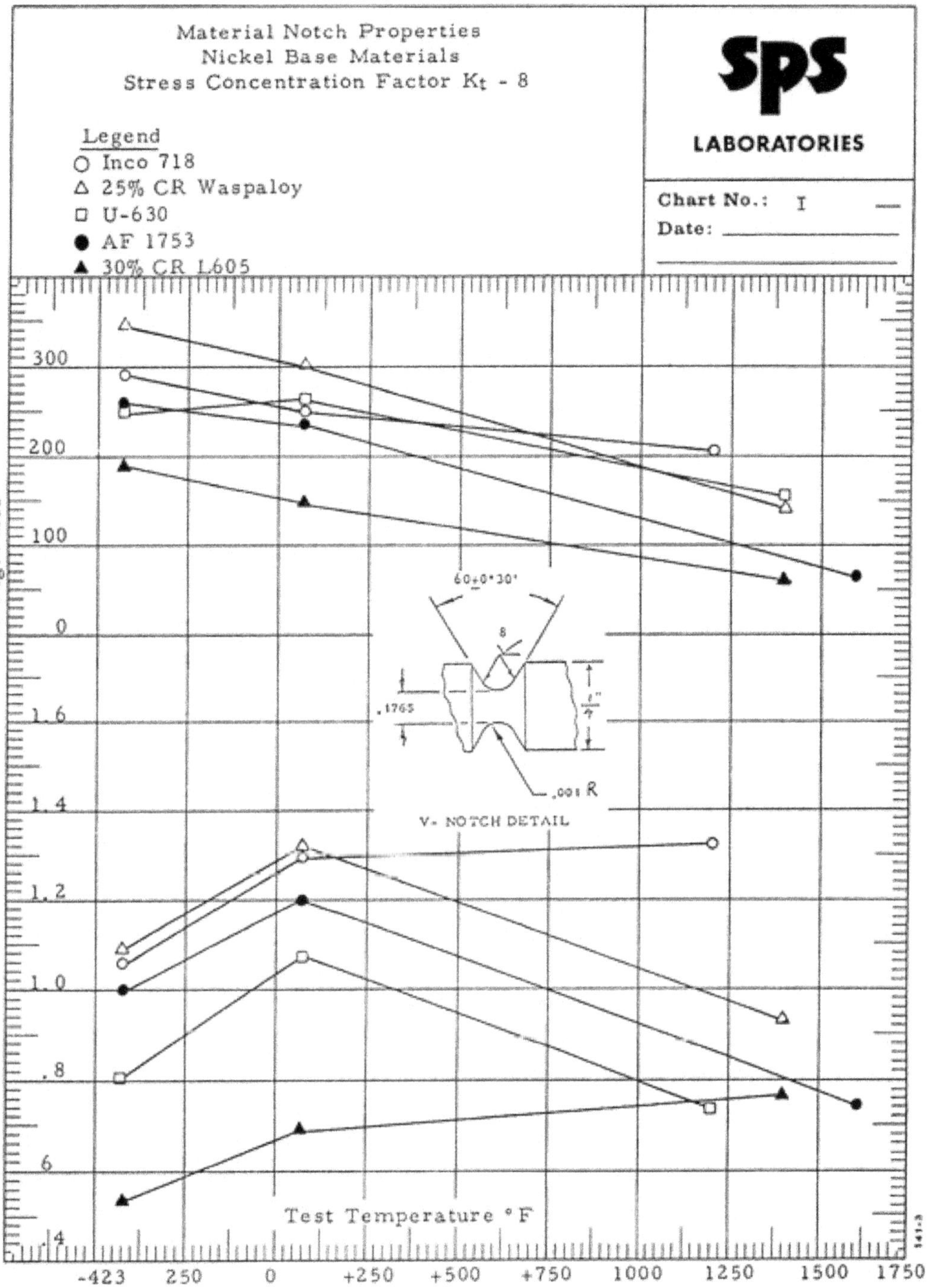

CHART 1.1 Material Notch Properties (From Ref. [31].)

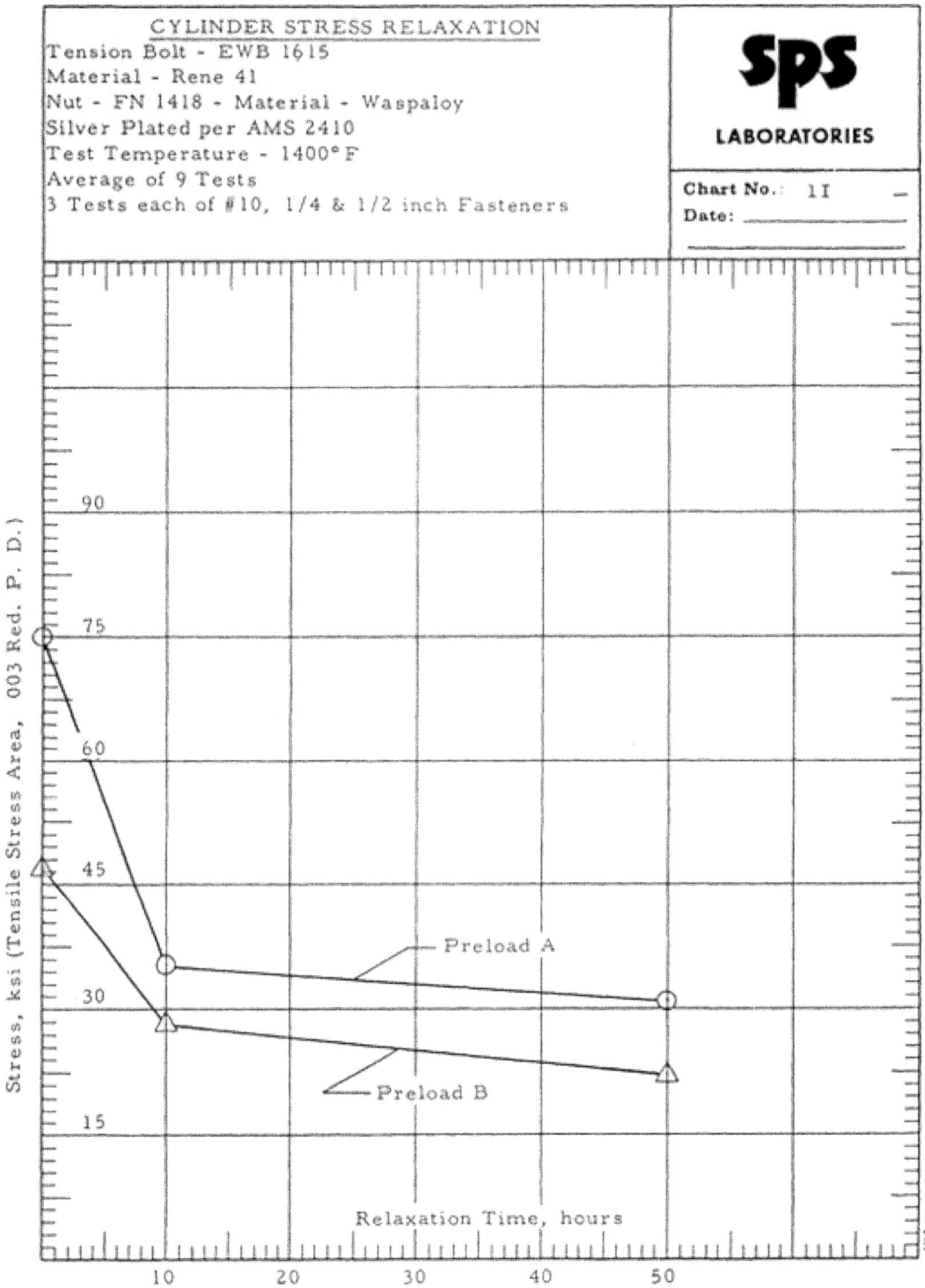

CHART 1.II Cylinder Stress Relaxation (From Ref. [32].)

TABLE 1.18
Mechanical Properties

	Bolt NAS 1275-48; Ti6Al-4V (160 ksi) Nut FN 1216-820; A-286 (160 ksi); 1/2-20		
Test no.	**Test Temperature (°F)**	**Ultimate Load (lb)**	**Ultimate Stress[a] (psi)**
		A. Axial Bolt Properties	
1	–423	32,500	203,000
2	–423	37,800	236,000
3	–423	32,400	202,600
4	–320	32,400	202,600
5	–320	30,000	187,600
6	–320	33,900	212,600
7	70	29,500	184,400
8	70	29,700	185,700
9	70	29,600	185,100
	B. Angle Block Bolt Properties (3°∠. at Nut Bearing Face)[b]		
10	–423	27,000	157,200
11	–423	26,600	155,000
12	–423	28,010	163,500
13	–320	19,000	118,800
14	–320	21,060	131,700
15	–320	20,300	127,000
16	70	25,800	161,400
17	70	22,000	137,600
18	70	21,000	131,300

Source: From Ref. [25].

[a] Stress calculated at the tensile stress area of 0.1599 in.^{2}.

[b] Bolt grip length 3.0 in.

1.11 SUMMARY

The selection of fastener materials starts with the user's need to have a fastener made from the same material as the joint or one compatible with it. The fastener material must have the capability to provide high tensile strength, notch strength, ductility, and yield strength—high enough to take tightening loads with minimum plastic deformation.

Resistance to corrosion and stress corrosion is important, as is the capability to provide high fatigue resistance. Certain manufacturing techniques can provide these characteristics. Temperature resistance to change in properties over the widest possible range is valuable, although fasteners have been developed for specific temperatures.

Finally, the fastener must be incorporated without difficulty into the overall structural design of a product. The performance of fasteners from different materials may not always be the same for the same material in each test. Part of this problem was recognized a number of years ago with the development of the MIL-STD-1312 series of fastener test procedures. This is documented in

Table 1.19 along with tests for magnetic particle inspection, plating thickness, and metallography. Standardization of tests is necessary for the presentation of any data on fastener materials and the fasteners made from them.

TABLE 1.19
Fasteners Test Procedures

	Hardness	
MIL-STD-1312-6	ASTM-E3	ASTM-E18
ASTM-A870	ASTM-E10	ASTM-E384
SAE-J417		
	Tensile	
MIL-STD-1312-8	ASTM-A470	SAE-J429
MIL-STD-1312-18	ASTM-F606	SAE-J995
MIL-STD-1312-23		
	Shear	
MIL-STD-1312-13	ASTM-F606	
MIL-STD-1312-20	ANSI/ASME-B18.8.2	
MIL-STD-1312-28		
	Fatigue	
MIL-STD-1312-11	NAS1069 (National Aerospace Standards)	
MIL-STD-1312-21		
	Stress Durability (Embrittlement Testing)	
MIL-STD-1312-5	ASTM-F606	
MIL-STD-1312-14		
	Vibration	
MIL-STD-1312-7		
	One of the Tests in Product Specifications NAS 1675	
Torque		
MIL-STD-1312-15	MIL-STD-1312-25	IFI-100/107
MIL-STD-1312-24	MIL-STD-1312-31	IFI-125
	AS1310 (SAE) Fastener Torque for Threaded Applications, Definitions of	
Torque-Tension/Preload		
MIL-STD-1312-15	SAE-J174	IFI-101
MIL-STD-1312-16	ANSI-B18.16.2	IFI-124
Corrosion		
MIL-STD-1312-1	MIL-STD-753	
MIL-STD-1312-3	ASTM-B117	
MIL-STD-1312-19		
Stress Rupture		
MIL-STD-1312-10		
Stress Relaxation		
MIL-STD-1312-17		
Magnetic Particle Inspection		
AMS 2640 (SAE)	ASTM-E709	ASTM-1444
MIL-STD-271	SAE-J420	

(Continued)

TABLE 1.19 (*Continued*)
Fasteners Test Procedures

	Federal Test Method Standard STD No. 1516	
Liquid Penetrant Inspection		
AMS 2645 (SAE)	MIL-STD-6866	
ASTM-E165	MIL-STD-271	
SAE-J426		
Plating Thickness		
MIL-STD-1312-12	ASTM-B487	ASTM-B499
ASTM-B504	ASTM-B530	ASTM-B568
ASTM-B659	ASTM-E376	
	Federal Test Method STD No. 151, Methods 520.1 and 523	
Metallography		
ASTM-E3	ASTM-E1077	SAE-J123
ASTM-E112	ASTM-F788	SAE-J418
ASTM-E340	ASTM-F812	SAE-J419
ASTM-E384	ASTM-J121	SAE-J423
ASTM-E407	SAE-J122	SAE-J1061
ASTM-E883		

Source: Ref. [32].

REFERENCES

1. SAE J429-83, Mechanical and Material Requirements for Externally Threaded Fasteners, SAE Handbook, Vol. 1, Society of Automotive Engineers, Warrendale, PA.
2. ASTM A 307-94, Standard Specification for Carbon Steel Bolts and Studs, 60,000 psi Tensile Strength, ASTM Standards, Vol. 15.08, American Society for Testing and Materials, Philadelphia.
3. ASTM A325-94, Standard Specification for Structural Bolts, Steel, Heat Treated 120/105 ksi Minimum Tensile Strength, ASTM Standards, Vol. 15.08.
4. ASTM A 490-93, Standard Specification for Heat Treated Steel Structural Bolts, 150 ksi Minimum Tensile Strength, ASTM Standards, Vol. 15.08.
5. ASTM A354-95, Standard Specification for Quenched and Tempered Alloy Steel Bolts, Studs, and Other Externally Threaded Fasteners, ASTM Standards, Vol. 15.08.
6. SAE J995-79, Mechanical and Material Requirements for Steel Nuts, SAE Handbook, Vol. 1.
7. ASTM A 563-94, Standard Specification for Carbon and Alloy Steel Nuts, ASTM Standards, Vol. 15.08.
8. SAE J 123c, Surface Discontinuities on Bolts, Screws, and Studs, SAE Handbook, Vol. 1.
9. Bickford, J. H., *An Introduction to the Design and Behavior of Bolted Joints*, 3rd ed., Marcel Dekker, New York, 1995.
10. Gill, F. L., Boron Steels for Mechanical Fasteners, SAE Technical Paper 680127, 1968.
11. Lin, C. S., J. J. Laurilliard, and A. C. Hood, Stress corrosion cracking of high strength bolting, In *Stress Corrosion Testing*, ASTM STP425, ASTM, Philadelphia, p. 84, 1967.
12. ASTM F467-93, Nonferrous Nuts for General Use, ASTM Standards, Vol. 15.08.
13. ASTM F468-93, Standard Specification for Nonferrous Bolts, Hex Cap Screws, and Studs for General Use, ASTM Standards, Vol. 15.08.
14. ASTM F593-95, Standard Specification for Stainless Steel Bolts, Hex Cap Screws, and Studs, ASTM Standards, Vol. 15.08.
15. ASTM F594-95, Standard Specification for Stainless Steel Nuts, ASTM Standards, Vol. 15.08.
16. Patel, S., and E. Taylor, New high strength fastener materials resist corrosion, *Metal Progress*, 100(3), p. 98, September 1971.

17. Hood, A. C., MP35N—A multiphase alloy for high strength fasteners, *Metal Progress*, 93(5), pp. 81–84, May 1968.
18. Taylor, E., Multiphase alloy Environmental Resistance, SPS Laboratory Report 5817 (Revised), SPS Technologies, January 1, 1980.
19. Taylor, E., Stress-Corrosion Cracking Evaluation of Aerospace Bolting Alloys, Special Technical Publication 610, ASTM, p. 243, 1976.
20. Taylor, E., Stress Corrosion Crack Protection from Coatings on High Strength H-11 Steel Aerospace Bolts, Special Technical Publication 518, ASTM, p. 131, 1972.
21. Hood, A. C., Protective Coating for Ferrous Metals, U.S. Patent No. 3,897,222, July 29, 1975.
22. Military Specification Coating, Metallic-Ceramic, MIL-C-81751B, Jan. 17, 1972.
23. Hood, A. C., Corrosion in threaded fasteners—causes and cures, *Machine Design*, 36, pp. 153–156, Dec. 17, 1964.
24. Hood, A. C., Preventing stress-corrosion cracking in threaded fasteners, *Metal Progress*, 92(3), pp. 85–88, September 1967.
25. Product Engineering Report, Fastener Seminar, Third Printing, SPS Technologies, 1980.
26. ASTM A-193-95, Alloy Steel and Stainless Steel Bolting Materials for High Temperature Service, ASTM Standards, Vol. 01.01.
27. ASTM A194-95, Carbon and Alloy Steel Nuts for High Pressure and High Temperature Service, ASTM Standards, Vol. 01.01.
28. ASTM A437-94, Alloy Steel Turbine-Type Bolting Material Specially Heat-Treated for High Temperature Service, ASTM Standards, Vol. 01.01.
29. ASTM A453-94, Bolting Materials, High Temperature, 50 to 120 ksi Yield Strength with Expansion Coefficients Comparable to Stainless Steels, ASTM Standards, Vol. 01.01.
30. Buzolits, S. R., and L. A. Kline, Bolting alloy fills high temperature gap, *Advanced Materials and Processes*, 147(2), pp. 33-35, February 1995.
31. Glackin, J. J., and E. F. Gowen, Jr., Evaluation of Fasteners and Fastener Materials for Space Vehicles, NASA Contract No. NASH-11125, Final Report, November 1963–November 1965, SPS Laboratories.
32. Borg, K., *Specifications for Fastener Testing*, Fastener Technology International, p. 44, February 1995, Akron, OH.

2 Joining with Aluminum Alloy Bolts and Nuts

Russell E. Mack

2.1 WHY USE ALUMINUM BOLTS?

The characteristics of aluminum that cause it to be chosen for a wide range of end product applications are also those that provide the same advantages for bolts and nuts.

Resistance to atmospheric corrosion; nonstaining: Aluminum forms a protective oxide coating upon exposure to the atmosphere. The oxides of aluminum are colorless, and such a corrosion product, if it occurs, will not stain and mar the appearance of the fastener or the assembled materials.
Weight/strength: Aluminum alloy fasteners have the highest strength-to-weight ratio of any metal in common use for fasteners. Their strength is comparable to that of low-carbon steel fasteners, and aluminum fasteners weigh about one-third as much as steel and stainless steel fasteners.
Electrical conductivity: The high rate of conductivity of aluminum makes it ideal for joints intended to carry electricity. Depending on the alloy choice, aluminum fasteners have a conductivity rating of 38%–43% International Annealed Copper Standard [at 68°F (20°C)]
Nonmagnetic: All aluminum alloy fasteners are nonmagnetic and nonsparking.
Compatibility/galvanic corrosion: The use of aluminum fasteners in aluminum joints minimizes the danger of galvanic corrosion.
Color identification/corrosion resistance: Inherent corrosion resistance can be increased by anodizing the aluminum fastener for applications that are considered to be highly corrosive. The electrolytic anodic coating can be dyed before it is sealed in a variety of colors suitable for identification or aesthetic reasons. Anodic coatings and the process are covered by Military Specification MIL-A-8625.

2.2 AVAILABILITY

Aluminum alloy bolts are domestically produced by a number of manufacturers in the types and sizes typically available in other materials. Several master distributors make it possible to obtain a wide range of popular sizes throughout the U.S. fastener distribution system.

While size range capability varies among manufacturers, the design engineer can expect to obtain bolts made to appropriate American National Standard specifications in types and sizes to meet any requirement.

Hex head bolts are available either as headed hex head bolts (sometimes termed indented hex head bolts) or finished hex head bolts (sometimes termed cap screws). A comment about the headed hex bolt is appropriate as there exist misconceptions about it. Since the head is forged complete in the initial headed blank (i.e., not sheared like a cap screw), it has slightly rounded edges on the top and underside of the hex as well as on the corners. This in fact offers certain advantages; the head will not mar the bearing surface, and engagement of a wrench or socket is facilitated.

DOI: 10.1201/9781003501183-3

2.2.1 Size Ranges

Diameters 7/8–9, 1–8, and larger are produced by several specialty fastener manufacturers. Fine pitch threads are also manufactured but are not readily available and may entail substantial production runs. Metric equivalents are also produced on special runs by a limited number of manufacturers. Lengths are typically ¼ in. increments, and lengths in the ranges shown in Table 2.1 are readily available. Longer lengths are also produced on special production runs.

The applicable standard for hex head bolts produced in the United States is ANSI B 18.2.1, which is included in the *Fastener Standards Handbook* published by Industrial Fastener Institute, Cleveland, Ohio.

Hex head bolts for structural bolting of aluminum transmission towers covered by ASTM Specification F 901 are produced in appropriate aluminum alloys by selected manufacturers on special manufacturing runs. They are head marked and color anodized for size identification.

The applicable standard for carriage bolts produced in the United States is ANSI B 18.5, which is also included in the *Fastener Standards Handbook*. Sizes tabulated in Table 2.2 reflect those that are most readily available; lengths are typically ¼ in. increments. Other sizes, fine pitch threads, and short neck carriage bolts are produced by some manufacturers on special production runs for which mill minimums are required.

The bolts discussed above, hex and carriage, are the ones most commonly used in applications where the characteristics of aluminum are needed and appropriate. However, aluminum alloy bolts can be and are produced in virtually all head types including heavy hex head, hex flange, hex socket, square head, and 12-point.

TABLE 2.1
Common Sizes of Hex Head Bolts

Size	Length (in.)
1/4–20	½–2½
5/6–18	¾–3
3/8–16	¾–3½
1/2–13	1–4
5/8–11	1¼–4
3/4–10	1¾–4

TABLE 2.2
Common Sizes of Round Head, Square Neck Carriage Bolts

Size	Length (in.)
10–24	½–2
1/4–20	½–2½
5/6–18	¾–3
3/8–16	¾–4
1/2–13	1–4

2.2.2 Nuts

Aluminum alloy hexagon nuts for all the bolt sizes listed are readily available through the U.S. fastener distribution network. They are manufactured in accordance with ANSI B 18.2.2. Generally, they are cold-forged and are chamfered and countersunk on both sides to facilitate assembly. Aluminum alloy nuts should always be lubricated.

2.3 ALLOY SELECTION

The choice of aluminum alloys for male and female fasteners is relatively limited. Unless the proper alloy is used and is correctly heat treated and aged, corrosion and strength problems can be catastrophic. Heat-treating procedures should adhere to the practices and controls of MIL-H-6088. Bolts that are roll threaded after heat treating and aging are found to have slightly greater tensile strength than bolts that have cut threads or are roll threaded before heat treating.

Bolts are most commonly produced in alloy 2024-T4, which has strength, toughness, and corrosion resistance suitable for most applications. Alloy 2024-T4 bolts are frequently anodized to enhance their inherent resistance to atmospheric corrosion.

Where the designer finds a need for slightly higher strength, bolts can be manufactured in alloys 7075-T73 or 7050-T73, the latter being very special, probably difficult to obtain, and very costly.

For limited applications where strength can be sacrificed to gain further resistance to general corrosion and to some chemical corrosion, alloy 6061-T6 can be specified, but it is also not as readily available as 2024-T4 and still entails special production runs.

To develop the full strength of aluminum alloy bolts and to avoid stress corrosion cracking of the aluminum nut, the alloy for nuts must be either 6061-T6 or 6262-T9. Cold-forged nuts are manufactured in alloy 6061-T6. Nuts machined from bar can be either 6061-T6 or 6262-T9. It can be noted that 6262-T9 nuts are not readily available, and since they are manufactured by machining, they will be more costly than 6061-T6 forged nuts.

2.4 MECHANICAL PROPERTIES

The minimum mechanical properties and raw material specifications for the aluminum alloys discussed here are given in Table 2.3. Chemical composition limits and minimum strength values are covered by ASTM specifications F 467 for nuts and F 468 for bolts. Typical tensile and shear strengths are listed in Table 2.4.

The tensile strength of an aluminum bolt can be approximated by multiplying the tensile strength of the stock material by the cross-sectional area of the bolt at the root of the threads. Data confirming this are shown in Table 2.5.

TABLE 2.3
Minimum Mechanical Properties

Alloy and Temper	Tensile Strength (psi)	Tensile Yield Strength (psi)	Elongation in 2 in. or 4*D* (%)	Shear Strength (psi)	ASTM Material Designation
2024-T4	62,000	40,000	10	37,000	B 316
6061-T6	42,000	35,000	10	25,000	B 316
6262-T9	52,000	48,000	5	30,000	B 211
7050-T73	70,000	58,000	10	39,000	B 316
7075-T73	68,000	56,000	10	41,000	B 316

TABLE 2.4
Typical Mechanical Properties

Alloy and Temper	Tensile Strength (ksi)	Shear Strength (ksi)
2024-T4	68	41
6061-T6	45	30
6062-T9	58	35
7050-T73	75	45
7075-T73	73	44
Mild steel	60	45

TABLE 2.5
Tensile Strength of 2024-T4 Bolts and Stock Material

Bolt Size	Lots Tested	Total Test	Tensile Strength of Nominal Root Area (psi)	Average Tensile Strength of Stock[a] (psi)	Ratio of Strength of Root Area to Strength of Stock
1/4-20	209	638	73,800	68,800	1.07
5/16-18	78	234	70,100	69,100	1.02
3/8-16	54	161	69,900	69,100	1.01
1/2-13	68	205	69,000	68,900	1.00
5/8-11	34	101	67,300	69,200	0.97
3/4-10	5	16	67,100	68,600	0.98

[a] Tensile strength of raw material used to manufacture bolts heat treated with bolts.

TABLE 2.6
Tensile Strength of Bolts in Axial and 10° Wedge Test[a]

Bolt		Tensile Breaking Load (lb)		Strength Reduction
Size	Alloy and Temper	Axial Test	10° wedge Test	Because of Wedge (%)
3/8-16	2024-T4	5,100	4,900	4
1/2-13	2024-T4	9,000	8,900	1
5/8-11	2024-T4	14,200	13,900	2
3/4-10	2024-T4	20,000	19,800	1
3/8-16	7075-T73	6,000	5,600	7
1/2-13	7075-T73	11,000	10,300	6
5/8-11	7075-T73	17,100	16,900	1
3/4-10	7075-T73	25,600	25,200	2

[a] Bolts tested had a minimum of 3/8 in. unthreaded shank.

Axial tension and 10° wedge tension testing (per ASTM F 606) give emphasis to the suitability of 2024-T4 and 7075-T73 for bolts from the standpoints of strength and toughness. Results are shown in Table 2.6.

The load-carry capacity in shear is calculated by multiplying the shear strength of the bolt alloy by the cross-sectional area of the bolt in shear. If the threads are in the shear plane, the area at the root of the thread should be based on the minimum root diameter. Calculated minimum tensile and single-shear loads are given in Table 2.7.

TABLE 2.7
Minimum Tensile and Single-Shear Loads for 2024-T4 and 7075-T73 Bolts

			Tensile Strength (lb)		Single-Shear Strength (lb): Threads in Shear Plane		Single-Shear Strength (lb): Shank in Shear Plane	
Nominal Size (in.)	**Basic Major Diameter (in.)**	**Threads per inch**	**2024-T4**	**7075-T73**	**2024-T4**	**7075-T73**	**2024-T4**	**7075-T73**
10	0.190	24	926	1,020	553	598	978	1,060
¼	0.250	20	1,710	1,880	1,020	1,110	1,720	1,860
5/16	0.3125	18	2,880	3,160	1,720	1,860	2,710	2,930
⅜	0.375	16	4,300	4,710	2,560	2,770	3,930	4,240
½	0.500	13	7,950	8,720	4,750	5,130	7,060	7,640
⅝	0.625	11	12,800	14,000	7,620	8,230	11,060	12,000
¾	0.750	10	19,100	20,900	11,400	12,300	16,000	17,300
⅞	0.875	9	26,400	29,000	15,800	17,100	21,800	23,600
1	1.000	8	34,700	38,100	20,700	22,400	28,500	30,800

TABLE 2.8
Breaking Torque for Bolts

			Breaking Torque (lb.–in.)	
Alloy and Temper	**Shear Strength[a] (psi)**	**Screw or Bolt Size**	**By Test[b]**	**Calc.[c]**
6061-T6	29,500	3/8-16	193	188
		5/8-11	966	973
2024-T4	39,000	3/8-16	246	250
		5/8-11	1,148	1,280
7075 T73	48,100	3/8-16	316	308
		5/8-11	1,598	1,580

[a] Determined by standard double-shear test of stock.
[b] Test made in torsion machine.
[c] Torque, $T=\pi f_s d3/13$, where f_s=shear strength of bolt material in psi and d=minor diameter of bolt thread in inches.

2.5 INSTALLATION PROCEDURES

Although installation methods for aluminum alloy bolts are similar to those used for bolts in other materials, it should nonetheless be recognized that, if overtightened, aluminum bolts are not as forgiving as are bolts in other materials. Consequently, more careful attention to the procedure used is recommended. Also, as with fasteners of all materials, a quantitative measure of tightness is difficult to obtain.

The torsion shear strength of the bolt thread area can be used as a guide in developing installation torque. The torsion shear strength can be calculated or found by pragmatically determining free breaking torque, the latter by straining the bolt shank and applying a calibrated torque wrench to the head of the bolt. Table 2.8 presents data for both methods for three common alloys.

Whether employing a predetermined torque value or "the turn of the nut" from a specified snug position, it should be noted that clamping load can vary widely due to variables including thread

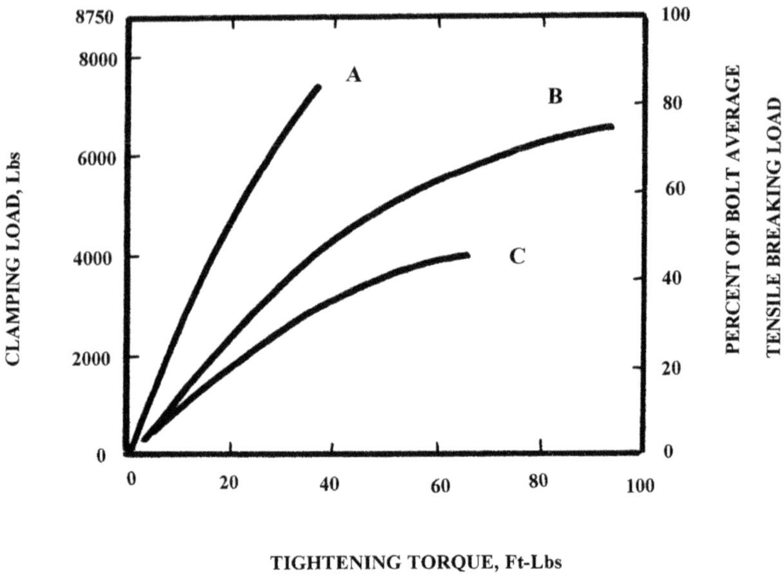

FIGURE 2.1 Effect of lubrication on clamping load of 1/2-13 2024-T4 bolts. A, the effect of a petroleum-based lubricant; B, the effect of a wax-based lubricant; C, the torque-load relationship obtained with a bare, unlubricated bolt.

fit, bearing surface condition, and lubrication. While this is true with all bolt materials, aluminum bolts are particularly sensitive to these variables, especially the presence or absence of lubrication. Consequently, experience has shown that installation torque is more appropriately determined by trial under actual installation conditions. A number of bolts can be installed and tightened until failure occurs, with a calibrated torque wrench used to record the failure torques. Subsequent job installation torque with a satisfactory safety factor would be 80% of the lowest recorded failure torque.

Since job site installation may be difficult using a calibrated torque wrench, a turn-of-a-nut method may be employed after the number of turns has been correlated either with a calculated or with empirical tightening torque as described above.

Proper lubrication in an aluminum bolted joint cannot be emphasized enough. Figure 2.1 graphically shows the salutary effect of lubricant on clamping load relative to bolt tensile breaking load. If tightening torque is applied to the nut, the common practice is to use a bulk lubricated nut so that the threads and the bearing face are lubricated. Most commercially available nuts are waxed by the manufacturer, but the user should be certain this has been done. If the bolt is to be tightened, then arrangements will have to be made to have the commercially available bare aluminum bolt lubricated, which again is best done in a bulk method to be sure the threads and head bearing surfaces are lubricated.

If aluminum materials are being joined, it is appropriate to use aluminum fasteners. The alloys for aluminum bolts and other fasteners are usually as strong as or even stronger than the aluminum structures being assembled. While several cautionary caveats have been mentioned, by no means should the designer be reluctant to specify aluminum bolts; they are easy to use. I have been acquainted with or aware of the trouble-free use of literally hundreds of millions of aluminum bolts and other fasteners.

BIBLIOGRAPHY

Aluminum, Vol. 3, Fabricating and Finishing, American Society of Metals, Metals Park, OH.

Design Considerations for Aluminum Fasteners, SAE Paper 80455, Society of Automotive Engineers, Warrendale, PA, 1980.

Fastener Standards, Industrial Fastener Institute, Cleveland, OH.

3 Titanium Screw Joint Systems

Dario Croccolo and Nicolò Vincenzi

3.1 INTRODUCTION

Why use titanium screws? In recent years, the increased demand for lighter components in different sectors (aerospace, automotive, motorsport, and automatic machines) has led to the widespread use of lightweight materials, such as titanium alloys, and the need for material-compatible mechanical fasteners, namely titanium threaded fasteners.

Titanium is available on the market as a pure material or as an alloy. Pure titanium has a hexagonal Alfa-type crystalline structure which increases its mechanical strength (the Ultimate Strength UTS) depending on its classification grade (grade 1 the lowest, grade 4 the highest). With the addition of alloyed elements such as aluminum, vanadium, and molybdenum, titanium alloys turn to cubic crystalline structures of the Alfa+Beta type. These kinds of alloys are commonly used in mechanical and aerospace applications even though the most interesting ones are Ti6A14V (grade 5) and Ti3A125V (grade 9) which are used for bolts, mechanical parts, and frames. The recent increase in using titanium and its alloys is mainly attributed to their desirable properties, which are listed below:

- high strength-to-weight ratio
- good corrosive resistance
- high-temperature resistance
- good electric resistivity.

The manufacturing process of titanium screws must be carefully designed and precisely controlled as well as frequently inspected for quality and repeatability. In particular, the fastener head should formed by a hot forging process (not cold forging), due to the known undesirable plastic behavior of the titanium material, stemming from the grinding and the stress-relieving treatment, whereas some compressive residual stresses are usually left in the underhead fillet radius using the cold deformation to increase the fatigue strength. Screws are finally undergone by the thread rolling process, but after the heat treatment: the surface of the thread is, therefore, more accurate and smoother since the grain flow in the threads is continuous, with its maximum density at the bottom of the root radius and without the presence of tool marks typical of the metal-cutting operations.

Titanium threaded fasteners are mainly advocated for their lightness more than their frictional characteristics. Their replacement often does not seem to be as obvious neither does appear to be non-immediate. It is common knowledge that frictional coefficients are key parameters in the initial design, and subsequent behavior, threaded systems bolted joints design, changes drastically when switching from steel to titanium material. Some results concerning the frictional behavior of bolted joints involving titanium screws are, therefore, presented: friction and torque coefficients were calculated according to ISO 16047 for joints made up of a hexagon socket head screw made of titanium alloy, a bush made of aluminum alloy and a steel nut. Data were collected by performing several tightening tests on ad-hoc designed specimens, which allowed the evaluation of the different tribological behavior of the same joint under three different conditions of lubrication (dry, Teflon® added oil, and ceramic paste). Twenty repeated tightenings (re-tightenings) have been analyzed in order to simulate some maintenance operations. The design of experiment (DOE) method was applied to manage the tests, while the results were analyzed by the analysis of variance (ANOVA) and *P*-value

DOI: 10.1201/9781003501183-4

methods. Out of the two lubricants examined, the ceramic paste showed the best results in terms of friction coefficient constancy throughout the re-tightening operations as well as the best protection of the thread and underhead surfaces against wear.

3.2 INITIAL PRE-LOAD VERSUS TIGHTENING TORQUE

The bolt's initial load is normally achieved by using a torque wrench applied to the head or the nut so that the tribological aspects of the tightening phase are critical to define the actual torque-preloading relationship [1–3]. The tightening torque is mostly consumed in overcoming two friction components: the underhead friction due to the sliding of the fastener head on the flanges and the thread friction between the male and female threads. The residual torque component produces the fastener tension by generating the joint clamping force. Inaccuracies in determining the friction components may lead to an overestimation or underestimation of the bolted joint performances. The torque-preloading relationship is often simplified by using a constant K, known as torque coefficient or nut factor.

$$T = K \cdot F_V \cdot d \tag{3.1}$$

The expression for the torque coefficient is reported in Eq. (3.1), where T [Nmm] is the input (total) tightening torque applied to the fastener, F_V [N] is the needed preloading force, and d [mm] is the nominal screw diameter.

Some authors [4,5] suggest that an approximated or mean value of $K=0.20$ could be used, which holds true only in the presence of screws, nuts, and flanges made of steel. For this reason, some recent European standards prescribe that the appropriate torque coefficient must be provided by the bolt manufacturer [6–8]. Because the scatter of the friction coefficients is too great to provide a reliable value, especially for critical joints, and, furthermore, only steel screws are considered. Motosh [9] and VDI 2230 [10] proposed the full torque-preload relationship, and the metric (ISO) thread profile is given as follows:

$$T = F_V \cdot \left[0.16 \cdot p + 0.58 \cdot \mu_t \cdot d_2 + 0.5 \cdot \mu_u \cdot d_u\right] \tag{3.2}$$

p [mm] is the thread pitch, μ_t is the friction coefficient between male and female threads, d_2 [mm] is the mean thread diameter, μ_u is the underhead friction coefficient and du [mm] is the underhead mean diameter. By considering a M8×1.25 bolt (d_2=7.19 mm, d_u=11 mm) and a mean value for friction coefficients μ_t and μ_u equal to 0.15, it is easy to verify that 88% of the total input torque T is consumed in overcoming friction [1]. Unless the majority of bolted joints in mechanical systems still comprise steel screws, many kinds of lightweight materials have been recently adopted for screws construction [11,12]. Since the use of threaded fasteners is worldwide considered as an easy and simple way of joining components, sometimes designers do not take into account the changed boundary conditions correctly when replacing conventional steel screws with lightweight ones or simply when replacing the surface finishing of bolted components. In [1], the authors examined several failures occurred on clamped joints (Figure 3.1) due to the wrong estimation of friction coefficients. Such components belonged to front motorbike suspensions [13–15] and were made up of one or two galvanized steel screws acting on aluminum alloy. The thread and the underhead regions could be anodized or spray-painted: the DOE methodology was applied to determine reliable values for the friction coefficients (and therefore for preloading forces) involved in the joints in order to avoid failures.

Here a similar methodology is applied to some M8×1.25 socket head screws made of titanium alloy: because of the favorable ratio between mechanical strength and density, titanium screws are eligible for applications that require low weights and high performances, e.g., sports cars, where titanium screws are used to connect the wheel rim to the wheel hub (Figure 3.2).

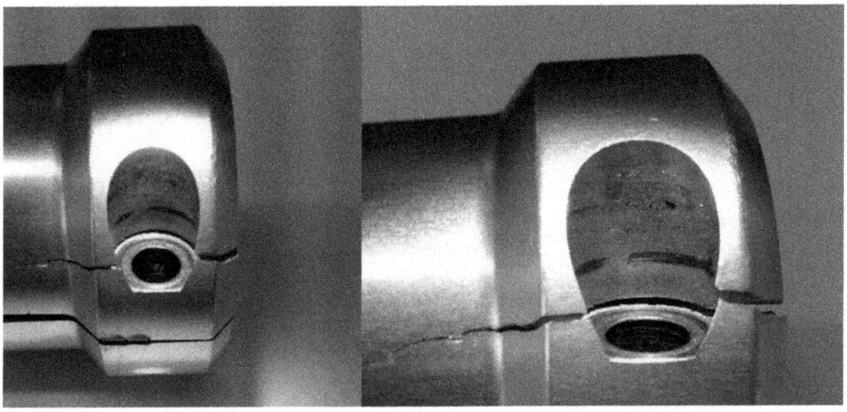

FIGURE 3.1 Failure occurred on a clamped joint connecting the front axle to the axle bracket of a motorbike.

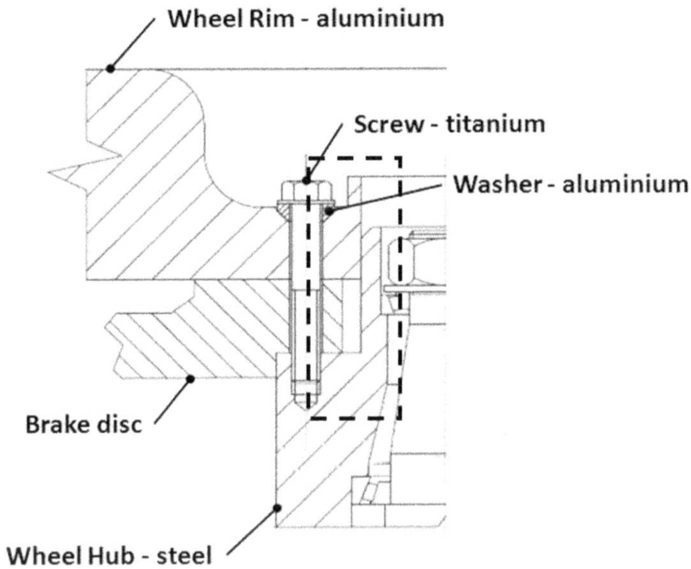

FIGURE 3.2 Car wheel assembly with titanium screw.

Moreover, titanium screws are often sold as aftermarket special parts, intended to replace steel screws on existing assemblies. Unfortunately, as well indicated in the next sections, the friction coefficients relevant to titanium screws are quite higher than those relevant to steel ones. Since the wrong estimation of friction coefficients may lead to catastrophic repercussions [15–18], the replacement of steel screws with titanium ones is not quite obvious and immediate because the same tightening torque prescribed for steel screws would produce a lower locking force if titanium bolts are adopted. To preserve the preloading force, it is, therefore, necessary to increase the torque applied during the tightening operation, even if the high values of friction coefficient, especially in the case of unlubricated surfaces, can lead to the socket head failure shown in Figure 3.3.

For this reason, the frictional behavior of a bolted joint made up of three elements, namely titanium screw, aluminum underhead, and steel threads, has been investigated. Such a combination approximates the behavior of the joint in Figure 3.2, and the results can be useful for car manufacturers. To fully exploit the capabilities of lightweight alloys and to avoid excessive wear,

(a)

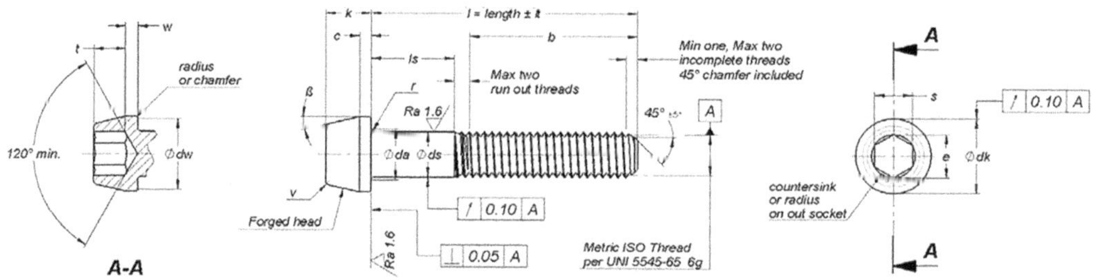

THREAD SIZE		b	β	c	Øda	Øds	Ødk	Ødw	ls	lt	e	k		r	v	s		t	w
EXT.	PITCH	SEE TAB 1B		+0.2 -0	MAX	+0 -0.1	+0 -0.15	MIN	MIN	±	MIN	MAX	MIN	MIN	MAX	MIN	MAX	MIN	MIN
M4	0.7	20 - 25	13°	0.8	4.7	4	7	6.53	ls = lenght – b – 2pitch	0.3	3.44	4	3.82	0.2	0.4	3.02	3.08	2.0	1.4
M5	0.8	20 - 25	13°	0.8	5.7	5	8.5	8.03		0.3	4.58	5	4.82	0.2	0.5	4.02	4.10	2.5	1.9
M6	1	22 - 25	13°	1.0	6.8	6	10	9.38		0.5	5.72	6	5.7	0.3	0.6	5.02	5.14	3.0	2.3
M8	1.25	26 - 30	13°	1.0	9.2	8	13	12.33		0.5	6.86	8	7.64	0.4	0.8	6.02	6.14	4.0	3.3
M10	1.5 - 1.25	26 - 30	13°	1.5	11.2	10	16	15.33		0.5	9.15	10	9.64	0.4	1	8.03	8.18	5.0	4.0
M12	1.75 - 1.5 - 1.25	26 - 30	10°	1.5	13.7	12	18	17.23		0.5	11.43	12	11.57	0.6	1.2	10.03	10.18	6.0	4.8
M14	2 - 1.75 - 1.5 - 1.25	26 - 30	10°	1.8	15.7	14	21	20.17		0.5	13.72	14	13.57	0.6	1.4	12.03	12.21	7.0	5.8
M16	2 – 1.75 – 1.5	26 - 30	10°	2.0	17.7	16	24	23.17		0.5	16	16	15.57	0.6	1.6	14.03	14.21	8.0	6.8

(b)

FIGURE 3.3 (a) Failure of a M8 hexagon socket conical head in a titanium screw and (b) screw dimensions.

lubrication during the assembly is suggested by standards [19,20]: therefore, the frictional behavior of the joint is assessed when different lubricants are applied to the mating parts before tightening. The evolution of friction coefficients over twenty assembly and disassembly operations was also observed during the experimentation in order to simulate the maintenance operations due to periodical tire substitution.

3.3 EVALUATION OF FRICTION COEFFICIENTS: THE INFLUENCE OF REPEATED TIGHTENINGS AND LUBRICATION

To evaluate the friction coefficients accurately, and to retrieve their mathematical expression, the DOE method was applied. After some screening analyses, a full factorial plane, characterized by two variables with three levels each, was prepared. To reduce the influence of noise (experimental error) and any non-investigated factors three replicas were carried out [1]. Then, a total of $3\times3^2=27$ experimental observations were performed and used in the DOE, whose parameters are indicated in Table 3.1. Due to the experimental observations needed in correspondence with the twentieth tightening operation for level 2 (see Table 3.1), a total of $27/3\times20=180$ different tests were performed.

$$T = F_V \cdot \left[0.16 \cdot p + 0.58 \cdot \mu_m \cdot d_2 + 0.5 \cdot \mu_m \cdot d_u\right]$$

$$\mu_m = \frac{T/F_V - 0.16 \cdot p}{0.58 \cdot d_2 + 0.5 \cdot d_u} \tag{3.3}$$

As already shown in [1] and according to VDI 2230 [10] and ISO16047 [20], Eq. (3.2) is rearranged into Eq. (3.3), considering both μ_t and μ_u equal to μ_m in order to derive the overall friction coefficient of the bolt/nut assemblies.

Each test planned by the DOE must provide a value of the overall friction coefficient μ_m (Eq. 3.3). To calculate μ_m, some specific specimens, sharing the key parameters of the actual components involved in the wheel-rim joint, were designed and realized (Figure 3.4).

The screw is a hexagon socket head M8×1.25 with a length $L=60$ mm, made up of Ti-6Al-4V alloy rolled after heat treatment and without any surface coating (roughness $R_a=0.8$ µm). Mechanical properties can be summarized by a yield strength in the range of 830–880 MPa, an ultimate strength in the range of 900–1.050 MPa, Young's modulus of 105 GPa, minimum elongation at break 10%, mass density 4.4 g/cm^3, and Vickers hardness 320. In terms of mechanical strength titanium screws can be considered similar to 10.9 steel screws. Ti-6Al-4V is the most commonly used titanium alloy, applied to many aerospace applications too [21]. The bush is made up of EN AW 7075 T6 aluminum alloy, without any surface coating (roughness $R_a=1.6$ µm). The nut is an ISO 4032—M8×1.25-8, without any surface coating. The specimen has two strain gauges applied 180° apart from each other, in order to clean the signal from undesired bending moments due to a possible misalignment between the bush and the screw axes. The lateral cuts allow clamping the specimen in a standard workbench vise: at the same time, they prevent the nut from rotating with respect to the specimen. Each specimen has two HBM 1-LY43 strain gauges with a 3 mm grid and 120 Ω resistance, arranged in a half bridge configuration. The total torque T is applied by means of the digital torque wrench with a measurement range between 10 and 120 Nm. The lubricants used for the experimentation are an industrial oil fortified with Teflon® (EP plus oil) and a ceramic paste suitable (also) for high-temperature applications (HT 1200). A picture of the experimental tools is reported in Figure 3.5.

TABLE 3.1
The Design of Experiment (DOE) Parameters: Variables and Levels

Variable	Level 0	Level 1	Level 2
A. Tightening number	First ($A=0$)	Tenth ($A=1$)	Twentieth ($A=2$)
B. Lubrication	Dry ($B=0$)	PTFE Oil ($B=1$)	Ceramic Paste ($B=2$)

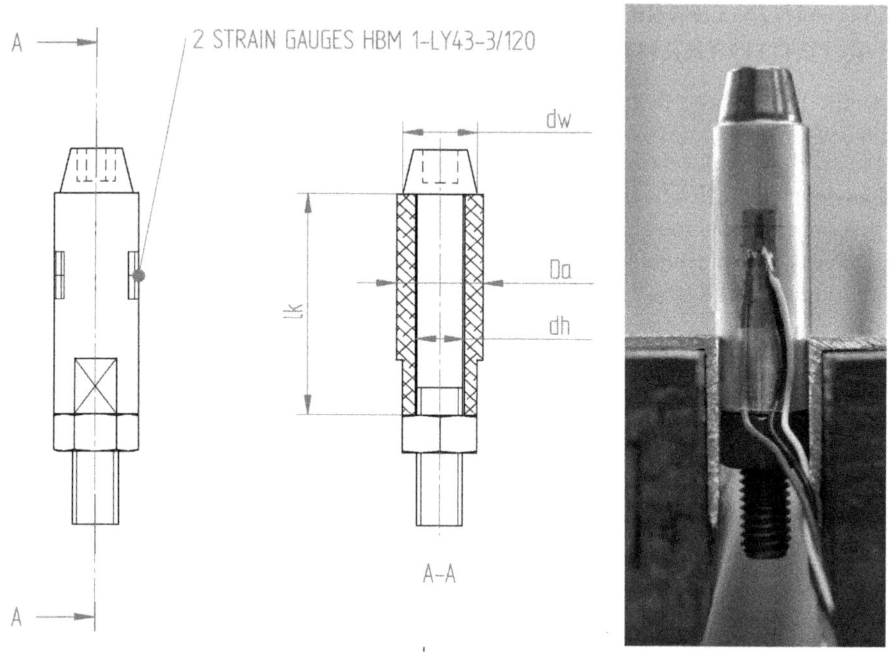

FIGURE 3.4 Test specimen.

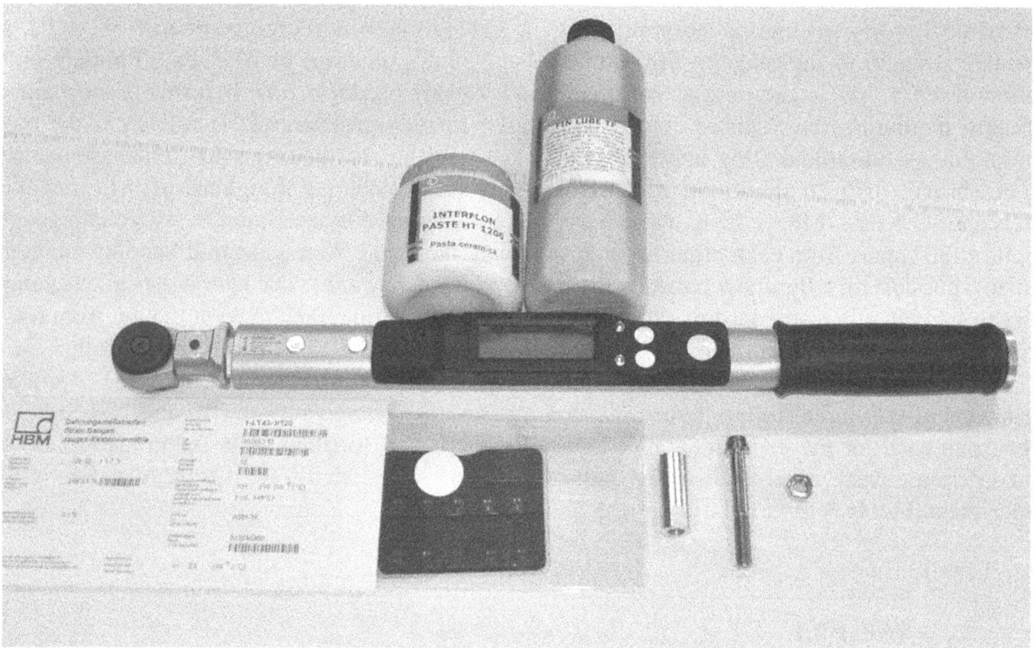

FIGURE 3.5 Experimental tools.

The preloading force F_V is evaluated using the strain gauges placed on the external surface of the specimen.

$$|F_C| = F_V = |\varepsilon_C| \cdot E \cdot A_{eq} \tag{3.4}$$

The compression force F_C (Eq. 3.4) acting on the specimen is equal, in magnitude, to the preloading force acting on the screw (the system works like a series of springs during the tightening phase).

E [MPa] is Young's modulus of the aluminum specimen (69,000 MPa) and A_{eq} [mm²] is the cylindrical equivalent cross section of the compressed region when the external diameter of the bush (D_A = 15 mm) exceeds the bolt head diameter (d_w = 13mm) as indicated in the drawing of Figure 3.4.

$$A_{eq} = \frac{\pi}{4}\cdot\left(d_w^2 - d_h^2\right) + \frac{\pi}{8}\cdot d_w\cdot\left(D_A - d_w\right)\cdot\left[\left(\sqrt[3]{\frac{l_k\cdot d_w}{D_A^2}} + 1\right)^2 - 1\right] = 120.83\ \text{mm}^2 \tag{3.5}$$

A_{eq} can be easily calculated using the formula proposed by [22] and reported in Eq. (3.5), as a function of the internal diameter of the specimen d_h=8.5 mm, the length of the specimen l_k=40 mm, and of the two parameters D_A and d_w.

$$A = \frac{\pi}{4}\cdot\left(D_A^2 - d_h^2\right) = 120\ \text{mm}^2 \approx A_{eq} \tag{3.6}$$

Since the specimen was expressly designed to be very similar to a thin bush, the equivalent cross section A_{eq} can be considered equal to the actual cross section A reported in Eq. (3.6): the difference is negligible.

During the screening tests, the ratio between T and F_V was checked to remain constant for tightening torque values T ranging between 10 and 50 Nm (see Figure 3.6)

Such a test was performed on a new specimen (first tightening) with dry surfaces. For subsequent tests, it was decided to set the tightening torque at T=25 Nm because this value is about in the middle of the measurement range of the torque wrench shown in Figure 3.6 and, moreover, it is low enough to avoid damaging the screw with an excessive axial load when lubricants are applied. According to the DOE methodology, the experimental tests were randomized [23,24]: the preloading force F_V was derived via strain gauges, whereas both the overall friction coefficient μ_m (Eq. 3.3) and the torque coefficient K (Eq. 3.1) were calculated as functions of the variables and levels reported in Table 3.1. Once the values of μ_m and K were calculated, their results were studied by applying the ANOVA [1,23,24]. Such a technique is based on a statistical approach and allows evaluating the effect of each variable and of their interactions, in order to identify the variables and interactions that are significant to define the friction coefficients value.

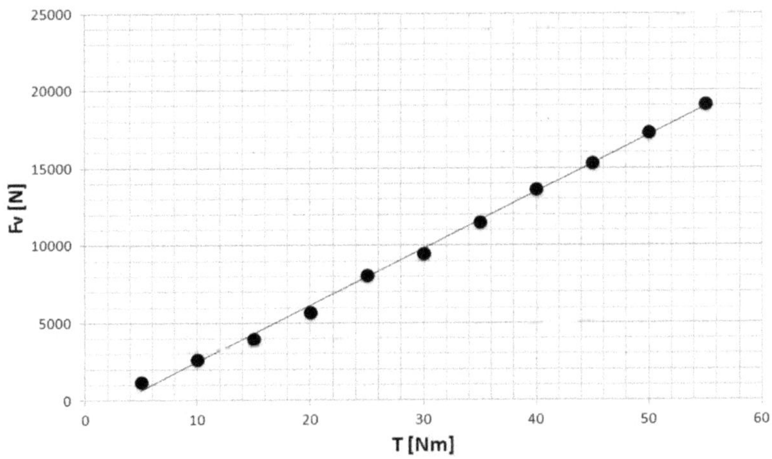

FIGURE 3.6 $T-F_V$ diagram for torque values between 10 and 50 Nm, unlubricated surfaces.

By means of the software Stagraphics®, the ANOVA table was generated, and the P-value test was done (see Table 3.2). All the five effects (singles and interactions) examined had P-values less than 0.05, indicating that they are significant in changing friction coefficients with a confidence level equal to 95%. Therefore, the mathematical models for μ_m and K were found: they are reported in Eqs. (3.7) and (3.8), respectively. In such expressions, each effect (or interaction) is weighted by an appropriate regression coefficient. A and B are the DOE parameters: they can be equal to 0, 1 or 2, depending on the number of tightenings and lubrication conditions (see Table 3.1).

$$\mu_m = 0.240 + 0.084.A - 0.156 \cdot B - 0.016 \cdot A^2 - 0.030 \cdot A \cdot B + 0.038 \cdot B^2 \quad (3.7)$$

$$\mu_m = 0.313 + 0.100.A - 0.184 \cdot B - 0.019 \cdot A^2 - 0.036 \cdot A \cdot B + 0.044 \cdot B^2 \quad (3.8)$$

For the sake of clarity, the μ_m and K values are calculated by means of Eqs. (3.7) and (3.8) depending on the different parameters are reported in Tables 3.3 and 3.4.

Looking at Tables 3.2–3.4 and at Eqs. (3.7) and (3.8), it can be appreciated how lubrication (B) is the most significant parameter. The evolution of the mean friction coefficient μ_m versus the tightening number, in the case of dry surfaces ($B=0$), is reported in Figure 3.7. In Figures 3.8 and 3.9, the same evolution is shown in the cases of EP oil (B=1) and HT1200 paste ($B=2$) lubricated surfaces, respectively. It is worth highlighting that each lubricant was only applied before the first tightening: the specimens were not re-lubricated during the twenty re-tightenings. In each diagram, a gray box highlights the mean value of μ_m, while the scatter bands show the extreme values recorded in the experimental campaign. The dashed line represents the values of

TABLE 3.2
Example of ANOVA Table Results for μ_m

Effect (Interaction)	Sum of Squares SS	DoF	Mean Square MS=SS/DoF	P-Value
A. Tightening number	0.0098	1	0.0098	0.000
B. Lubrication	0.2178	1	0.2178	0.000
AA	0.0015	1	0.0015	0.038
AB	0.0108	1	0.0108	0.000
BB	0.0086	1	0.0086	0.000
Error	0.0062	21	0.0003	

TABLE 3.3
μ_m Values for the Different Combination of Parameters

μ_m		$A=0$	$A=1$	$A=2$
Lubrication Condition		**Tightening Number**		
		1	10	20
$B=0$	DRY	0.240	0.308	0.344
$B=1$	EP Plus oil	0.122	0.160	0.166
$B=2$	HT 1200 paste	0.08	0.09	0.07

TABLE 3.4
K values for the Different Combination of Parameters

K		A=0	A=1	A=2
Lubrication Condition		**Tightening Number**		
		1	10	20
$B=0$	DRY	0.313	0.394	0.437
$B=1$	EP Plus oil	0.173	0.218	0.225
$B=2$	HT 1200 paste	0.121	0.130	0.101

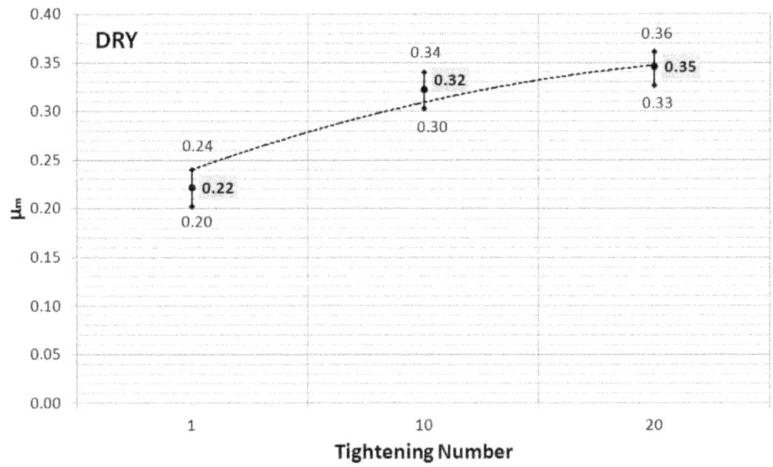

FIGURE 3.7 Friction coefficient μ_m versus tightening number for dry surfaces (B=0).

μ_m calculated by means of Eq. (3.7) once the lubrication parameter B is set equal to 0, 1, or 2 for dry, EP oil, and HT1200 paste, respectively. Equation (3.7) was verified to be effective for any arbitrary value of the tightening number parameter A between 1 and 20.

Referring to the left column of Figure 3.7 (first tightening), it can be noticed that μ_m takes a value of 0.22 ± 0.02 in the case of dry surfaces, which can be considered quite high. Moreover, Figure 3.7 shows a noticeable growth of μ_m along the whole series of twenty tightenings (+41% and +59% in correspondence of the tenth and twentieth tightening, respectively). Figures 3.8 and 3.9 show that the addition of a lubricant brings the mean friction coefficient down at first tightening: the lowest values of μ_m are found for the HT1200 paste. Throughout the re-tightening operations, the HT1200 paste shows no increase of μ_m while a sensible increase can still be observed in the case of EP Plus oil application (+33% and +42% in correspondence of the tenth and twentieth tightening, respectively). Of utmost importance is the fact that the scatter is negligible if the HT1200 paste is used: this condition allows getting always the same preloading force.

The thread friction coefficient (μ_t) and the underhead friction coefficient (μ_u), which combine to bring about the mean friction coefficient (μ_m), were measured as well, but they are not reported here for the sake of brevity. To retrieve the thread contribution, by nulling the underhead contribution of friction, a thrust bearing was applied under the screw head as indicated in Figure 3.10. The thread friction coefficient (μ_t) and underhead friction coefficient (μ_u) were then calculated from Eq. (3.2) following the indications of ISO16047.

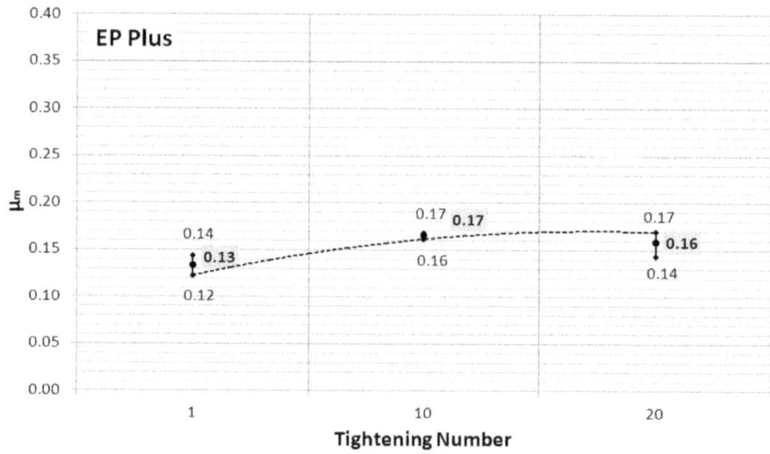

FIGURE 3.8 Friction coefficient μ_m versus tightening number for EP Plus oil lubricated surfaces ($B=1$).

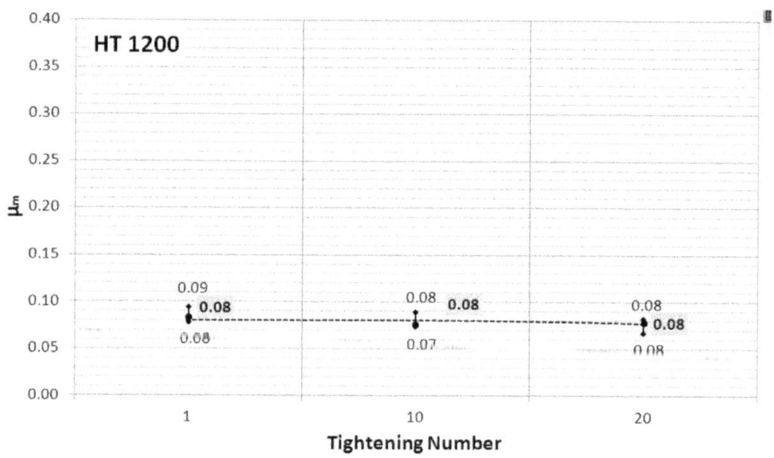

FIGURE 3.9 Friction coefficient μ_m versus tightening number for HT1200 paste lubricated surfaces ($B=2$).

During such a test, and only in the case of EP Plus oil addition, the stick-slip phenomenon was observed on about 20% of the performed tightenings so that, in this case, it is difficult to obtain a reliable value of μ_t. It is known from the literature [25] that stick-slip may appear in the presence of high pressure and low sliding velocity when the surfaces in contact have no coating and the joint is lubricated with mineral oil: such conditions match very well to the case under investigation. It is worth remarking that the tests performed without the thrust bearing never showed stick-slip phenomena and that the HT1200 paste avoided any stick-slip occurrence also when the thrust bearing was applied. As indicated in the diagram of Figure 3.11, the preloading force decreases drastically from the first to the twentieth tightening in the unlubricated specimens, provided that the tightening torque is constant.

This occurrence is due to the wearing and the spoiling of the thread and underhead surfaces when the re-tightening operations are carried out [1,26]. Both tested lubricants produce more leveled values of the preloading force since their protective film preserves the surfaces from the wearing. Such behavior is much more pronounced in the case of HT1200 paste, where the preloading force does even show a small increase during the twenty operations. The former statement can be

FIGURE 3.10 Thrust bearing under the screw head for μ_t evaluation.

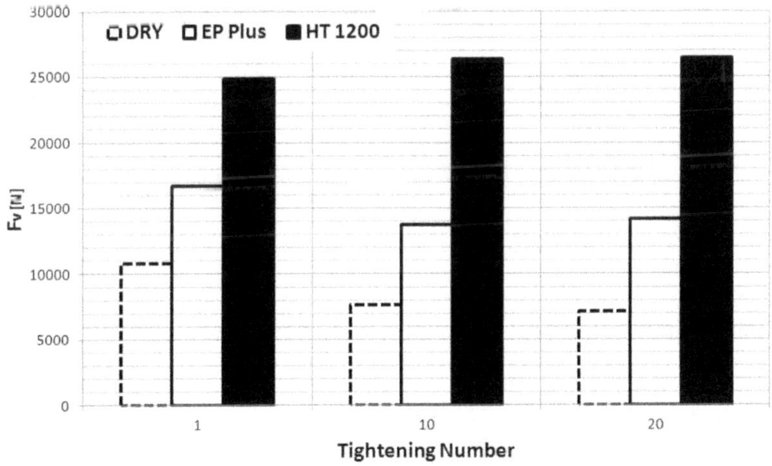

FIGURE 3.11 Preloading force F_V versus tightening number ($T = 25$ Nm).

easily confirmed by looking at Figure 3.12, where the different amount of wear on the screw threads after twenty re-tightening operations can be appreciated, by comparing the presence of HT1200 paste (a) with the absence of any lubrication (b).

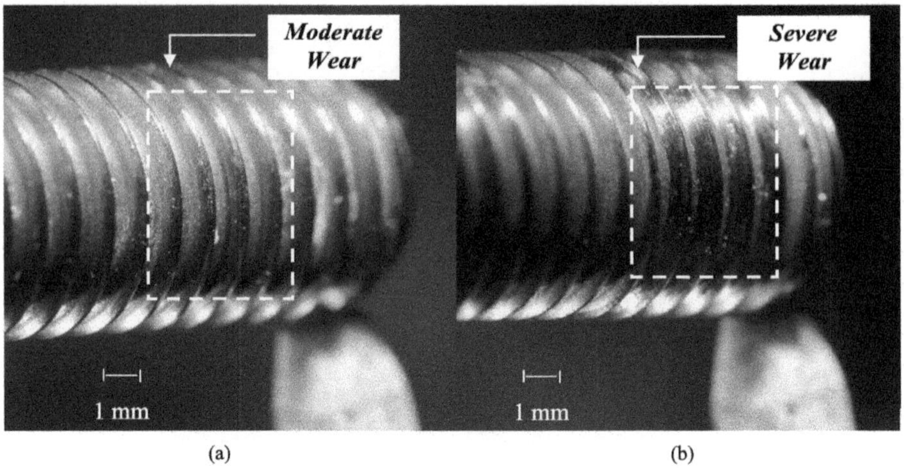

FIGURE 3.12 Effect of twenty re-tightenings on engaged threads: (a) screw lubricated with HT1200 paste and (b) dry screw.

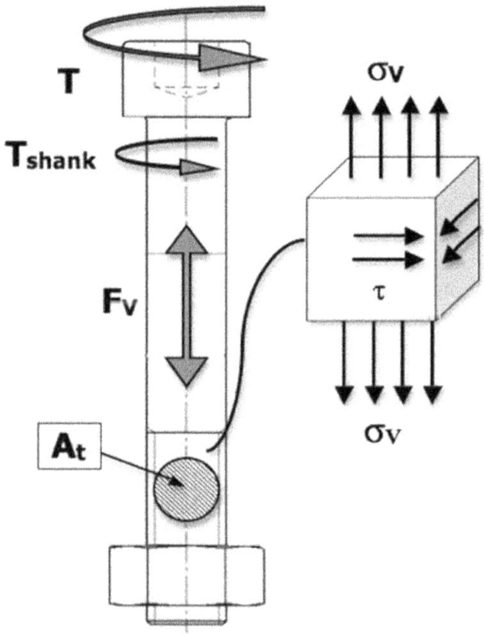

FIGURE 3.13 Tensile state on the screw shank during the tightening.

3.4 SCREW TENSION AFTER TIGHTENING

During the tightening operation of a screw, its required axial preloading is achieved by the applied tightening torque T. By analyzing the tensile state present on the screw shank during the tightening, it is possible to observe that besides the axial preloading stress σ_V [MPa] there is the additional shear stress τ [MPa] related to the portion of torque T_{shank} that crosses the underhead, as reported in Figure 3.13.

It is well known that upon the release of the torque wrench, the shank torsion reduces by a certain amount due to the elastic release of the screw-plates system. It is generally agreed that approximately half of the initial shank torque is released just a few seconds after the torque wrench removal. It is, therefore, advisable to calculate the stress state of the screw in two different ways: (1) at the end of the tightening with the total amount of the torsion and without considering the relaxation and the external load or (2) when the external loads are applied with a reduced torsion applied on the shank. In the second scenario, the external (axial) load increases the axial stress of the screw while the shear stress reduces by half of its initial value.

$$\sigma_{EQ} = \sqrt{\sigma_V^2 + 3\cdot\tau^2} = \sqrt{\left(\frac{F_V}{A_t}\right)^2 + 3\cdot\left(\frac{T_{shank}}{W_t}\right)^2}$$
$$= \sqrt{\left(\frac{4\cdot F_V}{\pi\cdot d_t^2}\right)^2 + 3\cdot\left(\frac{16\cdot F_V\cdot(0.16\cdot p + 0.58\cdot\mu_m\cdot d_2)}{\pi\cdot d_t^3}\right)^2} \quad (3.9)$$

It is, therefore, possible to calculate the equivalent axial stress on the shank at the end of the tightening σ_{EQ} [MPa], by applying the Von Mises criterion, according to Eq. (3.9), where d_t [mm] is the stress diameter related to the thread stress area A_t (Figure 3.13), and to compare such value with the screw yield strength.

For the sake of clarity, the coefficient of friction to be used in Eq. (3.9) is the thread coefficient μ_t instead of the mean one (μ_m) retrieved from the experimentation test; anyway, since the titanium screws are used in the presence of lubricant, the μ_m value is almost equal to μ_t and μ_u ones so that it can be adopted for calculating the equivalent stress.

To fully exploit the strength of the screw, the equivalent stress σ_{EQ} should be as close as possible to the axial preloading stress σ_V. In other words, the torsional (undesired) stress should be as low as possible: this condition is achieved by leveraging low values of friction coefficient. The three different lubrication conditions for titanium screws were, therefore, compared by calculating the stress ratio SR=σ_{EQ}/σ_V given by the ratio between the equivalent stress and the axial preloading stress; results are reported in Figure 3.14.

Automotive companies usually specify, for bolted joints, the acceptable coefficient of friction range to obtain a low preload scatter. For example, German manufacturers specify the range 0.09–0.14 [27] which implies a stress ratio SR limited to 1.28. Similarly, French manufacturers specify

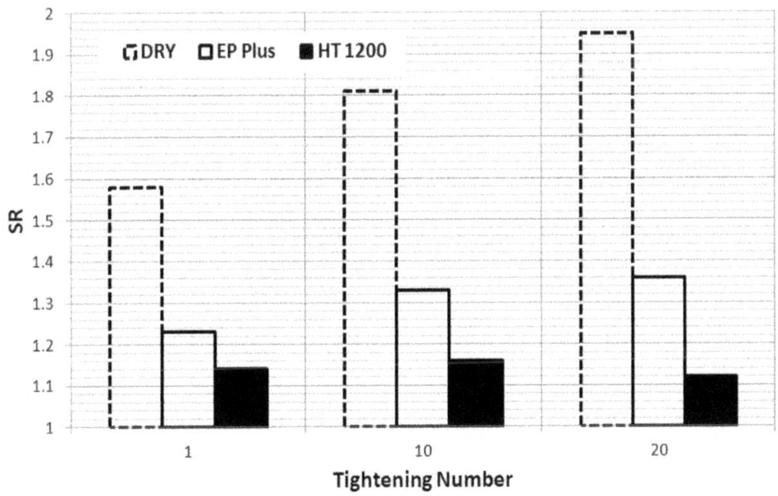

FIGURE 3.14 SR= σ_{EQ}/σ_V as a function of the number of tightenings, for the three different lubrication conditions.

the range 0.12–0.18 [26] which means having a stress ratio SR limited to 1.39. VDI 2230 [10] suggests the friction coefficient to be between 0.08 and 0.16 with a stress ratio SR up to the value of 1.33 represented by the dashed gray line indicated in Figure 3.14. The diagram in Figure 3.14 strongly advises against the use of titanium screws without lubrication (this advice can be found, for instance, in the Aerospace series standard EN 2491 [19]). By equating σ_{EQ} to the material yield stress or to the proof stress S_p [MPa] equal to 90% of yield stress S_y as suggested in VDI2230, it is possible to evaluate the maximum preloading force during the tightening phase as $F_{V_max}=(A_t{\cdot}S_p)/$SR. Engineering values of the parameter 1/SR for titanium screws are summarized in Table 3.5 (handbooks typically suggest tightening the value $F_V=0.70{\cdot}A_t{\cdot}S_p$). The results presented in Table 3.5 indicate that the torsional stress due to high coefficients of friction limits the preloading capability of dry screws: to reach the value $0.70,{\cdot}A_t{\cdot}S_p$ lubrication is mandatory.

It is worth noting that this kind of experimentation is quite different from the standard frictional tests, like pin-on-disk, which are usually based on low pressures and high sliding velocities between the tested materials, at least if compared to the bolt tightening phase. In the case of screws, a very different situation can be found: (1) the sliding velocity is limited to low values; (2) looking at the FEM contour plot of the tested titanium screws reported in Figure 3.15, the pressure distribution in the underhead or in the thread is far from a constant or regular distribution [26]. Up to now, a mean value of pressure is considered, and a mean value for friction coefficient is calculated, according to previous sections, whereas future investigations must take into account the local value of friction depending on the local pressure distribution.

TABLE 3.5
Maximum Preloading Forces F_{V_max} for Titanium Ti-6Al-4V Screws

	First Tightening	Tenth Tightening	Twentieth Tightening
Dry (unlubricated)	$0.63{\cdot}A_t{\cdot}S_p$	$0.55{\cdot}A_t{\cdot}S_p$	$0.51{\cdot}A_t{\cdot}S_p$
Lubricated with EP plus oil	$0.82{\cdot}A_t{\cdot}S_p$	$0.75{\cdot}A_t{\cdot}S_p$	$0.74{\cdot}A_t{\cdot}S_p$
Lubricated with HT1200 paste	$0.88{\cdot}A_t{\cdot}S_p$	$0.86{\cdot}A_t{\cdot}S_p$	$0.89{\cdot}A_t{\cdot}S_p$

Ti6Al4V: $S_{u\_min}=980$ MPa; $S_{y\_min}=860$ MPa; $S_{p\_min}=770$ MPa

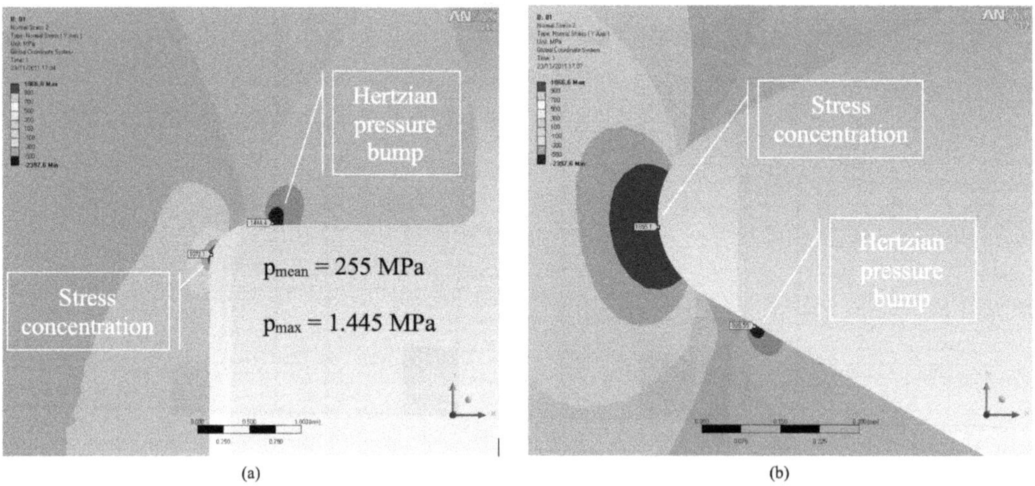

FIGURE 3.15 Stress and pressure distribution (F_V= 15kN, μ_m=0.08) in the surfaces in contact: (a) underhead and (b) first engaged thread.

3.5 EFFECT OF EXTERNAL TENSILE SERVICE LOADS F_V

In addition to the preloading of the threaded joint system at initial assembly, the system must also withstand subsequent external in-service loads F_v without failure. For this reason, it is important to carefully study the effects of the external (working) loads on the static and dynamic strength of the joint system. The well-known force-displacement diagram, reported in Figure 3.16, is useful to assist the reader.

During the initial tightening phase, the bolt (*B*) and the plates (*P*) share the same initial force F_i (the system works as a series of springs), to which corresponds different variations in length due to different stiffnesses. Once the components are clamped, an external load parallel to the bolt axis (R_n) produces an increase of the tension (ΔF_B) acting on the bolt as well as a reduction of the pressure (ΔF_P) acting on the plates: the amounts of ΔF_B and ΔF_P can be calculated as portion of the total external load R_n depending on the stiffnesses of two components (the clamped members work now as a parallel of springs, with the same displacement).

$$F_B = F_i + \Delta F_B = F_i + C \cdot R_n \tag{3.10}$$

The accurate definition (Eq. 3.10) of the maximum tensile load (F_B [*N*]) acting on the screw is important to select the appropriate screw.

$$C = \frac{k_b}{k_b + k_p} \tag{3.11}$$

The dimensionless parameter *C* used in Eq. (3.10) and defined by Eq. (3.11) is the so-called load factor, and it results as a function of the bolt (k_b [N/mm]) and plates (k_p [N/mm]) stiffnesses.

The use of titanium screws instead of the steel ones offers both the advantage of the weight and the maximum load (F_B) reduction. k_b is directly related to Young's modulus *E* [MPa] of the material: $E_{titanium}$~105GPa versus E_{steel}~210GPa. An example of stiffnesses and load factors © calculation is reported below for a M8(×1.25)×60 mm screw clamped on aluminum (large) flanges with a steel nut.

k_{b_steel} = 134 kN/mm
$k_{b\_titanium}$ = 71 kN/mm
$k_{f\_aluminum}$ = 277 kN/mm
$C_{steel\ screw/aluminum\ plates}$ = 0.33 vs $C_{titanium\ screw/aluminum\ plates}$ = 0.20

The ΔF_B in the case of the titanium screws is, therefore, 39% lower than the ΔF_B in the case of the steel ones.

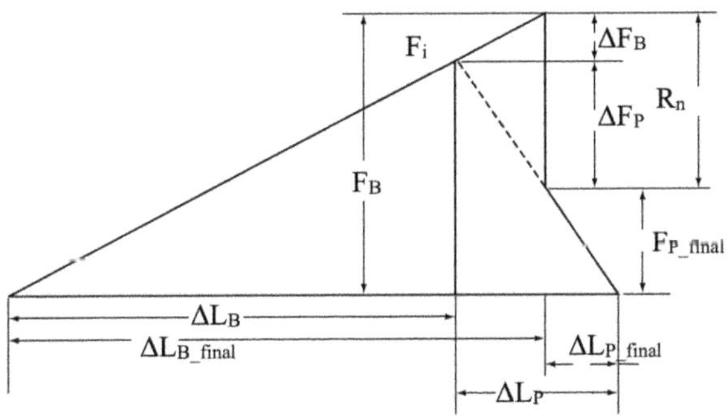

FIGURE 3.16 Force-displacement diagram: F_i is the initial (preloading force), F_B=F_i+ΔF_B is the actual load acting on the bolt during the service.

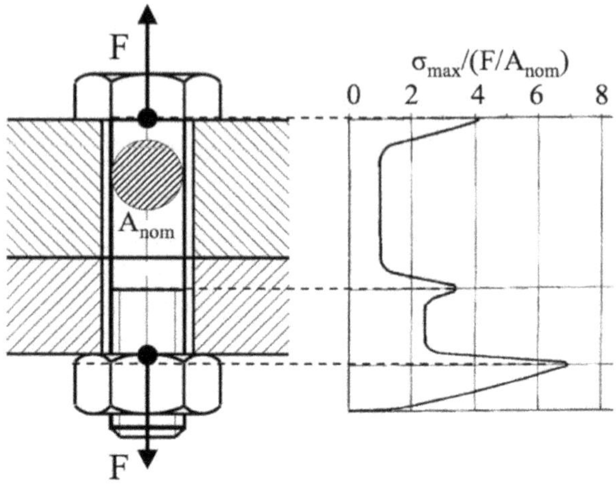

FIGURE 3.17 Stress trend along the screw shank.

On the other hand, the strength performances (UTS and Yield Point) of titanium screws made by Ti6Al4V are comparable to a 10.9 steel screw.

In many mechanical applications, the external load R_n represents a cyclic load for the clamped parts. In this case, the bolt fatigue cycle is composed of a minimum load F_{min} $[N]$ equal to F_i and a maximum load F_{max} $[N]$ equal to F_B (see Figure 3.16).

$$\begin{cases} F_{max} = F_i + \dfrac{\Delta F_B}{2} \\ F_{alt} = \dfrac{\Delta F_B}{2} \end{cases} \tag{3.12}$$

The mean F_{mean} $[N]$ and the alternate F_{alt} $[N]$ loads are, therefore, calculated in Eq. (3.12). The bolted connection has high-stress concentration factors, as reported in Figure 3.17, which limit its fatigue resistance [28].

To define the fatigue strength of the bolt, it is possible to apply the standard ISO3800, in which threaded fasteners are tested using the equipment represented in Figure 3.18 and are subjected to fluctuating tension-tension loads (tests with constant mean stress σ_{mean} or constant stress ratio $\sigma_{min}/\sigma_{max}$ may be used). Wiegand and Haas [29], in the early 1960s, performed a set of tests with a similar procedure on threaded fasteners made of different steels: the mean stress applied without plates between the bolt head and the nut (according to the recent ISO3800) was set equal to 350 MPa. The authors considered two different production processes even if the rolled process performed after the heat treatment is recommended: in this scenario, the alternate fatigue limit $\sigma_{a,SG}$ (MPa) is within the range ~60–100 MPa.

Some detailed and complete results useful for choosing steel bolt performances against high cycle fatigue ($N=2\times10^6$) are reported in the VDI-2230 [11], which prescribes to calculate the axial alternating stress $\sigma_{alt}=F_{alt}/A_t$ and compare it with a reference value that is valid when $\sigma_{mean}/\sigma_{p0.2}$ is within the range of 0.3–1.

$$\sigma_{altSV} = 0.85 \cdot \left(\frac{150}{d} + 45 \right) \tag{3.13}$$

$$\sigma_{altSG} = \sigma_{altSV} \cdot \left(2 - \frac{\sigma_{Mean}}{\sigma_p 0.2} \right) \tag{3.14}$$

FIGURE 3.18 Fatigue tests equipment for the threaded fasteners.

For bolt threads rolled before the heat treatment ($_{SV}$), the allowable stress amplitudes at 2×10^6 cycles are given by Eq. (3.13), whereas for bolts rolled after the heat treatment ($_{SG}$), the allowable stress amplitudes at 2×10^6 cycles are higher because the fatigue strength is positively influenced by the residual compressive stresses generated by plastic deformations; in this case, the reference values are given by Eq. (3.14). Equation (3.13) is independent of the initial load, and it is valid for bolt classes 8.8, 10.9, and 12.9; the factor 0.85 is inserted to take into account the scatter of the data. Equation (3.14), valid for the same bolt classes, depends on the mean (static) stress as the Haigh diagram well indicates.

Unfortunately, no systematic and standardized values can be found for screws made of lightweight material (e.g., aluminum or titanium alloys), but some consideration can be done about the titanium screws. In particular, according to the research on smooth and notched specimens (with particular reference to *V*-notched specimens, very similar to the thread geometry) made of titanium Ti6Al4V [30–32], it is possible to state that:

1. Titanium Ti6Al4V fatigue limit at 10^7 cycles is higher than the fatigue limit of steel (i.e., 34CrMo4 used for class 10.9) when the UTS is the same;
2. The fatigue stress concentration factor K_f of titanium alloy is lower than the fatigue stress concentration factor of steel in the presence of the same theoretical stress concentration factor K_t because the notch sensitivity $q=(K_f-1)/(K_t-1)$ is lower;
3. Threads on titanium screws are generally rolled after undergoing heat treatment so that the favorable grain structure strengthens the part against fatigue.

Based on the aforementioned considerations, the allowable fatigue stress amplitude (σ_{alt_SG}) for titanium screws rolled after the heat treatment is almost 20%–30% higher than the one reported in Eq. (3.14) for the steel ones.

In short, in the presence of high cycle fatigue stresses, titanium bolts present the double advantage of reducing both the load factor C and the fatigue stress concentration factor K_f if compared with the steel ones.

3.6 FINAL REMARKS

- Titanium Ti6Al4V has a Yield Point and UTS very close to class 10.9 steel and has, therefore, a better strength-to-weight ratio.
- The use of titanium screws without lubrication is not advisable due to the very high friction coefficients between the mating materials.
- In the presence of lubrication, it is possible to stabilize and align all the friction coefficients in the range of 0.07–0.17, considering also the repeated tightening, and to get an initial bolt load that is equal to 70%–90% of the proof load.
- The portion of external loads acting on the titanium screws is less than that applied to the steel ones due to their lower load factor.
- The fatigue strength of titanium screws is higher than that of steel ones due to their higher fatigue limit for both smooth and notched material.

BIBLIOGRAPHY

[1] D. Croccolo, M. De Agostinis, N. Vincenzi, Failure analysis of bolted joints: Effect of friction coefficients in torque–preloading relationship. *Engineering Failure Analysis* 18 (2011) 364–373.

[2] S.A. Nassar, H. El-Khiami, G.C. Barber, Q. Zou, T.S. Sun, An experimental study of bearing and thread friction in fasteners. *Journal of Tribology* 127 (2005) 263–272.

[3] S.A. Nassar, S. Ganeshmurthy, R.M. Ranganathan, G.C. Barber, Effect of tightening speed on the torque tension and wear pattern in bolted connections. *ASME Journal of Pressure Vessel Technology* 129 (2007) 426–440.

[4] R.C. Juvinall, K.M. Marshek, *Fundamentals of Machine Component Design*, 3rd ed., Wiley, New York (2000).

[5] S.A. Nassar, P.H. Matin, G.C. Barber, Thread friction torque in bolted joints, *ASME Journal of Pressure Vessel Technology* 127 (2004) 387–393.

[6] EN 1993-1-8, Design of steel structures Part 1–8: Design of joints (2005).

[7] EN 1090-2, Execution of steel structures and aluminium structures - Part 2: Technical requirements for steel structures (2008).

[8] EN 14399-1, High-strength structural bolting assemblies for preloading - Part 1: General requirements (2005).

[9] N. Motosh, Development of design charts for bolts preload up to the plastic range. *ASME Journal of Engineering for Industry* 98 (1976) 849–851.

[10] VDI 2230, Systematic calculation of high duty bolted joints—Joints with one cylindrical bolt (2003).

[11] C. Friedrich, Reliable light weight fastening of magnesium components in automotive applications, *2004 SAE World Congress*, Detroit, Michigan March 8–11 (2004).

[12] C. Friedrich, T. Gerhard, D. Koch, H. Kopfer, Threaded fastening systems with aluminium components—special characteristics for light weight design, *2009 European Aluminium Congress*, Düsseldorf, Germany, November 23–24 (2009).

[13] D. Croccolo, M. De Agostinis, N. Vincenzi, Recent improvements and design formulae applied to front motorbike suspensions. *Engineering Failure Analysis* 17 (2010) 1173–1187.

[14] D. Croccolo, N. Vincenzi, Tightening tests and friction coefficients definition in the steering shaft of front motorbike suspension. *Strain—An International Journal for Experimental Mechanics* 47 (2011) 337–342.

[15] D. Croccolo, M. De Agostinis, N. Vincenzi. An analytical approach to the structural design and optimization of motorbike forks. *Proceedings of the Institution of Mechanical Engineers, Part D: Journal of Automobile Engineering* 226 (2012) 158–168. doi: 10.1177/0954407011416178
[16] N. Eliaz, G. Gheorghiu, H. Sheinkopf, O. Levi, G. Shemesh, A. Ben-Mordechai, H. Artzi. Failures of bolts in helicopter main rotor drive plate assembly due to improper application of lubricant. *Engineering Failure Analysis* 10 (2003) 443–451.
[17] A.K. Jha, S.K. Singh, M.S. Kiranmayee, K. Sreekumar, P.P. Sinha, Failure analysis of titanium alloy (Ti6Al4V) fastener used in aerospace application. *Engineering Failure Analysis* 17 (2010) 1457–1465.
[18] M.E. Stevenson, J.L. McDougall, K.G. Cline, Metallurgical failure analysis of titanium wing attachment bolts. *Practical Failure Analysis* 3 (2003), 75–80.
[19] EN 2491, Molybdenum disulphide dry lubricants: Coating methods (1997).
[20] ISO 16047, Fasteners - torque/clamp force (2005).
[21] Y. Cui, Q. Meng, B. Zhang, F. Yan, Fracture analysis of Ti–6Al–4V bolts. *Engineering Failure Analysis* 13 (2006) 669–672.
[22] G. Niemann, H. Winter, B.R. Hohn, *Maschineelemente B.d. I.* Springer-Verlag, Berlin (2005).
[23] D.C. Montgomery, *Design and Analysis of Experiment*, 5th ed. Wiley, New York (2001).
[24] R. De Vor, T.H. Chang, J.W. Sutherland, *Statistical Quality Design and Control*, Maxwell Mcmillan International Edition, New York (1992).
[25] S. Baragetti, A. Terranova, M. Vimercati, Friction behaviour evaluation in beryllium–copper threaded connections. *International Journal of Mechanical Sciences* 51 (2009) 790–796.
[27] G. Dinger, C. Friedrich, Avoiding self-loosening failure of bolted joints with numerical assessment of local contact state. *Engineering Failure Analysis* 18 (2011) 2188–2200.
[26] W. Eccles, I. Sherrington, R.D. Arnell, Frictional changes during repeated tightening of zinc plated threaded fasteners. *Tribology International* 43 (2010) 700–707.
[28] H. Wiegand, K.H. Kloos, W. Thomala. *Schrauben-verbindungen.* Springer, Berlin Heidelberg, New York (2007).
[29] H. Wiegand, B. Haas. *Berechnung und gestaltung von schrauben-verbindungen.* Konstruktionsbucher, Springer, Berlin (1961).
[30] G. Owolabi, O. Okeyoyin, A. Olasumboye, H. Whitworth. A new approach to estimating the fatigue notch factor of Ti-6Al-4V components. *International Journal of Fatigue* 82 (2016) 29–34.
[31] S. Hosseini. Fatigue of ti-6al-4v. In *Biomedical Engineering - Technical Applications in Medicine* (2012). doi: 10.5772/45753
[32] Y. Okazaki. Comparison of fatigue properties and fatigue crack growth rates of various implantable metals. *Materials* 5 (2012) 2981–3005. doi: 10.3390/ma5122981

Part II

Processing of Fasteners

4 Fastener Manufacturing

Jesse A. Phebus and Peter F. Kasper

4.1 INTRODUCTION

To the design engineer, the primary concern is the specific unit under design, its function, and the components' ability to perform as intended, to achieve reliability, and meet and/or beat the total production cost parameters. Usually, the fasteners that hold the product and its components together are almost an afterthought.

When the design engineer gets to the final stages, the fasteners enter the picture. The designer develops the fastener package necessary to meet the needs of the completed unit and to install the unit so it will perform the intended function. Rarely is much thought given to where and how the fasteners are to be manufactured.

How important is the fastener manufacturing process? Knowing where and how fasteners are made can affect the strength, reliability, cost, and serviceability of the entire unit.

Fasteners are one the few universal components that must meet all the requirements of strength as they function in the ground, under water, in the air, and in outer space. They perform their intended function in any and all environmental conditions encountered.

A vast variety of materials, processes, and surface coatings come into play in fastener manufacturing processes. The manufacturer's knowledge and skills should be compatible with the fastener type, style, strength, and application, and the machinery and processes used in their manufacture should be suited to the specific fasteners required. Manufacturers do not all have the same capabilities, and few, if any, produce the full range of fastener products. Most are niche producers. By locating those producers best suited to your fastener design requirements, you can achieve the best quality, price, and service.

There are two categories of basic fastener manufacturing methods: forming and threading.

The primary forming methods are

1. Cold heading
2. Warm heading
3. Hot forging or forming
4. Turning or screw machining.

The primary threading methods are

1. Roll threading (forming or material displacement)
2. Cut threading
3. Ground threading.

4.2 FORMING METHODS

4.2.1 Cold Heading

Cold heading is a room temperature process. No temperature elevation device is used either internally or externally in the process. The forming process is a force, and the expended energy allows the material to flow into the shape of a fastener. The friction energy developed by the forming process generates sufficient heat to make the fastener feel hot to the touch.

DOI: 10.1201/9781003501183-6

4.2.2 Warm Heading

Usually, a header is equipped with an in-line heater, commonly an induction heater. The raw material or wire is preheated just prior to the forming process to a precalculated temperature that is dependent on the material alloy composition and the part configuration. This elevated material temperature is generally in the range of 425°F–1,100°F. Thus, the term "warm heading" has been applied to this process. The elevated temperature allows the header, with less force than is required for cold heading, to form the fastener material more easily and with less chance of head bursts or cracking. This process is commonly used when heading tougher materials such as stainless steels.

4.2.3 Hot Forging

With hot forging, the fastener product material is typically heated to 2,100°F–2,250°F. The heated material will develop a cherry red to cherry white appearance. This practice is used to form medium carbon to exotic high-temperature alloys. It allows the forging of fastener blanks with fewer forming stations and with smaller machines of less forming tonnage. The hot forging method is employed when more exotic alloys are used to impart high strength to fasteners for nuclear applications and aircraft or aerospace use and/or when extremely low and high-temperature fasteners are specified.

4.2.4 Turning or Screw Machining

A turning or screw machine uses full-size bar stock from which a fastener shape is generated through the removal of stock. Preformed shapes, such as hexagonal bar stock, are used from which hex head bolts and hex-shaped nuts are turned, bored, or drilled.

The advantages of this process are its compatibility with small runs, lower capital costs, and less variation in some product characteristics. The disadvantage is higher unit costs due to lower production rates per machine and higher material losses from the turnings removed from the full-size bar stock and a greater number of unusable bar ends.

4.3 THREADING METHODS

4.3.1 Roll Threading

Roll-formed threads are commonly formed on an extruded blank diameter close to the thread pitch diameter. Their strength is generally superior to that of other threads. During the roll forming process, the axial grain boundary flows and follows the individual thread contour, thus developing greater strength than a cut or ground thread. The thread surface finish is quite acceptable, better than that from cut threading but not as good as the surface attained with the ground threading method. The quality of the roll thread dies and the die alignment can result in thread flank, root, or crest discontinuities. High-strength critical parts can be threaded after the hardening operation, resulting in improved thread fatigue strength; however, this increases manufacturing cost.

4.3.2 Cut Threading

Cut threading is one of the oldest methods used to produce a thread. This process uses thread-cutting chaser sets or a turning operation such as a lathe and a cutting tool approach. This method is the least desirable from the point of view of thread finish quality and thread strength. Many small lot fasteners have cut threads. Cut threads are produced on full body diameter blanks where the major diameter of the thread and the full body diameter are the same.

4.3.3 Ground Threading

The ground threading method results in a high-quality thread with respect to dimensional control, thread strength, and surface finish. The ground thread is usually created on hardened and contour-ground fastener blanks. This method is more costly than roll forming and cut threading.

4.4 PRODUCTION SUPPORT ELEMENTS

The factors discussed in this section contribute to the manufacturer's ability to produce cost-effective quality fasteners.

4.4.1 Raw Materials

A wide range of materials are used to produce fastener products, both ferrous and nonferrous. The material strength requirements are specified for all products that are made to specific national, international, and/or industry standards. In addition, special fasteners can be designed to meet an application's requirements.

Generally, the fastener is specified to meet specific mechanical properties, and the manufacturer can use a range of material chemistry in achieving these properties.

4.4.1.1 Cold Heading Quality and Scrapless Nut Quality Material

Certain elements are necessary to produce cost-effective quality products. If a mill is to supply cold heading quality (CHQ) rod or scrapless nut quality (SNQ) material, this in itself defines a particular specification; in most cases, standard grades are purchased. The material specification controls the chemistry, surface discontinuities, grain size, and general quality for cold heading.

Wire processing plays a vital role in the material quality required for heading. The heading operation requires close tolerance wire size and a smooth coated surface finish. Hence the cleaned and annealed green rod from the mill will be cold drawn into wire suitable for the cold heading process. There are three key factors in the rod-to-wire conversion:

1. Mill scale is removed by acid pickling or mechanical descaling.
2. Annealing or in most cases spheroidizing annealing is required to soften the material prior to cold heading.
3. The wire drawing, usually done with carbon dies, results in size control and also provides the lubricated surface necessary to prevent galling in the header-forming tools.

4.4.1.2 Warm Heading

Warm heading materials are the same as those used for cold heading. Some exotic and high-end alloy materials that are not easily cold formed can be cold headed through the use of warm heading.

4.4.1.3 Hot Forging or Forming

For hot forging or hot forming, hot rolled rod, bar, or cold finished material is used in either cut-to-length bars or coils. There is a significant cost savings compared to CHQ/SNQ materials and screw machine bar stock.

4.4.1.4 Turning or Screw Machining

In most cases, cold-drawn finished bar material is selected from specific material grades that are easier to machine, sometimes called free machining materials.

4.4.2 Tool Control

In heading operations, the forming tools control most fastener dimensional characteristics. Tool changes contribute to most machine downtime; thus, a sound tool control program is important.

4.4.3 Machines and Equipment

The manufacture of fasteners has evolved over hundreds of years. As indicated below, there are a great variety of machine types. In addition, each narrow range of fastener diameters requires a machine of a specific size. There are other characteristics also that enter into the realm of the designer and purchaser for consideration. These are covered below for each of the primary forming methods.

4.4.3.1 Cold Heading

The equipment in use today for cold heading can range from a vintage date of 1918 to state-of-the-art machinery. Modern equipment can include load-sensing or load-monitoring devices, tool quick-change or automatic changing features, short feed detection, thread form monitoring devices, etc. Table 4.1 presents a list of equipment types and typical products for which each is used.

4.4.3.2 Warm Heading

Warm heading technology has advanced to temperature-sensing devices that automatically maintain the temperatures desired.

4.4.3.2.1 Heading Equipment

For warm heading, the equipment is the same as that used to cold head the part headed blank. Depending on the material alloy and the part configuration, a slower cold header may result in a

TABLE 4.1
Cold Heading Equipment

Equipment Type[a]	Typical Product
Single-die, single-blow	Nails, rivets
Single-die, two-blow	Blanks with simple round head with phillips, struck slot, etc.
Two-die, three-blow	Shank extrusion, square under head, more than 2.5 wire diameters head volume, large thin heads
Boltmaker (or similar machines), four or five dies and matching hammers with integrated machine point and roll thread capability	Cap screws or similarly shaped parts
Nutformer (or similar machines)	Hex, flange, and lock nut blanks
Progressive headers, three to eight dies with matching hammer stations	Complex multiple shank extrusions, backward extrusions; complex head shapes with up to 6 wire diameters of head volume material
Long-stroke headers	To run parts with approximately double the under-head length of standard or highspeed headers
Open-die and/or rod headers suitable for basic parts, an under head fin is present, caused by split heading dies	Very basic parts of longer lengths such as appliance tie rods
Parts former; generally not a threaded product. Short-stroke machines with high tonnage and multiple forming stations.	Run part blanks such as gears, bearing race, parts that may generate some rough finished size, and surfaces that are later machined or ground

[a] Boltmaker and Nutformer are trade names of National Machinery Company, Tiffin, Ohio.

more effective forming process, one that will produce better yields than high-speed headers. The heated raw material will generally respond better to the more gradual application of force, thus allowing the tougher metal alloys to flow without cracking and avoiding excessive tool breakage and wear.

4.4.3.2.2 *Heating Equipment*

There are a number of induction heating equipment suppliers. Most of the equipment generally performs satisfactorily. The equipment selection should be based on the specific needs and the overall warm heading requirements of the fastener producer.

4.4.3.3 Hot Forging and Hot Forming

Hot forging and hot forming started with the blacksmith's technology for heating metals. Today the metals are heated electrically by a series of coils or heater boxes depending on the specific hot forge manufacturing method employed. Bolts are generally forged by heating in induction coils, the area heated depending on the volume needed to form the head. Nuts are forged either from cut-to-length bars or coils of mill-supplied green rods. In either case, the nut material is heated as the material is fed into the machine. A few producers may be using gas-fired furnaces to heat the metals for forging.

4.4.3.3.1 *Hot Forging or Externally Threaded Fasteners*

The types of machines and processes in use range from one step above the blacksmith era to a semi-automatic in-line process linking individual operations.

Equipment Type	Typical Product
Single-die, two-blow—split or solid die, no point	Square and hex head bolts; cut thread may be pointed for thread start
Single-die, two-blow; solid die, extrusion for thread, pointer	• Cap screws; parts with washer face under head, pointed and roll thread start

The above equipment can be either horizontal headers or vertical presses. These are hand-fed or may be linked together with some degree of automation. The quality and cost can vary significantly from producer to producer.

More complex hot forged products require additional operations such as descaling of the hot-rolled bars for either surface peeling and turning or possible cold drawing. Sometimes the full body diameter and/or washer face is ground.

4.4.3.3.2 *Hot Forming of Internally Threaded Product (Nuts)*

Generally, nuts from ¾ to 1⅞ in. in diameter are produced on hot nut formers from either cut-to-length hot-rolled bars or hot-rolled coils. Some smaller diameter and larger diameter parts such as ⅝ in. and 2 in nuts are run on nut-forming machines. The manufacturing methods in use can be quite varied. Most producers are using hot nut formers equipped with in-line heaters. Some use vertical presses to form and pierce the nut blanks; large nuts, 2 in. and up in diameter, are formed on presses that may or may not have some degree of automation.

Internal threads are tapped. There are a number of OEMs who manufacture and supply this equipment to the industry. Some large nuts that are hot formed will have machined threads. Very large nuts, i.e., 6–12 in. in diameter, are machined complete including the threads.

4.4.3.4 Turning or Screw Machining

The turning or screw machine is used in the production of small quantities of special nuts, cap screws, or bolts—areas where traditional high-volume fastener producers cannot compete in price, service, or delivery.

The types of equipment employed to machine threaded fasteners from metal bar stock will depend on the type of product to be produced.

Machine Type	Typical Product
Turning lathe	Small quantities of very special nonrepetitive externally threaded parts
Turret lathe	Small quantities of very special nonrepetitive nuts
Screw machines	This family of machines can generally produce any fastener that is economically feasible

Depending on the quantity and complexity of the product, these machines are manually operated or computer-controlled.

4.4.4 Management Committed to Customer Satisfaction

Most cost-effective quality manufacturers subscribe to and practice a management system that promotes the attitude that customer satisfaction is of topmost importance to the company. For those seeking the best manufacturer of the fasteners required in their designs, the choice of a company certified to ISO/QS 9000 or an equivalent quality management philosophy is a good start in the selection process.

5 Fastener Coatings

John Laurilliard

5.1 INTRODUCTION

This chapter covers various standard platings normally applied to high-strength precision military and aerospace threaded fasteners and briefly cites other nonelectrodeposited coatings also commonly applied to this class of fasteners.

While the information supplied on coatings, techniques, process control, and testing applies to high-strength precision-threaded fasteners, it is also applicable to less costly commercial fasteners. The same problems encountered when coating precision fasteners also occur with the less precise variety but are often overlooked or not detected because of economics and their noncritical applications.

5.2 SELECTION OF COATINGS[1]

Threaded fasteners are such simple mechanical devices that they are often overlooked during the initial design stages of the complex mechanical equipment that they must hold together. Without adequate engineering information, the choice of coating is often given even less consideration and is normally made on the basis of corrosion resistance alone. This simplistic approach does not consider many critical physical, mechanical, and chemical properties threaded fasteners must have, especially the high-strength precision variety. Therefore, the selection of electrodeposited coatings requires specialized engineering knowledge.

Ideally, the fastener coating should provide useful functional properties not provided by the bare alloy, but it should also not cause significant degradation of the substrate or the joint structure.

Some factors to consider when choosing a fastener coating are listed in Table 5.1.

At least twenty-four metals and alloys (see Table 5.2) with widely different chemical and physical properties are commercially electrodeposited. The appearance and physical properties, and to a lesser extent the chemical properties, of fasteners may be varied by altering the plating bath formulation and the operating conditions such as temperature and current density. As an example, nickel may be plated as a bright, highly stressed decorative coating with a hardness of 650 Vickers or as a nonreflective, soft, low-stress or stress-free coating with a hardness of 130 Vickers.

Additionally, special properties may be produced by the codeposition of two metals that form alloys, some of which can be produced only by electrodeposition and not through standard metallurgical techniques. Nickel-zinc alloy coatings have chemical properties between those of the two-component metals. Although not as good as either metal alone, the alloy has better high-temperature oxidation resistance than zinc and better atmospheric corrosion resistance than nickel. However, nickel-tin deposits have properties unlike those of either of the alloying metals alone. This alloy has a pleasing bright pinkish appearance and approaches the noble metals in corrosion resistance.

The ideal fastener coating would have good atmospheric and marine corrosion resistance, a high melting point, and good lubricity, have no adverse effect on fatigue, and not cause undue galvanic corrosion of the joint material. The technique of application should produce a thin uniform adherent coating and be capable of coating both large and small complex shapes made of widely different metals and alloys. Most importantly, the process should cause little or no hydrogen embrittlement, and the product should experience minimal reduction of fatigue life due to surface preparation.

Just as no one alloy or material can meet every bolt application, no one coating will meet all the requirements of the ideal coating. But by knowing the fastener application and intended environment, an intelligent choice can be made (Table 5.3).

DOI: 10.1201/9781003501183-7

TABLE 5.1
Factors Affecting the Choice of Fastener Coatings

Coating Factors

1. Corrosion resistance—environment
2. Temperature limitation
3. Effect on fatigue life
4. Effect on torque vs. induced load—lubricity
5. Effect on locking torque
6. Stress corrosion resistance—environment
7. Galvanic compatibility with joint material
8. Ability to relieve hydrogen embrittlement
9. Abrasive resistance
10. Post-plate hydrogen embrittlement
11. Stress alloying
12. Cost

Application Technique Factors

1. Degree of induced hydrogen embrittlement
2. Effect on fatigue life
3. Dimensional change or degradation of basis metal during preparation
4. Thickness and distribution of coating
5. Adhesion of coating
6. Limitation of coating method because of fastener material, size, and shape complexity
7. Cost

TABLE 5.2
Fastener Coatings[a]—Methods of Application

Metallic Coatings				
Method of Application				
Electrochemical Electrodeposits		Chemical		
Unalloyed	Alloys	Electroless	Immersion	Physical (Various[b])
Cadmium	Cadmium-titanium	Nickel-phosphorus	Copper	Vacuum cadmium
Chromium	Copper-zinc	Nickel-boron	Gold	Ion vapor aluminum
Cobalt	Copper-tin	Cobalt-phosphorus	Tin	Hot dip aluminum
Copper	Nickel-zinc			Hot dip tin
Gold	Nickel-iron			Hot dip zinc
Indium	Nickel-cadmium (diffused)			Mechanical cadmium
Iron	Tin-cadmium			Mechanical cadmium-tin
Lead	Tin-lead			Mechanical tin
Nickel	Tin-nickel			Mechanical zinc-tin
Palladium	Tin-zinc			
Platinum				
Rhodium				
Silver				
Tin				
Zinc				

(Continued)

TABLE 5.2 (*Continued*)
Fastener Coatings[a]—Methods of Application

Metallic Coatings Method of Application					
Electrochemical Electrodeposits		Chemical			Physical (Various[b])
Unalloyed	Alloys	Electroless		Immersion	
Nonmetallic coatings					
	Inorganic	Method of application		Organic/inorganic	
	Chemical			Physical	
Electrochemical (anodize)	Oxide	Phosphate[c]	Chromate	Lubricants	Corrosion-resistant
Aluminum	Steel	Steel	Cadmium	Oils (various)	Paints (various)
Magnesium	Copper	Cadmium	Beryllium	Long chain, alcohols	Zinc chromate
Titanium	Stainless steel	Zinc	Aluminum	Molybdenum disulfide	Fluoropolymers
Zinc	(passivation)	Aluminum	Silver	Graphite-grease	Aluminum-ceramic
		Titanium	Zinc		Zinc flake-organic

[a] Consult ASTM Standards Vol. 02.05 [1] for a complete list and cross index of specifications governing metallic and inorganic coatings.
[b] Other physical methods include flame and plasma spray, pack cementation, and sputtering of various metals and alloys.
[c] Mn, Zn, Fe, F.

For greater detail on the properties and method of application of various finishes, consult Refs. [2–6].

5.3 PRINCIPLES OF ELECTROPLATING

Electroplating is a common coating technique used on both industrial and commercial products and may be defined as an electrochemical technique of applying a thin metal coating onto another metal. A short list of familiar items would include small appliance items such as toasters and electric coffee pots, automotive trim, plumbing fixtures, silverware, jewelry, copper-clad pots and pans, printing rolls, diamond drills, tin-plated steel, printed circuits, bearings, and the ubiquitous nuts and bolts.

To the uninformed individual, plating and metal finishing are often thought of as a low-tech dip-and-swish operation. But in reality, they are complex processes involving almost every field of the physical sciences: inorganic, organic, electro-, physical, and analytical chemistry; physics; metallurgy; mechanics; and electronics.

Platers often use and are exposed to some of the most deadly and hazardous chemicals known; cyanides, caustic soda, sulfuric, nitric, and hydrofluoric acids, and others, as shown in Table 5.4. A good plater must know detailed techniques, learned and acquired from years of on-the-job experience, especially for manual operations.

5.3.1 Mechanism of Electrodeposition

A simple example of an electroplating process is nickel plating. A standard nickel plating solution consists of approximately 48 oz/gal of nickel sulfate; 6 oz/gal of nickel chloride, and 5 oz/gal of boric acid, with the pH adjusted to 3.0–3.5 and operated at a temperature of 130°F.

Nickel sulfate and nickel chloride, when dissolved in water, form positively charged nickel ions and negatively charged sulfate and chloride ions, respectively. The article to be plated serves as the cathode, and the anode is normally a slab of nickel metal placed opposite the workpiece.

TABLE 5.3
Properties of Metallic Coatings

No.	Coating	Specification	Application Method	Type of Solution	Important Properties of Coating or Method	Limitations	Costs
1	Aluminum	MIL-C-83488	IVD	—	A. Atmospheric corrosion resistance B. High melting point, 1,220°F C. Galvanic compatibility with aluminum joints	D. Need lubrication E. High cost F. Special equipment G. Abrasive blast prior to coating	High
2	Cadmium	QQ-P-416; ASM-2400; ASTM-B-766; ISO-2082	Ed	Bright cyanide	A. 1A, C above B. Uniform deposits C. Coverage in recesses D. Precise control of dimensions E. Lubricity F. Thin coatings G. Solderability	H. Use below 450°F I. Low MP (610°F) J. H2 embrittlement relief difficult K. Vaporizes in vacuum L. Stress alloying in titanium at low temp. (125°F) M. Stress alloying in steel above 450°F N. Toxic; not for food use	Low
3	Cadmium	QQ-P-416; ASM-2401; MIL-STD-870	Ed	Unbrightened cyanide	A. Porous deposit allows H2 embrittlement relief B. 1A, C; 2E above	C. 1G; 21, K, L, M, N, above D. Carbon filtration for organics removal E. Causes H2 embrittlement during use from corrosion F. Less thickness uniformity than No. 2 G. Little or no coverage in recesses H. Must be rack plated; barrel plating peens deposit, closing porous deposit	Higher than No. 2
4	Cadmiumtitanium	MIL-STD-1500; AMS-2419	Ed	Unbrightened cyanide with titanium additive	A. Low H2 embrittlement B. Better corrosion resistance than No. 3 C. 1A, C; 2E above	D. 1G; 2H, I, J, L, M, N above E. Elaborate titanium additive system and controls F. Initial strike required for coverage G. Adhesion difficult on high chromium alloy steels	Higher than No. 3
5	Cadmium	NAS 672	Ed	Fluoborate	A. Low H2 embrittlement B. High current densities possible	C. 2H, I, K, L, M, N; 3F, G above	Medium
6	Cadmium	MIL-C-8837	Vac	—	A. No H2 embrittlement B. No post-plate bake	C. 2H, I, K, L, M, N; 3F, G above D. Costly vacuum equipment	High

(*Continued*)

TABLE 5.3 (*Continued*)
Properties of Metallic Coatings

No.	Coating	Specification	Application Method	Type of Solution	Important Properties of Coating or Method	Limitations	Costs
7	Cadmium	MIL-C-81562	Mech	—	A. Low H2 embrittlement	B. 2H, I, K, L, M, N above C. Thin deposits on sharp edges and recesses D. Possible nicking on large parts from tumbling	High
8	Chromium, hard (thick)	QQ-C-320; AMS-2406; ASTM-B650	Ed	Chromic acid	A. Hard (Rc70); wear-resistant B. Thick deposits, >0.001 in.	C. Heavy buildup on high current density areas. D. Often requires post-plate grinding for deposit uniformity E. Normally on bolt shank and under head F. Shot peening required to minimize fatigue loss G. Requires application of stop-off lacquers	High
9	Chromium, decorative	ASTM-B456	Ed	Chromic acid	A. Bright cosmetic appearance	B. Applied over copper/bright nickel plating C. Normally racked, but small parts can be barrel-plated	Medium
10	Copper	MIL-C-14550; AMS-2418; ASTM-B734	Ed	Cyanide or acid sulfate	A. High conductivity B. Flash plate for increased adhesion prior to silver C. Lubricity for extrusion D. Stop-off for nitriding	E. Can cause H2 embrittlement	Medium
11	Gold	MIL-G-45204; AMS-2425; ASTM-B488; ISO-4523	Ed	Cyanide	A. Oxidation-resistant B. High infrared reflectivity	C. 10E above	Very high
12	Nickel	QQ-N-290; AMS-2424; ASTM-B689	Ed	Sulfamate, chloride, sulfate/chloride	A. Oxidation resistance B. Low stressed deposits with sulfamate for minimal fatigue strength loss C. Used under silver plate for adhesion	D. 10E above E. Less deposit uniformity than cyanide cadmium	Medium

(*Continued*)

TABLE 5.3 (*Continued*)
Properties of Metallic Coatings

No.	Coating	Specification	Application Method	Type of Solution	Important Properties of Coating or Method	Limitations	Costs
13	Nickel-cadmium, diffused	AMS-2416	Ed	Sulfa mate nickel + cyanide cadmium + 630°F diffusion	A. Oxidation-resistant to 900°F B. Good atmospheric corrosion resistance	C. Minimum nickel required in recesses D. Control nickel deposit stress E. Loss of fatigue strength if diffusion is delayed beyond 1 h	High
14	Nickel-phosphorus (electroless nickel)	MIL-C-26074; AMS-2404; ASTM-B656	El	Hypophosphite, sulfate	A. No electric current required B. Uniform deposits in recesses and threads C. Deposits can be hardened to 1,000 Vickers by a 1 h, 750°F heat treatment	D. Solution control critical for deposit properties E. Low deposit ductility F. Short solution life	High
15	Silver	QQ-S-365; AMS-2410; AMS-2411; ASTM-B700	Ed	Cyanide	A. Lubricant for high-temperature alloys B. Uniform thickness distribution C. High melting point (1,761°F) D. High load carry capacity E. Heat and electrical conductivity	F. Requires nickel and silver strike prior to main deposit G. Prolonged exposure to temperatures above 450°F in the air reduces adhesion H. Oxygen penetrates silver at high temperature (>450°F)	High
16	Tin	QQ-T-435; MIL-T-10727; AMS-2408; ASTM-B545	Ed	Alkaline stannate	A. Lubricity B. Soft-seals thread C. Lower evaporation rate in vacuum than Cd, Zn, Pb, and Ag D. Uniform thickness distribution E. Good coverage in recesses F. Solderability	G. 450°F melting point H. Converts to brittle gray tin below 40°F	Moderate
17	Zinc	ASM-2402; ASTM-B633; ISO-2081	Ed	Cyanide; acid or neutral chloride	A. Good atmospheric corrosion resistance in industrial environment B. Higher melting point than cadmium (787°F) C. Good thickness distribution D. Good coverage in recesses	E. Poor corrosion resistance in marine environment F. Less lubricity than cadmium	Low (less than cadmium)

Ed, electrodeposition; El, electroless; IVD, ion vapor deposition; Mech, mechanical.

TABLE 5.4
Electrochemical Data for Plated Metals

Metal, Symbol	Atomic Weight[a]	Valence	Density[b]		Electrochemical Equivalents				Typical Plating Solution
			g/cm³	g/in.³	g/(A·h)[c]	oz/(A·h)	lb/(1000 A·h)	A·h/ft² to deposit 0.001 in.	
Cadmium, Cd	112.41	2	8.65	141.7	2.097	0.07400	4.62	9.74	Cyanide, acid sulfate; fluoborate
Chromium, Cr	51.996	6	7.19	117.8	0.3234	0.01141	0.813	52.47	Chromic acid
Cobalt, Co	58.933	2	8.90	145.8	1.099	0.03878	2.42	19.10	Acid sulfate
Copper, Cu	63.546	1	8.96	146.8	2.371	0.08364	5.23	8.92	Cyanide
		2	8.96	146.8	1.186	0.04182	2.61	17.83	Acid sulfate; pyrophosphate
Gold, Au	196.967	1	19.3	316.3	7.350	0.2592	16.2	6.20	Alkaline, neutral, or acid cyanide
Hydrogen, H	1.00794	1	0.0899[d]	—	0.03761	0.42[e]	—		Acid, alkaline, or neutral
Indium, In	114.82	3	7.31	119.8	1.428	0.0504	3.15	12.08	Cyanide; acid sulfate fluoborate
Iron, Fe	55.847	2	7.86	128.8	1.042	0.0368	2.30	17.80	Acid chloride
Lead, Pb	207.2	2	11.4	186.8	3.866	0.1364	8.52	6.96	Fluoborate
Nickel, Ni	58.693	2	8.90	145.8	1.095	0.0386	2.41	19.18	Acid sulfate; chloride; sulfamate
Palladium, Pd	106.42	2	12.0	196.6	1.985	0.0700	4.38	14.26	Alkaline amine dinitrite
Platinum, Pt	195.08	4	21.4	350.7	1.820	0.0642	4.01	27.75	Amine, nitrito acid chloride
Rhodium, Rh	102.906	3	12.4	203.2	1.280	0.0451	2.82	22.86	Acid sulfate; phosphate
Silver, Ag	107.868	1	10.5	172.1	4.025	0.1420	8.87	6.16	Cyanide
Tin, Sn	118.71	2	7.30	119.6	2.215	0.0781	4.88	7.78	Acid sulfate
		4	7.30	119.6	1.107	0.0391	2.44	15.56	Alkaline stannate
Zinc, Zn	65.39	2	7.14	117.0	1.220	0.0430	2.69	13.81	Cyanide acid chloride

[a] From International Table of Atomic Weights (1989).
[b] At 300 K, 27°C, 81°F. Density of deposited metal may differ significantly from the values shown because of porosity or occluded organic material. Values given in g/in.³ were calculated from values in g/cm³.
[c] Assumes 100% cathode efficiency.
[d] In g/L.
[e] In L/(A·h) at STP.

The workpiece or cathode is electrically connected to the negative terminal of a 6–12 V source of direct current electricity such as a silicon diode power supply. The anode is connected to the positive terminal.

When the electric circuit is closed, an electric field is created between the cathode and anode, with the cathode becoming charged with electrons. The positively charged nickel ions are attracted to, and migrate to, the cathode, where they plate out as nickel metal.

Meanwhile, at the nickel anode nickel atoms are each giving up two electrons and are going into solution as nickel ions. This essentially maintains a constant nickel concentration in the plating solution. The stream of electrons given up by the nickel atoms at the anode flow through the power supply to the cathode, where they are reunited with nickel ions plating out as nickel metal. If a nickel ion is not readily available for the electrons they may reduce water to hydrogen gas and a hydroxide ion. The following equations summarize the cathode and anode reactions.

Cathode reaction:

$$Ni^{2+}\,(\text{nickel ion}) + 2e\,(\text{electrons}) \rightarrow Ni\,(\text{deposited nickel}) \tag{5.1}$$

$$2H_2O\,(\text{water}) + 2e \rightarrow H_2\,(\text{hydrogen}) + 2OH^-\,(\text{hydroxide ions}) \tag{5.2}$$

$$H_2O + le \rightarrow H\ (\text{nascent hydro}) + OH^- \tag{5.3}$$

Anode reaction:

$$Ni - 2e \rightarrow Ni^{2+} + 2e \tag{5.4}$$

$$2H_2O + 4e \rightarrow O_2\ (\text{oxygen}) + 4H^+\,(\text{hydrogen inos}) \tag{5.5}$$

5.3.2 Faraday's Law

Equation (5.1) shows that each nickel ion, in order to deposit as a nickel atom, accepts two electrons at the cathode surface. Therefore, 1,000 nickel ions require 2,000 electrons, 1 million nickel ions require 2 million electrons, etc. In other words, the amount of nickel deposited is in direct proportion to the current that flows through the anode–cathode circuit. This relationship was enunciated in 1833 by Michael Faraday as one of two basic laws of electrolysis. These laws, incidentally, were the first quantitative demonstration of the electrical nature of matter and have long defined the unit quantity of electricity.

Faraday's first law states that the quantities of substances set free at the anode or cathode are directly proportional to the quantity of electricity that passes through the solution. The quantity of electricity is equal to the product of amperes and the time for current flows. Therefore, it is possible to deposit twice the thickness if either the time of plating is doubled or the amperes are doubled.

The second law, which today's plating industry uses to determine the required current and time to produce a given deposit thickness, states that the amount of different substances liberated by the same quantity of electricity is proportional to their chemical equivalent weights. To deposit one equivalent weight (atomic weight ÷ valence) requires 1 faraday (F) of electricity or 96,485 coulombs (C). One coulomb of electricity is equal to 1 ampere (A) flowing for 1 second (s). One faraday of electricity, therefore, is equal to 96,485 ampere-seconds (A·s), 1,608 ampere-minutes (A·min), or 26.8 ampere-hours (A·h). Platers normally use ampere-hours for large amperage processes such as barrel plating and ampere-minutes for small manual plating tank operations.

To calculate the amount of metal deposited for any amperage and time, Faraday's law is used in equation form:

$$\text{Wt} = \frac{\text{AW}}{V} \times \frac{ite}{F}$$

where

Wt = weight of metal deposited
AW = atomic weight of metal being deposited
V = valence (number of electrons required to deposit one atom)
i = current flowing through the plating solution
t = time that current flows (s, min, h)
e = cathode efficiency (% of current that deposits the metal)
F = Faraday's constant = 96,485 A·s, 1,608 A·min, or 26.8 A·h per equivalent weight

The thickness of the deposit (T) is then calculated as

$$T = \frac{\text{Wt}}{DA}$$

where

D = density of metal (g/in.^{3})
A = area of parts (in.^{2})

Combining Faraday's law equation and the thickness equation allows the plater to calculate the number of ampere-hours required to plate a given thickness on a load of parts:

$$\frac{\text{A·h}}{\text{Load}} = \text{metal factor} \times \frac{\text{area}}{\text{load}} \times \text{thickness} \times \frac{1}{e}$$

where

Metal	Metal Factor (A·h/(ft²·in.))
Cadmium	9,737
Silver	6,180
Nickel	19,180
Zinc	14,300

$$\frac{\text{Area}}{\text{Load}} = \frac{\text{in.}^2}{\text{piece}} \times \frac{\text{pieces}}{\text{load}} \times \frac{1\ \text{ft}^2}{114\ \text{in.}^2} = \frac{\text{ft}^2}{\text{load}}$$

Thickness = thickness of deposit (in.)

e = cathode efficiency (%)

5.3.3 Cathode Efficiency

During the plating process, the solution chemistry in the cathode film, the 0.005–0.010 in. thick film in immediate contact with the cathode, controls the plating process and the nature and properties of the metal deposited.

A distinguishing feature of the cathode film is its metal concentration gradient, which is the range in the bulk solution concentration from the furthest from the cathode down to 25%–60% of that value nearest the cathode depending on the current density. The nickel ions diffuse through the cathode film, accept two electrons each at the cathode, and deposit as nickel metal.

If too much current is applied to the cathode, there will be more electrons available than there are nickel ions diffusing through the cathode film to accept them. The excess electrons will then react with water molecules either to release nascent hydrogen, which can penetrate the nickel and basis metal crystalline structures, or form hydrogen gas bubbles, which are released into the plating solution.

As long as sufficient nickel ions are diffusing through the cathode film to accept the available electrons, very little hydrogen is evolved. When hydrogen is evolved, the amount of metal deposited is lesser than the amount that should theoretically have been deposited. The ratio of the amount of metal deposited to what theoretically should have been deposited is known as the cathode efficiency:

$$\text{Cathode efficiency} = \frac{\text{actual weight of metal deposited}}{\text{theoretocal weight of matel per Faraday's law}} \times 100$$

In our nickel plating example, the highest cathode efficiency and the least amount of hydrogen released are achieved by (1) keeping the nickel metal above a minimum concentration in the body of the solution, (2) heating the solution to increase the rate of diffusion of the nickel ions through the cathode film, (3) agitating the solution to decrease the thickness of the cathode film, and (4) keeping the current below the limiting current density for the particular solution chemistry involved so that only enough electrons are being supplied to the cathode to plate out the nickel ions arriving to accept them.

5.3.4 Current Density

The number of amperes applied to a plating solution through the cathode–anode circuit must be controlled and adjusted according to the solution chemistry and the surface area of the work being plated. Current density is the amperes flowing through the plating solution divided by the surface area being plated and is usually expressed as amperes per square foot.

$$\text{Current density} = \frac{\text{amperes}}{\text{unit area}} = \frac{\text{A}}{\text{ft}^2}$$

Each type of plating solution requires that the current density be within a particular range to produce a quality deposit. This range of current densities increases or decreases as the concentrations of the plating ingredients change. Usually, the metal concentration is the controlling factor. As the metal concentration increases, a higher current density can be used without deleterious effects. If the current density exceeds its limiting value for a given solution formulation, the amount of evolved hydrogen rapidly increases. In the case of nickel plating, the pH in the cathode film increases to the point where nickel hydroxide precipitates on the cathode, resulting in a "burned," granular, defective deposit.

5.3.5 Throwing Power

Ideally, the thickness of plating on a fastener should be uniform. Unfortunately, plating thicknesses are not uniform over the surfaces of a bolt or nut, or any other item for that matter. The plating tends to be thickest on prominent projecting surfaces such as the point and head of the bolt and the outside of a nut, and the thinnest deposits are found in recessed areas of a bolt such as the shank-to-head fillet or the internal threads of a nut. This ability or inability of a plating solution to plate a deposit of uniform thickness over the entire surface of a complex shape is known as its throwing power.

Cyanide and alkaline solutions tend to have better throwing power than acid solutions. Chromium plating solutions have the least throwing power of all plating solutions. The cause of good or bad throwing power is complex, but it is known to be influenced by the ohmic resistance of the plating solution, which determines the primary current distribution; the geometry of the plating system; the shape of the part; and the position of the part in the plating solution relative to other parts and to the anode.

Polarization, or a reverse EMF, occurs in the cathode film at high current density areas and shifts the plating current to more recessed areas, thereby depositing more metal in those areas than would have been deposited with the primary current distribution.

Throwing power is improved by using lower operating temperatures, lower metal concentrations, and lower current densities and by avoiding agitation of the solution.

Mechanically, throwing power may be improved by keeping the anode as far away from the work as possible, by arranging parts so that their high current density areas are adjacent to each other, and by using auxiliary "dummy" cathodes near high current density areas to accept some of the current.

5.3.6 Solution Composition

Plating solutions can be broadly classified as acid, neutral, or alkaline. Solutions with a pH of less than 5 can be considered acid; from 5 to 9 neutral; and above a pH of 9, alkaline.

Acid and neutral solutions, except in the case of acid cyanide gold, use either simple inorganic metal salts such as sulfates or chlorides or complex inorganic salts such as fluoborates, pyrophosphates, or sulfamates. The majority of alkaline solutions for plating cadmium, zinc, silver, copper, and gold contain cyanide complexes. Tin is plated from non-cyanide-containing alkaline stannate solutions.

Several variables must be controlled to produce a metallurgically sound deposit during electroplating: the solution chemistry, solution temperature, current density, and, to a lesser degree, the basis metal.

Chemicals added to a plating solution serve at least one of the following functions:

1. As a source of the metal being deposited.
2. To improve conductivity.
3. To provide anode corrosion.
4. To provide brightness or a smoother, finer grained deposit.
5. To buffer or maintain the proper pH.

In the case of our nickel solution example, the nickel sulfate supplies the nickel metal and also the conductivity because of its relatively high concentration. The nickel chloride, besides supplying some nickel, is added primarily to provide chloride ions to promote anode corrosion. It also contributes to conductivity. The boric acid acts as a buffer to minimize rapid pH changes. Frequently a wetting agent such as sodium lauryl sulfate is added to prevent pitting from hydrogen gas bubbles adhering to the plated surface.

To achieve a decorative bright nickel deposit, several proprietary organic chemicals would be added. Without these "brighteners," the nickel deposits are smooth and fine grained but matte gray in color.

For a very low tensile stressed and ductile nickel deposit, a sulfamate formulation is used instead of the sulfate solution. If a high rate of deposition is required, a fluoborate formulation can be used.

Although it is important that plating solutions have closely controlled concentrations of major ingredients, it is also important that they be free of certain impurities peculiar to each plating system. For example, all plating solutions, except chromium solutions, are sensitive to hexavalent chromium, even in the parts per million range. Hexavalent chromium drastically reduces cathode efficiency and in high enough concentrations will prevent deposition altogether. Other solutions such as the alkaline tin solution must have all their metal ions of the same valence. The tin must be present as the stannate (4+) without any stannite (2+) in order to produce smooth, fine-grained deposits.

Just as small amounts of impurities are deleterious in most plating solutions, small amounts of certain chemicals are necessary in some plating solutions to deliver sound deposits. Chromium plating solutions, both those used for decorative deposits and those used for thick deposits, must contain a small amount of sulfate, usually in a concentration of 1% of the chromic acid concentration. Above and below this range, cathode efficiency rapidly decreases. In bright acid copper plating solution, chloride in a range of 20–80 ppm is necessary to achieve the maximum effect of the organic brighteners.

5.3.7 Addition Agents

In addition to their major and minor inorganic ingredients, most plating solutions must also contain a small amount of organic material or chemical to produce an acceptable deposit. These materials are referred to as addition agents and sometimes as brighteners.

Even though these additives are added to plating solutions in small quantities and do not take part in the electrolysis (although they may be absorbed into the deposit, oxidized at the anode, or reduced at the cathode) they have a profound effect on the character of the deposit. They refine the crystal structure, resulting in denser, less porous, and brighter deposits. More importantly, especially in cyanide solutions, they greatly increase throwing power and allow the use of higher current densities.

As an example of the intense effect of minute amounts of addition agents, glue was once used to prevent treeing in acid copper plating at a concentration of only 3 mg/L (3 ppm). Although not in common use today, glue is still included in some lead fluoborate plating formulations, at a much higher concentration of 200 mg/L.

Addition agents or brighteners were serendipitously discovered by an accidental contamination of a silver cyanide solution with carbon disulfide in 1843. This chemical combination is still used today by some platers.

It is interesting to note that the early development of brighteners centered on the small additions of common organic chemicals found around the home, especially in the kitchen. As an example, the first brightener for cadmium cyanide solutions was a caustic solution of dissolved wool in 1922. During this period such common organics as dextrin, gelatin, milk sugars, Instant Postum, molasses, and sulfonated castor oil were often used by cadmium platers.

Currently used commercially available proprietary brighteners are materials whose active ingredients are identifiable chemicals with definite chemical structures and are manufactured or refined by closely controlled processes.

5.4 PLATING PROCEDURES

The essential features of a typical plating procedure include

1. Preplating preparation
2. Electroplating
3. Postplating treatment

The preplating preparation phase removes any soil, dirt, scale, or oxides that would interfere with the adhesion of the subsequent coating. The preplating step may consist of one or more, or all, of the following: vapor degreasing, chemical descaling, mechanical scale removal such as abrasive blasting, soak or electrolytic cleaning, and finally an activation acid drip.

The electroplating phase consists of immersing the parts to be plated into the electroplating solution, positioning them properly between the anodes for optimum plating distribution, adjusting the electric current, and measuring the time to deposit the required thickness. The parts are then removed and water rinsed to remove any residual plating solution drag-out.

The postplating phase may consist, in the case of cyanide cadmium plating, of a mild neutralizing/passivating dip in a weak chromic acid solution followed by one or more cold water rinses and a final hot water rinse to facilitate drying. The part is then baked at 375°F for 23 h to drive off any absorbed hydrogen. The chromate conversion coating, if required, and any other subsequent coating such as a wax or molybdenum disulfide dry film lubricant, are then applied. A typical cadmium plating cycle is outlined below.

Step 1. Vapor degreasing: Removes bulk of grease, oil, and shop dirt prior to subsequent treatments.
Step 2. Rust and scale removal—Abrasive blast with glass beads (no base metal removal) or fine aluminum oxide: Removes tenacious heat treating scale and corrosion products.
Step 3. Soak or anodic alkaline cleaning: Removes the last traces of soil. Produces chemically clean surface to promote good adhesion of the deposit.
Step 4. Cold water rinse: Removes alkaline chemicals from parts.
Step 5. Acid dip (when permitted by specification): Activates surface by removing oxides produced by alkaline cleaning. Low concentration, 30 s maximum. Promotes good adhesion.
Step 6. Cold water rinse: Removes acid from part.
Step 7. Cadmium plate: Depending on material strength, the solution is a cyanide solution either with or without brighteners.
Step 8. Cold water rinse: Removes plating solution from parts.
Step 9. Chromate dip: Removes or neutralizes residual plating solution.
Step 10. Cold water rinse: Removes chromate solution.
Step 11. Baking: Bake at 375°F for 4 or 23 h to remove hydrogen absorbed during plating.
Step 12. Chromate conversion coating: When required. Produces an iridescent or olive drab chromate coating that improves corrosion resistance.
Step 13. Cold water rinse: Removes residual chromate solution from part.
Step 14. Warm water rinse: Facilitates drying of part. Used with compressed air blow-off for large parts or a centrifugal spin dryer for small parts.
Step 15. Inspection: Check for blisters, pits, nodules, adhesion, thickness, and salt spray.

These steps may not all be necessary in all plating procedures.

The flowchart in Figure 5.1 illustrates this (water rinses not shown).

5.4.1 Rack or Barrel Plating?

Fasteners may be plated either in bulk in a rotating barrel or individually on a fixture or rack. The decision to barrel or rack plate may depend on the quantity of parts being plated, economics, part size, and the probability of thread damage.

Generally, precision high-strength bolts 3/8 in. in diameter or less by approximately 2 in. or less in length are barrel plated. Beyond these limits, there is buildup beyond dimensional limits on the ends before the minimum thickness is achieved at the center of the part. But a more critical reason for restricting barrel plating to small bolts is that the larger the diameter and length, the more likely it is that parts will suffer thread damage from handling and peening over of the thread crests from prolonged tumbling.

The above restrictions are not generally of concern with lower grade fasteners. The economics and vast volume indicate that they are barrel plated.

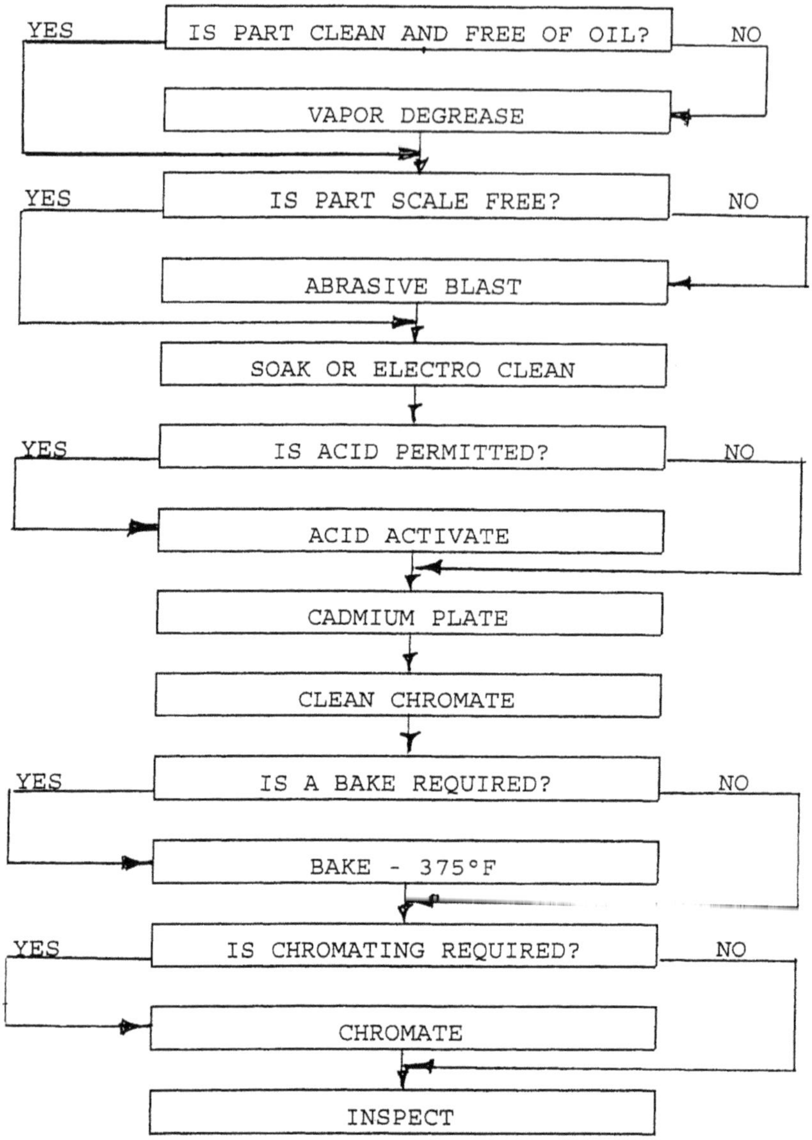

FIGURE 5.1 A flowchart for the cadmium plating cycle.

Rack plating, although costly, is necessary for larger diameter and longer precision fasteners with close dimensional tolerances. Plating buildup on the lead thread can easily be prevented by making electrical contact at the thread end for part of the plating cycle. Normally midway through the plating cycle, the bolts are repositioned on the rack by rotating 90° or 180° depending on the type of rack. This permits areas that are low current density areas, and that have minimum plate thickness, to become high current density areas, making the final thickness more uniform over the entire surface of the bolt.

Most people, including engineers, think that electroplating is a simple dip or dunk process. It definitely is not. It is a science that employs various aspects of chemistry, physics, electronics, and electrochemistry along with mechanical, plumbing, and electrical trades know-how, combined with a great deal of common sense and a bit of art.

Electroplating is a unique manufacturing operation. Along with heat treating, electroplating is one of two manufacturing techniques of special concern to users and purchasers of high-strength precision fasteners.

While other fastener manufacturing operations such as drilling, forging, tapping, grinding, and thread rolling impart obvious and easily verified shapes and dimensions, heat treating and electroplating induce intrinsic properties that allow a fastener to function as intended under specific conditions and environments.

5.4.2 Arc Burns

A common problem caused by electroplating is arc burns. An arc burn occurs, almost exclusively during barrel plating, when there is too little contact area between the electrical connectors inside the barrel and the bolts or nuts being plated.

In the case of steel fasteners, when arcing occurs, the temperature of the burned area momentarily reaches a temperature above the transition temperature at which martensite converts to austenite. As the austenite cools it transforms back to untempered martensite. These untempered martensite spots are areas that are conducive to fatigue or stress fractures. To minimize the possibility of arc burning, the electrical contacts inside the barrel should have as large a contact area as possible. Rather than use the standard fist-size knob contacts at the end of the limp electric cables, large diameter discs with a large surface area should be used. The more parts in contact with the discs, the less likelihood that arc burns will occur.

Arc burns can be detected by stripping the plating from a sample of barrel-plated parts and using the standard nitol test for untempered martensite.

5.5 DIMENSIONING FASTENERS FOR PLATING

There are few things as useless as a fastener that doesn't fit, being either oversized or undersized for the mating threads. Manufacturing unplated fasteners to close dimensional limits is not difficult, but if the fastener is to be plated and still meets the final blueprint dimensions, it must be manufactured unsize to allow for the buildup due to the added plating thickness.

5.5.1 Tolerance

Tolerance is defined as the amount of variation permitted in the size or dimensions of a part. If a part is to be plated, there is not only a tolerance allowed in the size of the part before plating, but because the thickness of plating cannot be deposited exactly, a plating thickness tolerance is also necessary. Plating thickness tolerances are necessary because the plating thickness will vary on the same part from point to point, the average thickness on each part will vary from part to part, and finally, the average thickness of each batch plated will vary from batch to batch.

If 0.0003 in. of cadmium plate is applied to only one side of a panel, then the dimension of the panel will increase by 0.0003 in.; but if 0.0003 in. of cadmium is plated on both sides of a panel, the panel's dimensions will increase twice 0.0003 in., or 0.0006 in. total. When a thread is plated with 0.0003 in. of cadmium, the pitch diameter increases by 4×0.0003 in., or 0.0012 in. These three examples are illustrated in Figure 5.2.

As shown in Figure 5.2, a plating thickness of 0.0003 in. increases the pitch diameter of the thread by 0.0012 in. because of the 30°/60°/90° triangle generated by the thickness of plating and the increased distance from the original unplated diameter to the pitch diameter of the plated thread. The hypotenuse, the increase from the unplated pitch diameter, is twice the side opposite the 30° angle, the thickness of the plating, or 0.0006 in. Since this increase also occurs on the opposite side of the thread, the total increase in pitch diameter is 4 times 0.0003, or 0.012 in.

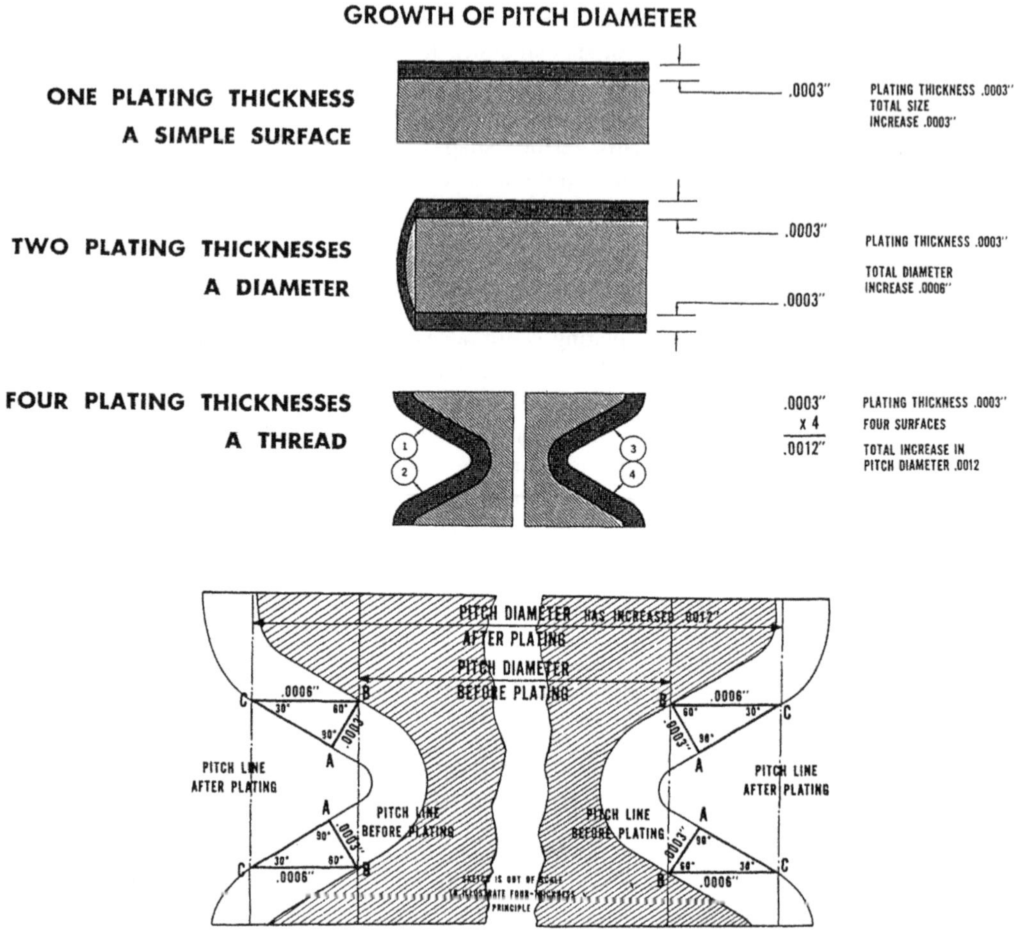

FIGURE 5.2 The effect of plating thickness on thread dimensions.

5.5.1.1 Establishing Tolerances

Tolerances for multiple manufacturing processes are additive. To go back to the example of plating a single surface of a panel, the total tolerance of the plated panel is

$$\text{After-plate tolerance} = \text{before-plate tolerance} + \text{plating tolerance}$$

or, transposing,

$$\text{Before-plate tolerance} = \text{after-plate tolerance} - \text{plating tolerance}$$

Example 5.1

If the before-plate dimensions of the panel are allowed to vary from 0.1000 to 0.1010 in., a tolerance of 0.001 in., and the plating thickness is held to 0.0002–0.0004 in., a tolerance of 0.0002 in., then the total tolerance of the plated panel is

$$0.001\text{ in.} + 0.0002\text{ in.} = 0.0012\text{ in.}$$

However, if the plated panel must be held to the original unplated tolerance of 0.001 in., the tolerance of the panel before plating must be held to 0.0008 in.

$$\text{Before} - \text{plate tolerance} = \text{after} - \text{plate tolerance} - \text{plate tolerance}$$
$$= 0.001 \text{ in.} - 0.0002 \text{ in.} = 0.0008 \text{ in.}$$

Example 5.2

If a bolt body is plated on both sides, with the before-plate and plating tolerances as given at the beginning of Example 1, the total after-plate tolerance is now

$$\text{After} - \text{plate tolerance} = \text{before} - \text{plate tolerance} + 2 \times \text{plating tolerance}$$
$$= 0.001 \text{ in.} + 2(0.0002 \text{ in.}) = 0.0014 \text{ in.}$$

However, if the after-plate tolerance must be maintained at 0.001 in. as in the first example, then the dimensional tolerance of the bolt body panel must be reduced to 0.0006 in. If the tolerance on the body diameter cannot be reduced that much, then both the preplate diameter tolerance and the plating thickness tolerance can be reduced by 0.0001 in. for a thickness of 0.0002–0.0003 in., or 0.00025–0.0003, 5 or 0.0003–0.0004 in. or some combination that will satisfy the final 0.001 in. tolerance. For example,

$$0.001 \text{ in.} = 0.0008 \text{ in.} + 2(0.0001 \text{ in.})$$

When plating threads, the tolerance equation is

$$\text{After-plate tolerance} = \text{before-plate tolerance} + 4 \times \text{plating tolerance}$$

or

$$\text{Before-plate tolerance} = \text{after-plate tolerance} - 4 \times \text{plating tolerance}$$

The amount of preplate tolerance for threads requiring plating increases as the nominal thread size Increases. This increased tolerance is needed because lead error and thread drunkenness, caused by long thread roll dies, is more prevalent as the thread size increases. Some of these tolerances are listed in Table 5.5.

TABLE 5.5
Final and Preplate Pitch Diameter Tolerances for Some Typical UNF Threads

Nominal Thread Size	Final PD[a] (in.)	Final Tolerance (in.)	Preplate PD[a] (in.)	Preplate Tolerance (in.)
#6-40	0.1198–0.1218	0.002	0.1109–0.1202	0.0012
1/4-28	0.2243–0.2268	0.0025	0.2235–0.2252	0.0017
3/8-24	0.3450–0.3479	0.0029	0.3442–0.3463	0.0021
1/2-20	0.4643–0.4675	0.0032	0.4635–0.4659	0.0024
3/4-16	0.7056–0.7094	0.0038	0.7048–0.7078	0.0030
1-12	0.9414–0.9559	0.0044	0.9407–0.9543	0.0036
1 1/2-12	1.4411–1.4459	0.0049	1.4403–1.4443	0.0040

[a] PD, pitch diameter. These dimensions may not be valid for all fastener manufacturers. The minimum or maximum preplate sizes may be greater or less to compensate for subsequent manufacturing operations. For example, the minimum preplate PD may be increased by 0.004 in. to compensate for 0.0001 in. of scale formed during heat treatment. When this scale is removed during the preplate cleaning and descaling operation the minimum PD just prior to plating will have decreased to that shown in the table.

5.5.1.2 External Dimensions

The maximum material condition or maximum after-plate dimension occurs on parts that are at the maximum before-plate dimension and receive the maximum specified plating thickness. Conversely, the minimum material condition or minimum after-plate dimension occurs on parts that are at the minimum before-plate dimension and receive the minimum specified plating thickness. The keywords above are "receives the … specified plating thickness." If parts are within the required before-plate dimensions and the part receives either more or less than the specified thickness, then the parts will be dimensionally out of tolerance.

The rules, then, for calculating the before-plate dimensions are similar to those cited above for determining tolerances, except that two calculations are made: one to determine the maximum and one to determine the minimum dimension.

For the bolt body maximum before-plate diameter, subtract twice the maximum specified thickness from the maximum after-plate diameter. For the minimum bolt body before-plate diameter, subtract twice the minimum specified thickness for the minimum plate diameter.

$$\text{Maximum before-plate diameter} = \text{maximum after-plate diameter} - 2 \times \text{maximum plate thickness}$$

$$\text{Minimum before-plate diameter} = \text{minimum after-plate diameter} - 2 \times \text{minimum plate thickness}$$

Similar calculations are made to determine the preplate pitch diameter (PD) dimensions of bolt threads, except that four times the plating thickness is subtracted from the after-plate pitch diameter dimensions.

$$\text{Maximum before-plate PD} = \text{maximum after-plate PD} - 4 \times \text{maximum plate thickness}$$

$$\text{Minimum before-plate PD} = \text{minimum after-plate PD} - 4 \times \text{minimum plate thickness}$$

5.5.1.3 Internal Preplate Dimensioning

Calculating preplate dimensions for internal surfaces is not just the reverse of that for external surfaces. The minimum metal condition after plating occurs when the hole or pitch diameter is the greatest (largest hole) and has the minimum plate thickness applied. The maximum metal condition exists when the hole or pitch diameter is the least (smallest hole) and has the maximum plate thickness.

For holes:

$$\text{Maximum before-plate diameter} = \text{Maximum after-plate diameter} + 2 \times \text{minimum plate thickness}$$

$$\text{Minimum before-plate diameter} = \text{minimum after-plate diameter} + 2 \times \text{maximum plate thickness}$$

For pitch diameters:

$$\begin{aligned}\text{Maximum before-plate pitch diameter} &= \text{maximum after-plate pitch diameter} \\ &\quad + 4 \times \text{minimum plate thickness}\end{aligned}$$

$$\begin{aligned}\text{Minimum before-plate pitch diameter} &= \text{minimum after-plate pitch diameter} \\ &\quad + 4 \times \text{maximum plate thickness}\end{aligned}$$

Note: Some manufacturers use a large maximum pitch diameter before plating to compensate for the thicker plating buildup on the first thread at each end of the nut. Plating thickness is measured on the bearing surface of the nut per MIL-STD-1312 method.

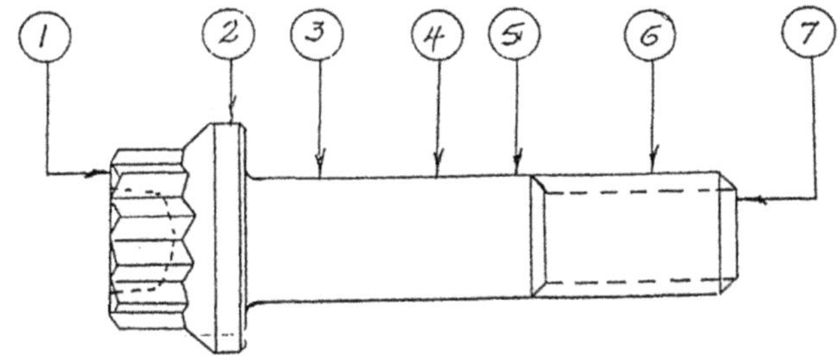

		THICKNESS - INCHES						
METHOD	SCREW SIZE	1	2	3	4	5	6	7
BARREL	1/4-28x1.072	.00048	.00042	.00033	.00033	.00035	.00043	.00062
BARREL	1/4-28x2.375	.00044	.00045	.00033	.00032	.00030	.00034	.00055
RACK	1/4-28x5.012	.00059	.00047	.00034	.00033	.00032	.00034	.00055
BARREL	3/8-24x1.645	.00050	.00048	.00029	.00028	.00033	.00029	.00040
RACK	3/8-24x4.270	.00045	.00044	.00027	.00026	.00026	.00025	.00052
RACK	1/2-20x2.038	.00035	.00033	.00022	.00021	.00022	.00024	.00037
RACK	1/2-20x4.375	.00031	.00032	.00020	.00025	.00023	.00037	.00034
RACK	3/4-16x7.431	.00050	.00052	.00028	.00030	.00028	.00030	.00036

FIGURE 5.3 Cadmium plating thickness distribution on 12-point fasteners.

5.5.2 Plate Thickness Distribution

Electrodeposited coatings vary in thickness from area to area on a bolt or a nut. In general, the greatest thickness is found on the head and point of a bolt and on the outside of a nut, and the thinnest deposit is found on the middle of a bolt and on the inside of a nut. Figure 5.3 illustrates typical plating thickness distributions on several sizes and lengths of bolts plated to 0.0002 in. minimum according to QQ-P-416.

Pitch diameter preplate dimensions for nonelectroplated coatings, such as vacuum deposition of aluminum or cadmium or mechanical deposition of zinc, cadmium, or tin, must be determined by trial because the amount of metal deposited on the pitch diameter may be only one-half of that deposited on the shank.

All of the above recommendations and equations for dimensioning bolts and nuts apply to parts with 60° threads. For parts with a 75° thread, the pitch diameter increases by almost 8 times the plating thickness or twice that of a 60° thread.

5.6 EFFECT OF COATINGS ON FASTENER PERFORMANCE

All fastener coatings affect fastener performance in some respect, although some effects may arise not from the physical presence of the coating but from the method of application. Some of these effects are discussed below.

5.6.1 Hydrogen Embrittlement

Electroplating is unique in its ability to deposit a variety of thin, adherent uniform metallic coatings to suit the fastener's intended application and environment. However, it has a serious side effect. The electroplating process can cause hydrogen embrittlement in high-strength fasteners.

TABLE 5.6
Factors Affecting the Degree of Hydrogen Embrittlement

Factors Prior to Plating

1. Type, quality, and strength level of the material
2. Microstructure of the material
3. Surface condition of the material
4. Mechanical, chemical, and electrochemical surface treatments prior to plating

Factors during Plating

1. The metal deposited
2. Nature of the deposit
3. Thickness of the deposit
4. Type and composition of the plating solution
5. Addition agents and impurities present in the plating solution
6. Temperature of the plating solution
7. Voltage and current density during plating
8. Length of plating time

Factors after Plating

1. Temperature and duration of the postplating embrittlement relief bake
2. Stripping of defective deposits
3. Corrosion while in service

Hydrogen embrittlement can be defined as delayed brittle failure during static loading at stress levels well below the tensile strength of the material due to the presence of absorbed hydrogen. The process of hydrogen embrittlement and the role of contributing factors are complex. The degree of embrittlement, whether slight, moderate, or severe, is related to the amount of hydrogen absorbed and retained in the basis metal during manufacturing and chemical processing.

The many factors affecting the degree of hydrogen embrittlement in high-strength fasteners are grouped in chronological order of occurrence in Table 5.6.

Much of the practical research into the causes of hydrogen embrittlement and its minimization was done during the period 1950–1970 by the military and the aerospace industry because of the increased use of high-strength and ultrahigh-strength materials during that period.

In light of today's sensitive embrittlement test techniques, some misleading conclusions were drawn from some of the earlier research programs because of the use of some relatively insensitive stress methods such as bend tests and reduction of area tensile tests, which could detect gross embrittlement but not slight embrittlement.

It was not until the universal adoption of notched tensile specimen testing that test data could be reliably compared with the results of other research. However, even notched tensile specimen data can vary widely because of differences in material batches or heats, the notch sharpness or K_t stress concentration factor, and the methods or techniques for the manufacture and preparation of the specimens.

From reported research on hydrogen embrittlement, the following general comments appear to be valid concerning the variables listed in Table 5.6. Most of the emphasis of these research efforts was concerned with cadmium plating because it is commonly used as a sacrificial corrosion-preventive metal coating by the aerospace industry and the military.

The degree to which a material can be embrittled depends on the intensity and longevity of the embrittling conditions, but the indicated degree of embrittlement is determined by the type and severity of the embrittlement test and the criteria chosen to indicate freedom from embrittlement.

5.6.2 Factors Prior to Plating

5.6.2.1 Type, Quality, and Material Strength

Most iron-base alloys and steel are susceptible to hydrogen embrittlement but to different degrees. The susceptibility to embrittlement increases as the strength of the material increases and generally manifests itself at strength levels at or above approximately Rockwell hardness C-30 [1].

Materials of the type AISI 4340 heat treated to high strength levels are the most susceptible to hydrogen embrittlement. This susceptibility varies among different production heats [2].

High-strength alloy steels containing high percentages of silicon [3,4] or tool steels similar to AISI H-11, which contains 5% chromium, show a relatively low degree of embrittlement [4–6].

Vacuum-melted steel has a lower susceptibility for hydrogen embrittlement than air-melted steel [7].

High-temperature alloys of the type Inco 718 (180 ksi), Waspalloy (190 ksi), René 41 (200 ksi), and U212 (180 ksi) have such a low susceptibility to hydrogen embrittlement that they do not fail when cathodically charged while being stressed to 80% of their yield strength [8].

5.6.2.2 Material Microstructure

Austenite has a higher solubility for hydrogen than either ferrite or martensite and therefore tends to concentrate in retained austenite in steel. Baking does not remove this hydrogen unless the baking also transforms the retained austenite. The hydrogen in the retained austenite does not cause embrittlement. However, if the austenite decomposes to untempered martensite in service and releases the hydrogen, the steel may be embrittled.

5.6.2.3 Surface Conditions of Steel

Rough surfaces, because of the greater surface area, will absorb more hydrogen than a smooth surface. Surface carburization produces a high surface hardness and high surface stresses that are conducive to embrittlement failure.

5.6.2.4 Mechanical, Chemical, and Electrochemical Surface Treatment Prior to Electroplating Mechanical

Mechanical surface treatments such as cold drawing, grinding, and polishing have a tendency to increase the susceptibility to embrittlement. These effects can be relieved by retempering within 50°F of the original tempering temperature or by anodic etching to remove the stressed surfaces.

Shot peening has been shown to be beneficial prior to cyanide cadmium plating.

Dry abrasive blasting, rather than wet vapor honing, should be used to remove the heat-treated scale because hydrogen is generated by the reaction of water with the oxide-free iron surface.

5.6.2.4.1 Chemical

The degree of embrittlement from acid treatment is influenced by the type of acid, concentration, temperature, and length of treatment.

Highly ionized acids such as hydrochloric (HCl), sulfuric (H_2SO_4), and hydrofluoric (HF) acids cause embrittlement because of increased absorption of hydrogen by materials. Less ionized acids such as phosphoric (H_3PO_4) cause less embrittlement.

Cadmium fluoborate plating according to NAS672 permits the use of a 10 s maximum dip in a 1.5% concentration of either nitric, sulfuric, or fluoboric acid for activation of bolts just prior to cadmium plating.

In general, the use of inhibitors in pickling solutions reduces the amount of hydrogen absorbed during pickling. However, they are not used for acid activation immediately prior to the plating operation because the tenacious inhibitor film will interfere with the adhesion of the subsequent deposit.

5.6.2.4.2 *Electrochemical*

Alkaline cleaning either by immersion or by anodic electrocleaning does not cause embrittlement, but cathodic electrocleaning is severely embrittling.

Thin (0.000125 in.) undercoats of bright cyanide cadmium, electroless nickel, gold, or pyrophosphate copper plus a bake have been reported to minimize embrittlement from cyanide cadmium plating.

5.6.3 Factors during Plating

5.6.3.1 Type of Metal Deposited

All electroplating processes appear to cause hydrogen embrittlement to some degree. Whether or not a particular plating process causes embrittlement depends on the ability of the hydrogen absorbed during the plating process to diffuse out of the basis metal and through the plated metal during the post-plate bake operation.

Because hydrogen has low solubility in cadmium, zinc, and gold, it diffuses so slowly through dense deposits of these metals that outgassing of hydrogen cannot occur during the post-plate bake to prevent hydrogen embrittlement. However, embrittlement due to chromium and nickel plating can be relieved with a suitable post-plate bake because hydrogen can diffuse through these metals.

5.6.3.2 Nature of the Deposit

A porous crystalline deposit of cadmium, plated from a cadmium cyanide solution without any organic brighteners, allows hydrogen to outgas, while a bright dense deposit plated from a similar solution containing organic brighteners causes severe embrittlement.

5.6.3.3 Deposit Thickness

The thinner the deposit, the shorter the time for hydrogen to diffuse out through the coating. A method based on this concept was proposed that consisted of plating a thin layer of cadmium, approximately 0.0001 in., followed by a short bake to outgas the hydrogen, and finally followed by plating to the required total thickness of cadmium. But more sensitive embrittlement tests showed that this method also caused embrittlement.

5.6.3.4 Type and Composition of the Plating Solution

Cadmium is severely embrittling when plated from a cyanide solution but much less so when plated from a fluoborate [7], sulfamate, or perchlorate solution. The cyanide ion acts as a catalyst, promoting the absorption of hydrogen into steel. Cadmium solutions with complexing agents such as pyridine, triethanolamine, methanol, and dimethyl formamide show a low tendency to cause embrittlement.

Although seemingly contradictory, acid plating solutions, with the exception of chromium, cause less embrittlement than alkaline solutions.

5.6.3.5 Plating Solution Addition Agents and Impurities

Oxidizing agents such as hydrogen peroxide, nitrates, and titanium compounds have been added to cadmium cyanide solutions to reduce the degree of embrittlement. These materials oxidize the nascent hydrogen before it can diffuse into the basis metal.

Organic brighteners or grain refiners and organic contaminants produce dense cadmium deposits through which hydrogen cannot diffuse during the post-plate bake.

Sulfide has been a suspected impurity in unbrightened cadmium cyanide solutions, while iron contamination nullifies the beneficial effect of titanium.

5.6.3.6 Plating Solution Temperature

Temperatures in the range of 60°–90° for both zinc and cadmium cyanide plating produce less hydrogen than elevated temperatures.

5.6.3.7 Voltage and Current Density

A reduction of over 1,000 times in the hydrogen input rate can be achieved by lowering the voltage across the plating tank to 1 V maximum. In unbrightened cyanide cadmium, plating current densities above 40 A/ft^2 produce a porous deposit, allowing for a more efficient out-gassing of hydrogen during the post-plate bake.

5.6.3.8 Plating Time Duration

A shorter cathodic treatment such as plating limits the amount of hydrogen migrating into the basis metal. Also, higher current densities result in shorter plating time for equal thickness because most hydrogen is absorbed during the initial part of the plating period.

5.6.3.9 Current Fluctuations

Low current ripple is beneficial in hydrogen absorption, and periodic reverse current is of no value.

5.6.4 Factors after Plating

5.6.4.1 Temperature and Duration of the Post-Plate Bake

Baking after plating relieves embrittlement by lowering the hydrogen concentration in the basis metal to a nonembrittling level in one or both of the following ways:

1. By outgassing the hydrogen through the plating
2. By dispersing the hydrogen uniformly throughout the basis metal volume

For materials extremely sensitive to hydrogen embrittlement, relief must occur in the first way, while for materials less sensitive or of a lower strength level, relief is achieved in the second way. In this mechanism, the initial high hydrogen concentration near the surface is reduced by diffusion of the hydrogen into the body of the basis metal to a concentration below the threshold value necessary to cause embrittlement.

As the strength level of steels increases, the post plate baking time must be increased. For materials below 190,000 psi tensile strength, a 3 h bake at 375°F ± 25°F is adequate. Above this level, a 23 h bake is necessary. At strength levels above 280,000 psi, no amount of baking gives full recovery.

The baking temperature should be as high as possible without affecting the strength of the material or the integrity of the coating. The baking temperature for cadmium is generally 375°F, whereas chromium can be baked at 700°F and nickel as high as 1,000°F.

5.6.4.2 Stripping of Plating

The chemical stripping of metallic coatings can be a source of hydrogen, especially if acid solutions are used. A baking operation should follow the stripping operation but precede replating.

5.6.4.3 Corrosion

When cadmium, zinc, or aluminum coatings on steel corrode in service, hydrogen is liberated on the steel surface at breaks in the coating where steel is exposed or at the steel/coating interface of porous deposits. This hydrogen can diffuse into the steel, and if the part is under stress, delayed brittle failure can occur. Since bright cyanide cadmium is less porous than dull unbrightened cadmium there is less of a tendency for this type of failure with the former.

5.6.5 Fatigue Strength

There are many factors other than plating that influence fatigue strength, and they are covered in another chapter of this book.

Most plated coatings and coating processes reduce the fatigue strength of steel bolts. The degree of decrease, or in rare cases increase, depends on the type and strength of the steel, the residual surface stresses in the steel, the mechanical and chemical properties of the coating, the testing environment, and the method of fatigue testing, whether by rotating–bending or tension–tension.

Any damage, no matter how slight, to the original thread rolled surface will decrease the fatigue strength of a high-strength bolt. A major cause of surface damage is grit blasting or acid pickling to remove heat treat scale. The loss of fatigue strength may be minimized by substituting alkaline descaling for acid pickling or by grit blasting using fine glass beads or fine aluminum oxide, such as 350 grit, at as low an air pressure as possible.

The need for blasting or chemical descaling can be eliminated if an inert atmosphere such as argon is used during heat treating and tempering; this will prevent scale formation. If blasting is necessary, especially on large diameter bolts, automatic individual blasting machines will give a consistency of blasting time and coverage that is impossible to achieve with a manual operation.

5.6.6 Stress Corrosion

It is important that high-strength steel fasteners be resistant to stress corrosion cracking. Electrodeposited nickel (0.0005 in.) or nickel (0.0003 in.) plus cadmium (0.0002 in.), either as-plated or diffused, was twice as effective as cadmium (0.0005 in.) alone in preventing stress corrosion failure in bolts of 290,000 psi tensile strength loaded to 90% of their proportional limit and exposed for more than 1,000 h of alternative immersion in 3.5% sodium chloride solution.

The nickel, nickel-cadmium coatings were far superior to such coatings as vacuum-deposited cadmium, electrodeposited gold, or zinc chromate primer. However, nickel plus SermeTel W, a phosphate chromate bonded aluminum coating, surpassed almost all other coatings tested.

If resistance to stress corrosion cracking is of prime importance as a bolt property, it is better not to rely on plated coatings to provide resistance but to consider the use of corrosion-resistant alloy bolts that require no protective system.

5.6.7 Stress Alloying

Cadmium-plated high-strength steel shows some reduction in tensile strength, stress rupture, and fatigue strength at temperatures as low as 500°F. But at the melting point of cadmium, 610°F, there is a sharp decline in these properties. Therefore cadmium plate is limited for use on steel parts that do not exceed 400°F in service. Cadmium plating is prohibited on titanium and titanium alloy fasteners because stress alloying or solid cadmium embrittlement can occur in service at temperatures as low as room temperature.

5.6.8 Corrosion Protection

Coatings that are electronegative or cathodic to steel, such as nickel, chromium, or silver, protect steel from corrosion by the barrier effect. These coatings, being more corrosion-resistant than steel, protect the steel by keeping the corrodant from contacting the steel. However, since coatings for fasteners are thin, in the range of 0.0003–0.0005 in., some porosity is present. It is at these pores, or in mechanical scratches or scrapes if they are present, that a galvanic cell is produced. These corrosion-resistant coatings, rather than protecting the steel, cause it to corrode because the steel is now the anode in the galvanic cell.

By contrast, coatings such as zinc, cadmium, and aluminum protect steel not only by the barrier effect but also by acting anodically to the steel. They will corrode in preference to steel at scratches and scrapes in the coating. However, this can be detrimental for high-strength steels; because the steel will be the cathode in any corrosion cell, hydrogen would be generated, which could diffuse into the steel and cause hydrogen embrittlement.

5.6.9 Lubricity

Torque tension is one of the most important fastener properties because it indicates the quantity of tightening or induced stress of the fastener and joint components for a given amount of wrenching effort or energy. However, the amount of torque required to develop a given tension or load in a fastener depends on the friction between mating surfaces of the bolt, joint, and nut. The fastener coating may either increase or decrease this friction. Coatings such as cadmium, copper, and silver will lower friction as is demonstrated in the use of these metals as lubricants during the manufacture of nuts and bolts in various sizing and shaping extrusion processes.

One of the two most popular plated coatings, cadmium requires less torque to develop a particular preload than zinc. However, the advantage of one over the other depends on whether the plating is on the bolt or nut or both.

The well-known torque–tension equation

$$T = \frac{KDW}{12}$$

where

T=torque, lb-ft
K=friction factor
D=bolt diameter, in.
W=bolt tension, lb.

cannot be relied on for accurate torque–tension determinations because the handbook values of K are only estimates. The true value of K must be determined for each particular fastener-coating-joint combination at hand.

5.7 PROCESS CONTROL

With electroplating, it is not obvious whether the applied coating has the specified properties of thickness, adhesion, corrosion resistance, and lubricity or whether the plating process has insidiously caused hydrogen embrittlement. These properties can be verified only by lengthy and costly and in some cases destructive tests with conclusions based on limited statistical data and sampling. Accordingly, it is readily recognized that if good plating results are to be consistently achieved, the plating process must be a precise and rigidly controlled operation. Many commercial job shop platers cannot be relied on to perform specified plating procedures to meet current critical fastener plating specifications. These specifications can be met only by well-trained platers who follow and adhere to established, proven, documented detailed procedures, operating with process and plating solutions under the proper controls of time, temperature, chemical composition, and current density.

With ordinary industrial fasteners, the plater's chief concern is just to plate the minimum thickness. If plating builds up on the lead thread there is still enough tolerance or allowance to meet the specification gauge fit. But with precision fasteners with narrow final tolerances, plating buildup is a problem. The plater must meet the minimum thickness on the body without building up the lead thread beyond the accepted gauge fit.

Close-tolerance plating can be accomplished by a system of tight process control. The difficulty with which plating thickness can be held to a 0.0002 in. tolerance increases as the minimum thickness increases. When the thickness requirement is 0.0002–0.0004 in., the maximum thickness is 100% higher than the minimum. But when the thickness range is 0.0003–0.0005 in., the maximum is only 66% greater than the minimum; and as the plating range moves to 0.0004–0.0006 in., the maximum is only 50% greater than the minimum and requires extremely close process control. These controls are outlined below.

5.7.1 Solution Control

The composition of all plating solutions should be controlled to within ±5% of the set concentration for each component. For example, a cadmium plating solution whose cadmium concentration is set at 4.0 oz/gal should not be allowed to vary beyond a range of 3.8–4.2 oz/gal.

All solutions should be analyzed as frequently as necessary to prevent the individual components from varying beyond the set limits. The time interval may be once a week for solutions where parts are plated on racks, the drag-in and drag-out is small, and production is low. However, the interval between analyses could be as short as a few hours in the case of the pH of a barrel nickel plating solution where the ampere-hours per gallon value is high, drag-in and drag-out of solution is high, and production is high. Chemical additions should be made promptly. Consistent filtration of the solution will minimize roughness from particles, and periodic solution purification using activated carbon will remove organic contamination and products of the organic breakdown of brighteners and wetting agents.

5.7.2 Process Control

Ammeters and voltmeters should be standardized every 6 months, and ampere-hour meters should be checked monthly. Temperature controls should be checked weekly. The post-plate bake oven temperatures are checked weekly.

Mandatory detailed plating procedures should be adhered to when manual plating is done. Computer-controlled automatic plating machines accomplish this rigidly.

A controlled plating load zone of 55 ft^2 is optimum, with 65 ft^2 a maximum for a 14 in.×30 in. hexagonal plating barrel.

The plater should use a computer or nomograph to determine the number of pieces per barrel load and the ampere-hours necessary for the specified thickness.

All plated parts must be in the post-plate oven within 1–4 h depending on the specification requirement.

5.7.3 Post-Plate Inspection

The plating operator will inspect plated parts for appearance, blisters, thickness dimensions, and gage fit at the plating facility before sending parts for the post-plate bake. Parts should be reinspected after the post-plate bake before they leave the plating department.

5.8 TERMINOLOGY RELATED TO ELECTROPLATING[2]

Addition agent: A material added in small quantities to a plating solution to alter the characteristics of the deposit such as grain size, smoothness, brighteners, hardness, and stress corrosion.

Ampere: Current that flows at the rate of 1 coulomb per second and is capable of depositing 0.0011180 g of silver in 1 s.
Anion: A negatively charged ion such as chloride (Cl^-) or sulfate (SO_4^{2-}).
Anode: The electrode, usually made of the metal being deposited, connected to the positive terminal of the electrical power supply. The anode dissolves to replenish the metal being deposited at the cathode.
Anode bar: The metal bar, normally a 1 in. round copper rod, from which anodes or anode baskets are hung.
Anode basket: A metal basket hung from the anode rod that is filled with small anodes shaped as balls, cones, chips, buttons, or rods to provide a large anode surface area.
Barrel plating: Bulk plating of small parts where the parts are plated, while in a rotating perforated cylinder.
Basis metal: The metal on which the electrodeposited metal is applied.
Brightener: An addition agent that causes the plating to deposit bright rather than dull.
Burning or burnt deposit: A rough, dark, powdery deposit caused by using too high a current density.
Bus bar: A copper bar, ¼×4 in. for carrying up to 1,000 A from the power supply to a process tank in the plating system.
Cathode: The electrode or metal on which the electroplating is deposited. It is connected to the negative side of the power supply.
Cathode efficiency: The weight of metal actually deposited divided by what should have theoretically been deposited according to Faraday's law. Expressed as a percent.
Cathode film: A film approximately 0.005–0.010 in. thick that forms adjacent to the cathode during plating and through which the metal ions pass to be deposited. The metal concentration in this film is significantly lower than in the body of the plating solution.
Cation: A positively charged ion such as sodium (Na^+) or nickel (Ni^{2+}).
Cathode bar: The metal bar, normally a 1 in. round copper rod, from which the cathode or work to be plated is hung.
Chemical equivalents: The weight of an element in grams equal to the atomic weight divided by the valence of that element for a specific reaction. The number of atoms in a chemical equivalent is equal to Avogadro's number (6.022×1023) divided by the valence.
Combined cyanide: The cyanide in the plating solution that exists in the cyanide complex. (*see* Cyanide complex.)
Coulomb: The quantity of electric current flowing through an electric circuit when 1 A flows for 1 s; abbreviated C.
Covering power. The ability of a plating solution to deposit metal in recessed areas of parts at very low current densities.
Current density: The current per unit area expressed as amperes per square foot (A/ft^2).
Cyanide complex: A negatively charged ionic group formed when a metal ion combines with two or more cyanide ions as in the reaction $Cd^{2+} + 4\,CN^- \rightarrow Cd(CN)_4^{2-}$.
Electrochemical equivalent: The weight of metal deposited or dissolved or of the gas evolved at either the cathode or anode during the passage of a unit quantity of electric current expressed in faradays, ampere-hours, or ampere-minutes.
Faraday: One faraday (1°F) of electricity is equal to 96,485°C or A·s. One faraday will deposit one chemical equivalent of a metal.
Flash: A very thin plating, normally 0.0001 in. or less (see also Strike).
Free cyanide: The amount of cyanide present in a plating solution that is not included in a cyanide complex.

High current density area: An area on a part that is closest to the anode or a protrusion that develops current densities much higher than those in other parts of the part.

Hydrogen embrittlement: Embrittlement of a metal due to absorption of hydrogen during cleaning, activation, and plating. Hydrogen embrittlement may lead to failure when the part is stressed at levels considerably below its tensile strength.

Ion: An atom or molecule that has gained a positive or negative charge by giving up or acquiring one or more electrons.

Limiting current density: The maximum current density at which a satisfactory deposit can still be obtained.

Low current density area: An area on a part that is farthest from the anode or a recessed area of a part such as a hole, which will develop current densities much lower than other areas of the part.

pH: The hydrogen ion concentration is expressed as a negative logarithm. The pH scale ranges from 1 to 14, with a pH of 7 being neutral, those below seven acid, and those above seven alkaline.

Pickling: The removal of scale and oxides from metal surfaces using an acid solution either alone or with other chemicals or acids.

Plating rack: A spline or frame that articles to be plated are attached to or suspended from so that several parts can be plated simultaneously.

Plating range: The current density range over which a satisfactory deposit is obtained.

Strike: A very thin deposit usually plated out at high current densities and short times to achieve coverage and adhesion.

Throwing power: The ability of a plating solution to deposit a uniform thickness of plating over the entire surface of a part.

Total cyanide: The total of the free and combined cyanide concentrations of a plating solution.

Work: The part or parts being plated.

NOTES

1 This section draws heavily on Refs. [1] and [2].

2 See ASTM B 374, "Standard Terminology Relating to Electroplating," for a complete list of terms used in plating and metal finishing.

REFERENCES

[1] Cross Index of Standard Documents for Metallic and Inorganic Coats, Annual Book of ASTM Standards, Vol. 02.05, Metallic and Inorganic Coatings; Metal Powders, Sintered P/M Structural Parts, 1991, p. 788.

[2] Lowenheim, F. (ed.), *Modern Electroplating*, 3rd ed., Wiley, New York, 1974.

[3] Durney, L. (ed.), *Electroplating Engineering Handbook*, 4th ed., Van Nostrand Reinhold, New York, 1984.

[4] Safranek, W. H. (ed.), *The Properties of Electrodeposited Metals and Alloys*, 2nd ed., American Electroplaters and Surface Finishers Society, Orlando, FL, 1986.

[5] Lowenhein, F. A., *Electroplating*, McGraw-Hill, New York, 1978.

[6] *Metal Finishing Guidebook and Directory Issue*, Metals and Plastics Publications, Inc., Hackensack, NJ, 1995.

[7] *Selection of Electrodeposited Coatings*, American Electroplaters and Metal Finishers Society, Orlando FL.

[8] Pearlstein, F., Selection and application of inorganic finishes: Part V, Metal deposits; Part VI, Barrier layer protective metal deposits; Part VII, Metal deposits for specialized characteristics, *Plating and Surface Finishing*, April 1979, p. 28; May 1979; June 1979, p. 42.

6 Thread Lubricants

Gary J. Novak and Tom Patel

6.1 INTRODUCTION

Lubricants are compounds that are applied to threaded and other contacting surfaces on fasteners during the assembly of a bolted joint. Thread lubrication has a significant effect on the threaded fastener during the assembly, in-service, and disassembly phases [1]. Thread lubricants facilitate the assembly of bolted joints by providing smooth assembly and reducing torque-tension scatter. Fastener tension level, fastener tension variation around the joint, and successful assembly without fastener failure are influenced by the thread lubricant. Without proper assembly of the bolted joints, powerplants, papermills, refineries, chemical processing plants, internal combustion engines, and numerous other machines and process equipment systems would not function properly. The results of improper fastener assembly can include fastener overload leading to fastener failure, loose fasteners, nonuniform flange loads, leaking joints, and crushed and overloaded gaskets. These types of failure lead to high plant maintenance cost and total system shutdown. Thread lubricants can help reduce the potential damage arising from improper assembly, which could range from customer inconvenience through loss of production, financial and bodily harm, hazardous emissions, and leakage of hazardous materials to loss of life. When removed or taken apart, fasteners that were assembled with thread lubricants are disassembled more readily. Some of the lubricant or a residue from the lubricant remains on the threads and provides a layer that lubricates and prevents galling during disassembly. During repairs, service, or disassembly, cost, time, and damage to the assembled parts including fasteners and threaded components are reduced through the use of lubricants.

Thread lubricants provide control of the thread friction level and prevent metallurgical bonding during the assembly of a fastener. Lubricants affect both the nominal level and the variability of the resultant tension that is developed in a fastener when it is assembled to a target torque level. A lubricant that is properly selected will prevent galling, fretting, and seizing and reduce friction and wear of the fastener compared to a dry-assembled fastener. Thread lubricants can also provide an antiseize function during the disassembly or removal of the fastener. The lubricant can protect the fastener from moisture or chemical exposure. Corrosion can be reduced and, in some cases, prevented during storage and shipping and in service through the use of lubricants. Thread lubricants have a significant effect on the torque-tension relationship of a fastener.

6.2 HOW DO THREAD LUBRICANTS WORK?

When a fastener is under high load, the high contact points on the mating surfaces of the threads deform elastically until the contact area is sufficient to support the load. If the surfaces are clean (unlubricated) or if the lubricant breaks down, the surfaces will gall. The threads may weld together at several sites due to the load and/or elevated temperature, and significant portions of the thread may be torn off as the fastener is assembled or disassembled. To prevent this galling and potential thread failure, the two surfaces must be kept separated. A good thread lubricant fills the spaces between contact points, thereby reducing the metal-to-metal contact and preventing galling and seizing. In effect, it acts like a microscopic ball bearing to keep the surfaces separated, ensuring quick and easy disassembly and reducing equipment downtime.

DOI: 10.1201/9781003501183-8

6.3 SELECTION OF A THREAD LUBRICANT

Thread lubricants are primarily selected to facilitate the assembly and disassembly of fasteners. However, the lubricant must be compatible with the environmental and functional requirements of the bolted assembly. Several important factors must be considered when specifying a thread lubricant. The nominal value of the nut factor, K, is an important parameter that is used to evaluate the effectiveness of a thread lube as an assembly aid. The nut factor is related to the coefficient of friction on the surfaces lubricated with the subject thread lubricant. Table 6.1 shows some typical nut factors for a variety of bolting materials and lubricants. Low nut factor values will result in greater bolt tension for a given level of torque. In addition to the nut factor, the type of application and environment are also important issues when selecting a thread lubricant. Table 6.2 provides a thread lubricant selection guide for various applications and environments. Some of the issues that should be considered when specifying a thread lubricant are environment, fluctuating loads, storage and handling, initial assembly or service application, and the assembly procedure.

The prevention of thread galling and the achievement of target thread friction or nut factor are also important issues to be considered. The specification of a low friction lubricant may result in a nut factor that is too low and overtensioning of the fastener. The specification of a lubricant

TABLE 6.1
Nut Factors K for Various Fastener Materials and Coatings[a]

	Reported Nut Factor		
Fastener Materials and Coatings[b]	**Min**	**Mean**	**Max**
Pure aluminum coating on AISI 8740 alloy steel [2]	0.42	0.52	0.62
Electroplated aluminum on AISI 8740 alloy steel [2]	—	0.52	—
Mild or alloy steel on steel, as received	0.058	0.2	0.267
Stainless steel on mild or alloy steel, as received	—	0.3	—
A490, 1 in. diam. [3]			
As received	—	0.179	—
Very rusty[c]	—	0.389	—
With Johnson 140 stick wax	—	0.275	—
Black oxided 7/8 A325 and A490, slightly rusty[d] [4]	0.15	—	0.22
Black oxide	0.109	0.179	0.279
Cadmium plate (dry)	0.106	0.2	0.328
Vacuum cadmium+chromate [2]	—	0.21	—
Copper-based antiseize	0.08	0.132	0.23
Cadmium plate (waxed)	0.17	0.187	0.198
Cadmium-plated A286 nuts and bolts [5]	0.15	—	0.23
Cadmium plate plus cetyl alcohol on A286 nuts and bolts [5]	0.11	—	0.16
Cadmium-plated nuts used with MP35N bolts [5]	0.18	—	0.29
Dag (graphite+binder)[e] [6]	0.16	—	0.28
Dicronite (tungsten carbide in lamellar form)[e,f]	0.045	—	0.075
Emralon (PTFE+resin)[e] [6]	0.10	—	0.15
Everlube 810 (MoS_2/graphite in silicon binder)[e] [7]	0.09	—	0.115
Everlube 811 (MoS_2/graphite in silicate binder)[e] [7]	0.09	—	0.115
Everlube 6108 (PTFE in phenolic binder)[e] [7]	0.105	—	0.13
Everlube 6109 (PTFE in epoxy binder)[e] [7]	0.115	—	0.14
Everlube 6122	0.069	0.086	0.103
Fel-Pro C54	0.08	0.132	0.23
Fel-Pro C-670	0.08	0.095	0.15
Fel-Pro N 5000 (paste)	0.13	0.15	0.27

(Continued)

TABLE 6.1 (*Continued*)
Nut Factors *K* for Various Fastener Materials and Coatings[a]

	Reported Nut Factor		
Fastener Materials and Coatings[b]	**Min**	**Mean**	**Max**
Mechanically galvanized A325 bolts [4]			
As received	0.35	—	0.49
Clean and dry	—	0.46	—
Slightly rusty[d]	0.36	—	0.39
Mechanically galvanized A325 bolts; lubed with 1 part water, 1 part Jon Cote 639 was [4]	0.11	—	0.26
Hot-dip galvanized 7/8 A325 [4]			
As received	0.14	—	0.31
Slightly rusty[d]	0.09	—	0.17
Clean and dry	0.09	—	0.37
Hot-dip galvanized 7/8 A325 lubed with 1 part water, 1 part Jon Cote 639 was [4]	0.10	—	0.16
Gold on stainless steel or beryllium copper [8]	—	0.4	—
Graphite coatings	0.09	—	0.28
Lube Lok 1000 (ceramic-bonded coating)[e,f]	0.275	—	0.31
Lube Lok 2006 (silicone resin bonded MoS_2+graphite)[e,f]	0.075	—	0.25
Lube Lok 4253 (high silver and indium content coating)[e,f]	0.21	—	0.24
Lube Lok 4856 (contains powdered tin+lead)[e,f]	0.21	—	0.25
Lube Lok 5306[e,f]	0.04	—	0.18
Machine oil	0.10	0.21	0.225
Microseal 100-1 (graphite plus inorganic binder)[e] [7]	0.09	—	0.115
Molydag (MoS_2+binder)[e] [6]	0.16	—	0.40
Moly paste or grease	0.10	0.13	0.18
Neolube	0.14	0.18	0.20
Never-Seize (paste)	0.11	0.17	0.21
N5000, threads only, not nut [9]	0.097	0.106	0.117
Pepcoat 6122 on nut, threads, and washer [9]	0.080	0.085	0.089
Pepcoa[f] [10]	0.09	—	0.11
Phos-Oil	0.15	0.19	0.23
PTFE plus binder[e] [6]	0.10	—	0.15
SermaGard (aluminum particles in ceramic binder)[f]			
#902 or 846; basecoat only	—	0.50	—
Basecoat+SermaLube 1000	0.09	0.15	0.40
Basecoat+751 wax	0.18	0.23	0.90
Misc. Sermagard systems with modified silicone topcoat	—	0.17	—
Serme Tel-W (phosphate-bonded, aluminum-rich coating) [2]	—	0.30	—
Solid film PTFE	0.09	0.12	0.16
Zinc plate (waxed)	0.071	0.288	0.52
Zinc plate (dry)	0.075	0.295	0.53

[a] See Chapter 18 for list of coating suppliers and their addresses.
[b] All tests involve steel bolts in steel joints, unless otherwise noted.
[c] Exposed outdoors for 2 weeks.
[d] Dipped in water, air-dried 24 h twice.
[e] Use with extra caution. These nut factors were computed from published coefficients of friction (μ) data using the equation $K=\mu(r_t/D \cos \beta)+r_n/D+P/2D\pi$. See Eq. (2) for terms and units. Threads ranging from 1/4-20 to 2-4-1/2 were assumed in the calculations.
[f] From manufacturer's literature.

TABLE 6.2
Lubricant Selection Guide

Thread Lubricant Type	General Purpose Anti-Sieze	Extreme High Temperature Resistance (to 2,400°F–2,600°F)	High Temperature Resistance (to 1,800°F)	Extreme Chemical Resistance	For Maximum Lubricity	Electrically Conductive	For Aluminum/ Soft Metals	For Stainless Steel	Copper-Free Formulation	Nuclear Certified	USDA Authorized	Solid Film	Typical Nut Factor
Copper and Graphite Flakes	•		•			•	•	•			•		0.16
Nickel and Graphite Flakes		•		•		•		•	•		•		0.15
Copper, Graphite, and Aluminum Flakes	•		•				•	•					0.17
Graphite Flakes Calcium Fluoride Powder		•					•	•	•				0.16
Molybdenum Disulfide Powder					•			•	•		•		0.11
Copper and Graphite Flakes-Nuclear Grade			•					•		•			0.17
Nickel and Graphite Flakes-Nuclear		•		•				•	•	•			0.15
Food Grade							•	•	•		•		0.163
Graphite Flakes	•			•		•	•		•				0.13
Zinc Dust							•						0.15
Molybdenum Disulfide Powder					•			•	•				0.133
Molybdenum Disulfide Powder, Graphite		•			•				•			•	0.11
Graphite Flakes Molybdenum Disulfide Powder		•			•				•			•	0.13

• Preferred Choice • Good Choice • Acceptable Choice

that yields a high nut factor will result in bolt tensions that are relatively low when the fastener is assembled to a specified torque. The selection of a thread lubricant for an existing application should include an evaluation of the assembly process and torque specifications to ensure that final preloads in the assembled fasteners are within the target tension range. The nut factor values in Table 6.1 include the minimum, mean, and maximum because there is significant variability in the resultant nut factor that is measured for each set of conditions. The mean value of the nut factor ranges from 0.086 to 0.52. An inspection of the nut factor minimums and maximums in Table 6.1 shows that some lubricants are more effective than others in reducing the variability in the nut factor.

The specification of a thread lubricant and the corresponding nut factor will have a significant effect on the state of stress in the fastener during assembly to (or disassembly from) a desired level of tension. The state of stress includes the fastener tensile stresses, a shear stress that develops due to the applied torque during assembly, and other stresses due to any external forces and moments applied to the bolted joint. The tensile and torsional shear stresses will be directly affected by the selection of lubricant. The amount of torsional stress retained in a tightened fastener depends on the fastener under-head friction and the presence of any locking devices. The selection or lack of thread lubricant can control bolt twist failures during assembly and disassembly as well as the state of residual stress in the fastener after assembly. The residual state of stress will affect other modes of failure such as direct tension, fatigue, or stress corrosion cracking (SCC).

Thread lubricant not only affects the tension of an individual fastener but it may also affect clamping load distribution across a bolted joint that has many fasteners. In their study, Chaaban et al. [11] concluded that for some situations the selection of a thread lubricant generally leads to more uniform bolt tensions. Also, this study points out that a joint assembled with molybdenum disulfide-based thread lubricant demonstrated the most uniform gasket stress distributions of the four lubricants of their study, the other three being a machine oil, a nickel-based antiseize compound, and a copper-based antiseize compound.

In-service operating conditions may expose the fastener and lubricant to elevated temperatures. The liquid contents of the lubricant may evaporate, but the solid materials will remain on the thread contact areas to provide release and antiseize function during disassembly. Stresses and deformation of the fastener and related components may arise from these thermal loads. Differential thermal expansion between the bolt, flanges, and attached structures of the joint can increase or decrease fastener preload, deformation, and stress. These loads should be considered in the analysis of fastener tension and preload. Oxidation of the fasteners can also occur and will generally accelerate at elevated temperatures. The lubricant should be chemically compatible with the fastener materials as well as any materials that may contact the lubricant during storage, handling, assembly, or in service. Corrosion can be reduced by using thread lubricants that prevent damage to the surface oxide or coating layers of the fastener during assembly and provide a protective coating.

SCC and hydrogen embrittlement can be reduced by using thread lubricants. Hydrogen embrittlement can be reduced using a lubricant such as molybdenum disulfide instead of electroplating. Both hydrogen embrittlement and SCC are less likely to occur when stresses at sites of potential failure are reduced through the use of a thread lubricant.

The chemistry of the lubricants should be compatible with the conditions of assembly and service. Molybdenum disulfide paste should be considered carefully when fasteners are operating at temperatures above 500°F. In the presence of humidity, the lubricant can release sulfur to form sulfuric acid, which can lead to corrosion or SCC failures. The copper-based antiseize compounds should also be specified with care. Copper-based lubricants should not be used in systems where they will be exposed to ammonia, where strong chemical reactions may take place. Copper should not be used in the processing of edible oils because it may accelerate spoilage. The copper in some lubricants will damage and may deactivate platinum catalysts used in some refining operations. These are a few of the many compatibility issues that should be considered prior to the selection of a thread lubricant. A less reactive nickel flake/graphite flake lubricant is available for applications where high temperatures or chemical compatibility are issues of concern. The nickel and graphite flake lubricants are significantly more expensive than copper or molybdenum disulfide lubricants.

Thread lubricants are used in regulated industries such as food processing, nuclear power generation, and military products. Lubricants that meet the specifications for performance and composition for these industries are available. Purity, particle size, and other quality factors are monitored and certified to conform to various specifications including military (MIL) specifications, the federal Meat and Poultry Products Inspection Program of the USDA, and nuclear equipment manufacturers' standards. Nuclear-grade lubricants are supplied in containers that identify the batch as well as the compositional analysis. The analysis includes levels of relevant impurities measured in parts per million. In the nuclear industry [1], the important elements to monitor include the halogens fluorine and chlorine, sulfur, and certain heavy metals such as lead, tin, zinc, cadmium, and mercury.

6.4 THREAD LUBRICANTS

6.4.1 Types of Thread Lubricants

Thread lubricants are essentially classified into three categories: liquid lubricants (oils), paste lubricants (combination of oil and/or grease and solid lubricants), and dry film lubricants (resin-bonded solid lubricants).

Oil-based liquid lubricants include petroleum-based oils (e.g., mineral oil or synthetic oil) that include silicone, esters, olefins, glycols, and polybutenes. These lubricants have limited applications at moderately high temperatures because oxidation occurs at temperatures around 350°F (higher oxidation temperature for silicone fluids), and the thread surfaces are left in a state of dry metal-to-metal contact. After exposure to these temperatures, threads may seize; moisture, chemicals, or gases may get between threads, and corrosion might occur. Oil-based liquid lubricants may squeeze out under extreme pressure, creating sufficient metal-to-metal contact to cause cold welding.

Paste lubricants are formulated with solid lubricants dispersed in liquid oil and/or grease. They are commercially known as antiseize lubricants. Most of the thread lubricants used for maintenance, repair, and operation (MRO) applications fall into this category. The solid lubricants used in these products are metal powders or flakes such as copper, nickel, aluminum, lead, or zinc. Nonmetallic lubricants include graphite, molybdenum disulfide, boron nitride, polytetrafluoroethylene, mica, talc, and metal oxides, hydroxides, and fluorides, such as calcium oxide, calcium fluoride, zinc oxide, titanium dioxide, magnesium oxide, calcium hydroxide, barium oxide, and tin oxide. Oil-based lubricants are based on mineral oil or synthetic oils from the silicone, ester, olefin, glycol, and polybutene families. Greases are mineral oil thickened with sodium, aluminum, calcium, or lithium, soaps or oil thickened with chemically treated bentonite clays, silicas, or polymer-thickened oils such as polyurea greases.

Paste lubricants are effective even when the oil boils off at elevated temperatures above 350°F. They leave a solid lubricating film on the threads, which allows fasteners to be removed without seizing or galling. Under heavy loads, they remain between metal surfaces and will not squeeze out like oil on the threads.

Dry film (or bonded) lubricants consist of dry lubricating solids dispersed in binders. The most commonly used lubricating solids are molybdenum disulfide, graphite, tungsten disulfide, polytetrafluoroethylene, boron nitride, calcium fluoride, and metallic oxides such as antimony oxide, calcium oxide, and magnesium oxide. Common binders are epoxies, silicones, silicates, phenolics, polyamides, polyimides, phosphates, and ceramics. Dry film lubricants can be cured at room temperature or at elevated temperatures depending on the formulations. These products perform as well as paste lubricants, but many carry hazardous solvents. Once the solvent evaporates, these products become a film that is dry to the touch, clean, and easy to handle. Dry film lubricants are used in original equipment manufacturing because fasteners can be coated in bulk before assembly.

6.4.2 Application of Thread Lubricants

Thread lubricants are applied to clean, dry surfaces. The lubricant may be applied with a brush, spray, dip, or other process depending on the specific application. The threaded parts are then assembled. In some cases, spray-applied lubricants are air-dried or oven-baked prior to assembly to provide a dry lubricating film. Thread lubricants are sometimes applied to the underside of the bolt head to reduce friction.

6.5 NUT FACTORS AND THE TORQUE–TENSION RELATIONSHIP

The applied torque and resulting bolt tension of a fastener are most often modeled as a linear function for the purpose of design. In most cases, the linear model is a very good approximation of the actual behavior. The equation determines bolt tension from applied torque, fastener diameter, and the nut factor as follows:

$$T = KDF \tag{6.1}$$

where

K = the nut factor (dimensionless)
D = the nominal diameter (in., m)
F = the bolt tension force (in.-lb, N·m)
T = the torque (in.-lb, N·m)

Nut factors are determined experimentally. Assembly torque and the corresponding bolt tension are measured during the assembly of a fastener. The torque–tension relationship is then established. Chaaban et al. [11] used a strain-gaged bolt to monitor tension and a torque wrench to apply bolt torque to prescribed values. In this study of six conditions that included four different thread lubricants, the six torque–tension curves were linear over the reported range of data. The slope, c, of the torque–tension curve is used to determine the nut factor K. The nut factor K is calculated from the torque–tension slope as

$$K = \frac{1}{cD} \tag{6.2}$$

The bolt geometry and the coefficients of friction of the bolt thread and bolt head contact areas have a major effect on the nut factor. The performance of a lubricant is a function of numerous factors including thread geometry, material, hardness, heat treatment, temperature, finish, alignment, and cleanliness. These factors vary from application to application. Reported nut factor values should not be considered as absolute values but should be used as reference values in selecting a lubricant. A torque–tension study using a strain-gaged bolt or ultrasonic tension measurements should be performed with actual fasteners under actual conditions to determine more accurate nut factor values for lubricants and fastener materials not specifically called out in the tables. The reported nut factors in Table 6.1 should be used only as a guide.

6.5.1 Coefficient of Friction

The assembly torque of a fastener is consumed by two mechanisms: friction-free bolt stretching and overcoming friction. Useful work is done when the contacting surfaces slide relative to each other up the inclined planes formed by the thread surfaces. A portion of the applied torque on the bolt head creates normal forces on the threads to stretch the bolt. The balance of the bolt torque overcomes thread friction and under-head friction. The assumed linear torque–tension relation is a simplification of the actual behavior. More complicated models reflect the effects of thread friction, bolt head friction, and the helix and thread flank angles. The following equations [12, pp. 758–762] characterize the external bolt torque, T, required to tighten or loosen a fastener:

$$T_t = \frac{F}{2}\left[d_n f_n + d_t\left(\frac{f_t + \cos\theta\tan\alpha}{\cos\theta - f_t\tan\alpha}\right)\right] \quad (6.3)$$

$$T_l = \frac{F}{2}\left[d_n f_n + d_t\left(\frac{f_t - \cos\theta\tan\alpha}{\cos\theta + f_t\tan\alpha}\right)\right] \quad (6.4)$$

where

F=the bolt tension (lb, N)
T_t=the torque to tighten (in.-lb, N · m)
T_l=the torque to loosen (in.-lb, N · m)
υ=the thread flank angle (deg)
α=the helix angle (deg)
f_n=the friction between the bolt head and the flange (dimensionless)
d_n=the mean diameter of the head bearing face (in., m)
f_t=the bolt thread coefficient of friction (dimensionless)
d_t=the mean thread contact diameter (in., m)

These equations can be rewritten in the following form:

$$T = \frac{FD}{2}\left[\frac{d_n f_n}{D} + \frac{d_t}{D}\left(\frac{f_t + \cos\theta\tan\alpha}{\cos\theta - f_t\tan\alpha}\right)\right] \quad (6.5)$$

A mathematical expression for the nut factor can then be written in the following form:

$$K = \frac{1}{2D}\left[d_n f_n + d_t\left(\frac{f_t + \cos\theta\tan\alpha}{\cos\theta - f_t\tan\alpha}\right)\right] \quad (6.6)$$

For small lead angle α and low thread friction, the nut factor can be expressed as [13, p. 224]

$$K = \frac{d_n f_n}{D} + \frac{d_t f_t}{2D\cos\theta} + \frac{d_t\tan\alpha}{2D} \quad (6.7)$$

The torque–tension equations show that the input torque is divided into three components. First, the applied torque must overcome the friction between the bolt head and the flange acting at diameter d_n with a coefficient of friction f_n. Second, the applied torque must also overcome the thread friction forces acting at mean thread diameter d_t with thread friction coefficient f_t. The balance of the applied torque does the useful work of stretching the bolt.

The efficiency of the screw, e, is defined [12, pp. 817–822] as the ratio of the torque required to develop a bolt load without friction to the torque required to develop the same bolt load with friction:

$$e = \frac{d_t\tan\alpha}{d_n f_n + d_t\left(\dfrac{f_t + \cos\theta\tan\alpha}{\cos\theta - f_t\tan\alpha}\right)} \quad (6.8)$$

The efficiency indicates how much torque is being used to stretch the bolt and how much is used to overcome friction. When the coefficients of friction at the bolt threads, f_t, and under the bolt head, f_n, are zero, the efficiency is 100%, and all of the torque is used to stretch the bolt. When f_t or f_n is nonzero, the screw efficiency is less than 100%. The use of thread lubricants with low coefficients of friction reduces the nut factor value and also increases the bolting efficiency *e*. An analysis of the equation describing the torque to loosen, T_l, shows that a low coefficient of friction at the bolt threads and under the bolt head will reduce the torque required to loosen. Fasteners that are assembled with low friction lubricants may be more liable to loosen than those with higher friction lubricants because the loosening moments have less frictional torque to overcome.

6.6 COMBINED TENSION AND TORSION STRESS IN THE FASTENER

The tightening of a fastener results in a state of stress that includes the uniaxial tensile stresses and shear stresses due to twisting. The combination of these stresses along with the stresses induced from external service loads should be used for failure analysis and design. In the case of a design evaluation against the maximum shear stress theory, the tensile stress and the shear stress are used to calculate the maximum shear stress due to this biaxial loading [12, p. 769].

$$\tau_{\max} = \left[\left(\frac{\sigma}{2} \right)^2 + \tau^2 \right]^{1/2} \tag{6.9}$$

$$\tau_{\max} = \left[\left(\frac{F}{2A} \right)^2 + \left(\frac{16Q}{\pi d_r^3} \right)^2 \right]^{1/2} \tag{6.10}$$

$$A = \frac{\pi}{4} \left(\frac{d_p + d_r}{2} \right)^2 \tag{6.11}$$

where

$\tau_{\max}$ = the maximum shear stress (psi, MPa)
σ = the shear stress due to pure tension (psi, MPa)
τ = the shear stress due to pure torsion (psi, MPa)
A = the tensile area (in.^{2}, m^2)
d_r = the thread root diameter (in., m)
d_p = the thread pitch diameter (in., m)
F = the bolt tension (lb, N)
Q = the twisting torque (in.-lb, N·m)

The twisting torque Q is the portion of the assembly torque T that is transmitted through the bolt shank. The lost torque overcomes the friction between the bolt head and the flange, and the transmitted torque overcomes thread friction and does useful work in stretching and tensioning the bolt. The twisting torque is less than the total applied bolt torque:

$$Q = \eta T \tag{6.12}$$

where η varies from 0 to 1 depending on the values of f_n and thread geometry.

When the torque–tension relationship is included, the resulting expression shows the effect that thread lubricant selection has on the maximum shear stress within the threaded fastener:

$$\tau_{max} = F\left[\left(\frac{1}{2A}\right)^2 + \left(\frac{16\eta KD}{\pi d_r^3}\right)^2\right]^{1/2} \tag{6.13}$$

In most situations, the coefficient of friction between the bolt head and the flange is about the same as the bolt thread friction. For typical lubricated and metal friction coefficients and thread geometries, it can be shown that approximately 50% of the total applied torque T is transmitted as twisting torque Q, and η is approximately 0.5.

When assembled to target levels of tension F, low nut factor lubricants will result in lower maximum shear stress values. Fasteners with lower maximum shear stress will function with a higher factor of safety when elevated against the maximum shear theory:

$$\tau_{max} = \frac{S_y}{N} \tag{6.14}$$

where S_y=the stress (psi, MPa) and N=the factor of safety (dimensionless).

6.7 FLUCTUATING LOADS

Gasket leakage, fastener overload, fastener loosening, and fastener fatigue failures can occur if the proper preload is not developed at initial assembly. The specification of thread lubricants should be consistent with the assembly and torque specification to yield a preload that is in the "proper" range to avoid these failures.

When external loads are applied to a joint, the fastener tensions and gasket loads will change. Enough preload should be provided to prevent leakage when fluctuating loads cause the gasket to unload. If the bolt preload is too high and if gasket compression is increasing during a condition of fluctuating loads, the gasket crushes and mechanical failure may occur. The selection of a thread lubricant must not change the initial preload out of the design range or else gasket overload or underload may lead to failure. For service or maintenance applications, thread lubricants should not be specified that have a nut factor significantly higher or lower than that specified by the factory without a detailed analysis of loading, assembly methods, and torque specification. New applications should have fastener size, lubricant, and assembly procedures specified to meet applicable code requirements and to provide suitable assembly bolt tension. The determination of initial bolt preload is discussed in Section VI of this handbook.

Mechanical systems are often exposed to cyclic loads that affect the performance of a fastener. These fluctuating loads can arise from sources such as rotary imbalance, reciprocating imbalance, cyclic external loads, and intermittent external loads that induce a structural ringing. When the frequency of cyclic loads is near any of the system's resonant frequencies, large cyclic fluctuations of the bolt loads, both axial and shear, can occur. These high cyclic loadings can affect the performance of the fastener if sufficient preload is not maintained. Two modes of failure that occur under these conditions are self-loosening and fastener fatigue failure.

6.7.1 SELF-LOOSENING

The fasteners in an assembled bolted joint may loosen over a period of time when subjected to fluctuating loads. There are many theories as to the mechanism of vibratory loosening. The topic of self-loosening is discussed in detail in Chapters 40 and 41 of this second edition of the handbook. The self-loosening condition can lead to other modes of failure as the preload is lost. Gasket failure or leakage, fastener fatigue failure, joint loosening, flange misalignment, fastener misalignment, or complete failure of the bolted joint can result from a condition of self-loosening. In general, the solution or preventive action for self-loosening is based on the concept that fasteners will not self-loosen if the friction between the threads is not reduced under the conditions of fluctuating external loads [13, pp. 527–563]. A reduced coefficient of friction and a steep thread lead angle will increase the tendency for vibratory loosening compared to a fastener with a high coefficient of friction and/or a shallow thread lead angle. Self-loosening can be controlled by proper preload and thread friction and by mechanical devices. In situations where significant fluctuating loads are anticipated, a thread lubricant with a higher friction coefficient may be specified. Devices that create a prevailing torque, such as nuts without round holes, nylon locking collars on nuts, nylon patches, or lock washers can reduce the concern regarding low nut factor thread lubricants and self-loosening. Devices that mechanically lock the fasteners, for example wiring or pinning, can be used to prevent self-loosening. See Section 6.8 of this chapter and *Handbook* Chapters 21–23.

6.7.2 Fatigue

Another issue concerning vibration and fluctuating loads is fatigue. Fasteners must have sufficient initial preload to minimize the portion of the fluctuating external load that is carried by the bolt and hence improve its fatigue performance. A widely used fatigue failure criterion, the Soderberg criterion, evaluates the minimum and maximum bolt stress to determine whether the fastener can be expected to fail by fatigue. For most metals, a state of stress that is fluctuating between S_{min} and S_{max} is more likely to result in a failure than a constant stress of S_{max} [12, p. 319]. The bolt preload and external loads are used to determine the mean stress. The fluctuating external loads on the bolted joint are used to determine the alternating stress level. Joint stiffness effects will split fluctuating external forces between the bolt and the flange. The fluctuating bolt force is combined with the bolt stress area and any relevant stress concentration factors to determine an alternating stress level. A fastener design is expected to satisfy the Soderberg criterion when the mean stress and the alternating stress fall below the Soderberg line. Some of the other criteria that are used for the design of fatigue-resistant fasteners and other structures include the Goodman, modified Goodman, and Gerber [14, p. 299]. See Chapters 36–38 for additional information on the fatigue failure of fasteners. When preloads are too low, the flanges can completely unload and cause all of the external load to be carried by the bolt, which results in excessive fluctuating stress. Too much preload can result in either a direct, noncyclic tensile overload or a fatigue failure due to the high mean stress. Either condition of improper preload, too high or too low, can result in failure [12, pp. 817–822]. The selection of a thread lubricant can have a significant impact on the initial preload applied to a fastener. Care must be taken to keep the preload at the proper level for both new design and repair and maintenance situations by specifying lubricants that result in proper preloads to meet fatigue criteria.

6.8 RELATED DEVICES

Mechanical thread-locking devices and liquid sealant thread adhesives create a prevailing torque condition that prevents the loss of load through self-loosening. These devices are important to the selection of a thread lubricant because the prevailing torque eliminates the self-loosening concerns when selecting a thread lubricant. Mechanical thread-locking devices include lock washers, nylon thread inserts, cotter pins, and other devices. Liquid sealants are generally not used in addition to thread lubricants because they are applied to the same threaded surfaces.

Liquid sealants include retaining compounds, thread-sealing adhesives, and thread-locking adhesives [15, pp. 130, 137]. Retaining compounds are normally anaerobic or cyanoacrylate adhesives. They fill the voids in the threaded assemblies and then chemically harden. In many cases, the assembled fasteners may become permanently bonded. The threads are sealed, and they become vibration-resistant, protected from corrosion, and leak-free. Thread-sealing adhesives also fill the gaps between the threads of an assembled fastener and provide a level of thread locking. These materials are often used to seal threads on pipe joints. Thread-locking adhesives are intended to provide a level of breakout torque. These adhesives come in a variety of strengths and formulations depending on the application. The popular formulations are one part solvent-based chemical curing, one part anaerobic, two parts curing, and "passivated" preapplied adhesives. The primary function of thread-locking adhesives is thread locking. Thread-sealing performance will vary with manufacturer and formulation. See Chapters 22 and 23 for more information on thread sealants.

REFERENCES

1. Schaefer, W. L., Thread Lubricants/Joint Compounds, Fel-Pro Internal Communication, April 25, 1985.
2. Taylor, E., SermeTel W (TM) aluminum coating for aerospace fasteners, SPS Laboratory Report No. 5392, SPS Technologies, Inc., Jenkintown, PA, Oct. 30, 1980.
3. Tests on 1 inch A490 conventional and twist-off high strength bolts and direct tension indicators conducted in the as-received condition and after weathering, Report No. 870702, Pittsburgh Testing Laboratory, Pittsburgh, PA, April 10, 1986.
4. Yura, J., K. Frank, and D. Polyzois, High Strength Bolts for Bridges, PMFSEL Report No. 8-3, University of Texas, May 1987.
5. Crispell, C., Torque vs. induced load of A-286 and MP35N nuts and bolts with cadmium, dry film and cetyl alcohol lubricants, Request No. 8-1-8-EH-07694-AP29-B, NASA Contract No. NAS8-32525, Nov. 14, 1978.
6. Emrich, M., Corrosion protection for fasteners, Part 1, *Assembly Eng.*, October 1982.
7. Gresham, R. M., Bonded solid-film lubricants for fastener coatings, Fastener Technol. Inc., April/May 1987.
8. Hung, N., Clamping force for electrically conductive fasteners, *Machine Design*, 56(23), 158–158, 1984.
9. Brenner, H. S., Results of special evaluation of Pepcoat 6122 and nuclear grade N5000 antiseize lubricant systems on torque–tension performance of ASTM A325 series hex-head bolts submitted by Power and Engineered Products Company, Inc. (PEPCO), South Plainfield, NJ, Report No. C 16034-1, Almay Research and Testing Corp, Los Angeles, Oct. 25, 1982.
10. *Pepcoat, A Study in Protection and Performance*, G* Chemical Corp., Wayne, NJ, undated manufacturer literature.
11. Chaaban, A., M. Derenne, A. Bouzid, and W. Schaefer, Evaluation of torque coefficients and gasket stress distributions in a bolted flanged joint using different types of lubricants. *Presented at CETIM 3rd International Symposium on Fluid Sealing*, Biarritz, France, Sept. 15–17, 1993.
12. Deutschman, A. D., W. J. Michels, and C. E. Wilson, *Machine Design Theory and Practice*, Macmillan, New York, 1975.
13. Bickford, J. H., *An Introduction to the Design and Behavior of Bolted Joints*, 3rd ed., Marcel Dekker, New York, 1995.
14. Shigley, J. E., and C. R. Mischke, *Mechanical Engineering Design*, 5th ed., McGraw-Hill, New York, 1989.
15. Brown, M. W., *Seals and Sealing Handbook*, 3rd ed., Elsevier Science, Oxford, UK, 1990.

7 Tribology of Threaded Joints
Sustainable Lubricants

Mattia Mele and Giorgio Olmi

7.1 ROLE OF TRIBOLOGY IN BOLTED JOINTS

Throughout the process of tightening a bolted joint, the application of preload arises from the relative motion between mating surfaces. Specifically, the interaction between the screw and the nut threads plays a pivotal role in the conversion of torque into axial force. Moreover, the rotational motion induces either the screw or nut head to slide against the adjoining surface. It is discernible that the intricate contact phenomena occurring at these interfaces significantly affect the ultimate correlation between the applied torque (T) and the resultant preload axial force (F) experienced by the screw. This relationship is encapsulated within the widely recognized Motosh formula, as reported in Eq. (7.1).

$$T = T_p + T_t + T_b \tag{7.1}$$

T_p is the pitch torque, namely the torque responsible for the clamping force. This depends only on the pitch of the thread, as highlighted in Eq. (7.2)

$$T_p = F \frac{p}{2\pi} \tag{7.2}$$

T_t is the thread friction torque, i.e., the torque needed to overcome the friction between the threads of the screw and those of the nut. This can be expressed as in Eq. (7.3)

$$T_t = F \frac{\mu_t R_t}{\cos \beta} \tag{7.3}$$

The angle β in Eq. (7.3) is half of the thread profile angle. The term μ_t refers to the coefficient of friction at the interface between the threads of the screw and those of the nut, while R_t is the effective radius of threads.

The last addendum (T_b) is the bearing torque, i.e., the torque required to overcome the friction between the head of the bolt (or the nut) and the surface of the underlying part. T_b can be calculated as in Eq. (7.4), where R_b and μ_b are, respectively, the effective bearing radius and the friction coefficient between the surfaces at the underhead.

$$T_b = F \mu_b R_b \tag{7.4}$$

It can be remarked that the terms in Eqs. (7.3) and (7.4) are directly proportional to the coefficients of friction μ_t and μ_b. This indicates that the tribological characteristics at the interfaces between mating surfaces significantly influence the correlation between the applied torque (T) and the resulting clamping force (F). Consequently, getting a comprehensive awareness of the variables impacting friction between the contact pairs emerges as a crucial point. Such an understanding is highly relevant to warrant the optimal functionality of the bolted joint within the intended design parameters.

DOI: 10.1201/9781003501183-9

7.2 FACTORS AFFECTING THE TRIBOLOGY OF BOLTED JOINTS

There are several internal and external factors that can affect the tribology of a threaded joint. Among these, the most relevant to practical applications are the following:

- Material
- Surface finishing
- Surface treatments
- Coatings
- Lubrication
- Number of tightenings

Their main effects are shortly described in the following. It is worth mentioning these factors are mutually influenced, so that none of them can be considered independently from the others. The brief introduction provided in this chapter does not pretend to give a complete insight into such complex relations, for which the reader is invited to consult specialized handbooks.

7.2.1 Material

The selection of materials composing the interconnected elements, such as the bolt and nut, demands meticulous attention when examining the tribological aspects of their contact surfaces. This is primarily because the forces operating between the mating surfaces can be perceived as electromagnetic forces originating from the interactions among the atoms or molecules of the two materials as they draw close to each other.

These intermolecular forces, often referred to as Van der Waals forces, hinge on the chemical composition of the molecules involved and their relative proximity. To promote relative motion between the contacting parts, it becomes imperative to supply sufficient energy, so that the Van der Waals forces can be overcome. In addition, the presence of charged particles within the materials can give rise to electrostatic forces, which may manifest as either attractive or repulsive forces depending on the polarity of the charges carried by the two components.

In certain instances, specific material pairs have the capacity to form chemical bonds upon contact, leading to adhesion forces that exhibit significantly higher strength than mere Van der Waals and electrostatic forces. Consequently, greater external energy is necessary to enable sliding movement between the mating components.

The chemical interactions between the materials at the contact interface can also prompt the creation of oxide layers or reaction products. These distinct elements, differing chemically from the materials constituting the mating parts, have the capability to modify both the coefficient of friction and the wear characteristics of the contacting surfaces.

The hardness of materials emerges as another pivotal material property influencing the tribological behavior of contacting elements. Indeed, material hardness affects its resistance to deformation, abrasion, and wear. Generally, harder mating materials exhibit superior resistance to surface wear in sliding contact scenarios. Conversely, significant discrepancies in hardness between contacting materials may result in substantial wear of the softer material. Moreover, if one component is considerably softer than its counterpart, it tends to fill the surface roughness valleys, thus consequently increasing the actual contact area and potentially elevating the coefficient of friction. Therefore, it is typically recommended to select materials with comparable hardness when selecting pairs of materials for contacting surfaces.

7.2.2 Surface Finishing

As well known to the reader, real parts are characterized by a certain surface roughness that deviates the actual surface from the nominal one. Considering the section of a profile, like that being

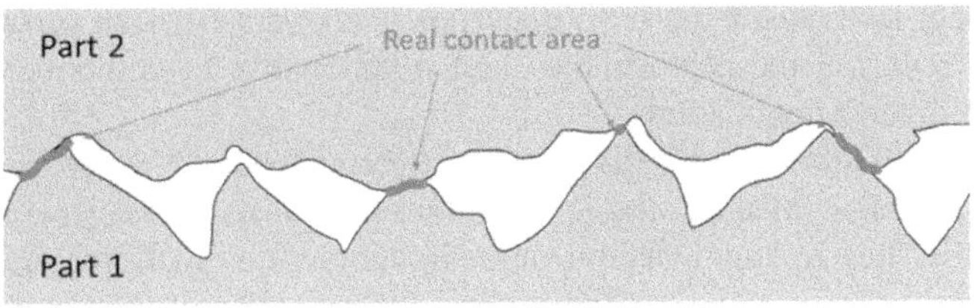

FIGURE 7.1 A scheme of surface roughness and real contact area.

schematically sketched in Figure 7.1, it is possible to distinguish a sequence of peaks and valleys. The relative distance between peaks and valleys is generally used to provide a measurement of roughness along a given direction.

The surface finishing of the two contacting surfaces holds paramount significance in dictating the tribological behavior of a contact pair, as it directly impacts the macroscopic coefficient of friction.

First, roughness plays a crucial role in determining the actual contact surface between the interfacing parts. As depicted in Figure 7.1, the areas where the two components are in contact constitute only a fraction of the nominal mating surface. These zones may also be related to the regions where the previously discussed chemical interactions between the materials take place (as detailed in Section 7.2.1). Higher roughness typically reduces the genuine contact surface between the components and, consequently, worsens the adhesion between the materials.

The latter assertion might suggest that higher surface roughness could lead to a reduction in the coefficient of friction. However, empirical observations frequently demonstrate an opposing trend, because additional detrimental effects are induced by surface roughness.

Elevated roughness, characterized by increased peak height, if compared to valleys, may foster interlocking between the asperities (microscopic irregularities) of the two mating surfaces. This phenomenon creates a mechanical hindrance to the relative sliding of the parts, thereby sharply increasing the apparent coefficient of friction. Furthermore, the energy required to surmount this resistance may result in the fracturing of these peaks, leading to surface wear and abrasion, which is particularly prominent in brittle materials.

Another consequential effect of surface roughness involves the presence of cavities that trap lubricants. This phenomenon will be further explored in subsequent sections to address its implications.

7.2.3 Surface Treatments

Numerous methodologies and techniques are available within industrial practices to selectively modify and enhance the physical as well as mechanical attributes of components. These methods encompass a wide spectrum, including mechanical, thermal, chemical, and thermochemical treatments. Their widespread use in industry is primarily aimed at the achievement of specific and desired properties on the external surfaces of these components. Notably, these treatments significantly affect the tribological characteristics of the overall system, and have the capability of yielding substantial alterations in most of the previously discussed properties, such as chemical composition, hardness, and surface finishing.

In many practical applications, surface treatments play a pivotal role in preparing the surfaces of individual parts intended for assembly, and often make it possible to accomplish the prescribed performance criteria. Therefore, it becomes essential for the engineer holding the responsibility of threaded joint design to possess a comprehensive awareness of these treatments. This understanding can support informed decisions and sufficiently accurate friction coefficient estimations, which are crucial in determining the joint's functionality and durability.

In addition, these surface treatment techniques can also be strategically applied to screws and nuts and endow these elements with distinct performance characteristics that may be considerably enhanced with respect to those of the base materials. This versatility in surface treatment applications also entails tailoring the performance and functionality of components to meet specific requirements within diverse applications.

Furthermore, it is essential to note that surface treatments serve a diverse array of purposes when applied to screws. For instance, the application of black-oxidation treatment to steel screws represents a preventive measure against environmental corrosion. Such treatments serve as protective measures, ensuring the prolonged integrity and functionality of the components within their operational environments.

7.2.4 Coating

Coating involves the application of an external layer onto the surface of a component, differing from the material of the base part. Similarly to surface treatments, this alteration impacts the chemical composition, mechanical attributes, and surface finishing of the external layer, thus implying related enhancements. However, coating introduces an additional concern associated with the potential detachment of the applied layer from the underlying base material.

The detachment issue arises due to variations in the chemical and mechanical properties between the coating and the base material. Depending on the type of coating and the intensity of contact forces, there is a likelihood that the applied layer might abrade or detach from the base material. This detachment significantly affects the local contact conditions because contact is substantially transferred to the base material. Furthermore, any debris entrapped between the contacting parts exacerbates the alteration of contact conditions, intensifying surface abrasion. This phenomenon is particularly pronounced when the coating is compromised, which leads to increased wear and potential degradation of the surfaces.

Like surface treatments, coating can be applied not only to the parts intended for clamping but also to the screws or nuts themselves. A relevant example is the electroplating of steel screws, such as zinc electroplating, aimed at preventing oxidation and enhancing corrosion resistance.

In summary, on the one hand, coatings offer several advantages such as altering surface properties and providing protection against environmental degradation; on the other hand, the potential for detachment and subsequent alteration of contact conditions requires careful consideration by designers and engineers when implementing coating techniques in bolted joint applications. Understanding the specific properties of coatings and their interactions with the base materials is crucial to ensure the desired functionality and durability of the assembled components.

7.2.5 Lubrication

Lubrication involves the insertion of a malleable film between surfaces in contact, aiming to lower friction. This lubricant can manifest in liquid or solid form. The thickness of this lubricating layer plays a crucial role in mitigating or completely preventing direct contact between the interfacing surfaces. Thus, the application of the lubricant must ensure a uniform spread across the contacting surfaces to effectively reduce friction.

Choosing an appropriate lubricant demands a comprehensive consideration of various factors, including the operating environment and the potential interactions between the lubricant and the base materials involved in the contact surfaces. It is essential to select a lubricant that is suitable both to the environment and to the interfacing materials.

However, while lubricants are typically used to decrease the coefficient of friction and improve the ease of movement between parts, they can potentially induce adverse effects on the self-relaxation and stability of threaded joints. These effects must be thoroughly characterized to ensure the required performance and the safety and durability of the connection following a suitable tightening.

Therefore, lubrication is indeed an effective means to reduce friction between contacting surfaces, but it needs a meticulous selection process and a thorough understanding of its potential impacts on threaded joints. The fulfillment of a balance between friction reduction and joint structural integrity preservation is crucial for ensuring the desired performance and durability of the connection.

7.2.6 Number of Tightenings

Up to this point, surface properties have been predominantly perceived as static, determined solely by the material's intrinsic characteristics and its manufacturing processes. However, it is crucial to remark that wear resulting from the sliding motion between parts is likely to significantly alter the morphology and properties of the involved contact surfaces. This holds particular importance in the context of bolts subjected to multiple cycles of tightening and loosening, wherein progressive surface degradation can lead to fluctuations in coefficients of friction. Consequently, this variation influences the relationship between tightening torque and preload force.

In real-world applications, accurately predicting these fluctuations poses a challenge. As a practical solution, keeping track of the number of re-tightenings becomes essential, because it makes it possible to adjust prescribed torque accordingly. However, in common practice, maintaining consistent friction coefficients across these cycles often holds greater significance than precisely determining their values and trends. Consequently, minimizing variations in contact conditions over multiple tightening cycles becomes a relevant issue.

The consistent behavior of coefficients of friction through multiple tightening cycles emerges as a critical factor, as it directly impacts the relationship between torque application and resulting preload force. Striving to minimize deviations in contact conditions over repeated cycles of tightening and loosening becomes imperative in ensuring the reliability and durability of bolted joints. This focus on maintaining consistent frictional behavior over time is a key consideration in the design and maintenance of these joints, ultimately contributing to their long-term performance and reliability.

7.3 EXPERIMENTAL METHODS FOR DETERMINING FRICTION COEFFICIENTS IN BOLTED JOINTS

The tribology of bolted joints is highly influenced by numerous correlated factors, making it exceedingly challenging to accurately estimate real friction coefficients through theoretical considerations only. Consequently, experimental measurement of friction coefficients at the interface between the mating surfaces becomes crucial for practical applications.

The Motosh formula provides a means to glean comprehensive information about tribological contact by measuring three out of the four torques in Eq. (7.1). It is worth mentioning that determining the preload force is akin to measuring the pitch torque (T_p), as a geometric constant allows for the calculation of one from the other. A comprehensive guide regarding the variables to be measured and corresponding calculation formulas for attaining specific information is outlined in the international standard ISO 16047:2005+A1:2012.

Section 7.2.1, based on the aforementioned standard, provides an example elucidating how the mean coefficient of friction can be determined using straightforward equipment. Meanwhile, Section 7.2.2 introduces a more sophisticated testing apparatus with the capability of determining all coefficients of friction, distinctly separating the contributions of bearing and thread friction.

7.3.1 Simplified Testing to Measure the Average Coefficient of Friction

The mean coefficient of friction μ_m can be defined as that value which allows satisfying Eq. (7.1) under the hypothesis that $\mu_b = \mu_t = \mu_m$. The value of μ_m can be easily measured by using a torque wrench and a strain gauge. A schematic representation of the experimental setup is shown in Figure 7.2a.

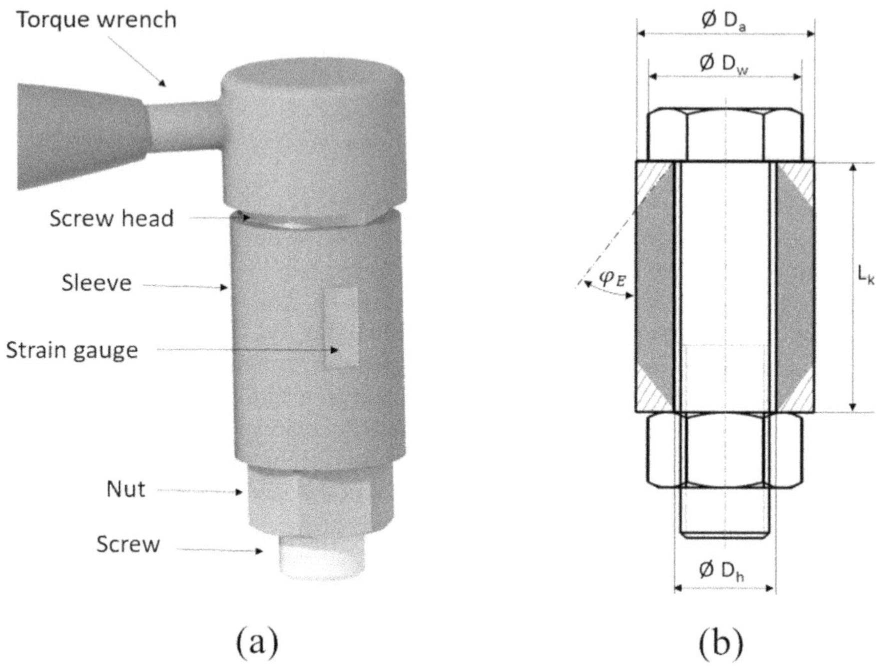

FIGURE 7.2 (a) Scheme of the test and (b) characteristic dimensions of the connection.

The bolt in Figure 7.2 is tightened onto a cylindrical sleeve, while a quarter-bridge strain gauge is bonded to the external surface of the sleeve, aligned with the cylinder's main axis. Careful consideration must be given to the dimensions of the sleeve and the strain gauge's positioning to ensure a consistent compression of the sleeve within the gauge's region. To validate this condition, the angle of the substitutional deformation cone must be computed as in Eq. (7.5) [1]:

$$\varphi_E = \arctan\left[0.348 + 0.013\ln\left(\frac{L_k}{D_w}\right) + 0.193\ln\left(\frac{D_A}{D_w}\right)\right] \tag{7.5}$$

where D_w is the external diameter of the screw head, D_A is the external diameter of the sleeve, D_h is the hole diameter and L_k is the length of the sleeve, as shown in Figure 7.2b.

The strain gauge allows for measuring the strain ε_S of the sleeve. Since the bolt and the sleeve are in series during the tightening phase, the preload F of the screw can be easily calculated by Eq. (7.6)

$$F = E\ A_{eq}\ \varepsilon_S \tag{7.6}$$

where the equivalent compressed area can be calculated, according to [2], as in Eq. (7.7)

$$A_{eq} = \frac{\pi}{4}\left(D_w^2 - D_h^2\right) + \frac{\pi}{8}\ D_w\ \left(D_A - D_w\right)\left[\left(\sqrt[3]{\frac{L_k\ D_w}{D_A^2}} + 1\right)^2 - 1\right] \tag{7.7}$$

As the tightening torque T is measured and controlled by the torque wrench, it is possible to derive μ_m from the Motosh formula; the solution becomes as shown in Eq. (7.8):

$$\mu_m = \frac{\dfrac{T}{F} - \dfrac{p}{2\ \pi}}{\dfrac{R_t}{\cos\beta} + R_b} \tag{7.8}$$

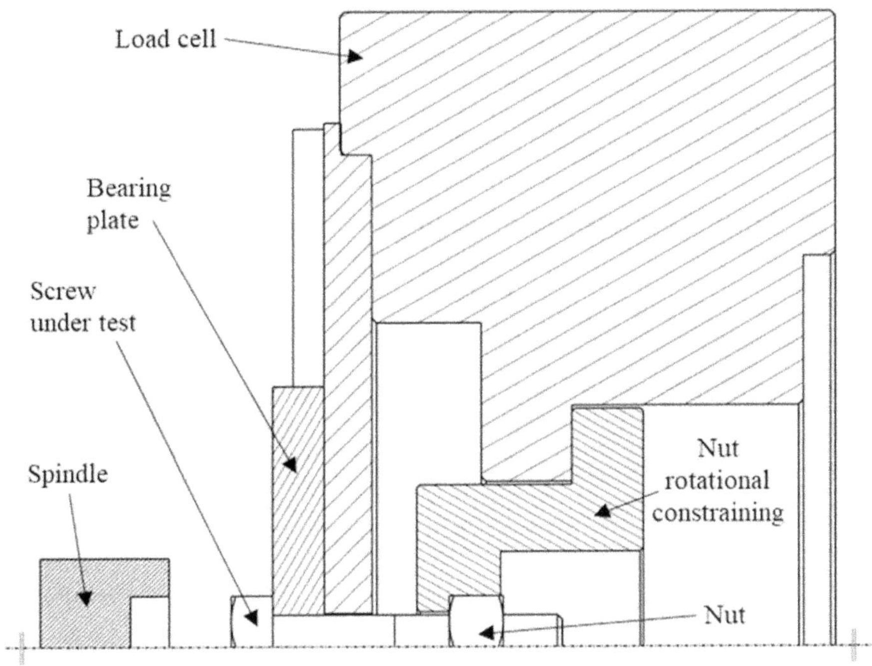

FIGURE 7.3 Schematic representation of a testing rig for measuring all the coefficients of friction during tightening.

7.3.2 Testing Rig to Measure All the Coefficients of Friction

Figure 7.3 [3] shows a schematic representation of a testing rig that can be used to fully characterize the tribological characteristics of the joint.

The screw under test is tightened on a bearing plate. The latter is coupled to a measuring head. A nut rotational constraining is used to prevent the rotation of the nut when the bolt is tightened. As shown in the scheme of Figure 7.3, the nut rotational constraining is free with respect to the load cell.

The screw is tightened by a spindle that can measure the tightening torque T applied to the screw. The measuring head comprises an axial and a ring torsional load cell. During tightening, the axial load cell is capable of measuring the preload force F applied to the screw, while the ring torsional load cell measures the bearing torque T_b. the shank torque, $T_t + T_p$, can then be immediately estimated by Eq. (7.1). As mentioned above, the pitch torque may be easily determined following the experimental measurement of the axial preload because it is directly proportional through thread geometrical parameters. Thus, it is possible to work out the T_t contribution, which allows for friction estimation. The coefficients of frictions μ_t and μ_b can be thus easily calculated using Eqs. (7.3) and (7.4), respectively.

The picture of a real testing rig is shown in Figure 7.4.

As mentioned, this approach allows for getting more detailed information by separating the effects of bearing from those of the thread, i.e., distinguishing between the two coefficients of friction. Another relevant advantage is the opportunity to have strict control of the tightening process, e.g. setting the tightening speed and possible waiting times. In addition, it is possible to choose between different input signals to control the test cycle; for example, the ending condition of the test can be set to be the tightening torque or the actual preload of the screw.

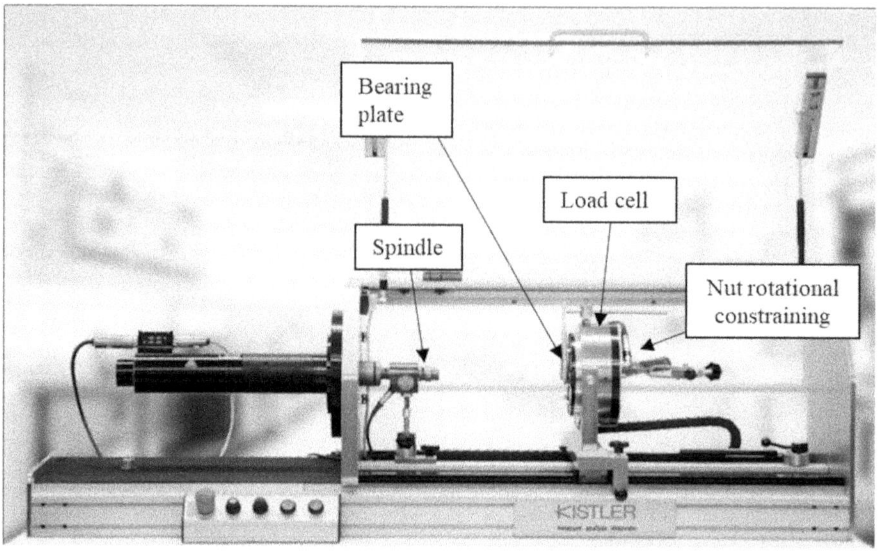

FIGURE 7.4 Picture of a testing rig capable of measuring all the coefficients of friction during tightening (courtesy of Kistler Group).

7.3.3 Testing for Real-World Applications

As highlighted in the previous sections, the tribological behavior of the system is remarkably influenced by the specific characteristics of the screw, nut, and the components intended for joining. Therefore, achieving an accurate estimation of the actual tribological response demands that the base part utilized for testing closely resembles the one intended for the final application. Depending on the component's geometry and complexity, a simplified testing approach using a torque wrench and a strain gauge can often be directly applied to the part designated for assembly.

In more complex scenarios, particularly when utilizing sophisticated testing rigs like the one outlined in Section 7.2.2, it becomes necessary to extract a bearing from the part where the bolt is intended to be applied. However, this extraction process can significantly contribute to the overall cost of the experimental campaign, especially when comparing the effects of several multiple treatments or manufacturing operations applied to the final part.

Therefore, the selection of appropriate testing methodologies together with the consideration of cost-effective approaches become pivotal factors when designing and running well-comprehensive tribological campaigns at affordable costs.

7.4 EFFECT OF LUBRICANTS AND SURFACE MODIFICATIONS

7.4.1 Effect of Lubrication

As discussed in Section 7.1.5, lubrication creates an intermediate layer between the parts in contact. This generally reduces the coefficient of friction at the interfaces during tightening. For example, it was shown that the adoption of a ceramic paste allows to halve the coefficient of friction in case of black oxidized 12.9 screws tightened on a cast iron flange.

Figure 7.5 shows the change in the mean coefficient of friction over multiple retightenings in dry and lubricated conditions [4].

As shown in Figure 7.5, in the case of dry bolts, the mean coefficient of friction tends to increase over tightenings. Conversely, it can be observed that the friction coefficient trend is barely flat when

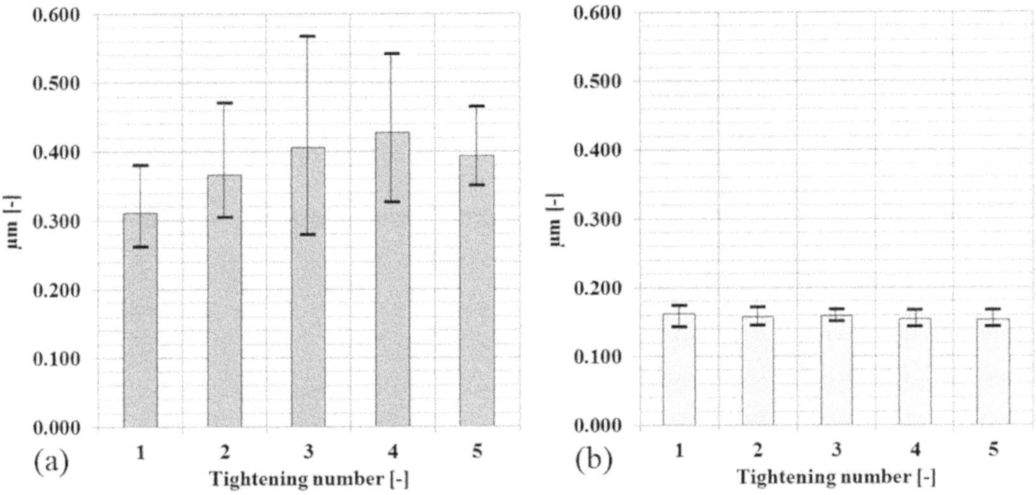

FIGURE 7.5 Mean friction coefficient evolution for (a) dry and (b) paste lubricated joint [4].

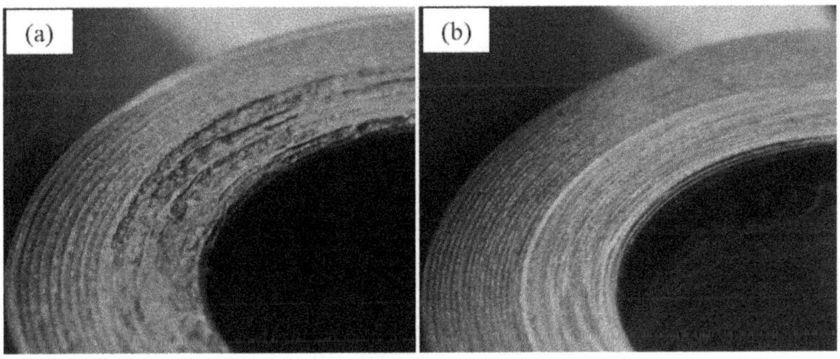

FIGURE 7.6 (a) Sleeve surface after dry tightening, (b) Sleeve surface after tightening with ceramic paste [4].

the lubricant paste is applied. This can be explained by considering the different wear of the underhead surfaces in dry and lubricated conditions.

Figure 7.6a and b depict two sleeve underhead surfaces after five re-tightenings in, respectively, dry and lubricated (treated with ceramic paste) conditions [4].

Across five re-tightening cycles, the ceramic paste-treated surface in Figure 7.6a exhibits notably reduced degradation compared to its dry-operated counterpart in Figure 7.6b. The prominent large pits that are visible in Figure 7.6a indicate significant material loss. Indeed, as far as dry condition is concerned, the wear of the underhead over multiple tightenings determines the detachment of debris from the cast iron part. Figure 7.7 shows the underhead of two screws after five tightenings in dry and lubricated conditions, respectively.

Based on the image in Figure 7.7a, it is evident that debris from the cast iron flange has been transferred onto the steel screw. Conversely, Figure 7.7b exhibits a clean and smooth surface without any visible debris. This discrepancy highlights the impact of lubrication, and, in turn, the resulting surface conditions on the tribological behavior between the components.

The reduction in the coefficient of friction not only lowers the stress on the bolt during tightening but also ensures consistency in achieving the desired preload of the screw, and, consequently joint safety, even after multiple disassembly and reassembly cycles.

FIGURE 7.7 Screw underhead after operation: (a) dry tightening and (b) tightening with ceramic paste [4].

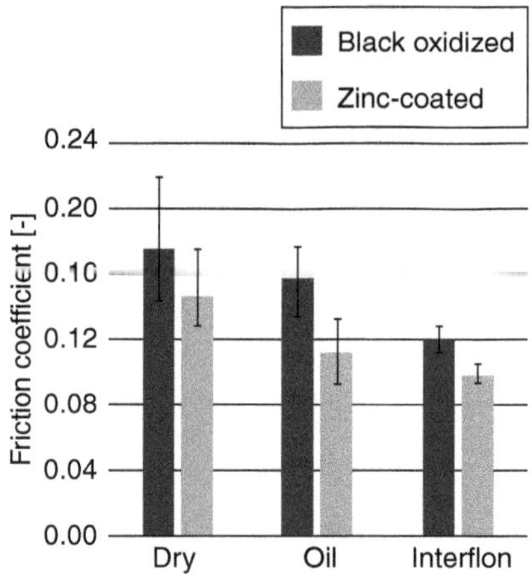

FIGURE 7.8 Bar graph indicating the effects of screw plating and of lubrication on friction [8].

The analysis of Figure 7.5 reveals a noteworthy observation: the standard variation of μ_m is significantly higher in the case of dry screws. This discrepancy is attributed to the random nature of wear phenomena. Therefore, it can be concluded that the adoption of lubricants makes the tribological behavior of bolted joints steady and well repeatable (provided the same lubricant is used complying with the same application procedure), particularly across numerous tightening cycles.

Various types of lubricants, including ceramic paste, grease, mineral oil, and polymeric layers, have been investigated in the relevant literature [5–7]. However, as mentioned above, the selection of the most proper lubricant depends on multiple factors such as materials, dimensions, operating temperatures, and applied tightening torque.

As previously mentioned, the factors influencing the tribological behavior of bolted joints are interrelated, and, therefore, their effects cannot be easily separated. For instance, the efficacy of lubrication greatly varies depending on the plating of the screw, a phenomenon being highlighted by the bar graph in Figure 7.8 [8].

This outcome exemplifies the intricate interplay among multiple factors that influence the tribological characteristics of bolted joints.

Based on the insights from Figure 7.8, both black oxidation and zinc coating yield a reduction in the coefficient of friction within the tested joint. In particular, the efficiency of zinc coating surpasses that of black oxidation owing to its capability to create a smoother treated surface. Notably, this enhanced smoothness plays a key role in reducing friction between the components.

Additionally, the comparison between solid lubricant (interflon) and oil reveals that the solid lubricant proves more effective in decreasing the coefficient of friction within the joint. This finding suggests that solid lubricants, such as interflon, exhibit superior performance in reducing friction compared to liquid lubricants like oil within the tested bolted joint. The solid lubricant proves to be efficient for both the two plating treatments, but frictional coefficient reduction appears to be greater for black-oxidized fasteners.

7.4.2 Surface Treatment

Several mechanical treatments can be applied to the surface of screws. The most common treatments are deep rolling and shot-peening. Experimental observations demonstrated that deep rolling applied at the fillet between the screw shank and the head does not have any significant influence on the tribological behavior of screws. In fact, the surface at the underhead is left almost untouched by this treatment that is usually applied to enhance the fatigue resistance of the bolt. Conversely, it was found that shot peening at the underhead surface can be effectively used to modify the bearing coefficient of friction μ_b [9]. Figure 7.9 shows the evolution of μ_b over-tightening cycles in a 36 NiCrMo grade 13.9 screw fastened on a steel plate [9].

Shot peening is a cold working method utilized to enhance the fatigue life and strength of metallic components, including screws. It involves bombarding a material's surface with small spherical media, known as shot, at high velocities.

When applied to screws, shot peening induces controlled plastic deformation in the surface layer, creating numerous small dimples due to the impact of the shot. These indentations lead to the generation of compressive residual stresses. Typically, screw threads are protected during shot peening to prevent damage, thus confining the tribological modifications induced by this treatment to the screw's underhead.

Based on the trends in Figure 7.9, shot peening treatments entail a reduction in the coefficient of friction at the bearing with respect to dry conditions. Moreover, shot-peened surfaces exhibit a less pronounced increase in friction over multiple tightening cycles.

The impact of shot peening on friction and wear characteristics can be explained in the light of the formation of dimples on the surface, inducing compressive residual stresses. These dimples play a crucial role in beneficially influencing both friction and wear characteristics. The porous texture and increased contact points resulting from these dimples contribute to friction reduction.

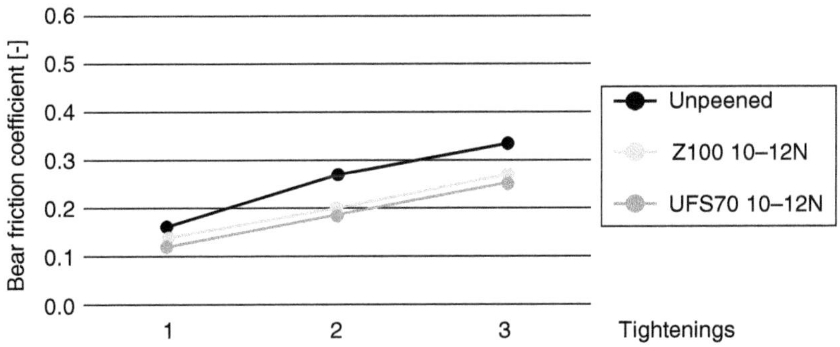

FIGURE 7.9 Bearing friction coefficient over multiple tightening for unpeened and shot-peened bolts [3].

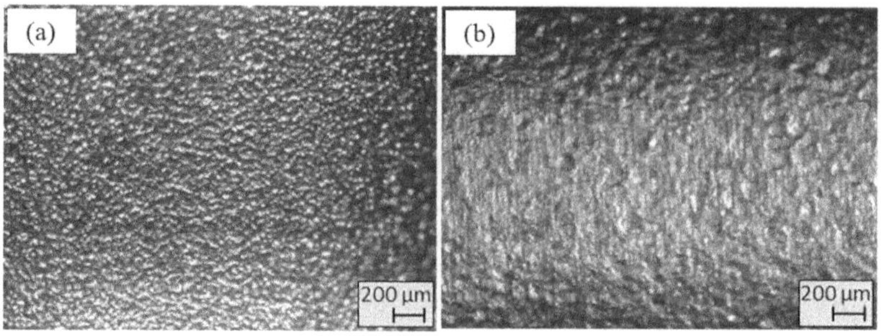

FIGURE 7.10 Bolt surface shot-peened with (a) UFS70 10-12N and (b) S110 8.8N treatments [3].

Moreover, the generated debris during wear tends to accumulate within these indentations, reducing its impact on progressive wear and friction during surface sliding [9].

Figure 7.8 illustrates diverse trends stemming from distinct shot-peening treatments, such as Z100 10-12N and UFS70 10-12N. The efficacy of shot peening may depend on various parameters, including the shot material and size, impact energy, and treatment coverage. The size, shape, and material composition of the shot peening beads influence treatment (Almen) intensity and coverage. Variations in shot size impact the depth of residual stresses induced, while different materials affect surface effects on treated components. Additionally, parameters like impact velocity, angle, and coverage significantly affect the distribution and entity of induced residual compressive stresses. Careful control of these parameters ensures uniform treatment across the surface area targeted for shot peening.

However, detailed surface morphologies post-shot-peening treatments, such as those shown in Figure 7.10 depicting UFS70 10-12N and S110 8.8N treatments, can further exemplify the alterations induced by these treatments on bolt surfaces [3].

Comparing the two pictures in Figure 7.10, it can be observed that the two treatments produce dimples with different dimensions. Specifically, the average size of dimples in UFS70 10-12N is 60.8 ± 13.3 µm, while the average size of S110 8.8N is 153.1 ± 15.4 µm. The two treatments are also characterized by different surface roughness Ra, that is Ra = 2.0 ÷ 2.1 µm for UFS70 10-12N and Ra = 0.5 ÷ 1.0 µm for S110 8.8N. These differences are due to the higher dimensions and the lower impact energy of the shots in the case of S110 8.8N treatment [3]. In particular, the latter treatment takes advantage of the reduced impact force that is spread over a wider contact area as an effect of the greater shot size. This results in a smoother surface, which turns into a remarkably low roughness level. At the same time, the aforementioned features arising from the textured surface are maintained, which, as highlighted below, reflects in a remarkably improved frictional response.

Figure 7.11 shows the different tribological behavior of 42CrMo screws shot-peened with these two treatments and tightened multiple times.

The observation of the outcomes in Figure 7.11 reveals a discernible reduction in the bearing coefficient of friction when comparing surface-treated screws to non-treated ones. Notably, both treatments exhibit lower coefficients of friction at the bearing interface if compared to those for untreated screws. However, in the case of the UFS70 10-12N treatment tested in dry conditions, there is an observed increase in μ_b over tightening cycles, albeit less pronounced. Remarkably, this treatment leads to a significant reduction in frictional coefficient when lubrication is applied.

The phenomenon of reduced frictional variations in lubricated scenarios after the peening treatment can be clarified, considering the hydrostatic effect of the lubricant confined within the dimples generated by the shot-peening treatment. The combination of a lubricant and shot-peening exhibits a synergistic effect, leading to a substantial reduction in surface wear, as evidenced in Figure 7.12 [9].

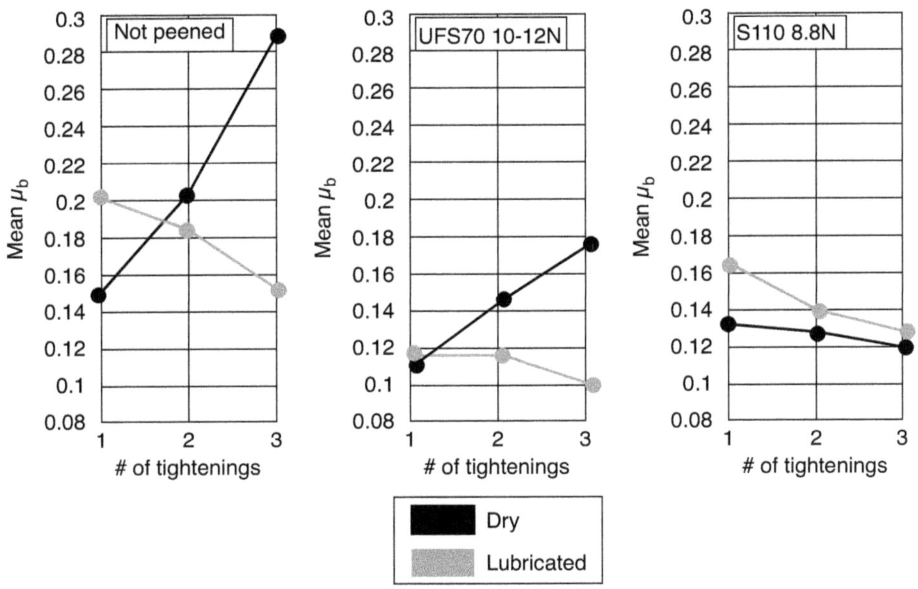

FIGURE 7.11 Bearing coefficient of friction of 42CrMoV with different surface treatments [3].

FIGURE 7.12 Microscope images of the surfaces at the underhead after tribological tests in lubricated condition with reference to (a) an unpeened and (b) a UFS70 10-12N peened screw [9].

In the case of S110 8.8N treatment, which is characterized by a larger size of the dimples, Figure 7.11 shows that the bearing coefficient of friction is stable over multiple tightening cycles. This is indeed related to the previously mentioned lower roughness of the surface. In addition, the residual compression stress induced by the shot-peening treatment entails a considerable reduction in the surface wear. This is clearly visible, even in dry conditions, in Figure 7.13 by comparing the underhead surface of the shot-peened screw to that of a regular screw.

The observations from Figure 7.11 highlight a crucial finding: the presence of lubrication leads to a decrease in the coefficient of friction over subsequent tightening cycles, which is even more relevant when larger dimples are present. However, this reduction effect poses challenges as it may introduce a bias in the relationship between tightening torque and screw preload.

In light of these findings, it can be inferred that shot-peening treatments characterized by smaller dimensions of dimples, typically achieved through the use of smaller shots and higher impact energy, are more preferable in scenarios where bolted joints operate under lubricated conditions. Conversely, treatments that result in larger beads and lower surface roughness should be preferred for bolted joints operating in dry conditions. It is worth mentioning that in the latter case

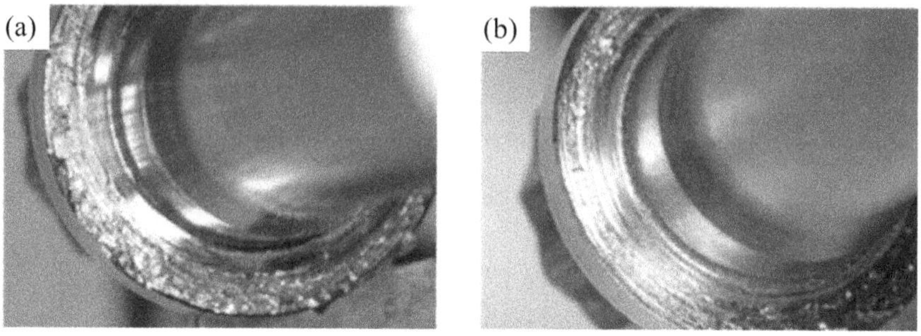

FIGURE 7.13 Underhead surface after three tightening cycles in dry conditions with reference to (a) a non-treated screw and (b) a screw shot-peened with S110 8.8N [9].

the achievable properties in dry conditions are particularly interesting in the framework of a set of industrial applications. The achievable friction coefficients, in the range between 0.1 and 0.15, even after repeated tightenings, are well aligned with those in lubricated conditions. Therefore, a proper choice of peening parameters can lead to a new trend involving the reduction of lubricant use with positive outcomes on eco-sustainability.

7.4.3 Effect of Tribology on Self-Loosening

Indeed, while much emphasis has been placed on understanding the tribological effects during the tightening phase of bolted joints, it is equally crucial to consider the impact of tribology during their operational phase. Although reducing the coefficient of friction brings about benefits such as minimizing energy dissipation and surface wear during tightening, its effects on the functioning of bolted joints in service are noteworthy.

So important is the impact of coefficients of friction on the phenomenon of self-loosening in threaded fasteners, that it has been the topic of several studies. In particular, this aspect has been investigated in an analytical model proposed by Housari and Nassar [10], which accounts for the influence of friction on fastener self-loosening due to vibrations and dynamic loading.

Potential self-loosening in a screw is highly influenced by coefficients of friction: regarding this point, a relationship is displayed in the graph shown in Figure 7.14 [10]. Being able to understand and control friction coefficients becomes crucial for mitigating self-loosening tendencies, thus ensuring the integrity and reliability of bolted joints during their operational phase.

The trend depicted in Figure 7.14 highlights the importance of not excessively reducing coefficients of friction, as low values could have adverse effects on the self-loosening behavior of the bolted joint.

As outlined in the preceding sections, the bearing friction coefficient significantly reduces energy dissipation during the tightening process, thus lowering energy consumption, which is a noticeable advantage. Hence, most recent trends involve reducing the bearing coefficient (μ_b) while maintaining a reasonably higher coefficient (μ_t) on the thread. This strategic balance is exemplified in treatments like the previously described shot peening, which specifically impacts the underhead of the screw. Such considerations are also relevant when applying selective lubrication. Studies have shown that targeted lubrication, focused solely on the screw head, can notably decrease wear across multiple tightening cycles without detrimentally impacting the self-loosening tendency of the bolted joint [8].

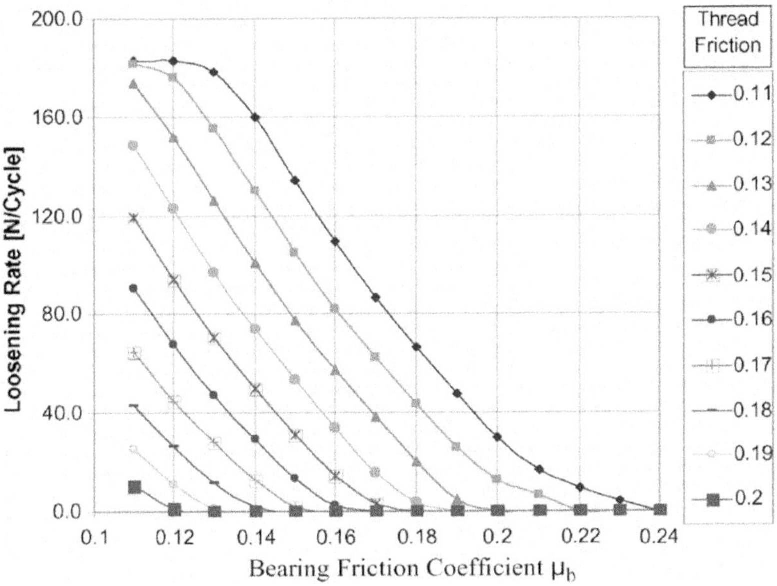

FIGURE 7.14 Effect of friction coefficients on self-loosening [10].

BIBLIOGRAPHY

[1] V. D. I. Richtlinie, "2230," *Bl.* 1" *Syst. Berechung hochbeanspruchter Schraubenverbindungen", VDI Düsseld.*, 2001.

[2] G. Niemann, H. Winter, and B. R. Höhn, "Maschinenelemente, Band 1," Konstr. und Berechnung von Verbindungen, 2005.

[3] D. Croccolo, M. De Agostinis, S. Fini, M. Mele, and G. Olmi, "Effect of different underhead shot-peening and lubrication conditions on high-strength screws undergoing multiple tightenings," *Tribol. Int.*, vol. 188, p. 108874, 2023.

[4] M. De Agostinis, S. Fini, and G. Olmi, "The influence of lubrication on the frictional characteristics of threaded joints for planetary gearboxes," *Proc. Inst. Mech. Eng. Part C J. Mech. Eng. Sci.*, vol. 230, no. 15, pp. 2553–2563, 2016, doi: 10.1177/0954406215605863.

[5] Q. Zou, T. S. Sun, S. A. Nassar, G. C. Barber, and A. K. Gumul, "Effect of lubrication on friction and torque-tension relationship in threaded fasteners," *Tribol.* Trans., vol. 50, no. 1, pp. 127–136, 2007, doi: 10.1080/10402000601105490.

[6] F. Robusto, M. Gerini-Romagnoli, M. De Agostinis, J. H. Lee, and S. A. Nassar, "Effect of using 3D printed parts on the torque-tension relationship of threaded joints," *ASME Int. Mech. Eng. Congr. Expo. Proc.*, vol. 2-A, pp. 1–8, 2022, doi: 10.1115/IMECE2022-95614.

[7] J. B. Aycock, J. Wilson, and A. Bausman, "Variables affecting nut factors in bolted flange connections," *Am. Soc. Mech. Eng. Press. Vessel. Pip. Div. PVP*, vol. 2, pp. 1–10, 2019, doi: 10.1115/PVP2019-93721.

[8] D. Croccolo, M. De Agostinis, S. Fini, and G. Olmi, "Tribological properties of bolts depending on different screw coatings and lubrications: An experimental study," *Tribol.* Int., vol. 107, pp. 199–205, 2017, doi: 10.1016/j.triboint.2016.11.028.

[9] D. Croccolo, M. De Agostinis, S. Fini, G. Olmi, L. Paiardini, and F. Robusto, "Tribological properties of connecting rod high strength screws improved by surface peening treatments," *Metals*, vol. 10, no. 3, 2020, doi: 10.3390/met10030344.

[10] B. A. Housari and S. A. Nassar, "Effect of thread and bearing friction coefficients on the vibration-induced loosening of threaded fasteners," *J. Vib. Acoust.*, vol. 129, no. 4, pp. 484–494, 2007, doi: 10.1115/1.2748473.

Part III

Threaded Fasteners

8 Computing the Strength of a Fastener

Joseph Barron

8.1 INTRODUCTION

The strength of a threaded fastener is the most basic consideration in the design or analysis of bolted joints. However, as with many engineering problems, the first step in the calculation procedure is to determine the actual or expected loading conditions. Once the loading conditions are defined, the threaded parts can be evaluated for acceptability. This chapter discusses the issues that have to be considered in evaluating the strength of a threaded fastener. The loading conditions are defined in Section IV of this handbook. The techniques presented here are applicable to threaded connections as well as threaded fasteners.

8.2 POSSIBLE FAILURE MODES

Most threaded fasteners that are installed and put into service never fail. However, many fasteners do fail. It is the designer's responsibility to select the correct material and size combination for a given application. The first step in the selection process is to understand the possible failure modes. Then it will be possible to develop calculation procedures to avoid these failures. There are several static failure modes for a threaded connection. The governing mode is determined by the combination of loading conditions, thread dimensions, and material properties. For connections subject to axial loading, failure can occur by

1. Stripping of external threads
2. Stripping of internal threads
3. Tensile failure of external threads

For connections subject to shear loading, failure can occur by

4. Shearing of external fastener

These four failure modes represent the catastrophic failure of the threaded parts. The basic premise of design is to preclude such failure. Moreover, there are usually criteria that define "failure" prior to the onset of actual part breakage. This could be thread yielding, which may make disassembly and subsequent reassembly impossible. Or it could be a maximum allowable stress imposed by a design specification.

8.3 BASIC DESIGN RULE

The preceding section discussed the various failure modes for threaded connections. Thread stripping failures should be avoided. In the worst case, such failures render the parts useless. In the best case, expensive and time-consuming repairs are usually required. Also, thread stripping failures

DOI: 10.1201/9781003501183-11

may not be readily apparent during installation or subsequent inspections. A broken bolt is much more noticeable than a bolt that is assembled in a stripped hole. Therefore, threaded connections should be designed using the following rule:

> Axially loaded threaded connections should fail by tensile failure of the external thread, not by thread stripping.

This rule requires that the stripping strength of both the internal and external threads exceed the tensile strength of the external thread. However, as with many rules, there are exceptions. Some threaded connections are not intended to develop the full strength of the externally threaded part. An example would be a bearing retaining nut used to secure a ball bearing on a shaft. Here, the nut must only withstand the maximum axial design load of the bearing, not the tensile load of the shaft.

8.4 TENSILE LOADING

The tensile strength of a threaded fastener could be computed like that of any other machine part. The cross-sectional area multiplied by the material strength would equal the strength of the fastener. The tensile area of the body or unthreaded portion of a bolt or screw would simply be its cross-sectional area.

$$A_B = \frac{\pi}{4} D_B^2 \tag{8.1}$$

where

A_B=body area (in.^{2}, mm^2)
D_B=diameter of body (in., mm).

However, the tensile area of an external thread is not readily apparent upon inspection. One could use the minor diameter. This cross section is defined as the root area (AR) and is calculated as follows:

$$\text{AR} = \frac{\pi}{4}\left(D - \frac{1.3}{n}\right)^2 \quad \text{(for inch series)} \tag{8.2}$$

$$\text{AR} = \frac{\pi}{4}(D - 1.3p)^2 \quad \text{(for metric series)} \tag{8.3}$$

where

AR=root area (in.^{2}, mm^2)
D=nominal size (in.)
d=nominal size (mm)
p=thread pitch (mm)
n=number of threads per inch (in.$^{-1}$).

Yet when threaded fasteners or threaded specimens are tensile tested, they exhibit strengths that are greater than that predicted by the material strength and the root area. These parts behave as if they had a larger cross-sectional area. An empirical formula was developed that accounts for this phenomenon. The tensile area is based on a diameter midway between the minor and pitch diameters and is calculated as follows [1,2]:

$$AT = \frac{\pi}{4}\left(D - \frac{0.9743}{n}\right)^2 \qquad \text{(for inch series)} \tag{8.4}$$

$$AT = \frac{\pi}{4}(d - 0.9382p)^2 \qquad \text{(for metric series)} \tag{8.5}$$

where AT = tensile area (in.^{2}, mm^2).

While the tensile area is the most appropriate for tensile loading conditions, some design standards and codes still specify that the root area be used. The use of this smaller area "builds in" an additional factor of safety. The above tensile area equations are applicable to the standard Unified and metric thread forms. Thread forms with a defined root radius, known as *J* threads, require different equations because they have larger external thread root diameters. The tensile area of *J* threads is calculated as follows [3,4]:

$$AT_J = \frac{\pi}{4}\left(D - \frac{0.6495}{n}\right)^2 \qquad \text{(for inch UNJ series)} \tag{8.6}$$

$$AT_J = \frac{\pi}{4}(d - 0.6495p)^2 \qquad \text{(for metric MJ series)} \tag{8.7}$$

where AT_J = *J* thread tensile area (in.^{2}, mm^2).

SAE standard MA 1520 [4] provides equations for the tensile area of MJ threads that differentiate between threads that are roll threaded before and those roll threaded after heat treatment. However, these values differ by only a few percent from those calculated using Eq. (8.7).

The preceding discussion presented equations for the tensile and root areas of an eternal thread. The equations presupposed the normal geometry of a solid externally threaded part. For parts that do not meet this definition, the cross-sectional areas would have to account for any holes, slots, flutes, or other changes in geometry.

The number of threads in the grip can affect the tensile strength of fasteners. There appears to be a length effect that must be considered. Most commercial bolts and screws are tested with six threads exposed in the grip. For this condition, the standard tensile area equation produces acceptable results. However, most aerospace bolts and screws are tested with two to three threads in the grip. This situation requires the use of a larger cross-sectional area as defined by Eqs. (8.6) and (8.7) even if the fastener does not have J threads.

8.5 SHEAR LOADING

Shear is loading transverse to the axis of the fastener. This is illustrated in Figure 8.1. The shear strength of a fastener would be equal to the shear area multiplied by the shear strength.

$$SS_s = (AR)(S) \qquad \text{(for single shear plane)} \tag{8.8}$$

where

SS_s = shear strength of fastener (lb, N)
S = material shear strength (lb/in.^{2}, N/mm^2).

As shown in Section 8.4, fastener shear area is smaller than the tensile area. Also, the shear strength of typical fastener metals is approximately 60% of the tensile strength. Therefore, shear loading

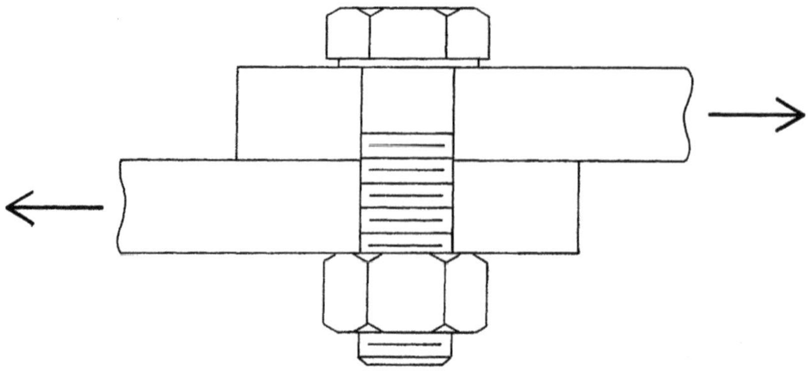

FIGURE 8.1 Bolt loaded in shear with a single shear plane passing through the threaded region.

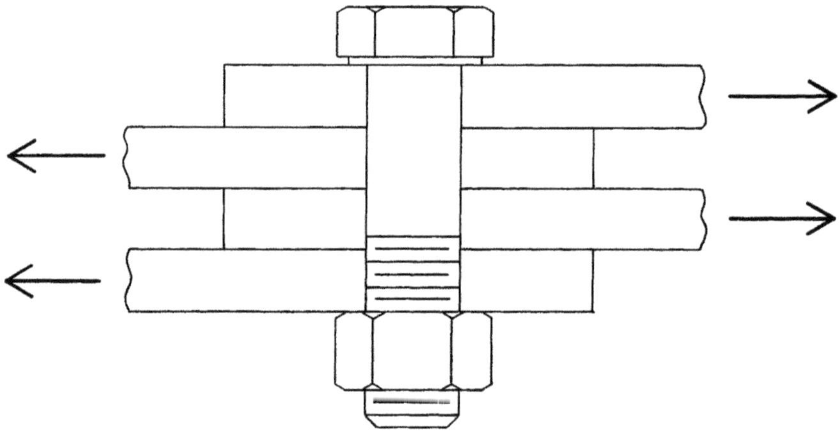

FIGURE 8.2 Bolt loaded in shear with two shear planes passing through the body and one passing through the threaded region.

is not an efficient use of material. To avoid subjecting a fastener to shear, it is recommended that shear pins or another design feature be used to resist the shear forces. The preceding assumed a single shear plane. For applications with more than one fastener shear plane, the shear strength of the fastener would be equal to the total shear area multiplied by the shear strength. An example of a fastener with three loaded shear planes is illustrated in Figure 8.2. The shear strength of this fastener would be

$$\mathrm{SS}_s = (2A_B + \mathrm{AR})S \qquad (\text{for three shear planes}) \tag{8.9}$$

8.6 TORSIONAL LOADING

Since most fasteners are preloaded by an applied torsional moment (torque), the externally threaded fastener is subject to torsional stress. Examining the long-form torque–preload equation described in Ref. [5], the input torque applied to the bolt head or nut (whichever is turned) is apportioned to three reaction torques. The three torques are bearing face friction, thread friction, and bolt stretch (preload). The proportions are typically defined as 50%, 40%, and 10%, respectively. Since the bolt "sees" the thread friction and stretch components, it is subject to 50% of the input torque. This torsion is resisted by the body and/or threads. The shear stress due to torsion can be calculated as follows:

$$\tau = \frac{T\rho}{J} \tag{8.10}$$

where

τ = shear stress (lb/in.², N/mm²)
T = torsional moment (lb-in., N·mm)
ρ = radius at point in question (in., mm)
J = polar second moment of inertia (in.⁴, mm⁴).

For fasteners with a central hole, the maximum shear stress is reduced to

$$\tau_{\max} = \frac{16\mathrm{TD}_O}{\pi\left(D_O^4 - D_I^4\right)} \tag{8.11}$$

where

D_o = outside diameter (in., mm) (use external minor diameter for threaded region)
D_I = diameter of central hole (in., mm).

For solid fasteners, this reduces further to

$$\tau_{\max} = \frac{16T}{\pi D_O^3} \tag{8.12}$$

Torsional failure is not uncommon. Many fasteners fail when they are "wrung off" at assembly or disassembly. Another troublesome instance of torsional failure is seen during the installation of interference fit threads. Here a stud that is slightly larger than normal is screwed or "driven" into a normal size tapped hole. There is a definite interference at the mating pitch diameters. The intent is to produce an assembly in which the stud becomes semi-permanently installed in the hole. Due to the thread interference, a driving torque is required to install the stud even though no preload is being produced. The shear stress developed is defined by Eq. (8.12) except that the torsional moment would be equal to the stud installation torque. Approximate installation torques for medium carbon steel inch size studs can be found in ANSI B1.12 [6]. Torsional failure of interference fit studs during installation can be especially problematic when the stud is made of a low-strength material or is a small nominal size.

8.7 COMBINED LOADING

So far, tensile and shear stresses have been discussed on an individual basis. Since most fasteners are preloaded using an applied torque, it is necessary to examine the effect of these combined stresses on the fastener. Tests show that the breaking strength of a bolt is reduced by 15%–20% during torque tightening [7]. This indicates that the torquing process applies additional stress on the fastener above that produced by the preload. Fortunately, when the torque is removed, it is generally assumed that the torsional stress dissipates quickly. Utilizing the distortion energy or von Mises–Hencky failure theory that is applicable to ductile metals, the combined or effective stress can be calculated as

$$\sigma_c = \sqrt{\sigma^2 + 3\tau^2} \tag{8.13}$$

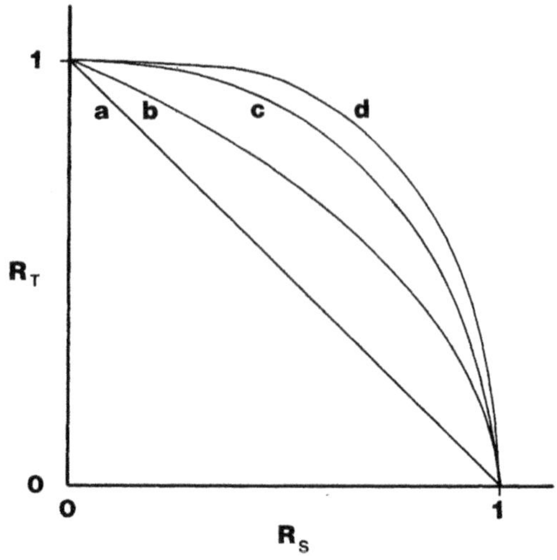

FIGURE 8.3 Plots of interaction equations for combined shear and tensile stresses. Equation (a) is the most conservative, (a) $R_S+R_T=1$; (b) $R_S+R_T^2=1$; (c) $R_S^2+R_T^2=1$; (d) $R_S^3+R_T^2=1$.

where

σ_c=combined stress (lb/in.², N/mm²)
σ=tensile stress (lb/in.², N/mm²)
τ=torsional stress (lb/in.², N/mm²).

This combined stress can then be compared to the material's yield strength or other defined allowable tensile stress. Another method of evaluating combined stresses is to use interaction equations. Here, the shear and tensile stresses are compared with their respective allowable values. These load ratios, R_S and R_T, respectively, are then combined in the interaction equation. Satisfying the equation predicts a safe design. A number of interaction equations have been proposed [8]. These are illustrated in Figure 8.3. By inspection, it is evident that some equations are more conservative than others. MIL-HDBK-5 [9] proposes the use of $R_S^3+R_T^2=1$. By manipulation of Eq. (8.13), the interaction equation for the distortion-energy theory incorporating a factor of safety is

$$R_S^2+R_T^2=\frac{1}{f} \tag{8.14}$$

where

R_S=shear stress ratio=τ/τ_{Allow}
R_T=tensile stress ratio=$\sigma/\sigma_{\text{Allow}}$
τ_{Allow}=allowable shear stress (lb/in.², N/mm²)
σ_{Allow}=allowable tensile stress (lb/in.², N/mm²)
f=factor of safety.

Regardless of which interaction equation is used, it is clear that an increase in the level of one stress must be accompanied by a decrease in the level of the other stress.

8.8 BENDING STRESS

In Section 8.7, it was shown how torsional moments increase the stress on a fastener. There is another source of additional stress—bending. Bending can be introduced into the grip of a fastener by a number of conditions such as

Joint bearing surface nonperpendicular to hole axis
Fastener bearing surface nonperpendicular to fastener axis
Joint members transversely misaligned
Fastener threads eccentric to body
Applied loads such as flange rotation or joint prying

The amount of stress generated by an applied bending moment can easily be calculated from the familiar flexure formula

$$\sigma = \frac{Mc}{I} \tag{8.15}$$

where

M = applied bending moment (lb-in., N·mm)
c = outer radius (in., mm)
I = second moment of area (in.^{4}, mm^4).

Inspection of this formula explains why bending should be avoided. Fasteners are typically slender. Therefore, the moment of inertia of a fastener's cross section is rather small. This situation produces a large stress for a given moment. The resulting bending stress must then be added directly to the preload or service load tensile stress in the fastener. Reference [5] contains a thorough treatment of eccentrically loaded joints and their effect on fastener stress. While it is possible to compute the additional tensile stress created by a bending moment, in practice the designer seldom has enough information to complete the calculations. Therefore, the most practical method to account for bending stress in design is to reduce the sources of bending and their severity.

8.9 MODIFIED BODIES

Most externally threaded fasteners (bolts, screws, studs) have solid, full-size bodies or shanks. It is sometimes advantageous to employ a reduced cross section. For instance, reducing the cross section will reduce the stiffness (see Chapter 11). This is sometimes useful to increase the fatigue life of the fastener. A reduced cross section can also be used to reduce the strength to a predetermined level. Therefore, the fastener would fail at a specific location. There are a number of ways to modify the body of a fastener. These include

1. Reducing the diameter of the body
2. Drilling an axial hole in the body
3. Machining flutes into the body
4. Machining slots into the body or threads.

In all these cases, the actual dimensions of the fastener need to be considered in order to calculate the strength of the fastener.

8.10 DYNAMIC LOADING

When the stress level in a part is held constant, it is said to be statically loaded. For tensile loading, failure occurs when the tensile strength of the material is exceeded. When the stress level varies as a function of time, the part is said to be dynamically loaded. Failure under dynamic conditions is termed fatigue failure or just fatigue. It is important to predict fatigue failures. Experience has shown that dynamically loaded parts fail at average stress levels well below the tensile strength or even the yield strength of the material. This section addresses why fasteners are especially susceptible to fatigue, methods to reduce this susceptibility, and how to determine whether a fastener is safe from fatigue failure. The reason that parts fail by fatigue while loaded to seemingly low stress levels is that stress concentrations are present. At these locations, which are typically at changes in cross section and sharp corners, the localized stress is many times the average stress. Therefore, the localized stress can easily exceed the yield strength. Over the course of many load cycles, a crack initiates where there is a stress concentration and grows under this high stress until failure occurs. The thread roots, thread runout, and head-to-body fillet areas on externally threaded fasteners are all areas of stress concentration. Standard threads have stress concentration factors between 4 and 10 [10]. The nonuniform nature of thread load distribution in engaged threads also contributes to fastener fatigue. The first thread at the nut bearing face is subjected to the majority of the thread load as well as the full cross-sectional load. With the combination of the highest average stress and the stress concentration of the thread root, the first thread is the location for most fastener failures. It would seem that threaded fasteners are fatigue failures waiting to happen. However, there are methods to improve the fatigue performance of fasteners. An obvious consideration is to reduce the stress concentrations. The standard UN and metric thread forms can have sharp corners at the thread roots. The UNJ and MJ thread forms have a specified root radius. These thread forms have three to four times greater fatigue life than the other forms. There are also proprietary thread forms that are even more effective [11]. Since fatigue failure is due to tensile stress, the presence of a residual compressive stress would reduce the magnitude of the tensile stress, thus improving fatigue performance. Most threads are produced by thread rolling. This cold working introduces a residual compressive stress in the thread roots, exactly where it is needed. However, many fasteners require heat treatment, which would reduce the residual stress. That is why most aerospace applications requiring long fatigue life use fasteners that are roll threaded after heat treatment. Nonperpendicularity of bearing surfaces is another source of fatigue problems. Test results show that an angularity of only 0.5° reduces fatigue life by almost one-half while 2° reduces it to almost zero [11].

Whether or not a fastener is safe from fatigue failure can be determined through the use of a fatigue diagram. Here, the mean stress (fastener preload) is plotted against the dynamic or alternating stress. Next it is necessary to define the failure criteria. There are several failure theories. The most commonly used is the Goodman criterion. Figure 8.4 is a fatigue diagram including the Goodman line. Only points below that line are safe. The Goodman line extends from the endurance limit at zero mean stress to the tensile strength at zero alternating stress. This indicates that, just as in Section 8.7 for combined tensile and torsional stresses, a decrease or increase in the alternating stress must accompany an increase or decrease, respectively, in the mean stress. The likelihood of

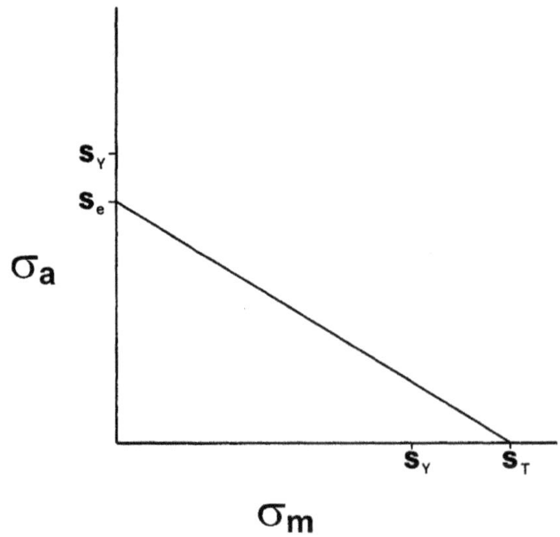

FIGURE 8.4 Fatigue diagram using the Goodman line for demarcation of safety. Points below that line are considered safe.

failure is evaluated by plotting the mean and alternating stress conditions for the fastener. The mean stress is the in-service fastener preload. Determination of this value is described in Chapters 18 and 36. Fastener alternating stress is not the full alternating load applied to the joint; it is the stress due to that portion of the alternating applied load that the fastener experiences. This value is determined by using joint diagrams as described in Chapter 36. The Goodman line can also be expressed using an interaction equation. Expressing the alternating and mean stresses as ratios of their respective allowable values, the interaction equation incorporating a factor of safety would be

$$R_a + R_m = \frac{1}{s} \tag{8.16}$$

where

R_a=alternating stress ratio=σ_a/S_e
R_m=mean stress ratio=σ_m/S_T
σ_a=alternating stress (lb/in.^{2}, N/mm^2)
S_e=material endurance limit (lb/in.^{2}, N/mm^2)
σ_m=mean stress (lb/in.^{2}, N/mm^2)
S_Y=material yield strength (lb/in.^{2}, N/mm^2).

See Chapter 37 for a further discussion of fatigue.

8.11 THREAD SHEAR AREA

Threaded connections transfer axial loads between the external and internal threads by contact of the thread flanks (see Figure 8.5). The thread shear area of a fastener is controlled mainly by the mating thread. The relative strengths of the two parts define where the shear force will be applied. When the strength of the internal thread material greatly exceeds that of the external thread, thread failure will occur by stripping of the external thread. The external threads shear at a diameter established by the minor diameter of the internal thread (see Figure 8.6a). It is possible to calculate the shear area of the external thread at this diameter as follows [1,2]:

$$AS_s = \frac{\pi \cdot LE \cdot d_{1\max}}{p}\left[\frac{p}{2} + 0.57735\left(d_{2\min} - D_{1\max}\right)\right] \tag{8.17}$$

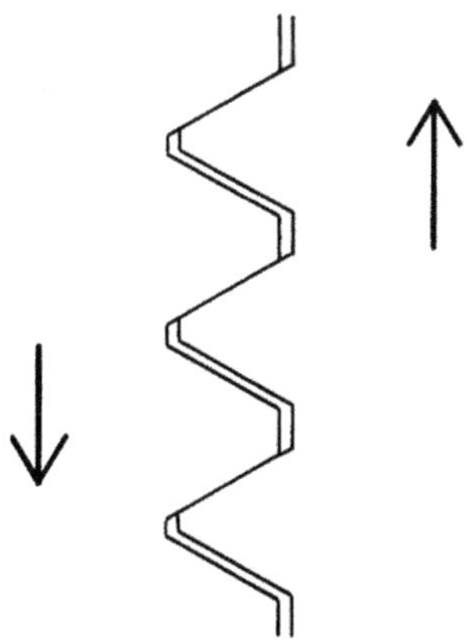

FIGURE 8.5 Cross section of assembled screw threads showing transfer of axial load by contact of mating thread flanks.

where

AS_s = external thread shear area (in.^{2}, mm^2)
LE = length of engagement (in., mm)
D_{1max} = maximum minor diameter of internal thread (in., mm)
d_{2min} = minimum pitch diameter of external thread (in., mm).

When the strength of the external thread material greatly exceeds that of the internal thread, the assembly will resist the axial load through shearing of the internal threads at a diameter established by the major diameter of the external thread (see Figure 8.6b). It is possible to calculate the shear area of the internal thread at this diameter as follows [1,2]

$$AS_n = \frac{\pi \cdot LE \cdot d_{min}}{p}\left[\frac{p}{2} + 0.57735\left(d_{min} - D_{2max}\right)\right] \tag{8.18}$$

where

AS_n = internal thread shear area (in.^{2}, mm^2)
d_{min} = minimum major diameter of external thread (in., mm)
D_{2max} = maximum pitch diameter of internal thread (in., mm).

Inspection of Figure 8.6 reveals an important condition. The shear areas are different at each of the failure locations. For example, given a 1/2-13UNC-2A/2B combination, the external and internal thread shear areas per unit length of engagement would be 0.779 and 1.123 in.^{2}/in., respectively. The internal thread shear area is over 40% larger than the external thread shear area. Regarding thread dimensions, it is possible to produce more shear area by decreasing the thread dimensional tolerances (i.e., using a higher class of fit). For example, the unit internal thread shear area for a 1/2-13UNC-3A/3B combination is 1.164 in.^{2}/in. But this is only 3.7% larger than the value for the previous 2A/2B example. Therefore, the use of higher thread classes is not an efficient method of producing more thread shear area. There is one application where altering thread dimensions does produce a significant increase in thread shear area. For the case of external thread area, the minor diameter of the internal thread is a prime factor. The minor diameter of a tapped hole is controlled by the tap drill or reamer used to prepare the stock for tapping. Minor diameters tabulated for standard Unified and metric screw threads will produce internal thread heights of 66%–83% of the height of a sharp V thread. The height of a sharp V thread, H, is equal to $0.8660254p$. Decreasing the tap drill or reamer size will decrease the minor diameter, thereby

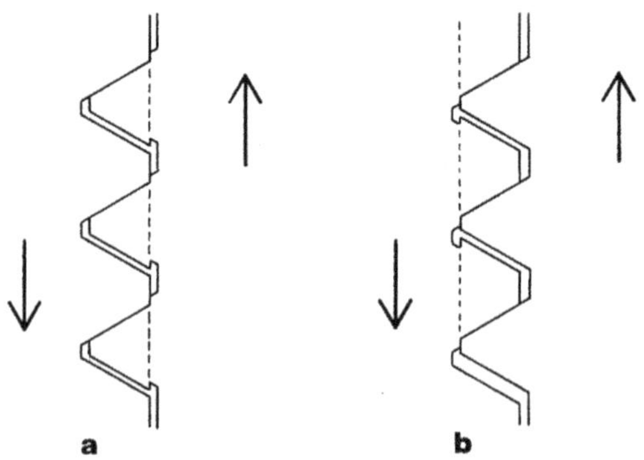

FIGURE 8.6 (a) External thread shearing at minor diameter of internal thread and (b) Internal thread shearing at major diameter of external thread. Dashed lines show the location of thread shear.

increasing the external thread shear area. Equations defining the percentage of thread height for a given tap drill hole size are as follows [12]:

$$Z = n\frac{D-h}{0.013} \qquad \text{(for inch series)} \tag{8.19}$$

$$Z = \frac{76.98}{p}(D-h) \qquad \text{(for metric series)} \tag{8.20}$$

where

Z=percentage of sharp V thread height (percent)
h=tap drill hole size (in., mm).

Also, by rearranging terms, the above equations can be used to calculate the required hole size for a given percentage of thread height. However, this procedure has disadvantages. Increasing the percentage of thread height requires more power to turn the tap because more material is being removed. Also, the tap driving torque increases dramatically. This increases the torsional stress in the tap, leading to tap breakage. For example, the Metal Cutting Tool Institute found that for a 3/8-16 tapped hole in AISI 1020 steel, the tapping torque doubled for an increase in thread height of 60%–72% and almost tripled at 80% [12]. Tables of tap drill sizes for inch and metric screw threads that produce standard minor diameters can be found in many sources [12,13].

8.12 LENGTH OF ENGAGEMENT

The previous section provided equations to calculate the amount of thread shear area for a given length of engagement. However, for most design applications, it is the minimum length of engagement that must be determined. Referring back to the basic design rule in Section 8.3, the following equation can be written:

$$(\text{AS})(S) \geq (\text{AT})(S_T) \tag{8.21}$$

where

AS=thread shear area (internal or external) (in.^{2}, mm^2)
S_T=material tensile strength (lb/in.^{2}, N/mm^2).

This equation states that the shear strength of the engaged threads must be equal to or greater than the tensile strength of the fastener. Therefore, by substituting the various equations for AS, the length of engagement can be determined. This was the approach used until a more comprehensive approach was published by Alexander [7]. This method provides equations to calculate three failure loads: bolt breaking load, bolt stripping load, and nut stripping load. The variables, of which length of engagement is one, are then selected until the basic design rule is satisfied (i.e., bolt stripping and nut stripping loads are greater than the bolt breaking load). The method considers such effects as internal thread dilation and bell-mouthing, thread bending, thread relative motion, and thread chamfering. The paper [7] also contains a Fortran computer program that computes the minimum nut height (length of engagement). One useful feature of this program is the use of statistical theory. Instead of using worst-case values (i.e., minimum material condition), the program generates random values consistent with typical production processes. This allows the calculation of nut heights that are shorter than those produced by worst-case analysis but still statistically acceptable.

The following non-computer-based treatment is adapted from the Alexander approach. It considers nut dilation, thread bending, and thread chamfers. Since no statistical theory is employed, the results are conservative with respect to answers generated by the program.

8.12.1 Internal Thread Dilation

Due to the 30° flank angle of standard threads, axial loads produce a dilation of the internally threaded part. An internally threaded part with thinner walls allows more dilation. This explains the strength advantage of heavy hex nuts over regular hex nuts. A dilation strength reduction factor is calculated as follows:

$$\mathrm{C1} = -\left(\frac{s}{D}\right)^2 + 3.8\left(\frac{s}{D}\right) - 2.61 \tag{8.22}$$

where

C1 = dilation strength reduction factor
s = width across flats for nuts (in., mm).

This equation is valid for values of *s/D* between 1.4 and 1.9 and is plotted in Figure 8.7. Note that no advantage is given for values of *s/D* greater than 1.9.

8.12.2 Thread Bending

The relative strengths of the internal and external threads determine which will bend. As the engaged threads bend, there is less effective thread shear area. Also, the bent thread has a decreased flank angle, creating more dilation. The strength ratio is defined as

$$R_S = \frac{S_n(\mathrm{AS}_n)}{S_s(\mathrm{AS}_s)} \tag{8.23}$$

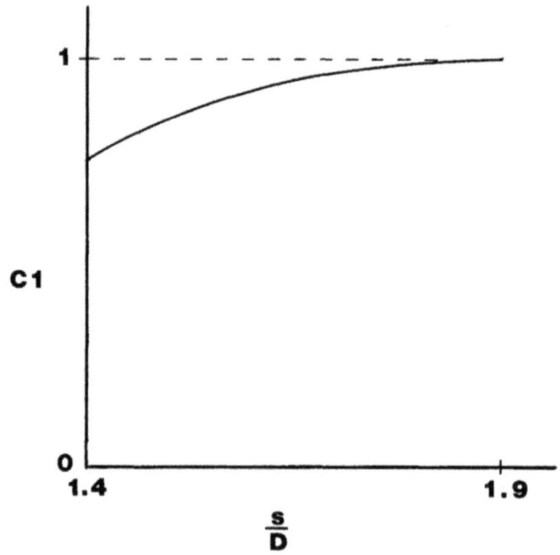

FIGURE 8.7 Plot of dilation strength reduction factor. The lower limit for *s/D* of 1.4 is established by the fact that the outside diameter must be larger than the nominal screw size in order to have an assembly.

where

R_S = strength ratio
S_n = internally threaded part tensile strength (lb/in.², N/mm²)
S_s = externally threaded part tensile strength (lb/in.², N/mm²).

Strength reduction factors C2 and C3 account for the bending of the external and internal threads, respectively. These factors are calculated as follows:

For $1 < R_S < 2.2$:

$$C2 = 5.594 - 13682 R_S + 14.107 R_S^2 - 6.057 R_S^3 + 0.9353 R_S^4 \quad (8.24a)$$

For $R_S \le 1$:

$$C2 = 0.897 \quad (8.24b)$$

For $0.4 < R_S < 1$:

$$C3 = 0.728 + 1.769 R_S - 2.896 R_S^2 + 1.296 R_S^3 \quad (8.25a)$$

For $R_S \ge 1$:

$$C3 = 0.897 \quad (8.25b)$$

where

C2 = external thread strength reduction factor
C3 = internal thread strength reduction factor

8.12.3 Tapped Hole Countersinks

Internal threads are usually chamfered or countersunk. While removing the feather edge of the thread reduces the possibility of cross-threading, it also reduces the amount of thread shear area in the length of engagement. Based on experimental results, Alexander found that for a nut with a 90° countersink, the length of threads in the countersink contributed only 40% of the strength of an equal length without the countersink. Therefore, countersinks can be accounted for by modifying the length of engagement used to calculate thread shear areas as follows:

$$LE' = LN_{Nom} - 0.6\left(D_{Cmax} - D_{1min}\right) \quad (8.26)$$

where

LE' = effective length of engagement (in., mm)
LE_{Nom} = nominal length of engagement (in., mm)
D_{Cmax} = maximum countersink diameter (in., mm)
D_{1min} = minimum minor diameter (in., mm).

It is now possible to compare the three failure loads.

$$BBL = (S_t)(A_S) \quad (8.27)$$

$$BSL = 0.6(S_s)(AS_s)(C1)(C2) \quad (8.28)$$

$$NSL = 0.6(S_n)(AS_n)(C1)(C3) \quad (8.29)$$

where

BBL = bolt breaking load (lb, N)
BSL = bolt stripping load (lb, N)
NSL = nut stripping load (lb, N).

The factor of 0.6 in the stripping load equations denotes the common ratio between shear strength and tensile strength for ductile metals. Comparison of the results of Eqs. (8.27)–(8.29) indicates which failure mode will prevail. The variables are then manipulated until the desired condition of bolt breakage is achieved.

8.13 DEPTH OF ENGAGEMENT

Thread depth of engagement is defined as the radial overlap of the mating threads. Mated screw threads are usually drawn with the axes of the internal and external threads coincident. However, given sufficient tolerance and/or allowance, it is possible for the mating threads to be eccentric enough to allow the threads to disengage on one side. This is important because the thread shear area equations of Section 8.11 assume concentrically located mating threads. A significant amount of eccentricity will reduce the stripping strength of both the internal and external threads. The amount that mated threads can be displaced is [14]

$$\text{ecc} = T_{D2} + T_{d2} + \text{es} \quad (8.30)$$

where

ecc = possible eccentricity of mated threads (in., mm)
T_{D2} = tolerance of internal pitch diameter (in., mm)
T_{d2} = tolerance of external pitch diameter (in., mm)
es = thread allowance (in., mm).

Maintaining sufficient depth of engagement is not a concern when standard thread dimensions are used. If it is necessary to specify a nonstandard thread, it is recommended that the following conditions be satisfied [14].

1. To provide for a minimum depth of engagement equal to 50% of thread height:

$$d_{min} - D_{1max} \geq 0.75H \quad (8.31)$$

2. To provide for a minimum depth of engagement equal to 50% of the thread height on one side and zero on the other when the assembly is eccentric:

$$D_{2max} - d_{2mix} \leq 0.625H \quad (8.32)$$

Loss of depth of engagement can also be caused by internal thread dilation and external thread contraction. Obviously, these effects are more pronounced in thin wall parts.

8.14 PROOF STRENGTH

The previous sections have discussed the calculation of strength for externally threaded parts as well as assemblies. These values are then compared with the expected joint loading conditions. Given that most joints are assembled using standard nuts and bolts, it is not necessary to calculate the strength of every fastener. For specified materials, nuts and bolts of standardized dimensions will have specified strengths. This fact gave rise to the concept of proof strength. Proof strength is a rating. The proof strength of a nut is not its axial stripping strength. The proof strength of a bolt is not its tensile strength. Proof strengths are specified loads or, in the case of proof stresses, specified stresses that the fastener must withstand without any permanent deformation. For nuts, the axial load is applied to the finished part using a hardened mandrel [15,16]. The load is equal to the proof stress multiplied by the tensile area of the same size external thread. The nut must not strip or rupture and must be removable from the mandrel with the fingers. Failure to meet either of these conditions indicates that permanent deformation has occurred and is cause for rejection. Since nuts come in various configurations, the proof load must be compatible with not only nominal size and material but also nut height, width across flats, slotting, and overtapping. Nut standards typically specify the following relative ratings for nut proof loads:

Regular hex nut	100%
Heavy hex nut	108%–110%
Slotted hex nut	80%
Overtapped hex nut	60%
Hex jam nut	60%

To provide a more severe proof load test to verify the integrity of hot-forged or heat-treated nuts, the cone proof test was developed. Here the load is applied to the bearing face of the nut with a truncated cone instead of a flat plate. The cone introduces a severe dilation of the nut. The test is used to verify that seams, laps, or cracks have not had a deleterious effect on the ability of the nut to carry its rated load without stripping or rupturing. For bolts, the axial load is applied to the finished part [15,16]. Again the proof load is equal to the specified proof stress multiplied by the tensile area of the bolt. The bolt must withstand the load without breaking or stripping. In addition, the length of the bolt is measured before and after loading. There can be no increase in length, which would indicate permanent deformation. Failure to meet either of these conditions is cause for rejection.

REFERENCES

1. ASME Standard B1.1, Unified Inch Screw Threads (UN and UNR Thread Form), 1989.
2. ASME Standard B1.13, Metric Screw Threads—M Profile, 1983.
3. National Aerospace Standard NAS 1348, Fasteners—Recommended Tensile Stress Areas for External Threaded, Aerospace Industries Association of America, May 30, 1979.
4. Metric Aerospace Standard MA 1520, Areas for Calculating Stress or Load Values for Metric MJ Externally Threaded Fasteners, Society of Automotive Engineers, Warrendale, PA, July 1987.
5. Bickford, J. H., *An Introduction to the Design and Behavior of Bolted Joints*, Marcel Dekker, New York, 1995.
6. ASME Standard B1.12, Class 5 Interference-Fit Thread, 1987.
7. Alexander, E. M. Analysis and Design of Threaded Assemblies, SAE Paper No. 770420, 1977.
8. Barrett, R. T., Fastener Design Manual, NASA Reference Publication 1228, March 1990.
9. Military Handbook MIL-HDBK-5, Metallic Materials and Elements for Aerospace Vehicle Structures, Revision F, Nov. 1, 1990.
10. VDI 2230, Systematic Calculation of High Duty Bolted Joints, Joints with One Cylindrical Bolt, VDI Society for Product Development, Design and Marketing, Committee for Bolted Joints, Verein Deutscher Ingenieure (VDI). Translation of German edition July 1986 by C. Junker and J. Newnham.

11. Product Engineering Report 4718, Fastener Seminar, SPS Technologies, Aerospace Products Division, 1974.
12. Drilled Holes for Tapping—A Guide to Selection of Drills for Tapping, Metal Cutting Tool Institute, 1986.
13. Federal Standard FED-STD-H28, Appendix A3, Screw-Thread Standards for Federal Services, Mar. 31, 1978.
14. Federal Standard FED-STD-H28/2B, Screw-Thread Standards for Federal Services, Aug. 20, 1991.
15. ASTM Standard F606, Determining the Mechanical Properties of Externally and Internally Threaded Fasteners, Washers, and Rivets, 1986.
16. ASTM Standard F606 M, Determining the Mechanical Properties of Externally and Internally Threaded Fasteners, Washers, and Rivets (Metric), 1995.

9 Computing the Stiffness of a Fastener

Joseph Barron

9.1 INTRODUCTION

When threaded fasteners are used to assemble a joint, they are usually preloaded so that there is a residual compressive load placed on the joint members. The fastener supplies this load by being stretched from its free-state length during the assembly process. This is comparable to stretching a helical spring. Although the typical fastener may not stretch as much as a spring, it does behave like a spring. Therefore, in order to characterize the behavior of the fastener, it is necessary to determine the most important property of a spring; its stiffness or spring rate. Also, it is possible to determine and control fastener load by measuring the length of the fastener. The relationship between fastener length and load is again determined by fastener stiffness.

9.2 EFFECTIVE LENGTH

The process of determining fastener stiffness is based on simple materials science principles. The standard tensile test provides the necessary information. When an axial tensile load is applied to the ends of a straight cylindrical rod, it stretches. A graph of the applied tensile load versus the resultant axial deflection for a given size specimen is shown in Figure 9.1. The linear portion of the curve can be expressed as

$$F = k\delta \tag{9.1}$$

where

F = applied load (lb, N)
k = spring rate (lb/in., N/mm)
δ = deflection (in., mm).

This shows that the deflection is directly proportional to the applied load. It should be noted that the load–deflection curve is not universally applicable to all parts due to geometrical differences.

By converting the data to stress and strain, parts of any size can be analyzed. Load and deflection are related to stress and strain as follows:

$$\sigma = \frac{F}{A} \tag{9.2}$$

$$\varepsilon = \frac{\delta}{L} \tag{9.3}$$

where

σ = applied stress (lb/in.^{2}, N/mm^2)
A = cross-sectional area (in.^{2}, mm^2)

DOI: 10.1201/9781003501183-12

FIGURE 9.1 Diagram of load–deflection curve for an elastic-plastic material. The slope of the linear portion is the stiffness of the part.

ε=engineering strain (in./in., mm/mm)
L=original length (in., mm).

A graph of applied tensile stress versus resultant axial strain is shown in Figure 9.2. Here, the linear portion of the curve can be expressed as

$$\sigma = E\varepsilon \tag{9.4}$$

where

E=elastic modulus (lb/in.^{2}, N/mm^2).

Of particular interest is the linear portion of the curve. Point A is defined as the proportional limit where the linear relationship between stress and strain stops. Point B is defined as the elastic limit. Beyond this point, the specimen has been plastically deformed (strained) and will not return to its original length when the stress is removed. Point C is defined as the yield point. This is typically characterized by a specified amount of plastic strain, usually 0.2%. While many fasteners are intentionally preloaded beyond the yield point, the determination of the stiffness of a fastener is confined to the elastic region. By combining Eqs. (9.1)–(9.4) and rearranging terms, an equation defining the deflection can be written:

$$\delta = \frac{\mathrm{FL}}{\mathrm{AE}} \tag{9.5}$$

Further manipulation produces the equation defining the stiffness:

$$k = \frac{\mathrm{AE}}{L} \tag{9.6}$$

However, the preceding equation is applicable only to parts with a uniform cross section.

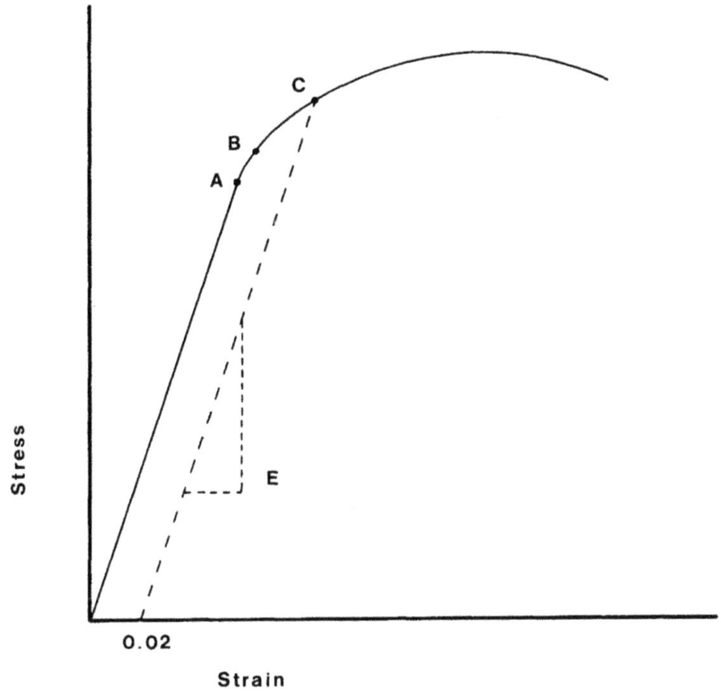

FIGURE 9.2 Diagram of stress–strain curve for an elastic-plastic material. The slope of the linear portion is the elastic modulus of the material. The yield strength at 0.02% offset is defined at point C.

Threaded fasteners typically have significant changes in geometry. Therefore, it is necessary to determine the deflection of each segment and then add the values to obtain the total deflection. For example, given a fastener with three segments, the total deflection would be

$$\begin{aligned}\delta &= \text{deflection}_1 + \text{deflection}_2 + \text{deflection}_3 \\ &= \frac{PL_1}{A_1E} + \frac{PL_2}{A_2E} + \frac{PL_3}{A_3E} \\ &= P\left[\frac{L_1}{A_1E} + \frac{L_2}{A_2E} + \frac{I_3}{A_3E}\right] \\ &= P\left[\frac{1}{k_1} + \frac{1}{k_2} + \frac{1}{k_3}\right]\end{aligned} \tag{9.7}$$

It is sometimes easier to work in terms of the inverse of the stiffness, or $1/k$. This is known as resilience or compliance and can be written

$$\frac{1}{k} = \frac{1}{k_1} + \frac{1}{k_2} + \frac{1}{k_3} \tag{9.8}$$

Therefore, the total resilience of the fastener can be written

$$f = f_1 + f_2 + f_3$$

where

f = resilience (in./lb, mm/N) = $1/k$.

It can be seen that the stiffness of a fastener is dependent on the number of segments, the area and length of each segment, and the elastic modulus. Assuming that the fastener is made of one homogeneous material, only the length and area are variables. Determination of the length of each segment is straightforward except for those sections including the head of the bolt or screw or the threaded area engaged with a nut or tapped hole. Here, the designer needs to know how much of the bolt head or depth of the tapped hole contributes to the length of the fastener being stretched. The amount of contribution of these areas to the geometrical length affects the effective length of the fastener.

Experiments have shown that fasteners exhibit stiffnesses that indicate an effective length longer than the grip length. VDI 2230 [1] specifies that a length equal to 0.4*D*, where *D* is the nominal diameter, be added to the grip length to account for the contribution of the fastener head. It also specifies that a length equal to 0.4*D* be added to account for the contribution of the engaged threads in nuts or tapped holes. Other sources recommend values from 0.3Z) to 0.6*D*. Bickford [2] recommends a value of 0.5*D* for both heads and engaged threads. The amount of cross-sectional area in each segment is established as follows:

$$A_{\text{HEAD}} = \frac{\pi d^2}{4} \tag{9.9}$$

$$A_{\text{SHANK}} = \frac{\pi d_1^2}{4} \tag{9.10}$$

$$A_{\text{THREAD}} = \frac{\pi}{4}\left(D - \frac{0.9743}{n}\right)^2 \quad (\text{for unified threads}) \tag{9.11}$$

$$A_{\text{THREAD}} = \frac{\pi}{4}(D - 09382p)^2 \quad (\text{for metric threads}) \tag{9.12}$$

where

D = nominal diameter (in., mm)
d_1 = shank diameter (in., mm)
n = number of threads per inch
p = thread pitch (mm).

It is also necessary to account for any axial holes in fasteners. These are typically used to allow the insertion of stud heaters or length measurement rods.

$$A_{\text{THREAD}} = \frac{\pi}{4}\left(d_1^2 - d_h^2\right) \tag{9.13}$$

where

d_h = diameter of hole (in., mm).

The calculation of fastener stiffness, while based in part on experimental work, is theoretical. Since the stiffness is based on dimensional data, dimensional tolerances will have an impact. For example, a ¼-13UNC×3 in. long hex head cap screw is assembled with a hex nut to give a grip length of 2¼ in. Standard tolerances for the overall length and the length of threads for this combination would produce a variance of 4% in the value of the stiffness. Another source of variation is in the elastic modulus. Many texts quote the elastic modulus of steel between 29×10^6 psi (200 GPa) and 30×10^6 psi (207 GPa). This is a 3% variation. Also, it has been found that fastener stiffness and hence deflection do not agree with experimental values when the fastener is "short." In regard to stiffness calculations, fasteners should be considered "short" if the length-to-diameter ratio (l/D) is less than 5. However, these variations can easily be reduced if greater accuracy is required. It is a simple matter to produce load–deflection data for each lot of fasteners or even for each fastener. Only you can determine if the added accuracy will offset the added costs.

9.3 EXAMPLE STIFFNESS CALCULATIONS

In these examples, a factor of 0.5 is used for the amount of contribution of bolt heads and threaded areas to the effective length of the fastener.

Example 9.1

Hex head bolt and nut (see Figure 9.3)

$$\frac{1}{k_{\text{BOLT}}} = \frac{1}{k_H} + \frac{1}{k_B} + \frac{1}{k_T} + \frac{1}{k_N}$$
$$= \frac{0.5D}{\pi D^2 E/4} + \frac{L_B}{\pi D^2 E/4} + \frac{L_B}{A_T E} + \frac{0.5D}{A_T E}$$

Example 9.2

Double-ended reduced body stud and nuts (see Figure 9.4)

$$\frac{1}{k_{\text{STUD}}} = \frac{1}{k_N} + \frac{1}{k_{T1}} + \frac{1}{k_B} + \frac{1}{k_{T2}} + \frac{1}{k_N}$$
$$= \frac{0.5D}{A_T E} + \frac{L_{T1}}{A_T E} + \frac{L_B}{\pi d^2/4} + \frac{L_{T2}}{A_T E} + \frac{0.5D}{A_T E}$$

Example 9.3

Socket head cap screw in a tapped hole (see Figure 9.5)

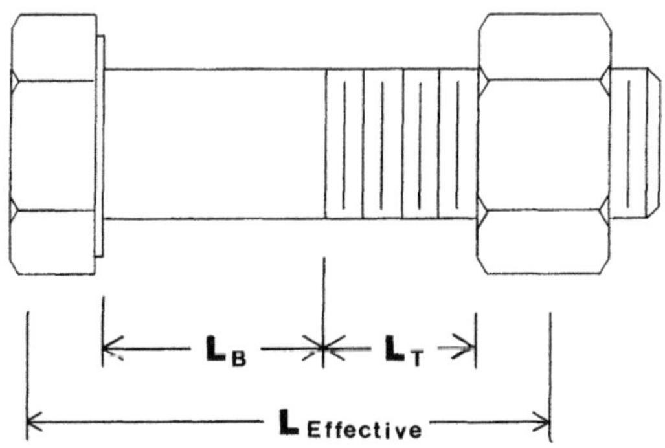

FIGURE 9.3 A hex head bolt assembled with a hex nut. The body and thread lengths are shown along with the effective length.

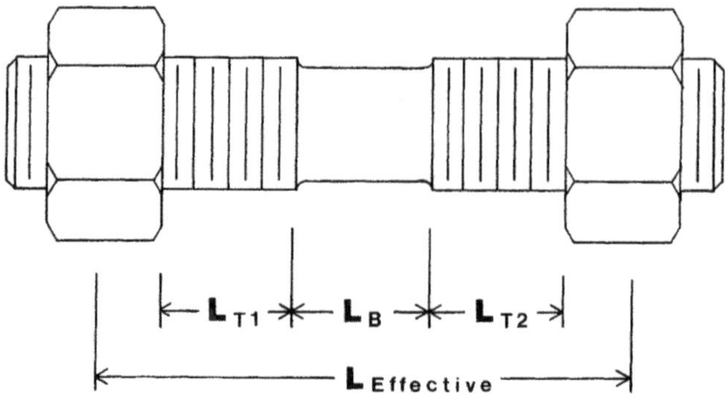

FIGURE 9.4 A reduced body stud assembled with a nut at each end. The body and thread lengths are shown along with the effective length.

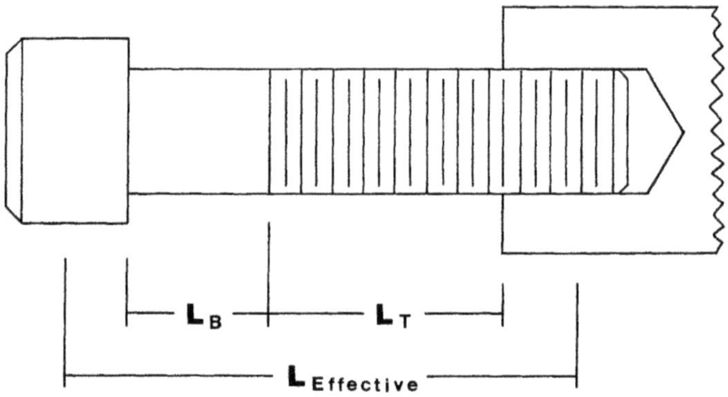

FIGURE 9.5 A socket head cap screw installed in a tapped hole. The body and thread lengths are shown along with the effective length.

$$\frac{1}{k_{SCREW}} = \frac{1}{k_H} + \frac{1}{k_B} + \frac{1}{k_T} + \frac{1}{k_N}$$

$$= \frac{0.5D}{\pi D^2 E/4} + \frac{L_B}{\pi D^2 E/4} + \frac{L_B}{A_T E} + \frac{0.5D}{A_T E}$$

Example 9.4

Set stud with heater hole (see Figure 9.6)

$$\frac{1}{k_{STUD}} = \frac{1}{k_N} + \frac{1}{k_T} + \frac{1}{k_B} + \frac{1}{k_h}$$

$$= \frac{0.5D}{(A_T - A_h)E} + \frac{L_T}{(A_T - A_h)E} + \frac{L_B}{(\pi d^2 - A_h)E} + \frac{0.5D}{A_T E}$$

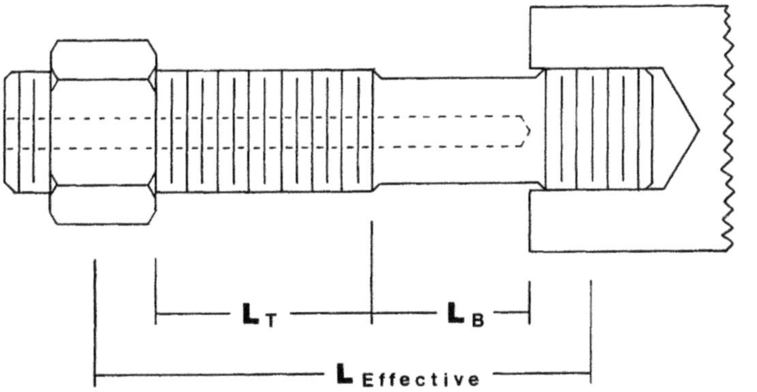

FIGURE 9.6 A stud with a heater hole installed in a tapped hole. The body and thread lengths are shown along with the effective length.

REFERENCES

1. VDI 2230, Systematic Calculation of High Duty Bolted Joints, Joints with One Cylindrical Bolt, VDI Society for Product Development, Design and Marketing, Committee for Bolted Joints, Verein Deutscher Ingenieure (VDI). Translation of German edition July 1986 by C. Junker and J. Newnham.
2. Bickford, J. H., *An Introduction to the Design and Behavior of Bolted Joints*, 3rd ed., Marcel Dekker, New York, 1995.

10 Computing Joint Stiffness

Sayed A. Nassar and Antoine G. Abboud

10.1 JOINT STIFFNESS K_C AND INTRODUCTION OF NEW STRESS CONE ANGLE α

10.1.1 Introduction

After outlining a procedure for calculating the bolt stiffness in Chapter 9, this chapter is dedicated to joint stiffness calculation. Needless to say that a more accurate calculation of respective bolt and joint stiffnesses k_b and k_c would significantly affect the torque-tension-turn signature of threaded joint systems during initial assembly as well as the distribution of subsequent external service loads [2-5]. This is a particularly important issue in critical and safety-related applications, where both the initial clamp load level as well as the sufficiency and stability of its residual level during operation would directly impact performance, reliability, and product safety. Figure 10.1 shows a typical bolted joint signature during initial assembly; bolt and joint stiffnesses determine the clamp load increase per degree of bolt head turn in the elastic zone.

Validated equations for joint stiffness are introduced in this section. Figure 10.2 illustrates a user-friendly Excel calculator for joint stiffness using equations provided in this chapter. Of particular importance is the implementation of a test-validated new value of the stress cone angle $\alpha=36°$, which improves both the accuracy and practicality of calculated joint stiffness k_c.

Section 10.1.2 briefly outlines the methodology that led to the improved equations for joint stiffness k_c, effective plane area A_c, and a newly developed angle $\alpha=36°$for the stress cone envelope [1].

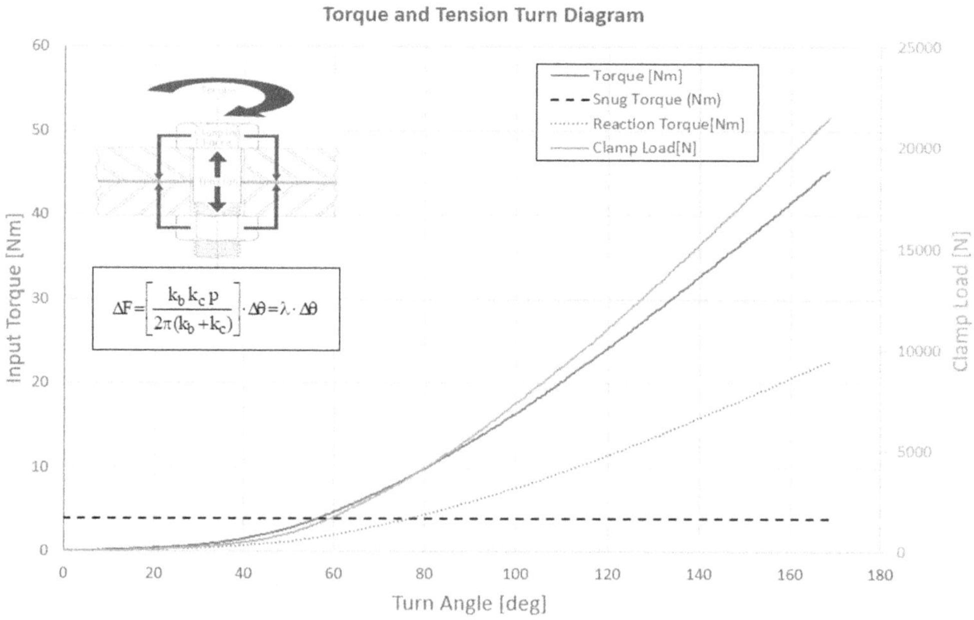

FIGURE 10.1 Typical bolted joint signature: torque-tension-angle.

DOI: 10.1201/9781003501183-13

Input	
Hole Diameter D_h(mm)	12
Joint Width D_o(mm)	120
Bearing Diameter D_w (mm)	18
Plate 1 thickness t_1(mm)	20
Plate 2 thickness t_2(mm)	30
Modulus of Elasticity "E_1" for Plate 1 (N/mm^2)	200000
Modulus of Elasticity "E_2" for Plate 2 (N/mm^2)	73000
Output	
Joint Stiffness (N/mm)	**996026**

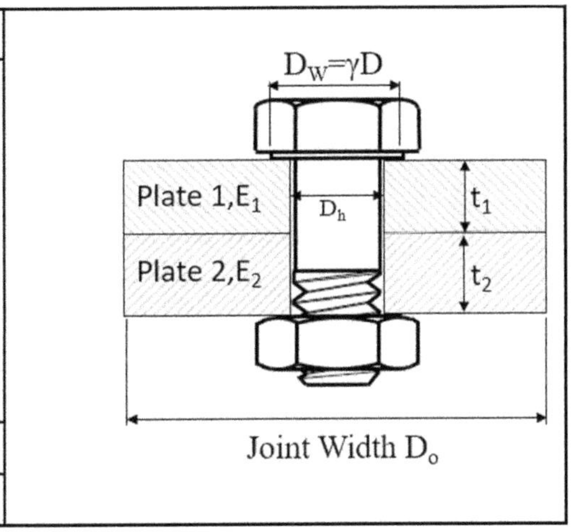

FIGURE 10.2 Sample calculation of joint stiffness k_c [1].

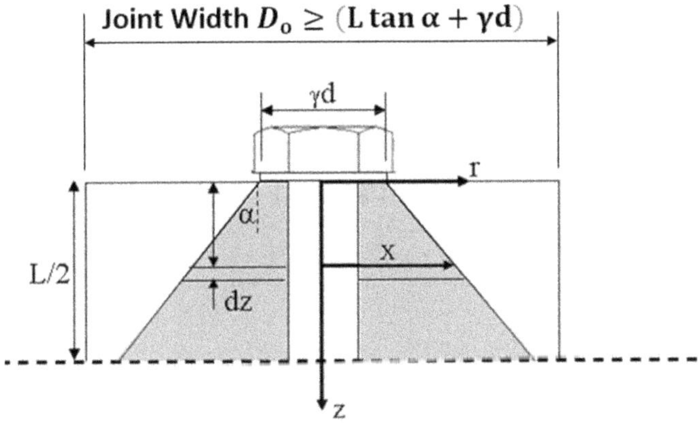

FIGURE 10.3 Fully developed stress cone in the upper plate.

Section 10.1.3 outlines Test and FEA validation of the new equations, while Section 10.1.4 illustrates the use of those stiffness equations for six joint model scenarios of geometry and plate materials.

10.1.2 Summary of Methodology

The new joint stiffness equations are simply obtained by linearly correlating the average compression of clamped axisymmetric joint model to the compressive clamp load F_c from tightened bolt [1]. As illustrated in Figures 10.3 and 10.4, all combinations of contact interface with the joint, angle α, joint thickness (grip length) L, hole diameter, and joint width D_o would determine the outer stress contour of the clamped joint. Depending on the value of joint width D_o, the stress contour would generally expand away from the hole edge. For the joint configuration in Figure 10.3, since the joint width $D_o \geq (L \tan \alpha + \gamma D)$, the stress contour would be fully developed with a maximum value of A_c. However, for a smaller joint width $\gamma\ d < D_o < (L.\tan \alpha + \gamma D)$, a smaller stress area A_c would develop (Figure 10.4).

Nassar and Abboud [1] introduced a model in which, at each z-location, the clamp load F_c creates a compressive stress with its peak value σ_o at the edge of bolt hole and reduces in a fourth order

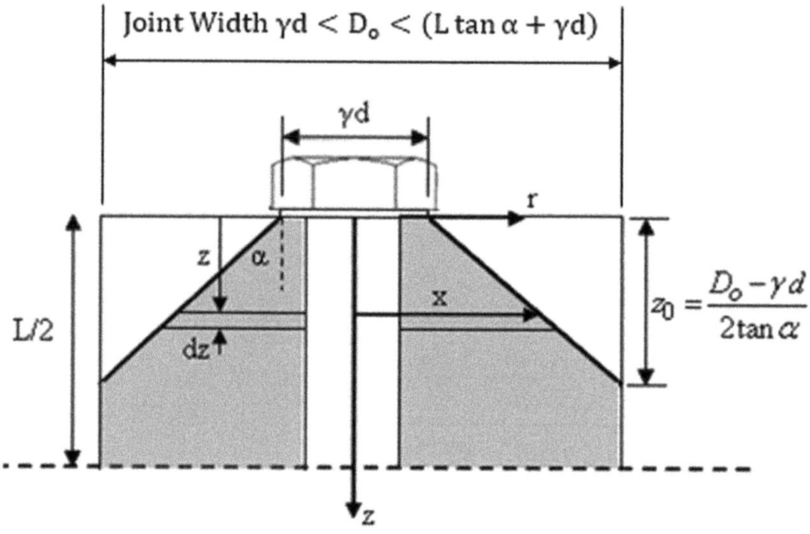

FIGURE 10.4 Partially developed stress cone.

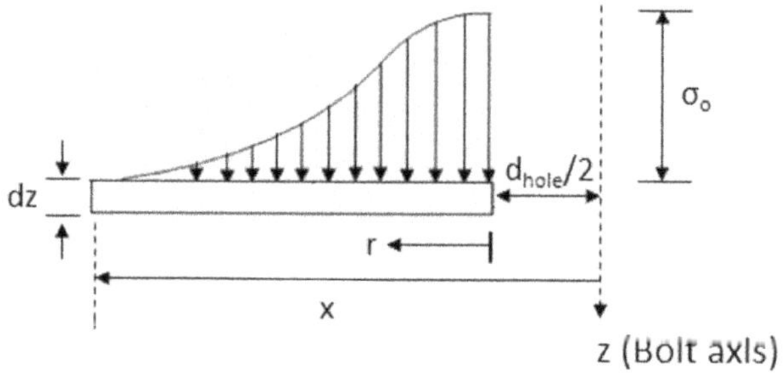

FIGURE 10.5 Stress distribution on joint element *dz* of the joint.

fashion outward from the edge of bolt hole (Figure 10.5). To obtain the peak compressive stress σ_o at each z location within the joint grip length, static equilibrium is enforced between the clamp load F_c and the integrated compressive stress over the corresponding pressure area perpendicular to the axis of the bolt hole at each location z. Thus, the joint compressive stress field on each infinitesimal joint disc of thickness dz (located at distance z along the bolt axis) is expressed in terms of the clamp load F_c, the location z, and the radial location r from the hole edge.

That compressive stress distribution on each disc is converted to an axial strain field (by using Hooke's law), which leads to an average reduction of thickness for that element dz at each location z within the grip length and shown stress contour lines. Finally, the total joint compression δ_c is obtained by summing up (integrating) the reduction of thicknesses of each disc between the limits $z=0$ and $z=$joint grip length L, which would be in terms of the clamp load F_c, material properties, joint grip length L and width D_o, angle α, and hole diameter. Direct correlation of the clamp load F_c with the resulting expression for joint compression δ_c gives the expression as follows:

$$k_c = \frac{F_c}{\delta_c} \tag{10.1}$$

Both compression test results and FEA simulation results [1] for joint stiffness k_c correlated best with those given by Eq. (10.1) by setting the stress cone angle α equal to 36°. Once the new joint

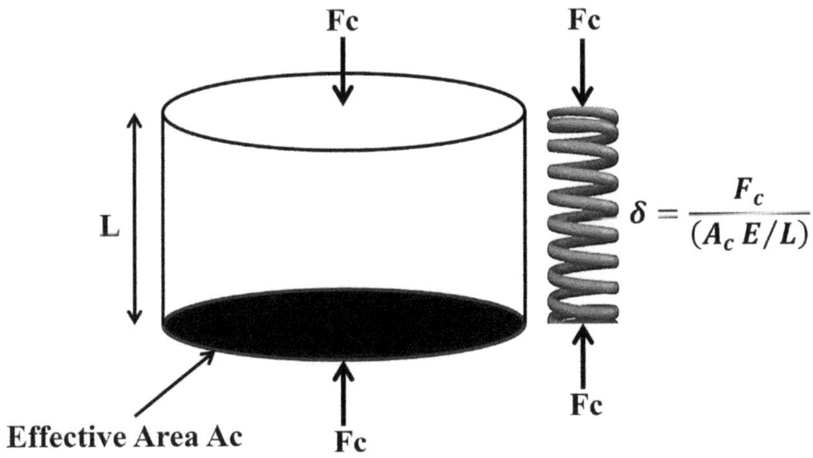

FIGURE 10.6 Virtual 1-*D* rod analogy of joint stiffness ($k_c \equiv A_c E/L$).

stiffness k_c expression was determined by Eqs. (10.2), (10.4) and experimentally validated, the revised angle value $\alpha = 36°$ is also used for getting a more accurate effective stress area A_c as given by Eqs. (10.3), (10.5), which is explained. Firstly, they used the analogy of a virtual one-dimensional rod in Figure 10.6, whose stiffness would simply be equal to $A_c E/L$, with E and L being those of the joint. Setting joint stiffness equation for k_c equal to that of virtual 1-*D* rod would give an expression of what the equivalent stress area A_c should be.

For example, applying this concept to the joint in Figures 10.3, 10.4 would give the joint stiffness k_c and the effective clamped area A_c for similar material as follows:

For the Joint model in Figure 10.3

$$k_c = \frac{F_c}{\delta} = \frac{d\pi E \tan\alpha}{2\left[\ln\left(\frac{(\gamma+3)}{\gamma d + L\tan\alpha + 3d}\right)\left(\frac{\gamma d + L\tan\alpha - d}{(\gamma - 1)}\right)\right]} \tag{10.2}$$

$$A_c = \frac{d_h \pi \tan\alpha L}{2\left[\ln\left(\frac{\gamma d + d_h}{\gamma d + L\tan\alpha + 3d_h}\right)\left(\frac{\gamma d + L\tan\alpha - d_h}{\gamma d - d_h}\right)\right]} \tag{10.3}$$

For the Joint model in Figure 10.4

$$k_c = \frac{F_c}{\delta} = \frac{\pi E \tan\alpha}{2\left[\frac{1}{d_h}\ln\left(\frac{\gamma d + 3d_h}{D_o + 3d_h}\right)\left(\frac{D_o - d_h}{\gamma d - d_h}\right) + \frac{4(L\tan\alpha - D_o + \gamma d)}{(D_o + 3d_h)(D_o - d_h)}\right]} \tag{10.4}$$

$$A_c = \frac{L\pi \tan\alpha}{2\left[\frac{1}{d_h}\ln\left(\frac{\gamma d + 3d_h}{D_o + 3d_h}\right)\left(\frac{D_o - d_h}{\gamma d - d_h}\right) + \frac{4(L\tan\alpha - D_o + \gamma d)}{(D_o + 3d_h)(D_o - d_h)}\right]} \tag{10.5}$$

It should be noted that for the case when the joint width D_o is smaller than or equal to the outer diameter of the bolt head ($D_o \leq \gamma d$), then the underhead compressive stress distribution can be considered constant and the effective area will be approximately equal to the net area of the joint.

10.1.2.1 Dissimilar Joint Materials

For joints made of dissimilar plates, the deflection of each clamped part is calculated, and the joint stiffness is obtained by dividing the clamped force over the sum of the deflections in both plates. The joint stiffness is slightly modified from [1] to account for the hole size and it is provided as follows:

$$k_c = \frac{E_1 E_2 d\pi \tan\alpha}{(E_1+E_2)\ln\left(\dfrac{\gamma d+3d_h}{\gamma d-d_h}\right)+E_1\ln\left(\dfrac{\gamma d+2t_2\tan\alpha-d_h}{\gamma d+2t_2\tan\alpha+3d_h}\right)+E_2\ln\left(\dfrac{\gamma d+2t_1\tan\alpha-d_h}{\gamma d+2t_1\tan\alpha+3d_h}\right)} \tag{10.6}$$

for $D_o \geq (L\tan\alpha+\gamma d)$ fully developed stress envelope

And

$$k_c = \frac{E_1 E_2 \pi \tan\alpha}{\dfrac{(E_1+E_2)}{d_h}\ln\left(\dfrac{\gamma d+3d_h}{D_o+3d_h}\right)\left(\dfrac{D_o-d_h}{\gamma d-d_h}\right)+\dfrac{4E_1(2t_2\tan\alpha-Do+\gamma d)+4E_2(2t_1\tan\alpha-D_o+\gamma d)}{(Do+3d_h)(Do-d_h)}} \tag{10.7}$$

for $\gamma d < D_o < (L\tan\alpha+\gamma d)$ partially developed stress envelope.

10.1.3 Validation of New Joint Stiffness Equations and Cone Angle Value

Figure 10.7 shows the setup for joint compression testing using an MTS Test System 810, which measures the compressive force F_c and the corresponding joint compression δ_c. Measured joint stiffness would be given by Eq. (10.1).

FIGURE 10.7 Joint compression testing.

The corresponding equivalent stress area $A_{c\ test}$ in Table 10.1 is obtained from joint stiffness as follows:

$$A_{c\ \text{test}} = \frac{k_{c\ \text{test}} \times L}{E} \tag{10.8}$$

Test validation data Tables 10.1 and 10.2, respectively, show a reasonable correlation between the calculated stress area A_c and those obtained by testing and FEA simulation for the various combinations of joint parameters. That included different ratios of the grip length to hole diameter, hole size, and various ratios of the joint width to hole diameter were used in the validation. Additional details and validation discussion are found in Ref. [1].

TABLE 10.1
Calculated Stress Area A_c vs. Test Validation

Joint Parameters				Stress Area A_c (mm²)		
Hole Diameter d_h (mm)	Grip Length L (mm)	Joint Width D_o (mm)	Cone Angle $\alpha°$	Calculated A_c with $\alpha=36°$ [1]	Test Data $A_{c\ test}$	Percent Difference (%)
12.7	30	25.4	36	279	280	0.5%
12.7	60	25.4	36	297	351	15.4%
12.7	30	40.5	36	373	444	15.9%
s12.7	60	40.5	36	520	550	5.5%
12.7	30	63.5	36	373	426	12.4%
12.7	60	63.5	36	581	585	0.6%

TABLE 10.2
Calculated Stress Area A_c vs. FEA Validation

Joint Parameters			Stress Area A_c (mm²)		
Grip Length/Hole Diam. (L/d_h)	Joint Width/Hole Diam. (D_o/d_h)	Hole Diameter d_h (mm)	FEA Validation (mm²)	Calculated A_c with $\alpha=36°$ [1]	Percent Difference (%)
2.4	1.9	8	105.5	101.8	3.4%
2.4	3.2	8	132.4	149.3	12.8%
2.4	5	8	138.8	149.3	7.6%
4.7	1.9	8	117.3	106.0	9.6%
4.7	3.2	8	202.1	206.3	2.1%
4.7	5	8	223.2	229.9	3.0%
2.4	1.9	12.7	280.1	278.6	0.5%
2.4	3.2	12.7	339.4	372.9	9.9%
2.4	5.0	12.7	351.8	373.0	6.0%
4.7	1.9	12.7	328.7	296.4	9.8%
4.7	3.2	12.7	515.0	520.0	1.0%
4.7	5.0	12.7	580.0	582.0	0.3%
2.4	1.9	16	417.7	407.3	2.5%
2.4	3.2	16	522.6	597.3	14.3%
2.4	5	16	535.5	597.4	11.5%

10.1.4 Calculation Procedure for Joint Stiffness

In this section, new equations are provided for calculating [more accurately] the joint stiffness k_c for various geometric and material combinations. A user-friendly Excel calculator is provided for stiffness calculations. Stiffness equations and calculation output for six different joint configurations are shown in Figures 10.8–10.13.

Out of the seven joint parameters, two were kept arbitrarily constant in all six joint configurations, namely hole dimeter (D_h=13 mm) and underhead contact diameter D_w ($D_w=\gamma D$=18 mm). The other five parameters were arbitrarily varied to illustrate the effect of plate thicknesses (t_1, t_2=30-t_1), joint multi-material aspects (E_1, E_2), and the joint width D_o that determine whether or not the stress area A_c would fully develop which is happening at D_o=41.3 mm since the joint grip length is the same for all these configurations (t_1+t_2=30 mm).

Below is a list of the analyzed six joint configurations that include:

I-a Joint Stiffness for scenario 1 is determined by using Eq. (10.2) [similar material; same plate thickness; fully developed stress envelope].

Input	
Hole Diameter D_h(mm)	13
Joint Width D_o(mm)	50
Bearing Diameter D_w (mm)	18
Plate 1 thickness t_1(mm)	15
Plate 2 thickness t_2(mm)	15
Modulus of Elasticity "E_1" for Plate 1 (N/mm^2)	200000
Modulus of Elasticity "E_2" for Plate 2 (N/mm^2)	200000
Output	
Joint Stiffness (N/mm)	**2189836**

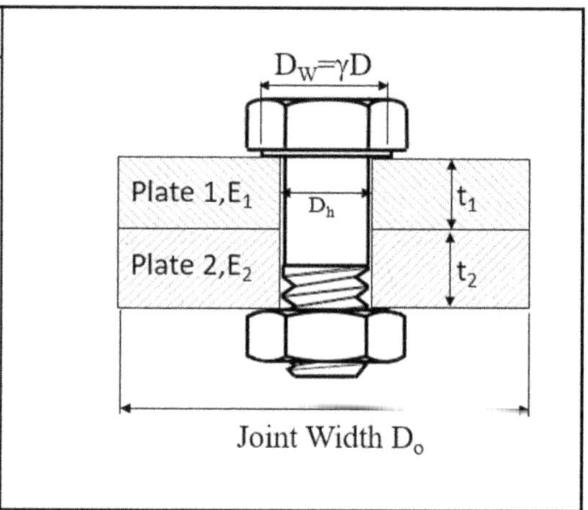

FIGURE 10.8 Scenario 1.

Input	
Hole Diameter D_h(mm)	13
Joint Width D_o(mm)	30
Bearing Diameter D_w (mm)	18
Plate 1 thickness t_1(mm)	15
Plate 2 thickness t_2(mm)	15
Modulus of Elasticity "E_1" for Plate 1 (N/mm^2)	200000
Modulus of Elasticity "E_2" for Plate 2 (N/mm^2)	200000
Output	
Joint Stiffness (N/mm)	**2022670**

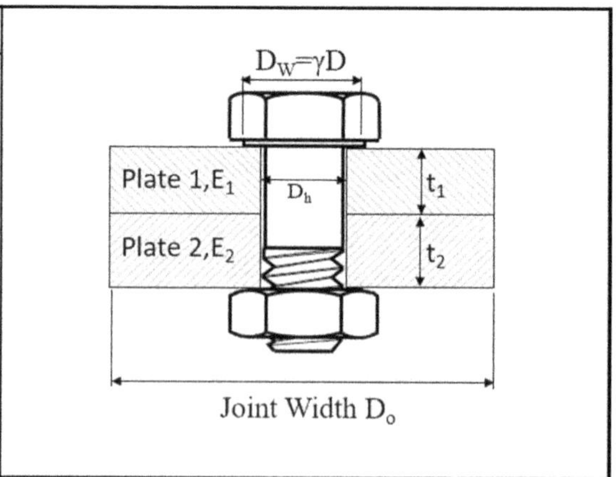

FIGURE 10.9 Scenario 2.

Input	
Hole Diameter D_h(mm)	13
Joint Width D_o(mm)	50
Bearing Diameter D_w (mm)	18
Plate 1 thickness t_1(mm)	15
Plate 2 thickness t_2(mm)	15
Modulus of Elasticity "E_1" for Plate 1 (N/mm^2)	200000
Modulus of Elasticity "E_2" for Plate 2 (N/mm^2)	73000
Output	
Joint Stiffness (N/mm)	**1171121**

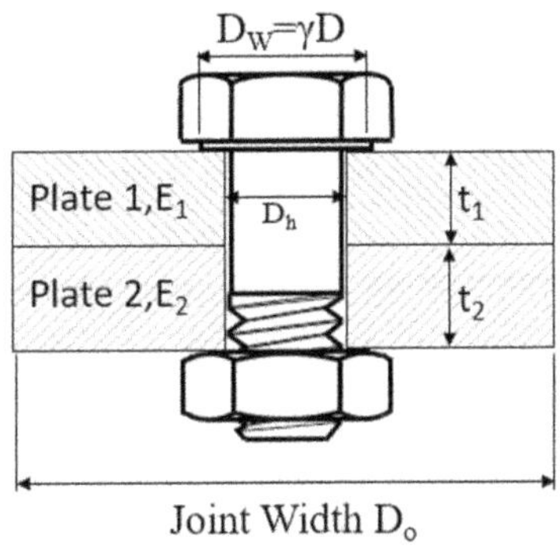

FIGURE 10.10 Scenario 3.

Input	
Hole Diameter D_h(mm)	13
Joint Width D_o(mm)	30
Bearing Diameter D_w (mm)	18
Plate 1 thickness t_1(mm)	15
Plate 2 thickness t_2(mm)	15
Modulus of Elasticity "E_1" for Plate 1 (N/mm^2)	200000
Modulus of Elasticity "E_2" for Plate 2 (N/mm^2)	73000
Output	
Joint Stiffness (N/mm)	**1081721**

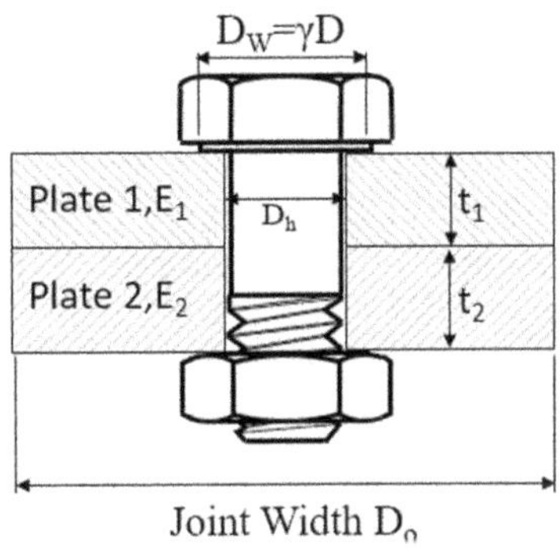

FIGURE 10.11 Scenario 4.

Input	
Hole Diameter D_h(mm)	13
Joint Width D_o(mm)	50
Bearing Diameter D_w (mm)	18
Plate 1 thickness t_1(mm)	20
Plate 2 thickness t_2(mm)	10
Modulus of Elasticity "E_1" for Plate 1 (N/mm^2)	200000
Modulus of Elasticity "E_2" for Plate 2 (N/mm^2)	73000
Output	
Joint Stiffness (N/mm)	**1285185**

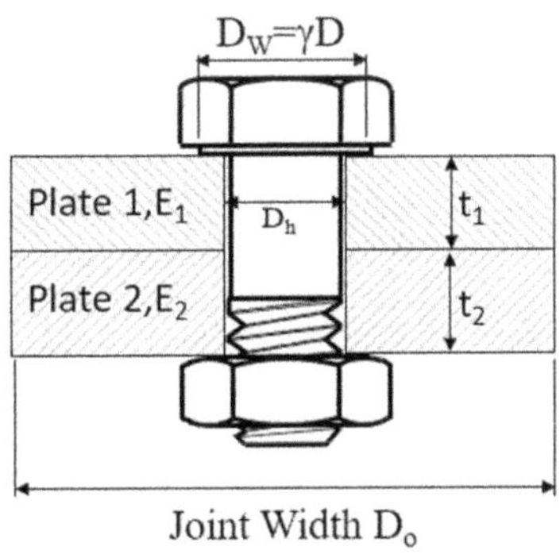

FIGURE 10.12 Scenario 5.

Input	
Hole Diameter D_h(mm)	13
Joint Width D_o(mm)	30
Bearing Diameter D_w (mm)	18
Plate 1 thickness t_1(mm)	20
Plate 2 thickness t_2(mm)	10
Modulus of Elasticity "E_1" for Plate 1 (N/mm^2)	200000
Modulus of Elasticity "E_2" for Plate 2 (N/mm^2)	73000
Output	
Joint Stiffness (N/mm)	**1204772**

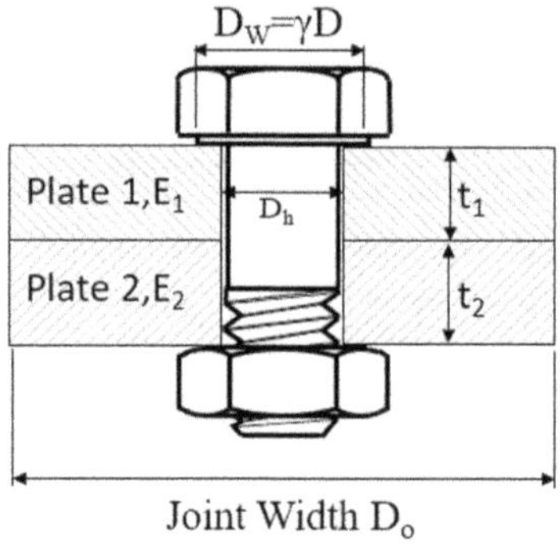

FIGURE 10.13 Scenario 6.

I-b Joint Stiffness for scenario 2 is determined by using Eq. (10.4) [similar material; same plate thickness; partially developed stress envelope].
II-a Joint Stiffness for scenario 3 is determined by using Eq. (10.6) [dissimilar material; same plate thickness; fully developed stress envelope].
II-b Joint Stiffness for scenario 4 is determined by using Eq. (10.7) [dissimilar material; same plate thickness; partially developed stress envelope].
III-a Joint Stiffness for scenario 5 is determined by using Eq. (10.6) [dissimilar material; different plate thickness; fully developed stress envelope].
III-b Joint Stiffness for scenario 6 is determined by using Eq. (10.7) [dissimilar material; different plate thickness; partially developed stress envelope].

REFERENCES

[1] Nassar, S.A. and Abboud, A., "An Improved Stiffness Model for Bolted Joints", *Journal of Mechanical Design- ASME Transactions,* vol. 131, no. 12, p. 121001, 2009.
[2] Juvinall, R.C. and Marshek, K. M., "Fundamentals of Machine Component Design," 2000, 3rd edition, Chapter 10, John Wiley & Sons, New York.
[3] Bickford, J.H., and Nassar, S.A, "Handbook of Bolts and Bolted Joints," Marcel Dekker, New York, 1998.
[4] Motosh, N., 1976, Determination of Joint Stiffness in Bolted Connections," *Journal of Engineering for Industry*, pp 858–861.
[5] Lehnhoff, T.F., Ko, K.II, Mckay, M.L., 1992, "Member Stiffness and Contact Pressure Distribution of Bolted Joints," *Journal of Mechanical Design*, vol.116, pp. 550–557.

11 Washers
A Guide to Their Function and Selection

Marc Levinson

11.1 INTRODUCTION

Washers play an essential role in generating and sustaining integrity in bolted assemblies. They are active contributing components from the moment torque is applied in the assembly process.

When the primary purpose of the fastener is to resist shear forces, a washer may be added to the assembly to

1. Prevent damage to the bearing surface adjacent to the fastener or to prevent possible embedment of the fastener by distributing the clamp load
2. Function as a locking, or backoff-resistant, device

When fasteners are used to provide a clamping force to ensure the integrity of an assembly, a washer may be added to

1. Prevent damage to the surface adjacent to the fastener
2. Prevent head embedment
3. Span oversize holes
4. Generate and/or maintain tension in the fastening system
5. Function as a locking, or backoff-resistant, device

Still other washers are used to compensate for tolerance variations and eliminate end play or rattles in assemblies. In many assembly applications, a combination of two or more of these functions may be required in a washer.

11.2 THE LOOSENING PHENOMENON

One of the important contributions that a washer makes to the integrity of an assembly is inhibiting the loosening phenomenon that leads to assembly failure.

The integrity of an assembly can be negatively affected in many ways. The fundamental nature of the threaded assembly, which involves the mating of inclined planes under load (the thread lead and flank angle of the screw or bolt and the angle of the mating thread in the nut plate), creates an inherent tendency for the mating pieces to slide "downhill" until there is no longer any load being applied to the inclined surface (see Figure 11.1).

This transverse sliding/loosening tendency is accelerated when joined materials are soft or ductile and yield under load rather than generating tension through resistance. Thermocycling, with its resulting expansion and contraction of materials, also dissipates needed assembly tension.

Only by generating and retaining needed tension in the assembly can equilibrium be sustained in the inclined surfaces of the mating threads to prevent transverse sliding and loosening. In the

DOI: 10.1201/9781003501183-14

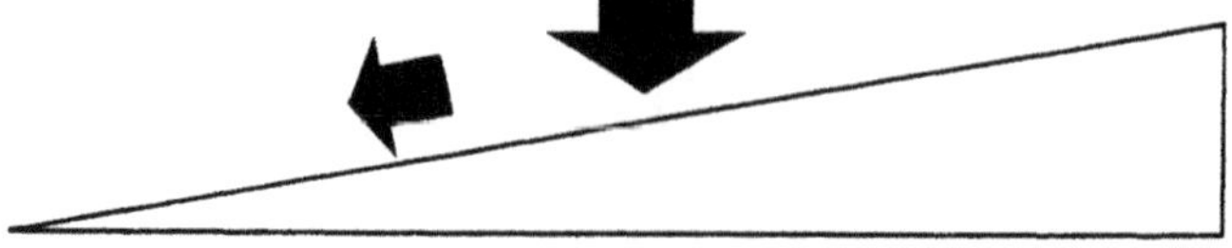

FIGURE 11.1 Transverse sliding of threaded assembly.

ideal assembly environment, a bolt will stretch enough, when engaged in a nut plate, to generate the tension needed to offset the effects of head embedment or any yielding of the clamped material and by doing so will sustain a secure assembly.

A steel bolt elongates approximately 0.002 in. per inch of effective length when it is stressed to 60,000 psi, assuming that the yield strength of the material has not been exceeded. A long bolt, properly torqued, can therefore serve as a very effective tension-generating spring.

If a bolt can be long enough and if sufficient load can be applied, the only reason for adding a washer to the fastening system would be to protect the bearing surface adjacent to the bolt or nut by distributing the load over a wider area. Unfortunately, in the majority of commercial applications, space and weight considerations limit acceptable bolt diameter and length, and, without a supplementary washer, loosening of an assembly can occur. For example, the head of a Grade 5 fastener with an effective length of 1/2 in. will have to embed only 0.001 in. due to burrs, paint, or dirt on the faying surfaces of the joint to lose 100% of a 60,000 psi tensile stress preload.

Assemblies are often subjected to cyclic loads (shock and vibration), and that too can produce the momentary release of applied tension and permit slippage on the inclined plane.

11.3 WASHER FUNCTION AND SELECTION

Needed tension in an assembly can be most efficiently generated by adding a washer component. A wide variety of washer types have been developed to offset specific loosening influences. In selecting a washer appropriate to an assembly application, the purpose and function of many washer options available must be understood.

11.3.1 Washer Option No. 1: Distribute Clamp Load to Prevent Screw and Bolt Head Embedment and Surface Damage and Span Oversized Holes

A flat washer is used to protect faying surfaces by distributing clamp load over the largest effective bearing area. It is placed under the fastener component that is rotated during assembly. The flat washer is also used to span oversized holes. Its only claims to contributing to needed tension in an assembly are that it prevents embedment of the bolt head or nut and it adds to effective bolt length by its thickness.

Flat washers are available in a wide variety of shapes, sizes, and materials. Let's examine the three basic styles of the widely used Type B flat washer as described in Industrial Fastener Institute (IFI) standards; they are listed in Table 11.1.

It might appear that if space limitations were not a factor, the flat washer with the largest outside diameter should always be selected for maximum distribution of clamp load over the largest effective bearing area. This is not always the case.

Load transmission is effective only through a washer for a diameter equal to the bearing surface of the bolt head plus twice the thickness of the washer. This is illustrated in Figure 11.2.

In Table 11.2 charts, we have applied the principle illustrated in Figure 11.2 to select the optimum size flat and bearing washers for typical hex head and hex washer head bolts or screws.

TABLE 11.1
Nominal Dimensional Comparison of Type B Flat Washer Standards

Screw diameter	Narrow		Regular		Wide	
	OD	Thickness	OD	Thickness	OD	Thickness
#6	0.312	0.032	0.438	0.040	0.562	0.040
#8	0.375	0.040	0.500	0.040	0.625	0.063
#10	0.406	0.040	0.562	0.040	0.734	0.063
#12	0.438	0.040	0.625	0.063	0.875	0.063
1/4	0.500	0.063	0.734	0.063	1.000	0.063
5/16	0.625	0.063	0.875	0.063	1.125	0.063
3/8	0.734	0.063	1.000	0.063	1.250	0.100
7/16	0.875	0.063	1.125	0.063	1.469	0.100
1/2	1.000	0.063	1.250	0.100	1.750	0.100

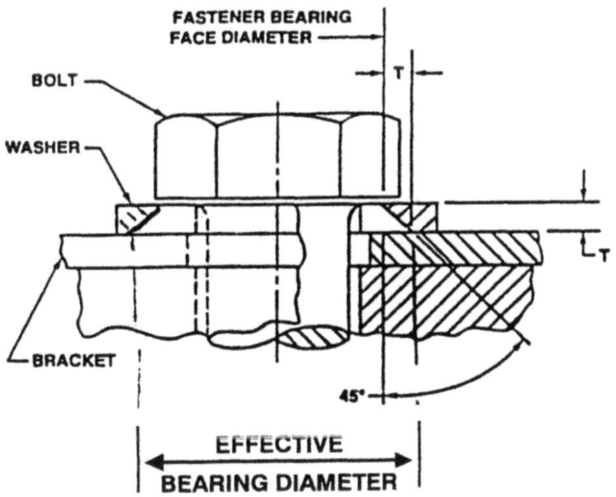

FIGURE 11.2 Flat washer—effective load transmission.

TABLE 11.2
Optimum Washer Diameters

Hex Heads								
Screw Diam.	Hex Across Flats	Type B Washer Thickness			Effective Bearing Diameter			Matching Washer
		Narrow	Regular	Wide	Narrow	Regular	Wide	
#8	0.250	0.040	0.040	0.063	0.330	0.330	0.380	Narrow
#12	0.312	0.040	0.063	0.063	0.390	0.440	0.440	Narrow
3/8	0.562	0.063	0.063	0.100	0.690	0.690	0.760	Narrow
Hex Washer Heads								
Screw Diam.	Hex Washer Flats	Type B Washer Thickness			Effective Bearing Diameter			Matching Washer
		Narrow	Regular	Wide	Narrow	Regular	Wide	
#8	0.348	0.040	0.040	0.063	0.430	0.430	0.470	Regular
#12	0.432	0.040	0.063	0.063	0.510	0.560	0.580	Regular
3/8	0.780	0.063	0.063	0.100	0.910	0.910	0.980	Regular

It is evident from the dimensional comparisons that the narrow series of flat washers is actually optimum for the hex heads and the regular series is optimum for the hex washer heads.

The question that logically follows is: Why is the wide series needed at all? If the bearing surface is a particularly soft or thin material that will tend to yield under load, the bearing stresses will be distributed toward the outside of the washer as the material under the head of the screw yields. In such a case, the wide diameter washers will contribute by reducing the bearing stresses in the soft or thin materials. In other cases, manufacturing considerations may override the purely engineering considerations and result in the selection of a larger diameter washer than the tables suggest.

A note of caution: Flat washers need to be truly flat to provide effective load distribution. IFI defines acceptable flatness for flat washers as being 0.005 in. in washers with diameters up to 0.875 in. and 0.010 in. for larger diameters. This level of flatness cannot be obtained in conventional pierce-pilot-pierce progressive stamping dies due to a tendency to crown the washer in the final piercing station. Manufacturers of flat washers should use a secondary flattening station or compound tooling to achieve acceptable flatness.

11.3.2 Washer Option No. 2: Generate and/or Sustain Tension in the Fastening System

The conical washer, with all of its variations, is one of the more widely used types of tensioning washers. Conical washers, while absorbing driving torques and sustaining needed tension, also span alignment holes and effectively distribute bearing loads over soft or thin materials.

11.3.2.1 Belleville Washers

The Belleville conical washer shows an almost linear rate of load vs. compression from its crown height to the flat condition (see also Chapter 15). The yield strength will not be exceeded even when the washer is compressed to flat, and the washer will exhibit 100% spring return when the compression force is removed. This optimum performance is achieved by maintaining very specific relationships between the washer's inside diameter, outside diameter, thickness, and crown height. The ratio of material thickness to rim width, for example, is held to about 1:5. Crown height should not exceed 40% of material thickness.

To obtain required load deflection characteristics, Belleville washers are often stacked as shown in Figure 11.3.

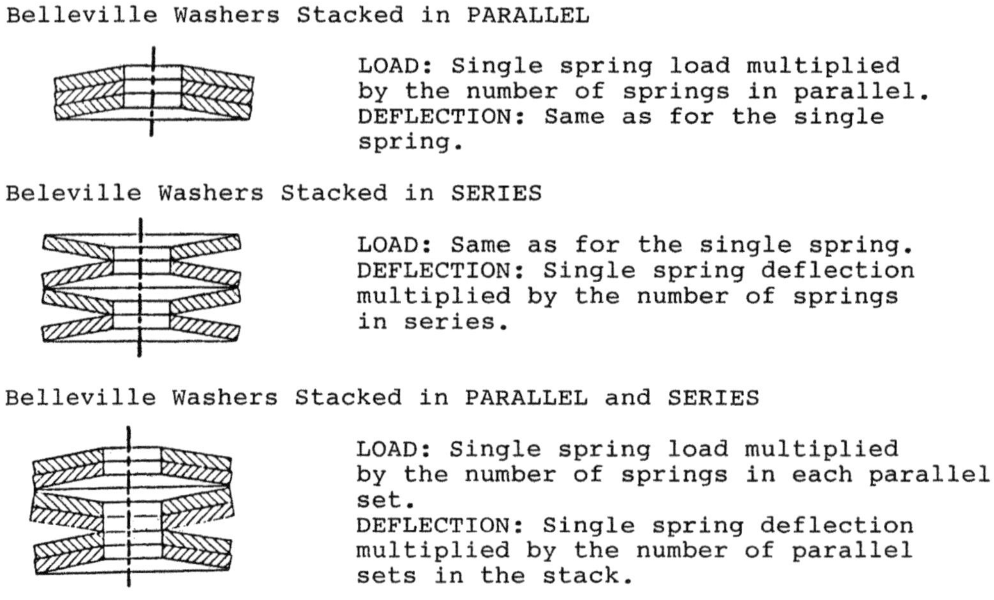

FIGURE 11.3 Alternative stacking arrangements for Belleville washers.

When stacked in parallel, the force required to flatten the washers is multiplied by the number of washers in parallel. If one washer flattens at 100 lb, three similar washers stacked in parallel will flatten at 300 lb.

When Belleville washers are stacked in series, it is the deflection that is multiplied by the number of washers. For example, if the washer that requires 100 lb to flatten has a crown height of 0.025 in., then four washers stacked in series will have a total deflection of 0.100 in. under a 100 lb load. By alternating parallel and series stacking, a wide range of load and deflection needs can be met.

11.3.2.2 Conical Washers

Although Belleville washers are produced to specifications designed to optimize load–deflection performance, there is a wide range of conical washers produced to less stringent specifications. Most of the conical washers used in preassembled screw and washer assemblies called "Sems," for example, have crown heights far greater than 40% of the washer thickness, yet these washers are intended to be flattened when used. In the case of conical washers for Sems fasteners, the increased crown height is for orientation during washer–screw assembly.

When a conventional conical washer is flattened the first time, the steel is stressed beyond the yield point. Consequently, the washer is permanently deformed and takes a set (see Figure 11.4).

As the washer is initially compressed, it exhibits a linear load deflection characteristic, but eventually the stresses in the washer become high enough for the material to permanently yield. When this occurs, the load begins to drop, and the crown height reduces to a height that is compatible with the induced stress, which the material is then capable of absorbing without further yielding.

It is probably more correct to say that the washer has been re-formed rather than deformed, for if it is unloaded and then reloaded, it will act in a consistent elastic manner over its reduced crown height.

Conical washers are frequently used in applications where they function entirely within the elastic range and are never fully compressed to a flat condition. It is quite common to see a conical washer that would yield if completely flattened used in an application where the deflection and induced stress never approach the yield point.

By varying thickness (t) and crown height (h) relationships, conical washers can be designed to meet a wide range of load–deflection requirements.

Reflecting the Belleville principle, when crown height (h) equals 40% or less of thickness ($0.4t$), an elastic condition is maintained. As h increases in relation to t, the washer exceeds the yield point of the material when flattened. When h is in the range $0.4t$–$0.8t$, a fairly constant spring rate is maintained. An h/t ratio of up to 1.4 will sustain a positive rate of increase of the load up to 100%

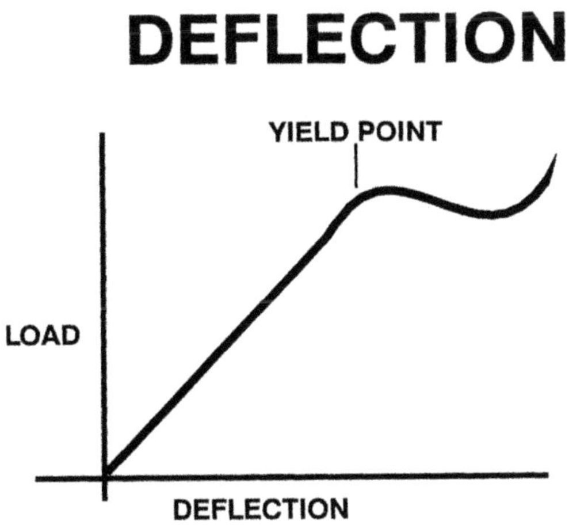

FIGURE 11.4 Stressing a conical washer beyond the yield point.

deflection. An h/t ratio of 1.5 sustains a constant load over a fairly large deflection, making washers of this type useful in applications where they must absorb an extreme wear condition. When the h/t ratio exceeds 1.5, yielding will occur; in extreme conditions, oil canning or even inverting can occur.

11.3.2.3 Square Cone Washer[1]

A significant improvement on the basic conical washer, the high performance Square Cone washer, was developed and introduced by ITW Shakeproof a number of years ago. This unique washer has two distinct conical sections and consequently two separate load deflection rates (see Figure 11.5).

The outer cone, or first stage, of the Square Cone washer flattens first and has a relatively low spring rate. Once the first stage has flattened, additional loading begins to flatten the inner cone, which, due to its geometry, has a significantly higher spring rate.

The Square Cone washer is capable of accepting a high tension load while maintaining a high spring action and generating a constant load, even with variations in installation torques. It provides a higher retained clamp load and improved compensation for thermal cycling and vibration.

The relationship between torque and tension in a fastened joint is very difficult to predict. Theoretical methods exist for calculating torque–tension relationships, but in the real world variations as high as 30% can occur in a typical joint design. In spite of this, in most applications, torque is used as a method of determining whether adequate clamp load has been generated in a joint. As can be seen in Figure 11.6, the Square Cone washer provides for more uniformity in the torque/tension ratio. With the Square Cone washer, torque variation is just over 10% and tension variation 14%. Torque variation on the conventional conical washer is 22%, and the variation in tension is 33%.

The reason for the more uniform relationship between torque and tension in a joint that incorporates the Square Cone washer is demonstrated in Figure 11.7, which compares the torque-time relationship during the driving of a screw with a 7/8 in. outside diameter Square Cone washer and a screw with a regular conical washer having the same 7/8 in. outside diameter.

In achieving the same clamp load, the standard conical washer required about 30% more torque, and the torque was developed over a much shorter time, which can transmit a shock load to the operator. Shock loading has been shown to cause carpal tunnel syndrome, a physical problem resulting in serious operator discomfort, lost time, and high medical costs.

The Square Cone washer produces a more uniform torque–tension relationship because so little change in tension occurs over a wide deflection range. This is apparent in Figure 11.8, which depicts the typical Square Cone tension–deflection curve. This characteristic provides a large "window" for accommodating screw gun torques, resulting in more predictable clamping force and reducing shock loading on the drive tool.

FIGURE 11.5 The Square Cone washer has two load deflection rates.

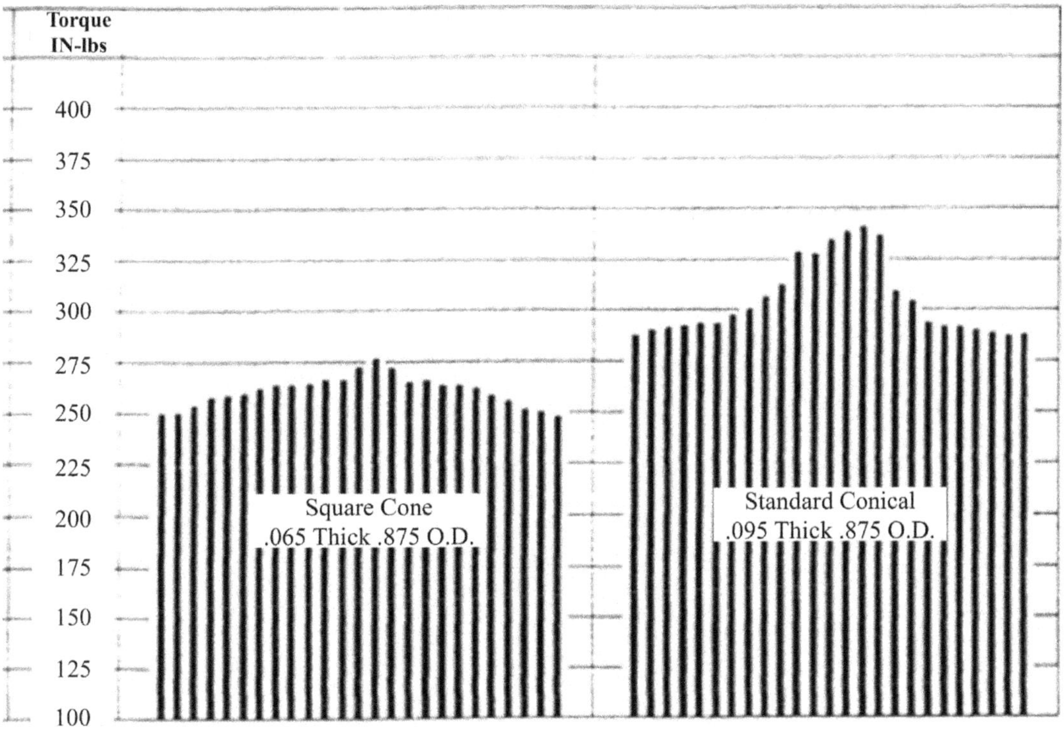

FIGURE 11.6 Torque distribution for the Square Cone washer compared to that of a standard conical washer.

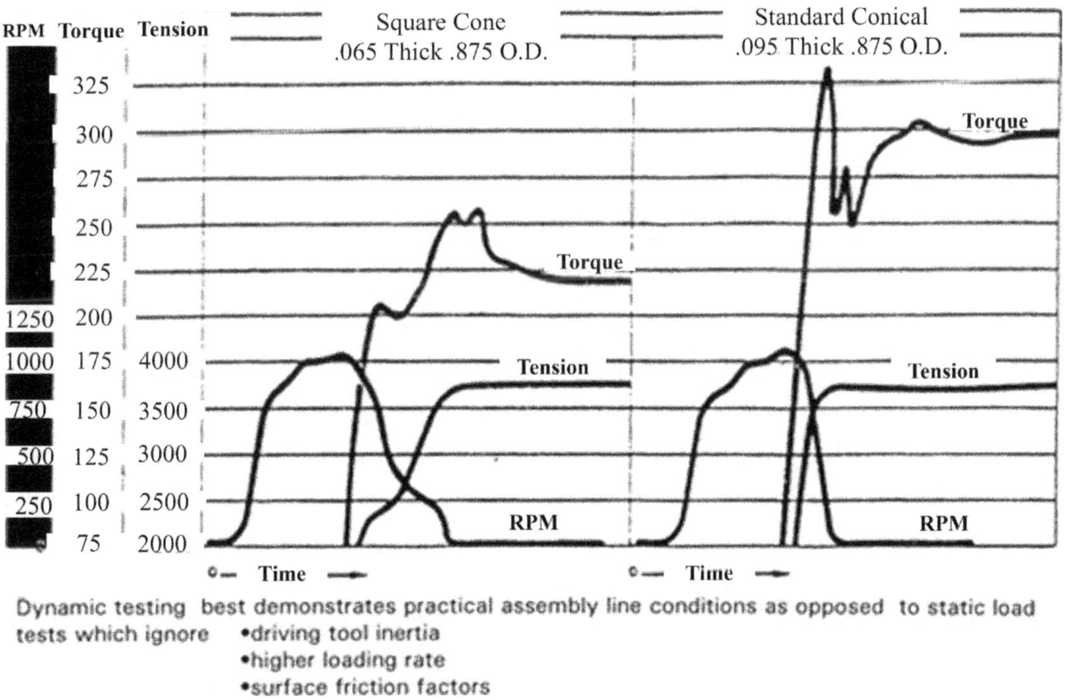

FIGURE 11.7 Torque–time relationship for the Square Cone and standard conical washers.

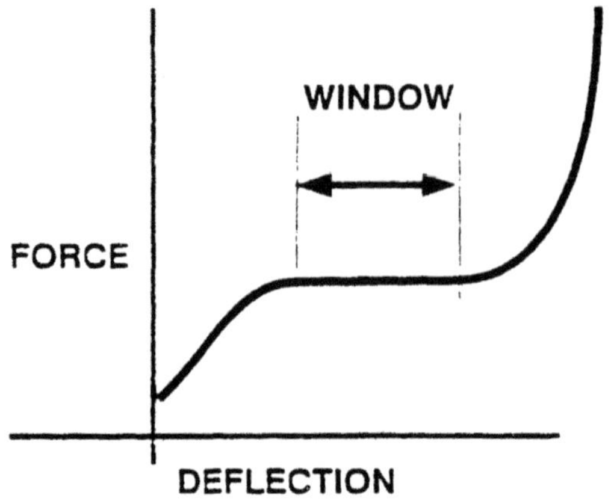

FIGURE 11.8 Tension–deflection relationship for the Square Cone washer.

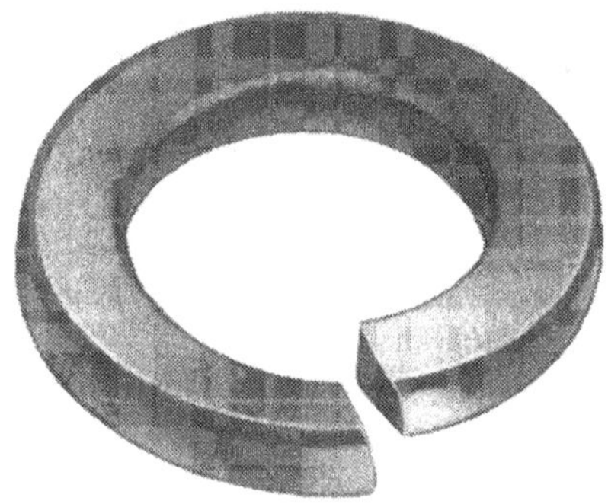

FIGURE 11.9 Helical spring lock washer.

11.3.3 Washer Option No. 3: Resist Loosening Effects of Shock and Vibration

Certain washers are designed specifically to resist loosening or backoff rotation while enhancing and sustaining tension in an assembly. Basically, these are the helical spring and tooth lock washers in all of their variations.

11.3.3.1 Helical Spring Lock Washers

In 1887, the helical spring lock washer was the first to be introduced as a loosening-resistant or locking washer (see Figure 11.9). It has proven its claim through the years. This device is essentially a single coil spring, manufactured from trapezoidal wire, with the maximum thickness being on the inside diameter. A single coil spring is not capable of providing very much reactive tension in a joint, even when flattened.

Figure 11.9 shows that the maximum tension achieved from a 7/16 in. washer is about 450 lb, which compares to a proof load of 9,050 lb and a proof load stress of 85,000 psi for a Grade 5 bolt of this size. This would seem to indicate that the helical spring washer does not contribute much to joint integrity. Yet it has performed the job expected of it for over 100 years. Why?

Tests conducted at Lawrence Technological University show that these washers have a definite spring rate that develop after the single coil is closed and the washer is visually flat. Further loading, it has been shown, results in additional deformation of the washer. The trapezoidal section apparently twists so that the face of the washer lies flat. The maximum outside diameter of the washer is actually increased (see Figure 11.10).

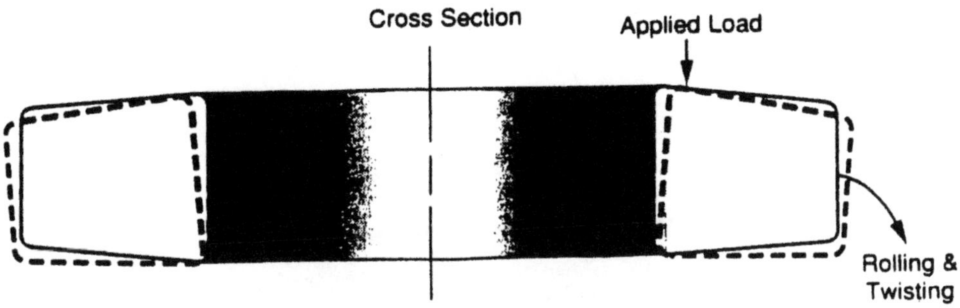

FIGURE 11.10 Helical spring lock washer under load.

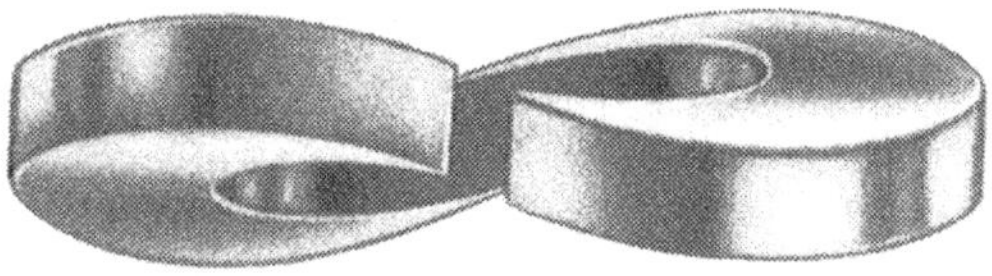

FIGURE 11.11 Spak "split-wave" washer.

On a 7/16 in. diameter helical spring lock washer, this deflection provides a spring rate of 6240 lb/0.001 in. at 60% of the proof load of the fastener. This is the equivalent of adding 0.72 in. to the effective length of the bolt plus the thickness of the washer for a total increase in effective bolt length of 0.85 in. As previously explained, the longer the effective length of the bolt, the less likely it is that embedment of the fastener will result in dropping the prestress level below the induced stress. Therefore, it is less likely that the frictional balance between the screw and the nut plate will lose its equilibrium.

The section width and thickness of helical spring lock washers can be modified to meet application needs. Standard sections include regular, heavy, extra duty, and light. They are also available in special types, such as the Hi-Collar washer for use with socket head cap screws and in other limited space applications and the Double Coil, which provides much greater reactive range needed in power line applications and in other wood-to-metal assemblies used in construction.

11.3.3.2 Spak Washers

The Spak washer is the newest variation of the split lock washer (see Figure 11.11). This unique washer adds the reactive tension and torque absorption qualities of the wave or spring washer to the sustained locking tension of the split lock washer. It is produced as a split lock washer and is then stamped to form a wave. It has the ability to absorb shock and vibration and sustain tension. Invented and patented in Germany and introduced in Japan in the 1960s, Spak washers, traditionally preassembled to screws in Sems form, are the fasteners of choice in Asia for electronic and automotive assemblies.

11.3.3.3 Twisted Tooth Lock Washers

The twisted tooth lock washer functions by biting into the head of the fastener and into the earing surface of the component being clamped. The twisted teeth are oriented so that as the screw is tightened, the teeth slide on the rotating surface and are flattened, thereby providing a flexing feature that adds tension to the joint.

Any tendency for the screw to loosen causes the teeth to try to return to their original form, resulting in the sharp corners biting into the mating surface. The teeth then function as struts, resisting further backoff rotation or loosening (see Figure 11.12).

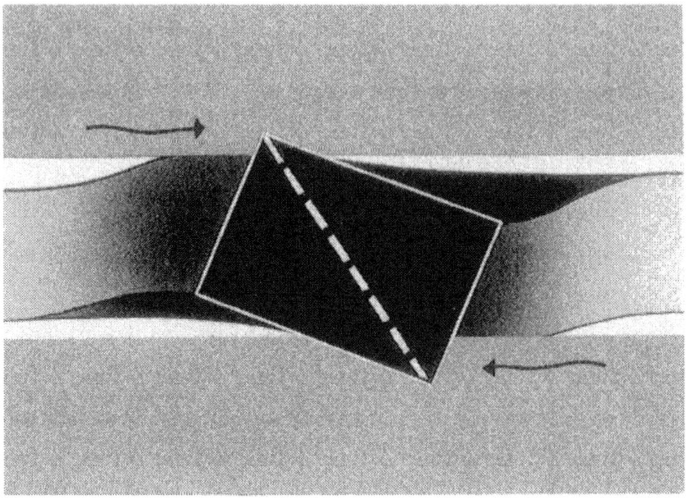

FIGURE 11.12 Tooth lock washer—resistant strut action.

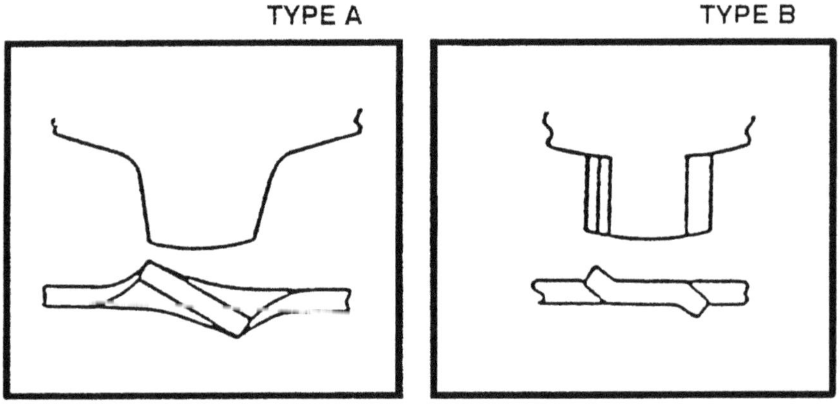

FIGURE 11.13 Type A (left) and type B (right) tooth forms.

Twisted tooth lock washers, which are hardened into the range (per ASME) of 40–50 HRC, work best when they are about 10 HRC points harder than the other materials in the joint.

Two basic tooth forms, Type A and Type B, are available in the United States (see Figure 11.13). Both of these tooth forms work well, with Type A having an advantage in light gage metal applications and Type B having an advantage in harder materials or casting. Type B tooth form also functions well when an electrical ground connection is required. Type A tooth form has a good spring return, while Type B, with its rigid tooth form, has almost no spring action.

Even with Type A tooth form, spring action is limited compared to a conical washer. Generally, tooth washers are smaller in diameter than conical washers and may be preferable in applications where space is limited or the joint is visible.

The basic external and internal tooth washers are the widely used standards in the majority of OEM applications (See Figure 11.14). The external tooth washer provides maximum resistance to backoff, as the tooth bite is positioned under the outer edge of the standard screw head.

Exposed applications are sometimes not well suited to the external tooth washer. It may be desirable to avoid visual marring around the circumference of the screw head, or there may be a danger of snagging something on the exposed teeth. The internal tooth washer was developed to accommodate these applications. The internal tooth washer also works well in close-fitting counterbored holes where the fastener head is recessed.

FIGURE 11.14 External tooth lock washer (left) and internal tooth lock washer (right).

FIGURE 11.15 External/internal tooth lock washer.

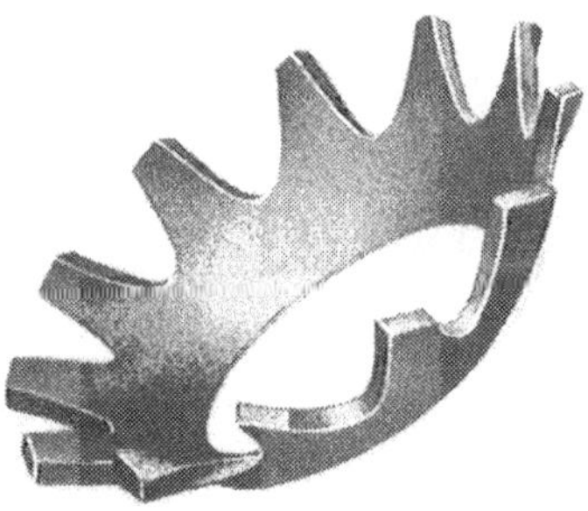

FIGURE 11.16 Countersunk tooth lock washer.

With the external/internal tooth lock washer, twisted teeth are formed on both the inside diameter and the outside diameter of the washer (See Figure 11.15). This type of washer is particularly effective between two components to prevent shifting and retain relative position. It is ideal for electrical grounding and is also used where a large bearing surface is required for oversized or elongated holes.

All of these tooth washers are available in varying rim widths and thicknesses to meet application requirements.

The design versatility inherent in the tooth washer principle has also made it possible to accommodate flush-fitting screws in countersunk holes. Today, the more popular sizes of countersunk tooth washers are used with 82° countersunk screw heads (see Figure 11.16).

11.3.3.4 Combination Locking/Tension Washers

With the inherent adaptability of the tooth form, it was inevitable that it would be combined with variations of the conical washer to provide optimum locking and tension in a single washer that could also span oversized holes and distribute load. Two types, the dished tooth lock washer and the domed tooth lock washer, were introduced many years ago.

The dished type tooth washer is resilient and is excellent for distributing a load over a soft or thin metal and for spanning oversized or elongated holes. The tooth periphery version (Figure 11.17) is designed to bite into an upper sheet to prevent shifting when an oversized hole is needed for adjustment purposes. A flat rim version is available for bearing against hard plastic surfaces or to avoid marring of the upper sheet.

The dome type tooth lock washer (Figure 11.18) is designed to function in the same type of application as the dished type of washer, but it provides added rigidity to support greater loads without oil canning or inverting even when spanning large holes.

The pyramidal type washer (Figure 11.19) is designed for applications that require very high tightening torques. It holds up under heavy loads while providing resiliency and optimum holding power. The washer is produced from heavy gage materials, and the crown is strengthened with six radial creases embossed into the material that run from the inside diameter to the corners of the hexagonal outer diameter. The sharp points at the outside corners of the washer prevent shifting, and the rigid crown forces the teeth on the inner diameter to bite deeply into the bolt or nut for maximum holding power.

FIGURE 11.17 Dished tooth periphery lock washer.

FIGURE 11.18 Domed plain periphery lock washer.

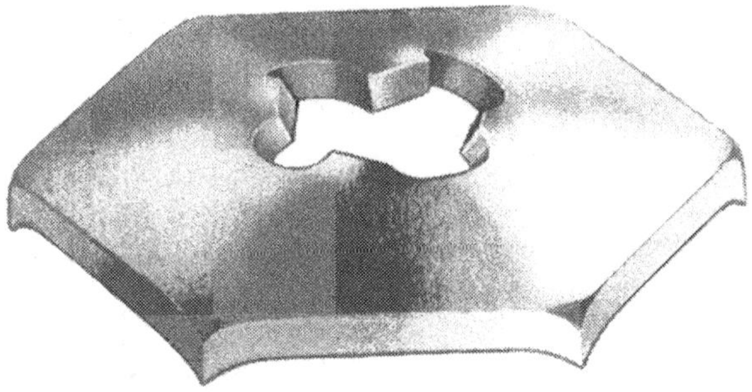

FIGURE 11.19 Pyramidal lock washer.

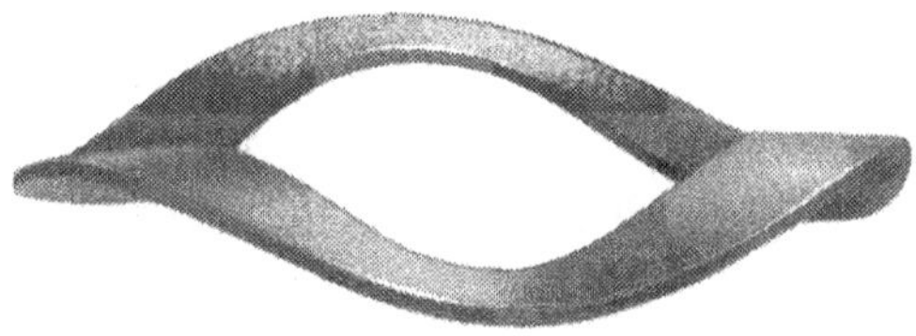

FIGURE 11.20 Multiple-wave spring washer.

11.3.4 Washer Option No. 4: Compensate for Tolerance Variations

11.3.4.1 Spring and Wave Washers

Spring or wave washers are specifically designed to provide a compensating force and sustain a load or absorb shock (see Figure 11.20). Many design variations have evolved to best serve one or the other of these functions or to optimize both functions in a single part within specific inside and outside diameter limits.

Variations in washer thickness, number of waves, and wave height directly affect load-bearing and deflection capabilities. Single-wave washers are generally best in applications involving low loads and requiring high deflection. The load–deflection curve for this type of washer is virtually linear. Typical applications would include the elimination of end play in electric motors, the take-up of large tolerance variations, the elimination of rattles, and the cushioning of light loads.

When considering multiple-wave washers (traditionally three or four waves) it should be remembered that the greater the number of waves the greater the load-bearing capability and the less the deflective capability. Multiple-wave washers have wide application as take-up springs.

In specifying spring washers, in addition to the amount of load that may be applied to the washer, the type of load (static or dynamic) must also be considered. In a static load environment, the basic function of the spring washer is to retain load. In this type of environment, the elastic limit of the material may actually be exceeded.

In a dynamic load environment, the washer performs as a regularly functioning spring, and the elastic limits of the material must therefore not be exceeded. Loading the washer beyond its yield strength results in permanent distortion of the wave height, and consequently its responsiveness to dynamic loads is reduced.

The type and magnitude of the load to which a spring washer will be subjected and the reactive range required are the primary factors in determining the type (single-wave, three-wave, four-wave, etc.) and physical characteristics (material thickness, wave height, etc.) that should be specified for the washer.

Formulas for determining load and deflection characteristics for single-wave and multiple-wave washers are given in Section 3.4.2.

Although their effectiveness is often overlooked, the proper specification and use of spanning, locking, tensioning, and compensating washers can make the difference between a reliable, secure assembly and an assembly that causes problems in field service.

11.3.4.2 Load and Deflection Formulas

11.3.4.2.1 Single-Wave Washers

Load equation:

$$P = \frac{4S(D-d)t^2}{D6} \tag{11.1}$$

Deflection equation:

$$f = \frac{SD^2}{6Et} \tag{11.2}$$

where

P=load (lbf)
E=modulus of elasticity of material (30,000,000 psi for steel)
t=material thickness (in.)
f=deflection (in.)
d=inside diameter (in.)
D=outside diameter (in.)
S=maximum allowable stress (200,000 psi for steel).

11.3.4.2.2 Multiple-Wave Washers

Load equation:

$$P = \frac{SN^2t^2(d-d)}{0.75(D+d)\pi} \tag{11.3}$$

Deflection equation:

$$f = \frac{S\pi^2D^2}{12EtN^2} \tag{11.4}$$

where

N=number of waves
P=load (lbf)
E=modulus of elasticity of material (30,000,000 psi for steel)
t=material thickness (in.)
f=deflection (in.)
d=inside diameter (in.)
D=outside diameter (in.)
S=maximum allowable stress (200,000 psi for steel).

Note: These equations are presented only as a guide. Hardness, for example, will influence the reaction of a spring washer to load but is not factored into these equations. They are presented to guide a specifier to general design parameters, and calculated values may vary by as much as 30% from actual values in some cases.

NOTE

1 "Square Cone" is a registered trademark of Illinois Tool Works, Inc.

12 Vibration-Resistant Fasteners

Joseph R. Dudley

12.1 INTRODUCTION

Most bolted joints are exposed to motion, vibration, or shock, which can cause a momentary loss of pressure between the thread flanks. This loss of pressure will cause the joint members to slip sideways and loosen. A wide variety of locking fasteners resist this loosening. Some fasteners use an accessory part to provide a mechanical lock. Others resist loosening through the use of prevailing torque.

This chapter presents data on vibration-resistant bolts and screws, which are fasteners with external (male) threads, and nuts, which have internal (female) threads. Prevailing torque and locking type fasteners are covered here; other types are discussed in Chapter 41.

Following some basic definitions of terms in Section 12.2 and a general description of types of vibration-resistant fasteners in Section 12.3, Section 12.4 presents data on fasteners with external threads whose locking mechanisms are nylon patches or chemical adhesives. Section 12.5 concludes with data on fasteners with internal threads. In addition to 180° and 360° patch and adhesive-locking nuts, Section 12.5 includes data for prevailing torque type steel hex locknuts. In all cases, tables are included for both inch and metric series fasteners.

12.2 DEFINITIONS

Prevailing torque: Force necessary to rotate a fastener relative to its mating component against a frictional resistance.

Initial installation torque (maximum): Force produced the first time the locking fastener is installed, measured during the first five revolutions after engaging the locking element.

Prevailing on-torque: Torque force read in the first 360° of rotation in the installation (on) direction after the component that contains the frictionally resistant member is fully engaged and after the initial installation torque has been determined.

Prevailing off-torque: Torque read during the first 360° of rotation in the off direction after the break-loose or breakaway torque has been measured. Both high and low torque should be recorded.

Seating torque: Tightened torque or clamp load. The maximum value in lb-ft or N·m that occurred at the final assembly of the joint, as the joint became fully seated in the tightening direction.

Break-loose torque: The force required to initiate reverse rotation to a fastener that has been fully seated, clamped, or tightened.

Breakaway torque or static torque: The force required to initiate reverse rotation to a fastener that has not been seated, clamped, or tightened.

12.3 TYPES OF VIBRATION-RESISTANT FASTENERS

Prevailing torque fasteners that use physical interference function on the wedge principle. Normally, a polymer compound, usually nylon in the form of a pellet, strip, or patch, is permanently attached either to an external or to internal thread. This nylon pellet, strip, or patch forms a wedge between the mating threads, creating friction, which is measured as torque. The nylon patch can be applied

DOI: 10.1201/9781003501183-15

circumferentially from partial coverage up to 360°, according to torque requirements, either to internal or to external threads. 180° patch coverage is normally specified. The nylon can be reused for a minimum of five times; it is nonreactive and can be used at a temperature of up to 250°F continuous. There is also a high temperature version that functions up to 430°F.[1] Nylon material is generally specified when a joint has to be disassembled and reassembled for service and repair (see Tables 12.1, 12.2, and 12.5–12.8).

Another type of prevailing torque fastener is the *all-metal deformed thread* internal threaded product. This product, since it uses thread interference to produce torque, has limited reuse. Its temperature range is the same as that of steel, and it is nonreactive. Its use is usually permanent, and when the bolted joint is disassembled, a new nut should be used for reassembly. See Tables 12.11 and 12.12.

External threads are also specified with deformed threads, usually a thread forming style configuration. There are many types of threads, and as the joint comes to clamp load the metal in the thread form is displaced or destroyed. Reuse is generally not recommended.

A third type of vibration-resistant uses a chemical adhesive on either the internal or external thread to provide a locking action. It is usually not reusable and is not used in a joint that wouldv normally be disassembled. It can be used at temperatures up to 340°F (see Tables 12.3, 12.4, 12.9, and 12.10).

The torque achieved with either a fine or coarse metric series or inch series thread is quite comparable within a size. There are many factors that will affect the friction and thus the torque in a bolted joint: the hardness of the parts, type of material plating, finish, lubricants, tightening speed, thread fit, clearance hole size, surface pressure, and many others. One engineer estimated that over 200 factors can affect friction.

12.4 FASTENERS WITH EXTERNAL (MALE) THREADS

The performance standards for various types of fasteners with external threads are presented in Tables 12.1–12.4.

The standard procedure in testing is that similar materials and finishes are preferred for the nut and bolt. The most common materials for the inch series is a Grade 5 bolt with phosphate coating and plain finish. Nuts are class B steel and zinc-plated. Metric fasteners use 10.9 bolts with phosphate coating and class 10 steel nuts with plain finish unless otherwise specified.

No table has been developed or published for externally threaded thread forming screws. Because of their vast variety and the fact that they are not a prevailing torque type of fastener, it is very difficult to determine or publish torque values for their installation.

The following rules apply in testing chemical locking externally threaded fasteners (Tables 12.3 and 12.4).

1. *Performance of test*: During the complete performance of the test, the nut shall be turned while the test bolt shall be restrained from turning. If the nut is restrained so that the torque reading is not adversely affected, it is permissible to turn the test bolt.
2. *Prevailing on-torque*: The test nut shall be assembled onto the preapplied adhesive-coated area of the bolt until the full thickness of the nut is engaged into the adhesive. The maximum prevailing on-torque shall be measured and recorded during the next 360° of rotation, while the fastener is in constant motion. The speed of rotation shall be approximately 10–20 rpm.
3. *Breakaway/prevailing off-torque*: Assemble the adhesive-coated bolts and nuts. The activated adhesive-bolt-nut assembly shall be allowed to cure at 25°C±2°C for 24 h. The fastener shall then be rotated in the off direction with a torque-measuring device of suitable capacity and sensitivity. Torque readings shall be measured and recorded at breakaway torque and within 360° of further rotation for the maximum prevailing off-torque.

TABLE 12.1
Performance Standards for Nonmetallic Resistant Element Type Prevailing Torque Screws—Nylon Patch—Inch Series[a]

Nominal Thread Diameter and Thread Flush	Clamp Load (lb) Strength Grade of Screw			Prevailing on-Torque, Maximum (lb-in.)	First Removal Prevailing Torque, Minimum (lb-in.)	Fifth Removal Prevailing Torque, Minimum (lb-in.)
	SAE Grade 5	SAE Grade 8	ASTM A574			
No. 2-56	234	330	390	2	3[b]	1.5[b]
No. 2-64	250	355	410			
No. 4-40	385	540	635	5	1	0.5
No. 4-48	420	595	690			
No. 6-32	580	820	950	8	2	1
No. 6-40	650	910	1,060			
No. 8-32	900	1,260	1,470	12	2.5	1.5
No. 8-36	940	1,330	1,540			
No. 10-24	1,120	1,580	1,840	18	3	2
No. 10-32	1,280	1,800	2,100			
1/4-20	2,020	2,850	3,220	40	5	3
1/4-28	2,320	3,260	3,680			
5/16-18	3,340	4,720	5,300	85	8	5
5/16-24	3,680	5,210	5,870			
3/8-16	4,950	6,980	7,880	110	14	9
3/8-24	5,590	7,880	8,900			
7/16-14	6,790	9,600	10,800	150	20	12
7/16-20	7,580	10,600	12,000			
1/2-13	9,080	12,800	14,400	220	26	16
1/2-20	10,200	14,400	16,200			
9/16-12	11,600	16,400	18,400	270	35	22
9/16-18	13,000	18,300	20,600			
5/8-11	14,400	20,300	22,900	350	45	30
5/8-18	16,400	23,000	26,000			
3/4-10	21,300	30,100	33,800	460	60	45
3/4-16	23,800	33,600	37,800			
7/8-9	29,500	41,600	46,800	700	95	65
7/8-14	32,500	45,800	51,500			
1-8	38,600	54,500	61,400	900	130	85
1-12	42,300	59,700	67,100			
11/8-7	42,400	68,700	77,200	1,050	150	110
11/8-12	47,500	77,000	87,000			
11/4-7	53,800	87,200	98,200	1,150	200	140
11/4-12	59,600	96,600	109,000			
13/8-6	64,400	104,000	117,000	1,300	240	150
13/8-12	73,000	118,000	134,000			
11/2-6	78,000	126,000	142,000	1,500	280	185
11/2-12	87,800	142,000	160,000			

Source: IFI Specification 124.

[a] The clamp torque applied in each case was equal to 75% of the proof load specified for the property class.

[b] oz-in.*Source*: IFI Specification 124.

TABLE 12.2
Performance Standards for Nonmetallic Resistant Element Type Prevailing Torque Screws—Nylon Patch—Metric Series[a]

Nominal Thread Diameter and Thread Pitch	Clamp Load 1 (Kn) Property Class Of Screw			Prevailing on-Torque, Maximum (N·m)	First Removal Prevailing Off-Torque, Minimum (N·m)	Fifth Removal Prevailing Off-Torque, Minimum (N·m)
	8.8	9.8	10.9			
M1.6×0.35	—	0.6	—	0.1	0.01	0.004
M2×0.4	—	1	—	0.2	0.02	0.01
M2.5×0.45	—	1.6	—	0.4	0.05	0.03
M3×0.5	—	2.4	—	0.6	0.14	0.06
M3.5×0.6	—	3.3	—	0.9	0.22	0.11
M4×0.7	—	4.3	—	1.2	0.26	0.16
M5×0.8	—	6.9	8.8	2.3	0.36	0.23
M6×1	—	9.8	12	3	0.45	0.3
M8×1.25	—	18	23	10	0.9	0.58
M10×1.5	—	28	36	14	1.8	1.1
M12×1.75	—	41	52	21	2.6	1.5
M14×2	—	56	72	30	3.6	2.3
M16×2	71	76	98	40	5	3.4
M20×2.5	110	—	150	60	8	5.5
M24×3	160	—	220	90	13	8.5
M30×3.5	250	—	350	120	19	13
M36×4	370	—	510	150	28	18

Source: IFI Specification 524.

[a] The clamp torque applied in each case was equal to 75% of the proof load specified for the property class.

TABLE 12.3
Performance Standards for Chemical Locking Externally Threaded Fasteners—Inch Series

Nominal Size And Thread Pitch	Prevailing on-Torque, Max (lb-in.)	Breakaway Torque (lb-in.)	Prevailing Off-Torque (lb-in.)
		Cured Min	Cured Min
1/4-20	15	20	12
5/16-18	20	40	15
3/8-16	45	100	45
7/16-14	60	150	60
1/2-13	80	200	90
9/16-12	100	250	120
5/8-11	125	350	150
3/4-10	150	450	200
7/8-9	180	550	250

Source: Data from Nylok Fastener test laboratory.

TABLE 12.4
Performance Standards for Chemical Locking Externally Threaded Fasteners—Metric Series

Nominal Size and Thread Pitch	Prevailing On-Torque, Max (N·m)	Breakaway Torque, Min (N·m)	Prevailing Off-Torque, Min (N·m)
M6×1	0.6	1	1.7
M8×1.25	1.5	8.5	5.6
M10×1.5	2	22	10.2
M12×1.75	3.5	27	14.1
M14×2	5.5	62	24.9
M16×2	8	88	37.6
M20×2.5	11	110	49.5
M24×3	14.5	137	61.9

Source: Data from Nylok Fastener test laboratory.

TABLE 12.5
Performance Standards for 180° Nylon Patch Nuts With Internal Threads—Inch Series[a]

Size	Initial Installation Maximum (lb-in.)	Prevailing 1st Off Minimum (lb-in.)	Prevailing 5th Off Minimum (lb-in.)
10-32	5.5	3	1.8
1/4-20	18	7	2.5
5/16-18	30	12.5	3.5
3/8-16	35	14	8.5
7/16-14	55	21	16
1/2-13	60	30	20.5
5/8-11	165	60	33
3/4-10	190	80	60
7/8-9	220	110	70
1-8	60[b]	30[b]	12[b]

Source: Data from Nylok Fastener test laboratory.

[a] Nut: Grade B steel HRC 28 max. Bolt: Grade 5 steel HRC 25-34. Test procedure per IFI 100/107.

[b] lb-ft.

12.5 FASTENERS WITH INTERNAL (FEMALE) THREADS

Performance standards for fasteners with internal threads are presented in Tables 12.5–12.12.

Caution is urged when selecting prevailing torque elements or locking elements in internally threaded products. Depending on the style of nut specified, especially when using standard finished or metric hex nuts, style 1 (no orientation required), there may not be enough threads inside the nut to coat with nylon or adhesive to achieve the amount of torque specified. Table 12.13 shows the limits and the number of threads available with standard nut thickness.

TABLE 12.6
Performance Standards for 180° Nylon Patch Nuts With Internal Threads—Metric Series[a]

Size	Initial Installation Maximum (N·m)	Prevailing 1st Off Minimum (N·m)	Prevailing 5th Off Minimum (N·m)
M4×0.7	0.6	0.2	0.1
M5×0.8	1.3	0.4	0.0
M6×1.0	2.2	0.8	0.4
M8×1.25	4.0	1.7	1.0
M10×1.5	12.0	5.0	2.5
M12×1.75	17.0	5.5	3.0
M14×2.0	24.0	11.0	4.5
M16×2.0	35.0	18.0	10.0
M20×2.5	70.0	35.0	20.5

Source: Data from Nylok Fastener test laboratory.
[a] Nuts: Class 10 zinc finish HRC 26–36. Bolts: Class 10.9 phos and oil finish HRC 33–39. Test procedure per ASME B 18.16.1 M. Program B, Table 12.4.

TABLE 12.7
Performance Standards for 360° Nylon Patch Nuts With Internal Threads—Inch Series[a]

Size	Initial Installation Maximum (lb in.)	Prevailing 1st Off Minimum (lb-in.)	Prevailing 5th Off Minimum (lb-in.)
10-32	11.5	5.0	2.5
1/4-20	18.0	9.0	3.5
5/16-18	30.0	16.0	5.0
3/8-16	42.0	15.0	10.0
7/16-14	60.0	21.0	14.0
1/2-13	65.0	24.0	17.0
5/8-11	95.0	55.0	20.0
3/4-10	200.0	95.0	45.0
7/8-9	240.0	80.0	55.0
r-8	44[b]	33[b]	15[b]

Source: Data from Nylok Fastener test laboratory.
[a] Nut: Grade B steel HRC 28 max. Bolt: Grade 5 steel HRC 25-34. Test procedure per IFI 100/107.
[b] lb-ft.

Adhesive patched nuts show low installation torque-on values as the compound is soft and mixed during installation. Breakaway torques can be as strong as the breaking point of some soft steels and can be modified to meet customer requirements. Prevailing off-torque of adhesive parts, once the bond is broken, is less than that of plastic patch parts. Some second time hardening occurs, as not all of the encapsulated beads crush and mix the first time, but the values obtainable are happenstance

TABLE 12.8
Performance Standards for 360° Nylon Patch Nuts With Internal Threads—Metric Series[a]

Size	Initial Installation Maximum (N·m)	Prevailing 1st Off Minimum (N·m)	Prevailing 5th Off Minimum (N·m)
M4×7	(1)	0.5	0.3
M5×0.8	(1)	0.7	0.4
M6×1.0	4.0	2.0	0.5
M8×1.25	6.0	2.0	1.5
M10×1.5	14.0	4.0	2.0
M12×1.75	23.0	7.0	3.5
M14 X 2.0	27.3	10.0	5.0
M16×2.0	33.0	17.0	8.0
M20×2.5	50.7	22.0	12.0
M24×3	55.0	25.1	15.0

Source: Data from Nylok Fastener test laboratory.
[a] Nuts: Class 10 zinc finish HRC 26–36. Bolts: Class 10.9 phos and oil HRC 33–39. Test procedure per ASME B 18.16.1 M.

TABLE 12.9
Performance Standards for Adhesive Locking Internally Threaded Fasteners—Inch Series, Coarse Threads[a]

Size (in.)	Initial Installation Maximum (lb-in.)	Breakaway Torque Minimum (lb-in.)	Prevailing Off-Torque 1st 360 after Breakaway Minimum (lb-in.)
1/4	3	20	15
5/16	5	28	20
3/8	7	40	28
7/16	9	90	120
1/2	16	125	145
9/16	18	151	163
5/8	20	260	255
3/4	45	470	405
7/8	50	500	480
1	72	560	530

Source: Data from Nylok Fastener test laboratories.
[a] Nuts: Grade B steel, commercial zinc-plated, HRC 28 maximum. Bolts: SAE Grade 5 steel, HRC 25–34.

numbers. Reuse of adhesive-coated fastener joints is generally not recommended. Breakaway values, as stated in Tables 12.9 or 12.10, could be quite high and, in the case of larger sizes (⅝ in. M16), make the parts difficult to remove. The time required for adhesive hardening might limit adjustment of the installed fastener. Heat and humidity can cause premature hardening of the adhesive, thereby limiting the shelf life and storage conditions of preapplied adhesive fasteners to inside areas away from furnaces and heaters. Because the adhesive is soft and fills the thread interspaces completely when smeared and mixed by the fasteners, long rundowns are not recommended.

TABLE 12.10
Performance Standards for Adhesive Locking Internally Threaded Fasteners—Metric Series, Coarse Threads[a]

Size	Initial Installation Maximum (N·m)	Breakaway Torque Minimum (N·m)	Prevailing Off-Torque 1st 360 after Breakaway Minimum (N·m)
M4×0.7	0.1	1.0	0.8
M×0.8	0.1	1.0	1.0
M6×1.0	0.2	2.5	2.0
M8×1.25	0.5	4.5	3.0
M10×1.5	0.7	6.0	4.5
M12×1.75	2.0	18.5	10.5
M14×2.0	3.0	30.0	13.0
M16×2.0	3.5	35.0	18.0

Source: Data from Nylok Fastener test laboratory.

[a] Nuts: Class 10 steel commercial zinc-plated, HRC 26–36. Bolts: Steel class 10.9, commercial zinc-plated, HRC 33–39.

The ideal condition is for the bolt end to be tightened to flush with the nut face so the adhesive coating is in full contact within the threads. It is suggested that no more than one- or two-thread protrusions be used for normal designs. Protrusion of four threads may reduce joint strength by 10%, due to the wipeout of the thread adhesive by bolt threads during installation. Another design consideration when using adhesive locking features is the number of threads available for adhesive coating. For small-size standard nuts, which would fewer internal threads available to apply the adhesive to. For example, a $^5/_{16}$in. (approx. size as M8) does have nut has four internal threads only, while a ¼ in. size nut (approx. M6) has only six internal threads. Since many specifications require one thread to be free of adhesive for assist in starting nut engagement with the bolt; only the two remaining the $^5/_{16}$ in nut threads would be available to apply the adhesive to: four threads for the ¼ in. nuts. That does not leave much room for properly applying the adhesive to achieve a high breakaway torques.

Nylon Patch Nuts, 180° patch coverage vs. 360° patch coverage compared to the collar nut. There are distinct differences in the applications of these three nuts. Whereas the various torque values are relatively the same, the 180° standard patch exposes 180° of bare metal to bare metal, while the 360° patch and the collar nut use a circumferential nylon surface against the external bare metal thread. The differences are shown in Table 12.14.

Commercial zinc plating is very smooth and presents a difficult base for adhesive bonding. When using adhesives on very smooth coatings (including chrome and nickel plates as well as zinc), consult Nylok for technical assistance. Smooth coatings may produce minimal torque values. Other finishes, such as dry phosphate, may double the locking forces due to its excellent bonding surface, and may generate very high breakaway torques.

Small internal threads will have low strengths due to limited area for adhesive adhesion.

TABLE 12.11

Prevailing Torque Type Steel Hex Locknuts—Inch Series—Coarse and Fine Thread Series

Nut Size and Threads per inch	Grade A Locknuts							Grade B Locknuts							Grade C Locknuts						
			Prevailing Torque							Prevailing Torque							Prevailing Torque				
				First Removal		Fifth Removal					First Removal		Fifth Removal					First Removal		Fifth Removal	
	Proof Load (lb)	Clamp Load (lb)	First Install., Max (lb-in.)	Highest Reading Min (lb-in.)	Lowest Reading Min (lb-in.)	Highest Reading Min (lb-in.)	Lowest Reading Min (lb-in.)	Proof Load (lb)	Clamp Load (lb)	First Install., Max (lb-in.)	Highest Reading, Min (lb-in.)	Lowest Reading, Min (lb-in.)	Highest Reading, Min (lb-in.)	Lowest Reading, Min (lb-in.)	Proof Load (lb)	Clamp Load (lb)	First Install., Max (lb-in.)	Highest Reading, Min (lb-in.)	Lowest READING, min (lb-in.)	Highest Reading, Min (lb-in.)	Lowest Reading, Min (lb-in.)
4.48	600	270	3	1	0.5	0.5	0.2	790	420	3	1	0.5	0.5	0.2	990	600	4	1	0.5	0.5	0.2
6.4	900	420	6	1.5	0.5	10	0.5	1,200	640	8	1.5	0.5	1	0.5	1,500	900	8	2	1	1	0.5
8.36	1,350	610	9	2	1	1.5	0.5	1,750	930	12	2	1	1.5	0.5	220	1,300	12	2.5	1	2	1
10.32	1,800	840	13	2.5	1	2	1	2,400	1,300	13	2.5	1.5	2	1	3,000	1,800	17	3.5	1.5	2.5	1
12.28	2,300	1,050	20	3.5	1.5	2.5	1	3,100	1,650	20	3.5	1.5	2.5	1	3,900	2,350	27	4.5	2	3	1.5
1/4-28	3,300	1,500	30	5	2.5	3.5	1.5	4,350	2,300	30	5	2.5	3.5	1.5	5,450	3,250	40	6	3	4.5	2
5/16-24	5,200	2,400	50	8	4	5.5	2.5	6,950	3,700	60	8	4	5.5	2.5	8,700	5,200	80	10.5	5	7.5	3
3/8-24	7,900	3,600	30	12	5	8.5	4	10,500	5,600	80	12	5	8.5	4	13,200	7,900	110	16	7.5	11.5	5
7/16-20	10,700	4,900	100	17	7.5	12	5	14,200	7,550	100	17	7.5	12	5	17,800	10,700	135	23	10	16	7.5
1/2-20	14,400	6,550	150	22	10	15	7.5	19,200	10,200	150	22	10	15	7.5	24,000	14,400	17[a]	30	15	20	10
9/16-18	18,300	8,350	17[a]	30	15	21	10	24,400	13,000	17[a]	30	15	21	10	30,400	18,300	25[a]	40	20	28	12.5
5/8-18	22,900	10,500	25[a]	39	17.5	27	12.5	30,700	16,300	25[a]	39	17.5	27	12.5	38,400	23,000	35[a]	52	25	36	15
3/4-16	33,600	15,400	35[a]	58	25	41	20	44,800	23,800	35[a]	58	25	41	20	56,000	33,600	45[a]	78	35	54	25
7/8-14	45,800	13,600	50[a]	88	40	62	30	61,100	32,400	50[a]	88	40	62	30	76,400	45,800	70[a]	117	50	82	40
-14	61,100	16,800	70[a]	120	60	84	40	81,500	43,300	70[a]	120	60	84	40	101,900	81,100	90[a]	160	80	112	50
-12	16,400	16,400	70[a]	120	60	84	40	79,600	42,300	70[a]	120	60	84	40	99,500	59,700	90[a]	160	80	112	50
1-1/8-12	21,200	21,200	75[a]	150	70	105	50	89,900	47,500	75[a]	150	70	105	50	128,000	76,800	100[a]	220	100	140	70
1-1/4-12	26,600	26,600	85[a]	188	90	132	60	113,000	59,700	85[a]	188	90	132	60	161,000	96,600	100[a]	250	120	176	90
1-3/8-12	32,500	32,500	100[a]	220	110	154	70	138,000	72,900	100[a]	220	110	154	70	197,000	118,000	135[a]	293	140	205	100
1-1/2-12	39,100	39,100	110[a]	260	130	182	90	16,600	87,700	110[a]	260	130	182	90	237,000	142,000	150[a]	346	170	242	120

Source: IFI Specification 100/107.

[a] Values are in lb/ft for these sizes.

TABLE 12.12
Prevailing Torque Type Steel Hex Locknuts—Metric Series

Nominal Nut Diam. and Thread Pitch	Proof Load (kN)			Clamp Load[a] (kN)			Prevailing Torque[b] (N·m)					
							Class 5 and 9			Class 10		
	Class 5	Class 9	Class 10	Class 5	Class 9	Class 10	First Installation	First Removal	Fifth Removal	First Installation	First Removal	Fifth Removal
M3×0.5	2.62	4.53	5.23	1.43	2.45	3.13	0.43	0.12	0.08	0.60	0.15	0.10
M4×0.7	4.57	7.90	9.13	2.50	4.25	5.47	0.90	0.18	0.12	1.20	0.22	0.15
M5×0.8	8.23	13.00	14.80	4.05	6.92	6.64	1.80	0.29	0.20	2.10	0.35	0.24
M6×1	11.70	18.40	20.90	5.73	9.80	12.50	3.00	0.45	0.30	4.00	0.55	0.40
M6×1.25	21.60	34.40	36.10	10.40	17.80	22.80	7.00	0.85	0.60	9.00	1.15	0.80
M10×1.5	34.20	54.50	60.30	18.50	26.30	36.10	10.50	1.50	1.00	14.00	2.00	1.40
M12 X 1.75	51.40	80.10	88.50	24.00	41.10	52.50	15.50	2.30	1.60	21.00	3.10	2.10
M14×2	70.20	100.00	121.00	32.80	56.10	71.60	24.00	3.30	2.30	31.00	4.40	3.00
M16×2	96.80	149.00	165.00	44.80	76.50	97.50	32.00	4.50	3.00	42.00	8.00	4.20
M20×2.5	154.00	225.00	280.00	69.80	110.00	152.00	54.00	7.50	5.30	72.00	10.50	7.00
M24 X 3	222.00	325.00	374.00	101.00	159.00	220.00	80.00	11.50	5.00	106.00	15.00	10.50
M30 X 3.5	363.00	516.00	595.00	94.50	252.00	349.00	108.00	16.00	12.00	140.00	19.00	14.00
M36 X 4	515.00	752.00	866.00	138.00	366.00	509.00	136.00	21.00	16.00	180.00	24.00	17.50

Source: ASME Specification B18.16.1M.

[a] The clamp loads for class 5 nuts are equal to 75% of the proof load of property class 5.8 bolts for diameters M3–M24 and 75% of the proof load of property class 4.8 bolts for diameters larger than M24. The clamp loads for class 9 nuts are equal to 75% of the proof load of property class 9.8 bolts for diameters M3–M16 and 75% of the proof load of property class 8.8 bolts for diameters larger than M16. The clamp loads for class 10 nuts are equal to 75% of the proof load of property class 10.9 bolts. Proof loads of bolts are given in SAE J1199.

[b] First installation torque is the torque occurring while the nut is being assembled on the test bolt for the first time. The torque is measured through the first 360° of nut rotation after two full bolt threads protrude through the nut. For the nut to be acceptable, the maximum torque reading shall not be greater than the tabulated value. First and fifth removal torques are the torques occurring during the first and fifth times the nut is disassembled from the last bolt. The torque is measured through the first 360° of nut removal rotation after (for first removal only) the clamp load has been fully relieved. For the nut to be acceptable, the maximum torque reading shall not be less than the tabulated value.

TABLE 12.13
Coverage of Locking or Prevailing Element in Standard Metric Hex Nut Threads[a]

Thread Size (Coarse)	Standard Hex, Style 1: Total Threads	Standard Hex, Style 1: Available Threads[c]	Standard Hex, Style 2[b]: Total Threads	Standard Hex, Style 2[b]: Available Threads[c]
M5×0.8	4	2	5	3
M6×1.0	4	2	5	3
M8×1.25	4	2	5	3
M10×1.5	5	3	6	4½
M12×1.75	6	4	6½	4½
M14×2	5	3	6	4
M16×2	5	3	6	4
M20×2	5	3	6	4

[a] One starting thread each side, no orientation.
[b] Style 2 nuts, not stocked, expensive, mostly by special order only.
[c] For hex flange nut, must be oriented, add two threads. For cap (acorn) nuts, add one thread.

TABLE 12.14
A Comparison of 180° Patch, 360° Patch, and Collar Nuts

	180° Patch	360° Patch	Nylon Insert Collar Nut
	A. External thread condition		
1. Thread chatter	Affects torque registry[a]	No effect	Affects torque registry
2. Plating finishes	Affects torque	No effect	Affects torque
3. Foreign material deposits	Affects torque	No effect	Affects torque
4. Long run-down	Good	Better	Good
	B. Nut		
1. Weight	Standard	Standard	Increases 15%–25%
2. Size	Standard	Standard	Increases by two threads minimum
C. Mating bolt	Standard	Standard	Increases by two threads minimum to overall height
D. Manufacturing effort	Good	Good	Extremely critical in skirt area and nylon placement
E. Cost	Standard	Standard	Increases up to 50% over patch nuts
F. Assembly forces	Horizontal against threads	Horizontal against threads	Vertical against nylon ring and crimped area
G. Plating	No effect	No effect	Plating cracks and splits after crimping

[a] The actual torque/tension on every automotive critical and/or safety joint is recorded on a computer. These printouts are the source for recalls. Chatter destroys the computer printout.

NOTE

1 NYTEMP, Nylok Fastener Corporation, Macomb, ML.

13 Aerospace Bolts

Shahriar M. Sadri

13.1 INTRODUCTION

Hundreds of different bolt designs with various sizes, strength levels, and materials are used in assembly of an aircraft. On average, for example, 2.4 million fasteners are used to assemble a Boeing 747 aircraft. Of this total, 22% are structural bolts (mostly titanium) and the rest are rivets.

Basic aircraft fastener types are

- Solid rivets
- Bolts
  - Threaded nut/collar
  - Swaged collar
- Upset button
- Recess drive screws/nutplates
- Blind rivets
- Blind bolts.

Introduction to all of these fastening systems, although a very interesting exercise, would be very extensive and beyond the scope of this chapter. An attempt is made, however, to cover the most commonly used aerospace structural bolts used in modern aircraft manufacturing.

Aerospace structural bolts in general are classified under two categories:

1. *Two-piece grooved pin/swage collar (lockbolt)*: This type of product, shown in Figure 13.1, and its mating collar component comes either in a pull version, installed with a manual pneudraulic or hydraulic tool, or in a stump version, which is mostly suitable for automatic installations.
2. *Two-piece threaded pin with threaded collar or locknuts*: As the name suggests, in this case, fastening is done by engagement of the threaded components (Figure 13.2).

FIGURE 13.1 Pull-type and stump-type lockbolts.

DOI: 10.1201/9781003501183-16

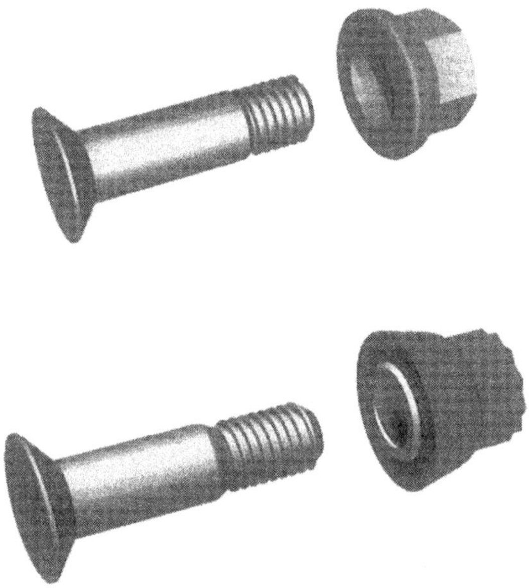

FIGURE 13.2 Threaded pin with hex or 12-point locknut.

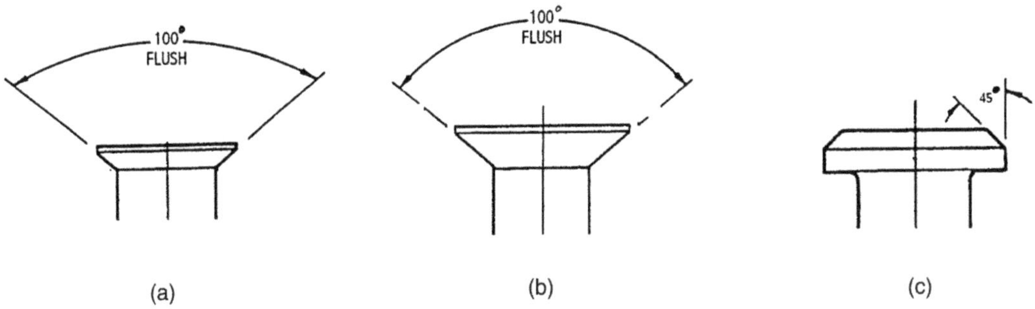

FIGURE 13.3 (a) 100° flush shear head; (b) 100° flush tension head; and (c) protruding head.

Normally, a hex recess is built into the bolt at the bolt end (opposite the head end). The parts are normally installed by using manual pneumatic installation tooling equipped with the proper nose assembly. The nose assembly consists of a stationary hex key and a 12-point rotating outer socket. The former engages the pin recess while the latter engages and rotates the outer hex (or 12-point) portion of the nut.

13.2 STRUCTURAL BOLT DESIGNS AND PROPERTIES

13.2.1 Head Styles

The most common bolt head styles used in aircraft structures are shown in Figure 13.3. The 100° flush head styles, for both the lockbolt type and threaded type, are used for aerodynamic flushness in areas such as skin attachments.

13.2.2 Two-Piece Grooved Pin–Swage Collar System (Lockbolt)

During installation of a swage-type fastener, a soft component called the collar is swaged onto the annular grooves of a pin by the action of an installation tool. Since the swage action involves

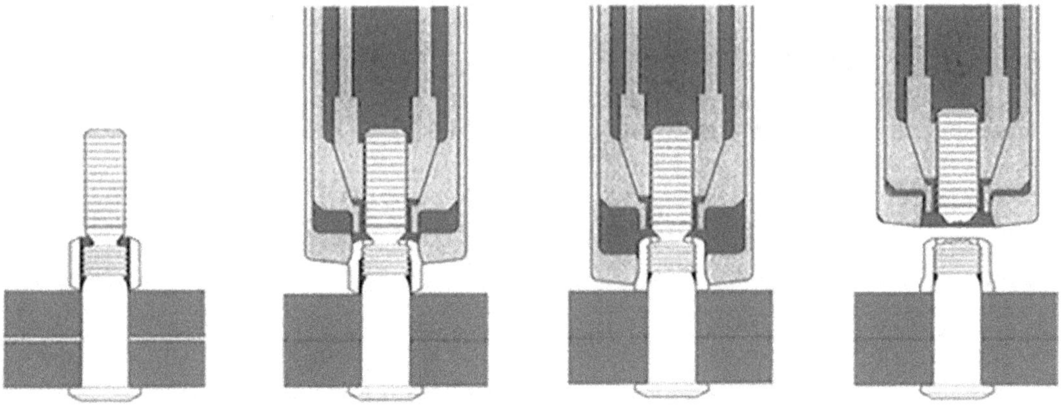

FIGURE 13.4 Installation sequence of a pull-type lockbolt.

pure tension-compression, as compared to the torque action in a threaded system, the final joint clamp variations are held to an absolute minimum. There are a number of different groove forms used in conjunction with dedicated collars, each intended for a specific application or design requirement. In general, the strength of a lockbolt is achieved through the control of a few factors:

1. Groove design and number of annular grooves.
2. Collar packing: the amount of interference between the swage cavity, the collar, and the pin.
3. Pin–collar material selection.

There are two methods used for the installation of lockbolt systems, depending on whether the lockbolt is a pull type or stump type.

A pull-type lockbolt has a tail portion, the pintail, which is broken off upon full setting of the fastener assembly. Figure 13.4 is a sketch of the installation sequence of a pull-type lockbolt. Today more than 60% of lockbolts are installed in this fashion.

The stump-type lockbolt does not have a pintail and is installed automatically by a modified drill riveter such as the one shown in Figure 13.5. The riveter will follow an automatic preprogrammed sequence for the installation of stump-type lockbolts. Typical steps taken during an automatic installation cycle are as follows:

1. The workpiece is held in position and securely clamped.
2. The hole is drilled.
3. A collar is advanced to a position directly under the hole.
4. A bucking bar in the upper ram pushes a pin, delivered by the pin feeder, into the drilled hole and through the collar.
5. The lower ram moves upward, forcing the swage anvil over the collar, thereby swaging the collar material into the grooves on the fastener.
6. The anvil is ejected from the collar, and the next cycle begins.

13.2.3 Two-Piece Threaded System

Typical threaded pins are shown in Figure 13.6. Like the lockbolt systems, the head styles and thread lengths are selected to meet the application requirements for shear, tension, etc. A sketch of an installed threaded pin and nut is shown in Figure 13.7.

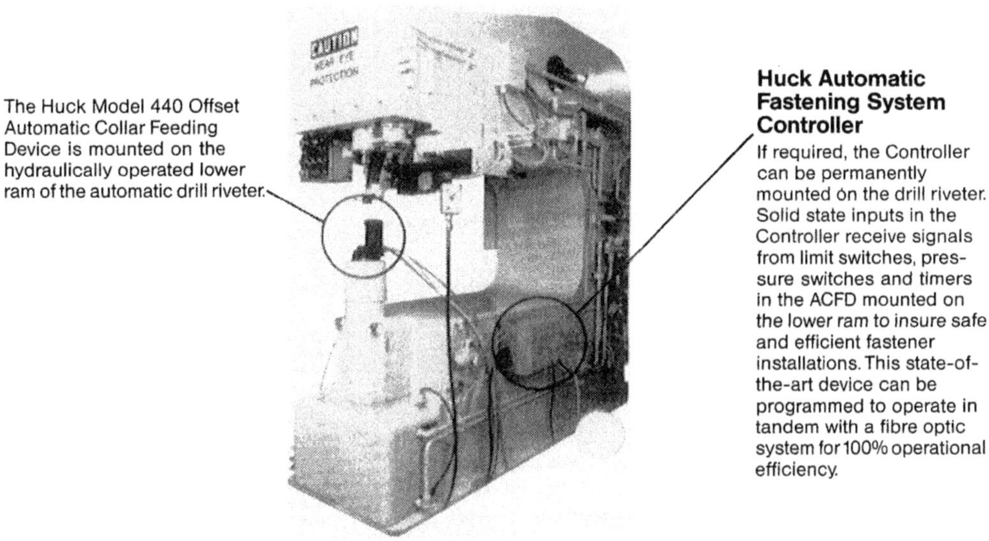

FIGURE 13.5 A modified drill riveter for automatic installation of lockbolts. ACFD = automatic collar feeding device. (Photo courtesy of Huck International, Inc.)

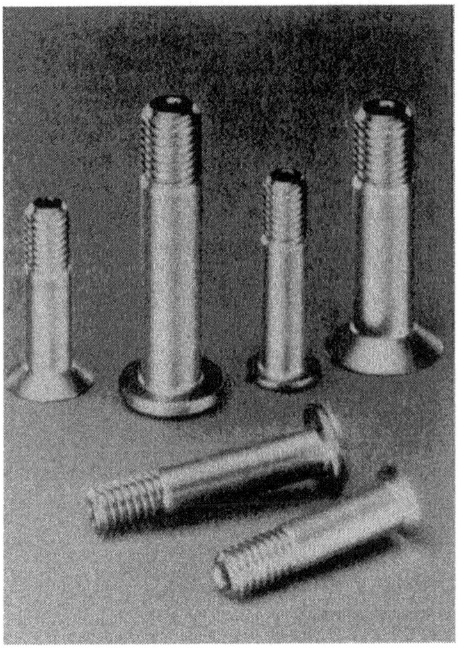

FIGURE 13.6 Threaded pins.

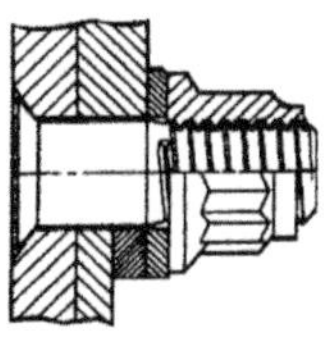

FIGURE 13.7 Installed threaded pin.

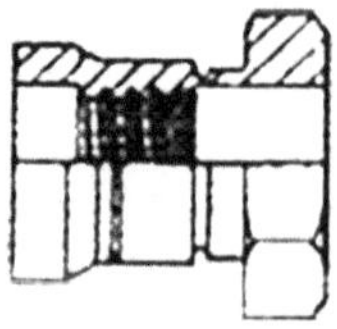

FIGURE 13.8 Frangible type nut and collar.

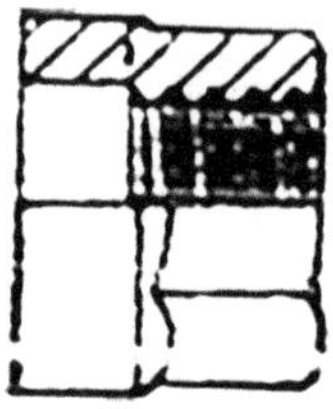

FIGURE 13.9 Twelve-point nut.

The pins are manufactured with a hexagonal drive recess. Generally, the pin threads are optimized so that a certain minimum strength in a predefined grip range (normally 1/16 in.) is achieved with the smallest number of full thread engagement. On some of the high-end systems, in order to maximize the strength-to-weight ratio, the pin threads are rolled using a special technique that would minimize the thread runout (partial thread) portion.

The mating nuts for threaded systems are categorized into two groups: frangible and nonfrangible. Full setting and installation of a frangible nut and collar (Figure 13.8) is achieved by shearing a hexagonal drive section of the nut at a certain torque established by a break groove.

A nonfrangible nut (Figure 13.9) is installed by using a torque-controlled tool. The installation tool is calibrated and will shut off when a certain predetermined torque level is reached.

Due to the high vibration environment of the application, the nuts used with the structural threaded bolts are required to have a locking feature. There are five methods commonly used for generation of locking torque for nut retention:

1. Deformed nut (oval, triangular crimp)
2. Deformed thread (localized crimp)
3. Installation swaging
4. Encapsulated adhesive (limited approval)
5. Lock wires or cotter pins (limited use).

13.3 BOLT MATERIALS AND COATINGS

The most widely used materials in the manufacture of aerospace bolts are listed in Tables 13.1 and 13.2. In these tables, the bolt (lockbolt or threaded) is listed together with the most common collar or nut material. For example, bolts made from titanium alloy are used with collars made of aluminum or titanium alloys. The most common bolt–collar–nut finishes are also listed in Tables 13.1 and 13.2.

13.4 APPLICATIONS

13.4.1 Basic Applications

Most common aerospace structural bolts, lockbolt or threaded, are made from 6AL-4V titanium heat treated to 95 ksi shear strength and coated with a low friction aluminum-pigmented epoxy coating.

TABLE 13.1
Lockbolt Materials and Finish Options

Lockbolt		Collar	
Pin Material	**Pin Finish**	**Collar Material**	**Collar Finish**
Titanium alloy 6AL-4V, BETA-C	Aluminum coating	Titanium alloy 3AL-2.5V, CP	Al coating, dry film lubricant
		Aluminum alloy 2024	Anodize
Aluminum alloy 7075	Anodize	Aluminum alloy 3000 series	Anodize
Stainless steel A-286	Passivate, cadmium plate	Stainless steel A-286	Dry film lubricant
		Aluminum alloy 2024	Anodize
Alloy steel AISI 8740, 4037	Cadmium plate or nickel-cadmîum	Stainless steel A-286	Dry film lubricant

TABLE 13.2
Threaded Bolt Materials/Finish Options

Threaded bolt		Collar	
Pin Material	**Pin Finish**	**Collar/Nut Material**	**Collar/Nut Finish**
Titanium alloy 6AL-4V	Aluminum coating	Stainless steel A-286, 300 series	Dry film lubricant
		Aluminum alloy 7075, 7050	Anodize
Stainless steel A-286	Passivate	Stainless steel A-286	Passivate
		Aluminum alloy 2024	Anodize
Alloy steel AISI 8740, 4037	Cadmium plate, nickel-cadmium	Stainless steel A-286	Passivate

The basic applications (or needs) for structural bolts are

1. Shear (predominate)
2. Tension
3. Fatigue (primarily wing)
4. Fuel tightness (wing)
5. High temperature
6. Corrosion control.

13.4.2 Aircraft Structural Materials and Bolt Selection

Mechanical fasteners are used in all areas of a commercial aircraft: primary, secondary, and nonstructural. Aluminum, in various alloy forms, is the predominant material used in the manufacture of commercial aircraft. For example, Table 13.3 lists the typical aluminum alloys used in commercial aircraft.

Many factors affect the selection of a specific type of bolt for a specific area or application in an aircraft. Some of the more important considerations are listed on the following page.

For the selection of primary fastening systems, the designer will look at all of the applicable factors and assign the appropriate weighting to them. The candidate systems are then compared, and a system is selected accordingly.

TABLE 13.3
Example of Aluminum Alloy Use on Commercial Aircraft

Structure	Portion	Material
Fuselage	Skin	2024
	Stringers	7075
Wing panels	Upper panel	7075
	Upper stringers	7075
	Lower panel	2024
	Lower stringers	2024
Wing spars	Web	7075
	Upper chord	7075
	Lower chord	2024
	Stiffeners	7075
Installed cost		Part pullup
Rate		Final clamp-up
Weight		Tooling requirements
Shear, allowable		Skin quality control
Skin thickness		Edge margin control
Access		Fuel tightness
Noise		Air leakage
Structural fatigue performance		Aerodynamic flushness
Corrosion		

BIBLIOGRAPHY

1. Huck International Inc., *Design Guide for Lockbolt Fasteners,* 1996, Carson, CA.
2. Huck International Inc., *Aerospace Fastener Manual,* 1996, California, USA.
3. Huck International Inc., *Installation Tooling Manual,* 1996, Carson, CA.
4. Huck International Inc., *VERI-LITE^{R} Fastening System Product Catalog,* 1992, Carson, CA.
5. Huck International Inc., *Fatigue Rated Fastening Systems*, 1994, pp. 7–95, Carson, CA.
6. VOI- SHANR, Division of Fairchild Corp., *Aerospace Fastener Design Manual,* 1975, Los Angeles, CA.

14 Thread Forming, Self-Tapping, and Thermoplastic Screws

Basil Housari and Ali Alkelani

14.1 INTRODUCTION

Machine thread screws are used in most applications where a solid and structural rigidity is required (Figure 14.1). In these applications, a nut of some type is required or a pre-tapped or pre-formed threads are needed in the mating part of the joint. However, there are certain applications where the fastener used can form the female threads in the mating component and attach the tightened part in one process (Figure 14.2). Instead of a nut, the screw requires a properly sized hole in the base material to form the female threads and join the part. These screws are perfect for cast parts made of aluminum, magnesium, or cast iron. These are thread forming screws.

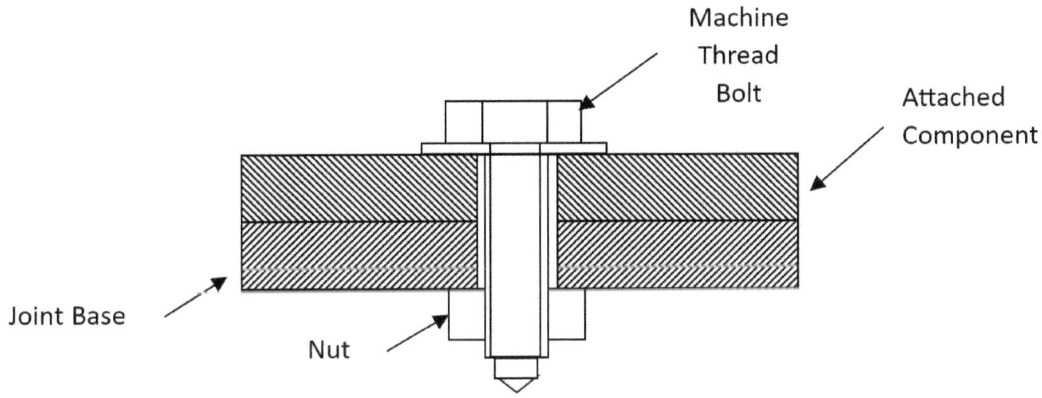

FIGURE 14.1 Typical joint with machine screws.

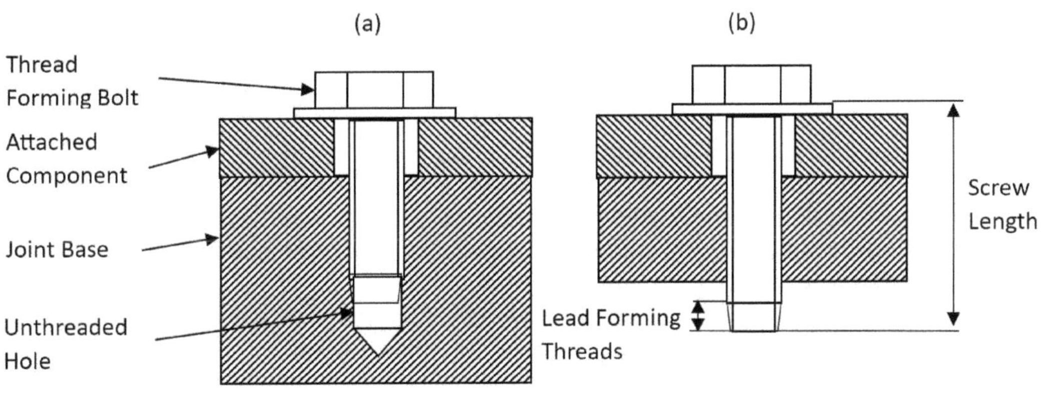

FIGURE 14.2 Joint with thread forming screw. (a) Blind hole; (b) Through hole.

DOI: 10.1201/9781003501183-17

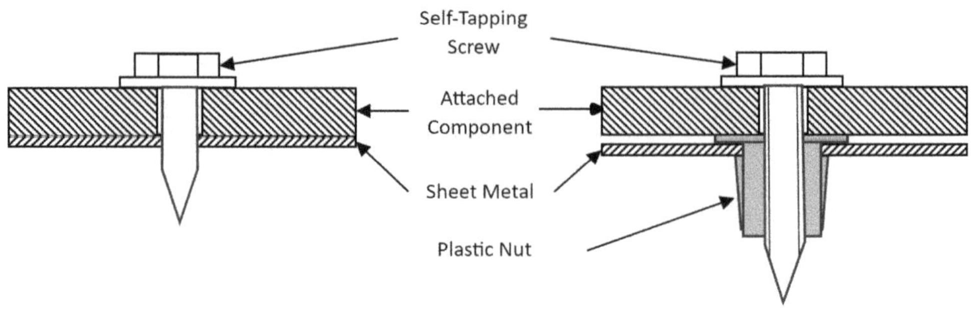

FIGURE 14.3 Joint with self-tapping screw.

On the other hand, in non-structural and light applications, the typical machine thread screws are considered an overkill in terms of strength, they add weight, complication, and cost to the product since the possibility of incorporating a compatible machine thread nut becomes difficult and expensive. The use of thread forming screws for these applications may not be possible due to the material and geometry of the joint base (Figure 14.3). The perfect solution in this case is self-tapping screws or thermoplastic screws depending on the joint base material and geometry. These screws can either cut threads during the tightening process or they can thread into a mating special type nut such as a C-clip or a plastic insert nut, which can be easily built into the joint base.

This chapter presents the key factors to consider when using thread forming screws, self-tapping screws, and thermoplastic screws. In addition, it provides a method to develop the proper tightening specification for these connections and highlights the key elements required to have a proper and stable fastening process in a mass production environment.

14.2 DEFINITIONS

Joint base: in thread forming applications, it is the joint member in which the thread forming screw forms the internal threads.

Thread forming screws: screws that form the internal threads into the joint base by displacing the joint base material to form the internal threads during the tightening process.

Self-tapping screws: screws designed to cut threads in a sheet metal joint base during the tightening process.

Thermoplastic screws: screws designed to cut threads in a plastic joint base during the tightening process.

Screw diameter (d): thread major diameter (outer diameter).

Bearing radius (d_b): the mean radius of the contact area under the head of the bolt.

Thread pitch (P): the distance between two consecutive thread peaks.

Thread profile: refers to the shape and geometry of the thread on a threaded fastener.

Ultimate torque (T_u): torque at which the screw fails.

Strip torque (T_s): the torque at which the threads in the joint base strip.

Target torque (T): torque to be used when tightening a screw in the joint.

Robust target torque window ($T_{min}-T_{max}$): the range available to select a robust target torque from, it is defined based on the minimum clamp load or torque requirements and the strip torque or ultimate torque.

Thread forming torque (T_{th}): the torque required to form the threads in a joint base throughout the entire length of thread engagement and fully seat the fastener.

Torque to failure test: joint test in which the tightened fastener is torqued until failure in the joint members or the fastener occurs.

Joint failure mode: refers to the type of failure observed during the torque to failure test, e.g., screw fracture, internal threads strip, or other joint member damage.

14.3 THREAD FORMING SCREWS

These screws are normally used to fasten components into a cast metal joint base, eliminating the need for tapping the female threads, or the use of standard nuts. This introduces significant cost saving and assembly simplification in the manufacturing processes.

Some of the advantages of using thread forming screws are:

- Eliminate pre-tapping or pre-forming of the female threads and all related processes such as chips removal and threads gaging and inspections.
- Eliminate the risk of cross threading and the need for hand starting during assembly.
- Provide good vibration loosening resistance due to zero tolerance between the threads; this leaves no room for transverse movement when the joint is subjected to external loads, hence it reduces the chance for self-loosening.

To successfully meet the final product performance requirements, the thread forming screw has to be carefully selected, the joint base has to be properly designed, and the final assembly torque has to be properly developed. The following sections provide insight on these three important topics.

14.3.1 Screw Selection

There are different designs of thread forming screws. The main differences between them are the thread profile, the design of the lead forming threads, and screw heat treatment. Each one of these parameters may vary based on the joint base that the screw needs to form the threads in, the company developing the thread forming screw, and the design objectives. For example, metric thread forming screws are design so that the formed threads are compatible with normal metric thread screw of the same size, even though the screw thread geometry may not match the typical metric thread profile. The design of the thread-lead helps guide the screw and provides a gradual thread forming in the base material as the screw advances during the tightening. Some thread forming screws are fully hardened, some have only the leading threads induction hardened, while others are case hardened. Due to the large variation in geometry, proprietary designs, and difficulty in measuring thread profile features, customers resort to requiring minimum performance specifications on the screw drawing instead of profile and dimensional specifications. The most important function requirements, in this case, are the thread forming capability and failure or strip off torque when tested with a reference base material with specific hole dimensions.

The best screw to use for an application is a screw that requires the lowest thread forming torque and highest possible strip or failure torque while satisfying the joint strength and function requirements. When selecting the screw for the application, the design engineer must identify the required screw type, material, size, length, coating, and lubrication.

14.3.1.1 Screw Type

The screw type is referring to thread profile, where some thread profiles are designed specifically to form threads into certain materials, such as magnesium, aluminum, or steel. The shape of the thread profile controls how the material flows during the thread forming process and may have some features that help prevent the screw from self-loosening. REMINC & CONTI is one of the known companies that developed different thread forming screw types for different materials. The selected screw must be suitable to form threads in the joint base material.

14.3.1.2 Screw Material Hardness and Strength

In order to be able to form the threads effectively, the screw threads must be harder than the joint base. Screw hardness more than three times the hardness of the joint base material is recommended. There are three common hardness configurations, either the whole screw is hardened (e.g., class 10.9 thread forming screws) or only the surface of the screw is hardened (case hardened screws), or only the leading threads are hardened (induction hardened screws). The type to use depends on the hardness of the joint base. Fully hardened screws are commonly used for cast Aluminum and Magnesium joint base materials. Induction hardened screws are recommended for harder joint base where the hardened leading tip cuts the threads and penetrates through the joint base, while the remaining softer body of the screw carries the clamp load. Case hardened screws are suitable for ductile materials, where the harder surface helps form the threads. For each application, the screw needs to be able to hold the required clamp load required for the joint.

14.3.1.3 Screw Material, Coating, and Lubrication

In addition to the hardness and strength of the screw, material compatibility, in terms of galvanic corrosion, should be taken into consideration. In some cases, a special washer material is used to eliminate this risk. The washer can also help control the friction under the head of the screw. In other cases, the screw coating serves as a sufficient barrier to prevent galvanic corrosion. The coating and the top coat are also important to prevent corrosion and to control the friction coefficient. Additional lubrication in the form of wax on the threads to facilitate the thread forming is often used; it also helps increase clamp load and prevent threads chipping during the forming process.

14.3.1.4 Screw Size

The required screw size must be determined based on the required clamp load for the joint. It is important to keep in mind that the weakest component in the joint may not be the screw as it is the case in most machine thread applications. This is highly dependent on the joint base material and length of thread engagement. For this reason, if long thread engagement is not feasible, a higher clamp load can be obtained by using a larger size screw with higher torque.

14.3.1.5 Screw Length

The screw length is dependent on the thickness of the joint members, the desired clamp load, and screw material used. In general, there are two conditions for the joint. The first case is when the screw does not penetrate through the joint base, Figure 14.2a, in this case, the required screw length needs to be greater or equal to the sum of the washer thickness (if the screw has a retained washer), the attached component or components thickness, the required thread engagement (defined in Section 1.2.2), and the leading thread forming section. The second case is when the screw goes completely through the joint base. In this case, the screw length must be greater or equal to the sum of the washer thickness, the thickness of all joint members, and the length of the leading thread forming section of the screw (see Figure 14.2b).

14.3.2 Joint Base

The main factors to consider in the joint base design are the joint base material, thread engagement, and hole size.

14.3.2.1 Joint Base Material

The joint base material and geometry are normally driven by the component functional requirements, however, the ability to have a robust joint must be kept in mind during the material selection in terms of material formability and hardness, and also in terms of geometrical configuration required for joining (thickness or special boss thickness). The material formability affects the ability to form the threads without chipping or cracking, and the joint base hardness is important for the selection of the screw hardness and heat treatment.

14.3.2.2 Thread Engagement

Thread engagement refers to the length of threads engaged in the joint base. The torque that can be applied, the clamp load that can be achieved in the joint, and the failure mode are directly affected by the length of thread engagement.

From a clamp load perspective, joints can be divided into:

- Light clamp load joint: these are joints subjected to low external forces; as a result, they don't require a high level of clamp load e.g. light brackets holding sensors or light components to an engine block. For light clamp load joints, the thread engagement length in the joint base can be limited ($< 1.5\times d$), therefore, the expected failure mode during a torque to failure test is stripping the threads formed in the joint base. For these joints, the torque to be used is limited by the threads strip torque.
- High clamp load joints: these are joints that require a high level of clamp load due to high external loads. They need to maintain a high level of clamp load throughout the life of the product. The recommended thread engagement for light metal joint base material (Aluminum and Magnesium) is between $2\times d$ and $3\times d$.

  When sufficient thread engagement is available, it is expected that the bolt breaks before stripping the formed threads in a torque to failure test. This makes it possible to get higher clamp load from the bolt used.
- For joints with thread engagement between ($1.5\times d$ and $2\times d$), the longer the thread engagement the higher the failure torque, which allows for higher tightening torque and higher clamp load. As the thread engagement increases, the failure mode changes from stripping the threads in the joint base to breaking the bolt.

14.3.2.3 Hole Size

The hole size is one of the critical factors in the design of a thread forming screw joint. A properly sized hole provides the space required for the displaced material to flow properly and form the proper thread profile during the tightening. A small hole will not provide room for the material to flow, which makes it hard or impossible to form the threads. On the other hand, a large hole size will not have enough material to push and form full threads, the result is incomplete weak threads that strip at a low torque.

In most cases, the hole is cast in the base material, where the screw forms the threads into a blind hole with a draft angle, as shown in Figure 14.4a. The 1°–1.5° degree draft angle is necessary to pull the part from the casting mold.

The hole size is the function of the size of the screw and the thread profile used for the application. The proper hole size is also affected by joint base material strength, length of thread engagement, and lubrication on screw threads. For this reason, it is not possible to provide a single formula for the hole size for all applications. It is best to determine the proper hole size experimentally using torque to failure test (see tightening torque development Section 1.3), however, as a starting initial hole configuration for thread forming screws into Aluminum and Magnesium alloys, a hole size at

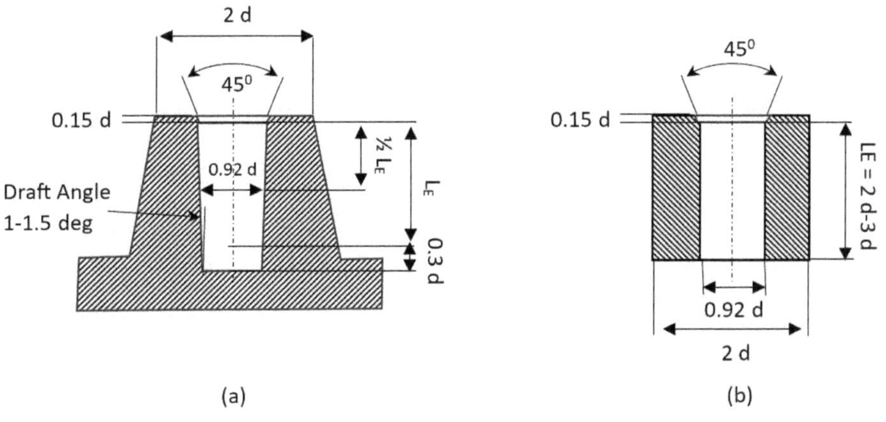

FIGURE 14.4 Joint base hole configuration for thread forming screws. (a) Blind hole; (b) Through hole.

50% of the thread engagement is 92% of the screw nominal diameter. For the remaining configurations of the cast hole the following dimensions are recommended:

- Boss diameter (in case of special joining boss) $= 2 \times d$
- Hole lead chamfer height $= 0.15 \times d$ at 45°
- Hole draft angle $= 1.2°$
- Clearance depth beyond end of screw $= 0.3 \times d$

Where d is the nominal diameter of the screw used.

After obtaining sample parts with the initial configurations above, a torque to failure test is performed and the hole size can be adjusted to obtain the optimal hole size based on the test results. The goal is to obtain the hole size that provides the lowest thread forming torque and the maximum strip torque to allow the use of a target torque that develops proper clamp load in the joint. For joint base with a drilled through hole, Figure 14.4b, the same ratio (92% of nominal diameter) can be used for the initial hole size.

14.3.2.4 Hole Size Tolerance

The thread forming torque and the strip torque are greatly affected by the hole size, therefore, once the proper hole size is obtained, it is important to keep the hole size consistent as much as possible, therefore, a tight hole size tolerance is required. It is recommended not to exceed $\pm 0.008 \times d$ for the hole tolerance. A tolerance may result in stripping the threads or not seating the screw during assembly.

Figure 14.4 shows a 45° lead in chamfer with $0.15 \times d$ depth. This lead chamfer helps accommodate the small bulge that may form at the hole entrance during the thread forming process resulting in flush contact surface between the two joined parts.

14.3.3 Clamp Load

One of the concerns when using thread forming screws is the consistency of the clamp load obtained in the joint after achieving the final tightening torque. The United States Council for Automotive Research LLC (USCAR) has issued a report investigating the clamp load variation in cast Aluminum and Magnesium castings using thread forming screws [1-2]. The tested material/screw combination was limited, where it included one Aluminum alloy and one Magnesium alloy, however, it considered the variation in screw lubrication and variation in hole size due to cast die wear. The report concluded that the clamp load variation obtained at a specific target torque are comparable to the clamp load variation results obtained for machine thread bolt and nut combination.

Even though the percentage of clamp load variation might be close to machine thread screws, it is important to note, that part of the torque applied is consumed by the thread forming process as shown in Figure 14.5a, and as a result, the component of torque converted into clamp load is reduced, hence the final clamp load obtained is less compared to a normal machine thread bolt and nut combination. The higher the thread forming torque, the lower the clamp load. For this reason, to provide a sufficient clamp load in the joint, the final target torque is recommended to be at least twice the maximum expected thread forming torque.

For joints with through hole joint base, when the lead thread penetrates through the hole, the thread forming torque drops close to zero, therefore, the final clamp load at the end of the tightening is not reduced by the thread forming torque, Figure 14.5b.

Due to the high number of variables affecting the clamp load, it is necessary to perform experimental testing for the specific joint to evaluate the level of clamp load achieved.

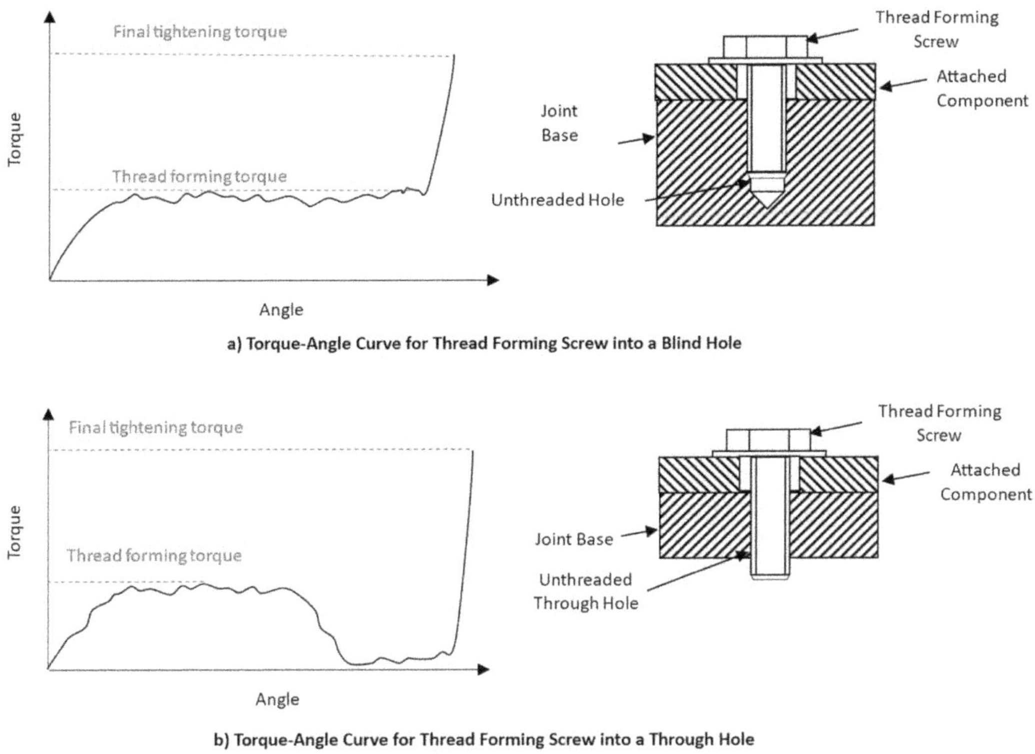

FIGURE 14.5 Torque-angle curve for thread forming screws. (a) Torque-angle curve for thread forming screw into a blind hole; (b) Torque-angle curve for thread forming screw into a through hole.

14.3.4 Development of Tightening Torque

It is not possible to have a theoretical calculation method to find the proper tightening spec for thread forming screw applications, due to many variables that are joint-specific. Instead, the tightening spec needs to be developed experimentally. The optimal tightening torque is the maximum safe torque that will always seat the screw and generate sufficient clamp load without damaging the joint. To identify the optimal torque, it is required to find the minimum failure torque at which the threads strip or the screw breaks. Once the lowest failure torque is identified, a safety margin is used to make sure that the tightening torque can be reached under all normal screw, hole, material, and friction variations. Because of all these variations, a 30%–40% safety margin is recommended. On the other hand, the target torque should be significantly higher than the maximum thread forming torque to generate sufficient clamp load.

After selecting the proper thread forming screw and designing the joint base according to the guidelines provided earlier, sample joints need to be tested to develop the proper tightening torque. Here are the steps for developing the proper tightening spec for a thread forming screw application:

1. Perform torque to failure test on a number of joints and collect the corresponding torque-angle curve for each joint.
2. From the torque-angle curve for each joint, identify the highest thread forming torque during the thread forming phase, and the failure torque at which the joint failure occurs.
3. Identify the minimum failure torque and the maximum thread forming torque for the tested samples.

4. Calculate ratio of the minimum strip torque to the maximum thread forming torque.
5. To have a stable robust thread forming tightening process, the calculated ratio needs to be equal or greater than 2.8. If this condition is met, the recommended tightening torque to use would be equal to 70% of the minimum failure torque obtained.

   If it is not possible to obtain a robust strip torque to forming torque, then a design change should be considered to either increase the failure torque or reduce the thread forming torque. To increase the failure torque, one or more of the following can be implemented:
   - Reduce the hole size
   - Use different screw with serrations under the screw head
   - Use a different screw type

   To reduce the thread forming torque, one or more of the following can be implemented:
   - Increase the holes size in the joint base
   - Use a screw with better lubrication
   - Use a screw that can provide better thread forming capability

14.4 SELF-TAPPING SCREWS

There are hundreds of kinds of self-tapping screws used in different industries based on the type of material the screw cuts the threads in. Some screws are equipped with a built-in drilling tip, others just have a pointy tip that allows them to penetrate through softer materials such as wood. Other screws require a pre-drilled or punched hole in the joint base material. The latter is the type that is used most in automotive, where the screw cuts through the sheet metal creating threads that allow the screw to pull the components together and create some clamp load. The fact that a nut is not needed makes this an economical method of joining components together, however, because there is a single thread or very few threads taking the load, the level of clamp load and strength of the joint are limited. The focus in this section is on ISO 1478 self-tapping screw threads and the most important factors to consider when these screws are used. These screws are available in different sizes ranging from 1.5 to 9.5 mm. The most common sizes used in automotive are ST-4.2 and ST-4.8. They are normally available in a flat leading tip (Form F tip), round leading tip (Form R tip), and sharp leading tip (Form C tip), see ISO 1478 for details (Figure 14.6).

14.4.1 Self-Tapping Screws Cutting Threads into Sheet Metal

In most applications, self-tapping screws are used to attach a component to a sheet metal part, where the screw cuts threads in an existing hole in the sheet metal part. The key factors for successful joining using self-tapping screws are:

14.4.1.1 Screw Size

The size of the screw to use depends on the joint strength required, e.g. pull strength before joint fails, and also depends on the level of clamp force needed to keep the attached component in place. The larger the screw size the higher the tap and strip torque are, which translates into a higher

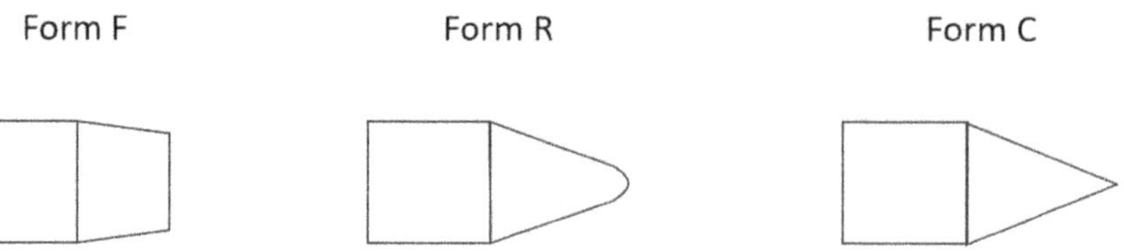

FIGURE 14.6 Types of leading tips for self-tapping screws.

TABLE 14.1
ISO 1478, Standard Self-Tapping Screw Sizes

Size	Pitch [mm]
ST 1,5	0,5
ST 1,9	0,6
ST 2,2	0,8
ST 2,6	0,9
ST 2,9	1,1
ST 3,3	1,3
ST 3,5	1,3
ST 3,9	1,3
ST 4,2	1,4
ST 4,8	1,6
ST 5,5	1.8
ST 6,3	1,8
ST 8	2,1
ST 9,5	2,1

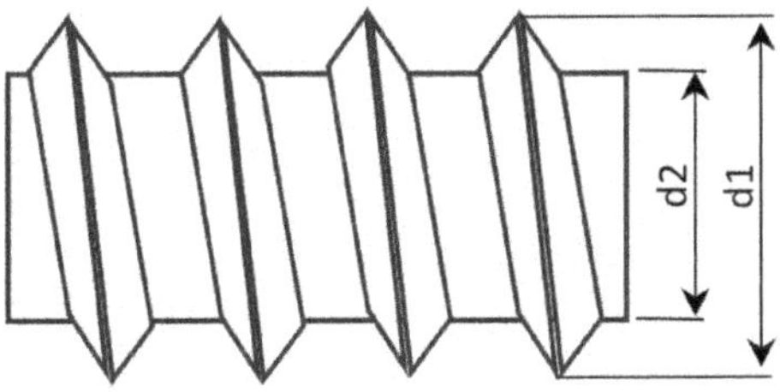

FIGURE 14.7 Inner and outer diameter for self-tapping screw.

tightening torque, higher clamp load, and higher joint strength. In general, it is recommended to use a screw with a thread pitch smaller or equal to the sheet metal part thickness. Table 14.1 provides the standard ISO 1478 sizes and corresponding thread pitch for each size.

14.4.1.2 Sheet Metal Material Strength

It is important that the screw has enough strength to cut through the sheet metal. Normally, these screws are not suitable to use with sheet metal strength greater than 500 MPa.

14.4.1.3 Hole Size in Sheet Metal

Determining the proper hole size is the most important factor in the design of a self-tapping screw joint. The torque required to cut the threads (T_t) and the torque at which the sheet metal strips (T_s). Both torque values are greatly affected by the hole size. A small hole size results in a higher thread cutting torque and a higher strip torque and vice versa. The optimal hole size is the hole diameter that provides the highest strip to tap ratio ($R = T_t/T_s$). Using a hole size smaller than the minimum screw diameter (d_2) shown in Figure 14.7 will result in a very high drive torque and will most likely cause screw or sheet metal failure before seating the screw.

The optimal hole size is affected by the sheet metal thickness and sheet metal strength. In most automotive applications, the used hole size ranges from 75% of screw nominal diameter for applications with low strength and thin sheet metal to 85% for applications with stronger and thicker sheet metal. An initial hole size equal to 80% of screw diameter is recommended as starting point, then the hole size can be optimized experimentally based on torque to failure test results.

Punched holes are subjected to cold work, therefore, the tap and strip torque values can vary for drilled holes versus punched holes. Holes in test samples used in the torque to failure test should be manufactured the same way planned for mass production manufacturing. Drilled hole samples can be used for the preliminary test, keeping in mind that the optimal hole size can be about 0.1 mm smaller if the hole is to be punched in mass production process.

It was reported that driving the screw from the hole punching side provides a better strip torque result, therefore, the side from which the screw is driven during testing should match the actual joint condition.

If a simple drilled or punched hole does not provide sufficient joint strength, an extruded hole or a stamped hole with conical impression can be used, see Figure 14.8.

14.4.1.4 Hole Clearance

The tap and strip torque values are sensitive to the hole size, therefore, using a tight tolerance for the hole size is a key for a stable self-tapping screw application. It is recommended not to exceed ±2% of the hole diameter.

14.4.1.5 Clearance Hole in Attached Component

The clearance hole in the attached part is important in providing support to the base sheet metal and preventing bending the base sheet metal into the clearance hole as shown in Figure 14.9. This bending of the sheet metal (i.e., getting sucked up into the clearance hole) weakens the tapped threads in the base sheet metal and results in a lower strip torque. The effect of the clearance hole is more pronounced in thinner sheet metal applications.

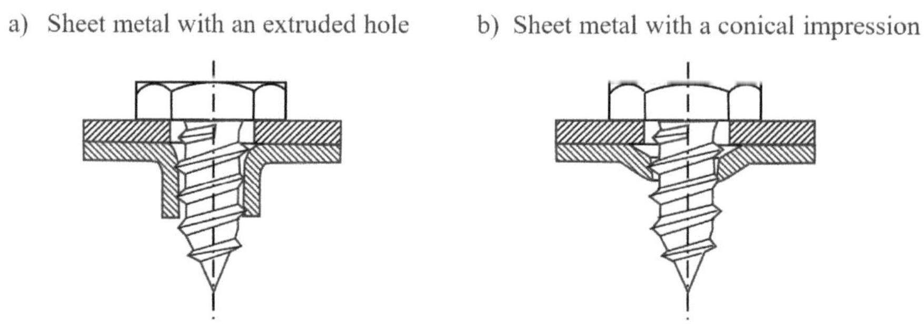

FIGURE 14.8 Self-tapping screw with alternative base sheet metal configuration. (a) Sheet metal with an extruded hole; (b) Sheet metal with a conical impression.

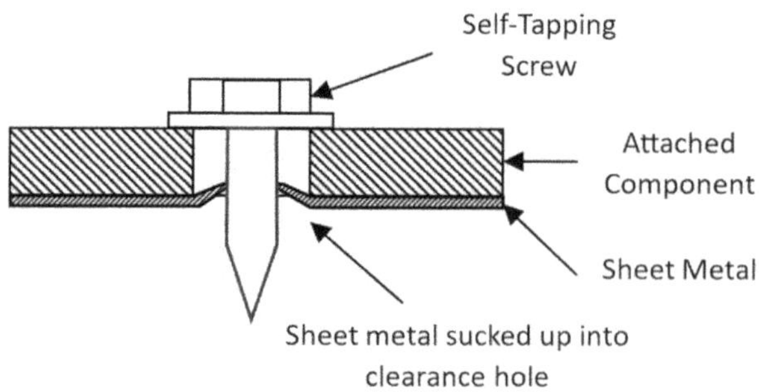

FIGURE 14.9 Effect of clearance hole in attached part on joint strength.

14.4.1.6 Tightening Torque

The desired tightening torque is the highest torque that will seat the screw and not strip the threads, regardless of normal parts and process variations. Figure 14.10 shows a typical torque-angle curve for a self-tapping screw application. At the beginning, the screw starts to cut the threads through the sheet metal; the torque required to cut through is called the tap torque. After penetrating through the sheet metal, the torque drops to almost zero until the screw is seated; after that, the tightening phase starts where parts are clamped together until the sheet metal or the screw strips and fails.

Similar to thread forming applications, it is important that the target tightening torque is high enough to make sure the screw is seated all the time regardless of hole size, material, friction, and screw variations. The target torque must also not strip the sheet metal as well. The same method used for thread forming screws can be used for self-tapping screws, where torque-angle curves are recorded for several samples made with holes equal to 80% of the nominal screw size, and then torqued to failure, the maximum tap torque and minimum strip torque values are identified and used to evaluate the possibility of having a proper tightening torque. If the ratio of the minimum strip torque to maximum tap torque is greater than 2.8, then it is possible to have a stable tightening process, and the target torque, to be used is equal to 70% of the minimum strip torque. If the 2.8 ratio is not achieved, then hole size optimization is needed to reduce the tap torque or increase the strip torque. If it is still not possible to obtain the proper ratio with hole size optimization, then changing the hole configuration to one of the alternative hole configurations shown in Figure 14.8 can be tried.

The main logic behind the strip to tap torque ratio is to account for dimensional and friction variations in the parts that may increase the tap torque to a level where the target torque is reached before the screw is seated, or reduce the strip torque to a level that causes the sheet metal to strip before reaching the target torque.

In some cases, if a DC tool is used for the tightening, it is possible to address the issue with high tap torque by setting a special two-step program. The first step is configured to cut through the sheet metal using angle control strategy, while the second is configured to go to the final target torque using torque control strategy, Figure 14.11. The target angle for the first step can be configured to start from a low threshold torque, while the target angle value must guarantee penetrating through the sheet metal thickness; it can be determined based on angles observed from the curves for the torque to failure test samples. The second step target torque is set to 70% of the minimum strip torque obtained from the torque to failure test. Although it is not recommended, with this strategy, it is possible to have the target torque set at values that are even lower than the tap torque as shown in Figure 14.11.

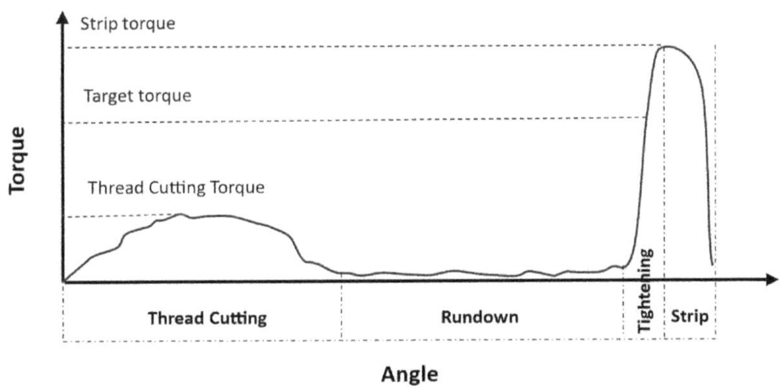

FIGURE 14.10 Typical Torque-Angle curve for self-tapping screw application.

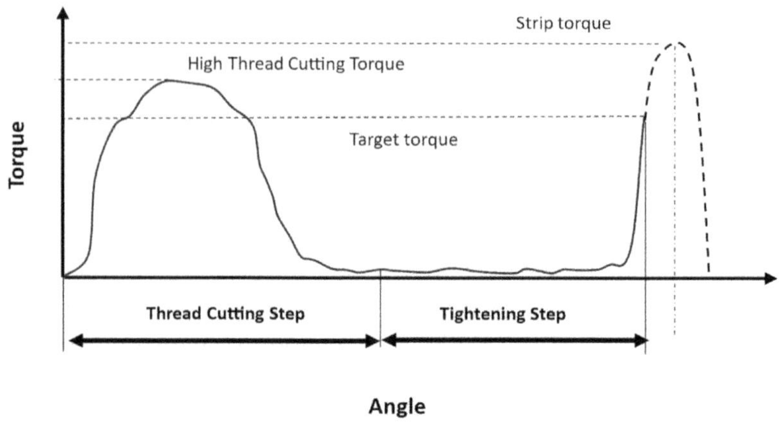

FIGURE 14.11 Self-tapping screw application with high tap torque.

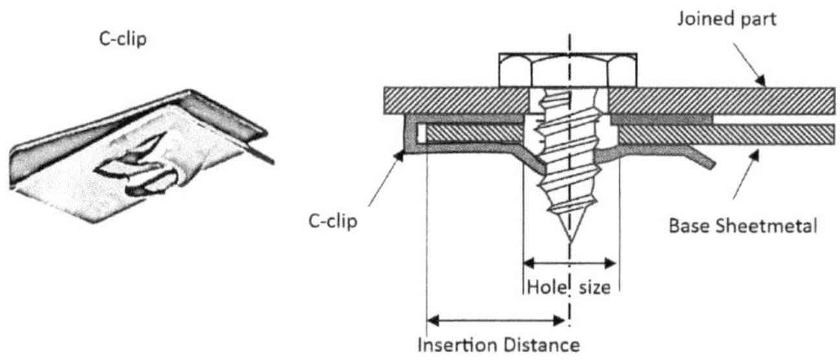

FIGURE 14.12 Self-tapping screw with C-clip.

14.4.2 Self-Tapping Screws with C-Clips

Self-tapping screws are used in some cases with a metal or plastic c-clip. C-clips are normally made of thin spring steel strips, that are formed into a U shape and have a single thread to match a specific screw size and pitch. The clip gets mounted to the edge of the base sheet metal or through a specific cut window with a pre-cut hole that aligns with the center of the C-clip, see Figure 14.12. The self-tapping screw does not cut threads into the base sheet metal, instead, it engages with the compatible pre-cut thread in the C-clip. The sheet metal must be made to accommodate the C-clip used. The main dimensions to consider when selecting the C-clip and designing the sheet metal configurations are:

- C-clip compatibility with the screw thread size and pitch
- C-clip compatibility with base sheet metal thickness
- Hole size in sheet metal is according to C-clip requirements
- Insertion distance (center of the hole to the edge of the base sheet metal) is according to C-clip requirements
- Tightening torque should be at least 20% lower than the minimum strip torque specifications provided by the C-clip manufacturer.

14.4.3 Self-Tapping Screws with Plastic Expanding Nut

Another option for joining parts using self-tapping screws is to use an expanding plastic nut that can be inserted into a cut square hole in the sheet metal to prevent the nut from rotating during the

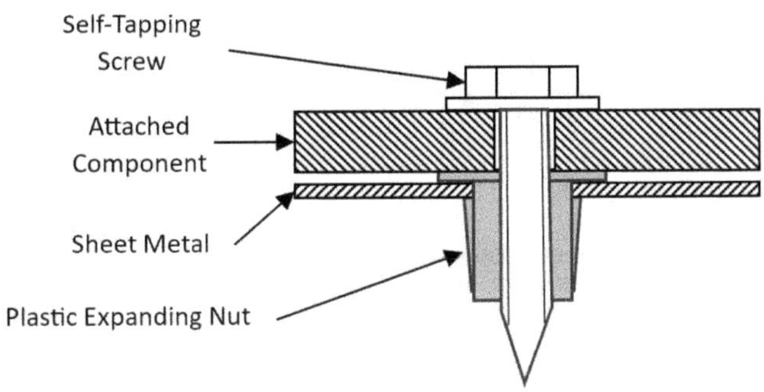

FIGURE 14.13 Self-tapping screw with expanding plastic nut.

FIGURE 14.14 Example thermoplastic screw.

tightening. The nut normally has a retaining feature that helps keep it attached to the sheet metal. When the self-tapping screw is tightened, it cuts threads into the center hole of the expanding plastic nut, see Figure 14.13. The screw causes the body of the plastic nut to expand radially creating a bulge on the back of the sheet metal, which increases the retention of the nut to the sheet metal. The key points to consider when using expanding plastic nuts are:

- Compatibility of the nut to the screw size
- Compatibility of the nut with the sheet metal thickness
- Recommended thread engagement length
- Sheet metal hole configuration requirements
- Recommended maximum tightening speed

The tightening torque to use should be at least 20% below the minimum strip torque specified by the nut manufacturer.

It is worth noting that the thread cutting torque and the plastic nut failure torque can be affected by the temperature of the plastic nuts since they become brittle and can fail easier when cold. For this reason, it is not uncommon to heat up expanding plastic nuts before use during winter.

14.5 THERMOPLASTIC SCREW APPLICATIONS

Thermoplastic screws are metal screws made with thin thread profile and course thread pitch, which are designed to cut threads into molded plastic parts, plastic C-clips, or plastic expanding nuts (Figure 14.14).

14.5.1 Thermoplastic Screws into Plastic Molded Parts

The following are the key factors that must be taken into consideration when designing a joint with thermoplastic screws.

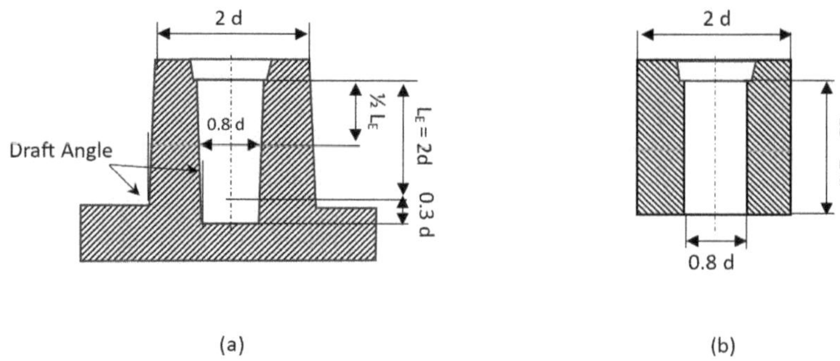

FIGURE 14.15 Generic initial hole configuration for plastic molded parts using thermoplastic screws. (a) Blind hole; (b) Through hole.

14.5.1.1 Screw Type

There are multiple thread profiles and different designs provided by different manufacturers. They can vary based on the plastic material they cut the threads in. The screw manufacturers normally provide some guidelines for the part design and materials the screw can be used for. It is important to select a screw that is compatible with the joint material.

14.5.1.2 Hole Configurations

When using thermoplastic screws, the hole configuration is one of the most important factors to consider in the design of the joint. The hole size, hole depth, and draft angle for molded components is function of the plastic material strength and the screw size. It is best to follow the hole configurations recommended by the screw manufacturer, however, if this information is not available from the screw manufacturer, the dimensions can be optimized experimentally, using an initial hole size of $0.8 \times d$, $2 \times d$ thread engagement, and a minimum outer boss diameter of $2 \times d$, see Figure 14.15.

The hole tolerance has to be kept tight for these components in order to have a stable and robust fastening process. Check recommendations from screw manufacturer. Typically recommended hole clearance is close to ±1% of the hole diameter.

14.5.1.3 Screw Length

The screw length needs to be equal to the washer thickness (if screw has a washer), plus the attached component thickness, plus the plastic boss leading hole, plus thread engagement (recommended $2 \times d$), plus distance to first full thread (normally 1–2 thread pitch).

14.5.1.4 Tightening Torque

The recommended tightening torque to use for thermoplastic screws has to be developed experimentally using torque to failure test. The method and rules presented earlier for thread forming screws can be applied for thermoplastic screws.

14.5.1.5 Effect of Parts Temperature

Parts temperature can have a significant effect on the performance of the thermoplastic screw. The tap and strip torque values can vary by up to 30% when the part temperature is changed between 5°C and 40°C.

14.5.1.6 Tightening Speed

High tightening speeds can generate significant amount of heat that can melt the plastic or make it weaker, hence having a low strip torque. In most applications, it is recommended not to exceed 400 RPM.

14.5.2 Thermoplastic Screws into C-Clips and Plastic Expanding Nuts

Thermoplastic screws can be used with expanding plastic nuts, plastic C-clips, or even compatible metal C-clips with compatible threads. The same design considerations presented for the use of self-tapping screws with C-clips and expanding plastic nuts apply when used with thermoplastic screws.

REFERENCES

1. Cleaver, R., Cleaver, T., & Talbott, R. (2012). Machining Elimination through Application of Thread Forming Fasteners in Net-Shaped Cast Holes. United States Council for Automotive Research LLC (USCAR), Project No.: FAS 1005.
2. ISO 1478: 1999, Tapping Screw Threads., International Organization for Standardization, Geneva.

Part IV

Design and Analysis of Bolted Joints

15 VDI Joint Design Procedures

Alex W. Heston

15.1 INTRODUCTION

VDI is the German society of engineers, Verein Deutscher Ingenieure. VDI 2230 is one of this society's published guidelines for bolted joints. The full title of this document has been translated into English as "Systematic Calculation of High Duty Bolted Joints: Joints with One Cylindrical Bolt."

The first step in predicting how a bolted structure will respond to external loads is to identify a singly bolted joint for analysis. Loads on the joint are presumed to be known or to be determinable from statics or other forms of analysis. These loads are shared by the bolt and the clamped (or jointed) material. The designer is permitted to use any analytic method, including finite element or boundary element methods, or experimental methods (strain gages, ultrasonics) to determine how the external loads are distributed. Equations based on beam theory are provided for the default method of analyzing the behavior of the joint under loading. Multibolted joints are handled by discretizing the system into several singly bolted joints and analyzing each of them separately.

15.2 UNITS

International (SI) units are used in VDI 2230 [1], as it is a European document, and are used exclusively here. There is nothing inherent to the analytic procedures to preclude the use of English units. Table 15.1 shows the most convenient metric and English units to use for most analyses. Note that the correlations given later in Eqs. (15.15), (15.25), and (15.26) must be calculated using metric units.

15.3 MODES OF FAILURE

The externally loaded joint is systematically checked against a list of criteria for failure. The five primary modes of failure that should be examined for all critical joints are

1. Insufficient preload
2. Tensile overload of the bolt
3. Fatigue failure of the bolt
4. Yielding of the joint material under the bolt head or nut
5. Vibration loosening of the bolt

TABLE 15.1
Preferred Units for Use in Joint Analysis

Feature	Metric Unit	English Unit	Multiplying Factor For Value In English Units
Length	mm	inch	25.40
Force	N	lb	4.448
Stiffness	N/mm	lb/in.	0.175
Compliance	mm/N	in./lb	5.710
Stress	N/mm^2 (MPa)	psi	6895

DOI: 10.1201/9781003501183-19

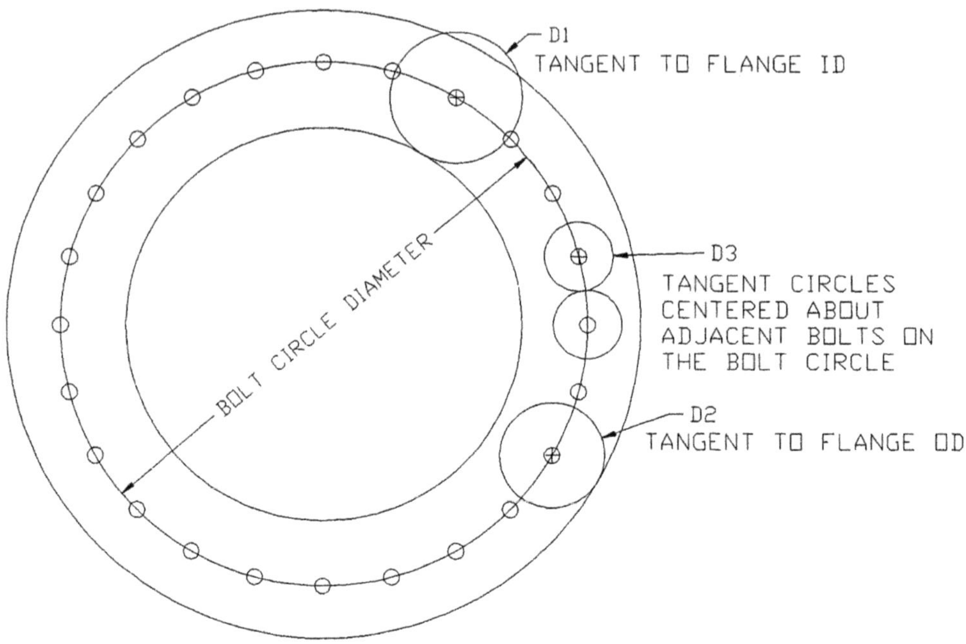

FIGURE 15.1 Characteristic circles for the joint diameter.

Guidelines are also given to guard against other design-specific modes of failure such as thread stripping resulting from insufficient thread engagement and head failures in socket head cap screws resulting from insufficient underbroach height.

15.4 IDENTIFYING THE BOLTED JOINT (CHARACTERISTIC DIAMETER)

When a bolt is tightened, the clamped material compresses, with the highest compression levels generally occurring near the bolt hole. The joint faces will remain in contact within a fixed zone around the bolt until either the preload is removed (preload loss or untightening) or external loads of sufficiently high levels are applied. Classical joint analysis generally isolates some axisymmetric section (usually conical or cylindrical) of the clamped material around the bolt for the purpose of making a stiffness estimate. The bolt and this clamped material make up the bolted joint. VDI 2230 presumes that the designer can readily identify the bolted joint and provides little in the way of guidelines for doing so.

Generally, a conservative approach for determining the joint diameter is taken by using the smallest characteristic circle for the joint. Figure 15.1 shows a flange connection and three characteristic circles concentric to the bolt hole. One circle is tangent to the flange inner diameter, the second is tangent to the flange outer diameter, and the third is tangent to a circle of the same diameter around the next nearest bolt hole on the bolt circle.

A basic assumption used for the joint stiffness is that the compressed region spreads out from under the bolt head, from the nut face, or from the mating threads at no more than a 45° angle. This leads to a second criterion for the smallest characteristic circle. Both criteria must be examined to determine the smallest characteristic circle diameter.

15.5 ESTIMATING LOADS ON THE BOLTED JOINT

VDI 2230 presumes that the designer can determine the applied loads for the joint. Typical loading is shown in Figure 15.2.

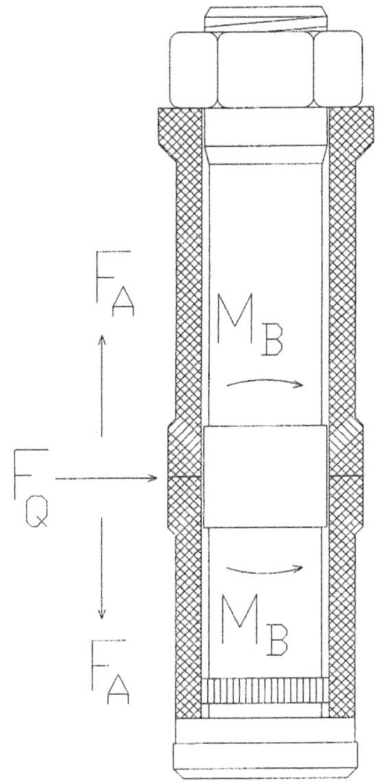

FIGURE 15.2 Free-body diagram for a joint.

There is an applied axial force F_A, an applied shear force F_Q, and an applied bending moment M_B. Most of the treatment is for the special case when $M_B=0$. Even for this special case, there may still be a bending moment M_b introduced into the analysis because the lines of action of the axial force and bolt clamp load are offset from the centroid of the joint face.

In some simple joints, the bending moment introduced by eccentric loading is statically determinant and can be determined from the axial load and local geometry. In many types of joints, the loading across the joint face is statically indeterminate. In the latter case, energy methods must be used to determine the bending moment and shear load associated with the axial loading. Energy method solutions for many types of geometries are published in standard handbooks. A finite element model or experimental test program may be the only recourse for some geometries. Figure 15.3 illustrates both statically determinate and statically indeterminate loading conditions. The righthand side of Figure 15.3 illustrates how bending is introduced into a conrod (connecting rod) joint.

Grotewhol [2] used energy methods to determine the ratio of the effective moment arm to the ring radius for a number of rings representing conrod geometries, a is used for the moment arm, and r_0 for the distance to the joint centroid (See Table 15.2).

Hagiwara [3] conducted additional research on the statically indeterminate conrod geometry.

Sometimes estimation of the external loads themselves is not straightforward. McNeill [4] developed and published a procedure for estimating the separating loads in a conrod based on the geometry, mass, and speed of the components.

$$F=\left[\left(M_{\text{REC}}+M_{\text{ROT}}\right)+\frac{M_{\text{REC}}r}{L}-\left(M_{\text{CAP}}+M_{\text{BOLT}}+M_{\text{SHELL}}\right)\right]r\omega^2 \tag{15.1}$$

where

M_{REC} = reciprocating mass

M_{ROT} = rotating mass

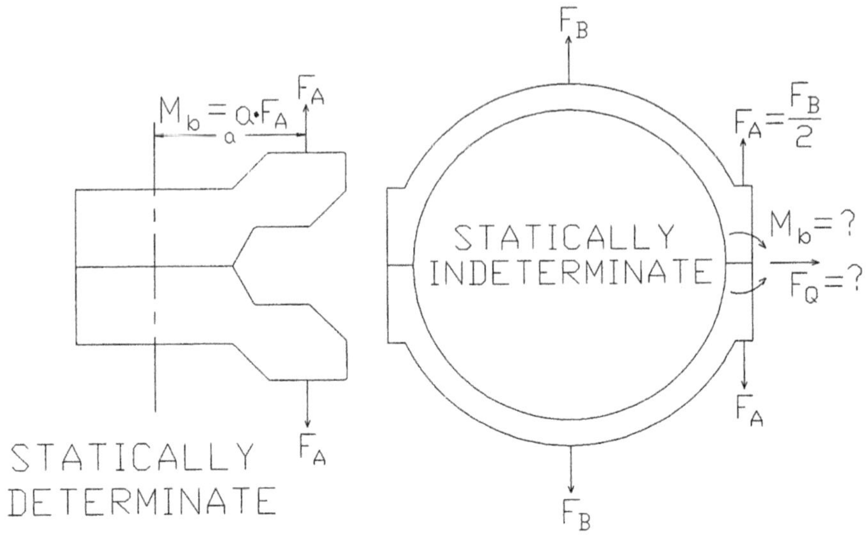

FIGURE 15.3 Statically determinate and indeterminate joints.

TABLE 15.2
Grotewhol Ring Solutions

Condition	Ratio a/r_o
Thin uniform ring	0.36
Bottom half twice as rigid	0.28
Bottom relatively rigid	–0.16
Bottom ring rigid between 225° and 315°	0.16
Actual conrod example	0.20

M_{CAP}=mass of the cap
M_{BOLT}=mass of one bolt
M_{SHELL}=mass of one shell
ω = engine speed in radians per second (1 rpm=0.1047 rad/s)
L=center-to-center distance from the crankshaft bore to the piston ring bore
r=crankshaft throw, i.e., eccentricity of crankpin axis to crankshaft axis. Half this load goes to each joint.

Often the external load can be determined more readily. In a cylinder head joint, adjacent cylinders do not fire simultaneously. Figure 15.4 shows a simplified cylinder head joint. Four bolts surround each cylinder of diameter D_{BORE}, and the combustion pressure is P_0. The external load is shared equally by all four bolted joints. The load per joint is given as

$$F_A = P_0\left(0.25\pi D_{BORE}^2\right)/4 \tag{15.2}$$

15.6 AXIALLY LOADED (CONCENTRICALLY LOADED) JOINTS

15.6.1 Springs in Series and Springs in Parallel

The classical theory of bolted concentrically loaded joints is based on a model of two linear springs representing the bolt and the clamped material. The springs may be in parallel, in series, or partially in parallel and partially in series, as discussed later in Section 15.6.6.

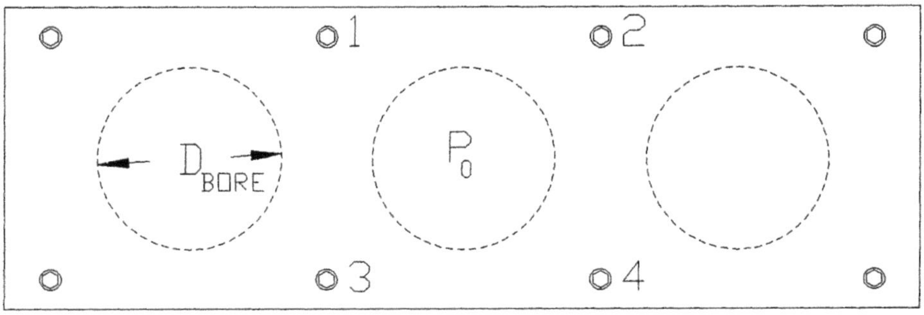

FIGURE 15.4 Axial loading in a cylinder head joint.

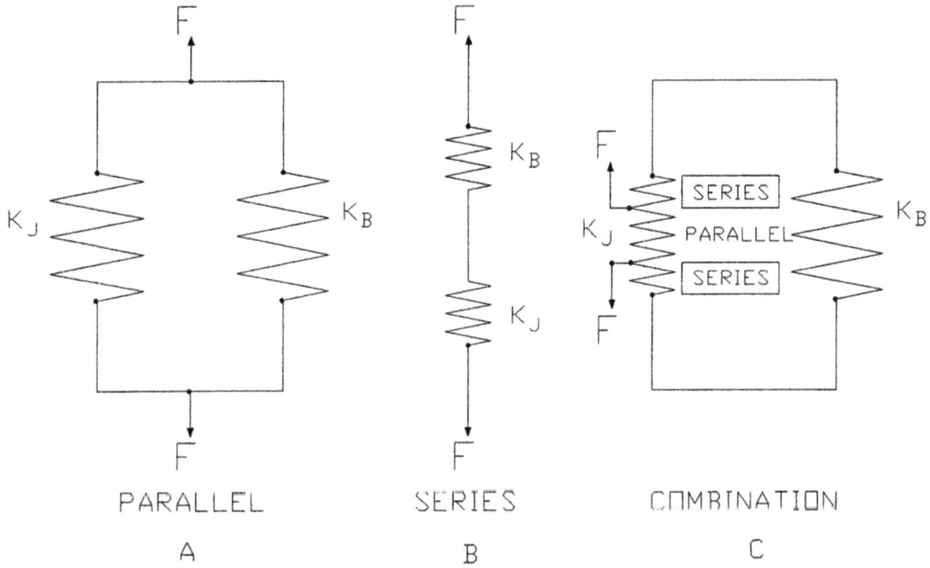

FIGURE 15.5 Spring models, (a) parallel; (b) series; and (c) combination.

Figures 15.5a and b illustrate the concepts of parallel and serial springs. Parallel springs stretch (or contract) by equal amounts, whereas serial springs carry equal loads. Parallel spring models are used to show how external load is shared by the bolt and the clamped material. Serial spring models are used to show how clamp load is affected by relative axial movement that often occurs between the bolt and the clamped materials at the interfaces during the initial tightening process.

Figure 15.5c shows a more generalized model where the spring representing the joint material is subdivided into three springs. The middle one of these is in parallel with the spring representing the bolt, and the other two springs are in series with the middle spring. This model is discussed in more detail in Section 15.6.6.

When a load F_A is applied to two springs in parallel, the portion of the load carried by each spring is directly proportional to its stiffness, K, Subscripts B and J correspond to bolt and clamped (joint) material.

$$F_B = \frac{K_B}{K_B + K_J} F_A, \qquad F_J = \frac{K_J}{K_B + K_J} F_A \tag{15.3}$$

Similarly, when springs in series are stretched, the portion of the elongation U_A carried by each spring is directly proportional to the inverse of its stiffness, which is known as its compliance, $\delta = 1/K$

$$U_B = \frac{\delta_B}{\delta_B + \delta_J} U_A, \qquad U_J = \frac{\delta_J}{\delta_B + \delta_J} U_A \tag{15.4}$$

15.6.2 Estimation of the Bolt Compliance δ_B

The bolt is modeled as a set of springs in series. The compliance of each section is calculated, and the sum of the compliances is computed. Only the portion of the bolt within the grip length is included in those calculations. The grip length l_k used for this computation is the distance from the first engaged thread to the underhead of the bolt (or screw) or to the other first engaged thread for a stud. Figure 15.6 shows a bolt passing through a 28 mm thick top plate and a 4 mm counterbore in the bottom plate before engaging in the tapped threads in the lower plate. The grip length and stiffness of the bolt depend not only on its dimensions but also on the dimensions of the joint, including the thicknesses of any washers used.

The compliance δ and stiffness K of a cylinder of diameter δ and length l made of a material with Young's modulus E are given by the equation

$$\delta = \frac{4l}{\pi d^2 E} = \frac{1}{K} \tag{15.5}$$

The general formula for the compliance of a bolt or screw is given in the equation

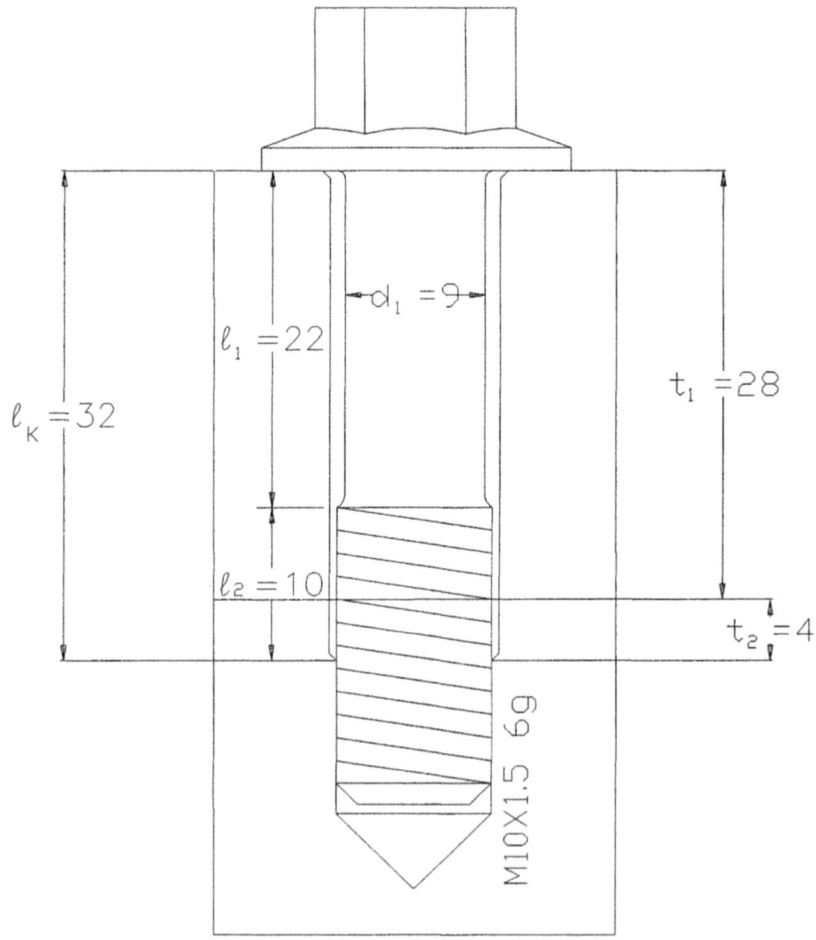

FIGURE 15.6 Bolt in joint for sample bolt compliance calculation.

$$\delta_B = \delta_K = \sum_{i=1}^{N} \frac{l_i}{E_B A_i} + \delta_G + \delta_M \tag{15.6}$$

where

$$\delta_K = \frac{0.4 d_n}{0.25\pi d_n^2 E_B} = \delta_G, \qquad \delta_M = \frac{0.5 d_n}{0.25\pi d_{\text{MINOR}}^2 E_B} \tag{15.7}$$

The factors 0.4 and 0.5 for the compliances of the head, δ_K, engaged external threads, δ_G, and engaged internal threads, δ_M, are empirical values for socket and hexagonal head bolts of nominal diameter d_n in steel joints, d_{MINOR} in Eq. (15.7) denotes the thread minor diameter. An additional δ_M would be added for a stud in which both threaded ends are stretched during tightening. The summation in Eq. (15.6) is over the N cylindrical sections of the fastener within the grip length (N can be 1 for a thread-to-thread part), with i being the dummy index.

Table 15.3 presents an example of compliance calculations.

15.6.3 Estimation of the Compliance of the Clamped Parts, δ_J

The determination of the characteristic joint diameter D_A in Section 15.4 plays an important role in determining the corresponding axial compliance. An equivalent area, A_{ers}, is defined so that a cylinder of length l_k with an area A_{ers} and material properties corresponding to those of the clamped parts also has the same compliance as the clamped parts.

The compressive region of the clamped material spreads radially from the bearing face of the bolt head or nut to the center of the joint at an assumed limiting angle of 45°. The characteristic joint diameter, D_A, is limited accordingly: $d_w \leq D_A \leq l_k + d_w$. d_w is taken as the nominal flange or underhead bearing diameter for the bolt or nut. In the event that there are two possible choices, such as a bearing diameter for a bolt and a bearing diameter for a nut, d_w should be the smaller of the two.

The calculation sequence is shown in the following two equations, which are evaluated in Table 15.4 for the joint shown in Figure 15.7.

TABLE 15.3
Compliance Calculations for Steel Bolt in Figure 15.6

Bolt Part	Length Used (mm)	Diameter Used (mm)	Compliance (mm/N)
Head	$0.4d_n=4$	10	2.462×10^{-7}
Shank	$l_1=22$	9	16.718×10^{-7}
Exposed threads	$l_2=10$	8.159	9.246×10^{-7}
Engaged threads	$0.4d_n=4$	10	2.462×10^{-7}
Mating threads	$0.5d_n=5$	8.159	4.623×10^{-7}
$K_B=1/\delta_B=280{,}200$ N/mm;		$\delta_B=35.511\times10^{-7}$ mm/N	

TABLE 15.4
Compliance and Stiffness Calculations for Figure 15.7

Data	$D_A=26$ mm	$d_w=20$ mm	$d_h=11$ mm	$l_k=32$ mm
Parameter	$x=0.982$	$A_{\text{ers}}=357.1$	$\delta_J=4.332\times10^{-7}$	$K_J=2.308\times10^{6}$
Units	—	mm²	mm/N	N/mm
Equation	(15.8)	(15.9)	(15.10)	(15.5)

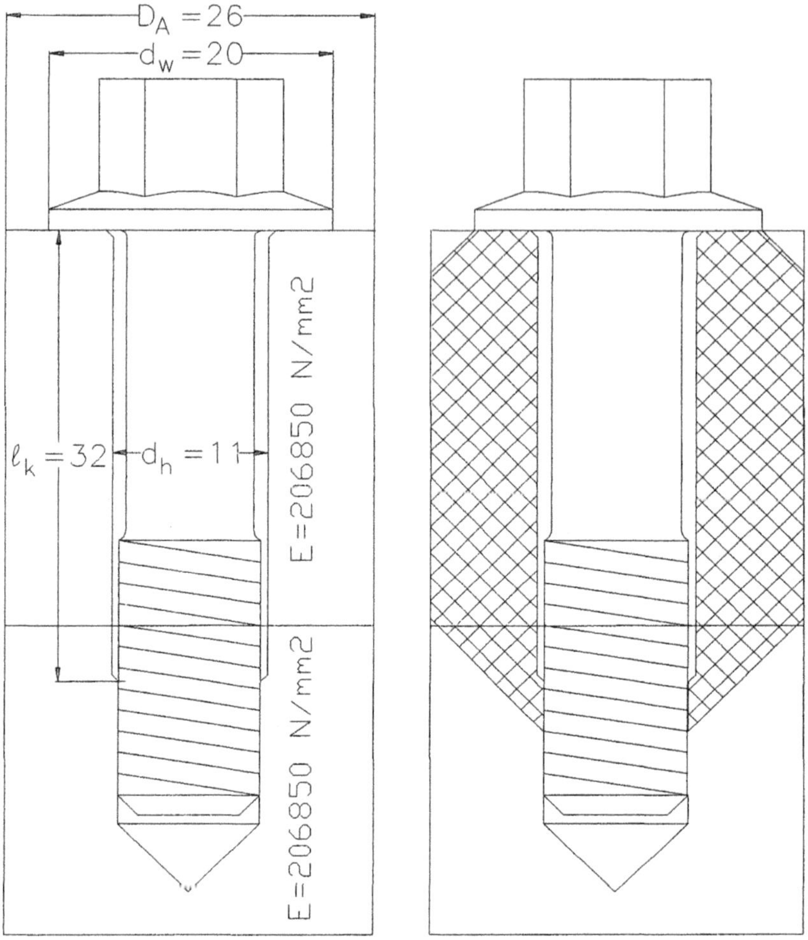

FIGURE 15.7 Bolt in joint for sample joint compliance calculation.

$$x \equiv \left(\frac{l_k d_w}{D_A^2} \right)^{1/3} \tag{15.8}$$

where x is a parameter used to interpolate between clamped shapes corresponding to a cylindrical sleeve and a 45° cone.

$$A_{\text{ers}} = \frac{\pi}{4}\left(d_w^2 - d_h^2\right) + \frac{\pi}{8} d_w \left(D_A - d_w\right)\left[(x+1)^2 - 1\right] \tag{15.9}$$

$$\delta_J = \frac{l_k}{E_J A_{\text{ers}}} \tag{15.10}$$

An equivalent average elastic modulus can be estimated when more than one material is clamped.

$$E_J = \sum \frac{E_{Ji} l_i}{l_k} \tag{15.11}$$

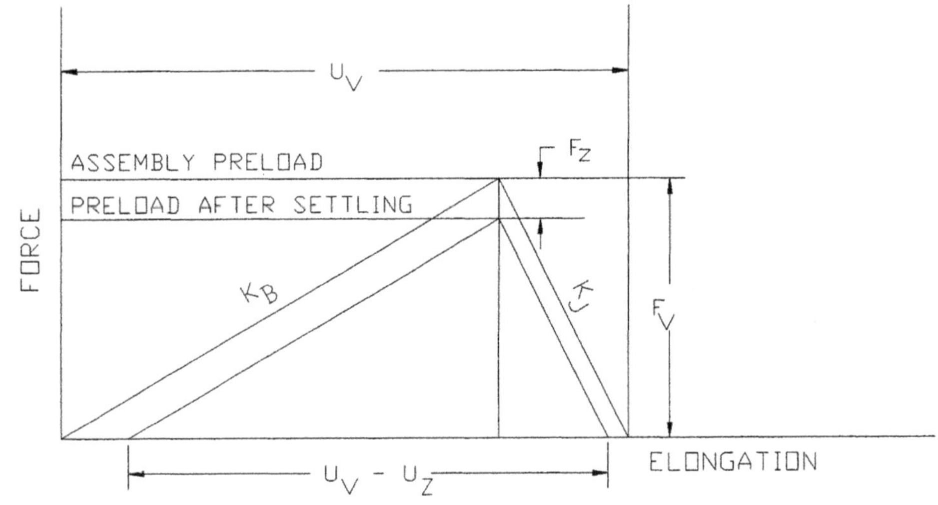

FIGURE 15.8 Tent diagram showing preload and settling losses.

15.6.4 Assembly and Settling Losses

When a fastener is tightened, the load–elongation curve is initially nonlinear because the mating surfaces are not fully in contact. Other sources of nonlinearity may be seating of gaskets, bearing shells, and nonparallelism. This nonlinearity is reflected in the preload angle curve. In many joints, the preload angle curve becomes linear and remains so until yielding takes place in the bolt or clamped parts. During this process, the bolt stretches and the joint material compresses. An idealized linear model is shown in Figure 15.8.

The bolt and clamped parts are seen to be acting like springs in series because the relative elongation, U_v, is the sum of the stretch of the bolt and the compression of the clamped parts. The clamp load in the bolt and in the clamped parts is the same.

$$F_V = F_B = K_B U_B = K_B \left(\frac{\delta_B}{\delta_B + \delta_J} \right) U_V \tag{15.12}$$

The stiffnesses, K_B and K_J, are in units of newtons per millimeter, and the displacements, U_B and U_z, are in millimeters. As the compliances, δ_B and δ_J, are the reciprocals of the stiffnesses, their values are in millimeters per newton.

The bolt threads turn 360° with respect to ground when they have advanced one pitch, p (mm), with respect to the mating internal threads. The relationship between the relative axial displacement between the external and internal threads, U_v, and the angle υ through which the bolt threads turn with respect to ground is given by the equation

$$U_V = \frac{\theta}{360} P \tag{15.13}$$

Substituting Eq. (15.13) into Eq. (15.12) and making use of the fact that the product of the bolt stiffness and bolt compliance is 1 ($K_B \delta_B = 1$) leads to the following relationship between clamp load and tightening angle:

$$F_V = \frac{(\theta/360)p}{\delta_B + \delta_J} = \frac{K_B K_J (\theta/360)p}{K_B + K_J} \tag{15.14}$$

Normally, some settling takes place in a bolted joint over time and results in a loss in clamp load. This amount may be determined by theoretical or experimental means. If no other information is available, the following VDI 2230 empirical correlation based on grip length l_K and nominal thread diameter d_n may be used to determine the settling in millimeters, U_z. Note that l_k and d_n must be in the same units (mm).

$$U_Z = 0.00329\left(\frac{l_K}{d_n}\right)^{0.34} \tag{15.15}$$

When settling occurs, the combined physical settling of the bolt and clamped parts, U_z, leads to an identical drop in the bolt clamp load and the joint compressive load, F_z. This is analogous to $F = F_z$ in the series spring model of Figure 15.5b, and so Eq. (15.4) applies.

$$F_Z = \frac{U_Z}{\delta_B + \delta_J} = \frac{U_Z(K_B K_J)}{K_B + K_J} \tag{15.16}$$

15.6.5 Thermal Loading and Expansion

The difference between the elongation of the clamped material and the elongation of the portion of the bolt within the grip length resulting from temperature variation is calculated, and then the rule for springs in series is applied to this differential elongation to determine the corresponding change in clamp load.

If the bolt material has a coefficient of thermal expansion ξ_B, the clamped material has a coefficient of thermal expansion $\xi_{J?}$ and both materials are heated (or cooled) from room temperature to a higher (or lower) operating temperature, the resulting change in clamp load is calculated according to the equation

$$F_{\text{THERMAL}} = \frac{(K_B K_J)(\xi_J - \zeta_B) l_K (T_{\text{OPERATING}} - T_{\text{ROOM}})}{K_B + K_J} \tag{15.17}$$

Heating of the bolted joint is not always uniform, in which case average values may have to be used. Steady-state temperatures for the bolt and clamped material may also differ. An average coefficient of thermal expansion can be determined using the approach of Eq. (15.11) when more than one material is clamped.

15.6.6 The Loading Plane Factor, *n*

The applied axial load is generally not introduced at the interfaces between the bolt and the clamped material. Figure 15.9 shows an idealized case of load being applied at two distinct points within the clamped material. The bolt is seen to be a spring acting in series with springs corresponding to the portions of the clamped material above and below the points where the load is applied. This combined spring is in parallel with a spring corresponding to the portion of the clamped material between the points where the load is applied. The load in the bolt spring, F_b, resulting from an applied load F is calculated using a formula similar to that for parallel springs, Eq. (15.3). Equation (15.18) has an additional cofactor, $K_J/K_{J2}=n$. n is referred to as the loading plane factor. Note that $0 \leq n \leq 1$.

$$F_B = \frac{K_J}{K_{J2}}\left(\frac{K_B}{K_B + K_J}\right)F = \frac{nK_B}{K_B + K_J}F = \frac{n\delta_J}{\delta_B + \delta_J}F \tag{15.18}$$

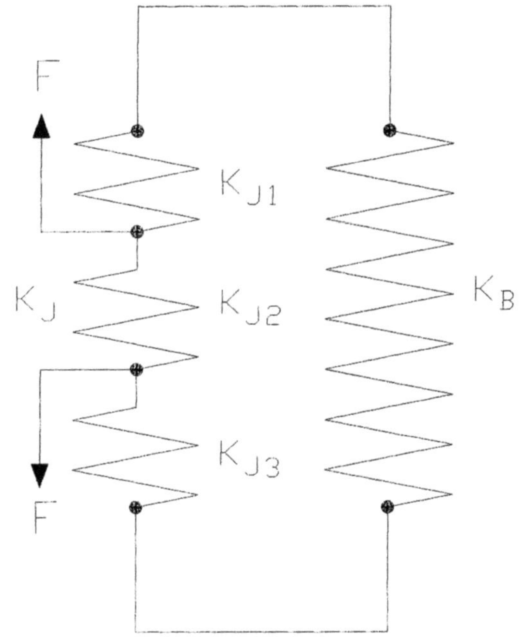

FIGURE 15.9 Illustration of the loading plane factor n.

For joints with relatively little clamped material extending beyond the bearing diameter d_w, the ratio K_J/K_{J2} is approximately the ratio of the length of the portion of the joint material between the points of applied loading to the grip length. This can be recognized easily by noting that A_{ers} has little sensitivity to l_k when $D_A \approx d_w$ because the second term on the right in Eq. (15.9) is then approximately zero.

Cylinder head joints are often assumed to have a loading plane factor of zero. In this case, none of the combustion load is carried by the cylinder head bolts.

15.6.7 The Force Ratio Φ

The force ratio ϕ is defined as the portion of the external load carried by the bolt. Mathematically, ϕ is the computed cofactor for F in Eq. (15.18):

$$\phi = \frac{n\delta_J}{\delta_B + \delta_J} = \frac{nK_B}{K_B + K_J} \tag{15.19}$$

If the applied load does not overcome the preload, then the portion of the applied load F_A carried by the bolt is ϕF_A and the clamp load at the joint face is reduced by $(1 - \phi)F_A$.

VDI 2230 uses several different subscripts for ϕ to indicate that the force ratio accounts for eccentricity, loading plane factors, and/or bending moments. This approach is not used here; instead, the different formulas for ϕ are kept distinct by subsection.

15.6.8 Determining Preload Requirements

The minimum possible preload developed by the assembly specification should leave sufficient clamp load to keep the joint closed after settling losses, thermal losses, and other miscellaneous losses have occurred. In the case of gasketed joints, keeping the joint closed may translate into maintaining a minimal average sealing pressure over a region. If the joint also experiences shear loading, then the remaining clamp load must maintain sufficient friction across the joint faces to prevent one from sliding past the other.

F_{KR} is used here to denote any special force required for sealing, crushing (or seating) of bearing shells into conrods or main bearings, etc. F_{KR} is zero unless some of the bolt clamp load is required to compress an element in parallel with the clamped material or the nominal compression across the joint face must always exceed zero for functional reasons such as sealing. The minimum preload required to withstand the axial loads is calculated as follows:

$$P_{MIN} = F_{KR} + F_Z + (1-\phi) F_A \tag{15.20}$$

where

F_z = settling loss
ϕ = force ratio (number between 0 and 1 representing the fraction of applied axial load carried by the bolt)
F_A = applied axial force

In cases of random or cyclical loading, the value used for F_A should correspond to the maximum possible axial load.

A similar calculation is made to determine the preload required to prevent sliding of the joint faces caused by a shear force F_Q.

$$P_{MIN} = \frac{F_Q}{\mu_J} + F_Z + (1-\phi) F_A \tag{15.21}$$

μ_J should be the minimum established coefficient of friction between the joint faces. The value of F_A used in Eq. (15.21) should correspond to the axial load at the time the shear force F_Q is applied. It will differ in value from the value of F_A used in Eq. (15.20) if the shear force, usually the maximum shear force, does not occur at the same time as the maximum axial load. In cases of random or cyclical loading, it may be necessary to evaluate Eq. (15.21) under more than one set of loads to determine the most severe condition. F_z is again the settling loss.

The minimum required preload is the larger of the two values of P_{min} calculated from Eqs. (15.20) and (15.21). If thermal loading can lead to a decrease in clamp load, that drop should be accounted for by including an additional term in Eqs. (15.20) and (15.21).

The maximum allowable preload is limited by the bolt strength, the thread strength, or the compressive strength under the bolt head or nut face. The maximum allowable preload is estimated assuming that no settling takes place.

$$P_{MAX} = F_{MAX} - \phi F_A \tag{15.22}$$

The value for F_{MAX} used in Eq. (15.22) should be the smallest of the three following values: the bolt yield strength, the stripping strength of the mating threads, and the embedment limit of the clamped material.

If the bolt is yield-tightened by a torque plus angle or gradient strategy, it will work harden, as illustrated in Figure 15.10. In my opinion, the value used for the tensile yield strength may be increased to a level a few percent higher than the maximum level specified on the bolt print. Ideally, tensile testing of yield-tightened parts will be used as a basis.

Generally, the bolt material is at least as strong as, if not stronger than, the material used for the internal threads. So thread stripping of the bolt material is usually not a concern. A check should be

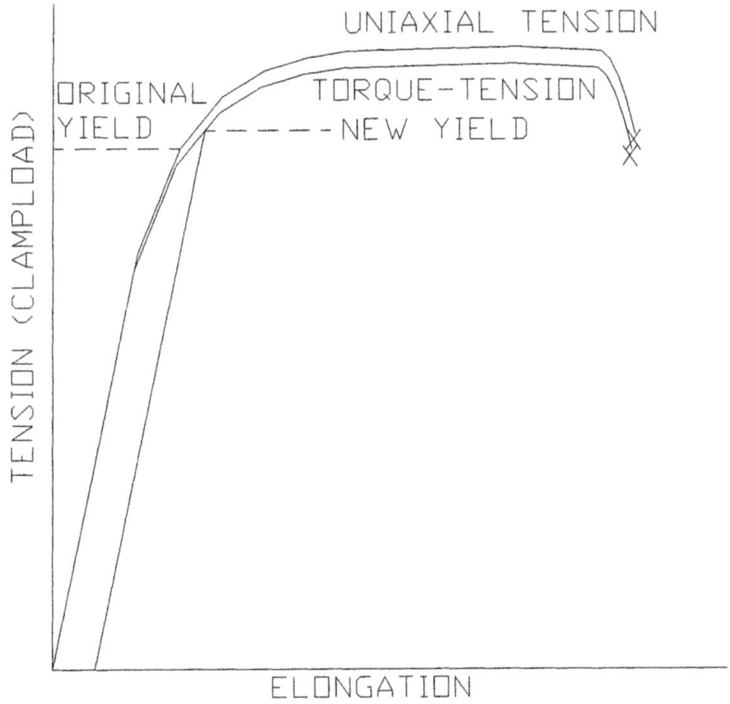

FIGURE 15.10 Increase in yield load from work hardening.

made to ensure that there is sufficient thread engagement for the internal threads to carry the full tensile strength of the bolt, preferably without significant yielding. Equation (15.23) gives the stripping strength of internal threads.

$$F_{STRIP} = \tau_U L_{ENGAGED} \pi D_{SMIN} \left(\frac{1}{2} + \frac{D_{SMIN} - E_{NMAX}}{\sqrt{3}p} \right) \quad (15.23)$$

where

- τ_U = ultimate shear strength of the mating material, about 50%–60% of its ultimate tensile strength
- p = thread pitch
- $L_{ENGAGED}$ = thread engagement length
- D_{SMIN} = minimum major diameter of the bolt thread
- E_{NMAX} = maximum pitch diameter of the mating thread

Values for D_{SMIN} and E_{NMAX} may be found in most reference books on threaded fasteners. The minimum value of $L_{ENGAGED}$ based on the bolt print and the print of the part with the mating threads should be used in Eq. (15.23).

Example: Determine the required 6H thread engagement in 310 aluminum for a 10.9 metric strength level M10×1.5−6g bolt. The maximum bolt strength for a 10.9 metric strength level bolt may be taken as the nominal strength of a 12.9 metric strength level bolt, 1,200 MPa. The tensile stress area of the threads is 58 mm^2. The maximum bolt load is 1,200×58 or 69,600 N. The ultimate shear strength of the aluminum is 130 MPa. Tabulated thread specifications show D_{SMIN} = 9.968 mm and E_{NMAX} = 9.206 mm. Substituting 69,600 for F_{STRIP} in Eq. (15.23) and using the data provided here leads to $L_{ENGAGED}$ = 21.6 mm.

Tests with hex nuts have shown that dilation may reduce the stripping strength of internal threads if there is insufficient wall thickness. The data have been correlated in terms of the ratio of the across-flats dimension w and the nominal thread diameter d_n. The following equation is applicable for w/d_n ratios between 1.4 and 1.9

$$\text{Dilation factor} = -\left(\frac{w}{d_n}\right)^2 + 3.8\frac{w}{d_n} - 2.61 \qquad (15.24)$$

This factor, which is between 0.75 and 1.0, is multiplied by the stripping strength calculated in Eq. (15.23) to calculate the reduced level for parts with thinner walls.

The bolt head and nut face, if any, bear on clamped material. If the clamp load is too high, the bearing surfaces will embed in the clamped materials, resulting in a loss in clamp load. The limiting clamp load is the product of the clamped material yield stress and the bearing area.

The underhead bearing area in Figure 15.11 is an annulus. The area is

$$0.25\pi\left(19^2 - 12^2\right) = 170.43 \text{ mm}^2$$

The underhead bearing area for football head and D-head bolts may be somewhat more difficult to evaluate. If a nut is used, the bearing area for the nut face must also be calculated. Work hardening of the clamped material during tightening and constraint by the unloaded material around it may increase the apparent yield strength above published levels. If operating conditions lead to an increase in clamp load resulting from thermal loading, then the right-hand side of Eq. (15.22) should be reduced by that amount.

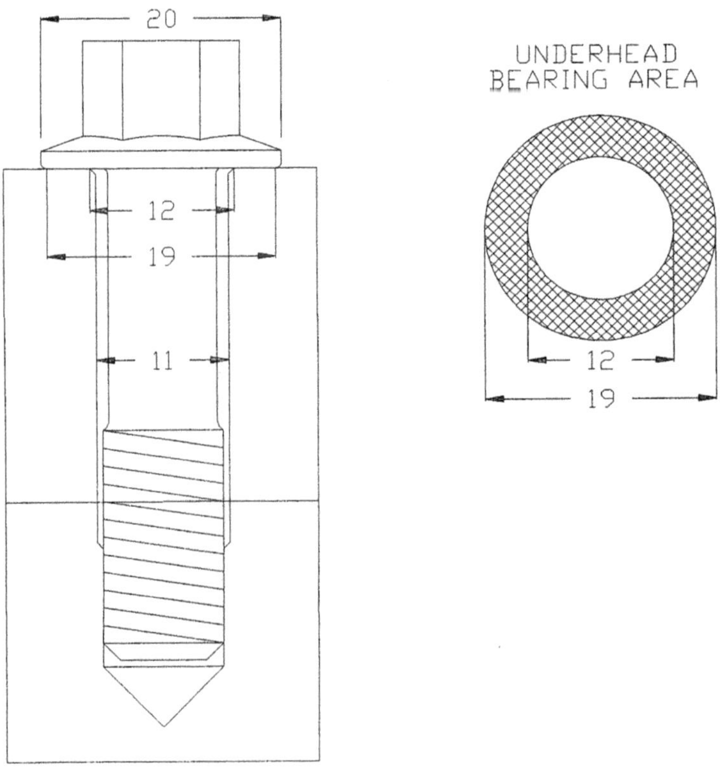

FIGURE 15.11 Definition of bearing area.

The range of preload from P_{MIN} to P_{MAX} dictates what type of tightening strategy is suitable. The tightening strategy will produce a maximum preload of F_{SP} and a minimum preload of F_{SP}/α_A, where the tightening factor α_A is about 1.0 for strain gage or ultrasonic measurements, 1.22 for yield tightening, and anywhere from 1.4 to 2.7 depending on coatings and equipment for other less accurate elastic-range tightening strategies such as torque control, torque plus angle control, and hydraulic tensioning.

15.6.9 Fatigue Analysis

VDI 2230 uses the approach that the applied fatigue cycle on the bolt can be characterized by a calculated applied alternating stress. This alternating stress is compared to the endurance limit for the bolt, which is the maximum alternating stress under comparable mean stress conditions that will not cause the bolt to fail.

The alternating force is one-half the difference between the maximum and minimum applied forces. The alternating stress is the alternating force divided by the thread diameter minor area, A_{MINOR}. The mean or average force is simply the average of the minimum and maximum applied axial forces. The mean stress is the mean force divided by the thread tensile stress area. (The traditional use of thread (TSA) for calculating the mean stress and of thread minor diameter area for calculating the alternating stress has been a cause of unresolved contention.)

The alternating stress on a bolt should not exceed its endurance limit. Figure 15.12 can be used to estimate the endurance limit of an alloy steel bolt, metric strength class of 8.8-12.9, with rolled threads.

The endurance limit σ_A for alloy steel bolts with threads rolled before heat treatment (RTBHT) is relatively independent of the mean load over a large range. This is a somewhat unexpected trend, but a theory has been proposed by Yoshimoto [5] to explain it. σ_A in megapascals (MPa=N/mm^2)

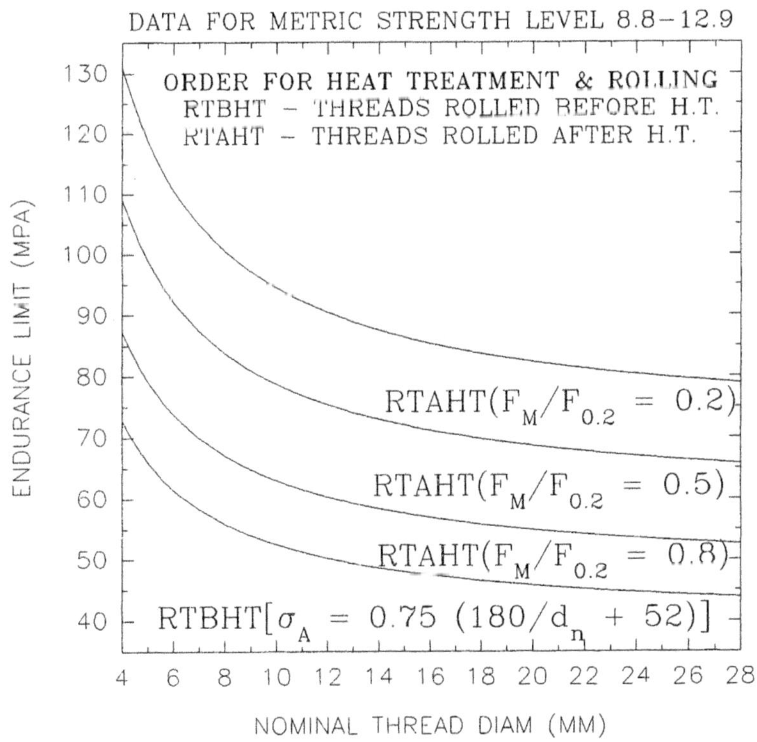

FIGURE 15.12 Fatigue diagram—endurance limit vs. bolt nominal diameter.

is correlated solely in terms of the nominal thread diameter d_n in millimeters and does not depend on the metric strength level.

$$\sigma_A = 0.75\left(\frac{180}{d_n} + 52\right) \tag{15.25}$$

The endurance limit of alloy steel bolts with threads rolled after heat treatment (RTAHT) is normally somewhat higher and depends on the ratio of the mean load F_m to the 0.2% yield load $F_{0.2}$ over most of the range.

$$\sigma_A = 0.75\left(\frac{180}{d_n} + 52\right)\left(2 - \frac{F_M}{F_{0.2}}\right) \tag{15.26}$$

Equations (15.25) and (15.26) are adequate for mean stresses ranging from 20% to 90% of the 0.2% yield stress, based on the tensile stress area of the threads. Note that Eq. (15.26) uses the ratio of the mean force to the yield strength, which eliminates any ambiguity regarding the appropriate stress area for calculating the mean stress.

Example: Predict the endurance limit of a 10.9 M10 bolt with threads rolled after heat treatment when the minimum load is 10% of the maximum load F_{MAX}. The mean load is $(F_{MAX}+0.1F_{MAX})/2=0.55F_{MAX}$, and the alternating load is $(F_{MAX}-0.1F_{MAX})/2=0.45F_{MAX}$. Substituting these expressions into Eq. (15.26) leads to the equation

$$\sigma_A = \frac{0.45F_{MAX}}{A_{MINOR}} = 0.75\left(\frac{180}{d_n} + 52\right)\left(2 - \frac{0.55F_{MAX}}{F_{0.2}}\right) \tag{15.27}$$

where

A_{MINOR} = thread minor diameter area
d_n = thread nominal diameter
$F_{0.2}$ = yield strength

Using $d_n = 10$ mm, $\sigma_{0.2} = 940$ MPa, TSA = 58 mm², $F_{0.2} = 940 \times 58 = 54.52$ kN, and $A_{MINOR} = 52.2$ mm² leads to a value for F_{MAX} of 11,480 N. The endurance limit under this condition is then 11,480/52.2 = 220 MPa.

Bolts may be designed such that the shank diameter is sufficiently necked down to the point where fatigue failures will occur in the shank rather than in the threads. Fatigue failures may also initiate within knurled sections, in the bolt underhead fillet, and through other regions of geometric transition. Figure 15.12 overestimates the endurance limit for these cases.

The endurance limit may also be increased by using larger radius thread roots, i.e., an MJ profile, using specialty materials such as MP159, and by various methods of adjusting the load distribution between mating threads. In contrast, the endurance limits of bolts with cut threads are lower than those for corresponding parts with rolled threads.

For these reasons, experimental determination of the endurance limit is preferable to the use of Figure 15.12 when time and resources allow.

15.7 ECCENTRICALLY AND NONLINEARLY LOADED JOINTS

The primary difference between the analytic methods for concentric joints and eccentric joints is that beam bending is also taken into account. Generally, the bending stiffness of the clamped parts is much greater than that of the bolt. Thus, a simplifying assumption can be made that the curvature of the loaded joint is a function of its geometry and the bending moment. This bending moment is the sum of the external bending moment applied to the joint, M_B, the bending moment caused by the applied axial load, M_b, and the bending moment produced by the bolt because its axis is offset from the centroid of the joint face. Closed-form solutions can be derived for these conditions.

The same approach can be used for nonlinear joints, but graphical methods or iterative computer techniques are necessary to arrive at a solution.

The behavior of the joint generally becomes nonlinear if sufficient loads are applied to cause localized separation of the clamped material at the joint face. While VDI 2230 does provide some equations for analysis of this condition, the primary recommendation is that there be sufficient preload to prevent the condition from occurring.

15.7.1 Geometric Properties of the Joint Face

The section modulus is used in beam bending problems for determining the maximum tensile stress in the body. The same approach is used in VDI 2230. The process of drilling the bolt hole through the joint face shifts its centroid, and the moment of inertia is calculated using the parallel axis theorem. Figure 15.13 shows a generic joint drawn as a rectangle for expediency. CG denotes the centroidal axis.

The bolt hole may be offset from the centroid of the joint face. *s* is used for this dimension, and the convention is to use a positive sign for *s* when the bolt preload counteracts the applied moment. Generally, the axial load is tensile, and *s* is taken as positive when the bolt axis is on the same side of the centroid as the line of action of the applied force.

The following geometric characteristics must be determined.

a Equivalent moment arm of the applied force (mm)
u Distance from the centroid to the separable side (mm)
v Distance from the centroid to the opposite side (mm)
s Vector (size + sign) from centroid to bolt axis (mm)
A_B Planar area of the predrilled joint face (mm^2)

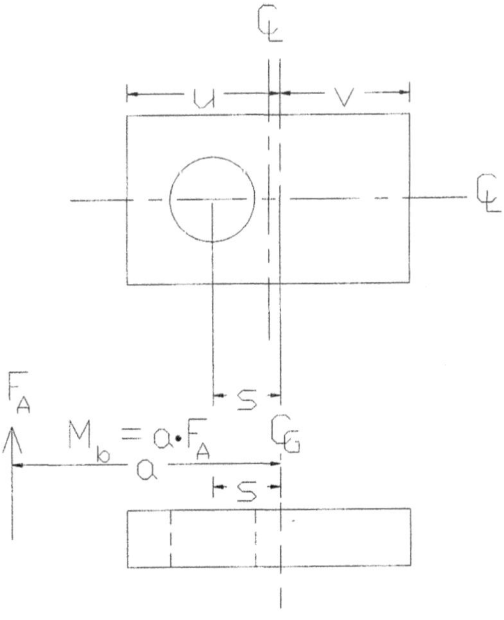

FIGURE 15.13 Definition of joint face geometric parameters.

A_D Planar area of the drilled joint face (mm²)
I_B Moment of inertia of the predrilled joint face (mm⁴)
I_{BT} Moment of inertia of the drilled joint face (mm⁴).

The values of A_B and I_B are not used directly but only in intermediate steps for calculating A_D and I_{BT}.

The moments of inertia are taken about centroidal axes that are parallel to the axis of the applied moment.

u is the distance from the centroid of the joint to the edge of the joint that will separate if the joint is overloaded, v is the distance from the centroid of the joint to the opposite end. The value of v is used only for prying joint calculations, which are not covered here.

If the applied axial force in Figure 15.13 were in the opposite direction (compressive), then the values used for u and v would be interchanged, and the sign of s would be negative.

15.7.2 External Loading—Conrod Example

A top view of the two joints is shown in Figure 15.14.

$A_B = 18 \times 20 = 360\,\mathrm{mm}^2$
$A_H = 0.25 \times \pi \times 10^2 = 78.54\,\mathrm{mm}^2$ (area of hole)
$A_D = A_B = A_H = 281.46\,\mathrm{mm}^2$
Offset = 1 mm (distance from hole to predrilled centroid)
$x_c = A_B/A_H - 1$ offset $= 0.28\,\mathrm{mm}$ (new centroid)
$u = 10 + 0.28 = 10.28\,\mathrm{mm}$; $v = 10 - 0.28 = 9.72\,\mathrm{mm}$
$s = 1 + 0.28 = 1.28\,\mathrm{mm}$ ($s > 0$ because $F_A > 0$ and on same side)
$I_B = (1/12) \times 18 \times 20^3 = 12{,}000\,\mathrm{mm}^4$
$I_H = (\pi/64) \times 10^4 = 490.87\,\mathrm{mm}^4$ (moment of inertia of hole).

$$I_{\mathrm{B1}} = I_{\mathrm{B}} + A_{\mathrm{B}}x_c^2 - I_{\mathrm{H}} - A_{\mathrm{H}}s^2 - 11{,}408.67\ \mathrm{mm}^2$$

$r_0 = 50/2 + u = 35.28\,\mathrm{mm}$
$a \cong .2r_0 \cong 7.06\,\mathrm{mm}$.

15.7.3 Vibration Loosening

VDI 2230 provides no quantitative calculations relating to vibration loosening. The general rules for eliminating vibration are long bolts, i.e., high l_k/d_n ratios, and high preload. If a locking feature is needed, the function must be properly understood.

Serrated bearing faces on the nut or bolt and adhesives are the only tabulated functional antirotational locking features. Deformed bolt threads, plastic patches or inserts, all-metal prevailing torque

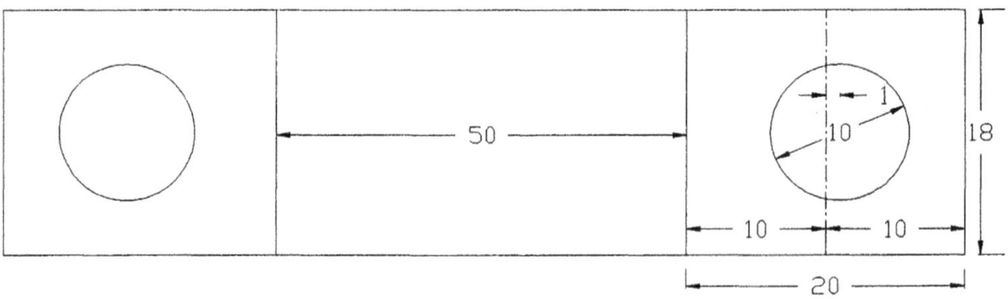

FIGURE 15.14 Geometry of joint face geometry for sample calculations.

nuts, castellated nuts, tab washers, and cotter pins (wire) are all listed as being primarily functional for preventing loosening following clamp load loss caused by embedment.

15.7.4 The Force Ratio Φ

The bending compliance β of a body is calculated using a formula similar to Eq. (15.5) with the moment of inertia used in the denominator rather than the area.

$$\beta = \frac{l}{EI} \tag{15.28}$$

As l is in mm, Young's modulus E is in N/mm^2 (MPa), and I is in mm^4, β is in units of (N·mm)$^{-1}$.

If the joint cross section is fairly constant, the value of I_{BT} calculated for the joint face may be substituted into Eq. (15.28) to determine the bending stiffness of the clamped material. In this case, the moment of inertia for the joint face is also the equivalent moment of inertia for the clamped material, I_{Bers} ($I_{BT}=I_{Bers}$).

When the cross section of the joint varies through the grip length, the equivalent moment of inertia is calculated by using an averaging scheme.

$$I_{Bers} = l_k \left[\int_0^{l_k} \frac{dx}{I(x)} \right]^{-1} \tag{15.29}$$

The following equation is the generalized formula for the force ratio. Note that VDI 2230 does not explicitly account for the hole in the moment of inertia, I_{Bers} for this equation. This was done in earlier revisions, however, and would appear to be more consistent with the overall beam bending formulation.

$$\phi = \frac{n\delta_J \left(1 + asA_{ers} / I_{Bers}\right)}{\delta_B + \delta_J \left(1 + s^2 A_{ers} / I_{Bers}\right)} \tag{15.30}$$

n again denotes the loading plane discussed in Section 15.6.6 and should be a value between 0 and 1. The ratio of the equivalent area for the clamped parts, A_{ers}, to the equivalent moment of inertia of the clamped parts, I_{Bers}, appears in both the numerator and the denominator of Eq. (15.30). VDI 2230 makes a point of relating this ratio to the radius of gyration [$(I_{Bers}/A_{ers})^{0.5}$] for the clamped parts. Note that Eq. (15.30) reduces to Eq. (15.19) when there is no geometric eccentricity, i.e., $s=0$.

15.7.5 Determining Preload Requirements

There is a certain amount of preload, F_{Kerf}, required to keep the joint from one-sided opening when the axial force F_A is applied.

$$F_{Kerf} = \frac{(a-s)uF_A}{I_{BT} / A_D + su} \tag{15.31}$$

This term is added to the others in Eq. (15.20).

$$P_{MIN} = F_{KR} + F_Z + (1-\phi)F_A + F_{Kerf} \tag{15.32}$$

Equation (15.21) remains unchanged and its calculated value is compared with that from Eq. (15.32) to pick the larger P_{MIN} as the minimum required preload. The limitations for the maximum allowable preload discussed in Section 15.6.8 are still applicable.

15.7.6 Fatigue Analysis for Combined Axial and Bending Loads

Fatigue data for combined tensile and bending stresses are almost always unavailable and are difficult to generate. The calculated largest alternating combined bending and tensile stress at the thread root, or other weakest section, is calculated and compared to the endurance limit for the bolt for the same mean load.

If M_B is constant (zero is a constant) and the bending stiffness of the bolt is much less than that of the joint, then the maximum alternating stress at the thread root corresponding to a range of axial loads F_{AU} and F_{A0} can be shown to have the form

$$\sigma_A = \left[1+\left(\frac{1}{\phi}-\frac{s}{a}\right)\left(\frac{l_k E_B a \pi s^3_{MINOR}}{l_{ers} E_J 8 I_{Bers}}\right)\right]\left(\frac{\phi(F_{AU}-F_{A0})}{A_{MINOR}}\right) \tag{15.33}$$

VDI 2230 explicitly accounts for the drilled hole in the calculation of I_{Bers} used for Eq. (15.33) and uses a bar over the symbol to distinguish this moment of inertia from the moment of inertia for a predrilled body.

l_{ers} is the equivalent bending length of the bolt based on the thread minor diameter. Procedures analogous to those used for determining the bolt axial compliance in Section 15.6.2 are used for determining its bending compliance. l_{ers} is the length that would be substituted into Eq. (15.28) with the moment of inertia corresponding to the thread minor diameter to arrive at the same bending stiffness or compliance as that of the bolt.

$$l_{ers} = 0.5+0.8\left(\frac{d_N}{d_{MINOR}}\right)^2+\sum_{i=1}^{N} l_i\left(\frac{d_i}{d_{MINOR}}\right)^2 \tag{15.34}$$

The factor 0.8 represents the sum of two factors of 0.4.

As discussed in Section 15.6.2, the factors 0.4 and 0.5 correspond to empirical values for hex head and socket head bolts and represent flexure in the head and mating threads. A second 0.5 term would be included for a stud in which both threaded ends are stretched during tightening.

The summation in Eq. (15.34) is over the N cylindrical sections of the fastener within the grip length (N can be 1 for a thread-to-head part), i is a dummy index for the summation.

15.8 CALCULATION OF THE FACTORS OF SAFETY

Factors of safety are calculated as ratios of acceptable levels of loading to calculated levels of loading.

$$SF_{PRELOAD} = \frac{F_{SP}/\alpha_A}{P_{MIN}} \tag{15.35}$$

where

F_{SP}=target clamp load
α=preload scatter factor representing the ratio of the maximum preload developed by the tightening strategy to the minimum preload developed by the tightening strategy
F_{MIN}=calculated minimum required preload for a functional joint

$$F_{FATIGUE} = \frac{\text{bolt endurance limit}}{\text{calculated alternating stress}} = \frac{\sigma_A}{\sigma_a} \tag{15.36}$$

$$SF_{YIELD} = \frac{\text{axial yield strength}}{\text{maximun axial load}} = \frac{F_{0.2}}{F_{SP}+\phi F_A} \tag{15.37}$$

$$FS_{SHEAR} = \frac{\text{min friction force}}{\text{shear force}} = \frac{\mu\left[F_{SP}/\alpha_A - F_Z - (1-\phi)F_A\right]}{F_Q} \tag{15.38}$$

$$SF_{PRESSURE} = \frac{\text{clamped material yield strength}}{\text{maximun bearing pressure}} = \frac{\sigma_{0.2\text{ for Clamped Parts}} \times \text{bearing area}}{F_{SP} + \phi F_A} \tag{15.39}$$

Again, the denominator in Eqs. (15.37) and (15.39) may have to be increased to reflect thermal loading. The numerator in Eq. (15.37) may also be increased based on experimental data relating to work hardening of the corresponding material.

15.9 FASTENER SIZING GUIDELINES

The entire process of designing a joint and calculating its safety factors may have to be repeated more than once to arrive at a satisfactory design. Table 15.5 provides a basis for the initial sizing of the fastener.

To use Table 15.5:

1. Pick the row with the lowest load that is greater than or equal to the applied load.
2. Estimate the minimum required preload.
   a. Move four rows down for shear loading.
   b. Move two rows down for eccentric fatigue loading.
   c. Move one row down for eccentric static or concentric fatigue loading.
3. Estimate the largest required preload.
   a. Move two rows down for straight torque tightening.
   b. Move one row down for elastic-range tightening using elongation measurements or torque plus angle monitoring or a calibrated torque wrench.
4. Pick the 8.8, 10.9, or 12.9 bolt size on that row.

TABLE 15.5
Table for Estimating the Appropriate Bolt Size

	Bolt Size for Corresponding Strength Level		
Load (kN)	**12.9**	**10.9**	**8.8**
0.25			
0.40			
0.63			
1.0			
1.6	3	3	3
2.5	3	3	4
4.0	4	4	5
6.3	4	5	5
10.0	5	6	8
16.0	6	8	8
25.0	8	10	10
40.0	10	12	14
63.0	12	14	16
100.0	16	16	20
160.0	20	20	24
250.0	24	27	30
400.0	30	36	
630.0	36		

Example: What size 10.9 bolt should be used for a shear load of 9 kN? Pick 10 kN as next biggest load in the leftmost column. Increase by four steps for shear to 63 kN. Increase by one step for a torque wrench to 100 kN. M16 would be used for 10.9 (or 12.9).

LIST OF SYMBOLS

A	Effective moment arm for axial force F_A; $M_b = aF_A$ (mm)
AB	Gross area of joint face before bolt hole is drilled (mm²)
AD	Net area of joint face after bolt hole is drilled (mm²)
A_{ers}	Equivalent area for clamped material (mm²)
A_i	Cross-sectional area of *z*th section (mm²)
AH	Hole area
A_{MINOR}	Cross-sectional area of thread minor diameter (mm²)
D	Generic diameter (mm) disthread nominal diameter (10 for M10, 12 for Ml2, etc.)
dh	Hole diameter (mm)
d_{MINOR}	Thread minor diameter (mm)
dn	Thread nominal diameter (mm)
dw	Bearing or washer diameter (mm)
DA	Characteristic joint diameter (mm)
D_{SMIN}	Minimum external thread major diameter (mm)
E	Generic elastic (Young's) modulus (MPa; N/mm²)
E_B	Elastic modulus of bolt material (N/mm²)
E_i	Elastic modulus of *i*th section of clamped material (N/mm²)
EJ	Elastic modulus of clamped material (N/mm²)
E_{NMAX}	Maximum pitch diameter of internal thread (mm)
F	Generic axial force (N)
$F_{0.2}$	Yield strength of bolt based on 0.2% yield stress (N)
fA	Applied axial force (N)
BB	Force in bolt (spring) arising from applied axial force (N)
fJ	Reduction in compressive force between clamped parts (spring) caused by the applied axial force (N)
F_{KR}	Additional force requirement for seating, sealing, etc. (N)
F_{Kerf}	Minimum clampload to prevent any separation of the joint face (N)
FM	Mean axial bolt load in fatigue cycle (N)
FQ	Applied shear force (N)
F_{STRIP}	Axial load that will strip internal threads (N)
FV	Preload or clampload (N)
i	Dummy summation index
I	Generic moment of inertia (mm⁴)
IB	Moment of inertia of predrilled joint face (mm⁴)
I_{Bers}	Effective moment of inertia of clamped parts (mm⁴)
I_{BT}	Moment of inertia of joint face with drilled bolt hole (mm⁴)
K	Generic axial stiffness (N/mm)
K_B	Axial stiffness of bolt (spring) (N/mm)
KJ	Axial stiffness of clamped parts (spring) (N/mm)
l	Generic length (mm)
l_{ers}	Equivalent bending length of bolt within clamped length based on thread minor diameter (mm)
lk	Grip length between bearing surface and first engaged thread or opposite bearing surface for studs (mm)
$L_{ENGAGED}$	Length of mating external and internal threads (mm)

Mb	Applied pure bending moment (N·mm)
Mh	Bending moment resulting from eccentricity of F_A (N·mm)
n	Loading plane factor ($0 \leq n \leq 1$)
N	Number of sections
P	Thread pitch (mm)
P_{MIN}	Minimum required preload for a functional joint (N)
P_{MAX}	Minimum required preload for a functional joint (N)
s	Distance from bolt axis to centroid of joint face (mm)
T	Temperature (°C)
T_{ROOM}	Room temperature
$T_{OPERATING}$	Operating temperature (°C)
TSA	Tensile stress area. Tabulated values can be found for standard threads. For metric threads $(\pi/4)(d - 0.9382p)^2$. For English threads $(\pi/4)(d - 0.9743p)^2$.
u	Distance from centroid of joint face to portion of clamped parts that will separate first when overloaded (mm)
UV	Initial combined bolt stretch and compression of clamped material developed when bolt is preloaded (mm)
UA	Applied axial displacement (elongation) to joint (mm)
UB	Elongation of bolt (spring) resulting from axial load (mm)
UJ	Decompression of clamped parts (spring) from axial load (mm)
v	Distance from centroid of joint face to edge in opposite direction to that used to measure *u* (mm)
w	Width across flats for hexagonal nuts (mm)
x	Interpolating parameter in Eqs. (15.7) and (15.8)
αA	Ratio of maximum to minimum preload developed in assembly
β	Generic bending compliance $[(N \cdot mm)^{-1}]$
δ	Generic axial compliance ($\delta = 1/K$) (mm/N)
δB	Axial compliance of bolt (spring) (mm/N)
δG	Axial compliance of engaged external threads (mm/N)
δJ	Axial compliance of bolt (spring) (mm/N)
δK	Axial compliance of bolt head (mm/N)
δM	Axial compliance of engaged internal threads (mm/N)
θ	Angle bolt turns with respect to internal threads (degrees)
μJ	Coefficient of friction between clamped materials at the joint face
ξB	Coefficient of thermal expansion for bolt material $[(°C)^{-1}]$
ξJ	Coefficient of thermal expansion for clamped material $[(°C)^{-1}]$
$\sigma_{0.2}$	Axial yield stress based on 0.2% offset (N/mm^2)
σA	Endurance limit of bolt for fatigue loading (N/mm^2)
σa	Peak alternating stress in bolt thread root (N/mm^2)
τU	Ultimate shear strength of internal thread material (N/mm^2)
ϕ	Force ratio, fraction of external axial load the bolt carries.

REFERENCES

1. VDI 2230 Bl.1, Systematische Berechung von hochbeanspruchten Schraubenverbidungen, zylindrische Einschraubenverbindungen, VDI, Düsseldorf, 1986.
2. Grotewhol, A., Calculation of a dynamically and eccentrically loaded bolted conrod connection according to VDI 2230, SAE Paper 750882, 1975.
3. Hagiwara, M., A method to simplify the strength design of bolted joints—Case of connecting-rod bolts, *Exp. Mech.*, 24, 28–32, March 1984.
4. McNeill, W., Dynamic analysis of connecting rod bolts, SAE Paper 881287, 1988.
5. Yoshimoto, I., A hypothesis concerning fatigue strength of bolt nut joints, *Bull. PME (TIT)*, 51: 43–46, 1982.

16 Design Guidelines for Automotive Bolted Joints

Ali Alkelani and Basil Housari

16.1 INTRODUCTION

Threaded fasteners are commonly used to form mechanical connections in many industries, from automotive to structural steel, primarily because they are inexpensive and easy to assemble and disassemble. The design of mechanical joints is often underestimated, which is the main reason for having frequent joint failures in the field. To properly utilize a threaded fastener, the development of the proper clamp load is critical to prevent joint failures such as leakage, joint separation, self-loosening, etc. Therefore, key design elements must be taken into consideration. In addition to obtaining the proper clamp load in the joint, it is important to make sure the fasteners selected meet all other design requirements, such as corrosion protection, loosening resistance, prevent joint component damage, etc.

The tightening strategy used during assembly is another important factor that contributes to the development of the proper level of clamp load and the desired quality of the joints. There are several tightening strategies that have been developed to control the amount of clamp load created at the end of the tightening process. These strategies include torque control, angle control, yield control, and stretch control. Torque control and angle control remain the most commonly used strategies in the automotive industry.

In this chapter, the selection of the fasteners including material, geometries, finish, special features, etc. are discussed, and practical guidelines to help engineers in the design of bolted joints are provided. In addition, a method for developing the right level of clamp load using both torque control and angle control strategies is covered in depth. Torque tables to aid the design engineers with the proper tightening torques for a variety of fastener sizes, materials, grades, etc. are provided including the resultant level of clamp loads. Also, a practical method for developing angle control tightening specifications and an easy method to calculate the final expected torque window for commonly used fastener sizes are explained. Finally, a method to detect joint quality issues for angle control tightening strategy is provided.

16.2 KEY DESIGN FACTORS FOR THE SELECTION OF THREADED FASTENERS

There are many types of fasteners that can be used to join different components of a machine or a device, such as normal machine thread bolts and nuts, thread forming screws, self-tapping screws, thermoplastic screws, rivet-nuts, rivet-studs, etc. In this chapter, the key factors to consider when selecting metric thread fasteners for an application are presented.

Before getting deep into the design and selection of bolted joints, it is important to cover some basic definitions related to threaded fasteners (Figure 16.1):

- *Thread profile*: the thread profile refers to the shape of the threads formed on the fastener (bolt or nut). There are many thread profiles used in different parts of the world for different applications; however, for machine thread bolts and nuts, the most common thread profile is the ISO Metric thread. The detailed dimensions for the metric thread profile are available in a number of ISO standards [1–7].

 DOI: 10.1201/9781003501183-20

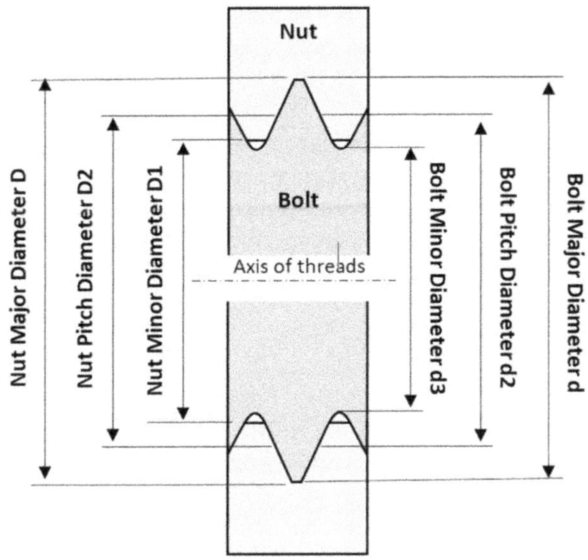

FIGURE 16.1 Basic thread diameters for metric thread bolt/nut (See ISO 68-1[1] for details).

- *Diameters*: when talking about the thread profile, there are three different diameters to be defined, namely, major diameter, pitch diameter, and minor diameter.
  - The major diameter is the diameter measured at the tip or peak of the external threads on a bolt or at the root of the inner threads in a nut.
    - The minor diameter is the diameter measured at the root or valley of the external thread on a screw or at the tip or peak of the inner thread in a nut.
    - The pitch diameter is the diameter of an imaginary cylinder that is equidistant from the peak and the valley of the threads on a bolt or in a nut.
  - *Thread pitch*: thread pitch refers to the distance between two consecutive threads on the same helix. For ISO metric fasteners, standard pitch values are used for each bolt size, which are referred to as coarse threads. A smaller fine thread pitch is also known and commonly used for larger size bolts (>M12).

There is a huge variety of bolts and studs used for bolted joints in different industries. In this chapter, the focus is on metric machine thread fasteners, which are commonly used in the automotive industry. The same design considerations can be applied to other fasteners made to other thread profiles; however, the calculations and numbers will be different.

The selection of metric thread bolt/nut combination depends mainly on the application requirements. Here are the major parameters that need to be considered during the fastener selection.

16.2.1 Material and Strength

The most common material used for metric machine thread fasteners is carbon steel. Aluminum fasteners are used for some applications for galvanic corrosion or weight-saving purposes. Stainless steel is also used for special applications, high-temperature applications, or for special corrosion requirements.

Steel bolts are available in different strength classes such as class 4.8, 6.8, 8.8, 9.8, 10.9, and 12.9. The most commonly used classes are 8.8 and 10.9. The first number in the classification refers to the ultimate strength in 100 MPa, and the second number refers to the minimum yield strength as a ratio of the ultimate strength, for example, class 8.8 indicates that the ultimate strength for the screw material is equivalent to 800 MPa and the yield strength is 0.8×800=640 MPa. The mechanical properties for standard metric bolts and nuts are available in DIN EN ISO 898-1 [8] for externally threaded fasteners (bolts and studs) and DIN EN ISO 898-2 [9] for internal threaded fasteners (nuts and threaded blocks).

Class 8.8 or 10.9 bolts are commonly used. Since Class 10 bolts/nuts are stronger, they can stand a higher tightening torque, hence providing higher clamp load in the joint. Class 12.9 and higher increase the risk of bolt failure due to hydrogen embrittlement and stress corrosion cracking; therefore, they are not commonly used. It is always recommended to use the same strength class for both the bolt and the mating nut; otherwise, the design should be based on the lower-class fastener.

16.2.2 Bolt Size and Thread Pitch

The bolt size refers to the bolt outer diameter size. For metric threads, the bolt or nut size is identified by the number after the letter M, for example, M8 refers to a metric eight-millimeter bolt/nut. The selection of the proper bolt size depends mainly on the clamp load requirements, external load requirements, and the geometrical restrictions on the maximum bolt size to be used.

It is important to make sure that the clamp load retained in the joint after tightening the joint is sufficient to prevent joint separation and joint members sliding when the joint is subjected to external service loads.

Once the required minimum clamp load is identified (based on joint analysis), a safety margin should be added to account for potential clamp load loss due to embedment, relaxation, and losses due to external loads.

If torque control tightening is to be used, then the variation in clamp load due to friction variations, and final tightening torque variations due to tool accuracy must be considered.

The required minimum bolt diameter can be calculated using the yield strength of the selected bolt material and the calculated clamp load using the equation below:

$$d = 0.64\ P + 1{,}000\sqrt{\frac{9\ F}{\pi\ \sigma y}} \tag{16.1}$$

where

- d is the bolt major diameter in mm
- P is the thread pitch in mm
- F is the minimum required clamp force in N
- σ_y is yield strength of bolt material in Pa (640×10^6 Pa for class 8.8, and 940×10^6 Pa for class 10.9).

The above equation takes into consideration tool tolerance and variation in coefficient of friction (0.09–0.15).

Once the minimum bolt diameter is calculated, the next larger standard bolt size is selected.

If the bolt is to be tightened beyond yield using yield control or angle control tightening, then the torsional stress component at the highest coefficient of friction must be considered. Thus, the equation for selecting the bolt size becomes:

$$d = 0.64\ P + 1{,}000\sqrt{\frac{5.33\ F}{\pi\ \sigma y}} \tag{16.2}$$

In automotive, both coarse and fine series threads are used; however, the coarse thread is most common. The pitch information is normally identified after the fastener size (e.g. M8x1); however, it is normally not listed in the designation when coarse thread pitch is used, see ISO 965-1 [5]. A coarse

TABLE 16.1
Coarse Pitch Values for Common Metric Coarse Thread Bolts

Bolt Size	**M5**	**M6**	**M8**	**M10**	**M12**	**M14**
Pitch [mm]	0.8	1.0	1.25	1.5	1.75	2.0

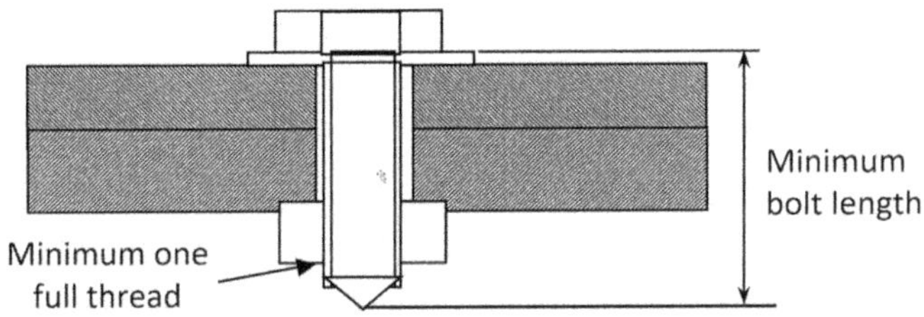

FIGURE 16.2 Minimum bolt length.

thread pitch bolt is less prone to cross threading and has shorter assembly time compared to the same size fine thread pitch bolt; however, it has a smaller cross section, hence it can stand a slightly lower torque/clamp load (about 7% less).

Table 16.1 provides the basic coarse pitch values for the most common metric course thread bolt sizes used in automotive:

16.2.3 Bolt Length

The bolt length is measured from the bearing surface under the bolt head to the tip of the bolt, including the leading tip except for bolts with a pilot point flat (PF) or with truncated cone (PC), where the leading tip is not included in the length, see DIN EN ISO 4753 [10].

A metric bolt is used either with a metric nut or with a threaded hole in a metal block. When the bolt is used with a metric nut, it is required to have at least one full thread protrude beyond the end of the nut to guarantee full thread engagement and proper load distribution on the threads.

The proper bolt length should be greater than the sum of the washer thickness (for bolts with captured washer), the joint members thickness, the nut thickness, one additional thread, and bolt lead feature length as shown in Figure 16.2.

When the bolt is used with a threaded block, the minimum thread engagement depends on the strength of the block material. The objective is to have sufficient thread engagement to support the maximum clamp load expected during the final assembly process. It is also desired that in case of over-tightening, the failure occurs in the bolt shank and not in stripping the internal threads in the block. This can be achieved by having sufficient thread engagement in the block and proper bolt strength selection. A general rule for aluminum blocks is to have a thread engagement of two times the bolt nominal diameter, and for steel one and a half the bolt diameter. The minimum bolt length in this case should be equal to the sum of the washer thickness, joint members thickness, minimum block thread engagement required, and the length of the thread lead feature, as shown in Figure 16.3.

The length of some lead types is not included in the declared nominal bolt length specified in the fastener designation. Therefore, special attention must be paid for having sufficient clearance for the screw lead to avoid collision. For more details, the lead dimensions and which type is included in the bolt length are given in DIN EN ISO 4753 [10].

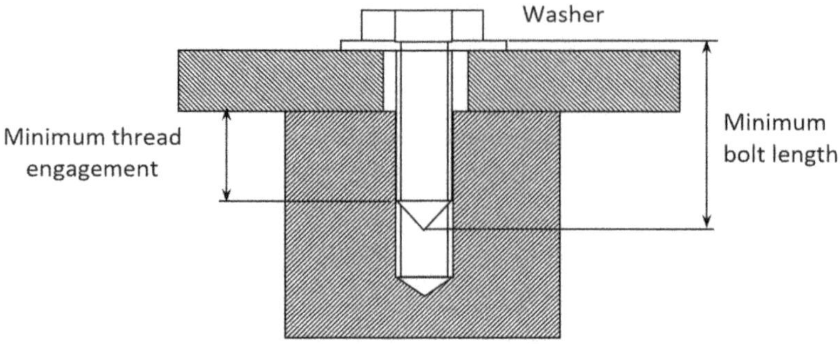

FIGURE 16.3 Minimum bolt length for blind hole joint.

TABLE 16.2
Recommended Minimum Salt Spray Resistance for Fasteners Coatings Based on Application Location in Automotive

Area	Minimum Requirement NSS (Hr)
Body in white	24
Interior	240
Under the hood	480
Exterior	720

16.2.4 Fastener Coating

One of the main objectives of fastener coatings is to provide sufficient protection against corrosion. In general, a standard salt spray test is used to measure the number of hours of protection before rust shows up when the coated fastener is continuously subjected to the salt spray cycles as described in ISO 9227 [11].

Table 16.2 provides the recommended minimum salt spray hours the fastener coating needs to provide based on the area the bolt is used in the car.

Some coatings or even washer materials are used to prevent galvanic corrosion concerns when composite or lightweight alloys are joined. There are also coatings used for decorative and appearance purposes, while other coatings are specifically used to provide better electrical conductivity when grounding through the fastener is required.

In addition to corrosion protection or other coating functions, the coating process is important for controlling the thread and bearing friction coefficient, which has a direct effect on the amount of clamp load obtained during the tightening. It is important that the friction coefficient range used in the calculations of clamp load obtained is specified for the final coated fastener. The range 0.09–0.15 is commonly used.

16.2.5 Head Diameter

The bolt head diameter must be large enough to provide enough bearing area that prevents high stress concentration that may damage the joint in contact with the bolt head. Otherwise, joint damage and clamp load loss may occur. When a large clearance hole is required in the joint, due to large variation in part position or parts tolerances stack up, the use of a large strong thick washer must be considered in order to prevent joint damage and bending the washer into the hole during the tightening. It is desired that the bolt head or washer cover the whole clearance hole regardless of bolt location within the clearance hole, especially if water intrusion is a concern depending on the joint location.

The bearing diameter must be sufficient to take the maximum clamp load expected after tightening the bolt (taking into account highest possible torque with lowest allowed friction coefficients).

The maximum surface pressure must not exceed the yield strength of the joint or fastener material. This is very important when metal spacers are used in plastic molded parts to prevent crushing the plastic part, where the washer material strength and thickness need to be designed properly to carry the maximum expected clamp load.

16.2.6 Washer Type, Material, Diameter, and Thickness

There are several sizes and shapes for captured washers. Washers are normally used for the following reasons:

- Enlarge bearing area, for example, for a joint with a large clearance hole
- To prevent joint surface damage due to bolt head friction against the joint during the tightening
- To prevent part rotation during the tightening due to bearing friction between the bolt and the joint
- To obtain more consistent clamp load due to more consistent friction between bolt and washer
- To reduce embedment by distributing the load under the bolt when joining softer material (e.g., plastic parts)
- To prevent galvanic corrosion when joining aluminum or magnesium parts
- Bevel washers can be used to help reduce/eliminate self-loosening, especially in very hard joints.

The washer size and thickness are selected based on the purpose for which the washer is used and the space available for the washer.

It is important to have the proper washer hardness; otherwise, the washer itself may deform during or after the tightening. Washers used in screw, and washer assemblies are recommended to have minimum 200 HV hardness when used with class 8.8 bolts, and minimum 300 HV for higher property class screws.

Special aluminum washers are used in some cases to prevent galvanic corrosion for tightening of magnesium components. Careful load calculations should be performed to prevent damaging the washers during the tightening.

16.2.7 Bolt Leading Tip

Bolt leading tips and lead specifications are provided in DIN EN ISO 4753 [10]. The objective of having a leading tip for bolts is to help prevent cross threading by aligning the bolt axis to the nut axis before the threads are engaged or simply to help find the hole during the blind hole assembly. If hand start is specified for the process, then only a chamfer tip is sufficient.

The most preferred lead to help prevent cross threading is the pilot point with truncated cone (PC). In addition to the leading tip, some fasteners have special shape threads at the leading tip to help reduce or prevent cross threading, such as short and long MAT thread lead.

16.2.8 Thread Locking

Thread locking is usually used where significant clamp load loss is expected due to self-loosening, embedment, or joint creep and relaxation. This happens often in joints with high external alternating loads (e.g., suspension dynamic joints) or with joints that contain soft components that can experience creep and relax after assembly (e.g., joints with a rubber gasket or a soft plastic component).

There are several methods of self-locking on metric thread bolts and nuts. The most common methods are locking coatings such as microencapsulated adhesive coatings (DIN 267-27) [12] and thermoplastic coatings (DIN 267-28) [13]. Another common locking feature is mechanical locking that normally uses an altered thread profile to create mechanical interference with the mating threads and as a result provide a locking function (e.g., powerlock screws). A good alternative option is to use a nut with self-locking feature such as crimped nuts or nuts with an embedded locking nylon ring. In all these cases, a rundown prevailing torque is expected during the initial rundown of the bolt/nut. This rundown varies based on the locking method used. In addition, the clamp load calculations when thread locking is used are affected and may not be easy to obtain without experimental testing to establish the friction coefficient range expected for the fastener with the locking feature.

16.2.9 Thread Sealing

Thread sealing can be achieved by applying a special teflon coating to the bolt threads to prevent water intrusion through the gaps between the threads. Locking coating applied to the threads can also serve as a thread sealant if applied on the whole thread circumference in the thread engagement area.

16.2.10 Paint Scraping Feature

Screws with paint scraping features are used where the nut threads get paint during the body painting process. The effectiveness of the paint scraping feature must be evaluated with worst case paint (i.e., vehicle body can be painted up to three times), where the rundown should not exceed 50% of the final released torque for the application. Nuts with paint scraping feature are also available for paint-coated studs.

16.2.11 Bolt Drive

The bolt drive refers to the shape on the head of the bolt that is used to drive the bolt during the tightening process. It can be in the shape of an external shape of the head (external drive), e.g. hex or external torx, or an internal recess in the head (internal drive), e.g. internal hex or torx. External head drives offer lower risk of slipping off the head of the fastener during the tightening process. External torx drives work well when the final torque is less than 60 Nm. For higher torque values, the socket or bit tends to come off (cam out) and slip off the bolt, unless the operator applies sufficient force to keep the driving socket engaged. In these cases, hex head drive is preferred.

Internal drives are good when lower head profile is needed due to space restriction or to reduce the weight in general; however, they have much higher risk of slip off and possibly damage the fastener.

The bolt head height is affected by the type of drive the bolt has. Internal torx drive normally provides the lowest head height. Availability of sufficient clearance for the bolt head and the final torque to use, and clearance for tool access might be the key factors for the selection of the bolt drive shape. It is important to make sure that the bolt drive is strong enough to stand the required final tightening torque and that it cannot get easily damaged during joint assembly or disassembly.

16.2.12 Tool Access

It is important to make sure that the fastening tool used for the process has easy, straight, and clear access to the bolt head based on the required assembly sequence. Even though wobble extensions are available in the market, they cause significant variation in the torque delivered to the fastener; therefore, they are not recommended to use.

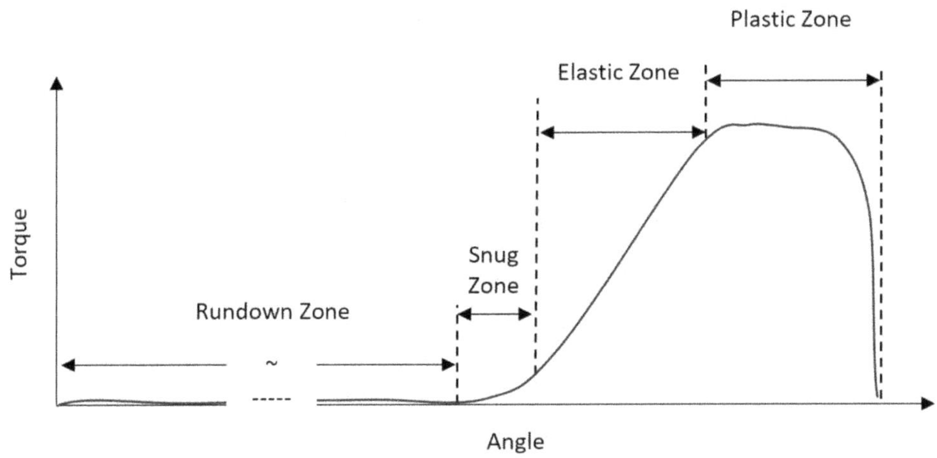

FIGURE 16.4 Torque-angle curve for a torque to failure tightening.

16.3 TIGHTENING STRATEGIES

Normally, the goal of all tightening strategies is to develop the required level of clamp load to keep the joint members together and prevent joint separation, slippage, squeak, rattle, self-loosening, leakage, and other types of failure. There are many tightening strategies employed over a wide range of bolted joint applications. The most common tightening strategy used is torque control, followed by angle control, then yield control.

16.3.1 Torque Control

Torque control is the most practical and commonly used tightening strategy as it requires the least expensive tooling and skill. There is a direct relationship between the tightening torque and clamp load which is known in the literature as the torque tension relationship [14]. When using torque control, the tool is set to stop the tightening when the configured target torque is reached within an allowed tolerance. The selected target torque and tolerances should always provide sufficient clamp load in the joint and stay below the load that may cause permanent deformation in the joint of the bolt. Due to tolerance required to address tool accuracy, and the wide variation of the friction coefficient, the clamp load obtained at the end of the tightening may vary by ±35%. Figure 16.4 shows a typical torque-angle curve for a torque to failure curve. For torque control tightening, the target torque including torque tolerance limits should be within the elastic zone area.

16.3.2 Angle Control

For joints where a higher clamp load level with less variation is needed, angle control tightening can be used. In angle control strategy, the bolt is tightened to a threshold torque followed by an additional specific angle taking the bolt slightly into the plastic zone shown in Figure 16.5. Since the clamp load does not increase much after yield, the result is a more consistent clamp load at the end of the tightening regardless of the friction coefficient variations. This provides a smaller variation in the resultant clamp load of about ±7%. Angle control tightening strategy maximizes the utilization of the fastener strength, which allows using a smaller fastener size to obtain the same clamp load. This can reduce the overall product weight and cost. Also, it can eliminate the need for expensive product design changes to fit a larger size fastener, when higher clamp load is required with space restriction. However, one of the disadvantages is that since the bolt is permanently stretched during

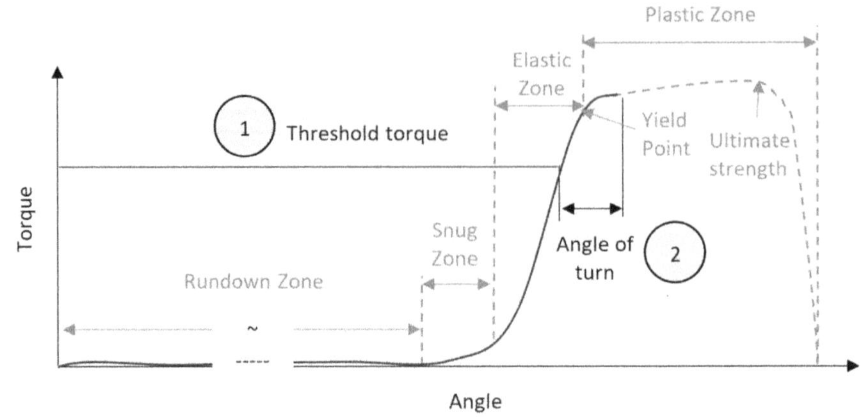

FIGURE 16.5 Angle control tightening strategy.

the tightening it cannot be reused multiple times, which means the bolt will need to be replaced when that connection is disassembled for whatever reason. In addition, angle control strategy complicates the assembly process and requires more expensive fastening systems.

16.3.3 Yield Control

Yield control is a special tightening strategy used for very few applications, where a very consistent high clamp load is required (e.g., cylinder head assembly). Normally, permanent bolt elongation limits are provided for yield control spec. In yield control tightening strategy, the torque to angle ratio (torque gradient) is monitored during the tightening, and when this gradient changes by a certain percentage (indicating the yield point), the fastening stops. In contrast to angle control, when yield control is used, the tightening stops at the beginning of the yield zone, which results in less permanent deformation in the bolt at the end of the tightening. Figure 16.5 shows the yield point and the change of torque angle slope that triggers the tool to stop. One of the main challenges is making sure that this change in torque gradient is due to bolt yielding and not something else. For this reason, final torque monitoring limits are used to help ensure that the bolt is tightened correctly. The expensive tooling required for yield control strategy and the difficulty of calibrating these tools make this strategy less common.

16.4 DETERMINATION OF TIGHTENING SPEC FOR TORQUE CONTROL TIGHTENING STRATEGY

16.4.1 Machine Thread Fasteners

This section provides a guideline for developing torque control tightening specifications for machine thread fasteners along with the expected clamp load level in the joint, taking into consideration the torsional stress developed in the fastener during the tightening, the friction coefficient variation, and the bolt head size.

Within the elastic zone shown in Figure 16.4, when the torque increases the clamp load in the joint increases proportionally. The relationship between the applied torque and the corresponding clamp load generated is expressed in the torque-tension relationship equation [14]:

$$T = \left(\frac{P}{2\pi} + \frac{\mu_t r_t}{\cos\alpha} + \mu_b r_b \right) F = K \; d \; F \tag{16.3}$$

The torque at which the fastener yield occurs (T_y) and the corresponding clamp load at yield (F_y) can be calculated using Eqs. (16.4) and (16.5) [15]

$$F_y = \frac{S_y}{\sqrt{\left(\frac{1}{A_t}\right)^2 + 3\left[\frac{16(0.15P + 1.16r_t\mu_t)}{\pi d_{\min}^{\;3}}\right]^2}} \tag{16.4}$$

$$T_y = \frac{\left(\frac{P}{2\pi} + \frac{\mu_t r_t}{\cos\alpha} + \mu_b r_b\right) d\ S_y}{\sqrt{\left(\frac{1}{A_t}\right)^2 + 3\left[\frac{16(0.15P + 1.16r_t\mu_t)}{\pi d_{\min}^{\;3}}\right]^2}} \tag{16.5}$$

where

T is the applied torque
P is the thread pitch
μ_t is the thread coefficient of friction
μ_b is the under-head coefficient of friction
α is half of thread profile angle
r_t is the effective contact radius of threads
r_b is the effective bearing radius under the bolt/nut head
F is the bolt tension (clamp force)
F_y is clamp force at yield
T_y is torque at yield
S_y is minimum yield strength of bolt material
K is the nut factor
d is bolt major diameter
A_t is the equivalent stress area of the bolt
$r_{\min}$ is the bolt threads minor radius
$d_{\min}$ is the bolt threads minor diameter.

All the dimensional features of the bolt, the friction coefficient limits, and the yield strength of the bolt material should be available from the bolt's drawing and applicable specifications. With that, the minimum torque, at which the fastener starts to yield, can be calculated using the lowest friction specification and the minimum yield strength of the bolt material. Since the desired tightening torque needs to be as high as possible without getting the bolt to yield, 90% of the calculated minimum yield torque would be recommended for the tightening torque to use.

Typically, after selecting the target tightening torque, the acceptable tolerance limit must be selected. The tolerance limits are necessary for the selection of the proper fastening system. Most fastening tools are capable of passing calibration requirements with ±10% tolerance limit.

Once the torque tolerance limits are defined, the minimum clamp load expected (F_{Min}) at the selected torque can be calculated from Eq. (16.1) using the highest friction coefficient limit and the lower end of the torque tolerance limit.

16.4.2 Fasteners with Self-Locking Features

Most locking features increase the rundown torque; therefore, the following should be noted when using fasteners with self-locking features:

- The torque value can be calculated as described in Section 4.1; however, the corresponding clamp load value is expected to be lower. This is because a portion of the input torque goes to overcome the resistance caused by the locking feature. It is common to have 20%–30% less clamp load when self-locking fasteners are used.

- The fastener specifications should be checked to obtain the maximum allowed rundown torque for the self-locking fastener; this information can help determine the amount of clamp load that can be lost due to the locking feature. The percentage of clamp load lost can be equivalent to the percentage of the maximum rundown torque compared to the final tightening torque used.

16.4.3 Self-Tapping and Thread Rolling Fasteners

Since self-tapping and thread rolling screws are very sensitive to many variables, such as thread type, hole size, coating, joint surface, and other factors, the appropriate tightening torque must be determined experimentally based on test data for the actual joint. A separate chapter is dedicated for self-tapping and thread rolling screw applications.

16.5 DETERMINATION OF TIGHTENING SPEC FOR ANGLE CONTROL TIGHTENING STRATEGY

Angle control or torque plus angle tightening strategy is often used when higher and more consistent clamp load in the joint is desired. Using angle control eliminates some of the uncertainty in clamp load caused by variations in friction that are inherent in torque control.

It is very common to use angle control tightening to take the fastener beyond its yield limit ensuring the maximum utilization of the strength of the fastener. Angle control tightening requires that the bolt be tightened to a specified threshold torque and then turned an additional angle. In this section, a method for determining the threshold and angle of turn parameters is provided.

16.5.1 Threshold Torque

The threshold torque is the torque from which the final angle count starts. It must be high enough to bring the joint members together and close any gap in the joint without taking the bolt to yield. The recommended highest torque that will not cause yield is the torque value provided in the last section for torque control tightening; therefore, the threshold torque to use for the angle control tightening can be obtained the same way the torque recommended for torque control strategy explained in Section 16.4.

16.5.2 Angle of Turn

When a bolt is run down, it advances by one pitch for every 360 degrees of rotation, however, once snug is reached, the bolt starts to stretch and the joint starts to compress. In this case, the sum of the stretch in the bolt and the compression in the joint after one full turn should be equivalent to one thread pitch. The relationship between the angle of turn (θ) and the clamp force (F) within the elastic region as function of bolt stiffness (K_b) and joint stiffness (K_c) is given as [14]:

$$\theta\ (\text{deg}) = \frac{360\ F}{P}\left(\frac{1}{K_b} + \frac{1}{K_c}\right) \qquad (16.6)$$

where the measurement of the angle of turn starts from snug. The angle at which yield starts (θ_y) can be calculated using the force at yield. From Eq. (16.2), the force at yield is affected by the friction coefficient and the yield strength of the fastener. This means that the minimum angle at which yield occurs (θ_{yMin}) can be calculated using the minimum force at which yield starts (F_{yMin}). This corresponds to a bolt with the minimum yield strength (S_{yMin}) and highest friction coefficient, while the maximum angle at which yield starts (θ_{yMax}) can be calculated using the maximum yield force at which yield starts (F_{yMax}). This corresponds to a bolt with maximum yield strength (S_{yMax}) and lowest friction coefficient.

Using Eq. (16.4) for both forces conditions and then dividing the two equations yields:

$$\frac{\theta_{\text{yMin}}}{\theta_{\text{yMax}}} = \frac{F_{\text{yMin}}}{F_{\text{yMax}}} \tag{16.7}$$

Since force measurement is not easy, we can use Eq. (16.1) to substitute the force by the corresponding torque:

$$\frac{\theta_{\text{yMin}}}{\theta_{\text{yMax}}} = \frac{T_{\text{yFMin}}\ K_{\text{Min}}}{T_{\text{yFMax}}\ K_{\text{Max}}} \tag{16.8}$$

where T_{yFMin} and T_{yFMax} can be calculated using Eq. (16.3), and K_{Min} and K_{Max} can be obtained from Eq. (16.1).

Similarly, the relationship between any angle at yield (θ_{yi}) for a sample joint compared to θ_{yMin} and θ_{yMax} is provided in Eqs. (16.7) and (16.8), respectively:

$$\frac{\theta_{\text{yMin}}}{\theta_{\text{yi}}} = \frac{T_{\text{yFMin}}\ K_{\text{i}}}{T_{\text{yi}}\ K_{\text{Max}}} \tag{16.8}$$

$$\frac{\theta_{\text{yMax}}}{\theta_{\text{yi}}} = \frac{T_{\text{yFMax}}\ K_{\text{i}}}{T_{\text{yi}}\ K_{\text{Min}}} \tag{16.9}$$

where K_i is the K factor for sample joint with θ_{yi} angle at yield. When using angle control tightening strategy, the fastener is tightened to a specific torque then an additional target angle (θ_t) is applied. The angle from snug to threshold is the threshold angle (θ_{th}), and the sum of the total angles is total angle applied from snug to the end of tightening (θ_s):

$$\theta_s = \theta_{\text{th}} + \theta_t \tag{16.10}$$

There are three different scenarios:

1. The applied torque+angle does not take the fastener to yield
   This occurs when $\theta_s < \theta_{\text{yMin}}$
2. The applied torque+angle does take the fastener to yield
   This occurs when $\theta_s > \theta_{\text{yMax}}$
3. The applied torque+angle may or may not take the fastener to yield (depending on the actual friction coefficient and material yield strength for the bolt in the joint)
   This occurs when $\theta_{\text{yMin}} < \theta_s < \theta_{\text{yMax}}$

The angle from snug to threshold torque varies a lot due to big variation in bolt and joint stiffness for the different applications and the unknown actual friction coefficient, however, it is possible to use the torque-angle curve for samples of the actual joint to develop a criterion for the angle to use for a torque plus angle tightening that will always take the fastener to yield, or will keep the fastener within the elastic range.

The actual tightening curve for sample joints will fall somewhere in the area between the two extreme curves. Since the actual friction coefficient values are not known, an assumption can be made to conservatively find the angle limits that will take the fastener to yield or to keep the fastener within the elastic range. The lowest allowed friction coefficient values μ_{tMin} & μ_{bMin} can be used to calculate K_i and use Eq. (16.7) to calculate the conservative θ_{yMax}, similarly, the highest allowed friction coefficient values μ_{tMax} & μ_{bMax} can be used in calculating K_i, and then use Eq. (16.8) in to calculate the conservative θ_{yMin}. These calculated conservative limits can then be used to find the tightening spec for the angle control tightening.

$$\theta_{yMin\ Conservative} = \frac{T_{FyMin}\ K_{Min}}{T_{yi}\ K_{Max}}\ \theta_{yi\ from\ sample\ curve} \quad (16.11)$$

$$\theta_{yMax\ Conservative} = \frac{T_{FyMax}\ K_{Max}}{T_{yi}\ K_{Min}}\ \theta_{yi\ from\ sample\ curve} \quad (16.12)$$

By introducing the ratio of the minimum K factor to the Maximum K factor as R_K, and using Eqs. (16.9), (16.10), and (16.11), the target angle to apply after reaching the threshold torque (θ_t) has two scenarios:

1. If it is desired to never take the fastener to yield

$$\theta_t <= R_K \frac{T_{yFMin}}{T_{yi}}\ \theta_{y\ from\ sample\ curve} - \theta_{th\ from\ sample\ curve} \quad (16.13)$$

   In this case, the minimum clamp load obtained is not known, but it is higher than the clamp load obtained at the threshold and always lower than the minimum clamp load at yield.

   If the calculated angle value is too small or negative, then the threshold torque needs to be reduced.

2. If it is desired that the applied torque+angle is guaranteed to take the fastener to yield

$$\theta_t >= \frac{1}{R_K} \frac{T_{yFMax}}{T_{yi}}\ \theta_{y\ from\ sample\ curve} - \theta_{th\ from\ sample\ curve} \quad (16.14)$$

It is recommended to round the calculated target angle value to the nearest multiple of 45° such as 45°, 90°, 135°, 180°, and 270°. This makes it easier for service later on where advanced digital tools may not be available.

Once the final target angle is found, it is important to make sure that this angle does not take the fastener too close to the ultimate point (this can happen normally with very hard joints). It is recommended that the angle from threshold to ultimate torque point is three times or more than the selected target angle of turn. For this evaluation torque to failure curves for the sample joints are required. For this reason, angle control tightening is not recommended for very hard joints (e.g., steel joint with short grip length) because, for these joints, the yield point is too close to the ultimate strength point after which the bolt fails.

For very soft joints (e.g. joints with soft gaskets), it is also not recommended to use angle control tightening in general. Finally, the method proposed here assumes that the bolt is the weakest component that will start to yield in the joint.

16.5.3 Torque Monitoring Limits and Selection of Tool Size

It is common to use DC electric tools, which are capable of measuring torque and angle, for angle control tightening processes. Torque monitoring limits are used to evaluate the torque reached at the end of the tightening. They are used to ensure proper installation of parts and detect any quality issues with the components being assembled. The torque monitoring limits can be calculated using Eq. (16.3). For the lower torque monitoring limit, $T_{y\ min}$, the lowest allowable coefficient of friction must be used. For the upper monitoring limit, the highest coefficient of friction must be used to determine the maximum torque at which yield may take place, $T_{y\,max}$. Since the torque can increase during tightening after the fastener reaches its yield limit, the upper monitoring torque limit can be a little higher.

The upper monitoring limit is needed for the selection of the proper tool size to use for the process.

Once the threshold torque and target angle are defined, a tolerance for both needs to be assigned. It is desired to have a tight tolerance on the threshold torque to have a precise starting point for the measurement of the target final angle. The tolerances for the angle limits should also be as precise as possible within a defined standard tolerance based on the available fastening system capabilities.

16.6 TORQUE AND CLAMP FORCE DESIGN TABLES

The use of the equations explained in Sections 16.4 and 16.5 to develop tightening specifications is time-consuming and makes it difficult for the engineering team to calculate the torque or torque plus angle specifications needed for an application. For this reason, standardization of bolt sizes, bolts geometry, allowed friction coefficient range, fasteners material class, torque tolerance used, etc. allows the creation of design tables that can easily provide the required information to design the joint, select the fastener size, and obtain the proper tightening specifications for both torque control and angle control applications.

There are several standards available in the industry, the most common are standards for metric bolts and nuts, such as ISO 4014 for hexagon head metric bolts, and DIN EN 1665 for Hexagon head bolts with flange. Once bolts standards are adopted, the geometrical specifications become available to use in Eqs. (16.1)–(16.13). Regarding the friction coefficients, a specification range is normally called out in the fastener coating specifications for the fastener used. Adopting a common range for all fasteners used can help standardize the calculations and design method. A common range used for both bearing friction and thread friction coefficients is 0.09–0.15.

Finally reducing the number of bolt/nut classes used is needed. The most common class used is class 10.9, followed by class 8.8.

Once all variables in the Eqs. (16.2) and (16.3) are defined, a design table can be developed and used. Here are the most important parameters to include in the design table for each common bolt size.

1. *Recommended tightening torque (T)*: this is the torque value recommended to use for torque control processes.
2. *Minimum clamp load at recommended torque (F_{Min})*: this value can be used for the joint design evaluation when torque control tightening is used. It is calculated using Eq. (16.1), with the lowest target torque limit and the highest allowed friction coefficient.
3. *Minimum torque at which yield may start (T_{yMin})*: this value is the minimum torque expected when angle control tightening beyond yield is used. It can be used for the minimum torque limit to set on the DC tool to ensure the quality of the fastening process. It is calculated with minimum friction and minimum yield strength values.
4. *Maximum torque at which yield may start (T_{yMax})*: it is useful for setting the maximum torque limit for tightening beyond yield using angle control strategy. It is also used for the tightening tool selection and equipment strength design. It is calculated using Eq. (16.3), with maximum friction and maximum yield strength values.
5. *Torque at maximum yield clamp load (T_{yFyMax})*: it is useful for calculating the target angle needed to guarantee taking the bolt to yield when angle control strategy is used (See Eq. 16.12). It is calculated using Eq. (16.3) with minimum friction and maximum yield strength.
6. *Torque at minimum yield clamp load (T_{yFyMin})*: it is useful for calculating the target angle needed to guarantee not taking the bolt to yield when angle control strategy without yield

TABLE 16.3
Friction and Strength Values to be Used for Design Table Calculations

Calculated Value	Friction Coefficient	Yield Strength
F_{Min}	Maximum	–
T_{yMax}	Maximum	Maximum
T_{yMin}	Minimum	Minimum
T_{yFMax}	Minimum	Maximum
T_{yFMin}	Maximum	Minimum
F_{yMax}	Minimum	Maximum
F_{yMin}	Maximum	Minimum

is desired (See Eq. 16.13). It is calculated using Eq. (16.3) with maximum friction and minimum yield strength.

7. *Minimum clamp load that can be obtained at yield* (F_{yMin}): this value can be used for the joint design evaluation when tightening beyond yield using angle control tightening. It is calculated using Eq. (16.2), with minimum friction and minimum yield strength.
8. *Maximum clamp load that may be obtained at yield* ($F_{y\,Max}$): this value can be used to make sure joint members can stand the maximum clamp load possible when beyond yield angle control tightening is used.
9. *The ratio of the minimum K factor to the maximum K factor* (R_K): this value is used for calculating the minimum angle to yield (θ_{yMin}) and the maximum angle to yield (θ_{yMax}) which are used to determine the target angle for angle control strategy. R_K is calculated using Eq. (16.1) and the minimum and maximum friction coefficient limits.

The bolts used should have a defined friction range and a defined strength specifications range. In some cases, the maximum yield strength is not explicitly defined, but it can be calculated based on the hardness spec range. Table 16.3 shows the friction coefficient limit and yield strength limit to be used when calculating the parameters listed above.

Example Design Table for hex head bolts made to ISO 4014, class 8.8 and class 10.9, with friction coefficient range 0.09–0.15:

ISO 4014, provides the hex head dimensions, while ISO 965-2 and ISO 68-1 provide the thread dimensions. ISO 898-1 provides the minimum yield strength for class 8.8 and 10.9 bolts (640 and 940 MPa, respectively). The maximum yield strength (S_{yMax}) can be calculated based on the minimum yield strength (S_{yMin}), minimum hardness (HRC_{Min}), and maximum hardness (HRC_{Max}), which are available from ISO 898-1 for both classes 8.8 and 10.9.

$$S_{yMax} = HRC_{Max} / HRC_{Min} \times S_{yMin}$$

For example, hardness range for class 8.8 is 22 to 32 HRC, minimum yield strength is 640 MPa, hence, the maximum yield strength is $S_{yMax} = 640 \times 32/22 = 931$ MPa.

Similarly, for class 10.9 $S_{yMax} = 39/32 * 940 = 1{,}146$ MPa.

With a friction coefficient range of 0.09–0.15, the parameters are calculated using Eqs. (16.2) and (16.3) with the values listed in Table 15.4. Based on this, Table 15.4 can be developed.

It must be noted that the calculations are based on the assumption that the fastener is the weakest component in the joint; meaning that the fastener is the first component to yield during the tightening process. This is true in most cases but should be confirmed based on the specific joint under consideration before using the values in Table 16.5. Similar tables for other standard machine thread

TABLE 16.4
Friction coefficient and yield strength values used in calculations for Table 16.5

Calculated Value	Bolt Class	Friction Coefficient	Yield Strength
F_{Min} (at $0.9\times T$)	8.8	0.15	---
	10.9	0.15	---
T_{yMax}	8.8	0.15	931
	10.9	0.15	1,146
T_{yMin}	8.8	0.09	640
	10.9	0.09	940
T_{yFMax}	8.8	0.09	931
	10.9	0.09	1,146
T_{yFMin}	8.8	0.15	640
	10.9	0.15	940
F_{yMax}	8.8	0.09	931
	10.9	0.09	1,146
F_{yMin}	8.8	0.15	640
	10.9	0.15	940

fasteners can be developed based on the geometrical configurations and friction range requirements for the fasteners uses.

16.6.1 Selecting Bolt Size Using Design Tables

Using the design table makes it very easy to find the bolt size to use based on the minimum clamp force required. From the proper design table, simply select the size that has a higher clamp force value in the F_{Min} column if torque control tightening is to be used. If beyond yield angle control tightening is to be used, then use the column F_{yM}.

16.6.2 Develop Tightening Spec Using Design Tables

For torque control, the target torque and tolerance are already given in the table in column T (±10%).

For Angle control, finding the proper tightening spec is done using the following steps:

1. Select threshold torque to use from the table based on the fastener size and class from the T (±10%) column, and assign tolerance based on standard tooling accuracy.
2. Collect sample torque-angle curves for few samples of the joint tightened to failure
3. Measure the angle from snug to threshold torque for each sample curves and calculate average angle (θ_{th}).
4. Measure the angle from snug to yield from sample curves and calculate average (θ_{yi}).
5. Measure the angle from threshold to ultimate torque for all samples
6. Get R_K, T_{FyMin}, and T_{FyMax} from the design table
7. Use Eqs. (16.12) or (16.13) to calculate the minimum angle to yield or the maximum angle to yield. The equation to use depends on whether it is desired to tighten the fastener beyond yield or only within the elastic region. In most cases, the intention to take the fastener to yield when angle control is used in order to maximize the clamp load.
8. If tightening beyond yield is intended, find the closest larger angle that is multiple of 45° and use it for the angle spec. If tightening below yield is intended, find the closest smaller angle that is multiple of 45° and use it for the angle spec.

TABLE 16.5
Recommended Torque, Min/Max Yield Torque and Yield Clamp Load for ISO Hex Head Bolts According to ISO 4014, with Friction Coefficient Range 0.09–0.15

Class	Fastener Size	T (±10%) [Nm]	F_{Min} [KN]	T_{ymin} [Nm]	T_{ymax} [Nm]	T_{FyMin} [Nm]	T_{FyMax} [Nm]	F_{yMin} [KN]	F_{yMax} [KN]	R_K
8.8	M4x0,7	2.4	4.1	2.6	5.3	4.5	3.8	5.6	7.3	0.65
	M5x0,8	4.7	6.6	5.2	10.6	9.0	7.6	9.1	11.9	
	M6x1	8.0	9.3	8.9	18.1	15.3	13.0	12.9	16.7	
	M8x1,25	19.4	17.0	21.6	43.9	36.9	31.3	23.4	30.6	
	M10x1,5	38.3	27.0	42.5	86.8	72.8	61.8	37.1	48.6	
	M12x1,75	66.5	39.4	73.9	151.2	126.8	107.5	53.9	70.7	
	M14x2	105.7	53.8	117.5	240.7	201.3	170.9	73.6	96.6	
10.9	M4x0,7	3.4	5.9	3.8	6.5	6.5	4.7	8.2	8.9	0.65
	M5x0,8	6.9	9.7	7.7	13.0	13.1	9.3	13.3	14.5	
	M6x1	11.8	13.6	13.1	22.2	22.4	15.9	18.8	20.5	
	M8x1,25	28.4	24.9	31.5	53.8	54.0	38.4	34.3	37.5	
	M10x1,5	55.9	39.6	62.2	106.3	106.5	75.8	54.3	59.5	
	M12x1,75	100.8	59.6	112.0	190.7	193.7	136.5	82.4	89.7	
	M14x2	154.7	78.7	171.8	295.0	294.2	209.4	107.5	118.3	

9. Compare the calculated value to the angle from threshold to ultimate torque. It is recommended that the target angle value be less than one-third of the smallest measured angle from threshold to ultimate torque.

16.6.3 Standardizing Torque Specification

It is common to have different standards for head configurations for the same thread size, such as flange hex head, flange external hexlobular head, hex head without flange, etc. It is possible to create a design table for each type, however, the end results will be close to each other with minor differences in the table values. In this case, it would be desirable to use one table that covers all types, or at least share the same recommended torque for all types of the same bolt size and class. In most cases, adopting the values for standard hex head, which normally have the smallest bearing diameter, works for all same size and same class fasteners. This helps reduce the number of torque values used and allows sharing the same tool for the same size and class fasteners regardless of the head drive configuration. This simplifies tool selection, tool setup, tool calibration, assembly process balancing, and rework processes.

REFERENCES

[1] DIN ISO 68-1 ISO general purpose screw threads – Basic profile – Part 1: Metric screw threads (ISO 68-1: 1998).
[2] DIN ISO 261 ISO general purpose metric screw threads – General plan (ISO 261: 1998).
[3] DIN ISO 262 ISO general purpose metric screw threads – Selected sizes for screws, bolts and nuts (ISO 262: 1998).
[4] DIN ISO 724 ISO general purpose metric screw threads – Basic dimensions (ISO 724: 1993).
[5] DIN ISO 965-1:1998, ISO general purpose metric screw threads — Tolerances — Part 1: Principles and basic data.
[6] DIN ISO 965-2 ISO general purpose metric screw threads – Tolerances – Part 2: Limits of sizes for general purpose bolt and nut threads – Medium quality (ISO 965-2: 1998).
[7] DIN ISO 965-3 ISO general purpose metric screw threads – Tolerances – Part 3: Deviations for constructional screw threads (ISO 965-3: 1998).
[8] ISO 898-1:2013, Mechanical properties of fasteners made of carbon steel and alloy steel Part 1: Bolts, screws and studs with specified property classes, Coarse thread and fine pitch thread.
[9] ISO 898-2:2022(en), Fasteners — Mechanical properties of fasteners made of carbon steel and alloy steel — Part 2: Nuts with specified property classes.
[10] DIN EN ISO 4753, 2011, Fasteners Ends of parts with external ISO metric thread.
[11] ISO 9227- Corrosion tests in artificial atmospheres; Salt spray tests.
[12] DIN 267-27, Fasteners - Part 27: Steel screws, bolts and studs with adhesive coating - Technical delivery conditions.
[13] DIN 267-28 Fasteners - Part 28: Steel screws, bolts and studs with locking coating - Technical delivery conditions.
[14] Bickford, J. H., and Nassar, S. A., 1998, *Handbook of Bolts and Bolted Joints*, Marcel Dekker, New York, New York.
[15] Housari, B., Alkelani, A., and Yocum, G., 2012, Development of tightening specification for post yield angle control tightening strategy. *Proceedings of Pressure Vessels and Piping Conference (PVP2012)*-American Society of Mechanical Engineers, vol. 55010, pp. 123–129.

17 Design of Gasketed Joints

Bernard S. Nau

17.1 INTRODUCTION

17.1.1 Why Gasketed Joints Are Special?

The reasons for treating the design of gasketed joints separately from that of other bolted connections are quite fundamental:

1. A pressurized gasketed joint is statically indeterminate.
2. Control of leakage is a crucial requirement.

The static indeterminacy arises from the fact that the gasket, the two flanges, and the bolting behave like an assemblage of coupled springs. The gasket resembles a compression spring (nonlinear), the bolting a tension spring, and the flange a rotational spring. Also, the lines of action of the "springs" are offset relative to one another, with the result that moments as well as forces act on the flanges. In general, the two flanges can differ in geometry and material.

In consequence of the static indeterminacy, iteration is needed to calculate the equilibrium displacements of the joint.

The general objective of the design process is to establish, for given gasket characteristics, safe economical dimensions of the flanges and bolting while ensuring adequate control of leakage.

17.1.2 Scope of This Chapter

Gasketed joints come in many geometries based on a variety of design concepts. Although most joints are circular, as in piping joints, examples of other shapes are a rectangular joint between a sump and cover plate; the complex joint between the two halves of a split-casing pump, valve, or turbine; and the cylinder head joint of an internal combustion engine. The contained fluid may be liquid or gas, pressure may range from vacuum to thousands of bar (applied internally or externally), and temperature may range from cryogenic to 500°C–1,000°C or more. Operating conditions may be steady or cyclic or transient or a combination of these. Life and leak rate requirements also vary widely.

Here we concentrate on internally pressurized circular bolted flanged gasketed joints to illustrate approaches to the design of joints more generally.

17.1.3 General Design Considerations

Regardless of geometry and operating conditions, the design of a gasketed joint must address two primary concerns [1-15]:

1. Stress throughout the structure must always be within safe limits.
2. Gasket stress must be sufficient to keep the leak rate below an acceptable limit (see Chapter 7).

These requirements must be met for all load conditions. Steady-state load conditions for consideration at the design stage include

DOI: 10.1201/9781003501183-21

1. Initial assembly, when bolts have been tightened.
2. Hydrostatic test at, for example, 1.5×normal working pressure.
3. Normal working conditions (e.g., elevated temperature and pressure).

Depending on the application it may also be necessary to consider

External loads, axial or bending, e.g., pipework weight and thermal expansion.
Bolt-tightening inaccuracy, depending on the method used (Chapters 23-35).
Transient conditions such as heating up or cooling down.
Fatigue due to cyclic operation.
Corrosion, which reduces metal thickness, necessitating the incorporation of a corrosion allowance on affected dimensions.
Creep of flanges, bolting, and gasket—all can reduce gasket compression stress.

Many applications can be met by choosing a standard joint design; this approach is discussed in Section 17.3. When a standard joint cannot be used, the next possibility to consider is the use of a "code" joint as described in Section 17.5.

To provide an understanding of the complexities of the design and behavior of bolted gasketed flanged joints, we first consider in some detail the workings of gasketed joints. This is sometimes done geometrically, using a rather complex load–deflection diagram (e.g., in the German draft standard DIN V2505 [1]); an algebraic approach is more explicit and is adopted here.

17.2 FORCES, MOMENTS, AND INTERACTIONS

17.2.1 Simplified Model of a Joint

Here we focus on joints using taper-hub weld-neck flanges, which are commonly chosen for high-duty applications. To understand the interactions of the loads acting on a joint and the resulting displacements, we can view the joint as a system of "spring" elements. In this simplified approach, the gasket, bolting, and flanges are each treated as some form of spring (linear), an approach first adopted by Wesstrom and Bergh in 1951 [2].

The spring rates can be written down directly for the gasket and the bolting.

Compressive stiffness of the gasket:

$$k_g = E_g A_g / t_g \tag{17.1}$$

where

E_g=compression modulus of the gasket, unloading condition (psi, N/mm²)
A_g=an appropriate gasket contact area (in.², mm²)
t_g=an appropriate gasket thickness (in., mm)
k_g=compressive stiffness (lb/in., N/mm)

Axial stiffness of the bolting:

$$k_b = E_b A_b / l_b \tag{17.2}$$

where

E_b=Young's modulus of the bolts (psi, N/mm²)
A_b=effective cross-sectional area of the set of bolts (in.², mm²)
l_b=effective clamping length of the bolts (in., mm)
k_b=axial stiffness of the bolting (lbf/in., N/mm).

In reality, the calculation of the deflections of the flange can be complicated by the existence of several factors affecting axisymmetric deflection:

1. Ring-mode rotation of the flange cross section about its centroid without change of shape of the cross section
2. Radial bending of the flange proper
3. Axial bending of the hub
4. Bending of the bolts
5. Radial stiffness of the wall of the adjoining piping or vessel
6. Angular stiffness of the wall of the adjoining piping or vessel.

A simple approximation is to model only the first of these, although a rigorous analysis requires consideration of all modes. Bending of the flange is important when the radial width is large compared with axial thickness; its neglect leads to underestimation of deflections. Neglect of bending of the hub also causes an underestimate. However, neglecting bolt bending is conservative, as it resists flange rotation.

The connection to the pipe or vessel wall constrains flange deflection but also introduces extra loads, due to geometric discontinuity, which appear as a moment and radial shear force at the junction.

Considering one flange of the pair, the major forces acting are shown in Figure 17.1, which also defines the coordinates x and ϕ, describing flange deflections relative to the mating flange. In Figure 17.1 the notation is as follows:

B=total bolt load (lbf, N)
G=gasket reaction (lbf, N)
P=fluid force due to pressure acting over the end of the pipe or vessel (lbf, N)
x=coordinate for axial translation of the flange centroid relative to that of the second flange (in., mm)
ϕ=coordinate for axisymmetric rotation of the flange as a ring (radians)
b=lever arm of the bolt force B about the centroid (in., mm)
g=lever arm of the gasket force G about the centroid (in., mm)
ρ=lever arm of the fluid end force P (initially zero) about the flange centroid (in., mm).

As operating conditions change, for example, due to pressurization of the fluid, the three forces B, G, and P interact due to changes in

1. Axial displacement x of the flange, as the gasket compresses or relaxes, and bolt stretch changes
2. Rotational displacement ϕ of the flange about its centroid, as moments vary with the changing forces.

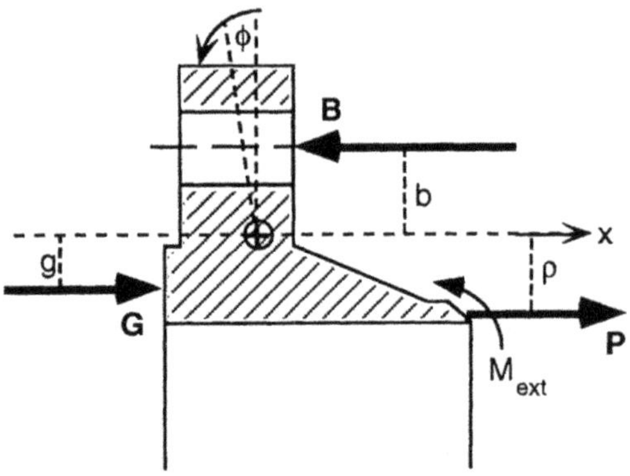

FIGURE 17.1 Major forces acting on a flange (from Ref. [3]).

These flange displacements are defined relative to the initial assembled condition. As we see later, these interactions also depend on three stiffness coefficients:

1. Compressive stiffness of the gasket, k_g (actually a nonlinear function of compressive stress)
2. Axial stiffness of the bolting, k_h
3. Rotational stiffness of the flange, k_ϕ (actually a nonlinear function of rotation if local yield occurs in the flange).

The mating flange might not be of the same design and material. Here for clarity, we assume that the mating flange is rigid. Regardless of this assumption, the values of B, G, and P acting on the second flange are the same as those on the first, but moments and flange stiffness may differ.

When the joint is assembled, before the fluid is pressurized (fluid pressure $p=0$), we can write an axial force equilibrium condition:

$$G(0) = B(0) \tag{17.3}$$

and after pressurization, fluid pressure $p>0$:

$$G(p) = B(p) - P(p) \tag{17.4}$$

Subtracting Eq. (17.3) from Eq. (17.4) gives an expression relating to changes in force:

$$\Delta G = \Delta B - \Delta P \tag{17.5}$$

These force changes are proportional to flange displacements (Δx, $\Delta\phi$):

$$\Delta G = -k_g\left(\Delta x + g\Delta\phi\right) \tag{17.6}$$

$$\Delta B = +k_b\left(\Delta x \quad b\Delta\phi\right) \tag{17.7}$$

Also, internal flange stresses are in equilibrium with the moments acting; therefore,

$$k_\phi\Delta\phi = \Delta G\ g + \Delta B\ b + \Delta P\ \rho + M_{\text{ext}}(p) \tag{17.8}$$

where $M_{\text{ext}}(p)$ represents pressure-dependent external moments (axisymmetric) due, e.g., to the adjoining pipe wall as it expands radially under the influence of the fluid pressure.

To simplify this equation define a coefficient C such that

$$C\ \Delta p = \Delta P\ \rho + M_{\text{ext}}(p) \tag{17.9}$$

where $\Delta p \equiv p$. Then

$$k_\phi\Delta\phi = \Delta G\ g + \Delta B\ b + C\Delta p \tag{17.10}$$

We now have four equations [(17.5), (17.6), (17.7), and (17.10)] to solve for four unknowns: ΔG, ΔB, Δx, and $\Delta\phi$. Rather than use algebraic manipulation, it is more convenient to write the equations in matrix form and solve using a computer matrix solver. The matrix equation takes the form

$$\begin{vmatrix} 1 & -1 & 0 & 0 \\ 1 & 0 & k_g & k_g g \\ 0 & 1 & -k_b & k_b b \\ g & b & 0 & -k_\phi \end{vmatrix} \begin{vmatrix} \Delta G \\ \Delta B \\ \Delta x \\ \Delta \phi \end{vmatrix} = \begin{vmatrix} -A_p\ \Delta p \\ 0 \\ 0 \\ -C\ \Delta p \end{vmatrix} \tag{17.11}$$

in which ΔP has been replaced by (A_p Δp), where A_p is area exposed to axial fluid pressure.

If the second flange is not rigid, Eqs. (17.6) and (17.7) have to be generalized by replacing

$$g\ \Delta\phi \text{ by } \{g_1\ \Delta\phi_1 + g_2\ \Delta\phi_2\}$$

and

$$\Delta x \text{ by } \{\Delta x_1 + \Delta x_2\}$$

with appropriate coordinate system, and writing Eqs. (17.8) separately for each flange. This gives five equations, which can be solved for the five unknowns.

17.2.2 Effects of Relative Stiffness of Gasket and Bolting

The special case of a joint having rotationally rigid flanges ($\Delta\phi$=0) gives useful insight into the significance of the *relative* stiffness of gasket and bolting in a joint. It can be shown, from Eqs. (17.5)–(17.7), that when fluid pressure is applied to the assembled joint, the gasket force and bolt force change as follows:

$$\Delta G = -k_g\ \Delta P/\left(k_b + k_g\right) \tag{17.12}$$

$$\Delta B = +k_g\ \Delta P/\left(k_b + k_g\right) \tag{17.13}$$

Thus the *gasket stress decreases* and the bolt force increases to maintain the axial force equilibrium as the fluid pressure is applied. There are two limiting cases of interest to consider.

Case (i): If the gasket is stiff compared with the bolting, $k_g >> k_b$, then

$$\Delta G = -\Delta P \text{ and } \Delta B = 0 \tag{17.14}$$

Case (ii): If the bolting is stiff compared with the gasket, $k_g << k_b$, then

$$\Delta B = -\Delta P \text{ and } \Delta G = 0 \tag{17.15}$$

Case (ii) gives constant gasket stress (ΔG=0), a desirable situation as the gasket stress maintains its design value. It might therefore seem that the gasket should be chosen to be very compliant; in practice, however, excellent gasket performance is obtained, at high pressure and temperature, when the gasket is stiff compared with the bolting. Why should this be? In fact, our simple model omits flange rotation, and when a flange rotates a stiff gasket applies a restoring moment to the flange, opposing rotation. A more complete model would bring out such effects. It might seem "obvious" that sheet gaskets and spiral-wound gaskets would be appreciably stiffer than the bolting in joints, but actually, this is usually not the case, especially for large sizes [4]. This slightly surprising fact is due to the combination of large gasket area with fairly high modulus (which increases at high compressive stresses),

TABLE 17.1
Examples of Specifications for Standard Steel Flanges

International Standard ISO 7005, *Metallic Flanges. Part 1: Steel Flanges.*
ASME B16.5—1988, *Steel Pipe Flanges and Flanged Fittings* (formerly ANSI B16.5 and then ASME/ANSI B16.5).
ASME B16.47—1990, *Large Diameter Steel Flanges:* NPS 26 Through NPS60.
European Draft Standard CEN prEN 1092-1, Flanges and Their Joints—Circular Flanges for Pipes, Valves and Fittings. Part 1. Steel Flanges—PN Designated.
European Draft Standard CEN prEN 1759-1, Flanges and Their Joints—Circular Flanges for Pipes, Valves, Fittings and Accessories, Class Designated. Part 1. Steel Flanges, DN 15 to DN 600 (NPS 1/2 to 25).

Source: Ref. [3].

TABLE 17.2
Examples of Standard Flange Types

Welded Construction	Nonwelded Construction
Weld-neck taper-hub flange	Threaded flange
Slip-on weld plate flange	Lap flange (engages abutment at pipe end)
Weld-socket plate flange	Blank flange ("end closure")

Source: Ref. [3].

which together give a high stiffness. If the flanges can also rotate, then it is possible for the bolt force to either increase or decrease. Either way, the gasket stress decreases.

17.3 STANDARD JOINTS

Many applications can be met by choosing a standard joint design. This is usually the quickest and most efficient approach. Standard flanges are commercially available, as are gasket sizes to match. Standard joints are specified in detail in national and international standards, such as those in Table 17.1.

17.3.1 Flange Types

Standard flange sizes range up to 600mm and more. Typical design types include those listed in Table 17.2 and illustrated schematically in Figure 17.2.

The weld-neck taper-hub flange (Figure 17.2a), with its relatively narrow raised face to accommodate the gasket, is a common choice for high duty applications. Its reinforcing hub gives it extra rotational stiffness, which reduces stresses in the metal and gives a better environment for the gasket. It is usually less expensive to extend or thicken the hub than to increase the axial thickness or diameter of the flange proper.

Flanges with full-face gaskets (e.g., Figure 17.2d) are regarded as light duty designs. This is because the compressive stress applied to the gasket by the given bolting is low compared with that for a narrower gasket. Threaded flanges (not illustrated) are also intended for light duty. In this case, because the threaded construction provides little reinforcement and the flange is essentially free to rotate under the moments generated by the bolting force, gasket reaction and fluid end-load.

The controlled compression gasket concept (Figure 17.2f) is not widely used, as its advantages are not generally appreciated. The chief of these is that the gasket does not suffer changes in compressive load as fluid pressure and temperature change, so it is maintained in an optimum state for

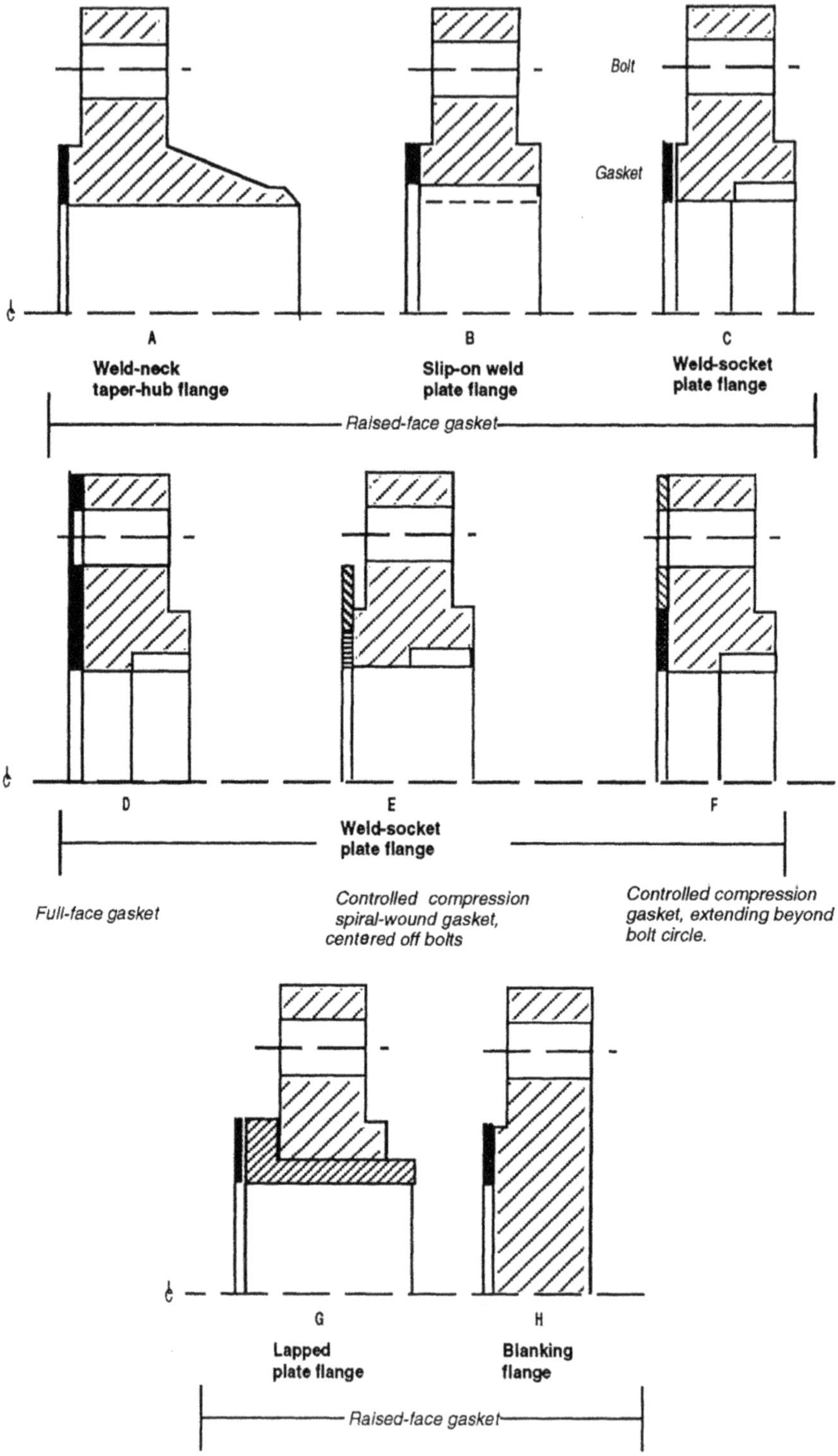

FIGURE 17.2 Examples of standard flanges (from Ref. [3]): (a) Weld-neck taper-hub flange; (b) slip-on weld plate flange; (c) Weld-socket plate flange; (d) Full-face gasket; (e) Controlled compression spiral-wound gasket, cantered off bolts; (f) Controlled compression gasket, extending beyond bolt circle; (g) Lapped plate flange; (h) Blanking flange.

leakage control. Also, provided that the bolts are tightened well beyond what is required to seat the flanges metal-to-metal, the bolts do not suffer cyclic loading, which reduces the risk of fatigue or loosening. Further advantages result from reduced flange rotation as a consequence of

1. The rotational constraint due to the pivot reaction force acting at the periphery
2. The direction of the bolting force moment is reversed due to the bolt circle being inboard of the flange pivot (unlike a raised-face joint).

The tendency of such flanges to bend about the bolt circle results in radial bending stresses here that have to be taken into account when evaluating design stresses. Consideration may also have to be given to possible stress corrosion problems resulting from the bolting being more highly stressed than in other types of joints.

Spiral-wound gaskets are commonly fitted with an outer ring (Figure 17.2e) to limit compression of the windings. This does not behave in the same way as the controlled compression concept just described. Because the limit ring is inboard of the bolt circle it is possible for overtightening of the bolts to cause the flange and ring to pivot on the outer edge of the flange's raised face. This unloads the inner region of the gasket and can actually *increase* leakage if the bolts are overtightened.

The lapped plate flange (Figure 17.2g) is another design that has not been fully exploited. With this arrangement, flange rotation is largely decoupled from the pipe-end hub and the gasket. This again benefits gasket performance.

The blanking flange (Figure 17.2h) is a standard design. In demanding applications, stresses can be reduced if the central disc is axially offset from the flange proper by inserting a length of pipe. This isolates the flange region from the large bending moment due to fluid pressure bowing the central region of the disc.

There may be limitations on the use of certain types of standard joints. For instance, socket-weld flanges and threaded flanges are not recommended for use at extreme temperatures when there is temperature cycling (below –40°C or above 260°C) or if large temperature gradients will exist in the joint.

17.3.2 Information in Standards

Flanged joint standards typically tabulate such details as

Dimensions of flanges and bolts.
Face details, including dimensions, finish, and permissible defects.
Bolt-hole templates.
Weld geometry and dimensions.
Material specifications.
"Ratings"; i.e., maximum-allowable non-shock pressure–temperature limits, for each size and duty class, for various groups of construction materials. The materials are grouped according to their stress–strain properties over relevant temperature ranges.

Gasket dimensions may be included in the main flange standard (e.g., ASME) or in separate gasket standards (e.g., ISO and CEN).

17.3.3 Terminology

17.3.3.1 Class and PN

Standard flange designs are grouped according to the severity of the duties for which they are suitable. For example, in ASME B16.5 [5], Class 150 joints are the lightest duty designs and Class 2500 are the heaviest duty designs. In ISO 7005 [6], designations range from PN 2.5, the lightest, to PN 420, which is the heaviest duty design.

The designation "Class" is used by ASME (and CEN in standards derived from ASME), and the designation "PN" is used by ISO and CEN. Both designations are simply numerical identification labels; they do not have physical units.

Note that ASME B16.5 also uses the numerical value of the class label in deriving its tabulated *p–T* ratings. Thus for Class 300, the primary rating pressure used in calculating ratings is 300 psi; for Class 600 it is 600 psi; and so on, with the exception of Class 150, for which it is only 115 psi. However, the primary rating pressure value does not appear explicitly in the rating tables.

17.3.3.2 NPS and DN

These are analogous designations to indicate diametral size. Thus ASME B16.5 uses NPS##, i.e., nominal pipe size; and ISO and CEN use DN### (CEN uses both for class-designated flanges). Again, such designations are intended only as an identification label, in this case to indicate a general diametral size category for use in identifying items of equipment of consistent size.

When items of equipment are combined together in a system, they are normally chosen so that all the items in the system have matching numerical designations for duty and similarly for size. For safety reasons one would not use a Class 300 flange on a Class 600 valve, for example, and for economic reasons, one would not do the reverse.

17.4 PROPRIETARY JOINTS

There are many special joint designs, and even more special gasket types, the latter being outside our scope. These are developed and/or marketed by specific companies and are usually the subject of patents. Examples of proprietary joints are given in Table 17.3 and Figure 17.3 to illustrate design features of interest. Most are more compact and weigh less than a standard flange while retaining or improving on *p–T* rating. They can be grouped according to whether they have axial bolting or a clamping arrangement.

TABLE 17.3
Some Proprietary Joint Designs[a]

Design Characteristics	Examples (Trade Names)
Axially bolted, wedge-seal	
Modified standard flanges (Figure 17.3a)	Destec (Destec Engineering Ltd, Lincoln, UK) Flangeplus (Techlok Ltd, Port Talbot, UK) Grayloc (Gray Tool Co., Houston, TX)
Special flanges (Figure 17.3b)	Taperlok (Taper-Lok Corp., Houston, TX)
Compact	
Lap joint (Figure 17.3c)	UKAEA (UKAEA, Warrington, UK)
Integral (Figure 17.3d)	Verax (Steel Products Offshore A/S, Drammen, Norway)
Clamped	
Hubbed pipe, controlled-compression wedge gasket (Figure 17.3e)	Destec G (Destec Engineering Ltd, Lincoln, UK) Flexitallic Clamp Joint (Flexitallic Ltd, Cleckheaton, UK) Grayloc (Gray Tool Co., Houston, TX) Techlok (Techlok Ltd, Port Talbot, UK)
Hubbed pipe, quick-disconnect, single bolt	Cefilac (Cefilac, Saint-Etienne, France)
Grooved pipe, elastomeric seal (Figure 17.3f)	Victaulic (The Victaulic Company, London, UK)
Plain pipe, ball swaged	Morgrip (Hydra Tight, Walsall, UK

Source: Ref. [3].
[a] See Figure 17.3.

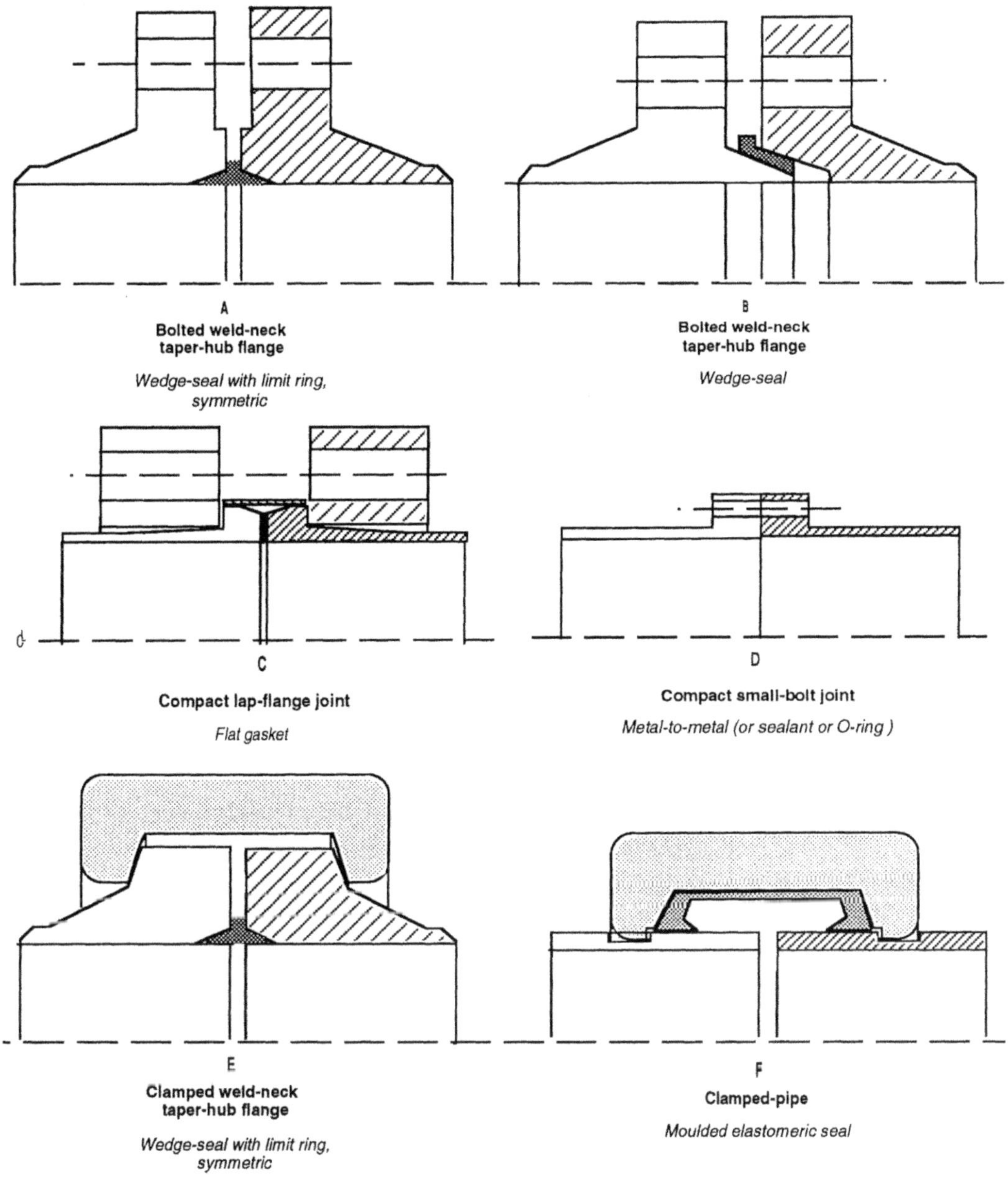

FIGURE 17.3 Proprietary joint designs in Table 17.3 (from Ref. [3]): (a) Bolted wedge-seal with limit ring, symmetric; (b) wedge-seal; (c) flat gasket; (d) metal-to-metal (or sealant or 0-ring); (e) clamped wedge-seal with limit ring, symmetric; (f) moulded elastomeric seal.

Wedge gaskets (designs a, b, and e in Figure 17.3) generate high stresses at the sealing contact without requiring very high bolt loads, giving the potential for good high duty performance. Clamp designs E and F have the advantage of fewer bolts, simplifying assembly and dismantling. On the other hand, some axially bolted designs (e.g., A) conveniently require only slight modification of standard flanges, avoiding costly specials. Design F uses a molded elas-tomeric seal rather than a more conventional gasket. Although this limits thermal and chemical compatibility, it avoids the need for any welding to the pipe. Design D aims at the ultimate in compactness by using small bolts located on a small pitch circle; it also minimizes flange rotation by arranging that pivoting is at the outer perimeter, and it does not require a gasket if the faces are suitably finished. However, like other compact flanges, it does not mate with standard flanges, which limits the scope of application. Design C achieves compactness by a different approach, again using small bolts on a small pitch circle but reducing flange rotation by using loose flanges. This decouples flange rotation from the gasket. The flanges are stiffened against rotation by virtue of their axial length, which is greater than in a standard loose flange; radial alignment of the pipes is ensured by a sleeve sliding over the end hubs.

17.5 CODE DESIGN JOINTS

17.5.1 What Is a Code?

The purpose of a joint design code is to provide a simple "recipe" or procedure that enables the user to design a gasketed joint that is structurally sound and meets leakage control requirements. Typically such codes comprise a series of formulas and charts. Key data, including dimensions, gasket characteristics, operating conditions, and limiting stresses of the materials of construction, are first collated. Forces and moments acting on the joint are then evaluated; this enables the bolting requirements, i.e., number and diameter of bolts, to be established. Next, the magnitudes of stresses that have to be checked are calculated and tested against designated maximum values.

The procedure requires that you first assume plausible flange dimensions, to be improved subsequently in light of the results of the code calculations. This sequence is inconvenient, but the nature of the problem is such that it is not possible to work in the preferred direction, i.e., to directly calculate the required flange thickness for given stress and leakage requirements.

Table 17.4 compares some features of major design codes.

TABLE 17.4
Features of Some Design Codes for Integral Raised-Face Flanged Joints

Subject	ASME Section VIII, Div 1, 1995[a]	DIN 2505, 1990[b]	CEN prEN 1591, 1994
Basis flange model	Taylor Forge annular plate	Ring	TGL Ring
Hub model	Cylindrical shell, variable thickness	Equivalent cylindrical shell	Equivalent cylindrical shell
Wall model	Cylindrical shell	—	—
Allowance for bolt holes	No	Yes	Yes
Reference state	Elastic	Elastic	Plastic collapse
Reference stress	$S_{\text{allowable}}(T)$	$\sigma_{\text{yield}}(T)$ or $\sigma_{\text{ult}}(T)$	$\sigma_{\text{yield}}(T)$
E_{gasket}	No	$E=E(S_g, T)$	$E=E(S_g, T)$
Gasket stress basis	m, y	m, σ_{vu}	options
Pressure Inflation	No	No	No
External moment	No	No	Yes
External thrust	No	Yes	Yes
Temperature differential	No	(Yes)	Yes
Flange rotation	No?	Yes	Yes
Load-deflection interaction	No	Yes	Yes
Gasket width	Arbitrary reduction	Empirical	Calculated
Bolt bending	No	No	No
Flange radial constraint	Hub junction	Natural	Natural
Reaction at hub/flange joint	Mid-hub	Mid-wall	Mid equivalent shell

Source: Ref. [3].
[a] Also BS 5500, CODAP, and others.
[b] (Yes) denotes partial treatment.

17.5.2 Stress Analysis

The longest established code is the ASME Boiler and Pressure Vessel Code for Unfired Pressure Vessels [7]. This originated in an analytical stress analysis of taper-hub flanges developed during the 1920s and 1930s [8], now called the Taylor Forge method. It is the basis of national codes in a number of other countries, including the British code BS 5500 and the French CODAP. The approach followed in the German code DIN 2505 is quite different.

An international code currently under development is the draft European code CEN/prEN 1591 [9]. This differs fundamentally from all those listed above, which use linear elastic analysis. The draft CEN code is based on elastoplastic limit load analysis. The basis of this code is one used successfully in the former German Democratic Republic (TGL 32903/13 [10]).

A limit load analysis is more realistic because it recognizes what has long been known, namely that many joints experience local yield and yet perform satisfactorily (see, e.g., Ref. [11]). Local yield allows high stresses to redistribute themselves safely, behavior that has long been used in the design of civil engineering structures and without which most bridges and skyscrapers could not exist. A drawback of this approach is that it introduces nonlinear behavior into an already complicated analysis; however, this is simply recognizing reality.

17.5.3 Flange Rotation

Flange rotation is of particular importance to the gasket and to leakage control. Until recently, ASME-type codes did not take account of this (except by a crude designation of an "effective width" of the gasket) even though the original theoretical development included an explicit formula for flange rotation [12]. The German DIN 2505 and draft CEN code do take direct account of flange rotation, the latter more rigorously.

17.5.4 Leakage

Leakage is another fundamental issue inadequately dealt with by ASME-type codes. Such codes involve the use of two gasket parameters (*m* and *y*) that together define the gasket stresses (and hence bolt load) required to "seal" the joint. In reality, leakage decreases progressively as gasket compressive stress is increased; there is no leak/seal divide as implied by the use of *m* and *y*. It is expected that future issues of the ASME code will include an improved treatment of this subject, based on a procedure developed by The Pressure Vessel Research Council. The draft European code can use either approach.

17.6 ANALYSIS OF GASKETED FLANGED JOINTS

17.6.1 Analytical Methods

The traditional approach is to consider a flange as an assemblage of interacting structural members with appropriate conditions for continuity of displacements, stresses, and moments imposed at the interfaces between them (Figure 17.4). The components are chosen on the basis that analytical solutions exist for the chosen substructure and loading system. This normally means using some combination of shells, plates, or rings. For instance, a taper-hub flange could be considered as a combination of

1. An annular ring that suffers rotation but not bending (this is the radial flange proper)
2. A variable-thickness cylindrical shell (the taper hub)
3. A cylindrical shell (wall of pipe or vessel)

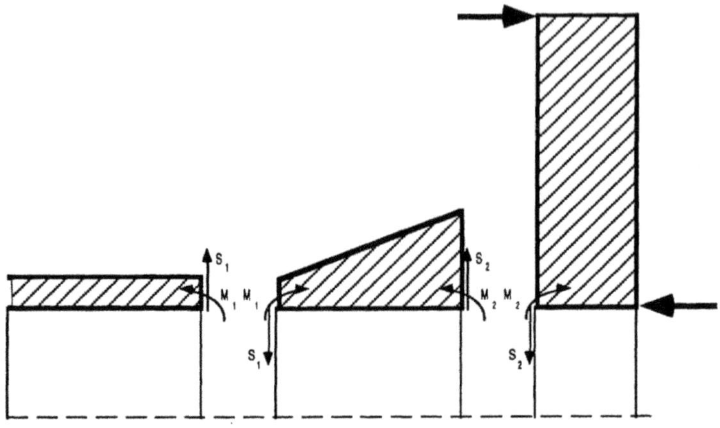

FIGURE 17.4 Substructuring model used by Taylor Forge method (from Ref. [3]).

TABLE 17.5
Validity of Flange Component Models

Model	Validity Condition[a]
Thin circular plate	$w/t > 4$
Ring	$1/4 < w/t < 4$
Beam	$w/t > 8$, negligible hoop stress

Source: Ref. [3].
[a] See Figure 17.5.

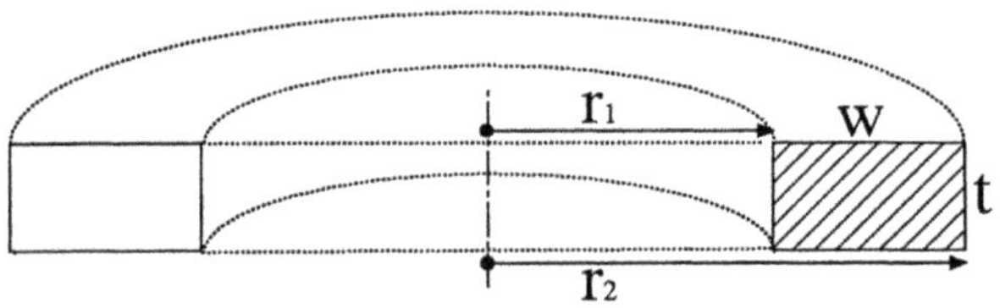

FIGURE 17.5 Notation for Table 17.5 (from Ref. [3]).

The assembled equations for the substructures comprise a set of simultaneous linear equations that can be solved by normal numerical methods. This approach can be implemented on a computer to provide a tailored model for joints of a given shape, with dimensions and materials as variable input data. Details of the formulation of such an approach are given by Rodabaugh and Moore [11].

When dividing the joint into substructures, the validity of simple models depends on the aspect ratio of the component, the ratio *w/t* in Table 17.5 and Figure 17.5. It is also assumed that the in-plane stress varies linearly through the thickness *t*. Design codes are based on this substructuring approach. Table 17.5 summarizes key features and approximates underlying several codes.

17.6.2 Finite Element Analysis

The substructuring method outlined above is essentially a finite element (FE) method with a single element for each "component." Each of these elements is a different type, chosen to model that specific component. Instead of simple polynomial interpolation within elements, as in conventional FE, more accurate analytical expressions are used. Interestingly, in the Taylor Forge analysis [12] the hub solution was actually obtained by an approximate numerical method involving minimization of energy, a method often used to develop conventional FE elements.

However, the use of a general-purpose finite element stress analysis package is an alternative to the use of a standard design code or an analytical approach. Advantages are

1. More accurate modeling of the geometry and material properties.
2. Comprehensive output of stress and deformation distributions.

On the other hand, perfection is not easily achieved for the following reasons:

1. Setting up an FE analysis can take longer than a complete code analysis.
2. Gasket stress–strain characteristics are nonlinear and have hysteresis.
3. Gasket stress distribution and bolt load both vary with flange rotation.
4. Elastoplastic analysis is required because flange stresses can exceed yield.
5. Static indeterminacy and material nonlinearities necessitate an iterative solution.
6. The procedure still uses the "cut and try" method to optimize flange and bolt dimensions.

REFERENCES

1. Flanged Joint Calculation, *German Draft Standard* DIN V2505, 1986.
2. Wesstrom, D. B., and S. E. Bergh, Effect of internal pressure on stresses and strains in bolted-flanged connections, *Trans. ASME* 73: 553–568, 1951.
3. Nau, B. S., *Fluid Sealing Lecture Notes: Joint Design*, 1995 (unpubl.)
4. Nau, B. S., On the design of bolted gasketed joints, Paper Dl, Proc. *12th International Conf. on Fluid Sealing,* BHRA, Cranfield, UK, 1989.
5. Steel Pipe Flanges and Flanged Fittings, *ASME Standard* B 16.5, 1988.
6. Metallic Flanges. Part 1: Steel Flanges, *International Standard* ISO 7005, 1990.
7. Boiler and Pressure Vessel Code for Unfired Pressure Vessels, *ASME Standard*, 1995.
8. Waters, E. O., D. B. Wesstrom, et al., Formulas for stresses in bolted flanged connections, *Trans. ASME* 59(3): 161–169, 1937.
9. Design Rules for Gasketed Circular Flange Connections, Draft *European Standard* CEN prEN 1591, 1994.
10. *German Democratic Republic Standard* TGL 32903/13, 1983.
11. Rodabaugh, E. C., and S. E. Moore, Evaluation of the Bolting and Flanges of ANSI B16.5 Flanged Joints—ASME Part A Design Rules, *Oak Ridge Nuclear Laboratory Report* ORNL/Sub/2913-3; NRC-5; PR. 115–7b, 1976.
12. Waters, E. O., D. B. Rossheim, et al., *Development of General Formulas for Bolted Flanges,* Taylor Forge & Pipe Works, Southfield, MI. Reprinted by Welding Research Council, New York, 1949.
13. Large Diameter Carbon Steel Flanges, *American Petroleum Institute Standard* API 605, 1980.
14. Flanges and Their Joints—Circular Flanges for Pipes, Valves and Fittings. Part 1. Steel Flanges—PN Designated, Draft *European Standard* CEN prEN 1092-1, 1994.
15. Flanges and Their Joints—Circular Flanges for Pipes, Valves, Fittings and Accessories, Class Designated. Part 1. Steel Flanges, DN 15 to DN 600 (NPS 1/2 to 24), Draft *European Standard* CEN prEN 1759-1, 1994.

18 ASME Boiler and Pressure Vessel Code Design Rules for Bolted Flanges

George Bibel

18.1 INTRODUCTION

This chapter presents a brief discussion giving insight into the ASME Code rules for design of bolted flanges [1]. However, it does not substitute for the detailed requirements of a Code analysis.

Bolted flanged joints are a very simple mechanical component used in many applications. In this context, a bolted flange is a pressure vessel and piping component.

Most pressurized vessels or piping components fail by excessive stress. Although flanges can also fail this way, failure is usually defined as leakage. Leakage usually occurs because of insufficient compressive stress on the gasket.

Bolted flange design can be broken down into three fundamental problems:

1. Selection of an appropriate gasket for the service conditions of pressure, temperature, and fluid compatibility.
2. Determination of required bolt loads to adequately compress the gasket and hold the joint tight under operating conditions.
3. Transmission of the bolt loading to the gasket without excessive flange stresses and deflections.

Gasket selection is dealt with in more detail in the sister volume *Gaskets and Gasketed Joints* [2] and elsewhere.

The current ASME gasket factors m and y determine the required bolt loading (see Section 18.3.3). New gasket factors G_b, G_s, and a have been developed for many gasket types. The new factors are based on extensive test data sponsored by the Pressure Vessel Research Council (PVRC). The new gasket factors will eventually become the basis of a new nonmandatory appendix in ASME Section VIII to be used as an option to m and y.

The discussion in this chapter is limited to ASME Section VIII, Division 1, Mandatory Appendix 2, based on gasket factors m and y. This is done for the following reasons:

1. The use of factors m and y as the basis for ASME bolted flange design is still mandatory and will be for the foreseeable future.
2. The current design rules have been used successfully for many years in most applications and will remain an important benchmark comparison.
3. The new proposed rules with gasket factors G_h, G_s, and a are unnecessary for our purposes in this chapter. For those who are interested, they are discussed at length in Refs. [2] and [3] and in many papers published by the ASME and the Welding Research Council (WRC).

DOI: 10.1201/9781003501183-22

It is important to note that structural analysis of the flange is essentially the same in the current ASME rules and in any proposed changes. What does change is the determination of bolt loading required to compress the gasket. Therefore, statements made in this chapter will also apply to the new proposed rules except for the calculation of required bolt loads.

18.2 FLANGE DESIGN: BACKGROUND AND THEORY

An increase in the operating temperatures and pressures of steam generation equipment created a need for flange design rules applicable over a range of conditions. In 1935, Waters and coworkers undertook the study of all existing flange stress formulas and the development of new rules over a practical range of applications. The results of their work [4,5] were adopted by the ASME in 1940 and are still in use today. A brief review will give insight into the format and complexity of these rules. Gasket factors *m* and *y*, first described by Rossheim and Markl [12], were included in the ASME Code in 1942 and are still used, with little change, in the present edition of the code [1].

The analysis of bolted flange structures is based on plate-and-shell analysis of three separate components as shown in Figure 18.1: (1) the shell, (2) the tapered hub, and (3) the ring flange. Each component is studied independently with unknown boundary conditions of displacement, rotation, shear, and moment. When the individual components are assembled, it is mathematically possible to solve all boundary and stress conditions in terms of external loads by equating the boundary conditions on either side of the junction.

This complex stress problem was reduced, with judicious approximations, to the point where manual calculations could be contemplated. The procedure was extended into a practical design procedure by precalculation of various functions (alternative design methods use different and possibly more accurate assumptions for the boundary conditions).

For example, the stress distribution in the shell and hub can be described in terms of M_{h1}, the moment on the large end of the hub. The calculations were done for a full range of hub and shell geometries and became flange factors F and V, given in the Code in formula and chart form. Different boundary conditions apply to loose types of flanges. Therefore, different charts and formulas are given for F_L and V_L [1,4,6, 7].

The flange ring is summarized by the precalculation of factors U, Y, Z, and T. The flange factors F, V, U, Y, and Z, with further adjustments given in the Code, multiply the flange moments caused by bolt loads, gasket reaction, and pressure thrust to form the flange stresses.

For a flange with a straight hub, the critical stresses are the radial and hoop stresses at the inside diameter of the ring and the axial hub stress at the surface of the junction with the ring. When the angle of the hub taper increases beyond a certain limit, the hub becomes more rigid and the location of the critical axial stress in the hub moves from the large end into the small end (see Figure 18.2).

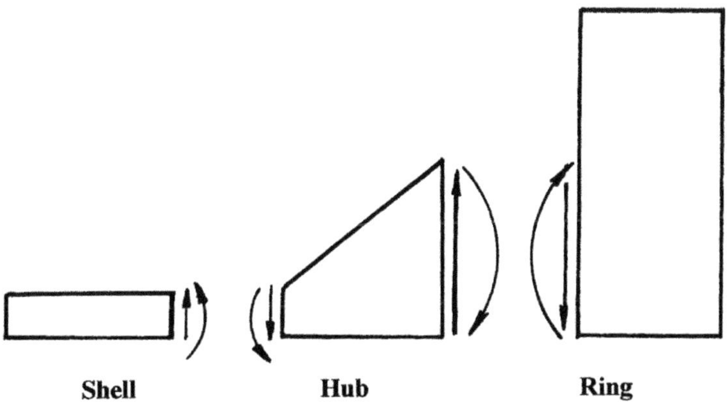

FIGURE 18.1 The three structural components of a flange.

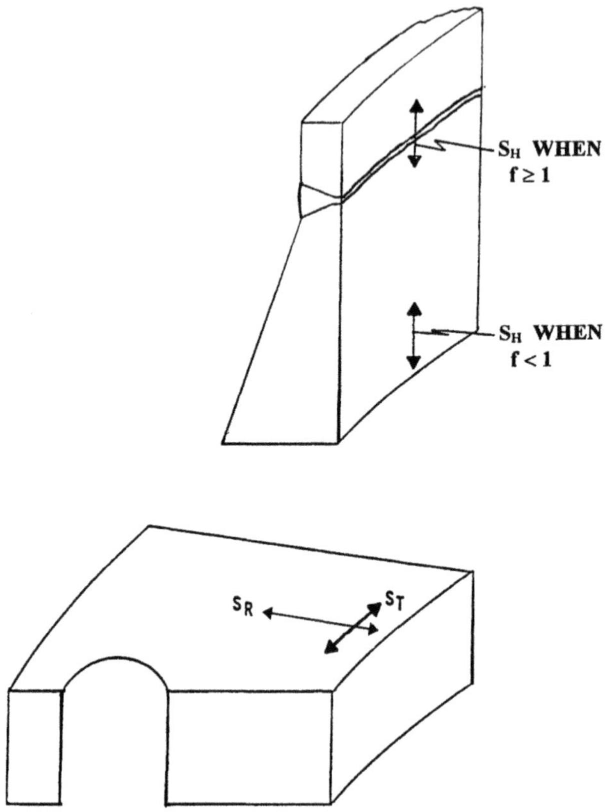

FIGURE 18.2 Flange stresses.

A hub stress correction factor *f* was introduced to obtain the "true" maximum stress. The stress at the hub ring junction multiplied by *f* is equal to the stress at the hub/shell junction. This factor is presented in graphical form in terms of the hub geometric parameters.

18.3 FLANGE DESIGN: APPLICATION

18.3.1 Scope

The rules in ASME Section VIII, Division 1, Appendix 2 apply specifically to "the design of bolted flange connections with gaskets that are entirely within the circle enclosed by the bolt holes, with no contact outside the circle." It is recommended that flanges conforming to the following standards be used when connecting to external piping:

ASME/ANSI B16.5, Pipe Flanges and Flanged Fittings [8]
ASME/ANSI B16.24, Cast Copper Alloy Pipe Flanges and Flange Fittings, Class 150, 300, 400, 600, 900, 1500, and 2500 [9]
ASME/ANSI B16.42, Ductile Iron Pipe Flanges and Flanged Fittings, Class 150 and 300 [10]
ASME/ANSI B16.47, Large Diameter Steel Flanges, NPS 26 Through PS 60 [11]

These standards may be used for other pressure vessel flange connections within their specified pressure–temperature ratings.

18.3.2 Procedure

The complex solution method has been reduced to a "cookbook" procedure in the Code. Numerous correction factors are used to adjust for the variety of possible geometric configurations. Once the flange geometry and gasket type have been identified, the procedure reduces to a four-part process.

1. Identify the required bolt loads for
   a. Gasket seating.
   b. Operating conditions.
2. Calculate the internal flange forces and moments for gasket seating and operating conditions.
3. Calculate the flange factors and stresses for gasket seating and operating conditions.
4. Compare to the Code allowable stresses.

The individual steps are reviewed below.

18.3.3 Required Bolt Load

Appendix 2 is based on two separate and parallel designs. First, the bolt loading required for gasket seating, W_{m2}, is evaluated.

$$W_{m2} = 3.14bGy \tag{18.1}$$

where

b=effective gasket width reduced to account for flange rotation about the gasket
G=gasket diameter at location of gasket load reaction
y=gasket seating load
W_{m2}=effective gasket area times a gasket seating factor y

W_{m2} reflects the bolt load required to compress the gasket a sufficient amount that it flows into the surface imperfections of the flange to provide an adequate seal. This calculation is done with no consideration given to design pressure.

After gasket seating, pressurization tries to separate the joint and reduce the gasket seating stress. The operating bolt pressure W_{m1} must resist the separating pressure and maintain a sufficient gasket load to retain a seal.

$$W_{m1} = 0.785G^2P + (2P \times 3.14Gmp) \tag{18.2}$$

where P is the design pressure. The first term on the right is the pressure thrust trying to separate the joint (the pressure times the area contained by the gasket). The second term is the residual load required on the gasket after pressurization and equals the gasket area times the operating pressure and a gasket factor m. Gasket factors m and y have little experimental basis but have proved adequate in the field for most applications (the PVRC test program sought at first to identify "better m and y factors").

The design bolt load W equals W_{m1} for operating conditions. The flange stresses created by W_{m1} must be evaluated against the Code allowable stresses at the design temperature.

The flange stresses created by gasket seating are evaluated by comparison to the Code allowable stresses for ambient conditions. Gasket seating and operating conditions must be separately evaluated to see which one "controls" the design.

The design bolt load W for gasket seating is not W_{m2} but an adjusted value. This adjustment is made to acknowledge the reality of field conditions. If there are any leakage problems it is common practice to torque the bolts excessively by whatever means. Therefore, the design bolt load W for gasket seating is increased to

$$W = (A_m + A_b)S_a/2 \tag{18.3}$$

This is the average of the sum of the required bolt area A_m and the actual bolt area A_b multiplied by the allowable bolt stress S_a at ambient temperatures. A_m is taken as the greater of W_{m1}/S_h or W_{m2}/S_a. S_b is the Code allowable bolt stress at operating temperatures. This average value increases the design gasket seating bolt load to acknowledge the possibility of overbolting.

18.3.4 Internal Flange Moments

For gasket seating the internal flange moment is given by

$$M_0 = W(C - G)/2 \tag{18.4}$$

where

$C.$=bolt circle diameter

G=location of the gasket load reaction.

For operating conditions, the loading forces and moment arms are as shown in Figure 18.2.

$$H_D = \text{hydrostatic end force inside the flange} = 0.758B^2P \tag{18.5}$$

$$H_T = \text{pressure force on the flange face} = 0.758G^2P - 0.758B^2P \tag{18.6}$$

$$H_G = \text{design bolt load minus total hydrostatic end force} = W - 0.758G^2P \tag{18.7}$$

These three loading forces are due to internal pressure and work to separate the flanges. They are balanced by the bolt load W.

The total internal moment for operating conditions is the sum of the product of each of the above forces and its moment arm as shown in Figure 18.3.

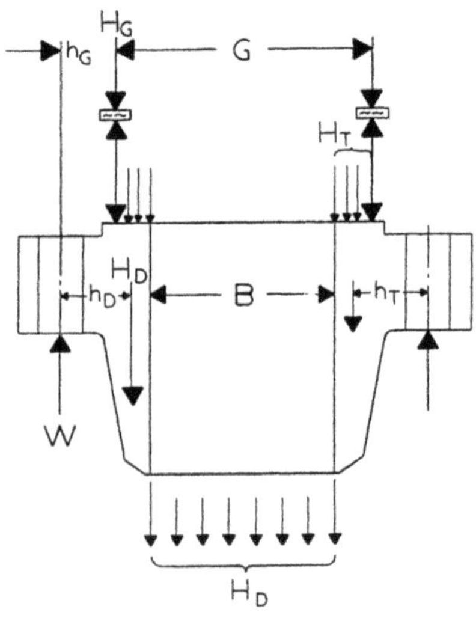

FIGURE 18.3 Forces and moment arms on a bolted flange.

$$M_0 = H_D h_D + H_T h_T + H_G h_G \tag{18.8}$$

where h_D, h_T, and h_G are the moment arms as shown in Figure 18.3.

18.3.5 Calculation of Flange Stresses

Once the internal flange moments are determined for gasket seating and independently for operating conditions, the flange stresses can be calculated.

Longitudinal hub stress:

$$S_H = fM_0/Lg_1^2B \tag{18.9}$$

Radial flange stress:

$$S_R = (1.33te + 1)M_0/Lt^2B \tag{18.10}$$

Tangential flange stress:

$$S_T = YM_0/t^2B - ZS_R \tag{18.11}$$

where t is the flange thickness and g_1 is the hub thickness at the back of the flange. Factors f, L, e, Y, and Z are correction factors given in the Code and part of the adjustments described earlier that reduce the numerical solution of a system of differential equations to a "cookbook" procedure.

For low pressure, small diameter flanges, seating conditions are expected to control the design, especially for gaskets with high y gasket factors. For high pressure, large diameter flanges, operating conditions are expected to control, especially for gaskets with high m gasket factors.

18.3.6 Flange Rotation and Rigidity

Large diameter, low pressure flanges tend to be flimsy with excessive rotation. The gasket serves as a pivot point for the rotation. It is even possible for leakage to increase with increased bolt loading as the flange rotates off the gasket.

The original analysis developed by Waters et al. [4] included a calculation of the hub ring junction rotation. Many designers have used this equation to place limits on flange rotation of 0.2° for loose type flanges and 0.3° on integral type flanges.

$$\theta = 52.14M_0V/LEg_0^2h_0 \text{ (degrees)} \tag{18.12}$$

This equation, in modified form, was recently added to the ASME Section VIII, Nonmandatory Appendix S, Design Considerations for Bolted Flange Connections. The flange rotation was changed to a rigidity index (J) because of concern that J may be interpreted as something more physically measurable than the accuracy of the theory permits. For integral flanges,

$$J = 52.14M_0V/LEg_0^2h_0K_1 \text{ (dimensionless)} \tag{18.13}$$

For loose type flanges with hubs:

$$J = 52.14M_0V_l/LEg_0^2h_0K_L \tag{18.14}$$

For loose type flanges without hubs (and optional flanges designed as loose type flanges), the rigidity index is modified as shown below.

$$J = 109.4M_0/Et^2 \ln(K)K_L \tag{18.15}$$

J must be equal to or less than 1. $K_1=0.3$ for integral or optional flanges, and $K_L=0.2$ for loose type flanges.

18.3.7 Step-by-Step Procedure

The Code procedure identifies 61 separate terms. Most designers have access to a computerized version of the design process. The standard for hand calculations remains sheet A from Taylor Forge's "Modern Flange Design," Bulletin 502 (Figure 18.4). The advantage of this format is that the terms are introduced as needed in the calculation process. An updated version of the form (Slight changes have occurred in nomenclature). The ASME Code is still required to determine the flange factors.

18.4 NOMENCLATURE USED IN FLANGE DESIGN FORMULAS

The symbols defined below are used in the formulas for the design of integral flanges are reprinted with permission from the 1995 ASME Boiler and Pressure Vessel Code, Section VIII, Division 1, Appendix 2. All references to figures, tables or paragraphs are from 1995 ASME Code (see ASME Code for loose type or lap type flanges).

A=outside diameter of flange or, where slotted holes extend to the outside of the flange, the diameter to the bottom of the slots (in.).
A_b=cross-sectional area of bolts using the root diameter of the thread or least diameter of unthreaded position, if less (in.^{2}).
A_m=total required cross-sectional area of bolts, taken as the greater of A_{m1} and A_{m2} (in.^{2}).
A_{m1}=total cross-sectional area of bolts at root of thread or section of least diameter under stress, required for the operating conditions (in.^{2}), $A_{m1}=W_{m1}/S_b$.
A_{m2}=total cross-sectional area of bolts at root of thread or section of least diameter under stress, required for gasket seating (in.^{2}). $A_{m2}=W_{m2}/S_a$.
B=inside diameter of flange (in.). When $B<20g_1$, it will be optional for the designer to substitute $B_\downarrow$ for B in the formula for longitudinal hub stress S_H.
$B_1=B+g_1$ for loose type flanges and for integral type flanges that have calculated values h/h_0 and g_1/g_0 which would indicate an f value of less than 1.0, although the minimum value of f permitted is 1.0.
$B_\lambda=B+g_0$ for integral type flanges when $f\geq 1$ (in.).
b=effective gasket or joint contact surface seating width (in.) [see Note 1, 2–5(c)(l)].
b_0=basic gasket seating width (in.) (from Table 2-5.2).
C=bolt circle diameter (in.).
c=basic dimension used for the minimum sizing of welds (in.), equal to t_n or t_x, whichever is less.
d=factor (in.^{3}) as follows: $h_0g_0^2$ for integral type flanges; $h_0g_0^2$ for loose type flanges.
e=factor (in.$^{-1}$) as follows: $e=F/h_0$ for integral type flanges; $e=F_1/h_0$ for loose type flanges.
F=factor for integral type flanges (from Fig. 2-7.2).
F_L=factor for loose type flanges (from Fig. 2-7.4).
f=hub stress correction factor for integral flanges from Fig. 2-7.6 (when greater than 1, this is the ratio of the stress in the small end of hub to the stress in the large end). For values below limit of figure, use $f=1$.
G=diameter at location of gasket load reaction. Except as noted in sketch (1) of Fig. 2-4, G is defined as follows (see Table 2-5.2): When $b_0\leq 1/4$ in., G=mean diameter of gasket contact face (in.); when $b_0>1/4$ in., G=outside diameter of gasket contact face less $2b$ (in.).
g_0=thickness of hub at small end (in.).

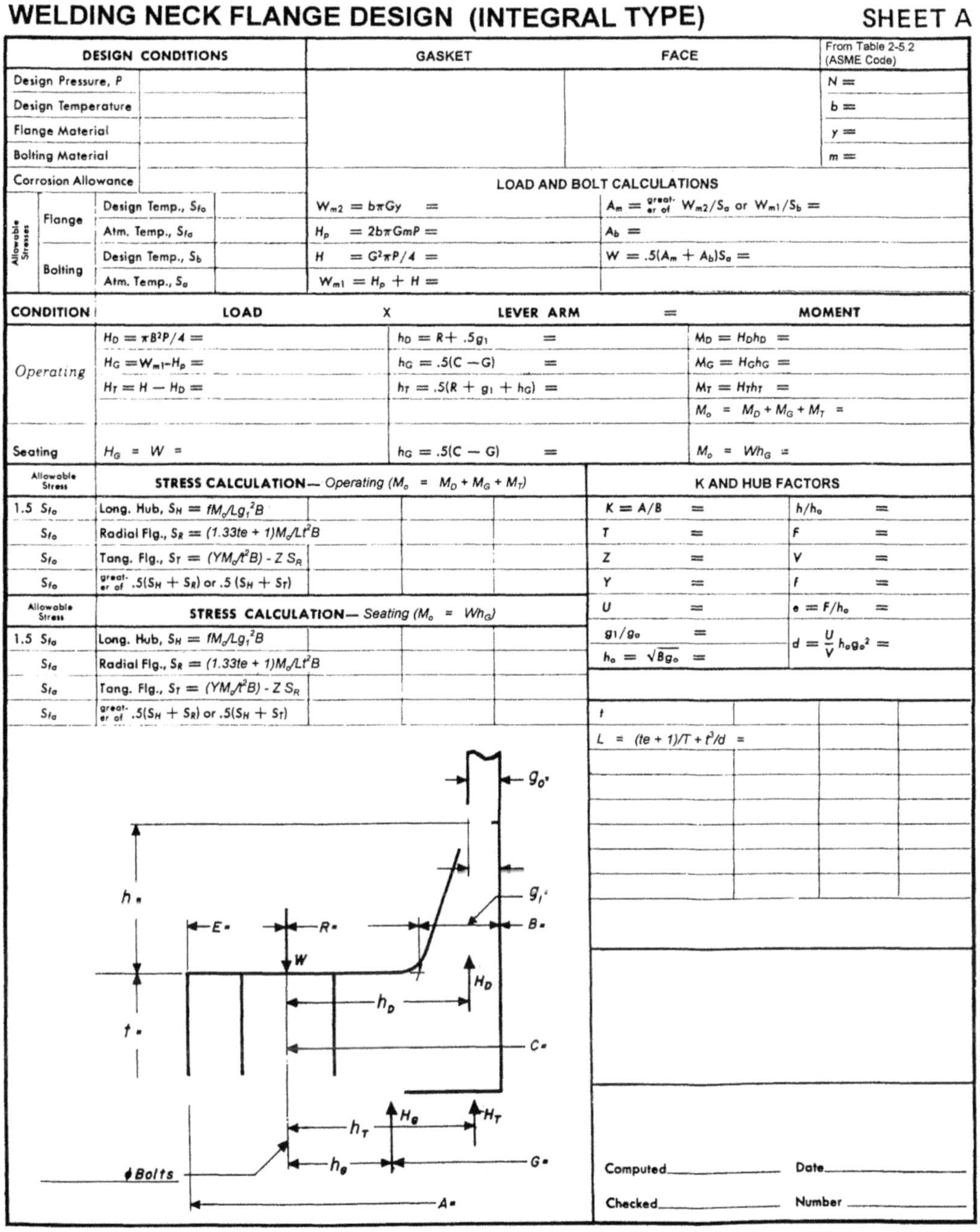

FIGURE 18.4 Welding neck flange design, integral type (see Code for loose or lap type flanges). Adopted from "Modern Flange Design: Bulletin 502" sheet A (Courtesy of Taylor Forge Company).

g_1 = thickness of hub at back of flange (in.).

H = total hydrostatic end force (lb); $H = 0.785G^2P$.

H_D = hydrostatic end force on area inside of flange (lb); $H_D = 0.785\ B^2P$.

H_G = gasket load (difference between flange design bolt load and total hydrostatic end force) (lb); $H_G = W - H$.

H_P = total joint contact surface compression load (lb); $H_P = 2b\ (3.14GmP)$.

H_T = difference between total hydrostatic end force and the hydrostatic end force on area inside of flange (lb); $H_T = H - H_D$.

h=hub length (in.).
h_D=radial distance from the bolt circle to the circle on which H_D acts, as prescribed in Table 2-6 (in.).
h_G=radial distance from gasket load reaction to the bolt circle (in.). $h_G=(C-G)/2$
h_0=factor (in.); h_0 $(Bg_0)^{1/2}$.
h_T=radial distance from the bolt circle to the circle on which H_T acts, as prescribed in Table 2-6 (in.).
K=ratio of outside diameter of flange to inside diameter of flange; $K=A/B$.
L=factor; $L=(te+1)/T+t^3/d$.
M_D=component of moment due to H_D (in.-lb); $M_D=H_Dh_D$.
M_G=component of moment due to H_G, in.-lb; $M_G=H_GH_G$.
M_0=total moment acting upon the flange, for the operating conditions or gasket seating, as may apply (in.-lb) (see 2-6).
M_T=component of moment due to H_T (in.-lb); $M_T=H_Th_T$.
m=gasket factor, obtained from Table 2-5.1 [see Note 1,2-5(c)(1)].
N=width (in.) used to determine the basic gasket seating width b_0 based on the possible contact width of the gasket (see Table 2-5.2).
P=internal design pressure (psi). For flanges subject to external pressure see 2-11.
R=radial distance from bolt circle to point of intersection of hub and back of flange (in.). For integral and hub flanges, $R=(C-B)/2-g_1$.
S_a=allowable bolt stress value at atmospheric temperature (psi) (see UG-23).
S_b=allowable bolt stress value at design temperature (psi) (see UG-23).
S_f=allowable design stress for material of flange at design temperature (operating condition) or atmospheric temperature (gasket seating), as may apply (psi) (see UG-23).
S_n=allowable design stress for material of nozzle, neck, vessel, or pipe wall, at design temperature (operating condition) or atmospheric temperature (gasket seating), as may apply (psi) (see UG-23).
S_H=calculated longitudinal stress in hub (psi).
S_R=calculated radial stress in flange (psi).
S_T=calculated tangential stress in flange (psi).
T=factor involving K (from Fig. 2-7.1).
t=flange thickness (in.).
t_n=nominal thickness of shell or nozzle wall to which flange or lap is attached (in.).
t_x=2×thickness g_0 when the design is calculated as an integral flange (in.) or two times the thickness of the shell or nozzle wall required for internal pressure when the design is calculated as a loose flange, but not less than ¼ in.
U=factor involving K (from Fig. 2-7.1).
V=factor for integral type flanges (from Fig. 2-7.3).
V_L=factor for loose type flanges (from Fig. 2-7.5).
W=. flange design bolt load, for the operating conditions or gasket seating, as may apply (lb) (see 2-5(c)).
W_{m1}=minimum required bolt load for the operating conditions (lb) (see 2-5(c)). For flange pairs used to contain a tubesheet for a floating head for a U-tube type of heat exchanger, or for any other similar design, W_{m1} shall be the larger of the values as individually calculated for each flange, and that value shall be used for both flanges.
W_{m2}=minimum required bolt load for gasket seating (lb) (see 2-5(c)).
w=width (in.) used to determine the basic gasket seating width b_0, based on the contact width between the flange facing and the gasket (see Table 2-5.2).
Y= factor involving K (from Fig. 2-7.1).
y=gasket or joint contact surface unit seating load (psi) [see Note 1, 2-5(c)].
Z=factor involving K (from Fig. 2-7.1).

REFERENCES

1. ASME Boiler and Pressure Vessel Code, Section VIII, Division 1, Appendixes 2 and S, 1995.
2. Bickford, J. H. (ed.), *Gaskets and Gasketed Joints*, Marcel Dekker, New York, 1997.
3. Bickford, J. H., *An Introduction to the Design and Behavior of Bolted Joints*, Marcel Dekker, New York, 1995.
4. Waters, E. O., D. B. Wesstrom, D. B. Rossheim, and F. S. G. Williams, Formulas for stresses in bolted flanged connections, *Trans. ASME* 59:161–169 (1937).
5. Waters, E. O., D. B. Wesstrom, D. B. Rossheim, and F. S. G. Williams, *Development of General Formulas for Bolted Flanges,* Taylor Forge and Pipe Works, Chicago, IL. Reprinted as a PVRC Monograph by the Welding Research Council in August 1979.
6. Gill, S. S., *The Stress Analysis of Pressure Vessels and Pressure Components,* Pergamon, New York, 1970.
7. *Modern Flange Design*. Taylor Forge Bulletin 502, Taylor Forge, Chicago, IL.
8. ASME/ANSI B16.5, Pipe Flanges and Flanged Fittings.
9. ASME/ANSI B16.24, Cast Copper Alloy Pipe Flanges and Flange Fittings, Class 150, 300, 400, 600, 900, 1500, and 2500.
10. ASME/ANSI B16.42, Ductile Iron Pipe Flanges and Flanged Fittings, Class 150 and 300.
11. ASME/ANSI B16.47, Large Diameter Steel Flanges, NPS 26 Through NPS 60.
12. Rossheim, D. B., and A. R. C. Markl, Gasket loading constants, *Mech. Eng.,* vol 9, pp. 647–648, 1943.

19 Fastener Selection for Non-Metallic Automotive Joints

Antoine G. Abboud

19.1 INTRODUCTION

In recent years, thc focus of the automotive industry is to reduce the weight of car and improve energy efficiency to meet government regulations. Thermoplastic ended up being an ideal solution for this goal and it is constantly growing in the automotive industry due to its light weight, strength, flexibility, and others. However, having thermoplastic component in bolted joint can be challenging due to its creep which will correlate to clamp load loss with time, and due to its lower tensile yield strength for some materials. In this chapter, a range of fastener selections are presented for joining two plastic components, and that is considered the worst scenario. If one of the two parts is metal, then the solutions provided in this chapter for the plastic part are still valid.

19.2 STRUCTURAL VS NON-STRUCTURAL JOINT

Before discussing the fastener selections, it is important to differentiate between structural and non-structural joint. A structural joint is a joint that is subjected to different types of service loads. Clamp load is critical for these joints and torque loss should be very minimum. Non-structural joint is a joint where clamp load loss is permissible since the joint is not subjected to any significant service load. In this scenario, the main function of the fastener is to hold the joint together, therefore, it is recommended for these types of joints to have either a nut with a prevailing torque feature or a bolt with a thread adhesive patch to prevent the bolt from falling off the joint upon torque loss due to creep. In this Chapter, different fastener solutions are presented for each of these two categories with the assumption that the nut member is the non-wrenchable fastener where the male fastener is the one being driven into the nut.

19.3 FASTENER SELECTIONS FOR STRUCTURAL JOINT

19.3.1 Threaded Insert

Threaded insert is a female thread that can be installed into a plastic part allowing the adequate assembly torque to be applied on the bolt without cracking the plastic. Figure 19.1 shows an example of joining two thermoplastic parts using threaded insert and compression limiter in the clamped component. A metal-to-metal contact is established between the insert nut flange and compression limiter to provide strength to the joint and minimize the torque loss due to creep. A plastic boss is required for threaded insert so it should be considered in the early design stage to incorporate that boss in the design. The boss dimension of the plastic should be provided by the insert supplier and varies from one to another depending on the insert design.

Different designs of threaded inserts are available (Figure 19.2) and that depends mainly on the installation process whether it is molded-in or post molding.

DOI: 10.1201/9781003501183-23

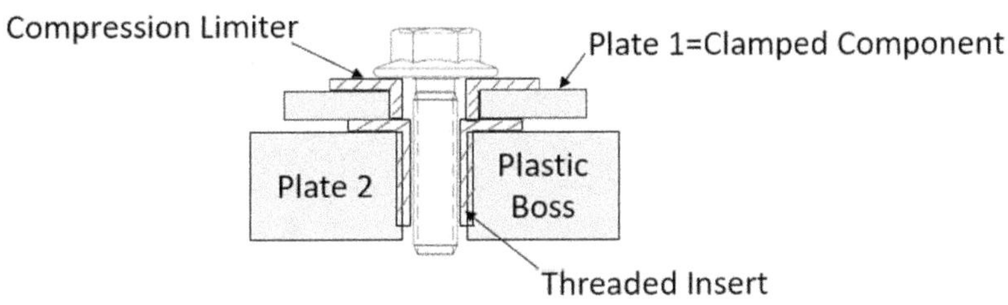

FIGURE 19.1 Joining of two thermoplastic materials using threaded insert and compression limiter.

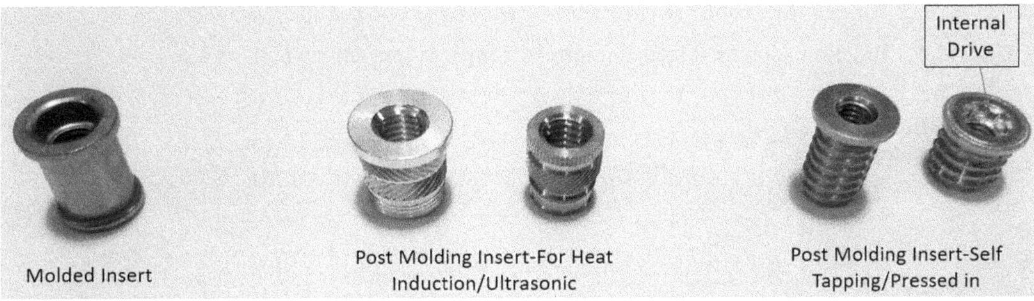

FIGURE 19.2 Different designs of threaded inserts.

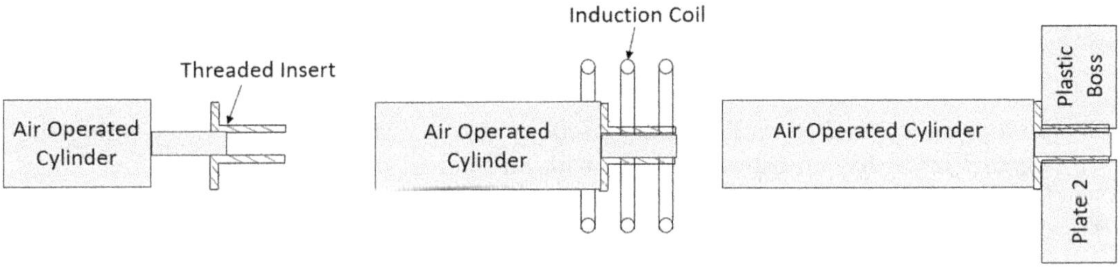

FIGURE 19.3 Heat induction installation steps (left to right).

Below is a brief description for different installation processes of threaded inserts:

- *Molded-in*: These types of inserts are installed during injection molding of Plate 2 by placing them over core pins in the mold cavity prior molding of Plate 2. Molded-in insert is in general superior in performance but if an application requires multiple inserts, then this option could be less desirable since it will affect the cycle time of the inject molding process.
- *Heat induction*: This is a post molding installation process. Figure 19.3 provides the sequence of this process. The insert is first installed into pin that is attached to air operator cylinder, then it will extend out to go through the induction coil to preheat the insert to a certain temperature depending on the material of Plate 2 before it is finally pressed into the mounting hole of the plastic part.
- *Ultrasonic*: It is similar to heat induction, but in this process, the heat portion is achieved by high-frequency vibration instead of induction coil. Inserts that are good for heat induction installation are very likely to be good for ultrasonic installation. However, the performance might be different.

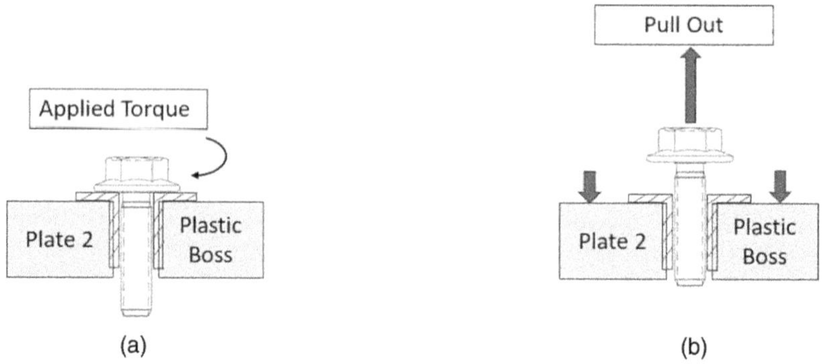

FIGURE 19.4 Test evaluation for threaded insert. (a) Torque to spin and (b) pull out test.

FIGURE 19.5 Contact vs no contact between insert and compression limiter. (a) Good contact between insert and compression limiter. (b) No contact between insert and compression limiter (bad design)

- *Self-tapping/pressed-in*: Self-tapping insert is also a post molding process where the threaded insert has an external thread with an internal drive feature on the top flange (Figure 19.2) so it can be torqued down into the plastic boss. Pressed-in insert has special design features that can be installed by pressing them down into the plastic boss.

Threaded insert material varies from one supplier to another depending on the manufacturing process. They are available in Steel, Aluminum, and brass.

The most common tests to evaluate threaded insert in any plastic material are Torque to Spin test and Pull out test (Figure 19.4). Torque to spin is the amount of torque that can be applied into the bolt before the insert starts to spin while the pull-out test is the amount of force that can be applied into the bolt before the insert is being pulled out. The performance of threaded insert depends on many variables such as: material of the plastic boss, installation process, knurling design, outer diameter and total length of threaded insert, and others. For soft plastic such as thermoplastic polyolefins or polypropylene, threaded insert might not be an ideal solution as the performance is going to be inferior.

Helpful Tips:

- Make sure to have a good contact between the threaded insert flange and the clamped component (Figure 19.5); otherwise, the torque applied to the bolt will be consumed to pull the insert out of the plastic boss instead of creating the clamp load.
- For heat induction, a lot of development is required at the early design stage to specify the temperature and speed of installation. Therefore, it is highly recommended to mold bosses made of the same thermoplastic material used in production at early design stage to evaluate the performance of threaded insert and the installation process.
- Make sure to follow the boss dimensions recommended by the supplier.

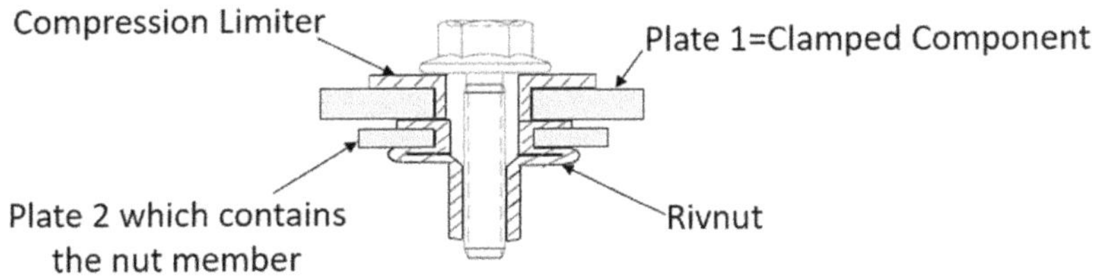

FIGURE 19.6 Joining two thermoplastic materials using rivet nut and compression limiter.

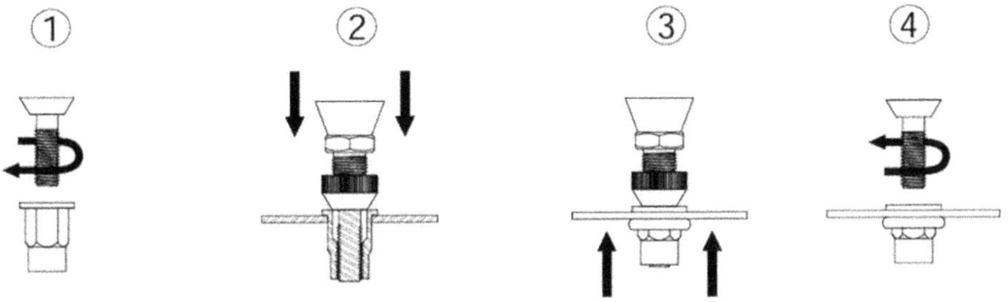

FIGURE 19.7 Rivet nut installation process (Acument Global Technologies, a Fontana Gruppo company).

19.3.2 Rivet Nut

Rivet nut is another female thread solution that is ideal for thermoplastic joint. Figure 19.6 shows an example of how to join two plastic components to create a structural joint by using a rivet nut and compression limiter. A metal-to-metal contact is established between the rivet nut head and compression limiter to provide strength to the joint and minimize the torque loss due to creep.

Rivet nut is available in different body styles such as hexagonal, round, and others. Choosing the suitable body for any application depends mainly on the parent material (Plate 2 in Figure 19.2). Rivet nut is available with various thread sizes and different grip range to accommodate the thickness of Plate 2.

One of the advantages of using a rivet nut in thermoplastic is that the installation is done post molding using pneumatic/hydraulic tool. Figure 19.7 shows the sequence of the rivet nut installation procedure as follows:

1. Spin the tool mandrel into the rivet nut thread.
2. Insert the rivet nut into the hole of Plate 2.
3. Pull the trigger and install the rivet nut.
4. Remove the tool mandrel.

Another advantage of using rivet nut is that it can be installed into thin panel with typical thickness for thermoplastic applications ranging from 2.0 to 4.0mm, and no need for plastic boss as in threaded insert. For soft plastic materials such as Thermoplastic Polyolefins or Polypropylene, rivet nut might not be an ideal solution since the maximum upset load will cause cracking in these materials and this will result in a low performance.

Once the selection of rivet nut is identified according to the material and thickness of Plate 2, then testing is necessary to achieve the optimum joint design. Torque to failure test (Figure 19.8a) is a key test that provides a good evaluation of the optimum assembly torque that can be applied on the bolt during production. Additional to torque to failure test, other tests (Figure 19.9) can be considered for evaluation depending on the application such as torque to spin, push out, and pull out. Special design features can be added to rivet nut such as knurling and serration under rivet nut head to improve the performance.

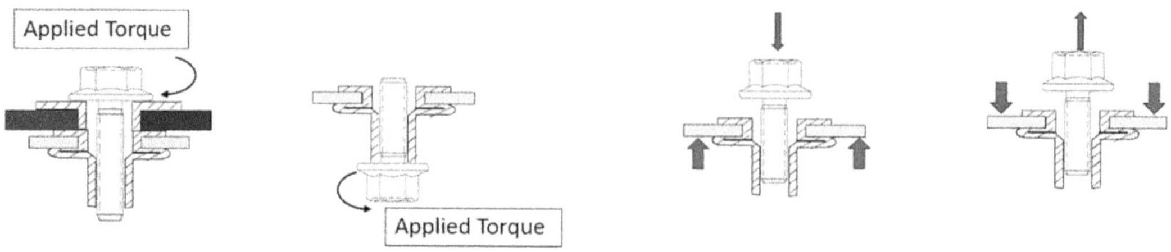

FIGURE 19.8 Different test types for rivet nut. (a) Torque to failure, (b) torque to spin, (c) push out, and (d) pull out.

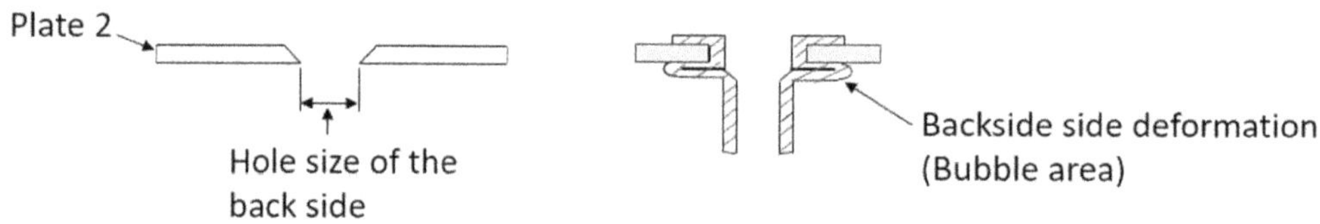

FIGURE 19.9 Dealing with draft.

Helpful Tips

- Similar to threaded insert, make sure to have enough contact between the rivet nut head and clamped component/compression limiter.
- Make sure that the hole size of Plate 2 is within the tolerance specified by supplier. Otherwise, the backside flange formation will not have enough contact with the parent material which will affect the overall performance of the rivet nut. If draft is an issue during molding, then make sure that the hole size of the backside of Plate 2 meets the specification given by the supplier (Figure 19.9)
- Most standard rivet nuts are designed to have a standard torque spec of Class 8.8 if the joint allows. For a torque spec of class 10.9, additional or hardened threads might be needed.
- Discuss with the rivet nut supplier for design recommendations if the hole size in Plate 2 is close to an edge.
- Evaluate the rivet nut at the early design stage by having coupon made of the same thermoplastic material and the same thickness of the parent material (Plate 2).

19.3.3 Compression Limiter/Shoulder Bolt

A compression limiter is a metal sleeve installed in the clamping part (Plate 1 in Figures 19.1 and 19.6) to carry most of the clamp load. They can be installed during molding process, but very likely they are installed post molding. For post molding installations, the two common types are split seam and the knurling design as shown in Figure 19.10. For split seam, it is made of carbon steel and it is installed by compressing the limiter inward before being released in the hole. The main disadvantage of this design is that the contact area between compression limiter and mating components is small since it is thin. Therefore, this design might not work for every application depending on the hole size of the mating components. For compression limiter with knurling, it is usually pressed into the hole of Plate 1. This design can be made of different materials such as steel, aluminum, or brass depending on the manufacturing process. The outer diameter of compression limiter will be then determined depending on the chosen material to make sure that it can carry the clamp load without any yield. It is recommended to have the yield strength of compression limiter equal to the yield strength of the bolt at minimum. Similar to threaded insert and rivet nut, application engineer needs to design the compression limiter in a such a way that there is a path of metal to metal

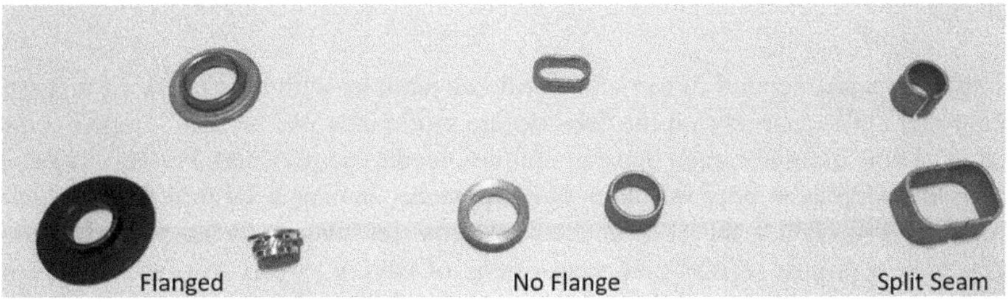

FIGURE 19.10 Different designs of compression limiter.

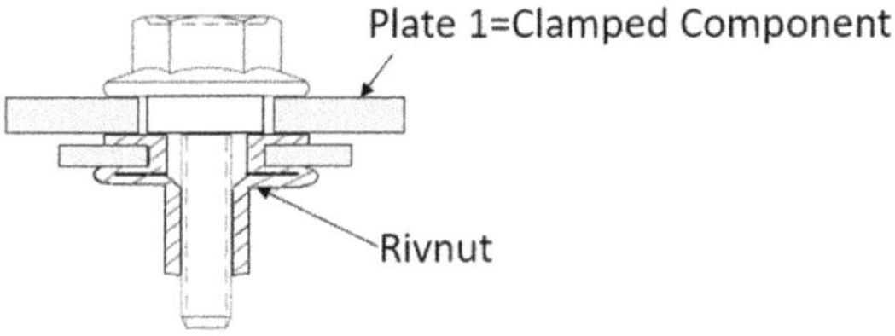

FIGURE 19.11 Shoulder bolt.

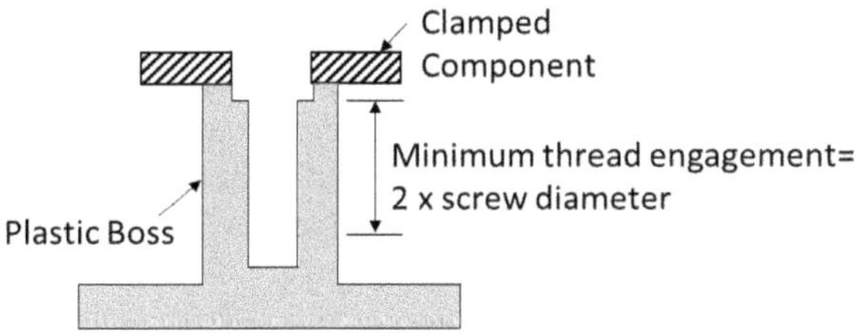

FIGURE 19.12 Plastic boss for thread forming screw.

contact between compression limiter and mating components otherwise the compression limiter is not going to function as it is supposed to be.

An alternative solution to compression limiter is shoulder bolt. The shoulder length will be the same as the length of the plastic component in the joint, but the shoulder diameter will be calculated to make sure that the contact area between shoulder and nut member is sufficient to withstand the torque specified for this application. In addition, the bolt head will be larger than the standard bolt so it can have sufficient bearing contact with the clamped component. Figure 19.11 illustrates the use of shoulder bolt.

19.4 FASTENER SELECTIONS FOR NON-STRUCTURAL JOINT

19.4.1 Thread Forming Screw into Plastic

Thread forming screw into plastic is considered one of the cheapest solutions for these types of joints since the nut member is replaced with the plastic boss, and screw is the only fastener needed. It can be used for all thermoplastic materials including thermoplastic polyolefins or polypropylene. Design engineer needs to make sure that there is enough space to package the plastic boss as shown in Figure 19.12. Since it is a thread forming screw, then there is no need to add an additional prevailing torque feature.

Helpful Tips:

- It is always recommended to run analytical calculations with the supplier for these types of joint and not to just rely on the boss design guidelines. When the clamped component has a large hole diameter, then these guidelines need to be revisited. For soft thermoplastic such as thermoplastic polyolefins or polypropylene, having a large hole in the clamped component will reduce the contact area between the boss and clamped material which might cause a failure around that area instead of having thread striping. As a result, the outer diameter of the boss might need to be bigger in these scenarios in order to increase the contact between the boss and clamping component.

19.4.2 Spring Steel U-Nut

Spring steel fasteners are made of high carbon steel and u-nut is one the most common clip that is used in plastic parts and it needs to be installed from the edge of the part. U-nut can be installed manually or through an automated process. There are other fastener clip solutions where they can be installed in the middle of the panel, but the focus on this section is only about u-nut. Key factors that need to be considered during design: Panel thickness, hole size, and edge distance (Figure 19.13), so it is important to follow the design recommendation provided by the supplier. U-nut is available in various thread sizes such as ST4.2/ST4.8 or standard thread sizes M4, M5, M6, M8 or M10. However, for thermoplastic it is not recommended to use a U-nut with a thread size larger than M6 since the contact between the barrel and plastic panel is not large enough to withstand a high torque regardless of the plastic material strength. For this reason, using u-nut in thermoplastic does not generate a high torque spec and that is due to the small contact area between barrel and plastic panel.

Spring steel nut can come with and without prevailing torque features as shown in Figure 19.14. As discussed earlier, it is recommended to have a prevailing torque feature for non-structural joint and U-nut can provide this feature in a very cost-effective way. Otherwise, a thread adhesive patch is required for the bolt.

Helpful Tips:

- For U-nut, make sure to follow supplier recommendation regarding edge distance, hole size, panel thickness (Figure 19.14)
- A Torque to Failure is required for these joints to develop the assembly torque specifications for the production.

19.4.3 Grommet Nut

Figure 19.15 shows different designs of Grommet. This is a quick solution where you can snap the grommet nut or just clip it into the plastic panel. Thread forming screws into plastic will be used with this nut. The performance of the grommet depends mainly on its material, and therefore testing is recommended to determine the torque specification of that joint.

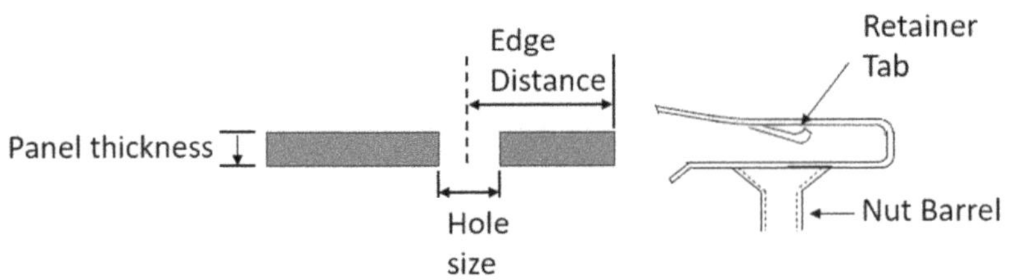

FIGURE 19.13 Design considerations for U-nut.

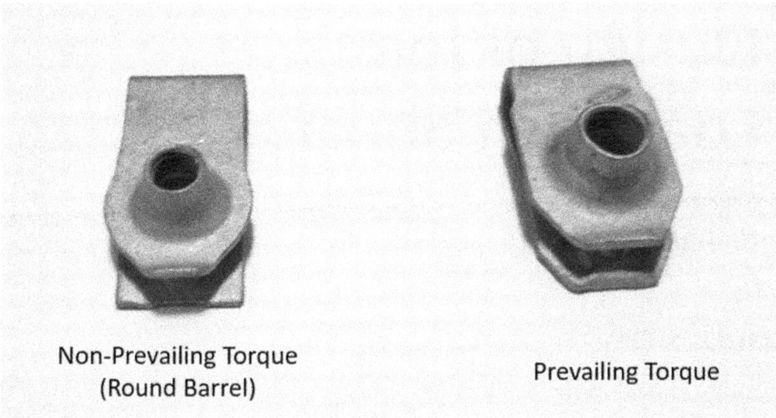

FIGURE 19.14 Non-prevailing vs prevailing torque U-nut.

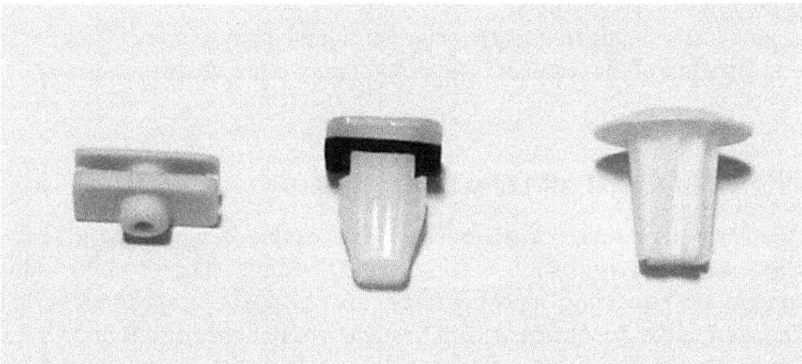

FIGURE 19.15 Different styles of grommet.

After discussing various fastener solutions for joining two thermoplastic parts, it should be noted that rivet nut and threaded insert can be used in the non-structural joint if compression limiter is not used in the clamped component, but in that scenario, these two solutions are not used effectively. In addition, rivet nut and threaded insert can be available as a male thread, and they are called rivet stud and insert stud if the application requires to have the nut as the driven fastener. Finally, the author would like to emphasize that testing is certainly necessary to validate any fastener solution in this chapter.

20 Design of Joints Loaded in Shear

Richard T. Barrett

20.1 INTRODUCTION

Much more attention is paid to the design of fasteners in tension than in shear, but shear design is just as important as tension. In particular, shear is the dominant fastener design factor in the aircraft and aerospace field, where structural components are thin and very elastic.

The two general types of shear fasteners are hole-filling (rivets) and clearance hole (bolts and screws). Both have important functions.

Shear loads and tension loads frequently occur simultaneously and must be combined and compared to the total strength of the fastener. These and many other design criteria are covered in this chapter.

20.2 DEFINITION OF SHEAR LOADING

Starting with the basic definition, a fastener shear joint is one in which the load is perpendicular to the fastener longitudinal axis as shown in Figures 20.1 (single shear) and 20.2 (double shear). In the single shear case, the equal and opposing forces are reacted to by shear stress on the fastener at plane A. In like manner, the double shear case has two reaction areas, at B and C, respectively. For riveted joints, there should be no gap between the material sheets. For a bolted joint in shear, it is imperative that the gap between material sheets be as close to zero as possible to prevent fastener bending. (Note that a single shear joint can have a small bending moment due to the eccentricity of the opposing forces.)

20.3 BOLTS AND SCREWS

20.3.1 Grip Length

The joint is weakened if the threads of the fastener lie in a shear plane between joint members or in the planes defined by the surfaces of the joint. To avoid this, only partially threaded fasteners are used, and a washer is placed under the nut. Thread runout occurs above the surface of the joint. The nut can be tightened fully, and only the smooth shank of the fastener lies within the hole.

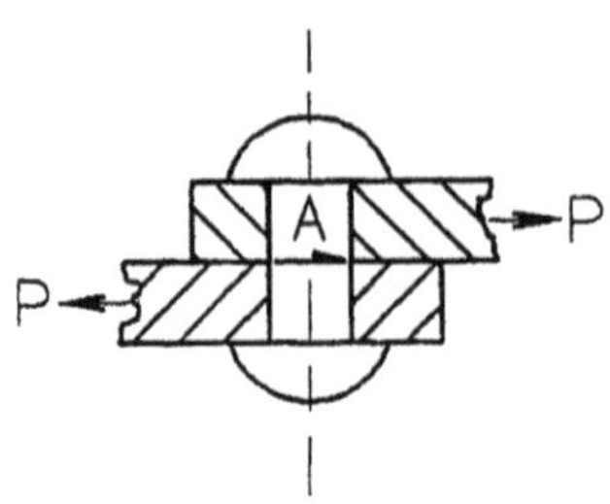

FIGURE 20.1 Single shear.

DOI: 10.1201/9781003501183-24

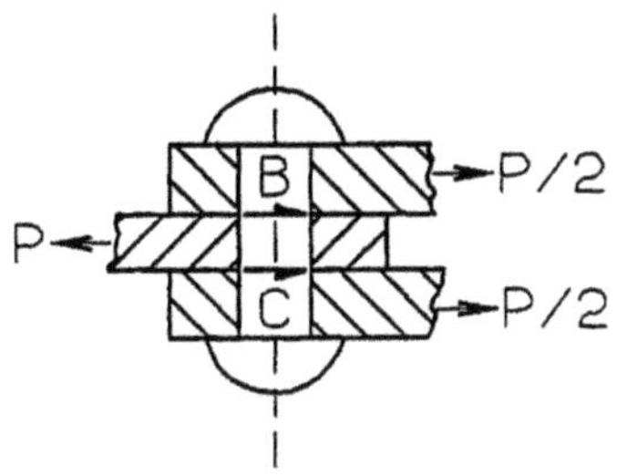

FIGURE 20.2 Double shear.

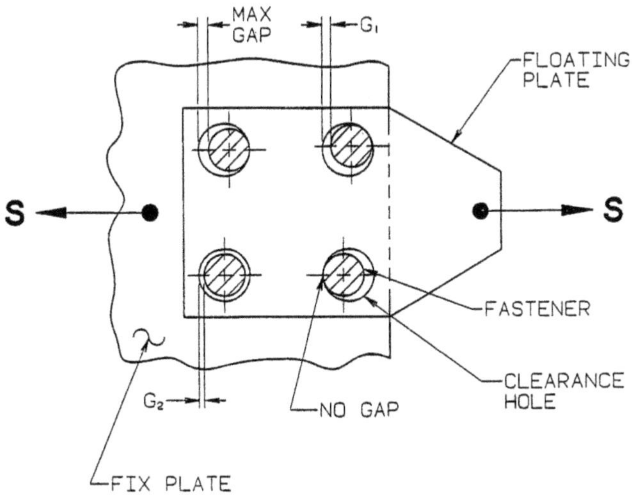

FIGURE 20.3 Clearance hole gaps on a fastener pattern (before loading).

All designs have their "weakest links." For fasteners, the ideal shear design would fail in the sheet or plate in bearing at a load near (but still less than) the shear yield of the fastener. Where a bolt pattern is being used, the only way to load up all of the bolts is for the bolt clearance holes to yield enough individually to close all the gaps between the bolts and the sheet (see Figure 20.3 for illustration). This means that the shear strength of the bolt must exceed the bearing yield strength of the sheet (the development of bearing stress allowables is covered in this chapter).

20.3.2 Fastener Edge Distance and Spacing

Although the actual edge distance and distance between fasteners can vary, a nominal value for edge distance is 2*D* with an absolute minimum of 1.5*D*, where *D* is the nominal diameter of the fastener. The nominal value for fastener spacing is 4*D*. These distances are measured from the hole centerlines, as shown in Figure 20.4.

The thickness of the bolted sheets and whether sealing is required between sheets are major factors in determining fastener spacing. Shear tearout calculations can be used to determine whether edge distances are adequate (shear tearout is discussed in Section 20.12.1).

20.3.3 Shear Heads and Nuts

If the design loads for a fastener are predominantly shear, it is possible to save weight and space by using fasteners with shear heads and/or nuts. The difference in thickness for comparable heads

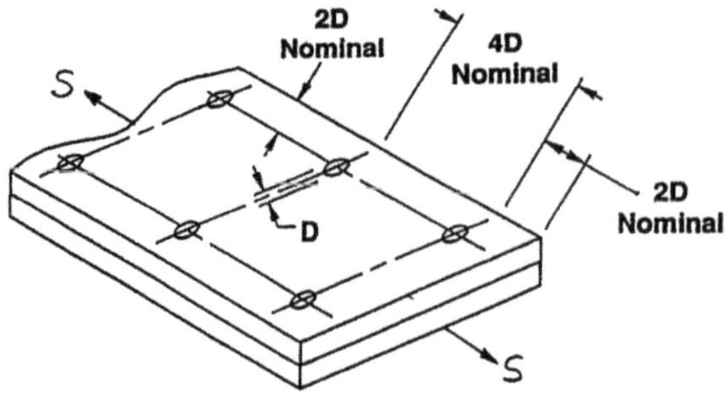

FIGURE 20.4 Fastener edge distance and spacing.

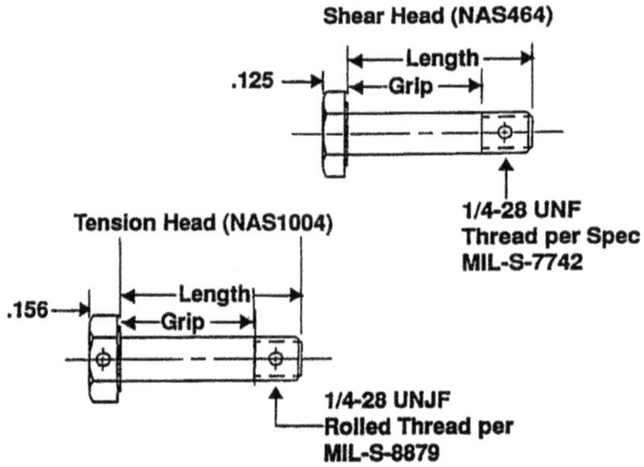

FIGURE 20.5 Shear head vs. tension head.

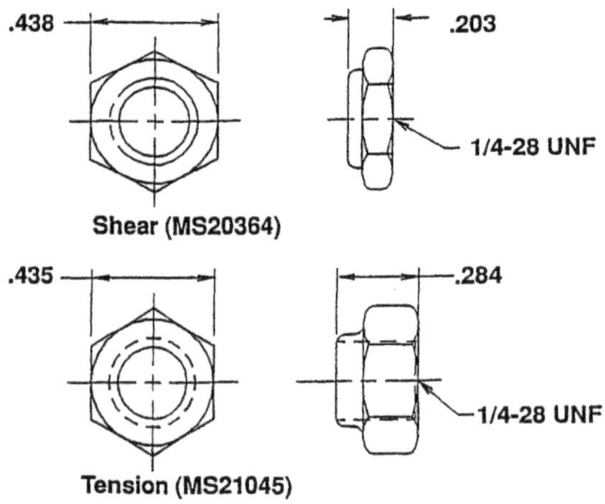

FIGURE 20.6 Shear nut vs. tension nut.

and nuts is shown in Figures 20.5 and 20.6, respectively. Note that the torque values must also be reduced by approximately 50% for fasteners with shear heads because the thin heads and nuts cannot develop the full tensile strengths of the fastener shanks.

20.3.4 Friction Effects

In most cases, friction forces between clamped (bolted) surfaces are not counted in determining load-carrying capability. The reason for not including friction forces is the inability to determine the coefficient of friction between clamped surfaces and accurate clamping forces generated by each fastener (Figure 20.7 illustrates the forces generated by joint friction). In some buildings and bridges, this contact friction is monitored closely enough to be used as a reaction force to prevent joint movement. However, the bolts are torqued to near yield levels to accomplish this goal.

20.4 RIVETS

The analysis of rivets is not much different from bolt analysis, except that rivets are normally interference fit in holes and have a lower tensile load capacity than bolts. The primary use of rivets is to carry shear loads in relatively thin sheet or plate (the maximum sheet thickness is usually less than the rivet diameter). The two main reasons for this requirement are to make the rivet critical in bearing rather than shear (sheet will yield before rivet yields) and to be able to form the shop head on the rivet (a more complete description of rivets is given in Ref. [1]).

20.4.1 Inter-Rivet Buckling

If the distance between rivets is too large, the sheet can buckle as a column between rivets. This principle can be illustrated by pulling a sheet of paper in a three-ring binder along the 11 in. border. The sheet will buckle between holes before the holes start tearing out. For a practical elastic buckling check of a riveted joint, as shown in Figure 20.8, use the buckling formula [2],

$$F_{\text{bcr}} = \text{KE}\left(\frac{t}{b}\right)^2 \tag{20.1}$$

where

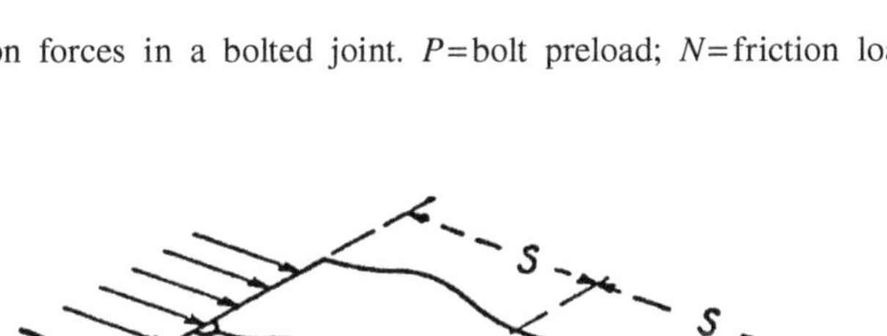

FIGURE 20.7 Friction forces in a bolted joint. P=bolt preload; N=friction load=$P\mu$. Shear load on bolt=$F - 2N$.

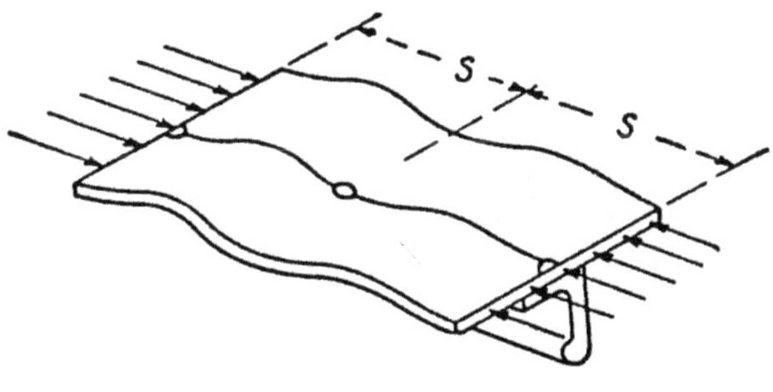

FIGURE 20.8 Sheet inter-rivet buckling (From Ref. [2]).

F_{bcr} = buckling stress (psi)
t = thickness of sheet (in.)
b = width of sheet (in.)
a = length of sheet (in.)
E = modulus of elasticity (psi)
$s = a$ = rivet spacing (in.)

and K is a factor taken from Figure 20.9.

Note that this formula is based on a Poisson's ratio of 0.3 and is for elastic buckling only. Since the ratio varies from 0.28 to 0.35 for all metals, a value of 0.3 should be satisfactory for most calculations. Euler's formula can be used for more exact calculations [2]. A modification of Eq. (20.1) using the tangent modulus Et must be used for plastic buckling.

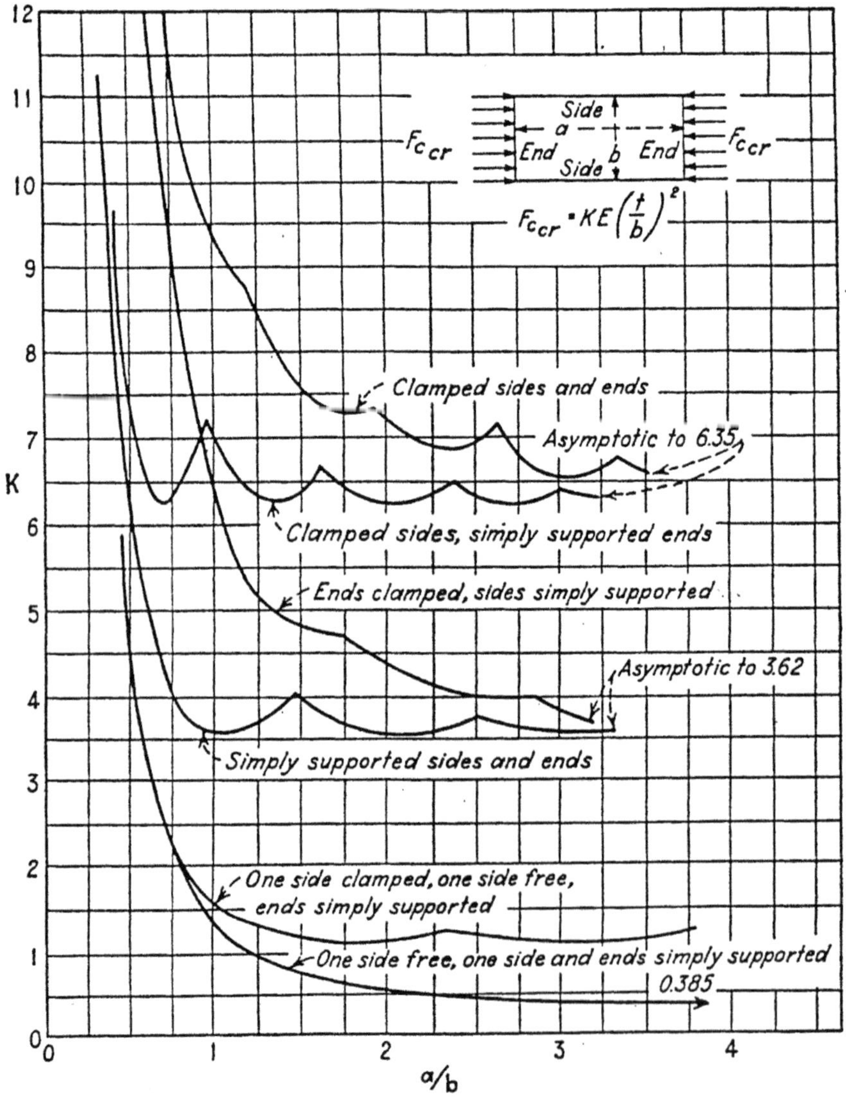

FIGURE 20.9 K vs. a/b for sheet buckling (From Ref. [2]).

20.4.2 Countersunk and Dimpled Holes

Countersunk (flush or flat head) rivets and screws are very common, particularly for attaching aerodynamic surfaces. The most common countersunk head angles are 82° and 100°, although special head angles can vary from 60° to 120°. A countersink should never extend all the way through the first sheet, since this would create a knife edge hazard, as explained below.

20.4.3 Knife Edge Hazards

Knife edges are stress risers and are to be avoided in countersunk sheet holes (see Figure 20.10). To avoid this problem, make the sheet thickness t greater than $1.5h$ as shown in the figure. If it is not possible to have $t \geq 1.5h$, then dimple the outside sheet and countersink the hole in the inner (thicker) sheet as shown in Figure 20.11. If neither sheet is thick enough to machine a countersink, both sheets can be dimpled to provide a flush outside surface as shown in Figure 20.12. However, the sheet must have sufficient ductility to dimple without cracking.

20.4.4 Hole-Filling versus Non-Hole-Filling Fasteners

Sometimes a designer will propose a combination of rivets and bolts or bolts and dowel pins to share a shear load. This is not an acceptable design. Rivets are hole-filling (interference fit). Dowel pins

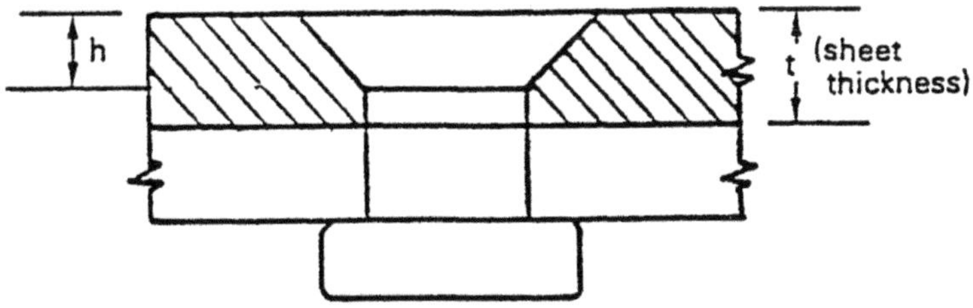

FIGURE 20.10 Knife-edge countersunk sheet.

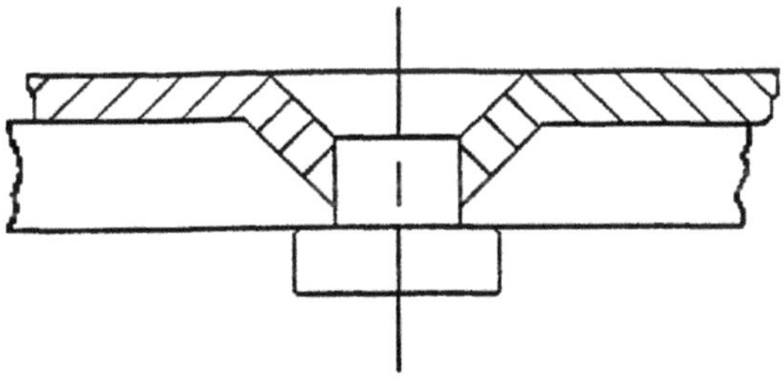

FIGURE 20.11 Dimpled and countersunk holes.

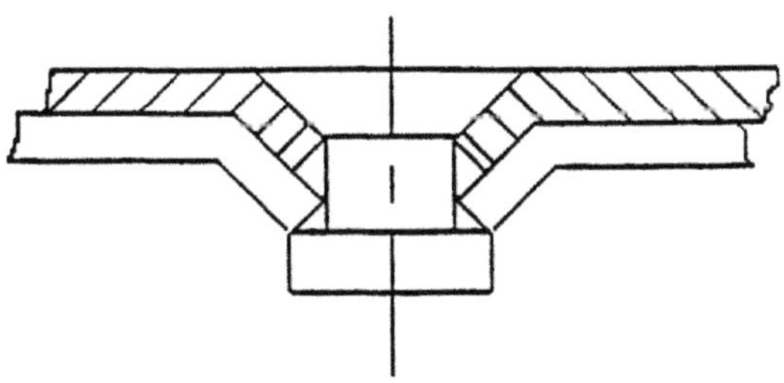

FIGURE 20.12 Both holes dimpled.

are slip or snug fit, and bolts have clearance holes. Therefore, any combination of these three types of shear fasteners will give uneven load distribution, with the tightest fasteners picking up most of the load first.

Note that this does not preclude the use of dowel pins for the alignment of a bolted joint as long as the bolts are designed to carry all of the structural loads.

20.5 MATERIAL ALLOWABLES

For fastener material selection, many aspects must be considered. Some of these are

1. Strength required
2. Operating temperature range
3. Corrosiveness of environment, including both chemical and galvanic corrosion
4. Availability of fasteners in the desired material
5. Fatigue resistance (and general ductility)

Aerospace fasteners are not available from the European or Asian markets. Therefore, metric aerospace fasteners are presently available only by special order from U.S. aerospace fastener manufacturers to AIA/SAE specifications.

It is best to pick a material ultimate tensile strength level under 200 ksi even for aerospace design, because of the stringent manufacturing and quality control procedures required for very high-strength materials. Typical aerospace fasteners are in the 140–190 ksi range, while industrial fasteners are in the 60–150 ksi range.

20.5.1 Shear Stress

Shear ultimate allowables for most aerospace materials are given in the handbook MIL-HDBK-5 [3]. Bearing yield and ultimate allowables are given in MIL-HDBK-5 for many of the common aerospace materials. However, commercial specifications such as those of SAE, ASTM, and AISC, rarely give more than tensile ultimate, tensile yield, and elongation.

In the absence of shear allowables, the following approximate ultimate values may be used (F_{su} = shear ultimate force and F_{tu} = tensile ultimate force):

For carbon steels, use $F_{su}=0.6F_{tu}$.
For stainless steels, use $F_{su}=0.55F_{tu}$.
For aluminum, use $F_{su}=0.55F_{tu}$.

Since a shear yield (F_{sy}) is a difficult thing to determine, my rule of thumb is to use $F_{sy}=0.75F_{su}$ for a shear yield allowable.

20.5.2 Bearing Stress

A bearing allowable is an empirical (or fictitious) value that is determined from a shear test (usually per MIL-STD-1312). The true bearing stress varies from 0 at the side of the hole (90° from the load direction) to a maximum at the center of the hole as shown in Figure 20.13a. This semicircular load distribution is equated to an equivalent stress, using a rectangular area equal to Dt, where D is the fastener diameter and t is the sheet thickness, as shown in Figure 20.13b. The bearing stress allowables such as those given in MIL-HDBK-5 are determined (per MIL-STD-1312) by testing a sheet to failure in shear. The failure load is divided by Dt to give the ultimate allowable bearing stress. In like manner, the yield load is used to calculate an allowable bearing yield stress. These allowables usually range from 1.5 to 2.0 times the actual compressive yield (F_{cy}) and compressive

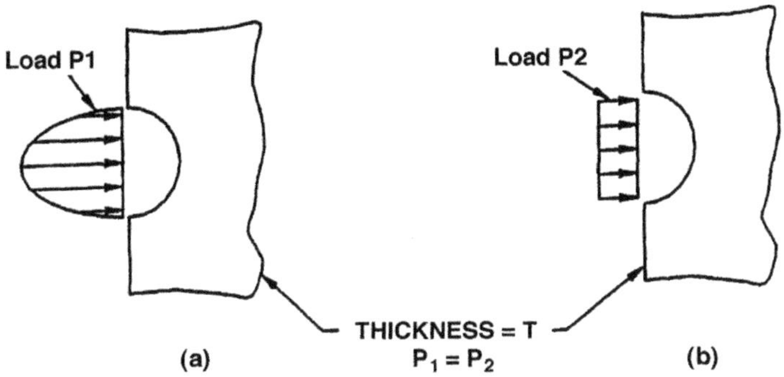

FIGURE 20.13 Development of bearing stress allowables.

TABLE 20.1
COSMOS/M1.7 Shear Load Summary

Bolt Diam. (in.)	Shear Load (lb.)	Plate Thickness (in.)	Hole Size (in.)	Max. Displacement (in.)
0.500	10,000	0.500	0.563	0.0225
0.375	7,500	0.500	0.438	0.0201
0.250	3,500	0.350	0.281	0.0117

ultimate (F_{cu}) of the material for an edge distance of *2D* on the sheet. In the absence of established bearing allowables, $F_{bru} = 1.5F_{cu}$ and $F_{bry} = 1.5F_{cy}$ may be used for design allowables for all common structural metals except magnesium (use 1.3 for magnesium).

20.5.3 Development of Bearing Stress Distribution in a Bolt Hole

C. Anthony Hermann of NASA Lewis and I wrote a computer program to determine the actual stresses and deformations from a bolt in bearing in a steel plate. We used the COSMOS/M1.70 Finite Element Model with plastic deformation. A summary of the cases run is provided in Table 20.1.

The bolts were assumed to be SAE Grade 8, and the plate was mild steel with a tensile yield of 37,500 psi. No fastener yield was reached in these calculations, and all of the displacements shown in Table 20.1 are in the plate material at the bottom of the clearance holes. The total contact angle was approximately 80° of the theoretical 180° for each of these holes. This analysis reinforces the design concept that a fastener pattern with clearance holes must be able to elongate the plate clearance holes to distribute the total load.

For details of the COSMOS/M program, contact C. Anthony Hermann at Lewis Research Center in Cleveland, Ohio.

20.6 COUNTERSUNK HEADS AND MATCHED DRILLING

A very common design error of countersunk fasteners in tapped holes is to make the mating plates on separate machines. Then an attempt is made to assemble two separate hole patterns, both of which are self-aligning (the countersink on one and the tapped hole on the other). In addition, the normal tolerances in true location of the holes cause further mismatch of the holes. Figures 20.14 and 20.15 show the problems caused by these various mismatches. The arrows in the countersunk areas indicate where head bending will start.

Tapped holes with countersunk heads should be avoided. A better design would be to use a clearance hole and a nut. At least this allows the countersunk head to do the alignment. If the back side of the plate is blind (not accessible for nut installation), a floating nutplate can be installed. However, a floating

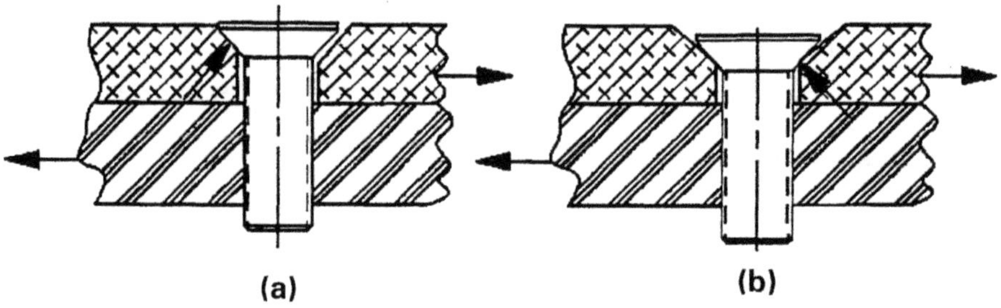

FIGURE 20.14 Shear loading of countersunk fasteners. (a) Holes match but countersink is not in line. (b) Wrong countersink used.

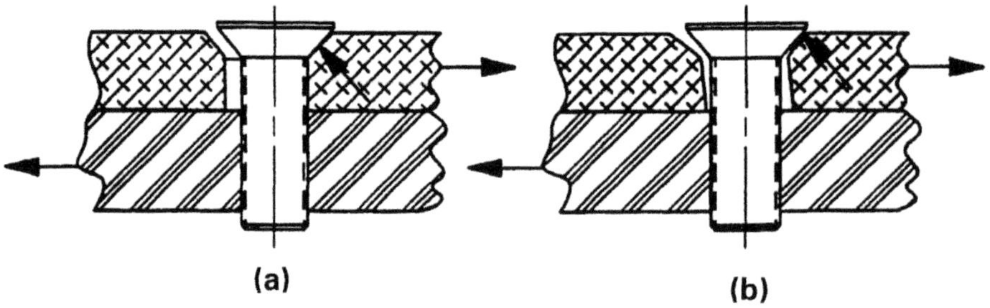

FIGURE 20.15 Shear loading of countersunk fasteners. (a) Holes parallel but not in line. (b) Holes not parallel.

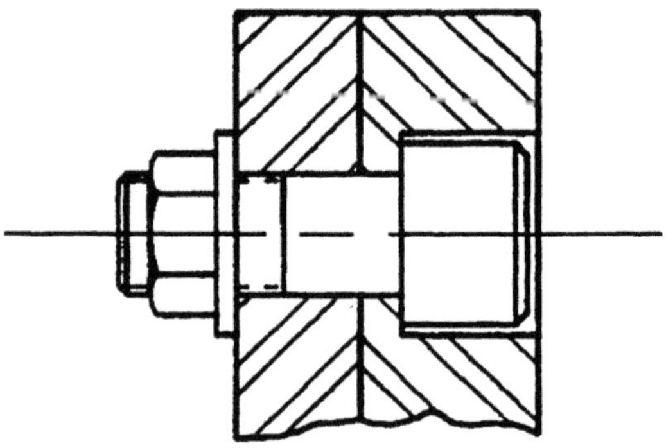

FIGURE 20.16 Counterbored hole.

nutplate only provides alignment adjustment parallel to the surface to which it is riveted. It does not twist to provide alignment for a bolt or screw that is not perpendicular to the mounting surface.

In summary, if countersunk holes are required, they should be line-drilled and countersunk while the mating sheets are clamped together to prevent misalignment problems.

20.7 COUNTERBORED HOLES

If weight is not critical, it is easier to design for flush surfaces with socket-head screws in counterbored holes. However, the plate must be thick enough to have adequate shear area left on the shank after cutting away a large portion of the total thickness (see Figure 20.16). A hex or 12-point head can also be installed in a counterbored hole, but the counterbored diameter must be large enough to allow a socket to go over the bolt head.

It is a little easier to get a counterbored surface parallel to the plate surface than to get a countersunk surface on line with the clearance hole axis.

A partial counterbore (which is also called a spotface) is sometimes used on sloped surfaces to give a top surface parallel to the bottom surface in a joint to prevent bending on the bolt head.

20.8 INSERTS

Inserts are normally used in relatively soft materials to provide strong threads to engage steel fasteners. There are two basic types of inserts for structural applications: wire thread (helicoils) and threaded (keenserts). Each has its advantages and disadvantages as given below. (Helicoils are made by Emhart Fastening Technologies of Shelton, CT; keenserts are a product of Fairchild Aerospace Fasteners, Torrance, CA.)

The wire thread insert is a wire spring with a diamond-shaped cross section that is installed in a tapped hole as shown in Figure 20.17. This insert can also have a self-locking feature as shown in Figure 20.18. The advantages of a wire thread insert are that it can be installed in a

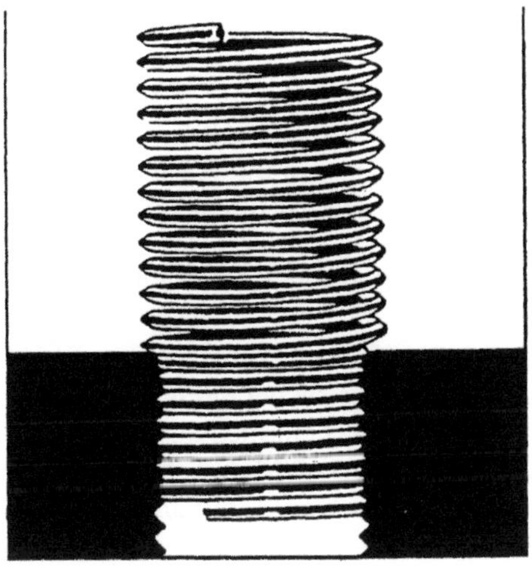

FIGURE 20.17 Wire thread insert installation.

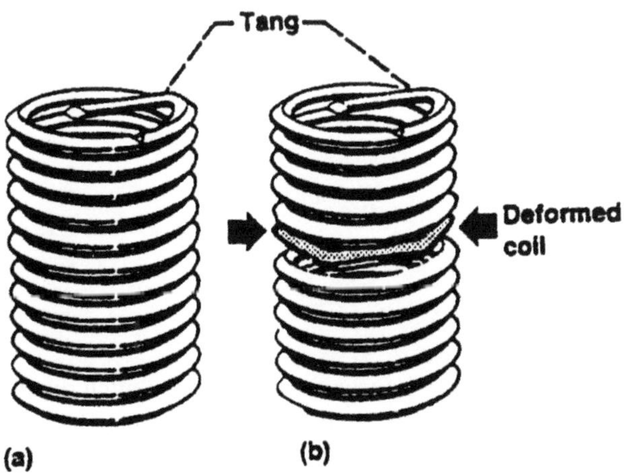

FIGURE 20.18 Wire thread insert types. (a) Free running; (b) locking.

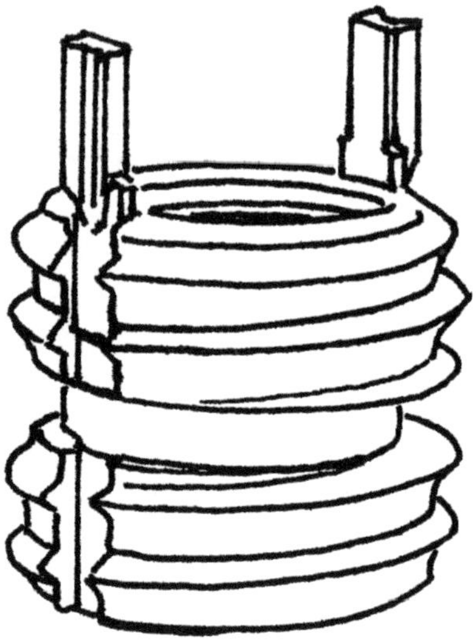

FIGURE 20.19 Keensert.

smaller tapped hole than a threaded insert (keensert), and it is much lighter than a keensert. On the negative side, the helicoil is harder to install, and it has a smaller outside diameter than a threaded insert (for the same diameter internal thread). This smaller diameter gives less thread shear strength in the softer material.

A helicoil does not lock as well as a threaded insert, so it will sometimes come out with the fastener when the fastener is removed.

The keensert is a solid threaded insert with both external and internal threads. It can have an external locking feature such as the prongs shown in Figure 20.19, or a nylon plug or a deformed thread. The internal locking can be done by deformed thread, plug, or pellet or by a split-beam (flat beam) locking nut on the bottom of the insert as shown in Figure 20.20b.

Helicoil inserts are available in lengths from 1*D* to 3*D* in order to develop the full bolt tensile strength in softer materials. However, keenserts have a fixed length for a given internal thread size, with the outside thread made larger for heavy-duty (higher load) applications.

Even if a keensert is not capable of developing the full tensile strength of a bolt, it is still satisfactory for shear applications if the bolt installation torque is kept low enough to prevent tensile pullout of the keensert. Remember that the edge distance (2*D*) must be based on the insert's external thread diameter.

Although insert manufacturers sometimes state pullout loads, the loads frequently have to be calculated. Pullout values for specific inserts are given in MIL-I-45914. Pullout allowables can be calculated by using the formula

$$P = \frac{\pi d_m F_S L}{3} \tag{20.2}$$

where

P = pullout load (lb)
d_m = mean diameter of threaded hole (approximately equal to pitch diameter) (in.)
F_S = material ultimate or yield shear stress allowable (psi) (note that aerospace design practices normally require both yield and ultimate margins of safety.)
L = length of full thread engagement (the last two or three threads at the bottom of the tapped hole are only partially formed).

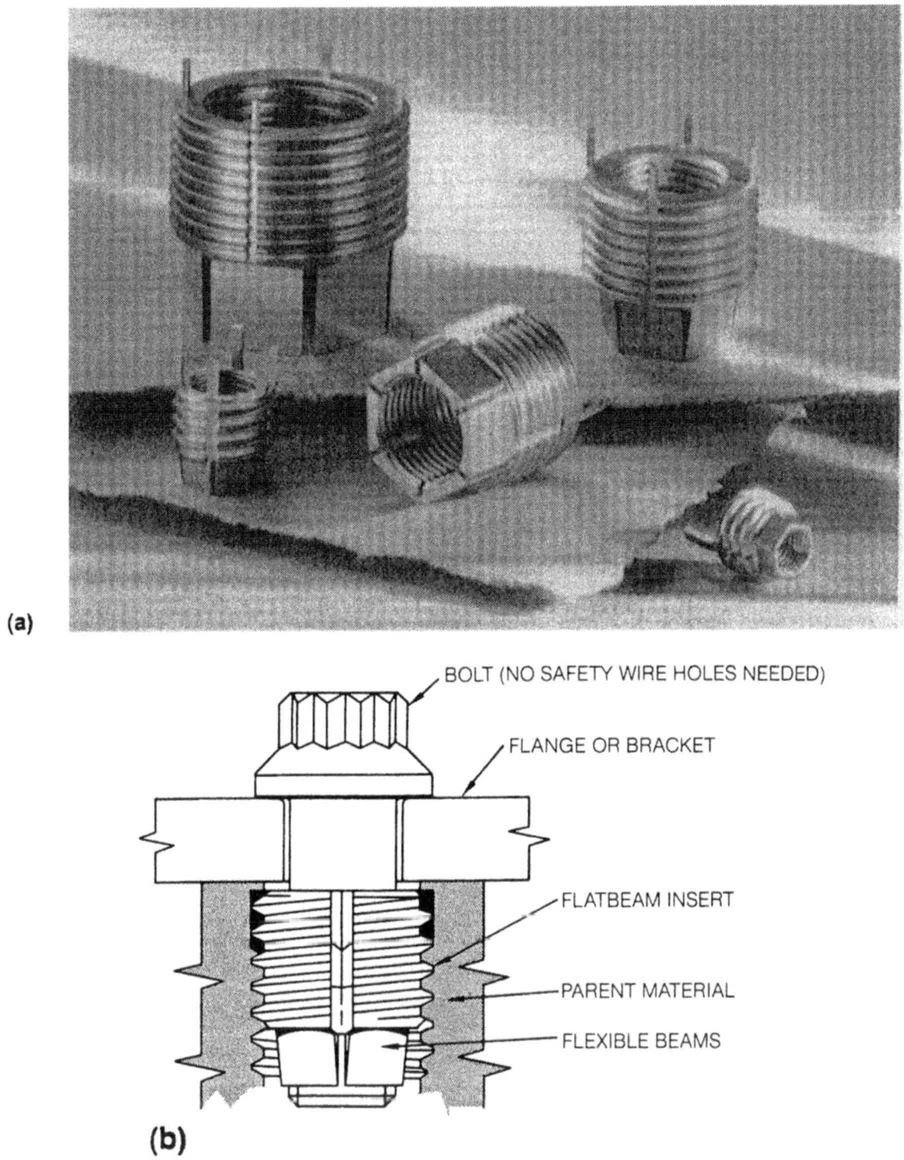

FIGURE 20.20 Flat beam insert manufactured by Tridair Fairchild. (Courtesy of Fairchild Fastener Division, Torrance, CA.)

Note that the 1/3 factor is empirical. If the threads were perfectly mated, this factor would be 1/2, since the total cylindrical shell area of the hole would be split equally between the bolt threads and the tapped hole threads. The 1/3 factor is used to allow for mismatch between threads.

20.9 NUTPLATES

A nutplate (or anchor nut) is a blind nut that is normally riveted to the back of a sheet that is too thin to tap. It requires the drilling of three holes: two for the installation rivets and a through hole for the fastener. The two basic types of nutplates are fixed and floating (see Figures 20.21 and 20.22). The floating nutplate is the most popular because the nut is mounted in a crimped "cage" to allow

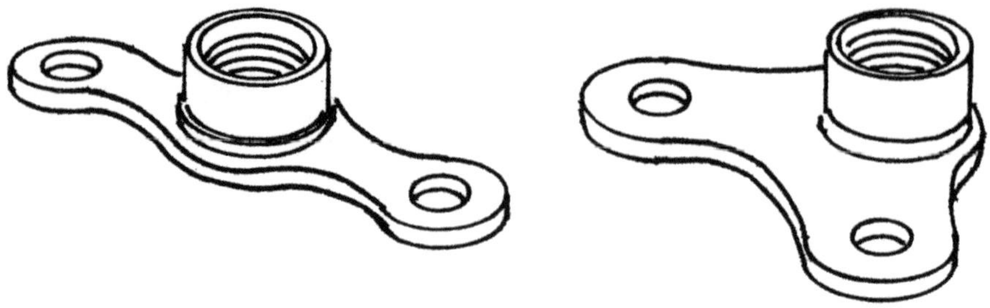

FIGURE 20.21 Fixed nutplates.

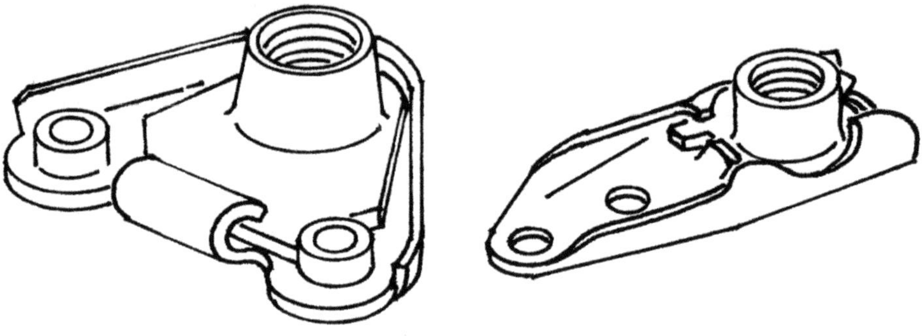

FIGURE 20.22 Floating nutplates.

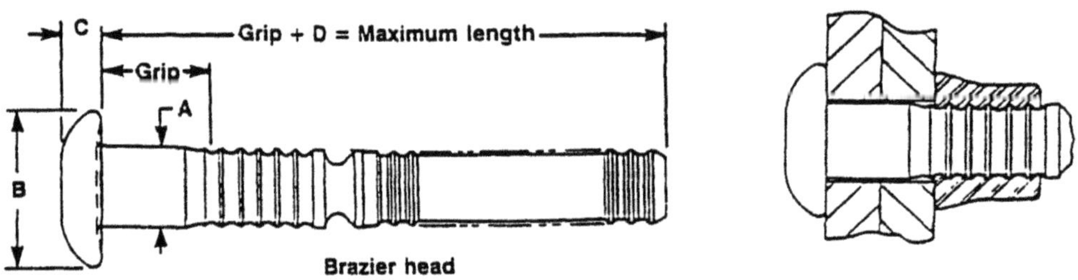

FIGURE 20.23 Installed huckbolt fastener. (Courtesy of Huck Manufacturing Co., Irvine, CA.)

the nut to align itself with the mating fastener. However, the nutplate only moves parallel to its mounting surface, so it will not correct for fasteners that are not perpendicular to the mounting surface. In general, nutplates can have the internal thread-locking capabilities of a regular nut. However, the most popular one in the aerospace field is the deformed thread type.

Nutplates are frequently used for shear and access panels. Once again, a good design should have the fasteners critical in bearing (sheet weaker than fastener) and the fastener grip length sized to have no threads in the through holes in the sheet.

20.10 LOCKBOLTS

In general, a lockbolt is a nonexpanding, high strength fastener that has either a swaged collar or a type of threaded collar to lock it in place. It is installed in a standard drilled hole with a snug fit but normally not an interference fit. A lockbolt is similar to an ordinary rivet in that the locking collar or nut is weak in tension loading and is difficult to remove once installed. An example is shown in Figure 20.23. Hence, lockbolts are normally used for shear designs.

The heavy truck industry uses lockbolts extensively for assembly of major structural components because they fit tighter than bolts and can be more quickly installed. Since these fasteners have clearances that are in between those of bolts and rivets, they should not be used in a shear design with rivets and/or bolts.

20.11 ANALYSIS OF A PATTERN OF FASTENERS IN SHEAR

For analyzing a pattern of fasteners, the first task is to find the centroid of the pattern. This is done by the standard static method of arbitrarily picking x and y axes and using a unit area times its distance from each axis to get the X and Y of the centroid (see the Appendix to this chapter). Although it is poor design practice to have different sizes of fasteners in a pattern, they can be analyzed by picking the most common one as a unit value. Then ratio the other areas to this one. For example, if eight bolts of a 12-bolt pattern are 0.375 in. in diameter and the other four are 0.313 in., the 0.375 in. diameter shank area, 0.1104 in.^{2}, would be the unit value of one and the value for the 0.313 in. diameter fasteners would be 0.0767/0.1104 or 0.69. The shank areas can be used for shear because it is not good design practice to have threads in bearing. However, for tension loads, the thread root area should be used for calculations.

In many cases, the fastener pattern may be symmetrical, as shown in Figure 20.24, so that the centroid is obvious and calculations are unnecessary. With the centroid located, the next step is to divide the load R by the number of fasteners n to get the direct shear load P_c on each fastener. To account for bending, the next step is to find $\sum r_n^2$ for the pattern of fasteners. The r_n's are the radial distances from the centroid to the center of each fastener. In the symmetrical example shown in Figure 20.25,

$$\sum r_n^2 = 4\left(r_1^2\right) + 4\left(r_2^2\right) \tag{20.3}$$

Now calculate the moment of the load R about the centroid ($M = Re$) also from Figure 20.25. The shear load (pounds) for a particular fastener due to this moment is

$$P_e = Mr / \sum r_n^2 \tag{20.4}$$

where r = distance from the centroid to the (usually) outermost fastener (in inches). Note that the expression for P_e is analogous to the formula for torsional stress (torsional stress = Tr/J) except that P_e is in pounds. The two shear loads (P_c and P_e) can now be added vectorially to give the resultant

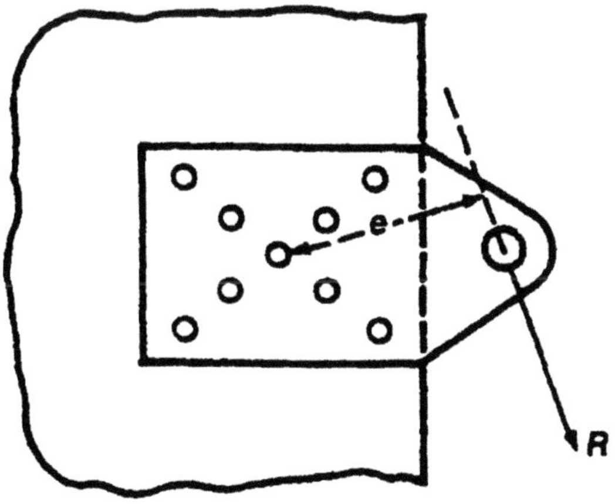

FIGURE 20.24 Symmetrical load pattern.

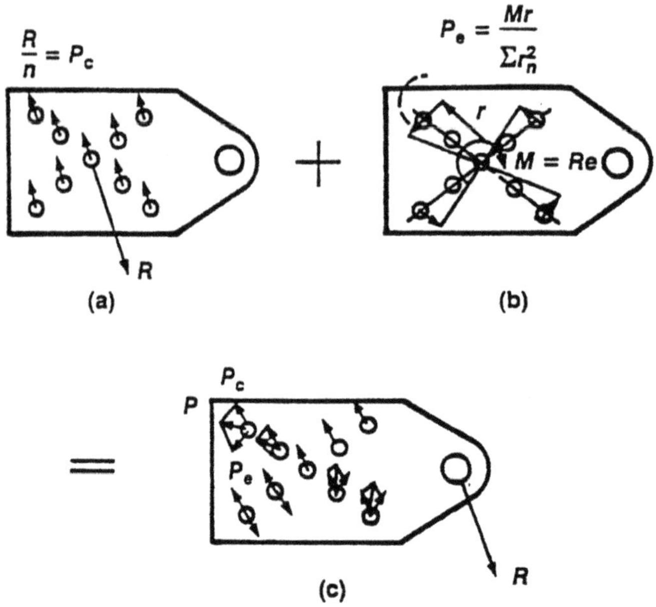

FIGURE 20.25 Combining of shear and moment loading.

shear load. This load must now be combined with the fastener tension load and compared to the total strength of the fastener, using load or stress ratios (see Section 20.13).

20.12 DEVELOPMENT OF FITTING STRESSES AND SHEAR TEAROUT STRESSES

20.12.1 Shear Tearout

Earlier in this chapter, we covered the development of bearing stress allowables. These allowables were predicted on an edge distance of $2D$, where $2D$ is the distance from the centerline of the fastener to the edge of the sheet (e.g., for a 0.250 in. diameter fastener, $2D=0.500$ in.). The adequacy of this assumption can be checked by taking two shear areas (see Figure 20.26) out of the wall thickness and comparing their capacity to the bearing allowable of the hole.

Since the hole is oversize, the actual length of each shear area is $2D - D_o/2$, where D is the nominal fastener diameter and D_o is the diameter of the clearance hole. Multiplying by the thickness t will give two shear areas of

$$A = \left(2D - \frac{D_o}{2}\right)t \tag{20.5}$$

Using as an example a 0.250 in. diameter bolt in a 0.281 in. diameter clearance hole in a 6061-T6 aluminum plate of 0.188 in. thickness:

$$A = \left[2(0.250) - \frac{0.281}{2}\right](0.188) = 0.0764 \text{ in.}^2$$

There are two of these areas, so

$$A_S = 2A = 2(0.0764) = 0.1528 \text{ in.}^2$$

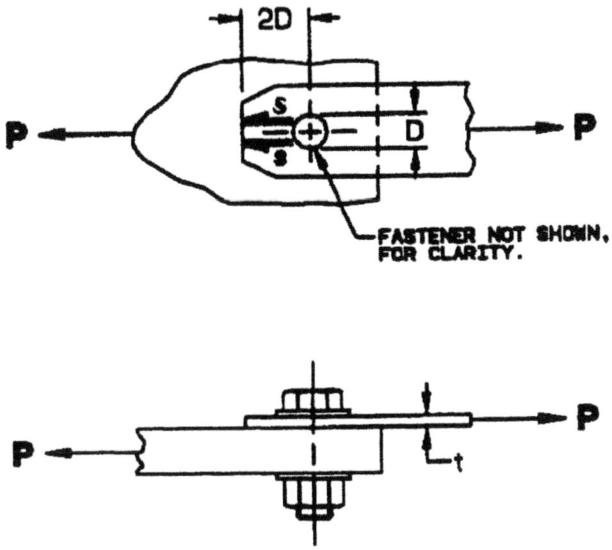

FIGURE 20.26 Shear tearout.

Using yield strengths for comparisons (although ultimate strength can also be used):

F_{sy} = 20 ksi (from Reynolds Aluminum)
F_{bry} = 58 ksi (e/d = 2.0) (from MIL-HDBK-5)

For shear,

P_{total} = (0.1528)(20,000) = 3,056 lb capacity in shear

For bearing,

$$P_{total} = (Dt)(58\text{ ksi}) = (0.250)(0.188)(58{,}000)$$
$$= 2{,}726\text{ lb, which is less than } 3{,}056\text{ lb}$$

This shows that the $2D$ edge distance is more than adequate in shear for this case. An edge distance of as little as $1.5D$ can be used in special cases, but it needs to be checked for shear tearout.

Shear tearout is more likely to occur along lines up to 40° from each side of the direction of the applied load, so taking two shear areas near the load line is slightly conservative.

20.12.2 Development of Fitting (Lug) Stresses

Lugs are usually used for rotating connections. They are usually thicker than an ordinary shear joint. Since lugs are thick, they normally have a double shear connection to prevent pin bending (see Figures 20.27 and 20.28). If the lug and clevis are relatively soft (e.g., aluminum) compared to the pin, they may also have bushings installed for wear and lubrication. The lug end is usually semicircular, which is desirable from both weight and clearance aspects.

The stress checks shown in Figure 20.27 must be made to verify the lug design. Remember that the hole diameter for the bushing must be used when determining the net wall thickness for stress checks. It is also imperative that the joint not have gaps sufficient to induce pin bending.

To illustrate the necessary stress checks, we will work through an example with the following dimensions (from Figure 20.27) and allowables:

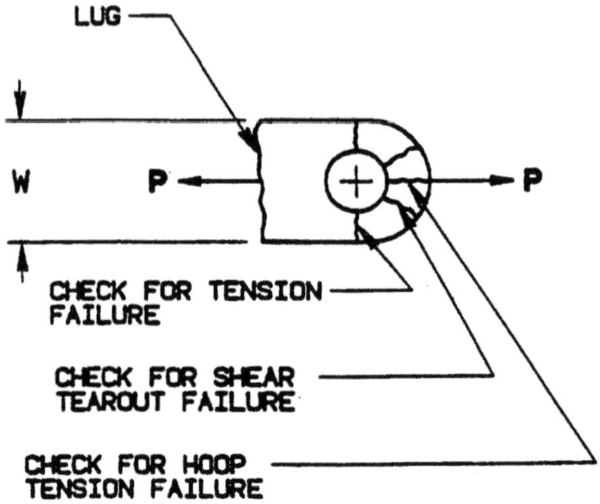

FIGURE 20.27 Semicircular lug (shown without clevis and pin for clarity).

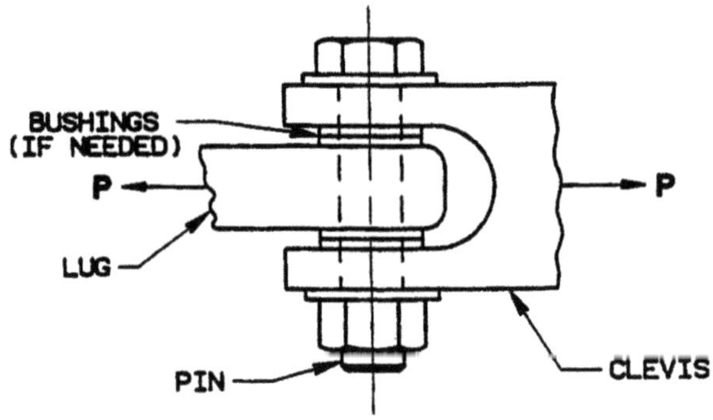

FIGURE 20.28 Semicircular lug assembly.

$W = 1.0$ in.
$D =$ pin diameter $= 0.5$ in.
$P = 3{,}000$ lb
$t =$ lug thickness $= 0.5$ in.

Material is 6061-T6 aluminum
From Ref. [3],

$F_{tu} = 42$ ksi, $F_{ty} = 35$ ksi, $F_{su} = 27$ ksi

From Ref. [4],

$F_{sy} = 20$ ksi

From Ref. [3],

$F_{bru} = 67$ ksi, $F_{bry} = 50$ ksi

For tension failure:

$$f_t = \frac{P}{A} = \frac{3{,}000}{2(0.25)(0.5)} = 12 \text{ ksi}$$

$$\text{Margin of safety (MS)} = \frac{\text{allowable load (stress)}}{\text{actual load (stress)} \times \text{safety factor}} - 1 \tag{20.6}$$

$$\text{MS} = \frac{F_{\text{tu}}}{f_t} - 1 = \frac{42}{12} - 1 = 2.5 \text{ (ultimate)}$$

$$\text{MS} = \frac{F_{\text{ty}}}{f_t} - 1 = \frac{35}{12} - 1 = 1.92 \text{ (yield)}$$

For shear tearout failure:

$$f_s = \frac{3{,}000}{2(0.25)(0.5)} = 12 \text{ ksi}$$

with margins of safety

$$\text{MS} = \frac{F_{\text{su}}}{f_s} - 1 = \frac{27}{12} - 1 = 1.25 \text{ (ultimate)}$$

$$\text{MS} = \frac{F_{\text{sy}}}{f_s} - 1 = \frac{20}{12} - 1 = 0.67 \text{ (yield)}$$

For bearing stress failure:

$$f_{\text{br}} = \frac{3{,}000}{(0.5)(0.5)} = 12 \text{ ksi}$$

with margins of safety

$$\text{MS} = \frac{F_{\text{bru}}}{f_{\text{br}}} - 1 = \frac{67}{12} - 1 = 4.58 \text{ (ultimate)}$$

$$\text{MS} = \frac{F_{\text{bry}}}{f_{\text{br}}} - 1 = \frac{50}{12} - 1 = 3.17 \text{ (yield)}$$

For hoop tension failure, the wall thickness is too high to use the hoop stress formula $f=Pr/t$, so a thick-wall formula [5] will be used:

$$f_t = \frac{Pr_i^2}{r_o^2 - r_i^2}\left(1 + \frac{r_o^2}{r^2}\right) \tag{20.7}$$

In this case,

r_o = outside radius = 0.5 in.
r_i = inside radius = 0.25 in.
r = radius to the point of stress desired
p = bearing stress = 12 ksi

Since we want the maximum stress (at the inside surface) we will let $r=r_i$. Then

$$f_t = \left(\frac{12{,}000(0.25)^2}{(0.5)^2 - (0.25)^2} \right) \left(1 + \frac{(0.5)^2}{(0.25)^2} \right)$$

$$= \frac{750}{0.1875} \times 5 = 20{,}000 \text{ psi}$$

$$\text{MS} = \frac{F_{\text{tu}}}{f_t} - 1 = \frac{42}{20} - 1 = 1.1 \text{ (ultimate)}$$

$$\text{MS} = \frac{F_{\text{ty}}}{f_t} - 1 = \frac{35}{20} - 1 = 0.75 \text{ (yield)}$$

Note that safety factors (SF) have not been addressed here. They can be included in the 3,000 lb load or they can be added to the MS calculation:

$$\text{MS} = \frac{F}{f \times \text{SF}} - 1 \tag{20.8}$$

In most aerospace applications the SF is different for yield and ultimate, so it would be added to the MS calculation. For example, a carrier-based plane will have higher safety factors than a land-based plane due to the impact loads from arresting hooks. Note also that stress concentration factors (CF) are to be included in the safety factors.

In summary, our lug is good for the 3,000 lb load, providing that no added safety factors are large enough to give negative margins of safety. Also note that rigorous lug design criteria are given in proprietary aircraft publications such as Lockheed's *Stress Memo Manual.*

20.13 COMBINING FASTENER SHEAR AND TENSION LOADS

When a fastener is subjected to both shear and tension loads simultaneously, the loads must be combined and compared to the total strength of the fastener. A good example of this type of loading is that of a fastener that has a tensile load created by torquing it (which also creates shear stress due to twisting) and an applied direct shear load.

The customary way to combine tensile and shear loading is to use load (stress) ratios [1]. The load ratios are

$$R_S \,(\text{or } R_1) = \frac{\text{actual shear load}}{\text{allowable shear load}} \tag{20.9}$$

$$R_T \,(\text{or } R_2) = \frac{\text{actual tension load}}{\text{allowable tension load}} \tag{20.10}$$

Note that the shear and tensile allowable loads can be obtained from publications such as MIL-HDBK-5 or Federal or SAE specifications for a given fastener size and strength.

The next step is to determine which interaction curve to use from Figure 20.29. The most unconservative is $R_1^3 + R_2^3 = 1$. (This series of curves is from an old edition of MIL-HDBK-5.) The current MIL-HDBK-5 [3] lists only the single formula, $R_S^3 + R_T^2 = 1$. If weight is not critical, I recommend the $R_1 + R_2 = 1$. However, the accepted aerospace curve is the $R_S^3 + R_T^2 = 1$. After calculating the R_S and R_T ratios, the margin of safety can be calculated as

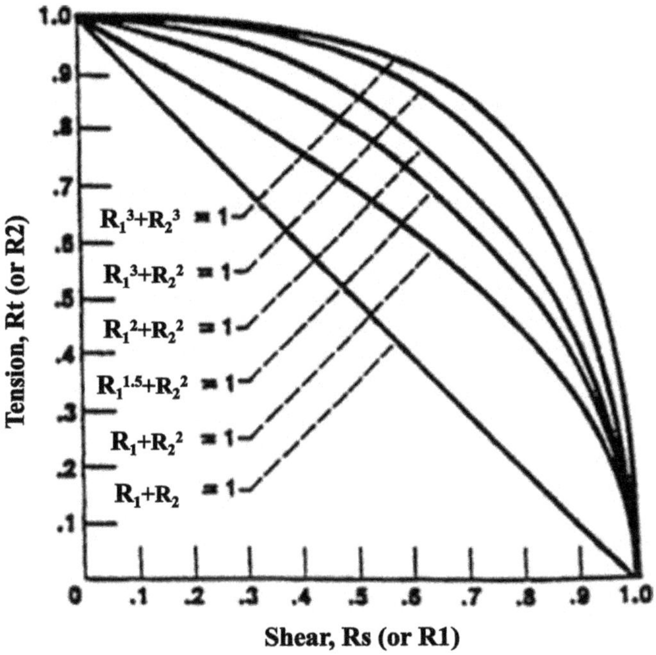

FIGURE 20.29 Interaction curves.

$$MS = \frac{1}{R_S^X - R_T^Y} - 1 \quad (20.11)$$

where *X* and *Y* are exponents chosen to suit your own degree of conservatism. Note that the fasteners should be made of ductile material with a minimum elongation of 10%.

20.14 DOWEL PINS

Dowel pins are usually solid close tolerance pins that are used for aligning two mating components. The pins are usually mounted in one half with a slight interference fit. The mating half has close tolerance clearance holes to slip over the protruding pins. Then the two halves are bolted together and the bolts are analyzed for the total shear load. However, the joint can be designed with enough dowel pins to carry the entire shear load. In this case, the bolts would only ensure that the joint was tight enough to prevent pin bending. Note that either the bolts or the pins (not both) carry the shear loads, owing to the differences in hole fits. The pins will load up first and yield their holes (or fail the pins) before the bolts start to pick up shear loads.

Dowel pins installed in blind holes with interference fits should be avoided if pin removal is desired. It is much better to have a through hole, which allows the pin to be driven out from the back side. The through hole also allows venting, which can be a problem on blind hole installations unless a longitudinal venting groove or flat is provided on the pin.

Shear allowables for dowel pins are usually determined by test by the manufacturer because the area at the shear plane may vary (the pin cross section may not be constant and/or a common geometric shape).

Some common dowel pins are shown in Figures 20.30 (straight), 20.31 (grooved), 20.32 (vented), and 20.33 (drilled and tapped with vent).

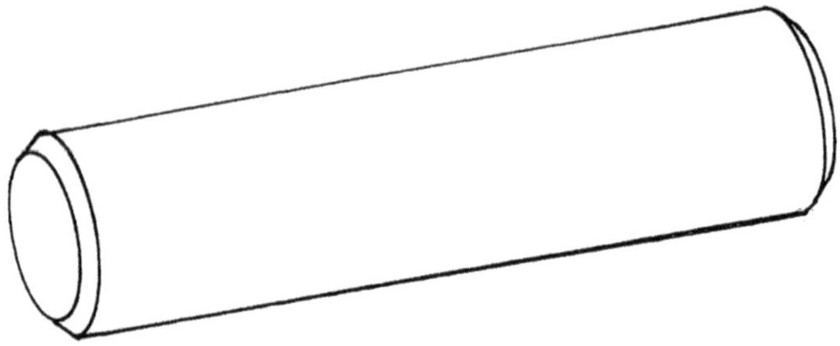

FIGURE 20.30 Straight (solid) dowel pin.

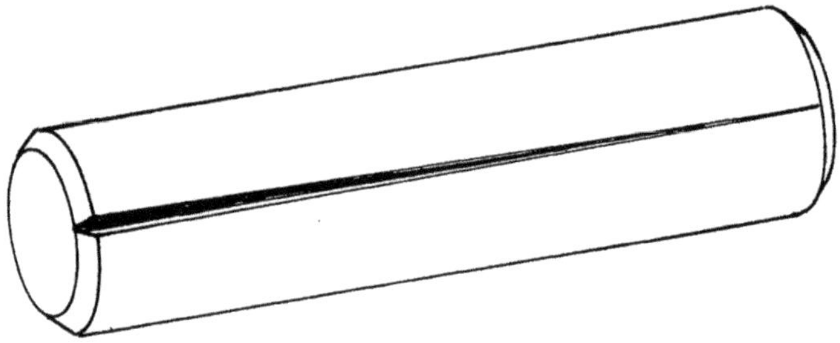

FIGURE 20.31 Grooved dowel pin.

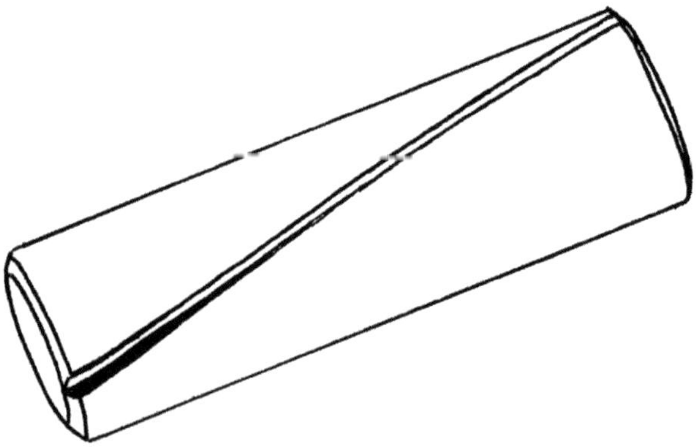

FIGURE 20.32 Vented dowel pin.

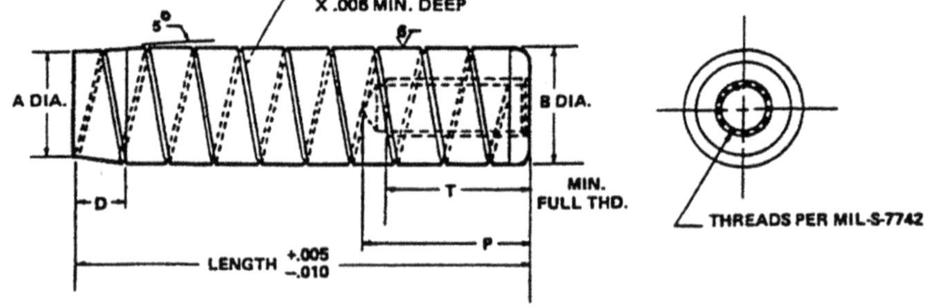

FIGURE 20.33 Drilled and tapped dowel pin with vent.

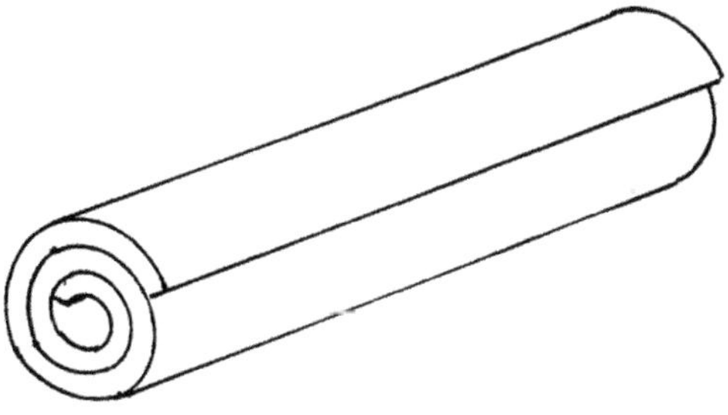

FIGURE 20.34 Roll pin.

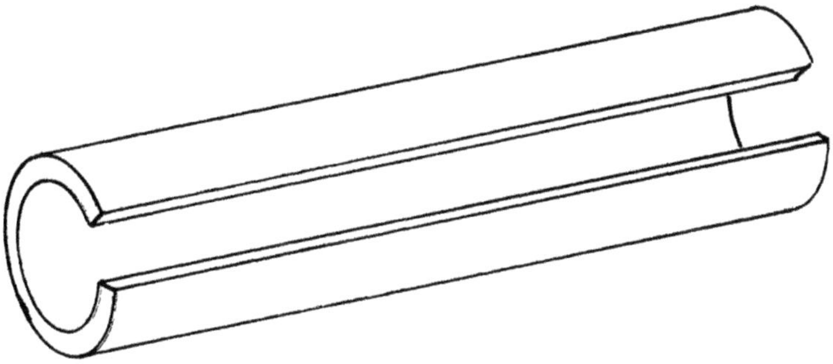

FIGURE 20.35 Slotted tubular pin.

20.15 ROLL PINS

Roll pins (also called spring pins) are usually made by rolling a piece of thin alloy sheet metal to a given diameter, with a chamfer on each end, and heat treating it to a high hardness. The chamfer on the ends allows it to be started and driven into a hole. The coiled cross section decreases in diameter during driving to give an interference fit. A typical roll pin is shown in Figure 20.34. However, if the cross section is not overlapped (Figure 20.35), the pin is called a slotted tubular pin.

Roll pins are used for quick-disconnect joints with interference fit, such as for attaching a handle to an actuating shaft. They can also be used as a fixed pin in a rotating joint where the rotating component has an oversize hole to allow free rotation around the roll pin.

Load-carrying capabilities for both roll pins and slotted tubular pins are usually determined by the pin manufacturer.

20.16 ANALYSIS OF COUNTERSUNK FASTENERS IN SHEAR

Referring back to Figure 20.10 we see that the only bearing area in contact with the rivet shank is $(t - h)D$ unless the holes are match drilled and countersunk. How well the countersunk rivet head can carry shear load depends solely on its fit in the countersunk hole. The most conservative analysis would use only $(t - h)D$ for the shear area. However, the head will carry some load if the rivet installation is closely controlled. Chapter 8 of MIL-HDBK-5 [3] gives shear allowables for countersunk rivets, based on test data. These values are acceptable for general design use, as long as installation tolerances are as good as those used for the MIL-HDBK-5 test articles. Rivet strengths must also be compared to those in the MIL-HDBK-5 table to ensure full compatibility of the tabulated values.

Since countersunk bolt heads have looser fits than countersunk rivet heads, the best design approach would be to use only the $(t - h)D$ bearing area for bearing stress calculations for countersunk bolts or screws.

20.17 HORIZONTAL SHEAR LOADS ON BEAM CONNECTION FASTENERS

If we revisit the strength of materials section, which deals with beam bending and horizontal shearing stresses $f_S = VQ/Ib$, we can determine the fastener sizes and spacing required for a composite beam. In this example case, the uniformly loaded beam cross section is symmetrical and consists of two flat plates that are bolted together as shown in Figure 20.36 (not very design-efficient but easy to work with).

Assign specific numbers for the dimensions shown in Figure 20.37 as follows:

$h = 2$ in. (two 1 in. plates)
$b = 6$ in.
$w = 400$ lb/in. of length
ℓ = length of beam = 50 in. (simply supported)

$$V_{max} = \frac{400 \text{ lb/in.} \times 50}{2} = 10,000 \text{ lb vertical shear}$$

$$M = \frac{W\ell^2}{8} = \frac{400(50)^2}{8} = 125,000 \text{ in.} - \text{lb}$$

$$Q = \frac{bh}{2}\left(\frac{h}{4}\right) = (6\times1)(0.5) = 3.0 \text{ in.}^3$$

$$I = \frac{bh^3}{12} = \frac{6(2)^3}{12} = 4.00 \text{ in.}^4$$

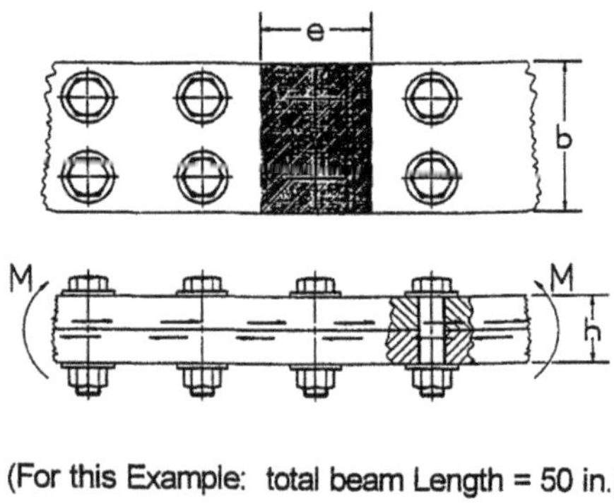

FIGURE 20.36 Horizontal shear load. For this example, total beam length = 50.

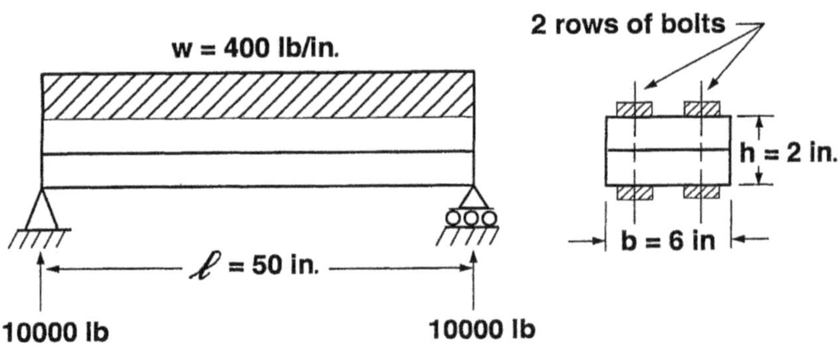

FIGURE 20.37 Beam loading for horizontal shear load.

Note: The moment of inertia I should be recalculated after the final bolt diameter is determined, as

$$I = \frac{(b-2D)h^3}{12}$$

where D is the bolt hole diameter. Then a final $f_S = VQ/Ib$ should be calculated.

For no bolt hole reduction,

$$f_S = \frac{10,000 \times 3.0}{4.00 \times 6.00} = 1,250 \text{ psi}$$

This stress will be acting over the entire shaded area of Figure 20.36, which is equal to $6e$.

Then $S(\text{lb}) = 1,250 \times 6e = 7,500e$. If we use 0.50 in. diameter Grade 5 bolts, two bolts are good for 10,500 lb each in single shear without yielding. Solving for e, the maximum distance between rows of bolts,

$$e = \frac{2 \times 10,500}{7,500} = 2.8 \text{ in.}$$

which is close for our plate thickness.

Go to 0.56 diameter bolts and recalculate a new moment of inertia with the hole areas deducted.

$$I = \frac{(b-2D)h^3}{12} = \frac{[6-2(0.63)](2^3)}{12} = 3.17 \text{ in.}^4$$

Now

$$f_S = \frac{\text{VQ}}{\text{IB}} = \frac{10,000(3.0)}{3.17(6.0)} = 1577 \text{ psi}$$

$$S - f_S \ A_S = 1,577(6e) = 9,464e \text{ lb}$$

Assume two Grade 5 (0.56 in. diameter) bolts with $F_{sy} = 13,900$ lb each. Then

$$e = \text{max. spacing} = \frac{2(13,900)}{9,464} = 2.94 \text{ in.}$$

Of course, bearing and bending stresses must be checked and the bolts must be critical in bearing to properly distribute their shear loads. Note also that bolt spacing can change as the vertical shear changes.

20.18 GALVANIC CORROSION AND STRESS CORROSION CRACKING

Although galvanic corrosion and stress corrosion cracking should be a major consideration in all structural design, we will touch on them here as they apply to fasteners.

20.18.1 Galvanic Corrosion

Galvanic corrosion is caused by contact between two dissimilar metals in the presence of an electrolyte. This electrolyte can be as simple as atmospheric air or water with other impurities present. A galvanic cell is created, and the more active (anode) of the two materials is eroded and deposited on the less active (cathode). The farther apart two materials are in the galvanic corrosion table, the greater the galvanic action between them [1].

To prevent or deter galvanic corrosion, one or both materials are coated (or plated) with a more inert material. For example, aerospace aluminum is anodized or iridited for protection. The fastener holes are coated with MIL-S-8802 sealant or zinc chromate paste to inhibit galvanic corrosion. Steel fasteners are also coated with cadmium or zinc. Aluminum rivets are usually painted with zinc chromate paint prior to installation.

It is good design practice to put hardened flat washers under both the head and nut of a bolt to minimize damage and corrosion to the coated surfaces around the holes.

20.18.2 Stress Corrosion

Stress corrosion occurs when a corrosion-susceptible material incurs tensile stress in a corrosive environment. An otherwise ductile material will fail at a stress much lower than its tensile yield strength due to surface imperfections (usually pits or cracks) caused by the corrosive environment. For a material that is susceptible to stress corrosion, the higher the strength (and the lower the ductility), the more sensitive it is to stress corrosion cracking.

Although no master list of stress corrosion-sensitive fasteners exists, two common ones are H-11 tool steel bolts and 5056 aluminum rivets. The H-11 bolts are usually high tensile strength (200–260 ksi), so they must have a protective coating to prevent corrosion. On the other hand, the 5056 aluminum has a low tensile strength (42 ksi) so it can be used for coldheaded rivets. It will sometimes crack from stress corrosion under "no load" from the tensile stresses induced during cold heading. However, 5056 aluminum is acceptable for riveting of magnesium because of their relative positions on the galvanic scale.

20.19 HELPFUL HINTS FOR FASTENERS USED IN SHEAR JOINTS

1. Avoid feather edges on countersunk or counterbored holes. Feather edges are prone to developing cracks when loaded in shear.
2. Shear allowables for many countersunk and protruding head rivets are given in Section 20.9 of MIL-HDBK-5 [3].
3. Single shear rivet joints produce some bending and some rivet tension due to the change in load path at the rivet. This load condition should be checked on critical designs because rivets have small tension capabilities compared to their shear capacities.
4. Do not combine rivets and bolts to share a shear load.
5. Do not combine dowel pins and bolts to share a shear load.
6. Size the grip length of a shear fastener so that no threads are in bearing (only smooth shank).
7. Do not use interference fit fasteners in composite materials, as the compressive stresses can cause decomposition of the holes.
8. Do not subject fasteners to bending loads.
9. A pattern of fasteners must be critical in bearing (rather than shear) to distribute a shear load properly.
10. Check fasteners for combined shear and tensile stress.
11. Use washers under both the head and nut of a bolt to avoid embedment in the joint material and give more predictable torque vs. tension values.
12. Avoid the use of countersunk holes unless they are match-drilled with their mating parts.
13. Perpendicularity of fastener holes to their mounting faces is critical (SAE allows ±2° tolerance) to avoid bending moments on the fastener heads.
14. Avoid the use of short "fat" fasteners in tension. Since a fastener is a tension beam with an elongation of $\Delta l = PL/AE$ (where P = tension load, L = length between nut and under side

of head, and A=cross-sectional area), we can see that a smaller A and larger L will give our fastener more elongation for a given load. The joint will then be easier to keep under compression and will be more fatigue-resistant.

15. Avoid the use of tapped holes if possible. A through hole with a nut is much cheaper and more convenient. Tapped holes have cut threads that are difficult to form and hard to inspect. If the tapped hole is in a much softer material than the fastener, it is usually necessary to use an insert. This adds further labor and cost to the design. Another disadvantage of tapped holes is that the elastic working length of the fastener (from hint 14 above) is shortened when part of the joint becomes the "nut."
16. Use a nut with a maximum hardness that is near the minimum hardness of the rolled external threads. This will allow the nut threads to yield enough to distribute the bolt load to adjacent threads (my thanks to my friend Dave Richardson of Lockheed Martin for this hint).

APPENDIX: FINDING THE CENTROID OF A BOLT GROUP[1]

We can usually find the centroid of a bolt group by inspection. If the bolts are not in neat rows, however, as in Figure 20.A1, we can use the following equations to locate it [7,8]:

$$\bar{X} = \frac{\sum A_i x_i}{\sum A_i} \tag{20.A1}$$

$$\bar{Y} = \frac{\sum A_i y_i}{\sum A_i} \tag{20.A2}$$

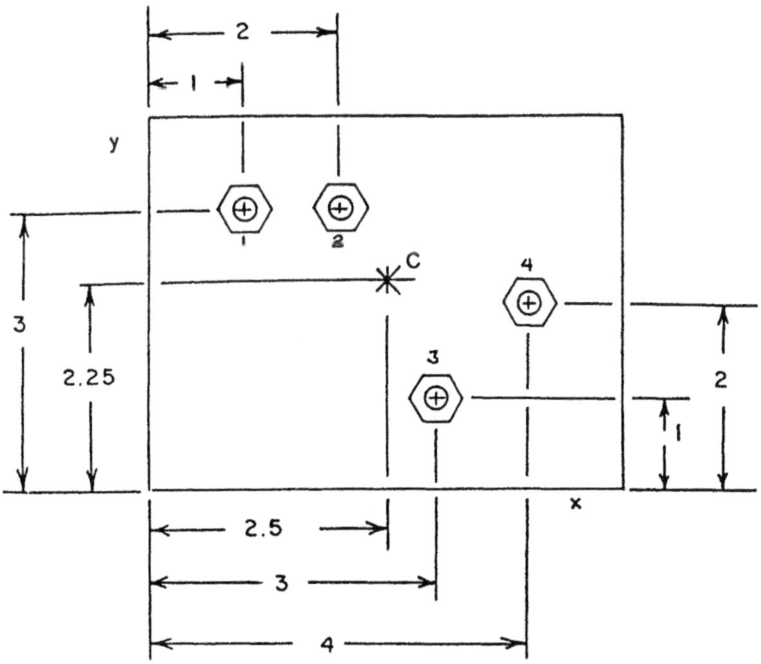

FIGURE 20.A1 This joint is used as an example in the text to compute the location of the centroid (C) of the bolt group. The bolts are ¼ in. in diameter. The x and y axes are arbitrarily located along the left and bottom sides of the plate.

where

A_i = cross-sectional area of the body of bolt *i* (in.^{2}, mm^2)
x_i = distance of bolt *i* from an arbitrarily located *y*-axis (in., mm)
y_i = distance of bolt *i* from an arbitrarily located *x*-axis (in., mm)
$\overline{X}$ = distance of the centroid from the *y*-axis (in., mm)
$\overline{Y}$ = distance of the centroid from the *x*-axis (in., mm)

With reference to Figure 20.A1, let us assume that the nominal diameter of the bolts is ¼ in. and that we have arbitrarily located *x* and *y* axes along the bottom and left edge of the splice plate as shown. The distance of bolt 1 from the *y*-axis is x_1; its distance from the *x*-axis is y_x; etc.

$$A_i = \frac{\pi\left(0.25^2\right)}{4} = 0.0491 \text{ in.}^2$$

$$\overline{X} = \frac{A_1x_1 + A_2x_2 + A_3x_3 + A_4x_4}{A_1 + A_2 + A_3 + A_4}$$

Since each bolt has the same cross-sectional area in this example,

$$\overline{X} = \frac{0.0491(1+2+3+4)}{4(0.0491)} = 2.5 \text{ in.}$$

Similarly,

$$\overline{Y} = \frac{0.0491(1+2+3+3)}{4(0.0491)} = 2.25 \text{ in.}$$

The centroid so located is labeled *C* in Figure 20.A1.

NOTE

1 This appendix is reprinted from Ref. [6] with minor changes. Reprinted with permission.

REFERENCES

1. Barrett, R. T., Fastener Design Manual, NASA RP-1228, NASA Lewis Research Center, Cleveland, OH, 1990.
2. Peery, D. J., *Aircraft Structures*, McGraw-Hill, New York, 1950.
3. Metallic Materials and Elements for Aerospace Vehicle Structures, MIL-HDBK-5G, Wright Patterson AFB, Dayton, OH, 1994.
4. *Reynolds Structural Aluminum Design*, Reynolds Metals Company, Richmond, VA, 1962.
5. Timoshenko, S., and G. H. MacCullough, *Elements of Strength of Materials*, Van Nostrand, New York, 1949.
6. Bickford, J. H., An *Introduction to the Design and Behavior of Bolted Joints*, 3rd ed., Marcel Dekker, New York, 1995.
7. Shigley, J. E., and C. R. Mischke (eds.), *Standard Handbook of Machine Design*, McGraw-Hill, New York, 1986, Section 23, Bolted and Riveted Joints, pp. 23.12–23.16.
8. Shigley, J. E., *Mechanical Engineering Design*, 3rd ed., McGraw-Hill, New York, 1977, p. 258.

21 Design Rules for Structural Steel Bolting

John H. Bickford and William A. Milek

21.1 INTRODUCTION

The Research Council of Structural Connections (RCSC), an organization of researchers, fastener manufacturers and suppliers, structural engineers, steel fabricators and erectors, and others first authored and adopted the Specification for Structural Joints Using ASTM A325 and A490 Bolts (hereinafter referred to as the Specification) in 1952 [1]. Since that time, the council has taken cognizance of the research developments and periodically updated the Specification based on the advances in the state of the art and usage. The council's Specifications of various adoption dates are reviewed, endorsed, and published by the American Institute of Steel Construction (AISC). The council also sponsored the writing of *Guide to Design Criteria for Bolted and Riveted Joints* [2], currently in its second edition (hereinafter referred to as the *Guide*). The *Guide* is based on approximately 800 references of experimental and theoretical studies dating from the late 1800s to recent times and is a comprehensive source of information on the state of the art of the behavior and strength of riveted and bolted joints. Additionally, the AISC implements the provisions of the council's Specifications in the form of extensive design aids and tables for a wide variety of typical structural applications in *Manual of Steel Construction*, Volume II [3]. Currently, the RCSC Specification and the implementing AISC Manuals Volume II are available in two formats, "Allowable Stress Design" (ASD) and "Load and Resistance Factor Design" (LRFD) [4].

These documents are of interest because, although focused on structural steel applications, they are the only bolting Specifications, of which we are aware, that not only provide authoritative criteria for strength of joints subject to various loading conditions but also tell the designer and user how much tension or preload is required (when preload is required) in installed bolts. To do this, the Specifications address all significant factors affecting the bolts, from their selection by the designer to their inspection upon receipt at the job site, handling and storage, verification of the suitability of fastener components (bolts, nuts, washers and lubrication thereof; each probably from more than one source) plus their suitability as an assembly and when installed by the method to be used in the work to achieve the specified pretension. They then specify how to achieve the required preload and how to inspect for it.

Because it is difficult to maintain control over many variables that affect bolt preload, this Specification has inspired the development of special fasteners that attempt to control or to indicate the attainment of the required preload in the fasteners during assembly. These documents address the inherent uncertainties of bolting and direct the designer's attention to the installation and inspection procedures necessary to ensure the performance of the connections assumed for the design. Other industries could benefit from such detailed attention to the selection and use of fasteners.

Another reliable structural steel bolt that should not be overlooked by designers is the ASTM A307. This bolt is a structural steel fastener for use primarily in shear-loaded applications, where the higher shear capacity or the slip resistance of tensioned high-strength bolts is not required. It is not covered by the Specifications that are the subject of this chapter because it is not considered high strength nor is it suitable for pretensioning to resist shear loads by friction between connected parts. On the other hand, it is a totally reliable bolt that was the standard common bolt for shear/

DOI: 10.1201/9781003501183-25

bearing applications prior to the development of the ASTM A325 and A490 bolts. The use of ASTM A307 bolts, wherever applicable, provides for economy not only through their lower cost but also because installation tensioning to a specified preload and inspection to ensure full pretensioning is not required, which greatly reduces field labor cost.

21.2 TWO VERSIONS OF ONE SPECIFICATION

The RCSC and the AISC provide two versions of the Specification based on different design philosophies. Both versions are supported by the same criteria for the strength and serviceability of fasteners and connected material extensively documented in the research literature. However, one version, subtitled "Allowable Stress Design" (ASD) [1] is based on the more traditional approach of applying a global factor of safety, based on experience and judgment, that is deemed adequate to provide for all the uncertainties in strength of material, tolerances in the manufacture of fasteners and connected material, improper storage and handling, variability of lubrication, human error in fabrication and field installation, reliability of estimation of applied loads and load combinations, reliability of engineering design formulas to predict actual connection performance, etc. The second version, subtitled "Load and Resistance Factor Design" (LRFD) [4] separates the load considerations from considerations of the resistance to load of the fastener and connected material. The resistance of bolts and the connected material to different potential modes of failure is developed by theory and test with resistance factors (ϕ factors) selected on the basis of statistical study of the data. The resistance, so determined, is required to be equal to or greater than the results of explicit consideration of the sum of loads in combinations (each load multiplied by a load factor 7, which may vary according to the combination). The general expression may be written as

$$\phi R_n \geq \gamma_D D + \gamma_L L + \gamma_W W + \ldots \tag{21.1}$$

where R_n=nominal tensile strength of the bolt (kip).

This deceptively simple formula is a truism that shrouds many problems. All it says is that the resistance (including all of its uncertainties) must be greater than the effects of the loads (including all of their uncertainties). The Greek letter ϕ (phi) is a coefficient less than unity reflecting the degree of uncertainty to which the strength may be overestimated, and the Greek letter 7 (gamma) is a coefficient greater than unity reflecting the degree to which the various load effects may be underestimated. With traditional ASD, the factors of safety for various modes of failure and load combinations varied widely. One basic objective in the development of the LRFD format was to calibrate the design concept to provide an essentially uniform "factor of safety" for the different failure modes. The safety factor 1.67 that had proven satisfactory in uncountable applications for the definitive case of a simple span beam subject to uniform loading was chosen as the target value for all cases of failure by yielding. A larger safety factor target value for the limit states of fracture was taken as approximately 2.0. A target central overall reliability factor β, encompassing both loads and resistance, was targeted at approximately 3.0. The result is that there are only small differences between the connections designed by the two procedures, with the more involved LRFD procedure providing slightly greater economy. The more important advantage of LRFD over ASD is in the more rational formulation, which is suited to logical evaluation of the effects that may result from advances in the state of knowledge of connection performance and to evaluating the effect of changes in any of the individual parameters on the load or resistance side or to evaluating any unique design problem.

A word of warning: The Specifications are subject to continuous scrutiny and updating by RCSC and AISC. While they are not legal documents in and of themselves, they become legal documents when incorporated in municipal building codes and in contract documents. Generally, designers and users must follow these Specifications if involved with steel structures such as buildings or bridges. If you deal with such structures, you should obtain the latest version of these documents from AISC and use them, rather than this chapter, to direct your activities. Hopefully, the following discussion will be helpful in imparting an understanding and use of the Specifications; but it should not be used as a substitute for the latest version of the documents on which this chapter is based (information on how to obtain these documents is included in the references at the end of the chapter).

This chapter alone may be helpful in the design of nonstructural steel joints and in the development of similar Specifications for other applications.

21.3 THE CONTENTS OF THE SPECIFICATIONS

As mentioned earlier, the provisions of the two versions of the Specifications are identical in stating requirements that are independent of the design calculation procedure to be used. Thus, it is possible to describe both documents with a single outline.

Section 1. Scope: The ASD and LRFD Specifications relate to the design for strength and slip resistance of structural steel joints using ASTM A325 or A490 bolts or equivalent fasteners, and to the real-world requirements for use to ensure that reliable performance is achieved.

Section 2. Bolts, nuts, washers, and paint: In both ASD and LRFD, each component of the fastener assembly is prescribed in detail by reference to applicable ASTM Specifications and/or American national standards. Insofar as paint applied to faying surfaces is concerned, the coefficient of friction of the cured paint coating is important only for joints specified as slip-critical. The Specifications require that to be used in slip-critical joints, the paint must be qualified. A test method to determine the slip coefficient for coating used in bolted joints, which was developed at the request of Steel Structures Painting Council and is based on research done at the University of Texas, is provided as an appendix to the Specification.

Section 3. Bolted parts: ASD and LRFD require that all material within the grip be steel. No material such as gaskets or insulation is permitted. Surface conditions that affect the serviceability limit state (slip resistance) are specified. Hole sizes and types—standard, oversized, short-slotted, and long slotted—are defined.

Section 4. Design for the strength of bolted connections and Section 5. Design check for slip resistance (different for ASD and LRFD) are addressed in detail subsequently.

Section 6. Loads in combination: ASD and LRFD have different wording but the same objective. More liberal stresses are permitted for the unlikely simultaneous occurrence of maximum loads of different types.

Section 7. Design details: This section of both ASD and LRFD specifies that standard holes may be used for any application but stipulates the conditions under which oversize, short-slotted, and long-slotted holes may be used and details how they are to be used when their use is permitted. The section also stipulates the conditions under which hardened washers must be used and where they must be located to prevent galling of the surface of softer connected material (which would cause high and misleading installation torques). Special attention is given to the use of special 5/16 in. plate washers or continuous bars with standard holes to completely cover long slotted holes and “bridge the gap” when they occur in an outer ply.

Section 8. Installation and tightening: This section is identical in ASD and LRFD. It covers the requirements for handling and storage to maintain the bolts and nuts in the clean, lubricated condition required for reliable installation. It specifies that a tension-measuring device must be on the job to be used daily where high-strength bolts are being installed and tightened (see Chapter 42). It then goes on to establish in detail the procedure that is to be followed for testing fastener assemblies, using the selected installation procedure to ensure that the specified installed tension requirements are achieved. Finally, it sets forth in detail the procedure to be followed for the installation and tightening of fasteners in the work using any one of four recognized methods (turn of the nut, calibrated wrench, alternative design bolts, or direct tension indicating devices).

Section 9. Inspection: Again, ASD and LRFD are identical. The most reliable method for ensuring the installation to a tension at least as great as the specified minimum is strict adherence to the recommended installation requirements of Section 8. Inspection should be directed primarily toward this goal. After-the-fact determination of the installed pretension to any degree of reliability is not possible with any known economical means. Testing by torque methods is subject to ±30% variation; thus, attempts to verify the tension in a bolt installed in accordance with specified procedures by torque-dependent methods is analogous to verifying the accuracy of a micrometer using a yardstick. For this reason, the section is primarily directed to inspector responsibilities to ensure proper installation in the first place. An arbitration inspection method is specified for use in cases of controversy.

21.4 DESIGN OF HIGH-STRENGTH CONNECTIONS

ASD and LRFD provisions in Section 21.4 are different in terms and formulations, each being unique to the design philosophy being presented; however, both may be related to the same criteria for the strength of the fastener or connected material as justified by the theories and test results documented in the *Guide*. The two philosophies of design lead to comparable connection designs. LRFD requires the detailed evaluation of the effects of multiple loads (dead loads, occupancy loads, snow loads, wind loads, and earthquake loads) in various combinations and each multiplied by a modifying factor, γ, that varies according to the combinations of loads being considered. A discussion of the load part of LRFD is beyond the scope of this chapter. Therefore, direct comparison of ASD and LRFD must be limited to the general statement that LRFD generally leads to a more economical design and a better understanding on the part of the designer of the effect of individual variables on the end result.

The performance of a bolt centered in the hole of a single bolt connection (low strength, high strength, fully or only partially tightened) subject to applied shear loading is shown diagrammatically in Figure 21.1. Note that as load is increased from zero, the deformation is small and proportional to the applied load. In this region, the bolt really does not "feel" the applied load. That is,

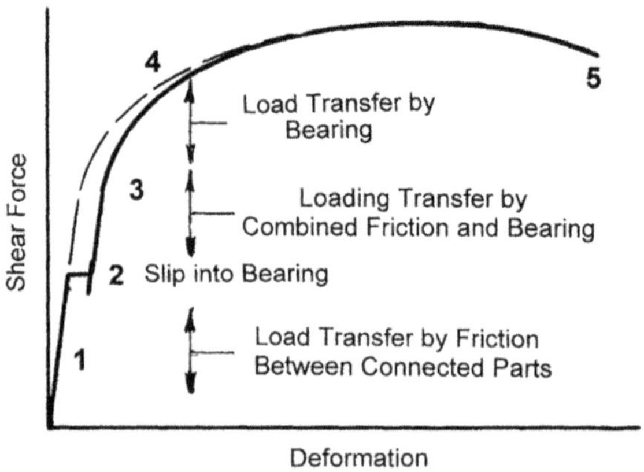

FIGURE 21.1 This graph illustrates the performance of a bolt centered in the hole of a single bolt connection if the bolt is subjected to an applied shear load. The bolt first deforms slightly (1) until the shear load causes major slip (2). The load transfer between joint members has been by friction up to this point; now and at higher loads it is transferred by bearing and friction. The bolt then experiences additional elastic deformation (3) until its yield strength is reached (4). If the shear load continues to increase, the bolt will experience plastic deformation until the bolt breaks (5).

the load is transferred by friction due to tension in the bolt between the bolted parts. At some level of loading, dependent on the coefficient of friction of the faying surfaces of the connected material and the preload in the bolt, a small well-defined displacement will occur, but nothing more. This displacement is due to the slippage of one part of the connected material relative to the other as the frictional capacity is exceeded. Slippage halts abruptly when the bolt comes into bearing against the sides of the holes. As the load is increased to higher levels, the deformations still remain essentially proportional to the load and on about the same slope as at low loads. In this region, load is probably transferred primarily by friction but with a little help from small bolt bearing forces against the sides of the hole—the relative proportions cannot be determined. As the load is increased still further, the proportion transferred by friction will reduce until eventually the connected material of the net section will begin to yield, with consequent reduction in thickness. At this point, the bolt clamp force reduces to zero, but the connection still does not fail. Even greater loads may be supported at the expense of deformation of the bolt and deformation due to bearing pressure on the connected material around the hole.

This graphical representation of the performance of a bolt in a hole is very instructive, and note should be taken of several points. Slip may occur early or late in the loading. The level at which it will occur depends entirely on the condition of the faying surfaces of the connected parts and the clamping force provided by bolt pretension. Slip is not a signal of imminent failure. Slip load is far below the ultimate strength of the connection and generally significantly below the level at which the bolt will be called upon to resist the load by direct shear alone. The strength of the connection (maximum supportable load) is totally independent of whether the bolt is finger tight or fully tightened. This typical performance has given rise to the classification of joints loaded in shear as either "slip-critical" (i.e., connection strength is not the primary concern; slip cannot be tolerated) or "shear/bearing" (i.e., slip will not occur, or if it does it is not important—strength is the primary concern). Because the predictable performance of slip-critical connections is dependent on a known minimum bolt pretension, the terminology provides convenient terms of reference to define the tightening requirements for connections that must be fully tightened even though slip per se at a clearly known value of pretension may not be a consideration. The design procedure for both ASD and LRFD is to first design the connection for the particular orientation of applied loading on the basis of strength and then check the connection for adequacy to prevent slip, if slip into bearing must be avoided.

21.4.1 Strength Criteria for Fasteners Subject to Applied Tension

The strength of a bolt subject to tension has been determined by theory and test to be accurately predicted by the product of the ultimate strength of the bolt material (F_u) times the stress area of the threaded part of the bolt (A_{sa}). Because different dimensional properties of the bolt (body area, tensile stress area, area at root of the threads) are the critical geometrical parameters for the failure modes for different types of loading, the pertinent geometrical property has been used in the theoretically correct formula for each failure mode. Then, for convenience of the user, the resulting formula is multiplied by the ratio of the theoretically correct dimensional parameter divided by the full body area and also multiplied by the body area to provide formulas for all failure modes in terms of the single dimensional property A_g. This procedure provides a sufficiently accurate formula but avoids the pitfall of inadvertently using the wrong area for a given case. The following formula is used to express the strength of a bolt subject to tension.

$$R_t = F_u A_g \left(A_{sa} / A_g\right) \tag{21.2}$$

where

R_t = strength of bolt subject to tension (ksi)
F_u = specified minimum tensile strength of the bolt (ksi)

A_{sa} = tension stress area of the bolt (in.²)
A_g = area corresponding to the nominal diameter of the bolt (in.²)

Because the ratio of tensile stress area to body area is essentially constant at approximately 0.75 for bolts with UNC threads of the usual diameter for structural applications (⅝ in. to about 1¼ in.), the tensile strength of a bolt in tension may be simply stated as

$$R_t = F_u \times 0.75 \times A_g \quad (21.3)$$

Thus, the strength of high-strength bolts subject to tension reduces to the following:

For ASTM A325 bolts, $R_t = 120 \times 0.75 = 90$ ksi (on the nominal diameter area).
For ASTM A490 bolts, $R_t = 150 \times 0.75 = 112.5$ ksi (on the nominal diameter area).

The ASD Specification establishes an allowable stress for A325 bolts in tension as 44 ksi, providing a factor of safety of approximately 90/44 = 2.05 to cover all uncertainties in load estimation, under-strength material, out-of-tolerance manufacturing, etc. Similarly, the allowable stress in tension for A490 bolts is established at 54 ksi, providing a factor of safety of approximately 112.5/54 = 2.08.

The LRFD Specification provision for bolts subject to tension derives from the same formula for strength. However, it is termed the nominal strength R_n, and it is required that it be multiplied by the reduction factor ϕ to cover all of the uncertainties (manufacturing tolerances including dimensions and heat treatment on bolts and nuts, material strength, concentricity of applied load, etc.) that bear on the accuracy with which the strength formula for R_n predicts the strength.

$$\phi R_n = \phi \times F_u \times 0.75 \times A_g \quad (21.4)$$

By means of statistical evaluation of test data, it has been determined that a value for ϕ of 0.75 is appropriate. The LRFD Specification specifies that

$$\phi R_n \geq \gamma_D D + \gamma_L L + \gamma_W W + \ldots$$

(as appropriate to the particular load case). Because specific requirements of loading cases are beyond the scope of this chapter, a discussion of LRFD is not pursued further for this case.

AISC has carried the process of simplification forward even further by implementing the Specification requirements in the form of allowable load tables in kilopounds (kip) for bolts of each size. Similarly, the nominal resistances for LRFD are presented in tabular form for each bolt size.

One important requirement applicable to bolts subject to tension loading that has not been touched on yet must be discussed briefly. Depending on the arrangement of connected parts and the stiffness of the parts, connections that subject the bolt to applied tension most often tend to deform in a manner that develops prying forces that increase the tension in the bolt beyond the level attributable to the externally applied load. Such prying forces, which are illustrated in Figure 21.2, may be small or as great as 50% of the applied load. Calculations require that a preliminary selection of connected fitting members be made and the number and size of bolts determined. The preliminary selection must then be analyzed to estimate the prying induced in the bolts. The connection must be modified accordingly to account for the increased forces and again reanalyzed. This can be tedious. The reader is directed to the AISC manuals of steel construction for reliable guidance and to minimize the effort required.

21.4.2 Shear on Fasteners and Bearing on Connected Material—General

In the second introductory paragraph of Section 21.4, the history of events that occur in the loading of a joint subject to shear was explained to introduce the difference between design for strength and

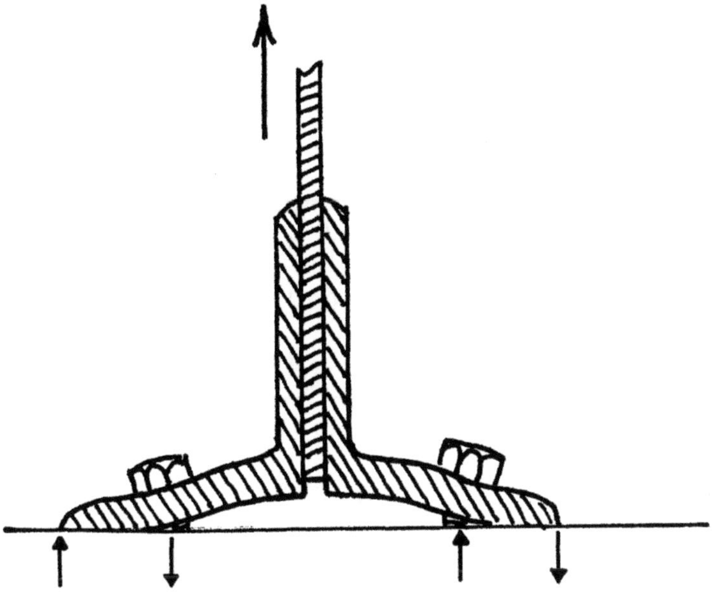

FIGURE 21.2 A bolt subjected to prying must support the external load L_X plus a reaction force created by the leverlike influence of that portion of the joint which lies outside of the bolt pattern. This phenomenon can significantly increase the loads borne by the bolts. Because the prying force is dependent upon the relative stiffness of the connected parts and the bolts, the magnitude of this "lever action" is difficult to estimate (from Ref. [5]).

design check for resistance to slip or serviceability. This section is devoted to the discussion of the strength per se of shear/bearing connections after the slip occurs.

Several interrelated parameters influence the shear and bearing strength of connections. These include such geometric parameters as the ratio of net area to gross area of the connected parts (see Figure 21.3), the ratio of the net area of the connected parts to the total shear-resisting area of the fasteners, and the ratio of transverse fastener spacing to fastener diameter and to the connected part thickness. In addition, the ratio of yield strength to tensile strength of the steel comprising the connected parts and the total distance between extreme fasteners measured parallel to the line of applied tensile force affect connection strength.

In the past, a balanced design concept was sought in developing criteria for mechanically fastened joints to resist shear between connected parts by having fasteners bear against the sides of the holes. This philosophy resulted in wide variations in the factor of safety for the fasteners because the ratio of yield to tensile strength increases significantly with increasingly stronger grades of steel. Also, it had no application at all in the case of very long joints used to transfer direct tension, because the end fasteners "unbutton" before the plate can attain its full strength or before the interior fasteners can be loaded to their rated shear capacity.

By using a mathematical model it has been possible to study the interrelationship of the previously mentioned parameters. It has been shown that the factor of safety against shear failure ranged from 3.3 for compact (short) joints to approximately 2.0 for joints with an overall length in excess of 50 in. It is of interest to note that the longest (and often the most important) joints had the lowest factor, indicating that a factor of safety of 2.0, as proven satisfactory in service, should be appropriate for all cases.

21.4.3 Strength Criteria for Fasteners Subject to Shear

The determination of the strength of a bolt in shear is based on the observation that the ratio of the shear strength through the body area of a single high-strength bolt to the tensile strength of the bolt is 0.62. Joints involving only a single bolt are rare, and it has been observed that with more than two

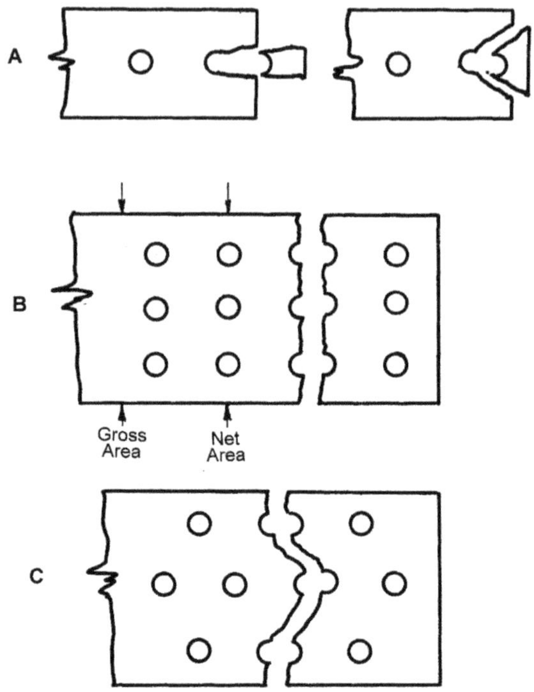

FIGURE 21.3 Illustration of joint tearout failure modes and of the definition of net and gross cross-sectional areas of the joint members. (a) End tear-out failures. (b) The net cross-sectional area is the area through a line of bolt holes; the gross area is that of the uninterrupted joint member. (c) Zigzag failure. (From Ref. [5].)

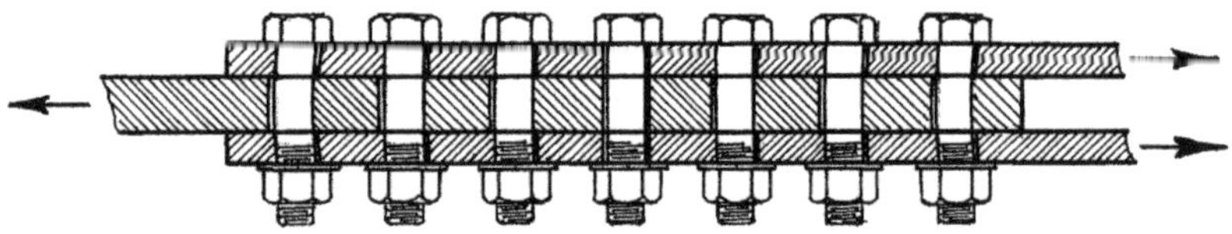

FIGURE 21.4 The bolts in a long joint loaded in shear do not see equal shear loads. Those at the outer ends of the bolt pattern are subjected to higher loads than those in the center of the pattern due to build up of stress in the plates along the length from opposite ends (from Ref. [5]).

bolts in the line of force the deformations of the connected material in bearing against the bolt and along the length of the joint cause nonuniform distribution of force to the bolts, with the bolts at the ends of the joint carrying a higher portion of the load than those at the center (see Figure 21.4). The strength of the connection in terms of the average force on the bolts decreases as the joint length increases. At the failure load of a joint with the length of about 50 in., the average force on a bolt is about 0.80 times the strength of a single bolt in shear. To avoid the need for a reduction formula based on joint length, a single reduction factor of 0.80 has been applied to the 0.62 factor. This permits the development of a formula that is applicable to joints up to 50 in. long that is not overly conservative for shorter joints. It must be noted that for joints longer than 50 in., the Specification requires that an additional 0.80 reduction be applied. Further, it has been observed in tests that the average value of the strength of a bolt loaded in shear through the threads is 0.833 times the strength of the same bolt loaded through the body area (see Figure 21.5). In the Specification, a conservative value of 0.80 has been applied to account for the shear strength of a fastener loaded through the threads in terms of the body area. Thus, the strength of a high-strength bolt in shear/bearing in connections up to 50 in. long may thus be reliably determined by the following formulas.

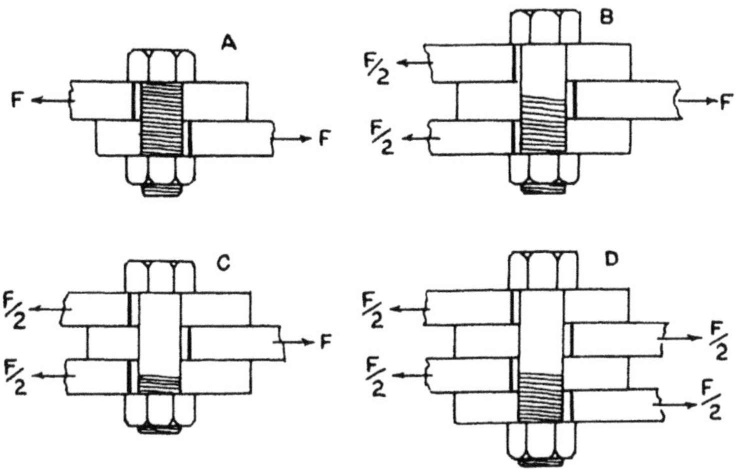

FIGURE 21.5 The shear strength of the bolt is reduced if bolt threads coincide with one or more of the shear planes of the joint, as they do here in examples a and b. The number of shear planes shown varies from one (a) to two (b and c), or three (d) (from Ref. [5]).

TABLE 21.1
Shear Strength Versus Allowable Stress for Bolts Loaded in Shear

Case Description	Shear Strength	Allowable Stress	Factor of Safety
A325: Threads not excluded from shear plane	48 ksi	21 ksi	2.29
A325: Threads excluded from shear plane	60 ksi	30 ksi	2.0
A490: Threads not excluded from shear plane	60 ksi	28 ksi	2.14
A490: Threads excluded from shear plane	75 ksi	40 ksi	1.87

With threads not excluded from the shear plane,

$$R_v = 0.62 \times 0.80 \times F_u \times 0.80 \times A_g = 0.40 \times F_u A_g \tag{21.5}$$

With threads excluded from the shear plane,

$$R_v = 0.62 \times 0.80 \times F_u \times A_g = 0.50 \times F_u A_g \tag{21.6}$$

The ASD Specification establishes allowable stresses for bolts loaded in shear. They are presented in Table 21.1 together with the strength as calculated above and with the apparent factors of safety provided.

The LRFD Specification uses the same formulation for the nominal strength, R_n, for bolts subject to shear but stipulates that R_n shall be multiplied by a reduction factor ϕ=0.75 to cover uncertainties in the determination of the resistance. For reasons explained above, the further treatment of the LRFD Specification is beyond the scope of this chapter.

21.4.4 Design Criteria for Combined Tension and Shear

The adequacy of bolts subject to tension combined with shear may be checked by means of the elliptical interaction formula

$$(f_v / F_v)^2 + (f_t / F_t)^2 \leq 1.0 \tag{21.7}$$

where

f_v=the applied shear stress (ksi)
F_v=the allowable shear stress (ksi)
f=the applied tensile stress (ksi)
F_t=the allowable tensile stress (ksi).

This concept is illustrated in Figure 21.6.

21.4.5 Strength Criteria for Connected Material Subject to Bearing Stress

Bearing stress produced by a high-strength bolt pressing against the side of the hole in a connected part is important only as an index to behavior of the connected part. It is of no significance to the bolt. The critical value can be derived from the case of a single bolt at the end of a tension member. (See Figure 21.7.)

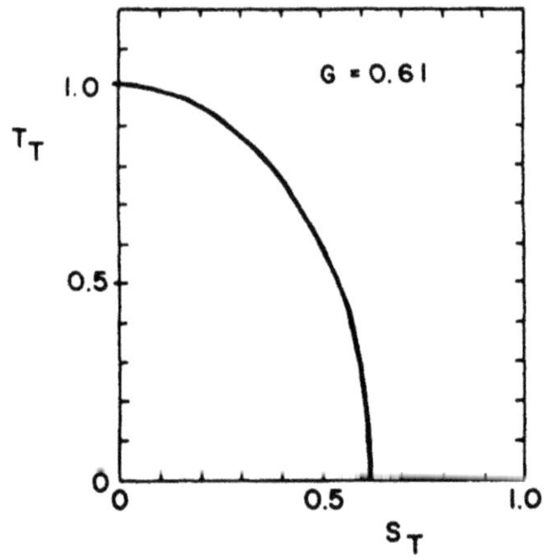

FIGURE 21.6 An elliptic curve can be used to relate the tensile capacity of a bolt to the shear stress imposed simultaneously on it, or vice versa. S_τ is the ratio of the shear stress in the bolt to the ultimate tensile strength of the bolt. T_τ is the ratio of tensile stress in the bolt to the same UTS. G is the ratio of shear strength to tensile strength of the bolt material. The curve shown here is for static loads on the thread stress area of ASTM A325 or A354 BD bolts (modified from Ref. [2]).

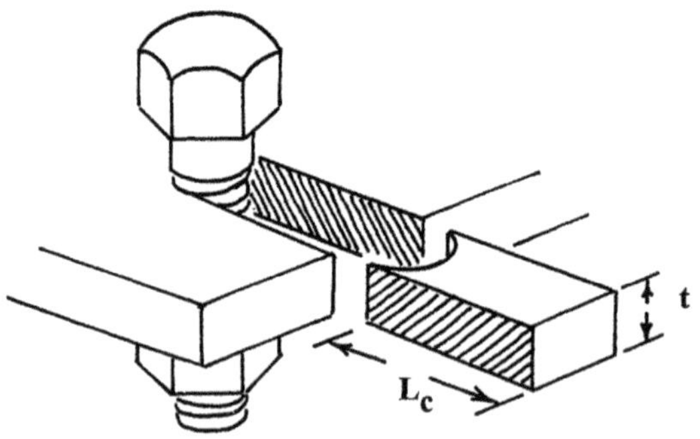

FIGURE 21.7 Many of the design rules discussed in this chapter are based on the assumption that the edge distance, L_c, from centerline of the bolt to the nearest edge of the joint in the direction of the applied load is at least 1.5d. To tear out of the joint, such a bolt would have to shear the joint area illustrated here (i.e., $2tL_c$).

It has been shown, using finger-tight bolts, that a connected plate will not fail by tearing through the free edge of the material if the following relationship is satisfied.

$$\frac{L}{d} \geq \frac{R_p}{F_u} \tag{21.8}$$

where

R_p = nominal bearing pressure (ksi)
F_u = specified minimum tensile strength of the connected part (ksi)
L = distance measured parallel to the line of applied force from center of a bolt to the free edge of the connected material (in.)
d = nominal diameter of the bolt (in.).

Although Eq. (21.8) has been proved by test to accurately predict strength, practical considerations must be observed, such as providing reasonable lower limits on edge distance and spacing to rule out excessive hole ovalization with excessively large edge distances and spacing as well as a lower limit to rule out ridiculously low allowable stresses at unpractically small end distances. By rearranging Eq. (21.8) and providing judgment-based practical limits, criteria with a factor of safety of 2.0 have been provided in the ASD Specification.

1. Connections using high-strength bolts in slotted holes with load applied in a direction other than approximately normal (between 80° and 100°) to the axis of the hole and connections in oversize holes shall be designed for resistance against slip.
2. Tabulated values (Table 21.2) apply when the distance L parallel to the line of force from the center of the hole to the edge of the connected part is not less than $1.5d$ and the distance from the center of a bolt to the center of an adjacent bolt is not less than $3d$ (see Figure 21.7). When either of these conditions is not satisfied, the allowable stress shall be determined according to the equation

$$F_p = \frac{LF_u}{2d} \leq 1.5\ F_u \tag{21.9}$$

3. For connections with more than a single bolt in the direction of force, the resistance may be taken as the sum of the resistances of the individual bolts.

The absence of allowable stress or strength provisions specifically for the case where a bolt in double shear has a nonthreaded shank in one shear plane and a threaded section in the other shear plane (Figure 21.5) is for purposes of simplification and because the implied accuracy would be inconsistent with practical experience and with test data relative to the performance of such connections. It also recognizes that knowledge of bolt placement (which might leave both shear planes in

TABLE 21.2
Allowable Working Stress on Fasteners Based on Connected Material

Loading Condition	Allowable Stress[a] (ksi)
Bearing on connected material with single bolt in line of force in a standard or short slotted hole	$1.0\ F_u$
Bearing on connected material with two or more bolts in line of force in a standard or short slotted hole	$1.2\ F_u$
Bearing on connected material in long slotted holes	$1.0\ F_u$

the threaded section in thin material) is not ordinarily available to the detailer. If threads occur in one shear plane, the conservative assumption is made that threads are in all shear planes.

21.4.6 Design Check for Slip Resistance

As discussed in the introductory paragraph of Section 21.4, a small amount of slip will occur in shear-loaded connections prior to reaching the maximum strength load. (This may not be true for the case of failure by bearing against connected material with very small edge distances.) Slip may also occur prior to reaching the allowable load (maximum strength divided by a factor of safety) as determined by strength considerations. In a great preponderance of connections, strength is the primary concern governing the design or slip cannot occur because in the real world joint assembly procedures generally ensure that the bolts will be in bearing with either the dead weight of the piece bearing on the loose erection bolts prior to final tightening or to the application of any other load or where, due to numerous bolts in large connections and the in-tolerance misalignments of holes, bolts are in bearing at all stages of loading. Reports of slip at design loads have rarely if ever been reported; however, there are situations where a high degree of assurance is required that slip will not occur. Slip must be prevented to avoid even minor changes in the geometrical relationship between members to ensure the integrity of the structure subject under static loads or to ensure that back-and-forth slippage of connections does not occur when they are subjected to significant reversal of loading or for other reasons. Such connections are advisedly identified in the Specification as "slip-critical" connections to call attention to the fact that they are intended to be used where special performance characteristics are important and to emphasize that their installation and tensioning require special control and inspection at additional cost. Joints considered to be slip-critical are those subject to significant load reversal or to cyclic fatigue loading, those with bolts installed in oversize holes, those with bolts installed in slotted holes when the line of action of the applied loads is more or less parallel to the long axis of the slots, those in which bolts and welds share the applied loads, and those that the engineer of record thinks would be adversely affected by slip.

For situations where slip into bearing must be prevented in order to ensure the connection's suitability for service (slip-critical connections), the Specification requires that, in addition to satisfying the requirements for strength, the slip resistance of the connection be checked. The allowable capacity to resist slip must be greater than the force tending to cause slip at working loads. The allowable slip resistance is expressed by the formula

$$P_s = F_s A_b N_b N_s \tag{21.10}$$

where

F_s=allowable load unit area of bolt based on slip resistance (ksi)
A_b=area corresponding to nominal body area of bolt (in.^{2})
N_b=number of bolts in the connection
N_s=number of slip planes.

The allowable slip load unit area of the bolt, F_s, is given in Table 21.3 for A325 and A490 bolts, for three classes of connected faying surface treatments and various bolt sizes. It is true from basic principles that slip load is solely dependent on the coefficient of friction and clamping force; however, out of concern for the ability to maintain installed clamping force when the high clamping force is applied to a reduced area at oversize and slotted holes, the Specification-allowable loads at such holes is reduced. For A325 bolts, the allowable slip loads vary by 10–28 ksi, with the highest allowed for bolts in standard holes with Class B surfaces and the lowest assigned to bolts in long-slotted holes parallel to the direction of loading with Class A surfaces. A490 bolts are treated in a similar manner, with allowable slip loads F_s ranging from 13 to 34 ksi.

TABLE 21.3
Allowable Slip Loads per Unit of Bolt Area for Slip-Critical Connections (Data in ksi)

	Hole Type And Direction Of Application Of Load							
	Any Direction				Transverse, Long Slots		Parallel, Long Slots	
	Standard		Oversize					
Surface Condition of Joint Members[a]	A325	A490	A325	A490	A325	A490	A325	A490
Class A	17	21	15	18	12	15	10	13
Class B	28	34	24	29	20	24	17	20
Class C	22	27	19	23	16	19	14	16

Note: Joint surfaces must be blast cleaned before Class A or B coatings are applied.

[a] **Class A**: Slip coefficient of 0.33 is expected if the joint surface consists of clean mill scale or blast-cleaned surface to which Class A coatings have been applied. Class A coatings are those that carefully defined tests show will provide a slip coefficient not less than 0.33.

Class B: Slip coefficient of 0.50; blast-cleaned surfaces or blast-cleaned surfaces with Class B coatings (which tests show will provide a coefficient not less than 0.5).

Class C: Slip coefficient of 0.40; hot dip galvanized and roughened surfaces.

In contrast to earlier versions of the Specification, which recognized a wide variety of surface treatments, and to prevent the chaos that could result from unlimited small differences in coefficients of friction claimed for various proprietary surface treatments, only three coefficients of friction are recognized by the current Specification, each corresponding to a group or class of surface treatment.

Class A—Slip coefficient 0.33: Appropriate to clean mill scale joint contact surfaces or blast-cleaned surfaces to which a Class A coating has been applied.

Class B—Slip coefficient 0.50: Appropriate to uncoated blast-cleaned surfaces or blast-cleaned surfaces to which a Class B coating has been applied.

Class C—Slip coefficient 0.40: Appropriate to hot-dip galvanized surfaces that have been roughened.

Note: Coatings for Class A and Class B surfaces must have been demonstrated by test to provide a coefficient of 0.33 or 0.50, respectively and must be applied only to blast-cleaned surfaces.

An extensive appendix to the Specification describes the tests to be used to determine the slip coefficient of a coating. The appendix stipulates long-term creep tests as well as short-term tests, such things as thickness of coating and curing time, and complete documentation to enable reproducibility in application.

A good practice followed by experienced designers and detailers on projects where some bolts are non-slip-critical and others are slip-critical or require full tightening for other reasons is to design all slip-critical connections using bolts of a single diameter that is different from that of all other bolts used in the job. In this way, all bolts that require special tightening and inspection are clearly identified on the drawings and in the material delivered to the job by their diameter. The drawings do not carry special notations for individual connections. The ironworkers and inspectors need not refer to drawings for special instructions.

21.4.7 Combined Axial Tension and Shear in Slip-Critical Joints

Consideration must be given to the situation in which the externally applied load is applied in such a manner that it reduces the clamping force in the connection that is provided by the bolt pretension. As illustrated in Figure 21.8, the oblique load in the brace is resisted by axial tension in the bolts as

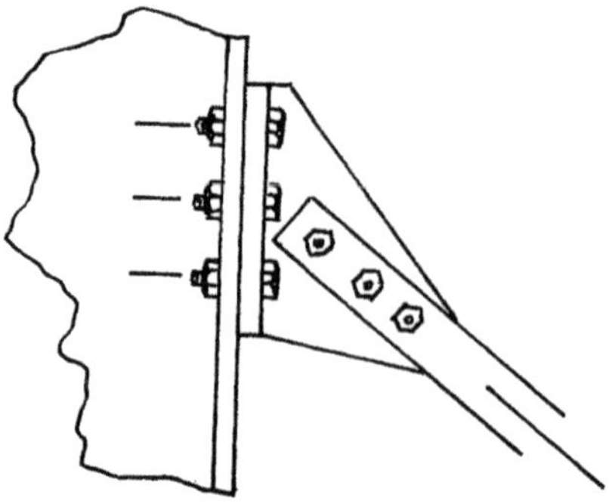

FIGURE 21.8 An oblique load of the sort shown here is resisted by axial tension in the bolts as well as by a shearing force at the faying surface. The tension in the bolts is not significantly increased by such a load, but the clamping force (and therefore slip resistance) between joint members is reduced.

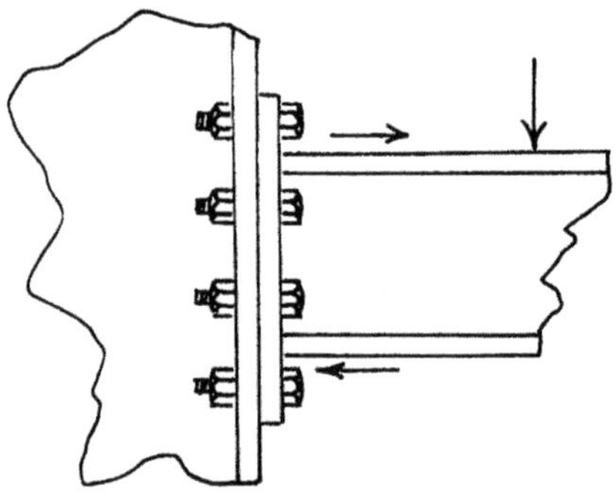

FIGURE 21.9 Combined axial and tensile loads do not always result in a loss of slip resistance. In the joint shown here, for example, any reduction in clamping force at the top bolts is matched by an equal increase around the bottom bolts, thanks to the bending moment on the beam.

well as by a shearing force at the faying surface. The axial tension from the external load does not increase the tension in the bolt to a level higher than the installed pretension; rather it reduces the clamping force at the faying surfaces, which are counted on to provide resistance to slip. That is, a portion of the clamping force that was provided by bolt pretension is taken over by the externally applied axial tension force. For such cases, the allowable design slip load, P_s [Eq. (21.10)], of the connection must be reduced to the degree that the clamping force at the faying surface is reduced by multiplying it by the factor

$$1-\frac{T_u}{T_m}$$

where

T_u = axial force per bolt induced by the externally applied load (kip)
T_m = specified installed prestress per bolt (kip).

The reduction in allowable slip load does not apply to all cases where the clamping force may be reduced around individual bolts by the externally applied load. As illustrated by Figure 21.9, when the externally applied load induces a bending moment as well as shear (a tendency to slip), the reduction in clamping force due to the applied load at the top bolts is exactly matched by an increase in the compression forces around the bottom bolts. The slip resistance is not affected.

21.5 SLIP RESISTANCE PER LRFD

Before discussing the LRFD requirements for slip resistance it should be repeated that it is not possible to present a truly usable summary of LRFD here because the load functions required for the calculations are well beyond the scope of this chapter. It might also be noted that there is considerable resistance to the LRFD Specification by practicing engineers. They claim that the work required to deal with the load aspects of this Specification is excessive. They cannot see the merit or advantage of spending the money required to change their current (often computerized) design procedures to achieve marginal economies for their customers. The academic community, however, believes that LRFD is a significant step forward and resists any further use or development of the ASD Specification. Since these people have considerable influence and may indeed succeed in replacing ASD with LRFD, it is felt that the introduction to LRFD contained in this chapter, even if unusable on a stand-alone basis, is necessary. With all that in mind, we now take a look at the LRFD approach to slip resistance.

As in the ASD version, the LRFD Specification distinguishes between slip-critical and shear/bearing connections, with slip resistance being of concern only for the first. Once again the frictional slip coefficient of the faying surfaces must be controlled; this is defined by three classes: A, B, and C, with coefficients of 0.33, 0.50, and 0.40, respectively, as defined in Section 21.4.4. And once again bolts must be installed and tightened to or above the minimum tensions listed in Table 21.1, following one of the tightening procedures recognized by both versions of this Specification (discussed in Section 21.6).

21.5.1 Slip Considerations for Nominal Load Levels

The LRFD Specification allows two different design criteria for slip-critical joints. In the first, nominal (service) loads are the design criterion and must not exceed the design slip resistance, which is given as ϕR_s, where

$$R_s = D\mu T_m N_b N_s \tag{21.11}$$

The factor D is the slip probability factor.

$D=0.81$ for Class A surfaces ($\mu=0.33$)
$D=0.86$ for Class B or C surfaces ($\mu=0.50$ or 0.40).

The factor μ is the coefficient of friction for Class A, B, or C surfaces (namely 0.33, 0.50, or 0.40) or as determined by the experiments discussed in Section 21.4.4.

This time the factor ϕ can take one of several values.

$\phi=1.0$ for standard holes
$\phi=0.85$ for oversized or short slotted holes
$\phi=0.70$ for long slotted holes transverse to the direction of the load
$\phi=0.60$ for long slotted holes parallel to the direction of the load.

This is not quite the end of the nominal load story. Modifications to Eq. (21.11) are often required, as discussed, for example, in Section 21.5.3. First, though, let us look at the second approach to slip resistance outlined in the LRFD Specification.

21.5.2 Slip Considerations at Factored Load Levels

When factored loads are the design criterion, they must not exceed the design slip resistance, which is defined as ϕR_{st}, where

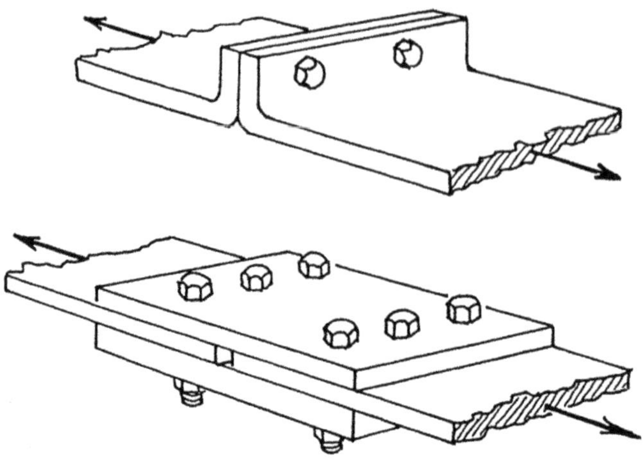

FIGURE 21.10 The upper sketch shows a joint loaded in axial tension, the line of action of the load being parallel to the axes of the bolts. The lower sketch shows a joint loaded in shear, with the line of action of the load perpendicular to the axes of the bolts (from Ref. [5]).

$$R_{st} = 1.13\mu T_m N_b N_s \tag{21.12}$$

If you run calculations comparing the results obtained with Eq. (21.12) to those obtained by use of the equivalent expressions in the ASD Specification, you will find, we think, that load and resistance factor design is somewhat less conservative than allowable stress design. This is why the above-mentioned "economies for their customers" can be achieved by the use of LRFD.

21.5.3 Modifications to Accommodate Tensile Loads on the Joint

If a tensile load, illustrated in Figure 21.10 (i.e., a load acting more or less parallel to the axes of the bolts) is applied to a joint, it tends to pull the joint apart, partially reducing the interface clamping force the bolts had created between joint members. This phenomenon is discussed at length in Chapter 42. It is the clamping force, however, that generates the friction forces that resist joint slip. As a result, it is important to take tensile loads, if any, into account when estimating the design slip resistance of the joint. The rather lengthy procedures for doing this are given in the LRFD Specification.

21.6 ASSEMBLY AND TIGHTENING

As already mentioned, both ASD and LRFD versions of the Specification incorporate the same criteria for the installation and tightening of high-strength bolts. The resulting rules depend on whether or not the joint is loaded in tension or shear, whether it is subject to cyclic (fatigue) loading, and whether a shear-loaded joint is considered to be slip-critical. Although different tightening requirements might have been justified for each type of loading or application that requires tightening, the single criterion is applied to all joints requiring tightening to minimize opportunities for errors in the field. The pretension required to be developed in the tightening procedure is upon which the allowable loads for slip-critical connections are based. The minimum installed fastener pretensions for ASTM A325 and A490 bolts of different sizes are as specified in the Specification (see Table 21.4).

Bolts in connections not classified as slip-critical or subject to tensile loads or cyclic fatigue loading need to be tightened only to a "snug tight" condition. This degree of tightness is defined as "tightened to the degree necessary to bring all joint members into firm contact." The Specification suggests that snug tightness can normally be achieved by a few impacts of an impact wrench or by the full effort of a man using an ordinary spud wrench.

TABLE 21.4
Minimum Fastener Tension (T_m) Required for Slip-Critical Connections and Connections Subject to Direct Tension Loads

Nominal Bolt Size (in.)	Minimum Tension (kip)	
	A325 Bolts	A490 Bolts
½	12	15
⅝	19	24
¾	28	35
⅞	39	49
1	51	64
1⅛	56	80
1¼	71	102
1⅜	85	121
1½	103	148

In contrast to the rules for snug-tightened joints, the installation requirements to ensure that the minimum pretensions equal to those specified in Table 21.4 (70% of the ultimate tensile strength of the bolts) are detailed in the requirements for proper receipt inspection, storage, job site testing capacity of the fastener assembly, as it will be used in the work, to achieve a tension 5% higher than required by the Specification and the use of one of four recognized methods for installation and tightening of bolts in the work.

It is informative to learn why 70% of ultimate strength was established as the required installation tension. Early versions of the Specification were limited to A325 bolts, and tightening was required to achieve the specified proof load of the bolt. Proof load is defined in the ASTM Specifications basically as the minimum required yield point as determined in a standard direct tension test (generally taken as the stress at 0.2% offset).

The Specification was extended to include the higher strength A490 bolts in 1964, and in the version of that adoption date, the installation tension requirement was established at the proof load. Proof load of the higher strength less ductile A490 bolts is specified by ASTM Specifications to be 80% of the ultimate tensile strength and is also determined in a standard direct tension test. In the practical application and use of bolts, however, the tension is induced by torquing a nut on the threaded end of the bolt, subjecting the area at the root of the threads to combined torsion and tension stresses. The failure load of bolts subject to combined torsion and tension is approximately 20% less than the failure load in direct tension as suggested in Figure 21.11; thus twist-off failures were soon encountered when bolts were being tensioned to meet the Specification's pretension requirements. Additionally, to minimize rejection rate, manufacturers must target production at, say, 10% above the minimum specified acceptance level. Thus the actual yield point approaches the ultimate tensile strength. This increased the potential for failures in service due to hydrogen embrittlement and stress corrosion cracking. To address these problems the specified minimum proof load in the ASTM Specification for A490 bolts was reduced by 10 ksi, and the RCSC installation pretension requirement was changed from proof load to 70% of the ultimate tensile strength, which just happens to be the percentage of ultimate strength for proof load with A325 bolts.

21.6.1 Preinstallation Check Testing

Specifying a minimum installation based on requirements to satisfy in-service fastener performance is easy. An assembly comprising a bolt, nut, washer, and connected material is deceptively simple, and for most applications tightening the assembly is a simple, straightforward operation.

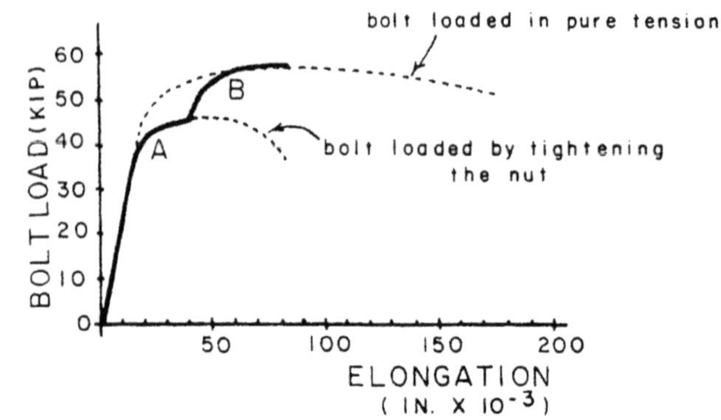

FIGURE 21.11 A fastener tightened with a torque tool, and therefore exposed to simultaneous torsion and tension, will yield at a slightly lower level of tensile stress (A) than will a fastener loaded in pure tension. This same fastener, as a result, will support an additional tensile load in service because the torsional stress component will, in general, disappear rapidly after the wrench is removed. Data are for a ⅞ A325 bolt with a 4⅛ grip length (Adapted from Ref. [2]).

However, when the tightening of the assembly must be controlled to ensure a specified minimum tension, the control becomes very difficult. It has challenged the best and most experienced minds and still does. Literally, hundreds of variables affect the results of developing tension in a bolt by tightening the nut. ASTM and ANSI Specifications establish geometrical, chemical, and physical property tolerances on the separate components (bolts, nuts, and washers) that are presumably right and proper for service; however, the user is not interested in a bolt or a nut or a washer. The only thing that is of importance is the complete assembly (including lubrication and connected material). The cumulative effect of tolerances on the individual parts, whether they were produced with new tooling or tools about to be retired; the conditions of handling and storage in the warehouse, during shipment, and at the job site; the adequacy and condition of the installation tools; the knowledge and diligence of the workmen; and many other factors all bear on the problem. Even in controlled laboratory environments, it has been demonstrated that torque–tension relationships vary by as much as ±30%–40% due to the effect of the numerous uncontrolled variables. Because it is not possible to adequately control these numerous individual variables, the Council Specification requires that their cumulative effect be compensated for by a testing and calibration procedure *prior to commencement of work* and repeated daily when the installation method might be dependent on tool operating condition, using the fastener components from each lot assembled as they will be assembled in the work and using the method that will be used by the crew that will do the work. The specified demonstration testing of fastener assemblies under the real conditions that will be used is essential for reliable results (see Chapter 52 for further discussion).

One point concerning which questions have been raised is the fact that the specified procedure requires testing to a level of tension 5% higher than the specified minimum pretension level. It has been argued that this does not constitute a fair and enforceable requirement because the minimum level for acceptance of the assembly by field test is 5% higher than the ASTM specified minimum proof load for A325 bolts. The argument is not valid. Rarely does the test identify bolts or nuts that do not comply with ASTM minimum requirements, but when it does, that's good: Such fasteners can be rejected before they are incorporated in the work rather than later when they are identified by postinstallation inspection. The proof load for high-strength bolts is comfortably lower than the ultimate tensile strength; thus proper bolts and nuts can easily achieve loads 5% above proof load. What the test does identify is improper use of the selected installation method, mismatch of components from different sources (although each component might be within tolerance for the individual part, they may not work well together), fasteners that have become contaminated by dust or rust from improper storage, dry or improperly lubricated fasteners that need relubrication to overcome the lowering of developed tensile strength when tension is induced by tightening the nut, overtapping of nuts in the case of galvanized fasteners, and other conditions. The **pretesting of fastener assemblies and installation method is the critical element** in ensuring proper tension in the work. It must not be given short shrift.

21.6.2 Tightening—General

Inspection testing to ensure that the fasteners meet requirements and the components will work together as an assembly is required for fastener assemblies that are to be tightened in slip-critical joints. Further testing of fastener assemblies is not specified in great detail. However, for assemblies requiring full tensioning, the Specification recognizes four tightening methods, any one of which will ensure that the specified tension is achieved if it is properly used. It can be said with equal assurance that unsatisfactory results and unknown tensions will result if any of the methods is misused.

21.6.3 Turn-of-Nut Tightening

Once the job site demonstration testing has shown that the fastener assemblies and procedure to be used will induce a tension 5% higher than the specified minimum tension, installation of fasteners may proceed. The method requires that bolts be installed in all holes of the connection and tightened to the snug tight condition starting with the fasteners in the most rigid area of the connection and proceeding systematically outward to the free edges. Generally, one or two impacts of an impact wrench will signal that a bolt is snug; however, with large joints in thick material more than a single cycle of snuggling bolts from the center outward may be required to bring all bolts to the snug condition. Once all bolts are essentially uniformly snug, final tightening may begin. Again starting with the bolts in the most rigid part of the connection and proceeding outward systematically to the free edges, the nuts (or bolt heads) are given not less than the required number of turns (see Table 21.5). In the case of large joints in thick material, it may be necessary to "touch up" the bolts that were tightened first as they may have been loosened by tightening of adjacent bolts.

The turn-of-nut method has been shown repeatedly to be one of the most reliable methods, when diligently applied, for attaining the required pretension in the bolts. This is true because none

TABLE 21.5
Nut Rotation from Snug Tight Condition for Turn-of-Nut Procedure[a,b]

	Disposition of Outer Face of Bolted Parts		
Bolt Length (Underside of Head to End of Bolt)	**Both Faces Normal to Bolt Axis**	**One Face Normal to Bolt Axis and Other Sloped Not More Than 1:20 (Beveled Washer Not Used)**	**Both Faces Sloped Not More Than 1:20 from Normal to the Bolt Axis (Beveled Washer Not Used)**
Up to and including 4 diameters	⅓ turn	½ turn	⅔ turn
Over 4 diameters but not exceeding 8 diam.	½ turn	⅔ turn	⅚ turn
Over 8 diameters but not exceeding 12 diam.	⅔ turn	⅚ turn	1 turn

Source: Refs. [1] and [4].

[a] Nut rotation is relative to bolt regardless of the element (nut or bolt) being turned. For bolts installed by ½ turn and less, the tolerance should be ±30°; for bolts installed by ⅔ turn and more, the tolerance should be ±45°.

[b] Applicable only to connections in which all material within the grip of the bolt is steel.

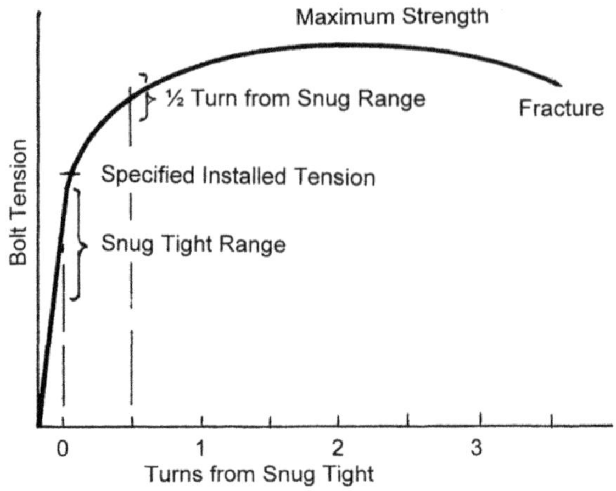

FIGURE 21.12 Illustration of the force–elongation curve generated when a bolt is tightened until it breaks. At first, the bolt deforms elastically; tension builds up in it, but there is little axial deformation. Finally the bolt yields, at the point shown by the knee of the curve. Any attempt to tighten it further is accompanied by significant deformation, until it breaks. In the turn-of-nut procedure, the bolt is first tightened to a minimum of 70% of its ultimate strength. This brings it to "proofload," which is just below the knee of the curve. The following, specified, turn of the nut (typically half a turn) will bring the bolt slightly farther into the inelastic region. As you can see from the curve, however, there is a large margin of safety (equal to about three turns) against overtightening (from Ref. [5]).

of the variables that affect torque-dependent methods affect the turn-of-nut method to the same degree. The tension developed in the bolt depends on the amount of bolt extension from the snug tight condition. Therein, however, also lies the source of the inaccuracy of results achieved by the method. Care must be exercised to ensure that the connected material is in contact throughout the area of the joint and that no bolts are loose at snug tight. Care must be exercised to apply required turns from the most rigid area to the free edges. Because the turns specified are adequate to bring the bolts to slightly more than the specified minimum, they will be slightly past the knee of the turns–tension curve (Figure 21.12) where small differences in the initial snug-tightness tension of individual bolts make even less difference in final tension. The main source of abuse and inaccuracy with the method is haste and lack of care on the part of the workmen. A frequently occurring abuse is to put bolts in all holes and then tighten them in any convenient order to final tightness, trying to note the orientation of the chuck on the fly as it begins to impact and then adding an estimated number of turns. This procedure will most assuredly render the first installed bolts loose to some unknown degree when adjacent bolts are later tightened. Another disadvantage of turn-of-nut, even when diligently applied, is that the inspector must monitor the work in progress to be able to assure the owner that the method was properly applied. He cannot come to the work after completion at his convenience and ascertain that the work was properly done.

21.6.4 Calibrated Wrench Tightening

The calibrated wrench tightening method is directly dependent on the torque–tension relationships of the fastener assemblies. Therefore, all of the numerous variables that affect the torque required to develop tension in the bolts by torquing the nut also affect the final result. Daily demonstration testing of the system is essential and must involve a representative sample of each lot, length, diameter, and grade of fastener assembly as they will be used on the job in a device capable of indicating tension in the bolt, using the actual tightening equipment that will be used (see Chapter 42). Retesting and readjustment of the system are required if any significant change in condition of the fasteners is noticed. In the testing, the required minimum tension must be developed without exceeding the bolt-head-nut rotation from snug tight specified in Table 21.5. If manual torque wrenches are to be used, the torque must be measured while the nut is being turned in the tightening direction. These are demanding but necessary requirements that must be met on a daily basis if reliable results are to be achieved.

It must be noted that the Council does not recognize any table of standard torque values or any formulas for the calculation of required torque for installation or inspection. Such tables and formulas are based on consideration of ideal thread helix angle, bolt diameter, and thread pitch but totally ignore the many variables that affect the results in the real world. They are rejected as unreliable and misleading.

Once the job site demonstration testing has shown that the fastener assemblies and procedure to be used will induce a tension 5% higher than the specified minimum tension, installation of fasteners in the work may proceed. Bolts shall be installed in all holes of the connections with hardened washers under the element to be turned in tightening, and all bolts brought to the snug tight condition. Snug tightening is required to start at the most rigid part of the connection and proceed systematically to the free edges until all bolts are uniformly snug tight and the mating parts of the joint are in firm but not necessarily continuous contact. Following this initial tightening, which may be accomplished using any wrench, the calibrated wrench shall be used in a systematic manner from the most rigid part of the connection to the free edges. The element not turned in tightening must be held to prevent rotation during final tightening operation and the turns observed to ensure that the threads are not stripped or damaged. More than a single cycle of tightening may be required to ensure that all the bolts are simultaneously at the same proven torque level.

21.6.5 Alternative Design Fastener Installation

Alternative design fasteners are those that meet all the geometrical, chemical, and physical properties specified in ASTM A325 or A490 Specifications as appropriate but have some additional feature such as a twist-off splined end and special installation wrenches to establish a specific level of torque at which the splined end will shear off and the tool will cease operation. Although such fasteners are in fact torque-controlled fasteners and therefore subject to the same effects of the many variables that affect the performance of conventional high-strength bolts installed by torque-controlled methods, they have a good track record and despite high cost are proving popular in the marketplace. This is probably because they are generally distributed through the producer's distribution representatives rather than through brokers who supply fastener components from many sources, because they are probably produced under conditions of greater quality control, because they are supplied as an assembly with the bolt and nut under the control of the same producer, and because they are probably lubricated, stored, and shipped under better control (their performance is strongly dependent on the lubrication provided). Even so, they are subject to the same preinstallation demonstration testing by lot, diameter, and length, to test their ability to provide a tension 5% higher than the specified installed tension requirements. If they fail to pass the test, the problem is usually one of failure to maintain proper cleanliness and lubrication and can often be corrected by cleaning and relubrication in accordance with the producer's requirements.

Alternative fastener installation can be abused and the good performance characteristics are defeated by installing bolts in the holes and carrying the tensioning forward in a single continuous operation. Only if the prescribed procedure is faithfully followed will the absence of the splined end give witness to a properly tensioned bolt. Only if the inspector monitors work in progress to ensure that the correct procedure is routinely used may it be assumed that the specified tension has been installed.

21.6.6 Direct Tension Indicator Devices

The typical direct tension indicator (DTI) is represented by devices that meet the requirement of ASTM A959 although others are not precluded (see Chapter 33). Although such devices do not provide the specified installed tension, they do constitute an integral part of the fastener assembly and are subject to the same demonstration testing for each lot, diameter, and length of fastener being used. It is sometimes argued that since they measure only the tensile stress developed in the bolt, repeated testing with fasteners of different lengths is excessive. This is not deemed to be a valid argument because the objective is to ensure that the fastener assemblies develop a tension 5% higher than the minimum specified tension. Instances have been encountered wherein DTIs that were targeted at a strength level slightly higher than the specified minimum so as to minimize the reject rate were actually as much as 15%–20% over strength. In application, such DTIs may be combined with bolts that are close to the minimum specified strength by direct tension testing. When tensioned in the joint by tightening the nut, the combined torque–tension effects may cause bolt failure because of the over-strength DTI. In such cases neither the bolts nor the DTIs should be summarily rejected; rather the solution is to be found in improving the fastener lubrication.

Once the job site demonstration testing has shown that the fastener assemblies including the DTIs, placed as they will be installed in the work and with the procedure to be used, will induce a tension 5% higher than the specified minimum tension, installation of fasteners in the work may proceed. In the work bolts with properly positioned DTIs must be installed in all holes of the connections and brought to the snug tight condition by the same systematic tightening from the most rigid area to the free edges but with care being taken to only partially compress the indicator protrusions. Final tightening may proceed when the mating parts are in firm contact and all bolts are uniformly tight. Tightening must, again, be accomplished by tightening in a systematic manner from the most rigid area to the free edges. In large connections, tightening should not be carried forward to final gap dimension in a single operation; connections may require more than a single cycle of tightening. Final compression of the DTIs to the specified gap should be accomplished after all bolts are uniformly tight.

Flat hardened washers must be placed under DTIs when the DTIs are installed in oversized or slotted holes in an exterior ply to avoid false indication of protrusion compression by dishing of the DTI into the large hole.

Direct tension indicator installation can be abused and the proper tensioning of the joints is defeated by installing bolts in the holes and carrying the tensioning forward in a single continuous operation. Only if the prescribed procedure is faithfully followed will the specified gap give witness to a properly tensioned bolt. The protrusions do not rebound once compressed if bolt tension is reduced; thus the tightening must be carried out in a manner that minimizes relaxation of previously tightened bolts. If a DTI is compressed to the final gap requirement before all bolts are uniformly tensioned, it should be removed and replaced. Only if the inspector monitors work in progress to ensure that the correct procedure is routinely used may it be assumed that the specified tension has been created.

21.7 INSPECTION REQUIREMENTS

Bolts in slip-critical connections or in connections that must support direct tension loads are subject to inspection, and such joints must be clearly identified on the working drawings.

Ideally, the inspector will be present at the job site to witness calibration tests, installation, and the actual tightening of some of the bolts. If not, or if the tightening of the bolts is challenged, the inspector must conduct a test in which a representative sample of five bolts of each diameter, length, and grade used on the job are tested in the tension-measuring calibrator described earlier and in Chapter 42. The bolts shall be first snugged to about 15% of the value shown in Table 21.4, then further tightened to the Table 21.4 amount. The torques used to create the final tension are recorded, and the middle three values are averaged to define a "job inspection torque," which is then applied to a small sample of the bolts that have been challenged. If none of them turn, the inspection is complete. If any one of them turns under the job inspection torque, all bolts shall be so tested.

Bolts that have been installed some considerable time before being challenged cannot be accurately inspected by the methods just described. Alternative test means must be defined and agreed upon. In a few cases, the ultrasonic equipment and techniques discussed in Chapter 34 have been used successfully for this purpose.

SYMBOLS AND UNITS

A_b Cross-sectional area of bolt corresponding to nominal diameter (in.^{2})
A_g Cross-sectional area of bolt corresponding to nominal diameter (in.^{2})
A_{sa} Tensile stress area of the threaded region of a bolt (in.^{2})
D Slip probability factor (dimensionless)
D Anticipated dead load on structure (kip)
d Bolt diameter (in.)
F_p Allowable bearing pressure on a bolt (ksi)
F_s Allowable slip load per unit area of bolt (ksi)
F_t Allowable tensile stress (ksi)
f_t Applied tensile stress (ksi)
F_u Specified minimum tensile strength of connected part (ksi)
F_v Allowable shear stress (ksi)
f_v Applied shear stress (ksi)
L Anticipated live load on a structure (kip)
L End distance from center of the nearest round bolt hole (or transverse slotted hole) to the edge of connected part, in the direction of the applied load; also the center-to-center distance between standard holes or transverse slots (in.)
N_b Number of bolts in joint
N_s Number of slip planes in joint
P_s Allowable slip resistance of a joint loaded in shear (kip)
R_n Nominal tensile strength of a bolt subjected to axial tension or shear; or to a combination of tension and shear (kip)
R_a Nominal bearing pressure exerted by bolt of joint or the bearing resistance of the connected material (ksi)
R_s Nominal slip resistance of the bolts in a joint for use at nominal loads (kip)
R_{st} Slip resistance of the bolts in a joint for use at factored load levels (kip)
R_t Strength of bolt subject to tension and shear (kip)
R_v Strength of bolt in a shear/bearing connection (kip)
T_m Specified minimum fastener tension (kip)
T_u Axial force per bolt induced by the externally applied load (kip)
W A load placed on the structure (kip)
y A coefficient less than unity reflecting the degree to which various load effects may be underestimated (dimensionless)
ϕ Resistance factor; a coefficient less than unity reflecting the degree of uncertainty to which the strength of the joint may be overestimated (dimensionless)
μ Mean slip coefficient of the joint's faying surfaces as listed in the Specification, or as established by tests (dimensionless).

NOTE

1 The AISC specifications can be obtained for a modest price from the American Institute of Steel Construction, Inc., One East Wacker Drive, Suite 3100, Chicago, IL 60601-2001. Publication sales can be arranged by telephone: 1-800-644-2400.

REFERENCES[1]

1. Research Council on Structural Connections, *Specification for Structural Joints Using ASTM A325 or A490 Bolts: Allowable Stress Design*, American Institute of Steel Construction (AISC), Chicago, IL, 1992.
2. Kulak, G. L., J. W. Fisher, and J. H. A. Struik, *Guide to Design Criteria for Bolted and Riveted Joints*, 2nd ed., Wiley, New York, 1987.
3. AISC, *Manual of Steel Construction, Vol.* 2, Connections (ASD, 9th ed.; LRFD, 1st ed.), AISC, Chicago, IL, 1992.
4. Research Council on Structural Connections, *Specification for Structural Joints Using ASTM A325 or A490 Bolts: Load and Resistance Factor Design,* American Institute of Steel Construction (AISC), Chicago, IL, 1992.
5. Bickford, J. H., *An Introduction to the Design and Behavior of Bolted Joints*, 3rd ed., Marcel Dekker, New York, 1995.

22 Quality Inspection of Automotive Joints

Ali Alkelani and Basil Housari

22.1 INTRODUCTION

Threaded fasteners are widely used in many industries mainly due to their relative ease of assembly and disassembly for repair and maintenance. Other advantages include low cost, ability to join dissimilar materials, and ability to develop the necessary clamp force required to prevent certain types of failure modes such as leakage.

There are hundreds of types of threaded fasteners used in the industry such as metric bolts and nuts, riv-nuts, weld nuts and studs, thermoplastic screws, thread rolling screws, and self-tapping screws. Also, there are several tightening strategies used such as torque control, angle control, yield control, and stretch control. The goal of all tightening strategies is to develop the required level of clamp load to keep the joint members together and to prevent joint separation, slippage, squeak, rattle, self-loosening, leakage, and other types of failure. Several types of tightening tools are available to perform the tightening. In this chapter, the joint classification, and the commonly used tool types along with their advantages and disadvantages are discussed. In addition, a practical guideline for tool selection and parameters optimization are provided in order to detect possible joint failures and improve the joint quality.

22.2 BOLTED JOINT CLASSIFICATIONS

Bolted joints can be classified in different ways. One way is based on load bearing. As the name implies, load bearing joints support high level of load and transfer forces between components. Non-load bearing joints, on the other hand, don't bear signification structural loads; rather they are used for alignment purposes and assembly of parts. Another classification is based on the type of load such as tension, shear, compression, or combined. Tension forces tend to pull the joint components apart, while shear loads tend to apply forces parallel to the joint component surfaces. Compression load causes the joint components to experience pressure while they are pushed together. In real world, the joint often experiences combined loading involving a combination of tension, shear, and compression forces. Of course, each type of loading requires specific considerations in bolted joint selection, design, and tightening to handle the applied loads and ensure the bolted joint reliability and integrity.

VDI 2862 classifies bolted joints based on a risk assigned to the joint [1]. The design engineer includes the joint category as a part of the joint release which serves an important purpose when selecting the tightening tool and planning the assembly process by the manufacturing engineer. VDI 2862 classification is commonly used in the automotive industry. Basically, there are three joint categories: A, B, and C based on the risk associated with the joint failure, functionality, and regulations.

- *Category A – High risk*: A bolted joint is considered as category A if the failure of this joint could lead to severe consequences in terms of people's safety or negative environmental impact. Examples of such joints could be airbag or some exhaust joints.

DOI: 10.1201/9781003501183-26

- *Category B – Medium risk*: A bolted joint is considered as category B if the failure of this joint may impact the functionality of the system resulting in malfunction or breakdown but does not pose a safety risk. Examples of such joints are the vehicle's starter to engine or alternator to engine joints.
- *Category C – Low risk*: A bolted joint is considered as category C when it does not fall under Category A or B. Typically, a failure of such joint could result in noise or rattle, and therefore, it is classified only as low risk joint with moderate reliability requirements.

This bolted joint category classification will be used to select the proper tool type for each joint in Section 22.4.

22.3 TIGHTENING TOOL TYPES

There are several tightening tool types available in the market with wide variety in terms of torque range, speed, cost, and accuracy. The following is an overview for the most commonly used types.

22.3.1 Pulse or Impact Tools

Pulse tools are used in the automotive industry and construction due to their durability, ability to deliver high torque with little reaction to the operator, and advantageous tightening speed which could easily exceed 4,000 RPM and may reach up to 10,500 RPM during free rundown. Tools of this type are available for a very wide range of torque going from few newton meters to thousands of newton meters. Some of these tools' drawback is the high noise and vibrations level, and the need for an air supply source which can be expensive.

Impact battery tools are similar in function, with an advantage in reducing the noise, and eliminating the need for an air supply. They are more limited in terms of speed and maximum torque capacity compared to air pulse tools.

For both pulse air tools and impact battery tools, precision, accuracy, and possibility of calibration are some of the main concerns that may limit the use of these tools.

22.3.2 Air Clutch Tools

These tools utilize a mechanical clutch mechanism to shut off the tool when the required torque is reached. They have a relatively high productivity due to their high rundown speed. The operator will have to take the full torque reaction torque during the tightening. They are more accurate and precise compared to pulse tools, and they are also more limited on maximum torque availability. Their high power-to-weight ratio allows for providing higher efficiency without sacrificing the user comfort. Since they require a compressed air source, this limits the flexibility in assembly lines for line balance and relocation purposes. The cost of compressed air and movement flexibility has affected their use in recent years. Additionally, they are not as accurate as some other tool types, and sometimes the high speed makes the calibration of these tools a challenge.

22.3.3 Battery Clutch Tools

As the battery technology improved significantly in the last two decades, battery clutch tools have become more and more the first choice for tightening tools. The torque range of battery tools could reach up to 100 Nm. These tools normally control the torque delivered by using a mechanical clutch to shut the tool off when the clutch activation torque is reached. The shut off torque is adjusted by mechanically adjusting the clutch to increase or decrease the clutch activation torque. In addition, with recent technology, some additional controls are possible by programming the settings in the electronic control board, such as speed in tightening and reverse, programming multiple steps with a different speed for each step and setting some control limits to detect some abnormalities during the tightening such as re-hit with audible and visual feedback to the operator. Low torque tools may have speeds up to 1,500 RPM or more. Higher torque tools are limited mostly to 100 RPM. Clutch battery tools are accurate even at high speeds. It is common to have different battery sizes for the same tool. The batteries with larger Amp-hour capacity provide longer working time before the battery must be recharged or replaced; however, they are less ergo-friendly due to their higher weight. The calibration and adjustment of torque or speed of battery tool is easy based on the joint requirements.

22.3.4 Mechanical Tools

Handheld mechanical torque wrenches are the simplest choice of tightening tools. They are widely used due to their low cost. However, since they are manually driven, the tightening takes much longer time which makes them not suitable for mass production. They are still used as backup for power tools or for rework purposes. They cover a very wide range of torque. They are available in different styles: preset click wrench, adjustable click wrenches, and cam-over wrench. Mechanical preset and adjustable wrenches are accurate in terms of clicking at the desired torque, however, the final torque delivered to the fastener is operator-dependent since the operator can continue to tighten the fastener even after the wrench clicks. Cam-Over wrenches address this issue and prevent the over-tightening, but they are difficult to use, this is because the cam mechanism suddenly releases when the target torque is reached, which is unpleasant to the operator.

22.3.5 DC Electric Tools

DC electric tools have a built-in transducer that monitors the torque and an angle encoder that monitors the changes in the torque and angle of rotation in real time during the tightening. They also have a controller that can analyze these readings and control the motor driving the tool to stop at the required torque or angle. This makes them capable of performing torque control, angle control, and yield control tightening processes. They are capable of setting monitoring limits to detect abnormalities in the torque, angle, or time. They can be programmed to have multi-step programs with different speeds and monitoring limits for each step. They can be used alone, or controlled by an external line PLC to activate different tightening programs based on an externally provided signal that is driven by the product type or product options. Even the number of required fastenings can be configured to complete a specific job, where the job is evaluated to be OK only when the required number of fastenings with OK results is performed. The tightening results such as torque, angle, timestamp, product identification number, and tightening curves can be stored and used for quality analysis of the tightening process and detection of quality issues.

They are the most precise tightening tools and the most expensive type as well. They are very reliable and repeatable. Regular calibration and maintenance are crucial for ensuring the accuracy and reliability of these tools as they are typically used for critical applications.

There are three main types of these tools, namely, cabled handheld tools, cabled fixtured tools, and battery-powered handheld DC tools.

22.3.6 Digital Tools

Manual digital wrenches are used as backup for DC electric tools. They are very precise yet dependent on the operator skills. Some have a built-in gyroscope, therefore, they are capable of measuring torque and angle; hence, they can be used for angle control tightening processes. They are available in a wide range of torque/sizes. They are typically equipped with a digital screen to give feedback to the operator when to stop the tightening process. In some cases, a special version of digital wrenches has built-in algorithms to measure the residual torque in the joint. Table 22.1 shows an overview comparison of previously mentioned types of fastening tools.

TABLE 22.1
Tightening Tools Overview and Comparison

	Air Tools	Battery Tools	Mechanical Wrenches	DC Electric Tools	Digital Tools
Tightening strategy	Torque	Torque	Torque	Torque Angle Yield	Torque Angle Yield
Quality monitoring	Low	Low	Low	High	High
Automation	Medium	Low	NA	High	NA
Torque range	Low to high	Low to Moderate	Low to high	Low to high	Low to high
Speed	Extremely high >1,500 RPM	Medium ->300–1,000 RPM	Manual	Medium however speed must be limited in the final tightening step to achieve accuracy ->50–150 RPM	Manual
Accuracy	Medium	Medium	Medium but operator dependable	High	High but operator dependable
Cost	Moderate to high	Moderate to high	Low	Very high	High
Applications	Automotive, construction,	Automotive, assembly	Rework, mechanic	Automotive, precision assembly	Precision rework

TABLE 22.2
Guide for Tightening Tool Selection

Tool Type	DC Electric Tools	Clutch Battery Tools	Air Tools	Mechanical Wrench	Digital Wrench
Category A/Angle Ctrl/Tight Tol.	X			Torque control joint rework	Angle control joint rework or applications, where power tools can't be used
Category B		X	X		
Category C		X	X		

22.4 TOOL SELECTION IN THE AUTOMOTIVE INDUSTRY

There are several factors that play a role when it comes to tool selection, namely, joint criticality, speed, accuracy, and cost. The maintenance of the tool is another important factor for mass production which cannot always be included in the tool cost as it will result in tool unavailability or downtime which is very critical on assembly lines. Finally, the choice of tool depends on the specific task, the required torque, and accessibility in the vehicle. Table 22.2 is a common practice used in the automotive industry to select the tool type based on the joint categories defined in VDI 2862, namely, A, B, or C categories. Category A joints are the most critical joints and require the use of special measures to guarantee the quality of the joint. In most of these cases, DC electric tools are used. In addition, angle control tightening processes require DC tools as well.

For Category B and C joints, either battery or air tools are normally used, however, due to the air hose connection, the air tools have limited reach and are more difficult to move to another line location when needed. This is why many companies switch to battery tools.

The mechanical tools can be used for any rework application as they are slow and not suitable for mass production.

Digital tools are also slow but they are a good choice for reworking angle control tightening processes.

22.5 TIGHTENING CURVES ANALYSIS

In safety-critical applications, monitoring the torque and angle changes during the tightening process is essential to ensure the joint meets the required quality and safety standards. DC tools provide the capability to do this, however, in order to be able to set the monitoring parameters correctly, one must understand the joint being tightened, and how the torque changes during the tightening process. This relationship is represented by the torque-angle curve. Each joint has its proper torque-angle signature, where a deviation from the typical shape indicates an anomaly in the joint. Here are two examples of a typical tightening curve for a joint using machine thread screw vs. joint with thread forming screw.

Figure 22.1 shows a typical torque-angle curve for a joint using a machine thread screw when tightened to failure. The curve can be divided into four different zones. The first zone is the rundown zone, in which the fastener turns freely without resistance or very little resistance (zero torque). The second zone starts when the head of the fastener touches the tightened joint member, and the parts are pulled snugly together. After that the tightening torque increases linearly as the fastener rotates until the fastener or joint starts to yield, this is called the elastic zone. If the tightening continues, plastic deformation in the bolt or in the joint continues until the fastener or the joint fails completely. This zone is called the plastic deformation zone.

When torque control tightening strategy is used, the tightening stops within the elastic zone close to the last quarter of that zone. This is typical for the majority of tightening applications. For applications where angle control tightening strategy is used, the tool is set to reach a specific torque value, called threshold torque, within the elastic zone, followed by an additional angle of rotation which normally ends within the plastic zone. This is done in order to maximize the clamp load obtained in that joint.

Figure 22.2 shows the torque-angle curve for a joint using a thread forming screw. During the rundown zone, the screw must cut or form the threads in the joint base, therefore, the torque is not zero. The torque required to make the threads (cut or form) in the unthreaded hole is usually called the tap torque. When the screw penetrates through the joint base before the screw is fully seated the torque drops down, if it does not penetrate through, the thread forming torque may keep increasing

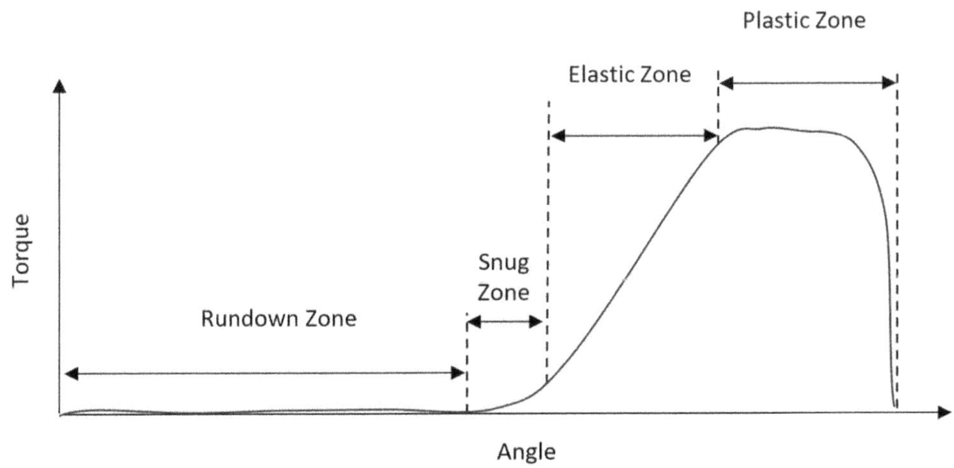

FIGURE 22.1 Typical tightening curve for machine thread fasteners.

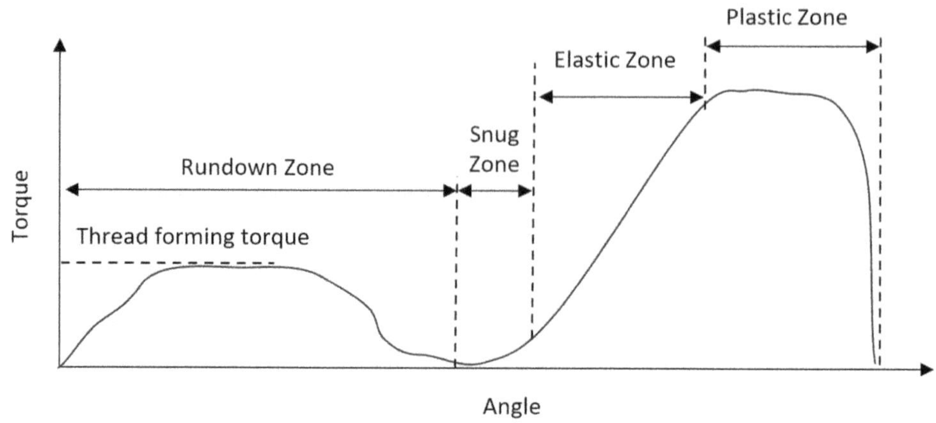

FIGURE 22.2 Typical tightening curve for self-tapping fasteners.

slightly until the screw is seated. After the screw is seated the torque increases again to clamp the parts together until the screw or joint starts to yield. If the tightening continues beyond this point, then joint failure takes place and both torque and clamp force drop to zero.

Each one of the zones has certain features that can be monitored, where deviations from the normal behavior can be used to abort the tightening and issue a warning to the operator that something is wrong in the joint. For example, an early rise in the torque during the rundown zone may indicate a cross threading condition, and stopping the tightening as soon as this condition is detected will protect the joint from extreme damage that could be very difficult and expensive to repair. Using the monitoring parameters for each stage during the tightening provides a real-time quality control. This is done normally by analyzing the tightening curves to understand the typical curve shape for good joints and then setting the proper parameters to detect deviations from the typical shape.

In addition, keeping a record of the torque angle curves, with the corresponding product ID, provides a powerful tool to analyze and understand the root cause for joint failure for warranty cases, or provide a proof that the joint was assembled correctly, in case of a lawsuit for example.

Figure 22.3 represents a torque-angle curve generated by a DC electric tool. A simple tightening program will include some rundown monitoring limits where a minimum number of degrees is expected before reaching some torque value (determined based on sample curves collected for good joints). It will also include a specific torque from which the final angle counting starts (typically 50% of target torque), and a minimum and maximum limit for the final torque achieved and the angle measured at that torque. Min and max torque limits are normally set based on the design allowed torque tolerance, while the angle limits are estimated based on the final angle values observed for a sample of good joints with enough tolerance to accommodate expected normal variation. The tool controller continuously evaluates the torque and angle data during the tightening, and it provides the overall tightening result when the tool stops; either OK if the good tightening conditions are met, or Not OK when any of the limits is violated.

The proper settings of the tightening system parameters are essential to detect quality issues that the joint may have. Figure 22.3 shows examples of different NOKs and lists some of the possible conditions that can cause these NOKs.

- It is important to establish the baseline or how a good curve looks like. A smooth, gradual increase is generally expected during a controlled tightening process. The tool is expected to stop inside the predefined torque-angle window shown in Figure 22.3, which means the curve aligns with the required torque and angle for the specific application.

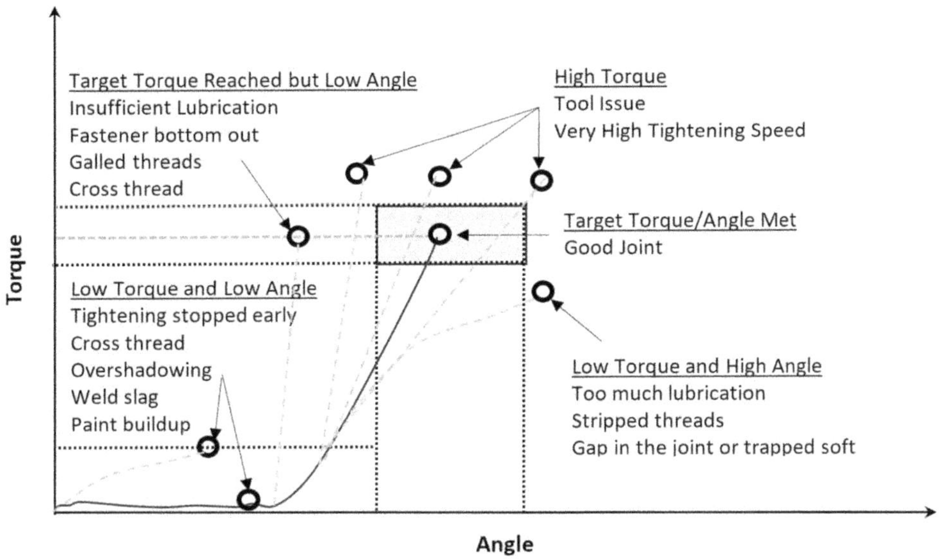

FIGURE 22.3 Common root causes for Not OK results based on tightening curve.

- Sudden changes (spikes or drops) in the curve are normally an indication of a problem in the tightening. A sudden increase during the beginning of the tightening might suggest a cross threading problem, while a sudden drop in the torque could signal slippage off or parts failure (crack or breakage).
- Plateau at a relatively high torque may indicate reaching the plastic zone of the bolt or the clamped joint. This could be normal for angle-control processes, since by design it is intended to reach the plastic zone by the end of the tightening. However, for torque control processes, this indicates an issue with the joint, such as incomplete threads, softer bolt or joint material, low friction coefficient, joint component failure, etc.
- If the tightening does not stop within the final predefined torque window for a torque control process, it is normally an indication of high final stage speed where the tool is not able to stop accurately at the target torque. Another possibility is an extremely hard joint or a stick slip condition. In all three cases, reducing the final tightening speed will help eliminate this issue.
- The final angle measurement normally starts from a threshold torque, which is typically set close to 50% of the target torque. However, in some cases if this torque is reached early due to high thread forming torque condition for example, the threshold torque could be reached early and as a result, the maximum angle limit is exceeded before reaching the final torque. In this case, the threshold torque can be increased to eliminate unnecessary Not OK results from the tool. On the other hand, in case of extremely hard joints, the final angle value measured is too low when the measurement starts from 50% of target torque. This makes it hard to distinguish a good tightening from a re-hit condition. Reducing the threshold torque may help in this case.

Tightening curves provide valuable insights into the fastening process, allowing for real-time quality control of the tightening process. Regular monitoring and analysis of the tightening curves can help maintain the reliability and safety of bolted joints, which is particularly important for critical applications. In the next section, guidelines for detecting quality issues are discussed through the setup of the tightening tool program.

22.6 GUIDELINE FOR SETUP AND OPTIMIZATION OF TIGHTENING PARAMETERS

DC electric tools, which are generally used for safety critical applications, are programable and when set properly can detect potential quality issues. The tool program shall be set to help flag NOK joints which is normally achieved by setting monitoring limits relevant to the joint characteristics. The following serves as a general guideline for the setup of DC tool tightening programs.

22.6.1 Programming DC Tools

The first step in developing a tightening program is to understand the tightening specifications provided by the product design group. These specifications include the tightening strategy, the target control value (Torque, or threshold torque + angle), and the corresponding tolerances allowed which need to be accommodated by the tool used.

The first step in setting up the tool program is to know the tightening strategy to use. The most commonly used tightening strategy is torque control, followed by angle control.

For either strategy, the tightening program can be one stage or multiple stages, where each stage has its own control parameter and monitoring limits. A two-stage program is normally used, where the first stage is utilized for rundown with relatively high speed to save process time; the second stage resumes the tightening up to the final target at a lower speed, which is necessary for the tool to accurately reach the target control value (torque or angle).

22.6.1.1 Torque Control Tightening Processes

For torque control tightening, the control parameter is the final torque set in the final tightening stage. The target torque and the upper and lower limits are set according to the design required specifications. In addition, angle limits are set based on analyzing sample tightening results for a number of good joints, taking into account potential normal variations. They are used to detect abnormalities in the joint.

Counting the final angle starts from a threshold torque after the fastener is fully seated, typically, 50% of the target torque is used. Figure 22.4 shows the torque tolerance limits and the angle monitoring limits that form the OK window.

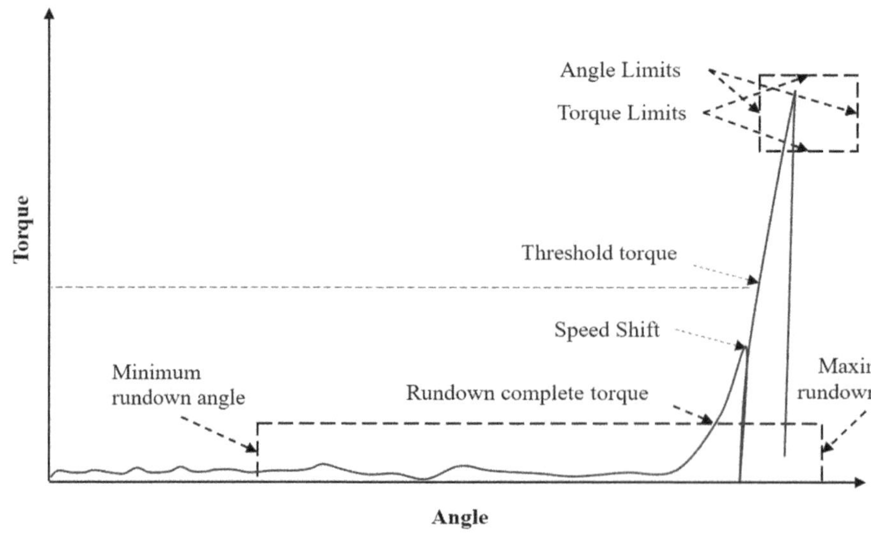

FIGURE 22.4 Torque-angle curve with rundown and final monitoring windows.

22.6.1.2 Angle Control Tightening Processes

The torque plus angle specification along with tolerances for both threshold torque and final angle are normally released by the Design department. Example: 56Nm (±10%)+90° (±15°).

The final expected torque limits are set based on analyzing sample tightening results for a number of good joints taking into account potential normal variations. Alternatively, the minimum final torque monitoring limit can be set to the minimum yield torque calculated for the fastener used (assuming the bolt is the weakest component in the joint), while the maximum final torque monitoring limit can be set based on the maximum expected fracture torque at the highest friction limit allowed and maximum material strength possible. Setting the proper final expected torque range is very important to detect possible quality defects. For example, one of the potential failures is tool slip-off of the fastener after reaching the threshold torque and before completing the final angle of turn; this can be detected if a proper minimum torque limit is used.

22.6.2 Tightening Stages

It is recommended to use minimum two stages in the tightening program, a rundown stage, and a final tightening stage. The rundown stage with additional monitoring parameters is used to detect errors caused by cross threading, weld slag, damaged threads, or other defects that are normally not detected by the final tightening stage parameters. The final stage involves tightening the fastener to the target tightening specification. This stage is monitored using a torque and angle window to capture possible stripping, component failure, slipping off, overshooting, or other defects.

22.6.2.1 Rundown Stage

Usually, the first stage has fast rundown speed (about 70% of maximum tool speed) and ends at a torque value that ensures that the fastener is fully seated; typically, 30% of final target torque is used unless the fastener has a prevailing torque or it is a self-tapping or a thread forming screw; in these cases, it must exceed the maximum expected prevailing torque or thread cutting/forming torque.

For thermoplastic screws going into a plastic nut or a plastic part, it is recommended that the rundown speed does not exceed 400 RPM. It is recommended to check the design recommendations and parts drawings for recommended maximum speed.

Rundown monitoring is used to detect cross threading, parts' overshadowing, and thread damage, and other quality issues. The concept is to make sure the fastener turns the number of turns (degrees) expected before torque start building up. To achieve this, three main parameters are set:

- *Rundown complete torque*: the torque at which rundown is complete. For applications using self-locking fasteners or self-tapping, thread rolling, and thermoplastic screws, the maximum expected prevailing torque or thread cutting/forming torque should be taken into consideration when setting up the rundown monitoring window parameters.
- *Minimum rundown angle*: minimum angle of rotation required before reaching the rundown complete torque. This angle is measured from the start of the tightening to the rundown complete torque or from a specific torque value.
- *Maximum rundown angle*: maximum angle of rotation that must not be exceeded before reaching the rundown complete torque.

These three parameters from the rundown monitoring window are shown in Figure 22.4. The values for rundown monitoring window are joint-specific and should be set based on the data collected for the actual joint. If the rundown monitoring window conditions are violated, the DC tool will give a NOK result indicating to the operator that the joint has a problem.

Here are a few quality issues that can be detected during the rundown stage:

- Parts misalignment (over shadowing) resulting in high rundown torque.
- Screw bottoming out due to use of longer fastener causing torque to build up quickly.
- Weld slag in threads causing a spike in torque during the rundown.
- Rehit (tightening an already tight fastener) causing torque to jump up immediately.

22.6.2.2 Final Stage

During the final stage of the tightening program, the speed is slowed down to improve the final torque accuracy (typically about 10%–30% of maximum tool speed, based on joint hardness and analysis of sample tightening results for torque overshooting). The tightening is supposed to stop when the target torque or angle is reached. If the final value is consistently greater than 3% of the target torque, then the final speed needs to be reduced.

The torque and angle monitoring limits form the acceptable monitoring window for the final torque and angle values at the end of the tightening. For torque control strategy, the torque monitoring limits are based on the torque tolerance provided by the design engineer. The final angle measurement starts from a threshold torque, which is normally set close to 50% of final target torque, as shown in Figure 22.4. In some cases, the threshold torque value must be increased or decreased depending on the joint hardness and characteristics. The final angle monitoring limits are calculated based on the results for a sample of good joints (25 or more samples are recommended) as follows:

$$\text{Minimum Final Angle Limit} = \text{average final angle} - 3 \times \text{standard deviation for collected samples}$$

$$\text{Maximum Final Angle Limit} = \text{average final angle} + 3 \times \text{standard deviation for collected samples}$$

In many cases, the minimum final angle calculated can be too small or negative, in these cases it is recommended to reduce the threshold torque and recalculate the limits. If the lower limit is still too low, it is recommended to set it to no less than 8 degrees. A lower minimum angle limit will reduce the effectiveness of detecting re-hit or other quality issues, however, if the joint is too hard then there is no option but to use a lower final angle limit without changing the joint design.

On the other hand, if the maximum final angle is too large (>200°), increasing the threshold torque will help reduce the average and the upper limit, especially for soft joints, such as clamp tightening applications, or thread forming applications, or self-locking fasteners.

In general, the more consistent the final angle results are, the tighter the final limits can be, which provides a better detection for any abnormalities in the joint. For angle control strategy, the angle monitoring limits are set according to the tolerance provided by the design engineer, while the final torque monitoring limits can be developed using the same method described above using the average final torque value and corresponding standard deviation for results obtained for a sample of good tightening results.

After setting all parameter limits, the tightening results of the process should be monitored for a period of time to check for abnormalities that need to be detected and look for any false “Not OK” results that may require further parameters adjustment.

22.6.3 Multi-Stage Tightening Programs

Most DC tools allow to add additional stages and may have some additional advanced monitoring steps or limits. Normally for automatic bolting stations or for special applications, a program with more than two stages is necessary to help align the fastener and stop the process before an extensive damage to the joint occurs when abnormal conditions are detected. For example, a find-step rotating the spindle one or two rotations counter-clockwise is to make sure the bolt is properly positioned before the tightening process is started. Another slow start step for one or two rotations would be also useful for proper thread engagement. Then the same recommendations listed above apply for the rundown and the final tightening stages; additional special stages can be used to detect any special defects based on the specific joint characteristics.

22.7 DETECTING TIGHTENING ANOMALIES USING MACHINE LEARNING

Monitoring and interpreting tightening curves is a very useful tool to ensure quality and reliability, however, this process is very time-consuming considering the high number of joints in manufacturing. Over the last few years, the advancement of machine learning science has allowed for building algorithms to detect abnormalities in the tightening curves. This involves training a model to recognize patterns associated with normal and anomalous tightening behavior. To build such a model:

- A diverse dataset of tightening curves, including both normal and abnormal cases must be collected. Any inconsistency or noise in the data must be removed. Additionally, the data must be stored in a format that is suitable for processing by machine learning algorithms.
- Key features from the tightening curves must be identified then data labeling must take place to label normal and abnormal curves.
- A machine learning model must be chosen, trained, and optimized to enhance its performance.
- Validation and evaluation of the model performance using known data set to determine if fine-tuning is necessary.
- Apply the model to real-time tightening data and monitor its performance. Feedback loops must be established to accommodate any scenarios that did not show up during the initial model training.
- Document the model architecture and training process for future enhancement and maintenance of the model.

Building and deploying a machine learning system for tightening curve anomaly detection requires collaboration between domain experts, data scientists, and fastening engineers. Regular updates and maintenance of the model are crucial to ensure its effectiveness over time. Many companies are moving in this direction to reduce manual work and improve productivity.

It is expected that DC tool manufacturers would build such models into their tools considering the most common joint behaviors and allow the customer to use the built-in artificial intelligence algorithms to monitor the tightening process and flag any abnormalities.

REFERENCE

[1] VDI/VDE 2862 Blatt 1, Minimum restrictions for application of fastening systems and tools - Applications in the automotive industry, 2012-04.

Part V

Assembly of Bolted Joints

23 Behavior of a Bolted Joint During Assembly

John H. Bickford

23.1 INTRODUCTION

This chapter covers bolted joint behvior during initial assembly using different assembly tools and procedures. Some of these will seem overly complicated to those who don't appreciate the complexity of bolted joints. In this chapter, we examine the behavior of the joint during assembly, introducing many of the factors that make it difficult to do a "perfect" job with even the most sophisticated tools and procedures.

That all-important clamping force that holds the joint together—and without which there would be no joint—is not created by a good joint designer or by high-quality parts. It is created by the mechanic on the assembly line or job site, using the tools, procedures, and working conditions he or she has been provided with. The force is brought into being as energy in the mechanic or power tool is converted to potential energy stored in the joint and bolt members. The correct amount of force cannot be created if the design is faulty or the parts don't fit together properly or if they break; but getting all this right, while necessary, is not enough. The final, essential creator of the force is the mechanic, and the time of creation is during assembly. So it is very important for us to understand the assembly process.

How do the bolts and the joint members respond as we tighten the bolts? We will see that the behavior of the parts during assembly is complex. This chapter takes a close look at several unseen, difficult-to-detect, difficult-to-quantify factors that can have a significant impact on the results, on the amount of clamping force developed in the joint. This will teach us that it isn't easy to control the buildup of clamping force in the joint and that mechanics need all the help we can give them, both as product designers and as production engineers.

In 11 of the chapters that follow, we will look at many of the options we have for control of the bolt-tightening process. Our knowledge of the behavior of the bolts and joints during assembly will help us evaluate the merits of these options.

Accurate control of the clamping force is often necessary; too much or too little clamp can degrade the behavior and life of the joint in service. If the bolts and joint members do not contain the correct amount of stored energy to create the correct amount of clamping force, there will be joint problems. In other words, proper assembly can be essential.

23.2 INITIAL VS. RESIDUAL PRELOAD

The clamping force a bolt exerts on the joint is usually called—or equated to—the "preload" in the bolt. This term is used in general in most of the literature on bolting to describe the tension in the bolt at any time, but this, in my opinion, is a mistake. I like to think of the preload created in an individual fastener when it is first tightened as "initial preload," even though that term may be redundant. As you'll see, the effects we're about to discuss will frequently modify this preload as the fastener relaxes and/or as we tighten other fasteners in the joint. I call the final preload in the bolts the "residual preload."

When the joint is put into service, a variety of things can act to modify the preload in individual fasteners still further. This could be called in-service tension in the bolts.

DOI: 10.1201/9781003501183-28

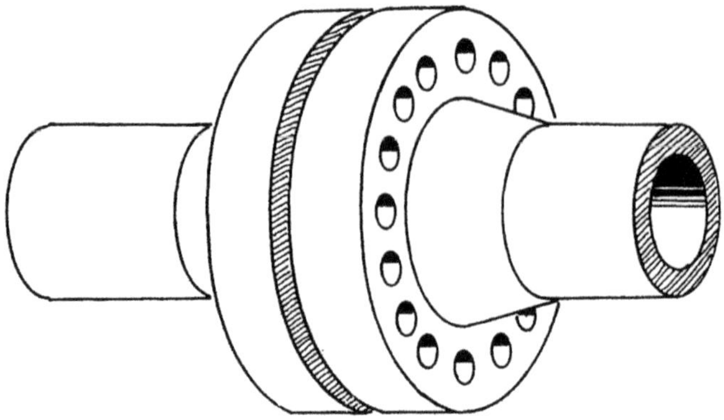

FIGURE 23.1 A sketch of the large-diameter pressure vessel joint used as an example in this chapter.

Each of these preloads or tensions is directly proportional to the amount of potential energy stored in the bolt as it is first tightened, after relaxation occurs, or in service. In most cases these preloads or tensions will also be directly proportional to the clamping force between joint members, but there are exceptions.

Now that we have these definitions under our belt, let's get on with it. What happens during assembly?

23.3 STARTING THE ASSEMBLY PROCESS

We are going to assemble a hypothetical joint, using as our example a round, gasketed, pipe flange joint held together by sixteen 1 1/8–8 ASTM A193 B7 bolts (see Figure 23.1). The large diameter and the presence of a gasket make this assembly a little more difficult than most but allow us to look at a more complete range of assembly problems than would a simpler example. Most of the discussion will apply to joints in general.

We are also going to measure the torque we apply to the nuts to control the buildup of initial preload in these bolts. This is the most common type of control and one of the simplest. It is the subject of Chapter 29, so we won't go into a lot of detail here about its pros and cons. We will just use it for now.

23.3.1 Assembling the Parts

We start by roughly aligning the flanges so that we can insert the bolts by hand. When we finish pushing and pulling on the flanges, their mating surfaces are not exactly parallel and the holes aren't aligned perfectly, so we have to tap a few of the bolts with a hammer to get them through their holes, and some of them stick out a little farther, on the nut end, than others. Now we're going to apply a preliminary "snugging torque" to run the nuts down and pull the flanges together.

23.3.2 Tightening the First Bolt

To load the joint (and gasket) evenly we apply the snugging torque in a cross or star pattern, as shown in Figure 23.2. We would use a similar pattern on a square or rectangular joint if the bolts were all around the edge. In a rectangular structural joint pattern with several rows of bolts, we would start snugging at the center of the bolt pattern and work our way out to the free edges.

We use 225 lb-ft of torque for this first, snugging pass. This is about one-third of the final torque we're planning to use, and we follow it with a second pass at two-thirds of final torque, and then a third and final pass at full torque. In a structural steel joint we would follow the snugging pass with

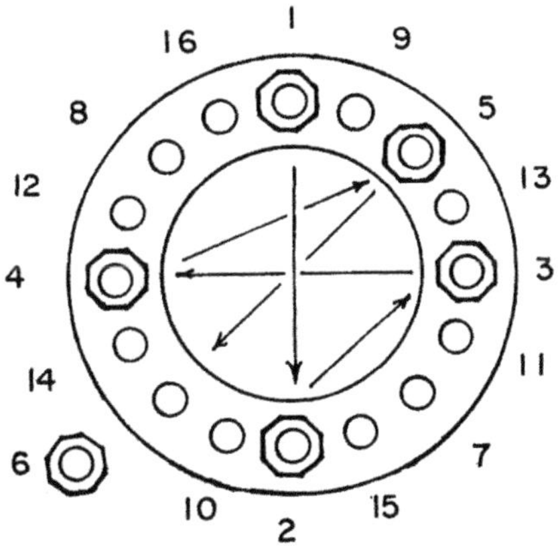

FIGURE 23.2 We'll tighten the bolts of our example joint in the "star pattern" sequence shown here. We will use three passes—at one-third, two-thirds, and final torque—following the same sequence on each pass.

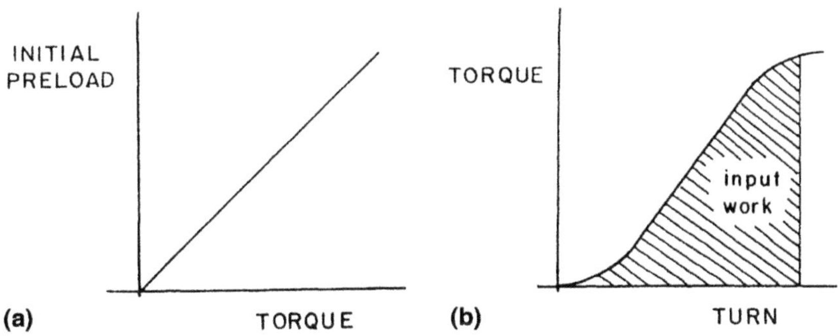

FIGURE 23.3 As we tighten the first bolt in our example joint we plot (a) the buildup of initial preload vs. applied torque and (b) applied torque vs. the angle through which the nut turns. The area under the torque-turn curve is equal to the amount of work we are doing on the nut and to the energy delivered to the fastener-joint system.

a second (last) pass at the final torque. Note that in each case we are following basically a two-step procedure: Pull the joint together and then tighten it. Because this is a learning experiment, we use ultrasonic equipment (Chapter 34) to measure the preload in each bolt as we tighten it. We also measure the angle through which the nut turns after it contacts the surface of the joint, and we measure the amount by which the bolt stretches and the amount by which the joint is compressed.

We now apply the snugging torque to the first bolt and use the resulting preload, torque, and turn data to plot the curves shown in Figure 23.3. We are doing work on this fastener as we tighten it. The amount of work is equal to the area under the torque–turn curve (measured in pound-feet or newton-meters times radians). Ideally, all of this work would be converted to potential energy in the bolt and in those portions of the joint members that surround it. If that were the case, all of the work we do on this fastener would end up contributing to the clamping force. Unfortunately and unavoidably, most of our input work is lost.

Typically about 90% of the work we do on a nut is converted to heat, owing to the frictional resistance between the face of the nut and the surface of the joint and between male and female threads. About 50% is lost under the nut and about 40% within the threads, as shown in Figure 23.4. Only 10% of the input work typically ends up as potential energy in the bolt, so only 10% ends up as bolt preload or as clamping force between joint members.

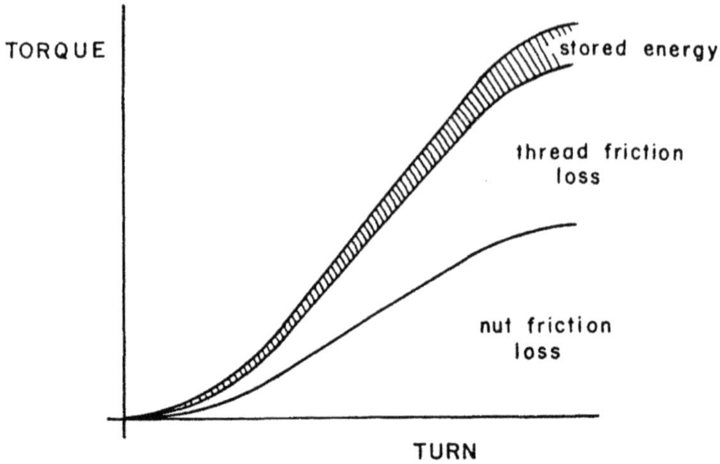

FIGURE 23.4 This diagram shows the approximate way in which the energy delivered to the fastener–joint system is absorbed by it. About 50% of the input is lost as friction generated heat between the face of the nut and the surface of the joint. Another 40% is lost as heat between male and female thread surfaces. Only about 10%, on average, ends up as potential energy stored in the bolt and joint "springs"; and only that 10%, therefore, ends up as preload in the bolt and clamping force on the joint.

We would like to apply a given torque to each bolt and create a given amount of initial preload—the same amount—in each bolt. But the fact that most of the work we do on the nuts is converted to heat makes this virtually impossible because these frictional losses are extremely difficult to predict or control. Let's assume, for example, that this first nut we're tightening is a little drier than average. As a result, let's assume that 52% of the input work is converted to heat at the nut/joint interface, rather than the "typical" 50%. A 4% increase in friction—from 50% to 52% of the input work—is easy to come by.

This 4% increase in friction loss—that extra 2% of the input work going into heat— means that 2% *less* of the input work will be converted to tension in the bolt. We started with the assumption that an average of only 10% of the input would be going into preload; now we've lost one-fifth of that. This bolt will therefore end up with only 80% as much preload as we expected it to. A 4% swing in friction has caused a 20% change in assembly preload, a very bad "leverage" situation. There are a lot of factors that can cause this kind of variation in friction.

Although in our learning experiment we're measuring both torque and bolt tension, we will not attempt to compensate for the frictional differences between bolts; we will apply the snugging torque of 225 lb-ft to the first bolt and let the initial preload end up where it may.

We also plot the deflections in the bolt and in the joint material surrounding the bolt versus the preload we create in the bolts and the presumably equal and opposite clamping force on the joint (see Figure 23.5a). We then combine these force–deflection curves, plotting the preload on a common axis, as shown in Figure 23.5b. This creates what the bolting world calls a joint diagram. You may complain that the preload in the bolt and the clamping force on the joint are equal and opposite action and reaction forces and that both should not be shown as positive values, but this joint diagram is a great convenience.

Since the diagram records the forces developed in bolt and joint and the deflection of each part, it also gives us a visual indication of the stiffness of the bolt and the stiffness of the joint. These are proportional to the slopes of the two straight lines and, as we will see in Chapter 36, can be computed as follows:

$$K_B = \frac{F_P}{\Delta L} \tag{23.1}$$

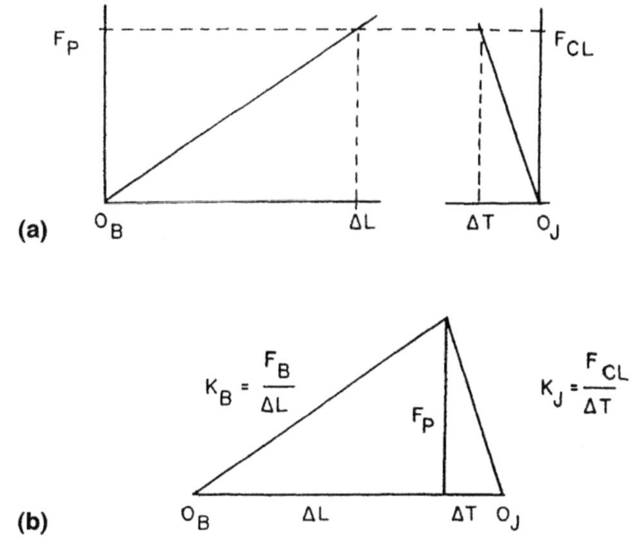

FIGURE 23.5 (a) As we tighten the first bolt we also plot the buildup of preload (F_P) in the bolt vs. the increase in length (ΔL) of the bolt, and the buildup of clamping force on the joint (F_{CL}) vs. the compression or change in thickness (ΔT) of the joint. At this point, we assume that the preload will be equal to the clamping force. (b) We then combine those two plots to start constructing a "joint diagram."

$$K_J = \frac{F_P}{\Delta L} \tag{23.2}$$

where

F_P=preload in the bolt and joint (lb, N)
ΔL=deflection (stretch) of the bolt (in., mm)
ΔT=deflection (compression) of the joint (in., mm)
K_B=stiffness of the bolt (lb/in., N/mm)
K_J=stiffness of the joint material being loaded by this bolt (lb/in., N/mm).

Note that the areas under the bolt and joint curves also equal the amount of energy stored in these parts. So this simple diagram (Figure 23.5b) contains a lot of useful information. This diagram is extended in Chapter 36 to add the effects of external loads on the joint. Joint diagrams are also used when we design joints. Now, however, we are merely interested in using the joint diagram to illustrate the preloading of bolts.

Our supervisor has used the torque–preload equation [Eq. (29.1), Chapter 29)] to compute the average preload he expected us to get in this first bolt when we applied 225 lb-ft of torque to it. He tells us that we should have created 12,000 lb of tension in the bolt. Because of the slightly higher than average friction loss described earlier, however, this bolt has ended up with only 80% of that preload or 9,600 lb.

Has this really created 9,600 lb of clamping force between joint members? Our joint diagram assumes it has, but the correct answer is "probably not"—at least as far as this first bolt is concerned. Remember that we had to tap some of those bolts into their holes? This implies that there was contact between the sides of those bolts and their holes—bolt-hole interference. Furthermore, the flange surfaces were not pulled into full contact when we first assembled the parts. They were slightly misaligned, as we could tell by the fact that some bolts stuck out farther on the threaded end than others. Before we go on to snug tighten the remaining 15 bolts in our example joint, let us take a look at how hole interference and/or nonparallel flanges might affect the buildup of clamping force during the assembly process.

23.4 BOLT PRELOAD VS. CLAMPING FORCE ON THE JOINT

The main purpose of the bolts is to clamp the joint members together. A common misconception is that there is always an equal and opposite action–reaction relationship between the tension in the bolts in a joint and the clamping force between the joint members. If there are eight bolts in the joint and an average tension of 10,000 lb in each of the eight bolts, then, simplistically, the joint is clamped together with an interface force equal to eight times 10,000, or 80,000 lb. That's usually true, but there can be some significant exceptions.

23.4.1 Effects of Hole Interference

Consider the situation shown in Figure 23.6. We are tightening this stud by turning the upper nut. Our goal is to clamp the joint members together. The hole in the upper joint member is undersized, so the stud is a press fit in this member.

Because of frictional and/or embedment constraints between the sides of the bolt and the walls of the hole, it will take some positive force to pull the bolt through the hole and then to stretch it within the hole. Where does this force come from?

The force is created, obviously, when we turn the nut, creating tension in the bolt. Thanks to hole interference some of this tension will not end up as clamping force between joint members. Part of it will be lost as the bolt is forced past the walls of the hole.

Envision the extreme case in which the holes are grossly undersized and the torque normally specified for a bolt of this diameter insufficient to even bring the joint members into contact, much less provide any real clamping force. Note that misalignment between the holes of upper and lower joint members could create a similar hole/bolt interface problem. This is obviously not a good situation, but hole misalignment, undersized holes, press-fit fasteners, etc. are relatively common in the bolting world.

Hole interference is used on purpose by the airframe industry, for example, to reduce the possibility of fatigue failure of the shear-loaded joints used in airframe structures (compressive stress built up in the walls of the bolt holes opposes the formation or growth of fatigue cracks).

The holes are purposely drilled smaller than the diameter of the bolts to create this interference. There is, of course, a manufacturing tolerance on the diameters of both holes and bolts, so the

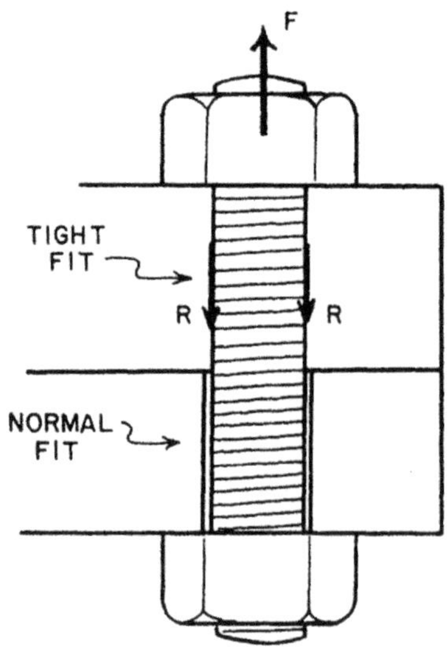

FIGURE 23.6 If there is interference between the bolt and the hole, the clamping force between joint members may be less than the tension in the bolt, changing the torque-clamping force relationship.

amount of interference varies. The greater the interference, the greater the force required to pull the bolt through its hole. It also requires more force to pull a bolt through a thick plate than through a thinner one, for a given amount of interference.

One airframe manufacturer measured the amount of force required to pull bolts of a given nominal diameter through holes drilled in plates of varying thickness. Some of the results are shown in Figures 23.7 and 23.8. The bolt and hole diameters used in the experiment varied through the full

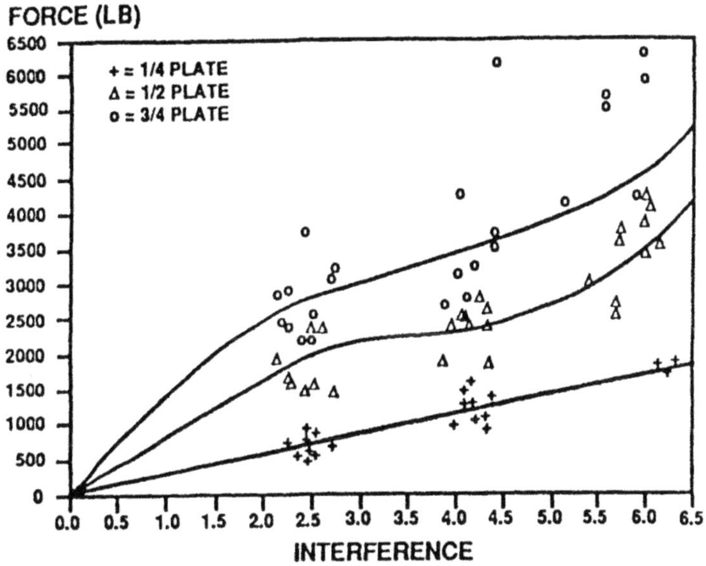

FIGURE 23.7 Chart showing the force required to push $^{5}/_{16}$-in. fasteners through interference fit holes in aluminum plates varying in thickness from ¼ to ¾ in. Bolt diameters are larger than the hole diameters by the amounts shown on the horizontal axis. Note the wide scatter in the results, which are summarized in Figure 23.8.

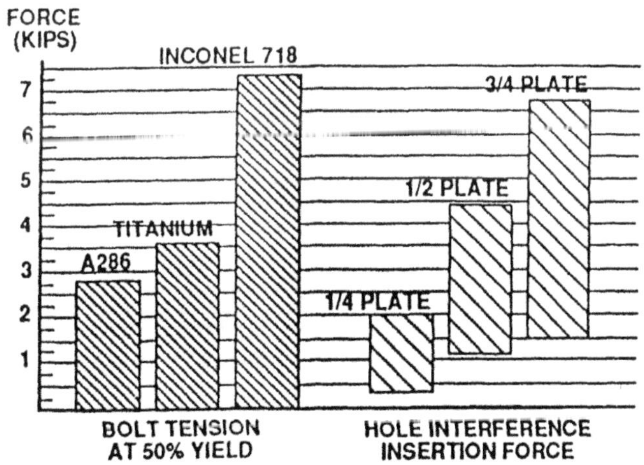

FIGURE 23.8 The three vertical bars at the right in this chart show the range of force required to thrust $^{5}/_{16}$in. fasteners through interference fit holes in aluminum plates of various thicknesses, summarizing some of the data shown in Figure 23.7. The three bars at the left show the nominal preload that would be generated by $^{5}/_{16}$in. bolts tightened to 50% of yield. Three bolt materials are shown. Note that only the Inconel bolts would have enough preload to overcome the worst-case interference forces and still provide some interface clamping force on the joint if ½ or ¾ in. plates were involved.

range of the manufacturing tolerance. It was found that the force required to pull some of these bolts through their holes exceeded the average preload that would be developed by the specified torque in a bolt of that diameter even if the bolts were used in regular holes with normal clearance. In other words, the specified torque could not be counted on to pull all of the bolts through their holes, much less go on to develop any clamping force between joint members. In spite of this, the same torque is specified by the airframe manufacturer for all bolts of a given diameter, without regard to the amount of hole interference seen by a given bolt or the thickness of the plates in which the bolt is used.

As a result, torque is *not* used in this application to pull the bolts through the holes. A bolt puller does that job, and temporary clamps are used to hold the joint members together until the bolts have been installed and tightened. But in some joints, the act of tightening the bolts is supposed to create some clamping force between upper and lower joint members, and the amount of this clamping force must vary widely.

23.4.2 Resistance from Joint Members

Another factor that can rob from the clamping force between joint members is shown in Figure 23.9. A heavy cover is being lifted up against a flange on a pressure vessel. As shown in the figure, the joint members have not yet been brought into contact, but the nut is being turned on the stud to bring them into contact. At the moment shown in the figure, there is already tension in the study equal to the weight of the cover. As a result, it will take torque to advance the cover up against the flange, due to the normal frictional constraints between male and female threads and between the nut and the cover.

So, at the point shown, there is torque, there is tension, there is friction loss, and there probably will be torsion in the stud, but there is zero clamping force between joint members.

Eventually the two joint members will be brought into contact. Further torque will be required at this point to create a clamping force between joint members to load the gasket. That torque should presumably be added to the torque required to pull the joint members together in the first place, but in practice it rarely is.

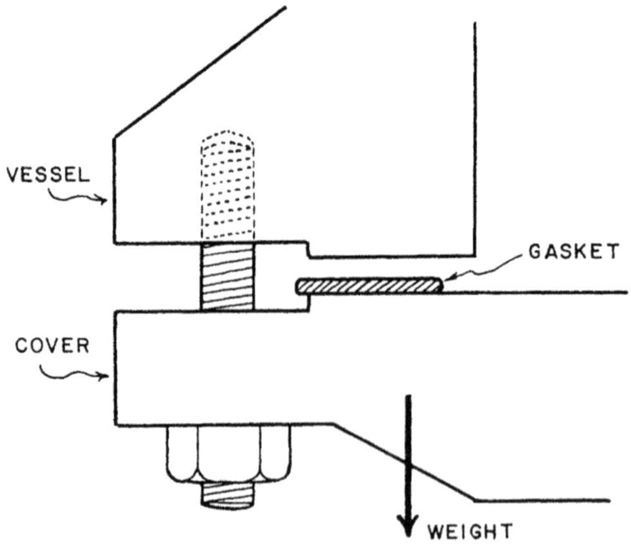

FIGURE 23.9 Heavy and/or misaligned or warped joint members can also affect the relationship between the torque applied to the fastener, the tension in the fastener, and the clamping force between joint members.

There aren't many applications in which a heavy weight is raised against a mating joint member by tightening the bolts, of course. It is, however, common for large joint members to be misaligned or nonparallel as the assembly process starts. Getting a pipe flange, for example, to mate with the flange of a pump or valve often requires a lot of motion in the flange members. The forces required to align such systems will have the same effect that the force created by the weight in Figure 23.9 has on that joint.

I wish I could tell you how much extra torque to add for misalignment or the like, but I can't. I have never seen anything in the literature on this subject, either. Yet I'm convinced that this factor can seriously degrade the relationship between tension in the bolts and clamping force between joint members, especially in large joints, as one example. Warped or nonflat (e.g., wavy) joint members, incidentally, could create the same sort of problem.

I once met a maintenance supervisor in a large petrochemical plant who took this problem seriously. He insisted that his crews align gasketed flanges within 12 mils before bolting them together. Bolting nonparallel flanges, he said, was a waste of time; "they'll always leak."

The Puget Sound Naval Shipyard also studied this problem [18]. They made theoretical calculations and conducted experiments to determine the forces, stresses, and moments in pipes and flanges when the flanges are misaligned. They used a hydraulic tensioner, for example (see Chapter 26), to measure the force required to pull misaligned flanges together, as shown in Figure 23.10. The chart in Figure 23.11 shows some of the results of their work: the stresses in the pipe adjacent to the flange as a function of nominal pipe diameter and with flange gaps of 0.015, 0.020, and 0.025 in. The forces (and torques) required to pull the flanges together would be proportional to these stresses.

In each case documented in Figure 23.11 it is assumed that there is a length of pipe equal to 100 times the nominal pipe diameter between the flange and the first rigid pipe support (a rod hanger or intersection with a larger pipe).

As a result of these studies, the shipyard developed some flange parallelism criteria, designed, as I understand it, to keep pipe stress below 3 ksi near any flange that is connected to turbines or other rotating equipment. Pipe stresses beside flanges located 50 or more pipe diameters away from rotating equipment were allowed to go slightly higher. The resulting criteria are shown in Table 23.1.

As a final note, if misalignments and stresses are too high they are sometimes reduced by the use of bellows-type expansion joints at some point in the pipeline.

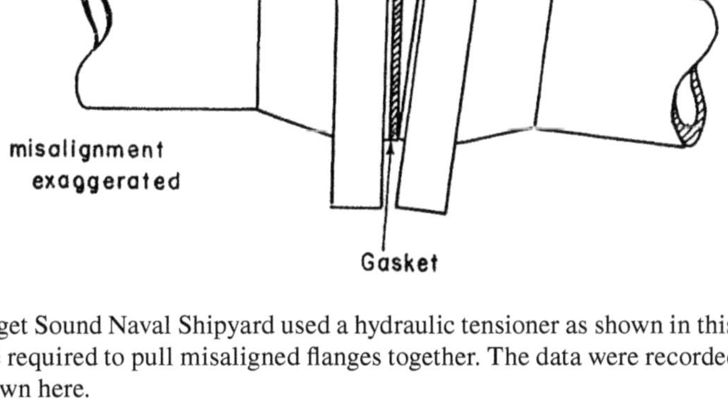

FIGURE 23.10 Puget Sound Naval Shipyard used a hydraulic tensioner as shown in this sketch to learn how much force would be required to pull misaligned flanges together. The data were recorded as a function of the size of the "gap" shown here.

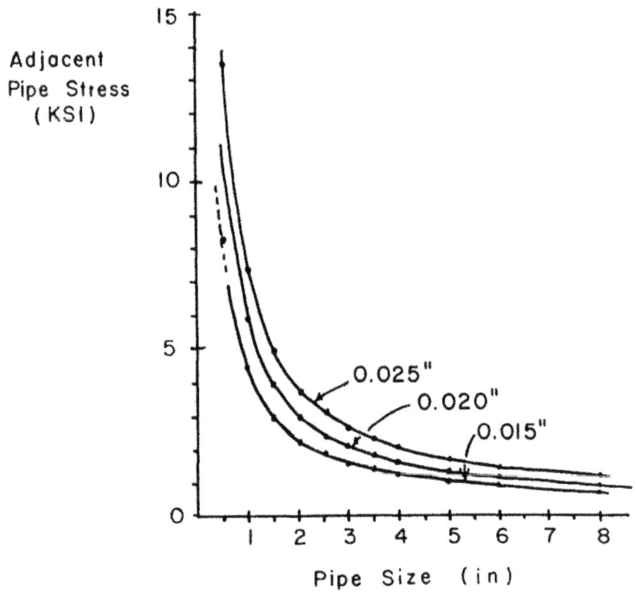

FIGURE 23.11 A plot of the stress in the pipe adjacent to a misaligned flange vs. the nominal diameter of the pipe, for misalignment gaps of 0.015, 0.020, and 0.025 in. It is assumed that the nearest rigid support for the pipe is located 100 pipe diameters away, along the pipe, from the misaligned flange. The moment on the flange and the force required to pull the two halves into full contact would be proportional to the pipe stress.

TABLE 23.1
Flange Parallelism Criteria[a]

	Flanges Adjacent to Rotating Equipment			Line Flanges
Pipe size	1 in. or less	1 1/4–8 in.		All sizes
Wall thickness	All	Sch 40 or below	Over 40 through Sch 80	All thicknesses
Raised face	0.10 in.	0.025 in.	0.020 in.	0.005 in./in. of contact diameter
Flat face	0.020 in.	0.035 in.	0.030 in.	0.005 in./in. of contact diameter

[a] If the pipe diameter is over 10 in. or wall thickness greater than Schedule 80, the specifications say that a special analysis is required.

23.5 CONTINUING THE SNUGGING PASS

We are now going to continue tightening the 16-bolt gasketed flange joint we are taking as an example in this chapter. We apply the snugging torque of 225 lb-ft to each of those bolts, following the cross-bolting pattern shown in Figure 23.2. We measure the preload, turn, and deflection in and around each bolt as we tighten it, and we record these data. We find that, on average, we have created the anticipated preload in these bolts but that individual bolt results vary from the average by ±30%. That, our supervisor assures us, is a typical result of torque-tightening a group of unlubricated, as-received steel bolts and nuts against steel joint surfaces. Everyone's happy, so we now go out and have lunch.

When we come back from lunch we find our boss's boss on the job site, reviewing our data. He wants to see how we managed to measure the tension or preload in these bolts, so we plug in our ultrasonic instrument and remeasure the preload in bolt number 1, the first bolt we tightened to a preload of 9,600 lb. To our embarrassment, we find only 900 lb of tension in that bolt. And we find a wide range of residual preloads in the other 15 bolts. We think that our instrument is misbehaving, but the boss disagrees. He gives us a lecture on bolt relaxation, which some people call "torque loss." Here's what he says.

23.6 SHORT-TERM RELAXATION OF INDIVIDUAL BOLTS

Whether or not there's a one-to-one relationship between bolt tension and interface clamping force, there will often be some initial loss of tension in individual bolts after they are initially tightened. Call this short-term relaxation to distinguish it from other effects such as stress relaxation [22] that will cause further loss of tension over a longer period of time.

In general, short-term relaxation occurs in a bolted joint because something has been loaded past its yield point and will creep and flow to get out from under the excessive load. This can be a component such as a soft bolt or a gasket; more commonly it is only a portion of a component, such as the first threads in a nut. Let's look at some examples.

23.6.1 Sources of Short-Term Relaxation

Here are some things that can cause relatively short-term relaxation, starting with the most common of all, embedment.

The surfaces of the threads in the nut, the bolt, and the faying surfaces of the structural members, washers, etc. are never perfectly flat, even if such parts are given a high polish— which is rarely the case with industrial structures and fasteners. Under a microscope, they are a series of hills and valleys.

When such parts are first loaded, they contact each other only through high spots on the metal surfaces. Even a small bolt, however, is able to exert extremely high surface pressures on structural members or on its own threads. Thread dimensions have been selected to support these high loads, but only if a significant percentage of the total thread surface shares that load.

Since initial contact areas are relatively small, the metal at the contact points cannot stand the pressure. Plastic deformation occurs until enough of the total thread surface has been brought into play to stabilize the situation and support the load without further deformation. The same thing happens in the faying surfaces of the structures, though perhaps to a lesser extent because larger surfaces are involved and initial contact areas are larger. Embedment is illustrated in Figure 23.12.

Many of these surface high spots are smoothed away during the tightening process [1]; at least they will be if the fasteners are torqued. Hydraulic tensioners don't load the active threads, or even

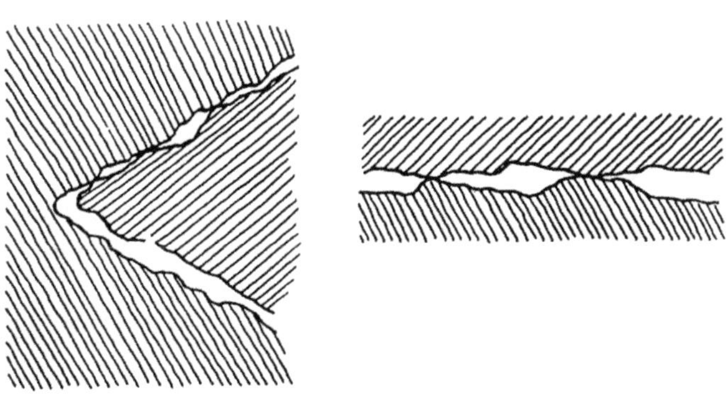

FIGURE 23.12 High spots on thread and other contact surfaces will yield and creep under initial contact forces. As a result, the surfaces will settle into each other until enough surface area has been brought into contact to stabilize the joint. The process is called embedment.

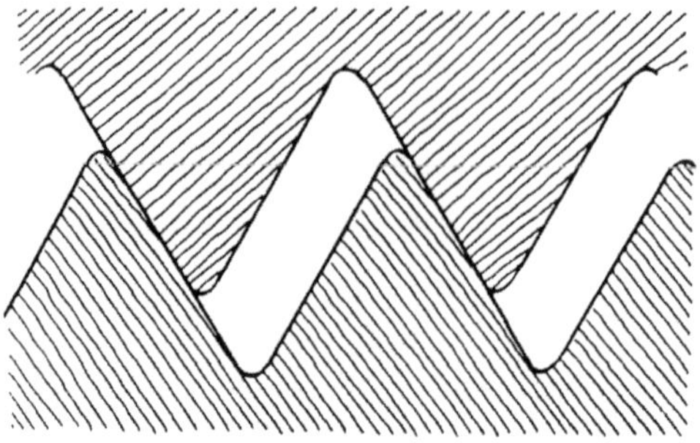

FIGURE 23.13 Poor thread engagement may be a major source of plastic deformation and therefore joint relaxation.

the joint surfaces underneath the nut or washer until they let go of the bolt. As a result, there is often more embedment relaxation after tensioning than there is after torquing.

Embedment is worse on new parts than on reused ones. In critical applications, it can be minimized by tightening, loosening, and retightening the fasteners several times. This is done on camera mounts in space satellites, for example [2].

23.6.1.1 Poor Thread Engagement

If the bolt is undersized, or the nut oversized, thread contact areas will be less than those planned by the designer, and substantial plastic deformation may occur, as shown in Figure 23.13 [3].

23.6.1.2 Thread Engagement Too Short

The length of thread engagement for steel fasteners should be at least 0.8 times the nominal diameter of the fastener. If the engagement length is too short (too few threads support the load), thread contact areas are again smaller than those intended by the fastener manufacturer and excessive relaxation can result. One author claims that if thread engagement length is greater than 1.25 times the nominal diameter, then "permanent set is negligibly small" [1].

23.6.1.3 Soft Parts

If parts are softer than intended by the designer—perhaps because of improper heat treatment or incorrect material—they may creep and relax substantially even if the geometry is correct and loads are normal.

23.6.1.4 Bending

If the fastener is bent as it is tightened, it will see higher stresses along one side than along the opposite side, as we have seen in Chapter 10. These higher stresses mean more plastic flow and therefore greater than normal embedment or thread relaxation.

23.6.1.5 Nonperpendicular Nuts or Bolt Heads

The contact faces of nuts and bolt heads are never exactly perpendicular to the axis of the threads or to the axis of the bolt hole. This means that only a portion of the contact surface of the nut or bolt head is loaded when we first tighten the fastener. These abnormally loaded surfaces will creep until enough additional contact area has been involved to reduce contact pressures and stabilize the joint.

23.6.1.6 Fillets or Undersized Holes

If the head-to-body fillet contacts the edge of the bolt hole as shown in Figure 23.14, the edge of the hole will break down under initial contact pressures. This may result in a complete loss of preload,

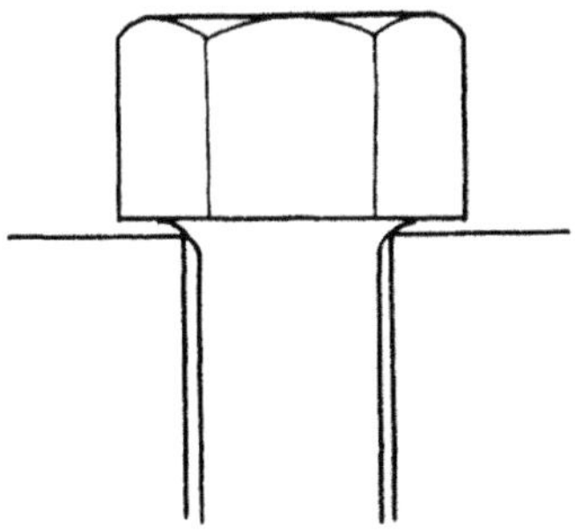

FIGURE 23.14 Oversized fillets and/or undersized holes may result in total relaxation of a preloaded fastener.

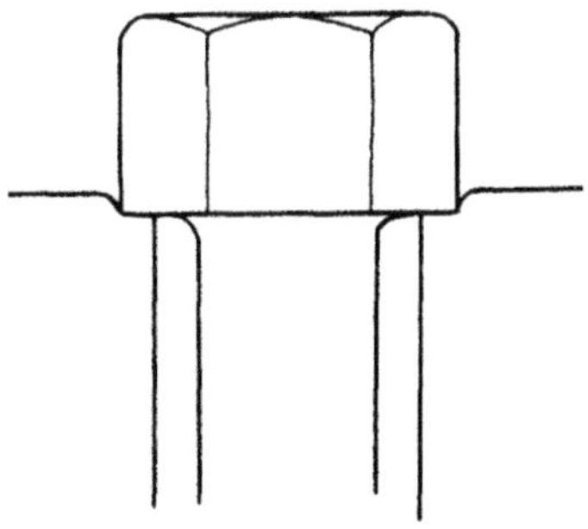

FIGURE 23.15 Oversized holes may also increase contact stress levels and therefore increase embedment relaxation.

since such effects are usually large compared to the amount by which the bolt was stretched when it was initially tightened [4].

23.6.1.7 Oversized Holes

Oversized holes can also be a problem. Now there is too little contact between nut and joint surface or between bolt head and joint surface. Unless a washer or something is used to distribute contact pressures and limit contact stresses, the head and/or nut will embed itself in the joint surfaces, as suggested in Figure 23.15 [3]. The amount of relaxation will, of course, depend on the strength of the surface supporting the nut or washer. The oversized or slotted holes used to aid the assembly of structural joints, for example, do not increase relaxation appreciably [5].

23.6.1.8 Conical Makeups

Surface irregularities exist on conical joint surfaces as well as flat ones. The effect on axial tension in the fastener, however, is magnified if the embedment occurs on a conical surface. A given amount of relaxation perpendicular to the surface may mean substantially greater relaxation in the axial direction, as suggested in Figure 23.16 and Eq. (23.3).

$$r = \frac{e}{\sin\theta} \tag{3}$$

where

e=embedment relaxation perpendicular to the surface of a conical joint member (in., mm)
r=resulting relaxation parallel to the axis of the fastener (in., mm)
θ=half-angle of the cone (deg).

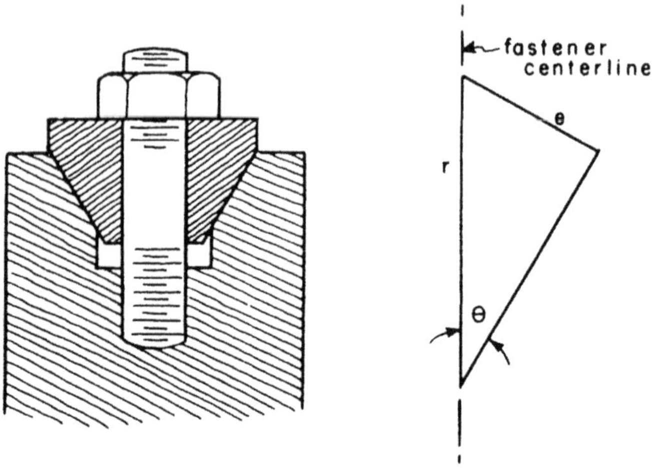

FIGURE 23.16 Conical or tapered joints usually relax more than flat ones for reasons given in the text.

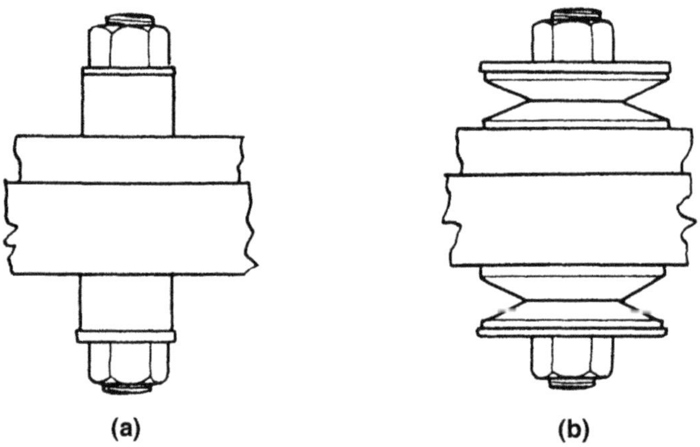

FIGURE 23.17 Since long, thin bolts will relax by a smaller percentage than short, stubby ones, many people use bushings as shown in (a) to reduce percentage relaxation in a given joint. Stacks of Belleville washers (b) are also effective.

23.6.2 Factors Affecting Short-Term Relaxation

Overstressed parts relax. Overstressing may be created in a number of different ways, as we have seen. The amount of relaxation that over stressing causes in a given bolt and joint, however, can depend on a number of secondary factors.

23.6.2.1 Bolt Length

Long, thin bolts will relax by a smaller percentage than short, stubby ones. The total embedment relaxation or the like will be the same for a given initial preload, but that embedment will be a different percentage of the total length of the bolt and therefore account for a different percentage loss in length. Preload loss will be proportional to the change in length (see p. 61 of Ref. [5]).

Many people take advantage of this fact. They add bushings above and below flange surfaces, for example, as shown in Figure 23.17a. This makes it possible for them to use longer bolts on a given joint.

23.6.2.2 Belleville Washers

Another common way to reduce the change in clamping force produced by a given amount of embedment is to use Belleville washers (Figure 23.17b). These springs have a very flat rate compared to the stiffness of either bolt or joint members. They will therefore determine (limit) the preload and clamping force in the system. Because of their flat rate, a small deformation in the bolt or joint will not make an appreciable difference in force levels. For the same reason—and more commonly—Bellevilles are used to compensate for the effects of differential temperature expansion. Spring manufacturers have trouble controlling the stiffness of Bellevilles, however, so this solution to a temperature or relaxation problem can increase the basic scatter in preload.

23.6.2.3 Number of Joint Members

Increasing the number of surfaces in a joint may increase relaxation effects because there are now more high spots to embed and settle in together. Doubling the number of contact surfaces, for example, will almost double relaxation in many cases [6].

23.6.2.4 Tightening Speed

Creep and flow take time. Fasteners that are tightened very rapidly do not have time to settle in together during the tightening process and relax more after tightening [7]. This is one of the advantages of hesitation tightening or torque-recovery tightening. Tighten with high-speed tools, but pause to give the parts time to relax (see Chapter 28).

Tightening bolts in a series of passes rather than applying full torque on the first pass allows time for relaxation. This procedure also pulls the joint together uniformly. For both of these reasons, progressive tightening is a virtual necessity on large gasketed joints.

23.6.2.5 Simultaneous Tightening of Many Fasteners

Some experiments [8] have suggested that tightening a group of fasteners one at a time results in more relaxation in a given fastener than does tightening several or all of them at once. Presumably, a fastener tightened before its neighbor is subjected to higher stress concentrations than if it is tightened simultaneously with the rest and all share the developing load.

23.6.2.6 Bent Joint Members

If joint members are soft or warped or bent, tightening one fastener can cause relaxation (or additional stress) in other fasteners. This sort of "cross-talk" between fasteners is very common, although it is not usually seen or recognized. More about this in Chapter 24.

23.6.3 Amount of Relaxation to Expect

The factors that cause and contribute to relaxation are many and difficult to predict. Although attempts have been made to write equations for the amount of relaxation to expect [6,8,9], in most cases the amount must be determined experimentally. And, as usual when dealing with bolted joints, it won't be "an" amount but rather a distribution of values around some anticipated mean.

In general, fasteners relax rapidly following initial tightening, then relax at a slower rate, following the pattern shown in Figure 23.18. The amount of relaxation varies greatly, depending on the condition of the parts, finishes, initial and local tension levels, fit of parts, and all of the other factors discussed earlier. Here are some of the relaxation amounts and times found in the references. Fisher and Struik report [5, p. 61] that tests of A325 and A354 Grade BD bolts in A7 structural steel showed a loss of 2%–11% of preload immediately after tightening, followed by another 3.6% in the next 21 days, followed by another 2% over the next 11.4 years. Bethlehem Steel reports [10] that only 5% of the initial tension will be lost in structural bolts set by turn-of-nut techniques—5% over the total life of the structure. Chesson and Munse report [11] a variety of results on a variety of structural

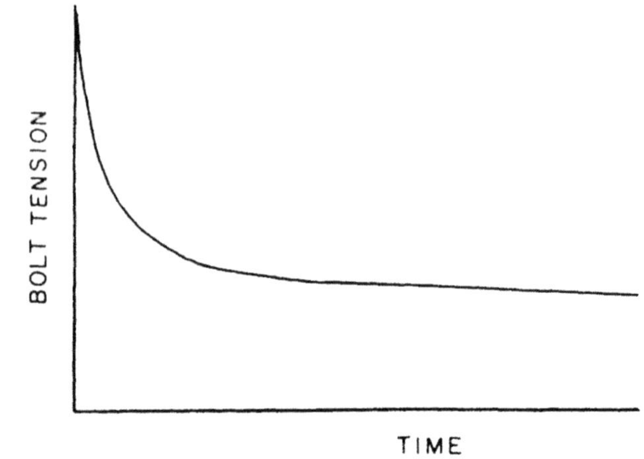

FIGURE 23.18 Most short-term relaxation occurs in the first few seconds or minutes following initial tightening but continues at a lesser rate for a long period of time.

bolts with different types of bolt heads, different nuts, with and without washers, etc. As one example, an A325 bolt with a regular (not heavy) head, a flanged nut, and no washer relaxed 2.6% in the first minute after tightening (most of this in the first 15–20 s). It had relaxed by 6.5% after 5 days. Mayer [12] and Hardiman [13] reports that most relaxation occurs in the first few seconds but that relaxation, usually, never stops. Southwest Research Institute [14] suggests that fasteners lose an average of 5% of preload right after tightening, "because of elastic recovery." The 2 3/4–8 × 12 3/4 Nitronic 50 top guide studs in a boiling water reactor (BWR) relaxed an average of 43% after tensioning. Grip length was 4.75 in.; studs were hydraulically tensioned to 160,000 lb; nuts were run down with a measured torque of 500 in.-lb. This is not all embedment, as we will see in Section 23.7.

Gasketed joints will relax substantially whether the bolts are torqued or tensioned. This is especially true during preliminary passes, when loss of as much as 80%–100% of initial tension is not at all uncommon for reasons to be discussed soon. Gaskets will eventually stabilize, however, and will retain the tension introduced in final tightening operations.

23.6.4 Torsional Relaxation

Preload or tension relaxation is of prime importance to us because of the general importance of correct preload. We must not forget, however, that torsional stress is also built up in a fastener as it is tightened and that this stress is also subject to varying amounts of relaxation. Many people, in fact, insist that torsional stress disappears immediately and completely when the wrench is removed from the fastener. Others find that it doesn't disappear until a breakaway torque is applied [15]. Our experience indicates that, like tension relaxation, torsional relaxation depends on many factors; the amount and rate of torsional relaxation will vary substantially from bolt to bolt as well as from application to application.

Figure 23.19 shows the tension and torsion relaxation measured in an experiment with a 2 1/4–8 × 12 B16 stud that had been lubricated with molybdenum disulfide (Moly-lube). Torsion relaxed 50% when the wrench was removed; tension actually *increased* 1–2% during this period. I have since seen this phenomenon on many other types and sizes of bolts. My guess is that, as embedment allows relaxation of both tension and torsion stress to occur, some of the torsional stress is turned into a little more tension stress. The twisted bolt screws itself farther into its own nut. The exchange is encouraged if you lubricate the threads but do not lubricate the face of the nut.

Sizable relaxation of tension occurring as the torsional stress disappears can, of course, mask this exchange of torsion for tension. Thus I have observed this exchange only on hard joints. But it

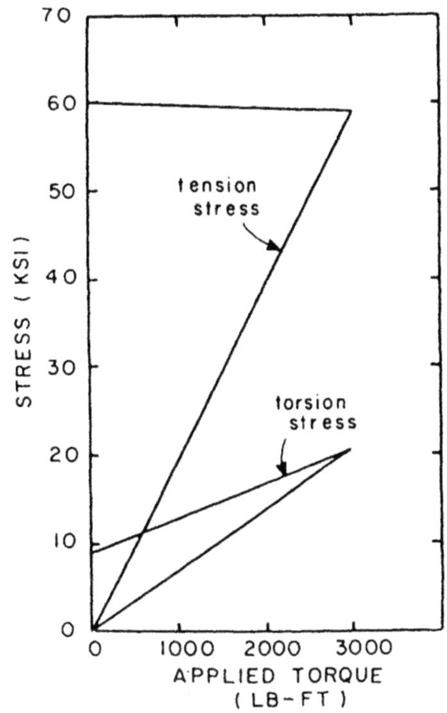

FIGURE 23.19 Relaxation of torsional stress in a bolt can be accompanied by an actual increase in tension. The bolt screws itself into its nut. Data shown were taken in tests on a 2 1/4–8×12 B16 stud.

is relatively easy to reproduce, and I think it is common. Many bolts in large joints mysteriously "grow" a little between passes, for example, even when there is no temperature change or the like to explain the growth and even when neighboring bolts remain at constant length. Presumably, elastic interactions play a role here as well.

A torque wrench appears to respond to torsional stress levels in the bolt as well as to preload levels. The torque required to restart a nut can be less than that required to tighten it in the first place, even if there has been no loss in preload in the meantime, *if* torsional stress has disappeared or been reduced. Repeated "torque recovery" can, as a result, gradually increase the tension in a fastener until it is substantially above initial anticipated levels. In one set of measurements on tank tread end connector bolts, for example, we first tightened the 5/8–18×1 3/4 Grade 8 bolts with a torque of 150 lb-ft. This stretched the bolt 0.0015 in. This was a tapered joint, so the bolt now relaxed to a stretch of only 0.001 in. We reapplied 150 lb-ft of torque, and the stretch returned to 0.0014 in. Incidentally, it took only 100 lb-ft to restart the nut.

The bolt now relaxed again, was tightened again, relaxed some more, etc., as shown in Figure 23.20. Final preload (stretch) was 33% greater than that achieved in the first pass, even though we never applied more than the initial 150 lb-ft of torque. The final restarting torque was still only 87% of the rated torque. Many mechanics would conclude from this that preload was still 13% below the initial value. This possible interaction between torsional and tension stress further complicates the task of predicting how much a given fastener will relax, of course. This is another complex situation.

23.7 ELASTIC INTERACTIONS BETWEEN BOLTS

Even if we can avoid the problems cited above and can count on achieving a certain amount of preload in the bolts we tighten at assembly one by one, there are going to be times when that preload

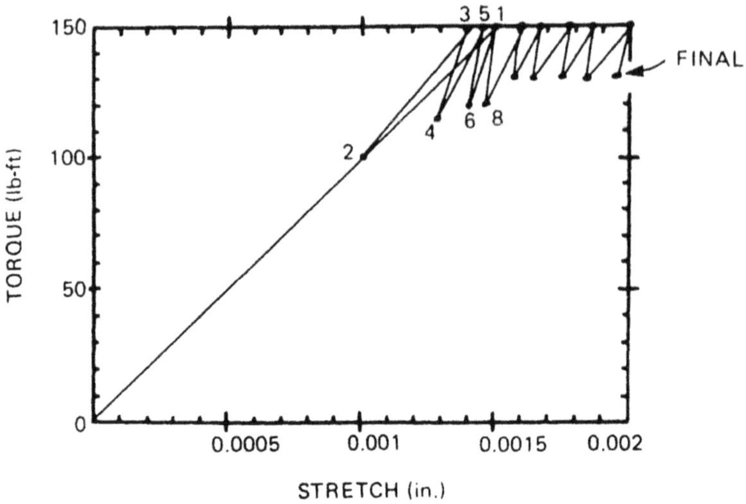

FIGURE 23.20 Torque–stretch–relaxation history of a 5/8–18 × 1 3/4 Grade 5 bolt. A torque of 150 lb-ft was applied repeatedly to this fastener, with a pause for relaxation between passes. Final preload was 33% greater than that achieved on the first pass.

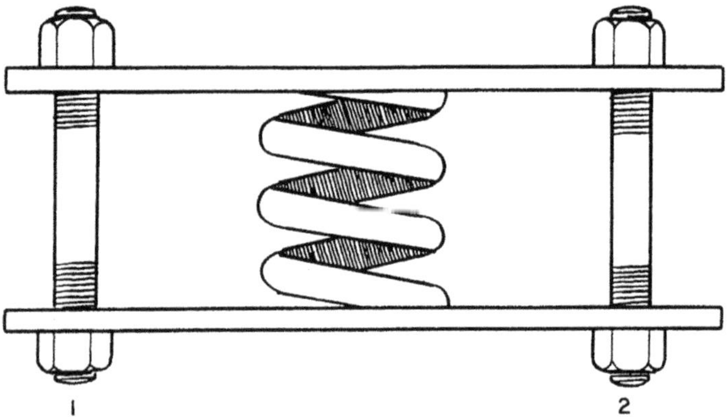

FIGURE 23.21 A simulated model of a bolted joint in which the joint members are represented by a large spring, here "loaded" by the first two bolts to be tightened.

will be significantly modified as we tighten other bolts in the same joint, owing to "elastic interactions" between bolts. Let's look at an example.

Assume that we are planning to tighten a circular flanged joint that contains eight bolts. We are going to use ganged hydraulic wrenches or hydraulic tensioners to tighten these bolts two at a time. The bolts we tighten simultaneously, of course, will be opposite each other on the flange, 180° apart.

To explain the process of elastic interactions, think of the joint as a large spring connected by rigid top and bottom plates to the bolts (smaller springs) that are going to be used to clamp it. This arrangement is suggested in Figure 23.21, which shows the first two bolts to be tightened.

Now, let us assume that we have tightened bolts 1 and 2 in this joint and that magically we have achieved exactly the initial preload we wanted in each bolt. Let's say that this preload is 10,000 lb of tension in each bolt.

The 20,000 lb of force that these two bolts are creating on the joint partially compresses the joint. We now go on to tighten bolts 3 and 4 located 90° away from bolts 1 and 2. Again, our tools work magic for us, and we create exactly 10,000 lb of initial preload in bolts 3 and 4 when they are tightened (Figure 23.22).

We now have four small springs (bolts) compressing the joint spring rather than the two small springs we had a moment ago. If bolts 1 and 2 had retained their full preload, we would now have 40,000 lb of force on this joint instead of 20,000. Doubling the compressive force on the joint spring would, of course, double the amount by which it is compressed. But what happens in bolts 1 and 2 when we tighten 3 and 4? Bolts 1 and 2 are allowed to relax a little as the joint is compressed by bolts 3 and 4.

At this point in the process, therefore, bolts 1 and 2 have a slightly lower preload in them than bolts 3 and 4—even though each of the four bolts started with the same initial tension of 10,000 lb. When we now go on and tighten bolts 5 and 6 in a third step, bolts 3 and 4 will relax a little, and bolts 1 and 2 will relax further. Tightening bolts 7 and 8 to complete the assembly will create relaxation in each of the six bolts tightened earlier.

The result is four different levels of residual preload in the eight bolts when they are tightened two at a time, even though the initial preload in each one was identical. And this is not just a theoretical possibility; it is a very common occurrence. Most people are not aware of this interaction,

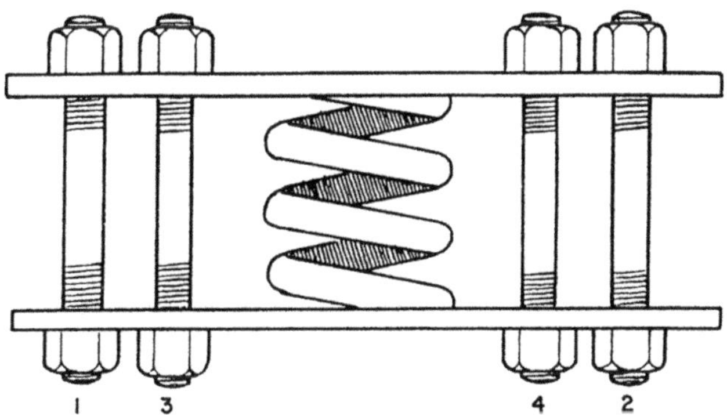

FIGURE 23.22 The joint model of Figure 23.21, but now with four bolts to be tightened.

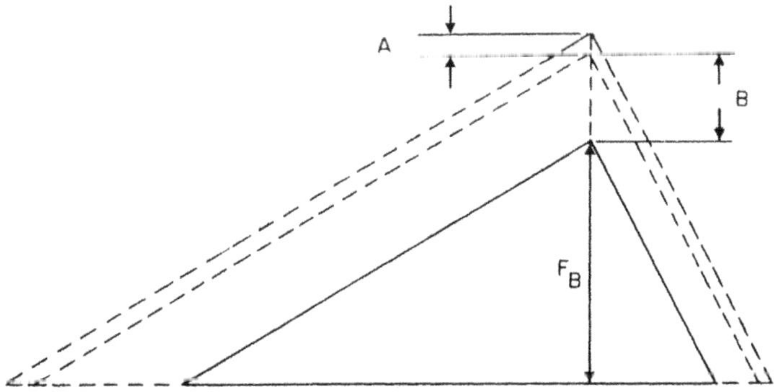

FIGURE 23.23 The joint diagram of Figure 23.5 modified to show the effects of embedment and elastic interaction on initial preload. The largest triangle reflects the initial, "just tightened" situation. Embedment reduces the initial preload by an amount shown as *A*. The average loss in a group of bolts caused by elastic interactions is shown as loss *B*. We use the average loss instead of the individual loss because joint behavior will more likely be determined by average loss than by worst case loss. The final average residual preload in this group of bolts is represented by the solid triangle.

however, which is visible only if you use ultrasonics or strain gages or something to monitor the tension in the bolts. This process, called elastic interaction, is the subject of Chapter 24 [16,17,19–21].

We can now modify the joint diagram of Figure 23.5b to show the loss in preload created by embedment and by the average elastic interaction effect, as in Figure 23.23. The dashed lines show the situation—bolt and joint forces and deflections—before relaxation. The solid line shows the residual preload situation at the end of the assembly operation.

23.8 THE ASSEMBLY PROCESS REVIEWED

After learning about elastic interactions, we make one more pass around our 16-bolt joint, in the reverse order, to reduce the scatter in residual preload. We remeasure the preloads in each bolt after that pass and find that each has changed. We accept the final results and then draw the block diagram shown in Figure 23.24 to summarize the things we now know can affect the amount of clamping force created on a multibolt joint when it is tightened. We include a few things learned elsewhere, the fact that some bolts will be bent slightly if tightened against nonparallel joint surfaces, and some minor factors not discussed, such as the very small heat loss in the tools and drive bars we're using. We also include the possibility of heat loss through "prevailing torque"—when interference of some sort between male and female threads makes it necessary to use a wrench simply to run a nut down before the nut contacts the joint.

Because of our interest in stored energy, we also list the many ways the input work we do on the nuts is absorbed by the bolts and joint members. The work

Heats the parts and tools
Enlarges interference fit holes
Bends and twists the bolts
Dilates the nuts
Pulls the joint members together
Overcomes prevailing torque
Deforms the parts plastically (embedment and gasket creep)

It also deforms the bolts and joint members elastically. It stretches the bolts and compresses the joint. Only the work done deforming the parts elastically ends up as stored energy, creating clamping force, but then some of this work is lost as the parts embed and relax.

We now realize why accurate control of the assembly process is so difficult. We would have to predict or control such unpredictable things as the friction forces between parts, the effort required to pull the joint together, the amount of hole interference, and the elastic interactions between bolts to achieve exactly the same amount of residual preload in each bolt. I think that it's safe to say that we'll never be able to afford the effort required to predict or control the many variables involved. This does not mean, however, that all is lost.

23.9 OPTIMIZING ASSEMBLY RESULTS

A great deal of work has been and is still being done on assembly tools and techniques. There are ways to get better results than we achieved with our example joint in this chapter. Most of these techniques are discussed in this section of the handbook. This information will allow you to pick the most appropriate assembly methods for your own applications.

Regardless of the type of control you adopt, furthermore, there are a number of relatively simple things you can do to improve your chances of success during assembly operations. In fact, the things listed below can often make more improvement than the use of elaborate or expensive preload control equipment. As you can see, most of the items on the list would tend to make your assembly practices and procedures more consistent. Although you cannot predict or control the many variables affecting results, you can reduce their variation by being consistent.

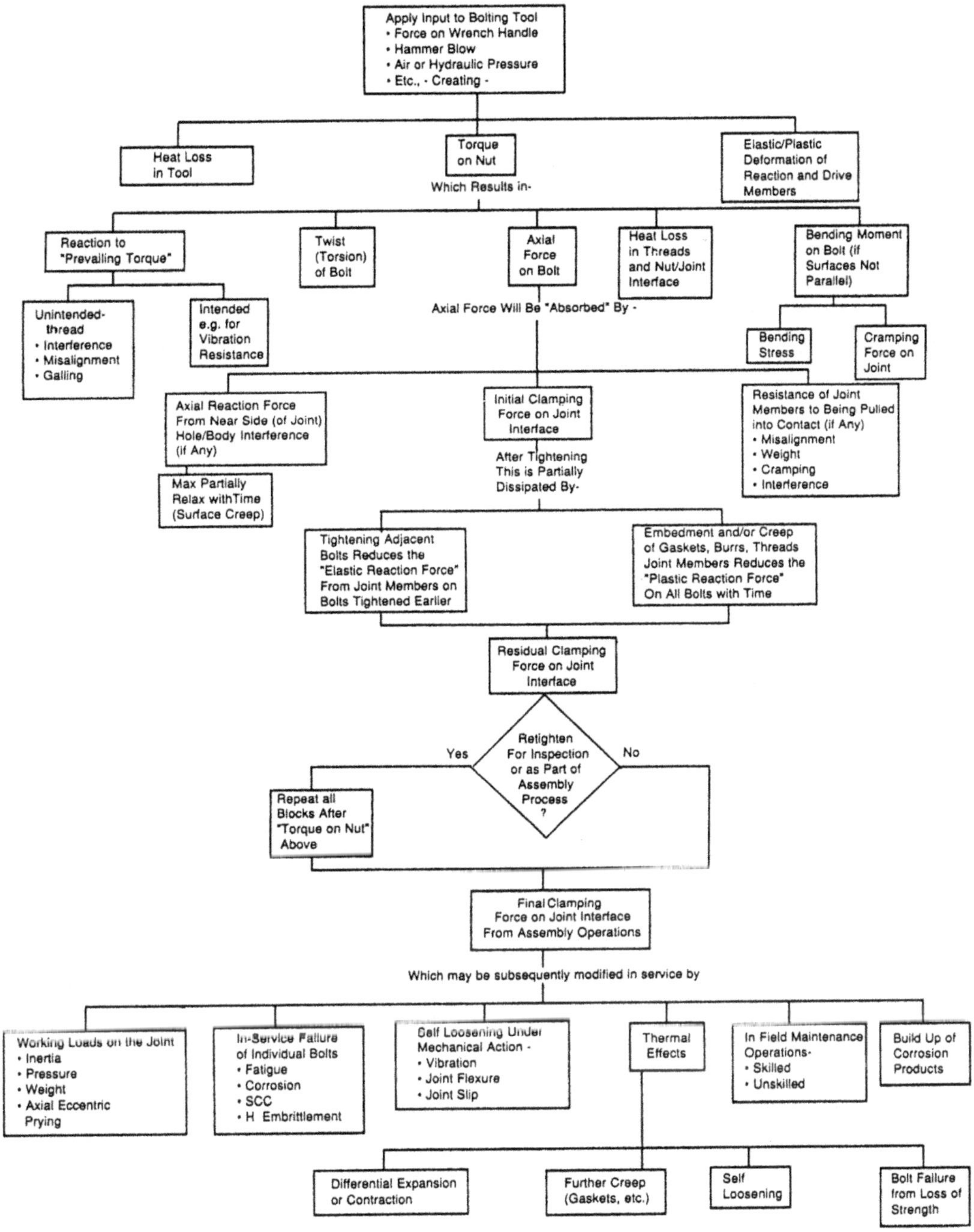

FIGURE 23.24 Block diagram showing most of the factors that affect the relationship between the input work done by the tool on the bolt or nut and the subsequent inservice clamping force between joint members.

1. Be as consistent as possible in your choice and use of tools, procedures, calibration frequency, etc.
2. Train and supervise the bolting crews. This can help far more than a "better" lubricant or more expensive tool. Let the people know how important good results are and how difficult they are to obtain. Enlist their help.

3. Make sure the fasteners are in reasonable shape. Wire-brush the threads if they're dirty and rusted. Use stainless steel bristles on alloy steel materials. Chase threads with a tap or die if they're damaged.
4. Use hardened washers between the turned element (nut or bolt head) and joint members.
5. If lubricants are to be used, make sure they are clean and fresh. Apply them consistently, applying the same amount to the same surfaces by using the same procedure. Preload scatter will be minimized if lubricants are used on both thread and nut (or head) contact surfaces.
6. Run the nuts down by hand. If you can't, the threads may need to be cleaned or chased.
7. Hold wrenches perpendicular to the axes of the bolts.
8. Apply torque at a smooth and uniform rate. Avoid a "stick-slip" situation as you approach the final torque. If necessary, back the nut off a little so the specified torque can be reached with the wrench in motion.
9. If hydraulically powered wrenches are used, be sure that adequate reaction points are used and that the tools aren't twisting or cramping as a result of cocked or yielding reaction surfaces.
10. Snug the joint first, with a modest torque; then tighten it. Try to align the joint members before tightening the bolts.
11. Tighten from the center of bolt patterns toward the free edges if the bolt pattern is rectangular (as in a structural steel joint). Work in a cross-bolting pattern on circular or oval joints.
12. Keep good records of the tools, operators, procedures, torques, and lubricants used. Bolting is an empirical "art" at present. If you record details of your "experiment," you'll have the information you need to make a controlled modification in the procedure the next time if the first procedure doesn't give the results you want.
13. If possible, develop your own nut factors experimentally rather than relying on a table. Perform the experiment under actual job conditions if possible, though a lab test on your joint (or a simulation of your joint) is better than nothing.

REFERENCES

1. Landt, R., *Preload Loss and Vibration Loosening*, SPS Technologies, Jenkintown, PA, 1979.
2. Conversation with Carl Osgood, consultant, Cranbury, NJ, 1974.
3. Friesth, E. R., Performance characteristics of bolted joints in design and assembly, SME Technical Paper AD77-715, 1977.
4. High strength bolted joints, *SPS Fastener Facts,* SPS Technologies, Jenkintown, PA, Sec. IV-C-4.
5. Fisher, J. W., and J. H. A. Struik, *Guide to Design Criteria for Bolted and Riveted Joints*, Wiley, New York, 1974, p. 179.
6. Meyer, G., and D. Strelow, How to calculate preload loss due to permanent set in bolted joints, *Assembly Eng.*, February and March 1972.
7. Ehrhart, K. F., Preload—the answer to fastener security, *Assembly and Fastener Eng.,* 9(6): 19–23 (1971).
8. Prodan, V. D., More precise load-relief factor for tightened screw connections, *Vest. Mashin.*, 54(1): 27–28 (1974).
9. Junker, G., *Principles of the Calculation of High Duty Bolted Connections—Interpretation of Guideline VDI 2230*, VDI Berichte, No. 220, 1974, published as an Unbrako technical thesis by SPS, Jenkintown, PA.
10. *High Strength Bolting for Structural Joints,* Booklet 2867, Bethlehem Steel Co., Bethlehem, PA, p. 9.
11. Chesson, E., Jr., and W. H. Munse, *Studies of the Behavior of High Strength Bolts and Bolted Joints*, University of Illinois College of Engineering, Engineering Experimental Station Bulletin 469, 1964.
12. Mayer, K. H., Relaxation tests on high temperature screw fastenings, *Wire*, 23: 1–5 (1973).
13. Hardiman, R., Vibrational loosening—causes and cures, *Presented at the Using Threaded Fasteners Seminar at the University of Wisconsin—Extension,* Madison, May 8, 1978.

14. Investigation of Threaded Fastener Structural Integrity, Report prepared by Southwest Research Institute, San Antonio, TX, under Contract NAS9-15140, DRL Tii90, CLIN 3, DRD MA 129 TA, SWRI Project 15-4665, October 1977.
15. Blake, T. C., and H. J. Kurtz, The uncertainties of measuring fastener preload, *Mach. Des.*, 37: 128–131 (1965).
16. Van Campen, D. H., A systematic bolt-tightening procedure for reactor pressure vessel flanges, *ASME First International Conference on Pressure Vessel Technology,* Delft, Sept. 29–Oct. 2, 1969, Part 1, pp. 131–141.
17. Rumyantsev, D. V., V. D. Prodan, and A. F. Pershin, A tightening procedure for fastening bolts applicable to high-pressure apparatus, *Khim. Neftyanoe Mashinostroenie*, 2: 3–5 (1973).
18. Tripp, O. S., and E. D. Stevens, Flange parallelism criteria, *Presented to the PVRC Subcommittee on Bolted Flanged Connections*, San Francisco, CA, January 1983.
19. Bibel, G., and R. Ezell, An improved flange bolt-up procedure using experimentally determined elastic interaction coefficients, *J. Pressure Vessel Technol.*, 114: 439 (1992).
20. Bibel, G., and R. Ezell, Bolted flange assembly: Preliminary elastic interaction data and improved bolt-up procedures, Draft of paper submitted to the Pressure Vessel Research Committee, May 18, 1993.
21. Bibel, G. D., and D. L. Goddard, Preload variation of torqued fasteners, a comparison of frictional and elastic interaction effects, *Fastener Technol Inter.*, Feb. 17, 1994.
22. Bickford, J. H., *Introduction to the Design and Behavior of Bolted Joints*, 3rd ed., Marcel Dekker, New York, 1995.

24 Tightening Groups of Fasteners in a Structure and the Resulting Elastic Interaction

George Bibel

24.1 INTRODUCTION

The relationship between fastener torque and resulting preload is a mathematical equation based on the summation of forces on the thread. This equation is described in detail elsewhere in this handbook. Other than statistical variation of dimensions due to manufacturing, all the terms in the equation are precisely known except for friction coefficients.

Many factors affect the coefficient of friction (again this is a topic discussed in detail elsewhere in this handbook). These factors turn the mathematical relationship between torque and preload into a statistical phenomenon best described with experimental data. The purpose of this chapter is to describe another phenomenon that results in significant preload reduction. This reduction occurs when groups of fasteners interact within a structure, and the phenomenon is known as elastic interaction or bolt cross talk.

In the sections that follow, elastic interaction is defined, data is presented that demonstrates greater preload variation than caused by friction, and elastic interaction is shown to depend on the ratio of joint stiffness to fastener stiffness. Finally, an improved flange bolt-up procedure that compensates for elastic interaction is described.

24.2 ELASTIC INTERACTION DEFINED

When a group of fasteners are tightened in a joint, the elongation of each individual bolt causes it to structurally interact with the bolted joint being compressed. (The tightened bolt acts like springs in parallel with the joint being compressed.) Subsequent tightening compresses the joint further and reduces the preload in fasteners previously tightened. This structural interaction is best explained with a series of simple illustrations.

Figure 24.1a shows bolt 2 initially tightened. In this case, the joint consists of a bolted flange and gasket. Local compression of the joint occurs under bolt 2. When the adjacent bolt 1 is tightened, additional compression of the joint is shown in Figure 24.1b. The preload is reduced in bolt 2 as shown. When bolt 3 is tightened, additional compression of the joint occurs, with additional loss in the preload of bolts 1 and 2 as shown in Figure 24.1c.

Elastic interaction always results in a loss of bolt preload. This differs from preload variation due to friction. Frictional variation results in a random preload, higher or lower, about some mean value. If the fasteners are torqued, frictional variation will be superimposed on the elastic interaction preload reduction. If the fasteners are loaded to a precise value with hydraulic tensioners or torqued to a precise preload value by monitoring bolt strains with ultrasonics, micrometers, strain gages, etc., then only preload reduction due to elastic interaction occurs.

 DOI: 10.1201/9781003501183-29

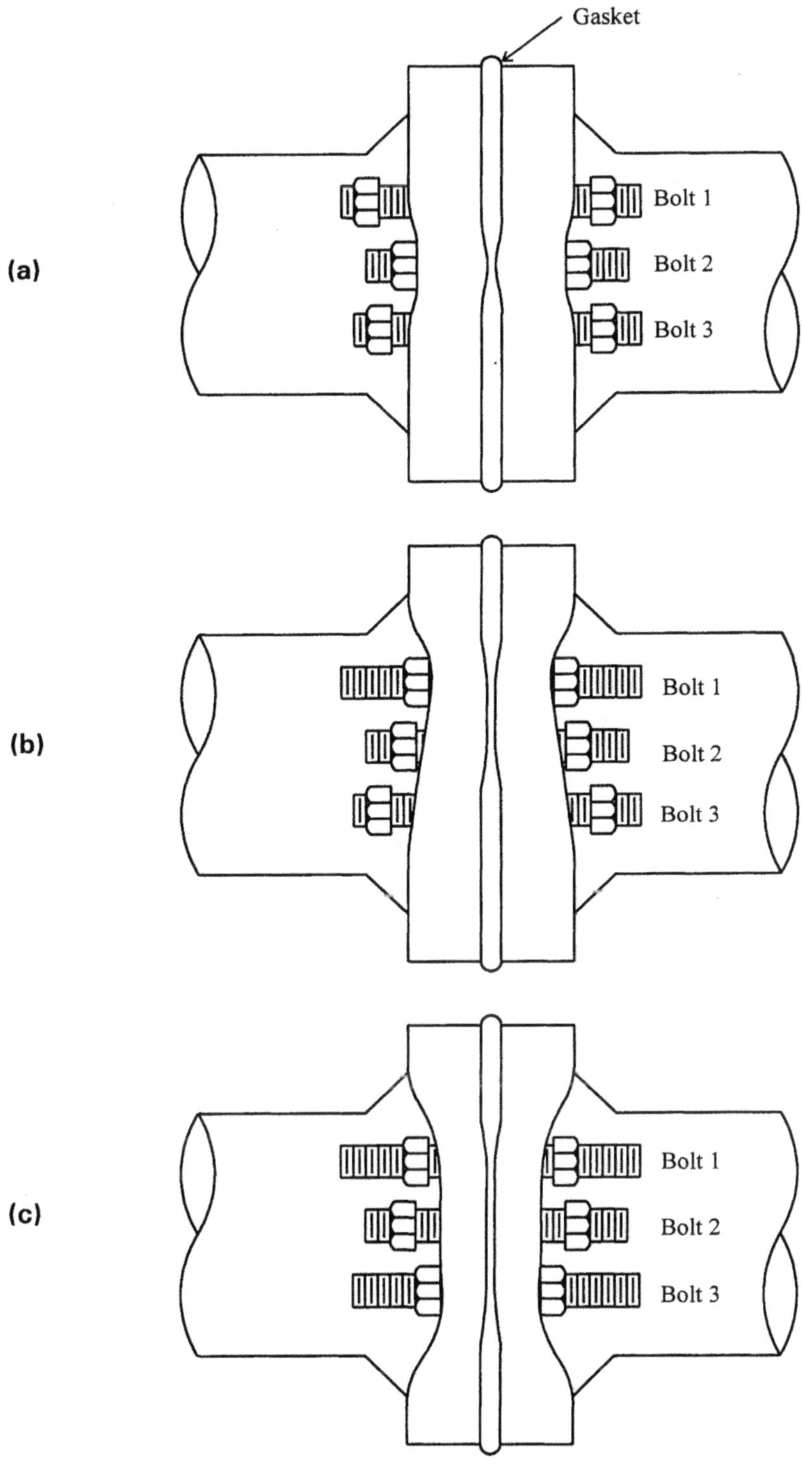

FIGURE 24.1 (a) Tighten bolt 2. (b) Tighten bolt 1; preload in bolt 2 is reduced. (c) Tighten bolt 3; preload is reduced in bolts 1 and 2.

24.3 ELASTIC INTERACTION DATA

24.3.1 Procedure for Obtaining Data

The following procedure can be used to obtain elastic interaction data.

1. Define a method to accurately measure bolt tension. This can be done with strain gages mounted on the fasteners, ultrasonics, digital micrometers, or strain-gaged washers.
2. Define the final preload desired in the bolts, in terms of microstrain, stress, or tension, whichever is appropriate for the means used to measure tension.
3. Tighten the bolts one at a time to this preload, following the normal sequence. The bolts in a circular pressure vessel flange, for example, are usually tightened in a cross or star pattern sequence. Alternatively, the bolts can be tightened in several passes, each at a higher torque than the last, to probe the interactions that occur during the normal assembly operations.
4. Measure the tension in each bolt when it is first tightened to the full or preliminary preload and after you have tightened each of the other bolts. Record the data in a data file. This data can later be displayed, perhaps following the format of Table 24.1, to reveal the effects of the elastic interactions between bolts.

TABLE 24.1
Elastic Interaction Data for Passes 1 and 2 of a 20-Bolt Flange Tightened in a Star Pattern (see Figure 24.2) as Follows: Bolts 1, 11, 16, 6, 2, 12, 17, 7, 3, 13, 18, 8, 4, 14, 19, 9, 5, 15, 20, 10

Pass 1

Bolts 1–6 Strain Data

493.00	0.00	0.00	0.00	0.00	501.00
243.00	485.00	0.00	0.00	0.00	469.00
251.00	492.00	0.00	0.00	0.00	485.00
219.00	508.00	0.00	0.00	0.00	501.00
235.00	532.00	0.00	0.00	0.00	282.00
172.00	305.00	493.00	0.00	0.00	219.00
180.00	305.00	493.00	0.00	0.00	219.00
110.00	289.00	516.00	0.00	0.00	235.00
110.00	305.00	532.00	0.00	0.00	195.00
118.00	250.00	360.00	501.00	0.00	117.00
125.00	250.00	352.00	501.00	0.00	117.00
71.00	211.00	360.00	516.00	0.00	133.00
71.00	219.00	367.00	532.00	0.00	133.00
71.00	211.00	297.00	376.00	492.00	39.00
79.00	219.00	305.00	383.00	492.00	47.00
24.00	140.00	266.00	383.00	508.00	47.00
32.00	140.00	273.00	383.00	524.00	47.00

Bolts 7-12 Strain Data

8.00	0.00	0.00	0.00	492.00	0.00
8.00	0.00	0.00	0.00	508.00	0.00
8.00	0.00	0.00	0.00	297.00	493.00
0.00	0.00	0.00	0.00	305.00	501.00
509.00	0.00	0.00	0.00	281.00	517.00
509.00	0.00	0.00	0.00	297.00	524.00

(Continued)

TABLE 24.1 (*Continued*)
Elastic Interaction Data for Passes 1 and 2 of a 20-Bolt Flange Tightened in a Star Pattern (see Figure 24.2) as Follows: Bolts 1, 11, 16, 6, 2, 12, 17, 7, 3, 13, 18, 8, 4, 14, 19, 9, 5, 15, 20, 10

524.00	0.00	0.00	0.00	258.00	337.00
532.00	0.00	0.00	0.00	258.00	337.00
329.00	493.00	0.00	0.00	211.00	329.00
297.00	501.00	0.00	0.00	211.00	337.00
297.00	517.00	10.00	0.00	211.00	282.00
305.00	517.00	0.00	0.00	219.00	282.00
251.00	329.00	501.00	0.00	133.00	258.00
180.00	313.00	501.00	0.00	140.00	258.00
188.00	313.00	509.00	0.00	156.00	251.00
188.00	321.00	517.00	0.00	148.00	251.00
180.00	274.00	321.00	516.00	54.00	188.00
		Bolts 13-18 Strain Data			
0.00	0.00	0.00	501.00	0.00	0.00
0.00	0.00	0.00	524.00	0.00	0.00
0.00	0.00	0.00	516.00	0.00	0.00
0.00	0.00	0.00	297.00	508.00	0.00
0.00	0.00	0.00	305.00	508.00	0.00
0.00	0.00	0.00	313.00	524.00	0.00
492.00	0.00	0.00	266.00	532.00	0.00
508.00	0.00	0.00	211.00	289.00	501.00
516.00	0.00	0.00	219.00	297.00	509.00
524.00	0.00	0.00	227.00	297.00	516.00
320.00	477.00	0.00	157.00	266.00	532.00
328.00	493.00	0.00	157.00	211.00	313.00
328.00	508.00	0.00	164.00	211.00	313.00
336.00	508.00	0.00	164.00	219.00	321.00
281.00	305.00	493.00	94.00	156.00	297.00
281.00	313.00	501.00	102.00	141.00	235.00
258.00	313.00	509.00	102.00	141.00	235.00
		Bolts 19-20 Strain Data			
0.00	0.00				
0.00	0.00				
0.00	0.00				
0.00	0.00				
0.00	0.00				
0.00	0.00				
0.00	0.00				
0.00	0.00				
0.00	0.00				
0.00	0.00				
0.00	0.00				
486.00	0.00				
486.00	0.00				
509.00	0.00				
509.00	0.00				
290.00	493.00				
290.00	485.00				

(*Continued*)

TABLE 24.1 (*Continued*)
Elastic Interaction Data for Passes 1 and 2 of a 20-Bolt Flange Tightened in a Star Pattern (see Figure 24.2) as Follows: Bolts 1, 11, 16, 6, 2, 12, 17, 7, 3, 13, 18, 8, 4, 14, 19, 9, 5, 15, 20, 10

Pass 2

		Bolts 1–6 Strain Data			
994.00	39.00	86.00	329.00	516.00	55.00
1,002.00	39.00	86.00	337.00	532.00	62.00
1,017.00	47.00	94.00	337.00	563.00	70.0
1,033.00	39.00	78.00	266.00	164.00	1,001.00
532.00	1,001.00	0.00	172.00	93.00	970.00
540.00	1,009.00	0.00	180.00	93.00	1,001.00
517.00	1,048.00	0.00	180.00	101.00	1,009.00
524.00	1,087.00	0.00	180.00	70.00	563.00
423.00	618.00	993.00	86.00	31.00	422.00
431.00	618.00	1,001.00	86.00	23.00	438.00
321.00	610.00	1,032.00	86.00	31.00	446.00
321.00	626.00	1,072.00	94.00	31.00	360.00
345.00	539.00	696.00	1,002.00	15.00	188.00
352.00	539.00	704.00	994.00	7.00	188.00
165.00	500.00	719.00	1,025.00	7.00	195.00
165.00	500.00	727.00	1,056.00	15.00	211.00
188.00	500.00	626.00	704.00	1,017.00	39.00
204.00	516.00	626.00	689.00	1,017.00	47.00
47.00	344.00	594.00	704.00	1,048.00	47.00
47.00	344.00	602.00	712.00	1,071.00	55.00
		Bolts 7–12 Strain Data			
196.00	274.00	321.00	516.00	54.00	196.00
211.00	274.00	235.00	188.00	985.00	24.00
219.00	274.00	235.00	195.00	1,001.00	24.00
39.00	157.00	188.00	195.00	1,017.00	24.00
39.00	165.00	204.00	219.00	1,024.00	24.00
47.00	180.00	204.00	133.00	547.00	1,002.00
39.00	180.00	204.00	141.00	555.00	1,017.00
994.00	71.00	94.00	102.00	555.00	1,041.00
986.00	79.00	110.00	109.00	555.00	1,041.00
1,009.00	79.00	126.00	117.00	461.00	603.00
1,017.00	79.00	118.00	117.00	469.00	618.00
626.00	994.00	32.00	39.00	391.00	618.00
579.00	1,010.00	32.00	47.00	399.00	618.00
579.00	1,025.00	40.00	55.00	414.00	524.00
587.00	1,033.00	40.00	55.00	414.00	524.00
501.00	650.00	994.00	8.00	250.00	493.00
352.00	611.00	1,010.00	8.00	258.00	493.00
344.00	611.00	1,025.00	8.00	281.00	493.00
360.00	618.00	1,025.00	8.00	289.00	493.00
360.00	525.00	673.00	1,001.00	78.00	384.00

(*Continued*)

TABLE 24.1 (*Continued*)
Elastic Interaction Data for Passes 1 and 2 of a 20-Bolt Flange Tightened in a Star Pattern (see Figure 24.2) as Follows: Bolts 1, 11, 16, 6, 2, 12, 17, 7, 3, 13, 18, 8, 4, 14, 19, 9, 5, 15, 20, 10

Bolts 13–18 Strain Data

266.00	320.00	524.00	110.00	149.00	211.00
156.00	266.00	516.00	110.00	164.00	211.00
133.00	156.00	172.00	1,002.00	31.00	70.00
140.00	164.00	180.00	1,009.00	23.00	78.00
140.00	172.00	188.00	1,041.00	31.00	102.00
39.00	70.00	125.00	1,017.00	31.00	102.00
39.00	62.00	86.00	485.00	1,001.00	8.00
39.00	70.00	94.00	493.00	1,009.00	8.00
39.00	62.00	94.00	501.00	1,041.00	8.00
1,001.00	7.00	39.00	399.00	1,033.00	8.00
1,025.00	7.00	47.00	305.00	563.00	1,009.00
1,048.00	15.00	47.00	313.00	563.00	1,017.00
1,056.00	7.00	47.00	321.00	571.00	1,033.00
618.00	985.00	8.00	172.00	508.00	1,064.00
625.00	1,017.00	8.00	172.00	415.00	642.00
641.00	1,048.00	8.00	180.00	422.00	642.00
649.00	1,048.00	8.00	180.00	422.00	649.00
547.00	626.00	1,002.00	63.00	282.00	610.00
555.00	633.00	1,025.00	70.00	282.00	524.00
532.00	649.00	1,041.00	78.00	289.00	524.00

Bolts 19–20 Strain Data

180.00	141.00
188.00	141.00
149.00	141.00
149.00	149.00
141.00	102.00
149.00	102.00
47.00	71.00
47.00	63.00
63.00	78.00
71.00	78.00
8.00	31.00
0.00	31.00
8.00	39.00
0.00	39.00
1,002.00	16.00
1,002.00	16.00
1,018.00	8.00
1,049.00	8.00
642.00	1,017.00
634.00	1,017.00

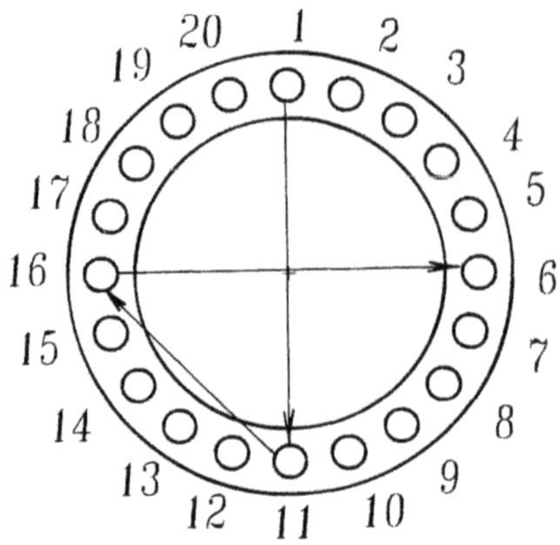

FIGURE 24.2 Bolt number identification. Bolts were tightened in the following sequence: 1, 11, 16, 6, 2, 12, 17, 7, 3, 13, 18, 8, 4, 14, 19, 9, 5, 15, 20, and 10 (four-point "star" pattern).

This procedure results in many more bolt tension readings than are normally taken. A joint with six bolts would have a total of 21 readings. The first bolt would be measured once when it is first torqued to its target load and five additional times, once after each of the other five bolts is torqued. The second bolt would be measured once as it reaches its target and four additional times once for each bolt tightened afterward. The subsequent bolts would be measured four times, three times, twice, and once, for a total of 21 readings for the six bolts.

There may be obvious patterns in the interactions that may reduce the number of readings required. For example, for the data presented here, significant interactions occurred only in the adjacent bolts.

24.3.2 An Example

Elastic interactions were measured on several standard ANSI pipe flanges (24 in. Class 150 and two 16 in. Class 300 [1]). Testing was done with increased joint flexibility (by varying the stiffness of the gasket) and increased flexibility of the bolts (by using longer studs with spacers.)

The target bolt strain was either 850 or 1,010 microstrain. This converts to about 25,000 psi or 30,000 psi bolt stress by multiplying stud strain, measured with strain gages, by the modulus, 29,300,000 psi [2]. The studs were tightened in a "standard" star pattern as shown in Figure 24.2.

Typical elastic interaction data are shown in Figures 24.3 and 24.4. The final bolt strain is plotted after all of the studs are tightened to an initial value of approximately 850 μstrain. Figures 24.3 and 24.4 represent extremes in joint flexibility: Figure 24.3 data are for a stiff joint with no gasket and little interaction, whereas Figure 24.4 is for a flexible joint (same flange with a soft gasket) with large interaction.

Another way to describe interaction is with a defined bolting efficiency, which is equal to the average bolt stress (or strain) divided by the design bolt stress. For Figure 24.3, even without a gasket, the bolting efficiency is calculated to be 84%. For the flexible joint elastic interaction data graphed in Figure 24.4, the bolting efficiency is seen to be only 50% after the first pass (i.e., tightening all of the bolts once) and 69% after the second pass.

Preload reduction due to elastic interaction occurs in patterns based on bolt tightening sequence. The first bolts tightened have the most preload reduction. The last bolts tightened have none. Bolts tightened between first and last suffer an intermediate amount of elastic interaction. Therefore, changing the sequence of the bolt tightening pattern or the number of bolts tightened simultaneously will change the amount of preload reduction. Figure 24.5 shows the results of a "scrambled" bolting sequence. The cyclic pattern of interaction can also be expected to be disturbed whenever the star

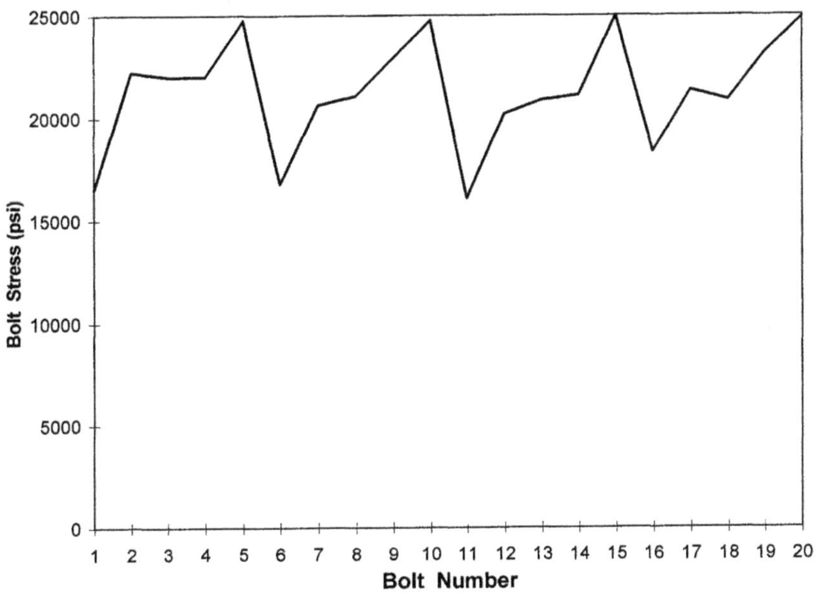

FIGURE 24.3 Elastic interaction data for a stiff joint. A 24 in. Class 150 ANSI pipe flange with no gasket. Bolting efficiency is 84%.

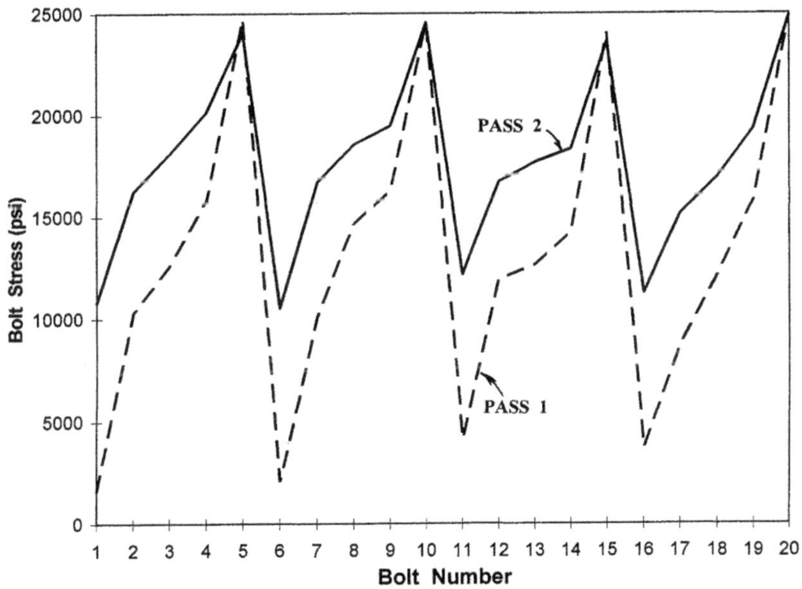

FIGURE 24.4 Elastic interaction data for a flexible joint. Same flange as in Figure 24.3 except that a flexible spiral wound gasket was used. Bolting efficiency is 50% after the first pass and only 60% after tightening all of the bolts twice to 100% of target load.

pattern is applied to a group of bolts that is not a multiple of four. (Of course, other patterns can be envisioned. The star pattern can be applied in groups of three with bolts being tightened every 120°.)

Another way to present the data is to tabulate the history of each stud throughout the bolt-up procedure, as shown in Table 24.1. The data presented this way show the proximity of interaction. Another mechanism, described later as rocking, can also be observed.

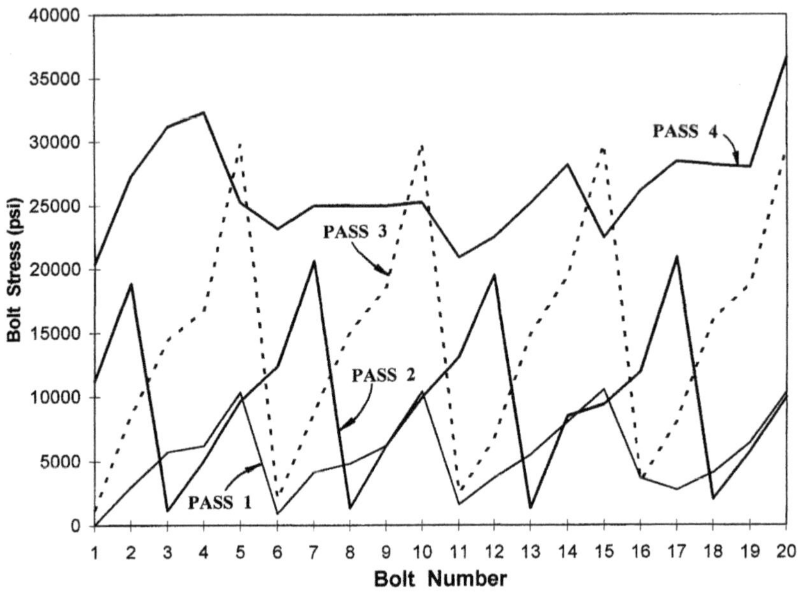

FIGURE 24.5 Elastic interaction for a modified four-point star sequence. Bolts were tightened as follows. Pass 1: One-third of 30,000 psi target stress with four-point star pattern starting at bolt 1. Pass 2: Two-thirds of 30,000 psi target with four-point star pattern starting at bolt 3. Pass 3: 30,000 psi target starting at bolt 6. Pass 4: 37,500 psi on consecutive bolts starting at bolt 1, moving clockwise.

Every time a bolt is tightened, the strain in all 20 bolts is measured and recorded to a data file. A one-pass procedure would create 20×20 or 400 strain readings, 19 of which are zero on the first pass only. (Tightening the first bolt results in one actual reading and 19 zero entries. Tightening the second bolt gives a strain gage reading for the second bolt, an interaction reading for the first bolt and 18 zero readings, etc.)

Pass 2 for Table 24.1 is read as follows: The first 20 rows of column 1 are the strain history of bolt 1 while tightening all other bolts according to the star pattern 1, 11, 16, 6, 2, 12, 17, 3, 13, 18, 8, 4, 14, 19, 9, 5, 15, 20, 10. Row 5 shows the first significant interaction when adjacent bolt 2 is tightened and bolt 1 strain drops from 1,033 to 532. Additional interaction is seen in bolt 1 when the other adjacent bolt (#20) is tightened (row 19) and the strain is seen to drop in bolt 1 from 204 to 47 μstrain.

24.3.3 What the Data Reveals

Presenting the strain history for each bolt demonstrates that most of the interaction is local (at least for the pipe flanges tested) in that only the two adjacent bolts have an appreciable effect. It is suggested that this is true in general, but one should be careful to extrapolate data of this type to other structures and bolting configurations.

In addition to frictional variation and elastic interaction, a third effect can be noticed during the bolt-up of pipe flanges. A rigid-body motion or rocking of the flanges can occur. Although the following bolting pattern was never used, it best illustrates the condition.

Consider 12 bolts located like numbers on a clock. If the 12 o'clock bolt is tightened first, followed by positions 5, 6, and 7, then the 12 o'clock bolt will in fact increase in preload. This occurs because bolts 5, 6, and 7 try to open the flange as shown in Figure 24.6.

Using the star pattern on a 20-bolt flange with tightening sequence 1, 11, 16, and 6, the tension in bolt 1 did, in fact, increase by up to 48% in some cases. This worst-case condition occurred when the bolts were tightened to the target load on the first pass (no incremental tightening).

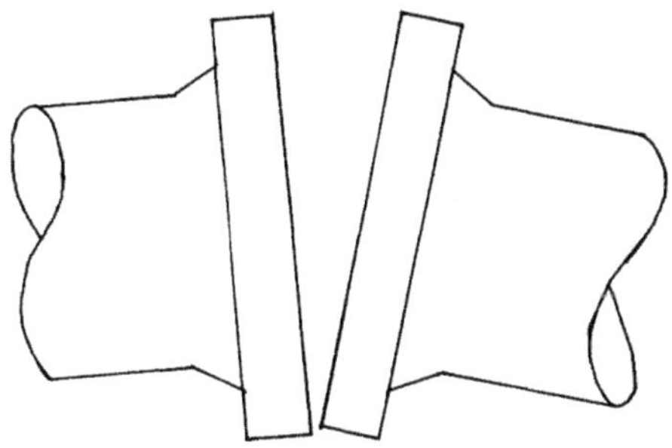

FIGURE 24.6 Flange opens up if several bolts are tightened on the bottom of the flange. This can increase the preload by as much as 48% on the top bolt.

This shows that the potential exists for a bolt tightened near yield to yield. This increase in bolt load also decreases the reduction of preload in bolt 1 during testing. Bolt 1 did consistently have less reduction than the other bolts. Other bolting patterns as extreme as the case described above could be expected to create a significant increase in preload in the first bolt tightened.

The rocking disappears rapidly or after enough bolts are tightened to make the incremental loading associated with a single bolt insignificant. Other ways to deal with rocking are to

1. Use incremental tightening. This seemed to eliminate the problem.
2. Use symmetrical tightening. For example, use four hydraulic wrenches, one every 90°.
3. Bring the flange faces into contact by any available means before tightening the bolts. The simplest way may be to apply a small amount of torque, by hand, to each of the bolts.

Bringing the flanges flush may create a different bolting problem. In a stiff piping system with misalignment, a significant portion of the bolt load may in fact be used up to correct any flange/piping misalignment. This may leave the actual bolt load available for clamping the joint or sealing a gasket far below design.

24.3.4 The Effect of Flange Surface Gaps and Misalignment

Two "identical" pairs of 16 in. Class 300 flanges were tested, with results as shown in Figures 24.7 and 24.8. Some preliminary conclusions can be drawn about flange surface gaps and misalignment from comparing test data for the flange pairs.

Data from bolts 1, 11, 16, and 6 on each pair of 16 in. flanges show "soft" spots. Stress reductions of up to 67% can be seen in bolts 6 and 16 for flange 1 (see Figure 24.7). A reduction of 87% is present in bolt 11 of flange 2 (see Figure 24.8). Differences among apparently identical flange pairs are presumably due to surface gaps and misalignments.

Flange surface misalignment apparently results in gaps of unknown size and location. The flange is not as stiff when the gap is open but becomes stiffer as the gap is closed. As the joint becomes more rigid with increased bolt loading and gap closure, the interaction is reduced. An infinitely stiff flange without a gasket would have no interaction, conversely, very flexible bolts, made for example out of rubber, would also have no interaction (the rubber bolts would be incapable of compressing the joint).

The grip length of the bolts used in this example was 4.86 in. This results in a bolt stretch of about 0.005 in. It is easy to imagine flange distortion due to welding and/or machining resulting in gaps of this order or a significant fraction thereof.

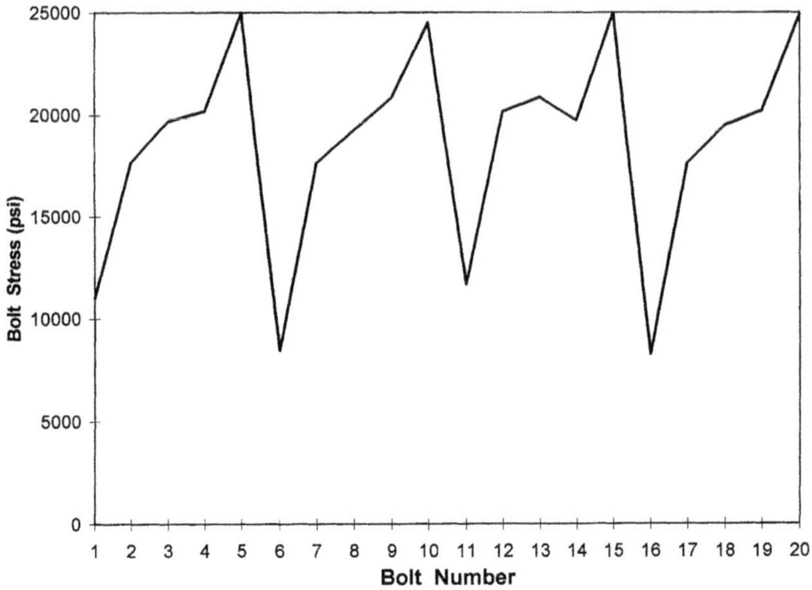

FIGURE 24.7 Single pass to 25,000 psi stress on a 16 in. Class 300 ANSI flange without a gasket.

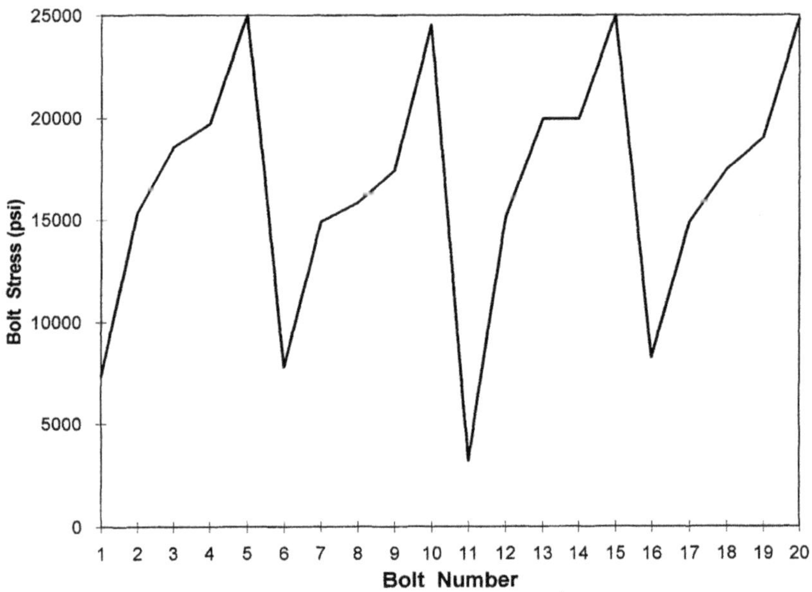

FIGURE 24.8 Identical to Figure 24.7 except that data are for a second pair of 16 in. flanges. Differences between Figures 24.7 and 24.8 are presumably due to nonuniform contact in the flange faces.

24.3.5 Finite-Element Analysis of Elastic Interaction

Discrepancies between finite-element (FE) modeling of interaction and experimental data can also be explained with surface misalignment. An axisymmetric model with nonaxisymmetric loading was used for the finite-element analysis. The method used is described in Ref. [3]. The model does not account for any flange misalignment.

The model predicted elastic interaction much less than that observed experimentally when only four bolts were tightened. However, with 19 of the 20 bolts tightened, the FE model predicted elastic interaction very close to the experimental value. With 19 bolts tightened, any gaps that exist are closed and the flange is much stiffer.

In Ref. [3], the finite-element model gave very good results with experimentally measured interaction data for a gasketed joint. In a gasketed joint, the gap is immediately filled by the relatively soft gasket material and is therefore less of a factor.

Reportedly FE analysis has been used with modest success to predict elastic interaction in a nuclear reactor. The flange of a nuclear reactor is essentially metal-to-metal contact. (Although there is a gasket, it is structurally insignificant.) I believe that FE analysis was of limited success because of the inability to accurately model any misalignments. Later in this chapter, an algorithm is presented to compensate for elastic interaction. Since the algorithm is based on experimental (instead of analytical) determination of interaction coefficients, there is automatic adjustment for any misalignments that occur.

24.3.6 More Data Are Available

Detailed elastic interaction data are presented in Ref. [4] for 24 tests on pipe flanges described above.

Data for a more complex flange, a heat exchanger with 48 1¾ in. diameter studs (effective length of 22.7 in.) and a spiral-wound gasket are given in Ref. [5], which is a case study of a problem flange. The required torque was initially evaluated with well-known torque-preload equations. Measurements with ultrasonics found the actual bolt stretch to be much less than expected. This decrease in bolt preload was attributed to elastic interaction. Retorquing to compensate for elastic interaction increased bolt stretch by 44%. Subsequently, the exchanger operated for 4 years without leading. Before the change in bolt-up procedure, leakage was expected every few weeks.

Interaction data for an automotive engine block are given in Ref. [6] (this test is discussed in more detail in the final section of this chapter).

24.3.7 Design to Minimize Elastic Interaction

By varying the joint stiffness and the stiffness of the bolts during testing, a pattern in the effect of the ratio of the stiffness of the joint to the stiffness of the fasteners was determined. Elastic interaction depends on the relative stiffness of the joint compared to the stiffness of the fastener. For very stiff joints or very flexible bolts, there is little interaction (i.e., rubber bolts have no interaction because no joint compression occurs). For very flexible joints or stiff bolts, there is a great deal of interaction. This gives some insight into the effect of joint design on elastic interaction and assembly.

In this context, the joint refers to the structure and the gasket, if any. Although any sort of gasket is expected to be significantly more flexible than a steel pipe flange, the data show much interaction even without a gasket. The flange will dimple and bend under the contact load. Although it is hard to imagine much deflection of the flange compared to that of a gasket, the flange deflections are relative to the equally small deflections of the bolt.

The stiffness of the gasket is not necessarily insignificant compared to that of the bolts. The axial stiffness of an elastic rod is AE/L, where A is the cross-sectional area, E is the modulus of elasticity, and L is the length of the rod. Although this is a very simplistic model to describe the axial stiffness of a nonlinear gasket, it does correctly illustrate the effect of gasket thickness and area. L for the gasket is much smaller than that of the bolts. Also, the cross-sectional area of the gasket is typically 2–3 times greater than that of the bolts. This somewhat compensates for the fact that the modulus of the usually nonmetallic gasket material is much less than the modulus of the steel bolts. Gasket stiffness data are given in Ref. [7].

24.4 COMPARISON OF FRICTIONAL VARIATION OF PRELOAD WITH ELASTIC INTERACTION

After testing with strain-controlled loading to isolate elastic interactions, follow-up tests were conducted to focus on frictional scatter in the same set of bolts.

In the first test, the preload scatter created by repeatedly torquing an individual stud was observed. For this type of test, the same stud–washer–nut configuration was tightened, loosened, tightened, loosened, etc. This was done without any relubrication or any other changes. Torque was measured with a torsional load cell using internal strain gages. This type of testing eliminates the effects of elastic interaction and proved to be quite repeatable. Preload scatter was ±5%. Typical scatter bands for two bolts are presented in Figure 24.9. Although each bolt exhibited about the same low amount of preload scatter, the slope of the torque–strain curve is noticeably different for each bolt.

The test was then repeated for 14 different studs. The scatter bands are shown in Figure 24.10. The 14 bolts when plotted together had scatter bands of about ±13%. This result is considered consistent with work done by others who have studied the statistical scatter of preload vs. torque. Note that care must be taken to separate out the elastic interaction from the frictional scatter.

It is hypothesized that minute misalignments of all of the surfaces (i.e., threads, nut face, washer, flange, etc.) due to manufacturing variation cause greater scatter in a group of bolts than in a single bolt. A single bolt tested repeatedly always sees the same misalignments. For groups of bolts, the misalignments are random and nonrepeatable.

Table 24.2 summarizes a comparison of preload variation for friction only and elastic interaction only for various degrees of joint flexibility. Joint flexibility is varied by (1) using no gasket in the pipe flange; (2) using a thin, stiff sheet gasket; (3) using a thicker, flexible spiral-wound gasket; and (4) using a spiral-wound gasket and longer studs with 4 in. spacers. The addition of the gasket makes the joint more flexible. Adding spacers to the studs makes the studs more flexible.

A few trends are seen that may affect bolt-up strategy.

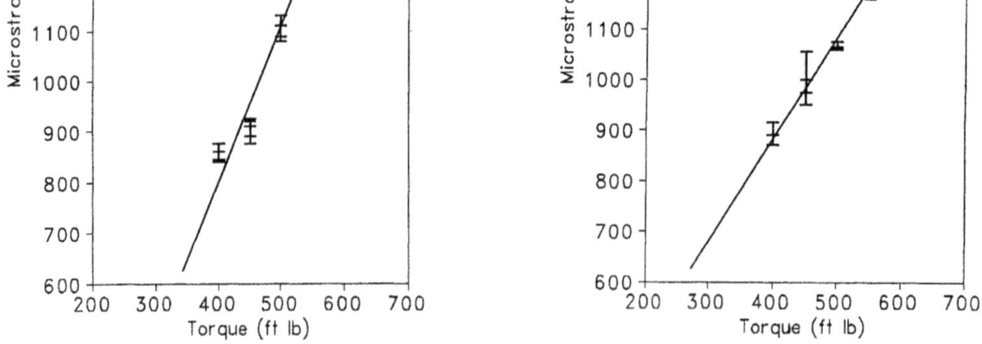

FIGURE 24.9 Typical scatter bands for two individual bolts torqued repeatedly. This testing was done to compare the preload variation due to friction to that due to elastic interaction. Although each bolt is quite repeatable, note how the strain vs. torque line has a different slope for each bolt.

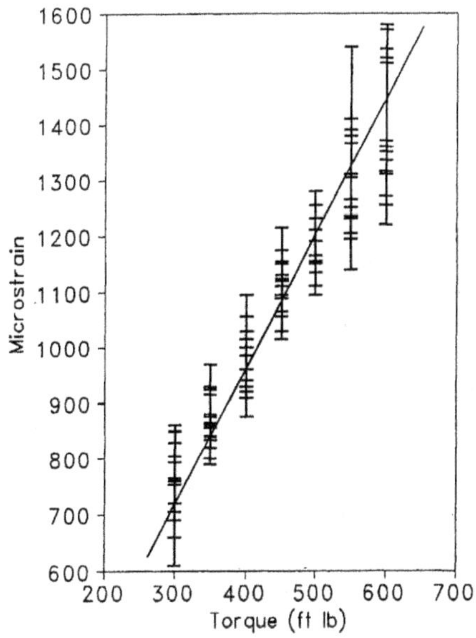

FIGURE 24.10 Scatter bands for a group of 14 bolts that were torqued. This testing was done to evaluate the repeatability of an individual bolt compared to a group of bolts and to compare frictional variation in preload to elastic interaction variation.

TABLE 24.2
Preload Variation

Case[a]	Preload Variation
I	±5%
II	±13%
III	+0-35%
IV	+0-59%
V	+0-95%
VI	+0-75%

Note: All testing in this data set was done with the same studs and same structure (i.e., 24 in. pipe flange).

[a] Case I: Preload variation in a single stud torqued repeatedly (no interaction).
Case II: Preload variation in a group of 14 studs torqued repeatedly one at a time (no interaction).
Case III: Preload variation of 20 studs in a stiff structure without a gasket (24 in. ANSI Class 150 pipe flange). Torquing is done to precise strain values by using strain gages mounted on the studs (no frictional variation).
Case IV: Same as above except joint is made more flexible by adding a 0.060 in. thick sheet gasket.
Case V: Same as above except joint is made more flexible by adding a 0.180 in. thick spiral wound gasket.
Case VI: Same as above except studs are made more flexible by using 4 in. spacers and longer studs. This approximately doubled the grip length.

1. Frictional variation is a random phenomenon with equal probability of plus or minus variation.
2. Elastic interaction can result only in the reduction of individual bolt preloads. Care was always taken to separate frictional preload variation from elastic interaction variation. In most real-world applications, the two effects would, of course, superimpose.
3. Adding a gasket makes the joint more flexible and results in more elastic interaction.
4. Making the studs longer (i.e., making the studs more flexible) reduces elastic interaction.

24.5 IMPROVED FLANGE BOLT-UP PROCEDURES

24.5.1 The Procedure Summarized

The data given in Table 24.1 can be used in a matrix equation to compute elastic interaction coefficients. These coefficients define the relationship between the initial preload in a given bolt, the final load in that bolt (after interaction), and the initial loads of the other bolts that cause interaction. This relationship can be put into equation form and expanded to include all the bolts of any size flange. Each bolt is represented by a linear equation that defines its relation to the other bolts in the flange. These equations can be easily solved with readily available engineering software.

Any bolted connection will have a set of equations that can be written in matrix form as

$$\{S_f\} = [A]\{S_i\} \tag{24.1}$$

where

$\{S_f\}$ is an $n \times 1$ matrix corresponding to the final bolt load
$\{S_i\}$ is an $n \times 1$ column matrix describing the initial bolt load
$[A]$ is an $n \times n$ matrix of interaction coefficients that correctly transforms S_i into S_f; n equals the number of bolts in the flange.

Coefficients of the $[A]$ matrix can be found experimentally by tightening the bolts once and measuring the interactions. Once the coefficient matrix is found, the matrix equation can be solved for the required initial bolt loads that result in any desired final bolt loads as shown.

$$\{S_i\} = [A]^{-1}\{S_f\} \tag{24.2}$$

Typically, each bolt must be tightened to a different initial load to create uniform final loads in each bolt.

24.5.2 Determining Interaction Coefficients

Consider the hypothetical elastic interaction data shown in Table 24.3 for a three-bolt system. Each bolt is initially tightened to 40,000 psi. The history of bolt 1 is as follows: Initially tighten to 40,000 psi. Bolt 1 drops to 25,000 psi after bolt 2 is tightened. Bolt 1 then drops to 22,000 psi due to interaction when bolt 3 is tightened.

Bolt 1:

$$a_{11} \times 40{,}000 = 40{,}000$$

$$a_{11} = 1.0$$

TABLE 24.3
Hypothetical Interaction Data for a Three-Bolt System[a]

	Resulting Load (psi)		
Tighten Bolt	**Bolt 1**	**Bolt 2**	**Bolt 3**
1	40,000	—	—
2	25,000	40,000	—
3	22,000	27,000	40,000

[a] The data are used in the text to demonstrate how the interaction coefficients are calculated.

where a_{11} describes the interaction of bolt 1 on itself.

$$a_{11} \times 40{,}000 + a_{12} \times 40{,}000 = \text{stress in bolt 1 after bolt 2 is tightened}$$

$$(1.0)(40{,}000) + a_{12} \times 40{,}000 = 25{,}000$$

$$a_{12} = -0.375$$

where a_{12} is the interaction of bolt 2 on bolt 1.

$$a_{11} \times 40{,}000 + a_{12} \times 40{,}000 + a_{13} \times 40{,}000 = \text{stress in bolt 1 after bolt 2 is tightened}$$

$$(1.0)(40{,}000) - (0.375)(40{,}000) + a_{13} \times 40{,}000 = 22{,}000$$

$$a_{13} = -0.075$$

where a_{13} is the interaction of bolt 3 on bolt 1.

Bolt 2:

$$a_{21} = 0.0$$

where a_{21} is the interaction of bolt 1 on bolt 2.

Since bolt 1 is already tightened, it does not affect bolt 2.

$$a_{22} = 1.0$$

where a_{22} is the interaction of bolt 2 on itself.

$$a_{21} \times 40{,}000 + a_{22} \times 40{,}000 + a_{23} \times 40{,}000 = \text{stress in bolt 2 after bolt 3 is tightened}$$

$$(0.0)(40{,}000) + (1.0)(40{,}000) + a_{23} \times 40{,}000 = 27{,}000$$

$$a_{23} = -0.325$$

where a_{23} is the interaction of bolt 3 on bolt 2.

Bolt 3:

$$a_{31} = a_{32} = 0.0; \; a_{33} = 1.0$$

Equation (24.1) becomes

$$\begin{bmatrix} 1.0 & -0.375 & -0.075 \\ 0.0 & 1.0 & -0.325 \\ 0.0 & 0.0 & 1.0 \end{bmatrix} \cdot \begin{bmatrix} S_i \\ S_{2i} \\ S_{3i} \end{bmatrix} = \begin{bmatrix} S_{1f} \\ S_{2f} \\ S_{3f} \end{bmatrix}$$

24.5.3 Procedures for Large Flanges

The procedure described above can be extended for use on large conventional flanges. A 20-bolt flange, for example, would be represented by a system of 20 equations with 20 unknowns. The method was first described by Van Campen [8], who developed interaction equations for reactor vessels using hydraulic jack pressures and bolt strain. He also assumed flange symmetry, which reduced the number of equations.

Van Campen assumed the interaction coefficients remain constant and can be determined by a single test. I have developed a more accurate procedure that involves two single-pass tests and results in interaction coefficients that are precise enough to allow the flange to be subsequently tightened with a final preload scatter of ±2%. This procedure is as follows.

1. *Test* 1: Tighten all bolts to initial preload or stress equal to that ultimately desired for the final preload. Let us use, by way of example, a stress goal of 25,000 psi. The final average bolt stress during this test is, of course, much less than 25,000 psi.
2. Use the data from Test 1 to compute preliminary interaction coefficients (i.e., establishing a preliminary [A] matrix).
3. Using Eq. (24.2), predict the initial bolt loads presumably required to obtain a uniform final bolt load of 25,000 psi.
4. *Test* 2: Loosen all the bolts and then retighten them to the initial stress values predicted in step 3.
5. Use the data obtained in Test 2 to update the [A] matrix and then recompute the initial stress values required to create a final stress of 25,000 psi in this set of bolts with Eq. (24.2).

The initial stress values can now be used on subsequent tightening of these bolts, with a resulting scatter in stress of only ±2% of 25,000 psi. Note that if the interaction coefficients were constant, as assumed by Van Campen, or if greater scatter could be tolerated, then only one test would be needed to determine the correct initial loads. Because the coefficients change some, a second test is required to improve the accuracy of the coefficients and ensure ±2% uniformity in future bolt-ups.

24.5.4 An Example

Experimental testing was completed on a 16 in. Class 300 RFWN (raised face welding neck) flange with 20 1¼ in. (31.75 mm) diameter bolts. The flanges and instrumentation are as described earlier.

Initial testing was done without a gasket (simulating a nuclear reactor head where the gasket is structurally insignificant). Subsequent testing was completed with a flat sheet gasket on a second pair of 16 in. flanges and on a pair of 24 in. flanges. Each test consisted of one tightening pass using the four-point star tightening sequence (see Figure 24.2). The first four bolts (1, 11, 16, 6) were torqued together, simulating simultaneous tensioning. This was done to eliminate flange rocking and excessive stress on the first bolt. Other bolting patterns could be used to eliminate this problem (e.g., hand tightening of all bolts to about 25 ft-lb). Tightening four bolts simultaneously eliminated the rocking problem by bringing the two flange surfaces rigidly against each other. Technically this resulted in what might be called a two-pass procedure; however, this first pass could be done without torquing equipment, without any time-consuming torque or bolt stretch monitoring, and without a special pattern. If the flange is brought flush with the first four bolts tightened simultaneously as described above, then perhaps the process should more accurately be described as a 1.2-pass procedure. Using multiple tensioners would result in a true one-pass procedure.

The procedure used was the one described earlier—a series of two tests designed to experimentally obtain and improve the coefficients in the [*A*] matrix, which were then confirmed in a third test. The initial bolt stress for each of the three tests is shown in Table 24.4.

24.5.5 Discussion of the Results

A three-test sequence completed with metal-to-metal contact (no gasket installed) is shown in Figure 24.11. The graph shows the final bolt stress for each one-pass procedure. In test 1 all bolts were tightened to approximately 25,500 psi (172.4 MPa). In tests 2 and 3, the bolts were tightened as shown in Table 24.4.

TABLE 24.4
Initial Bolt Stress in a Three-Test Series[a]

	Test Results (psi)		
Bolt Tightening Sequence	**Test 1**	**Test 2[b]**	**Test 3[c]**
1	25,500	42,090	42,210
11	25,500	41,190	40,500
16	25,500	47,190	41,820
6	25,500	46,620	43,110
2	25,500	34,170	34,620
12	25,500	31,110	32,280
17	25,500	34,170	34,080
7	25,500	34,260	34,230
3	25,500	31,650	32,310
13	25,500	30,210	30,810
18	25,500	30,570	32,070
8	25,500	32,040	32,820
4	25,500	30,630	30,690
14	25,500	30,630	31,230
19	25,500	30,630	30,630
9	25,500	29,790	30,510
5	25,500	25,500	25,500
15	25,500	25,500	25,500
20	25,500	25,500	25,500
10	25,500	25,500	25,500

Source: Reprinted with permission, from Welding Rsearch Council Bulletin 408, Jan. 1996.

[a] Final stress values are plotted in Figure 24.11.

[b] Calculated from Test 1.

[c] Calculated from Test 2.

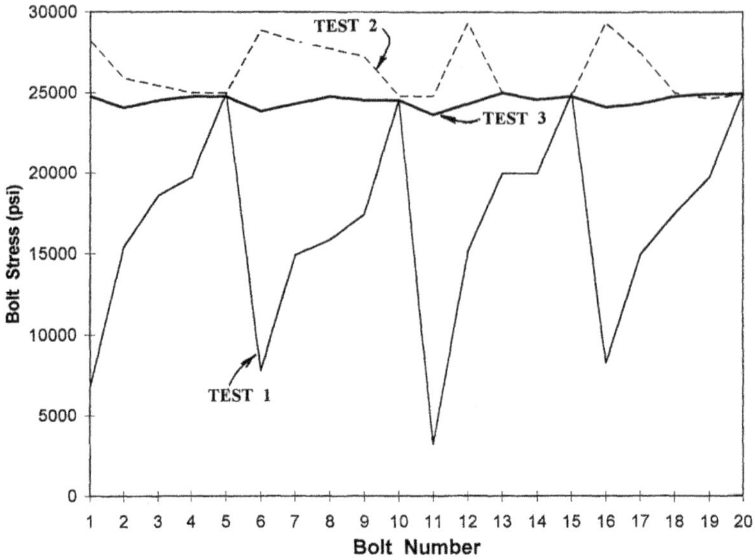

FIGURE 24.11 Final stresses for a three-test sequence (single-pass tests) on a set of 16 in. flanges without a gasket.

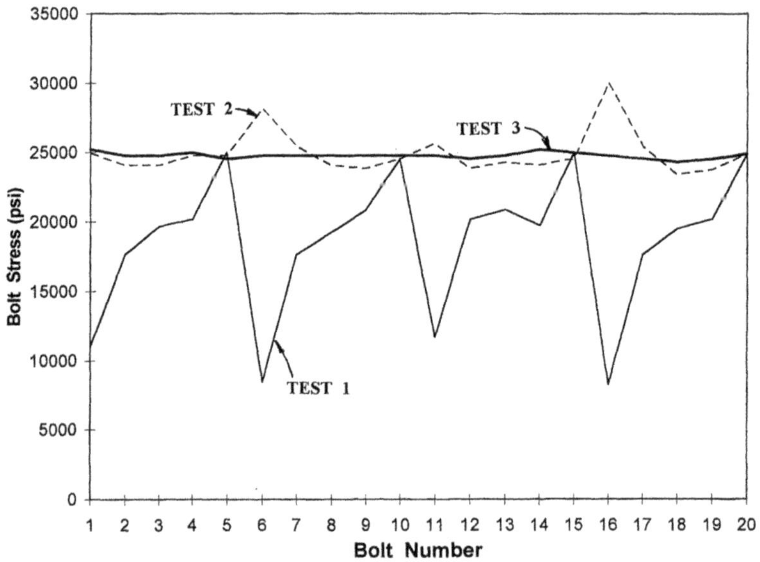

FIGURE 24.12 Final stresses for a three-test sequence (single-pass tests) on a second set of 16 in. flanges without a gasket.

The first test shows large variations in bolt stress with an average value of only 18,500 psi (127.6 MPa). Stress load reductions of up to 67% occurred. The second test in the series has an average stress load of 25,500 psi (172.4 MPa) with variations of +17 to −6.5%. The average bolt load for the final test is at the target stress of 25,500 psi (172.4 MPa) with scatter of only ±2%. This small amount of scatter can be attributed to instrumentation noise (±0.9%) and the inability to hand tighten to precise values (±1.8%).

Figure 24.12 shows similar results for the second pair of 16 in. flanges. Figure 24.13 shows results for a three-test sequence conducted with a sheet gasket 0.060 in. (1.52 mm) thick.

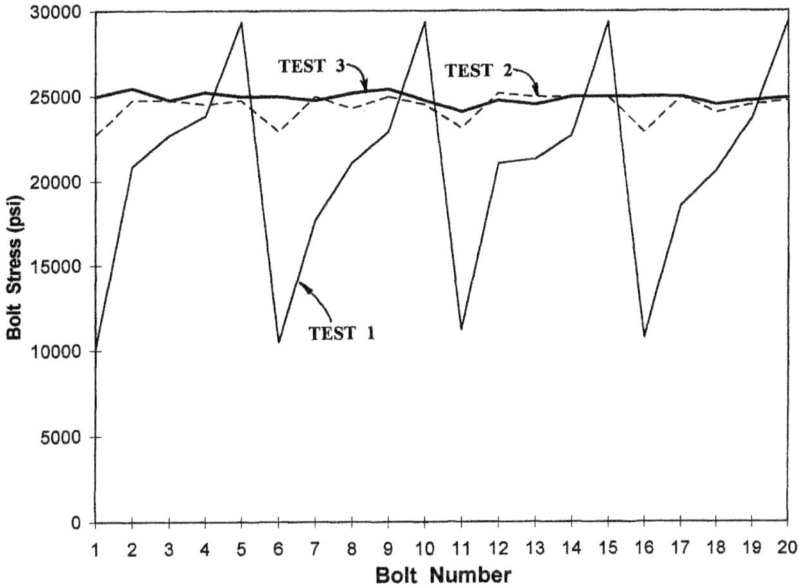

FIGURE 24.13 Final stress for a three-test sequence (single-pass tests) on a 16 in. flange with a sheet gasket.

Analysis of the coefficient matrix shows that the majority of interaction is caused by the adjacent two bolts. However, the other coefficients were not negligible. Collectively they also contribute to the final bolt load. A complete 20×20 interaction coefficient matrix is shown in Table 24.5 as an example. Table 24.5 is the [*A*] matrix for test 3 of a test series done on a 16 in. flange without a gasket.

The amount of interaction changes for different loading conditions. High clamping force creates less interaction than bolting with low clamping force.

(*Note*: The coefficients used for the procedure are ordered by tightening sequence number rather than bolt number. $A_{1,5}$ is the interaction coefficient for bolt 1 when the fifth bolt of the sequence is tightened. The fifth bolt is the adjacent bolt 2. Ordering in this manner makes the matrix inversion easier. The matrix becomes upper triangular with all 1s on the diagonal.)

As seen from Table 24.5, primary coefficients average 0.334 for the first four bolts tightened. The averages for the second, third, and fourth groups are 0.290, 0.254, and 0.215, respectively. This decreasing pattern is also consistent with the theory that interaction decreases with increased clamping force.

Additional testing was done to illustrate the stability of the coefficients. For example, if bolt A is tightened to 30,000 psi (206.9 MPa) and adjacent bolt B is tightened to 36,000 psi (248.2 MPa), the coefficient is

$$30{,}000 + a \times 36{,}000 = 16{,}500$$

$$a = -0.375$$

If bolt A is increased to 36,000 psi (248.2 MPa) or decreased to 24,000 psi (165.5 MPa), the coefficient remains almost constant:

$$36{,}000 + a \times 36{,}000 = 22{,}500$$

$$a = -0.375$$

$$24{,}000 + a \times 36{,}000 = 11{,}250$$

$$a = -0.375$$

TABLE 24.5
20×20 Coefficient Matrix for Test 3 of the Three-Test Series Without a Gasket

	1	11	16	6	2	12	17	7	3	13
1	1.0000	0.0000	0.0000	0.0000	–0.3292	0.0075	–0.0191	0.0000	0.0000	0.0000
11	0.0000	1.0000	0.0000	0.0000	0.0000	–0.3657	0.0066	–0.0262	0.0000	0.0079
16	0.0000	0.0000	1.0000	0.0000	0.0140	0.0075	–0.3120	0.0000	0.0000	–0.0531
6	0.0000	0.0000	0.0000	1.0000	0.0140	0.0149	0.0000	–0.3316	–0.0436	0.0069
2	0.0000	0.0000	0.0000	0.0000	1.0000	–0.0224	0.0066	0.0203	–0.2670	0.0000
12	0.0000	0.0000	0.0000	0.0000	0.0000	1.0000	0.0191	0.0271	–0.0076	–0.2616
17	0.0000	0.0000	0.0000	0.0000	0.0000	0.0000	1.0000	0.0000	0.0076	0.0026
7	0.0000	0.0000	0.0000	0.0000	0.0000	0.0000	0.0000	1.0000	0.0218	0.0157
3	0.0000	0.0000	0.0000	0.0000	0.0000	0.0000	0.0000	0.0000	1.0000	–0.0079
13	0.0000	0.0000	0.0000	0.0000	0.0000	0.0000	0.0000	0.0000	0.0000	1.0000
18	0.0000	0.0000	0.0000	0.0000	0.0000	0.0000	0.0000	0.0000	0.0000	0.0000
8	0.0000	0.0000	0.0000	0.0000	0.0000	0.0000	0.0000	0.0000	0.0000	0.0000
4	0.0000	0.0000	0.0000	0.0000	0.0000	0.0000	0.0000	0.0000	0.0000	0.0000
14	0.0000	0.0000	0.0000	0.0000	0.0000	0.0000	0.0000	0.0000	0.0000	0.0000
19	0.0000	0.0000	0.0000	0.0000	0.0000	0.0000	0.0000	0.0000	0.0000	0.0000
9	0.0000	0.0000	0.0000	0.0000	0.0000	0.0000	0.0000	0.0000	0.0000	0.0000
5	0.0000	0.0000	0.0000	0.0000	0.0000	0.0000	0.0000	0.0000	0.0000	0.0000
15	0.0000	0.0000	0.0000	0.0000	0.0000	0.0000	0.0000	0.0000	0.0000	0.0000
20	0.0000	0.0000	0.0000	0.0000	0.0000	0.0000	0.0000	0.0000	0.0000	0.0000
10	0.0000	0.0000	0.0000	0.0000	0.0000	0.0000	0.0000	0.0000	0.0000	0.0000

Source: Reprinted with permission from Welding Research Council Bulletin 408, January 1996.

Even though bolt A was varied by ±20%, the coefficients varied only from 0.0 to –5.6%. A –5.6% variation in the interaction coefficient results in only a 4.4% error in the predicted stress of bolt A after interaction. The method works because the coefficients become stable as the correct solution is approached.

24.5.6 Tests on a Noncircular Joint

The question often asked is: Is a circular symmetrical flange needed for the matrix method to work? The matrix method was applied to an automotive cylinder head and engine block [6], a highly irregular and noncircular shape. (I do not believe the circular shape of a pipe flange is of any significance to the matrix method. There is nothing symmetrical about tightening one bolt at a time in a circular flange.)

The bolts were threaded into the engine. This made it very difficult to mount strain gages on the fasteners. An alternative method was chosen to apply the matrix method. A plastic sensor about 0.005 in. thick (described in Ref. [9]) was placed on top of the gasket to monitor gasket contact stresses. Fifteen sensors were paired up with 15 bolts. Instead of strain gage readings on the bolt before and after elastic interaction, gasket contact stress before and after interaction was used in the matrix equations.

One sensor consistently gave a very low value. The matrix method predicted an excessive torque—enough to yield, if not break, the bolt. The assumed cause was a lack of flatness of the engine block and/or head. Instead of aiming to achieve uniformity, the goal was changed to obtaining a calculated distribution of sensor readings in a single pass by using the matrix algorithm. This test was considered successful and is described in detail in Ref. 5. All of this suggests that one application of the matrix method is to predict or "dial in" any favorable head gasket load distribution by varying bolt load, size, torque, or location.

24.6 TWO-PASS MATRIX ALGORITHM

The one-pass procedure worked very well for rigid flanges (metal-to-metal contact and flat sheet gaskets). It did not work as well with flexible systems such as spiral-wound gaskets. A two-pass matrix procedure was developed next.

$$[A]\{S_i\}+[B]\{S_{\text{inc}}\}=\{S_f\} \tag{24.3}$$

where

[A] = interaction coefficient matrix for the first pass
{S_i} = the initial value to which each bolt is tightened in the first pass
[B] = interaction coefficient matrix for the second pass.

18	8	4	14	19	9	5	15	20	10
−0.0617	0.0071	0.0016	0.0079	−0.1156	0.0079	0.0095	−0.0095	−0.1669	0.0000
0.0000	−0.0769	0.0080	0.0149	0.0000	−0.1308	0.0095	0.0095	0.0000	−0.2151
0.0000	0.0063	−0.0070	−0.0782	0.0313	0.0000	0.0000	−0.1858	0.0094	0.0000
0.0000	−0.0134	−0.0858	0.0000	0.0000	0.0226	−0.2130	0.0095	0.0000	0.0096
0.0139	−0.0071	−0.0240	0.0000	0.0000	0.0000	0.0095	−0.0095	−0.0271	−0.0096
0.0000	0.0134	0.0000	−0.0228	0.0000	0.0079	0.0000	0.0035	0.0000	−0.0382
−0.3128	0.0071	−0.0080	0.0079	−0.0382	0.0079	−0.0095	−0.0284	0.0094	0.0000
0.0000	−0.3217	0.0000	−0.0079	0.0078	−0.0268	−0.0367	0.0095	0.0000	0.0096
0.0139	0.0143	−0.2345	−0.0079	0.0157	0.0000	−0.0284	0.0000	0.0094	0.0000
0.0139	0.0206	0.0000	−0.2010	0.0000	0.0147	0.0000	−0.0272	0.0000	0.0096
1.0000	−0.0071	0.0080	0.0149	−0.3066	0.0000	0.0000	0.0095	−0.0271	0.0084
0.0000	1.0000	0.0230	0.0000	0.0000	−0.2763	0.0178	−0.0083	0.0000	−0.0287
0.0000	0.0000	1.0000	0.0000	0.0157	0.0069	−0.2130	−0.0083	0.0082	0.0000
0.0000	0.0000	0.0000	1.0000	0.0069	0.0079	0.0000	−0.2308	0.0094	0.0000
0.0000	0.0000	0.0000	0.0000	1.0000	0.0000	0.0000	0.0000	−0.2012	0.0000
0.0000	0.0000	0.0000	0.0000	0.0000	1.0000	0.0000	0.0000	0.0000	−0.2151
0.0000	0.0000	0.0000	0.0000	0.0000	0.0000	1.0000	0.0000	−0.0082	0.0084
0.0000	0.0000	0.0000	0.0000	0.0000	0.0000	0.0000	1.0000	0.0000	0.0000
0.0000	0.0000	0.0000	0.0000	0.0000	0.0000	0.0000	0.0000	1.0000	0.0000
0.0000	0.0000	0.0000	0.0000	0.0000	0.0000	0.0000	0.0000	0.0000	1.0000

{S_{inc}} = the incremental load added to each bolt in the second pass
[S_f] = the final bolt load after interaction in the first and second passes.

However, since [A]{S_i} equals the bolt load in each bolt at the end of the first pass, this equation can be simplified to

$$\{S_{\text{P1}}\}+[B]\{S_{\text{inc}}\}=\{S_f\} \tag{24.4}$$

where {S_{P1}} is the bolt load in each bolt at the end of the first pass.

Once the [B] matrix is found experimentally, the following matrix equation can be solved for the required increment in the second pass to give any final bolt load.

$$\{S_{\text{inc}}\}=[B]^{-1}\left(\{S_f\}-\{S_{\text{P1}}\}\right) \tag{24.5}$$

What follows is an example of how the coefficients are calculated and a description of experimental results.

TABLE 24.6
Hypothetical Interaction Data for a Two-Pass, Four-Bolt System[a]

Final Values at End of First Pass				
	Bolt 1	**Bolt 2**	**Bolt 3**	**Bolt 4**
	450	550	700	850
Second Pass Interaction Data				
	Bolt 1	**Bolt 2**	**Bolt 3**	**Bolt 4**
Bolt 1	850	400	650	600
Bolt 2	700	850	575	550
Bolt 3	650	725	850	475
Bolt 4	550	650	700	850

[a] Used in text to illustrate calculation of two-pass interaction coefficients.

24.6.1 Determining Two-Pass Interaction Coefficients

Hypothetical interaction data for a two-pass procedure are given in Table 24.6. In the first pass all the bolts are tightened to 850 µstrain, but interaction results in final values of 450, 550, 700, and 850 µstrain in bolts 1, 2, 3, and 4, respectively, as shown. There are no blank entries in the second-pass data matrix because all the bolts are tight at the end of the first pass and receive interaction in the second pass.

The two-pass matrix procedure is based on increments. Therefore the increments used in the second pass must be calculated. Bolt 1 ends the first pass at 450 µstrain. At the start of the second pass, bolt 1 is incremented by 400 to 850. This results in bolt 2 dropping to 400. Bolt 2 is incremented by 450 to 850 µstrain during the second pass. Similarly, bolt 3 is incremented 275 and bolt 3 is incremented by 375.

The change in bolt load is divided by the initial load of the bolt being tightened. For the [B] matrix coefficients, the denominator becomes the increment added to the bolt instead of the initial load used to determine the [A] matrix.

$$b_{ij} = \frac{\text{current load in bolt } i - \text{previous load}}{\text{increment added to bolt } j \text{ being tightened}}$$

Using this information the [B] matrix can be calculated as follows:

For bolt 1:

$$b_{11} = (850 - 450)/400 = 1.0$$

$$b_{12} = (700 - 850)/450 = -0.333$$

$$b_{13} = (650 - 700)/275 = -0.182$$

$$b_{14} = (550 - 650)/375 = -0.267$$

For bolt 2:

$$b_{21} = (400 - 550)/400 = -0.375$$

$$b_{22} = (850 - 400)/450 = 1.0$$

$$b_{23} = (725 - 850)/275 = -0.455$$

$$b_{24} = (650 - 725)/375 = -0.200$$

And so on for bolts 3 and 4.

Correct calculation of the [B] matrix can be verified by substituting the values into Eq. (24.4). The final values obtained in the first pass and the increments used in the second pass should, along with the correct [B] matrix, correctly predict the final bolt stress values. The complete substitution of the hypothetical data into Eq. (24.4) results in the matrix equation

$$\begin{bmatrix} 450 \\ 550 \\ 700 \\ 850 \end{bmatrix} + \begin{bmatrix} 1.0 & -0.33 & -0.18 & -0.27 \\ -0.38 & 1.0 & -0.46 & -0.20 \\ -0.13 & -0.17 & 1.0 & -0.40 \\ -0.63 & -0.11 & -0.27 & 1.0 \end{bmatrix} \cdot \begin{bmatrix} 400 \\ 450 \\ 275 \\ 375 \end{bmatrix} = \begin{bmatrix} 550.75 \\ 646.5 \\ 696.5 \\ 849.25 \end{bmatrix}$$

24.6.2 Experimental Data

Figure 24.14 shows interaction data for a two-pass procedure. In the first pass, each bolt is tightened to 500 μstrain. In the second pass, each bolt is tightened to 850 μstrain. The bolting efficiencies are 0.39 and 0.81, respectively. This data is for a 16 in. ANSI pipe flange without a gasket.

Figure 24.15 shows data for a two-pass procedure using the two-pass matrix technique. The first-pass target was 500 μstrain and was obtained with good accuracy by using Eqs. (24.1) and (24.2). The second-pass target of 1,000 μstrain was obtained by using Eq. (24.5). This test was done on a 16 in. flange without a gasket. The bolting efficiencies are seen to be nearly perfect.

Figure 24.16 shows the results of using the two-pass matrix technique on a spiral-wound gasket. The target strain is 500 and 850 μstrain for the first and second passes, respectively. The bolting efficiencies are 0.60 and 0.88, respectively. Although not as good as the results shown in Figure 24.15, these values compare favorably with those of a non-two-pass matrix procedure on a spiral-wound gasket with much lower bolting efficiencies of 0.26 and 0.53.

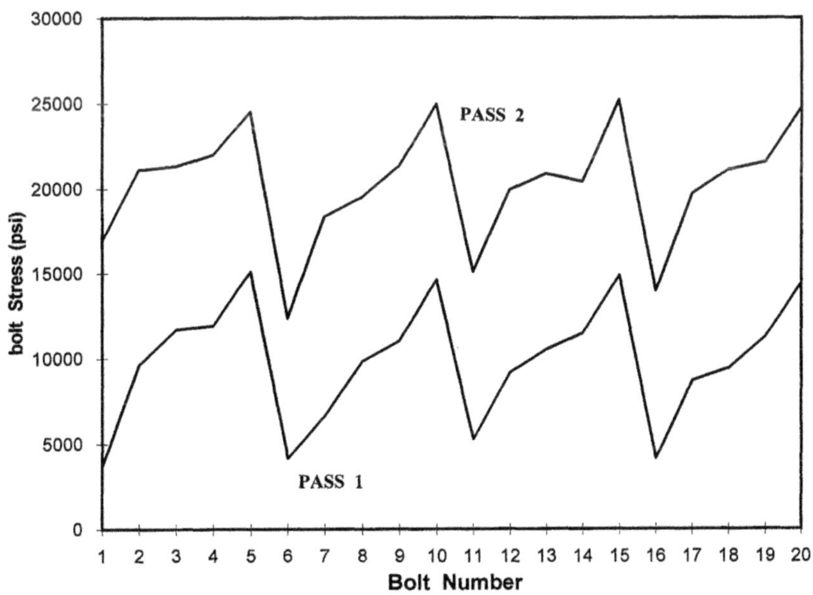

FIGURE 24.14 Data for a two-pass procedure (16 in. flange, no gasket). First pass to 15,000 psi stress; second pass to 25,000 psi.

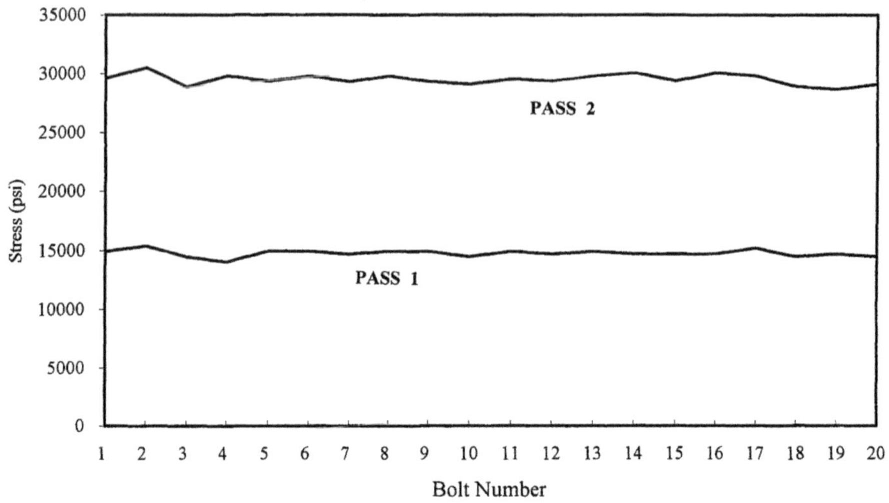

FIGURE 24.15 Final bolt stress for a 16 in. flange (without a gasket) using two-pass matrix technique. Second pass target is 30,000 psi stress.

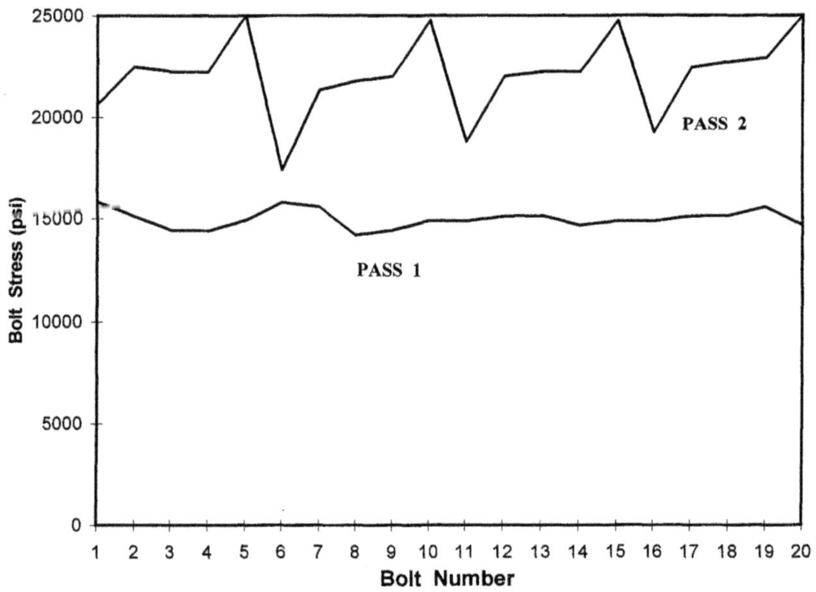

FIGURE 24.16 Final bolt stress for a 16 in. flange with a spiral-wound gasket using two-pass matrix technique. Second pass target is 25,000 psi stress.

24.6.3 Results and Discussion

The two-pass matrix procedure worked very well for a rigid system without a gasket. It did not work as well with a flexible spiral-wound gasket. The thicker, more flexible spiral-wound gasket has interaction coefficients that are not as stable as those for a rigid system. However, the bolting efficiencies for a two-pass matrix procedure were much higher than for any other spiral-wound bolt-up procedure after two passes.

The nonlinear bolted flange–gasket interactions are approximated with a system of linear equations. Performing a series of tests and updating the coefficients after each test is equivalent to experimentally iterating to the correct nonlinear solution with a series of linear approximations of the elastic interactions. If the spiral-wound gasket has less constant interaction coefficients, improvements can be expected with more iterations (i.e., more tests). The prediction should become better as the correct solution is approached and the coefficients become more accurate.

24.7 APPLICATIONS

Any critical joint is a good candidate for a matrix bolt-up procedure. A bolted connection with a specific bolt-up procedure or one that requires bolt load monitoring or stud tensioning equipment and/or receives additional engineering attention for whatever reason (i.e., history of leakage and/or critical nature of the joint) is a good candidate for a one-pass matrix bolt-up procedure.

For new equipment, it makes sense to require that a series of tests be done in the shop immediately after fabrication to determine the [*A*] matrix. Critical flanges should be purchased with their interaction data supplied.

It may be difficult to run a series of tests on operating equipment. However, most of the data can be taken during the normal bolt-up procedure. The [*A*] matrix can be determined without disrupting the existing procedure. The algorithm can then be used during the next bolt-up.

A modified two-pass matrix procedure could be used as an alternative approach in the field. Data from the first pass could be used to predict overbolting required for the second pass.

Data specific to each application must be developed. However, this does not prevent a matrix procedure from being used on ordinary flanges that are torqued. The accuracy for torqued flanges will never be ±2%. However, by virtue of torquing, the operator is accepting tremendous scatter in bolt-to-bolt loading. A generic torquing procedure could be developed, for example, for all 10 in. Class 300 flanges. The matrix method would give less scatter in fewer passes.

REFERENCES

1. Ezell, R., The analytical and experimental study of elastic interaction in a bolted flanged gasket joint, Master's Thesis, University of Akron, December 1992.
2. *1995 ASME Boiler and Pressure Vessel Code, Section II, Materials, Part D—Properties*, The American Society of Mechanical Engineers, New York, 1995.
3. Bibel, G. D., Experimental and analytical study of elastic interaction in a pipe flange, *International Conference on Pressure Vessel Technology,* Duesseldorf, 1992.
4. Bibel, G. D., and R. Ezell, Bolted flange assembly: preliminary elastic interaction data and improved bolt-up procedures, Welding Research Council Bulletin 408, May 1996.
5. Bickford, J. H., K. Hayashi, A. T. Chang, and J. R. Winter, A preliminary evaluation of the elevated behavior of a bolted flanged connection, Welding Research Council Bulletin 341, February 1989.
6. Goddard, D. L., and G. D. Bibel, Achieving a selected load distribution in the bolted joint of a cylinder head of highly variable stiffness and contact geometry, SAE Tech. Paper Series 940693.
7. Bazergui, A., and L. Marchand, PVRC milestone gasket tests—first results, Welding Research Council Bulletin 292, February 1984.
8. Van Campen, D. H., *A Systematic Bolt-Tightening Procedure for Reactor Vessel Flanges*, University of Technology, Delft, 1969.
9. Czernik, D. E., and F. L. Misczk, *A New Technique to Measure Real Time Static and Dynamic Gasket Stresses*, SAE International Congress and Exposition, Detroit, MI, Feb. 25, 1991.

25 Hydraulic Stud Tensioning

William L. Biach

25.1 INTRODUCTION

There are two requirements that need to be served when using a threaded fastener: (1) stretching the fastener and (2) maintaining the stretch. For most applications, the nut is used for both purposes. Rotating the nut against the fastener thread uses the inclined plane of the thread for mechanical advantage to stretch the fastener. It is then left in place to maintain the stretch. In hydraulic stud tensioning, a hydraulic cylinder stretches the fastener; then a nut is used to maintain the stretch. The standard process is as follows (refer to Figure 25.1):

1. The tensioner is attached to the fastener above the nut (usually 50%–75% of a fastener diameter).
2. Hydraulic pressure (6,000–25,000 psi) is applied to the tensioner, which pulls on the fastener and reacts on the flange.
3. This stretches the fastener, loads the flange, and compresses the gasket (if a gasket is being used), leaving a space between the flange and the bottom of the nut.
4. The nut is then rotated under a no-load condition to take up the space.
5. The hydraulic pressure is released, and the nut holds the load. There is usually a small drop in load as the previously unloaded fastener and nut threads deflect as they assume the load.
6. The tensioner is then removed.

Generally, more than one tensioner is used at a time, with a common hydraulic line. This ensures equal, concurrent, symmetric loading of the flange. Tensioners have been used for studs from as small as ¼ in. with a capacity of 2500 lb up to 20 in. with a capacity of 15,000,000 lb.

25.2 APPROPRIATE APPLICATIONS

25.2.1 Large Fasteners

Tensioners are most commonly used on fasteners 2 in. in diameter and up. In the range of 2–4 in., they are an alternative to torquing. In this range, tensioners are smaller, less cumbersome, and safer.

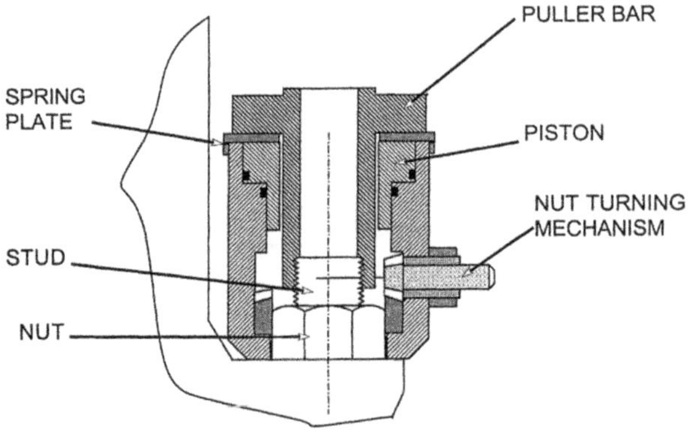

FIGURE 25.1 Tensioner cutaway.

DOI: 10.1201/9781003501183-30

Where fasteners are greater than 4 in. in diameter, hydraulic tensioning is the method of choice, as tensioning provides an even tightening plane that cannot be achieved with other methods.

Examples: Nuclear pressure vessels, large autoclaves, large press tie rods

25.2.2 Gasketed Flanges

The key to effective closure of a gasket flange is uniformity of gasket pressure. Multiple tensioners provide simultaneous loading at several points around a flange. It is common to tighten half the studs at one time in refinery applications, which provides extremely even gasket loading.

Examples: Pipe flanges, pressure vessel covers, manway covers

25.2.3 Controlled Bolt Group Centroid

When a structure is being anchored with a bolt group, the structure must be designed for the most conservative case for the loading of the bolts. If the bolts are torqued, the structure must be designed with the assumption that the bolts will be tighter on one side than on the other side. This puts the centroid near the edge and increases the structural requirements. Hydraulic tensioning ensures bolting uniformity, which permits the assumption that the centroid will be centrally located. This reduces the demands on the structure.

Example: Water tower designed for seismic conditions

25.2.4 Materials Subject to Galling

Since tensioning doesn't require sliding surfaces against each other under high load, galling is virtually eliminated.

Examples: Stainless steel, Inconel

25.2.5 Fasteners Operating Close to Yield

Axial stress and torsional stress are the primary components of the total stress in a fastener during tightening. Torsional stress becomes a significant component at high loads because the frictional force between the threads is high. Therefore a stud may yield while being torqued to an axial stress of only 75% of yield. As there is no torsional stress with hydraulic tensioning, this problem is eliminated. Although tensioning introduces the residual loss of tensioners, the loss may be reduced to 5% with the proper procedures and equipment.

Examples: Aerospace fasteners

25.2.6 Space Constraints

When fasteners are placed where access is limited, readily available tensioning tools may be used.

Examples: Box joints, bearings

25.2.7 Limited Reaction Surfaces

When there are no reaction surfaces or the available surfaces cannot be used, tensioning tools can be used as they require no torque.

Examples: Bridge suspension cables, underwater or outer space applications

25.3 TENSIONER SIZING

Characteristic of the tensioning process is a partial loss of load in the fastener when the hydraulic pressure is released and the nut retains the load. For example, if the fastener is stressed to 50 ksi while the tensioner is pressurized, the residual stress in the fastener may be 40 ksi when the nut is tightened and the tensioner is removed. The following are contributing elements to residual loss:

Nut and stud thread deformation
Change in effective grip length from where the tensioner is connected to the nut
Nut expansion under load
Compression of the flange material under the nut when it assumes load.

Residual loss can be anywhere from 5% to 50% of the fastener tension. However, under controlled circumstances, the loss is repeatable and predictable. Therefore, even when the characteristic loss of the application is high, the loading is still very uniform.

Since thread deformation accounts for at least 90% of the tension drop in the fastener, addressing this issue is the most effective way of reducing the loss. While the hydraulic tensioner has the fastener under load, the threads of the fastener above the nut are carrying the load, whereas the threads of the nut and the complementary threads of the fastener are carrying no load. When the nut takes over from the tensioner, the threads deform, permitting the stud to contract, which reduces the elongation strain. The amount of thread deflection is dependent on the design of the fastener (thread form, material, etc.) and the load. The loss in strain is essentially this deflection divided by the length of the portion of the fastener being loaded (effective length).

Example: Assume that the threads of a 3 in. diameter stud under a tension load of 325,000 lb would deflect 0.005 in. The joint length is 20 in., and we're seeking 285,000 lb of load in the stud. The calculation is as follows.

$$\text{Effective length of stud} = 20.4 + 1.25 \times 3 = 23.75 \text{ in.}$$

(The 1.25 factor accounts for the additional grip length in the threads at both ends of the joint.)

$$\text{Stress area of 3 in. stud} = 6.51 \text{ in.}^2$$

$$\text{Residual stress} = 285{,}000 \text{ lb}/6.51 \text{ in.}^2 = 43{,}778 \text{ psi}$$

$$\text{Residual strain} = \frac{43{,}778 \text{ psi}}{3 \times 10^7 \text{psi}} = 0.00146$$

(3×10^7 psi = Young's modulus for steel.)

$$\text{Residual elongation} = 0.00146 \times 23.75 \text{ in.} = 0.0347 \text{ in.}$$

$$\text{Tension elongation} = 0.0347 \text{ in.} + 0.005 \text{ in.} = 0.0397 \text{ in.}$$

(0.005 in. = thread deflection.)

$$\text{Tension strain} = 0.0397 \text{ in.}/23.75 \text{ in.} = 0.00167$$

$$\text{Tension stress} = 0.00167 \times 3 \times 10^7 \text{psi} = 50{,}100 \text{ psi}$$

$$\text{Tension load} = 50{,}100 \text{ psi} \times 6.51 \text{ in.}^2 = 326{,}151 \text{ lb}$$

$$\text{Tension design factor} = \frac{326{,}151 \text{ lb}}{285{,}000 \text{ lb}} = 1.14$$

Therefore, in order to achieve 285,000 lb residual load under these circumstances, we would need a tensioner capable of producing 14% more load to account for residual loss.

Factors affecting residual loss include

Nut seating torque: By torquing the nut sufficiently while the hydraulic tensioner is pulling on the stud, the threads can be deformed before the tensioner is released, reducing the load loss.

Coatings: Thick coatings such as paint and galvanized bolts crush when the nut takes over the load. This permits further loss of stud load.

Lubricants: Lubricants permit more "pre" deformation of the nut threads during nut rundown, thereby reducing the residual loss. This also makes the residual load more sensitive to the amount of nut seating torque applied. This can cause increased variation in the final loading.

Figure 25.2 can be used as a guideline to determine the tensioning factor. The x-axis shows the ratio of the effective length of the joint to the stud diameter. Each curve is for a different nut seating torque. The y-axis is the tensioner design factor.

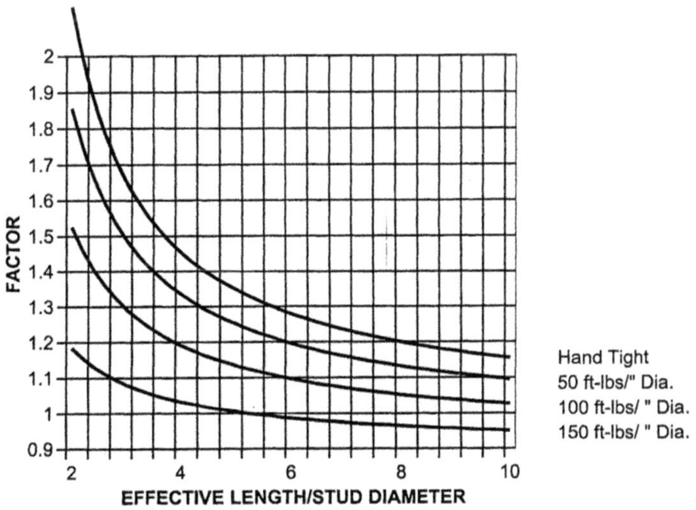

FIGURE 25.2 Tensioner design factor vs. joint length ratio for different nut seat torques.

Example: A 3 in. stud on a 20 in. joint length is required to maintain 285,000 lb load. We can put 150 ft-lb of nut seating torque on the nut. What capacity tensioner is required?

$$\text{Effective length of stud} = 20\ \text{in.} + 1.25 \times 3\ \text{in.} = 23.75\ \text{in.}$$

$$\text{Joint ratio} = 23.75\ \text{in.}/3\ \text{in.} = 7.92$$

$$\text{Seating torque per inch of stud diameter} = \frac{150\ \text{ft} - \text{lb}}{3\ \text{in.}} = 50\ \text{ft} - \text{lb/in.}$$

To determine the factor,

1. Find 7.92 on the x-axis of Figure 25.2. Go up to the curve for 50 ft-lb/in.
2. Go across to the y-axis to determine the tensioner design factor.

$$\text{Factor} = 1.15$$

$$\text{Tensioner capacity} = 285{,}000\ \text{lb} \times 1.15 = 327{,}750\ \text{lb}$$

Caution: This figure should only be used as a guideline. Accurate determination of the factor can only be made through testing.

26 Torque Control of Bolt Tightening

John H. Bickford

26.1 INTRODUCTION

The behavior and life of a bolted joint are, in large part, determined by the magnitude of the force that clamps the joint members together, a force created when we tighten the bolts. Too much clamping force can cause problems. So, more commonly, can be too little. As a result, it is often desirable to measure and/or control the buildup of tension or preload in bolts when they are tightened. The most practical and common way to do this is to use a torque wrench to control the amount of torque applied to the nut or to the head of the bolt. A few of the many types of torque wrenches available are shown in Figure 26.1. In other situations, we might use a more elaborate or powered tool of some sort to generate the torque. A few of these possibilities are illustrated in Figure 26.2. Further discussions of some of these tools can be found in Chapters 25 and 28.

We apply torque to create preload to create a clamping force. We measure torque to control the amount of preload. This means that we must know the relationship between these variables. Regardless of which type of tool we use, the relationship is the same: The preload developed in the bolt is related to the torque produced by the tool by the equation

$$T = KDF_P \tag{26.1}$$

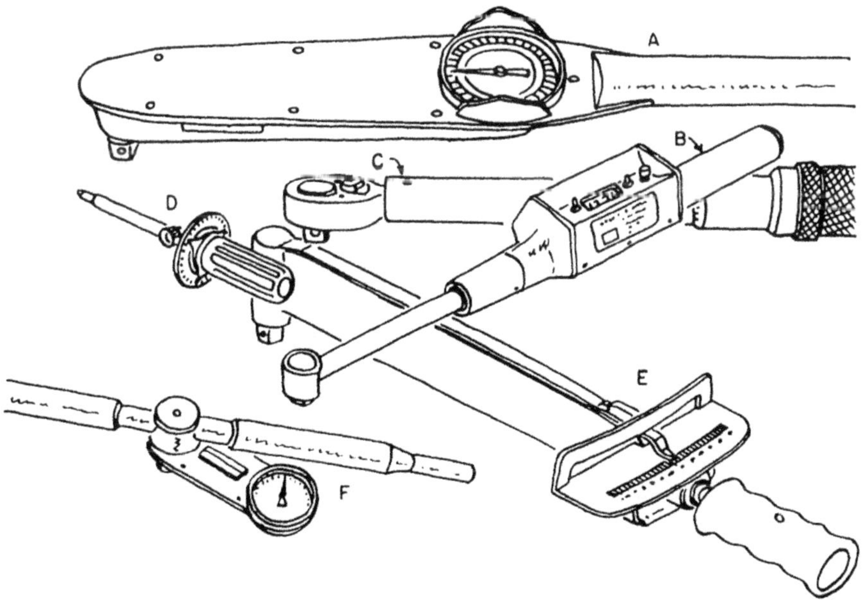

FIGURE 26.1 Manual torque tools come in a wide variety of styles, types, shapes, and sizes, only a few of which are shown here. As far as torque output is concerned, they range from tools such as the "torque screwdriver" (D), which can produce only a few inch-ounces of torque, to wrenches several feet in length (A), which can be used to create torques of many hundreds of foot-pounds (from Ref. [1]).

DOI: 10.1201/9781003501183-31

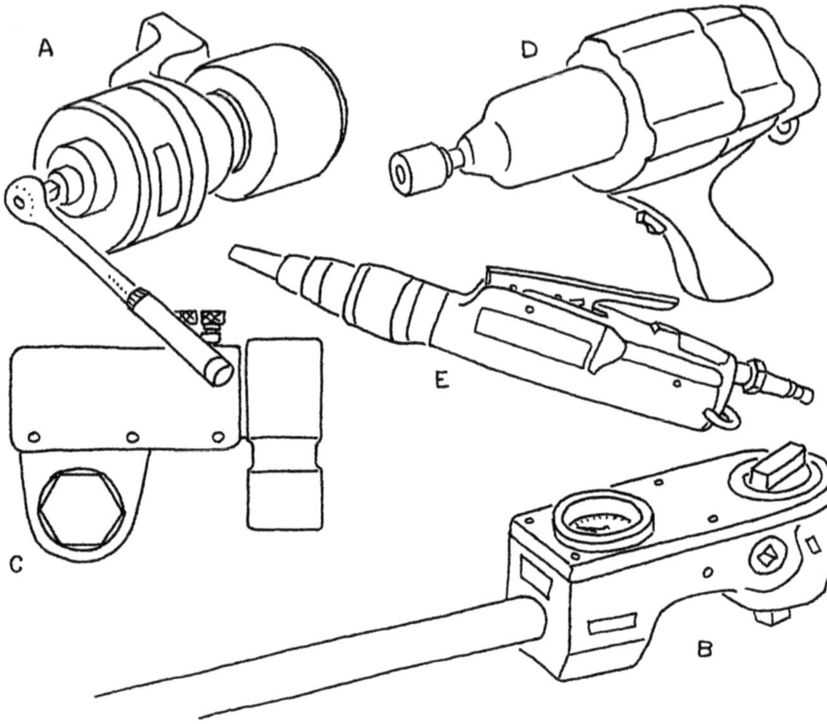

FIGURE 26.2 Many unconventional manual torquing tools are also available, as are a wide variety of powered tools. Those shown here include a "torque multiplier" (a), a geared wrench (b), a hydraulically powered wrench (c), an air-powered impact wrench (d), and an air-powered nutrunner (e). Tools like each of these are available from a variety of sources. Electrically powered tools, using brushless DC motors, are not shown here but have also become popular in recent years.

where

T = the applied torque (in.-lb, N·m)
D = the nominal diameter of the bolt (in., mm)
F_P = the preload developed in the bolt (lb, N)
K = a dimensionless constant called the nut factor.

Equation (26.1) defines a straight-line relationship between torque and preload, as suggested in Figure 26.3. At least the lower part of the relationship is linear. The line curves at the upper end when the heavily preloaded fastener yields. For our purposes, however, the assumption that we're dealing with a straight line is acceptable.

If we knew exactly what value to assign to the nut factor for a given fastener, then we would be able to predict with perfect confidence the slope of the torque–preload line and the amount of preload to be developed in that fastener by a given torque. Unfortunately, as we shall see, the torque–preload relationship is affected by a large number of variables, and this results in a good deal of uncertainty in the nut factor.

26.2 THE NUT FACTOR

The nut factor, K, is an experimentally determined constant. Its value depends primarily on the lubricity of the fastener, i.e., on whether or not a lubricant has been applied and if so, which lube and to which surfaces. The value of the nut factor for a given fastener is also affected by a number

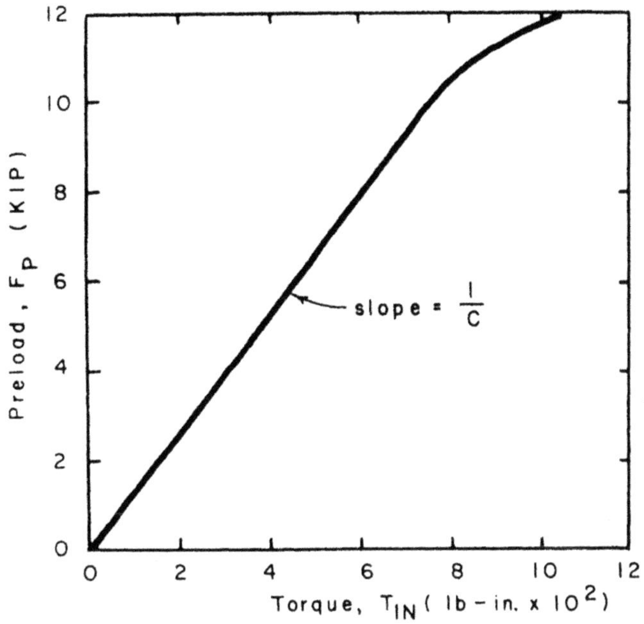

FIGURE 26.3 We can and almost always will assume that the relationship between the torque applied to a fastener and the preload achieved in it is a straight line, as shown by the lower portion of the curve shown here. This portion of the curve is defined by the torque–preload equation $T=KDF_P$. In practice, the "straight-line portion" of the curve can bend upward or downward a bit, for a variety of reasons, but we are rarely able to predict such nuances for a given set of parts (from Ref. [1]).

of other things that can alter the torque–preload relationship, as we see later. Because of this, and because even lubricity will vary from one bolt to another, there is a good deal of variability in the nut factor. For example, although a nut factor of 0.2 is usually assumed for as-received steel-on-steel fasteners, the amount of preload produced by a given torque when we tighten a large number of such fasteners, typically varies by ±30%, and the variation can often be worse than this. This means that the actual nut factor has ranged from 0.14 to 0.23. By "as-received fasteners," I mean those lubricated with only the residue of oil picked up during the manufacturing process.

Fasteners that have been cleaned and dried after receipt have still less lubricious surfaces than the as-received ones and would almost certainly have an average nut factor greater than 0.2. We would also expect them to exhibit more than ±30% scatter in the torque–preload relationship. On the other hand, if a good, high pressure thread lubricant is applied to the nut face and thread surfaces before assembly, then the average nut factor might be as low as 0.10 or 0.15 and the bolt-to-bolt variation in the preload achieved for a given torque might be ±10% or so rather than ±30%. Table 26.1 lists the average nut factors and reported variation in nut factor for a wide variety of thread lubricants. These are reported for various lubes, because the type of lubrication will usually dominate the situation, but these nut factors must also reflect any other factors that influenced the results of the various experiments used to quantify them. In fact, although the nut factor is determined primarily by lubricity, lubricity itself is only partly determined by the lubricant. More generally it is determined by friction, and that is only partially controlled by a lubricant.

Before we leave the subject of lubricants, it is worth noting that even they are subject to variation. A given type of lubricant does not always produce the same coefficient of friction between carefully tailored test surfaces. Brookhaven National Laboratories used heavily loaded metal blocks to test a number of molybdenum disulfide lubricants obtained from many different suppliers, for example, and found that the coefficient of friction (which strongly affects but is not the same as the nut factor) varied in magnitude by something like 10:1 [2]. The NRC tested graphite-based lubricants from a number of sources and found a variation in the magnitude of the coefficient of friction of 3:1 [3]. The range of values given in Table 26.1 in part reflects such findings. See Chapter 3 for a more detailed discussion of thread lubricants and nut factors.

TABLE 26.1
Nut Factors (*K*)[a]

	Reported Nut Factors		
Fastener Materials and Coatings	**Min**	**Mean**	**Max**
Pure aluminum coating on AISI 8740 alloy steel	0.42	0.52	0.62
Electroplated aluminum on AISI 8740 alloy steel	—	0.52	—
As received, mild or alloy steel on steel	0.158	0.2	0.267
As received, stainless steel on mild or alloy steel	—	0.3	—
As received, 1-in.-diam. A490			
as received	—	0.179	—
very rusty	—	0.389	—
with Johnson 140 stick wax	—	0.275	—
Black oxided ⅞ A325 and A490, slightly rusty	0.15	—	0.22
Black oxide	0.109	0.179	0.279
Cadmium plate (dry)	0.106	0.2	0.328
Vacuum cadmium+chromate	—	0.21	—
Copper-based antiseize	0.08	0.132	0.23
Cadmium plate (waxed)	0.17	0.187	0.198
Cadmium-plated A286 nuts and bolts	0.15	—	0.23
Cadmium plate plus cetyl alcohol on A286 nuts and bolts	0.11	—	0.16
Cadmium-plated nuts used with MP35N bolts	0.18	—	0.29
Dag (graphite+binder)	0.16	—	0.28
Dicronite (tungsten carbide in lamellar form)	0.045	—	0.075
Emralon (PTFE+resin)	0.10	—	0.15
Everlube 810 (MoS_2/graphîte in silicone binder)	0.09	—	0.115
Everlube 811 (MoS_2/graphite in silicate binder)	0.09	—	0.115
Everlube 6108 (PTFE in phenolic binder)	0.105	—	0.13
Everlube 6109 (PTFE in epoxy binder)	0.115	—	0.14
Everlube 6122	0.069	0.086	0.103
Fel-Pro C54	0.08	0.132	0.23
Fel-Pro C-670	0.08	0.095	0.15
Fel-Pro N 5000 (paste)	0.13	0.15	0.27
Mechanically galvanized A325 bolts			
as received	0.35	—	0.49
clean and dry	—	0.46	—
slightly rusty	0.36	—	0.39
Mechanically galvanized A325 bolts; lubed with 1 part water, 1 part Jon Cote 639 wax	0.11	—	0.26
Hot-dip galvanized ⅞ A325			
as received	0.14	—	0.31
slightly rusty	0.09	—	0.17
clean and dry	0.09	—	0.37
Hot-dip galvanized ⅞ A325 lubed with 1 part water, 1 part Jon Cote 639 wax	0.10	—	0.16

Source: Ref. [1].
[a] See Chapter 3, pages 45 and 46, for complete table.

How the fastener is lubricated has a major influence on friction and hence on the nut factor. Friction is not the only factor that affects and/or causes uncertainty in the torque–preload relationship, as we shall see, but it is probably the most important factor. Let's take a closer look at it and then look at some of the other things that can affect results when we tighten a bunch of bolts with a torque wrench. Then we shall go on to see what we can do to reduce the problems somewhat and to get the most out of torque control.

26.3 FRICTION: THE BOLT WON'T WORK WITHOUT IT, BUT ...

Contact pressures as high as 250,000 psi are generated between male and female thread surfaces when we tighten a bolt. Similar pressures are exerted by the bolt head and the nut on the joint. These pressures build up as the bolt is tightened, and they in turn result in large friction forces that fight further rotation of the nut or bolt. It has been reported, in fact, that typically about 10% of the torque applied to the fastener is converted to the thing we're after—preload in the fastener. The relationship between the input torque available to create preload and preload itself depends on the geometry of the helical thread and is quite predictable. The rest of the input, however, is absorbed primarily by friction. To predict the amount of preload we will get if we apply a given torque to the fastener, we must predict or at least estimate the amount absorbed by friction.

Anyone who has tried to design a machine component whose performance depends on high friction forces, such as a friction clutch, or a component that will work correctly only if friction forces are minimized, like a linear slide, knows that friction is a very unreliable ally. The friction forces developed between two surfaces depend on such things as the roughness of the surfaces, their hardness, their flatness, the magnitude of contact pressure, the relative velocity between the surfaces, surface temperatures, lubrication conditions, and dozens of other factors. Small, undetectable differences can often make a big difference in the forces, "this time, between this set of parts."

And all of this is true for the threaded fastener. It wouldn't work, wouldn't be able to create a sustained clamping force on the joint, without that all important friction restraint between thread surfaces that keeps the bolt from backing itself out of the nut. On the other hand, the bolt cannot be tightened to a precise preload with a torque wrench because of those same friction forces. Friction is an "essential troublemaker" in bolting.

A study made some years ago, sponsored by the Wright Patterson Air Force Base, identified several dozen variables that each had a statistically significant impact on the torque–preload relationship [4]. Most of these variables, including those few listed below, probably affect friction in one way or another.

The hardness of bolt, nut, joint members, washers
Surface finishes of each part
The type of materials used for each
The thickness, condition, consistency, and type of plating used on each
The manufacturing process used to form the threads (e.g., cut vs. rolled threads)
The thread form
Hole finish, alignment, concentricity
Burrs around edges of holes
Whether or not the holes have been countersunk
Type, number, size, location, and hardness of washers used
The type of tool used to tighten the fastener
The speed of the tightening operation
The fit of the wrench on the nut or bolt
Which was torqued—nut or bolt
Number of times the fastener has been used.

And we must not forget the most important factor: whether or not the thread, nut, and joint surfaces have been lubricated. It is not just "whether or not," either. Wright Patterson found that friction and

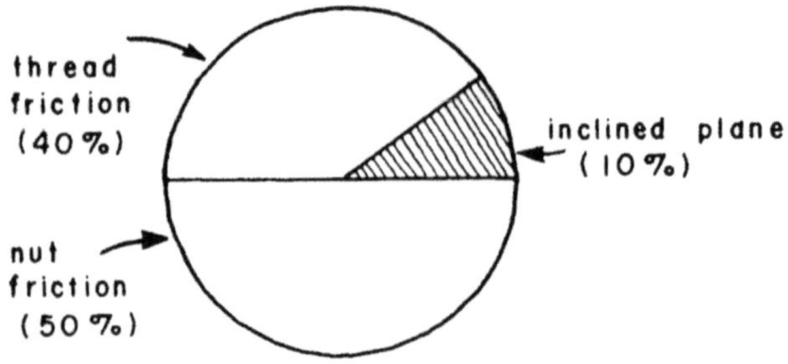

FIGURE 26.4 This diagram illustrates the relative magnitudes of the three reaction torques that oppose the input torque we apply to the nut or bolt. Friction between the face of the nut and the face of the joint creates a torque typically equal to 50% of the input torque. Friction between male and female threads creates a reaction torque equal to 40% of the input torque. Only about 10% of the input torque is converted to tension in the bolt (from Ref. [1]).

lubricity are affected by a long list of lubrication variables—the type of lubricant used, the quantity used, which surfaces it was applied to, the temperature of the lubricant, whether or not it was contaminated, how it was applied, how much was applied, and others.

As a result of all this, we cannot now and probably never will be able to predict the exact magnitude of the friction forces at work on a given set of parts. And this creates a bad "leverage" situation. Consider the situation diagrammed in Figure 26.4, which is based on the results of many studies of threaded fasteners. It suggests that "typically" 50% of the torque we apply to a nut is absorbed by friction between the nut and the joint. Another 40% is absorbed between nut and bolt thread surfaces. Only 10% is converted into preload.

Now see what happens if the friction between nut and joint is 10% greater than "typical" in a given set of parts, perhaps because their surfaces are a little rougher than normal. The torque absorbed at the nut/joint interface increases by 10%, from 50% of the applied torque to 55%. That extra 5% loss doesn't affect the thread friction loss; it only reduces the torque available for preload from 10% of the applied torque to 5% of the applied torque. And that means that the resulting preload, in this fastener, will be only half of that in the hypothetical "typical" fastener. So a 10% increase in friction has caused a 50% drop in preload. Unfortunately, variations in friction far exceeding 10% are commonplace.

26.4 GEOMETRIC FACTORS

There are a number of geometric factors that can also affect friction forces and therefore the torque–preload relationship. Here are some of them.

The fit between threads: Contact pressures between thread surfaces will be determined in part by the area of contact. Tight-fitting threads will create larger contact areas than will loose-fitting threads. Higher contact pressure would mean higher friction force.

Hole clearance: Larger holes with more clearance will reduce the contact area between nut face and bolt head and the joint, increasing contact pressures at those points.

Hole perpendicularity: If the bolt holes are not drilled exactly parallel to the surface of the joint, only the outer edges of the nuts and the bolt heads will contact the surface. This increases the radius of contact; it can also cause the nut, for example, to dig into the joint. Either or both can increase the torque loss at this point. A similar problem is caused if the axis of the nut threads is not perpendicular to the face of the nut.

Included angle of the threads: The faces of unified or metric threads are angled at 60° to the axis of the bolt. If nut and bolt threads are not at exactly the same angle, the area of contact between them will be reduced.

Hole interference: If the holes in mating joint surfaces are misaligned, then the bolt can be in hard contact with at least one wall of one of the holes. As the nut is tightened it must drag the bolt across this wall. The work required to do this reduces the amount available to stretch the bolt to create preload. Incidentally, most hole interference is inadvertent, but airframe manufacturers use interference fit holes on purpose to improve the fatigue resistance of joint members loaded in shear. In these applications the bolts are first pulled through the holes by a special tool before the nuts are torqued.

26.5 OTHER FACTORS THAT AFFECT RESULTS

The factors discussed so far affect primarily the friction forces developed when bolts are tightened. As a result they are the factors that most often cause variation between the applied torque and achieved preload even in well-managed assembly operations. They are not, however, the only factors that can affect the results obtained with a torque wrench. This section discusses a few other possibilities.

26.5.1 Nonparallel or Warped Joint Surfaces

Sometimes bolts are called upon to pull misaligned or nonparallel or heavy joint members together as the bolts are tightened. Preload or tension must be developed in the bolts to do this, but that preload does not create the normal amount of clamping force or pressure between joint members. The resulting joint can slip or leak even though the applied torque created typical preloads in the bolts.

26.5.2 Tool Accuracy

Torque wrenches are equipped with gages or preset devices that theoretically allow us to control the torque being applied to the fastener. But they will not do that correctly unless the gages or presets are of suitable quality, have been maintained in good condition, and have been properly and recently calibrated. Even then all such gages and presets are themselves subject to some inevitable variation or tolerance. Many hand torque wrenches, for example, are sold with specified gage accuracies; ±5% of midscale reading is a common specification, for example. Such a wrench will produce torque—not preload, but torque—with a ±5% accuracy of scale reading at midscale; it will produce torques with a variation greater than ±5% of scale reading at the lower end of the gage scale. The tolerance will be less than ±5% of scale reading at the upper end of the scale. These variations must be added, statistically, to the ±30% or whatever variation in the amount of preload achieved by a given torque.

26.5.3 The Operator

There are also human variables to be considered. Are the mechanics well trained? Are they well supervised? Do they care about what they're doing? Do they know that what they're doing is important and will be affected by their actions? The best thread lubricant coupled with the world's most accurate torque wrench can't compensate for gross errors on the part of the mechanic.

26.5.4 Job Conditions

Job conditions can have a major impact on assembly preloads. Are the bolts easy to get at? It has been said that the preload in most bolts is inversely proportional to their accessibility. Can the operator see what he's doing? Does he have to cope with high or low temperatures or rain or other difficult environmental problems? If so, preload control can be seriously affected.

26.5.5 Postassembly Relaxation

The tightened bolt is an energy storage device. We stretch it a bit when we tighten it, and the friction forces between bolt and nut threads retain that stretch. It is this locked-in stretch that maintains the clamping force with which the bolt grips the joint. In effect, the clamping force depends on potential energy stored in the bolts. Unlike a welded or bonded joint, the bolted joint is "alive," filled with energy. But it's not happy in this condition. It would like to relax, to shed that energy, and there are several ways in which it can do this. These include embedment as heavily loaded metal parts settle into each other, the creep of metal parts or gaskets, differential thermal expansion and stress relaxation, all of which are described in Chapter 23. Suffice it to say at this point that the final results we get with a torque wrench will also be affected by some or all of these relaxation mechanisms, all of which contribute to what is often called torque loss.

26.5.6 Interactions between Bolts

Everything discussed to this point affects the results we get when we tighten a given bolt because our subject is the relationship between the torque applied to the fastener and the resulting preload. To be complete, however, it is important to realize that results are further scattered when we tighten a *group* of bolts in a given joint. Let's say, for example, that we've tightened one bolt in a pressure vessel flange and that we created, in this bolt, the exact amount of preload predicted with the torque–preload equation and a "typical" nut factor. The bolt has been stretched a few thousandths of an inch as we tightened the nut. By tightening this bolt we have also pulled the joint together at this point on the flange and have partially compressed the joint members and the gasket that lies between them. When we tighten the bolts adjacent to this first bolt, however, we further compress the joint members and gasket. This allows the first bolt to relax a little. This process, called elastic interaction, can result in final preload scatter not of the ±30% or so noticed when we tighten as-received bolts one by one, but in max-min preload ranges of 4:1 or 10:1 or 20.1 or worse in a multibolt joint, regardless of the quality our torquing tools and procedures. Elastic interactions are discussed at length in Chapter 24.

26.6 WHAT TORQUE SHOULD WE USE?

Perhaps the most popular question in bolting is, What torque should I use on this fastener? There are many possible answers to that question.

26.6.1 Simple Procedures for Relatively Unimportant Joints

If this is a run-of-the-mill joint and occasional joint failure would be acceptable, then the easiest way to pick a torque is to use Eq. (26.1) and a suitable nut factor. I suggest that you use a factor of 0.2 unless you're planning to apply lubricant to the bolts. If you are going to lubricate them, use the average value of nut factor given in Table 26.1 or a value from Chapter 3.

As an alternative, use a torque taken from one of the many "torque tables" published by fastener or wrench manufacturers. Some of these tables are based on special experiments run by the manufacturer, but most of them, I believe, have been computed by using Eq. (26.1) and a K value of 0.2.

26.6.2 Creating a "Better" Nut Factor for a More Important Job

If you've had some problems with this joint and/or would prefer not to have any joint failures, then it would be appropriate for you to determine the "correct" nut factor for your application. Remember that the nut factor is not a "constant" for a given type of lube. It varies from supplier to supplier. And it is also affected by all the other variables we've considered: tools, operators, job conditions, etc. It is usually possible, therefore, to come up with a nut factor that is more accurate for your job than one obtained from a table. To do this you must use strain gaged bolts or load washers or ultrasonics or micrometers to measure the actual preloads achieved in a sample of bolts when you tighten them (see Chapters 30, 34, and 45 for details of these techniques). If possible, tighten them using your normal production tools, people, and conditions. Tighten as many as you can afford, but at least 25. Use a pocket calculator or a computer to find the mean value and the population standard deviation (σ_n) of the resulting data. The mean will be your average nut factor. The scatter can usually be assumed to equal ±3 times the standard deviation. If you must settle for a sample size significantly less than 25, then it is better to compute the scatter using the sample standard deviation (σ_{n-1}). One note for the statistically inclined: The distribution of preload obtained for a given torque is usually skewed right; this is not a normal distribution.

26.6.3 Basing Your Choice on Prior Experience

A really "good" torque is very difficult to predict because of the large number of variables that affect the results during assembly and the even larger number that act to alter assembly preloads when the joint is placed in service. One of the best ways to select a torque, therefore, is through prior experience. If you know the torque you have used in the past, and the results have been satisfactory, continue to use it. No amount of theory can lead you to a better value.

If you know the torque and results have been unsatisfactory, then you should either increase or decrease the torque by a small, recorded amount. For example, try 10% more torque next time if your problem has been vibration loosening, bolt fatigue, leakage from a gasketed joint, or joint slip. Use 10% less torque next time if your problem has been a crushed gasket, galled joint members, or fasteners that break at elevated service temperatures. Repeat these incremental changes until you experimentally find an appropriate torque.

If you don't have historic torque values to start this process, contact the designer or equipment supplier for a recommendation based presumably on the prior experience of others. If no recommendation is available—and that is often the case—use Eq. (26.1) or a torque table for starters.

26.6.4 When Success Is a Must

If you are dealing with a safety-related or economically critical joint, then it is unwise to use any of the above methods—with the exception of good, well-documented, prior experience— to pick a torque. The question must now start with the deeper question, What preload and in-service tension do we want in these bolts? This is a design question and should be answered by a qualified engineer (there is more on this in Chapter 36). Hopefully, the person who answers this question will also recommend a torque, will specify whether or not a lubricant should be applied to the bolts (and, if so, which one), and will even suggest the sequence in which the bolts should be tightened. He may, of course, also suggest some way to control assembly preloads other than by torque.

26.7 OPTIMIZING THE RESULTS OBTAINED WITH TORQUE CONTROL

Although many factors affect the preload results we obtain with a torque tool of some sort, torque remains the most practical type of control for most applications. It is less expensive than most other types of control, it is relatively simple to apply, and a wide variety of torque tools are available. Since most bolted joints are assembled either with no form of preload control or with torque control of preload, the joints have been designed to tolerate the inevitable uncertainty in preload. Based on past experience with similar joints, or on "gut feel," designers have used larger bolts and/or more bolts than would be required if preload could be controlled with greater precision. Joint members have also been "overdesigned" to accommodate the larger or more numerous fasteners. As a result, we'll get enough clamping force to hold this joint—this part of our machine or product—to hold it together in spite of the uncertainties.

There are also things that we as assemblers can do to reduce the uncertainties a bit and thereby make it more likely that our joints will behave themselves when put in service. The following discussion is taken from Ref. [1].

The things listed below can often make more improvement than the use of elaborate or expensive preload control equipment. As you can see, most of the items on the list would tend to make your assembly practices and procedures more consistent. Although you can't predict or control the many variables affecting results, you can reduce their variation by being consistent.

1. Be as consistent as possible in your choice and use of tools, procedures, calibration frequency, etc.
2. Train and supervise the bolting crews. This can help far more than a "better" lubricant or more expensive tool. Let the people know how important good results are and how difficult they are to obtain. Enlist their help.
3. Make sure the fasteners are in reasonable shape. Wire-brush the threads if they're dirty and rusted. Use stainless steel bristles on alloy steel materials. Chase threads with a tap or die if they're damaged.
4. Use hardened washers between the turned element (nut or bolt head) and joint members.
5. If lubricants are to be used, make sure they're clean and fresh. Apply them consistently, the same amount to the same surfaces by the same procedure. Preload scatter will be minimized if lubes are used on both thread and nut (or head) contact surfaces.
6. Run the nuts down by hand. If you can't, the threads may need to be cleaned or chased.
7. Hold wrenches perpendicular to the axes of the bolts.
8. Apply torque at a smooth and uniform rate. Avoid a "stick-slip" situation as you approach the final torque. If necessary, back the nut off a little so that the specified torque can be reached with the wrench in motion.
9. If hydraulically powered wrenches are used, be sure that adequate reaction points are used and that the tools aren't twisting or cramping as a result of cocked or yielding reaction surfaces.
10. Snug the joint first, with a modest torque; then tighten it. Try to align the joint members before tightening the bolts.
11. Tighten from the center of bolt patterns toward the free edges if the bolt pattern is rectangular (as in a structural steel joint). Work in a cross-bolting pattern on circular or oval joints.
12. Keep good records of the tools, operators, procedures, torques, and lubricants used. Bolting is an empirical "art" at present. If you record details of your "experiment" you'll have the information you need to make a controlled modification in the procedure the next time if the first procedure doesn't give the results you want.
13. If possible, develop your own nut factors, experimentally, rather than relying on a table. Perform the experiment under actual job conditions if possible, though a lab test on your joint (or a simulation of your joint) is better than nothing.

REFERENCES

1. Bickford, J. H., *An Introduction to the Design and Behavior of Bolted Joints*, 3rd ed., Marcel Dekker, New York, 1995.
2. Czajkowski, C. J., Investigations of corrosion and stress corrosion cracking in bolting materials on light water reactors, *Int. J. Pressure Vessel Piping*, 26: 87–96 (1986).
3. Anderson, W., and P Steiner, Degradation of threaded fasteners in reactor coolant pressure boundary of pressurized water reactor plants, U.S. Nuclear Regulatory Commission, NUREG-1095, May 1985.
4. Stewart, R., Torque/Tension Variables, List prepared at Wright-Patterson Air Force Base, April 16, 1973.

27 Torque–Angle Control of Bolt Tightening

Ralph S. Shoberg

27.1 INTRODUCTION

When torque control is used as the method for tightening a fastener, there is no way to be 100% certain that the desired tension will be created [1–8]. Using installation torque alone to control the process always introduces an element of "statistical gambling" into the assembly process. Installation torque measurements that are not backed up with simultaneous angle-of-turn measurements cannot be totally relied upon to ensure that proper fastener installation has been accomplished.

For bolted joints where safety and reliable performance are dependent on proper initial tension, both torque and angle of turn must be monitored and controlled during the tightening process. As each fastener is installed, the torque–angle tightening signature of the bolted joint should be compared to established assembly process limits to ensure that the specified assembly preload has been achieved.

The fundamental tightening procedure for torque–angle tension control is simply defined as follows:

1. Torque is applied until a specified "threshold" level is attained.
2. An additional angle of turn is then applied to finish the installation.

The engineering analysis, test measurements, and installation methods outlined in this chapter can be applied to virtually any bolted joint. A unified approach, starting with basic assumptions, is used to develop a reliable torque–angle control procedure for achieving specified clamp loads.

The test measurements used to establish the torque–angle process parameters can also be used as the basis for additional analysis. Torque–angle signatures can also be used to help verify joint strength, safety factors, and fatigue strength. They provide an extremely practical means to verify bolted joint design and engineering calculations.

The M-alpha diagram,[1] presented in detail in Section 27.8, illustrates the effect of friction coefficients on the torque–angle signature. The effect of the thread friction coefficient on the clamping force developed at the onset of bolt yield is also clearly illustrated by the use of the M-alpha diagram.

27.2 CLAMPING FORCE/STRENGTH CONSIDERATIONS

The clamping force and preload requirements for a bolted joint are determined by the static and dynamic loads that the assembly is expected to see in service. The bolted joint design must be completely engineered with regard to the axial (concentric), eccentric, and side shear loads to which the assembly will be subjected. This is the first step in any fastener engineering project.

After the external working loads have been defined, the necessary bolt preload can be calculated. Next, the safety factors against embedment and thread stripping must be checked to ensure that yielding in the bearing areas or threads will not limit the preload to less than the required amount.

The safety factors for embedment and thread stripping are important both for the initial installation of the fastener and for long-term reliability with regard to both loosening and fatigue resistance. Figure 27.1 shows three of the strength factors that should be evaluated with regard to expected service loads and preload requirements.

DOI: 10.1201/9781003501183-32

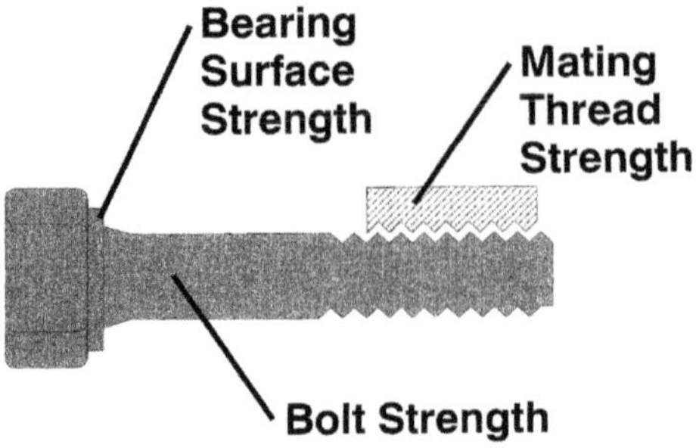

FIGURE 27.1 Basic geometric and material strength factors.

27.3 TORQUE–ANGLE SIGNATURE

The basic torque–angle tightening signature is used as a starting point for all analyses. The influence of underhead and thread friction on the tightening process is easily illustrated by this model curve. An increase in friction, in either the thread or underhead regions, results in a proportional increase in the slope of the torque–angle signature. The tension–angle coefficient for each bolted joint must be determined in order to establish the control parameters for torque–angle tension control. The R_{FM} slope from the FM-alpha (clamping force angle) diagram corresponds to the tension–angle coefficient for the tightening curve.

If the assembly tool is shut off at a specified angle of turn after the threshold torque is attained, the scatter in achieved tension will be much less than the scatter observed for the same fasteners tightened with torque-only control. For this process to work reliably, it is necessary that the threshold torque level for starting angle counting be set at a level that is above the alignment zone of the tightening process.

The curves in Figure 27.2 illustrate the basic principle of torque–turn tension control. A key element in understanding how bolted joints function is the concept of the "elastic origin."

The "elastic origin" is located by the projection of the tangent to the torque–angle signature curve to the zero torque or prevailing torque level. In the elastic tightening of the assembly process, bolt tension, or clamping force, is directly proportional to the angle of turn from the elastic origin.

A number of different methods can be used to determine the angle–tension coefficient for the bolted joint. A basic assumption is that as the fastener is turned to develop tension in the joint, the fastener stretches and the clamped parts compress elastically according to the effective spring rates of the fastener and the clamped parts.

After the tension–angle coefficient is determined for elastic clamping, it is relatively easy to estimate the tension achieved when tightening beyond the bolt yield point. The yield point must be analyzed to determine if the yield detected is in the bolt, or in the joint, or is possibly a result of thread stripping or underhead embedment.

After the installation process has been defined and implemented, methods must be specified to audit the results in order to verify that the process has achieved the desired fastener preload. Process audit procedures including the release angle measurement method and hand torque breakaway audits are presented in Chapter 43 of this book.

27.4 MODELING THE TIGHTENING PROCESS

The process of tightening a fastener involves turning (advancing the lead screw) and the application of torque (turning moment) so that preload (tension) is produced in the fastener. The desired result is a clamping force to hold the components together. The most general model of the fastener tightening process has four distinct zones as illustrated in Figure 27.3.

Zone 1 is the rundown or prevailing torque zone, which occurs before the fastener or nut contacts the bearing surface. Prevailing torque due to thread-locking features such as nylon inserts or

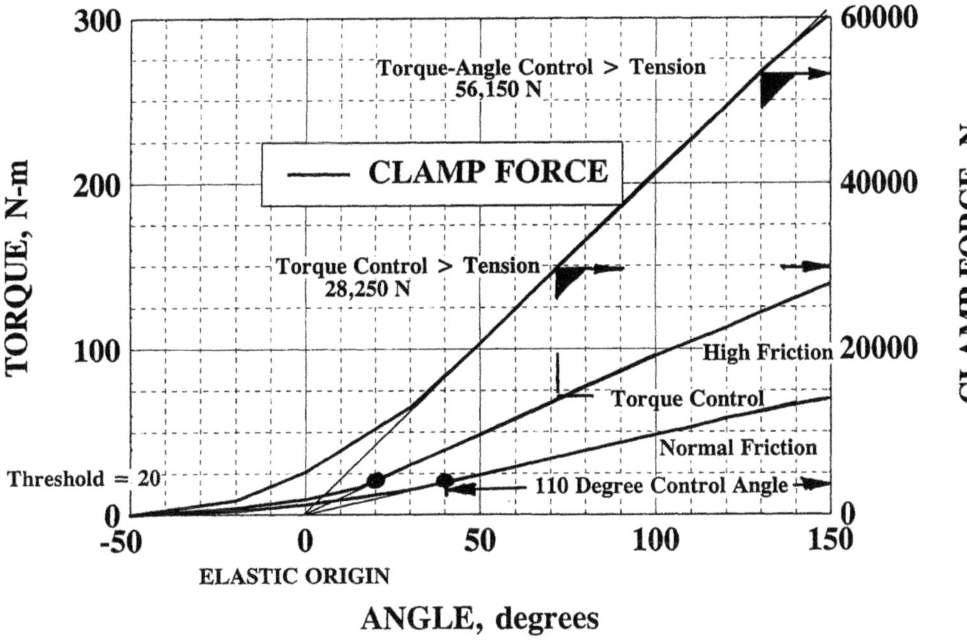

FIGURE 27.2 Torque–turn tension control principles.

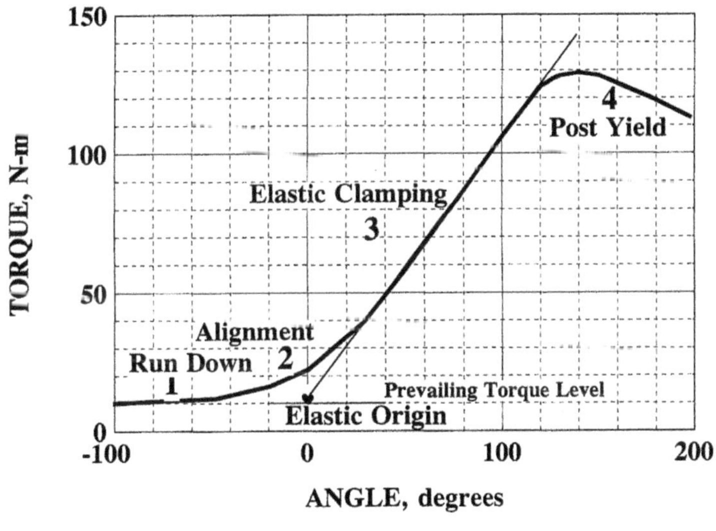

FIGURE 27.3 Four zones of the tightening process.

deformed threads will show up in the rundown zone. Frictional drag on the shank or threads due to misalignment of parts, chips, or foreign material in the threads as well as unintended interference due to out-of-tolerance threads are additional causes of prevailing torque in the rundown zone.

Zone 2 is the alignment or snugging zone, wherein the fastener and joint mating surfaces are drawn into alignment, or a stable clamped condition. The nonlinear alignment zone is a complex function of the process of drawing together the mating parts and bending of the fastener as a result of nonparallelism of the bearing surface and fastener underhead surface. In addition to the macro effects related to the alignment of parts, there are micro effects within the alignment zone. The micro effects include contact stress induced deformations of plating and coatings as well as local surface roughness and thread deformation.

Zone 3 is the elastic clamping zone, wherein the slope of the torque–angle signature curve is constant. The elastic clamping zone torque vs. angle slope is a very important characteristic of each bolted joint. This slope can be projected backward to locate the elastic origin. Angle of turn from the elastic origin is multiplied by the tension–angle coefficient to calculate the tension that has been created by the tightening process.

The elastic origin is located at the intersection of the prevailing torque level and the backward projection of the elastic clamping zone.[2] To further illustrate the concept of the elastic origin, the torque–angle signatures in Figure 27.4 show the increased slope, induced by increased friction, in the elastic tightening zone. Note that as friction increases, the torque required to bring the bolt to yield is also increased.

The curves in Figure 27.5 show that as friction increases the clamping force at the yield point is reduced, while the torque required to reach the yield point increases. This illustrates the fact that for a given fastener size, the torque required to yield the bolt is a function of the material yield strength and the thread friction coefficient.

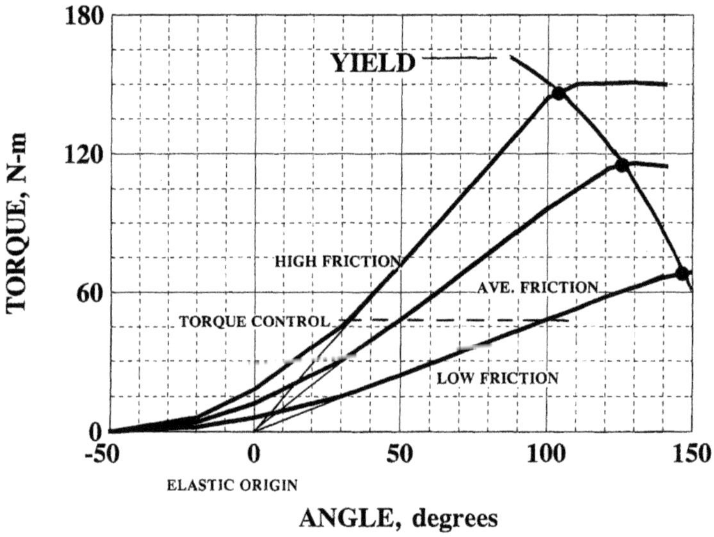

FIGURE 27.4 Effect of friction on yield point.

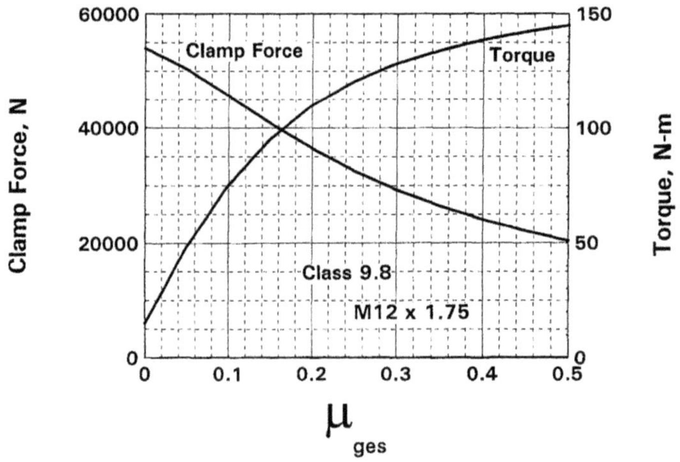

FIGURE 27.5 Effect of friction on applied torque and clamping force.

Once the torquing effort has been stopped in the elastic clamping zone, a line tangent to the straight-line portion of curve can be projected backward. The point where the projection line strikes zero torque, provided there is no prevailing torque, is called the elastic origin. If the angle of turn is measured from the elastic origin to the point where torquing was stopped in the elastic clamping zone, it has been found that the tension in the fastener is directly proportional to that angle of turn.

What is happening is that the compression of the parts and the stretching of the fastener are occurring in a linear fashion from the projected elastic origin. Even if friction between threads or in the underhead region of the fastener is varied, it still will be found that within the elastic region the tension generated is always proportional to the angle of turn from the elastic origin.

When there is prevailing torque in the initial rundown of a fastener, the line projected back along the elastic curve must stop at the prevailing level, and the elastic origin will be set there rather than at zero torque. Then the tightening angle is the angle from this elastic origin to the point where tightening was stopped in the elastic region.

Zone 4 in Figure 27.4 is the post-yield zone, which begins with an inflection point at the end of the elastic clamping range. Yielding can occur in the bolt or in the joint assembly as a result of underhead embedment or as thread stripping in the bolt or mating threads. The yield point can be used to establish or verify the tension–angle coefficient for the torque–angle–tension tightening process.

27.5 TORQUE–TENSION COEFFICIENT: NUT FACTOR, *K*

The basic equation, $T=KdF$, can be applied to the linear elastic clamping zone of the assembly tightening process. The prevailing torque during initial rundown must be subtracted from the total applied torque, a procedure that is usually accurate for all practical purposes. If insufficient torque is applied to tighten beyond the alignment zone, the tension estimated for a given assembly will still be correctly predicted if the overall K factor has been accurately estimated.

There are published tables of K factors for various combinations of materials, surface finishes, plating, coatings, and lubricants. However, experience has shown that it is highly unreliable to assume that any given K value applies to a specific fastener being assembled. The importance of the torque–angle approach to assembly cannot be overemphasized when you are responsible for the safety and reliability of critical bolted assemblies. It is not uncommon to see variations in friction coefficients of 2:1 or more as the same fastener is repeatedly tightened and loosened. The most practical way to minimize the variation in clamping force for bolted assemblies is to use both torque and angle of turn in your assembly process.

The factor K, often referred to as the "nut factor," can be expressed as a combination of three factors:

1. K_1—a geometric factor, a function of diameter d and pitch P
2. K_2—a thread friction factor, a function of μ_G and pitch diameter d_2
3. K_3—a bearing friction factor, a function of μ_K and D_{km}

The curve shown in Figure 27.6 illustrates the value of the nut factor K as a function of the average friction coefficient. Both the underhead friction coefficient μ_K and the thread friction coefficient μ_G are assumed to be equal to μ_{ges} for this example. See Glossary of Important Terms for definitions of variables.

27.6 EXPERIMENTAL DETERMINATION OF FRICTION COEFFICIENTS

Just as tabulated values of K are not reliable for application with any assurance of predictable results for a given bolt or nut assembly, it is equally true that friction coefficients determined experimentally

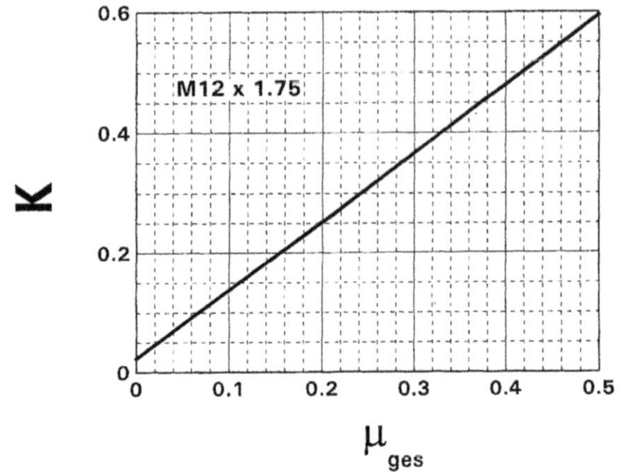

FIGURE 27.6 Nut factor, *K*, as a function of friction.

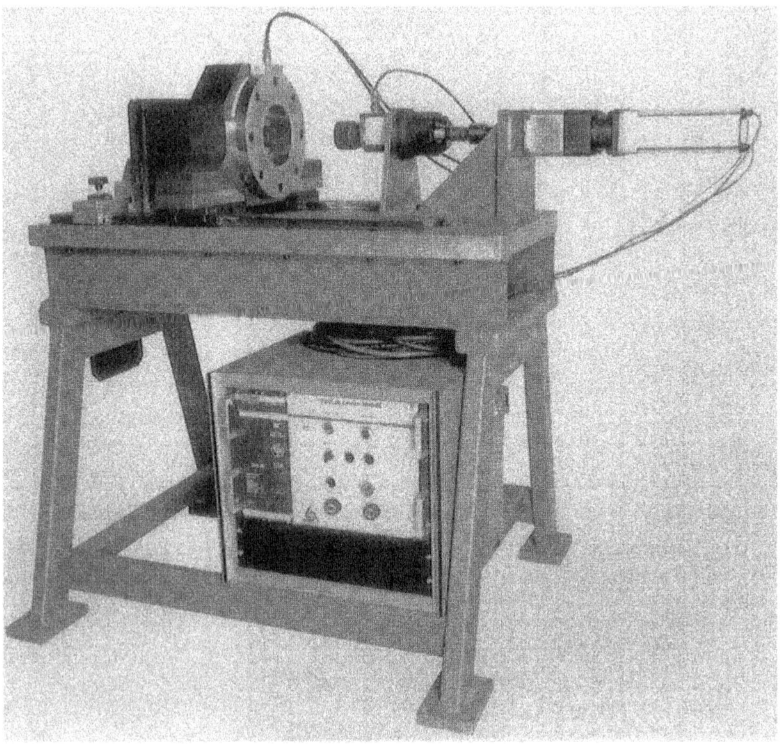

FIGURE 27.7 A typical laboratory fastener test machine.

must be used for comparison purposes and supplemented by torque–angle assembly monitoring to ensure proper assembly clamping forces.

The German norm DIN 946 (Determination of coefficient of friction of bolt/nut assemblies under specified conditions, October 1991) provides a practical standard method for the evaluation of friction coefficients on screw threads and underhead bearing surfaces. Fastener laboratory testing machines are available for the determination of friction coefficients according to DIN 946.

A fastener laboratory test machine for testing fasteners up to M36 size is shown in Figure 27.7. The largest existing machines of this type are capable of testing fasteners up to M100 with clamp

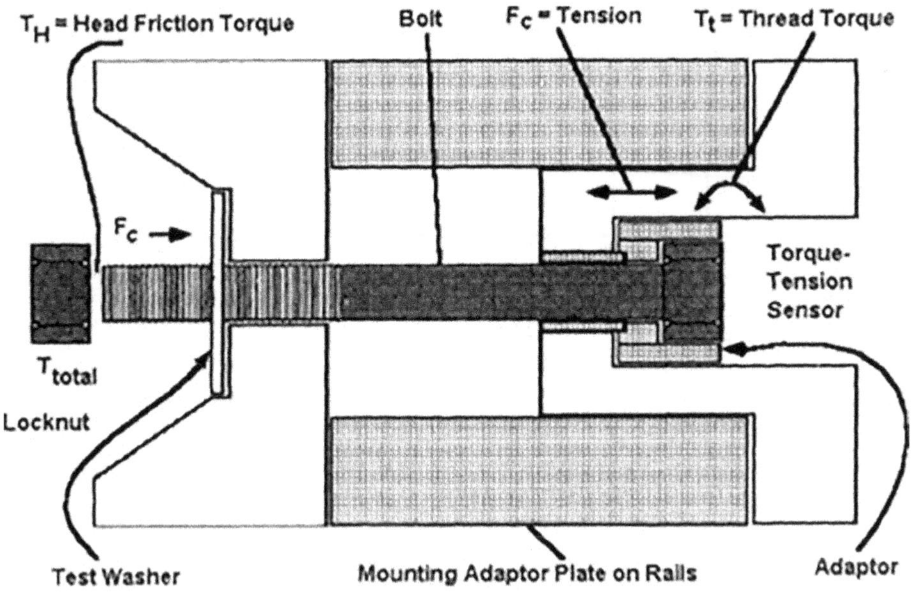

FIGURE 27.8 Torque–tension research head

load capacities of up to 500,000 N and torque capacities of 35,000 N·m. Small table-top units are available for testing fasteners of sizes below Ml (English fasteners down to 0–80).

27.7 THREAD AND UNDERHEAD FRICTION COEFFICIENT MEASUREMENTS

In the development of threaded fastener materials, surface finishes, plating, coatings, and thread-locking adhesives, it is necessary to measure and control the friction coefficients on both the threads and underhead regions of the fasteners.

A torque–tension research head (Figure 27.8) is a special load cell constructed to simultaneously measure both thread torque (pitch torque plus thread friction torque) and the tension created as the tightening torque is applied. Special devices such as locknuts, serrated underheads, thread-locking adhesives, and friction patches can be tested, and friction performance standards can be established and maintained through measurements made with the torque–tension research heads. The research head has the capability to fully test for friction coefficients according to DIN 946.

The average coefficient of friction, μ_{ges}, can be roughly estimated by measuring the change in length of a screw or bolt after application of a given tightening torque. This method can be applied only if there is no yield occurring in the bolt, threads, or elsewhere in the joint.

27.8 M-ALPHA DIAGRAM

The M-alpha (torque angle) diagram (Figure 27.9) is the straight-line projection of the tangent to the torque–angle assembly curve as projected backward from the final elastic tightening point to zero torque, or the prevailing torque level. This tangent projection is used to locate the elastic origin. Within the elastic tightening zone of the assembly process, the clamping force developed is directly proportional to the angle of turn from the elastic origin.

Since the M-alpha diagram was first developed as an extension of the SR1, VDI 2230 bolted joint design software, we have chosen to retain the terminology of the VDI standard glossary of terms (i.e., *M* for torque). Thus, we have torque–angle measurements from testing corresponding to the *M*-alpha diagram calculated as part of the SR1 VDI 2230 design software.

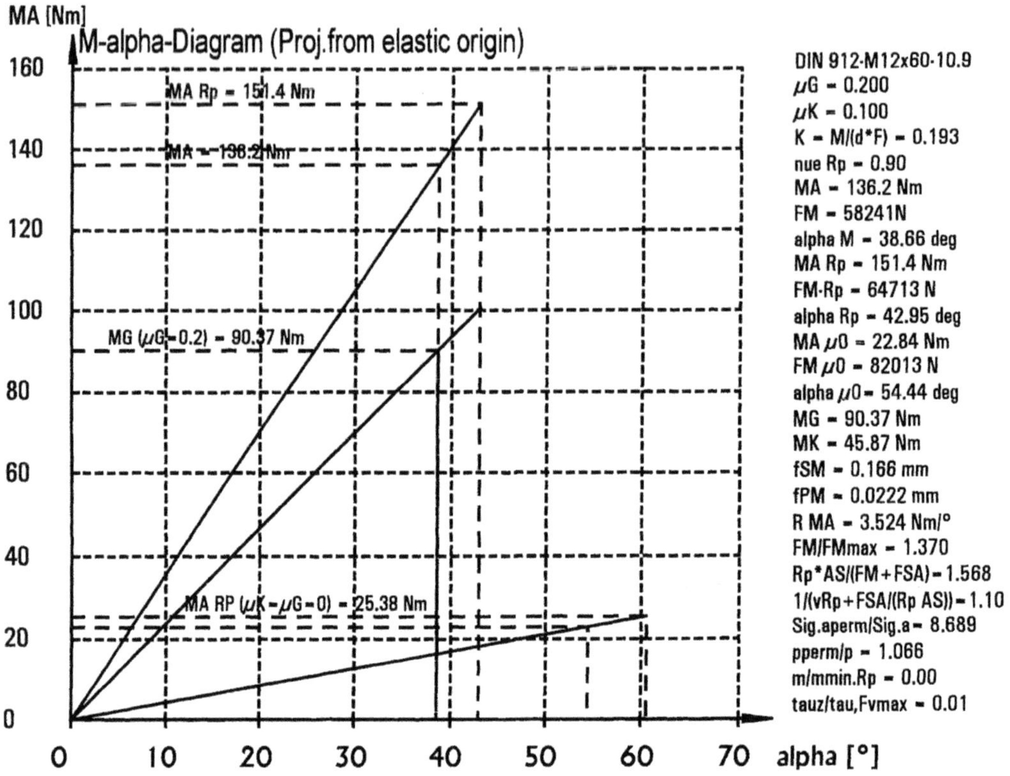

FIGURE 27.9 M-alpha diagram (from SR1 Bolted Joint Design Software © Hexagon, 1992–97.)

In addition to the applied torque MA, the M-alpha diagram has projections from the elastic origin for both the thread torque MG and the pitch torque MA_{RP} (where $\mu = 0$). A very useful feature of the *M*-alpha diagram is the easy manner in which the diagram clearly illustrates the distribution of the torque in a tightening process. With MA showing the total input torque, MG represents the thread torque, which is the thread friction plus the pitch torque that creates the clamping force. The difference between the MA and MG curves represents the underhead friction torque. The difference between the pitch torque curve and the MG curve represents the thread friction torque.

The SR1 M-alpha diagram is a straight line projected from the elastic origin to the yield point. By changing the thread friction (μ_G) and underhead friction (μ_K) coefficients assumed for the VDI 2230 analysis, the effect of friction on the tightening process can be clearly seen on the *M*-alpha diagram.

27.9 FM-ALPHA DIAGRAM

The FM-alpha (clamping force angle) diagram (Figure 27.10) is a new feature that was added to the SR1-VDI 2230 design software. This diagram illustrates the clamping force (F_m) vs. angle of turn (α) from the elastic origin. The slope of this curve in newtons per degree establishes the tension–angle coefficient for a specific bolted joint design.

The upper limit of the elastic clamping range is the yield point, which is calculated on the basis of the specified thread friction coefficient (μ_G) and the material tensile yield strength, i.e., M10.9, 940 MPa. The maximum shear strength Mohr circle calculation is used to determine the combined tension-torsional yield load.

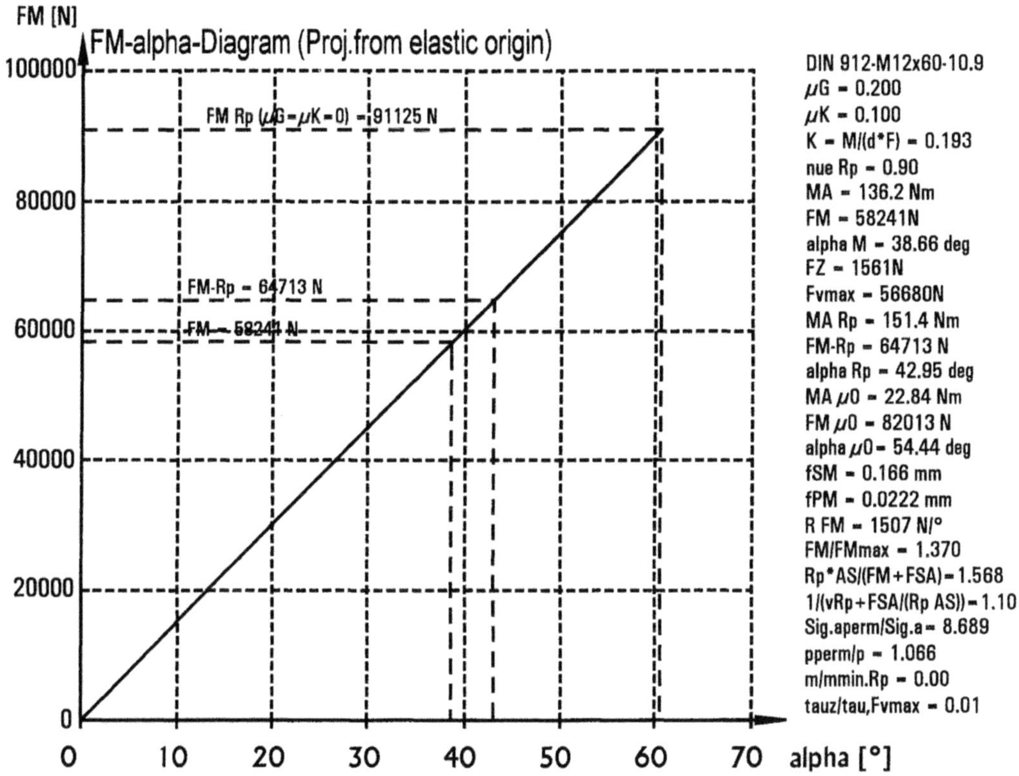

FIGURE 27.10 FM-alpha diagram (from SR1 Bolted Joint Design Software © Hexagon, 1992–97.)

27.10 ESTIMATING THE TENSION–ANGLE COEFFICIENT

The tension–angle coefficient can be estimated by several methods, each of which can be used to confirm the estimate made by one of the others. The methods are

1. Ultrasonic stretch
2. Strain gaged bolt
3. Force washer
4. Model calculation
5. Material property—Yield

Basic guidelines for these methods are given in the following sections.

27.10.1 Ultrasonic Stretch

The ultrasonic stretch method starts with ultrasonic stretch or tension calibration of the test bolt. After the bolt is calibrated, it is installed in the assembly and the torque–angle signature is recorded. The ultrasonic measurement is then made to determine the stretch or tension developed by the installation torque. Provided that the fastener was not stressed beyond the yield point and that no yield was induced in the threads or underhead, the tension reading can be used to calculate the tension–angle coefficient.

The elastic origin is located on the torque–angle signature curve recorded when the fastener was tightened. The clamping force measured by the ultrasonic stretch measurement is divided by the angle of turn from the elastic origin to determine the slope of the FM-alpha curve. Once this information is available, it is possible to confirm the original bolted joint design with detailed finite element analysis calculations or through the use of the SR1 design software.

27.10.2 Strain Gaged Bolt

The strain gaged bolt method of determining the tension–angle coefficient starts with the calibration of the bolt prior to its installation in the assembly. The calibrated bolt is tightened in the assembly, and the torque–angle signature is recorded. Provided that only elastic tightening has occurred, i.e., no yield is detected, the tension reading from the test bolt can be used to determine the tension–angle coefficient.

The torque–angle signature is evaluated to locate the elastic origin. The elastic tightening angle from the elastic origin to the final tightening torque is divided into the measured clamp load to determine the tension–angle coefficient.

If a two-channel transient recorder, such as the RS Technologies Model 960 (Figure 27.11) is used to record the torque–angle signature, you should be able to directly plot torque–tension, torque–angle, and force–angle curves. M-alpha and FM-alpha curves are standard plots generated by the Model 960 based on the recorded torque–angle signatures.

The use of a strain gaged bolt will result in a modification of the bolt stiffness unless the bolt is dimensionally the same as the ungaged bolt. Calculation of the tension–angle coefficient should take into account any estimated change in stiffness for the bolt from the fasteners to be assembled in the normal assembly process.

FIGURE 27.11 Two-channel transient recorder.

The friction coefficients for the thread and underhead regions of the fastener will most likely not duplicate the conditions on the normal production parts. This should not be seen as a major problem as long as we are focusing on obtaining the tension–angle coefficient. If we were trying to obtain a torque–tension coefficient, the futility of the effort should be obvious to anyone familiar with how fastener friction affects the torque–tension relationship.

27.10.3 Force Washer

The use of a calibrated force washer to determine the tension–angle coefficient for a bolted joint requires some extra calculations to estimate the effect of the increased bolt length and the spring rate of the force washer in parallel with the clamped parts. The increase in angle of turn due to the longer bolt and softer joint when using a force washer means that the actual tension–angle coefficient will be larger than when tightening the assemblies with no force washer present in the joint.

The torque–angle signature as well as the torque–tension curves can be recorded. The tension vs. angle curves and FM-alpha curves can be directly plotted from the transient recorder. Using the SR1 design software, the increased length of the bolt and the stiffness of the force washer can be introduced into the design calculation model.

By adjusting the shape and stiffness of the clamped elements in the SR1 design software, the slope of the calculated FM-alpha plot can be matched to the actual measured curve as plotted by the transient recorder. Since clamp load is proportional to angle of turn from the elastic origin, calculations for the FM-alpha diagram are independent of the friction coefficients up to the bolt yield point.

Once we have refined the joint model with the force washer in place so that the measured FM-alpha curve is matched by the model, we can shorten the bolt and remove the force washer element from the model. The final model will now allow us to directly calculate the tension–angle coefficient. The resulting M-alpha diagram will now show the effect of friction coefficient variations on the slope of the elastic tightening region of the torque–angle signature. The effect of thread friction on the yield load can also be demonstrated.

Of course, once the bolted joint is fully modeled by a combination of experimental and theoretical calculations, the SR1 design software automatically calculates all of the safety factors according to the VDI 2230 guide.

27.10.4 Model Calculation: Estimate of Tension–Angle Coefficient

The model calculation method in its simplest form involves selecting a value for the nut factor, K, which is used to calculate an initial estimate of the clamping force expected for a given torque. While it is acceptable to simply pick a value for K to be used in preliminary calculations, a more precise model can be developed if coefficients are estimated for thread and underhead friction and K is calculated based on these assumptions. The formula for the calculation of K is

$$K = K_1 + K_2 + K_3$$

where

K_1 = the geometric function of diameter d and thread pitch P
K_2 = thread friction factor, a function of the thread friction coefficient μ_G and the pitch diameter d_2
K_3 = bearing friction factor, a function of the underhead friction coefficient μ_K and the mean bearing or underhead diameter D_{km}.

Using an effective length of the bolt equal to the grip length plus one diameter, the bolt stretch due to the estimated clamping force is calculated. For a given thread pitch, the angle of turn required to provide the desired stretch is calculated.

The torque–angle signature is obtained while tightening to the specified torque. After locating the elastic origin on the signature plot, it is possible to determine the total tightening angle, which includes bolt stretch and compression of the clamped parts.

The tension–angle coefficient is estimated as the clamping force derived from the basic $T = KdF$ formula using the assumed value of K, with the angle of turn from the elastic origin obtained from the test signature curve.

From this information, it is possible to make a preliminary estimate of the relative stiffness of the clamped parts compared to the bolt. The stretch of the bolt is translated to terms of angular rotation based on the simple relationship between the stretch due to the estimated load and the pitch P of the thread.

Using finite element analysis (FEA) modeling methods, it is also possible to model the bolt and joint spring rates so that an estimate of the tension–angle coefficient can be made. Since this method involves estimating a proper K value, it must always be backed up by one or more of the other methods. A combination of torque signature analysis with the SR1 design software is perhaps the most efficient way to calculate the joint and bolt spring rates necessary for a complete bolted joint analysis.

27.10.4.1 Stress–Strain and Torque–Tension Diagrams

In the development of bolted joint tightening models, it is helpful to introduce the concept of the approximate equivalency of the stress–strain and corresponding torque–angle diagrams, as shown in Figure 27.12, for the elastic clamping point on the tightening curves.

27.10.4.2 Calculation of Nut Factor, *K*

27.10.4.2.1 Experimental Determination of Friction Coefficients

Just as tabulated values of K are not reliable for application with any assurance of predictable results for a given bolt or nut assembly, it is equally true that friction coefficients determined experimentally must be used for comparison purposes and supplemented by torque–angle assembly monitoring to ensure proper assembly clamping forces.

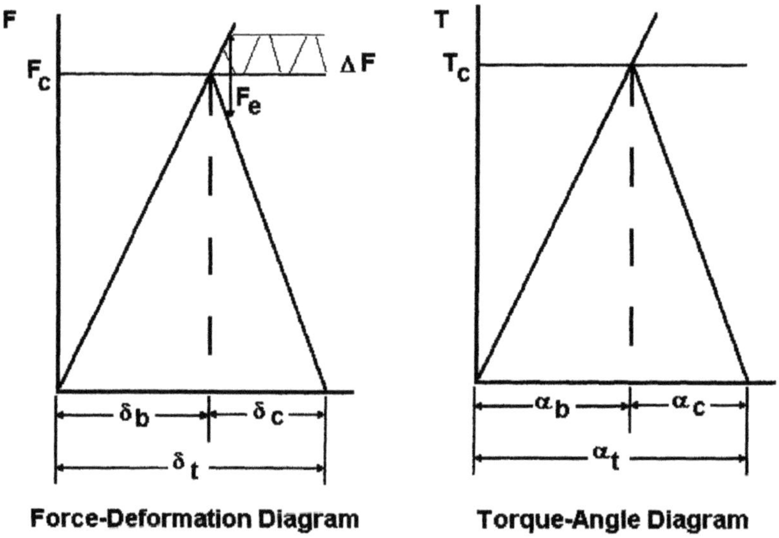

FIGURE 27.12 (a) Force–deformation diagram and (b) Torque–angle diagram.

The German norm DIN 946 provides a practical standard method of evaluation of friction coefficients on screw threads and underhead bearing surfaces. Fastener laboratory testing machines are available for the determination of friction coefficients according to DIN 946.

27.10.4.2.2 Torque–Tension Coefficient, Nut Factor K

As described in Section 27.5, the basic equation, $T=KdF$, can be applied to the linear elastic clamping zone of the assembly tightening process and used to estimate the K factor. There are published tables of K factors for various combinations of materials, surface finishes, plating, coatings, and lubricants. However, actual experience has shown that it is highly unreliable to assume that any given K value applies to a specific fastener being assembled.

27.10.4.3 Laboratory Measurement of Friction Coefficients

A torque–tension research head (Figure 27.13) is used to make the measurements necessary to determine the friction coefficients for the underhead and thread regions. The equations for this analysis are directly derived from the force–torque free-body diagrams for the 60° Sellers thread form. Mechanical engineering design textbooks such as those by Robert C. Juvinall and Joseph Shigley [1,2], provide a detailed derivation of the torque–tension relationships for threaded fasteners. The equations presented in the German standard DIN 946 can be verified following the derivations presented by Juvinall and Shigley. Using the terminology of DIN 946, which also corresponds directly with the terminology of VDI 2230, it is possible to calculate the nut factor, K from the friction coefficients, bolt pitch diameter, and thread pitch. The formula for the calculation of K is based on the simplified torque–tension equation $T=KdF$ as follows:

$$K = \frac{0.159P + 0.578 d_2 \mu_G + \left(\frac{D_{KM}}{2}\right)\mu_K}{d}$$

where

P=thread pitch
d_2=pitch diameter
μ_G=thread friction coefficient

FIGURE 27.13 Torque–tension research head.

D_{km}=mean bearing diameter or underhead
μ_K=underhead friction coefficient
d=the nominal diameter of the bolt.

With properly designed test fixtures, the research head[3] is used to test a variety of fastener configurations, i.e., (1) tightening the bolt, (2) tightening the nut, (3) tightening self-tapping screws, and (4) tightening pipe thread fittings. For example, for testing conducted according to ANJ-514 for high-pressure pipe fittings, the use of special tooling permitted a ¼ in. grip length to produce significant information that was used to compare cadmium plating with various zinc formulations.

27.10.5 Material Property—Yield

27.10.5.1 Yield Signature Analysis

When a fastener is taken beyond the yield point, the torque–angle signature method is first used to find the elastic origin. Then the yield point is located at the upper end of the elastic tightening curve where the linear portion of the signature rolls over into the post-yield tightening zone.

The added tension after yield is the amount of angle relative to the straight-line elastic clamping portion to that maximum torque point and is not proportional to the overall angle of turn (Figure 27.14). First, project a line along the elastic curve above the yield point. Then project the maximum torque level back to that line. This procedure works properly only if the observed yield point is due to the bolt yielding. Joints where thread stripping or underhead embedment occurs before bolt yield cannot be evaluated by this method.

It is generally true that once threads strip or embedment starts, the clamp load no longer increases at the same rate relative to angular turn. In most cases, particularly in response to thread stripping, the clamp load does not increase at all although the torque may increase 25%–35% after the onset of yielding. Failure to properly recognize this phenomenon has been the downfall of a number of attempts to use "yield-sensing" tightening strategies.

The clamp load at the bolt yield torque is primarily a function of the bolt material properties and the thread friction coefficient, μ_G.

Underhead embedment resulting from compressive yield in the bearing area depends on the strength of the bolt or nut relative to the clamped surface strength.

Thread yield, or thread stripping, is a function of the length of thread engagement and the relative strengths of the bolt and mating thread materials.

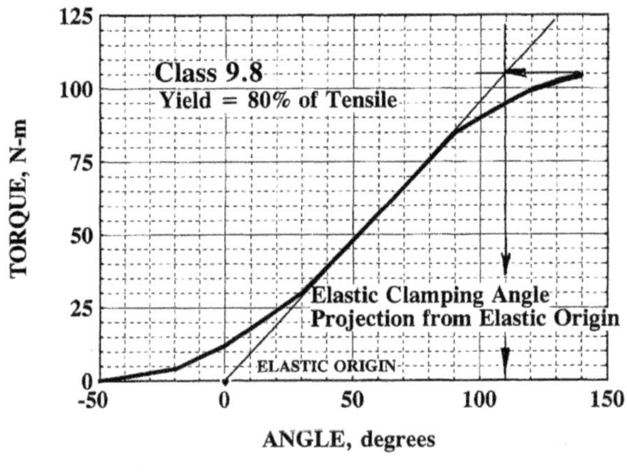

FIGURE 27.14 Determining clamping force beyond the yield point.

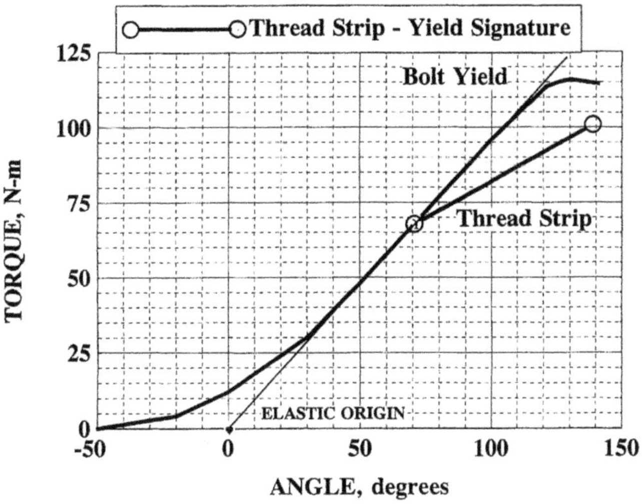

FIGURE 27.15 Yield signature curves, (o—o) thread stripping-yield signature.

Figure 27.15 illustrates the characteristic "yield signature" curves for both bolt yield and thread stripping or embedment. Note that for bolt yield the maximum increase in torque after the yield point is only 10%–20% due to the typical ratio between yield and tensile strengths for commonly used bolt materials; for a metric class 10.9 bolt, the difference is 10%.

Also illustrated is the torque–angle signature beyond the yield point for a tightening process where thread stripping or underhead embedment has occurred. Torque increases of 25%–35% or more after the underhead yield or thread stripping point are often observed. The difference between the two types of yield phenomena is usually very clear. However, this again can be used as a reminder to try to calculate the relative safety factors for bolt yield, thread stripping, and underhead embedment.

When a bolt is torqued to the yield point, the yield clamp load is always less than the simple tension test yield load. This is due to the combined tensile and torsional stresses on the shank of the bolt resulting from the torquing process.

Figure 27.5, presented earlier in this chapter, illustrates the effect of thread friction on the torque and clamping force at the yield point.

27.10.5.2 Use of Yield Point to Estimate Angle–Tension Coefficient

With the torque–angle signature, it is possible to measure the elastic tightening angle from the elastic origin to the yield point. If a nominal friction coefficient of 0.1 is assumed for the thread and underhead, then the clamp load at the bolt yield torque will be about 80% of the tensile test yield load.

The tension–angle coefficient is then calculated as the yield clamp load divided by the angle of turn from the elastic origin to the yield point. The tension–angle coefficient is the slope of the FM-alpha diagram created by the SR1 design software.

If thread stripping or underhead embedment occurs before bolt yield, it is necessary to estimate the load at this yield point and divide the load by the elastic tightening angle to the tension–angle coefficient.

It is sometimes possible or necessary to cold work a bearing surface by repeated tightening beyond the yield point so that adequate clamping force can be achieved on the bolt. However, it is poor practice to design bolted joints where thread stripping or embedment occurs before the bolt yields.

27.11 TORQUE–ANGLE TENSION CONTROL SUMMARY

The proper application of any assembly process includes the requirement that limitations and safety factors associated with the specified procedures be understood and well documented. This philosophy is in accordance with the principles defined by ISO 9000 and QS-9000 documentation requirements, which have become the guiding methods for manufacturing worldwide.

A relatively simple two-step process of torque–angle tightening was defined at the beginning of this chapter with a few short sentences. It should now be clear to the reader who has taken the time to go into the details that followed the introduction that this "simple" assembly procedure is potentially far more complicated than it initially appears to be. The benefits gained through understanding how fasteners really work can be enormous. In terms of cost, both material cost and the cost of human life and suffering, the results of poor bolted joint design and failed assembly procedures have been reported at many millions of dollars per year.

For a total engineering approach to assembly, an understanding of the principles of mechanics and materials science underlying the performance of threaded fasteners is essential. The effects of thread friction, underhead friction, bearing areas, material strengths, thread strength, and joint strength must all be accounted for when a tightening process is specified.

In the final analysis, only actual experience in the field will tell us if a design or assembly is successful. However, starting with a proper understanding of the assembly process through torque–angle signature analysis, we can develop significant insight into the assembly process and the potential reliability of any bolted joint.

GLOSSARY OF IMPORTANT TERMS

D_{km}	Effective diameter for the friction in the bolt head or nut bearing area (in., mm)
F	Force, general (lb, N)
F_M	Initial clamping load (assembly preload) (lb, N)
F_v	Preload, general (lb, N)
K	Nut factor for basic equation $T=KdF$
M_A	tightening torque for the assembly until preloading of bolt reaches F_M (lb-ft, N·m)
P	Pitch of the bolt thread (in., mm)
T	Torque, general (lb-ft, N·m)
D	Bolt diameter (=outside diameter of thread; nominal diameter) (in., mm)
d_h	Bore diameter of the clamped parts; inner diameter of the substitution cylinder (in., mm)
d_w	Outer diameter of the plane head bearing surface (at the inlet of the transition radius of the head) (in., mm)
d_2	Pitch diameter of bolt thread (in., mm)
μ	Coefficient of friction, general
μ_G	Coefficient of friction in thread
μ_{ges}	Average coefficient of friction for thread and bolt head bearing surface
μ_K	Coefficient of friction for bolt head bearing surface.

NOTES

1 The M-alpha diagram was created in 1996 by Fritz Ruoss of Hexagon and Ralph Shoberg of RS Technologies as an extension of the SR1-VDI 2230 Bolted Joint Design Software. See glossary of terms used for SR1-VDI 2230 at end of chapter.

2 A line drawn tangent to any point on the torque–angle curve prior to the yield point can be used to directly estimate the relative spring rate of the clamped assembly at that point.

3 RS Technologies Torque–Tension Research Head with capacities of 1,000 kN (225,000 lb) and 4,000 N·m (3,000 lb-ft).

REFERENCES

1. Juvinall, R. G, *Fundamentals of Machine Component Design*, John Wiley & Sons, New York, 1983.
2. Shigley, J. E., *Mechanical Engineering Design*, McGraw-Hill Book Company, New York, 1977.
3. Juvinall, R. G, *Engineering Consideration of Stress, Strain, and Strength*, McGraw-Hill Book Company, New York, 1967.
4. Shoberg, R. S., Torque-Angle Signature Analysis, *Fastener Technology International Magazine*, February 1996.
5. Shoberg, R. S., *Engineering Fundamentals of Tonque-Tum Tightening, Clemson University "Fastening Technology'* Conference, Cleveland, OH, 1996.
6. VDI 2230, Part 1: Systematic Calculation of High Duty Bolted Joints, English Language Version, Verein Deutscher Ingenieure, Düsseldorf, 1986.
7. DIN 946, Determination of Coefficient of Friction of Bolt/Nut Assemblies under Specified Conditions, Deutsches Institut fur Normung, Berlin, 1990.
8. Ruoss, F., *Sri-Bolted Joint Calculation Manual, Hexagon Industriesoftware GmbH,* Kircheim/Teck, 1996–97.

28 Torque-to-Yield Control of Bolt Tightening

Paul W. Wallace and Sayed A. Nassar

28.1 INTRODUCTION

The practice of tightening critical bolts in automotive applications to clamp loads that are at or beyond the torque–tension yield point is becoming more widespread. The result is closer control and higher initial clamp load in the joint, generated by a process controlled by bolt strength properties (yield tightening) rather than friction (torque tightening). Since the strength of bolts is much easier to control than their friction properties, the scatter in clamp load from yield or beyond-yield tightening methods is typically less than one-fourth of that in conventional torque-only control of bolt tightening.

The designer of a bolted joint is concerned with achieving the minimum clamp load necessary to provide in-service durability. If scatter in the tightening process is large, then the range of clamp loads to be accommodated in the design must also be large. Large scatter also requires that a larger size bolt be used, which usually means that the joint must also be larger to accommodate the bigger bolt. An additional consequence of large clamp load scatter in bearing-type joints (connecting rod, main bearing) is that variable clamp load causes inconsistent distortion of the bearing, which may lead to reduced service life.

Therefore, designers are drawn more and more toward tightening processes that minimize preload variation in order to obtain the benefits of more predictable service life and better power-to-weight ratios. This leads them to examine yield-controlled tightening or torque plus angle tightening beyond yield, as the only high volume production methods that will give them relatively tight control of preloads (around 10% total scatter).

28.2 EFFECT OF CLAMP LOAD ON JOINT STRENGTH

The strength of a bolted joint is almost proportional to the clamping force exerted by the fastener. For a joint to separate or for a bolt to fail in fatigue, it is necessary for the original clamping effect of the bolt to be overcome [1,2].

Figure 28.1 shows the characteristics of a concentric joint. The bolt was preloaded to three different loads, representing the lower and upper bolt stresses expected for torque-controlled tightening (450 and 750 N/mm^2) and the stress expected for yield-controlled tightening (1,050 N/mm^2). The joint components were then externally loaded by different amounts and the bolt stress was measured. It can be seen from Figure 28.1 that the bolt initially felt about 4% of the external load. After further loading, until about 80% of the preload, the bolt stress increased nonlinearly until the bolt was receiving the full external load.

This joint was then tested in fatigue. Figure 28.2 shows the *alternating load* cycles to failure (*S–N*) curve. It can be seen that the *dynamic load* at the endurance limit is greater than the bolt preload.

This means that the joint had to be open, and open significantly, for fatigue failure to occur. For the bolt preloaded to 26,100 N, the maximum dynamic external joint load at the endurance limit was 31,000 N. For the bolt preloaded to 43,500 N, the maximum load was 48,000 N at the endurance

 DOI: 10.1201/9781003501183-33

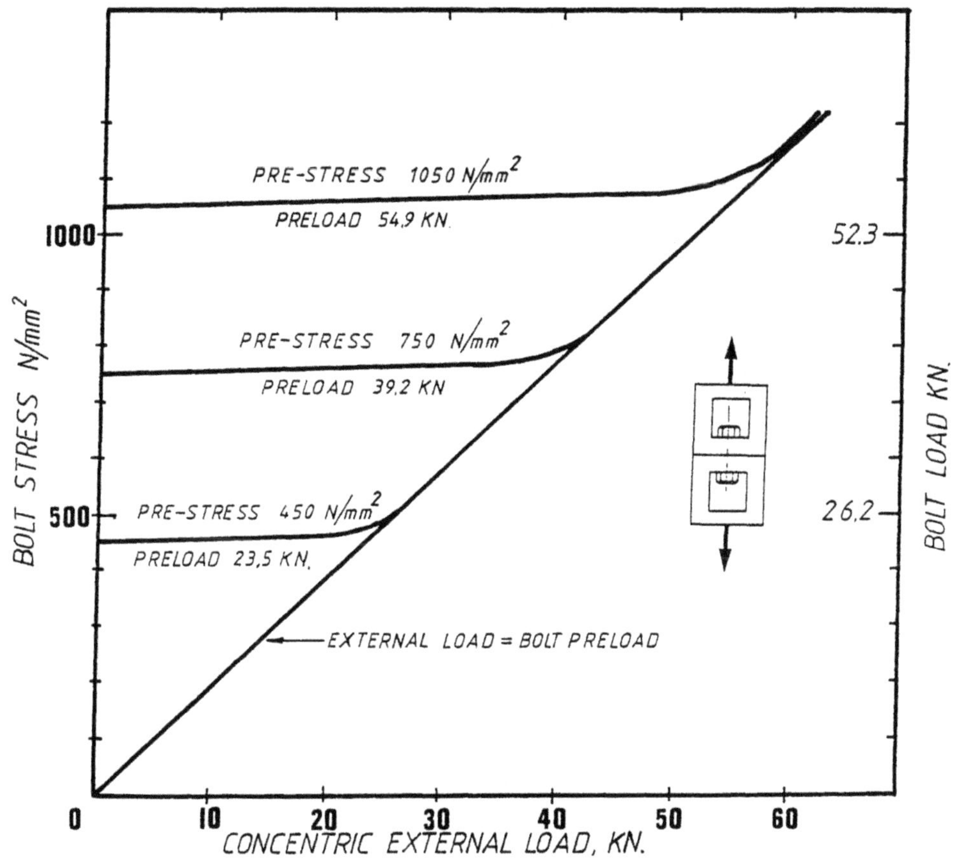

FIGURE 28.1 Effect of external load on the total joint on bolt stress. The application of external load to a joint causes little change to the bolt load initially. Only when the external load approaches the bolt preload and the joint begins to separate does the bolt load increase significantly. Separation occurs at loads greater than the bolt preload. Joint concentrically loaded with a single M10 Grade bolt, 12.9 (from Ref. [1]).

limit. In both cases, therefore, fatigue occurred when the maximum load was some 5,000 N higher than the bolt preload. This clearly indicates that in a hard concentric joint, preload is the controlling factor governing fatigue.

The response of eccentrically loaded joints is more severe than in the concentric case. In Figure 28.1 we saw that the load felt by the bolt was linear with external load until the external load reached 80% of the preload or more. In an eccentric joint, the linear portion is much less. Figure 28.3 shows the response of an eccentrically loaded joint similar in other respects to the concentric joint discussed above and preloaded again to three different levels. It can be seen that the load felt by the bolt starts from zero but rapidly increases to 40% or more of the external load. This is critically dependent on the preload: When 6,000 N was externally applied, the most highly loaded bolt experienced about 490 N additional load whereas the most lightly loaded bolt experienced 15,200 N additional load. This dramatic difference indicates that in an eccentrically loaded joint, the bolt preload is the major factor controlling the additional bolt load. This is reflected in the fatigue performance of this joint, which shows the fatigue endurance limit increasing in proportion to preload (Figure 28.4).

It should be noted that the vast majority of joints are eccentrically loaded in that the bolt axis and external load axis do not coincide.

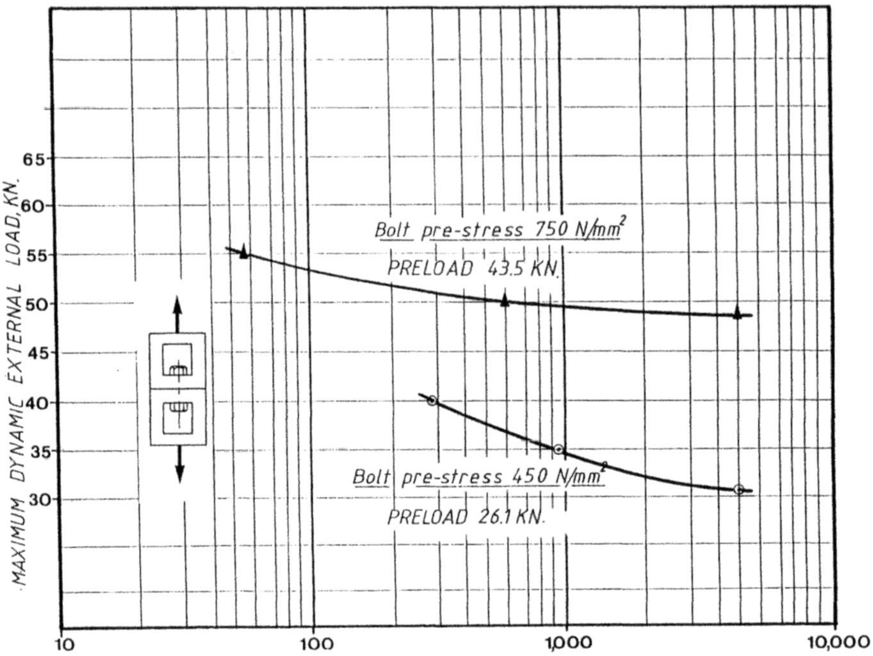

FIGURE 28.2 When the complete concentric joint in Figure 1 was subjected to alternating external loads, it exhibited these fatigue curves. The maximum external load applied at the endurance limit is greater than the bolt preload, indicating that joint separation occurred before failure (from Ref. [1]).

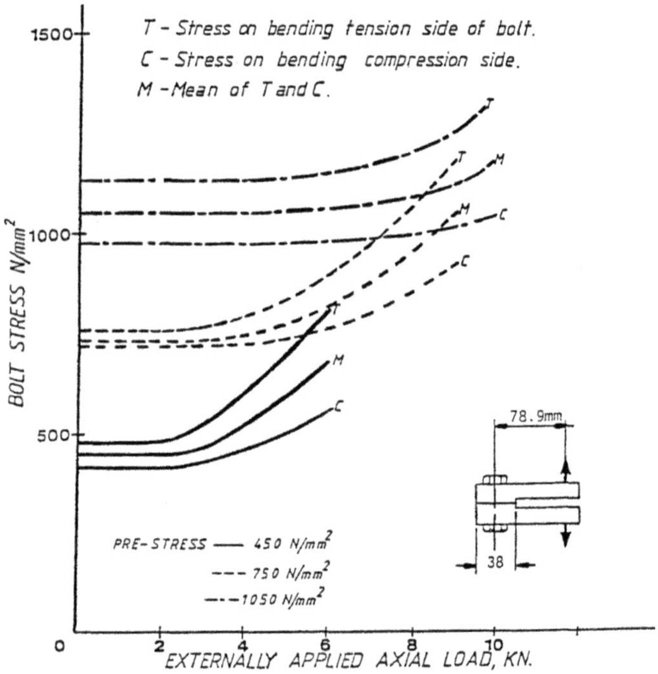

FIGURE 28.3 In an eccentrically loaded joint, bolt load and stress increase nonlinearly with total external load. Joints with higher preload withstand higher external load before the bolt stress begins to increase. (From Ref. [1].)

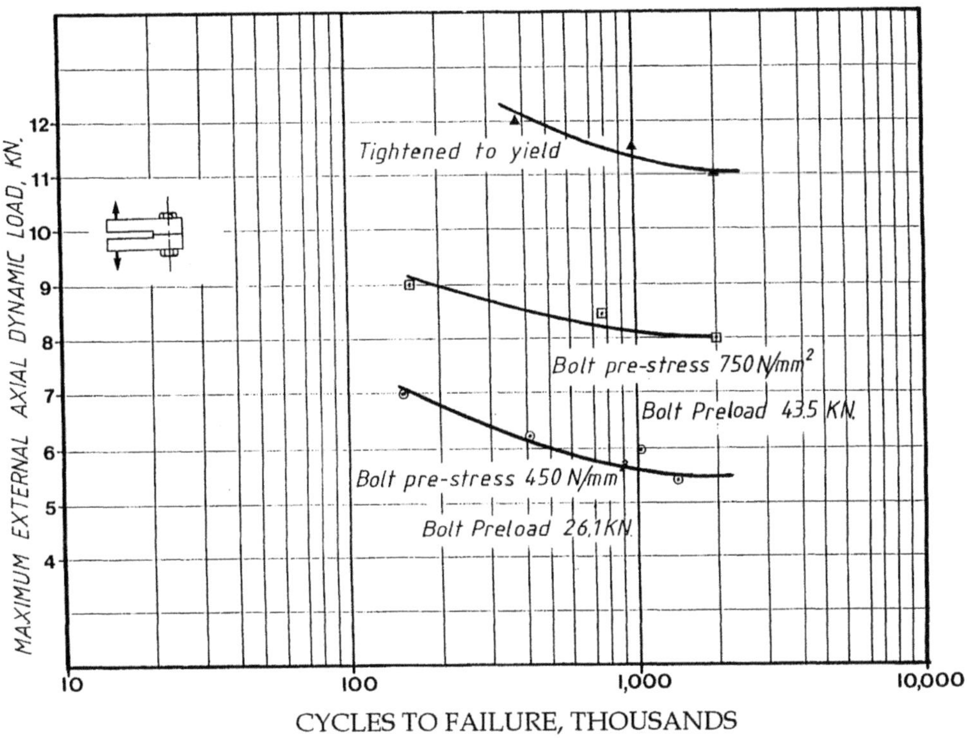

FIGURE 28.4 When an eccentric joint is subjected to alternating load, its fatigue performance also depends on bolt preload. The maximum external load at the endurance limits correlated with bolt stress increase in Figure 28.3 (from Ref. [1]).

28.3 TIGHTENING TO YIELD

Since higher clamp load is the key to higher performance bolted joints, much effort has been put into achieving the highest possible value. In practice, this occurs when the bolt is tightened to a point where it takes on a small permanent set. The load at this point is less than the yield strength of the fastener when loaded in pure tension. The applied tightening torque creates torsional stresses that reduce the tensile axial load required to cause permanent stretch in the bolt [1-5].

This characteristic is shown by the bolt load–extension diagram in Figure 28.5. The solid line represents pure tensile loading while the dashed line represents the behavior during tightening by the application of torque. At low loads, the two curves are coincident. When tightening, a yield point occurs at a load below the pure tensile yield, normally between 70% and 80% of the tensile yield, depending on the thread fiction coefficient. Note that friction is responsible for most of the torsional stress.

If the fastener is tightened beyond yield and the joint is subsequently loaded, the characteristic of the bolt returns to that under pure tension. Consequently, significantly more load must be applied before the bolt reaches its uniaxial yield point and suffers additional permanent extension.

Tightening bolted joints by yield control generates the highest level of clamp load without the risk of overtightening. Joints show no premature bolt yielding in service, even under external loads. There is a strength reserve (i.e., margin of safety) between the as-tightened tensile stress in the bolt and its uniaxial yield strength. Therefore, even when the bolt has been tightened to yield, further yield will not immediately occur when an external load is applied. In most cases, the full external load required to separate the joint can be applied before the bolt yields further in tension.

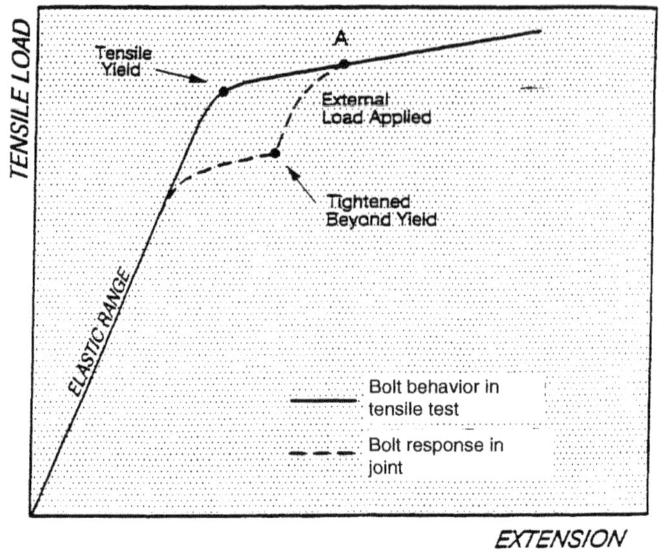

FIGURE 28.5 The load–extension curve for a bolt in pure tension follows the classic curve (solid line). For a bolt tightened with a wrench, the torsional stresses cause yielding at a lower tensile load (dashed line). This explains the additional load capacity of yield-tightened bolts when the joint is subjected to an external load (from Ref. [1]).

The maximum level of in-service clamp load is important in many types of bolted joints. Initial preload is not the only factor determining the clamping force between the joint surfaces under operating conditions. A specific example involves aluminum joints clamped by steel bolts when the structure is exposed to elevated temperatures. The expansion rate of aluminum is almost twice that of steel, so that as temperature increases, additional tensile load is imposed on the bolts. This additional tensile load is absorbed by the strength reserve between the as-tightened and the uniaxial yield strength.

A conservative calculation of the clamp loads generated by tightening to yield for Grade 12.9 fasteners [1] is shown in Table 28.1. The minimum clamp loads that can be ensured with yield tightening are 75% of bolt uniaxial tensile yield strength. Although this may appear low, it is about 40% higher than the best that can be achieved with controlled torque tightening by accurate power tools.

TABLE 28.1
Yield Strength and Clamp Loads for Grade 12.9 Metric Fasteners (Standard Threads)[a]

			Torque Control			Yield Control	
Size	Stress Area (mm²)	Min. 0.2% Tensile Yield Strength at 1,080 N/mm²[b] (N)	Applied Torque[c] (N·m)	Min. Clamp Load, 56% of Y^d (N)	Avg. Clamp Load, 73% of Y^e (N)	Min. Clamp Load, 75% of Y^f (N)	Avg. Clamp Load, 80% of Y^g (N)
M4	8.78	9,500	4.6	5,200	6,750	6,900	7,500
M5	14.2	15,300	9.5	8,550	11,100	11,300	12,300
M6	20.1	21,700	16	12,000	15,600	15,900	17,300
M8	36.3	39,500	39	22,100	28,700	29,300	32,900
M10	58.0	62,500	77	35,200	45,700	46,700	50,800
M12	84.3	91,000	135	51,400	66,700	68,200	74,100
M14	115	124,000	215	70,300	91,300	93,300	101,400
M16	157	170,000	330	97,000	126,000	128,800	140,000

(Continued)

TABLE 28.1 (*Continued*)
Yield Strength and Clamp Loads for Grade 12.9 Metric Fasteners (Standard Threads)[a]

Size	Stress Area (mm²)	Min. 0.2% Tensile Yield Strength at 1,080 N/mm²[b] (N)	Torque Control: Applied Torque[c] (N·m)	Torque Control: Min. Clamp Load, 56% of Y[d] (N)	Torque Control: Avg. Clamp Load, 73% of Y[e] (N)	Yield Control: Min. Clamp Load, 75% of Y[f] (N)	Yield Control: Avg. Clamp Load, 80% of Y[g] (N)
M18	192	207,000	455	118,000	153,000	156,400	170,000
M20	245	265,000	650	152,000	197,000	201,500	219,000
M22	303	327,000	870	189,000	245,000	250,200	272,000
M24	353	381,000	1,100	219,000	284,000	290,300	315,600
M27	459	496,700	1,650	288,000	374,000	382,300	415,600
M30	561	606,000	2,250	350,000	454,000	464,000	504,400

[a] These are conservative values based on the calculations summarized in footnotes b–g.

[b] Minimum tensile yield strength calculated as 0.2% yield stress of 1,080 N/mm² divided by stress area. The 12.9 grade has a yield stress range from 1,080 to 1,260 N/mm².

[c] This torque value is calculated at a friction coefficient of 0.125 to give a combined effective stress $\left(\sqrt{\sigma^2 + 3\tau^2}\right)$ equal to 90% of the yield stress (1,080 N/mm²).

[d] Minimum clamp load, using torque control, is based on 11% of the average clamp load value, i.e., assuming ±23% variation from the average. It is approximately 56% of the bolt uniaxial yield strength (see footnote b).

[e] Average clamp load, using torque control, is calculated at 90% of the combined stress yield. Torque values were taken from Ref. [3], VDI 2230, Table 28.2, at a friction coefficient of 0.125.

[f] Mmimum yield control clamp load is calculated as 8% less than the average value and is 75% of bolt yield strength.

[g] Average yield control clamp load is calculated to give a combined stress yield $\left(\sqrt{\sigma^2 + 3\tau^2}\right)$ equal to 1,080 N/mm² and is approximately 80% of the bolt yield strength.

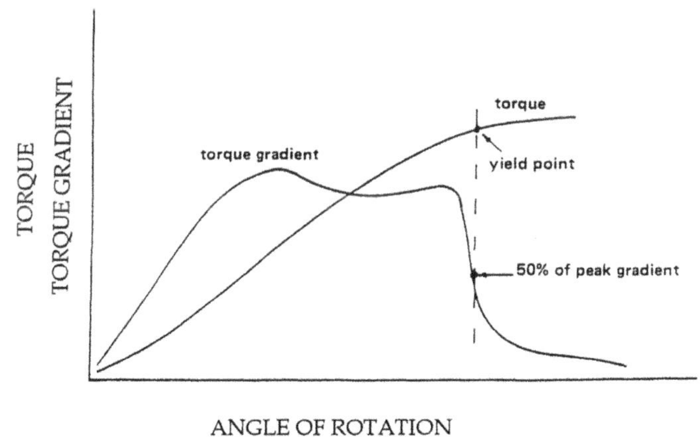

FIGURE 28.6 Tightening curve showing the relationship between torque and fastener rotation angle. The gradient of the torque–angle curve remains roughly constant until yield occurs when it drops rapidly (from Ref. [4]).

28.4 METHODS OF TIGHTENING TO YIELD

Tightening to yield requires the measurement of torque and rotation during tightening. The relationship between torque and rotation for a typical fastener in a joint is shown in Figure 28.6. After an initial transition, torque increases linearly with rotation until the yield process described above begins.

At that point, the slope of the torque–angle curve decreases. Eventually, the slope approaches zero and the torque becomes almost constant.

Measuring torque and rotation is the first step; deciding when yield has occurred is the second. The simplest method of achieving yield is to tighten through a fixed rotation angle known to take the fastener safely past its yield point. This is frequently done by selecting an easily measured angle (e.g., 90°, 180°) and then establishing an appropriate snug torque from which to initiate angle measurements. It is not difficult to establish safe values for this process if excessive permanent extension is acceptable. It simply requires experimentation to determine the torque and angle values required to permanently extend the fastener, say 0.010–0.015 in. (0.25–0.4 mm).

This method is used in both structural steel and automotive engine assembly where fastener reuse is not required (see also Chapter 25). Most fasteners cannot take more than two or three applications of torque–angle tightening into yield without the risk of thread distortion or necking. The uncertain state of the fastener in terms of both elongation and the number of tightenings it has experienced usually forces users to specify a new fastener for every tightening. In practice, therefore, torque–angle tightening into yield is not recommended where reassembly of the product using the same fasteners is required. (See Chapter 24 for a further discussion of structural steel practices.)

By contrast, the torque gradient characteristics can be used to ensure the yield of the fastener with minimum elongation. The gradient is the slope of the torque–rotation curve shown in Figure 28.6. The shape of the gradient curve is remarkably constant over a wide range of fasteners and joints. Essentially, the gradient rises sharply as soon as the fastener snugs the parts together, then plateaus during tightening and drops rapidly when the fastener yields.

The gradient drop is so rapid that the permanent extension of the fastener is little more than 0.002 in. (0.05 mm) from the top to the bottom of the gradient drop. Detecting and stopping the tightening at the gradient drop provides a method of achieving the high, accurate clamp loads attainable at yield combined with low plastic elongation; the result is the potential for several reuses of the fasteners.

The most effective yield tightening method employs the gradient curve to detect the yield point. Torque values are sampled for every degree of rotation, and the gradient is computed over a fixed angular measurement. The gradient value is monitored by an algorithm at every step. The present value is then compared with the stored peak, and when it declines to two-thirds or one-half of the peak, the tightening is stopped.

This yield tightening method can be applied, with little change in parameters, to fasteners in a range of strengths and sizes. No special calibration is required. The adaptive nature of the method enables the yield point to be identified for each fastener from its own characteristic curve.

28.5 QUALITY CONTROL ASPECTS OF THE PROCESS

The yield tightening process has not only the ability to tighten bolts but also the ability to measure process quality. Variations in joint characteristics, bolt function, and thread quality can all be detected by placing quality control limits on the torque and angle of rotation at the point of yielding. Figure 28.7 illustrates the way in which defects can be identified.

By measuring torque and rotation, additional quality assurance can be achieved through the "acceptance window," which has the preset maximum and minimum values for torque and angle. If the shutoff point does not lie in the window zone between these values, an error condition exists.

With this tightening system, nearly 100% control can be exerted on the quality of the assembly. The signals can be used with warning devices and relays that control the production line. A permanent record can be kept as evidence of the condition for each assembly. In the light of experience in using the system, many advantages in the area of product reliability have been found, and their number should increase in the future.

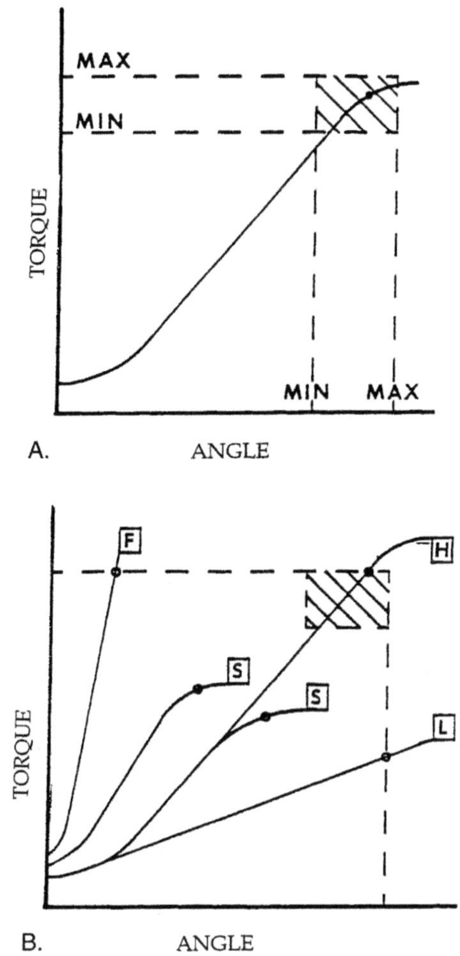

FIGURE 28.7 (a) Limits can be placed on the allowable torque and angle values measured at the detected yield point. This gives the potential of measuring the quality of the total fastening process. (b) Various faults can be detected by using torque and angle limits. Examples of these are illustrated by curves H, S, F, and L as follows. H: Bolt too hard. The high yield point is caused by a heat treatment or material problem. Could lead to delayed fracture. S: Bolt too soft. Heat treatment or material problem. Could lead to breakage in assembly or in service. F: Fault in the male or female thread causing high torque. May lead to inadequate clamp load. L: Low friction caused by wrong surface condition. May lead to overtightening and breakage.

28.6 EQUIPMENT FOR YIELD TIGHTENING

The essential requirements for yield tightening are

1. A tool with torque and angle measurement capabilities
2. A controller with yield-computing capabilities

Majority of power tools with these capabilities are driven by DC brushless motors either in fixtured or in hand-held versions. Figure 28.8 shows an example of these tools, which have seen growing use in automotive power train and vehicle assembly in both Europe and the United States, since the early 1980s.

For manual tightening, specialized tools have been developed with an integrated process and special-purpose angle sensors. The process integrates the torque–angle relationship through

FIGURE 28.8 Power tool systems for yield-controlled tightening typically consist of a control unit and a power tool with torque and angle measurement capabilities. Pictured here is a DC electric angle wrench with a controller that can be programmed for different tightening control methods. (Photo courtesy of Ingersoll-Rand Co., Tool and Hoist Division).

the ratcheting cycle, which permits the operator to tighten the fastener in a normal manner. A well-known tool for this purpose is shown in Figure 28.9. It is widely used for experimental tightening in the laboratory environment, in testing fasteners, in torque audit, and in the reassembly of equipment originally assembled by yield tightening.

28.7 TORQUE-TO-YIELD: ANALYSIS OF BOLT STRESSES

This section provides an insight into the torque-to-yield control method of bolt tightening with emphasis on the state of bolt stress. This control method relies on real-time monitoring of the constant slope of joint torque-turn signature and to stop the tightening process as soon as the slope is reduced signifying reaching the yield limit. At that point in time, the maximum principal stress in the bolt (from the combined effect of elongation and torsion) has just reached the yield stress of bolt material. However, that does not mean that the bolt tension and corresponding clamp load became equal to yield load. Additionally, subsequent disengagement of the tool socket from fastener reduces the torsional stresses significantly to a reduced level the principal stress below the yield stress; the bolt axial stress (from elongation remains unchanged by socket removal.

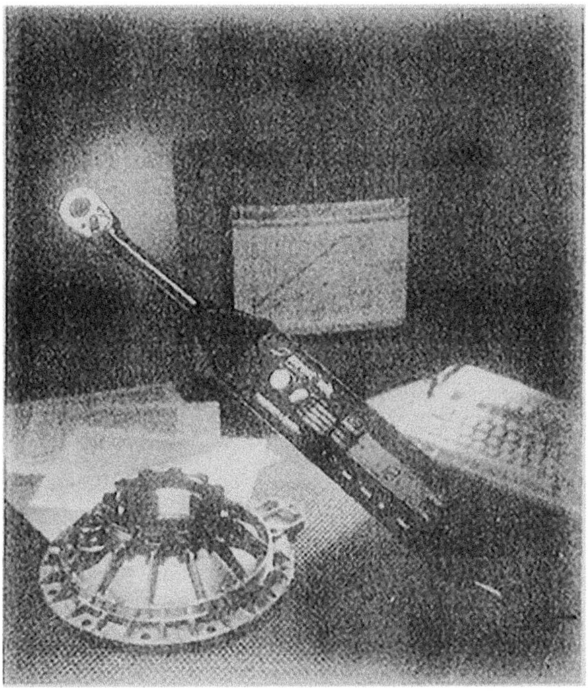

FIGURE 28.9 Hand-operated tools such as this Sensor I wrench are also available with a built-in yield-tightening program. This enables the user to tighten bolts of any size and grade to yield without complex adjustments. Tools come in four sizes covering 10–1,000 Nm. (Photo courtesy of Ingersoll-Rand Co., Tool and Hoist Division).

The main objective of this section is to analyze the state of bolt stresses at the end of torque-to-yield control of bolt tightening and to show that all stresses go back into the elastic zone upon disengaging the tool from the fastener.

As mentioned earlier in this chapter, the tool is programmed to automatically stop as soon as it detects nonlinearity in its acquisition of torque-angle data (past a specified threshold torque for part seating and alignment past free spinning run down) For a properly designed and engineered threaded joint, such tool stoppage means that the bolt material has reached its yield limit, under the combined effect of elongating and twisting the bolt simultaneously. The combined effect of bolt axial and torsional stresses contributes to the bolt principal stress that controls the torque-to-yield control of bolt tightening.

Accordingly, the maximum principal stress in the bolt σ_{1b} before socket removal is given by

$$\sigma_{1b}\left(\text{before removing tool socket}\right) = \frac{\sigma}{2} + \sqrt{\left(\frac{\sigma}{2}\right)^2 + \tau^2} = S_y \tag{28.1}$$

where σ is the bolt axial stress that is equal to bolt preload F divided by its tensile area A_t, τ is the torsional stress due to bolt twisting torque T_{tp}, which is the sum of thread friction and pitch torque components T_t and T_p, respectively. Equation (28.1) gives the bolt axial stress σ as follows

$$\sigma = S_y - \frac{\tau^2}{Sy} \tag{28.2}$$

The twisting torque T_{tp} is approximately equal to one half (50%) of the peak tightening torque T; the corresponding bolt torsional (shear) stress τ is given by

$$\tau = \frac{T_{\text{tp}}\,(D_t/2)}{\frac{\pi}{2}\left(\frac{D_t}{2}\right)^4} = \frac{4T_{\text{tp}}}{A_t D_t} \approx \frac{2T}{A_t D_t} \tag{28.3}$$

where T_{tp} is reaction torque which is the sum of friction and pitch torque components. That peak torque T may be expressed as follows:

$$T = \text{KDF} = \text{KD}(\sigma A_t) \tag{28.4}$$

where the nut factor K is either determined experimentally from torque-tension testing of the actual system or estimated by the Motosh model in terms of the bearing and thread friction coefficients μ_b and μ_t, respectively, as follows

$$K = \left(p/2\pi + \mu_b r_b + \mu_t r_t / \cos 30°\right)/D \tag{28.5}$$

where p is the thread pitch, r_b and r_t are, respectively, the effective bearing and thread friction radii that are approximated by [6] as follows:

$$r_b \approx 0.65D \text{ for standard hexagonal fastners} \tag{28.6}$$

$$r_t \approx (0.55 - 0.055p)\,D \tag{28.7}$$

If Eqs. (28.6) and (28.7) are substituted in Eq. (28.5), the following approximation is obtained for the nut factor of standard hexagonal-head fasteners

$$K \approx \; 0.16p/D + 0.65\mu_b + (0.6 - 0.06p)\,\mu_t \tag{28.8}$$

For illustration, Eq. (28.8) gives $K \sim 0.25$ for a hex-head M12×1.25 bolt in a joint system with $\mu_b = \mu_t = 0.2$. Substituting Eq. (28.4) into Eq. (28.3) gives the peal torsional stress τ on the bolt cross section as follows:

$$\tau \;(\text{before removing tool socket}) \approx \frac{2KD}{D_t}\sigma \tag{28.9}$$

The diameter ratio D/D_t in Eq. (28.9) is obtainable from the bolt tensile area A_t as follows:

$$D/D_t \approx 1.2 / \left(\frac{1}{A_t}\right)^{0.5} \tag{28.10}$$

The bolt tensile stress σ is obtained by substituting Eq. (28.9) into Eq (28.1) and solving the following quadratic equation

$$\sigma^2 \cdot \frac{1}{S_y}\left(\frac{2\text{KD}}{D_t}\right)^2 + \sigma - S_y = 0 \tag{28.11}$$

Solution to Eq. (28.11) gives bolt preload stress σ, which results from the bolt elongation achieved by the peak tightening torque T in a torque-to-yield control method

$$\sigma = S_y \cdot \frac{-1 + \sqrt{1 + 4\left(\frac{2\text{KD}}{D_t}\right)^2}}{2\left(\frac{2\text{KD}}{D_t}\right)^2} \tag{28.12}$$

Accordingly, prior to removing the tool socket, the state of bolt stress is (σ, τ) as respectively given by Eqs. (28.12) and (28.9); the corresponding combined stress σ_{1b} is equal to yield stress S_y of bolt material per Eq. (28.1). Upon the removal of took socket, however, the state of bolt stress changes to (σ, $\tau/2$) due to a 50% decrease of reaction torque from its peak prior to removing the tool socket, as evidence by torque-tension tests. Obviously, the corresponding principal stress σ_{1a} becomes less than the yield stress (i.e. $\sigma_{1a}<S_y$). The issue at hand is to assess how far into the elastic zone does the bolt principal stress σ_{1a} go back upon dis-engaging the tool from the fastener in a torque-to-yield control of bolt tightening. The reduced principal stress σ_{1a} is given by

$$\sigma_{1a}\left(\text{after removing socket}\right) = \frac{\sigma}{2} + \sqrt{\left(\frac{\sigma}{2}\right)^2 + (\tau/2)^2} < S_y \tag{28.13}$$

Finally, substituting Eq. (28.9) and (28.12) into Eq. (28.13) and comparing to Eq. (28.1), gives the bolt principal stress ratio $\sigma_{1a}/(\sigma_{1b}=S_y)$ after and before socket removal before and after removing the tool socket in a torque-to-yield control method as follows:

$$\frac{\sigma_{1a}}{S_y} = \frac{1+\sqrt{1+\left(\dfrac{2KD}{D_t}\right)^2}}{1+\sqrt{1+\left(\dfrac{4KD}{D_t}\right)^2}} \tag{28.14}$$

The corresponding percent reduction of bolt principal stress [upon removing the tool socket] is given by

$$\%\ \text{reduction of bolt principal stress below the yield limit} = 100\left[1-\left(\sigma_{1a}/S_y\right)\right] \tag{28.15}$$

Solved Example 28.1

Metric bolts M8, M10, M12, M14 Class 10.9 with fine and coarse threads are used for clamping bolted joints using the torque-to-yield control method. Use Equations (28.1)–(28.14) to generate user-friendly data and graphs to prove that bolt principal stress drops below the yield strength of bolt material upon the removal of tool socket at the end yield tightening of the bolt. (Note: Eq. (28.1) shows that σ_{1b} (before socket removal) is equal S_y, while Eq. (28.13) shows that σ_{1a} (after removing socket) is less than the yield stress.

TABLE 28.2
Fastener Geometric Data

Fastener Size	Pitch p [mm]	Tensile Area A_t [mm^2]	Section Polar Moment of Area J [mm^4]
M8×1	1.00	39.20	244.56
M8×1.25	1.25	36.60	213.20
M10×1.25	1.25	61.20	596.11
M10×1.5	1.50	58.00	535.40
M12×1.25	1.25	92.10	1,350.02
M12×1.75	1.75	84.30	1,131.03
M14×1.5	1.50	125.00	2,486.80
M14×2	2.00	115.00	2,104.82

TABLE 28.3
Bolt Stresses upon Removing Tool Socket: Torque-to-Yield (Class 10.9- *Fine Threads*)

Fastener Size	Peak Torque T Eq. (4) (K=0.2) [Nm]	Peak Torsional Stress τ Eq. (3) [MPa]	Bolt Axial Stress σ Eq. (12) [MPa]	Reduced Principal Stress $\sigma_{1a}<S_y$ Eq. (13) [MPa]	Bolt Principal Stress Ratio $\sigma_{1a}/(\sigma_{1b}=S_y)$ Eq (14)
M8×1	50.19	362.47	800.23	839.36	89%
M10×1.25	97.94	362.58	800.15	839.31	89%
M12×1.25	177.86	356.67	804.66	842.42	90%
M14×1.5	281.53	357.06	804.37	842.22	90%

TABLE 28.4
Bolt Stresses upon Removing Tool Socket: Torque-to-Yield (Class 10.9- *Coarse Threads*)

	Peak Torque T Eq. (4) (K=0.2) [Nm]	Peak Torsional Stress τ Eq. (3) [MPa]	Bolt Axial Stress σ Eq. (12) [MPa]	Reduced Principal Stress $\sigma_{1a}<S_y$ Eq. (13) [MPa]	Bolt Principal Stress Ratio $\sigma_{1a}/(\sigma_{1b}=S_y)$ Eq (14)
M8×1.25	46.44	371.73	793.00	834.40	89%
M10×1.5	92.16	369.82	794.51	835.43	89%
M12×1.75	160.94	368.56	795.49	836.11	89%
M14×2	256.23	368.26	795.73	836.27	89%

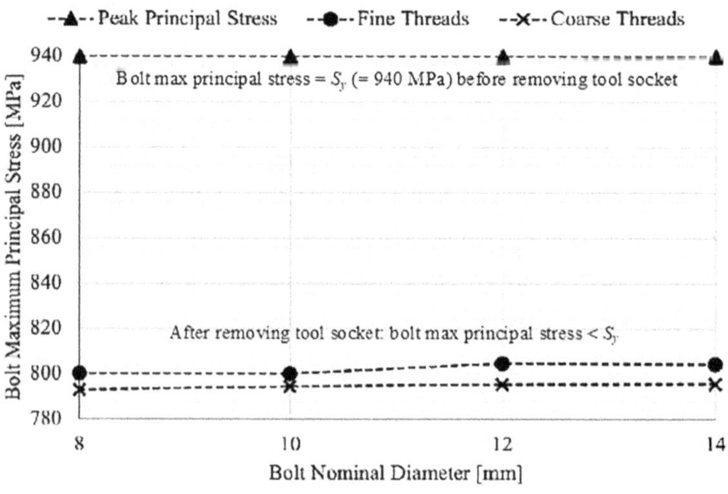

FIGURE 28.10 Bolt maximum principal stress before and after removing the tool socket (Class 10.9).

28.8 SUMMARY

Bolt tightening to yield is a primary method of ensuring high, accurate clamp loads. A step-by-step procedure is provided for calculating the reduction of bolt principal stress upon the removal of tool socket after its torque-to-yield controlled tightening. This control method is more reliable in producing repeatable bolt preload as compared to the Torque only or torque-turn control methods. Bolt stress analysis provided in Section 28.7 clearly demonstrates that the bolt axial stress given by

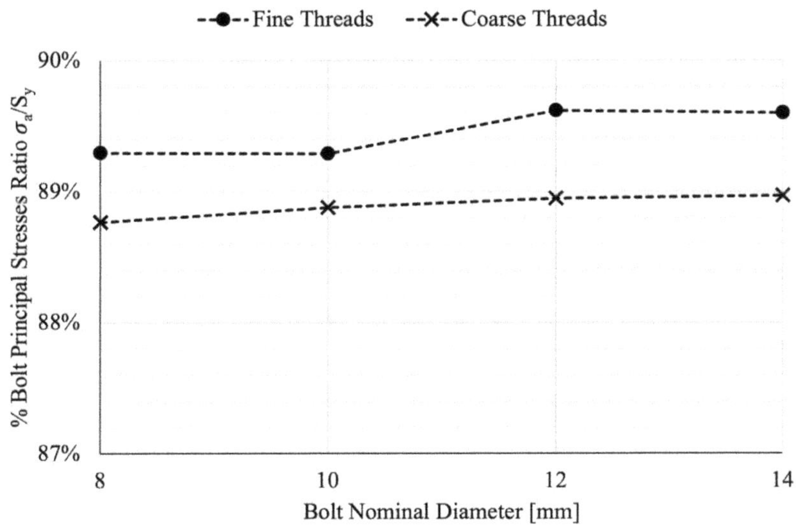

FIGURE 28.11 Bolt principal stress ratio upon removing tool socket from torqued-to-yield fasteners (Class 10.9).

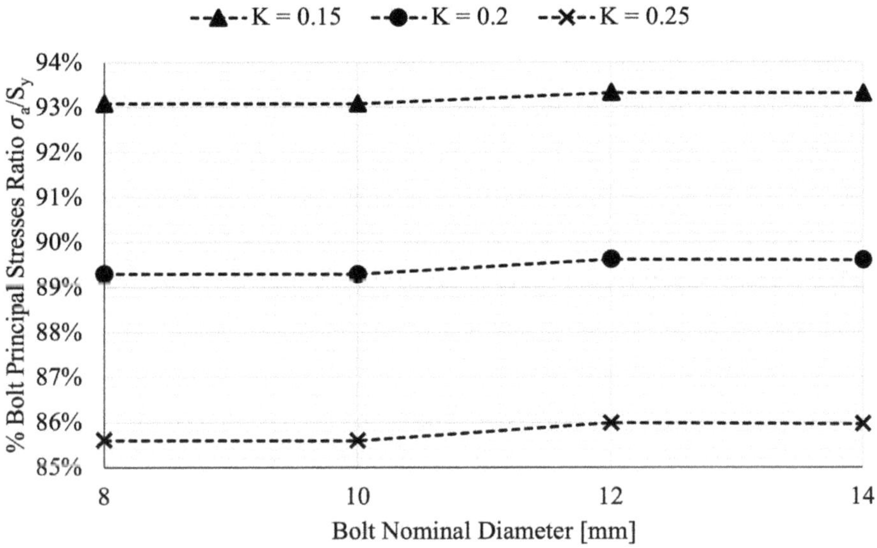

FIGURE 28.12 Effect of nut factor K on bolt principal stress ratio (Eq. 14) upon removing socket from torqued-to-yield threaded fasteners (Class 10.9).

Eq. (28.12) – after its torqueing-to-yield process – remains always less than 90% of its yield stress. Torque-to-yield control is widely used in highly stressed machines such as internal combustion engines and other critical applications. Mass production equipment is commercially available to implement the yield-tightening process

Chapter Problems

1. Comment on the assumptions and validity of major assumptions made in Solved Example 28.1 in Section 28.7.

2. Follow the steps in Solved Example 28.1 to generate similar graphs using nut factor values that range between $K=0.1$ to $K=0.3$ with increments of 0.05.
3. Redo sample problem 1 for SAE Grade 8 bolts size ¼″, 3/8″, 1/2″, and 5/8″ with fine and coarse threads.
4. Generate similar data/tables/graphs same tables//data/graphs similar to those in solved Example 28.1 for all commercial sizes of metric fasteners in material Class 8.8, 9.8, 10.9, and 12.9 with fine and coarse threads.
5. Repeat problem 4 for SAE Grades 2, 5, and 8 fasteners in all commercially available SAE sizes.

ACKNOWLEDGMENT

Marco Gerini Romagnoli and Daniel Nappi for generating and graphing the numerical data for the solved Example.

GLOSSARY

Angle: The angle of rotation of a fastener during tightening, measured from a starting point called snug torque.

Bolt preload; bolt prestress: The axial load and average tensile stress, respectively, are produced in a bolt by the tightening process.

Bolt load; bolt stress: The axial load and average tensile stress, respectively, in a bolt when stress by tightening or by subsequent external loading of the joint.

Clamp load: The force applied to the clamped surfaces of a joint.

Concentric joint; concentric external load: A joint with a single bolt in which the external load tending to separate the clamped surfaces is applied concentric with the bolt axis.

Eccentric joint; eccentric load: A joint with a single bolt in which the external load is applied away from the bolt axis but parallel to it.

External load: A tensile force applied to the clamped parts, acting in the same direction as the bolt axis.

Maximum external dynamic load: The peak value of an alternative external load applied to a joint in fatigue testing.

Preload: The initial clamping force applied to the joint by the tightening of the bolt.

Principal Stress: Combined axial and torsional stress due to elongating and twisting the bolt.

Tensile load: The axial load applied to a bolt by tightening and/or by the action of an external load on the joint.

Rotation: See *Angle.*

REFERENCES

1. Chapman, I., J. Newnham, and P. Wallace, The tightening of bolts to yield and their performance under load; *Trans. ASME, J. Vibration, Acoust., Stress Reliability Des.,* 108:213 (1986).
2. Newnham, J., and L. Curley, Bolted joints under tensile service loads: The effect of torsion in the bolt on in-service clamping loads, *Trans. ASME, J. Mech. Design,* 115:36 (1993).
3. Nassar, S. A. and P. H. Matin, Nonlinear strain hardening model for predicting clamp load loss in bolted joints. *ASME J. Mech. Des.*, (2006): 1328–1336.
4. Nassar, S. A. and P. H. Matin, Clamp load loss due to fastener elongation beyond its elastic limit. *ASME. J. Pressure Vessel Technol.*, (2006): 379–387.
5. Boys, J. T., and P. W. Wallace, Design and performance of an automatic control system for fastener tightening, *Proc. Inst. Mech. Eng.,* 191 (1977):311 .
6. Nassar, S. A., P. H. Matin, and G. C. Barber, Thread friction torque in bolted joints, *ASME. J. Pressure Vessel Technol,* 127, 4 (2005): 387–393.

29 Stretch Control

Rod Corbett

29.1 INTRODUCTION

There are three basic factors that determine the reliability of a bolted connection; namely, joint design, bolt and component quality, and achieving design bolt tension on installation. Investment in design and quality assurance in recent years has been substantial, as borne out by the development of ISO 9001/9002. In a bolted joint, this effort and investment has one objective: to ensure installed bolt tension will equal that intended by the designer, that the clamping force created withstands the working loads, thereby providing reliability and safety. Despite this fact, the effort and investment the industry spends on ensuring that this fundamental design objective is met on installation is negligible.

The vast majority of bolted joints in engineering are tightened in an uncontrolled manner; i.e., the achieved bolt tension is unknown. This is truly remarkable when you consider that 90% of all bolted joint failures can be attributed to incorrect bolt tightening.

Industry is demanding greater safety, reliability, and cost effectiveness from plant, equipment, and structures. It correspondingly spends millions of dollars on controlling and monitoring process parameters such as pressure, speed, flow rates, and temperatures. However, it is reluctant to control and monitor the one parameter that holds it all together—bolt tension.

Perhaps the source of this reluctance can be explained by a lack of understanding of bolted joint technology and the fundamental importance of bolt tension and its effect on safety and reliability. Many operators are unaware of the control limitations of their tightening equipment, especially hydraulic tensioners. In cases of persistent joint failure, the bolt-tightening situation is commonly perceived to be satisfactory, with the cause of failure lying elsewhere, e.g., the gasket.

Whatever the reason, design bolt tension objectives can be controlled and measured reliably. Those designers and operators who do address this fundamental mechanical concept inevitably benefit from increased safety, reliability, and reduced costs.

29.2 BOLT TENSION MEASUREMENT

Metallic fasteners are elastic bodies. In the elastic range, their behavior under axial load is governed by Hooke's law, which states that stress is proportional to strain:

$$\text{Stress} = E \times \text{strain} \tag{29.1}$$

The proportionality factor is the elasticity modulus for the alloy involved (E).

$$\text{Stress} = \frac{\text{load}}{\text{area}}$$

The only way to reliably measure bolt tension is to measure directly stress or strain in the bolt itself. Strain in a bolt is the elongation per unit length (E), and it is the bolt elongation or extension, ΔL (for a specified bolt, tension), that provides the most realistic parameter for measurement.

$$\text{Stress} = \frac{\Delta L}{L_0} \tag{29.2}$$

DOI: 10.1201/9781003501183-34

where

L_0 = the initial effective length of the fastener as defined below (in., mm)
ΔL = the change in length of the bolt (in., mm).

The extension for uniform cross section all-thread stud bolting can be calculated with the above formulas. However, many fasteners have a relatively complex shape. Elements such as head, shank, and thread will have their own load–extension relationship. The following formula breaks down the extension situation for each element and therefore provides a more realistic assessment.

$$\Delta L = F\left(\frac{0.4D}{EA_D} + \frac{L_1}{EA_1} + \frac{L_2}{EA_2} + \frac{0.4D}{EA_D}\right)$$

where

ΔL = overall bolt extension (in., mm)
F = bolt tension (lb, N)
D = nominal bolt diameter (in., mm)
E = Young's modulus (psi, GPa)
L_1/A_1 = length/area ratio of shank (in./in.^{2}, mm/mm^2)
L_2/A_2 = length/area ratio of grip length thread (in./in.^{2}, mm/mm^2)
A_D = area at nominal diameter (in.^{2}, mm^2)
$0.4D$ = the effective length (for stretch measurement purposes) of the bolt head and also the effective length of the threads engaging the nut or tapped hole (in., mm). Hence this term is repeated in Eq. (29.2).

Note that the effective length, L_0, of the fastener is equal to $0.4D + L_1 + L_2 + 0.4D$.

Calculated stretch values, however, can give only a "ballpark" figure useful for operational planning. Assured accuracy comes only from load test calibration on the fastener.

There are several extension measurement devices available that are capable of measuring very small displacements in fastener length under favorable conditions.

The amount of bolt extension to be measured can vary substantially. Bolt extension depends on bolt load magnitude and the bolt length or, to be more accurate, joint grip length or gage length over which extension is measured. Bolt extension can vary from micrometers for short, lightly loaded bolts to several millimeters for longer, highly loaded fasteners. Although there are applications where this is not true, engineering logic often dictates that a measuring system must be capable of measuring to within 10% of the main unit under consideration.

At these very small levels of displacement, it is understandable that variation in the engineering environment can be very significant to the reliability of any reading or indication.

29.3 FACTORS AFFECTING THE RELIABILITY OF EXTENSION MEASUREMENT

Under favorable conditions, traditional extension measurement with micrometers, dial gages, and ultrasound are all capable of measuring the levels of displacement associated with large diameter bolting. When it comes to shorter, smaller diameters involving bolt loads below 4.0 tons, users have to question the validity of correlating minute bolt stretch with tension. Reliability might be attainable in the lab, but in the engineering environment, it is often not realistic. Variations caused by workplace conditions, the operator, and the bolt itself tend to be of greater significance than the capability of the instrument.

The main factors affecting the reliability of measuring bolt stretch are the subject of this section.

29.3.1 Operating Environment

The in-service environment affects not only the tension-measuring system but also the operator. Deepsea, nuclear reactor, and long-wall mining applications are obvious examples where an operator's capability to use a control system would be impaired, thereby affecting the reliability and accuracy of any reading.

Even the location, i.e., the accessibility, of the bolted joint can affect procedural capability.

29.3.2 Temperature

Variations in operating temperatures usually cause variations in extension measurements. Temperature affects both bolt behavior and the measuring system; for example, ultrasound velocity varies with temperature. Many bolt-tightening procedures recognize the effect of temperature on correlation reliability. Generally, they call for recalibration of the measuring equipment and/or procedure if temperatures vary by more than ±2°C. With the more complex ultrasound techniques, compensation for the speed of sound is required.

In practice, significant temperature changes occur during round-the-clock assembly operations. Recalibration is, of course, time-consuming and expensive. Variations are perhaps most significant where regular bolt tension monitoring is part of planned maintenance procedures.

29.3.3 Measurement Datum Point

The integrity and consistency of the measurement datum point are crucial to extension accuracy. For optimum accuracy, each of the traditional extension measurement systems must be used on fasteners that have flat, perpendicular, parallel, and smooth ends. Bolts need to be purchased with identification or grade marks located away from intended datum points. Removal of these marks during datum face preparation creates quality assurance (QA) traceability problems.

In recent years, some suppliers of extension-measuring systems have claimed that datum point surface quality has no effect on accuracy. Operators should be wary of such claims.

Once the datum point has been established, its quality needs to be protected for the service life of the bolt. This is vital if in-service monitoring is an ongoing requirement. Corrosion, contamination, and oxidation can all change the datum point, thereby affecting the accuracy of future readings. The use of plastic protective end caps can be effective in providing long-term protection (see Figure 29.1).

Operator skill or "feel" is an important factor in establishing a consistent, reliable contact point.

Ultrasound techniques are particularly sensitive to datum point variation. The placement of the transducer, instrument movement, couplant layer thickness, and uniformity all cause readout variation. Some users find readout variation from one transducer to another. To minimize these effects, users try to use the same transducer for a particular set of bolts (for more on ultrasonic measurements, see Chapter 34).

29.3.4 Operator Skill

There is little doubt that operators need to understand the equipment they are using and be skilled in its use. On more than one occasion, users have chosen to purchase a particular system because of claims of "ease of use" and then experienced unreliability. The supplier of the equipment then cites lack of operator skill as the reason for poor performance. In reality, the end user has been shortchanged as he often ends up hiring the supplier's technician to operate the equipment he has purchased.

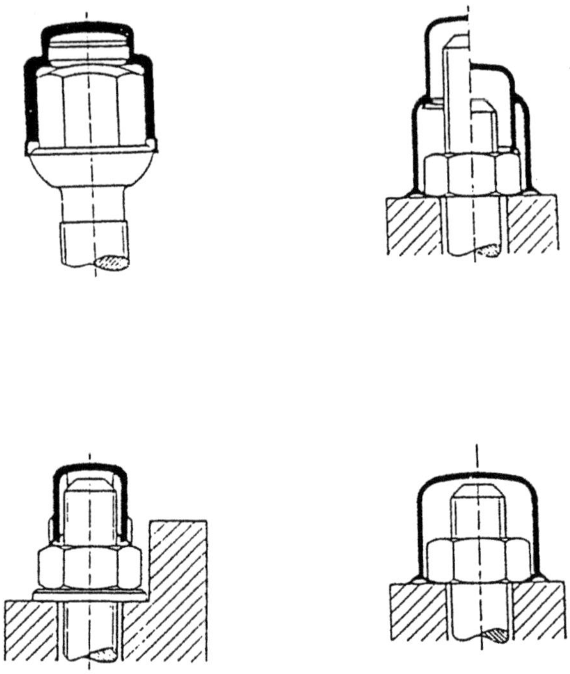

FIGURE 29.1 Plastic caps for protecting exposed fastener surfaces and extension measurement datum faces.

Some end users abandon control altogether, leaving expensive equipment on the shelf, and return to living with less effective bolted connections. This waste of resources can be avoided by using skilled operators from the start.

Experiments have shown that different stretch readings can be obtained on the same bolt by trained operators [1, p. 257]. Similar variations have been observed in the field; major power stations in the United Kingdom allow bolt stretch measurement to be done by only one operator; if that operator is not at work, it doesn't get done.

Any doubt about the integrity of a reading leads to increased time and cost. Readings are open to interpretation; the more skilled the operator, the more accurate the readings. Much of this interpretation centers on datum point selection and uniformity and therefore has a pronounced effect on accuracy.

Organizations that have used extension measurement techniques involving mechanical extensometers or ultrasound over a long period of time have been known to question the cost effectiveness of tension measurement. Improvement in the reliability of these techniques can be achieved through a better understanding of the practical limitations. By taking positive steps to minimize the impact of the operating environment, operator, and fastener quality, reliability will be improved. The greatest improvement, however, will be achieved by applying realistic levels of calibration. This has been recognized by the more technologically advanced extension measurement companies [1, p. 345].

29.3.5 Calibration

Calibration levels used by direct length measurement users, at least up to now, have often been inadequate, resulting in insufficient improvement in joint reliability. Optimum levels of assurance can be achieved only by 100% calibration of all fasteners in a joint. Experience has shown that a number of factors significantly affect the tension–extension relationship. The following factors have been identified by my company, Rotabolt, over the last decade or so. During this time, over 1 million bolts of all types, materials, and strength grades have been load tested. Very early on in the company's life, Rotabolt managers concluded that 100% calibration was essential if any assurance of commercial accuracy could be offered and validated.

Rotabolt converts ordinary fasteners into precision stretch-measuring "instruments" (see Section 29.4.5). The inherent stability of the product has allowed validation/QA procedures not previously applied to direct length measurement.

One result of all this fastener testing has been a better appreciation of how much the fastener load–extension relationship can vary and the causes for this variation. Once the variations are known, their effects can be minimized.

29.3.6 Bolt Variation–Load Extension Correlation

Hooke's law says that stress is proportional to strain. Therefore, bolt load is proportional to bolt stretch. From this, designers assume that once the required bolt extension has been established, the tightening of all future bolts to this stretch value ensures joint reliability.

A homogeneous test specimen of elastic material with a uniform cross section and clearly defined length will exhibit a straightforward stress–strain relationship predicted by Hooke's law as long as it is loaded to a stress level below the elastic limit. Unfortunately, industrial fasteners, although made of elastic material, vary from end to end and from fastener to fastener in metallurgical homogeneity, in dimensions, and often significantly in terms of thermal and stress history.

The heavy engineering and energy industries, among others, obtain their fasteners from a wide variety of sources that use many different manufacturing procedures. These variations can have a significant impact on the stress–strain behavior of an individual bolt. Small-sample statistical sampling of bolts will often result in a misleading definition of a "standard stretch value." In fact, these variations mean that no two apparently similar bolts can be relied upon to stretch the same. One hundred percent calibration is needed to ensure an optimum level of accuracy. This, of course, can have a significant impact on length-measuring systems in terms of time and cost. Property variant techniques, such as ultrasound, are very much affected by internal fastener variations such as composition, structure, and defect levels.

Rotabolt experienced periodic problems with statistically sampled air gaps (bolt extension vs. bolt load). The following examples typify the variations found and justify the decision to 100% calibrate all production indicators and load test all bolts.

Example A

Consider two M36×400 Grade X bridge cable clamp studs for which the following data are obtained [2].

Bolt	Extension	Load
1	4 mm	57.0 tons
2	4 mm	70.0 tons

These results were obtained by independent laboratory load extension tests (Figures 29.2 and 29.3). The tests were commissioned by consulting engineers to provide Rotabolt with specified bolt load values. The decision was made to replace the original length-measuring system procedure with Rotabolt control (described in Section 29.4.5). Improvements in bolting procedures were a significant part of a bridge strengthening project needed because of the increase in traffic levels since the bridge was originally built. The variation tabulated here, obtained under clean controlled conditions by skilled personnel, highlights the problems with load extension correlation.

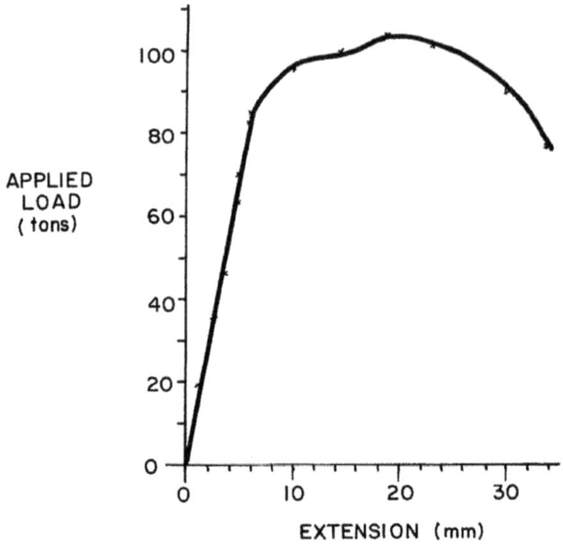

FIGURE 29.2 Axial load vs. extension for M36×482 stud bolt No. 1 (Rotabolt test).

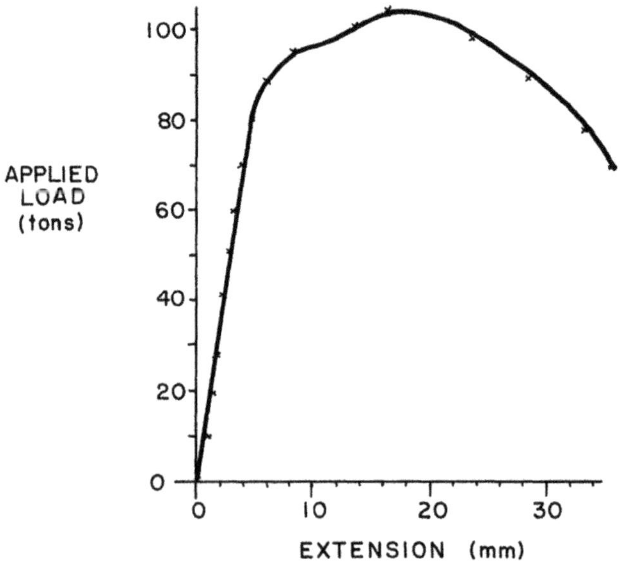

FIGURE 29.3 Axial load vs. extension for M36×482 stud bolt No. 2 (Rotabolt test).

Example B

This example concerns a crane structural bolting application using M30×180 mm hex head bolts of strength grade 10.9 and 65 mm gauge length [3]. The data obtained on 98 bolts were

Bolt Group	Mean Stretch (in.)	Load (tons)
A	0.0040	21.4
B	0.0087	21.4

Five bolts (group B) out of 98 sample bolts tested at Rotabolt experienced double the extension of the rest of the batch under a test load of 21.4 tons. Basic evaluation at Rotabolt involving hardness and dimensional testing revealed no explanation for this substantial variation.

29.3.6.1 Effect of Nut Mass

Load extension calibration should ideally be carried out using the actual nuts to be used in the installation. If this is not possible, a nut of the same style and dimensions must be used. If a different nut mass is used, variations will occur in the load–extension relationship. The factor responsible for most variation is the relationship between circumferential mass and nut height.

This was first noticed during calibration of indicators for conrod studs on high-speed marine diesel engines. The indicators were calibrated in the factory using standard nuts. The studs were installed on an experimental engine where all important components including the fasteners were strain gaged. On installing the fasteners, discrepancies were observed between the strain gage readings and the bolt tension indicators. Investigations finally uncovered the reason for the apparent indicator inaccuracy: nut mass. Subsequent resetting involving the bottom end of the conrod rather than a standard nut resulted in identical correlation between installed bolt tension indication and the strain gaging (Figure 29.4).

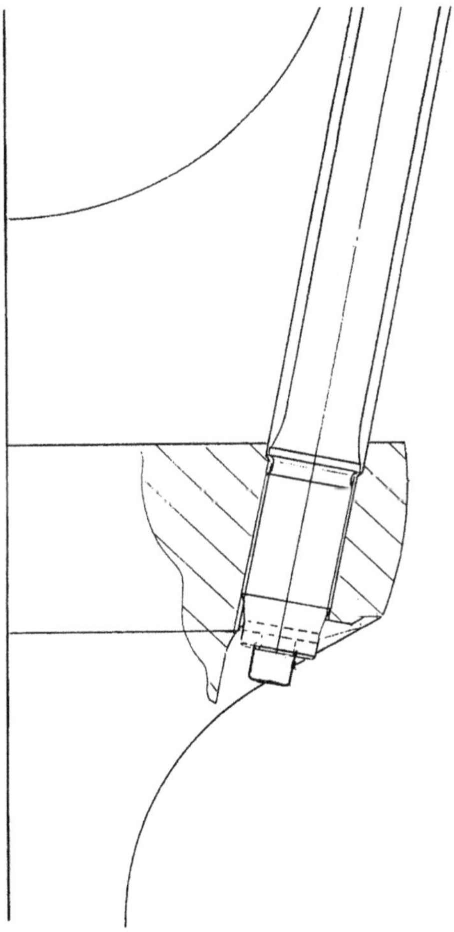

FIGURE 29.4 Conrod tapped thread nut mass for M22×2p V Grade Rota stud.

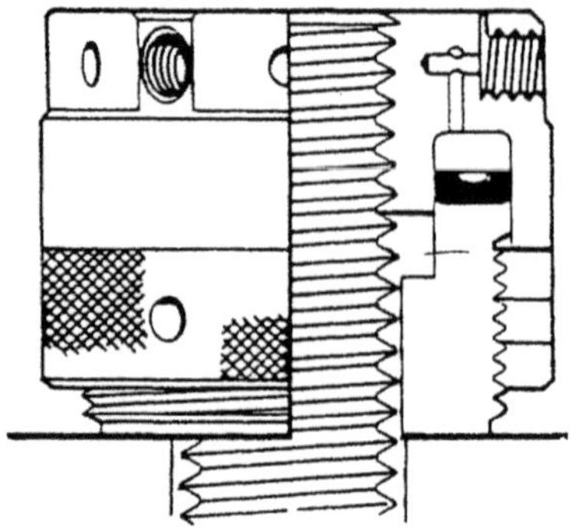

FIGURE 29.5 Schematic drawing showing a typical hydraulic tensioning nut.

Fastener description	M22 ×2p× 350 mm BS 970 817 M40 'V'
Nut used	BS 3692 M22 nut AF 32 mm/AC 37 mm/18.0 thick
Conrod mass	45 mm height
Set load	16.0 tons

29.3.6.2 Hydraulic Nut Differences

A popular tightening method for bolt applications requiring high tightening power in confined space is the hydraulic nut. This is a hydraulic tensioning nut (see Figure 29.5) that stays with the bolt after tightening. Load transfer is initiated by an integral collar between the hydraulic puller and the nut's bearing face. As with conventional hydraulic tensioners, load transfer relaxation is unpredictable, so subsequent residual loads are unknown and uncontrollable. Hence the need for a tension-measuring system to control the installation tightening.

It has been noticed that differences in extension can occur with nuts from different manufacturers. In this case, nut mass does not explain the variation as the nuts are apparently dimensionally the same. Whatever the cause, the following variances are typical.

Nut	M30×3.5
Bolt	M30×235 mm long 10.9 grade allthread stud
Bolt load variation	
Nut A	23.2 tons
Nut B	20.3 tons

29.3.6.3 Residual Stresses

Bolts of certain materials when subjected to loads within their elastic limit do not return immediately to their original condition after release of load. One popular material in particular, ASTM A193 B7, is prone to this phenomenon [4]. The problem was initially identified during the preparation of tension control indicators for offshore fasteners. When the indicator bolts were checked 2 days after production prior to an independent Lloyd's load test witness, the indicators were found to be inaccurate, despite 100% quality load tests on the finished product. Further investigation showed that the fastener load setting had actually increased.

The technology of the tension control indicator was such that an increase in load could be created only if the bolt had effectively shrunk over a certain time period. This meant that the bolt did not return to its original unstressed condition immediately after its initial load test. This delayed spring-back effect is thought to be due to residual stresses induced by the fastener's manufacturing deformation and thermal history. Similar experiences have been observed with some fasteners manufactured from certain high alloy content steels. The occurrence is frequent and would suggest that B7 studs produced via a particular production route are more prone to this phenomenon. Unfortunately, B7 studs can be produced by three or four different production routes depending purely on the manufacturer's preference.

Analysis of 73 batches of B7 bolts processed by Rotabolt over a 1-year period showed that nearly 70% exhibited delayed spring-back effects. There was no apparent correlation between load level, diameter, length, or source.

Rotabolt load test experience suggests that the greater the level of cold work and residual stress level, the greater the elastic instability. Load testing of SAE Grade 2 railcar wheel bearing cap screw bolting showed significant residual stress effects [5]. The fasteners were manufactured by cold heading and cold thread rolling with no subsequent heat treatment. Apparently, most fasteners of this type are supplied in this way despite the fact that the specification calls for a stress relieve option.

Delayed spring-back effects caused up to 22% variation in the load–extension relationship. Indeed, with this type of fastener, it was virtually impossible to obtain a stable, consistent load–extension relationship. Subsequent heat treatment trials demonstrated that stress relieving reduced the instability. Bolts in the fully heat-treated condition similar to SAE Grade 5 were completely stable.

29.3.6.4 Bolt Yielding Effects

Fasteners are designed to operate in their elastic range. Bolt extension measurement is feasible only in this range. If it is loaded into the nonlinear plastic zone, the fastener becomes permanently deformed and elongated. Any subsequent extension measurement will always give an erroneous higher bolt load indication than actually exists.

A designer should always ensure that maximum bolt loading is within a bolt's elastic limit. Traditionally, designers use 0.1% or 0.2% of permanent elongation to define the bolt's yield strength. Unfortunately, load testing on a wide range of bolt sizes, alloys, and selected maximum load conditions has shown that 0.2% permanent set value is an unreliable parameter for fastener yield.

The common 0.2% permanent elongation is obtained from standard tensile tests (ISO 6892) carried out on machined test pieces. The offset based on 0.1% or 0.2% of the test gage length is obtained by drawing a line parallel to the elastic linear position of the stress–strain curve (see Figure 29.6). The point at which the offset line cuts the stress–strain curve is the 0.2% value.

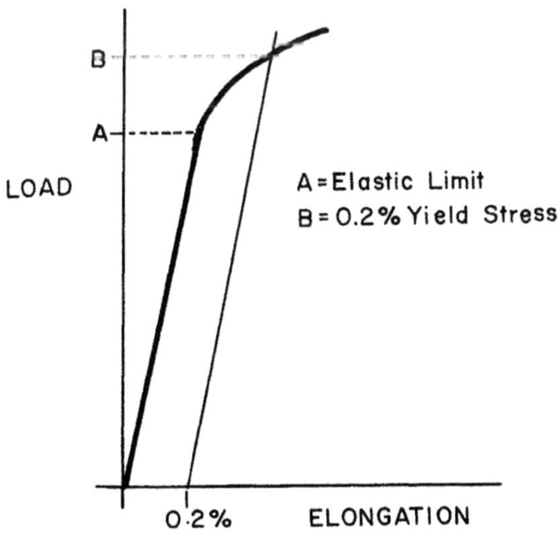

FIGURE 29.6 Typical load–extension curve for a small diameter, medium carbon low alloy steel bolt, showing the 0.2% permanent elongation point compared to the elastic limit.

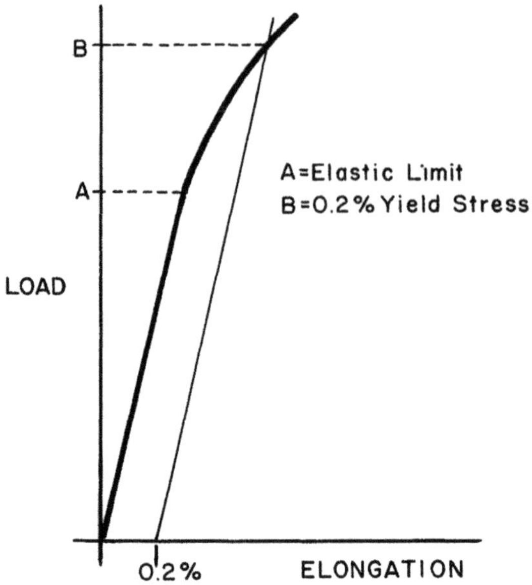

FIGURE 29.7 Typical load–extension curve for nonferrous alloys and large diameter alloy steels, showing the 0.2% permanent elongation point compared to the elastic limit.

TABLE 29.1
Elastic Limit of some non-ferrous materials

Alloy	0.2% Stress at Permanent Elongation (N/mm²)	Stress at True Elastic Limit (N/mm²)
Monel K500	655	475
Ferralium 255	649	442
Marinel 21A	720	487
Titanium Beta C	834	679
Titanium 6ALV4 ELI	830	593

The stress–strain curve characterized in Figure 29.6 has a 0.2% value only marginally higher than the bolt's actual elastic limit. The stress–strain curve in Figure 29.7 exhibits the same 0.2% value as Figure 29.6, but the bolt's true elastic limit is substantially lower.

Experience has shown that the reliability of 0.2% permanent elongation as a yield indicator worsens with increasing diameter, higher alloy content, and decreasing strength grades for medium carbon low alloy steels. The problem is exaggerated with nonferrous stainless alloys, as can be seen in Table 29.1.

Any fastener selection involving bolt loads in excess of 50% of the 0.2% permanent elongation runs the risk of yielding the fastener. If the most accurate extension measurement tension control is to be used, it is vital that all fasteners in the batch be calibrated. If not, the risk of erroneous results and subsequent joint reliability problems is significantly increased. In addition, bolt yielding during installation will be avoided if loads are below the 50% value.

Some significant differences between the 0.2% stress and true elastic limit have been observed even with B7/L7, the standard energy industry bolt [6]. Process variations at the raw material bar straightening stage can cause elastic limits to be 35% lower than the minimum 725 N/mm^2 0.2% stress specified. In this situation, any bolt load greater than 65% of 0.2% stress will cause the fastener to yield and adversely affect the reliability of the measuring system.

Risks of overtightening are particularly high when hydraulic tensioning is used. The hydraulic overload needed to compensate for load transfer relaxation is often a problem. Any load extension calibration needs to incorporate loading to the anticipated overload values so as to ensure that the fastener will not yield. The following example is typical of the problems that can arise in controlling hydraulic tensioner tightening of nonferrous fasteners [7].

Application	Structural clamps
Fastener	M33 Monel K 500 studs
0.2% stress	655 N/mm^2
Design tension	470 N/mm^2
Hydraulic overload	572 N/mm^2
Actual yield	475 N/mm^2

This was one of the first cases highlighting problems in basing bolt strength capability on 0.2% yield stress.

This problem can be eliminated at the fastener procurement stage by calling for a proof load test [8] on a full-sized bolt. This involves loading the fasteners to a minimum specified value. The fastener lengths before and after the test are compared for any permanent set. In effect, this test determines a fastener's minimum elastic limit or true yield.

Standard specifications such as SAE J429, BS 3692, ISO 898, and BS 1768 do incorporate a load test. Unfortunately, the energy industry bolting specs such as ASTM A193 B7 and BS 4882 do not.

Bolt yielding on the finished or supplied fastener often creates problems for the end user. Invariably bolts are the last item to be purchased during the last few weeks of a project's critical path. There is often insufficient time to repurchase. This leads to some surprising decisions:

1. Reduce bolt loads to avoid yielding.
2. Take bolts and use them, regardless of the situation, without tension control.
3. Select an extension-measuring system that will not detect fastener yielding.

Another option often tried is to recalibrate using the theoretical new elastic line created by yielding. In theory, the fastener should have work hardened to a higher yield point (see Figure 29.8), thereby enabling the original load to be achieved elastically. This procedure can work, particularly with the medium carbon low alloy steels, but its success cannot always be assured. Assurance is even less with nonferrous bolting alloys. Work hardening to a higher value often takes several load cycles before the new elastic line is established. The subsequent elastic load–extension relationship can

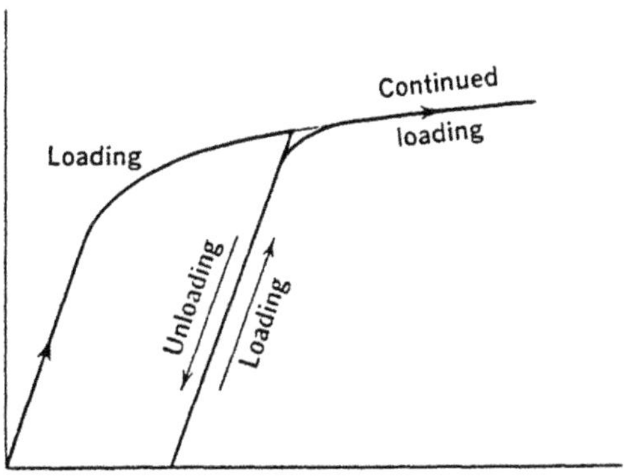

FIGURE 29.8 Effect of loading, unloading, and reloading in the plastic range.

then change with time with the fastener appearing to take up its original yielded length position. These effects have been noted with Monel K500 [6] and martensitic stainless steel.

It is far better to avoid fastener yielding altogether. Bolt yielding is an uncontrollable situation in terms of load–extension relationships. It is also potentially detrimental to the fastener and the reliability of the joint.

29.3.6.5 Service Life

Generally, the use of tension control extends the life of the fastener. However, repeated tightening can mechanically degrade the bolt unless great care is taken with lubrication, etc.

Where fasteners are exposed to corrosive environments, their functionality at the end of the operating period is suspect. The same is true of bolts exposed to temperatures in the creep/oxidation range of bolting alloys. Recalibration of the tension measuring is possible, but often the fastener itself has lost its functionality—nuts seize to studs, etc.

29.3.7 Summary

The use of bolt extension measurement can significantly improve the reliability of the bolted connections. Most direct length-measurement systems will ensure accurate bolt tensioning provided the user pays attention to the fundamentals:

- 100% calibration
- Assured datum points
- Good records
- Ensured grip/gage length
- Trained personnel
- Environmental impact.

Where traditional direct length-measuring systems and procedures have not been successful, one or more of the above points have not been addressed.

Calibration is absolutely vital. Reduced calibration levels cannot compensate for variations in bolt and nut metallurgical and dimensional detail.

29.4 REVIEW OF CURRENT EXTENSION METHODS

29.4.1 C Micrometers

These days the use of C micrometers tends to be limited. Because measurement is based on the total length of the fastener, the system is applicable only to through joints that provide access to both ends of the fastener. C mics cannot be used on blind hole joints.

29.4.2 Depth Micrometers

The use of depth micrometers is more common. Basically, there are two variations on the same theme:

1. A rod runs down the complete length of the fastener.
2. The rod runs along part of the stud. This is more appropriate for longer fasteners. These short rods are often called gage pins.

The measuring rods or gage pins are anchored by tapped holes in the parent bolt or by the use of threaded end plugs. As the bolt stretches under load, the rod, which is not stretched, appears to move into the drilled hole. The depth micrometer measures this displacement. This type of measurement is less prone to operator variances than measurements made with ordinary C micrometers.

The main disadvantage compared to a C micrometer is the expense of fastener preparation. Holes have to be gun drilled to ensure concentric holes, and gun drilling is expensive.

The depth of the hole or gage length also has to be accurately controlled, which is difficult. As a result, accuracy can be a problem. Accuracy will improve, however, with 100% calibration.

Even without 100% calibration, micrometer control can be more accurate than many of the alternatives such as torque control and is therefore widely used on large diameter fasteners, especially in the process chemical and power industries.

29.4.3 Gage Pin Bolts

The use of gage pins inserted into the fastener enables direct bolt extension measurement. The system is prone to error, especially if the reference or datum point happens to be on the joint or flange surface, because during tightening these surfaces deform, deflect, or embed. For optimum accuracy, therefore, it is vital that the extensometer or dial gage be mounted to the fastener itself.

An interesting variation on the gage pin method is the use of an extensometer with its own built-in age pin (see Figure 29.9). In operation, the gage pin is inserted into the fastener through

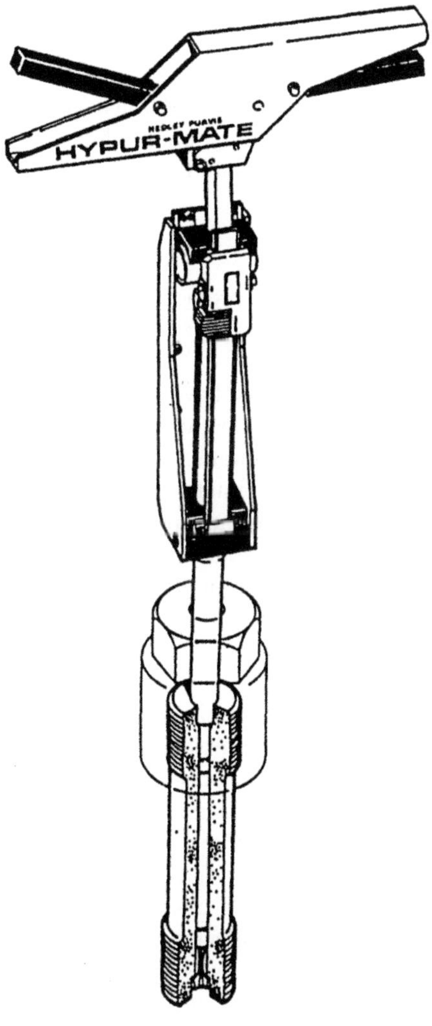

FIGURE 29.9 Hypur-Mate bolt extensometer by Hedley Purvis (UK).

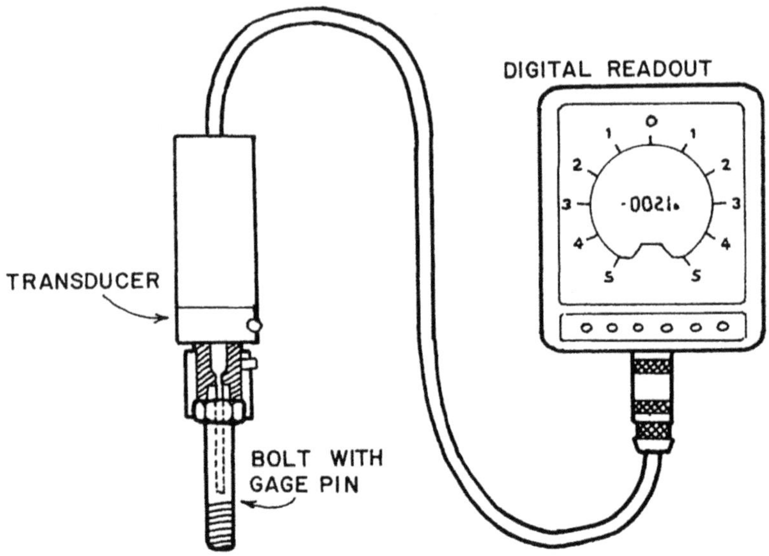

FIGURE 29.10 SPS monitor bolt.

a hole until an upper stop comes into contact with the fastener surface. A spring-loaded collet is opened that locks this position and anchors the pin in the hole. The digital readout is then zeroed. Any subsequent stretch in the bolt is then measured.

There are two advantages to this system:

1. Operator variations are minimized.
2. The datum point remains consistent provided fastener surfaces are not allowed to deteriorate.

The main disadvantage is that the fastener must have a gun-drilled through hole. This significantly increases the fastener cost for the majority of applications. End-stop fastener surfaces must still be flat and square. As with all other systems, calibration of all fasteners is needed for the best accuracy.

Another gage pin style of bolt made available in the mid-1980s was the SPS monitor bolt (Figure 29.10). The displacement between the gage pin and the bolt head was measured electronically off the machined gage pin head. The system is applicable to most fastener forms. Once again, reliability is dependent on the quality of the datum point and level of calibration.

29.4.4 Ultrasound

A fully detailed description of an ultrasound system is given in Ref. [1]. The technique is a property variant method and uses a transducer to send ultrasonic pulses through the fastener. The wave will echo off the far end and return. The time of the journey can be measured and is proportional to bolt length. As bolt length increases, the speed of sound decreases proportionally, further increasing journey time. The machine can process this information in terms of extension, stress, or tension. See Chapters 34 and 43 for further discussion.

There is no doubt that ultrasound is capable of measuring very small changes in bolt length. However, the engineering environment has an adverse effect on the system's effectiveness. Variations in the environment, along with calibration problems, can reduce the technique's capability. General accuracies of ±5% in the field and ±2% in the lab have been quoted. Such accuracies are possible in both environments only if 100% bolt calibration is included.

To put system effectiveness into perspective, the UK nuclear power generation industry has concluded that current ultrasound systems are not accurate enough to control important bolted connections [9]. This conclusion probably reflects the operator problems caused by the power generation environments rather than the potential accuracy of the equipment. Insufficient calibration would also have contributed to its reduced effectiveness.

This preference of the UK nuclear power industry for more mechanical methods is also supported by some on-site tightening companies in the United Kingdom. They feel that bolt installation is more effective using extensometry and/or micrometry. The latter techniques are not so adversely affected by the operating environment.

In spite of the difficulties, however, properly calibrated ultrasonic equipment used by well-trained mechanics can provide more accurate bolt preload control than do such options as torque control and has found widespread use as a QA tool or on critical joints. See Chapters 42 and 43 for further discussion.

29.4.5 Rotabolts

Rotabolt is a new direct tension indicating system that currently reflects state-of-the-art stress measurement in terms of accuracy, reliability, durability, and ease of use. Rotabolt is a gage pin style of bolt with tension indication given by simple finger feel. Indicator reusability provides quick and easy checkability and condition monitoring in service (Figure 29.11).

Rotabolt indicators come in two forms: Rotabolt 1 and Rotabolt 2. The original Rotabolt 1 provides a single tension indicator, whereas Rotabolt 2 offers two tension values that can be set at the required operational range.

The technology involves converting standard bolts and studs to Rotabolts. Initially, the fastener end is machined to provide a good quality square datum point. A hole is then drilled down the center of the fastener to create a gage length over which bolt stretch is going to be measured. There is no need to drill the complete length of the fastener. Drill depths are sufficient to provide the accuracy and sensitivity needed at the lowest cost. Although the Rotabolt system is based on Hooke's law and measures extension for a particular bolt load, all calibration and finished product load testing is done against bolt tension, not stretch; it is therefore a direct tension indicator as opposed to an extension-measuring device.

A headed gage pin is then inserted and eventually permanently fixed into the core of the fastener. The headed pin retains a Rota disc. The indicator is calibrated by physically loading the bolt to the required setting and measuring the stretch developed over the gage pin length. An air gap is then

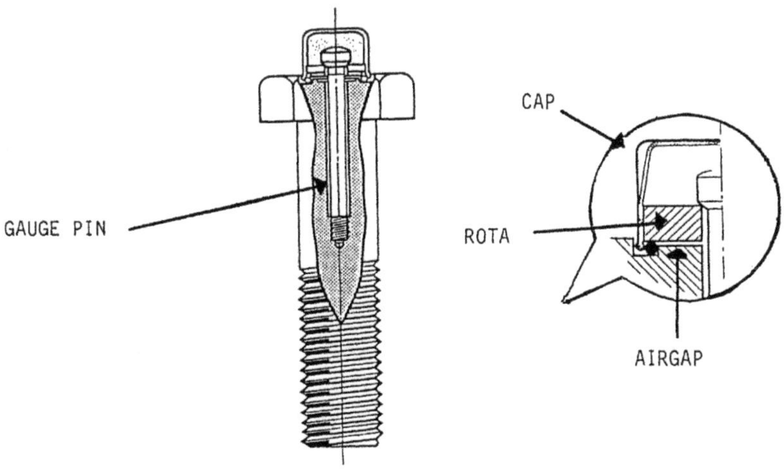

FIGURE 29.11 Rotabolt 1 indicator in a converted hex head bolt.

created between the machined datum point on the fastener end and the retained Rota. At no load, the Rota is able to spin in the set air gap. As the bolt is tightened, it starts to stretch, and the air gap decreases. At the required load the bolt stretches the set amount, and the Rota comes up against the machined datum point; the air gap is closed. At this point, the Rota can no longer be turned. The operator knows that the correct tension has been achieved.

If tension is lost after tightening, the air gap opens and the Rota can be turned. Reusability enables the operator to retighten until the Rota locks again. Instant condition monitoring is also provided.

Rotabolt indicators are finished with stainless steel caps that crimp onto the outer knurl of the Rota. They become a permanent part of the assembly and provide a comfortable finger checking facility.

Rotabolt 2 is based on the same technology. This time, however, a dual indicating cap encases a normal "main" gage pin and Rota. A second smaller pin that retains a second Rota is inserted into the main pin. Once again, respective air gaps and load settings are established by physical load tests.

What makes Rotabolt state of the art is the load test calibration procedure and product design.

The fastener variations highlighted previously make it very difficult to asssure load extension because of load–length correlation, elastic instability, and unexpected bolt yielding. These effects are identified and compensated for in the Rotabolt conversion process. Every Rotabolt is 100% load tested so as to ensure ±5% general accuracy. Fasteners showing residual stress effects are load tested 24 h after initial calibration. Those fasteners that are found to yield under the required loading conditions are immediately identified, and remedial action is taken. This usually means bolt replacement, although sometimes, when times do not allow for a bolt replacement, users will reduce the bolt load requirement if the design allows.

Invariably, time is a problem when a bolt is found to be out of specification. The fastener is the last item purchased despite being on the critical path. It is only when a problem occurs with the bolt that the engineer finally recognizes how important the component is.

Another course of action tried in such situations is to yield the bolt several times to work to harden the fastener to a higher elastic limit such that the required design load is achievable elastically. This practice is known as scragging. This drastic action can work, but its effectiveness is not certain or assured. Medium carbon low alloy steel fasteners are best suited to the practice.

With stainless steels and nonferrous bolting alloys, the effects of scragging or yield/work hardening are totally unpredictable. The creation of a new higher elastic limit can take up to five or six load cycles. The new limit is not immediately created as suggested in the schematic of Figure 29.6.

Experience with Monel and some heat-treatable stainless steels has highlighted some eccentric behavior. Load–extension data taken from the new higher value elastic line appears to revert to the original yield situation when reloaded 24 h later. This unpredictable behavior means that the nonferrous bolting alloys should not be loaded anywhere near their elastic limit if reliable, consistent load–extension behavior is to be achieved. Indeed, with alloy steels, the scragging practice should be regarded only as a last gasp desperate action. Great care must be taken. Tension control and procedural accuracy must be of the highest order, and scragging should not be attempted on-site.

Residual stress effects and fastener yielding change the Rotabolt indicator and are immediately identified by the load test. Where hydraulic tensioning is involved, the anticipated hydraulic overload to be used on installation is incorporated into the production load test so as to ensure that the parent bolt is not embarrassed.

Environmental conditions can have a significant effect on the accuracy and reliability of extension measurement techniques. A common problem is the effect on the measurement datum point. The Rotabolt design protects the air gap datum point by use of the protective cap, labyrinth seal and O-ring, and internal grease barrier. Ingress of a wide range of media—water, dust, paint, etc.—is prevented.

Accelerated corrosion testing involving combined maximum severity salt spray to MIL STD 07–55 and battery acid did not cause indicator seizure over a 28-day test period [10].

The Rotabolt design is such that the datum point is always the same.

One disadvantage Rotabolt has compared to other direct length measuring methods is that it provides a single load (i.e., yes/no) or an operational tension range (high load-low load). Direct length measurement systems can provide the load at that particular time. This disadvantage is probably more significant in R&D-type applications.

29.5 CONCLUSION

Wherever possible, tension measurement systems should be used if assurance of safety and reliability is needed. The use of tension control will improve the reliability of any connection.

There are a range of systems available that can be used to control installation. In-service monitoring can also be provided; however, the effects of the operating environment must be considered by the user.

The systems and products mentioned in this chapter do not represent every option in existence. They do, however, represent those that are commercially available and in regular use.

Over the last 10 years gage pin bolts based on Poisson's ratio have been experimented with, where the fastener section reduction is assessed as opposed to its stretch at a particular load. Other systems use a preloaded hair grip type of pin that deflects when the bolt stretches and relieves the preset stress in the pin. Systems such as these, regardless of their future potential, at present do not offer a viable option to industrial users from a technical or commercial viewpoint.

If the user recognizes limitations in service life and plans accordingly, the benefits of accurate control will follow.

REFERENCES

1. Bickford, J. H. (ed.), *An Introduction to the Design Behavior of Bolted Joints*, 2nd ed., Marcel Dekker, New York, 1990.
2. Rotabolt Ltd, Contract No. R2043. Dudley, 1987.
3. Rotabolt Ltd, Contract No. R5285. Dudley, 1994.
4. Corbett, R. F., The importance of high integrity bolt tension control, *2nd International Symposium on Fluid Sealing (CETIM)*, Labaoule, 1990.
5. Corbett, R. F., The effect of fastener manufacturing routes and material selection on the performance of bolted connections, *3rd International Symposium on Fluid Sealing (CETIM)*, Biarritz, 1993.
6. Corbett, R. F., The importance of high integrity bolted joints for engineering pipelines, *Proceedings of the Institution of Mechanical Engineers, Piping Engineering and Operation Conference C45CI/030*, 1993, pp. 53–64.
7. Scott, J. A., Studbolting experiences in the underwater environments of the North Sea, *Paper OTC 5050 presented at 17th Offshore Technology Conference*, Houston, TX, 1985, pp. 269–275.
8. BS EN 20 898-1, Mechanical properties of fasteners. PTI Bolts, Screws and Studs, 1991.
9. Nuclear Electric Operations, Memorandum NEOM 102, the Selection, Care and Replacement of Threaded Fasteners Operating at Temperatures Above 375°C, 1994.
10. Lakeman, J. B. Evaluation of "Rotabolts" for submarine main battery terminals (U), Defence Research Agency PRA/AWS TM AWME 1365 (L), 1993.

30 Use of Ultrasonics in Bolted Joints

Sayed A. Nassar

30.1 INTRODUCTION

Ultrasonics is used for high precision measurement of fastener elongation during the tightening process, but may also subsequently be used in service for measuring the residual bolt elongation. Round-trip transit time of longitudinal sound waves is measured and converted to distance equal to twice the bolt length, using the speed of sound traveling in the fastener material. Transit time is measured and recorded once before the tightening. Subsequent measurements of the transit time are directly proportional to the fastener elongation. The fastener preload is then determined by multiplying the elongation times the spring rate of the fastener.

Most control methods for bolt tightening have drawbacks and limitations. But each of these methods is good enough for many applications, thanks to the fact that most bolted joints are overdesigned and/or to the fact that we are usually not too concerned about the consequences of an occasional failure in less critical applications. In more and more applications, however, we are becoming concerned and would like to find a better way to control bolt tension and/or clamping force. Fortunately, a better way to control bolt tension, at least, is emerging: the use of ultrasonic techniques. These techniques allow us to get past dozens of the variables that affect the results we achieve with torque and/or turn control. They allow us to see and compensate for the elastic interactions and other factors that limit the accuracy with which we can tighten bolts with hydraulic tensioners. In short, they give us a new, more accurate, and often more convenient way to get the advantages of stretch control or strain gage control.

30.2 FASTENER ELONGATION (BOLT STRETCH Δ_B)

Fastener tension and the resulting clamping force in a bolted joint are directly proportional to the fastener elongation created by the tightening of the fastener at assembly. Similarly, the residual tension in a fastener remains directly proportional to the residual elongation. In contrast with the torque–tension relationship, friction plays no role in the correlation between bolt stretch and bolt tension. Consequently, the fastener elongation is a much more reliable predictor of the bolt preload and joint clamping force, provided that the spring rate of the fastener is accurately determined. In the elastic range, the relation between fastener tension F_b and the corresponding elongation Δ_b is

$$F_b = k_b \Delta_b \text{ (lb, N)} \tag{30.1}$$

where k_b is the spring rate of the fastener (lb/in., N/mm). This relationship is illustrated in Figure 30.1.

For a desired fastener tension F_b, the fastener must be sufficiently tightened that it stretches by

$$\Delta_b = F_b / k_b \text{ (in., mm)} \tag{30.2}$$

In some automotive applications, the fastener elongation Δ_b may be as small as 0.001 in. This situation requires measuring the elongation using instrumentation with high precision and high resolution such as strain gages or ultrasonics.

DOI: 10.1201/9781003501183-35

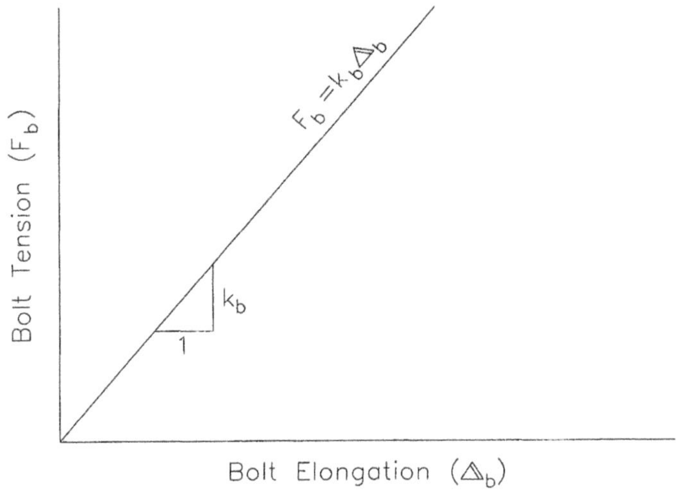

FIGURE 30.1 Elastic relation between fastener tension and elongation.

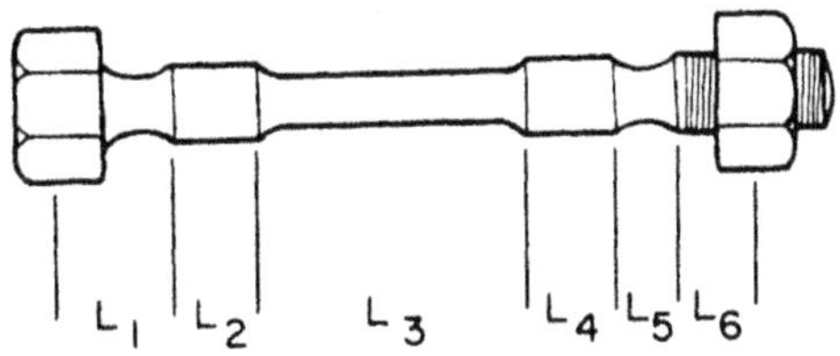

FIGURE 30.2 The stiffness k_i for each segment must be considered (from Ref. [1] with permission).

30.3 FASTENER SPRING RATE K_B

Industrial and research experience shows that the most reliable method to determine the spring rate k_b (bolt stiffness) of a given fastener is by testing. The test fixture must resemble the actual fastener in the actual joint and with the same grip length L. The fastener is then pulled in a tensile test, and the tensile load in the fastener F_b is plotted against the corresponding fastener elongation Δ_b as shown in Figure 30.1. The slope of the straight line is the spring rate of the fastener, k_b.

An estimate of the fastener spring rate k_b is given by

$$k_b = \frac{A_b E_b}{L} \left(\text{lb/in., N/mm}\right) \tag{30.3}$$

provided that the entire grip length L has the cross-sectional area A_b. The fastener material is represented by its Young's modulus E_b

$$\frac{1}{k_b} = \frac{1}{k_1} + \frac{1}{k_2} + \frac{1}{k_3} + \frac{1}{k_4} + \ldots \tag{30.4}$$

where

$$k_i = \frac{A_i E_b}{L_i},\ i = 2,\ 2,\ 3,\ 4,\ \ldots \tag{30.5}$$

is the stiffness of each segment of the fastener and A_i is the cross section for each segment L_i of the fastener.

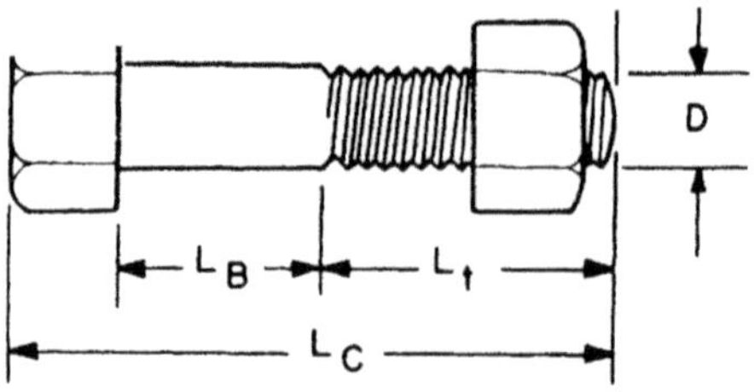

FIGURE 30.3 Example of fastener spring rate k_b.

Example[1]

Let us compute the stiffness and change in length of a ⅜″–16×1½ SAE Grade 8 hex bolt (shown in Figure 30.3) in a joint having a 1 in. grip length L_G. We can get the nominal dimensions we need by measuring a sample bolt or, more safely, by referring to the pertinent specifications as follows:

Specification	Dimension
SAE J 104 (nut)	Height of nut (T_N)=0.3285 in.
SAE J 105 (bolt)	Height of head (T_H)=0.2345 in.
	Thread length (L_T)=1.000 in.
ANSI Bl.1-1974 (thread)	Tensile stress area of threads (A_S)=0.0775 in.^{2}

From the description of the bolt we already know, of course, that

Nominal body diameter (D)=0.375 in.
Nominal shank length−1.500 in.

We must now compute the nominal cross-sectional area (A_b):

$$A_b = \frac{\pi D^2}{4} = \frac{\pi(0.375)^2}{4} = 0.1104 \text{ in.}^2$$

The nominal length of the body (L_B) is

$$L_B = L - L_t = 1.5 - 1.0 = 0.5 \text{ in.}$$

and the effective lengths of body and threaded sections, remembering that we are interested only in the threads that are actively engaged in carrying load, are

$$L_{be} = L_B + T_H/2 = 0.5 + 0.1173 = 0.6173 \text{ in.}$$

and

$$L_{se} = L_G - L_B + T_N/2 = 1.0 - 0.5 + 0.1643 = 0.6643 \text{ in.}$$

We are now ready to compute the reciprocal of the spring constant of this bolt in this joint:

$$\frac{1}{k_b} = \frac{L_{be}}{EA_b} + \frac{L_{se}}{EA_S} = \frac{0.6173}{30\times10^6\times0.1104} + \frac{0.6643}{30\times10^6\times0.0775}$$

$$= 0.4708\times10^{-6} = 04708\times10^{-6} \text{ in./lb}$$

$$k_b = 2.124\times10^6 \text{ in./lb}$$

Once we have this, we can compute the change in length for a given force. What force should we use? A typical tightening specification may be "60% of yield strength," which, for an SAE Grade 8 bolt, would be $0.6 \times 130{,}000$, or 78,000 psi.

To convert this desired stress level to axial force, we multiply the stress value by the tensile stress area of the threads:

$$F_b = \sigma \times A_S = 78{,}000 \times 0.0775 = 6{,}045 \text{ lb}$$

We can now compute the fastener change in length that this tensile force would create.

$$\Delta_b = F_P\left(1/K_b\right) = 6{,}045 \times 0.4708 \times 10^{-6} = 0.00285 \text{ in.}$$

30.4 BASIC CONCEPT OF ULTRASONIC MEASUREMENTS[2]

In bolted joint applications, the type of ultrasonic instruments most commonly used to measure fastener elongation is a "pulse-echo" or "transit time" instrument [1]. A small acoustic transducer of some sort is placed against one end of the bolt being controlled (see Figure 30.4). An electronic instrument delivers a voltage pulse to the transducer, which emits a very brief burst of ultrasound (typically two or three cycles). This burst passes down through the bolt, echoes off the far end, and returns to the transducer. The electronic instrument measures very precisely the amount of time required for this burst of sound to make its round trip in the bolt.

As the bolt is tightened, the amount of time required for the ultrasound to make its round trip increases for two reasons:

1. The bolt stretches as it is tightened, so the path length increases.
2. The average velocity of sound within the bolt decreases because the average stress level has increased.

Both of these changes are linear functions of the preload in the fastener, so the total change in transit time is also a linear function of preload. As a result, ultrasonic equipment is used primarily in applications involving relatively few bolts in critically important joints, or it is used for quality control audits or to set up and calibrate other types of assembly equipment (torque, torque–turn, etc.) or to conduct laboratory or field experiments.

Currently, available ultrasonic equipment is ideal for QC and laboratory roles because it can monitor bolt tension not only as it is being developed in a bolt during assembly but also at any time afterward. It can measure residual preloads after embedment or elastic interactions; it can show

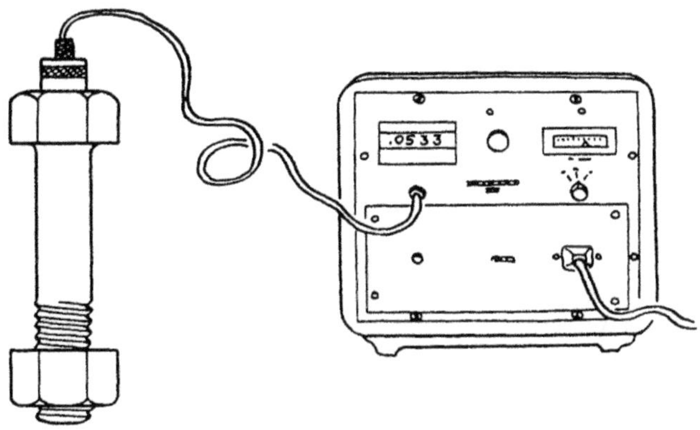

FIGURE 30.4 An acoustic transducer is held against, or mounted on, one end (either will do) of the fastener to measure the fastener's change in length as it is tightened.

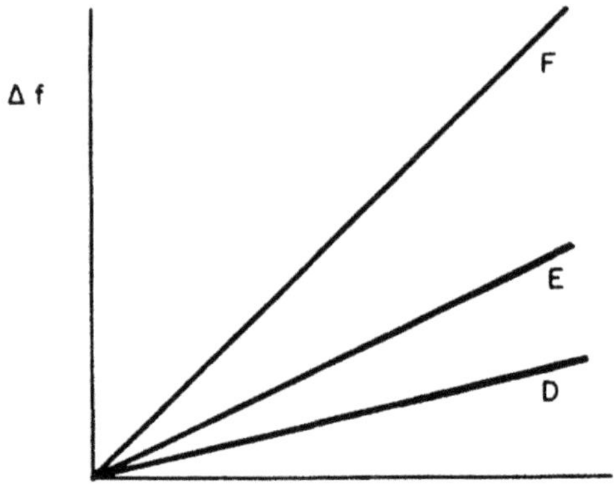

FIGURE 30.5 A transit time ultrasonic instrument responds to a change in bolt length (curve D) and a change in the velocity of sound through the bolt (curve E). Both changes increase the transit time linearly, as stress (or preload) in the bolt increases. Curve F is the total change in transit time as the fastener is loaded (from Ref. [1] with permission).

changes in bolt tension as external loads are applied; it can monitor the effects of temperature change. Thus, ultrasonics provides an alternative to strain-gaged bolts.

Currently available instruments can be used on virtually any kind of bolt material, including steel, aluminum, titanium, Inconel, MP35N, and other exotic materials. There are, however, some practical limitations.

The instruments cannot be used on bolts whose surface temperatures exceed 400°F because higher temperature transducers are not available. Both ends of the fastener must have relatively large flat surfaces on which to mount the transducer and/or against which to generate an echo. Socket head screws can be a problem, therefore, unless they are large enough to mount the transducer down within the socket. Furthermore, the bottom of the socket hole must be machined flat.

The instrument has been designed to measure the change in transit time that occurs during tightening and to interpret and report the results as either a change in length of the fastener, a change in stress level within the threaded region of the fastener, or a change in the tension within the fastener.

The preceding description of "how it works" is obviously simplified. We will take a more detailed look when we discuss currently available instruments and the accuracy with which we can measure bolt stress or strain at the present state of the art. In fact, we'll take a much closer look at how ultrasonic instruments work than we did at torque or tensioning equipment because ultrasonic techniques are only "emerging" at the present time, and relatively little information is available on how they work and on the factors that affect their accuracy. A detailed knowledge of their operating principles and behavior is often required for best results (Figure 30.5).

30.5 CURRENT USE OF ULTRASONICS IN BOLTED JOINTS

Although recent research efforts at Lawrence Technological University have led to the integration of ultrasonics to control the tightening of a fastener using a prototype single-spindle assembly tool, the use of ultrasonics to measure or control bolt stress or strain is not yet common. The equipment and technique are relatively new, and operator training is required. In some applications, an oscilloscope must be used in conjunction with the ultrasonics (at least with the simpler instruments). In any event, operator involvement is required, so the equipment cannot be used "automatically" on a production line. Typical applications currently include such things as

Measuring the residual tension in pressure vessel, reactor, heat exchanger, and other bolts in petrochemical plants.

Controlling the tightening of nuclear steam generator manway studs.
Calibrating and monitoring the performance of multispindle, torque–turn air tools used to assemble automotive engines (e.g., head bolts, main bearing bolts, and connecting rods).
Controlling the tightening of bolted joints in the engines of the space shuttle.
Monitoring the tension of the 40 ft long tie rods in large die-casting machines.
Monitoring the tension in the studs that support the vertical cables on suspension bridges.
Conducting field studies to determine the residual tension in structural steel bolts tightened by calibrated torque or torn-of-nut procedures.
Controlling the tightening of 6 ft long studs used to brace the legs of offshore oil platforms. (This work was done at depths of up to 200 ft, in the North Sea, in February.)
Conducting routine qualification tests on fasteners purchased for automotive production applications (eliminating the need to strain gage a sample of the fasteners).

Note, however, that the use of ultrasonics to monitor or control bolt stretch or tension is an emerging technology; designs and concepts are still evolving. Even if newer equipment has become available by the time you read this, knowledge of earlier products can help you understand the basic techniques and problems of this technology.

30.6 CURRENT TRANSIT TIME ULTRASONIC INSTRUMENTS[3]

30.6.1 Current Transit Time Instruments

30.6.1.1 Change-of-Length Measurement

Figure 30.6 shows two of the most recently introduced transit time instruments. These are digital electronic systems designed to measure the transit time of ultrasonic signals with great precision and then use those measurements to compute the change in length of the bolt, the stress in the threaded region, and/or the tension in the bolt. To do this, the instrument must also be fed certain other information concerning the geometry or stiffness of the bolt and the properties of the material from which it is made. This information is entered manually, by keyboard, or fed in electronically from a data disc or directly from another computer.

The basic principle of operation is as described earlier: The electronics generates brief bursts of ultrasound, which travel through the fastener, echo off the far end, and return to the sending/receiving transducer (which can be mounted on either end of the bolt). Transit time is measured either from the main bang to the first echo or between the first and some subsequent (e.g., the fifth) echo. The main bang technique involves larger signals, which can make life easier for the operator, but bolt ends must often be premachined to eliminate dishing, nonparallel surfaces, raised grade markings, and the like. This is still the most common way to measure larger fasteners, however (those over ½ in. in diameter and over ¾ in. in length).

Measuring transit time from the first to a later echo requires the instrument to detect and respond to weaker signals but allows it to be used on as-received and/or very short fasteners (including those too short to strain gage). It cannot be used on very long bolts (several feet in length).

In any case, a preliminary measurement of transit time is made before the bolt is tightened. The instrument interprets this as an "initial length" for that bolt. This information is displayed by the instrument and is also stored in its memory (the instruments can be programmed for any tightening sequence or strategy of interest to the user).

The bolt is now tightened, in some cases while the transducer is still mounted on one end (either the far end or inside the drive socket). In other cases, the transducer is removed, the bolt is tightened, and then the transducer is replaced. The transit time in the tightened bolt is, typically, 0.3% greater than it was in the untightened bolt.

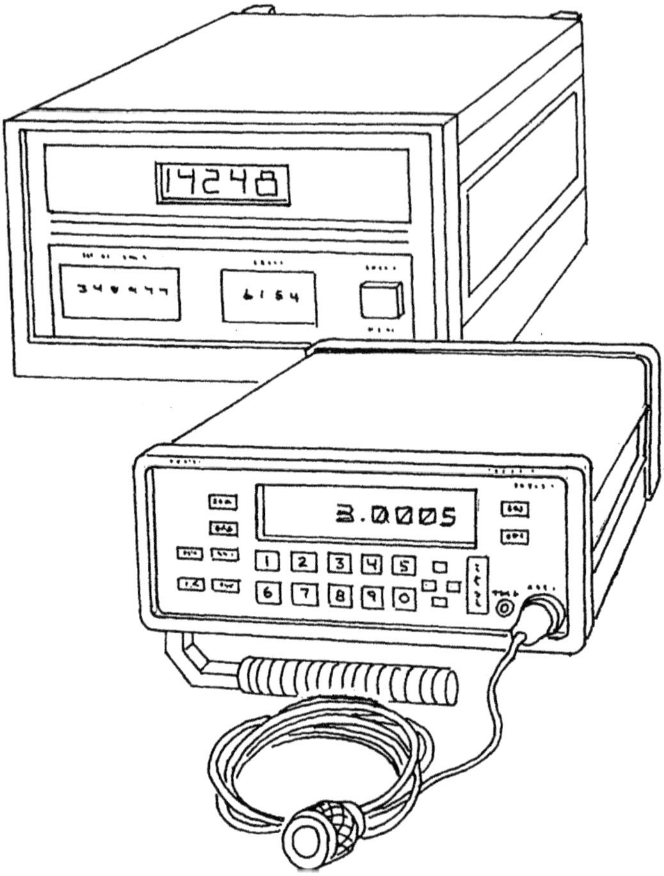

FIGURE 30.6 Two currently available digital ultrasonic instruments from Raymond Engineering. The nearer one includes a built-in oscilloscope for monitoring pulse and echo signals and for displaying digital information. The farther one is designed specifically for use with high-speed production air tools.

A microcomputer in the instrument, or a computer connected to it, now compares the transit time in the unloaded bolt to the final transit time. It then uses the following algorithm to compute the change in length of the bolt that occurred as the bolt was tightened:

$$\Delta L_S = \left[\mathrm{MV}\frac{T_2}{2} - \Delta t \left(\mathrm{MV}\frac{T_2}{2} \right) \mathrm{TF} - \mathrm{MV}\frac{T_1}{2} \right] \mathrm{SF} \tag{30.6}$$

where

ΔL_S = the change in length of the bolt that has been created by the tension in the bolt (i.e., ignoring temperature-induced change) (in., mm)
MV = the velocity of sound in the unloaded bolt (in./s, mm/s)
T_2 = transit time in the tensioned bolt (s)
T_1 = transit time in the unloaded bolt (s)
Δt = the change in temperature of the bolt, if any, between transit time measurements (°F, °C)
TF = a material "temperature factor" ($°F^{-1}$, $°C^{-1}$)
SF = the stress factor.

This equation involves three calibration constants that must be determined beforehand and fed to the instrument, as previously described. These are MV, SF, and TF. Let's take a closer look at them.

The material velocity (MV) is the velocity of sound in the unstressed bolt at room temperature. It is typically about 250,000 in./s. This velocity is used to calculate the new length of the bolt under tension, even though the actual velocity at that point will have decreased by something like 0.2%. This would introduce a slight error if it weren't for the fact that MV is always multiplied by the important stress factor (SF), to be discussed soon. SF becomes a catchall calibration constant that automatically takes any change in actual sonic velocity into account.

The temperature factor (TF) is a "per °F" (or per °C) multiplier that introduces the two effects that a change in temperature will have on the transit time of the ultrasonic signals. One change involves the linear coefficient of expansion; if the temperature increases, the bolt will get longer. We are interested only in changes in length that have been accompanied or caused by a change in tension within the bolt, so we want to factor out changes caused solely by temperature.

An increase in temperature will also cause a decrease in the velocity of sound within the bolt [1] (Figure 30.7). Like thermal expansion, this phenomenon will cause an increase in transit time that is unrelated to a change in tension in the bolt, so it must be factored out. Both these changes are essentially linear functions of the change in temperature, and they can be combined into one constant that we determine during a calibration exercise (to be determined later).

We can, and do, usually use a "typical velocity" for MV, and we will need a temperature factor only if a significant time elapses between the measurement of initial and final transit times, so a "typical" or "theoretical" TF is often acceptable. But there's no avoiding the final calibration constant, the stress factor (SF).

The SF is the ratio between the change in transit time created by the increase in length of the bolt and the total change in transit time created by the change in stress and the change in length. As illustrated in Figure 30.5, therefore, it could be expressed as the slope of line A divided by the slope of line C.

As already mentioned, the transit time in a bolt increases by something like 0.3% when the bolt is tightened; two-thirds of it has been contributed by a reduction in ultrasonic velocity, which in turn was caused by the increase in stress within the bolt. This means that the tension-induced increase in the length of the bolt has contributed 0.1%, and the stress factor for this bolt is

$$\text{SF} = \frac{\Delta T_e}{\Delta T_T} = \frac{0.1}{0.3} = 0.33$$

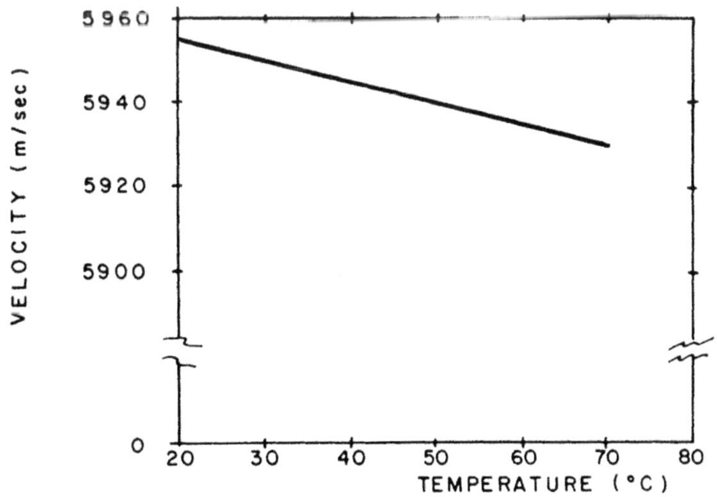

FIGURE 30.7 Velocity of ultrasound in alloy steel as a function of bolt temperature (modified from Ref. [2]).

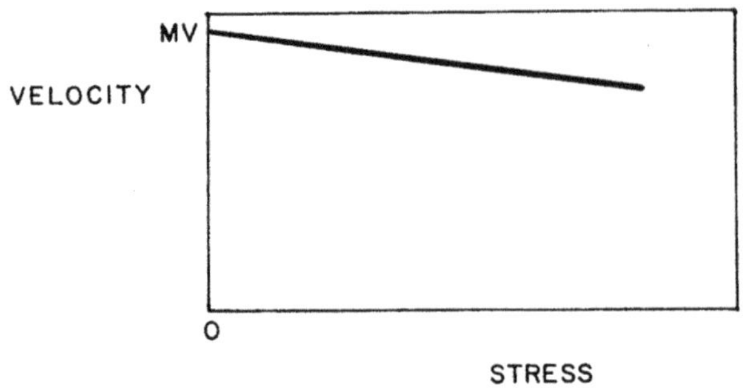

FIGURE 30.8 Velocity of the ultrasonic wave through a bolt as a function of the stress in the bolt. We are primarily interested in the slope of the curve, which is typically -8×10^{-3} for a low alloy bolting material.

where

SF = stress factor
ΔT_e = change in transit time created by the strain (i.e., change in length) of the bolt
ΔT_T = total change in transit time (caused by both the velocity and length changes).

The terms ΔT_e and ΔT_T can be expressed as absolute changes in transit time or in seconds.

$$\mathrm{SF} = \frac{\mathrm{MV}}{\mathrm{MV} - K_T E} \tag{30.7}$$

where

MV = the velocity of sound in the unstressed bolt (in./s, mm/s)
E = the modulus of elasticity of the material (psi, GPa)
K_T = the (negative) slope of the velocity-stress curve (Figure 30.8) expressed as a decimal (in./[(s/in.)/(s/psi)], mm/[(s/mm)/(s/kPa)]).

The stress factor usually ranges between 0.25 and 0.35 for most common bolts.

Theoretically, the stress factor is a material constant whose value depends on three material properties as follows:

Equation (30.7) gives the "theoretical" expression for SF. In practice, we select SF during a calibration exercise, and we use it, in part, to compensate for the change in sonic velocity that occurs as the bolt is stretched (because there is no independent way for the bolt gage to detect or estimate the velocity shift). As a result, the SF we use in practice could more properly be called a "calibration constant" and, although there's some debate on this, it sometimes appears to be a function of fastener geometry. In any event, and unfortunately, it is clearly affected by residual manufacturing stresses within the fastener, since these affect the no-stress sonic velocity (MV).

30.7 ACCURACY OF TRANSIT TIME ULTRASONIC INSTRUMENTS

Bickford [1] addresses the issue of accuracy as follows.

30.7.1 General

How accurately can a transit time instrument measure or control the change in length of a bolt or the tension in it? As we consider the answer to this question, we'll see that ultrasonic control or measurement, while often giving us a significant advantage over torque, turn, or some of the other techniques we have already considered, is far from perfect. Many variables still affect the accuracy of the result. Fortunately, none of these variables are subject to as much scatter as, for example, the coefficient of friction between thread surfaces. But they are scattered. One big advantage of ultrasonics, however, is that we can usually use the calibration exercise to reduce the uncertainties. We can calibrate the instrument using a small sample of our bolts, then use it with reasonable accuracy on the rest of the bolts—at least those in the same lot. This sort of thing is not possible with torque control, where, by comparison, we would have to "calibrate" each individual bolt in many cases.

In general, ultrasonic equipment is capable of measuring the change in length of a fastener to the nearest ±1% or so. In some situations, it can also measure the tension in a bolt to a similar accuracy. It is not easy or economical, however, to achieve this level of measurement accuracy in most applications, and it is even more difficult to control preloads with this precision. In automotive or aerospace work we will usually settle for preload control of ±5%–10% when using ultrasonics. In field applications involving large bolts tightened under difficult circumstances, we will often settle for preload (not measurement) accuracies of ±20%–30%. This frequently leads to the charge that "ultrasonic control is no more accurate than torque." But that's not true. Ultrasonics, even if used in a "quick and dirty fashion," can give us ±30% of control of *residual* preload or tension in the bolts after such things as embedded relaxation, elastic interactions, thermal effects, and the like. Torque control under ideal conditions may give us a scatter of ±30% of *initial* preload in individual bolts. However, the max-to-min scatter in residual tension can be as great as 4:1 or 10:1 or even worse in some cases.

30.7.2 Basic Principle of Operation

Four basic steps are always involved when we use an ultrasonic instrument to measure the tension in a bolt.

1. We calibrate the instrument for that particular type of bolt in that particular application.
2. The instrument is now used to measure the time required for a burst of ultrasound to make a round trip through the unloaded bolt and the slightly longer time required for a burst of sound to make its round trip in that same bolt after it has been tightened.
3. A microprocessor within the instrument now uses the difference between the unloaded and loaded transit times to estimate the change in length of the fastener as it was tightened. In making this calibration, the instrument also relies on the "calibration constants" that we input to the instrument as a result of our calibration exercise. Thanks to the calibration exercise there is usually a very good relationship between the measured change in transit time and the computed change in length. The variables involved here are reliable and easy to define. We usually have a good deal of confidence in the change in length displayed by the instrument.
4. The microprocessor or a computer associated with the instrument now uses the change in length of the bolt to estimate the tension in it, or the stresses in the threaded region, whichever we are interested in. To do this, it must again use calibration information plus, this time, information about the geometry of the bolt and the joint. More variables are involved here than in the estimation of change in length, and several of these variables are more difficult to pin down than those we used earlier. As a result, the accuracy with which the instrument can compute tension is usually less than the accuracy with which it can compute a change in length.

To understand the accuracy—or the limitations—of ultrasonic measurements, we must (1) look at anything that affects the measurement of transit time itself, (2) look at anything that affects the use of a change in transit time to estimate the change in length, and (3) look at anything that affects the use of the estimated change in length to compute tension. Such analysis will give us the "theoretical" accuracy of the instrument. In addition, we must recognize that we're living in the real world, and so we must also look at potential operator and equipment problems, things that could prevent us from realizing the full potential of the instrument.

30.7.3 Measurement of Transit Time

Sound travels through an alloy steel bolt at approximately 250,000 in./s. If we wish to measure a change in length of that bolt to the nearest 0.001 in. we must be able to resolve transit time to the equivalent of the nearest 0.00001 in. From a transit time point of view, we must then be able to resolve.

$$2\times\frac{1\times10^{-5}}{250,000}=0.8\times10^{-10}\,\text{s}$$

The "2" is introduced because we are dealing with round-trip transit times.

Because we are using a digital instrument, to resolve 0.00001 in. in a single round trip we need a system that responds at a 12.5 GHz rate.

$$\frac{1}{0.8\times10^{-10}}=12.5\times10^{9}\ \text{Hz}$$

That is not practical at the present time so, once again, we must resort to statistical sampling techniques. In one example, the instrument "interrogates" the bolt 160 times. It measures the length of the bolt "crudely" each time. Each answer is stored in memory; then the 160 answers are averaged to determine the most likely mean of the measurements. The central limit theorem tells us how many measurements are required to achieve a desired level of accuracy. With a basic clock frequency of 100 MHz and 160 measurements, we can estimate the actual transit time with a standard deviation on the order of 10^{-8} s.

Incidentally, the other numbers used in this discussion—100 MHz, 160 measurements, etc.—were for example only. The actual values depend, again, on the instrument used, the bolts involved, and the application.

30.7.4 Transit Time vs. Change in Length

Several factors affect the relationship between transit time—or, more correctly, a change in transit time—and the estimated change in length of the bolt. Let us take a look at these factors. The round trip transit time is related to bolt length as follows:

$$T_T=\frac{2L_C}{\text{MV}} \tag{30.8}$$

where

MV = velocity of sound through the unloaded, unstressed fastener (in./s, mm/s)
L_C = length of the bolt (in., mm)
T_T = round trip transit time (s).

Depending on the material used and its condition, the velocity of sound in a bolt will range from about 150,000 to over 300,000 in./s (3,910–7,260 m/s). We are primarily interested in how it varies for our bolts. Theoretically, the velocity of sound in a rod or bar is [3]

$$\text{MV}=\left(\frac{E}{\rho}\right)^{1/2} \tag{30.9}$$

where

MV = the velocity of sound (in./s, m/s)
ρ = the density of the rod or bar (slug/in.^{3}, kg/m^3)
E = modulus of elasticity (psi, Pa).

Simplistic handbook data would suggest that E and p will be constants for a given material. But this is not true; some experts say that the modulus of as-received bolts varies from 29×10^6 to 30×10^6 psi. Others say that even wider variations are possible. Density may vary as well. Certainly, both modulus and density will be functions of bolt temperature. And the density will be affected by stress, creating the stress-vs.-velocity effect seen in Figure 30.8.

As mentioned earlier, the estimated room temperature velocity of sound in the unloaded bolt is used in the algorithm to estimate the change in length or preload. Variations in the modulus of elasticity and the density of as-received bolts should not cause too much concern, because manufacturing operations such as thread cutting, thread rolling, cold heading, forging, and heat treatment leave a variety of residual stresses in the bolts. Although the average residual stress level is low, there can be significant concentrations at thread roots, head-to-body fillet, etc. The ultrasound will travel more slowly through the concentrations than in unstressed regions of the bolt. The average velocity, therefore, tends to vary somewhat from bolt to bolt in a given lot and will often vary substantially between bolts made by different vendors.

As a result, several bolts having identical physical lengths will usually have different "acoustic lengths" if these lengths are measured by a bolt gage set at a given stress factor and material velocity. Variations in average velocity, bolt to bolt, will result in apparent variations in length.

As an example, we once compared physical and acoustic lengths in a group of A193-B7 studs that were nominally 18 in. long. The indicated acoustic lengths differed from actual lengths by a minimum of about 50 mils. This doesn't sound like a large difference, but a bolt with a 14 in. grip length tightened to 50% of yield would be stretched only 21 mils or so. As a result, variations in acoustic length (caused by variations in physical length and in velocity) become very important. We're forced, in practice, to measure and store or log the initial acoustic lengths of individual bolts before tightening them.

30.7.5 Velocity vs. Stress

Stress affects velocity; that's one of the principal attractions of ultrasonic measurements. We use shifts in velocity (determined by measuring transit times) to estimate stress levels. The exact relationship between stress and velocity, however, is also subject to bolt-to-bolt and lot-to-lot scatter. We cope with this variation by using the previously described techniques to determine the mean and scatter of the stress factor calibration constant.

30.7.6 Length Change Not Caused by Tension

If something changes the length of the bolt without changing its stress level, the bolt gage could interpret the change in transit time incorrectly. Such changes are not uncommon, but, fortunately, they can be accommodated.

Assume that our transit time instrument has been calibrated to read, correctly, the change of length in a particular bolt when this bolt is stretched. One-third of the change in transit time is created by an increase in path length (ΔL); the rest of the change is created by the stress-induced reduction in the velocity of sound.

We measure the length of the bolt with the instrument, with the bolt unstressed. Then we remove 0.015 in. of material from one end and use the bolt gage to measure the change in length. The bolt is still unstressed. Since the instrument has been calibrated to interpret only one-third of any change in transit time as a change in dimension, it will tell us that the bolt is now 0.015/3 or 0.005 in. shorter. This is the sort of "wrong answer" it gets when it measures steps in gage blocks.

In practice, we can encounter dimensional changes that are unrelated to stress change when the temperature of the bolt changes.

$$\Delta L = \alpha(\Delta t) L_T$$

where

ΔL = change in length (in., mm)
α = coefficient of linear expansion (in./(in./°F), mm/(mm/°C))
Δt = change in ambient temperature (°F, °C)
L_T = total (not effective) length of the bolt (in., mm).

The length will increase or decrease, of course, as the temperature goes up or down. The coefficient of linear expansion of steel might be 6.48×10^{-6} in./(in./°F) or 11.7×10^{-6} mm/ (mm/°C) (the coefficient varies somewhat with alloy content, etc.).

We can also encounter a change in length unaccompanied by a change in stress if there's a long time between initial and final readings and one end of the bolt has corroded. We can't make good acoustic coupling to the bolt through rust, so we must remove the rust with a file or sandpaper. If we shorten the bolt by 0.006 in. in this process, the instrument will think it sees a decrease in length of 0.002 in.

The temperature factor calibration constant (TF) allows the instrument to compensate, automatically, for changes in length caused by thermal expansion or contraction. Other dimensional changes, caused by mechanical means, are much less common and must be dealt with "manually." One trick is to measure the reduction in length of a bolt as it is loosened, then take the unloaded reading of its length and compare this to the length logged before the bolt was first tightened. The log can now be corrected, and the bolt can be retightened. All of this is rarely necessary, however.

30.7.7 Stress Factor Accuracy

The stress factor calibration constant (SF) has entered the discussion repeatedly. By now you must realize that it plays a key role in determining the accuracy of the change-in-length measurements. It defines the basic ratio between velocity and change-in-length effects in the fastener material, and it can be, and usually is, used to tidy up such things as variations in initial velocity caused by residual stress in the bolts. If we don't input the correct value for SF, the instrument will not be able to report the actual change in length.

The accuracy of the SF will depend entirely on the accuracy of the equipment and techniques used to determine the true change in length of or tension in the sample bolts when we calibrate the bolt gage. Statistical techniques can be used to derive SF with acceptable accuracy; manufacturers of ultrasonic equipment normally provide technical support on how to do it. In any event, however, instruments used to calibrate the ultrasonics must, themselves, be properly (and frequently) calibrated, maintained, and used. If this is done, and the statistical work is up to par, all other room temperature variations and questions can be ignored. The ultrasonic equipment can be likened to a "rubber ruler" that can be adjusted to give any reasonable answer. If we program it with the right SF, it will give the correct answer, at least for room temperature measurements. Adding the other important calibration constant, the temperature factor (TF), will lead to correct answers at all temperatures. It's certainly useful for you to know the factors that cause uncertainty in the ultrasonic instrument's estimates of change in length—the factors discussed in this chapter—because there may be times when you'll have to reduce the bolt-to-bolt uncertainties by controlling or measuring velocities or moduli more carefully. In most cases, however, your measurements will be accurate enough if you follow the manufacturer's instructions to calibrate the instruments properly.

30.7.8 Resulting Accuracy in Change-in-Length Measurements

Results depend on calibration techniques, bolt size, the care taken in using the instrument, bolt condition and heat treatment, temperature measurements, and all the other factors we've discussed but, in general, it's probably safe to say the following. Most present-day transit time instruments are capable of measuring the change in length of fasteners to the nearest 0.0001 in. Special instruments are also available that can detect change to the nearest 0.00001 in. These absolute accuracies get less as fastener length increases, if for no other reason than that the digital circuits and display can handle only a fixed number of significant digits. The stretch of threaded rods up to 24 ft in length, however, has been measured with accuracies that are believed to have approached the nearest 0.001 in.

The equipment accuracies cited above are not always achieved in practice, especially when the equipment is used under difficult field conditions. Since there are few ways to confirm or challenge measurements of this accuracy in the field, data are generally unavailable. Bolt stretch can probably be measured to the nearest 5%. Production line or laboratory measurements will more likely approach the accuracy for which the instruments are capable. In any case, the accuracy of the calibration exercise and the way the instruments are used, rather than the capability of the instruments, determine the accuracy of the results.

30.7.9 Accuracy of Preload Estimates

We now use the change-in-length algorithms of Eqs. (30.6) and (30.7) (or equivalent) to estimate bolt preload. The problems in doing this, in relating change of length to tension, are described in detail in Chapter 5 and involve uncertainty in the geometry of individual features more than in material properties. The most significant uncertainty, at least in relatively short bolts ($L/D<4{:}1$) concerns the way the male threads engage the female threads, since this determines the "effective length" of the fastener and that in turn determines its stiffness. Thread runout vs. body length is also significant in short bolts.

As mentioned earlier, an accurate stress factor tells us what percentage of the total change in transit time measured by the instrument was caused by the stress related velocity change in the bolt. In fully threaded bolts, this can give us a relatively clear picture of tensile stress and preload. All we need to know in addition to the change in transit time is the effective length of the bolt. In a partially threaded bolt we must also know, separately, the effective length of the threads. Both dimensions are subject to bolt-to-bolt and lot-to-lot scatter. They can, and often do, change as a bolt is reused (because of localized plastic deformation and creep in the threads). The length of time a bolt has been in service can make a difference too because creep is a time based phenomenon.

What prevents all this from creating serious problems in most applications is calibration. In fact, measuring "actual" forces produced by a bolt on a load cell is easier than measuring "actual stretch," partially offsetting the additional variables. And the bolt-to-bolt scatter in stretch vs. preload is usually reasonably small, at last within a given lot. As a result, we feel that the equipment is capable of residual preload measurements with an accuracy approaching ±5% in most field situations, and can approach ±2% in laboratory work. There are even well-documented examples where ±1% was achieved consistently in aerospace applications, but this was accomplished only by calibrating the instrument several times a day, by controlling bolt temperatures very carefully, by careful design and treatment of the bolts (with special attention to end surfaces flatness and parallelism), etc. I think it's worth noting that the instruments are capable of such accuracy but that economics make this precision impossible in most applications.

As a final note, the accuracies cited above for stretch and preload measurements should be considered typical but not definitive. Detailed studies have been made of the theoretical precision with which present instruments can measure stress or strain in bolts. Uncertainties were estimated using first-order, second-moment methods, lognormal approximations, and Monte Carlo simulation. Eleven variables were found to play key roles in the measurements, and it was found that in most cases one or two of these variables contribute most of the uncertainty in the final estimates. Furthermore, different variables dominate different applications. For example, if the bolts are very long, variations in the modulus of elasticity can be one of the main problems. In very short bolts, uncertainty in grip and thread lengths can dominate. One value of the study is that it allows us to identify one or two variables that, if more carefully measured or controlled, can produce a significant improvement in overall accuracy. We needn't control them all. We needn't, in fact, "control" any of them; measuring and knowing them is sufficient if the instrument is programmed to accept changes in these factors.

30.8 USAGE FACTORS AFFECTING ACCURACY OF MEASUREMENT AND CONTROL[4]

We've analyzed the accuracy with which current transit instruments are able to measure bolt stretch or tension. Several times we learned that there can be a significant difference between the accuracy of the instrument and the accuracy we actually achieve in the field. We often settle for cruder measurements, and we almost never try to use the instrument to control preload with instrument precision. Let's look next at some of the factors that lead us to "settle for less."

30.8.1 Transducers

The transducer plays an important role. It consists of a crystal mounted in a protective housing with a wear plate on the business end to protect the crystal from mechanical damage or abrasion. Since it's important that the transducer remain in a fixed position during a measurement, the protective housing is usually provided with strong magnets to hold the transducer in place. Transducers used on nonmagnetic bolt materials are held in place by mechanical means. Sometimes, for example, they're threaded onto the end of a stud or bolt.

30.8.1.1 Selection of a Transducer

Transducers come in a variety of diameters, heights, resonant frequencies, operating temperatures, etc. They range in diameter from about in. to 1 in. In general, one should always use the largest diameter that will fit on the end of the bolt without extending past the root diameter of the threads. One reason for selecting transducer with a larger diameter is that it will pump more acoustic energy into the bolt than one with a smaller diameter. This means stronger echo signals and usually greater reliability and accuracy. Furthermore, a transducer whose diameter basically matches that of the bolt is easy to position or reposition on the bolt (see below).

The resonant frequency of the transducer is another important consideration. The frequencies used for bolt measurements usually range from 2 to 10 MHz, with 5 MHz being the most popular. A lower frequency will penetrate the bolt with less loss but results in a broader "cone" of transmitted energy, which can mean unwanted echos from nearby threads, etc. As a result, higher frequencies usually work more reliably on longer, thinner bolts.

Most transducers are unable to function at operating temperatures in excess of 150°F (302°C), but special ones are available that can tolerate temperatures up to 400°F (752°C).

In the past, it was often necessary to try several transducers before finding one that would work most reliably on a given application, but advances in the ultrasonic equipment (e.g., more "punch" and the ability to respond to weaker echos) make trial and error much less common now.

30.8.1.2 Placement of the Transducer

The way the transducer is placed on the bolt can affect results. If the transducer and the bolt have the same or similar diameters, placement is simple: The transducer is roughly centered on the end of the bolt, with no part of the transducer overhanging the edge of the bolt.

On large diameter studs, the transducer can also be placed at the center. It is then usually moved around slightly while the operator observes the "initial length" displayed by the instrument. He searches for the minimum displayed length and leaves the transducer in that spot. If the transducer is left in place as the fastener is tightened, no further action is necessary. If repeat measurements are to be made in the future, with the transducer removed in between, it is also desirable at this point to mark the selected location of the transducer with a prick punch, or spot of paint or the like, so that it can be replaced in the same location.

This procedure is also required if the transducer is used on a stud having a heater or gage rod hole gun-drilled along the axis. A large diameter transducer can be centered over a small diameter hole, but more commonly the transducer will be located to one side of the hole, between the hole and the outer diameter of the bolt. Again the operator tries several spots until he finds the minimum initial length. That location is then marked for future reference, if required.

30.8.2 Acoustic Couplant

30.8.2.1 The Role of the Couplant

The strength of the ultrasonic signal affects the accuracy with which the instrument can measure the bolt. In general, the stronger the signal, the more precise and dependable the measurement. A number of things influence this signal strength.

As one example, the strength of the ultrasonic signal is dependent on the amount of energy that can be pumped across the gap bridging the transducer and the fastener. If this is an air gap, the signal will be almost totally reflected back to the transducer before it reaches the bolt, because there is a poor acoustic impedance match between the transducer and air. In practice, therefore, the gap is always filled with an acoustic coupling material of some sort; glycerine, water, various oils, and greases have all been used. Each provides a better impedance match between transducer and fastener than would air alone. The choice of couplant for a given application depends on such things as roughness of the end of the fastener, the ambient temperature, and fastener location and orientation. Sometimes (e.g., in nuclear applications), contamination or corrosion problems limit our choice.

30.8.2.2 Voids in the Couplant

Each spot on the end of the transducer functions as a tiny "antenna" to propagate an acoustic wave. These small waves all combine to create the wave front that travels down the bolt, echoes off the far end, and returns to the transducer.

If there are voids in the film couplant between transducer and bolt, then some of these tiny spot antennas will be unable to contribute much energy to the combined wave front. The wave front becomes distorted. The shape and/or polarity of the various portions of the echo signal are distorted, and this will affect the measurement of round-trip transit time. As a result, it is important to make sure that the gap between transducer and bolt is completely filled with a uniform layer of couplant.

30.8.2.3 Gap Errors

Transit time instruments will measure the distance between the face of the transducer element and the far end of the fastener. This distance may include not only the length of the bolt—which we are interested in—but also the thickness of the gap between the operating face of the transducer and the bolt. Since we are making very precise measurements of the change in length of the fastener, it is important that this gap be controlled. We don't care what it is. We want it to remain the same as the bolt is tightened or be the same before and after tightening. This means that the transducers must be held securely against the bolt by magnetic or mechanical means during the bolt-tightening operation or must be consistently mounted if used before and after. Changes in thickness of the acoustic couplant must be minimized. There is also a wear plate between the working face of the transducer element and the outer surface of the transducer housing assembly. We must be sure always to use the same transducer on a given bolt or to calibrate out changes in wear plate thickness if we change transducers on that bolt (and are looking for the ultimate in accuracy).

None of the above is a problem, of course, if the transit time instrument is designed to measure the time between the first and second echoes rather than between main bang and first echo (or between first and tenth echoes as with the Harrisonics system). In this case, however, there is the offsetting difficulty that the extensometer must work correctly with weaker signals.

30.8.3 Condition of Fastener Ends

30.8.3.1 Roughness, Grade Markings, Socket Heads

A very rough surface on either end of the bolt can create problems in two ways. Roughness between the transducer and the input end of the bolt can reduce the amount of ultrasonic energy pumped across the gap, creating weak, uncertain echoes. Extreme roughness on the far or echo surface of the fastener can lead to interference effects and can distort the signal in the same way that near-field interference effects do. This, again, can lead to improper timing of the echo signal. This is not to say that bolt ends must be ground and polished in every application. In most cases, roughness affects accuracy rather than ability to do a job.

Raised grade markings, dished ends, hex sockets, etc. can also cause coupling problems (but are not a problem, usually, if they're on the echo-surface end of the bolt). If the transducer can't be placed on a reasonably flat surface, inside or beside surface holes or embossings, the bolt cannot be read ultrasonically.

Grade markings can be removed with a file. The resulting surface is usable from an ultrasonic point of view, but removing such markings may not be acceptable to quality control people.

Fasteners with special end treatments can be expensive, forcing many people to decide against ultrasonic measurements for this reason. A few people, however, have decided to machine the ends or remove grade markings on a small sample of the fasteners they use, painting one end red or something, so that these can be readily identified in the final assembly.

These special fasteners are fed into the assembly operation in a normal fashion, and conventional assembly tools—torque or turn, or whatever—use them along with the regular fasteners. Quality control inspectors then watch for "red bolts" in the final assembly. Each one of these is measured ultrasonically to monitor the performance of the basic torque–turn or other control system being used on a production basis. If each of the special fasteners has approximately the same acoustic length to begin with (which generally equates to approximately the same physical length), then this system is a very effective way to monitor the behavior of assembly tooling and/or to follow subsequent working loads or relaxation effects.

30.8.3.2 Nonparallel End Surfaces

If the opposite ends of the fasteners are not sufficiently parallel with each other, the acoustic signal will be weakened or may be lost altogether. This is not usually a problem with as-received fasteners (although the ends are never exactly parallel to each other), but it is a frequent problem when the bolt bends as it is tightened.

Nonparallelism can cause a problem simply because the "length" of the bolt is not uniform from one point to another. This is no problem if the diameter of the transducer is nearly as large as or as large as the diameter of the bolt; under these circumstances, the ultrasonics will average the length and/or change in length of the fastener.

If we're using a ½ in. diameter transducer on a 4 in. diameter bolt, however, the instrument will see a different length when it's mounted in a 12 o'clock position than it would in a 6 o'clock position or when mounted on the centerline of the fastener (mounting on the centerline is not always possible because of heater holes and the like). Under these conditions, the location of the transducer must be

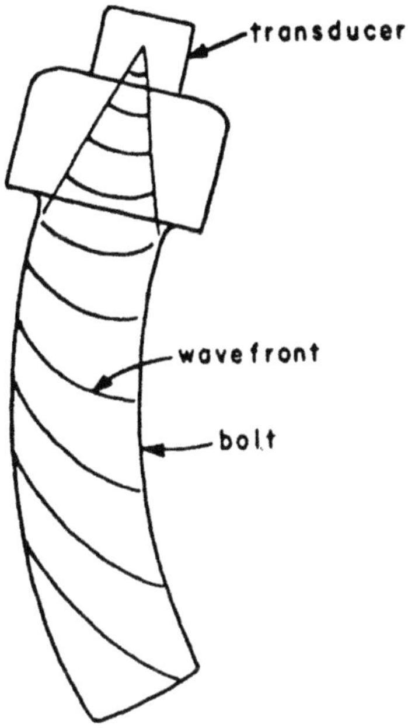

FIGURE 30.9 Tensile stress is higher on one side of the bent bolt than on the opposite side. This will rotate the wave front of the acoustic signal as shown.

marked with dye or a prick punch when the zero stress reading is taken. It must then be returned to approximately the same position for the aftertightening reading.

If the nonparallelism is crated by or made worse by bending, an additional problem arises. Remember that the velocity of the ultrasound is inversely proportional to the stress level within the fastener. The tension stress down one side of a bent bolt is, of course, higher than the tension stress down the opposite side. The ultrasound wave front will therefore travel more slowly down one side of the bolt than the other. The bolt acts as sort of a lens, gradually warping the ultrasonic wave front, as suggested in Figure 30.9.

If the difference in stress level from one side of the fastener to the other is too great, the ultrasonic signal can end up echoing off the side of the bolt rather than the end and will, in effect, be lost. This is an extreme case, but it does happen, especially in bolts that have heater holes in them (the hole adds an additional level of complication, since the bolt can no longer act as a waveguide for the signal).

Nonparallelism, regardless of its sources, can also distort the shape of an echo signal. Each portion of the echo surface acts as a tiny reflector. All such reflections are combined to produce the echo signal. If, because of nonparallelism, these small reflective surfaces are out of phase with each other, they can partially cancel each other. These interference effects distort the signal in the same way that the near-field effect distorts it, as suggested in Figure 30.9.

30.8.4 Controlling Preload

The problems we have just discussed can introduce error into our measurements, but that still doesn't explain why we sometimes settle for preload scatter of 20% or worse when the ultrasonic instruments are capable of a relatively easy 5%. The reason for the discrepancy: Even though we can *measure* preload during assembly, or after assembly, with an accuracy of 5% or better, we cannot always afford to *control* it with that precision. The normal example of this is the large pressure vessel flange. We tighten the bolts one or a few at a time, in a crossbolting pattern. Ultrasonic measurements are made on a few of the bolts during the early passes—the first two passes, typically.

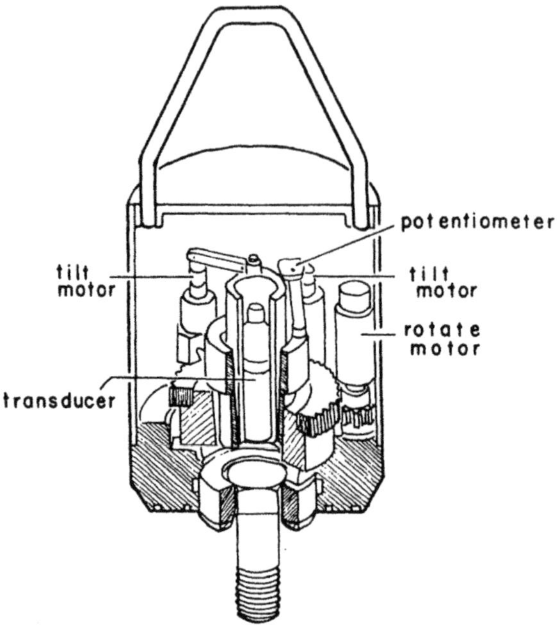

FIGURE 30.10 Interference patterns produced by nonparallel transmission and echo surfaces can seriously distort, or even destroy, an echo signal.

Only on the final routine pass is the tension in each bolt measured. Meanwhile, elastic interactions, embedment, and variations in the torque-pre-load relationship have led to the results illustrated by Figure 6.25, a "saw-toothed curve" with residual preloads varying by 4:1 or more in most cases. And we'll get these results whether or not the initial preloads achieved in each bolt are monitored ultrasonically as the bolts are tightened individually. Even perfect ultrasonic measurement at this point cannot prevent subsequent scatter.

When we first used ultrasonics on such joints, we assumed that it was only proper to end up with residual preload scatter comparable to the accuracy of the instrument, ±1%–2% or so. And we found that this was certainly possible—if one could spend a day or two tightening an individual joint. But, thank goodness, we also found ourselves forced to give up when scatter had been reduced from 4:1 to ±10% or 20% because every other part of the plant was ready to go back on line. As a result of being forced to stop, we learned that ±20% or ±30% or "worse" was a big improvement over 4:1 and was acceptable on all but the most stubborn leakers (Figure 30.10).

On hard joints, or on joints where most or all of the bolts are tightened simultaneously, scatter in the ±5%–10% range is probably the accepted norm, because interactions are fewer or are nonexistent. Even here, however, the final variations are determined more by the nature of the joint and by the tools used to generate bolt loads than by the accuracy of the ultrasonic measurements.

30.9 CASE STUDY: AUTOMOTIVE TRANSMISSION JOINT (CONTROL VALVE-TO-CASE)

30.9.1 Description of the Bolted Joint

In this particular joint, a cast steel valve body was to be bolted to the cast aluminum case of an automatic transmission as shown in Figures 30.11 and 30.12; the joint is gasketed. Ten bolts, numbered 1–10 (Figure 30.12), were used; seven were 5/16″-18 SAE Grade 2 (locations 1, 2, 5, 6, 8, 9, 10), and the remaining three were 1/4″-20, SAE Grade 5 (locations 3, 4, 7). No design changes were to be

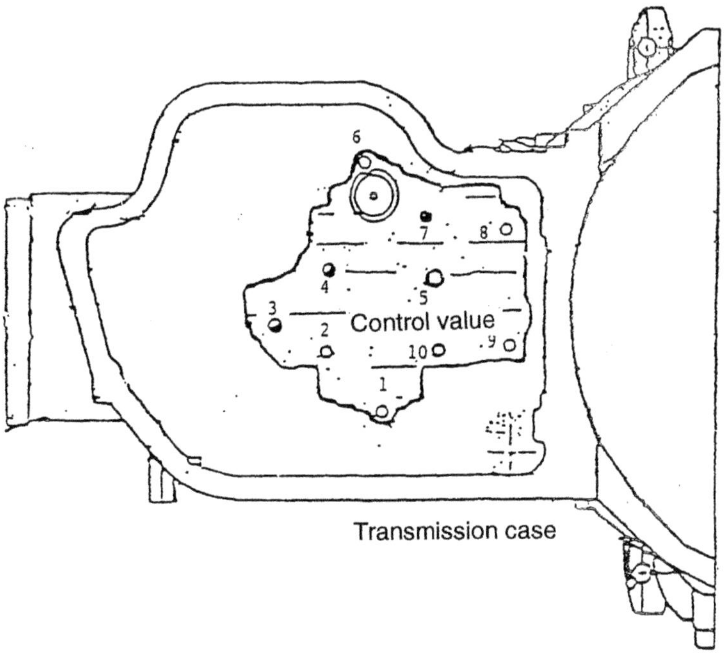

FIGURE 30.11 Automatic transmission joint.

recommended; it was only necessary to determine the assembly parameters needed for a uniform preload of 2,000 lb.

The goal was to develop a tightening strategy and a set of parameters to achieve simultaneous uniform preload of 2,000 lb in each of the joint's ten fasteners.

The assembly tool to be used was a ten-spindle nut runner with brushless DC motors. Ultrasonics was used to measure bolt stretch.

30.9.2 Procedure

1. Determine the spring rate k_b of each bolt by tensile testing each fastener in the joint. The experimental values of the spring rates are given in Table 30.1 (see Section 9.3). To obtain the spring rate, the ultrasonic stretch is plotted against tensile load for each fastener; the slope of the plot is k_b.
2. Determine the desired bolt stretch to achieve a uniform 2,000 lb preload in each fastener by using $\Delta = 2{,}000/k$, as shown in Table 30.1.
3. By trial and error, program each spindle by using various values for torque and angle of turn, until the desired bolt stretch is reached.
4. Record the spindle control parameters that achieved the desired preload of 2,000 lb.
5. Program the control parameters for each spindle, and use the multispindle to assemble "fresh" joints using fasteners with machined ends. Then use ultrasonic measurement of bolt stretch at the end of assembly to verify whether the desired stretches of Table 30.1 were achieved.
6. Repeat steps 3–5 until the control parameters of "torque" and "turn" produce the desired bolt stretch in a repeatable and reliable fashion. Then report recommendations on tightening parameters (Table 30.2, Section 9.3.2).
7. For comparison purposes, use ultrasonics to audit bolt stretch and tension on the assembly line, using premachined bolts.

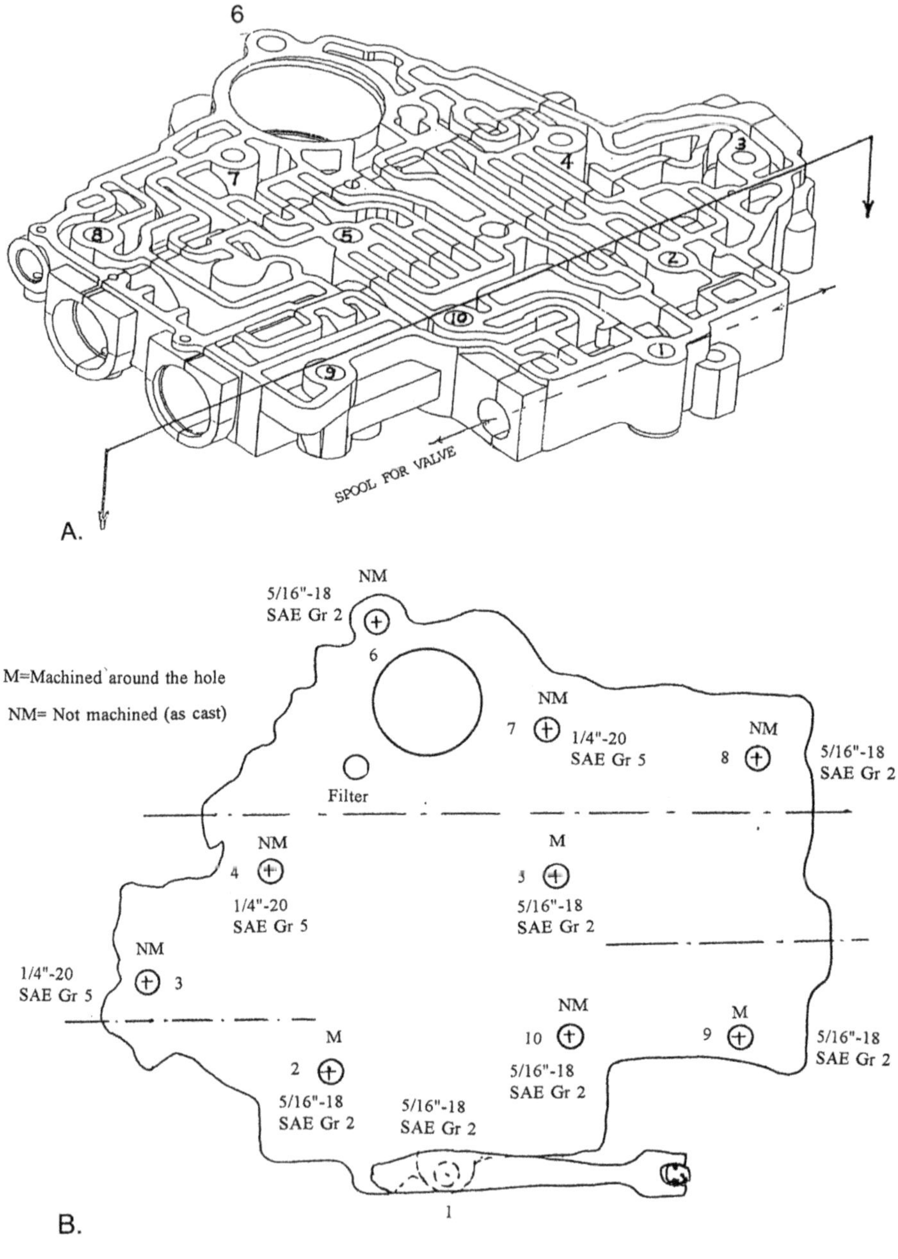

FIGURE 30.12 (a) Control valve casting. (b) Fastener layout.

30.9.3 Test Results

30.9.3.1 Measurement of Spring Rate

The actual spring rate for each of the ten bolts in the transmission joint was experimentally determined by plotting the tensile load versus the ultrasonic stretch. The actual grip length of each bolt, as it was in the actual joint, was duplicated for the tensile test machine. Table 30.1 provides the actual spring rates and the corresponding bolt stretch necessary to achieve 2,000 lb preload in each of the ten bolts in the transmission joint.

TABLE 30.1
Bolt Spring Rates and Desired Stretch

Bolt Location (Figure 30.12)	Bolt Size (in.)	SAE Grade	Spring Rate *k* (lb/in.) (Multiply by 10^5)	Desired Bolt Stretch (in.) for 2,000 lb Preload
1	5/16	2	9.643	0.00207
2	5/16	2	9.417	0.00212
3	1/4	5	7.846	0.00255
4	1/4	5	7.659	0.00261
5	5/16	2	9.242	0.00216
6	5/16	2	9.225	0.00217
7	1/4	5	8.447	0.00237
8	5/16	2	9.067	0.00221
9	5/16	2	10.698	0.00187
10	5/16	2	10.039	0.00199

TABLE 30.2
Recommended Tightening Parameters for a Uniform Preload of 2,000 lb

Bolt Location (Figure 30.12)	Threshold (snug) Torque T_t (N·m)	Angle of Turn (Deg) from T_t	High Torque Limit (N·m)
1	8.5	21	16
2	6.0	25	16
3	7.0	29	15
4	7.9	25	14
5	5.7	25	16
6	9.4	22	18
7	5.2	34	13
8	8.6	20	18
9	7.4	21	16
10	7.4	22	16

30.9.3.2 Tightening Parameters

Using ultrasonic measurements of bolt stretch, a set of torque–turn parameters was developed to achieve a uniform preload of 2,000 lb in each of the ten bolts in the joint at initial assembly. Table 30.2 provides the control parameters to which each spindle of the assembly line tool must be programmed. Note that each bolt has its own set of parameters that achieves the desired 2,000 lb preload. Variation in the control parameters for bolts of the same size is mainly contributed to the variation in underhead surface conditions (machined vs. as-cast) and to the variation in bolt and/or joint stiffness.

As an example (see Figure 30.12), the valve casting was machined around the hole for bolt 5, while the surface condition around the hole for bolt ten was "as-cast" or "not machined." Both bolts were the same size (5/16″–18, SAE Grade 2). As a result, the underhead friction conditions for the two bolts are significantly different, requiring different sets of control parameters to achieve the same level of fastener preload. Later audit of bolt tensions produced by a torque-only control method (Figure 30.13) clearly demonstrates the fact that a significantly smaller preload was achieved in bolt 10 than in bolt 5 of the same size. Both bolts were tightened to the same torque using a ten-spindle nut runner with DC brushless motors.

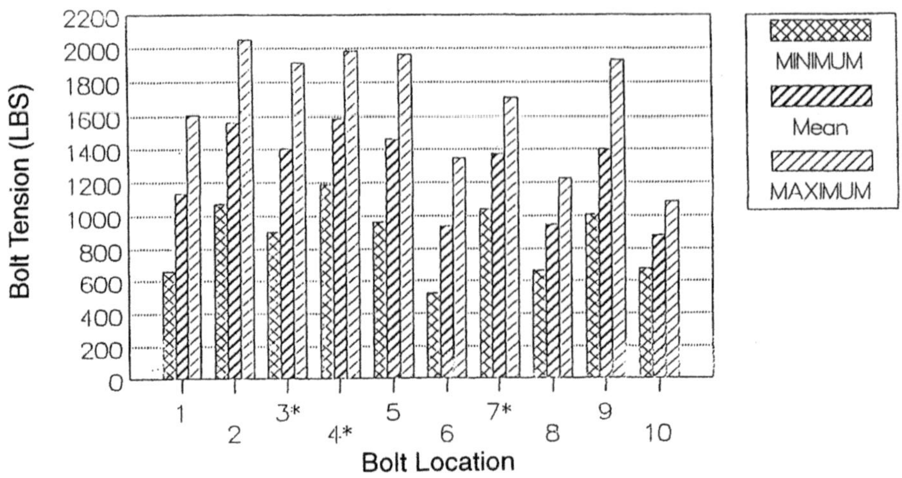

FIGURE 30.13 Ultrasonically measured bolt tension by location in control valve body.

30.9.3.3 Evaluation of Torque Control Using Ultrasonic Audit

Motivated by the fact that each spindle was to be programmed differently in order to achieve a uniform preload in the joint (Table 30.2), an experimental evaluation of the torque control method was performed on seven transmission cases as they were assembled on the line. A total of 70 bolts were specially machined at both ends in preparation for the ultrasonic measurements of bolt stretch at the end of the assembly process in an actual plant environment. During the experiment, the transmission cases were intercepted after the assembly station for the ultrasonic audit of bolt stretch, after which the transmission joints were immediately disassembled.

Figure 30.13 shows bolt preloads at the end of the initial assembly using a torque control process. The mean values were determined using seven transmission cases. The effect of the underhead surface conditions on the achieved preload is clearly depicted by comparing results for bolts 5 and 10: Both are the same size, but bolt 5 had a machined underhead surface whereas the underhead surface for bolt 10 was not machined (as-cast). The scatter shown for the maximum and minimum bolt tension corresponds to a 95% statistical confidence in test results (1.96 standard deviations from the mean for a sample size of seven transmission cases).

Figure 30.14 presents the same test results separately for the 5/16″-GR 2 bolts (sample size of 49) and for the 1/4″–GR 5 bolts (sample size of 21). The fastener preload was 1,200 lb with a scatter of ±40% for the larger bolts (5/16 in.) and 1,450 lb with a scatter of ±15% for the smaller fasteners (1/4 in.).

30.9.3.4 Repeatability of Ultrasonic Stretch

At this time, operator skills are a major factor in obtaining repeatable readings of the ultrasonic stretch of automotive fasteners, where the bolt stretch is normally small. Accuracy and repeatability of the measurements are significantly affected by the placement of the ultrasonic transducer on the end of the fastener, by the temperature as it creeps up while the operator holds the fastener head by hand, and even by the amount of pressure the operator's hand exerts on the transducer while it is mounted on the fastener.

During this project, a simple technique was followed to ensure the repeatability of ultrasonic stretch readings. In this technique, the operator moves the transducer around on the bolt head or end until he or she achieves the "lowest" reading of bolt stretch; then the operator must repeat the "same lowest reading" three consecutive times by removing the transducer and randomly placing it again on the bolt head or end until the "absolute lowest" reading has been repeated for three consecutive

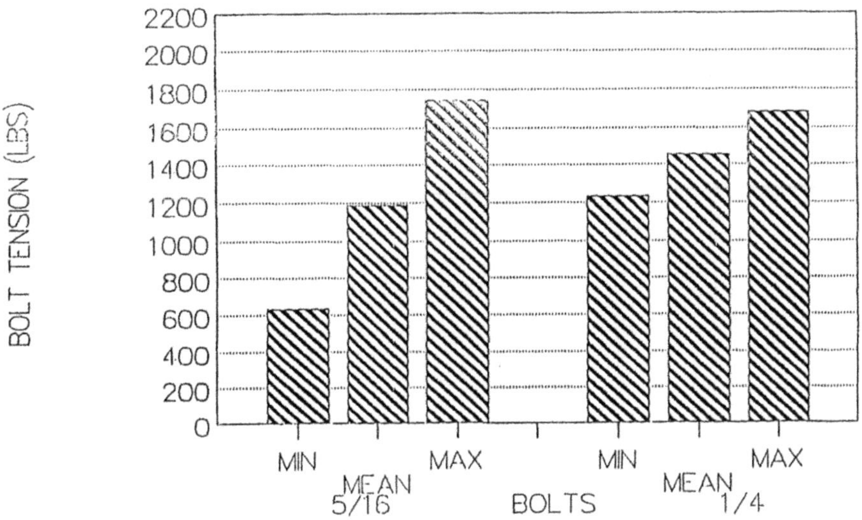

FIGURE 30.14 Ultrasonically measured bolt tension by size in control valve.

times or until a "lower" reading has been obtained and repeated for three consecutive times. This procedure is often time-consuming and tedious, but it reduces the reliance on the operation, to a certain degree.

30.9.3.5 Recent Advances in Ultrasonics

During this project the idea of continuous (dynamic) ultrasonic measurement of bolt stretch, as the bolt is being tightened by an automated assembly tool, grew and was initially well received by the automotive industry. A consortium was formed to sponsor the integration of ultrasonics and use it as a control method in automated assembly tools. This research was done at Lawrence Technological University. In the first phase, the ultrasonic stretch control capability was added to a single-spindle computer-controlled research tool that was originally capable of controlling the tightening of a fastener according to one of the following control modes: torque control, torque–turn control, yield control, or clamping force control.

Figure 30.15 provides a typical graph from the single-spindle ultrasonic prototype. Line 5 shows the ultrasonic stretch achieved with an ultrasonic transducer built in the socket. The transducer was in turn connected to an ultrasonic "box," which fed the measured ultrasonic stretch to a computer that had software that could control the tightening process according to the selected control parameter.

Full integration of ultrasonics into production tools has not been accomplished yet for a variety of reasons. First, current ultrasonic technology makes it possible to multiplex several spindles, with ultrasonic transducers mounted in them, to one ultrasonic box, which significantly reduces the sampling rate of bolt stretch. A slow sampling rate coupled with a high-speed assembly tool may result in "overshoots" in bolt stretch beyond desired values. If one chooses to dedicate one ultrasonic box to each spindle, this may not be an economical proposition. Low volume production of ultrasonic instrumentation still keeps its price high.

Second, the need for special preparation of bolt ends and the need to use acoustic couplant are added steps that may not be easy to implement in a plant environment. However, there are some reports about an emerging "ultrasonic fastener" technology that integrates the transducer into the fastener itself and hence eliminates the need for special preparation of the bolt ends. Third, at the present time, the mechanical connection between the ultrasonic transducer in the socket and the fastener itself is a weak link, especially when tightening at high rpm.

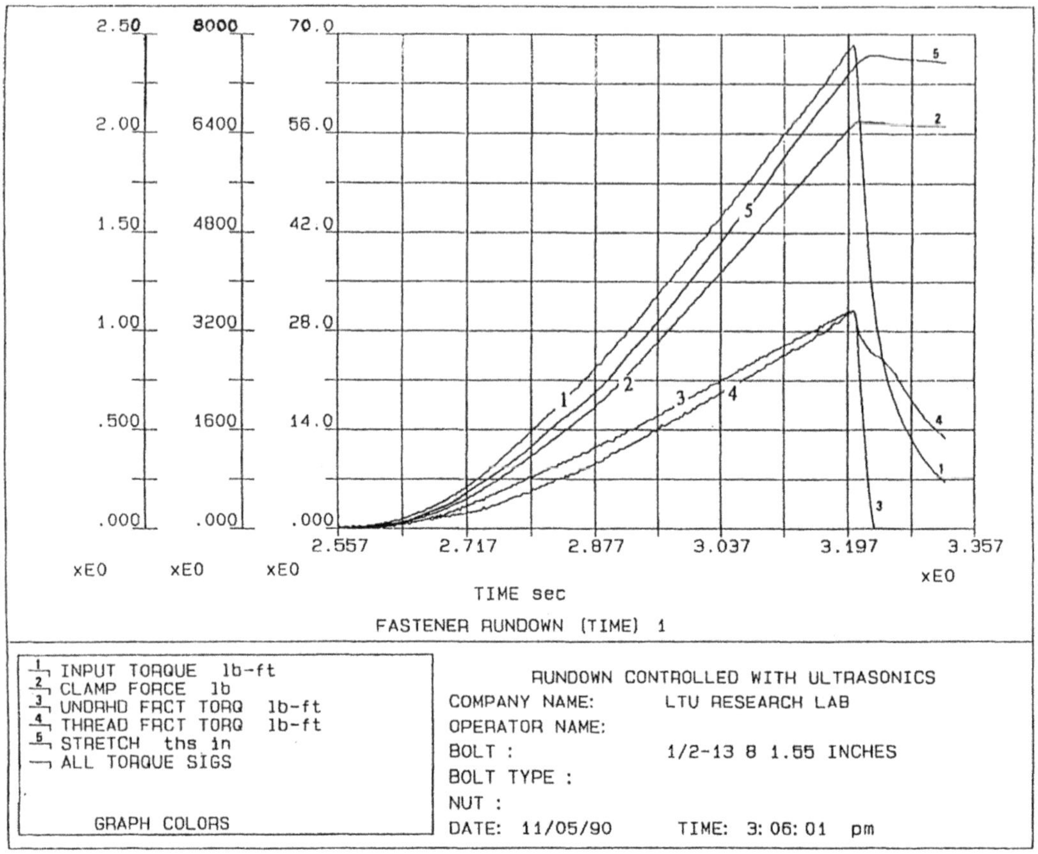

FIGURE 30.15 Ultrasonic stretch control of bolt tightening.

NOMENCLATURE

A, Ab	Nominal cross-sectional area of fastener (in.^{2}, mm^2)
AS	Stress area of the fastener (in.^{2}, mm^2)
D	Nominal diameter of the fastener (in., mm)
E, Eb	Young's modulus of elasticity for fastener material (psi, Pa)
Fb	Fastener tension (lb, N)
Fp	Fastener preload (lb, N)
KT	Negative slope of velocity–stress curve, Eq. (30.7)
kb	Fastener spring rate or stiffness (lb/in., N/mm)
L, *LG*	Fastener grip length (in., mm)
LC, LT	Total length of the fastener
MV	Velocity of sound in fastener material (in./s, m/s)
SF	Stress factor
TF	Temperature factor (°F^{-1}, °C^{-1})
TT	Round trip transit time (s)
α	Coefficient of thermal expansion [in./(in.-°F), mm/(mm°C)]
Δt	Change in ambient temperature (°F, °C)
Δb, Δle, ΔL	Fastener elongation or stretch (in., mm)
p	Density of fastener material (slug/in.^{3}, kg/m^3)
σ	Tensile stress (psi, MPa).

NOTES

1 Taken from Ref. [1], with permission.
2 This section closely follows a discussion in Ref. [1].
3 This section is adapted from Ref. [1], with permission.
4 This section is a condensed version of material in Ref. [1].

REFERENCES

1. Bickford, J. H., *An Introduction to the Design and Behavior of Bolted Joints*, 3rd ed., Marcel Dekker, New York, 1995.
2. McFaul, H. J., An ultrasonic device to measure high-strength bolt preloading, *Mater. Eval.*, 32(11): 244–248 (1974).
3. Sears, F. W., *Mechanics Heat and Sound*, Addison-Wesley, Reading, MA, 1950, p. 497.

31 Control with Direct Tension Indicators

David Sharp

31.1 WHAT IS A DIRECT TENSION INDICATOR?

A direct tension indicator (DTI) is a steel washer-shaped device with protrusions projecting from one face and corresponding depressions or pockets on the opposite face (Figure 31.1). DTIs have been used in the fabricated structural steel industry for over 35 years as a means to ensure achievement of proper tension in high strength structural steel bolts in slip-critical or tensioned connections (see Chapters 35 and 36 of this handbook). More recently, the use of DTIs has been adopted in the petrochemical, off-highway heavy equipment, and other markets.

The most common method of installing a DTI is to place it on a bolt with its protrusions under and against the bolt head. Before tightening, there are noticeable gaps in the spaces between the protrusions (Figure 31.2a). As torque is applied to the nut and the bolt begins to clamp the joint, the protrusions start to flatten. When the protrusions are flattened such that the height of the gaps between the protrusions has been reduced to a specified residual gap, required bolt tension has been accurately achieved (Figure 31.2b).

31.2 USE OF DTIs IN STRUCTURAL STEEL

In structural applications, it is the purpose of the DTI to ensure that the bolts are tensioned (preloaded) in accordance with the requirements of the Research Council on Structural Connections (RCSC) specification. The RCSC specification requires tensions of at least 70% of the minimum specified tensile strength of the bolts on which the DTIs are used. Tables 31.1 and 31.2 depict the specified target load ranges for inch series and metric series DTIs to be used with structural fasteners in accordance with RCSC requirements. Achieving this minimum required level of tension using uncoated DTIs is gaged by the refusal of a 0.015 in. feeler gage (0.400 mm for metric bolts and DTIs) in half or more of the spaces around the perimeter of the bolt head. The manufacturing and testing requirements for DTIs used in structural applications are covered in ASTM F959 for inch series and ASTM F959M for metric series.

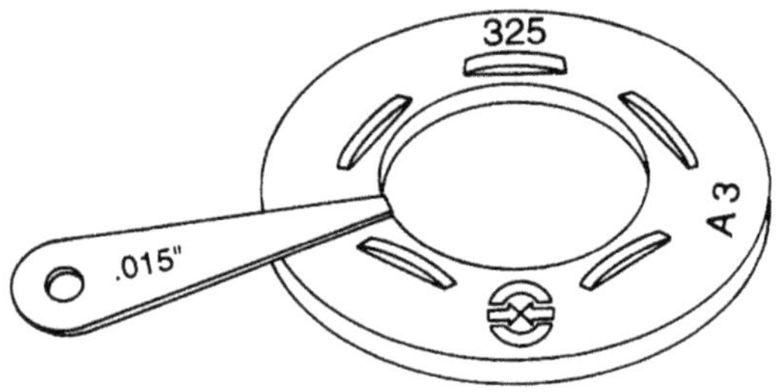

FIGURE 31.1 The direct tension indicator (DTI), a steel washer-shaped device with raised projections on one face. These projections are partially flattened when a bolt is tightened. A feeler gage, such as that shown here, is used to determine how much the projections have been flattened and, as a consequence, how much preload has been developed in the bolt.

DOI: 10.1201/9781003501183-36

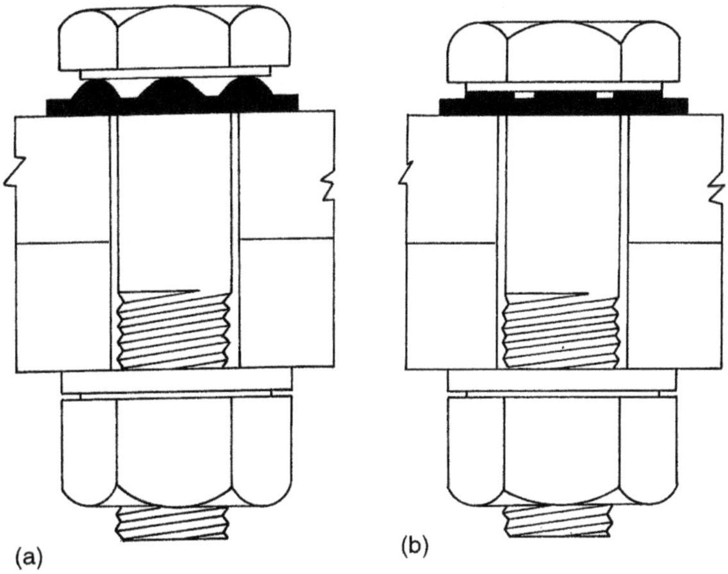

FIGURE 31.2 (a) A DTI installed under the head of an as yet untightened bolt. (b) The DTI after the projections on it have been partially flattened when the bolt was tightened.

TABLE 31.1
Acceptable Range of Compression Loads—Inch Series

	Compression Load Range (kips)	
DTI Size (bolt diameter, in.)	**Type 325**	**Type 490**
½	12–14	15–18
⅝	19–23	24–29
¾	28–34	35–42
⅞	39–47	49–59
1	51–61	64–77
1⅛	56–67	80–96
1¼	71–85	102–122
1⅜	85–102	121–145
1½	103–124	148–178

TABLE 31.2
Acceptable Range of Compression Loads—Metric Series

	Compression Load Range (kN)	
DTI Size (bolt diameter, mm)	**Type 8.8**	**Type 10.9**
M16	91–109	114–131
M20	142–170	179–206
M22	176–211	221–254
M24	205–246	257–296
M27	267–320	334–348
M30	326–391	408–469
M36	475–570	595–684

Figure 31.2 also depicts the most common installation arrangement; the DTI is placed under the head of the bolt, and torque is applied to the nut. This is the preferred method and is to be used whenever possible. Under certain conditions, it may not be possible to assemble the joint as depicted in Figure 31.2 due to inaccessibility of the installation tool or other factors. In such cases, the DTI can be installed under the nut with the nut end turned, or under the bolt head with the bolt head turned, provided that a hardened washer is used between the DTI and the turned element. The criterion for inspection when the DTI is used under the turned element is refusal of a 0.005 in. gage in half or more of the spaces.

Galvanized DTIs are frequently used with galvanized bolts, nuts, and washers on bridges and other structures subject to adverse weather and environmental conditions. Galvanized DTIs are used like plain finish DTIs, except that a 0.005 in. (0.0125 mm) feeler gage is used as the no-go gage for inspection purposes.

31.3 STRUCTURAL INSTALLATION

To successfully construct a slip-critical or tension joint of any kind in structural steel, it is first necessary to follow the RCSC procedure to ensure that the joint is properly snug and the bolts are not trapped prior to tensioning. "Trapped bolts" are bolts that are pinched in shear between the steel plies, usually the results of premature removal of the drift pins that are used to align the bolt holes. Trapped bolts may be difficult or impossible to tighten to required tension. A DTI will detect the inability to tighten a trapped bolt because the DTI is under the head of the bolt, the effort (torque) to tighten the bolt is applied to the nut, and the only way the DTI protrusions will flatten is if sufficient tension is transmitted through the full length of the bolt.

The first step in constructing a slip-critical joint is to bring the members of the joint together and align the holes with drift pins as depicted in Figure 31.3. Bolts should not be used as drift pins to achieve alignment. The second step is to fill the remaining holes with bolt assemblies that include DTIs. The third step is known as "snugging" the joint. Snugging is the process by which all of the joint members or plies are drawn into firm contact with one another. "Snug" is defined by the RCSC as "the tightness that exists when all plies in a joint are in firm but not necessarily continuous contact." Snugging should be started at the center of the bolted connection and then continued toward the free edges. Application of torque to the nuts partially tightens the bolts, partially compresses the DTI protrusions, and snugs the plies. Proper snugging[1] of a joint using DTIs has occurred when a noticeable initial flattening of the DTI arches has begun (Figure 31.4).

Only after proper snugging has been completed can full and proper tensioning be achieved. Once again, torque is applied to the nut side of each bolt assembly in the same pattern used during the

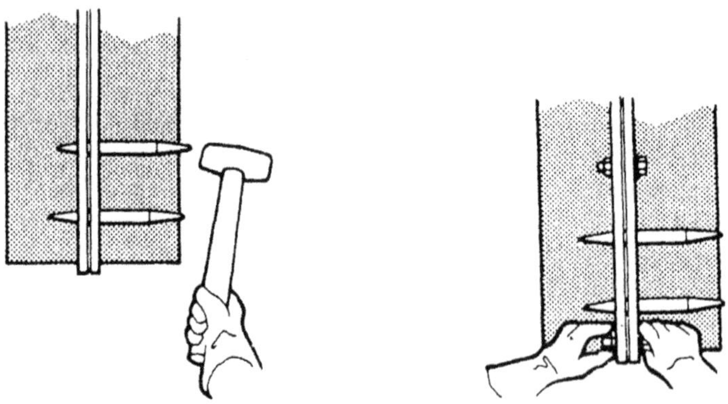

FIGURE 31.3 The bolt holes in a structural steel connection are first aligned by the use of drift pins as shown here. Interference between the bolts and the walls of the bolt holes will make it impossible for the bolts to be properly tightened if they, instead of drift pins, are used to pull the joint plates into alignment.

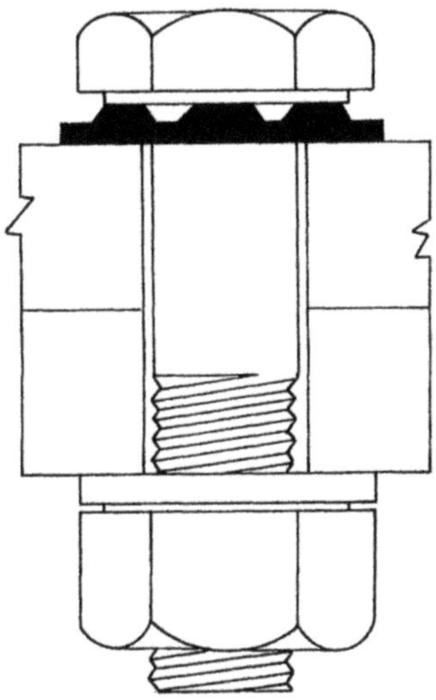

FIGURE 31.4 Specifications for structural bolting mandate that all bolts in a joint be "snugged" before any of them are fully tightened (see Chapter 26). Snugging partially flattens the projections on the DTI, as shown here.

snugging operation, from the center to the free edges of the bolted connection. Torque is applied to each nut until sufficient tension is achieved in the bolt as evidenced by reduction of the DTI gap below the specified value (0.015 in. or less for plain finish DTIs and 0.005 in. or less for mechanically galvanized DTIs).

The amount of torque required to create a desired tension in a fastener varies widely, and it greatly depends on many factors including thread condition, coatings, and surface finish. It is not uncommon to experience variation on the order of ±35% in the amount of torque required to create a desired bolt tension in a given joint. Use of a tool capable of delivering sufficient torque for the "worst-case scenario" is required. To minimize the torque required to install dry or rusty bolts, lubricants such as stick wax are often used to extract more tension out of a given torque, making it easier on the construction crew. The use of a lubricant on dry or rusty bolts also minimizes the chance of seizing or breaking a bolt during tightening.

Drift pins should be left in place until all of the assemblies in the connection are tensioned, as premature removal of drift pins could cause joint slippage and "trapping" of the bolts. Finally, the drift pins can be knocked out and replaced with bolt assemblies, which are then tensioned until the DTIs are likewise sufficiently flattened.

Bolting crews learn to use DTIs by practicing on about a dozen DTIs that they judge by eye to have been sufficiently flattened. They simply install the DTI until it is "almost flattened" and then check their own work with a feeler gage. In a short period of time, they are able to accurately tighten the fasteners without using the feeler gage at all. The bolting crew may later check their own work with a feeler gage, although it is not necessary. If an occasional inspection shows insufficient flattening, additional torque can be applied as necessary to sufficiently flatten the DTI. If for any reason the crew is unable to sufficiently flatten the DTI, minimum required tension has not yet been achieved. The DTI is merely a sentinel reporting tension.

Unlike torque-controlled tensioning methods, which require all fasteners to be fully tightened on the day in which they are removed from protected storage and installed, a DTI can be installed 1 day, snugged the following week, and fully tensioned any time thereafter. The DTI is unaffected by lubricity or other factors that detrimentally affect the torque tension methods such as the use of a calibrated wrench or a twist-off bolt. Valid final inspection of DTIs can take place at any time after tensioning and can be performed efficiently at any time convenient to the construction crew and the inspection team. Inspection requires only one person with a feeler gage.

Some of the advantages cited by DTI users include the following:

1. Savings in inspection costs exceed the cost of DTI.
2. Inspection of DTIs is considered a "value-added" operation because only tension can flatten the protrusions on installed DTIs; many of the other methods measure factors correlated to tension, not tension directly.
3. Slipped joints and trapped bolts are immediately detected because it is impossible to flatten the DTIs.
4. Rusty or dry bolts may be lubricated and used with DTIs rather than discarded or returned to the manufacturer (as is required with twist-off style torque control bolts).

Some projects have an additional requirement that "overtensioning" be avoided. DTIs provide for this requirement in that they may be installed and inspected to ensure that at least a minimal visible gap remains after tensioning. This indicates that the DTI has not been completely flattened or tightened beyond the completely flattened condition, thus providing visual proof that overtensioning has not occurred.

31.4 STRUCTURAL INSTALLATION IN SEISMIC ENERGY DISSIPATERS

Following pioneering work performed at the University of California at Berkeley, the use of DTIs in slotted bolted connections (SBCs) has become a safety-critical application of DTIs. SBCs are modified bolted connections that dissipate energy through friction during rectilinear tension and compression cycles. For example, SBCs address situations where it is necessary to limit the forces transmitted to brittle elements within an existing structure's lateral force-resisting path.

An SBC is a type of friction grip or slip-critical joint that is designed to slip and dissipate energy through frictional resistance at predetermined limits. Therefore, the mere achievement of a *minimum* target load is insufficient for tensioning of SBC joints. Engineers of SBCs design for a target load with minimal variation in either direction expressed in terms of upper and lower limits (usually no more than 10% total variation). This total variation is statistically analyzed for each and presented graphically. Figure 31.5 is constructed by taking the standard deviation in pounds and expressing it as a percentage of the target load in order to predict the total variation within a given lot. For example, a lot of ¾ in. Type 325 DTIs with a standard deviation of 280 lb at the target load of 28,000 lb could be expressed as a 1% standard deviation and could be included in the 1% column of a histogram like Figure 31.5.

DTIs used in SBCs are calibrated full scale including the target load. Therefore, DTIs used in these applications are installed to a specific gap directly corresponding to the specified target load. Figure 31.6 depicts a typical full-scale calibration graph for a specific lot of DTIs. It uses a compression-stress curve to show the relationship between compression (the size of the DTI gap) and stress (the force applied). As the graph suggests, the compression-stress relationship is essentially linear, with minimal variation between pieces from a given lot.

The original work with SBCs performed by Grigorian et al. [1] and Lizundia [2] was performed using DTIs provided by J&M Turner, Inc. that are unique in design and calibration features developed for such applications. These features are not required by the applicable ASTM standard for DTIs, although they are incorporated into all of that producer's structural DTIs. Figure 31.7 depicts a standard SBC connection.

31.5 PRESSURE VESSEL FLANGE INSTALLATION

The use of DTIs with bolts and studs in flanged connections is a comparatively recent development. Controlling specifications for bolt tensioning using DTIs in flanges are being developed. Bolt tension levels are chosen by the design or process/assembly engineers based on calculations and experience.

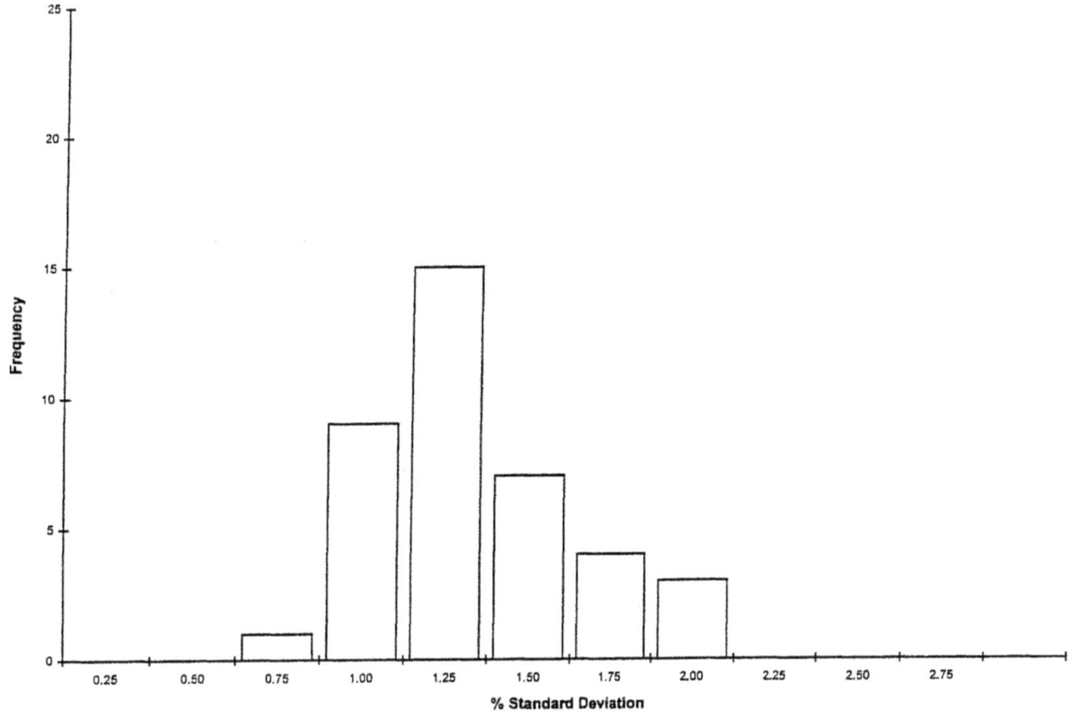

FIGURE 31.5 Histogram showing the lot history of a group of ASTM F959-94a DTIs used with ASTM A325 bolts. The standard deviation data are based on "percentage of load."

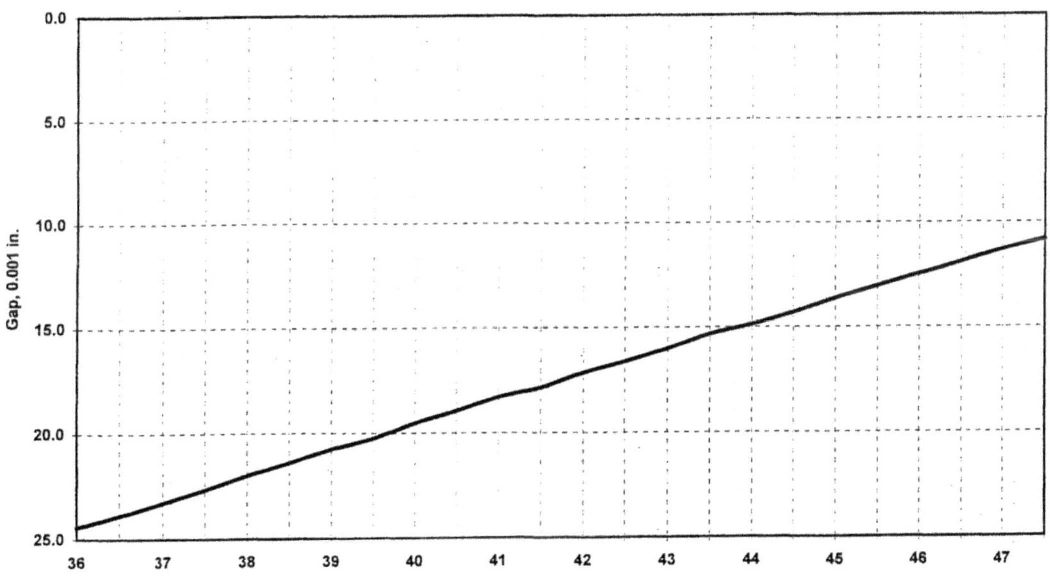

FIGURE 31.6 Graph showing the linear relationship between preload in a bolt and the resulting compression of the projections on the DTI. The latter is determined by measuring the height of the "gap" between projections (i.e., the remaining height of the projections). The data plotted are for the average of 27 samples.

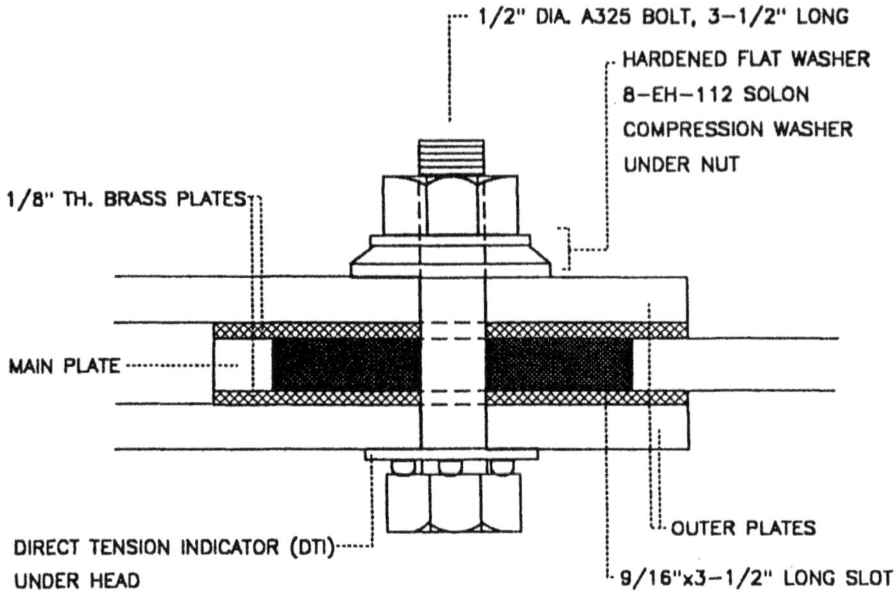

FIGURE 31.7 Sketch of a slotted bolted energy-dissipating connection. Friction between members in this joint is used to dissipate energy during seismic events. DTIs are used in such connections to guarantee that clamping forces, and therefore the interface friction forces, will be as large as intended by the designer of the structure.

In general, the target tensions in such applications are a much lower percentage of usable bolt strength than in structural applications, due in part to the nature of bolting of gasketed joints. Typical target loads range from 40% to 60% of the yield strength of the fasteners on which the DTIs will be used. Most often, such joints are fastened using ASTM A193 Grade B7 studs or bolts in combination with ASTM A194 Grade 2H nuts, or similar fasteners. DTIs are used in these applications as an alternative to the rather expensive ultrasonic inspection methods often used or as a replacement for more troublesome torque inspection methods.

Like seismic DTIs, the DTIs used in the flange market are calibrated for a range of tensions. The DTI manufacturer plots a compression–stress curve similar to Figure 31.6 for each lot of DTIs. compression–stress curves are created from at least 25 samples on a lot-by-lot basis. The compression–stress curve (and accompanying certification data) for each lot is used by the installation crew as a means to select the appropriate go and no-go feeler gages for the specific application and target tension required.

Installation of DTIs with B7 studs and 2H nuts in flanges proceeds similarly to structural applications, except that several passes are used to bring the joint up to tension rather than the two-step (snug and tension) process used in structural joints. The most significant difference between the use of DTIs in this market and in the structural market is that the DTIs are installed and inspected to a specific target gap to achieve a target tension rather than flattened anywhere below a specified gap to achieve at least a minimum tension. Other differences between these B7 DTIs and those used in structural joints include the following:

1. B7 DTIs have reduced outside diameters to accommodate the reduced bearing area and clearances available on many flanges.
2. B7 DTIs have much lower hardnesses, typically HRC 22 or less.
3. B7 DTIs are rarely used without some corrosion-inhibiting finish such as black oxide.

31.6 CAP SCREW INSTALLATION

The most recent development in DTI use pertains to the use of DTIs as a means to ensure tension in cap screws. Hex cap screws in both inch series and metric series have much smaller bearing surfaces than equivalent structural bolts. Consequently, DTIs were not historically available for cap screws due to technological limitations, which include the inability to transmit sufficient loads through these much smaller bearing areas without using excessively hard DTIs (i.e., HRC 38 and above).

Recent technological and design breakthroughs [3] have enabled the development of DTIs for use on SAE Grade 5 and Grade 8 fasteners as well as their metric Class 8.8, 9.8, and 10.9 counterparts. As of this publication, specifications for the use and installation of these DTIs are being coordinated with original equipment manufacturers of off-highway heavy equipment and will be submitted to the Fastener Committee of the Society of Automotive Engineers.

31.7 SUMMARY

Many changes have occurred since the invention of the DTI in 1962. The original DTIs offered reasonable predictability compared to other installation and inspection methods. Up until the early 1990s, DTIs were only modestly improved over their original designs. However, better DTIs were required to satisfy the needs of designers of seismic energy dissipation systems and for other uses. Fortunately, sufficient funding and exhaustive research led to a number of breakthroughs, which not only resulted in dramatic improvements in the standard DTIs and allowed production of DTIs for energy dissipaters but also piqued the interest of users in new markets. The humble DTI, which is often introduced into use as the "problem solver" for troublesome joints, is likely to find greater use as the device specified to help installers avoid problems in the first place.

NOTE

1 Never use torque or "bedding torque" to snug joints with DTIs. Research clearly demonstrates that torque is a poor predictor of the snug condition, particularly with large diameter bolts through multiple plies. A proper procedure to snug a joint in accordance with the RCSC specification follows the criterion that "snug tight is defined by partial compression of the direct tension indicator protrusions."

REFERENCES

1. Grigorian, C. E., T.-S. Tang, and E. P. Popov, Slotted bolted connection energy dissipators, Report No. UCB/EERC 92/10, College of Engineering, Univ. California at Berkeley, 1993.
2. Lizundia, B., *Seismic Strengthening with Steel Slotted Bolt Connections*, Rutherford and Chekene Consulting Engineers, San Francisco, CA, 1994.
3. J&M Turner, Inc., Patents 5,370,483, 5,487,632, and 5,667,346 and patents pending.

32 Selecting Preload for an Existing Joint

John H. Bickford

32.1 INTRODUCTION

This chapter will help you decide how much preload you should use in an existing joint, or "what torque?" as the question is usually worded. To answer this question, we must briefly review many of the topics covered in this handbook. Hopefully, this summary will help you decide which of these many factors will affect your results. And it will give you a reasonable way, I think, to make acceptable assembly decisions without having all the hard and fast data you would like to have. Such data cost money—often a lot of money—and are rarely available to those who must deal with existing joints. In fact, in many situations, you won't even need most of the procedures described in this chapter. You'll be able to use the "simple ways to select preload" described in Section 32.3. If you've had problems with a joint, however, or have failure or economic concerns, then you'll need something practical but a little more elaborate; you'll need the procedures that make up the bulk of this chapter. For safety-related joints, you may want to start with these procedures but use the information in prior chapters and/or the many references to go well beyond them. So, here's how we would start our search for a "better torque" if we had had problems or wanted to do a better job with an existing joint.

In picking preload or torque for such a joint we should answer two questions: How much initial, assembly clamping force do we *want* in this joint, considering the service loads and conditions the joint will face? and How much clamping force—and scatter in clamping force—can we *expect* from the assembly torques, tools, procedures, etc. we plan to use? Let's see how we might answer these questions in an economically acceptable way.

32.2 HOW MUCH CLAMPING FORCE DO WE WANT?

32.2.1 Factors to Consider

We start by considering the in-service clamping force needs of the joint. What loads and service conditions must the clamping force between joint members resist? Note carefully that this is the in-service clamping force we're talking about, not just the clamping force theoretically created by the initial assembly preloads. That initial clamp must be high enough to compensate for all of the mechanisms that may reduce the clamping force to the in-service level, including

Embedment relaxation
Elastic interactions
Creep of metal parts, gaskets, etc.
External tensile loads
Hole interference
Resistance of joint members to being pulled together
Prevailing torque
Differential thermal expansion

DOI: 10.1201/9781003501183-37

We must start by deciding how much clamp the joint will need in service, then try to estimate how much clamping force each of the above factors will have robbed from the initial clamp. Only then can we decide how much initial clamp—and therefore initial assembly preload—we need. OK—here's a checklist of the main factors that the in-service clamping force might have to resist.

32.2.1.1 Joint Slip

Joint slip is a key factor because slip can cause a lot of problems. It can cause unfortunate stress concentrations in a slip-distorted structure. It can cause fretting corrosion or fatigue of joint members. It can cause self-loosening or misalign and cramp bearings, etc. It's relatively easy to calculate the clamping force to reduce slip, however, if we know the magnitude of the shear loads imposed on the joint.

Let's assume that a shear load L_x is to be imposed on this joint. To avoid slip, the frictional forces created by the clamping force must exceed the external load as follows:

$$F \geq L_X \mu \tag{32.1}$$

where

F = clamping force on the joint (lb, N)
L_x = external load (lb, N)
μ = coefficient of friction (typically 0.15–0.30).

32.2.1.2 Self-Loosening

As we'll learn in Chapter 40, self-loosening will occur when transverse loads cause slip between joint members and thread surfaces. Although it is often difficult to quantify the vibratory or other forces creating self-loosening, the joint slip equation above could be used to estimate the point at which the joint members will slip if the external loads are known.

If external loads cannot be estimated and the application is likely to involve vibration or other forms of self-loosening, then good practice would be to plan for the "maximum clamping force the parts can stand." The more the better, up to, but probably not exceeding, yield.

32.2.1.3 Pressure Loads

The influence of pressure loads on bolted joints is explained in some detail in Chapter 19. One way to quantify the required initial and in-service clamping forces is to use the equations of the Boiler and Pressure Vessel Code. These are currently (early 1998) being modified to use new PVRC gasket factors, but the equations found in the Code at present are

$$W_{M1} = \frac{\pi G^2}{4} P + 2\pi GbmP \tag{2}$$

$$W_{M2} = \pi bGy \tag{3}$$

where

W_{M1} = tension in the bolts in service (lb, N)
W_{M2} = initial tension in the bolts at assembly (lb, N)
G = diameter of gasket (in., mm)
P = contained pressure (psi, Pa)
b = effective width of the gasket (in., mm)
m = gasket maintenance factor
y = initial gasket stress at assembly.

These equations assume that the entire pressure load will be seen by the bolts (i.e., ignoring the implications of the joint diagram), but these equations or their equivalent could presumably be used for an approximate (and conservative) estimate of the clamping force required in most gasketed joint situations.

32.2.1.4 Joint Separation

In some applications—a noncritical foundation bolt is an example—gravity holds the joint in place and it is sufficient for the bolts merely to maintain alignment. In this situation, clamping force could be as low as 0 without risk of joint failure. Even here, however, some clamp would be useful merely to retain the nuts. This little clamp would also be acceptable in those structural steel joints that are not slip-critical. In most tension joints, however, it's as important to avoid separation as it is to avoid joint slip—maybe even more important. Separation can lead to such horrors as gross leakage and low-cycle fatigue of bolts. To avoid separation we must be sure that the initial preloads are high enough to compensate for all of the clamp loss factors listed a minute ago and still leave some residual in-service clamp. A margin of safety, if you will.

32.2.1.5 Fatigue

Although joint separation can reduce the fatigue life of bolts by a substantial amount, it is not, as some handbooks imply, the only cause of fatigue failure. Too little clamping force or, less commonly, too high a mean bolt tension can also cause problems. This time, however, we have no simple equation to compute the amount of in-service bolt tension or the related interface clamping force required. A fairly complex analysis is required, as suggested by Chapter 37 and in the references cited at the end of that chapter.

32.2.2 Placing an Upper Limit on the Clamping Force

When determining the amount of clamping force required to combat separation, self-loosening, slip, or a leak, we are interested in establishing the essential minimum of force. In each of those situations, additional clamping force is usually desirable (for added safety) or is at least acceptable. You might remember that another early theme was that we always want the maximum clamping force the parts can stand. There is, however, always some upper limit on that clamping force. If that weren't the case, we could simply "tighten them a lot more" to avoid failures. Instead, we must define an upper limit for our application.

The maximum tension or stress in the bolts will usually, though not always, be the thing that places an upper limit on the amount of clamp we want. We might *like* more, but not if it means broken or threatened bolts. So, here are some bolt factors that limit clamping force.

32.2.2.1 The Yield Strength of the Bolt

There is a good deal of debate about the yield strength of the bolt in the bolting world at the present time, but most people feel that it is unwise to tighten bolts past yield in most applications. There are many exceptions, with structural steel being the most obvious. As we discussed in Chapter 31, torque–turn equipment that tightens the fastener to or past yield is growing in popularity for automotive and similar applications. Usually, however, we won't want to tighten them past yield during initial assembly. Bolt yield, then, is one easy-to-estimate, "worst-case," upper limit on the in-service clamping force.

32.2.2.2 Thread-Stripping Strength

Obviously, we will never want to tighten the fasteners past the point at which their threads will strip. This then provides another, simplistic, worst-case upper limit. If we didn't consider this limit when selecting the bolts, we should do so now (unless one of the limiting factors listed below will obviously dominate our decision).

32.2.2.3 Design Allowable Bolt Stress and Assembly Stress Limits

We always want to identify any limits placed on bolt stress by codes, company policies, standard practices, personal bias, or the like. Both structural steel and pressure vessel codes, for example, define maximum design allowable stresses for bolts. It is necessary, however, to distinguish between a maximum design stress and the maximum stress that may be allowed in the fastener during assembly. This will differ from the maximum design allowables if a design safety factor is involved. In the structural steel world, for example, bolts are frequently tightened well past yield, even though design allowables are only 35%–58% of yield. Pressure vessel bolts are commonly tightened to twice the design allowable. Aerospace, auto, and other industries may also impose more stringent limits on design stresses than on actual stresses to force the designer to use more or larger bolts than he might otherwise select (and therefore to introduce safety factors in the design).

32.2.2.4 Torsional Stress Factor

If the bolts are to be tightened by turning the nut or the head, then they will experience some torsional stress as well as tensile stress during assembly. If tightened to yield, they will yield under a combination of tensile and torsional stress. If we plan to tighten them to or near yield, it's pertinent to reduce the maximum tensile stresses allowed at assembly by a "torquing factor" which makes room for the torsional stress. If as-received steel-on-steel bolts are used, then a reduction in the allowable tensile stress of 10% is probably reasonable. If the fasteners are to be lubricated, you might use 5%. The true amount of strength absorbed by torsion will be determined by all of the variables that affect the torque–preload relationships, so the torsional effect on the tensile capacity of the bolts may be much greater than 5% or 10% (and is as difficult to predict as the specific torque–tension relationship for a given set of parts). Anyway, 5%–10% will probably be acceptable in most applications. Most people don't tighten bolts to the yield point anyway.

32.2.2.5 Shear Stress Allowance

If the bolts will also be exposed to shear stress we must take that into account in defining maximum assembly preloads and the resulting in-service clamping force, since shear stress will reduce the amount of bolt strength capacity available for the tensile stress.

The shear reduction is explained and illustrated in Chapter 10 (Eqs. 32.13 and 32.14).

32.2.2.6 Stress Cracking

As we'll learn in Chapter 39, stress corrosion cracking (SCC) or hydrogen embrittlement is encouraged by excessive tension in the bolts and we are often told to be alert if service loads exceed 50% of yield, at least for low alloy quenched and tempered steels. See Chapter 39 for a more complete definition of this important upper limit.

32.2.2.7 Combined Loads

As we try to determine the maximum stress the parts can stand, to avoid damage to the parts, SCC, or other problems, we must evaluate our estimates in light of the anticipated service loads on the bolts. Most of these service loads would add to the bolt stresses introduced during assembly (the preloads) even as they reduce interface clamping force.

The joint diagrams of Chapter 36 can be used to add external loads to preloads. The effects of a temperature change can be estimated using the procedure described in Chapter 18. The bolts must be able to withstand worst-case combinations of these preservice and service loads.

All of the limiting factors discussed so far deal with bolt strength, which in turn limits the clamping force available for the joints the bolts are used in. There are also a few joint factors that can limit clamping force. Here are some of the most common ones.

32.2.2.8 Damage to Joint Members

Too much tension in a bolt can cause its head and nut to embed themselves into the surfaces of the joint, not just by a "normal" amount of a few mils, but enough to cause visible damage to the joint surfaces. The VDI Directive 2230, for example, says that the "boundary surface pressure" (P_G) of joint materials is usually slightly greater than the yield strength of the material, and they recommend that bolt tension not exceed the value suggested by the equation

$$P_G A_P \geq 0.9 \max F_B \tag{32.4}$$

where

A_P=contact area (e.g., between nut face and joint) (in.^{2}, mm^2)
F_B=tension in the bolt (lb, N)
P_G=boundary surface pressure (psi, N/mm^2).

The 0.9 is added as a safety factor.

32.2.2.9 Distortion of Joint Members

Joint members can sometimes be distorted by excessive bolt loads. For example, the outer ends of raised face flanges can be pulled toward each other—perhaps bent—by too much preload. This can unload the inner diameter of the gasket, opening a leak path.

32.2.2.10 Gasket Crush

Excessive preload can so compress a gasket that it will not be able to recover when internal pressure or a thermal cycle partially unloads it. Contact the gasket manufacturer for upper limits, which will be a function of service temperatures.

32.2.3 Summarizing Clamping Force Limits

Let's try to summarize the range of clamping force we want. Not too high—that might cause joint problems or, more commonly, bolt problems. Not too low—that might cause joint slip or separation or leakage.

Figure 32.1 illustrates one possible scenario based on some of the factors just discussed. Each factor is described as a percentage of the ultimate strength of the bolt, to give us a common vertical axis. The figure shows that 100% of ultimate would break the bolts. It suggests that bolt tension above about 80% of ultimate would take the bolts past yield, and assumes that we don't want to do that in our application; this region is considered "unusable."

The chart suggests that we subtract about 5% from the tensile yield strength of the bolts to account for torsional stress introduced when torque is applied. It suggests that we might have reduced the maximum allowable bolt stress still further to accommodate a Code or other specification limit. And it further reduces the limit by a 5% "safety factor" to acknowledge that our data are limited and/or that we haven't included factors such as external loads or differential expansion that can add to bolt tension. The final result is a suggested upper limit on the desired clamping force equal to 62% of the ultimate strength of the bolts.

The chart also defines an acceptable lower limit on clamping force, suggesting that we want a clamping force equivalent to at least 30% of ultimate to make sure the nuts don't fall off and that, more seriously, we want a clamping force equivalent to as much as 42% of ultimate to prevent the failure mode (or modes) we're most concerned about—self-loosening, leakage, slip, fatigue, or combination of these. Since the exact amount of load or vibration to be seen by a joint is often hard to calculate, and since "more is generally better," we throw in a larger safety factor on the low end than

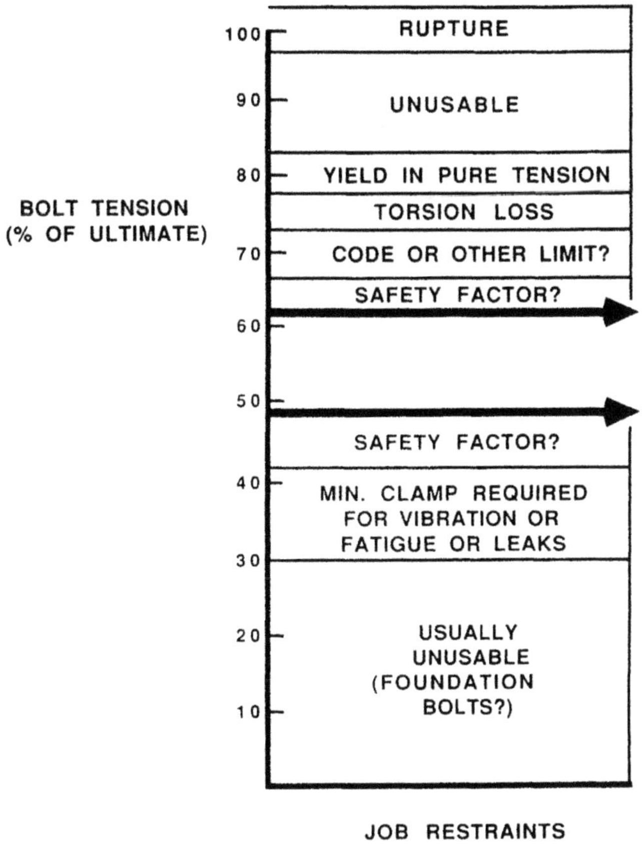

FIGURE 32.1 Chart summarizing the clamp load decisions made for a hypothetical joint. See the text for a detailed discussion.

we did at the high. The result: a lower limit on desired clamp that is equivalent to 48% of the bolt's ultimate strength (or about 60% of yield if the bolt yields at 80% of ultimate).

All of which is very broad-brush and seat-of-the-pants, but that's the way most assembly preload or torque decisions are made in the field. The principal advantage of the procedure is that it forces us to consider separately, and assign at least gut-feel values to, those factors we think might affect our joint. The chart shown in Figure 32.1, incidentally, can also be constructed in terms of yield strength or torque, for single bolts or the whole group.

OK. We have now roughly defined the range of clamping force we'd like to see in our joint, a range defined in part by limitations in the strength of the bolts. Now we want to pick a target preload to specify for assembly purposes. If we are dealing with a problem joint, we also want to estimate the range of clamping force we can expect to achieve during assembly, taking our choice of preload plus our assembly tools and procedures into account. It will often be necessary, too, to estimate how much that range might be increased by postassembly and in-service conditions. With all this information in hand, we can then compare the desired range of clamping force to the anticipated range. If they agree, we're done. If they don't agree, we'll have to take steps to correct our assembly procedures—perhaps pick a better torque—to make them agree.

Before proceeding, let us acknowledge, once again, that it usually won't be necessary for us to go through the "full" procedure described in this chapter. If service conditions are reasonable and we have no safety concerns, we can choose a preload or torque by some of the very simple ways described next. We'll want to go beyond these things only if we're concerned about results or—more commonly—have had problems with this joint in the past.

32.3 SIMPLE WAYS TO SELECT ASSEMBLY PRELOADS

In most situations, as the bulk of prior experience tells us, we can pretend we never read this book and can "pick a preload" by one of several simple, time-proven methods that have worked on most joints in the past. We don't want to forget this point, because if we do we can waste time and money solving problems that don't exist. So, before considering a procedure to use when it is necessary to "do it right," let's look quickly at some of the simpler ways that will often be sufficient.

32.3.1 Best Guide: Past Experience

If you've had previous experience assembling this joint, and the results have been acceptable, don't change your procedures or tools. As we've seen, bolting involves more variables than we can cope with. We can *never* predict results with perfect accuracy; so prior experience, if satisfactory, is better than all of our theories. Leave it alone! Note that the tools and procedures you've used have, by default, "selected" a preload range for you. It worked. It's acceptable.

Note that you won't usually be able to identify a successful preload per se in situations such as those we're discussing. You'll know what torque the mechanics have used, or which size of tool, or which procedure.

Note that even if we know exactly which torque has been used on this joint we don't know "the" preload created by that torque. A given torque—any torque—will create a range of preloads in a group of bolts. Variations in friction can scatter the preload by as much as ±30%. Elastic interactions can leave residual preloads scattered by max/min ratios of 20:1 or more. Embedment, tool error, operator problems, etc. can also enter the picture, as we have seen. In spite of all this, even though we haven't measured, or even been aware of, such factors, if the tools, procedures, torques, etc. we've used in the past have *worked*, we should keep using them.

32.3.2 Second Best: Ask the Designer

If you've had no prior experience with this joint, or have had some problems, your best source of information is the man who designed the joint. As we've seen, many factors must be considered in the design of a successful joint. It's a rare customer or user or assembler who knows enough about the materials and configurations of—or loads on—a joint to duplicate the designer's expectations when it comes to specifying the correct clamping force. If the selection of preload matters, the designer should know what the desired range of values should be. Presumably, he will have gone through a design procedure equivalent to that discussed in Chapter 18.

There will be many cases, however, in which the designer has done no such thing, but even here he may have been guided by the past experience of other customers or by general experience with this type of product. Even if he hasn't made an analysis, therefore, he may be able to recommend a preload—again, specified as a torque in most situations. Note that the designer's recommendations can sometimes be found in the operating or maintenance manuals that came with the equipment. If not, give the designer a call!

32.3.3 Unimportant Joint: No Prior Experience

In another common situation, you'll have had no prior experience with this joint, and there is no way to reach the designer (or he can't help), so you must pick the preload (probably a torque) yourself. If the joint is a "common" one, and you have no real concern about the consequences of failure, then the normal procedure is to pick a torque from a suitable table. Most torque tool manufacturers can provide such a table. Picking a torque (and a tool and a procedure) determines the range of preload you'll achieve at assembly, even if you won't be able to quantify this range with any precision. Most joints are overdesigned; most joints will behave themselves when assembled with a torque selected this way.

Again, picking a torque will, in effect, pick a range of preload, as suggested in Section 32.3.1.

TABLE 32.1
Typical Target Preloads as a Percentage of the Yield Strength of the Bolts

This Percentage of Yield	Might be Appropriate for Applications Such as These
25	Unimportant nongasketed joints exposed to static loads, foundation and anchor bolts under static load, also joints where there have been serious stress corrosion cracking problems
40	Gasketed joints in routine service, including those covered by the ASME Code that have not given problems
50–60	Average nongasketed joint, with "normal" safety or performance concerns, where past experience does not suggest higher or lower preloads; a good place to start a search for the optimum preload when some trial and error is acceptable
	Probably the maximum acceptable preload for gasketed joints designed to ASME Code rules (although there will be a few exceptions)
70–75	Upper limit for nongasketed joints with which you've had "low preload" problems in the past (leaks, self-loosening, fatigue, etc.) and where torque control will be used at assembly
85–95	Joints that have had consistent "low preload" problems in the past; where the need to avoid failure dictates the use of special techniques such as stretch or ultrasonic control; and where service loads (or ignorance of service loads) make it unwise to take the fasteners any closer to the yield point
100	Structural steel bolts tightened by turn-of-nut procedures; also high performance or problem joints facing self-loosening, fatigue, or other "low preload" problems, but where service loads can be predicted with sufficient accuracy to guarantee that the fasteners won't be "ratcheted" to rupture later on; sometimes used in gasketed joints in automotive applications, for example; requires special assembly techniques

32.3.4 When More Care Is Indicated

If your past experience with the joint could be better and/or you're a little concerned about the consequences of failure, you might want to compute an appropriate torque instead of picking one from a table. One way to do this:

1. Determine the yield strength (S_y) of the bolt material at the operating temperature of the joint (data on yield strengths will be found in Chapters 1 and 2).
2. Pick a target "percentage of yield" (P) from Table 32.1.
3. Determine the tensile stress area of the bolt (A_s):

$$A_s = 0.7854\left(D - \frac{0.9743}{n}\right)^2 \quad \left(\text{See p. 135 for units.}\right)$$

4. Estimate the lubricity of the fastener by picking a nut factor (K) from Table 3.1 in Chapter 3.
5. Now use the following version of the short-form, torque–preload equation to compute an assembly torque (T) in in.-lb.

$$T = KDPS_yA_s \tag{32.5}$$

where D is the nominal diameter of the bolt in inches.

Why is a computed torque "better" than a torque taken from a table? The computation takes specific job constraints into consideration; a table is based on "normal" assumptions, which may or may not be valid for your application. In the procedure above, for example, we made application-specific decisions based on

1. Percentage of yield strength as a preload (not torque) target
2. The lubricant we're planning to use
3. The operating temperature of the joint.

In addition, or instead, you could base the desired preload on such things as the fatigue endurance limit of the bolts or on the limitations suggested by stress corrosion cracking concerns instead of basing it on a percentage of yield. But the end result at this level of the preload selection process is still a torque, with all the subsequent uncertainties of the torque–tension relationship. Neither this process nor a torque table should be used on critical joints.

Again we end up with a specified torque and with the unspecified but unavoidable range of preload, it will create.

32.3.5 If Improvements Are Required

If you've used a torque based on past experience, or picked from a table, or computed with the short-form equation, and have had nuisance problems, and can't get any help from the people who designed the equipment, you might want to try a different torque.

Bolting is an empirical art at present; experiments are often the most cost-effective way to make improvements. They may, in fact, be your only choice. (Again, you shouldn't use this or any other "simple" preload selection technique if the joint is critically important and failure would be disastrous.)

What torque should you try? If the problems you've experienced included a leak, self-loosening, joint slip, joint separation under load, or fatigue, then more preload is indicated. Try increasing it (e.g., try increasing assembly torques) by 10% or so.

If the problem has involved stress corrosion, stripped threads, crushed gaskets, or excessive flange rotation, then try 10% less preload or torque the next time.

In either case, keep good records so that you can benefit fully from your "experiments" and ultimately find the optimum preload, torque, procedure, etc.

Note that "picking a better torque" probably won't reduce the range of preload created in the bolts; it will merely move the average or mean preload to a higher or lower level. But that will frequently solve a problem.

32.3.6 Selecting Preload for Critical Joints

The procedures described above can be used in most applications. Obviously they can't be applied, except perhaps as a first step, to critical or safety-related joints. A good design engineer must then be called upon to select a preload and an assembly procedure (to control the range). Either he or a well-trained bolting engineer might be able to use the more rigorous procedures described in Section 32.7, hopefully with the help of well-founded data on the factors that can affect results in a given application.

Let's assume that we've used the quick-and-dirty procedures described above to pick a target preload we will use to specify an assembly torque. We know that this preload will result in a range of residual preload and that range will be further modified by service conditions to give us a still wider in-service range. We're very interested in estimating the size and limits of this range because that range in preload will determine whether or not we have created at least the minimum clamping force we decided we needed in Section 32.2. Let's see how we might estimate the in-service range of clamping force.

32.4 ESTIMATING THE IN-SERVICE CLAMPING FORCE

Placing exact limits on the in-service clamping force will usually be impossible. There are too many variables, and we'll rarely have good data with which to define their exact effect on the overall results. We can, however, often use the procedure I'm about to describe, plus general knowledge of "typical" values, to give us some idea as to the possible range. The answers we get will be good enough for most applications.

The procedure we'll follow is this:

1. We'll list the variables, or groups of variables, that we think will affect the preload achieved at assembly and/or the stability of that preload in service. We might identify possible variables by looking at the table of contents or the index to this book. In most situations, incidentally, you'll be primarily concerned with a half-dozen or so key factors. You don't have to consider every little thing that could affect the outcome.
2. Next we'll estimate the contribution that each key variable or group of variables might make to the preload scatter experienced during assembly or might contribute to the subsequent instability. We'll deal with the key variables one at a time, for one group at a time, as we make these estimates.
3. Then we'll use the chart shown in Figure 32.2 to combine these estimates and so arrive at final answers to our questions.

The procedure is simple and is easy to use, in part because it's heavily based on assumptions, estimates, and "typical values." In most situations, this is all we'll have. Accurate data will be rare, and only in the most critical situations will we be able to afford the time and money to make the

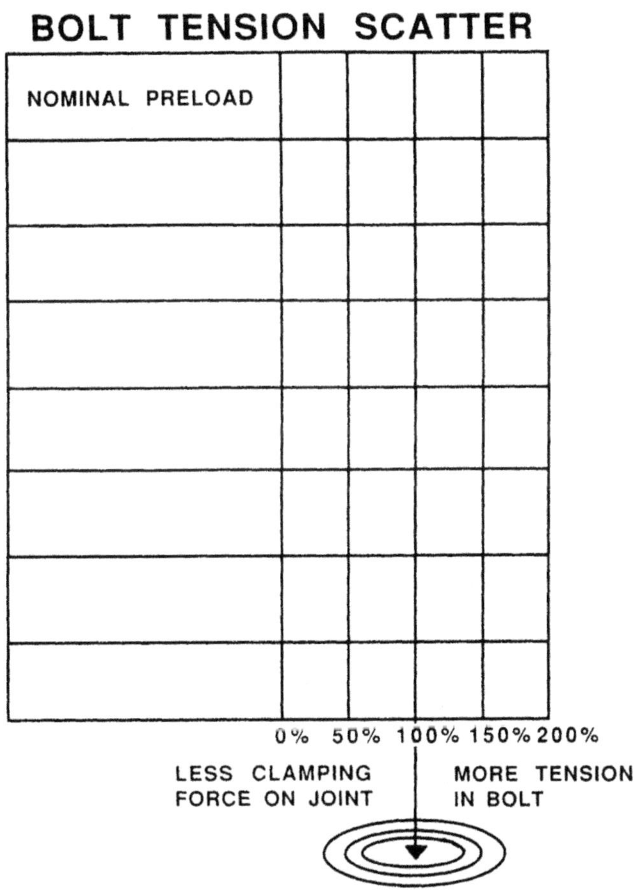

FIGURE 32.2 Chart used to summarize the results of scatter in the major variables that affect the amount of bolt tension and interface clamping force expected during assembly and when the joint is put into service.

experiments or analyses required to obtain hard data. Nevertheless, the proposed process will take us well beyond the "usual procedure" for estimating in-service bolt loads or clamping force on the joint.

What is the "usual procedure"? Most people seem to base their estimates of assembly accuracy—or in-service preload scatter—entirely on the process used to tighten individual bolts. If torque control is used at assembly, for example, it is usually assumed that final assembly preloads and in-service bolt tensions will be scattered ±30% around a mean or target value of tension. But the group of variables that create the uncertainty in the torque–tension relationship for single bolts is only one of the many groups of variables that can affect assembly or in-service results. Elastic interactions, relaxation effects, thermal effects, and many other things can also play a significant role. The procedure I'm about to describe allows us to consider all or most of the factors that will make a significant difference in a given application.

As mentioned, Figure 32.2 is the worksheet we will use to estimate the combined effect of the factors we think will affect the outcome in a particular application. Let's work an example to see how it's used.

32.4.1 Basic Assumptions

Assume that we will use a torque wrench to tighten the bolts on a straightforward, nongasketed joint. Also assume that we have used the procedure described in Section 32.2 to define an acceptable range or preload (48%–62% of the bolt's ultimate tensile strength as in Figure 32.1). Now, let's further assume that we've decided that five factors, each of which involves a different group of variables, will determine assembly results and subsequent changes of bolt tension in service. At the end of the example, we'll briefly review some of the other factors we might have considered.

We've picked the five factors listed below. We could have picked more—or fewer. There's nothing magic about five. The following list, however, includes several different types of problems or variables: some assembly, some in-service; some introducing plus or minus scatter, others biased in only one direction; some uncertain, others inescapable; etc. So this is a good list with which to study the procedure.

- Tool accuracy
- Operator accuracy
- Control accuracy
- Short-term relaxation
- External loads.

32.4.2 Combining the Scatter Effects

Now let's use the worksheet to combine these assembly and service factors. Then we'll be able to compare anticipated results with the desired results established in Section 32.2.

We start with the concept that there is an "ideal" in-service tension for this joint. If we could always introduce that tension at assembly, and if subsequent load and environmental factors never altered it, we'd have achieved perfection. We would have "hit the bull's-eye" 100% of the time. We represent this ideal tension by a vertical line drawn from the top of our worksheet to the center of the target at the bottom. As a first cut, we'll assume that this tension is at the midpoint of the 48%–62% range established earlier—i.e., our target, in-service bolt tension corresponds to 55% of the ultimate tensile strength of the bolt. This ideal bolt tension, of course, will create an ideal clamping force between joint members.

Now we introduce our first "factor" of concern—the accuracy of the tools to be used at assembly. Note that anything that reduces the preload or service tension in the bolts also reduces the all-important clamping force on the joint, making it less than our ideal or target value. Anything that increases the preload or tension means more tensile stress in the bolt than is "ideal." Let's see what our first variable, tool accuracy, does to us.

By "tool accuracy" I mean the accuracy with which the tool produces the thing it's supposed to produce. In this example, the tool produces "torque," and we'll assume that it does this with an accuracy of ±5%. (We'll consider the accuracy with which that torque creates preload in a minute.) We'll assume that this 5% (and the other percentages to follow) are "3 sigma" values and represent the worst-case errors we can expect to encounter in nearly all such assemblies. The worksheet could, instead, be based on one standard deviation (68% of the population), or on 2 sigma (2σ) values (95%), to be less conservative. But we'll assume 3 sigma (99%) for now.

In the real world, would we really consider tool accuracy to be a key variable? Sure. It's not uncommon to believe (sometimes correctly) that a more accurate tool can make a significant difference. For our example, let's pretend that a torque wrench salesman has heard that we've had vibration loosening problems, and he says that our problems will be solved if we use his 5% wrench instead of the 15% multiplier arrangement we've been using. We can evaluate his claims as we make our analysis; we can decide if a better, more accurate, torque tool would be likely, or unlikely, to reduce our problems.

We said that we expected the torque accuracy of this tool to be ±5%. Since torque is linearly related to preload (see the short-form equation in Chapter 29), this means that tool inaccuracies, worst case, will introduce as much as 5% more tension in some of our bolts than we'd like to see, or create, in other assemblies, as much as 5% less clamping force in the joints. These are both undesirable results and therefore are of concern to us. We don't want to break the bolts (too much tension), and we don't want the clamping force to be too low (remember that we always want the maximum clamping force those parts can stand). So, as we plot tool accuracy on our worksheet, as in Figure 32.3, we label plus errors as "more tension in the bolt" and minus errors as "less clamping force on the joint." The accumulated 3 sigma scatter at this point is, of course, just ±5%.

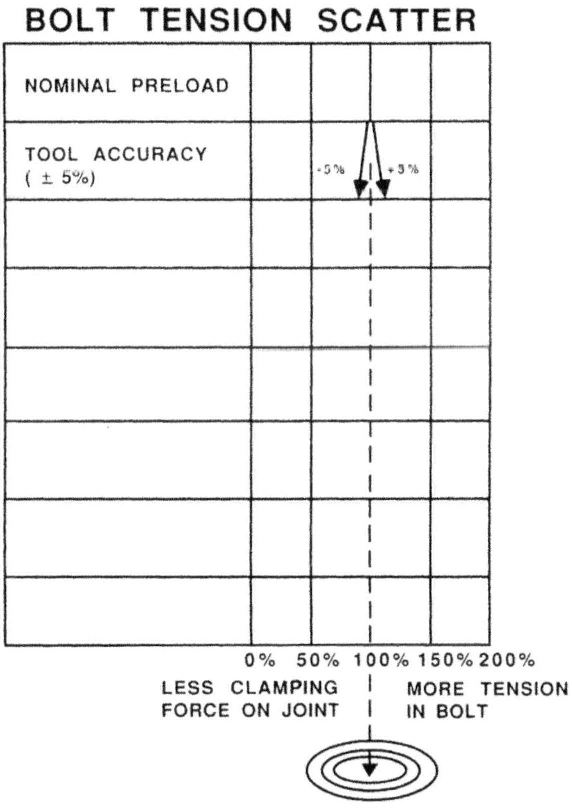

FIGURE 32.3 The first variable, tool accuracy, is assumed to contribute ±5% of scatter to the bolt tension or clamping force.

Now we introduce the second factor of concern: operator accuracy. (By "operators" I mean the mechanics or assemblers or others who "operate" the tools, use the wrenches, and tighten the bolts.) This factor defines the amount of scatter introduced by operator errors, operator carelessness, poor accessibility (which makes it difficult for the operator to do a "perfect" or "ideal" job), etc. Let's assume that this factor alone will contribute ±10% of scatter.

Do operators contribute this much uncertainty to bolted joint results? They certainly do. Some people claim, in fact, that *all* in-service problems result from improper assembly or maintenance practices carried out by operators who don't care or lack skill. We'll assume that we're dealing with well-trained mechanics who know that bolting is important. They'll "contribute" ±10% to the scatter primarily because accessibility is poor and working conditions are difficult. Even with a perfectly accurate torque wrench, they won't always be able to apply the "ideal" or target torque to each nut.

Note that this contribution to the scatter will be especially difficult to quantify. Our procedure, however, makes our guess more realistic by isolating it from the other variables we must deal with. Anyway, let's assume ± 10% for the example.

We don't just add this ±10% to the tool's ±5%. Probability theory tells us that it's unlikely that we'll experience a worst-case tool error and a worst-case operator error in the same assembly. We may see that once in a while, but in 99% of the assemblies (3 sigma again) the combined effect of these two errors will be less. We can compute the probable combined error by taking the square root of the sum of the squares of the two error percentages, as follows:

$$\pm V_T = \pm\left(V_{\text{TL}}^2 + V_{\text{OP}}^2\right)^{1/2} \tag{32.6}$$

where

V_T = the total 3 sigma scatter
V_{TL} = the 3 sigma scatter contributed by the tool
V_{OP} = the 3 sigma scatter contributed by the operator.

So the combined effect of these two errors is only ±11%, as shown in Figure 32.4.

Now we add the third factor, control accuracy. This is the accuracy with which the selected control variable—torque in this example—produces the thing we're interested in, which is always bolt tension. Table 32.2, incidentally, lists typical control accuracies for a variety of assembly tools and procedures. In our example, we'll use the conventional wisdom and say that the torque–tension scatter will be ±30%. We could introduce elastic interactions as an additional variable affecting the bolt tension achieved at assembly, but we won't for this first example. We'll assume a fairly rigid joint with metal-to-metal contact and insignificant interactions.

We take the square root of the sum of the squares again, taking three variables into account this time, and now we find a cumulative scatter of ±32% (Figure 32.5).

Our next variable, short-term relaxation, is handled the same way, but now we have to compute positive and negative errors separately, since they're different. Let's assume embedment relaxation of –10%. Relaxation will never increase bolt tension, so the plus scatter is zero. Cumulative errors now stand at, worst case, +32% and –34%. This tells us that over a long period of time and many assembly operations (i.e., a large statistical sample), we could expect to see, worst case, 32% more tension in some bolts than the "ideal" hoped for and that in other cases we'd see 34% less than the ideal clamping force in some of the joints (Figure 32.6).

This summarizes the situation at the end of assembly. Note that the ±30% torque–tension variable has dominated the results so far (justifying those who often assume that it's the only variable!). It wouldn't have dominated results if we had included elastic interactions in our analysis or had assigned a large scatter to the operators.

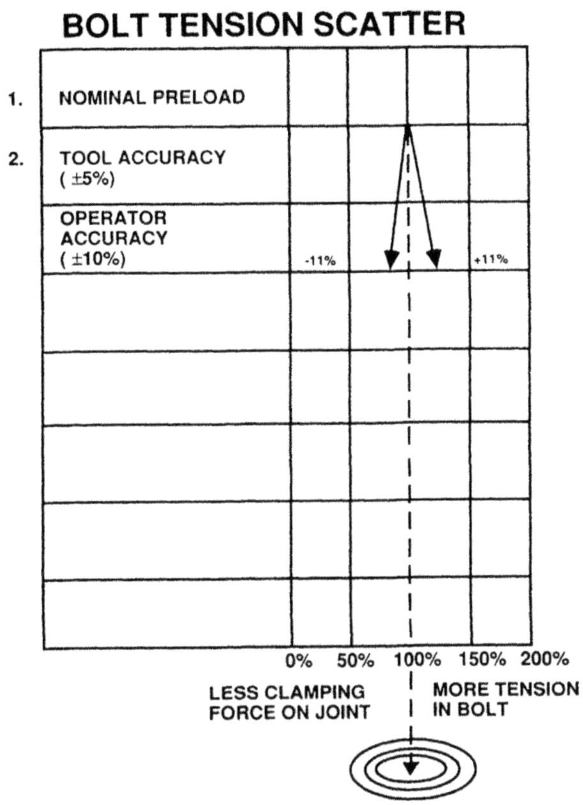

FIGURE 32.4 The second variable studied in the example given in the text is operator accuracy, which is assumed to contribute ±10% of scatter. The accumulated scatter at this point is ±11%. (See text for that calculation.)

TABLE 32.2
Preload Scatter Reported for a Variety of Bolting Tools or Procedures[a]

Torque control with hand wrench	±30
Small torque air tool	±35
Click type torque wrench	±60 to 80
Torque wrench plus multiplier	–70 to +150
Turn–of–nut (structural steel)	±15
Computer–controlled air tool to yield point	±3 to 10
Rockwell International's LRM (torque–angle) system	±3 to 10
Strain gaged load washers	±15
Strain gaged bolts	±1
Swaged lockbolts	±5
Air–powered impact wrench	–100 to +150
Hydraulic tensioners with vernier gage readout	±20
Operator feel	±35
Bolt stretch (micrometers)	±3 to 15
Ultrasonic control	±1 to 10

[a] All values are in percent and come from the author's own experiences or from Ref. [1].

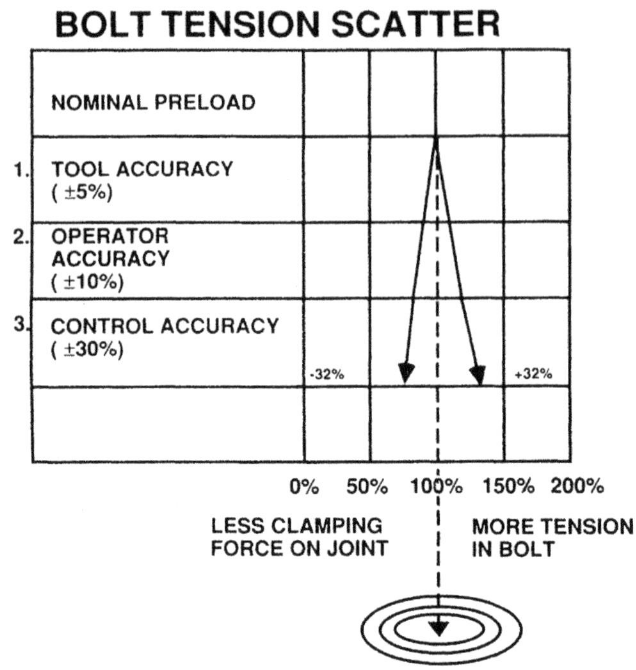

FIGURE 32.5 Control accuracy (torque vs. tension) scatter of ±30% raised the accumulated scatter to ±32%.

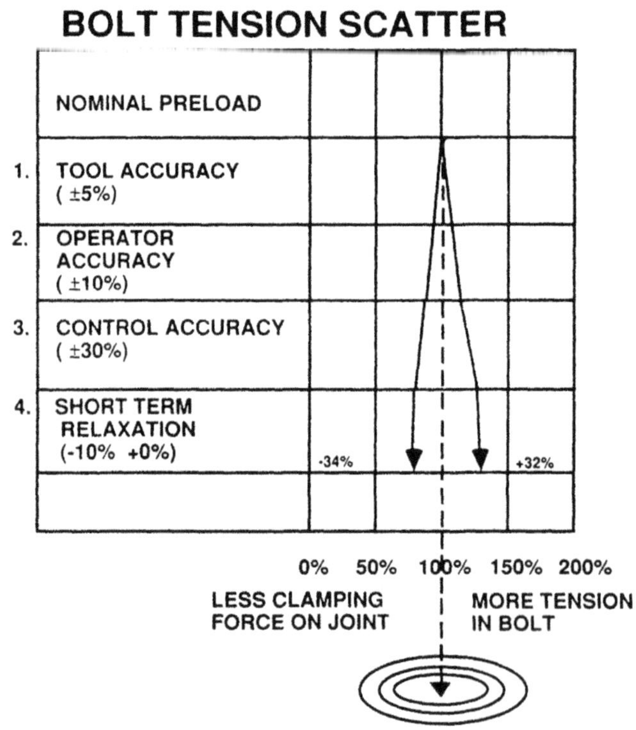

FIGURE 32.6 Short-term relaxation can decrease the clamping force but cannot increase bolt tension. The cumulative scatter is now +32% and –34%.

The final variable we've selected involves an in-service condition. We'll handle it in the same way, however. When we put the joint into service it sees an external tensile load. This load will both increase the tension in the bolts and decrease the clamping force on the joint (refer to the joint diagrams in Chapter 36). If we assume a 5:1 joint-to-bolt stiffness ratio and an external tensile load equal to 25% of preload, we might see a change of +7% in bolt tension and –20% in clamping force on the joint when the load is applied.

This factor differs from the others we've considered in several ways. First, it simultaneously affects both bolt tension and clamping force. Most of the previous variables would affect both, but not in the same assembly. Second, the external load may be relatively easy to quantify, compared at least to such factors as operator accuracy or torque–tension scatter. Third, this effect will probably be unavoidable. Tools and operators may perform perfectly, in some assemblies at least, but all assemblies have been designed to carry an external load. When we combine this "certain" variation with the earlier "probabilistic" ones, therefore, we're adding apples to oranges and will undoubtedly distress our neighborhood expert on probability. But I think the procedure is useful.

We're trying to estimate—roughly, simply, inexpensively—the combined effect of many variables. Most of the "data" we're using are "soft," to say the least. This latest violation of probability mathematics (if that's what it is) won't, in my opinion, make things much worse.

Adding the first service factor of concern, the effect of tensile loads doesn't change the picture much, thanks to the "smoothing" effect of the square root of the sum of squares equation. The cumulative worst-case errors now stand at +34% and –39%, as shown in Figure 32.7.

Let's pause at this point and see what those errors suggest. Plus 34% means that some of our bolts will experience 134% of ideal or optimum tension because of the combined assembly and

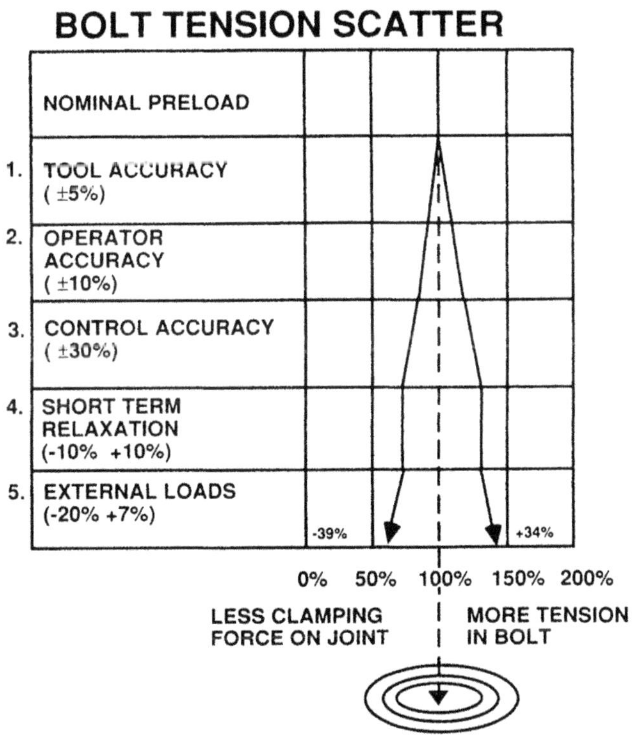

FIGURE 32.7 An external (tensile) load will simultaneously increase bolt tension by 7% and reduce clamping force by 20% if the bolt/joint stiffness ratio is 5:1 and the external load is 25% of preload. Note that the cumulative scatter can now be defined as –39% to +34% of the desired or target tension, or it can be defined as 61%–134% of the desired tension.

TABLE 32.3
Joint Overdesign Factor

Ratio between the Amount of Tension the Bolts Must Support and the Amount of Clamping Force We Can Count on (Worst Case)	That Ratio would Require a Bolt of the Diameter Shown Below (a Perfectly Tightened ¾ in Diameter Bolt Could Provide the Same Clamping Force)
2:1	1⅛ in.
5:1	1¾ in.
10:1	2½ in.
16:1	3 in.

service factors considered so far. Minus 39% tells us that, worst case, we can count on only 61% (100%–39%) of the desired clamping force in some of those joints. The ratio, therefore, between the tension the bolts must be able to support without breaking and the clamping force we can count on in 99% (3 sigma) of our assemblies in 134%/61%; in other words, the ratio is greater than 2:1.

Let's assume that past experience has suggested to the designer that a 1⅛ in. diameter bolt will probably be required for this joint. This means, presumably, that a 1⅛ in. bolt can withstand the maximum 134% of ideal stress experienced by bolts in this application. But we can count on only 61% of the ideal clamping force. We could get that much clamp from a perfectly tightened ¾ in. bolt.

This demonstrates a phenomenon that should be mentioned. Bolted joints are usually "overdesigned" to compensate for uncertainties in assembly results and/or in-service conditions. Designers know that they have to use lots of big bolts to avoid problems. If the assembly and service variations could be reduced, less overdesign would be required. Smaller bolts, lighter joint members, and smaller tools could do the job now done by the heavy joints and large bolts. All of this would mean economic advantages—which would be partially offset by the added costs of better control at assembly. At present the benefits offset the costs only in applications where weight, size, efficiency, etc., are also significant factors.

Even though we've considered only five variables, the ratio between the force we can count on and the force the parts must support is over 2:1. Many additional variables will have to be included in some applications, often increasing the ratio well beyond 2:1. Table 32.3 shows how the ratio affects the amount of overdesign required in a joint.

In Section 32.6 we'll try to answer the question, which variables should we include in our analysis? The answer will have a large impact on the results we get and on the estimated ratio between the maximum bolt tension and the minimum per-bolt clamping force in the joint. The ratio between "must support" and "can count on" will often be much greater than 2:1. Before deciding which variables to include, however, let's see how we might use the resulting estimate.

32.5 RELATING DESIRED TO ANTICIPATED BOLT TENSION

We estimated the range of bolt tension we wanted in Section 32.2 and the range we can expect, in Section 32.4. Now we will combine our two analyses. To do this we turn the target of our last worksheet on its side and hold it up against the "desired tension" summary of Figure 32.1. The results are shown in Figure 32.8. In doing this we have to convert the scatter percentages of Figure 32.7 to percentages of ultimate strength. The target or ideal preload was 55% of ultimate. On the downside, then, we'd expect a worst-case 61% of 55%, or 34% of ultimate. On the upper end, we'd expect 134% of 55%, or 74% of ultimate.

We have said that we want bolt tension to lie between 48% and 62% of ultimate tensile strength. The assembly/behavior analysis suggests the actual range will be 34%–74%, outside both the high and low ends of the desired range. How can we respond to this?

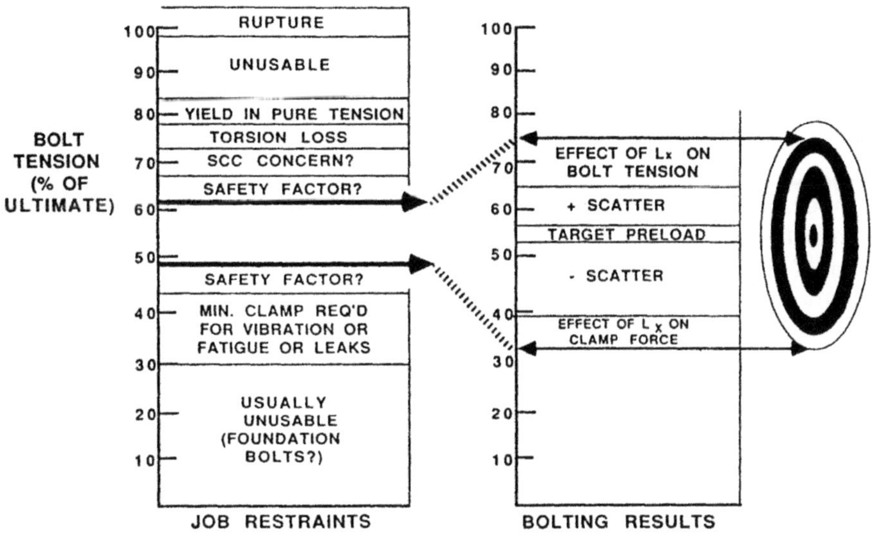

FIGURE 32.8 Chart combining the estimates of Figures 32.1 and 32.7. The range in bolt tension desired is shown on the left; the range expected is summarized on the right.

Remember that the anticipated range was based on a preselected type of assembly tool, a torque wrench in our example. Note that we were using a "better" torque wrench with ±5% scatter instead of ±15%. This undoubtedly got us closer to the desired results, but many assemblies will still lie outside the desired range. Going to a still better torque tool (±2% is also available) obviously wouldn't provide sufficient improvement.

The scatter in anticipated results could be narrowed considerably by going to a computerized torque–turn system and reduced even further by ultrasonic control of assembly. A second analysis would show whether or not such refinements would make the anticipated range fall within the desired range. I'm sure, in this example, that they would. But this much improvement in assembly accuracy may be economically unattractive. Fortunately, there are other options.

If the consequences of failure aren't too great, you can shrug your shoulders and accept the fact that some assembled joints will fail—will require field repair or maintenance, for example. The ranges we've estimated are 3 sigma ranges, covering the worst case with 99% probability. Most of the preloads introduced at assembly will be nearer the ideal than the extremes we've estimated. And, of course, we *are* making estimates, and not very accurate ones at that. We may luck out! So, let some fail. Keep good records. Build on your experience to modify your original selection of the target preload and/or assembly procedures to minimize failure even if you can't eliminate it.

Note that your estimates of the bolt tension desired may involve as much uncertainty as your estimates of expected tension. In our example we estimated max and min tensions desired, then a couple of safety factors. Maybe that was unnecessarily conservative. So, all is not lost if "expected" exceeds "desired."

If the analysis is correct, however, and it matters, another option would be to use a different fastener. A stronger one would increase ultimate strength, raising the upper end of the desired range. The target preload could now be shifted upward as well. The lower end of the desired range would fall (would become a smaller percentage of the higher ultimate strength of the new bolt). So a stronger bolt will open up the desired range.

The desired range can also be opened up by other fastener-related changes. You may have selected a relatively high minimum tension to fight vibration loosening. Perhaps you can use an anaerobic adhesive or a prevailing torque fastener to reduce minimum tension requirements.

The maximum desired (acceptable) tension might be increased by choosing a different fastener material if stress corrosion cracking is the concern, for example. There are many other possibilities of this sort that can be used to make the bolts, and the joint, better able to tolerate a wide variation in assembly and/or behavioral results. Think "overdesign"!

In some situations, you'll find it easy to fit the anticipated range into the required range. In this case, you could ask yourself whether less accurate assembly procedures would be acceptable. Presumably, that would reduce assembly costs, and there's no economic sense in keeping the range of anticipated results under tighter control than the joint requires. Such results might also indicate excessive overdesign.

All of this will sound too casual to the purists reading this book. But we're dealing with a "chaotic" number of variables. A rigorously accurate analysis may be impossible and is often going to be prohibitively expensive. It will rarely be worth the effort. But the estimating procedure I've described, even though crude, will take you well beyond the "accuracy" implied by codes and other documents that merely suggest that you "tighten them correctly." And in some cases, the accuracy of our rough estimates can be quite good.

That accuracy, of course, will depend on the accuracy with which we estimate the scatter of the individual key variables. Sometimes an experiment can be made to refine our guess for one or a couple of the variables that contribute the most to the overall scatter. The larger the number of variables you pin down this way, the more accurate your estimate, of course. But you'll never—or rarely—be able to pin them all down.

In any event, I think you'll find the analysis summarized in Figure 32.8 to be an efficient and relatively easy way to get an overview of the many different aspects of bolted joint we have discussed in this book.

32.6 WHICH VARIABLES SHOULD WE INCLUDE IN THE ANALYSIS?

32.6.1 In General

Which variables should we include when making an analysis of this sort to pick preload or torque for an existing joint? The answer to this question will obviously have a major impact on the results we get, and the answer depends upon what it is we are trying to do. Why are we trying to pick preload or torque or other control variables for this particular joint? Picking preload or a torque is something the designer should have done if a particular value is important. If we're doing it, it can only be because we think a choice is more important than he did, or his choice wasn't communicated to us, or because—most commonly—we've been having troubles with this joint (whether or not the designer specified a preload).

32.6.2 Possible Factors to Include

The VDI analysis (see Chapter 18) includes a long list of factors with which to start our own lists when picking a torque or other type of preload control. In their procedure they consider

Magnitude of the external load
Stiffness ratio ("load factor")
Eccentricity, if present
Embedment
Bending stress in the bolts
Tool scatter (e.g., torque-to-preload variations)
The effect of torsional stress on the tensile stress capacity of the bolt
The fatigue endurance limit of the bolts
The bearing stresses between nut or bolt head and joint surface
The clamping loads required to combat self-loosening, fatigue, or leakage
Differential expansion.

As far as I know, however, the VDI Directive 2230 doesn't include any of the following effects.

Gasket creep
Elastic interactions
Hole interference
Resistance of joint members to being pulled together ("weight effect")
Operator errors caused by lack of training, carelessness, or accessibility problems
Tool calibration errors
Safety factors.

32.6.3 Which Factors Should We Include?

Almost all of these factors are present and influence results when we tighten a group of bolts. Only differential expansion, gasket creep, hole interference, and the weight effect are application-specific and may be absent. And we won't care about the fatigue endurance limit if the applied loads are noncyclical. But the others, including such major factors as tool scatter and elastic interactions (and often operator problems) are essentially always there. If that is so, why leave any of them out?

One big reason most of these factors have always been left out is that we have been generally unaware that they exist. Another reason is that even if we know they exist we have had few or no data with which to support specific values. The VDI Directive is the only publicly available document I'm aware of that includes things like embedment relaxation and tool scatter in a formal design procedure and gives tables of typical values. Gasket manufacturers publish some gasket creep data, but it is usually limited to one thickness of a given material tested only at room temperature. In any event, the available data will often not apply to your situations.

How can we leave many factors out and still design or assemble most joints successfully? Easy; we have traditionally overdesigned bolted joints and included healthy "safety factors" to cover our ignorance. The overdesign gives us room to include the factors now, whenever we want a clearer understanding of why a joint has caused problems and want to pick a better torque or equivalent. If this is a safety-related joint, however, in an industry where designers *do* consider many of these factors, we should use the analysis only to probe our problems. Only the designer should pick a better torque.

Even though we'll usually include factors the designer overlooked and/or use different assumptions than he did about things like the stiffness ratio and the magnitude of external loads, it's still useful to make the analysis described earlier—or the more rigorous one to be described in Section 32.7. Either analysis is relatively easy to make, so we can repeat them, using several different sets of factors and using the "typical" data found in this book if we have no application-specific data of our own. Such analyses can often reveal design flaws and/or suggest root causes for a troublesome problem.

But don't overdo it. Start by including only the three or four factors you believe will have the most impact on the behavior of your joint, based on previous experience with that joint whenever possible. Include a safety factor or "tolerance" to cover uncertainties. Revise your list, or the values you're using, if the results don't seem reasonable or don't agree with prior experience. Add or subtract percentages from some of the variables to see if the results are more realistic. Try adding or subtracting 10% from the target torque, as suggested earlier in this chapter, and see what effect that has on your estimated results. *Play* with this stuff; don't let it dominate your life. We know too little to lock it in concrete. Remember: Past experience is a better guide than *any* theory or procedure or equation. If it ain't broke, don't fix it!

32.7 A MORE RIGOROUS PROCEDURE

The procedure we've looked at is very useful for making seat-of-the-pants decisions about preload and other matters, using our best guess about which factors might most influence the results and estimating the possible magnitude of each effect. We use rough statistics to make the results less intimidating than they might be if we assumed the worst case for each effect. We ignored many factors we might have included and assumed typical values for such difficult-to-estimate factors as the magnitude of the external load or the joint stiffness ratio. In spite of all this, and although we certainly wouldn't want to use it to make final decisions on safety-related or critical joints, I think the procedure is valid for most situations.

32.7.1 Experiments Required for True Accuracy

When truly accurate answers are required, we'd have to make carefully controlled experiments on the actual joint. Finite element analysis might help, but at the present time experiments are the only way to get the accurate application-specific data we need about such things as elastic interactions, stiffness ratios, and the response of the joint to service loads. Just picking a torque or other assembly control parameter isn't enough, either. We'd also have to impose strict quality standards on such things as assembly procedures, bolt quality, and lubricity to improve the chances that the results of our experiments accurately and consistently reflect production assembly results. The auto industry, for one, does all of this now and has for years.

32.7.2 The Equations

Whether or not we have experimental data with which to make our analysis, we will often want a more rigorous way to analyze the possibilities than the seat-of-the-pants method described earlier. Here's a set of equations roughly based on those found in the VDI Directive 2230 (see Chapter 18) but including elastic interactions. The equations can easily be extended to include more factors, or reduced to include fewer, to fit your own applications. The discussion in Section 32.6 is just as pertinent for this rigorous analysis as for the simpler one described earlier.

Maximum Bolt Load

$$\max F_B = (1+s)F_{Pa} - \Delta F_m - \Delta F_{E1} + \Phi_{en} L_X \pm \Phi_{en} K_J''(\Delta L_J - \Delta L_B) \qquad (32.7)$$

Minimum Per-Bolt Clamping Force on the Joint

$$\min F_J = (1+s)F_{Pa} - \Delta F_m - \Delta F_{EI} + (1-\Phi_{en})L_X \pm \Phi_{en} K_J''(\Delta L_J - \Delta L_B) \qquad (32.8)$$

Total Minimum Clamping Force on a Joint Containing N Bolts

$$\text{Min total } F_J = N \times \text{per} - \text{bolt Min } F_J \qquad (32.9)$$

where

F_{Pa}=the average or target assembly preload (lb, N)
ΔF_m=the change in preload created by embedment relaxation (lb, N); $\Delta F_m = e_m F_{Pa}$
e_m=percentage of average, initial preload (F_{Pa}) lost as a result of embedment, expressed as a decimal
ΔF_{EI}=the reduction in average, initial, assembly preload caused by elastic interactions (lb, N); $\Delta F_{EI} = e_{EI} F_{Pa}$
e_{EI}=the percentage of average, initial preload (F_{Pa}) lost as a result of elastic interactions, expressed as a decimal
ΔL_B=the change in length of the grip length portion of a loose bolt created by a change of in temperature, *At* (°*F*,°*C*) (in., mm);
ΔL_J=the change in thickness of the joint members, before assembly, if exposed to the same Δt (in., mm)

s=half the anticipated scatter in preload during assembly, expressed as a decimal fraction of the average preload

K_J''=the stiffness of a joint in which both the axes of the bolts and the line of application of a tensile force are offset from the axis of gyration of the joint and in which the tensile load is applied along loading planes located within the joint members (lb/in., N/mm); K_J'' is the reciprocal of the resilience of such a joint

Φ_{en}=the load factor for the joint whose stiffness is K_J''

N=number of bolts in the joint

L_X=external tensile load (lb, N).

32.7.3 Minimum Clamping Force—Some Examples

32.7.3.1 First Example—Using Worst-Case Values

Let's feed these equations some numbers to see how important it is to use the best available numbers. Let's start with Eq. (32.8) and compute the anticipated minimum clamping force on the joint. For a first round, we'll assume the following parametric values:

Average assembly preload (F_{Pa})=1430 lb.

Tool scatter (s): Assume a full 30% below target preload, so s=0.3.

Embedment: Assume a "typical" loss of 10%, so e_m=0.1.

Elastic interactions: Assume an "average" loss of 18% for a two-piece, metal-on-metal joint, using results obtained in Dr. George Bibel's experiments, so e_{EI}=0.18. (Note that Dr. Bibel tells us that the average loss would be 30% if a sheet gasket were included in the joint, or 46% if the gasket is a spiral-wound type. (See Chapter 24.)

External load effect: Assume that the tensile load reduced the per-bolt clamping force on the joint by 269 lb, based on the assumption that the external load was equal to 25% of the preload we were trying to compute (talk about winging it!) and on the "eccentrically and internally loaded" load factor Φ_{en} we're using.

Differential expansion loss: Assume a per-bolt clamping force loss of 208 lb based on the assumptions that Inconel bolts are being used in a mild steel joint and that the operating temperature would be 400°F.

These assumptions give us the results shown in Figure 32.9: a minimum, per-bolt clamping force on the joint of 123 lb. The ratio between average, initial preload and residual clamping force is about 12:1. A ratio of 15:1 was calculated in Chapter 13 of Ref. [2] but was based on maximum, not average, preload and also included a gasket creep factor that I've omitted here.

Whether the ratio is 12:1 or 15:1, this is not a pretty picture! And it doesn't include many factors that might have made it worse, such as gasket creep or hole interference or the weight effect (see Chapter 23). But how valid is this calculation?

It purports to define the minimum tension we can expect to see in some of the bolts. To do that, however, we would have to assume a worst-case elastic interaction loss of 36% instead of an average 18%. I tried that—and computed a residual, minimum, per-bolt clamping force of minus 135 lb. Since my goal was to teach you how to use Eq. (32.8), not to define and analyze an actual example, I used 18%. But it means that the computed 123 lb is an "almost worst case" answer.

Equation (32.9) says that we can compute the minimum anticipated clamping force on a joint containing N bolts by multiplying 123 by N. But this assumes that *every one* of the N bolts in some joints will only contribute an "almost worst case" amount of clamping force to the joint. I think that that's highly unlikely. I think that Eq. (32.8) can reasonably predict the worst-case clamp contributed by an occasional bolt, but not by each bolt in a given joint.

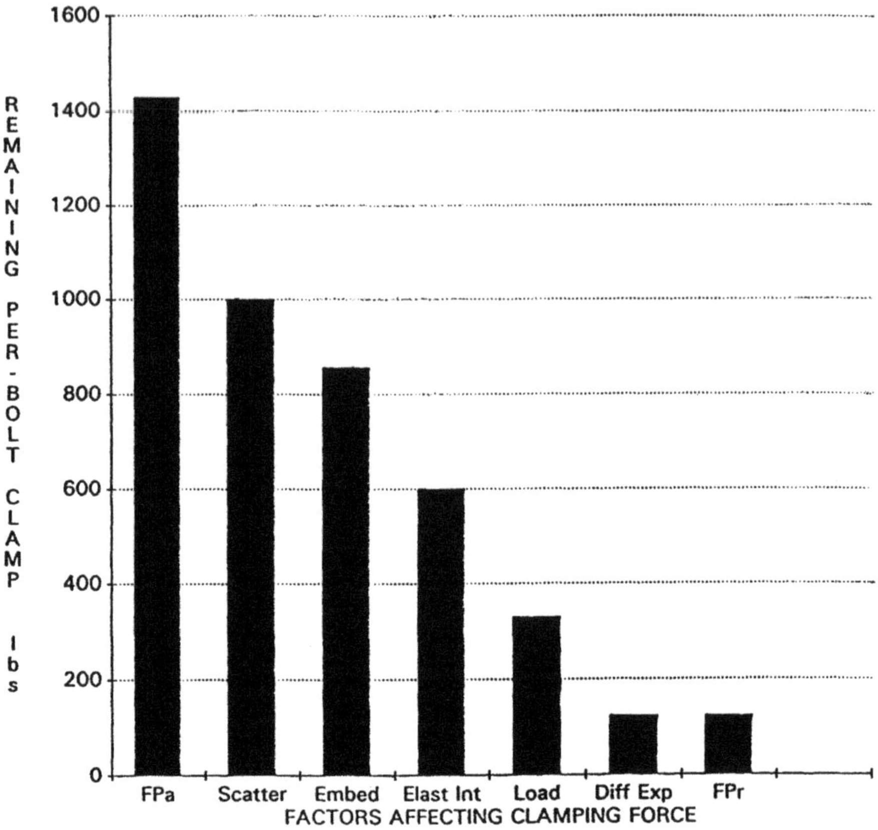

FIGURE 32.9 Chart showing the large difference expected between the average, initial, per-bolt clamping force applied to a joint during assembly (F_{Pa}) and the residual, inservice clamping force (F_{Pr}) when we assume worst-case tool scatter and embedment losses in bolt tension, plus an average elastic interaction loss, plus the unavoidable losses in clamping force caused by the external tensile load and differential expansion between bolts and joints. This is a plot of the worst-case calculations made in Ref. [2].

32.7.3.2 Second Example—Using Statistically Combined Values

In the quick-and-dirty procedure we discussed earlier in this chapter we used the square root of the sum of the squares of the variances to reduce the combined effect of tool scatter, embedment, and elastic interactions to get a more cheerful—and more realistic—picture. This introduces the assumption that no one bolt will simultaneously be exposed to the worst-case minimum assembly preload, maximum embedment, and maximum interaction loss. The results, shown in Figure 32.10, are based, therefore, on the square root of the sum of squares with the following assumptions:

Preload scatter: ±30%
Embedment: –10% max
Elastic interactions: –36% max.

The load and differential expansion effects remain unchanged at 269 and 208 lb, respectively. Combining the three factors tabulated above suggests a 3 sigma maximum loss of only 47.9% of average assembly preload (from which we further subtract the load and differential losses). The resulting per-bolt clamping force is now 268 lb, more than twice what it was under our earlier assumptions. The preload-to-residual-clamp ratio is about 5:1, not something to celebrate but a definite improvement.

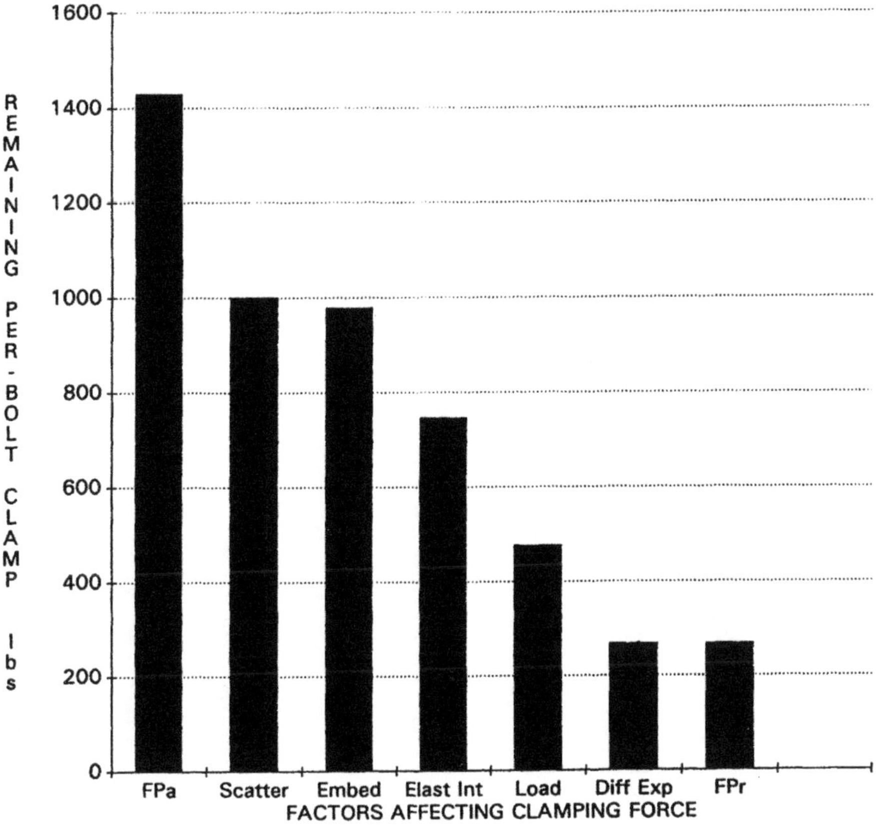

FIGURE 32.10 Chart showing the anticipated difference between initial and residual clamping force when we assume that worst-case effects will never (or rarely) occur simultaneously, that the expected deviations in individual factors can be combined by the square root of squares Eq. (32.6).

But again, I think it would be very conservative to assume that each of the *N* bolts in a given joint ended up with 47.9% less tension than anticipated by the target preload. What we really need is a series of experiments, a resulting list of "combined losses," and the mean and scatter for the combined data. Then we could use Eq. (32.9) with more confidence. Such data, unfortunately, do not exist.

32.7.3.3 Third Example—Using Average Values

Could we resolve this by feeding only "average" data into Eq. (32.8)? That would suggest the following:

Tool scatter: 0% average loss!
Embedment: 5% average
Interaction loss: 18% average
Load and differential expansion losses unchanged

These assumptions would result in a *big* improvement in predicted clamping force, thanks to the fact that the average of a ±30% tool scatter factor is 0%. Although the ±30% figure appears again and again in the literature, I think that it's unwise to assume that the plus scatter will equal the minus scatter on a given joint. We know that more factors lead to less preload than to more preload. Therefore, although your own experience may well lead you to use a different value, I'm going to use –10% for the average, single-joint, per-bolt, tool scatter.

The results are shown in Figure 32.11: an average worst-case (if there is such a thing) per-bolt, in-service, clamping force of 495 lb, and a preload-to-clamping force ratio of only 2.9: 1. I would

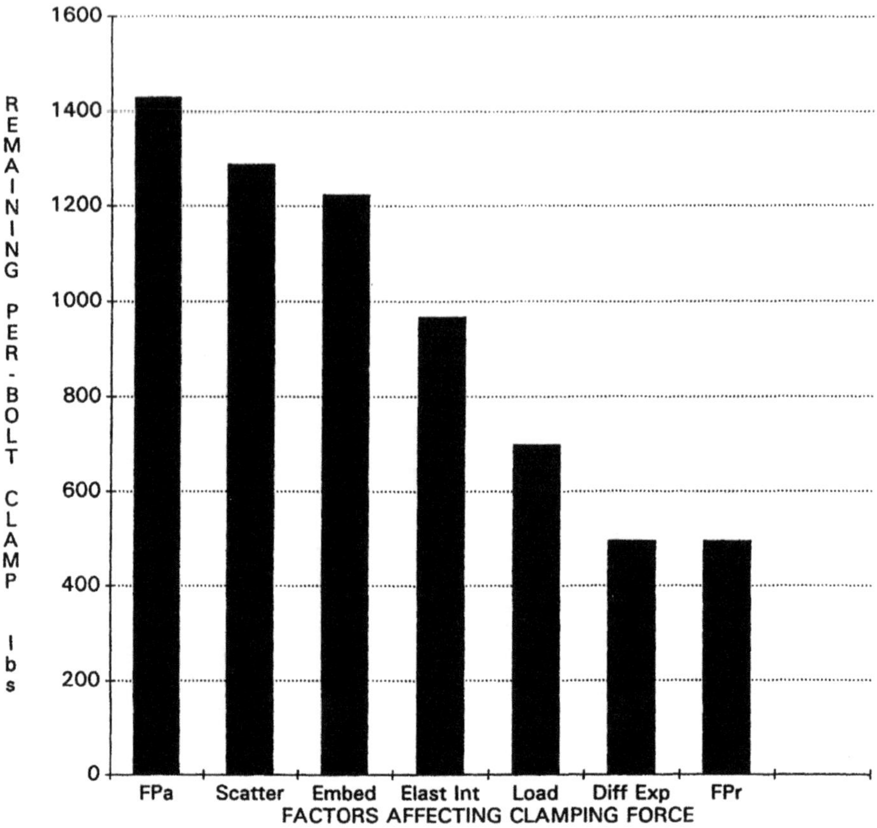

FIGURE 32.11 Chart showing the difference between initial and "average" residual per-bolt clamping force when we assume an average deviation for the various loss factors. This is probably a more realistic view of the results than those shown in Figures 32.9 and 32.10, especially if our goal is to compute the total clamping force applied to the joint by *N* bolts. See the text for a detailed discussion.

feel comfortable multiplying this number by *N* to estimate the total force on the joint. To repeat, however, every answer we get depends upon which factors we've included. Only experiments on the actual joint will give us a truly accurate answer.

32.7.3.4 Fourth Example—Using Feedback Control Values

If we use strain gage, ultrasonic, or other "feedback" control of final, assembly preload we can reduce the combined effects of tool scatter, embedment, and elastic interactions to ±10%. The results are shown in Figure 32.12: a ratio between target preload and final, per-bolt clamping force of only 1.8 to 1. Since the load effect and differential expansion losses can be accurately predicted if we know joint materials and geometry—and, the tough part, the magnitude of the loads we will see in service—this would also eliminate the need for joint experiments. The feedback gives us the "experimental" data we need as we assemble each joint. No more need to estimate assembly loss factors or even to decide which to include in our analysis. But this type of control is expensive, and we'll usually have to settle for estimates which we'll use with either the quick-and-dirty procedures described first in this chapter or in a more rigorous analysis such as that described in this section.

32.7.4 Maximum Bolt Tension

We have focused this "more rigorous" discussion entirely on the use of Eqs. (32.8) and (32.9). We should never accept the assembly preloads suggested by the use of these equations without also

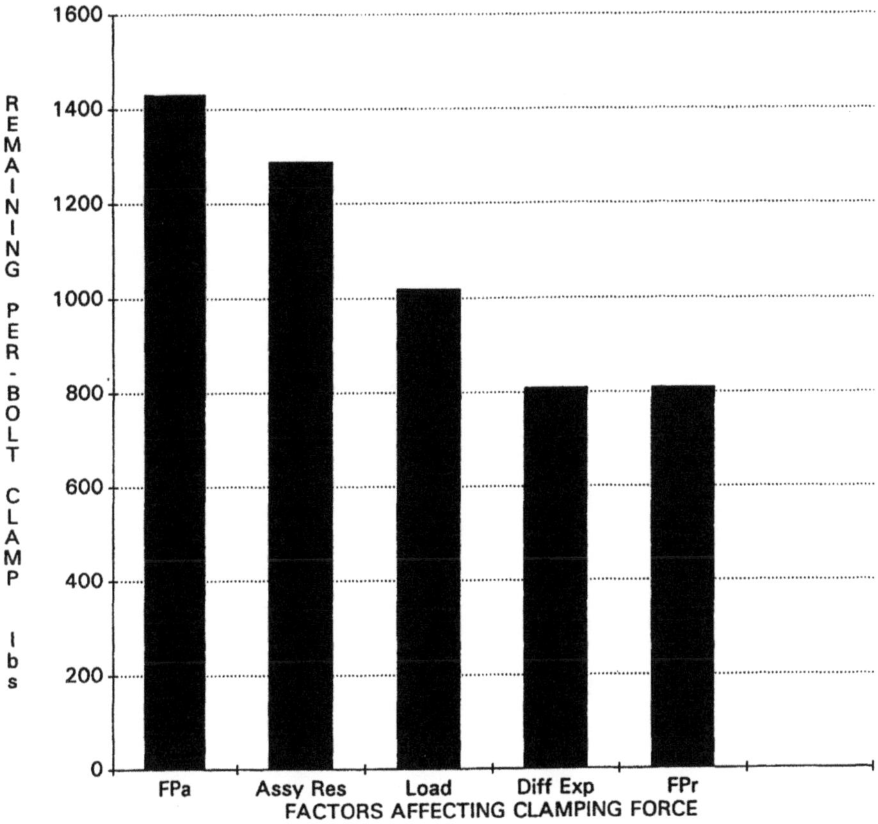

FIGURE 32.12 Chart showing the anticipated difference between initial and residual clamping force when feedback control is used during assembly to correct for tool scatter and compensate for embedment and elastic interaction loss, giving a practical, worst-case deviation of assembly preload of only 10% of average assembly preload.

using Eq. (32.7) to estimate the worst-case stresses our choice would create in some of the bolts. Note that it's perfectly reasonable to use worst-case deviations when estimating maximum bolt loads because we don't want any bolt to break and we don't multiply the results by N or something. Even here, however, our inputs to the equation can distort the results and penalize our design. If we're planning to use torque, is it reasonable to assume a worst case +30% scatter, for example, or will that be reduced when we lubricate the bolts?

Regardless of which procedure you use, I think that each of those we've examined in this chapter has its place and can be useful—can help you to estimate, however roughly, the assembly and in-service clamping forces in your joints and thereby help you to analyze and solve many bolting problems. They can certainly give you the insight you need to use good engineering judgment when picking a better torque—or when picking a more appropriate assembly control variable.

REFERENCES

1. Shigley, J. R., and C. R. Mischke (eds.), *Standard Handbook of Machine Design*, McGraw-Hill, New York, 1986.
2. Bickford, J. H., *An Introduction to the Design and Behavior of Bolted Joints*, 3rd ed., Marcel Dekker, New York, 1995.

Part VI

The Joint in Service

33 Clamp Load Decay in Multi-Material Joints due to Cyclic Heating and Cooling

Sayed A. Nassar, Andrew Q. Smail, and Amir Kazemi

NOMENCLATURE

$A(T)$	Temperature-dependent creep parameter
B	Material constant
e_b	Bolt thermal growths
K_b	Bolt stiffness
K_c	Joint stiffness
L_b	Bolt free length at room temperature
L_c	Joint free length at room temperature
n	Stress power law exponent
Q	Activation energy
R	Universal gas constant
T	Temperature
α_b	Bolt thermal expansion coefficient
α_c	Joint thermal expansion coefficient
$\dot{\varepsilon}$	Steady-state creep rate
Δ_b	Bolt initial preload deformation at room temperature
Δ_c	Joint initial preload deformation at room temperature
ΔT_b	Bolt temperature changes
ΔT_c	Joint temperature changes
δ	Shift in assembly position due to temperature change
δ_b	Bolt deformations from free length at temperature
δ_c	Joint deformations from free length at temperature
σ	Stress

33.1 INTRODUCTION

Bolted joints are widely used in assemblies due to their advantages such as the capability of creating clamp force, the ease of disassembly for repair, tolerance to environmental effects, relative low cost, and higher fatigue resistance compared to other joining methods. On the other hand, combining different materials enables greater design possibilities and material selection for superior product performance, lower cost, and lower mass, which play an important role in the performance of mechanical joints, especially in the aerospace and automotive industries. Multi-material joints can optimize the weight of the assembly by use of a specific material for a specific purpose and make the design more efficient.

Several research studies on lightweight design represent a tendency toward a greater combination of materials in the automotive industry and predict over 300 projected changes in vehicle by 2020 [1,2]. The joint performance depends on different parameters such as applied loading condition,

DOI: 10.1201/9781003501183-39

material properties of the bolt and substrates, joint geometry, and environmental condition. Therefore, bolted joints should be carefully designed for high responsibility. Although the common bolt and joint would normally tolerate linear elastic deformation at room temperature, they have nonlinear behavior at high and low temperature due to the material changes and thermal expansion difference in the joint parts which cause permanent clamp load decay. It may lead to part separation that would significantly decrease the joint performance and increase the cost of replacement and maintenance.

Design analysis and assembly procedure presented by ASME [3] don't consider the effect of elevated/lowered cyclic temperatures on the clamp load decay in bolted joint design so that a join can withstand cyclic temperatures without leakage or loosening of assembly. Kumano et al. [4] studied the mechanical behavior of bolted joints under steady heat conduction and analyzed temperature distributions and axial bolt variation of the joints using the finite difference method, but they didn't consider the creep behavior under cyclic temperature loading. Bouzid et al. [5] evaluated the effect of steady-state thermal loading on the deflections of a flange joint and showed that the change in bolt load due to temperature change requires the knowledge of both the radial and axial displacements of individual flange components. Nechache and Bouzid [6] determined the load changes in bolted gasketed joints subjected to elevated temperature. Sato and Kado [7,8] carried out the heat transfer and elastic plastic FE analysis to investigate the behavior of dissimilar material flanges with metal ring gasket at elevated temperature, but they didn't consider the effect of creep.

Limited literature is available on temperature cycling effect on dissimilar-material bolted joints. Hanus et al. [9] conducted experimental tests on the mechanical behavior of Grade 8.8 bolts under heating and cooling phases and demonstrated that the presence of tensile forces in axially restrained beams during the cooling phase of a fire could lead to bolt failures. Caccese et al. [10] determined the influence of stress relaxation on clamp-up force in hybrid composite-to-metal bolted joints and briefly addressed environmental effects including temperature cycling. They showed that the initial clamp-up load and the load at the start of the temperature cycling are given along with the peak load recorded during each temperature cycle but after the second cycle the load achieved by the temperature cycling leveled out and the percent of initial clamp-up load remained relatively constant.

Ekh and Schon [11] studied the effect of temperature in load transfer in multirow composite-to-aluminum single-shear lap joints and demonstrated that temperature changes could significantly affect the load distribution for member plates with different thermal expansion properties. Yang et al. [12] investigated the clamp load decay due to the elevated cyclic temperature of an automotive exhaust threaded fitting with a collapsible metal gasket and proposed the nonlinear analytical models for the thermo-mechanical behavior of the bolted joints. They developed a good creep-relevant model for the clamp load variation under temperature cycling.

None of the previous studies has considered all of the various combinations of the material coupon thickness, and preload level under cyclic temperature loading on the clamp load in bolted joints. Moreover, most of the studies are either analytical or finite element simulations, so more experimental data is needed.

In bolted joints, creep effects are observed in the through-thickness direction. Chen and Kung [13] have shown that the preload in composite bolted joints is highly sensitive to changes in temperature and humidity. Shivakumar and Crews [14] presented relaxation in simple T300/5208 graphite/epoxy bolted connections. They showed that significant relaxation could happen even at room temperature; elevated temperature and moisture increased the relaxation rate. However, the sensitivity of clamp load in different dissimilar metal-to-metal bolted joint is not addressed.

In this chapter, experimental and finite element models are presented for the investigation of the effect of cyclic thermal loading on clamp load decay on dissimilar material joints due to creep and thermal expansion mismatch. Material combinations include steel A36, aluminum 6061, and magnesium AZ31 with different thicknesses. The range of cyclic temperature profile varies between –20°C and +150°C. Tests continued until the rate of clamp load decay per cycle has become relatively small; that corresponded to ten consecutive 10-h cycles. The effect of preload level on clamp load decay is illustrated for various joint combinations of material and thickness. Thermal analysis is performed using 3-D finite element simulation (FEA) with ABAQUS software [15]. Temperature-dependent mechanical properties are used in the FEA model. The correlation between FEA model predictions and the experimental data is discussed.

33.2 THERMOMECHANICAL MODEL

Bolt and clamped members in a preloaded bolted joint can be considered as bar-pair as shown in Figure 33.1. Temperature changes cause a new equilibrium condition and change the internal forces in the bolt and joint.

Subsequent temperature changes would change the bolt length as follows:

$$e_b = \alpha_b \Delta T_b L_b \tag{33.1}$$

where e_b is bolt thermal growths, α_b is bolt thermal expansion coefficients, ΔT_b is bolt temperature changes; and L_b is bolt free length at room temperature. Similarly, the change e_c in the joint thickness is given by

$$e_c = \alpha_c \Delta T_c L_c \tag{33.2}$$

where α_c is the coefficient of thermal expansion of the joint material, ΔT_c is joint temperature changes, and L_c is the free length of the joint at room temperature. Combined changes in length for the bolt and clamped joint due to preload and elevated temperature are

$$\delta_b = (\Delta_b + \delta) - e_b \, \delta_c = (\Delta_c + e_c) - \delta \tag{33.3}$$

where δ_b and δ_c are bolt and joint deformations from free length at temperature, Δ_b and Δ_c are bolt and joint initial preload deformation at room temperature and δ is shift in assembly position due to temperature change. Initial preload F_i at room temperature is

$$F_i = K_b \Lambda_b = -K_c \Delta_c \tag{33.4}$$

where K_b and K_c are bolt and joint stiffnesses, respectively.

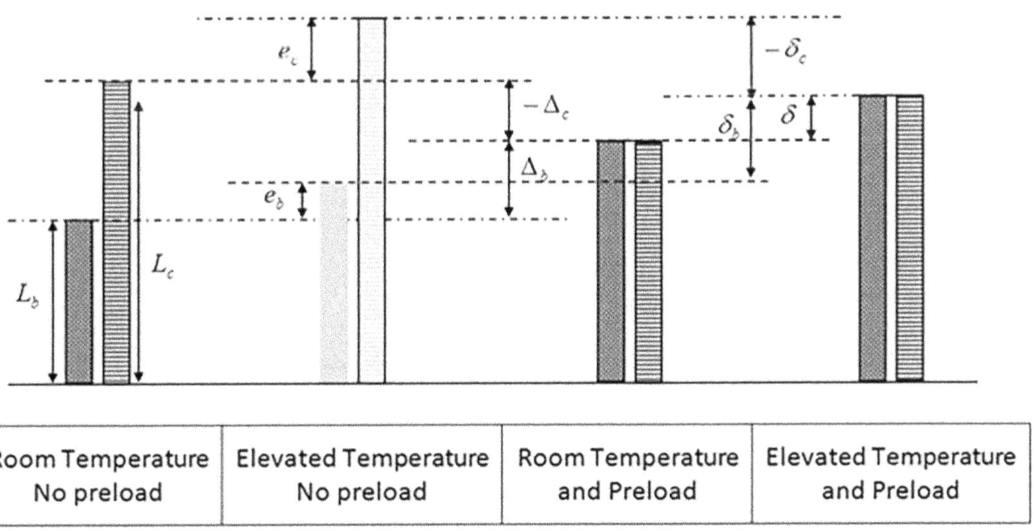

FIGURE 33.1 Schematic of changes in length due to thermal and/or preloading effects.

The bolt force F_b at elevated temperature is given by

$$F_b = K_b\delta_b = K_b\left(\Delta_b + \delta\right) - K_b\alpha_b\Delta T_b L_b \quad (33.5)$$

The corresponding clamp load F_c at elevated temperature would be given by

$$-F_c = K_c\delta_c = K_c\left(\Delta_c - \delta\right) + K_c\alpha_c\Delta T_c L_c \quad (33.6)$$

Since there is no external force applied to the joint, the following static equilibrium condition must be satisfied

$$F_b + F_c = 0 \quad (33.7)$$

Substituting Eqs. (33.5 and 33.6) into Eq. (33.7) gives the shift in assembly position δ as follows:

$$\delta = \left(\frac{1}{K_b + K_c}\right)\left(K_b\alpha_b\Delta T_b L_b + K_c\alpha_c\Delta T_c L_c\right). \quad (33.8)$$

Substituting Eq. (33.8) into Eq. (33.5), gives the bolt force F_b after the temperature change as follows:

$$F_b = F_i + \left(e_c - e_b\right).K \quad (33.9)$$

where the stiffness parameter K is given by

$$K = \frac{K_b.K_c}{K_b + K_c} \quad (33.10)$$

33.3 TEMPERATURE-DEPENDENT CREEP OF JOINT MATERIALS

Several methods have been used to model the creep behavior of materials. The most common expression to describe the stress, temperature, and microstructure dependence of the steady-state creep rate is based on Norton power law [16] equation

$$\dot{\varepsilon} = B\sigma^n \exp[-Q / RT] \quad (33.11)$$

where $\dot{\varepsilon}$ is the steady state creep rate, B is a material constant, σ is the applied stress, n is the stress power law exponent, Q is the activation energy for the creep process, R is the universal gas constant and T is the temperature (Kelvin). In conjunction with Eq. (33.11), the following Norton creep law [16] is used

$$\dot{\varepsilon} = A\sigma^n \quad (33.12)$$

where A is a temperature-dependent creep parameter of the material.

Since the value of the stress power law exponent n and the activation energy Q in Eq. (33.11 and 33.12) are not readily available, typical values for n and Q are used from literature [17–21] for calculating creep material parameter$A$ in this study.

33.4 EXPERIMENTAL PROCEDURE AND TEST SETUP

In this section, the experimental procedure and test setup are outlined. Test joints are instrumented with high-temperature load cells for monitoring the clamp load variation in real time during temperature cycling. All load cells were calibrated at 10V bridge excitation using Model A2

Signal conditioning Amplifier system and material test system MTS-810. A computer-controlled Thermotron SE-300 environmental chamber is used for applying precision cyclic temperature environment to all instrumented test joints. Corresponding real-time data on clamp load variations is collected using Agilent DAQ- U2331A software. Figure 33.2 shows the test equipment and instrumentation.

Hexagonal head 3/8"-16 SAE Grade 5 bolts and nuts with suitable types of washers are used in all test specimens (single-lap bolted joints). Each test specimen is made of two 2″×2″ metal coupons. A total of 12 combinations of coupon material and thickness are used, while keeping a constant 1″ overall joint thickness (Figure 33.3). Material combinations include Steel A36, Aluminum 6061, and Magnesium AZ31B. Coupon thickness combinations are 3/4″, ½″ and ¼″. Tables 33.1a and 33.1b includes all test combinations in this study. Initial clamp load is created by tightening the threaded

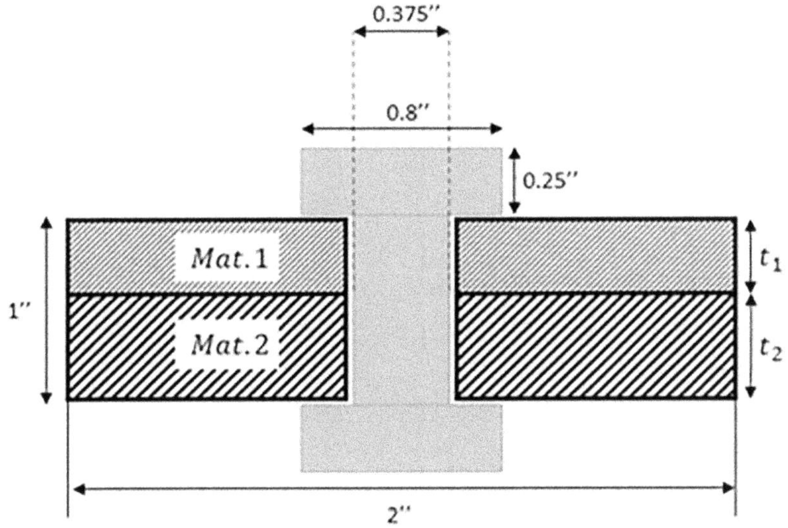

FIGURE 33.2 Coupon thickness and material combinations for FEA and experimental setup.

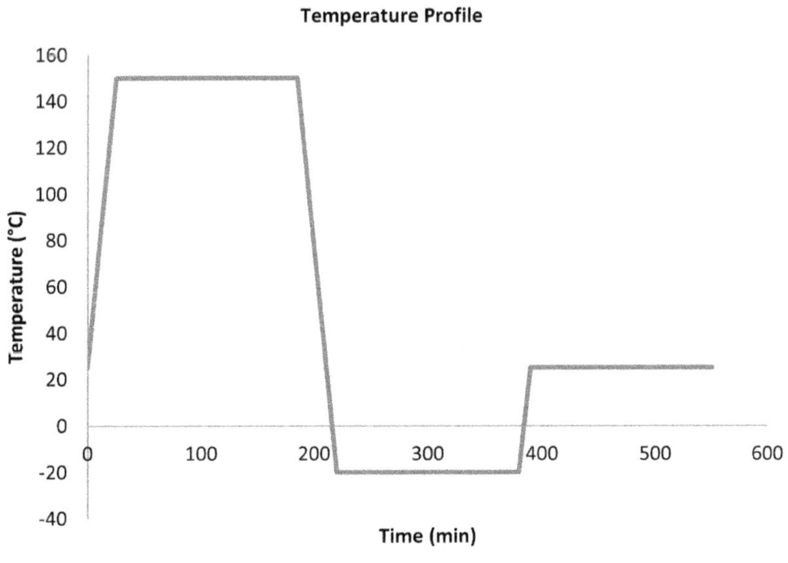

FIGURE 33.3 Illustration of the cyclic temperature profile.

TABLE 33.1a
Joint Material Combinations of Steel (St), Aluminum(Al), and Magnesium (Mg)

	Comb. 1.	Comb. 2	Comb. 3	Comb 4	Comb 5	Comb 6
Material 1	St	St	St	Al	Al	Mg
Material 2	St	Al	Mg	Mg	Al	Mg

TABLE 33.1b
The Coupon Thickness Combinations

t_1	¼″	½″	¾″
t_2	¾″	½″	¼″

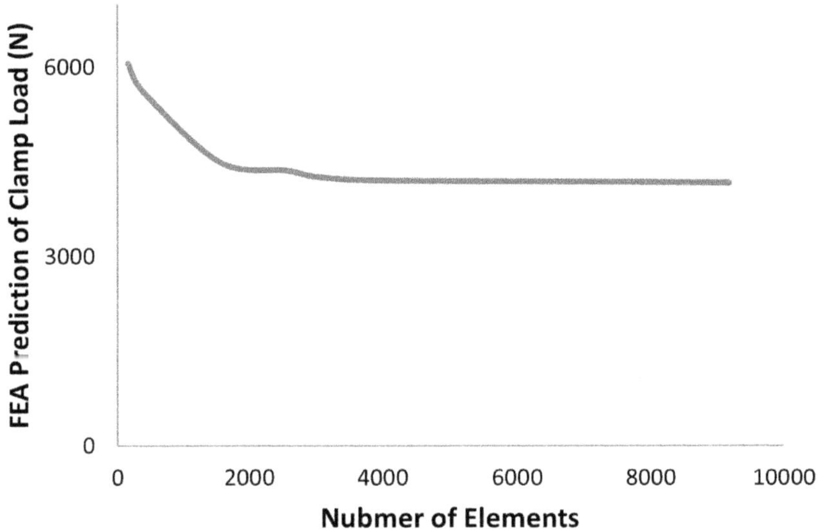

FIGURE 33.4 Illustration of the effect of mesh size on FEA results.

fastener until the desired load is reached as indicated by the load cell. Both high and low preload levels are separately investigated; namely, 70% and 30% of the bolt proof load, respectively. Preloaded joint specimens are subsequently placed in an environmental chamber for thermal cycling between –20°C and +150°C, as shown in Figure 33.4. Real-time clamp load data are collected over ten thermal cycles where the rate of clamp load decay has become fairly small.

33.5 FINITE ELEMENT SIMULATION

ABAQUS commercial FEA software [15] is used in this study. Various combinations of material and coupon thickness are modeled for investigating clamp load variation due to material creep during thermal cycling. Values of the temperature-dependent creep parameter A are used in Tables 33.2 to 33.4.

TABLE 33.2
Steel Material Properties

	Material Properties			
Temp. (°C)	Young's Modulus E (GPA)	Passion's Ratio (ν)	Coefficient of Thermal Expansion (α) (μm/m-°C)	Creep Parameter A $1/\left(s.\left(\frac{N}{m^2}\right)^6\right)$
–20	212	0.33	1.2E-5	2.5E-72
25	210	0.33	1.2E-5	2.6E-72
150	203.1	0.33	1.2E-5	3.5E-71

TABLE 33.3
Aluminum Material Properties

	Material Properties			
Temp. (°C)	Young's Modulus E (GPA)	Passion's Ratio (ν)	Coefficient of Thermal Expansion (α) (μm/m-°C)	Creep Parameter A $1/\left(s.\left(\frac{N}{m^2}\right)^6\right)$
–20	70.60	0.28	2.2E-5	4E-70
25	68.9	0.28	2.2E-5	3E-69
150	62.67	0.28	2.2E-5	2E-60

TABLE 33.4
Magnesium Material Properties

	Material Properties			
Temp. (°C)	Young's Modulus E (GPA)	Passion's Ratio (ν)	Coefficient of Thermal Expansion (α) (μm/m-°C)	Creep Parameter A $1/\left(s.\left(\frac{N}{m^2}\right)^6\right)$
–20	43.5	0.28	2.68E-5	3E-62
25	41.3	0.28	2.68E-5	3E-60
150	32.8	0.28	2.68E-5	3E-54

Using ABAQUS sketch tools, a 3/8″ Grade 5 bolt and two plates of 1/4″ and 3/4″ are modeled for creep analysis using material properties from Tables 33.2 to 33.4.

Contact conditions are specified between the two joint coupons and between the bolt and plate hole surfaces. Bolt load was applied initially and kept fixed at current length to allow the clamp force to freely change within the plate's expansion and shrinkage during the simulation steps.

Six steps were repeated for each thickness and material combination. The clamp load change was conducted using the interaction contact force in the history field output in ABAQUS solution.

Mesh size was refined until the FEA results would no longer be affected by further refinement as shown in Figure 33.5.

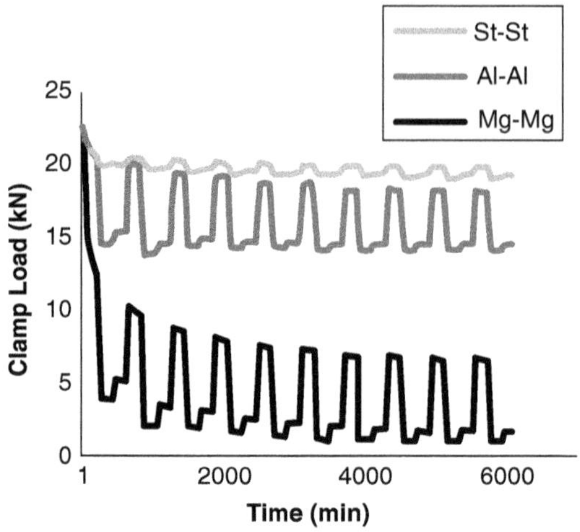

FIGURE 33.5 Clamp load variation in similar-material joints versus thermal cycling time for high initial clamp load.

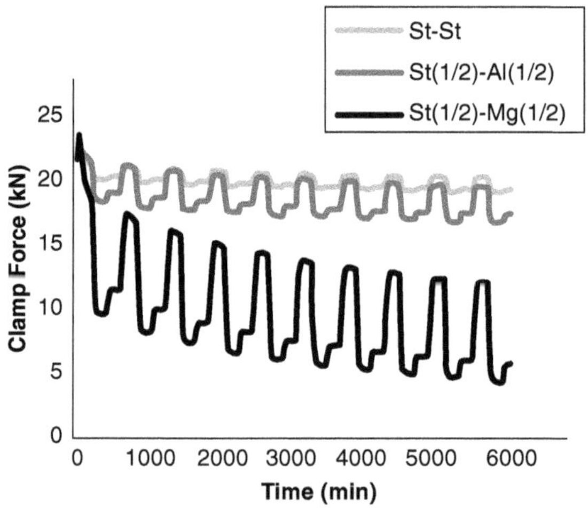

FIGURE 33.6 Clamp load variation in dissimilar-material joints versus thermal cycling time for high initial clamp load.

33.6 RESULTS AND DISCUSSION

Figures 33.6 and 33.7 provide sample real-time clamp load variation with thermal cycling between −20°C and +150°C for high and low initial preload levels, respectively. For all joint material combinations, the most significant clamp load drop was observed during the first thermal cycle. The rate of the clamp load decay decreases by with subsequent cycles; clamp load decay becomes insignificant after ten cycles.

Figure 33.6 displays the experimental clamp load variation data under the cyclic temperature for joints without material mismatch (i.e. Steel/Steel, Aluminum/Aluminum and Magnesium/Magnesium joints,) in high initial clamp load level at 70% of bolt proof load. It shows that for aluminum and magnesium joints the clamp load increases slightly first due to the thermal expansion mismatch between the joint and bolt, but it subsequently decreases rapidly due to creep. It indicates

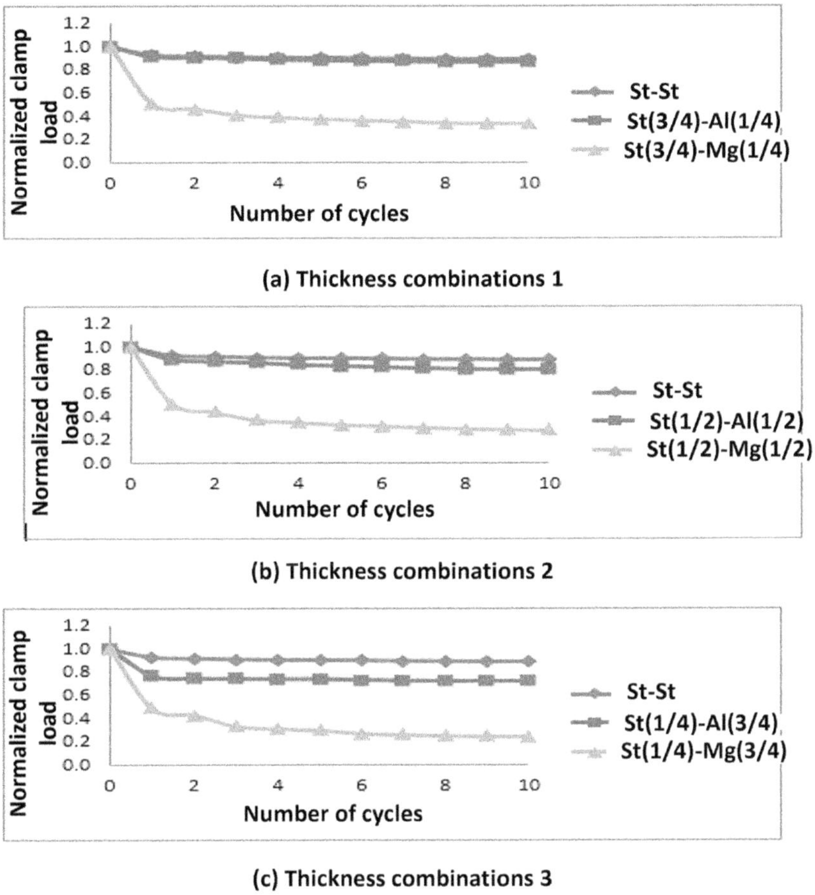

FIGURE 33.7 Normalized clamp load versus number of thermal cycles for steel-based joints.

that the dwell time at the peak temperature (+150°C) continuously reduces the clamp load with time due to creep stress relaxation. When the temperature goes down to –20°C, the clamp load naturally decreases due to the thermal contraction mismatch between the joint and bolt. Low temperature the load decay rate due to creep is very small and dwell time has no effect on clamp load decay. Figure 33.4 also indicates how creep causes the significant load decay in magnesium joints as compared to steel and aluminum.

Figure 33.7 displays the experimental clamp load data versus time under the cyclic temperature for dissimilar material joint when half of the joint remains steel and the second half is replaced by aluminum and magnesium for low initial clamp load level (30% of proof load). It shows how clam load decreases by substitution of half of the steel joint with aluminum and magnesium; this load decay increases significantly by replacing magnesium.

Figures 33.8–33.10 present the normalized clamp load level versus the number of thermal cycles for various steel-based, aluminum-based and magnesium-based material combinations, respectively. The most significant preload drop happens during the first cycle. The rate of the clamp load decay is reduced with the number of the cycles until becomes relatively small.

Figures 33.8a–c illustrates that in steel-based joints, the use of aluminum or magnesium coupons would cause to more clamp load decay as compared to pure steel-on-steel joints. The use of magnesium coupon causes more significant decay as compared to aluminum. Although substitution

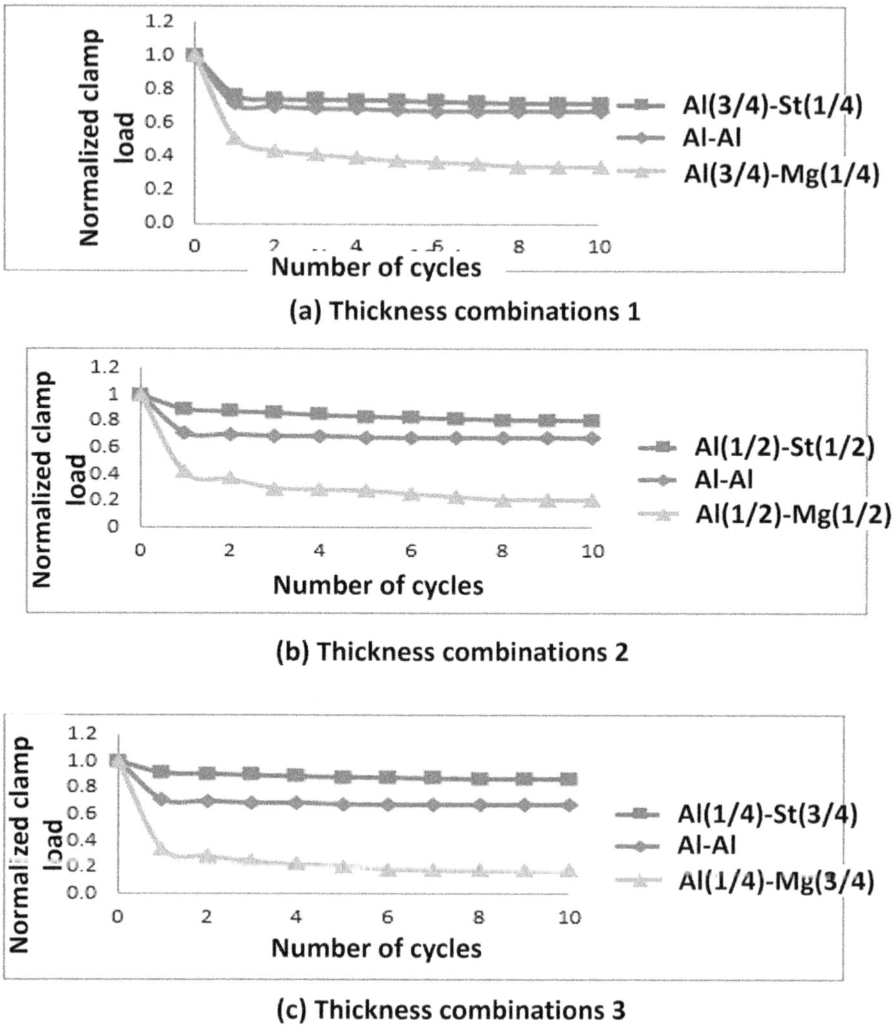

FIGURE 33.8 Normalized clamp load versus number of thermal cycles for aluminum-based joints.

of magnesium significantly increases the clamp load decay, increasing the magnesium thickness doesn't have substantial effect.

Figure 33.8a indicates that when the main part of the joint is steel (3/4″), replacing the second part (1/4″) by aluminum doesn't have a significant effect in clamp load decay. However, remarkable clamp load decay is observed for magnesium substitution. Figures 33.8b and c display a similar behavior as Figure 33.8a for magnesium, however replacing thicker secondary material increases the clamp load decay. Figure 33.9a shows that when the main part of the joint is aluminum (3/4″), replacing 1/4″ of steel doesn't have a significant effect in clamp load decay. However, magnesium significantly increases the clamp load decay load decay is observed for magnesium substitution.

Figures 33.9b and c show a similar behavior as Figure 33.9a for magnesium.

Figure 33.10a indicates that when the main part of the joint is magnesium (3/4″), replacing 1/4″ of steel or aluminum doesn't have much difference in clamp load decay and magnesium plays the main role. Figure 33.10b shows the similar behavior as Figure 33.10a and confirms the

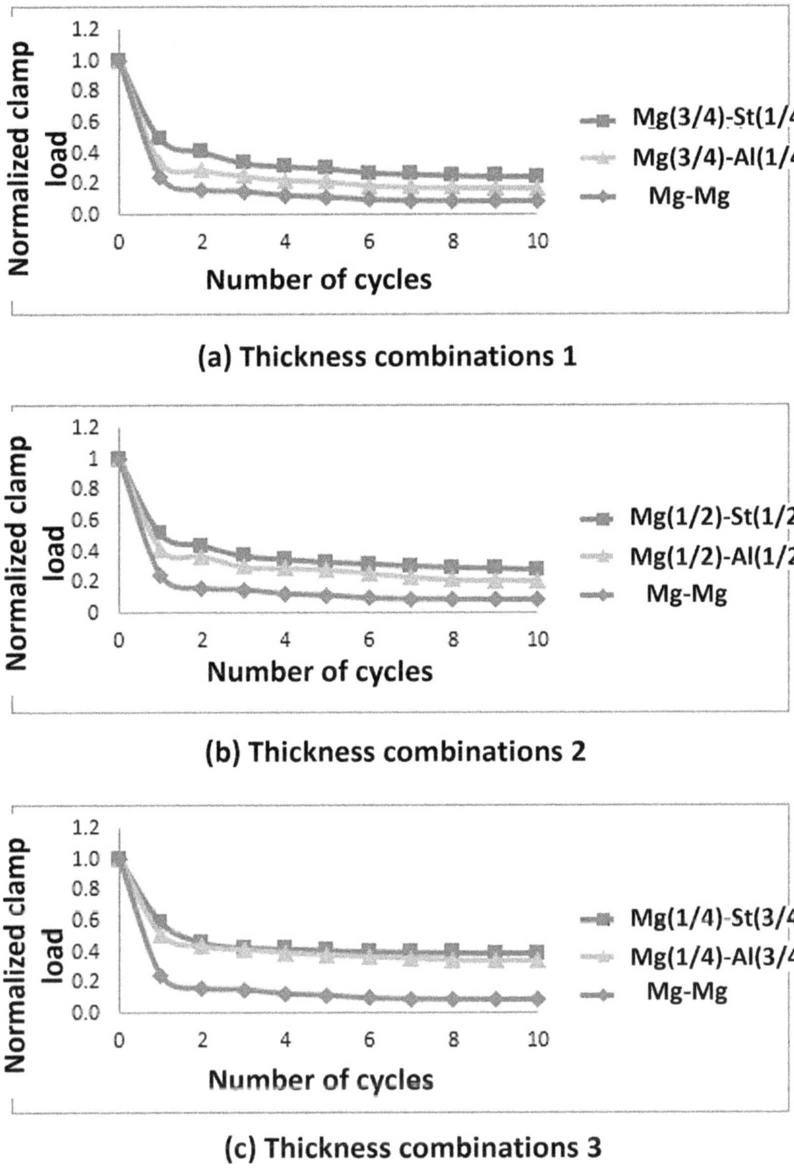

FIGURE 33.9 Normalized clamp load versus number of thermal cycles for magnesium-based joints.

importance of magnesium role in the load decay. Figure 33.10c shows that although substitution of 3/4″ steel or aluminum can decrease the amount of load decay, huge load decay still happens in the joint due to magnesium existence.

33.7 EFFECT OF PRELOAD LEVEL

Figure 33.11 show the effect of initial bolt preload level on the normalized clamp load decay with thermal cycles. Results show insignificant effect on the clamp load decay for various coupons thickness combinations. They also illustrate that the effect of initial load on the load decay percentage is not related to the material and thickness combination.

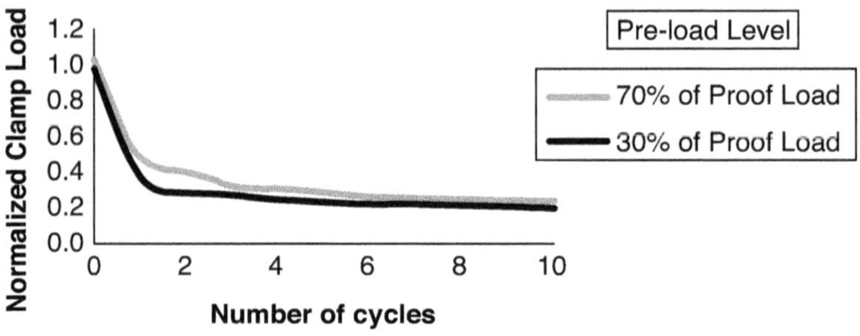

(a) St(1/4) –Mg(3/4)

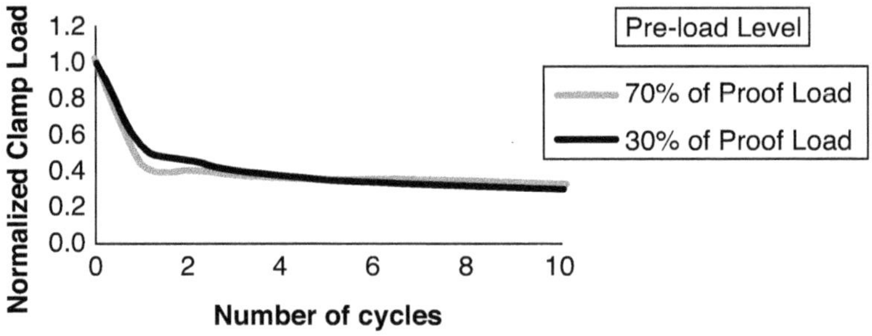

(b) St(1/2) –Mg(1/2)

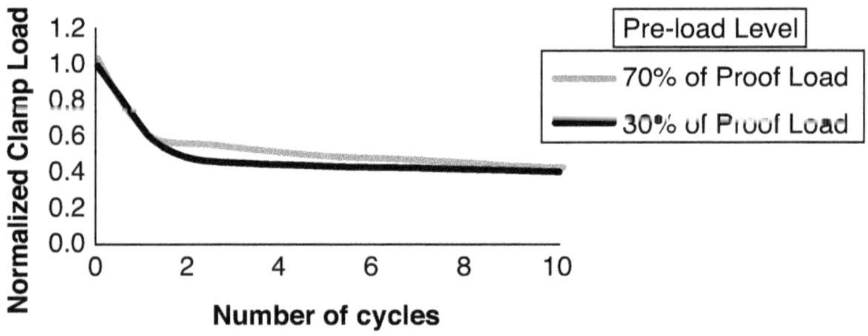

(c) St(3/4) –Mg(1/4)

FIGURE 33.10 The effect of initial clamp load level on the clamp load decay due to creep.

33.8 COMPARISON OF FEA AND EXPERIMENTAL RESULTS

Figure 33.11 shows the comparison between finite element and experimental data on clamp load decay due to cyclic heating and cooling of clamped dissimilar-material joints. Data shows a slight difference between the finite element and experimental results, but they are still in reasonable agreement. FEA results are obtained from several different combinations of joint materials coupon thickness and bolt preload levels. It should be noted that since temperature-dependent creep parameter A is sensitive to stress and activation energy at different temperatures, a significant difference could be observed between the experimental and finite element results. Additionally, the finite element model did not account for the bolt twist deformation or thread geometry; that could also contribute to the difference between the Experimental data and the FEA results.

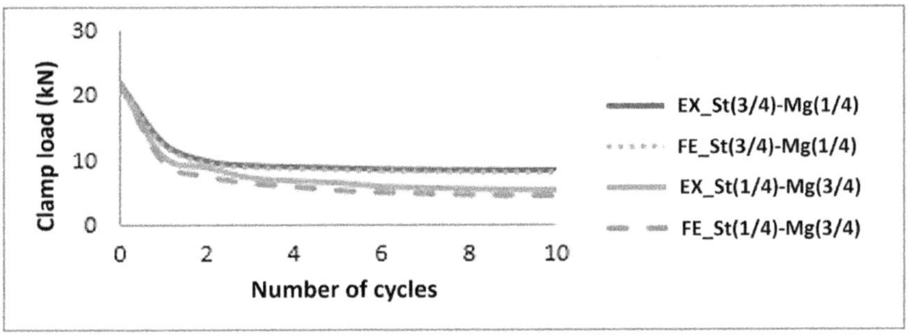

(a) Steel-Magnesium combinations at high preload level

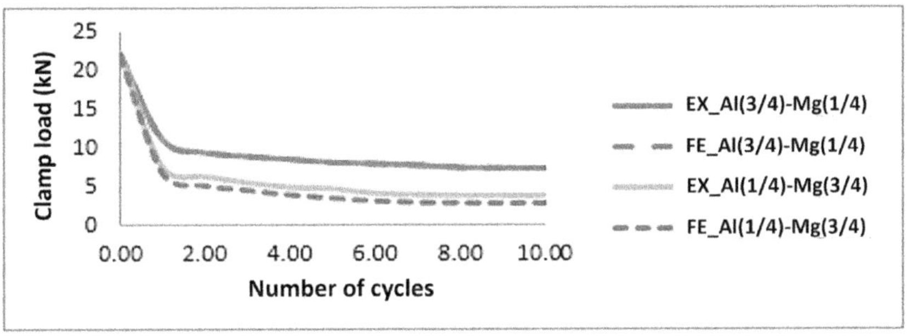

(b) Aluminium-Magnesium combinations at high preload level

FIGURE 33.11 Experimental and finite element data comparison.

33.9 CONCLUSIONS

FEA and experimental methods are used for investigating the effect of cyclic thermal cycling on the clamp load decay in bolted dissimilar-material lap joints due to material creep. Joint material combinations include A36 steel alloy and two other lightweight materials; namely, Aluminum 6061, and Magnesium AZ31B. Clamp load decay prediction by the FEA due to creep model is in reasonable agreement with the experimental data. The following observations and conclusions are made.

1. For all combinations of joint materials, the most significant clamp load drop occurs after the first full thermal cycle.
2. The rate of the clamp load decay decreases with the number of thermal cycles until the clamp load decay becomes insignificant after the tenth cycles, approximately.
3. Magnesium joints have the highest rate of clamp load decay with thermal cycling.
4. Initial clamp load level has insignificant effect on clamp load decay with thermal cycling.

ACKNOWLEDGMENT

US Army TARDEC for grant W56HZV-07-2-0001 for collaborative multi-material joining research (MMJ) between the Fastening and Joining Research Institute (FAJRI) at Oakland University and the Composite Vehicle Research Center (CVRC) at Michigan State University; Mohamed Dyab for the FEA modeling and data.

REFERENCES

[1] Lutsey N., 2010, *Review of Technical Literature and Trends Related to Automobile Mass-Reduction Technology*, University of California, Davis,Davis, California.

[2] Lotus Engineering, 2010, "An Assessment of Mass Reduction Opportunities for a 2017-2020 Model Year Vehicle Program.", Consulting Report, March 30, 20210, The International Council on Clean Transportation (icct), Consulting Report, Lotus Engineering (https://theicct.org/publication/an-assessment-of-mass-reduction-opportunities-for-a-2017-2020-model-year-vehicle-program/).

[3] ASME Boiler and Pressure Vessel Code, 2001, "Rules for Bolted Flange Connections with Ring Type Gaskets.", Section VIII, Division 2, Appendix 2, American Society of Mechanical Engineers, New York, New York.

[4] Kumano H., Sawa T., Hirose T., 1994, "Mechanical Behavior of Bolted Joints under Steady Heat Conduction," *ASME Journal of Pressure Vessel Technology*, 116, 42–48.

[5] Bouzid A., Nechache A., Brown W., 2002, "The Effect of Steady State Thermal Loading on the Deflections of a Flange Joint with a Cover Plate," In *Proceedings of the 2002 ASME-PVP Conference*, Vancouver, Canada, PVP-Vol.433, Analysis of Bolted Joints, PVP2002-1093, pp. 153–162.

[6] Nechache A., Bouzid A., 2003, "The Determination of the Load Changes in Bolted Gasketed Joints Subjected to Elevated Temperature", In *Proceedings of the 2003 ASME-PVP Conference, PVP2003-1883,* Cleveland, Ohio, pp. 139–148.

[7] Sato T., Kado K., 2006, "Inelastic Analysis of Sealing Characteristics of Flanges with Metal Ring Joint Gasket at Elevated Temperature", In *Proceedings of the 2006 ASME-PVP Conference*, July 23–27, 2006, Vancouver, Canada, PVP2006-ICPVT-11-93288.

[8] Sato T., Kado K., 2005, "Inelastic Analysis of Dissimilar Material Flanges with Metal Ring Gaskets at Elevated Temperature", In *Proceedings of the 2005 ASME-PVP Conference*, July 17–21, 2005, Denver, Colorado, USA, PVP2005-71131.

[9] Hanus F., Zilli G., Franssen J.-M., 2011, "Behaviour of Grade 8.8 Bolts under Natural Fire Conditions Tests and Model," *Journal of Constructional Steel Research*, 67, 1292–1298.

[10] Caccese V., Berube K. A., Fernandez M., Melo J. D., Kabche J. P., 2009, "Influence of Stress Relaxation on Clamp-Up Force in Hybrid Composite-to-Metal Bolted Joints", *Journal of Composite Structures*, 89, 285–293.

[11] Ekh J., Schön J., 2006, "Load Transfer in Multirow, Single Shear, Composite-to-Aluminum Lap Joints," *Journal of Composites Science and Technology*, 66, 875–885.

[12] Yang X., Nassar S. A., Wu Z., 2008, "Thermo-Mechanical Behavior of a Stainless Steel Threaded Fitting with a Precompressed Gastet," In *Proceedings of PVP2008 ASME Pressure Vessels and Piping Division Conference*, Chicago, Illinois, USA.

[13] Chen H. S., Kung H. K., 2002, "A Hygrothermal Sensitivity Evaluation on the Clamp Torque of Bolted Composite Joint," In *17th Annual Technical Conference of the American Society for Composites.*

[14] Shivakumar K. N., Crews J. H., 1982, Bolt Clamp-Up Relaxation in a Graphite/Epoxy Laminate. NASA TM-83268, Langley Research Center, 30 pp.

[15] ABAQUS 6.13, commercial FEA software and documentation, Dassault Systems, USA.

[16] Norton F. H., 1929, *The Creep of Steels at High Temperatures*, McGraw-HilI, New York.

[17] Spigarelli S., Mehtedi M. El, 2010, "Creep as an Extension of Hot Working: A Unified Approach to High Temperature Deformation of AZ31 Alloy," *Journal of Materials Science and Engineering, A*, 527, 5708–5714.

[18] Lou B. Y., Huang J. C., Wang T. D., Langdon T. G., 2002, "Flow Behaviour of Aluminum-Based Materials at Ultrahigh Temperatures in the Presence of Liquid Phase," *Journal of Material Transactions*, 43(3), 501–509.

[19] Mahmudi R., Ansary Sh., Esfandyarpour M. J., 2012, "Indentation Creep of the Wrought AZ31 Magnesium Alloy," *Journal of Material Science*, 47, 7181–7188.

[20] Yang X. J., Nassar S. A., 2007, "Elasto-Plastic Clamp Load Analysis of Bolted Joint for Bolt Strain Hardening Material under Separating Loading", *Journal of Materials and Manufacturing-SAE Transactions*, 116, 672–680.

[21] Nassar S. A., Kazemi A., Dyab M., 2014, "Clamp Load Decay in Preloaded Dissimilar Lightweight-Material Joints due to Cyclic Temperature", In *ASME Proceedings of the Pressure Vessels and Piping Conference*, July 21–26, Anaheim, CA.

34 Joint Diagrams

John H. Bickford

34.1 INTRODUCTION

The bolted joint consists of an assembly of parts—bolts and joint members—that deform when the bolts are tightened. In effect, the bolts act as "small springs" that when tightened, are stretched to a length slightly greater than their original length. This deformation will usually be elastic; the bolts will return to their original lengths when they are loosened unless they have been tightened past their yield point. When the bolts are tightened, the joint members are compressed to some new, slightly reduced thickness. The joint, in effect, acts like a stiff spring, and again the deformation will be primarily elastic. If the joint contains a gasket, it too is compressed when the bolts are tightened, although this time the compression is only partially elastic.

When the bolts have been tightened, the joint is said to be "preloaded." The tensile forces in the bolts oppose the equal and opposite compressive forces in the joint members. These forces may be reduced a bit by various relaxation mechanisms—embedment, metal creep, gasket creep, and/or differential thermal expansion (which can either increase or decrease the forces established during assembly)—but in general they will still be acting when the joint is put in service. If tensile service loads are then placed on the joint, they can either increase or decrease the tensile and compressive forces in bolts and joint members in ways that are not at all obvious. By tensile loads, I mean loads that act more or less parallel to the axes of the bolts. Shear loads, acting at right angles to the axes of the bolts, will generally *not* change the preloads in bolts and joint members.

In general, the behavior of the joint during assembly and in service can be complex. As a result, engineers have found it useful to use joint diagrams to illustrate the forces on and the deflections of the various parts and to help analyze the changes that occur when the bolts relax and/or external service loads are applied to the joint-bolt system. Several chapters of this handbook include joint diagrams to illustrate various design procedures or joint behavior or failure modes. It's useful, therefore, to devote a chapter to the diagrams themselves.

34.2 CONSTRUCTING THE BASIC JOINT DIAGRAM

If we apply a slowly increasing tensile load to a bolt, it will stretch elastically until the load exceeds its yield strength. We can draw a force–deflection diagram of this action on a single bolt as shown on the left-hand side of Figure 34.1. This diagram looks much like a stress–strain curve for the material, but it is not.

Note that the slope of the force-deflection line reflects the stiffness of the bolt "spring," K_B.

$$K_B = F_P/\Delta L \tag{34.1}$$

where

K_B = the stiffness of the bolt (lb/in., N/mm)

F_P = the preload (tensile force) in the bolt (lb, N)

ΔL = the corresponding change in the length of the bolt caused by that tensile force (in., mm).

We can construct a similar force-–deflection curve for the joint members as on the right-hand side of Figure 34.1. Because we want to use this information in conjunction with the information we've plotted for the bolt, we plot the clamping force created by a single bolt on the joint and the deflection of the joint in the vicinity of that bolt.

DOI: 10.1201/9781003501183-40

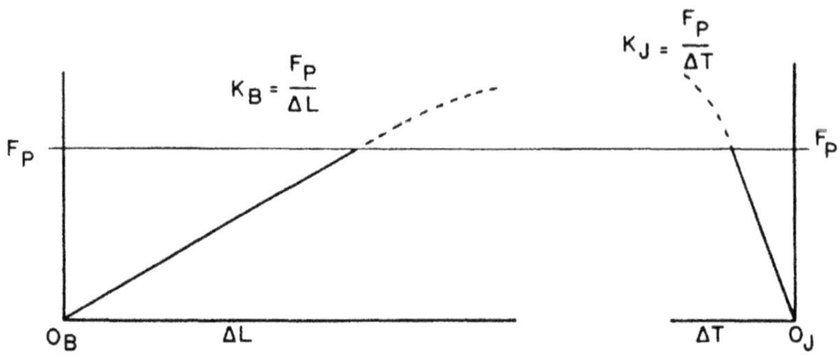

FIGURE 34.1 Elastic curves for bolt and joint members.

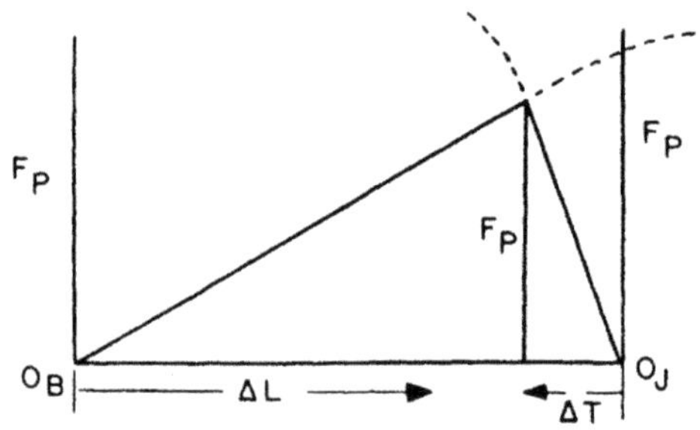

FIGURE 34.2 The elastic curves for bolt and joint can be combined to construct a joint diagram. O_B is the reference point for bolt length at zero stress. O_J is the reference point for joint thickness at zero stress.

Since the forces on the joint are equal and opposite to those on the bolts, it would be appropriate to call them negative and draw them below the deflection axis. For our purposes, however, we will find it more useful to ignore this sign convention and draw the force–deflection curve for the joint as shown in Figure 34.1.

Once again the slope of the line represents stiffness, this time, the portion of the joint members that is being loaded by a single bolt.

$$K_J = F_P/\Delta T \tag{34.2}$$

where

K_J=stiffness of the combined joint members (lb/in., N/mm)

F_P=the clamping force applied to the joint by the bolts, which is equal to the preload in the bolts (lb, N)

ΔT=the corresponding change in the thickness of the joint created by that clamping force (in., mm).

The force–deflection curve of the joint is shown in Figure 34.1 as a straight line, suggesting that joint deflection is fully elastic. This is often not the case, but we assume full elasticity when constructing the basic joint diagram.

Since the force exerted by the bolts on the joint is equal and opposite to that exerted by the joint on the bolts, we can plot the two on the same axes, as long as we ignore the sign convention. This allows us to combine the charts shown on the left- and right-hand sides of Figure 34.1, giving us the initial joint diagram shown in Figure 34.2. This could be called the "joint diagram during assembly." It provides us a preliminary look at the force vs. deflection and stiffness characteristics of the bolt–joint system. The initial preload in the joint at this point is F_P.

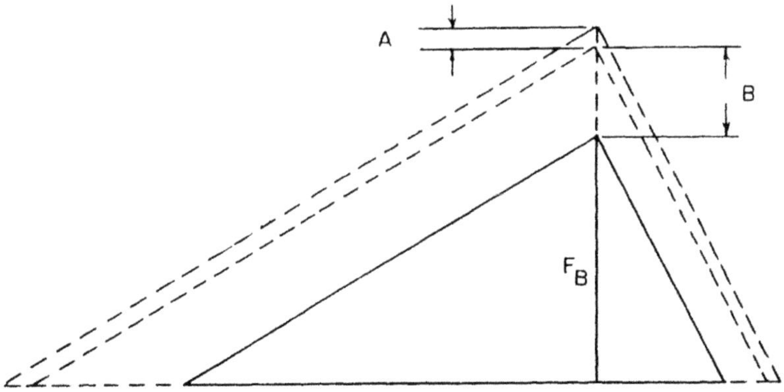

FIGURE 34.3 A joint diagram that shows the loss of initial preload caused by embedment (a) and the average loss caused by elastic interactions (b) on the preloaded joint. The largest dashed triangle represents the initial condition, before relaxation. The residual preload in this bolt is represented by the smaller triangle with side F_B.

As mentioned earlier, some of this initial preload is lost because of embedment, relaxation, and/or elastic interactions between bolts. The joint diagram shrinks in size without changing its basic shape, as shown in Figure 34.3, to reflect these losses. I'll refer to the residual tension in the bolts as F_B to distinguish it from the initial *pre*load F_P, but most people, including several authors in this handbook, prefer to use "preload" for the final, assembly tension in the bolts, which is certainly acceptable.

34.3 ADDING THE EFFECTS OF EXTERNAL LOADS

Assume that the residual preload in the bolt is 150,000 lb. This means that the bolt is exerting that much force on the joint—*and the joint is "pushing back" with an equal force of 150,000 lb.* Now let's apply a hypothetical separating load (L_x) of 2,400 lb to this bolt-joint system of springs by pulling in opposite directions on the head and nut ends of the bolt. At first glance, one might think that the tension in the bolt would increase by 2,400 lb. But as we pull on the bolt, it stretches a bit more, and in so doing it partially pulls away from the joint. This allows the joint to relax a little. It no longer pushes back with a force of 150,000 lb. Part of its job is now being done by the external tensile load. In practice, we say that "the joint has absorbed a portion of the external load." The bolt absorbs the rest. In effect, by pulling on the bolt we have applied a tensile load not just to the bolt but to a group of springs, the bolt-joint system, and the forces in and deflections of each spring in the system are affected by the load. The bolt stretches a little, to a new deflection we will call $\Delta L'$, and the thickness of the joint increases, by the same amount, to a new value we will call $\Delta T'$. The tensile force in the bolt increases, though not by 2,400 lb, and the clamping force it exerts on the joint is relaxed a little.

The resulting joint diagram is shown in Figure 34.4, where the external load, L_x, is plotted in such a way that it reveals the changes in force and deflection in both bolt and joint springs. The tension in the bolt has been increased by ΔF_B above the preload (F_P) to a new value of F_B, and the bolt has, as a result, stretched to the new value $\Delta L'$. The rest of the external load has allowed the clamping force on the joint to relax by an amount we will call ΔF_J, leaving a residual clamping force of F_J and a final joint deflection of only $\Delta T'$. The loss of deflection in the joint equals the increase in deflection of the bolt; the joint follows and maintains contact with the bolt as the bolt is pulled away from the joint by the external load.

Note that the final force in the bolt, F_B, is equal to the residual force by which the joint still pushes on the bolt, F_J, plus the external load, L_x.

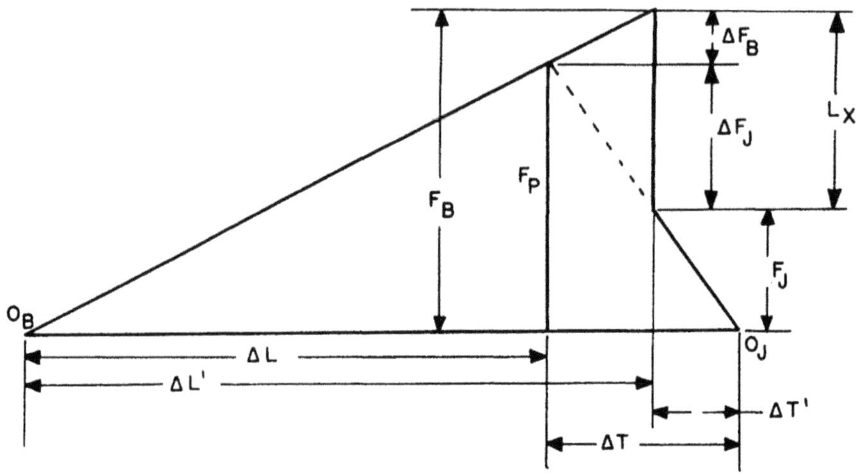

FIGURE 34.4 Summary of the discussion of the joint diagram. F_P=initial preload; F_B=present bolt load; F_J=present joint load, L_X=external tension load applied to bolt.

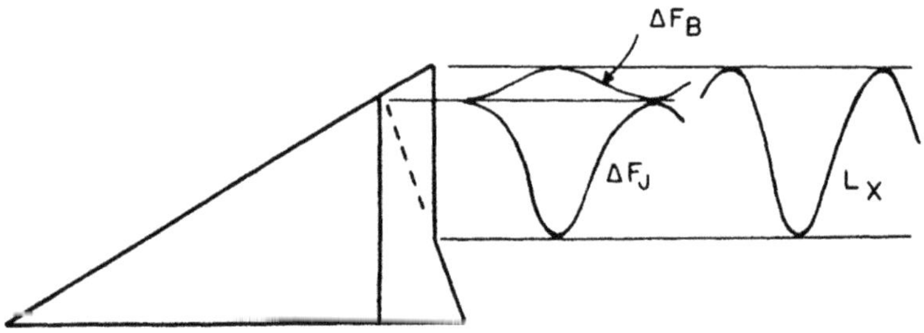

FIGURE 34.5 Bolt and joint loads when the external load fluctuates.

If the external load placed on the joint is cyclical and varies from zero to some maximum value and back to zero again, periodically, then the tension variation in the bolt and the clamping force on the joint will also be cyclical, as suggested in Figure 34.5.

34.4 THE "CRITICAL" EXTERNAL LOAD

The tension in the bolt increases as the external separating load is increased. At the same time, the clamping force on the joint decreases; the bolt is being pulled away from the joint. At some point, the clamping force will have fallen to zero. The external load required to pull the bolt completely away from the joint is called the critical load. This situation is illustrated by the joint diagram of Figure 34.6.

Note that any external load increment beyond the critical load is carried entirely by the bolt. The joint "spring" has effectively been taken out of the system of springs to which we first applied the load. The joint spring can no longer absorb part of the load. Any fluctuations in these large loads will be absorbed entirely by the bolt, too. This can cause special problems if the loads are cyclic because large excursions in load can lead to rapid fatigue failure.

34.5 THE EFFECT OF THE STIFFNESS RATIO

The behavior of a bolted joint is affected in several ways by the ratio between the stiffness of the bolt K_B and the combined stiffness of the joint members K_J, as is suggested in many other chapters of this handbook. Some of these effects are illustrated by the joint diagram. For example,

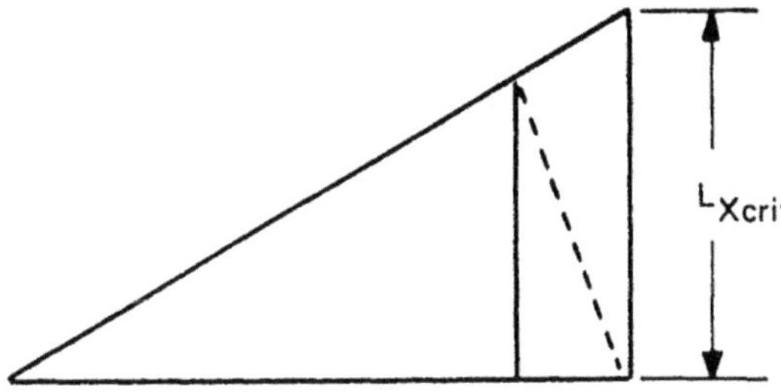

FIGURE 34.6 A critical external load (L_{xcrit}) fully unloads the joint (but not the bolt).

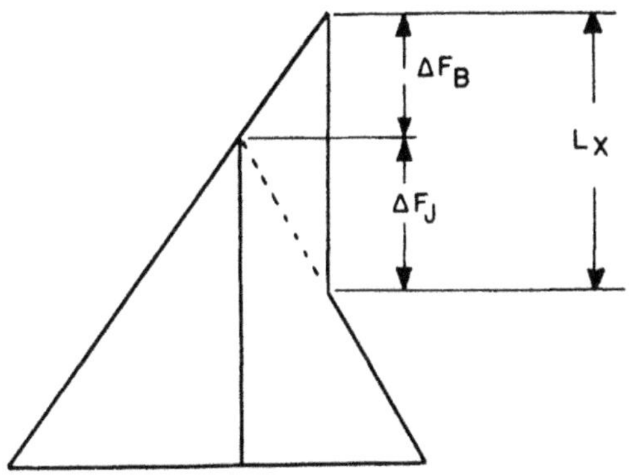

FIGURE 34.7 Joint diagram when the stiffness of the bolt nearly equals that of the joint.

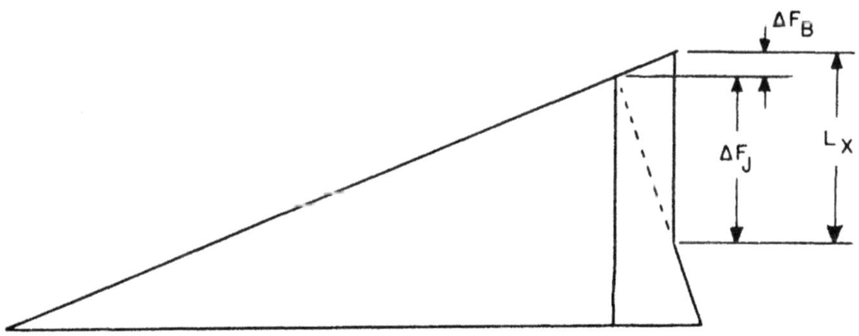

FIGURE 34.8 Joint diagram with softer bolt and stiff joint.

in Figure 34.7 if the stiffness of bolt and joint member are the same, about half of the external load is absorbed by the bolt, the other half by the joint, which could spell trouble in a fatigue loading situation but might be favorable for some types of gasketed joints. Compare the Figure 34.7 example to that shown in Figure 34.4, which shows a system with about a 4:1 ratio between joint stiffness and bolt stiffness. In this case, the bolt has to absorb only one-fifth of the external load until that load reaches the critical value.

If there is still a greater ratio between joint and bolt stiffness—a "softer" bolt and/or "stiffer" joint—the bolt will carry a smaller fraction of the external load, again until that load exceeds the critical value. Such a ratio, illustrated in Figure 34.8, would be good for fatigue loading situations.

34.6 OTHER TYPES OF LOADING

34.6.1 Joint Loaded at the Interface

So far we have only considered situations in which the external tensile load was applied to the head and nut ends of the bolt. Most joint diagrams are drawn with this assumption because it's a worst-case situation. The external load does two things we don't like: It adds tension to the bolt, and it partially or wholly unloads the joint. The mathematics of the diagram we've been using is also straightforward.

There are times, however, when we want a closer look at the "actual" effect of an external load on the joint and it has to be acknowledged that external loads applied to the ends of the bolts are rare. Here's a quick look at some of the less common joint diagrams that are used to illustrate other types of loads.

If the external tensile load is somehow applied at the interface between joint members, it helps one joint member to push on the other, and vice versa, until the critical load has been reached. At that point, the joint members have been pulled apart by the external load. Any external load greater than this will further compress the joint members and will increase the tension in the bolt. Until the critical load is exceeded there is no increase in bolt tension or joint compression; the external load is absorbed entirely as a decrease in the clamping force between joint members. This type of loading is illustrated in Figures 34.9, which shows the general situation, and 34.10, which shows the way the joint diagram changes after the critical load has been exceeded.

34.6.2 Joint Loaded from "Loading Planes"

Most tensile loads are applied to the bolt-joint system in ways other than the two described so far. They appear—from the way they affect bolt and joint deformations—to be applied from somewhere within the joint members. It would take a finite element analysis to determine just where within and just how the "springs" react to the load. Conventions have been established, however, that allow us to draw a modified joint diagram that, in most situations, is closer to the truth than either the diagram shown in Figure 34.4 or the one in Figure 34.9.

We start by assuming that equal and opposite tensile forces, parallel to the axis of the bolts, are applied along hypothetical "loading planes" within the joint members as shown in Figure 34.11. The separation between these loading planes is defined by a ratio (n) as

$$n = T_2/T_1 \tag{34.3}$$

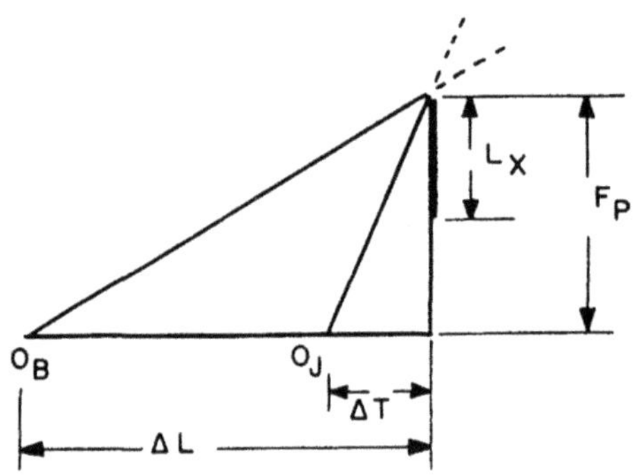

FIGURE 34.9 Joint diagram when an external tension load is applied to the joint interface. ΔL=elongation of bolt (in., mm), ΔT=compression of joint in., mm), F_P=original preload (lb, N).

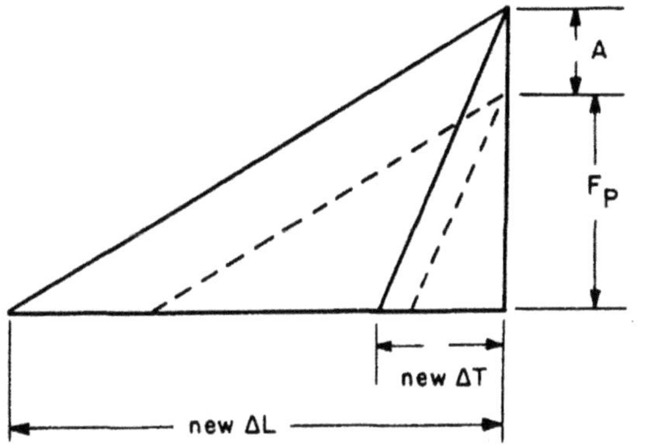

FIGURE 34.10 The external load applied to the joint interface has exceeded the critical load by amount A.

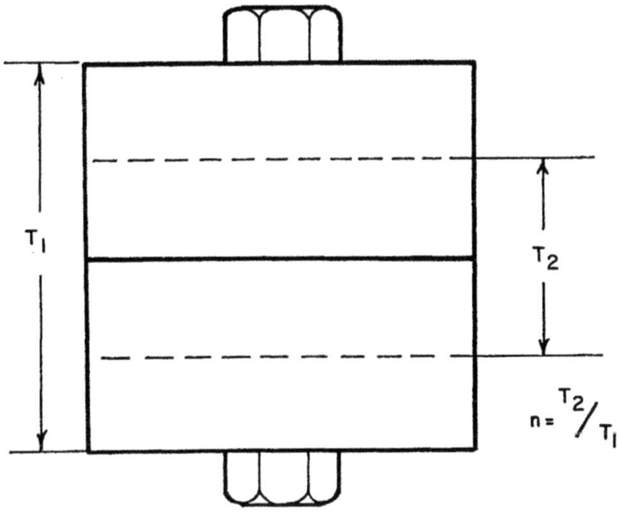

FIGURE 34.11 The "loading plane factor" discussed in the text is defined as the ratio between that portion of the joint that is unloaded by an external tensile load, T_2, and the total thickness of the joint, T_1.

where

n = the loading plane factor
T_1 = the combined thickness (grip length) of the joint (in., mm)
T_2 = the distance between loading planes (in., mm)

We can now modify the standard joint diagram of Figure 34.4, creating the diagram shown by the solid lines in Figure 34.12.

Unfortunately, we will rarely know exactly what value to assign to factor n. It has been suggested, however, that we draw a joint diagram, which is more accurate than that of Figure 34.4, if we assume that $n = 0.5$ and construct the diagram similar to the one shown in Figure 34.12.

34.6.3 Compressive Loads

Compressive loading is rare, as far as bolted joints are concerned, but it is not unheard of. You might find it applied to the piston of hydraulic cylinder, for example. Compressive loads will simultaneously reduce the tension in the bolts and increase the clamping force between joint members, as suggested in Figure 34.13.

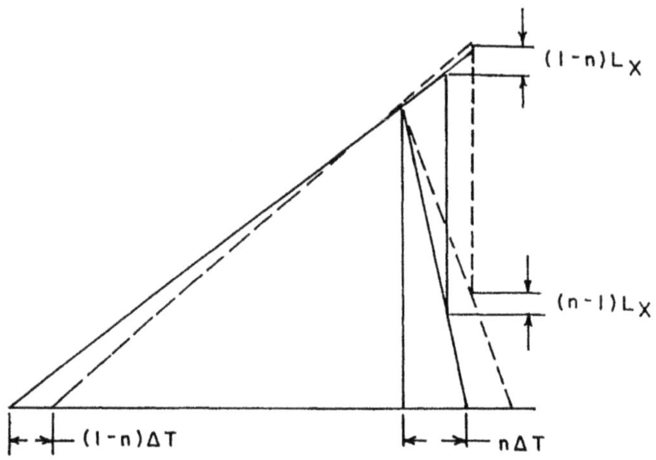

FIGURE 34.12 The dashed lines show the loading condition illustrated in Figure 34.4. The solid lines show how the diagram is modified to accommodate loading planes within the joint members. The location of these planes is defined by the loading plane factor, *n*, explained in Figure 34.11.

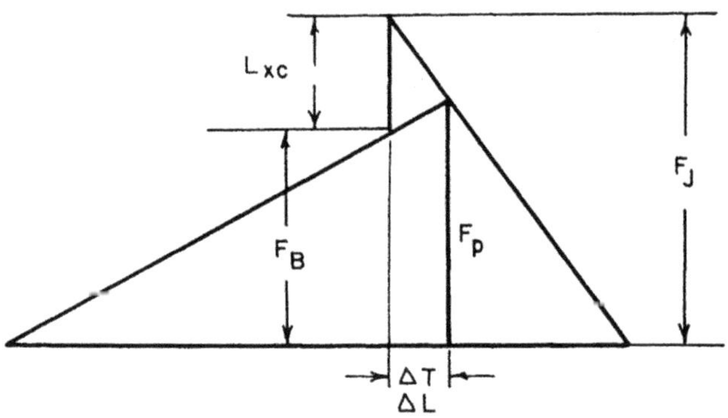

FIGURE 34.13 If a hydraulic cylinder is used in a push-pull mode, the joint between piston and rod will be subject to alternating tensile and compressive loads.

34.7 AN ALTERNATIVE TYPE OF JOINT DIAGRAM

The joint diagram we've been working with is very popular. Another type of diagram, however, that is sometimes encountered offers some advantages. This time we follow proper sign convention and plot bolt tension and clamping force on opposite sides of the horizontal axis, as shown in Figure 34.14. The bolts have initially been preloaded to a value of $+F_P$, and this has created an equal and opposite clamping force of $-F_P$ in the joint. In Figure 34.14, a small external tensile load, L_{X1}, which is plotted along the horizontal axis, has been applied to the preloaded bolt–joint system. As before, this load has slightly increased the tension in the bolt and has partially reduced the clamping force between joint members. The force change that occurs in each of these "springs" is reflected in the slopes of the load lines connecting $+F_P$ to the new bolt tension F_B and connecting $-F_P$ to the new clamping force F_J.

As before, the joint will be completely unloaded by any external load equal to or in excess of the critical load. Any additional tensile load beyond critical is born solely by the bolt, as shown in Figure 34.15. This type of diagram illustrates, more clearly than does the earlier type, the effects of loads that exceed the critical value. That makes this alternative diagram popular with those trying to analyze joints that will be placed in fatigue loading situations. One example of such analysis is shown in Figure 34.16.

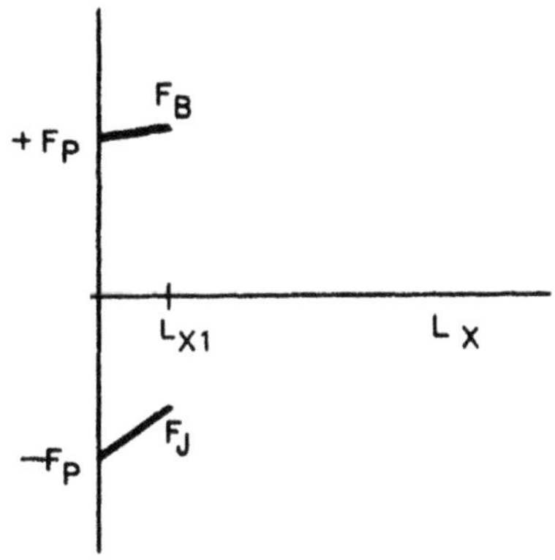

FIGURE 34.14 The initial preloads in bolt and joint are $\pm F_P$. The bolt load (F_B) increases and the joint load (F_J) decreases when we apply an external load L_{X1}.

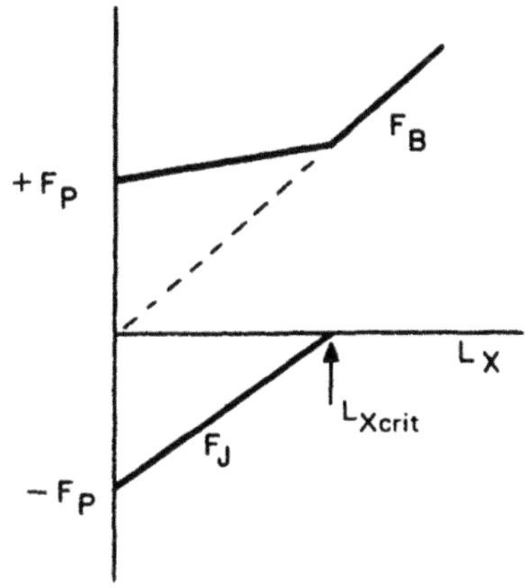

FIGURE 34.15 When the external load is increased until it exceeds the critical load, a sudden increase in the rate at which the bolt absorbs further external load is observed.

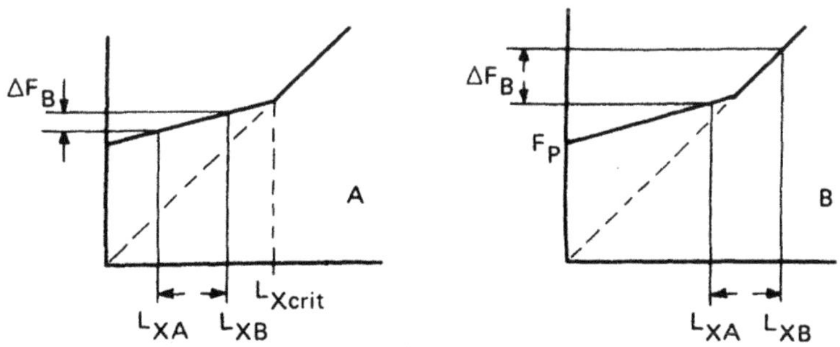

FIGURE 34.16 This alternative joint diagram gives us a dramatic picture of the effects of a large external load on bolt tension, a more accurate picture than that presented by the conventional joint diagram. Note that the bolt sees a far greater variation in tension (ΔF_B) if the external load exceeds the critical load required for joint separation (b) than it does when external loads are less than critical load (a). Note too that the initial preload and joint stiffness ratio are the same in each case.

34.8 NONLINEAR BEHAVIOR OF BOLTED JOINTS

Most joints are analyzed with the assumption that bolts and joint members deform elastically under tensile loads. In some situations, however, this is not the case. If a tensile load is applied to the joint at some distance from the bolts and the joint members are thin enough to flex, "prying loads" will be applied to the bolts. In addition, the contact pressure between joint members along one edge of the joint can actually be increased as the tensile load pulls the other side of the joint apart. This situation is illustrated in Figure 34.17. The resulting joint diagrams are shown in Figures 34.18 and 34.19.

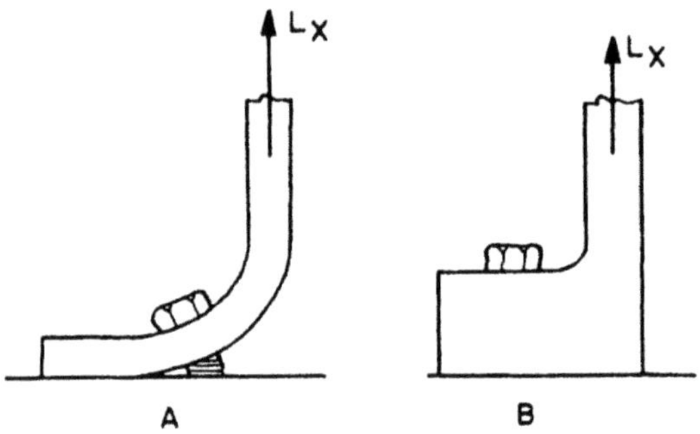

FIGURE 34.17 Joint members can act as levers to pry the bolt. The action is far more severe if the joint members are flexible (a) than if they are not (b).

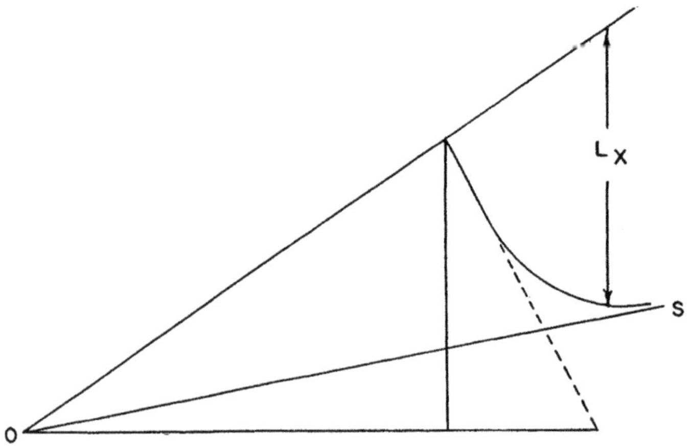

FIGURE 34.18 Joint diagram for a prying load. If the flange is flexible enough, the tension load cannot unload it completely. The joint experiences one-sided liftoff, starting near the point of application of the tensile load. After liftoff, the joint unloading line approaches the line O-S asymptotically.

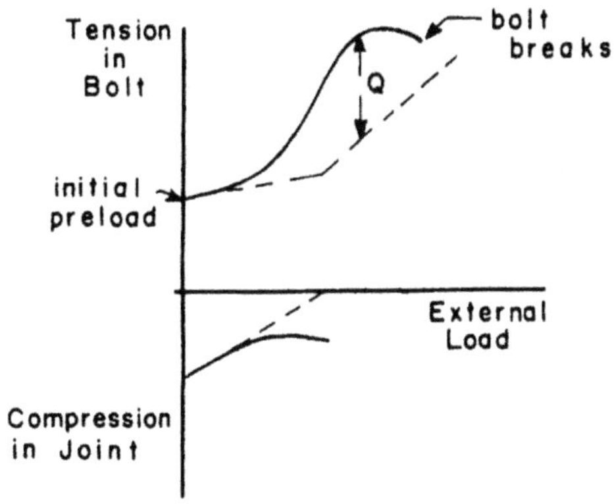

FIGURE 34.19 Alternative joint diagram when the bolt is subjected to a prying load.

ACKNOWLEDGMENT

The material in this chapter was adapted from portions of Chapters 6, 12, and 13 of my earlier text, *An Introduction to the Design and Behavior of Bolted Joints*, 3rd ed. (Marcel Dekker, New York, 1994).

35 Loss of Clamp Load in Post-Yielded Joint Systems Subjected to Axial Service Loading

Marco Gerini-Romagnoli

35.1 INTRODUCTION

This chapter is mainly focused on summarizing the work by Nassar and Matin [1,2] on clamp load in post-yielded joint systems that are subsequently subjected to axial service loads. That focus is motivated by the fact that joint reliability is primarily determined by both the level of joint preloading at initial assembly, and by the residual level of joint clamp load during joint life (or at least the time between scheduled maintenance and/or quality inspections). Needless to say, several other variable groups would also impact the entire joint system. Groups of variables may be related to the fastener, the joint, assembly tool, control method, post-assembly loads, and environmental factors. For more than three decades, systems approach to threaded joints has been steadily gaining momentum and popularity; especially in critical and/or safety-related applications. Material properties and the nature and magnitude of any static or cyclic external loads F_e (Figure 35.1) have the most significant effect on the stability of the clamp load, when the joint is tightened at or past its Yield strength and subsequently subjected to axial loading. The resulting partial loss of preload is irreversible, undetectable without costly structural health monitoring (SHM) instrumentation. This chapter provides an experimentally validated theoretical model that predicts the clamp load loss due to the plastic deformation of joint components under externally applied in-service axial loads.

35.2 ELASTO-PLASTIC BEHAVIOR OF THREADED FASTENERS: YIELD

The great majority of threaded fasteners are made of either carbon or stainless steel. Stainless steel fasteners are characterized by a distinct stress-strain signature, with a markedly nonlinear behavior starting from the upper end of the elastic range [3,4], as shown in Figure 35.2. The design of stainless steel bolted joints is further complicated by issues including but not limited to: (1) predisposition to galling, (2) time-dependent stress relaxation, and (3) dry bearing friction tribological characteristics [4]. Although tightening stainless steel joints at or beyond the yield strength of their components is commonly advised against, the model discussed in this chapter is applicable—with the appropriate assumptions—to stainless steel fasteners, especially from the austenitic and duplex families.

In contrast, the uniaxial tension stress-strain characteristic of carbon steel fasteners is fundamentally well understood, with a linear elastic region and a distinct yield point, determined by the bolt's grade (SAE fasteners) or class (ISO fasteners). A strain hardening phase follows, with values of ultimate strength commonly 10%–20% higher than the yield point. However, threaded fasteners are only subjected to conditions of uniaxial tension in the laboratory, when loaded in a tensile test unit with the appropriate test fixtures: in a bolted joint, the tightening and resisting (frictional) torques cause a significant shear stress distribution to appear on the fastener's cross section.

 DOI: 10.1201/9781003501183-41

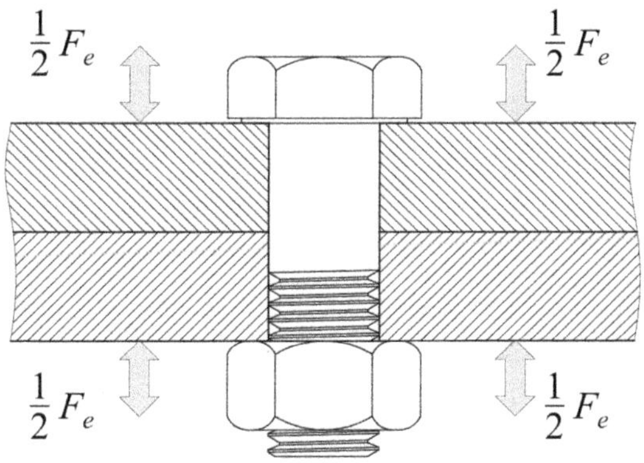

FIGURE 35.1 Example of bolted joint loaded with an axial force F_e.

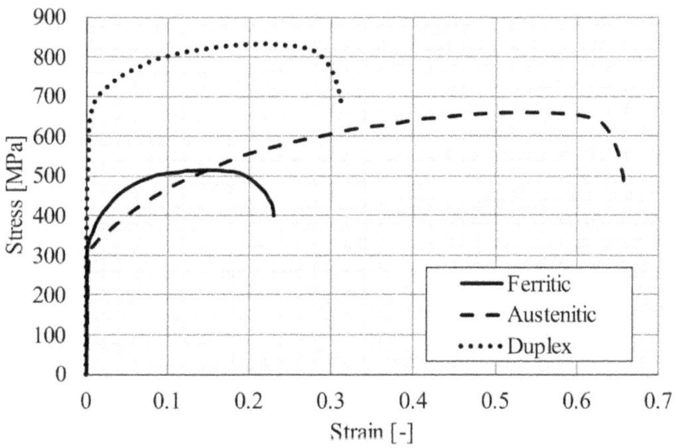

FIGURE 35.2 Stress-strain curves of different stainless-steel families [4].

35.2.1 Effect of Combined Axial-Torsional Loading on Bolt Preload-Elongation Curve

Figure 35.3 shows simplified torsional free body diagrams for a bolt. Figure 35.3a shows the torsional loads on the fastener during the tightening process: the tightening torque T is resisted by the sum of the bearing and thread friction torques (T_b and T_t, respectively), and by the pitch torque T_p. Figure 35.3b shows the torsional loads on a bolt *in service* (i.e. once the tightening process has stopped): bolt loosening (driven by the pitch torque T_p) is resisted by the bearing and thread friction torques T_b and T_t, respectively. The bolt shank experiences lower shear stresses than in the tightening phase, due to the reversal of the friction torques; however, the shear stress component is still significant, amounting to about ~10% of the tensile (normal) stress for fasteners of common sizes. Lastly, it is important to note that the diagram shown in Figure 35.3b is not statically determinate, i.e. the individual contributions of each friction torque component are undefined and depend on several parameters such as bolt/nut geometry, tribology of the bearing surfaces, etc.

The preload-elongation curve of typical threaded fasteners is therefore affected by the torsional shear stresses during the tightening process [5,6], especially near the elastic limit, as shown in Figure 35.4. The removal of the tightening tool causes a reduction in the shear stresses on the cross section (as explained in Figure 35.3), and a partial elastic recovery of the fastener. This phenomenon is exploited in Torque-to-Yield fasteners (see Chapter 28).

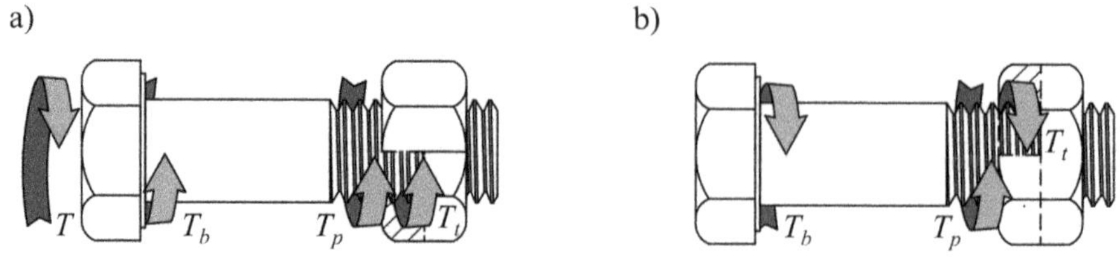

FIGURE 35.3 Torsional free body diagrams of bolt under tightening (a) and after the tool removal (b).

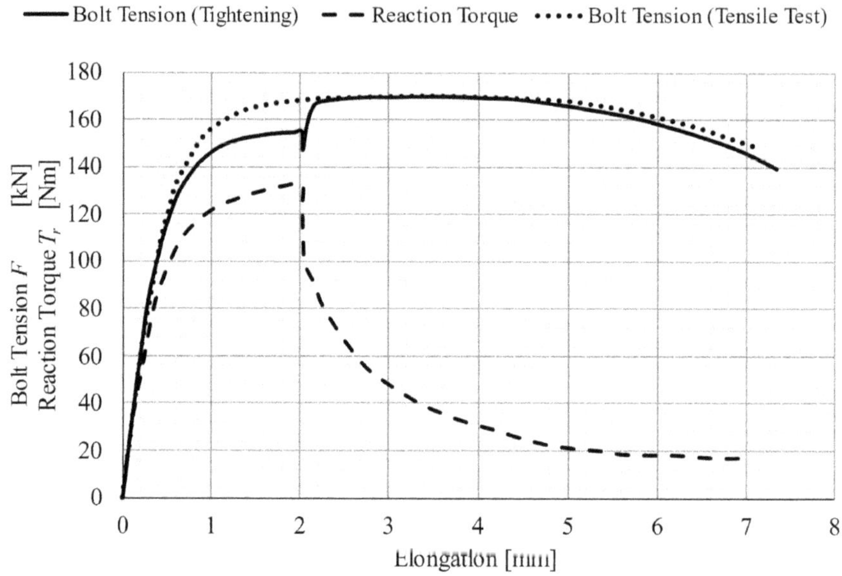

FIGURE 35.4 Fastener Load-Elongation curves during tightening vs. uniaxial tensile tests.

35.3 NONLINEAR JOINT DIAGRAM

Joint diagrams are important tools in the fastener engineer's arsenal. In a joint diagram, the force-displacement curve of the bolt is coupled to the curve of the clamped parts—which is mirrored about the vertical axis—and their intersection identifies the point of equilibrium achieved at the end of the tightening process. Consequently, each joint diagram is associated with a specific preload level. The application of an external separating (or compressive) load can be visualized as a vertical line shifting the bolt tension and clamp loads to the right (or to the left) of the initial equilibrium point.

The nonlinear behavior of the bolt (in tension) and of the clamped parts (in compression) can be easily implemented graphically in a joint diagram. It is useful to distinguish between three different scenarios, as shown in Figure 35.5: yielded fastener in an elastic joint (Figure 35.5a), yielded joint with an elastic fastener (Figure 35.5b), and yielded clamped parts and fastener (Figure 35.5c). The first scenario is common for torque-to-yield and, more generally, high preload fasteners, while the second scenario is common with gasketed joints.

The causes of bolt and joint yielding are often overlooked and simplistically reduced to overtightening. In the design phase, it is important to check if there are any conditions under which a bolted joint will be deformed plastically, and how that would occur. An important note is that although in Figure 35.5 all the joints were tightened *beyond* the material's yield strength, this is not the only case in which the plastic deformation of the fastener or of the clamped parts affects the joint

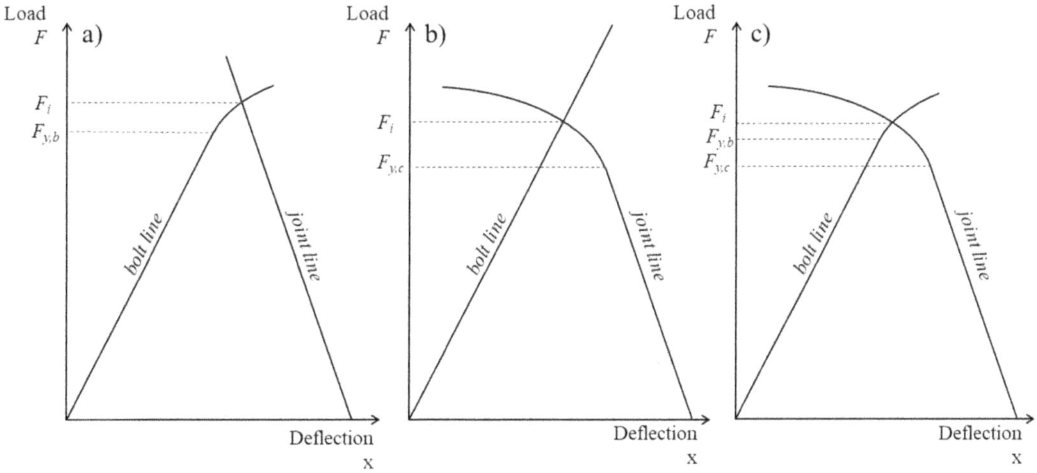

FIGURE 35.5 Nonlinear joint diagrams for joints tightened beyond Yield (scale exaggerated for clarity).

behavior. When a bolt is tightened to a sufficiently high preload—such as 90% of the proof load, well within the elastic range—moderate external separating loads can shift the equilibrium point to the nonlinear range (point *C* in Figure 35.6a). Similarly, if the initial preload is sufficiently close to the clamped parts' yield point, a moderate external compressive load can cause joint yielding (point *D* in Figure 35.6b).

Crucially, the yielding mechanism differs slightly between joints that are *directly tightened* beyond the elastic limit of the bolt, and joints that are *loaded* across the plastic threshold. In the first case, the material is subjected to gradually increasing *combined* normal and shear stresses, as described in Figure 35.4. In the second case, the yielding of the bolt is dominated by the increase in normal stress, while the shear stress on the cross section is less significant (but still increasing, due to the increase in pitch torque T_p), therefore following the uniaxial curve more closely.

Figure 35.7 shows a more complete joint diagram for a bolt that is tightened beyond its elastic limit, with two sets of strain hardening coefficients [5] and a 23.5 kN external separating load, represented as a vertical line linking the curves of the bolt and of the clamped parts. The first set of strain hardening coefficients applies to the bolt during the tightening process, which is characterized by larger shear stresses. The second set applies to the bolt under operating conditions. Should the experimental determination of the two sets of strain hardening coefficients be impractical or unfeasible, the Yield strength of the fastener can be lowered by 10% in the tightening phase, and the uniaxial curve can be used for the loading stage [2,5]. It is worth noting in Figure 35.7 that when a compressive load is applied to the joint, the bolt unloads along the dashed elastic line.

35.4 CLAMP LOAD LOSS

35.4.1 Scenario 1: Bolt Tension Beyond the Yield Point, while the Joint Remains Elastic

Regardless of whether a fastener is *tightened* (Figure 35.5a) or *loaded* (Figure 35.6a) beyond its elastic limit, the initial clamp load is defined by the intersection of the bolt and clamped parts curves on the joint diagram, and it remains stable until an external axial load is applied. When a separating load is applied to the joint, the bolt tension increases along its nonlinear curve, while the clamp load decreases along the elastic unloading line of the clamped parts, as shown in Figure 35.7. At this stage,

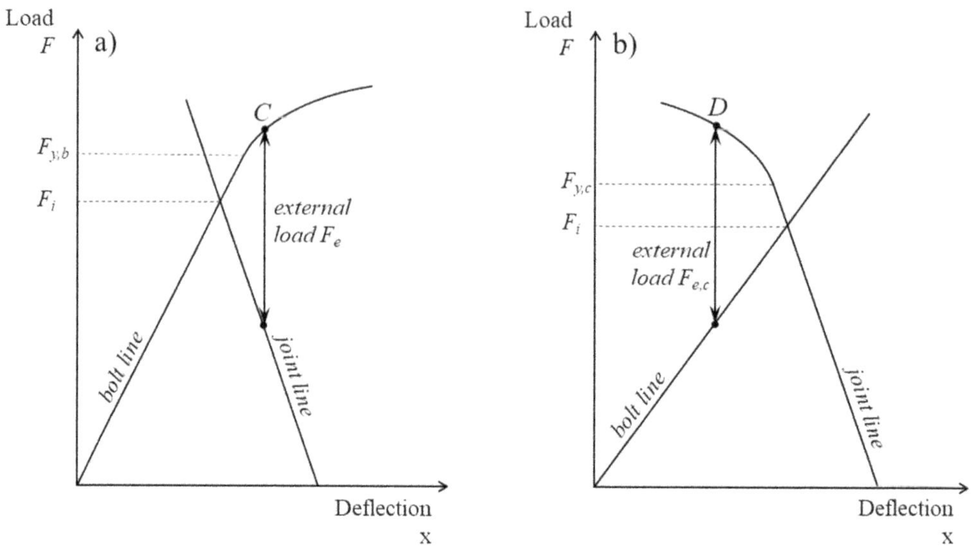

FIGURE 35.6 Nonlinear joint diagrams for joints loaded beyond Yield (scale exaggerated for clarity).

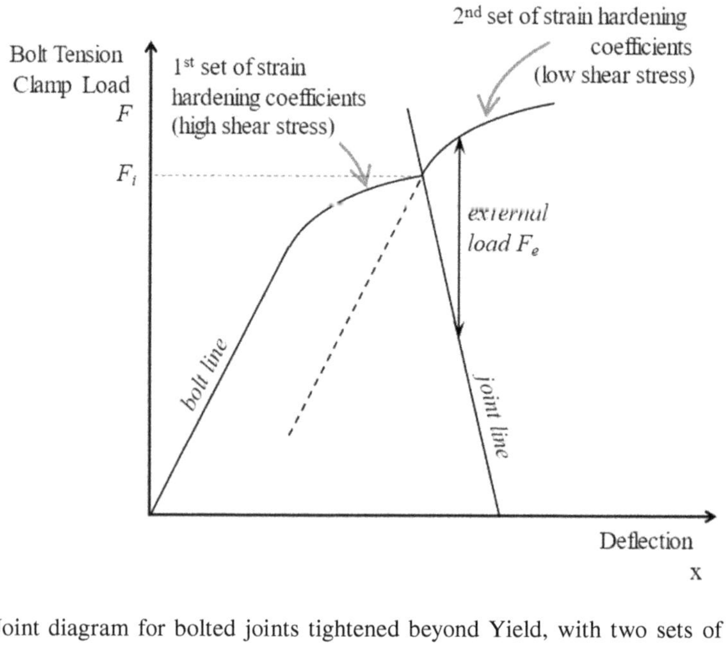

FIGURE 35.7 Joint diagram for bolted joints tightened beyond Yield, with two sets of strain hardening coefficients.

the reduced slope of the bolt curve causes a larger decrease in clamp load when compared to the linear elastic case. It is during the unloading phase, however, that the permanent loss of clamp load occurs. As shown in Figure 35.8, the bolt unloads along the dashed line *DH*, while the clamp load increases linearly along *OC*, and a new equilibrium point *H*, lower than the original point *C*, is established. Following the classical strain hardening theory [2], subsequent loading of the joint—up to the original external load F_e—does not cause further permanent clamp load loss, because the bolt loads and unloads along the elastic line *DH*. This latter point is important when cyclic (fatigue)

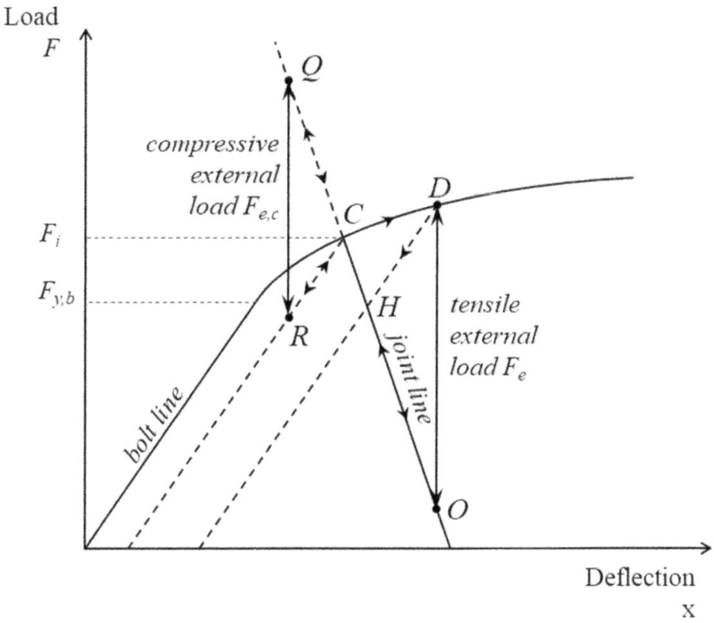

FIGURE 35.8 Joint diagram with nonlinear bolt behavior [2].

loading of the bolted joint is expected: in that case, the permanent loss of clamp load can be estimated by using the largest external load that is applied throughout the service life of the fastener.

A quantitative estimation of the permanent loss of clamp load can be obtained either graphically (i.e. from direct measurement of a sufficiently accurate joint diagram), or via the first-order approximation introduced by Nassar and Matin [1]:

$$\Delta F_c = \frac{\left(\dfrac{k_b + k_c}{k_c + \dfrac{nF_i^{1-1/n}}{L\left(\dfrac{1}{A\,K}\right)^{1/n}}} - 1 \right)}{\left(1 + \dfrac{k_b}{k_c}\right)} F_e \tag{35.1}$$

where ΔF_c is the clamp load loss, k_b and k_c are the bolt and joint stiffnesses, respectively, K and n are the strain hardening coefficients, L is the grip length, A is the tabulated tensile area of the fastener, F_i is the initial bolt tension, and F_e is the external separating load. It is important to note that when the applied separating external load F_e approaches the load that would cause joint separation (denoted as $F_{e,\max}$), this first order approximation underestimates the permanent loss of clamp load by 10%–15%.

Although less common, a compressive external load $F_{e,c}$ may be applied to the joint, causing the bolt to unload along the elastic line CR (Figure 35.8), and the clamped parts to compress further along the line CQ (provided that the initial preload is far enough from their elastic limit). In this case, both the bolt and the clamped parts follow the same elastic lines during the loading and unloading phases, and no permanent loss of clamp load occurs. In common applications, for non-gasketed carbon steel joints, it is reasonable to assume that the plastic deformation of the clamped members is negligible when compared to the bolt.

35.4.2 Scenario 2: Joint Is Initially Clamped beyond the Yield Point, while the Fastener Remains Elastic

In some cases, the point of initial tightening brings the clamped parts past their compressive Yield strength, while the bolt remains in the elastic range. The joint diagram for such a scenario is shown in Figure 35.9. Provided that point *C* is sufficiently far from the bolt's Yield strength, the application of an external *separating* load F_e does not cause any permanent loss of clamp load: the bolt loads and unloads linearly, while the joint follows its elastic spring back line (in accordance with classical strain hardening theory) even as it re-compresses. When the external load is fully removed, the system returns to the original equilibrium point *C*.

As previously mentioned, gasketed joints (sometimes referred to as "soft joints") are prone to yielding well before the fastener. *Compressive* service loads $F_{e,c}$ are reasonably common in multi-bolt gasketed flanged joints: when a bending moment is applied to a flanged joint (e.g. in a pipeline connection, or in a manufacturing robot end effector), parts of the flange experience compression. With reference to the diagram in Figure 35.9, the loading and unloading mechanisms are analogous to the case of a *separating* load: the bolt unloads along its elastic line, while the joint follows its spring back. However, the joint yields *further* during the loading stage, and a new equilibrium point *H* is reached upon removal of the *compressive* load $F_{e,c}$.

The clamp load loss is numerically equivalent to the previously analyzed scenario of a nonlinear bolt in an elastic joint subjected to a *separating* external load, leading to the following first-order approximation:

$$\Delta F_c = \frac{\left(\dfrac{k_b + k_c}{k_b + \dfrac{nF_i^{1-1/n}}{L\left(\dfrac{1}{A\ K}\right)^{1/n}}} - 1\right)}{\left(1 + \dfrac{k_c}{k_b}\right)} F_e \tag{35.2}$$

where *K* and *n* are the strain hardening coefficients of the clamped members.

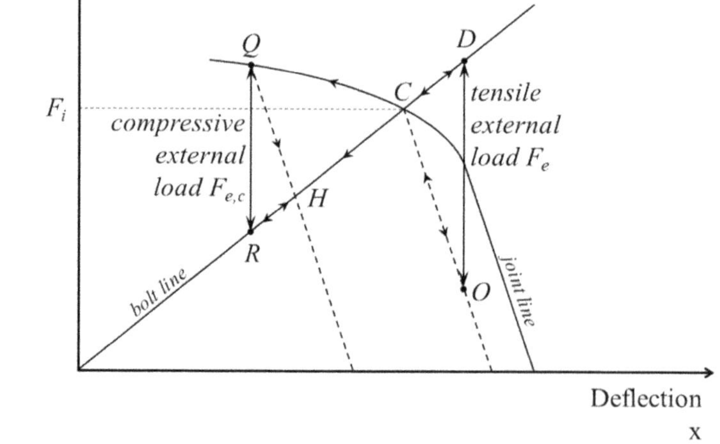

FIGURE 35.9 Joint diagram with nonlinear joint behavior and linear elastic bolt line.

35.4.3 Scenario 3: Both Fastener and Joint Elastic Limits Are Overcome during Tightening

Finally, the joint may be initially tightened to a preload that exceeds both the bolt's and the clamped members' elastic limit. The joint diagram for this scenario is shown in Figure 35.10. With reference to what was discussed for Scenarios 1 and 2, the effect of the external load on the permanent loss of clamp load can be summarized as follows:

- When a *separating* external load F_e is applied, the nonlinear behavior of the *clamped members* has no effect on the clamp load loss. Upon removal of the external load, the system returns to the equilibrium point H (Figure 35.10), and Eq. (35.1) can be applied directly.
- When a *compressive* external load $F_{e,c}$ is applied, the nonlinear behavior of the *bolt* has no effect on the clamp load loss. A new equilibrium point H' is reached after the external load is removed, and Eq. (35.2) can be applied.

35.5 CLAMP LOAD LOSS DUE TO CYCLIC SERVICE LOADS

In this section, the case of a joint subjected to a cyclic external load is analyzed, using the most general scenario of yielded bolt and clamped members, as shown by the joint diagrams in Figure 35.11. If a *reversed* cyclic or fatigue load is applied (i.e., if the external load cycles between *separating* and *compressive* stages, with the *separating* stage occurring first), the cumulative loss of clamp load can be estimated as follows:

- First, a joint diagram is generated with the largest *separating* and *compressive* loads experienced by the joint throughout its life (Figure 35.11).
- *Figure 35.11a:* if the *compressive* load is large enough to generate further plastic deformation of the clamped parts, then the cumulative clamp load loss $\left(\Delta F_c' + \Delta F_c''\right)$ can be obtained adding the result of Eq. (35.1) to Eq. (35.2). Point H' (corresponding to time instant 3 in the fatigue load diagram) is taken as the initial preload of the *compressive* stage.
- *Figure 35.11b:* If the *compressive* load is not large enough to generate further plastic deformation of the clamped parts, then the nonlinear behavior of the clamped parts has no effect on the clamp load loss.

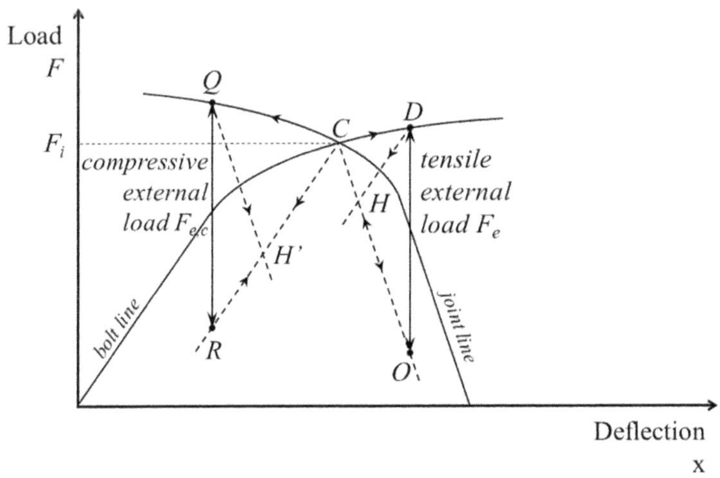

FIGURE 35.10 Joint diagram with nonlinear bolt and joint behavior, with loading and unloading stages.

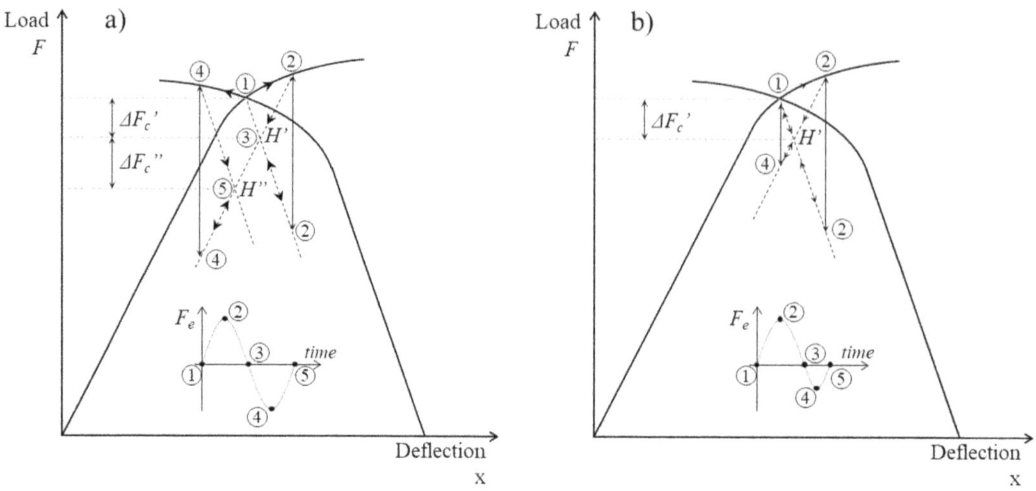

FIGURE 35.11 Reversed external load: cumulative loss of clamp load.

If the largest *compressive* stage occurs first, then the procedure is reversed: the result of Eq. (35.2) is added to Eq. (35.1), with point *H'* as the initial preload of the *separating* stage—i.e., $F_i = F_{H'}$ in Eq. (35.1).

35.6 JOINT TIGHTENED NEAR THE ELASTIC LIMIT

High performance bolted joints are often tightened near the elastic limit of the fastener or of the clamped members, without exceeding it. In these cases, loss of clamp load can still occur, if a sufficiently large external load is applied, and either the fastener or the joint are deformed plastically.

The case of a fastener tightened beyond its elastic limit (Scenario 1) applies to the case of torque-to-yield bolts, and more generally to any fasteners that are tightened at or very near their tensile Yield strength. Equation (35.1) can be used to estimate the loss of clamp load.

If the *initial* bolt preload is significantly lower than its elastic limit, however, direct application of Scenario 1 and Eq. (35.1) could lead to an overestimation of the loss of clamp load. In this case, a graphical approach, with an accurate joint diagram, is suggested, for a less conservative estimate. For *non-critical* applications, Eq. (35.1) may be used by subtracting the portion of the external load that does not cause fastener yield.

35.7 EFFECT OF JOINT AND BOLT STIFFNESS ON CLAMP LOAD LOSS OF JOINTS WITH YIELDED FASTENERS

The joint stiffness ratio, i.e. the ratio between the joint stiffness k_c and the bolt stiffness k_b, is the main parameter affecting the joint clamp load loss caused by a separating service load F_e. Joints with very large clamped parts, flange head bolts, and/or stainless steel bolts clamping carbon steel plates are characterized by high stiffness ratios, while aluminum or gasketed bolted joints are typically on the opposite end of the spectrum. Figure 35.12 shows the clamp load loss ΔF_c (as a ratio of the service load F_e, that causes joint separatin), is reduced rapidly with increasing the stiffness ratio $k_c/k_{b.}$. That is due to two separate factors; namely (1) the smaller increase in bolt tension from the external service load, and (2) a reduced level of external force $F_{e,sep}$ that would cause separation of very stiff joints with $k_{c}.>>k_b$.

In other words, it can be concluded that yielded soft threaded joint systems and subsequently subjected to a service tensile load are more prone to larger permanent loss of clamp upon the removal

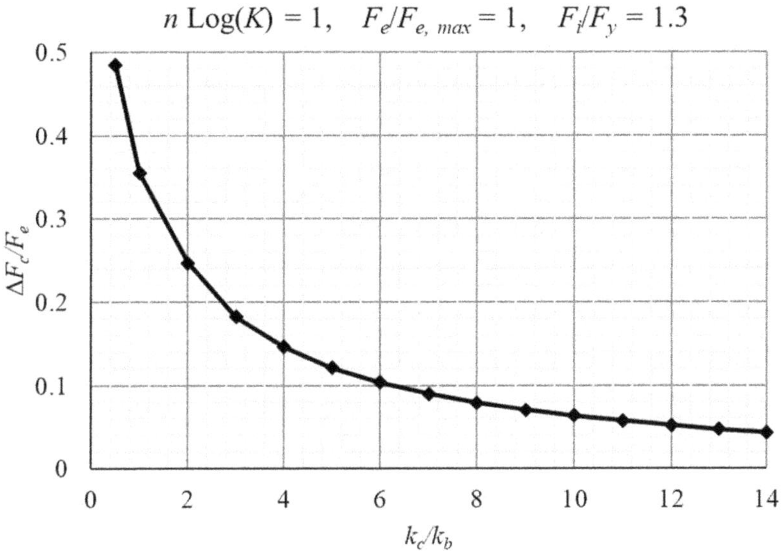

FIGURE 35.12 Clamp load loss vs stiffness ratio for stainless steel fasteners [1].

of the external load. That clamp load loss is caused by either one or two scenarios; namely (a) bolt yielding at initial preloading and subsequent application/removal of a *separating* service loads F_e, and/or (b)joint yielding during initial clamping and subsequent additional compression and removal of a compressive service loads $F_{e,c}$. Particular care, must be exercised in the design analysis and validation phases of soft or gasketed threaded connections.

35.8 SOLVED NUMERICAL EXAMPLES

35.8.1 Example 1: Yielded Bolt at Initial Preloading

The steel joint in Figure 35.13a is initially clamped to 120% of the bolt elastic limit using an M12×1.75 stainless steel fastener with the engineering stress-strain properties shown in Figure 35.13b. The joint remained in the elastic range at initial tightening. Subsequently, the joint is subjected to an external in-service separating load F_e=60 kN (during operation). Answer the following questions:

a. Use graph paper to draw the joint diagram to scale
b. Use the joint diagram to determine the bolt tension and the corresponding joint clamp load under an external tensile load F_e
c. Calculate the amount of clamp load loss after the external tensile load F_e has been completely removed

Solution

The tensile area of M12×1.75 fasteners is 84.3 mm², and the bering contact radii ratio γ is assumed to be 1.5. From the stress-strain diagram, the elastic limit is 760 MPa. The yield load is then

$$F_y = S_y \cdot A = 760 \cdot 84.3 = 64{,}000 \text{ N} = 64 \text{ kN}$$

The bolt is tightened to 120% of its elastic limit, corresponding to 912 MPa and 76.9 kN. The joint is then loaded with a 60 kN tensile service load. The service load does not cause joint separation because it is lower than the initial preload. The coordinates of the yield point and of a second point

TABLE 35.1
Respective Engineering Stress/Strain (S/e) and True Stress/Strain (σ/ε) for the Bolt in Example 1

Yield Point				**Point 2**			
S (MPa)	760	σ (MPa)	762.9	S (MPa)	1,130	σ (MPa)	1,220.4
e (−)	0.0038	ε (−)	0.00379	e (−)	0.08	ε (−)	0.077

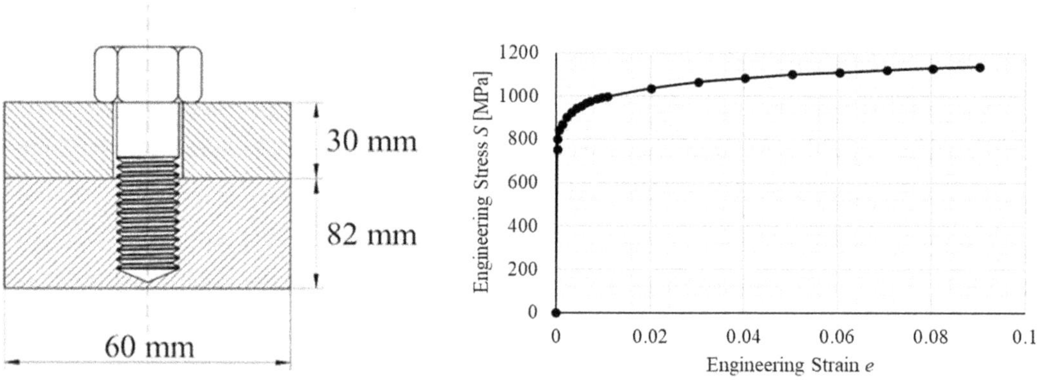

FIGURE 35.13 Joint dimensions (left) and bolt material properties (right) for Example 1.

on are retrieved from the stress-strain curve (any two points in the nonlinear region would be sufficient, **but using the elastic limit as first point is highly recommended**):

True stress and strain values are obtained from the respective engineering quantities as:

$$\sigma = S(1+e)$$
$$\varepsilon = \ln(1+e) \tag{35.3}$$

The strain hardening coefficients can be found solving a system of two equations, two unknowns:

$$\begin{cases} \sigma_1 = K \cdot \varepsilon_1^n \\ \sigma_2 = K \cdot \varepsilon_2^n \end{cases} \rightarrow n = \frac{\ln(\sigma_1/\sigma_2)}{\ln(\varepsilon_1/\varepsilon_2)} = 0.156,\ K = \frac{\sigma_1}{\varepsilon_1^n} = 1820.61 \text{ MPa} \tag{35.4}$$

The stress envelope in the joint is fully developed, and the joint stiffness can be calculated:

$$L \cdot \tan(36°) + \gamma D = 39.8 \text{ mm} < 60 \text{ mm},\ \gamma = 1.5 \text{ and } L = 30 \text{ mm}$$

$$k_c = \frac{D\,\pi\,E \tan 36°}{2\ln\left[\left(\frac{\gamma+3}{\gamma D + L\tan 36° + 3D}\right)\left(\frac{\gamma D + L\tan 36° - D}{\gamma - 1}\right)\right]} = 2{,}294 \text{ kN/mm} \tag{35.5}$$

The bolt stiffness can be calculated approximating the unthreaded length as 30 mm:

$$k_b = \left[\frac{L}{E\frac{\pi D^2}{4}} + 2 \cdot \frac{0.5D}{E\frac{\pi D^2}{4}}\right]^{-1} = 539 \text{ kN/mm} \tag{35.6}$$

The bolt line in the joint diagram can then be constructed by calculating the bolt tension F_b load in the elastic zone as follows

$$F_b = k_b \delta \tag{35.7}$$

In the nonlinear zone, however, the bolt tension F_b would be given by

$$F_b = A \cdot \frac{\sigma}{1+e} = A \cdot \frac{K\left[\ln\left(\frac{\delta}{L}+1\right)\right]}{1+\frac{\delta}{L}} \tag{35.8}$$

where δ is the bolt elongation. The joint line has slope $-k_c$ and meets the bolt line at initial preload F_i=76.9 kN. If an equation for the joint line is needed, a system of two equations and two unknowns can be iteratively solved:

$$\begin{cases} F_i = -k_j \delta_i + \mathrm{q} \\ F_i = A \cdot \dfrac{K\left[\ln\left(\frac{\delta_i}{L}+1\right)\right]^n}{1+\frac{\delta_i}{L}} \end{cases} \tag{35.9}$$

where the value of δ_i is found iteratively from the second equation; the corresponding clamp load F_c s is given by

$$F_c = F_i + k_c(\delta_i - \delta),\ \delta_i = 0.39 \text{ mm} \tag{35.10}$$

To draw the vertical line representing the application of the external service load (F_e=60 kN), it is sufficient to draw a vertical line on the joint diagram such that its length equals is equal to the external tensile service load F_e,. The bolt s elongation δ can be determined graphically from the joint diagram, or by solving the following equation interatively

$$F_b - F_c = F_e \rightarrow A \cdot \frac{K\left[\ln\left(\frac{\delta}{L}+1\right)\right]^n}{1+\frac{\delta}{L}} - F_i - k_c(\delta_i - \delta) = F_e \rightarrow \delta = 0.4159 \text{ mn} \tag{35.11}$$

When the external load is removed, the bolt relaxes following a force/displacement line that is parallel to the linear (elastic) portion of the bolt line. The joint is re-loaded following its elastic line. This brings the system to a new equilibrium point, causing clamp load loss (Figure 35.14). The amount of clamp load loss can either be evaluated graphically from the joint diagram or it can be calculated using a one-term expansion formula for the clamp load loss:

$$\Delta F_c = \frac{\frac{k_b - \lambda}{k_c + \lambda}}{\left(1+\frac{k_b}{k_c}\right)} F_e = 10.54 \text{ kN, with } \lambda = \frac{nF_i^{1-1/n}}{L\left(\frac{1}{A\ K}\right)^{1/n}} \tag{35.12}$$

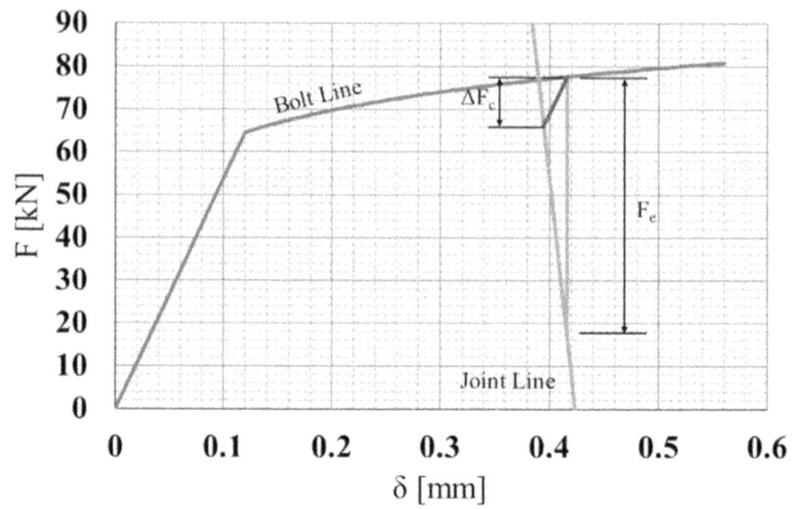

FIGURE 35.14 Joint diagram for solved Example 1.

35.8.2 Example 2: Yielded Joint at Initial Bolt-up

A Class 12.9 M14×1.5 bolt is used to clamp two mild steel plates with compressive force-deformation properties in the diagram in Figure 35.15. The joint is preloaded to 55 kN, and the bolt stiffness is k_b=565 kN/mm. An external tensile load of 40 kN is applied to the joint. Using the joint diagram, determine the bolt tension and joint compression:

a. Before the application of the service load.
b. When the service load is applied.
c. After the removal of the service load.

Solution

The joint is preloaded to 55 kN: without any external forces applied to the components (i.e. before the application of the service load), bolt tension and joint compression are in equilibrium:

$$F_b = 55 \text{ kN}, \; F_c = 55 \text{ kN} \tag{35.13}$$

The yield load of a Class 12.9 M14×1.5 fastener can be calculated using the tabulated values of yield strength (S_y=1,100 MPa for Class 12.9 bolts) and tensile area (A=125 mm²).

$$F_y = A \cdot S_y = 137.5 \text{ kN} \tag{35.14}$$

The joint diagram can be constructed: the slope of the bolt line is the bolt stiffness k_c, until the yield point (137.5 kN), and the joint force-deformation curve is mirrored and positioned in such a way to intersect the bolt line in correspondence to the initial preload level (55 kN, or point C in the diagram below).

The joint yields significantly during tightening, while the bolt remains elastic. In the limit case of an infinitely stiff fastener, the bolt tension after the application of the service load is equivalent to the sum of initial preload and external load. With F_i=55 kN and F_e=40 kN, the bolt does not approach the yield point (Figure 35.16).

The external load is represented by a vertical line in the joint diagram; after its application, the bolt tension rises to point D. The joint unloads along the elastic spring back line, and the compressive

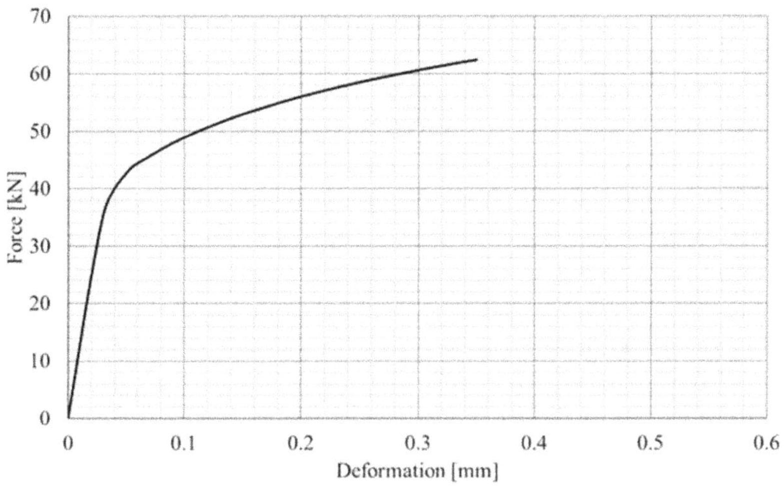

FIGURE 35.15 Force-deformation curve for the clamped parts in Example 2.

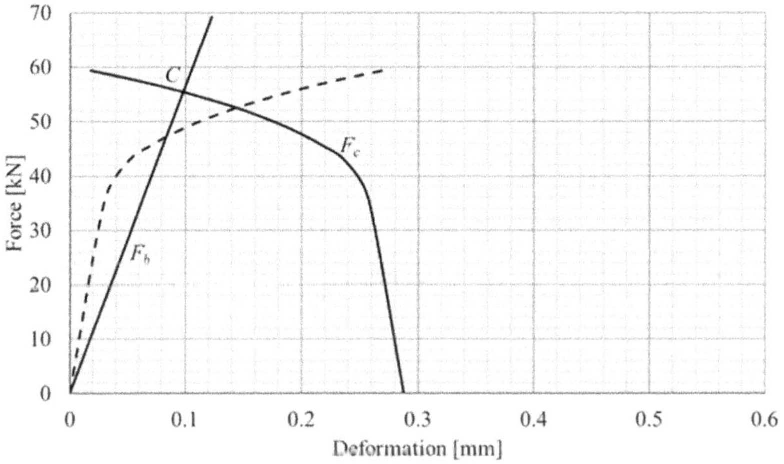

FIGURE 35.16 Construction of the joint diagram for solved Example 2.

force decreases to point *O* in the diagram. After the application of the external load, bolt tension and joint compression are

$$F_b = 68.5 \text{ kN}, \ F_c = 28.5 \text{ kN} \tag{35.15}$$

When the external load is removed, the bolt unloads along its elastic line, while the joint follows the spring back, until equilibrium is reached once again at point *C* (Figure 35.17). Yielding of the clamped members does not cause permanent loss of clamp load, and after the removal of the external load, bolt tension and joint compression are

$$F_b = 55 \text{ kN}, \ F_c = 55 \text{ kN} \tag{35.16}$$

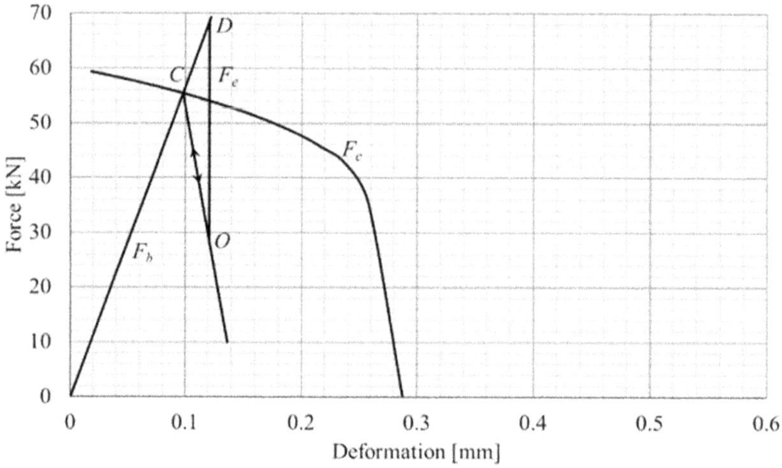

FIGURE 35.17 Complete joint diagram for solved Example 2.

BIBLIOGRAPHY

[1] Nassar, Sayed A., and Payam H. Matin. "Nonlinear strain hardening model for predicting clamp load loss in bolted joints." *ASME J. Mech. Des.* (2006): 1328–1336.

[2] Nassar, Sayed A., and Payam H. Matin. "Clamp load loss due to fastener elongation beyond its elastic limit." *ASME J. Pressure Vessel Technol.* (2006): 379–387.

[3] Afshan, Sheida, Itsaso Arrayago Luquin, Leroy Gardner, Graham Gedge, Michal Jandera, Esther Real Saladrigas, Barbara Rossi, Natalie Stranghoner, and Ou Zhao. *Design manual for structural stainless steel.* Ascot: Steel Construction Institute publications, 2017. Barcelona.

[4] Arrayago, I., E. Real, and Leroy Gardner. "Description of stress–strain curves for stainless steel alloys." *Mater. Des.* 87 (2015): 540–552.

[5] Nagata, Satoshi, Shinobu Kaneda, Hirokazu Tsuji, and Toshiyuki Sawa. "Three dimensional elastic-plastic finite element simulation on bolt tightening beyond yield." In *ASME Pressure Vessels and Piping Conference*, vol. 43659, pp. 201–206. 2009.

[6] William F. Smith, and Javad Hashemi. *Foundations of Materials Science and Engineering*. Mcgraw-Hill Publishing, 2006. Mcgraw-Hill Series in Materials Science and Engineering, New York.

36 Fatigue Performance of Ultra-High Strength Metric Bolts

Sayed A. Nassar and Tianwu Li

NOMENCLATURE

CA	Commercially available
HCF	High cycle fatigue
K_{I_c}	Critical stress intensity factor (mode I)
$\Delta\bar{K}_{I_c}$	Load ratio effect adjusted K_{I_c}
N_f	Fatigue life (cycles)
PZ	Plastic Zone
R^2	Linear regression coefficient
UHS	Ultra-high strength
S_a	Alternating stress
S_{a0}	Zero-mean-stress alternating stress
S_e	Fatigue limit with zero mean stress
S_{em}	Fatigue limit with non-zero mean stress
S_m	Mean stress
$S_{t\max}$	Maximum nominal tensile stress
S_y	Yield strength
S_u	Ultimate tensile strength
a_c	Critical crack length
k	Constant in Marin's equation
r	Load ratio
α	Constant in Marin's equation
β	Constant in Marin's equation
γ	Constant, being 0.42 for steel
σ_f	Fracture strength

36.1 INTRODUCTION

Tensile mean stress is known to be detrimental to the fatigue performance of preloaded threaded fasteners [1]. There are some commonly used models for mean stress effect by Goodman, Gerber, Morrow, Soderberg, and Elliptic; these models may be expressed by Marin's equation [2] as follows:

$$\left(\frac{S_a}{S_{a0}}\right)^{\alpha} + \left(k\frac{S_m}{S_u}\right)^{\beta} = 1 \tag{36.1}$$

DOI: 10.1201/9781003501183-42

	Model				
	Goodman	Gerber	Morrow	Soderberg	Elliptic (ASME)
α	1	1	1	1	2
β	1	2	1	1	2
k	1	1	S_u/σ_f	S_u/S_y	1

FIGURE 36.1 MTS 810 testing system.

where S_a is alternating stress for targeted mean stress level and a specific fatigue life; S_m is mean stress; S_{a0} is corresponding zero-mean-stress alternating stress; S_u is ultimate tensile strength; S_y is yield strength; σ_f is fracture strength; α, β, and k are constants and the values of them are listed. The above models are commonly used for high cycle fatigue (HCF) of steel components [1,3]. Some of these models may not be sufficiently accurate to use for threaded fasteners [4–11], especially in critical and safety-related applications. One main reason is that the mean stress had very little effect on the fatigue performance of the threaded bolts at a certain range of mean stress [8,12].

Essentially, the idea of Marin's equation is that the fatigue damage is the weighted sum of the damage caused by mean stress and by alternating stress, and the indices α and β are the indicators of the damage weights. In all the above criteria, α and β are constant values. But these criteria did not consider that α and β may also depend on fatigue life. From the S-N curves by Burguete and Patterson [8], it is observed that at different cycles, the influence of mean stress on

TABLE 36.1
Microstructure and Hardness

Bolt Size			M10X1.5		
Material Class	10.9	12.9	14.8	15.8	16.8
Microstructure	Martensite			Bainite	
Hardness (HRC)	38-39	40-42	44-46	47-49	50-52
Hardness (HV)	361	393	448	490	531

non-zero-mean-stress fatigue limit varied. It could be regarded as evidence that α and β depend on fatigue cycles.

In general, higher ultimate tensile strength S_u and hardness would lead to higher fatigue limit up to a certain level [13–16], beyond which the benefit may diminish [17–19], or even become detrimental to fatigue performance [20]. Budynas and Nisbett [21] suggested the limit to be 1,400 MPa; once S_u exceeds such value, the benefit of higher S_u to fatigue limit is no longer valid. Microstructure composition is considered to be the key to such a phenomenon by some researchers [22–25]. Other researchers suggested that fracture toughness, rather than S_u or hardness, would affect the fatigue limit. Lin et al. [26,27] observed a positive correlation between toughness and fatigue limit.

In this paper, a novel fatigue-life-dependent mean stress model is proposed for preloaded threaded fasteners; model results are compared with Goodman, Gerber, Soderberg, Morrow and Elliptic mean stress criteria using commercially available (CA) and ultra-high-strength (UHS) threaded fasteners. The effect of the microstructure and toughness on non-zero-mean-stress fatigue limit is discussed.

36.2 EXPERIMENTAL PROCEDURE AND TEST SETUP

Quasi-static tensile tests as well as fatigue tests are respectively conducted in accordance with ASTM-E8/E8M-09 and ASTM-E739-10, using an MTS 810 Testing System (Figure 36.1) at room temperature (20°C). All fastener test samples for this study were specially manufactured and supplied by Acument, along with relevant data shown in Table 36.1. Quasi-static tensile tests were performed at a displacement rate of 0.06 mm/s to obtain yield and ultimate tensile strength data by testing four samples from each fastener class (10.9, 12.9, 14.8, 15.8, and 16.8). Axial fatigue testing was conducted on fasteners that were pre-tensioned by the MTS system to a mean stress level of 50%, 70%, or 90% of their respective yield strength. The frequency of fatigue loading was 20 Hz. For each level of mean stress, respective fatigue life is recorded at four levels of amplitude with a sample size of 3 for each test.

36.3 DATA ANALYSIS

Following ASTM-E8/E8M-09 [28], Stress-Strain curves are generated by using the offset method. Respective averages of yield and ultimate tensile strength are shown in Table 36.2.

Normalized mean-stress-adjusted S-N curves are generated from fatigue tests following ASTM-E739-10 [28] and guided by Ali and Nassar [29,30]. Figures 36.2–36.4 show the normalized mean-stress-adjusted S-N curves for various mean stress levels; namely, 50%, 70%, and 90% of respective yield strength, S_y. Test data show that the UHS fasteners (Class 14.8, 15.8, and 16.8) have lower normalized alternating stress (S_a / S_u), as compared to that of CA fasteners (Class 10.9 and 12.9). Generally, among the UHS fasteners, Class 14.8 has the highest normalized alternating stress, while Class 16.8 has the lowest; the only exception corresponds to a mean stress $S_m = 0.5S_y$ (Figure 36.2), where Class 14.8 is lower than that of Class 15.8.

TABLE 36.2
Static Tensile Test Data

Bolt Dimension			**M10X1.5**		
Material class	10.9	12.9	14.8	15.8	16.8
S_y (MPa)	1,213	1,250	1,461	1,479	1,573
S_u (MPa)	1,271	1,370	1,539	1,634	1,749

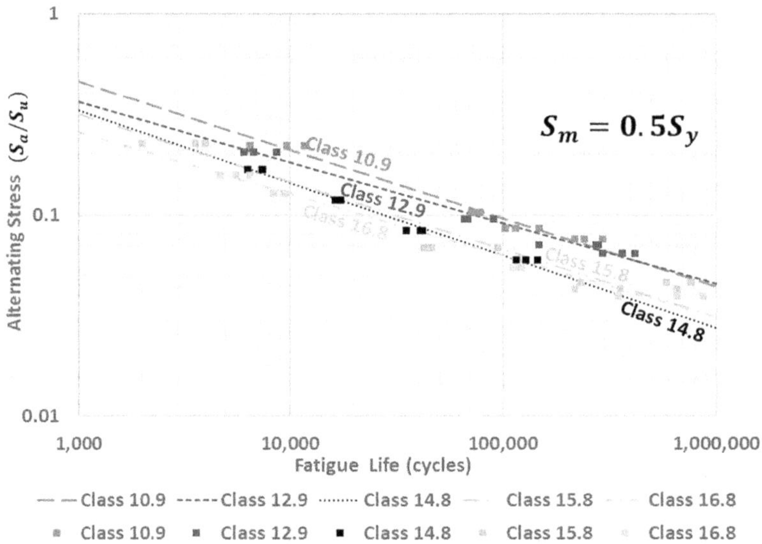

FIGURE 36.2 Normalized S-N curves for various material classes $(S_m = 0.5S_y)$.

Linear regression of fatigue test data on the S-N curves (Figures 36.2–36.4) has reasonable fitting coefficient R^2 values that range from 0.933 to 0.994 (Table 36.3).

36.3.1 Proposed Model for Mean Stress Effect

The following model is proposed for preloaded threaded fasteners:

$$\left(\frac{S_a}{S_{a0}}\right)^{\log_{10} N_f} + \frac{S_m}{S_u} = 1 \tag{36.2}$$

where N_f is fatigue life; S_m is mean stress; S_a is the corresponding alternating stress; S_u is ultimate tensile strength; S_{a0} is the zero-mean-stress alternating stress for the specific fatigue life. Because of $S_m = 40\%\ S_u$ being the starting point, the proposed model is represented by:

$$S_a = S_a' * \left(\frac{S_u - S_m}{0.6S_u}\right)^{\frac{1}{\log_{10} N_f}} \tag{36.3}$$

where S_a' is the alternating stress at mean stress being 40% S_u. The alternating stress value at $S_m = 40\%\ S_u$, S_a', is obtained by linearly curve fitting the data on the Haigh Diagram. Figure 36.5 and 36.6 shows the model prediction of the alternating stresses as compared to fatigue test data from Figures 36.2–36.4.

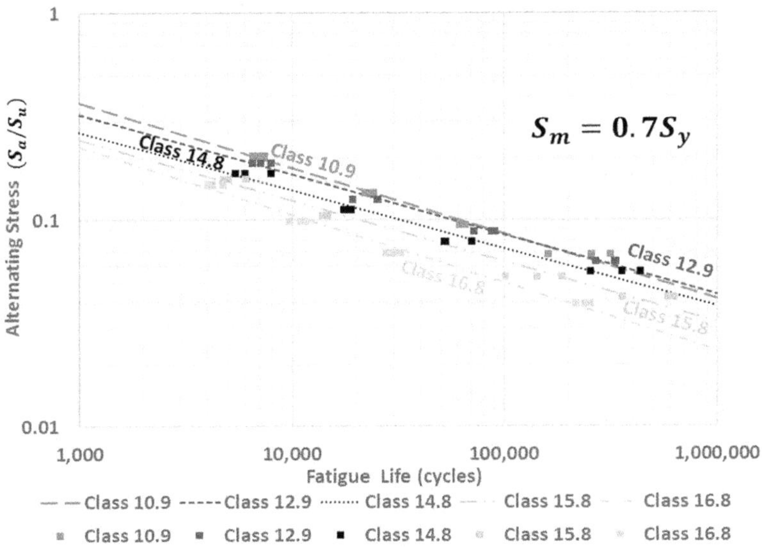

FIGURE 36.3 Normalized S-N curves for various material classes $(S_m = 0.7S_y)$.

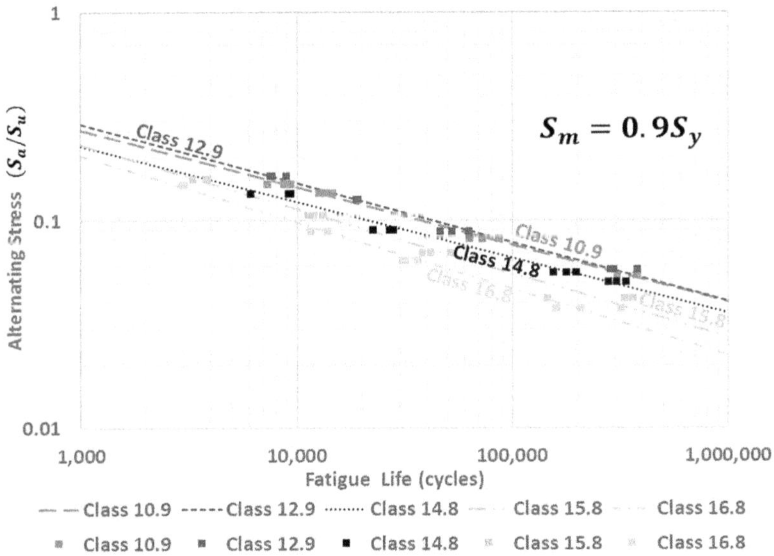

FIGURE 36.4 Normalized S-N curves for various material classes $(S_m = 0.9S_y)$.

TABLE 36.3
Linear Regression (R^2 Values)

R^2		Class				
		10.9	12.9	14.8	15.8	16.8
Mean	50%Sy	0.970	0.980	0.988	0.933	0.976
Stress	70%Sy	0.983	0.990	0.961	0.974	0.984
	90%Sy	0.994	0.987	0.984	0.979	0.987

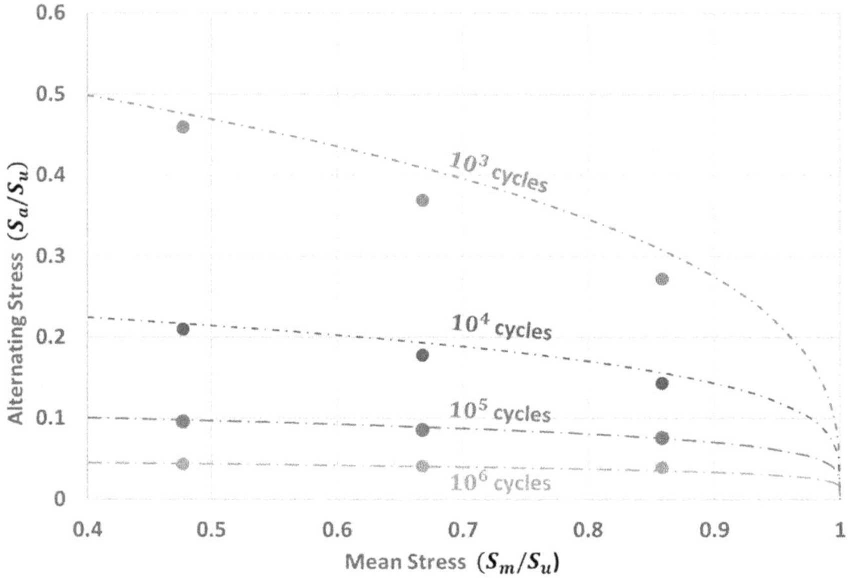

FIGURE 36.5 Sample model prediction of mean stress effect on alternating stress [Eqs. (36.2) and (36.3)] and data points from experimental S-N curves, CA class 10.9.

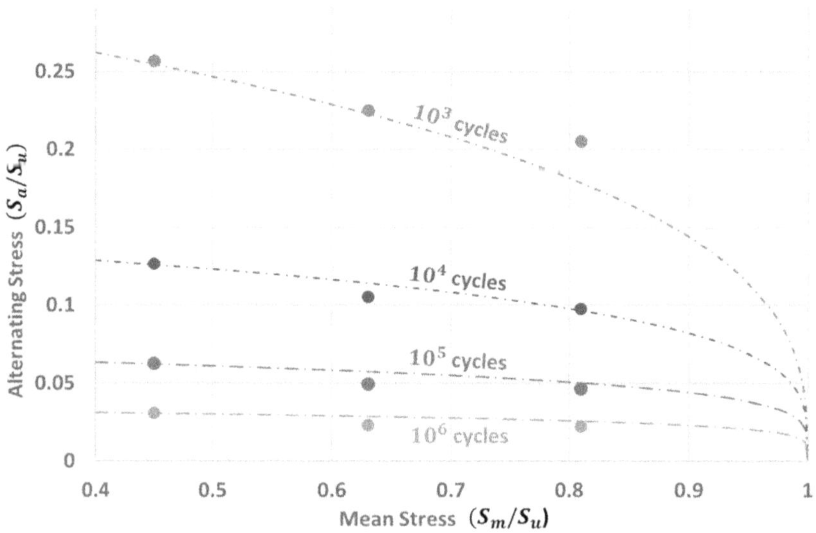

FIGURE 36.6 Sample model prediction of mean stress effect on alternating stress [Eqs. (36.2) and (36.3)] and data points from experimental S-N curves, UHS class 16.8.

The percent difference between model prediction and test data is defined as follows:

$$\%\text{Difference} = \left|\frac{\text{Model predicted } S_a - \text{ Experimental } S_a}{\text{Experimental } S_a}\right| \times 100 \tag{36.4}$$

The differences of each criterion at 10^3, 10^4, 10^5, and 10^6 cycles of fatigue life for each mean stress level are calculated. The averaged differences over the four fatigue lives at each mean stress level as well as those over the mean stress for each fatigue cycle are listed (Tables 36.4 and 36.5).

TABLE 36.4
Averaged %Differences between Model Prediction and Test Data for Various Mean Stress Levels over the Four Fatigue Lives

Mean Stress (S_m/S_y)	Goodman	Gerber	ASME	Soderberg	Morrow	Novel
50%	7.6%	7.5%	2.1%	48.5%	43.1%	1.2%
70%	32.4%	18.3%	9.3%	65.6%	57.2%	8.5%
90%	63.9%	47.7%	14.5%	87.1%	75.0%	9.9%

TABLE 36.5
Averaged %Differences between Model Prediction and Test Data for Various Fatigue Lives over the Three Mean Stress Levels

Fatigue Life (Cycles)	Goodman	Gerber	ASME	Soderberg	Morrow	Novel
10^3	30.9%	43.9%	6.9%	65.3%	56.0%	6.0%
10^4	33.6%	14.5%	6.4%	66.6%	57.7%	4.6%
10^5	36.0%	18.0%	8.8%	67.7%	59.3%	5.8%
10^6	38.0%	21.6%	12.5%	68.7%	60.7%	9.7%

Obviously, the novel mean stress criterion has better fitting to the experimental data than the five mean stress criteria.

The prediction given by the novel mean stress model is outstanding compared to the other five models, which indicates that the fatigue damage weights of alternating stress and mean stress vary with fatigue life. For threaded fasteners, plastic zone (PZ) is formed around the crack tip at the first engaged thread when the external load is applied. Increasing mean stress can only contribute to enlarging the plastic zone and alternating stress causes fatigue damage by cyclically opening and closing the crack tip [31]. Only if the size of the PZ reaches a threshold, the damage can be caused by alternating stress, while mean stress can only contribute to enlarging the PZ with causing minor damage. So at high mean stress level, even if the mean stress increases in a certain range, the alternating stress almost remains for specific fatigue damage, in other words, specific fatigue cycle. As fatigue life increases, the threshold size of the PZ for crack propagation gets lower, which means the mean stress' enlarging the PZ plays a smaller role in terms of fatigue damage. Correspondingly, the weight of the alternating stress gets larger, resulting in the weight index, α, increasing as the fatigue life increases ($\log_{10} N_f$).

36.3.2 Microstructure and Non-Zero-Mean-Stress Fatigue Limit

Approximate microstructure composition of each class (Table 36.6) is estimated by optical inspection of the microstructure for each class of fasteners (Figure 36.7). Figures 36.2–36.4 and Table 36.6 indicate that more bainite and less martensite will lead to lower normalized alternating stress.

In order to distinguish from the zero-mean-stress fatigue limit, S_e, the non-zero-mean-stress fatigue limits are denoted by S_{em}, which refer to alternating stress corresponding to 10^6-cycle life at non-zero mean stress levels. Non-zero-mean-stress fatigue limits for each class of bolts at different mean stress levels in this study would be analyzed together because their dependency on mean stress at high mean stress levels can be ignored (Figure 36.8). Statistical analysis (normality check

TABLE 36.6
Estimated Microstructure Gradients for Each Class

Class	10.9	12.9	14.8	15.8	16.8
Bainite	30%~40%	30%~40%	40%~50%	90%+	90%+
Martensite	60%~70%	60%~70%	50%~60%	5%~10%	5%~10%

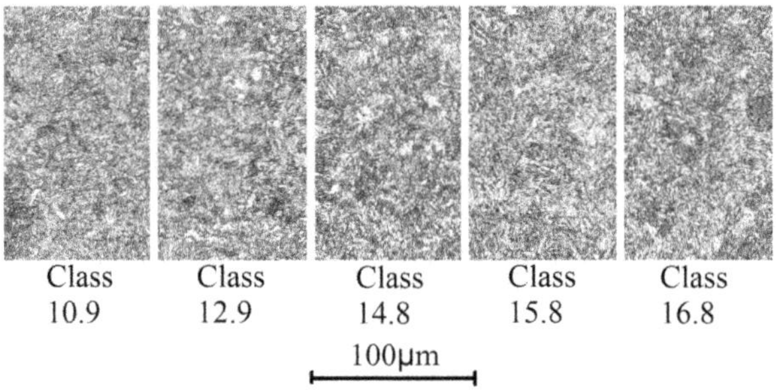

FIGURE 36.7 Microstructure of various fastener material classes.

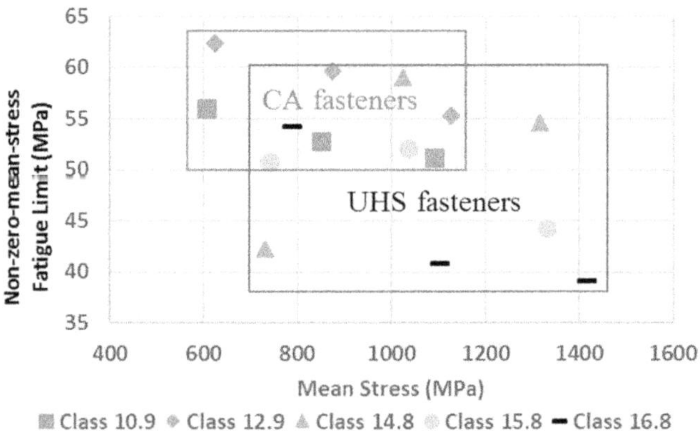

FIGURE 36.8 Non-zero-mean-stress Fatigue limits (test data).

and *t*-test) is adopted to verify the influence of microstructure on non-zero-mean-stress fatigue limits: the difference of the non-zero-mean-stress fatigue limits between the two groups, CA and UHS fasteners, should be confirmed, and meanwhile, other arbitrary groupings, rather than CA vs. UHS, should show no such difference (groupings are shown in Table 36.7).

Normality tests indicate that the *t*-test is valid for all groupings: *p*-values range from 0.111 to 0.973 for the three normality check methods, Anderson-Darling, Shapiro-Wilk, and Jarque-Bera [32,33]. In the results of *t*-tests (Table 36.7), the *p*-values of Grouping 0 and 9 are less than 0.05, which means the averages of the tested groups are statistically significantly different, while the other groupings do not show such statistically significant differences. So, it is clear that the difference of S_{em} couples with the difference of microstructure composition. As to Grouping 9 showing significance, it may be caused by that the microstructure composition of Class 14.8 falls in between Classes 10.9 & 12.9 and Classes 15.8 & 16.8 (Table 36.6), so in terms of non-zero-mean-stress fatigue limit, Class 14.8 is a little close to Classes 10.9 and 12.9. This result also fit for normalized

TABLE 36.7
***t*-Tests Result of Difference of Non-Zero-Mean-Stress Fatigue Limits between Groups**

			Result of *t* test			
			Absolute		Normalized	
Grouping Class			$P(T \leq t)$ Two-Tail	Significant?	$P(T \leq t)$ Two-Tail	Significant?
0	10.9&12.9	14.8&15.8&16.8	0.021	Yes	0.000	Yes
1	10.9&14.8	12.9&15.8&16.8	0.643	No	0.204	No
2	10.9&15.8	12.9&14.8&16.8	0.833	No	0.654	No
3	10.9&16.8	12.9&14.8&16.8	0.269	No	0.686	No
4	12.9&14.8	10.9&15.8&16.8	0.096	No	0.137	No
5	12.9&15.8	10.9&14.8&16.8	0.279	No	0.503	No
6	12.9&16.8	10.9&14.8&15.8	0.920	No	0.853	No
7	14.8&15.8	10.9&12.9&16.8	0.616	No	0.178	No
8	14.8&16.8	10.9&12.9&15.8	0.196	No	0.028	Yes
9	15.8&16.8	10.9&12.9&14.8	0.034	Yes	0.000	Yes

non-zero-mean-stress fatigue limit (Table 36.7 and Figure 36.9). So it is confirmed that at high mean stress levels, more bainite and less martensite will lead to lower non-zero-mean-stress fatigue limit.

36.3.3 Fracture Toughness

Figure 36.10 shows the fatigue fracture surfaces of the tested fasteners made of CA low carbon steel Classes 10.9 and 12.9, and UHS steel Classes 14.8, 15.8, and 16.8. For all of them, fatigue fracture occurred at the first fully engaged thread, which is the common fatigue failure mode of threaded fasteners [29,30,34]. The border, between the area for the fatigue crack propagation and the other area for the last cycle brittle fracture, is very clear in the view of optical microscope (Figure 36.10). Such final fracture is considered to follow the "Static Mode Mechanism" and characterized by the occurrence of inter-granular fracture (Figure 36.11) [25]. The corresponding critical crack lengths of typical samples are measured for toughness analysis below (Figure 36.10).

For each specific cyclic load condition, the critical crack length, a_c, of 1 sample (60 samples in total: five classes, three mean stress levels, and four alternating stress levels) is measured by optical microscope (Figure 36.10). The equations below are adopted to calculate the fracture toughness [35]:

$$K_{I_c} = 0.728 S_{t\max} \sqrt{\pi a_c} \tag{36.5}$$

$$\Delta \bar{K}_{I_c} = K_{I_c} (1-r)^{\gamma} \tag{36.6}$$

where K_{I_c} is critical mode I stress intensity factor, $S_{t\max}$ is maximum nominal tensile stress, $\Delta \bar{K}_{I_c}$ is fracture toughness, r is load ratio ($r = \frac{S_m - \sigma_a}{S_m + \sigma_a}$), γ is a constant (0.42 for steel). Figure 36.12 shows a linear relationship between the class-averaged non-zero-mean-stress fatigue limit S_{em} and class-averaged toughness $\Delta \bar{K}_{I_c}$. In the linear regression, the value of the intercept is nearly ignorable compared to the data scale, which is the reason to believe that there is a proportional relationship between the non-zero-mean-stress fatigue limit and toughness.

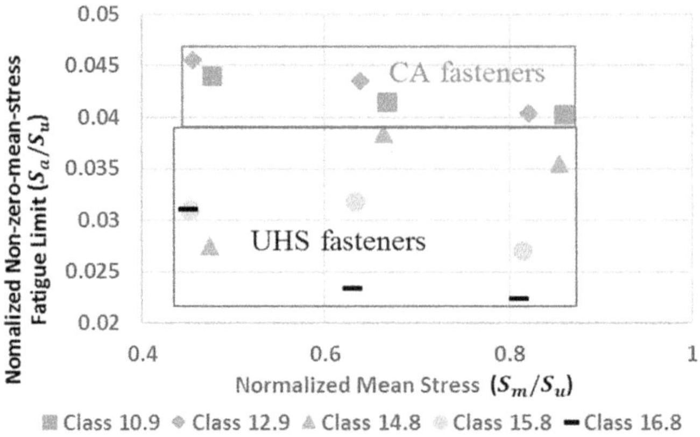

FIGURE 36.9 Normalized Non-zero-mean-stress Fatigue limits (test data).

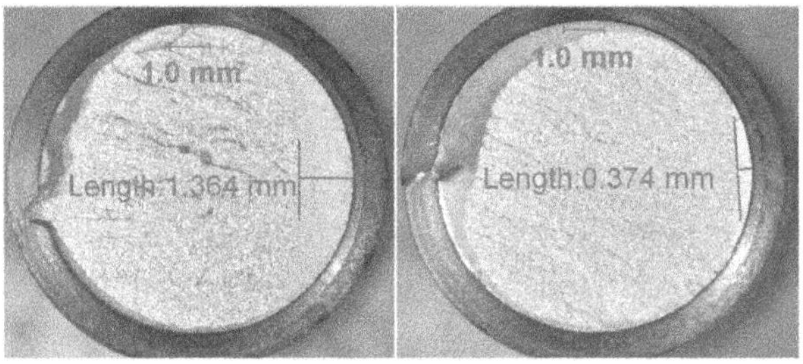

FIGURE 36.10 Sample fracture surfaces of CA (left) and UHS (right) fasteners

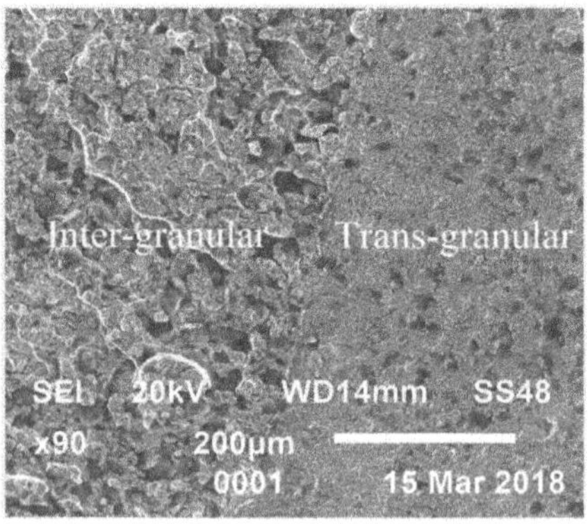

FIGURE 36.11 Border of Inter-granular and trans-granular areas by SEM.

36.4 CONCLUSIONS

A novel life-dependent model for mean stress effect on high cycle fatigue performance of preloaded threaded fastener is proposed [36-37]. The proposed model is compared with other existing models such as Goodman, Gerber, Morrow, Soderberg, and Elliptic (ASME) models; the new model's prediction is closer to fatigue test data. For all selected levels of mean stress, the normalized alternating stress of ultra-high strength fasteners is generally lower than that of commercially available fasteners. Bainite microstructure has a detrimental effect on fastener normalized fatigue performance, non-zero-mean-stress fatigue limit, and toughness, compared to martensite microstructure at high

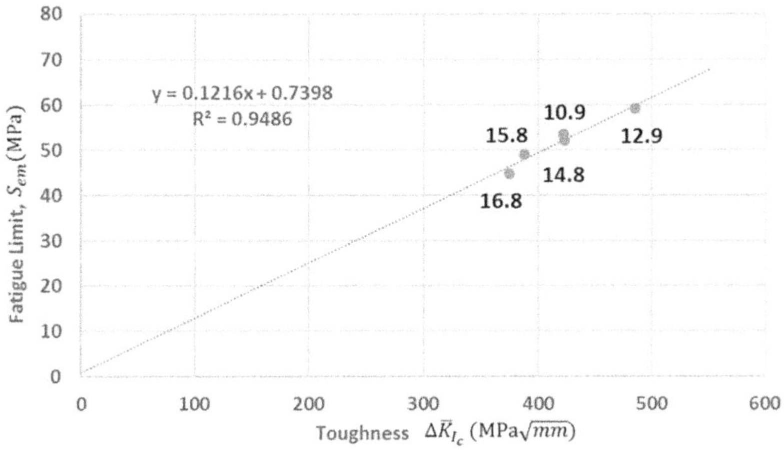

FIGURE 36.12 Fatigue limit, S_{em}, vs. toughness, $\Delta\bar{K}_{I_c}$, for various fastener material classes.

levels of mean stress. A proportional relationship is speculated between non-zero-mean-stress fatigue limit and fracture toughness.

ACKNOWLEDGMENT

The authors would like to thank Bill Turnmire and Rob Niec from Acument Global Technologies, Michigan-USA for providing all test samples and Kamel El-Agamy from Stanley Engineered Fastener for the help with metallurgical and optical inspection of microstructure.

REFERENCES

[1] A. Ince and G. Glinka, "A Modification of Morrow and Smith–Watson–Topper Mean Stress Correction Models," *Fatigue & Fracture of Engineering Materials & Structures,* vol. 34, pp. 854–867, 2011.
[2] J. Marin, "Interpretation of Fatigue Strengths for Combined Stresses," In *The International Conference on Fatigue of Metals*, London, 1956.
[3] A. Ince, "A Mean Stress Correction Model for Tensile and Compressive Mean Stress Fatigue Loadings," *Fatigue & Fracture of Engineering Materials & Structures,* vol. 40, pp. 939–948, 2017.
[4] N. Dowling, C. Calhoun and A. Arcari, "Mean Stress Effects in Stress-Life Fatigue and the Walker Equation," *Fatigue and Fracture of Engineering Materials and Structures,* vol. 32, pp. 163–179, 2009.
[5] N. Dowling, "Mean Stress Effects in Stress-Life and Strain-Life Fatigue," In *2nd SAE Brasil International Conference on Fatigue*, São Paulo, 2004.
[6] R. Burger and Y. Lee, "Assessment of the Mean-Stress Sensitivity Factor Method in Stress-life Fatigue Predictions," *Journal of Testing and Evaluation,* vol. 41, pp. 200–206, 2013.
[7] S. Wang, M. Dixon, C. Huey Jr and S. Chen, "The Clemson Limit Stress Diagram for Ductile Parts Subjected to Positive Mean Fatigue Loading," *Journal of Mechanical Design,* vol. 122, pp. 143–146, 2000.
[8] R. L. Burguete and E. A. Patterson, "The Effect of Mean Stress on the Fatigue Limit of High Tensile Bolts," *Proceedings of the Institution of Mechanical Engineers, Part C: Journal of Mechanical Engineering Science*, vol. 209, pp. 257–262, 1995.
[9] T. Wehner and A. Fatemi, "Effects of Mean Stress on Fatigue Behaviour of a Hardened Carbon Steel," *International Journal of Fatigue,* vol. 13, pp. 241–248, 1991.
[10] L. Pallarés-Santasmartas, J. Albizuri, A. Avilés and R. Avilés, "Mean Stress Effect on the Axial Fatigue Strength of DIN 34CrNiMo6 Quenched and Tempered Steel," *Metals,* vol. 8, no. 4, p. 213, 2018.
[11] A. Woodward, K. Gunn and G. Forrest, "The Effect of Mean Stress on the Fatigue of Aluminum Alloys," In *International Conference on Fatigue of Metals*, 1956.
[12] S. Cho, H. Chang and K. Lee, "Dependence of Fatigue Limit of High-Tension Bolts on Mean Stress and Ultimate Tensile Strength," *International Journal of Automotive Technology,* vol. 10, no. 4, pp. 475–479, 2009.

[13] N. Shamsaei and A. Fatemi, "Effect of Hardness on Multiaxial Fatigue Behaviour and Some Simple Approximations for Steels," *Fatigue & Fracture of Engineering Materials & Structures,* vol. 32, pp. 631–646, 2009.
[14] A. Casagrande, G. Cammarota and L. Micele, "Relationship between Fatigue Limit and Vickers Hardness in Steels," *Materials Science and Engineering A,* vol. 528, pp. 3465–3473, 2011.
[15] Q. Bader and E. Kadum, "Mean Stress Correction Effects on the Fatigue Life Behavior of Steel Alloys by Using Stress Life Approach Theories," *International Journal of Engineering & Technology,* vol. 14, no. 4, pp. 50–58, 2014.
[16] A. Kubit, M. Bucior, W. Zielecki and F. Stachowicz, "The Impact of Heat Treatment and Shot Peening on the Fatigue Strength of 51CrV4 Steel," In *21st European Conference on Fracture,* 2016.
[17] R. C. Juvinall and K. M. Marshek, *Fundamentals of Machine Component Design*, 5th ed., John Wiley & Sons, Inc., 2012, p. 325.
[18] K. Gillner and S. Münstermann, "Numerically Predicted High Cycle Fatigue Properties through Representative Volume Elements of the Microstructure," *International Journal of Fatigue,* vol. 105, pp. 219–234, 2017.
[19] A. Marcelo, A. Uehara, R. Utiyama and I. Ferreira, "Fatigue Properties of High Strength Bolts," *Procedia Engineering,* vol. 10, pp. 1297–1302, 2011.
[20] A. Yadollahi, N. Shamsaei, S. M. Thompson, A. Elwany and L. Bian, "Effects of Building Orientation and Heat Treatment on Fatigue Behavior of Selective Laser Melted 17-4 PH Stainless Steel," *International Journal of Fatigue,* vol. 94, pp. 218–235, 2017.
[21] R. Budynas and K. Nisbett, *Shigley's Mechanical Engineering Design*, 8th ed., McGraw–Hill, New York, 2006, pp. 277–283.
[22] R. Liu, Y. Tian, Z. Zhang, P. Zhang and Z. Zhang, "Fatigue Strength Plateau Induced by Microstructure Inhomogeneity," *Materials Science & Engineering,* vol. 702, pp. 259–264, 2017.
[23] V. Shamanth, P. Kumar and K. S. Ravishankar, "Effect of Heat Treatment on the High Cycle Fatigue Behaviour of S2205 Duplex Stainless Steels," *Materials Today,* vol. 4(10), pp. 10798–10802, 2017.
[24] Y. Qiu, J. Pang, C. Zou, M. Zhang, S. Li and Z. F. Zhang, "Fatigue Strength Model Based on Microstructures and Damage Mechanism of Compacted Graphite Iron," *Materials Science & Engineering A,* vol. 724, pp. 324–329, 2018.
[25] R. O. Ritchie, "Near-Threshold Fatigue Crack Propagation in Ultra-High Strength Steel: Influence of Load Ratio and Cyclic Strength," *Journal of Engineering Materials and Technology,* vol. 99, no. 3, pp. 195–204, 1977.
[26] C.-K. Lin and J.-Y. Wei, "High-Cycle Fatigue of Austempered Ductile Irons in Various-Sized Y-Block Castings," *Materials Transactions,* vol. 38, no. 8, pp. 682–691, 1997.
[27] C.-K. Lin, P.-K. Lai and T.-S. Shih, "Influence of Microstructure on the Fatigue Properties of Austempered Ductile Irons 1. High-Cycle Fatigue," *International Journal of Fatigue,* vol. 18, no. 5, pp. 297–307, 1996.
[28] American Society for Testing Materials. 2009. *Annual Book of ASTM Standards.* Section 3, *Metals Test Methods and Analytical Procedures. Vol. 03.01, Metals-Mechanical Testing, Elevated and Low-Temperature Tests, Metallography.* West Conshohocken: ASTM International.
[29] S. A. Nassar and R. Ali, "Fatigue Performance of Preloaded Threaded Fasteners," 2012.
[30] S. Nassar and R. Ali, "An Improved Cumulative Damage Criterion for Preloaded Threaded Fasteners," *Journal of Mechanical Design,* vol. 136, pp. 7–12, 2014.
[31] J. Schijve, *Fatigue of Structures and Materials*, 2nd ed., Springer, New York, 2009, p. 40.
[32] K. R. Das and A. H. M. R. Imon, "A Brief Review of Tests for Normality," *American Journal of Theoretical and Applied Statistics,* vol. 5, no. 1, pp. 5–12, 2016.
[33] D. Öztuna, A. H. Elhan, and E. Tüccar, "Investigation of Four Different Normality Tests in Terms of Type 1 Error Rate and Power under Different Distributions," *Turkish Journal of Medical Sciences,* vol. 36, no. 3, pp. 117–116, 2006.
[34] A. Kazemi, Z. Wu and S. A. Nassar, "Multi-Axial Fatigue of Preloaded Threaded Fasteners," In *ASME 2015 Pressure Vessels and Piping Conference,* vol. 2, 2017.
[35] Y.-L. Lee, J. Pan, R. Hathaway and M. Barkey, *Fatigue Testing and Analysis*, Elsevier Inc., 2005, pp. 256–257, 271–277, Philadelphia, Pennsylvania, USA.
[36] T. Li, S. A. Nassar and M. El-Zein, "Novel Model for Mean Stress Effect on High-Cycle Fatigue Performance of Threaded Fasteners", 2020, Special Issue of the Journal of Advanced Joining Processes, Springer.
[37] T. Li, Fatigue of Performance of Ultra High Strength Bolts, 2020, Ph.D. Dissertation, Department of Mechanical Engineering, Oakland University, Rochester, Michigan.

37 Fatigue Testing and Analysis of A286 Steel Bolts

Sayed A. Nassar, Michael J. Dube, and Rania Mansour

NOMENCLATURE

a	Y-intercept of the logarithmic $S-N$ line
A_t	Bolt tensile area
b	Slope of the $S-N$ curve
b^*	Slope of the $S'-N'$ curve after partial damage D
C_D	Specimen Size Factor
C_L	Load factor
C_m	Mean stress factor
C_R	Reliability factor
C_S	Surface factor
C_T	Temperature factor
F_u	Ultimate load
k_b	Bolt stiffness
k_c	Joint stiffness
k_o	Material constant
M	Mean stress sensitivity factor
N_j	Fatigue life corresponding to S_j (zero-mean)
q	Marin's exponent
R	Minimum-to-maximum bolt stress ratio
S	Fatigue strength at zero mean stress
S'	Mean stress-adjusted fatigue strength
S_a	Alternating stress amplitude (above and below the mean value)
$S_e\|_{R=-1}$	Endurance limit for $R=-1$
$S_e\|_{R=0}$	Endurance limit for $R=0$
S'_f	Fatigue strength coefficient at one cycle with zero mean stress
S_f	Fatigue strength coefficient at one cycle with mean stress level S_m
S_i	Initial preload stress
S_{max}	Maximum bolt stress
S_{min}	Minimum bolt stress
$S_j\|_{max}$	Maximum alternating stress amplitude in the loading history
S_j	Alternating stress amplitude corresponding to fatigue life N_j at zero mean stress
S'_j	Alternating stress amplitude corresponding to fatigue life N'_j at a mean stress level S_m
S_m	Bolt mean stress
S_u	Static ultimate strength of bolt material
S_y	Yield strength of bolt material

37.1 INTRODUCTION

Fatigue is a frequent failure mode to be considered in a mechanical design. Fatigue failure is often catastrophic and may have significant economic impact and cause injury or loss of human life.

DOI: 10.1201/9781003501183-43

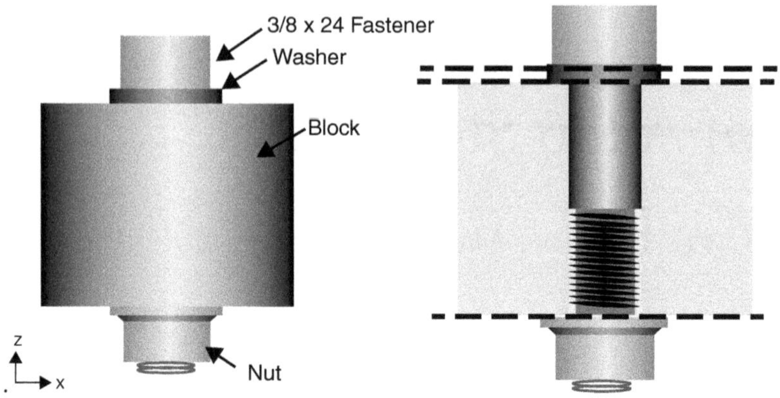

FIGURE 37.1 Schematic of the FEA model.

The level of axial mean stress can significantly affect the fatigue performance of a preloaded threaded fastener. On the one hand, a high mean stress level can accelerate the propagation of fatigue microcracks and shorten the fatigue life of a preloaded fastener. On the other hand, a low fastener mean stress is problematic as well; it could result in a localized joint separation under axial fatigue load. Localized joint separation would dramatically increase the amplitude of fatigue load on the bolt during the intermittent localized separation; that too would dramatically shorten bolt fatigue life.

This chapter generates mean stress-adjusted fatigue strength S' and fatigue life N' for preloaded A286 steel fasteners. Traditional $S-N$ data is normally generated from extensive fatigue testing of standard unnotched dog-bone specimens with zero mean stress, where the minimum-to-maximum stress ratio R is equal to -1. For preloaded threaded fasteners, however, both the effect of bolt preload as well as the notch would play a significant role in bolt fatigue performance. The $S-N$ curve of a smooth sample is modified for notch effect using two methods, Heywood [1] and Collins [2]. Based on Heywood's model, as shown in Figure 37.1, fatigue stress concentration factors K'_f and K_f are, respectively, used at 10^3 and 10^6 cycles. In the absence of experimental data, the 10^6 fatigue stress concentration factor K_f is 3 for rolled threads and 3.8 for cut threads [3] for SAE Grade 4 and above and ISO Class 8.8 and above. In this report, the theoretical stress concentration factor K_t is obtained from a three-dimensional finite element analysis (FEA) model using ABAQUS commercial software package. The FEA model and results are shown in Figures 37.1 and 37.2, respectively.

The FEA model uses A286 steel bolts with thread root equal to 0.144p (MIL-B-7838 for rolled threads), size 3/8″-24; the fastener material is assumed to have Young's modulus of 2.9×10^7 psi, Poisson's ratio of 0.3, and a density of 7.8 g/cm^3. Joint material is also assumed to have the same linear properties as the bolt material.

Table 37.1 and Figures 37.2 and 37.3 illustrate the procedure for calculating the fatigue stress concentration factor K_f at each engaged thread location for two standard values of thread fillet radius, namely 0.144*thread pitch and 0.18*thread pitch p. As anticipated, the maximum fatigue stress concentration factor is $K_f = 3.85$.

Reference [4] gives the value of K'_f (for 10^3 cycles) s as follows:

$$K'_f = \left(K_f - 1\right) q'_{1000} + 1 \tag{37.1}$$

where q'_{1000} is the notch sensitivity factor at 10^3 cycles that depends on the ultimate strength S_u [4]. The 10^3 cycle fatigue strength $S_{1000}\big|_{\text{notched}}$ would be given by

$$S_{1000}\big|_{\text{notched}} = \frac{S_{1000}}{K'_f} \tag{37.2}$$

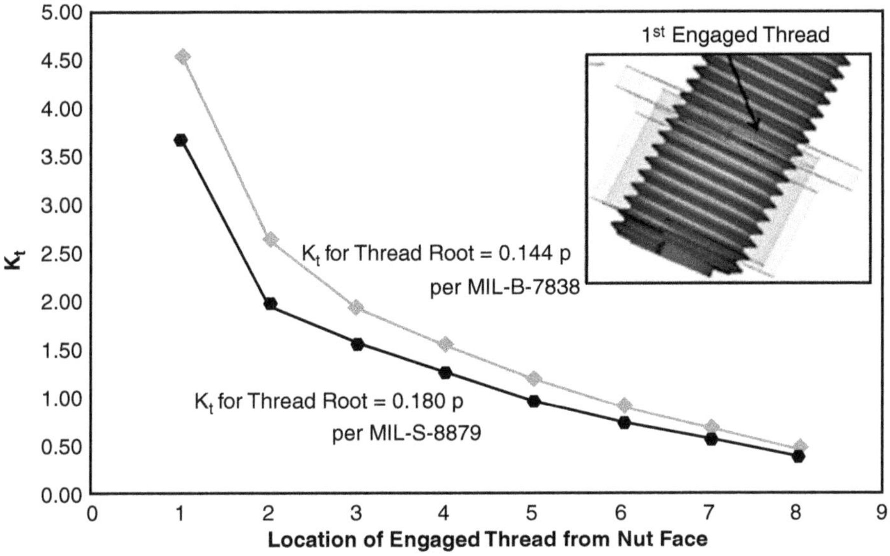

FIGURE 37.2 FEA stress concentration factor K_t vs. Thread Location.

TABLE 37.1
Calculation of Fatigue Stress Concentration Factor K_f

$K_f = 1 + (K_t - 1)*q = 1 + (K_t - 1)*0.8$
$(q = 0.8)$

<table>
<tr><th>Configuration</th><th>Thread Root Radius 0.144* P = 0.1523 mm Underhead Fillet Radius = 0.1 mm</th><th>Thread Root Radius 0.144* P = 0.1523mm Underhead Fillet Radius = 0.15 mm</th><th>Thread Root Radius 0.180* P = 0.1904 mm Underhead Fillet Radius = 0.1 mm</th><th>Thread Root Radius 0.180* P = 0.1904 mm Underhead Fillet Radius = 0.15 mm</th></tr>
<tr><td>First engaged thread</td><td colspan="2">3.85</td><td colspan="2">3.17</td></tr>
<tr><td>Second engaged thread</td><td colspan="2">2.31</td><td colspan="2">1.79</td></tr>
<tr><td>Third engaged thread</td><td colspan="2">1.77</td><td colspan="2">1.46</td></tr>
<tr><td>Fourth engaged thread</td><td colspan="2">1.45</td><td colspan="2">1.22</td></tr>
<tr><td>Fifth engaged thread</td><td colspan="2">1.16</td><td colspan="2">0.98</td></tr>
<tr><td>Sixth engaged thread</td><td colspan="2">0.93</td><td colspan="2">0.80</td></tr>
<tr><td>Seventh engaged thread</td><td colspan="2">0.75</td><td colspan="2">0.66</td></tr>
<tr><td>Eighth engaged thread</td><td colspan="2">0.58</td><td colspan="2">0.52</td></tr>
</table>

Theoretical stress concentration K_f = Actual FEA stress at thread root /Average tensile stress from bolt preload.
Fatigue stress concentration $K_f = 1 + (K_t - 1)*q = 1 + (K_t - 1)*0.8$ (note: used notch sensitivity factor $q = 0.8$).

where S_{1000} is the fatigue strength at 10^3 for an un-notched specimen. Heywood also gives the endurance limit S_e for a threaded fastener with zero mean stress as follows:

$$S_e\Big|_{\text{notched}} = \frac{0.54 S_u C_L C_s C_R C_D C_T}{K_f} \tag{37.3}$$

where S_u is the ultimate tensile strength, factors C_L, C_S, C_R, C_D, and C_T, are respectively the load, surface finish, reliability, specimen size, and temperature factors [3]. Collins [2] recommends the use of a one-line segment connecting the fatigue limit at 10^3 cycles and the fatigue strength

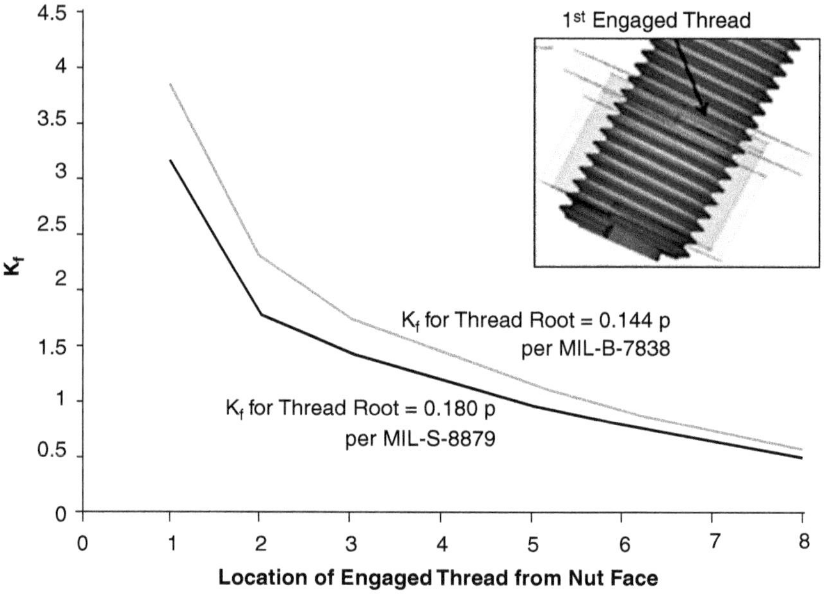

FIGURE 37.3 Fatigue stress concentration factor K_f vs. Thread Location.

coefficient S_f' at one cycle as illustrated by Figure 37.2. Collins' model simply removes some of the conservatism in the low-cycle fatigue (LCF) region as compared to Heywood's three-segment $S-N$ curve in Figure 37.1.

High tensile stress is the main factor in opening material voids and microcracks which accelerates the rate of fatigue crack propagation. Accordingly, a reliable method must be used for generating mean stress-adjusted fatigue curves to use for fatigue analysis of preloaded bolts.

Early empirical models were proposed by Gerber [5], Goodman [6], Haigh [7], and Soderberg [8] to account for tensile normal mean stress effects on high-cycle fatigue strength. These models can be graphically represented by constant life diagrams of alternating stress S_a versus mean stress S_m. The constant life diagram can be determined from $S-N$ curves generated at a specific combination of S_a and S_m.

Gerber [5] proposed a parabolic representation of Wohler's fatigue data on a plot of maximum stress as a ratio of ultimate strength S_{max} / S_u versus minimum stress as a ratio of ultimate stress S_{min} / S_u. Gerber's parabolic decrease of the fatigue strength S' by increasing the mean stress S_m is given by

$$S' = S\left(1-\left(\frac{S_m}{S_u}\right)^2\right) \tag{37.4}$$

Gerber method works well for ductile materials. Goodman [6] assumes a linear decrease of the mean stress-adjusted fatigue strength S' with increasing the mean stress S_m as follows:

$$S' = S\left(1-\frac{S_m}{S_u}\right) \tag{37.5}$$

Haigh [7] first represented fatigue test data for brasses on a diagram of alternating stress S_a versus mean stress S_m. Haigh's plot of Gerber [5] or Goodman [6] or mean stress connection is constructed by connecting the fatigue stress with zero mean stress S at a desired life and the ultimate strength S_u. Soderberg [8] suggested using the yield strength S_y to account for mean stress effect as follows:

$$S' = S\left(1 - \frac{S_m}{S_y}\right) \tag{37.6}$$

Wilson and Haigh [9] introduced the line of constant yield strength as an additional constraint for ductile materials on the safe design stress region, named as safe design region. Instead of using the static ultimate strength S_u by Gerber [1] Goodman [2], Morrow [10] uses the fatigue strength coefficient S'_f which is a fatigue property as follows:

$$S' = S\left(1 - \frac{S_m}{S'_f}\right) \tag{37.7}$$

An empirical model was introduced for relatively large mean stress, based on the concept of the mean stress sensitivity factor M as follows:

$$S' = S\left(\frac{M^* + 1}{M + 1} - \frac{M^* S_m}{S}\right) \tag{37.8}$$

For low mean stress$(-1 \le R < 0)$, M^* is equal to M. For high mean stress level$(0 \le R \le 1)$, M^* is equal to $M/3$. M is defined as [11]

$$M = \frac{S_{e,R=-1} - S_{e,R=0}}{S_{e,R=0}} \tag{37.9}$$

where $S_{e,R=-1}$, and $S_{e,R=0}$ are the endurance limit for $R = -1$ and $R = 0$ respectively. The mean stress sensitivity factor can be calculated based on the value of the ultimate tensile strength [12]. This model is widely used in spot welding, where the specimens are subjected to $R = 0$ loading to prevent any compression that leads to buckling of the metal.

In the finite life region as shown in Figure 37.4, Henry assumed that the virgin $S - N$ curve could be represented by the following equation of an equilateral hyperbola

$$N = \frac{k_o}{S - S_e} \tag{37.10}$$

where k_o is a material constant, S is the fatigue strength with zero mean stress, and S_e is the original endurance limit. Henry used the fatigue test data by Bennett [19] to compare it with his damage theory. Fatigue test data showed that the proposed Henry method [20] was a significant overestimation of the partial damage (Figure 37.5). Henry concluded that the significant difference between

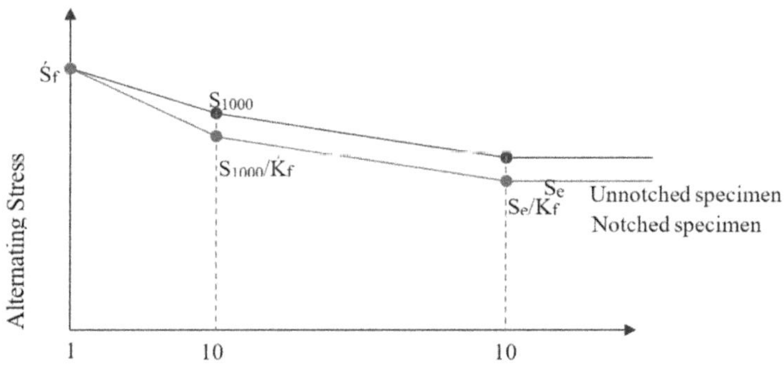

FIGURE 37.4 Heywood model for the effect of notch.

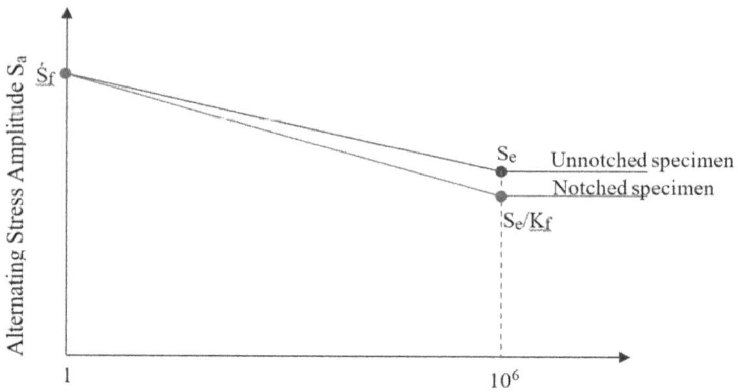

FIGURE 37.5 Collins model for notch effect.

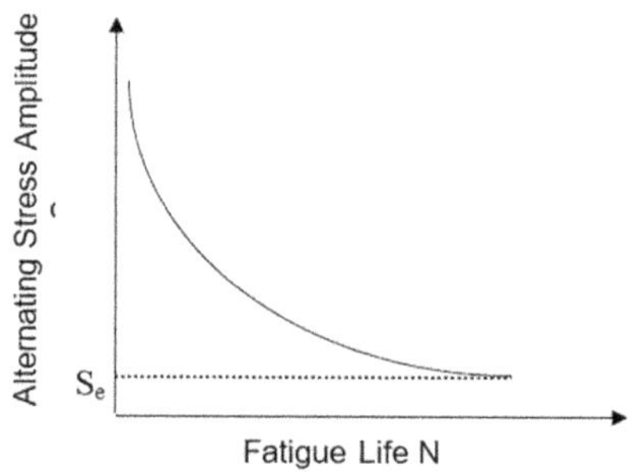

FIGURE 37.6 $S-N$ curve approximation proposed by Henry [20].

his model and the experimental data was due to the inaccurate representation of the $S-N$ as an equilateral hyperbola (Figure 37.6). He instead recommended the use of the actual logarithmic approximation given by

$$S = aN^b \tag{37.11}$$

37.2 MEAN STRESS-ADJUSTED FATIGUE CURVES

37.2.1 Experimental Data

Experimental fatigue data is generated using a standard 810 MTS fatigue testing system (Figure 37.7). Two sizes of A286 steel bolts with rolled threads are used; namely, #10-32 and 3/8″-24 with actual ultimate tensile strength of 194 ksi and 191 ksi, respectively. Test data is generated for the various combinations of mean stress and cyclic amplitudes shown in Table 37.2, and compared to the various theoretical predictions for mean stress effect according to Eqs. (37.4)–(37.9).

Two sets of fatigue data are generated. In the first set, test samples are cycled to failure at eleven different combination levels of mean and alternating stresses which are shown in Table 37.2. Sample size of 3 is used for each of the 11 load combinations. Test data is shown on the respective mean stress-adjusted fatigue $S'-N'$ curves as theoretically predicted by Gerber [5], Goodman [6], and Morrow [10] as shown in Figures 37.8–37.18.

In the second testing plan, the $S'-N'$ curves of the preloaded threaded fasteners are experimentally generated according to the ASTM E739-91 standard [22] at four levels of bolt mean stress

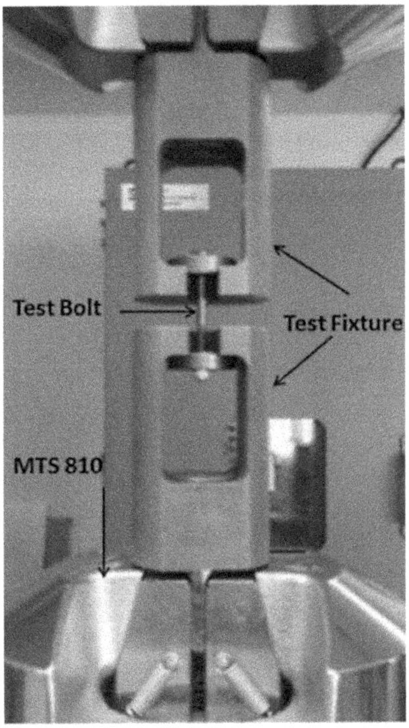

FIGURE 37.7 MTS 810 fatigue testing system and test fixture.

TABLE 37.2
Fatigue Test Plan

	Loading Combination ID										
	1	2	3	4	5	6	7	8	9	10	11
Cyclic stress amplitude S_a/S_u	0.09	0.075	0.06	0.175	0.13	0.08	0.24	0.15	0.09	0.125	0.175
Mean stress S_m/S_u	0.81	0.675	0.54	0.68	0.52	0.32	0.56	0.35	0.21	0.125	0.175

and then compared to the theoretically-generated $S' - N'$ curves provided by [5], [6], and [10]. Each experimentally generated $S' - N'$ curve is generated from testing nine bolt samples at three different stress levels with the same mean stress (Figures 37.19–37.22).

Based on the results from the two testing plans of A286 steel bolts, it is observed that for most load cases, the predictions of mean stress effect on bolt fatigue strength by Goodman [6] and Morrow [10] are both closer to the experimentally generated data $S' - N'$, as compared to the Gerber prediction [5].

37.3 FAILURE ANALYSIS

Fatigue failure will only occur if the following essential conditions are present: cyclic tensile loads, stress levels above a threshold value, fatigue-susceptible material, and a flaw in the material. If these conditions are present, then three events can occur leading to final fatigue rupture; Crack initiation, Crack growth (beach marks), and Crack propagation leading to final [brittle] fracture.

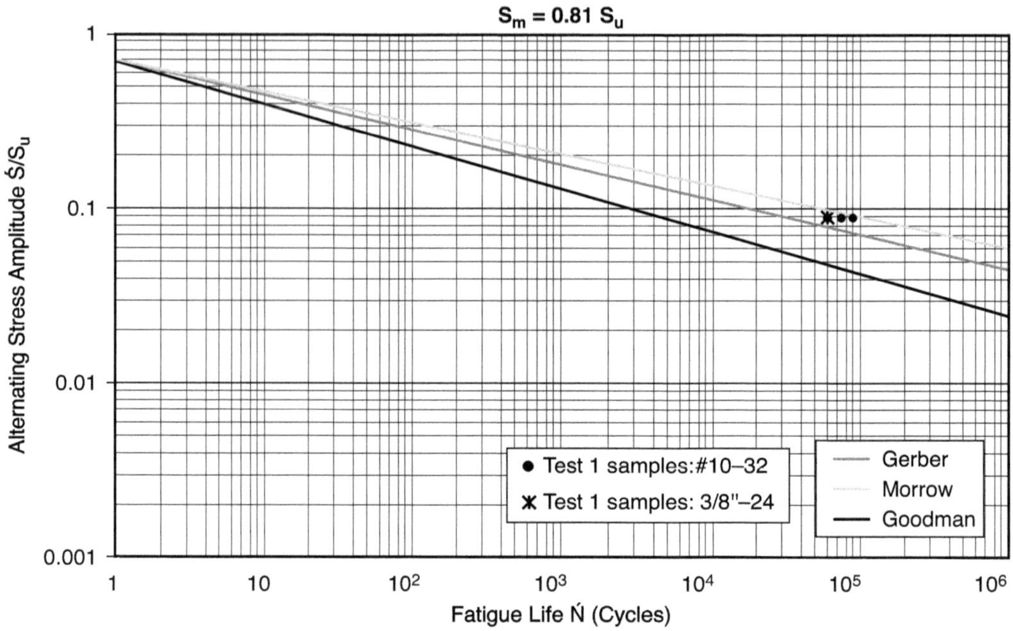

FIGURE 37.8 Various predictions of $S' - N'$ curves: $S_m = 0.81S_u$.

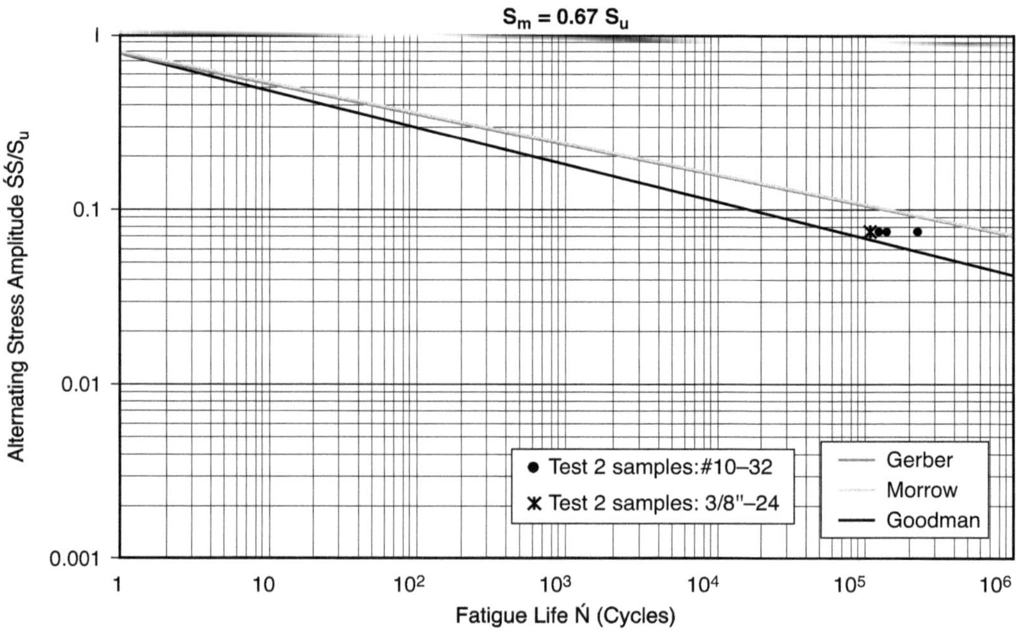

FIGURE 37.9 Various predictions of $S' - N'$ curves: $S_m = 0.67S_u$.

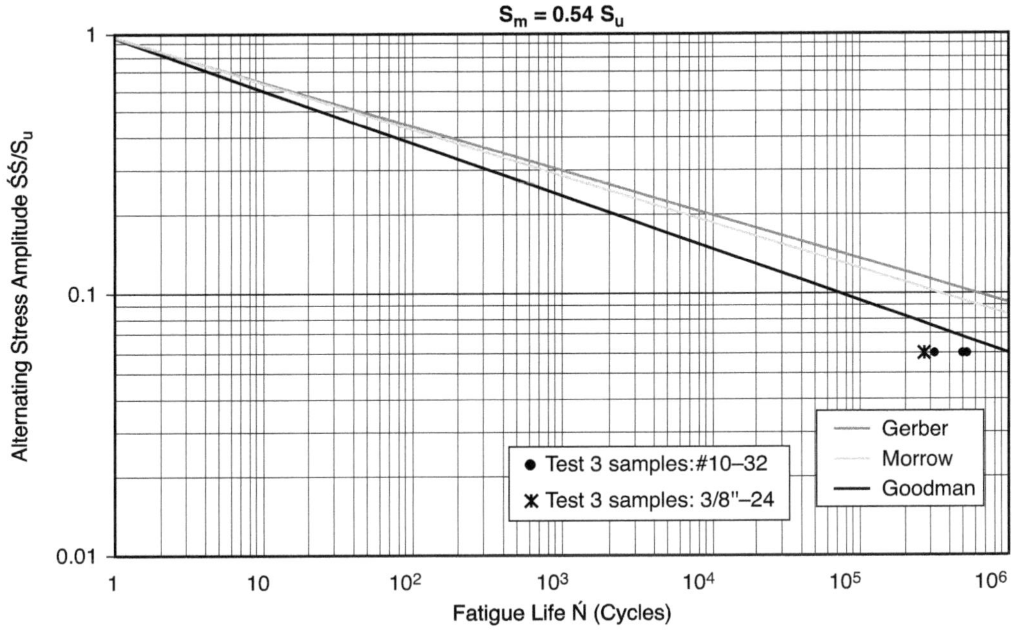

FIGURE 37.10 Various predictions of $S' - N'$ curves: $S_m = 0.54S_u$.

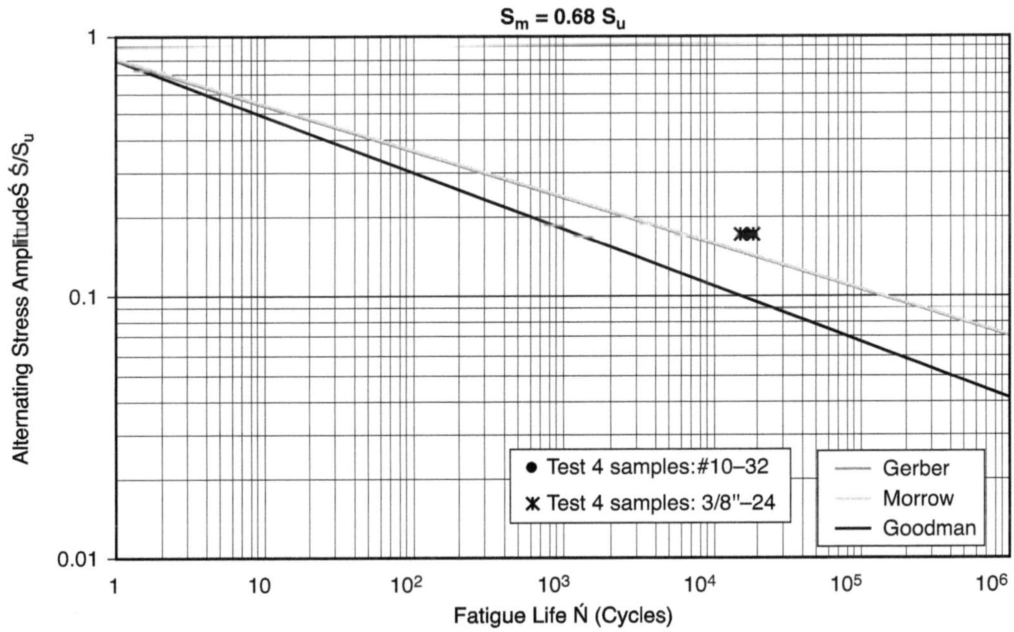

FIGURE 37.11 Various predictions of $S' - N'$ curves: $S_m = 0.68S_u$.

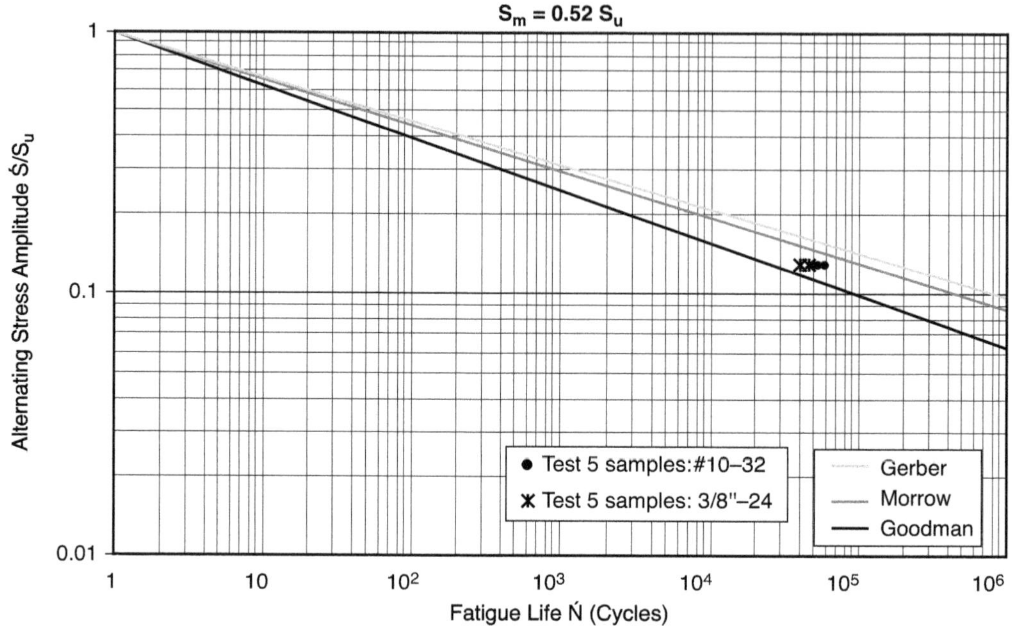

FIGURE 37.12 Various predictions of $S' - N'$ curves: $S_m = 0.52S_u$.

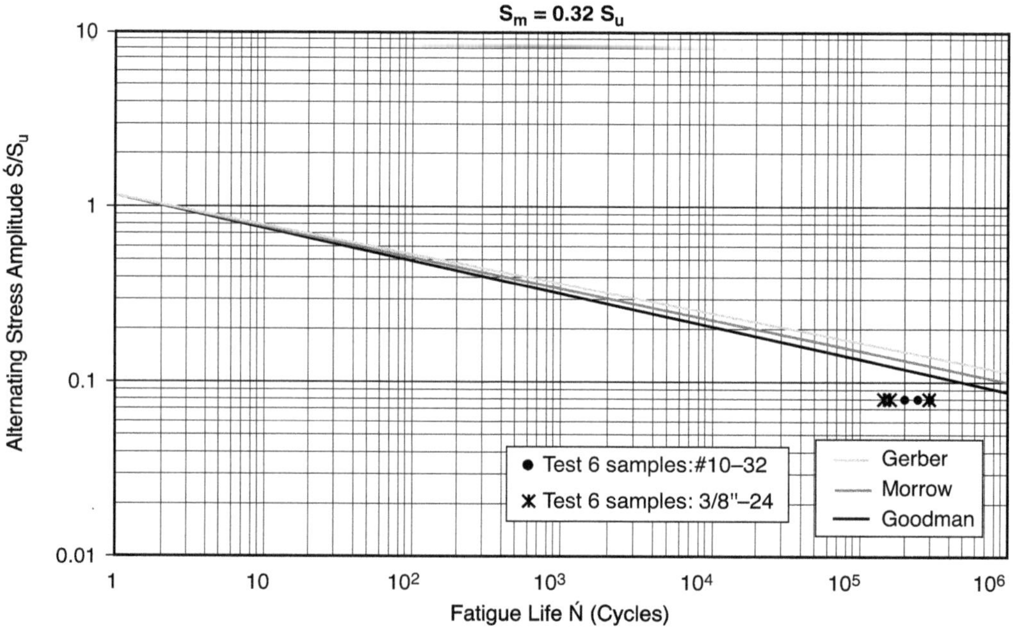

FIGURE 37.13 Various predictions of $S' - N'$ curves: $S_m = 0.32S_u$.

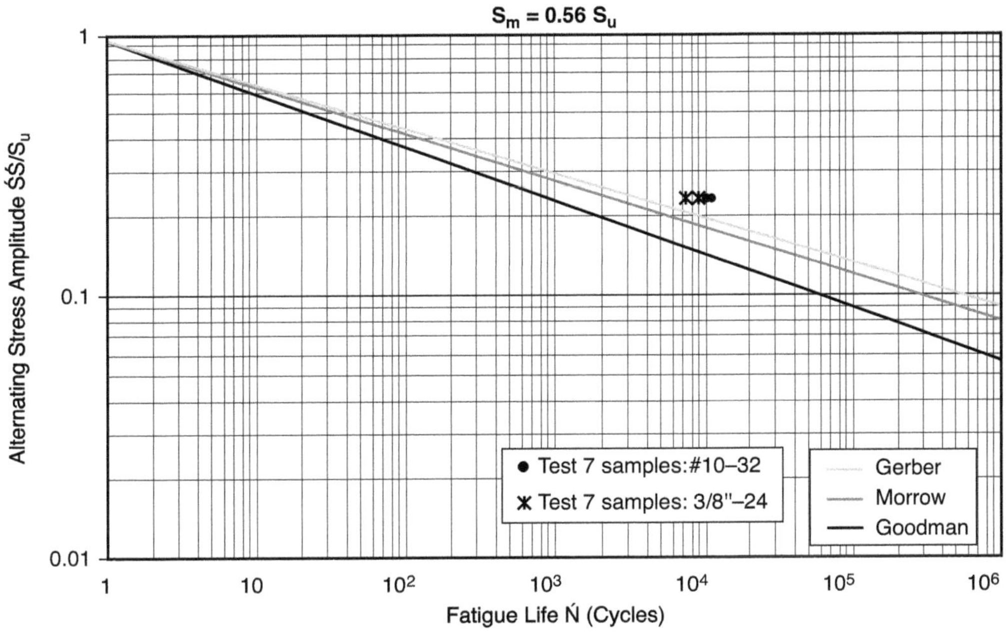

FIGURE 37.14 Various predictions of $S' - N'$ curves: $S_m = 0.56S_u$.

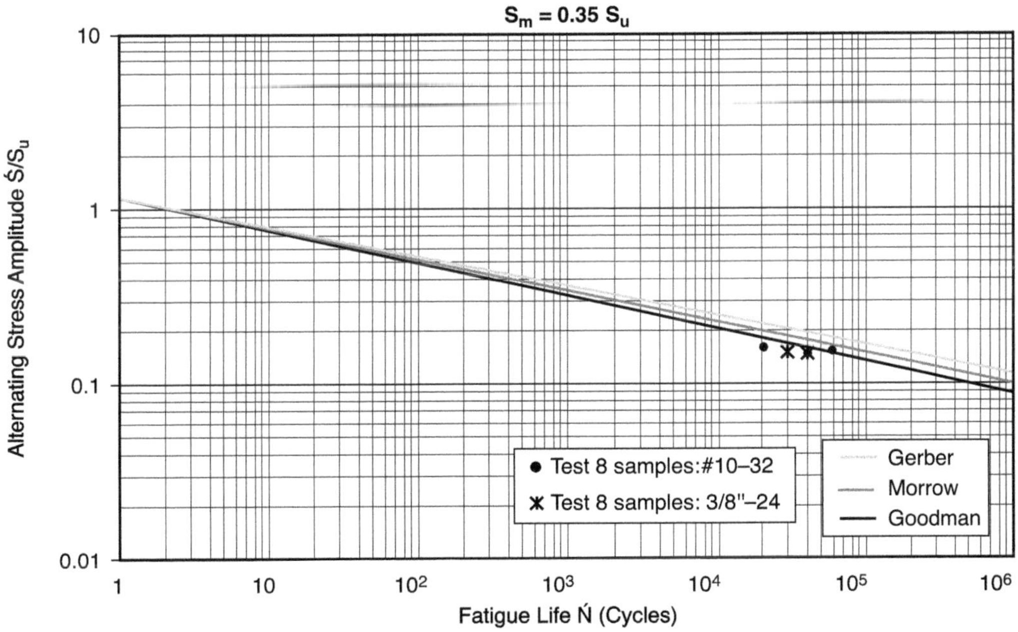

FIGURE 37.15 Various predictions of $S' - N'$ curves: $S_m = 0.35S_u$.

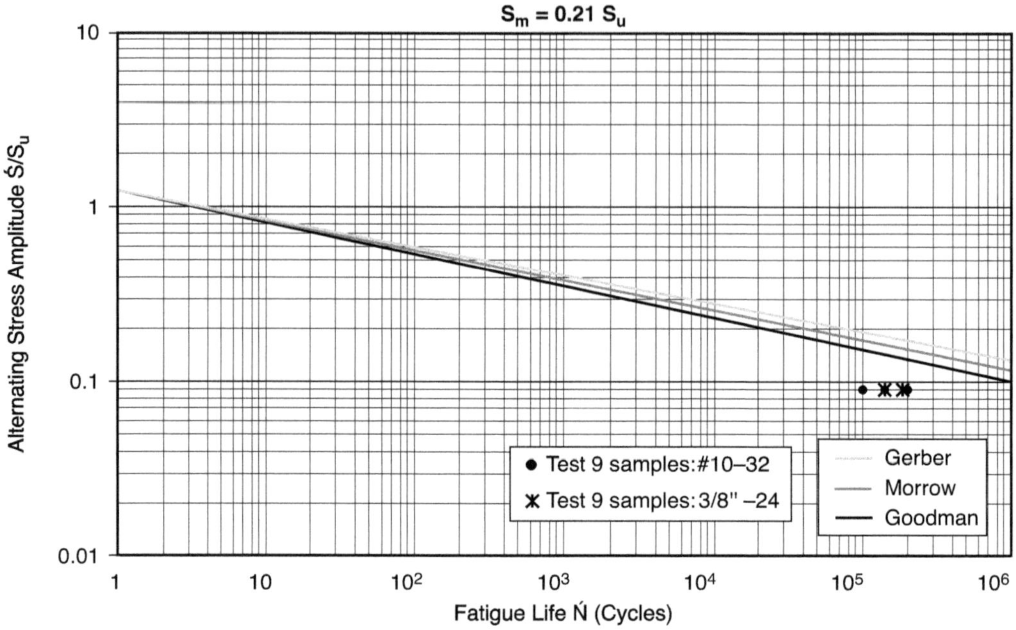

FIGURE 37.16 Various predictions of $S' - N'$ curves: $S_m = 0.21S_u$.

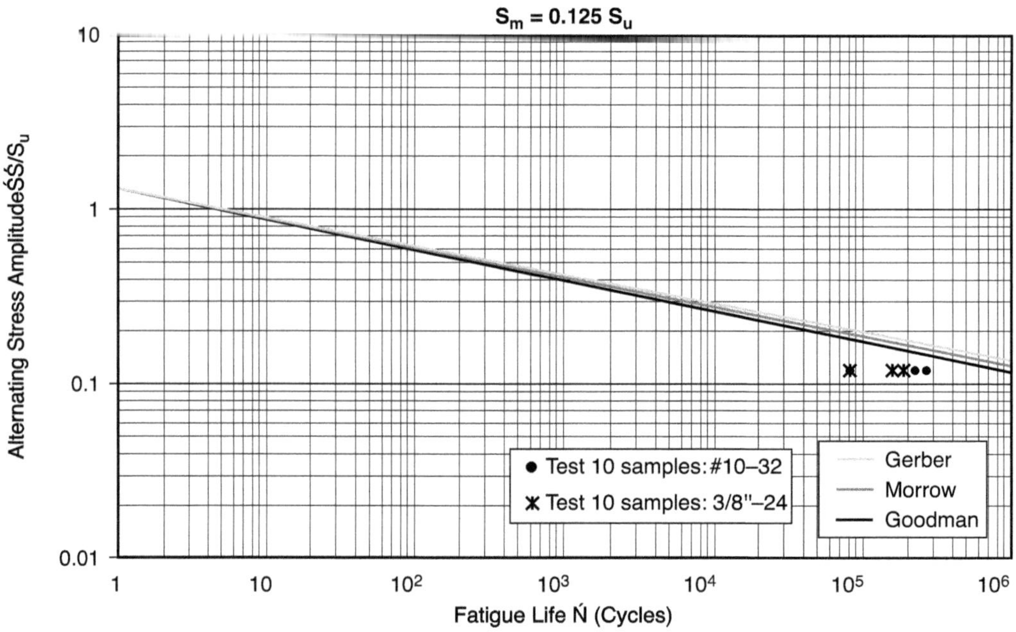

FIGURE 37.17 Various predictions of $S' - N'$ curves: $S_m = 0.125S_u$.

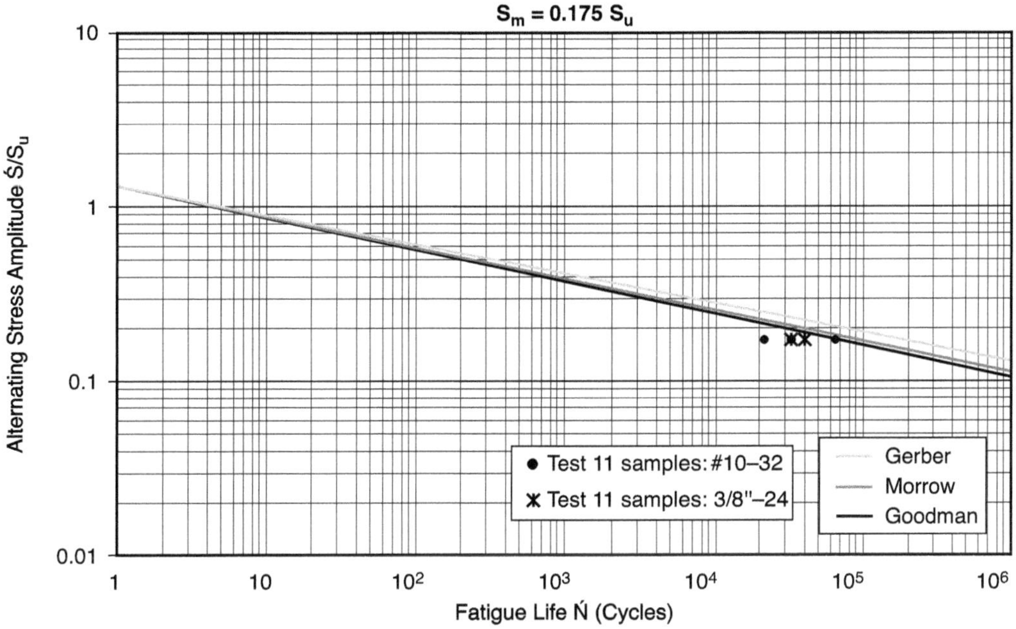

FIGURE 37.18 Various predictions of $S' - N'$ curves: $S_m = 0.175S_u$.

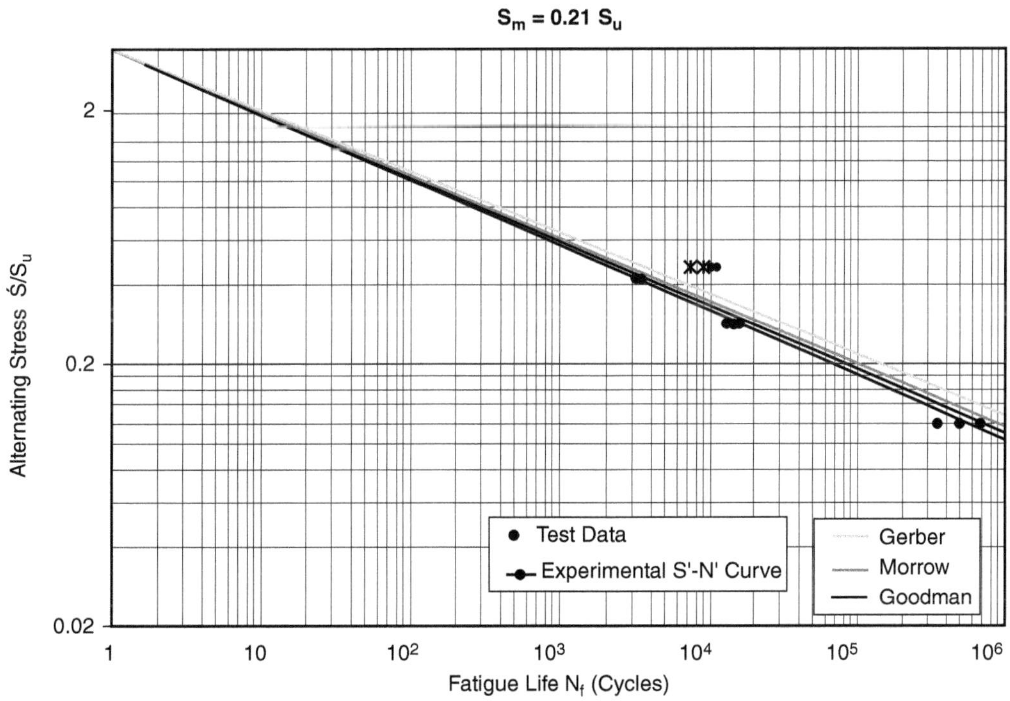

FIGURE 37.19 Comparison of experimental $S' - N'$ Curve with theoretical models: $S_m = 0.21S_u$.

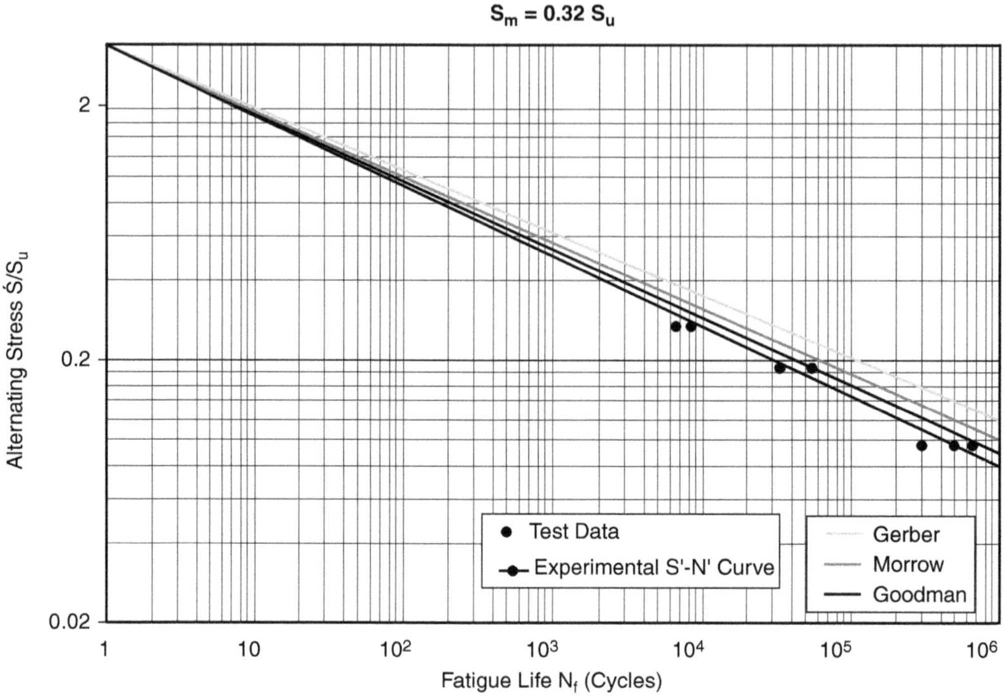

FIGURE 37.20 Comparison of experimental $S' - N'$Curve with theoretical models: $S_m = 0.32S_u$.

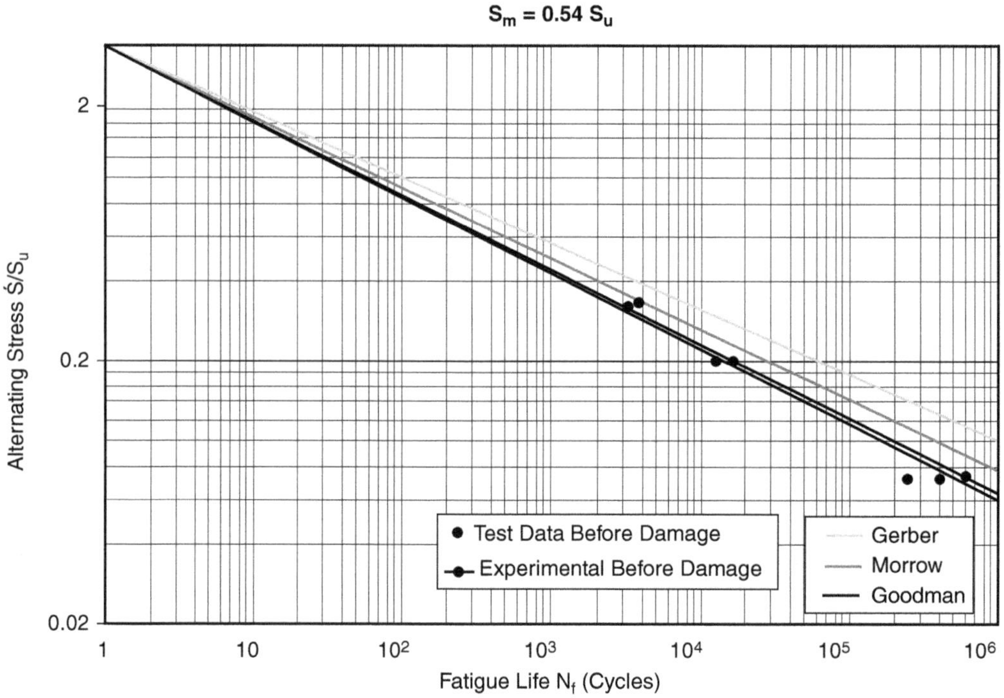

FIGURE 37.21 Comparison of experimental $S' - N'$ Curve with theoretical models: $S_m = 0.54S_u$.

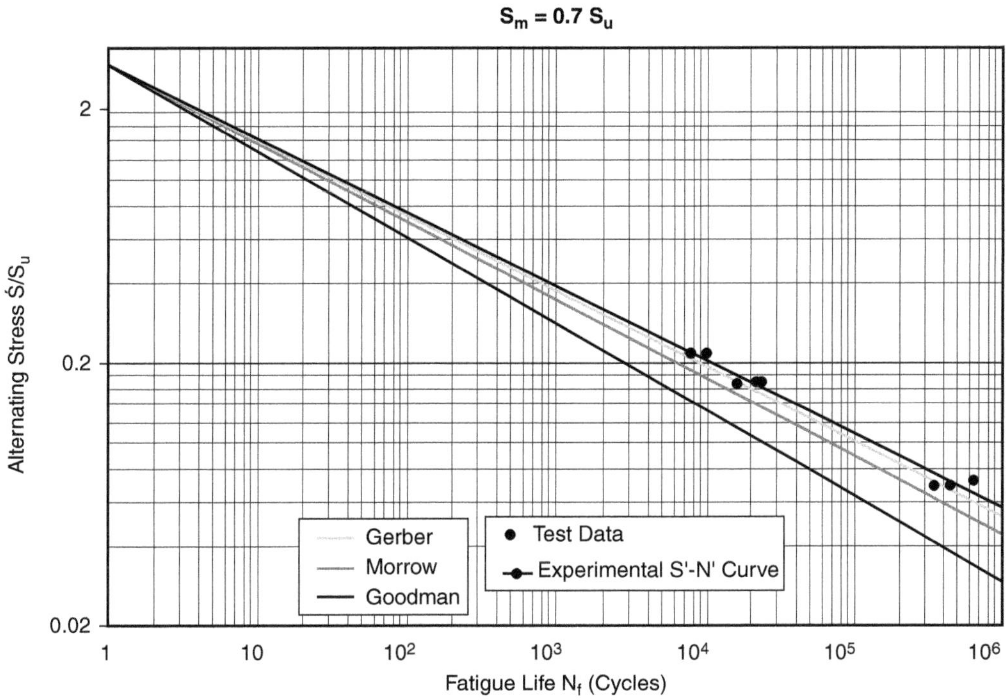

FIGURE 37.22 Comparison of experimental $S' - N'$ Curve with theoretical models.

FIGURE 37.23 Potential locations of fatigue cracks.

Fatigue cracks will originate on a free surface at a point of high stress concentration in the material. This may often be due to a pre-existing flaw in the material or perhaps a human-made discontinuity, such as the root of a thread, a tool mark, corrosion, or any point at which there is a sharp change in size or shape of the material.

As the part is subjected to cyclic tensile loading, the material around the stress concentrations undergoes a slip mechanism. The crack will grow along the slip plane, but will eventually change direction until it is growing in a plane perpendicular to the principle tensile stress. The crack grows more rapidly as stress levels increase. The process of crack growth is repeated until the crack reaches some critical length, at which time a sudden failure occurs, in either a ductile or brittle manner, depending on the characteristics of the material.

A bolt and nut loaded axially has three possible fatigue crack initiation locations as shown in Figure 37.23. The most common crack initiation location is at the first engaged thread (location a), then at the head-shank transition (location b), and the least common initiation location is at the runout of the threads (location c).

37.3.1 Scanning Electron Microscopy (SEM) of Bolts Fracture Surface

Twelve out of one hundred thirty-two test samples were inspected and analyzed using Hitachi S-4500 Field Emission Scanning Electron Microscope (SEM). For A286 3/8″-24 rolled threads bolts, three samples have failed at the same testing conditions given in Table 37.3. Visual inspection showed the test bolts were fractured in a classic cup and cone ductile failure as shown in Figure 37.24. It was concluded that the bolt had gone through ductile overload fracture beyond its yield limit because of the high level of max stress ($S_{max} = 0.8S_u$) which is very close to the yield strength $S_y = 0.84S_u$.

For A286 #10-24 rolled thread bolts, using visual inspection, it can be seen that nine bolts failed with the same fracture surface. Small "cup-like" features appear on the fracture surface, and the shape of the fracture face, as viewed from the side, is not flat and is curved as shown in Figure 37.25. Using SEM, many ratchet marks, carbon element of size 60 μm, dimple fracture appearance, and no overload zone was found under the bolt head as shown in Figures 37.26 and 37.27.

From the visual inspection and the SEM, it can be concluded that the curved side view of the fracture face indicates that the stress concentration played an important role in causing the failure. The presence of many ratchet marks and a very small overload zone indicates that the load was light, but there were high stress concentrations. The large carbon element (60μm) acts as a stress

TABLE 37.3
Fatigue Loading Conditions

	Loading Combination (Fraction of Ultimate Strength S_u)			
Stress Ratio *R*	Mean Load	Cyclic Amplitude	Max. Load	Min. Load
0.4	0.56	0.24	0.8	0.32

FIGURE 37.24 Cup and cone ductile failure.

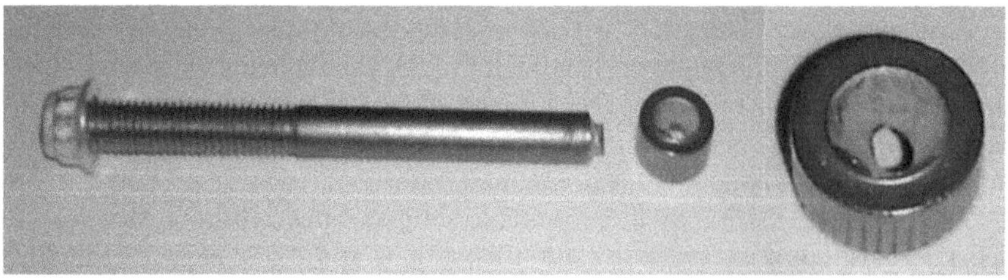

FIGURE 37.25 A286 #10-32 bolt: head fracture surface.

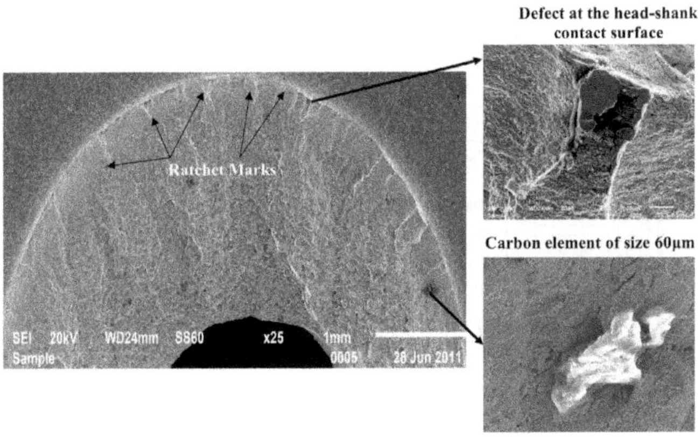

FIGURE 37.26 Ratchet marks and inclusion: A286 #10-32 under head fracture surface).

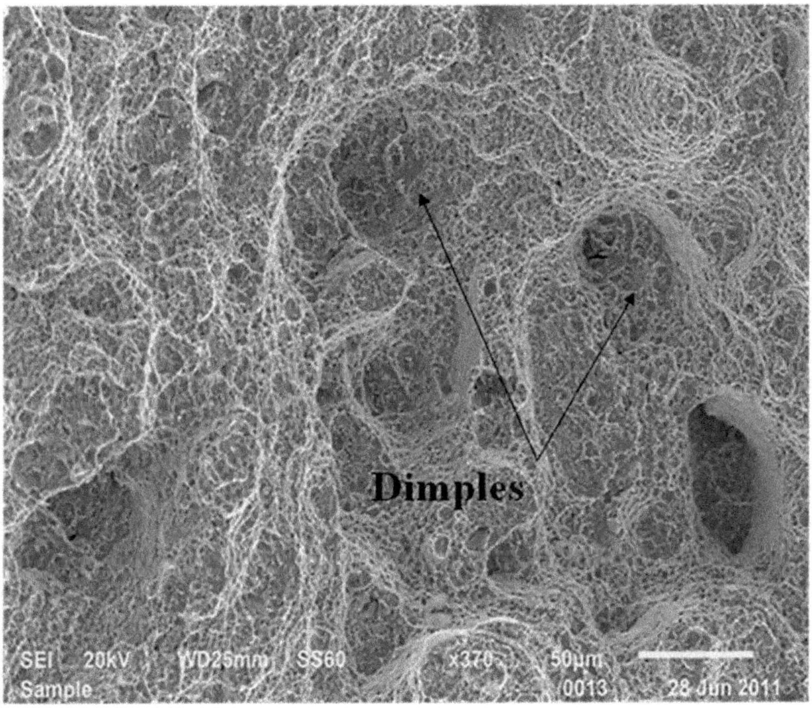

FIGURE 37.27 Tensile overload region (A286 #10-32 under head bolt surface).

riser and tends to increase the stress concentration. Finally, the under-head stress concentration was higher than it should have been due to the presence of large carbon element.

For both 3/8″-24 and #10-32 bolts with rolled threads, the underheard stress concentration was higher than that at the threads. Upon further inspection of socket head bolts, it was found that the depth *T* (Figure 37.28) was greater than it should be per applicable manufacturing standards. That is likely to be the reason for fatigue failure at that location. Table 37.4 shows a comparison between the manufacturing requirement standard for the depths *T* and *G* and versus their actual [measured] values.

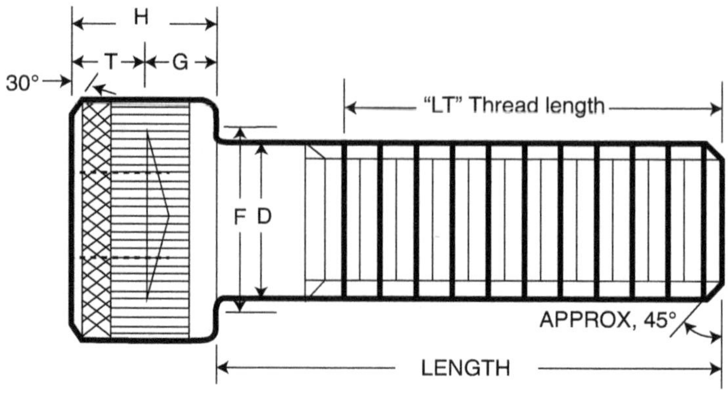

FIGURE 37.28 Standard design of A286 socket head bolts.

TABLE 37.4
Standard and Actual Measured Values for Socket Head Bolt Depth

Standard		Actual	
T	*G* (min)	*T*	*G*
0.182	0.143	0.25	0.1

37.4 CONCLUSIONS

In this chapter, an experimental procedure is developed for evaluating and comparing the fatigue performance of preloaded A286 steel bolts with published generic methods for accounting for the effect mean stress [5, 6, 10, 11-19, 21-24]. Generated fatigue data of preloaded A286 steel bolts correlates reasonably well with the predictions made by both Goodman [6] and Morrow [10] for the mean stress effect on fatigue strength. However, the Gerber prediction [5] of the effect of mean stress on the fatigue strength of A286 steel bolts [5] did not correlate well with the experimental data. Failure mode analysis of test bolts showed the potential locations of fatigue crack initiation and final fracture; namely, those localized geometric stress risers and/or material impurities. Thread roots, head-to-shank fillets, sizable material voids, dislocations, and/or alloy material additives in carbon steel are prime locations for fatigue fracture.

ACKNOWLEDGMENT

NASA Engineering and Safety Center (NESC) and the Threaded Fastener System Standard team (TFSS) for their support and helpful discussions of this work, which was part of an MS Thesis [24] by Ms. Rania A. Mansour at Oakland University; Saravanan Ganeshmurthy for generating the FEA data necessary for calculating the fatigue stress concentration at thread roots.

REFERENCES

[1] Heywood, R. B., 1962, *Designing Against Failure*, Chapman & Hall, London.
[2] Collins, J. A., 1993, *Failure of Materials in Mechanical Design—Analysis, Prediction, and Prevention*, 2nd ed., Wiley, New York.

[3] Juvinall, R., and Marshek, M., 2000, *Fundamentals of Machine and Component Design*, 3rd ed., John Wiley & Sons, New York.
[4] Lee, Y., Pan, J., Hathaway, R., and Barkey, M., 2005, *Fatigue Testing and Analysis (Theory and Practice)*, Elsevier, Burlington, MA, pp. 103–176.
[5] Gerber, W. Z., 1874, "Calculation of the Allowable Stresses in Iron Structures," *Z. Bayer Archit Ing Ver*, 6 (6), pp. 101–110.
[6] Goodman, J., 1899, *Mechanics Applied to Engineering*, Longmans Green, London, pp. 438–442.
[7] Haigh, B. P., 1917, "Experiments on the Fatigue of Brasses," *Journal of the Institute of Metals*, 18, pp. 55–86.
[8] Soderberg, C. R., 1930, "Fatigue of Safety and Working Stress," *Transactions of the American Society of Mechanical Engineers*, 52(2), pp. 13–28.
[9] Wilson, J. S. and Haigh, B. P., 1923, *Stresses in Bridges*, Engineering, London, pp. 446–448.
[10] Morrow, J., 1968, *Fatigue Design Handbook, Advances in Engineering*, Society of Automotive Engineers, Warrendale, PA, Vol. 6, pp. 21–29.
[11] Radaj, D. and Sonsino, C. M., 1998, *Fatigue Assessment of Welded Joints by Local Approaches*, Abington Publishing, Cambridge.
[12] Schütz, W., 1968, "View Points of Material Selection for Fatigue Loaded Structures (in German)," Laboratorium für Betriebsfestigkeit LBF, Darmstadt, Bericht Nr. TB-80.
[13] Pålmgren, A., 1924, "Die Lebensdauer von Kugellagern, Zeitschrift des Vereinesdeutscher Ingenierure," *Vereines Deutscher Ingenieure*, 68(14), pp. 339–341.
[14] Miner, M. A., 1945, "Cumulative Damage in Fatigue," *Journal of Applied Mechanics*, 67, pp. A159–A164.
[15] Kramer, I. R., 1976, "Prediction of Fatigue Damage," In *Proceedings of the 2nd International Conference on the Mechanical Behavior of Materials*, Boston, MA, American Society for Metals, Metals Park, OH, pp. 812–816.
[16] Marin, J., 1962, *Mechanical Behavior of Engineering Materials*, Prentice-Hall, Englewood Cliffs, NJ, pp. 142–144.
[17] Hashine, Z. and Rotem, A., 1978, "A Cumulative Damage Theory of Fatigue Failure," *Material Science and Engineering*, 34, pp. 147–160.
[18] Kommers, J. B., 1943, "The Effect of Overstress in Fatigue on the Endurance Life of Steel," *Proceedings of American Society for Testing and Materials*, 43, pp. 749–762.
[19] Bennett, J. A., 1946, "A Study of the Damaging Effect of Fatigue Stressing on X4130 Steel," *Proceedings of American Society for Testing and Materials*, 46, pp. 693–714.
[20] Henry, D. L., 1955, "Theory of Fatigue Damage Accumulation in Steel," *ASME Transactions*, 77, pp. 913–918.
[21] Baumel, Jr A. and Seeger, T., 1990, "*Metal Data for Cyclic Loading*," Supplement I, Elsevier Science Publisher, Amsterdam.
[22] ASTM E739-91, 1998, "Standard Practice for Statistical Analysis of Linear or Linearized Stress-Life (S - N) and Strain-Life (ε - N) Fatigue Data", American Society for Testing and Materials. www.astm.org.
[23] Nassar, S.A. and Ganganala, M., 2011, "Effect of Separating Load Eccentricity on the Clamp Load Loss in a Bolted Joint Using a Strain Hardening Model", *Journal of Pressure Vessel Technology*, 133, p. 021206.
[24] Mansour, R.A., "Fatigue of Preloaded Threaded Fasteners", 2012, MS Thesis, Department of Mechanical Engineering, Oakland University, Rochester, MI.

38 An Improved Criterion for Cumulative Fatigue Damage of Preloaded Steel Bolts

Sayed A. Nassar and Rania Mansour

38.1 INTRODUCTION

Cumulative fatigue damage analysis is critical for structural components that are subjected to variable-amplitude cyclic loads. Most fatigue cumulative damage theories use the baseline fatigue $S-N$ curves that are generated from a fully reversed cyclic loading with a minimum-to-maximum stress ratio $R=-1$.

Pålmgren [1] developed the early hypothesis for cumulative fatigue damage, known today as linear cumulative damage. Later, Miner [2] was the first to present a formula for Pålmgren hypothesis of cumulative damage D as follows:

$$D=\sum_{j}\left(\frac{n_j}{N_j}\right)<1 \tag{38.1}$$

where n_j is the actual spectrum cycles at stress level S_j and N_j is the corresponding fatigue life. Miner's rule is widely used for its simplicity; however, the method does not account for the stress amplitudes that are below the endurance limit, and neither does it account for the loading sequence.

For a fully reversed cyclic loading with zero mean stress, Kramer [3] proposed a linear damage rule that accounts for the damage history as follows:

$$D=\frac{n_1S_1^p}{\beta}+\frac{n_2S_2^p}{\beta}\left(\frac{S_1}{S_2}\right)^{pf_1}+\frac{n_3S_3^p}{\beta}\left(\frac{S_2}{S_3}\right)^{pf_2}\left(\frac{S_1}{S_2}\right)^{pf_1f_2}+\ldots\ldots<1 \tag{38.2}$$

where $P=-1/b$, b is the slope of the logarithmic $S-N$ curve, $\beta=a^P$, a is the y-intercept of the logarithmic $S-N$ curve, $f_1=n_1S_1^P/\beta$, and $f_1=\left(n_1S_1^P/\beta\right)\left(S_1/S_2\right)^{Pf_1}$. Later, Marin [4] proposed a nonlinear cumulative damage criterion that accounts for both the cycle ratio n_j/N_j and the loading sequence as follows:

$$D=\sum_{j}\frac{n_j}{N_j}\left(\frac{S_j}{S_j|_{\max}}\right)^q<1 \tag{38.3}$$

where $S_j|_{\max}$ is the maximum alternating stress amplitude in the loading history and q is an exponent given by

$$q=d+\frac{1}{b} \tag{38.4}$$

DOI: 10.1201/9781003501183-44

where $6.2 \leq d \leq 6.9$ for steel [4]. Hashine and Rotem [5] proposed a cumulative damage criterion given by

$$D = \left[\left(\frac{n_1}{N_1} \right)^{\log S_2 / \log S_1} + \cdots + \left(\frac{n_{j-2}}{N_{j-2}} \right)^{\log S_{j-1} / \log S_{j-2}} + \frac{n_{j-1}}{N_{j-1}} \right]^{\log S_j / \log S_{j-1}} + \frac{n_j}{N_j} < 1 \tag{38.5}$$

The concept of incremental lowering of the $S - N$ curve due to partial damage at each stress level was successfully introduced by Kommers [6] and Bennett [7], using a two-level step loading method. Henry [8] proposed a theoretical method for predicting the reduced endurance limit S_e^* of steel specimens that are subjected to a specific number of cycles at each stress level. He defined fatigue damage as the reduction in the endurance limit relative to that of the virgin material specimen. In the finite life region, Henry [8] assumed that the virgin $S - N$ curve could be represented by the following equation of an equilateral hyperbola

$$N = \frac{k_\circ}{S - S_e} \tag{38.6}$$

where $k_\circ$ is a material constant, S is the fatigue strength with zero mean stress, S_e and is the original endurance limit. Comparison with fatigue test data showed that the proposed Henry method [8] significantly overestimated partial damage. Henry concluded that the difference was due to the inaccurate representation of the $S - N$ curve as an equilateral hyperbola shown in Eq. (38.6). Instead, he recommended the commonly used logarithmic approximation given by

$$S = aN^b \tag{38.7}$$

This chapter uses Eq. (38.7) to develop an improved cumulative damage criterion that is experimentally validated by testing preloaded A286 steel fasteners. Damage analysis takes into account the non-zero mean stress condition ($R \neq -1$) by using mean stress-adjusted fatigue curves ($S' - N'$). Further, this paper contributes a way of constructing the whole $S' - N'$ curve at each damage level.

The proposed cumulative damage method necessitates the generation and utilization of fatigue curves at various levels of the mean stress (i.e., the mean stress-adjusted $S' - N'$ curves). Theoretically, those curves may be generated using various methods such as Goodman [9], Gerber [10], or Morrow [11]. Goodman [9] assumed a linear decrease of the mean stress-adjusted fatigue strength S' with increasing the mean stress S_m as follows:

$$S' = S\left(1 - \frac{S_m}{S_u}\right) \tag{38.8}$$

where S_u is the static ultimate strength. The Goodman method is more suitable for brittle materials and is too conservative for ductile materials. Gerber [10] revised the linear Goodman method to parabolic decrease of the fatigue strength S' with increasing the mean stress S_m as follows:

$$S' = S\left(1 - \left(\frac{S_m}{S_u}\right)^2\right) \tag{38.9}$$

The Gerber method works well for ductile materials. Instead of using the static ultimate strength S_u by Goodman [9] and Gerber [10], Morrow [11] uses the fatigue strength coefficient S_f' which is considered as fatigue properties as follows:

$$S' = S\left(1 - \frac{S_m}{S_f'}\right) \tag{38.10}$$

38.2 AN IMPROVED CUMULATIVE DAMAGE MODEL

An improved cumulative damage criterion is proposed for preloaded threaded fasteners that are subjected to loading spectra of multi-level stress amplitudes arising from external axial cyclic loads applied to a clamped joint. The proposed criterion [13] and [14] revises Henry method [8] by using the logarithmic relationship between the mean stress-adjusted fatigue strength S' and the corresponding fatigue life N' as follows:

$$S' = a(N')^{b} \tag{38.11}$$

From Eq. (38.11), the fatigue life N' is given by

$$N' = \left(\frac{S'}{a}\right)^{1/b} \tag{38.12}$$

Assuming that all $S' - N'$ curves before and after damage will have the same y-intercept a; the corresponding partial damage caused by n_j cycles at an alternating stress level gives a lower fatigue line for the partially damaged specimen that is represented by an equation similar to Eq. (38.11) as follows:

$$S' = a(N_r')^{b^*} \tag{38.13}$$

where $N_r' = N' - n$ is the remaining fatigue life after partial damage D, and b^* is the slope of the new $S' - N'$ curve after partial damage. Figure 38.1 illustrates fatigue lines for undamaged and partially damaged specimens.

Substituting the definition of N_r' into Eq. (38.13) and using Eq. (38.12) yield the following relationship for a partially damaged specimen

$$S' = a\left(\left(\frac{S'}{a}\right)^{1/b} - n\right)^{b^*} \tag{38.14}$$

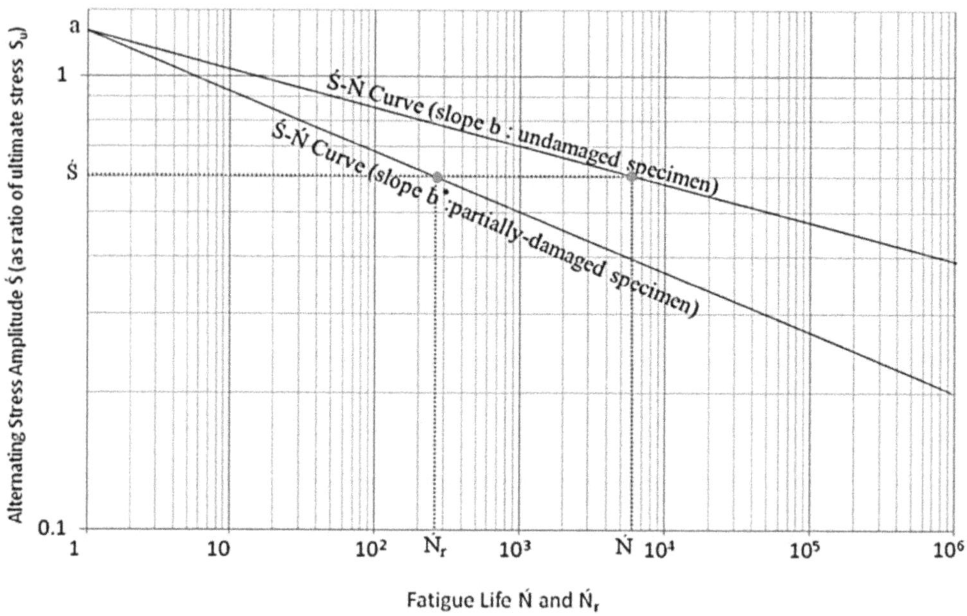

FIGURE 38.1 Schematic fatigue curves for undamaged and partially-damaged specimens.

Equation (38.14) gives the slope b^* of the damaged specimen fatigue line as follows:

$$b^* = \frac{\log\left(S'/a\right)}{\log\left(\left(\frac{S'}{a}\right)^{1/b} - n\right)} \tag{38.15}$$

After partial damage, the reduced endurance limit S_e^* at 10^6 cycles is given by

$$S_e^* = a\left(10^6\right)^{b^*} \tag{38.16}$$

Similar to the Henry procedure [8], the corresponding partial fatigue damage D is given by

$$D = \frac{S_e - S_e^*}{S_e} = \left(1 - \frac{S_e^*}{S_e}\right) \tag{38.17}$$

Likewise, the proposed cumulative damage procedure would still account for the loading sequence in the same manner as the Henry method [8] did.

38.3 EXPERIMENTAL VALIDATION

Experimental validation of the novel cumulative damage criterion is performed using a standard MTS 810 fatigue testing system (Figure 38.2), following the ASTM standards {12]. Data is generated for two sizes of A286 steel bolts with rolled threads.

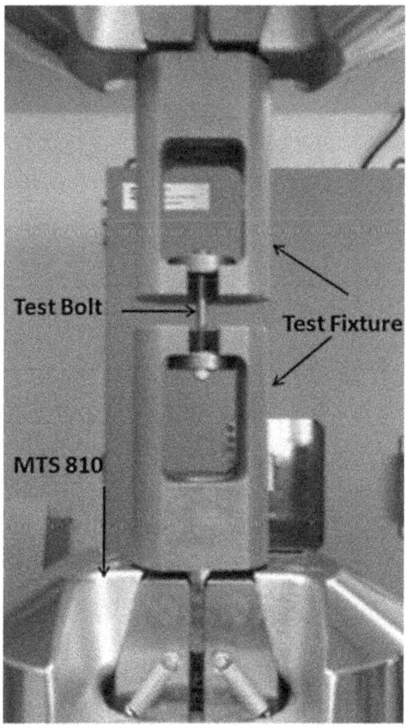

FIGURE 38.2 MTS 810 fatigue testing system and test fixture.

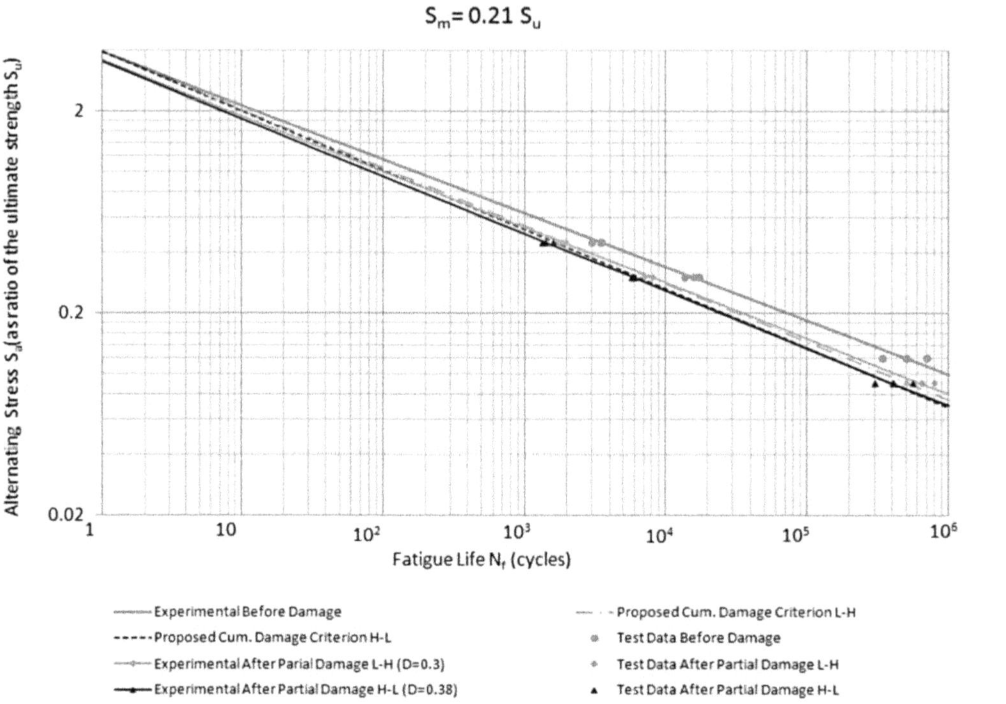

FIGURE 38.3 Comparison of proposed cumulative damage predictions with experimental data: $Sm = 0.21\ Su$.

Figures 38.3–38.6 show both the experimental and theoretical data. The upper line in each figure represents the mean stress-adjusted $S' - N'$ curve for undamaged specimens. The lower $S' - N'$ lines show the comparison between the experimental data and the proposed cumulative damage criterion after partial damage.

The original (undamaged) $S' - N'$ curves of the bolts are constructed according to the ASTM E739-91 standard [14] for various levels of bolt mean stress. Each $S' - N'$ curve is generated by testing nine bolt samples at three different stress levels with the same mean stress (upper line in Figures 38.3–38.6). The cumulative damage verification data is generated using three sets of three bolt samples; each is subjected to an arbitrary three-stage loading spectrum (Figure 38.7).

For the low-to-high (L-H) loading spectrum (Table 38.1), each preloaded bolt is first cycled for n_1 cycles at stress amplitude S_1', followed by cycling at amplitudes S_2' and S_3' for n_2 and n_3 cycles, respectively. Following the ASTM E739–91 standard a lower $S' - N'$ line is generated by testing the nine partially damaged (by S_1', S_2', S_3') bolts at three different stress levels.

To investigate the effect of the spectrum loading sequence on cumulative fatigue damage, a reverse high-to-low (H-L) loading sequence is also used (Table 38.2) for generating theoretical and experimental data following the same procedure outlined for the L-H loading spectrum as shown in Figures 38.3–38.6.

Each of those figures has five $S' - N'$ lines for the indicated level of bolt mean stress. Three of those lines are experimentally generated following the ASTM E739-91 standard; the upper most line is the fatigue line of undamaged bolts, while the two are for the partially-damaged bolts due to the L-H and H-L spectra shown in Tables 38.1 and 38.2. The remaining two fatigue lines are the corresponding fatigue lines of the partially damaged bolts as predicted by the cumulative damage criterion proposed by this study which is represented by Eqs. (38.11)–(38.17).

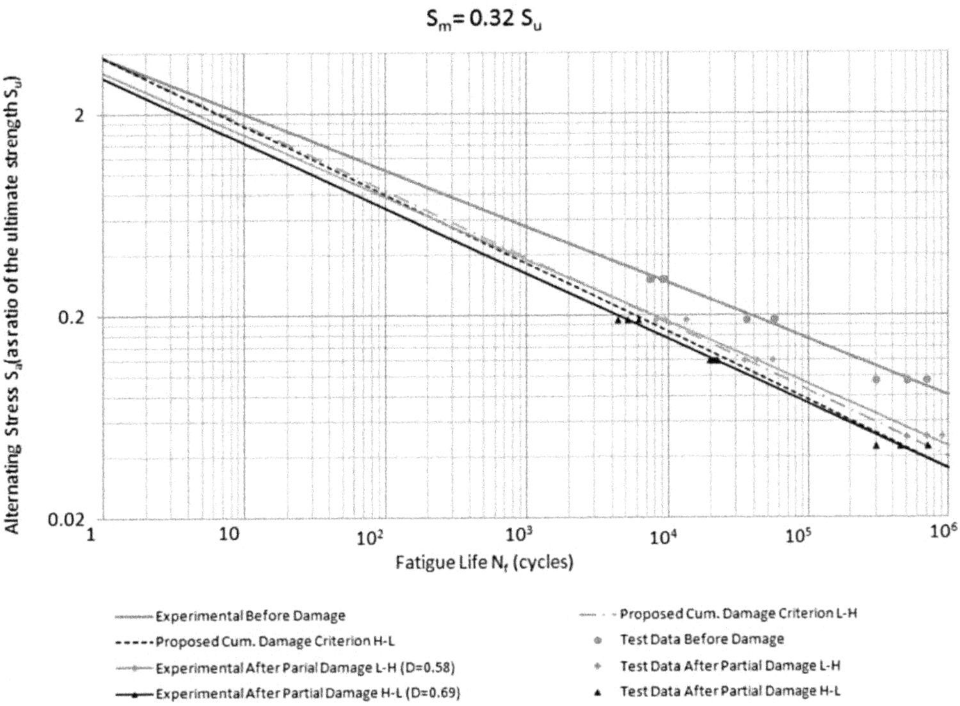

FIGURE 38.4 Comparison of proposed cumulative damage predictions with experimental data: $Sm = 0.32\ Su$.

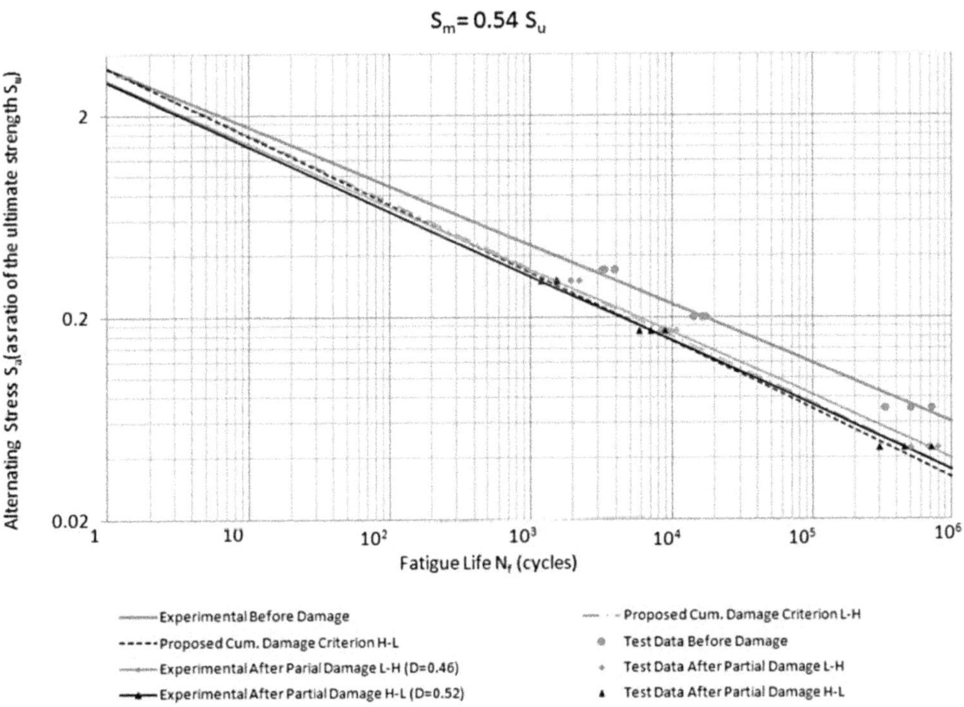

FIGURE 38.5 Comparison of proposed cumulative damage predictions with experimental data: $Sm = 0.54\ Su$.

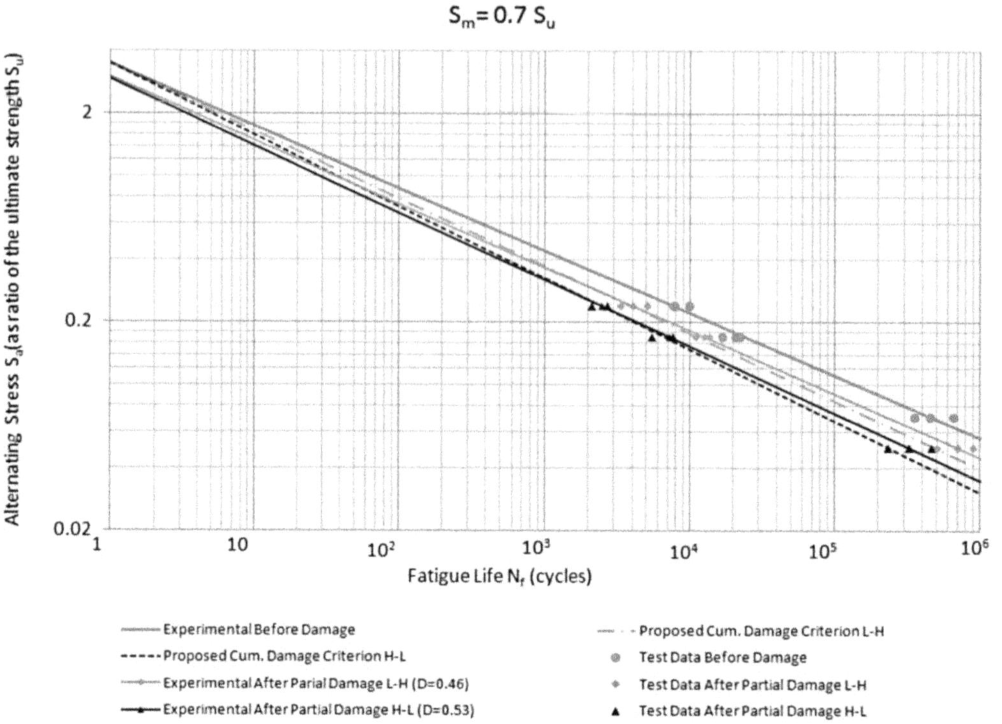

FIGURE 38.6 Comparison of proposed cumulative damage predictions with experimental data: $Sm = 0.7\ Su$.

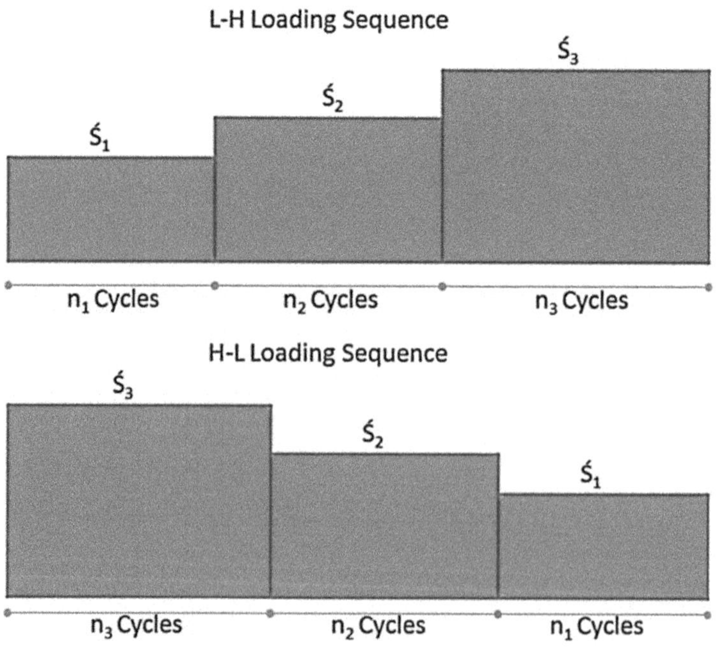

FIGURE 38.7 Schematic for loading sequences L-H and H-L.

TABLE 38.1
Low-to-High (L-H) Loading Spectrum

	Mean Stress (Ratio of Ultimate Stress *Su*)											
	0.21			0.32			0.54			0.7		
Applied stress amplitude (ratio of ultimate stress *Su*)	0.11	0.16	0.2	0.13	0.17	0.2	0.11	0.16	0.19	0.11	0.16	0.2
No of cycles *n*	100,000	40,000	27,000	30,000	20,000	14,000	14,000	7,000	6,000	12,000	4,000	6,000

TABLE 38.2
High-to-Low (H-L) Loading Spectrum

	Mean Stress (Ratio of Ultimate Stress *Su*)											
	0.21			0.32			0.54			0.7		
Applied stress amplitude (ratio of ultimate stress *Su*)	0.2	0.16	0.11	0.2	0.17	0.13	0.19	0.16	0.11	0.2	0.16	0.11
No of cycles *n*	27,000	40,000	100,000	14,000	20,000	30,000	6,000	7,000	14,000	6,000	4,000	12,000

Data represented by Figures 38.3–38.6 show a very good correlation between the experimental data and the proposed cumulative damage criteria in the high-cycle fatigue (HCF) range (fatigue life $10^3 \leq N' \leq 10^6$ cycles). In the low-cycle fatigue (LCF), there still is a reasonable correlation, but the difference is relatively greater than that in the HCF range. The higher difference is mainly attributed to the assumption that the y-intercept a used for the proposed (theoretical) damage criteria (Figure 38.1) is the same for both the undamaged and partially damaged bolt specimens.

38.4 CONCLUSIONS

Cumulative fatigue damage analysis is critical for preloaded threaded fasteners that are subjected to variable-amplitude cyclic loads. The authors would remark on the following points:

- An improved cumulative damage criterion is introduced for preloaded threaded fasteners; using the logarithmic relationship between fatigue life and the corresponding alternating stress amplitude.
- An experimental verification using preloaded A286 steel bolts under various values of mean stress is introduced.
- The new criterion demonstrates the influence of the loading sequence on the cumulative damage for all levels of mean stress. High-to-low (H-L) loading sequence causes more cumulative damage than low-to-high (L-H) loading sequence.
- The mean stress-adjusted $S' - N'$ curve is experimentally constructed for undamaged and partially damaged specimens to validate the suggested criterion.
- Cumulative damage prediction by the proposed criterion is in very good agreement with the experimental data for the high-cycle fatigue range (HCF). In low-cycle fatigue (LCF), there still is a reasonable correlation, but the difference is relatively greater than that in

the HCF range. The higher difference is mainly attributed to the assumption that the y-intercept a used for the proposed (theoretical) damage criterion is the same for both the undamaged and partially damaged bolt specimens.

ACKNOWLEDGMENT

The authors would like to acknowledge the NASA Engineering and Safety Center (NESC) and the Threaded Fastener System Standard team (TFSS) for their support and helpful feedback.

REFERENCES

[1] Pålmgren, A., Die lebensdauer von kugellagern, *Zeitschrift des Vereinesdeutscher Ingenierure*, Vol. 68, No. 14, 1924, pp. 339–341.
[2] Miner, M. A., Cumulative damage in fatigue, *Journal of Applied Mechanics*, Vol. 67, 1945, pp. A159–A164.
[3] Kramer, I. R., Prediction of fatigue damage, In *Proceedings of the 2nd International Conference on the Mechanical Behavior of Materials*, Boston, Massachusetts, American Society for Metals, Metals Park, Ohio, 1976, pp. 812–816.
[4] Marin, J., *Mechanical Behavior of Engineering Materials*, Prentice-Hall, Englewood Cliffs, New Jersey, 1962, pp. 142–144.
[5] Hashine, Z., and Rotem, A., A cumulative damage theory of fatigue failure, *Material Science and Engineering*, Vol. 34, 1978, pp.147–160.
[6] Kommers, J. B., The effect of overstress in fatigue on the endurance life of steel, *Proceedings of American Society for Testing and Materials*, Vol. 43, 1943, pp. 749–762.
[7] Bennett, J. A., A study of the damaging effect of fatigue stressing on X4130 steel, *Proceedings of American Society for Testing and Materials*, Vol. 46, 1946, pp. 693–714.
[8] Henry, D. L., Theory of fatigue damage accumulation in steel, *ASME Transactions*, Vol. 77, 1955, pp. 913–918.
[9] Goodman, J., *Mechanics Applied to Engineering*, Longmans Green, London, 1899, pp. 438–442.
[10] Gerber, W. Z., Calculation of the allowable stresses in iron structures, *Z. Bayer Archit Ing Ver*, Vol. 6, No. 6, 1874, pp. 101–110.
[11] Morrow, J., *Fatigue Design Handbook, Advances in Engineering*, Society of Automotive Engineers, Warrendale, PA, Vol. 6, 1968, pp. 21–29.
[12] ASTM E739-91, Standard practice for statistical analysis of linear or linearized stress-life (S - N) and strain-life (ε - N) fatigue data, American Society for Testing and Materials, 1998, www.astm.org.
[13] Ali, R., Fatigue Performance of Preloaded Threaded Fasteners, 2012, MS Thesis, Department of Mechanical Engineering, Oakland University, Rochester, Michigan.
[14] Nassar, S. A. and Ali, R., An improved cumulative damage criterion for preloaded threaded fasteners, *Journal of Mechanical Design- ASME Transactions*, vol. 136, No. 7, 2014, p. 074502.

39 The Susceptibility of Fasteners to Hydrogen Embrittlement and Stress Corrosion Cracking

Louis Raymond

39.1 INTRODUCTION

A threaded fastener is nothing more than a smooth round bar with a series of notches, which is analytically equivalent to a round bar with a series of built-in defects. For the round bar to be tolerant of these defects, it must have sufficient ductility or toughness that will not degrade the strength of the fastener enough to break below the inherent strength of the material, i.e., the bolt must not be notch-sensitive like glass [1-13]. Also, it cannot be sensitive to time-dependent slow crack growth by hydrogen embrittlement from the plating process or by stress corrosion cracking (SCC) from the environment.

39.2 THE FASTENER: A HIGHLY STRESSED PART WITH BUILT-IN DEFECTS

What is the worst design practice for a structural element? Answer: Putting a notch into a smooth bar to produce a local stress concentration and then applying a high stress. But that is exactly how tensile fasteners are made and used. The threads provide a series of notches, and, once installed, the fastener is under a high tensile force with the state of stress further complicated by bending, shear, and mechanical constraint.

Now the highly stressed notched bar (fastener) is placed in an environment with the hope that it will last forever. However, to improve our odds and protect the bolt against the environment, we will charge it with hydrogen while electroplating it with a dissimilar metal coating that creates conditions for galvanic coupling in a moist service environment.

Of course, the fastener has to be produced at a very low price, and if one bolt breaks in service, the failure of the entire assembly is the fastener manufacturer's fault. Now I ask you, Is this a challenge, or what?

This is the world of the fastener manufacturer—provide a low-cost product in large quantities that withstands the most rigorous design constraints of stress and environment and accept complete responsibility for the entire assembly if one of your components fails after years of service.

39.3 TO MEET THE CHALLENGE: A TECHNOLOGICAL BREAKTHROUGH

Fortunately, we are at a time of technological breakthroughs in our ability to understand the similarities and differences between hydrogen embrittlement and SCC, which are significant factors in the life of fasteners. Over the years, an enormous amount of research has been performed in attempting to understand these mechanisms, but for the first time, we are undergoing a procedural revolution because we now have equipment that conducts an accelerated test to make precise, real-time measurements of the threshold stress for fasteners or standard test specimens.

This equipment combines electrochemistry with mechanical testing as a Rising Step Load® (RSL)[1] under displacement control that permits precise measurement of the onset of slow or subcritical

DOI: 10.1201/9781003501183-45

crack growth in pure bending, which allows the testing and analysis of irregularly shaped coupons (see Appendixes for a complete list of the abbreviations used in this chapter).

The RSL machine is a powerful tool; in that, it has the profound advantage of measuring the threshold for subcritical crack growth in a small fraction of the time required by earlier methods (e.g., the cantilever bend sustained load test carried out according to ASTM E1681 can often take years). As a result, that threshold can be used to quantify the susceptibility of a part to hydrogen embrittlement or SCC with a parameter, H_{sr}. as has never been done before with the meaningless hold time technique with the plate test.

The RSL method has shown the threshold to be sensitive to microstructural changes beyond the detection limit of metallographic or other mechanical rupture load test methods such as ASTM STD E8. It combines the best features of all the other stress corrosion tests, including the slow strain rate and cantilever bend tests. The RSL test procedure has been standardized as ASTM STD 1624.

39.3.1 Revolutionary Findings

Significant findings have shown the erroneousness of the assumption that the large variance of results in the plate test is attributable to variable amounts of residual hydrogen having been introduced during barrel plating. In reality, the large variance of results in the plate test is due to large variances in the stress applied to the bolts by the same installation torque.

Barrel-plated zinc bolts evaluated by the plate test are claimed to have a bell-shaped or Gaussian distribution of residual hydrogen due to a variation in the throwing power because of the variation in distance of the bolts from the electrode during the barrel-plating process. The Gaussian distribution of the amount of residual hydrogen due to the barrel plating process is presumably verified by a plate test, wherein it is assumed that the tensile load is a constant for a given applied torque. Therefore, it is concluded that any evaluation of the degree of embrittlement must involve risk analysis based on a statistical sampling.

But the assumption of a Gaussian distribution of residual hydrogen due to plating is erroneous. The Gaussian distribution of the results is due to the variances in the applied load for a given torque in a plate test, not to the variation in the residual hydrogen in the plated parts.

It has been shown by torque tension tests of 6–32×¾ in. screws that the presumed constant 0.2 from the relationship $T=0.2\ P \times D$ can vary by much as ±0.1.

With the instrumented RSL system, the applied stress is precisely known, and it has been shown that the amount of residual hydrogen in each bolt of a barrel-plating process is the same, and the distribution of results in a plate test is due to the distribution of load for a given torque in that test. This conclusion was verified by barrel plating zinc on ten square bar specimens at 52 HRC-54 HRC at the same time as a large lot of screws were being plated. Using the precise stress-measuring RSL testing system, where the stress is unequivocally established, the threshold stress on all ten specimens was found to be ±2%, i.e., there was no variation whatsoever in residual hydrogen in the barrel-plating process.

The significant conclusions from these results are that there is no need for a large statistical sample size for testing for embrittlement in a plating lot.

Just as important, the barrel zinc-plated specimens at 52–54 HRC showed no degradation in bend strength. All ten specimens attained 98±2% of the bend strength; therefore, there is no need for this lot of screws to be baked, contrary to ASTM STD B849, which specifies that the bake time is a function of hardness, independent of the coating—another erroneous assumption.

The only information the data from a plate test provides once the threshold stress is precisely known is the actual distribution of the stresses on the bolt; only five of the 150 bolts had an installation stress above threshold. All the others, which did not fail, were below the threshold stress.

These results open entirely new avenues of thinking with regard to the sampling size and significance of quality assurance testing. By conducting one to three RSL tests, the same results are obtained as with 150–300 plate-tested bolts.

39.3.2 Future Uses

The RSL equipment and method enable us to

1. Perform a failure analysis by
   a. Identifying the actual source of the hydrogen, from manufacturing through use
   b. Quantitatively identifying the amount of degradation and interaction between the two mechanisms of internal hydrogen embrittlement (IHE) and environmentally induced hydrogen embrittlement (EHE), i.e., SCC.

2. Implement measures for failure prevention in manufacturing by
   a. Quantitatively comparing the quality of production hardware by measuring the hydrogen susceptibility ratio, $H_{sr\text{-}IHE}$, on coupons cut from bolts.
   b. Quantitatively identifying steps in the manufacture and processing of fasteners that degrade the product with hydrogen by measuring $H_{sr\text{-}IHE}$ on standard specimens.
   c. Quantitatively identifying the maximum hold time prior to baking—and the bake time for a plating process—by measuring $H_{sr\text{-}IHE}$.

3. Properly design a fastener system for failure prevention by
   a. Using fracture mechanics to select the proper material and coating for a bolt of given size and strength in a specific environment.
   b. Quantitatively selecting the maximum installation torque to prevent environmentally assisted hydrogen stress cracking for different coatings for a given size bolt by measuring $H_{sr\text{-}IHE}$ as a function of potential.
   c. Quantitatively identifying the severity of a service environment on a finished fastener by measuring $H_{sr\text{-}IHE}$ as a function of environment.

All of the above are possible through the development of a real-time, cost effective test method that can precisely measure the threshold stress for the onset of hydrogen stress cracking. Normally, the measurement of the threshold takes a long time, as much as 4 years for a Grade 8 steel fastener with a sustained load cantilever bend test. Obviously, this method is time and cost-prohibitive as a quality control test for hydrogen embrittlement. Now with the RSL technique, the threshold can be measured in less than a week, most often in less than 24 h. For years, the threshold as a test parameter has been a research tool, but now it can be used as a production and design tool.

39.4 THRESHOLD STRESS

39.4.1 What Is Threshold?

The threshold for the onset of hydrogen stress cracking has been described by Troiano and Hehemann [10] as follows:

> There is an applied stress below which failure will not occur. This critical stress is primarily dependent on the yield strength (material), although it is sensitive to other parameters, such as temperature and stress state.

Stated differently, there is some level of stress (threshold) below which no (time-dependent) cracking will occur, but above which subcritical cracking leading to time delay fracture will occur.

Technically, we can protect ourselves against or prevent the time delay fracture of fasteners in service due to stress corrosion or hydrogen embrittlement by keeping the sum of the applied and residual tensile stresses below the threshold stress for a given bolt geometry and loading condition or stress state.

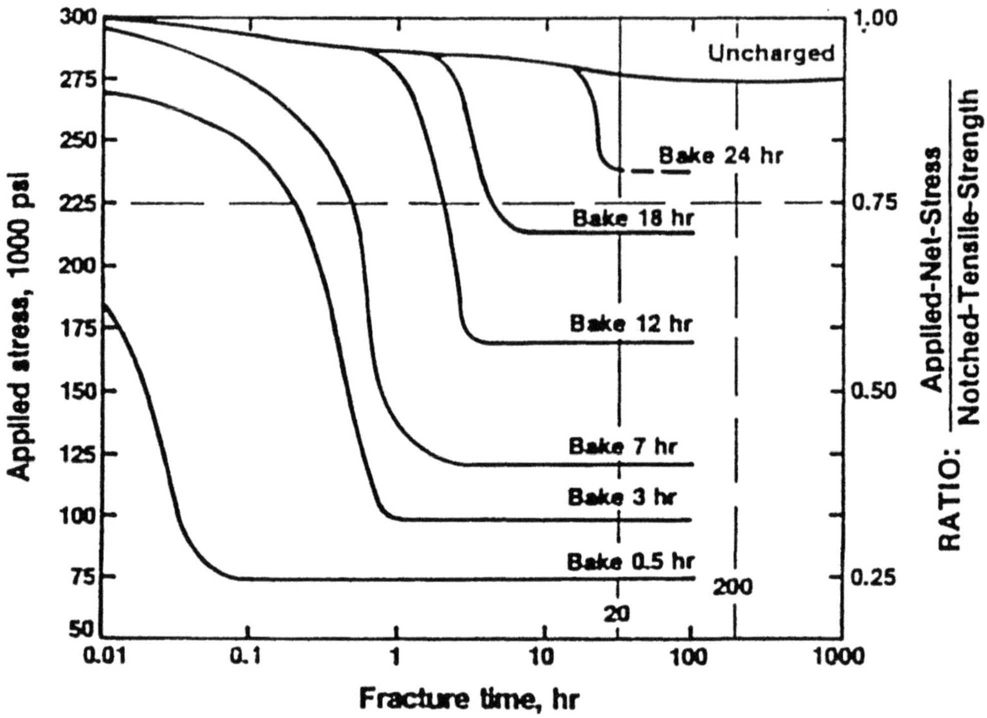

FIGURE 39.1 The effect of baking at 300°F (149°C) on time to failure at a given level of applied stress and stress ratio of notched specimens of 4340 steel heat treated to a strength level of 230 ksi (1586 MPa) (after Ref. 11).

39.4.2 How Is the Stress Threshold Measured?

Troiano [11] first defined and demonstrated the concept of threshold in hydrogen embrittlement by showing with notched round tensile bars that the stress threshold is a function of the hydrogen concentration (see Figure 39.1). AISI 4340 steel heat treated to 230 ksi ultimate tensile strength (UTS) per ASTM STD E8 produced a notched tensile strength (NTS) based on the net area or fracture strength in tension, FS(T), of 300 ksi. It should be noted that the fracture strength of the notched bar is 1.3 times greater than the tensile strength of the smooth bar:

$$\frac{\text{NTS}}{\text{UTS}} = \frac{\text{FS(T)}}{\text{UTS}} = \frac{300}{230} = 1.3$$

The upper limit or fracture strength should be independent of any residual hydrogen in the part because it is tested at ASTM E**8** loading rates, which are too fast to permit the diffusion of hydrogen. Even if residual hydrogen existed in the specimen, the results would not be affected. Therefore, the fracture strength provides a baseline that is independent of hydrogen concentration.

Loads were then applied to produce stresses below the fracture strength. These loads were maintained constant until fracture of the specimen occurred. The lower the stress, the longer the time to fracture, until an applied stress was reached at which fracture never occurred. This lower limit stress was defined as the *threshold stress* or the applied stress below which failure will not occur (infinite life). At applied stresses above this threshold but below the fracture strength, time-delayed failure occurs (finite life).

39.5 THRESHOLD STRESS SEPARATES "FINITE LIFE" FROM "INFINITE LIFE"

The stress essential for crack propagation is the tensile stress. Although we have other conditions of compression and shear, which can be resolved into a principal stress that does not have a shear component, it is the tensile stress that drives the crack perpendicular to the direction of the tensile stress.

The presence of a notch in a smooth bar, or a series of notches as with a bolt, requires additional consideration. It is generally accepted that if a notch is present in a smooth bar, the bar will break at a lower stress. But this is only true in brittle materials, which are then considered to be "notch-sensitive."

In ductile materials, the opposite can be true. Putting a notch in a smooth bar of ductile material will actually result in the bar requiring a higher stress to break—i.e., the material is "notch-strengthened." Evidence of this behavior is given by Troiano's notched tensile bar test results (Figure 39.1), where NTS/TS(T) = 1.3. This is true for the net stress, which is calculated from the minimum diameter.

Similarly, when the specimen or fastener is tested in a bending mode, which would appear to create a more vulnerable stress situation, the tensile strength is even higher. These effects can be explained by the theories of fracture mechanics, which are dealt with later in this chapter. What is known is that the fasteners will very likely see not only a tensile stress in service but also, more than likely, varying levels of bending stresses.

39.5.1 Load vs. Stress

Load and stress cannot be used interchangeably. The local stress produced by a given load on a bolt depends on where the load is applied. Generally, when a torque is applied, it is assumed that the resulting load is axial and that load and stress can be used interchangeably, just by dividing by the cross-sectional area of the bolt. As shown in Appendix A1, if a rigid arm is used to hold a load, tension and bending stress can be produced, but if a flex arm or wire and pulley is used to hold the load, a uniform tensile load is produced.

This simple relationship between load and stress is not true if the same load is applied with a slight offset on the head of the bolt. Then a condition of combined tension and bending will result. Since bending greatly amplifies the stress, the slightest offset will produce a large difference in the local maximum tensile stress. This condition significantly affects sustained load time-to-failure test results as demonstrated in Appendix A1. That is why great care must be taken in conducting a tensile test, especially when subjecting notched round tensile bars or bolts in tandem to hydrogen embrittlement testing.

The *state of stress* takes into account the conditions of stress and strain at a point. The boundary conditions in three mutually perpendicular directions define the triaxial constraint for plastic flow or yielding at the root of a notch, which manifests itself as an adjustment to the yield strength. In other words, the yield strength of a notched round bar under triaxial constraint in bending is greater than the yield strength of a notched round bar under triaxial constraint in tension, which in turn is greater than the yield strength of a smooth round bar in tension.

39.5.2 Importance of State of Stress

A smooth round bar tested in accordance with the geometrical configurations and loading rates of ASTM STD E8 produces the characteristically shaped curve for steel that defines a yield strength in tension, YS(T), which is about 0.85–0.90 times the ultimate tensile strength (UTS) or tensile strength in tension, TS(T), as shown in Figure 39.2.

A notched round bar tested in tension at ASTM E8 loading rates will have a fracture strength that can be either higher or lower than the UTS or TS(T), depending on the strength and toughness of the material.

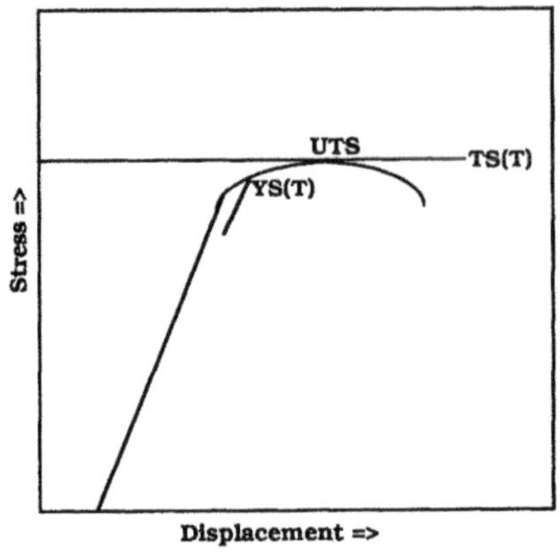

FIGURE 39.2 A typical stress–strain curve for steel tested as a smooth bar in tension according to ASTM STD E8.

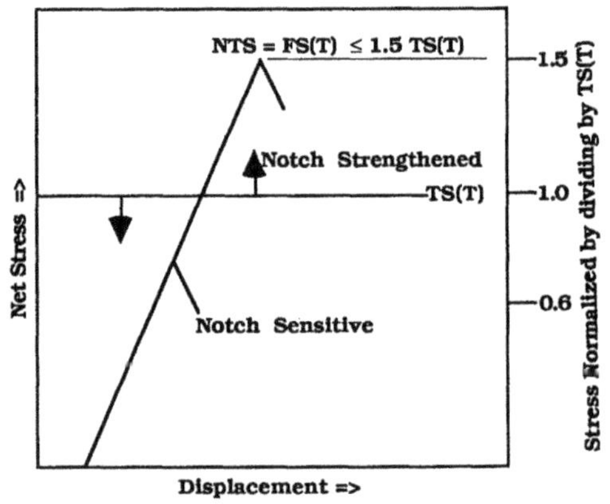

FIGURE 39.3 The effect of a circumferential notch on a round tensile bar depends on the strength and ductility of the material. The fracture strength in tension, FS(T), can either be lower (notch-sensitive) or higher (notch-strengthened) than the tensile strength of a smooth tensile bar, TS(T). The tensile strength normalized to TS(T) is shown on the right vertical axis.

Figure 39.3 shows the relationship between "notch-sensitive" and "notch-strengthened." When fracture occurs below TS(T), the material is said to be notch-sensitive. When fracture occurs above TS(T), the material is said to be notch-strengthened. The maximum value or limit of NTS/TS(T) or the fracture strength in tension, FS(T), of a notched bar to the tensile strength of a smooth bar is about 1.6 times the yield strength in tension or about 1.5 times the tensile strength in tension. That is, $FS(T) \leq 1.5\ TS(T)$.

Note: Because of notch strengthening effects in ductile materials that are used in fasteners, a specified "stress area" greater than the net area is often used to make the FS(T) more closely approach the TS(T).

For fasteners, the tensile strength in tension, designated as TS(T), which is equivalent to the UTS from a smooth bar tensile test per ASTM STD E8, is used to normalize the strength measurements. This means that all strength measurements are made dimensionless and referenced to the TS(T) by dividing them by TS(T), as shown in Figure 39.3.

39.6 FASTENER STRENGTH IN TENSION

The minimum tensile load is often specified for fasteners according to SAE J1237 or SAE J81. The minimum specified fastener strength in tension can be calculated by dividing this load by the net area, which is based on the minimum diameter of the fastener, *d*. The net area is used to calculate the minimum specified net tensile strength of the fastener, which by bolt design is equal to TS(T).

Since the minimum specified tensile load, and therefore TS(T), is not always given, as an alternative it can be measured according to ASTM STD F606 by machining a tensile specimen from the bolt and testing it at ASTM STD E8 loading rates. As a last resort, TS(T) can be estimated from the measured Rockwell hardness of the bolt, or it can be estimated from the mean of the specified hardness range selected and converted to the tensile strength in tension, TS(T), in accordance with ASTM E140.

39.6.1 Proof Testing against IHE

To ensure that a lot of bolts have not been embrittled by the plating process, they are proof tested. This is not to be confused with the "proof torque," which is a continuously increasing load application. The concept of a proof test for hydrogen embrittlement is to expose the plated bolt to a sustained stress higher than the maximum expected operating load, P_{max} (MEOP), that it will be exposed to in service.

Based on Figure 39.1, a bake time of 24 h at 300°F should be adequate to prevent a sustained load failure at 75% of the NTS. Passing this test does not mean that the bolts are free of residual hydrogen, only that there is not sufficient hydrogen to cause failure at 75% NTS. A higher stress would produce a time-delayed hydrogen embrittlement failure.

It should be noted that 75% NTS is equal to 225 ksi or about 100% TS(T), which was measured as 230 ksi from Troiano's tests. The usefulness of this information for actual industrial plating processes is limited because Troiano's tests were conducted on bare specimens. For plated specimens, it is only logical that the efficiency of the baking process would depend on the thickness and type of plating being deposited.

Surprisingly, Troiano's data are currently being used by the industry. It is further concluded that if the bolts do not fail in the 24 h test time, then they will not fail in service. Also, contrary to logic, ASTM STD B849 further assumes that the choice of a 4 or 24 h baking treatment depends solely on the hardness of the steel, independent of the type or thickness of the plating.

For fasteners, an inexpensive plate test is generally used to approximate tensile loads, or a more expensive MIL-STD-1312-5a test to ensure tensile loads by requiring that the bearing surfaces are parallel within ±1°.

As shown in the following example, the use of the plate test has been, and remains, the cause of so much embrittled product getting into service for the simple reason that the proof loads do not even attain the tensile strength of the material.

Example

The proof stress from the specified torque per SAE J81 for a hydrogen embrittlement test with a plate for a ¼–20 zinc-plated screws is 132 in. lb. From the relationship Torque $= 0.2 \times P_\tau \times D$, the load P_τ is

$$P_\tau = \frac{\text{torque}}{0.2D} = 5 \times \frac{\text{torque}}{D}$$

and the correspondence tensile stress is equal to P_τ/A_{net}.

Also, the minimum specified tensile strength, TS(T), is

$$TS(T) = P_{\text{m intensile}} / A_{\text{net}}$$

The percent of TS(T) generated by the torque is

$$\frac{P_\tau / A_{\text{net}}}{P_{\text{mintensile}} / A_{\text{net}}}$$

or

$$\%TS(T) = \frac{P_\tau}{P_{\text{mintensile}}} = \frac{5 \times \text{torque}}{D \times P_{\text{mintensile}}}$$

$$= \frac{5 \times 132 \text{ in. lb}}{0.25 \text{ in.} \times 4{,}300 \text{ lb}} = 60\%$$

Therefore, the specified proof stress in a plate test for a zinc-plated ¼–20 bolt is only a maximum of 60% TS(T). However, in service, if bending is involved, the bolt can actually experience local tensile stresses of 2×TS(T). For cadmium-plated screws, the conditions for embrittlement testing are even less discriminating because only 114 in.-lb of torque is specified, or a maximum IHE proof stress of only 52%.

Therefore, the assumption that $P_{\text{proof}} > P_{\text{MEOP}}$ is not met for a plate test. In fact, the service stress can be greater than three times the proof stress of a plate test—totally negating the intent of the proof test for hydrogen embrittlement. In conclusion, the plate test is a waste of effort, time, and money. Bolts will fail in service even though they pass the plate hydrogen embrittlement proof test. The only reason some bolts might fail in a plate test is that the Gaussian distribution or randomness of the load from torque and any bending might produce tensile stresses higher than those calculated.

39.6.2 When and How Should Bending Be Used?

Bending should be used in testing fasteners because it provides more constraint than a notched bar tensile test and allows the fastener to attain a higher maximum local surface tensile stress than it could ever attain in a tension test (Figure 39.4). For this reason, many failures occur in service in fasteners that have passed a proof test in tension prior to use. During installation, the fasteners can experience a bending stress, and then the maximum local surface tensile stress exceeds the proof stress in tension. For this reason, fasteners should be proof tested in bending. This test would provide the upper limit or limit load; i.e., no higher stress could ever be generated in the fastener.

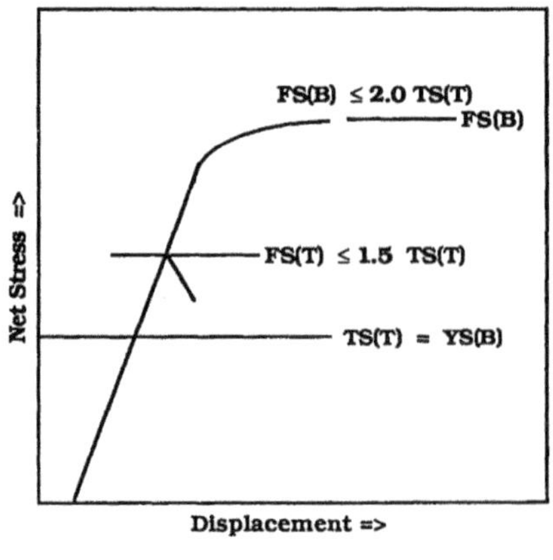

FIGURE 39.4 Relative positions of fracture strength in bending, FS(B), and in tension, FS(T), to the tensile strength in tension, TS(T), and yield strength in bending, YS(B).

The limit load for a ductile steel in bending is 2.3 times the yield strength in tension, which is about $(0.85-0.90)\times TS(T)$, or the limit load in bending is about 2.0 times the tensile strength in tension:

$$FS(B) \leq 2.0\ TS(T)$$

To maintain a consistent baseline for a fastener, the yield strength in bending, YS(B), is defined as being equal to the ASTM E8 TS(T); i.e.,

$$YS(B) = TS(T)$$

Figure 39.4 summarizes these relationships.

Previously it was shown that the maximum value or limit of the fracture strength in tension is 1.5 times that of the tensile strength of a smooth bar, or

$$FS(T) \leq 1.5\ TS(T)$$

39.7 SPECIMEN NOTCHED STRENGTH RATIO, R_{NS}

A convenient dimensionless parameter that can be used to describe the susceptibility to hydrogen stress cracking is the specimen strength ratio, R_{sb}, in accordance with ASTM E812. This parameter is the calculated maximum nominal fracture stress at the root of a sharp crack or fatigue crack in a square bar bent specimen, which is normalized by dividing by the yield strength of the material.

For bolts, the specimen strength ratio is modified to take into account a notch instead of a sharp crack and normalized by dividing by the tensile strength, TS(T), instead of the yield strength. Therefore, for bolts, this stress parameter will be defined as the notched strength ratio in tension, $R_{ns(t)}$, or in bending, $R_{ns(b)}$, which provides a dimensionless upper limit on the capacity of a fastener to carry a load or the fracture strength in tension or bending.

Our discussions regarding the state of stress and its limits can now be summarized as follows:

In tension,

$$R_{ns(t)} \equiv FS(T)/TS(T) \leq 1.5$$

In bending,

$$R_{ns(b)} = FS(B)/TS(T) \leq 2.0$$

The significance of these notched specimen strength ratio parameters is that they provide a method for comparing the actual tensile stress component under different loading conditions or states of stress. In all cases, the stress has been normalized by dividing by the tensile strength. Therefore, when the parameters are equal, they can be considered as having equivalent stresses, independent of the type of loading. If $R_{ns(b)}=1.7$, then $R_{ns(t)}=1.7$, and the maximum tensile stress in both cases is $1.7\times TS(T)$. Of note is the fact that the value 1.7 cannot be measured in a notched round bar tensile test because it exceeds the limit of 1.5. Bending provides a method of testing to a higher level of stress than tension of up to 2.0 TS(T).

Also to be noted is that these parameters identify the fact that the local surface stress in bending can be twice as much as the ultimate tensile strength in a smooth bar tension test and $2.0/1.5=1.3$ times greater than the net tensile strength of a circumferentially notched round bar, which in itself can produce a maximum local tensile stress at the notch that is 1.5 times the ultimate tensile strength of a smooth bar in a tension test. It should be noted that these upper limits are observed only in materials with high fracture toughness.

Appendix A2 summarizes the definitions and notations developed in this section and used throughout this chapter for testing in both tension and bend at ASTM E8 loading rates.

39.8 THE EFFECT OF HYDROGEN

Hydrogen in steel can cause a time-delayed or slow strain rate fracture of a fastener at applied stresses below the fracture strength of the bolt FS(T), i.e., at an applied stress less than $R_{ns(b/t)}$. In essence, the strength of the bolt is degraded with time.

This degradation was first demonstrated by Troiano with sustained load tests on circumferentially notched round bars in tension. Most threshold tests are now conducted in bending because a higher local maximum stress can be attained, as shown in Figure 39.5. The U.S. Navy was the first to conduct these tests extensively as sustained load cantilever bend tests; this practice has now been standardized as ASTM E1681.

Figure 39.6 illustrates typical results from using the cantilever bend test method of measuring the threshold stress, which again is a measure of degradation in strength (time-delayed) due to the presence of residual hydrogen in the steel.

39.8.1 Hydrogen Susceptibility Ratio, H_{SR}

Analogous to the notched specimen strength ratio for rupture or fracture strength that is measured at ASTM E8 loading rates, $R_{ns(b)}$, a parameter called the hydrogen susceptibility ratio, $H_{sr(b)}$, will be used as a measure of the degradation in strength due to the presence of residual hydrogen generated from either the manufacturing process or a service environment.

39.8.2 When Does Crack Growth Begin?

Hydrogen-induced crack growth occurs when a critical combination of applied and residual tensile stress and hydrogen concentration is attained. The greater the hydrogen concentration, the lower the critical stress; the lower the hydrogen concentration, the higher the critical stress. From either perspective, this definition gives a new meaning to the concept of threshold. What has been shown by

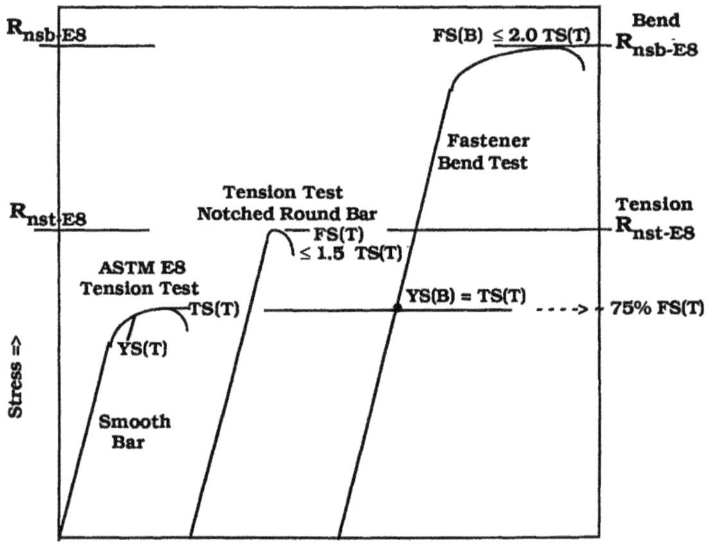

FIGURE 39.5 Expanded view of Figure 39.4 that shows the complete load-displacement curve at ASTM E8 loading rates for smooth and notched round bars in tension compared to a circumferentially notched round bar in bending. Note: Most hydrogen embrittlement tests are conducted on a circumferentially notched round bar at 75% FS(T), which is equal to about 100% of the TS(T) of a smooth bar or about 50% FS(B).

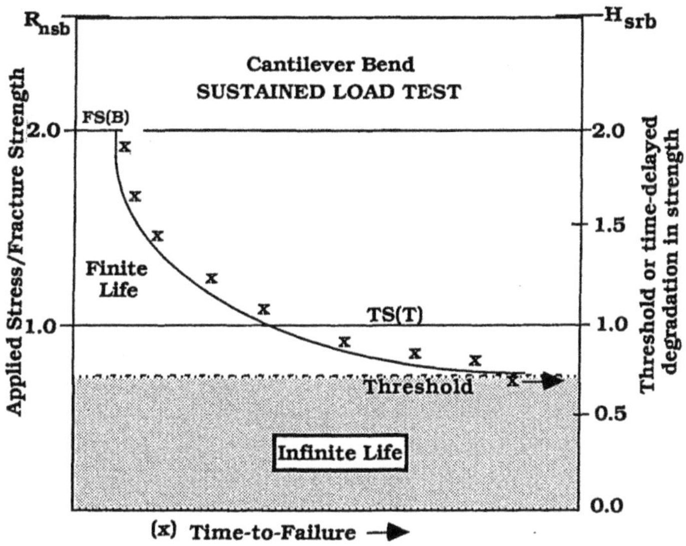

FIGURE 39.6 A conventional long-term method such as the sustained load cantilever bend test ASTM E1681 uses a large number of specimens and many machines with minimum runout times of as much as 10,000 h (over 1 year) to measure the threshold stress or threshold stress intensity below which no failure occurs in a specific environment if the service stresses are maintained below the threshold (infinite life).

the RSL test method is that physically the threshold is the instant the critical combination of stress and hydrogen concentration is attained at a sufficiently slow loading rate to initiate crack growth. Therefore, proper attention must be given to both stress and hydrogen concentration.

The RSL method allows us to precisely measure the threshold stress for the onset of hydrogen stress cracking in a real time, cost-effective manner. It combines the best of a slow strain rate test and a sustained load cantilever bend test to permit measurement of the threshold. The loads are applied at a modified slow displacement rate but in incrementally increasing steps in accordance with ASTM F1624. As shown in Figure 39.7, the threshold is compared to the fracture strength measured at ASTM E8 loading rates, to quantify the degradation of the part due to hydrogen embrittlement from processing or environmental exposure.

39.9 WHAT IS A SUFFICIENTLY SLOW LOADING RATE?

The RSL method is described in more detail in ASTM F1624. As shown in Figure 39.8, the rate is progressively decreased until no further change is detected in the measured threshold stress. The minimum loading rate at which the threshold measurement becomes an invariant is the "sufficiently slow loading rate."

As shown in Figure 39.9 by a schematic of Troiano and Heheman's Figure 39.1 [11], the greater the hydrogen concentration, the lower the threshold stress, $H_{sr(b)\text{-}F1624}$. The limits and notation in terms of stress and the dimensionless parameters H_{sr} and $R_{ns(b)}$ are illustrated.

The threshold in bending, $\sigma_{th(b)\text{-}[H^+]}$ $\sigma_{th(b)\text{-}[H+]}$, increases as hydrogen concentration, [H$^+$], decreases. Limiting stresses for electroplating are the fracture strength in bending, FS(B), and the threshold under an aggressive, worst-case, hydrogen charging environment, $\sigma_{th(b)\text{-}[H^+]}$. The upper limit of fracture strength in bending, FS(B), can be as much as 2.0 times the tensile strength in tension, TS(T), per ASTM E8; i.e., FS(B) $\leq$ 2.0 TS(T). Conversely, TS(T) $\geq$ 50% FS(B).

It should be noted that most ASTM F519 notched round bar tensile test are conducted at 75% FS(T). This stress is about equal to 100% of the ASTM E8 TS(T). This sustained stress is only about

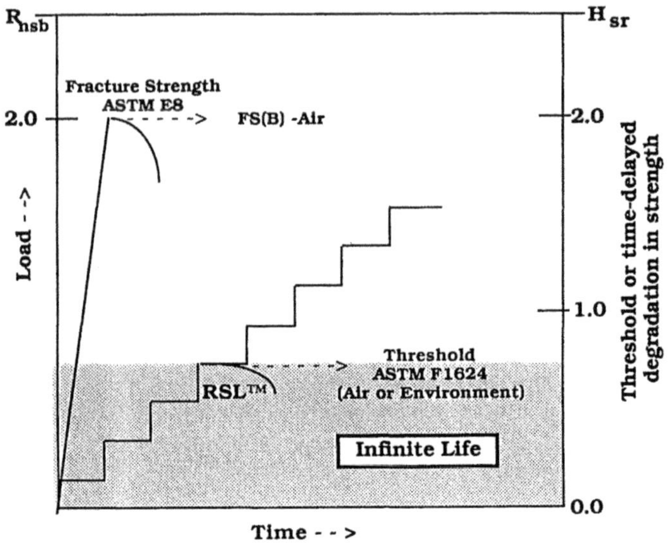

FIGURE 39.7 The new short-term or accelerated method to measure the threshold is the Rising Step Load (RSL) technique carried out according to ASTM F1624. Only a small number of specimens are used, and the maximum runout time can be as little as 24h (less than 1 day) to measure the threshold stress or threshold stress intensity to prevent hydrogen embrittlement service failures. The ASTM F1624 loading rates are much slower than the ASTM E8 loading rate. This difference allows time for the hydrogen to diffuse and cause degradation of the fracture strength.

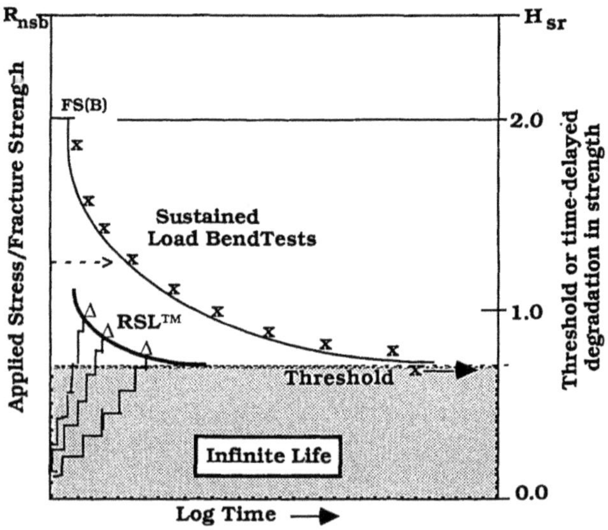

FIGURE 39.8 A comparison of the RSL incrementally increasing load (ASTM F1624) to the sustained load cantilever bend testing technique (ASTM E1681) shows that the threshold stress or stress intensity design parameter can now be measured in a shorter time, with fewer specimens.

50% of FS(B); i.e., bending can produce conditions such that the strength on a circumferentially notched bar in bending is about twice the fracture strength in tension. Therefore, even though bolts passed an ASTM F519 or MIL-STD-1312-5a sustained load tension test, they could possibly experience a service failure if the service stresses were in bending because they could very easily exceed 75% FS(T), even at a low installation torque.

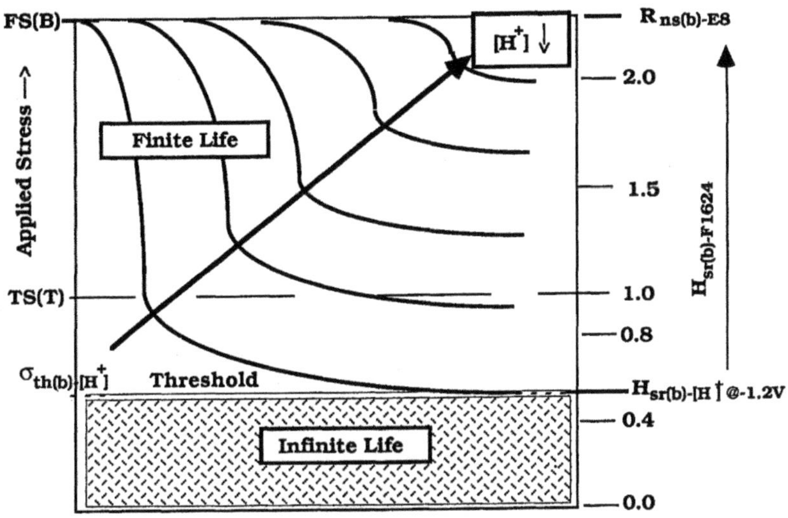

FIGURE 39.9 Schematic of a range of threshold curves of a circumferentially notched round bar or bolt in bending used to identify the various parameters and the effect of a changing hydrogen concentration (after Fig. 1 of Ref. [11]).

Figure 39.9 schematically illustrates conditions where the upper limit of the fracture strength of a circumferentially notched round bar or bolt in bending, FS(B), can be as high as 2.3 times the yield or 2.0 times the smooth bar tensile strength as measured per ASTM E8; i.e., FS(B)≤2.0×TS(T). Expressed in terms of the notched bend specimen strength ratio, $R_{ns(b)\text{-E8}} \leq 2.0$.

The lower limit produced by an aggressive hydrogen charging environment is identified as $\sigma_{th(t)\text{-}[H^+]}$ and designated as $H_{sr(b)\text{-}[H^+]}$. In the schematically drawn Figure 39.9, $H_{sr(b)\text{-}[H^+]} \approx 0.5$ or $\sigma_{th(t)\text{-}[H^+]} \approx 0.5\ \text{TS(T)}$.

The degradation of fracture strength in bending, $\text{FS(B)} = H_{sr(b)\text{-}[H^+]}/R_{ns(b)\text{-E8}} \approx 0.5/2.0 = 0.25$, or the fracture strength in bending has been degraded by 75% in the presence of an aggressive hydrogen charging environment. This factor is a measure of the susceptibility of the material to hydrogen-assisted stress cracking.

Appendixes A3.A and A3.B summarize the definitions and notations developed in this section and used throughout this chapter for RSL testing in both tension and bend at ASTM F1624 loading rates.

Now we must revise our thought process. Analogous to the specimen strength ratio in tension or bending at ASTM E8 loading rates,

$$\text{In tension, } R_{ns(t)\text{-E8}} \equiv \text{FS(T)/TS(T)} \leq 1.5$$

$$\text{In bending, } R_{ns(b)\text{-E8}} \equiv \text{FS(B)/TS(T)} \leq 2.0$$

The threshold stress in tension or bending at ASTM F1624 loading rates, σ_{th}, is normalized by dividing by the tensile strength in tension, TS(T), to provide similar parameters:

$$\text{In tension, } H_{sr(t)\text{-F1624}} \equiv \sigma_{th(T)}/\text{TS(T)} \leq 1.5$$

$$\text{In bending, } H_{sr(b)\text{-F1624}} \equiv \sigma_{th(B)}/\text{TS(T)} \leq 2.0$$

The lower limit of the threshold stress measured in an aggressive hydrogen charging environment, $\sigma_{th\text{-}[H^+]}$, is used to define the hydrogen susceptibility ratio, H_{sr}, which is calculated by dividing the threshold stress measured in an aggressive hydrogen charging environment, [H+], by the tensile strength:

$$H_{sr(t/b)\text{-}[H^+]F1624} \equiv \sigma_{th[H^+]}/TS(T)$$

The significance of this parameter is that it is considered to be a measure of the material's susceptibility to hydrogen-assisted stress cracking whether the source is hydrogen embrittlement (IHE) or environmentally induced hydrogen-assisted stress cracking, EHE (SCC).

As shown in Figure 39.9, the threshold stress, in terms of the specimen strength ratio, can range from a maximum of $R_{ns(b)\text{-}E8}$ to a minimum of $H_{sr(t/b)\text{-}[H^+]}$. It should be remembered that in bending, local maximum stresses as much as twice the tensile strength, TS(T)—i.e., ≤2.0 TS(T)—can be attained.

39.10 WHEN IS A PLATED PART "FREE" FROM SUSCEPTIBILITY TO HYDROGEN EMBRITTLEMENT AND SCC?

With hydrogen contamination either from IHE from the plating process (IHE) or environmentally induced hydrogen stress cracking (EHE, or SCC), the H_{sr} threshold values will fall between the two limits $R_{ns(b)}$ and $H_{sr(b)}$.

The fastener would be free of or immune to hydrogen-assisted stress cracking from either IHE or EHE if

$$R_{ns(b)\text{-}E8} = H_{sr(b)\text{-}[H^+]\text{-}F1624}$$

If $H_{sr(b)\text{-}[H^+]\text{-}F1624} < R_{ns(b)\text{-}E8}$, then all other H_{sr} threshold measurements will fall between the two extremes, increasing as the hydrogen concentration decreases.

A similar situation could be developed for pure tension. A fastener will be unsusceptible to either IHE or EHE up to $R_{nst} \approx 1.5$ if

$$R_{ns(t)\text{-}E8} = H_{sr(t)\text{-}F1624}$$

Working with tensile parameters is of limited value because the local stresses that can be attained in bending prior to rupture are much higher than those in pure tension.

The tools are now available to conduct quantitative analysis on the degree of embrittlement in terms of the specimen strength ratio and hydrogen susceptibility ratio. Now all that needs to be considered is the source of the hydrogen.

39.11 INTERNAL/ENVIRONMENTAL HYDROGEN EMBRITTLEMENT VS. SCC

Other than fatigue, time-delayed fractures in fasteners are thought to occur because of IHE from the plating process or environmentally induced SCC.

SCC occurs when a stressed fastener is exposed to an environment that generates a corrosion reaction on the surface of the fastener that results in an intergranular fracture surface that shows evidence of anodic dissolution, oxidation, or pitting on the surface. Such a surface is often used to differentiate SCC from IHE.

When a failed fastener is found to have a clean intergranular fracture surface, IHE is immediately blamed instead of SCC. However, a clean intergranular fracture surface could just as easily be found in a fractured plated steel bolt exposed to the environment. In the environment, the plating does its job by sacrificially corroding to cathodically protect the steel bolt from corrosion. As this occurs, hydrogen is evolved at the steel cathode, inducing hydrogen stress cracking from the environment, which results in a clean intergranular fracture surface.

ASTM STP 543 differentiates these two mechanisms that produce a clean intergranular fracture surface as IHE and environmental hydrogen embrittlement (EHE) (Appendix A4 summarizes the definitions developed in STP 543 and used in this chapter).

The issue then becomes one of finding a method to sort out the specific cause of failure. Scanning electron micrography (SEM) is generally used, but with limited success. The incrementally increasing load or RSL technique provides a method of precisely resolving this issue by measuring the threshold under a variety of test conditions.

39.11.1 SEM Fractography

If a fractured bolt is examined with the scanning electron microscope, inevitably one observes an intergranular fracture surface regardless of whether the failure mode is hydrogen embrittlement (IHE) or SCC (EHE). The subtle difference often used to discriminate between the two mechanisms is that a hydrogen embrittlement failure has clean faceted grains classically described as "rock candy," whereas stress corrosion would also exhibit faceted grains but with a pitted or oxidized surface.

If a bolt failed by hydrogen embrittlement and was subsequently exposed to a moist environment, however, it would also exhibit a pitted faceted surface. Conversely, if a zinc-plated steel bolt failed due to SCC in a moist environment, the zinc would sacrificially corrode and prevent the fracture surface of the steel from corroding, again resulting in clean intergranular grains. In conclusion, SEM examination of a fracture surface cannot be used to determine with certainty whether the mechanism is IHE or EHE.

If we compare the two mechanisms as shown in Appendix A5, the conclusion can be drawn that mechanistically IHE and EHE (SCC) are the same, the only difference being the sequence of exposure to stress and the hydrogen-generating environment. In IHE, the hydrogen is introduced during plating with no stress in the part, and the part is then exposed to stress. In SCC, the part is first stressed and then exposed to the environment. In either case, once the critical stress is attained for a given hydrogen concentration, slow crack growth begins.

39.11.2 Sorting Out the Cause

Figure 39.10 is a summary of the range of threshold values that can be obtained from different sources of hydrogen. From electroplating, a degradation in strength can be measured by the threshold stress $\sigma_{th(b)\text{-IHE}}$ or $H_{sr(b)\text{-IHE}}$. In general, the IHE threshold is higher than that in a hydrogen charging environment (EHE; SCC), where the degradation in strength can be measured as the threshold stress $\sigma_{th(b)\text{-EHE}}$ or $H_{sr(b)\text{-EHE}}$.

39.11.3 Specimen Notched Bend Strength Ratio, R_{NSB}, and the Hydrogen Susceptibility Ratio, H_{SRB}

The tensile strength in an ASTM E8 tension test, TS(T), is the normalizing factor used to calculate the bolt notched bend strength ratio, consistent with ASTM E812 for both the fracture bend strength, $R_{nsb\text{-}E8}$, and the threshold stress in bending, $H_{sr(b)\text{-F1624}}$.

The upper limit on the bolt notched strength ratio in bending, $R_{ns(b)\text{-}E8}$ can be as much as 2.0. The lower limit is $H_{sr(b)\text{-}[H^+]@\text{-}1.2v}$ in accordance with ASTM F1624. It is defined as the hydrogen susceptibility ratio, $H_{sr(b)\text{-}[H^+]}$. All other values of the threshold measured with a notched bend specimen or bolt fall in between these two values.

For service tensile stresses below $H_{sr(b)\text{-}[H^+]}$, the bolt should have infinite life in service. For service tensile stresses above $H_{sr(b)\text{-}[H^+]}$, the bolt will have a finite service life.

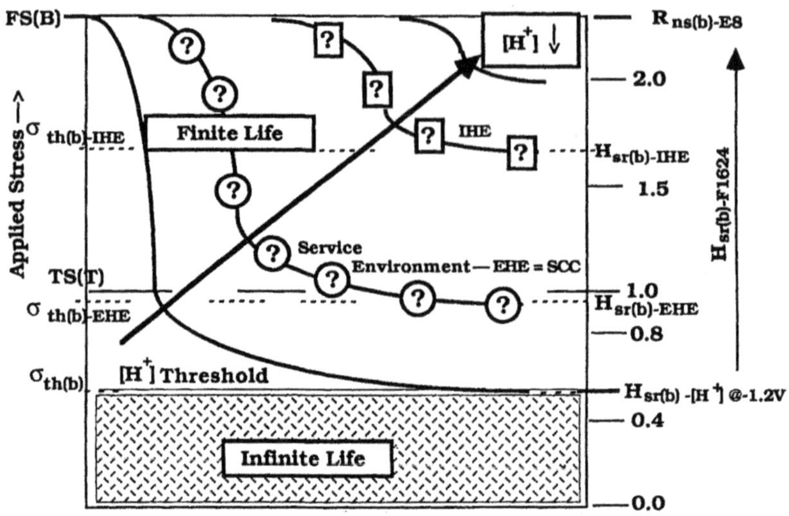

FIGURE 39.10 Schematic of a range of threshold curves of a circumferentially notched round bar or bolt in bending, illustrating that the threshold due to processing (IHE) or that due to a service environment (EHE, or SCC), falls between the two extremes of $R_{ns(b)\text{-}E8}$ and $H_{sr(b)@\text{-}1.2V\text{-}F1624}$. Note: TS(T) ≈ 75% FS(T) ≈ 50% FS(B).

When $H_{sr(b)@\text{-}1.2v} = R_{nsb\text{-}E8}$, the bolt is immune to hydrogen stress cracking from either processing or the environment.

The threshold tensile stresses measured in a bend test should be directly related to the threshold tensile stresses measured in a notched round bar tensile test, i.e., $H_{ns(t)} = H_{ns(b)}$. It should be noted that bending can induce about 1.5 times the maximum stress that is produced for a notched bar in tension. Therefore, to ensure structural reliability under worst-case loading conditions, all hydrogen embrittlement tests should be conducted in bending.

In summary, Figure 39.10 schematically illustrates conditions where the threshold due to classical hydrogen embrittlement (IHE) from an RSL test conducted in the air is significantly higher than the smooth bar tensile strength as measured per ASTM E8; i.e., $\sigma_{th(b)\text{-}IHE} \approx 1.7$ TS(T). Expressed in terms of the notched specimen hydrogen susceptibility ratio, $H_{sr(b)\text{-}IHE\text{-}F1624} \approx 1.7$.

Likewise, the threshold due to environmental hydrogen embrittlement (EHE or SCC) from an RSL test conducted in the environment is slightly less than the smooth bar tensile strength as measured per ASTM E8; i.e., $\sigma_{th(t)\text{-}EHE} \approx 0.95$ TS(T). Expressed in terms of the notched specimen hydrogen susceptibility ratio, $H_{sr(b)\text{-}EHE\text{-}F1624} \approx 0.95$.

The degradation of fracture strength in bending is found from the ratio $H_{sr(b)\text{-}IHE\text{-}F1624}/R_{ns(b)\text{-}E8} \approx 1.7/2.3 = 0.7$, or which indicates that the fracture strength has been degraded to 70% of its original value due to processing.

Likewise, in a service environment, the degradation of fracture strength in bending is found from $H_{sr(b)\text{-}EHE\text{-}F1624}/R_{ns(b)\text{-}E8} \approx 0.95/2.3 = 0.4$, which tells us that the fracture strength has been degraded to 40% of its original value.

39.12 USING OUR NEW TOOL

39.12.1 Use in Failure Analysis

In failure analysis, the RSL threshold can be used to identify the actual source of the hydrogen, from manufacturing through application. The amount of degradation and interaction between the two mechanisms of IHE and environmentally induced hydrogen embrittlement (EHE or SCC) can be quantitatively identified.

Common failure analysis questions can now be answered.

Is the failure an isolated case?
Is the failure related to only one lot, or to all lots?
If lot-related, can the remaining bolts be salvaged?

The answers to these questions reflect on issues of legal responsibility:

Were the bolts defectively manufactured (improper application of rolled threads; incorrect heat treatment; embrittling plating process; inadequate baking treatment; etc.)?
Was the bolting system defectively designed (improper choice of machined threads; wrong plating selection; galvanic coupling from dissimilar metals; choice of lubricant; etc.)?
Were the bolts abusively used? For example, was excessive installation torque applied, or were the bolts exposed to unexpected, aggressive service environments?

Answering the following questions illustrates the use of our ability to measure the threshold stress in an accelerated manner to solve these problems.

Question 1: If we have a steel bolt with an intergranular fracture surface, how does one determine if the cause is due to plating (IHE) or to EHE, often referred to as stress corrosion cracking (SCC)?

Answer: As mentioned earlier, SEM is not sufficient to discriminate between the two mechanisms. The only way to identify the root cause is to measure the threshold in air from other unused bolts from the failed lot. The ASTM F1624 RSL test will provide the required information in the shortest time.

The bolts should have originally been qualified by a tensile test and a hydrogen embrittlement test according to some applicable standard. The fracture strength test uses a measurement of bending in air as a baseline, $R_{ns(b)}$, and the RSL test in air measures the threshold due to the residual hydrogen in the steel bolt from processing, $H_{sr\text{-}IHE}$.

The RSL extends the testing range beyond 75% to 150% TS(T) if a notched round bar tensile specimen is used or to as much as 200% TS(T) if a notched bend specimen is used. The higher values could represent actual service conditions, perhaps unintentionally, due to bending during installation.

If the RSL bend test results in the air give $H_{sr\text{-}IHE}$ equal to $R_{ns(b)}$ or about 2.0, then failure could not occur from hydrogen embrittlement due to processing and therefore could be due only to environmentally induced hydrogen-assisted stress cracking.

Mathematically speaking, for

$$H_{sr(b)\text{-}IHE\text{-}F1624} = R_{ns(b)\text{-}E8} \approx 2.0$$

the bolt is immune to IHE under any loading conditions.

If a value of $H_{sr\text{-}IHE}$ much less than 2.0 is obtained from the RSL test in air, then an excessive amount of residual hydrogen is in the plated bolts, and failure by IHE is likely.

The logic tree is presented in the flowchart of Figure 39.11.

Question 2: If a bolt has a low value of $H_{sr\text{-}IHE}$, implying that it has a large amount of residual hydrogen due to processing, how can it be determined if the remaining bolts from the lot can be salvaged?

Answer: First, an unused bolt from the same lot that has not seen service should be pulled to rupture at ASTM E8 loading rates. If the test results do not meet or exceed the minimum specified fracture strength, TS(T), it can be concluded that the fastener has microcracks and has been embrittled and therefore cannot be salvaged.

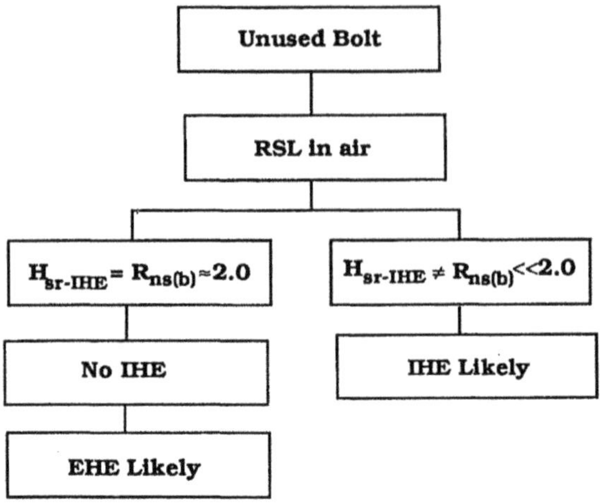

FIGURE 39.11 Summary flowchart for answer to Question 1.

If the test results do meet or exceed the minimum specified fracture strength, TS(T), it can be concluded that the fastener has no microcracks or has not been embrittled and can be salvaged because the process of removing hydrogen is reversible (see Appendix A4).

It should be recognized that the fastener contains residual hydrogen that could induce a time-delayed hydrogen embrittlement service failure. Hydrogen needs time to diffuse, and therefore testing at ASTM STD E8 loading rates would be insensitive to any residual hydrogen. All that has been established is that no permanent damage in the form of microcracks exists in the fastener or that hydrogen-induced cracking has not already occurred.

Additional baking and retesting according to ASTM F1624 should be performed until $H_{sr(b)\text{-IHE}} = R_{ns(b)} \approx 2$. Only then will the remaining bolts from the lot be free of hydrogen and available for service.

The logic tree for this answer is presented in Figure 39.12.

Question 3: What information can be obtained by RSL testing in air another fastener that has been exposed to the same service environment as the failed bolt?

Answer: From the failed bolt or an intact bolt from the same lot that has also seen service, the value of $H_{srb\text{-IHE}}$ is a measure of the extent to which the installed fasteners have picked up residual hydrogen from service in combination with the residual hydrogen from processing.

If $H_{srb\text{-IHE}} \approx 2.0$ for the unused bolt from the failed lot, then none of the degradation is due to residual hydrogen from processing.

If the $H_{sr\text{-IHE}}$ value of the failed bolt is the same as that of the unused bolt, then it is possible that the problem is isolated to a single bolt, and one should examine the failed bolt for pitting or local corrosion. If the $H_{sr\text{-IHE}}$ value of the failed bolt is lower than that of the unused bolt, then it is possible that the bolt picked up hydrogen from the service environment.

If the $H_{sr\text{-IHE}}$ value of another intact service bolt is the same as that of the unused bolt, then it is possible that the problem is isolated to a single bolt, and one should examine the failed bolt for pitting or local corrosion. If the $H_{sr\text{-IHE}}$ value of the service bolt is lower than that of the unused bolt, then it is possible that the failure is an isolated case.

If the value of $H_{sr(b)\text{-IHE}}$ for an intact bolt is much less than that for an unused bolt, then the problem could be related to all bolts from that lot that had service exposure.

This logic tree is presented in Figure 39.13.

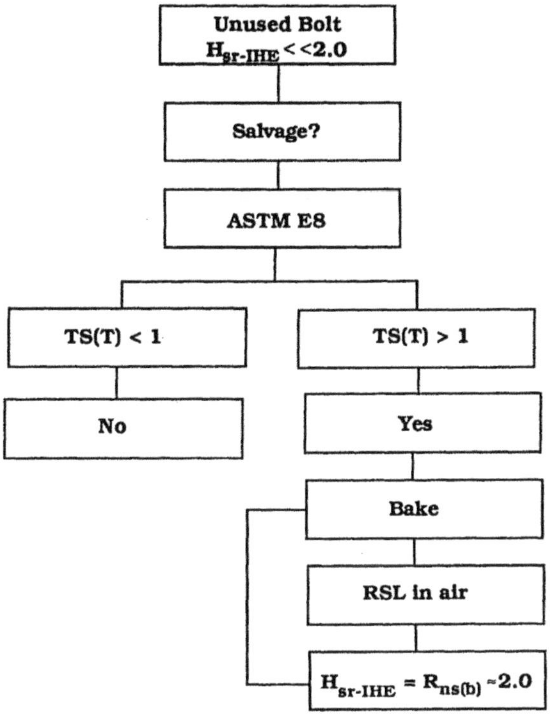

FIGURE 39.12 Summary flowchart for answer to Question 2.

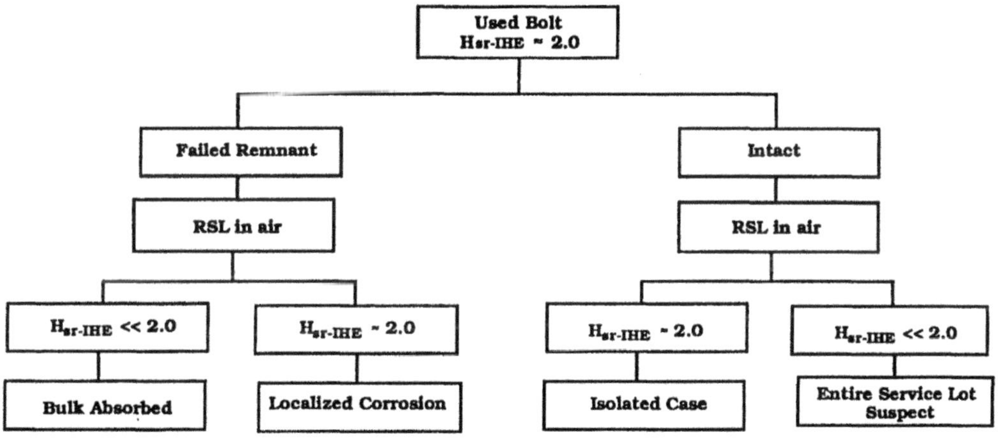

FIGURE 39.13 Summary flowchart for Question 3.

> *Question 4:* What information can be obtained by RSL testing an unused fastener in a hydrogen charging environment?

> Answer: Test for the threshold in a hydrogen charging environment is a measure of the material's susceptibility to hydrogen-induced stress cracking, regardless of the source of the hydrogen commonly found in bolting applications (see Appendix A5).

A bolt that has no residual hydrogen from processing should be used to conduct environmental susceptibility tests. This could be either an unplated bolt from the lot or a plated bolt. A hydrogen

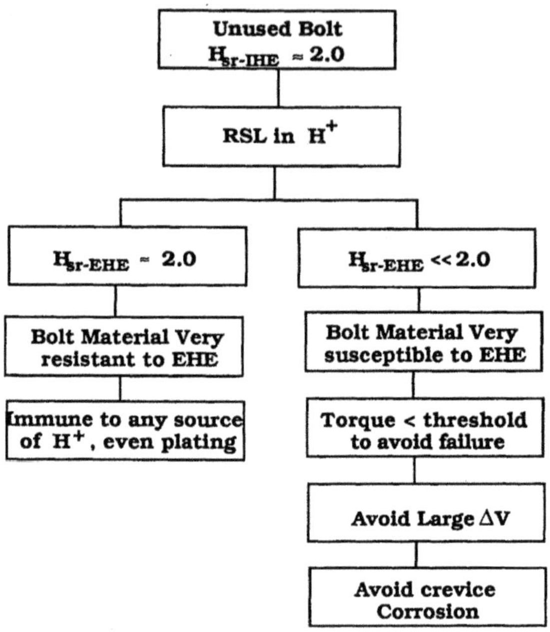

FIGURE 39.14 Summary flowchart for Question 4.

charging potential of –1.2V vs. a saturated calomel electrode (SCE) in a 3.5% salt water solution is recommended as the source of hydrogen, which should represent a worst-case situation under hydrogen charging conditions.

If the threshold $H_{sr(b)}$ is equal to $R_{ns(b)}$ or 2.0, then the material is very resistant to hydrogen-induced stress cracking. That lot of material would be essentially immune to any source of hydrogen during either plating or coating.

If the threshold $H_{sr(b)}$ is much less than $R_{ns(b)}$ or << 2.0, then the material is very susceptible to hydrogen-induced stress cracking. Special precautions must be taken when using that lot of bolts. Specifically, the installation torque must be controlled to limit the load to below the threshold stress, and care must be taken to minimize coupling to any materials that would produce a galvanic couple that would generate hydrogen in the presence of moisture.

The logic tree for this situation is presented in Figure 39.14.

39.12.2 Use in Failure Prevention

In failure prevention, the RSL threshold can be used in manufacturing to

1. Quantitatively compare the quality of production hardware by measuring the hydrogen susceptibility ratio $H_{sr\text{-}IHE}$ on coupons cut from bolts.
2. Quantitatively identify steps in the manufacture and processing of fasteners that degrade the product with hydrogen by measuring $H_{sr\text{-}IHE}$ on standard specimens.
3. Quantitatively identify the maximum hold time prior to baking and the bake time for a plating process by measuring $H_{sr\text{-}IHE}$.

Once the source of the problem has been identified, the problem can be prevented in the manufacture of new lots of bolts by applying appropriate processing or quality controls.

Critical manufacturing and processing variables must be identified and steps are taken to avoid repeating previous failures caused by improper application of rolled threads, incorrect heat treatment, embrittling plating process, inadequate baking treatment, and similar errors. To this end,

ASTM Subcommittee F07.04 is working on methods to provide test specimens or coupons cut from the actual bolts that will isolate and control the potential damage from any of the processing steps.

ASTM STD F519 uses specimens of prescribed geometry tested under worst-case conditions under the premise that if the specimen passes the hydrogen embrittlement test, then the plated parts of lower hardness and better material quality are certain to pass the hydrogen embrittlement test. Since for some applications of low hardness steels (Grade 8 or lower), this standard is too extreme, the F07.04 Subcommittee is working on a separate standard that permits the use of coupons manufactured from the actual bolts that can be used for quality assurance. As a baseline, a notched square bar specimen has been standardized and is available for testing in bending under displacement control that allows an upper limit stress of 2×TS(T) to be attained.

Question 5: How can notched square bar specimens be used to isolate specific sources of infusion of hydrogen into the bolts during the precleaning and plating process?

Answer: A group of specimens are processed along with the bolts. These specimens are selectively removed at different steps in the precleaning and plating process. $H_{sr\text{-}IHE}$ is measured in air in less than 24 h, and the degree of embrittlement introduced by each step along the way can be quantified. As an example, acid cleaning has been shown to decrease $H_{sr\text{-}IHE}$ from 0.7 to 0.5.

Question 6: How can notched square bar specimens be used to quantitatively identify the time required prior to and during baking for a plating process?

Answer: Since bake time depends on many variables other than the hardness of the bolt—e.g., plating thickness, bath efficiency, porosity of the plating—the appropriate bake time must be measured for each plating process and monitored within each plating house.

Use a group of test specimens that are cleaned and plated as a group. Selectively remove specimens one at a time at various delay times and bake times. Measure $R_{ns(b)}$ with RSL in the air; this gives a measure of residual hydrogen removed at each step of the baking process.

Using square bar specimens, a bake time of 4 h has been shown to be inadequate to remove hydrogen from bolts plated in a Ni-Zn plating bath. From Figure 39.15 it can be seen that at least 8 h is necessary to remove all of the hydrogen.

Question 7: How can notched square bar coupons be used to isolate the contribution of residual stresses from rolled threads improve the resistance in terms of threshold stress to hydrogen stress cracking?

Answer: Coupons cut from bolts with rolled threads can be compared to threaded coupons of the same geometry that have been machined. The difference in threshold at a hydrogen charging potential of −1.2 V vs. SCE in a 3.5% sodium chloride solution would be a measure of the residual compressive stress produced by thread rolling. The irregularly shaped coupons with the threads on one surface can be tested by using four-point bend loading or testing in pure bending.

39.12.3 Fastener System Design

In fastener system design, the RSL threshold can be used for failure prevention by

1. Selecting the proper material and coating for a bolt of given size and strength in a specific environment.
2. Quantitatively selecting the maximum installation torque to prevent environmentally assisted hydrogen stress cracking for different coatings for a given size bolt by measuring $H_{sr\text{-}EHE}$ as a function of potential.
3. Quantitatively identifying the severity of a service environment on a finished fastener by measuring $H_{sr\text{-}EHE}$ as a function of environment.

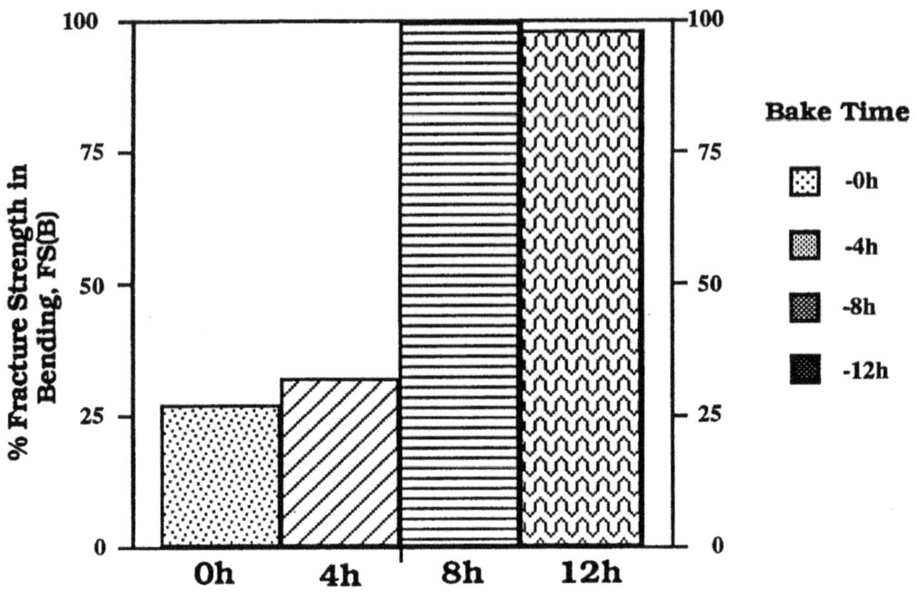

FIGURE 39.15 The effect of bake time after plating on the degradation of fracture strength in bending.

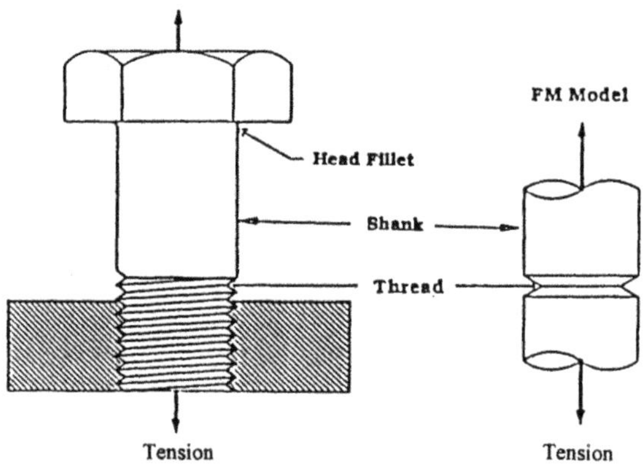

FIGURE 39.16 A circumferentially cracked round bar is used as the fracture mechanics model for threaded tensile fasteners.

To properly design a fastener system, material selection must be made on the basis of bolt size, strength requirements, coating, and environmental or service conditions.

39.12.3.1 Material

The material requirements for a given size bolt can be specified in terms of the fracture toughness-to-strength ratio as a function of the diameter. To mathematically relate these parameters, fracture mechanics must be used. As shown in Figure 39.16, a circumferentially cracked round bar is used as a model for a threaded tensile fastener.

The equation relating the tensile mode stress intensity factor K_I and the net stress or stress based on the minimum diameter, d, as shown in Figure 39.17 is found to be a function of the major diameter, D:

$$K_I/\sigma_{net} = \sqrt{D} \times f(d/D)$$

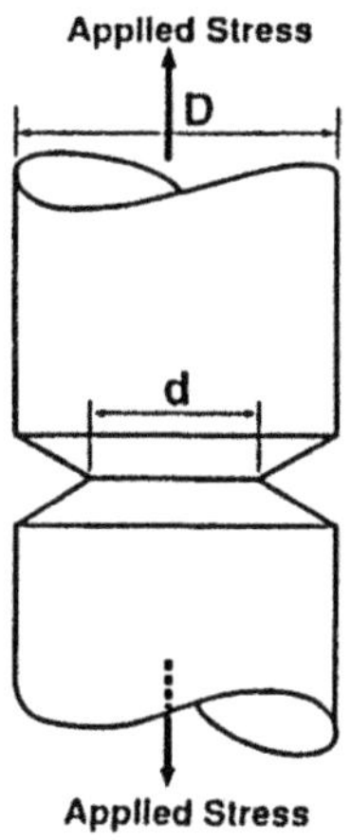

$$K_I/\sigma_{net} = \sqrt{D} \times f(d/D).$$

FIGURE 39.17 The parameters of a circumferentially notched round bar that are used in fracture mechanics to calculate the tensile mode stress intensity K_I.

TABLE 39.1
Ratio of Stress Intensity Factors to Net Section Stress for Tension and Bending

Round Bar in Tension with a Circumferential Crack $K_{It}/\sigma_{net\text{-}tension} = \sqrt{D} \times f_t(d/D)$		Round Bar in Bending with a Circumferential Crack $K_{Ib}/\sigma_{net\text{-}bend} = \sqrt{D} \times f_b(d/D)$	
d/D	$f_t(d/D)$	*d/D*	$f_b(d/D)$
0.60	0.422	0.60	0.364
0.65	0.425	0.65	0.377
0.70	0.425	0.70	0.387
0.75	0.420	0.75	0.393
0.80	0.413	0.80	0.393
0.85	0.399	0.85	0.381
0.90	0.363	0.90	0.349
0.95	0.287	0.95	0.279
0.97	0.230	0.97	0.227

Since the minor diameter is specified for each major diameter, the ratio of *d/D* can easily be calculated for each bolt size. Once *d/D* is known, the function *f(d/D)* can be calculated for both tension and bend as indicated in Table 39.1.

These values are plotted in Figure 39.18 to delineate the material property requirements in terms of the fracture toughness K_{Ic} and the yield strength YS to successfully make a threaded tensile fastener of a given diameter *D*. K_{Ic} is measured according to ASTM STD E399 or E1290. The yield strength is measured according to ASTM STD E8 (Figure 39.2). For K_{Ic}/YS values above the line, bolts of the indicated sizes can be successfully manufactured, exhibiting ductile fracture above the specified strength in a tensile test. For K_{IC}/YS values below the line, bolts of the

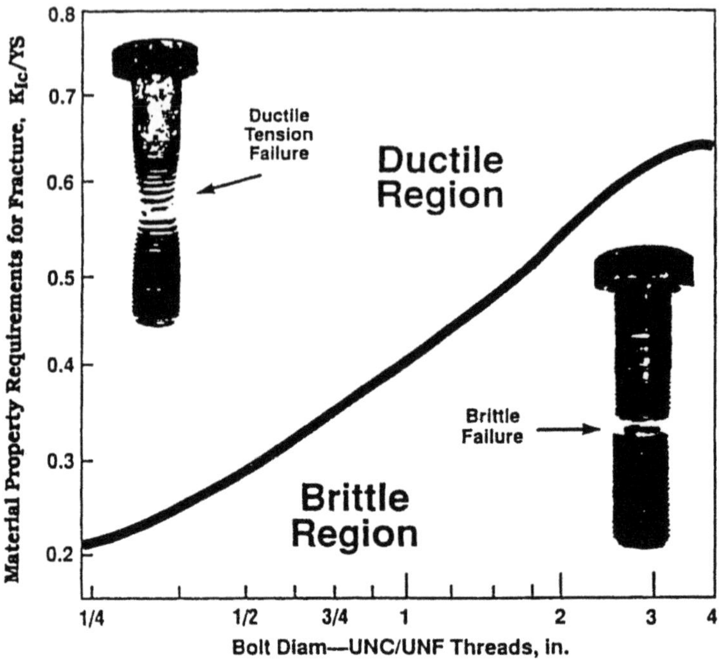

FIGURE 39.18 Material requirements in terms of the fracture toughness and yield strength are shown to increase as the bolt diameter increases. In other words, larger diameter bolts require more toughness to be successfully manufactured as a tensile fastener.

indicated sizes cannot be successfully manufactured, exhibiting brittle fracture below the specified strength in a tensile test.

39.12.3.2 Effect of Coating

A plating or coating is always of a different material than the steel bolt. If ever the coating is damaged, say on installation, and the bolt is exposed to an aqueous environment, a galvanic couple is set up between the coating and the steel. As a general guideline, the potential difference between the coating and the substrate should be minimized, but the substrate or bolt material should always remain cathodic relative to the coating. This ensures that the coating will sacrificially corrode to protect the steel bolt from rusting. The other chemical process that is taking place while the steel bolt is being protected is the evolution of hydrogen on the exposed steel surface. The greater the corrosion potential of the coating, the greater the amount of hydrogen that is being evolved.

When a bolt is installed or under stress, it acts as a sponge and can easily absorb hydrogen, creating an in situ service condition for hydrogen embrittlement. In the presence of stress, hydrogen is more easily absorbed into the steel than when the bolt is stress-free.

Environmentally induced hydrogen stress cracking is often referred to as stress corrosion cracking (SCC), although corrosion of the steel is not necessarily observed because the bolt is being protected by the coating. Since hydrogen is the cause of the embrittlement, the term EHE is used in preference to SCC, because all SCC in steel can be classified as EHE, but not all EHE can be classified as SCC.

This corrosion protection mechanism can be the source of a service-induced hydrogen embrittlement mechanism. The most appropriate coating depends on the degree of degradation in strength caused by the galvanic coupling between the anodic coating and the cathodic substrate.

The degradation in fracture toughness K_{Ic} due to hydrogen embrittlement from the plating process, $K_{Iscc\text{-}IHE}$, and that from EHE, $K_{Iscc\text{-}EHE}$, is shown in Figure 39.19 for three different steels

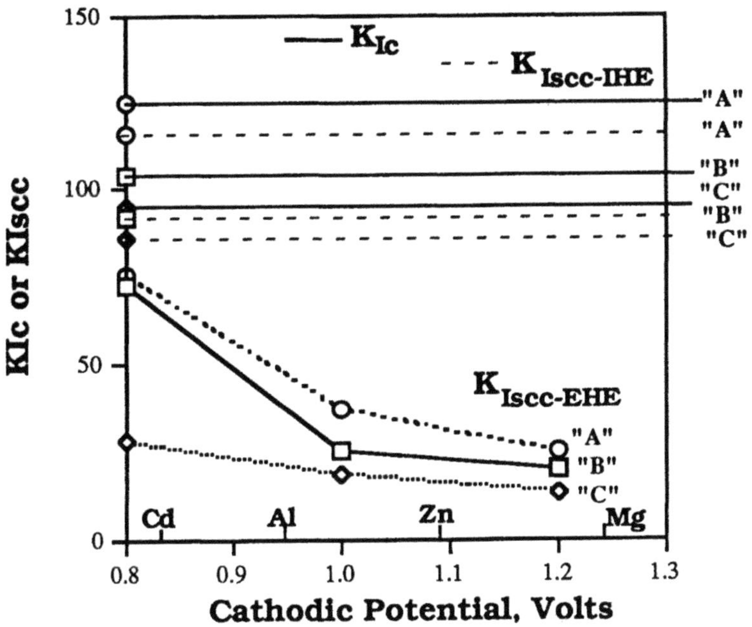

FIGURE 39.19 The effect of plating or coating a steel bolt with regard to hydrogen embrittlement from the plating process (IHE) vs. environmental embrittlement (EHE).

(denoted A, B, and C) for a range of platings from cadmium (Cd, or –0.85 V), through aluminum (Al) and zinc (Zn) to magnesium (Mg, or –1.25 V). Note that as the corrosion potential of the coating becomes more negative, the amount of hydrogen being generated increases, as does the amount of degradation in toughness. From the data presented in Figure 39.19, it can be concluded that the bolts were fabricated free of hydrogen embrittlement, $K_{\text{Iscc-IHE}} \approx K_{\text{Ic}}$, but are susceptible to environmentally induced hydrogen embrittlement, $K_{\text{Iscc-EHE}} < K_{\text{Ic}}$.

39.12.3.3 Effect of Hardness

It is generally known that as its hardness increases, a steel bolt becomes more susceptible to hydrogen embrittlement. Once a bolt is plated free of hydrogen, as shown in the previous section, it can still be susceptible to EHE. The extent to which a steel bolt can be affected by a galvanic couple between the plating or coating and the steel bolt substrate depends on the hardness or TS(T) as shown in Figure 39.20.

> *Question 8:* How can this information be used to prevent failures?
>
> Answer: Once the threshold stress has been measured for a given bolt size, processing conditions, specific plating, and in a specified environment, a chart similar to Figure 39.19 can be used to determine the maximum applied tensile stress (AS) to prevent a service failure. If the stress is maintained below threshold, a time-delayed service failure due to IHE or EHE will not occur.

Figure 39.21 delineates a series of regions for various ratios of applied stress to yield strength of the material, AS/YS. Once the ratio $K_{\text{Iscc-EHE}}$/YS has been measured, the maximum value of the AS/YS ratio can be identified for a given bolt diameter. The stress AS is the tensile stress and must be correlated to the applied torque from the torque–tension measurements to calculate the maximum applied installation torque. If a higher stress results because of bending or the scatter in the torque–tension correlation, failure will occur, and that will result in a brittle fracture. This design criterion is considered to be "safe life" as compared to "fail safe."

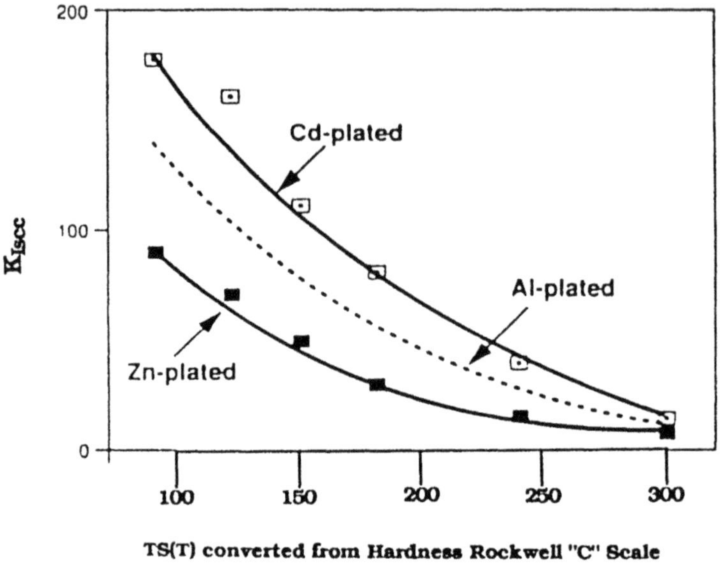

FIGURE 39.20 The amount of degradation in EHE is shown to increase as the TS(T) converted from the hardness on the Rockwell C scale (HRC) increases. Also to be noted is that for a given hardness the amount of degradation increases as the coating or plating varies from cadmium through zinc, consistent with Figure 39.19.

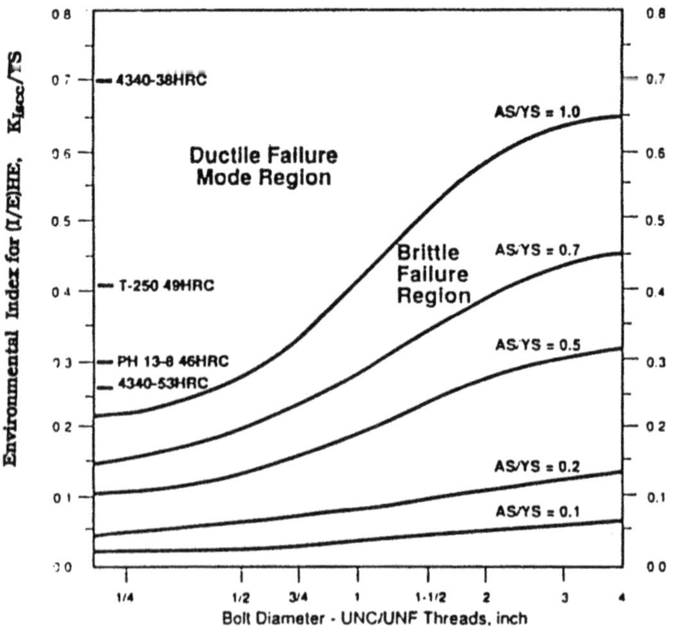

FIGURE 39.21 Knowing the value of the ratio K_{Iscc}/YS, the maximum installation torque for fracture prevention can be calculated from the stress ratio (AS/YS) for a given bolt diameter D for either internal or environmental hydrogen embrittlement.

The "fail safe" design criterion comes into effect when the ratio AS/YS is greater than or equal to 1. Then the only possible failure mode is failure by yielding. The bolt will be immune to any degradation in strength due to IHE or EHE.

Question 9: How can the state of stress be taken into account?

Answer: Ever important is the recognition that the slightest misalignment or bending under the head can greatly magnify the local tensile stress at the root of the thread. Designing with the assumption that only pure tension will be applied to a threaded bolt via a torque wrench is rather naive. If this avenue is pursued, then every precaution (such as facing the hole) should be taken to ensure that the bolt will be installed normal to the hole. Otherwise, a hydrogen-induced service failure is inevitable unless extremely low strength, ductile steel bolts are used. Again, using low installation torques or stress is not always the solution, because they can be significantly magnified by bending.

Question 10: What instrumentation can be used to avoid bending stresses that produce higher than anticipated local tensile stresses?

Answer: Even with expensive calibration instruments such as a strain gaged washer under the head of a bolt or ultrasonic displacement-measuring devices, bending could still be taking place and not be detected because these devices just measure the average axial load under the head. That is, all stress- or strain-measuring devices assume perfectly axial loads. All that is necessary to introduce bending is a slight misalignment of the head relative to the bearing surface of the bolt.

Question 11: Can installation stresses higher than anticipated in design ever be avoided?

Answer: No. Not without exorbitant expense; therefore, the only solution is to design with bending in mind. Test the bolt in bending, up to a maximum of FS(B), which can be as high as 2.3 YS or 2.0 TS(T). Testing in tension only can attain 1.5 TS(T) and might be misleading unless perfectly axial loads can be ensured during installation. Even bolting plates with predrilled and tapped holes can lead to bending just by a slight misalignment of the holes.

Question 12: Can the severity of a service environment affect the service life of a finished fastener?

Answer: Yes. Since water in the form of moisture or condensate cannot be avoided, it is most common to evaluate the performance of bolts in a saltwater environment. Throughout this chapter, this environment has been considered as the "worst case" under a cathodic hydrogen charging potential of –1.2 V. This is not to say that other special environments such as sour gas hydrogen sulfide or high-pressure hydrogen gas environments or a hydrotest fluid cannot be more aggressive. It has most recently been found that OSHA-approved environmentally compatible water-soluble dye penetrants used to find cracks can actually induce cracking in nickel-base superalloys. What is primarily considered in this chapter is the most common environment, that of water and salt.

The procedure to evaluate the severity of another environment is to measure the open-circuit corrosion potential of a coating in the anticipated environment. With the RSL, measure threshold at the imposed corrosion potential of coating in the anticipated environment. This threshold value will indicate the effect of a damaged coating with respect to environmentally induced hydrogen stress cracking in the anticipated environment.

ACKNOWLEDGMENT

Most of the material presented in this chapter has been presented as a series of articles in the *American Fastener Journal* and *The Fastening Journal*. I thank Mr. Fred Hunt of Fracture Diagnostics Inc. for his editorial and technical contributions.

Appendix A1: State of Stress

A1.A LOAD DISTRIBUTION

Load and stress are not always the same. The same load can result in different amounts of stress, depending on how the load is distributed.

Tension and Bending (Figure 39.A1, left). Bolting a rigid bar to hold a weight would produce a higher local surface tension stress on bolt 1 than on bolt 2 because bolt 1 would carry more load and therefore experience more stress than bolt 2.

Pure tension (Figure 39.A1, right). Using a pulley produces a pure axial tensile stress and therefore, for the same weight, would place less stress on bolt 1 than on bolt 2 when a rigid bar is used to hold the weight because each bolt would experience one-half the load.

A1.B WEIGHT THRESHOLD

Load and stress are not always the same. The same load can result in different stress, as demonstrated in Appendix A1.A and therefore different weight thresholds.

Plotting the time-to-failure test results as in Figure 39.A2 would produce two different threshold values. For a given weight, the rigid arm produces a higher local stress than the pulley and therefore would fracture in less time. Correspondingly, for the same weight, the pulley produces lower local stresses than a rigid arm, and therefore it would take longer for time-delayed fracture to occur under the same weight.

A1.C STRESS THRESHOLD

Load and stress are not always the same. The same load can result in different stress and therefore different points on the stress threshold (see Figure 39.A3).

The time to failure increases as the stress decreases. For the same weight, the bolts holding the rigid arm experience a higher local stress than those holding the pulley and therefore would fracture in less time.

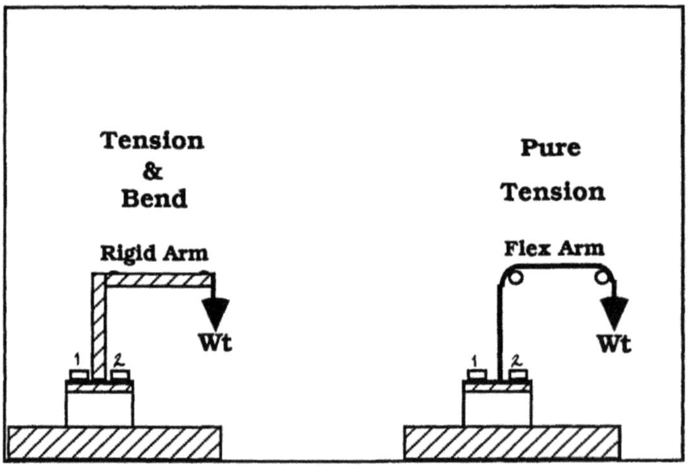

FIGURE 39.A1 Identical loads can produce different stresses if bending is involved.

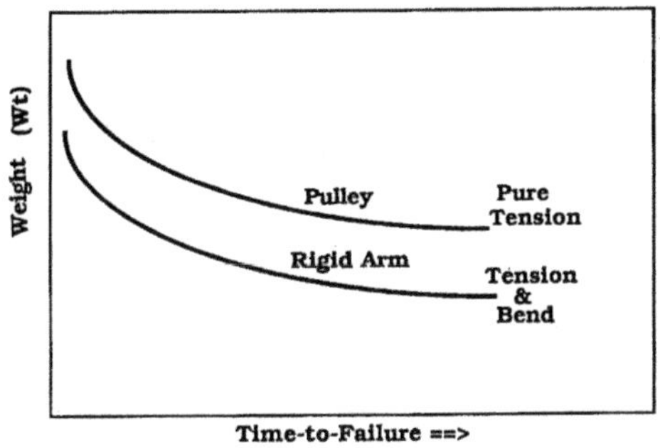

FIGURE 39.A2 Identical loads can result in different failure times and threshold loads if bending is involved.

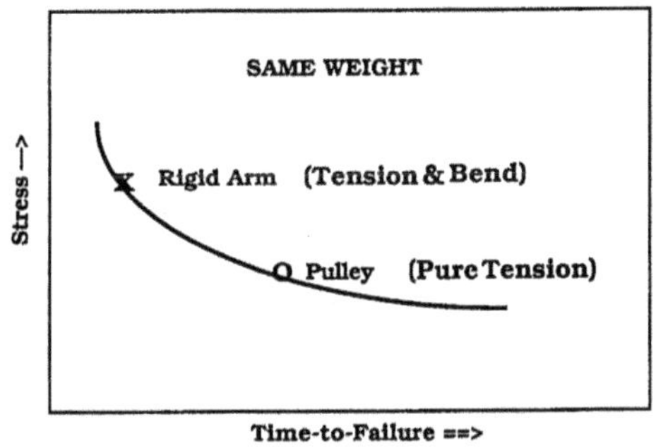

FIGURE 39.A3 Test data, including bending, can be normalized by calculating the maximum local stress that results in one stress threshold stress value.

Appendix A2: Definition of Terms: ASTM STD E8 Loading Rates

These results are independent of any residual hydrogen concentration because the tests are performed at a rate that does not allow sufficient time for the diffusion of hydrogen to occur.

A2.A TENSILE TEST

TS(T) ≡ Tensile strength (tension), ksi. Calculated from the minimum specified tensile strength (mst) and minor diameter of the fastener $= 4P_{\text{mst}}/\pi d_{\text{minor}}^2$

Example

SAE J1237/M10:

$$\text{TS}(\text{T}) = 1\left[\left(P_{\text{mst}} = 12{,}120\right)/\pi\right]\left(d_{\text{minor}} = 0.3285\right)^2 = 143 \text{ ksi}$$

FS(T) = Fracture strength (tension), ksi. Calculated from the measured fracture or ultimate tensile load of the fastener or notched or precracked test sample $= P_{\text{ult}}$/net area

$$\text{FS}(\text{T}) = 4\left(P_{\text{ult}}/\pi\right)d_{\text{minor}}^2 \geq \text{TS}(\text{T})$$

$R_{\text{ns(t)}}$ = Notched strength ratio in tension. Calculated from FS(T)/TS(T) $= R_{\text{nst}} \leq 1.5$.

A2.B BEND TEST

$$\text{YS}(\text{B}) \equiv \text{TS}(\text{T}), \text{ which for M10} = 143 \text{ ksi}$$

$$P_{ys(b)} \Rightarrow \left[\text{TS}(\text{T}) \times \pi d^3\right]/256 = \left[143 \times \pi\left(0.3285^3\right)\right]/256 = 62 \text{ lb}$$

$$\Rightarrow d \times P_{\text{TS(T)}}/64 = (0.3285) \times (12{,}120 \text{ lb})/64 = 62 \text{ lb}$$

FS(B) = Fracture strength (bend), ksi. Calculated from the measured fracture or ultimate bend of the fastener or notched or precracked test sample.

$$\text{FS}(\text{B}) = 256 \times \left(P_{\text{measured bend}}/\pi\right) d^3_{\text{minor}}$$

$R_{ns(b)}$ = Notched strength ratio in bending (ASTM E812). Calculated from FS(B)/TS(T) = $R_{nsb} \leq 2.3$.

Appendix A3: Definition of Terms: RSL – ASTM F1624 Loading Rates

A3.A IN TERMS OF STRESS

These results are independent of any residual hydrogen concentration because the tests are performed at a rate that does allow sufficient time for the diffusion of hydrogen to occur.

RSL = Rising Step Load test to measure the threshold stress per ASTM F1624, which is slow enough to allow for the diffusion of hydrogen to occur.

$\sigma_{th\text{-}air(t/b)}$ = Threshold stress at a given loading rate measured in air per ASTM F1624 in either tension or bend.

$\sigma_{th\text{-}@V(t/b)}$ = Threshold stress at a given loading rate measured at a given cathodic potential per ASTM F1624 in either tension or bend.

$\sigma_{th\text{-}H^+(t/b)}$ = Lower limit of the threshold stress at an invariant loading rate measured in the most aggressive hydrogen charging environment of –1.2V vs. SCE per ASTM F1624 in either tension or bend.

$H_{sr(t/b)}$ = Hydrogen susceptibility ratio ≡ $\sigma_{th\text{-}H^+(t/b)}$. A material property that is a function of composition, melting practice, thermomechanical processing, and heat treatment and is much more sensitive to microstructure than ASTM E**8** mechanical properties.

Evaluation Parameter

%FS(T/B) = Degradation factor = % fracture strength calculated from

$$\begin{aligned}\%\text{FS}(\text{T/B}) &= H_{sr(t/b)}/R_{ns(t/b)} \times 100\% \\ &= \sigma_{th\text{-}H^+(t/b)}/\text{TS}(\text{T})/\text{FS}(\text{T/B})/\text{TS}(\text{T}) \\ &= \sigma_{th\text{-}H^+(t/b)}/\text{FS}(\text{T/B})\end{aligned}$$

or the percentage degradation in strength as measured by the ratio of the worst case or minimum threshold value divided by the maximum attainable strength in air.

A3.B DIMENSIONLESS PARAMETERS

The test results are in terms of the notched strength ratio, $R_{ns(t/b)}$, which is the threshold stress normalized to TS(T). These results are independent of any residual hydrogen concentration because the tests are performed at a rate that does allow sufficient time for the diffusion of hydrogen to occur.

RSL = Rising Step Load test to measure the threshold per ASTM F1624, which is slow enough to allow for the diffusion of hydrogen to occur.

$H_{sr(t/b)\text{-IHE}}$ = Threshold notched hydrogen susceptibility ratio in either tension or bend for IHE which is given by

$H_{sr(t/b)\text{-IHE}} \equiv \sigma_{th\text{-air}(t/b)}/TS(T)$

$H_{sr(t/b)\text{-EHE}}$ = Threshold notched hydrogen susceptibility ratio in either tension or bend at a given cathodic potential for external or environmental hydrogen embrittlement, (EHE), which is given by

$H_{sr(t/b)\text{-EHE}} \equiv \sigma_{th\text{-@V}(t/b)}/TS(T)$

$H_{sr(t/b)@\text{-}1.2V}$ or $H_{sr(T/B)}$ ≡ Lower limit threshold notched hydrogen susceptibility ratio in either tension or bend for hydrogen-induced stress cracking in the most aggressive hydrogen charging environment of –1.2 V vs. SCE, which is given by

$H_{sr(t/b)@\text{-}1.2V} = \sigma_{th\text{-}H^+(t/b)}/TS(T) = H_{sr(t/b)}$

Appendix A4: Three Types of Hydrogen Embrittlement per ASTM STP 543 and Stress Corrosion Cracking

IHE is reversible (ductility can be restored provided that microscopic cracks have not yet initiated); also termed slow strain rate embrittlement and delayed failure. This type of embrittlement is referred to as the classic type of hydrogen embrittlement in steel although IHE has also been observed in a wide variety of other materials including nickel-base alloys and austenitic stainless steels provided that they are severely charged with hydrogen. For steels, IHE is most severe at room temperature. The problem results primarily from electroplating. Other sources of hydrogen are the processing treatments such as melting and pickling.

Environmental hydrogen embrittlement (EHE) is found in plated parts that cathodically protect the metal from corroding. Generates hydrogen at the surface of the metal. Produces a clean, intergranular fracture surface. Not reversible. Microcracking initiates on the outer surface. Other external sources of hydrogen, such as high-pressure hydrogen gas or hydrogen sulfide gas, can also cause EHE.

Stress corrosion cracking (SCC): Corrosion-produced hydrogen that is absorbed by metal under stress. Considered to occur while under anodic polarization. Not reversible. Produces an oxidized, intergranular fracture surface.

Reaction hydrogen embrittlement (RHE) is quite distinct from the other two types of embrittlement in that the hydrogen may react near the surface or diffuse a substantial distance before it reacts. Hydrogen can react with itself, with the matrix, or with a foreign element in the matrix. The new phases that form are usually quite stable, and embrittlement is not reversible. A hydride phase can be formed that is either spontaneous or strain-induced. Molecular hydrogen can be formed to produce flaking or "fisheyes." Methane gas can be formed in low alloy steels. Hydrogen can react with copper to form steam, resulting in blistering and a porous metal component.

Appendix A5: Comparison Chart

Similarities and Differences between Hydrogen Embrittlement and Stress Corrosion Cracking

IHE (Hydrogen Embrittlement)	EHE (Stress Corrosion Cracking)
Time-delay fracture when stressed in the air after exposure to environment.	Time-delay fracture in environment after exposure to stress in the air.
Requires $[H^+]_{critical}$ from processing (plating, acid cleaning) for a given stress above hydrogen stress cracking threshold.	Requires $[H^+]_{critical}$ from environment (corrosion reaction) for a given stress above hydrogen stress cracking threshold
Plating, acid cleaning, etc., considered environment in which "corrosion reaction" takes place to generate H^+, while stress is below threshold.	"Corrosion reaction" takes place in environment to generate H^+ or other environmental sources of H^+, while stress is above threshold.

The only difference between hydrogen embrittlement and stress corrosion cracking is the sequence of stress application and exposure to hydrogen. The phenomenon is the same. The mechanism is the same.

IHE (hydrogen embrittlement): $[H^+] + \text{stress} \rightarrow \text{rupture}$
EHE (stress corrosion cracking): $\text{Stress} + [H^+] \rightarrow \text{rupture}$

Note: If the part was bent or plastically deformed to introduce residual stress before being put into a plating bath, it would have cracked during the plating process, and the process would have been identified as hydrogen embrittlement although it had all the ingredients of stress corrosion cracking.

NOTE

1 Rising Step Load (RSL)® is a registered trademark of Fractured Diagnostics Incorporated.

REFERENCES

1. Raymond, L. (ed.), *Hydrogen Embrittlement: Prevention and Control*, ASTM STP-962, ASTM, Philadelphia, PA, 1988, pp. 10–16.
2. Dull, D. L., and L. Raymond, Electrochemical techniques, In *Hydrogen Embrittlement Testing* (L. Raymond, ed.), ASTM STP-543, ASTM, Philadelphia, PA, 1974, pp. 20–33.
3. Raymond, L., Hydrogen embrittlement control, *ASTM Standardization News*, December 1985, pp. 46–49.
4. Raymond, L., The many faces of hydrogen embrittlement, *ASTM Standardization News*, October 1996, pp. 40–43.
5. Raymond, L., Tests for hydrogen embrittlement, In *Metals Handbook*, Vol. 8, 9th ed., ASM, Metals Park, OH, 1985, p. 537–543.
6. Raymond, L., Evaluation of hydrogen embrittlement, In *Metals Handbook*, Vol. 13, 9th ed., ASM, Metals Park, OH, 1987, pp. 284–289.
7. National Materials Advisory Board. *Characterization of Environmentally Assisted Cracking for Design*, NMAB No. 386, National Academy of Sciences, Washington, DC, 1982.
8. Dull, D. L., and L. Raymond, Stress history effect on incubation time for stress corrosion crack growth in AISI 4340 steel, *Met. Trans.* 3:2943–2947 (1972).
9. Raymond, L., and W. R. Crumly, Accelerated, low-cost test method for measuring the susceptibility of HY-steels to hydrogen embrittlement, In *Current Solutions to Hydrogen Embrittlement Problems in Steels*, ASM, Metals Park, OH, 1982, pp. 477–480.
10. Troiano, A. R., and R. F. Hehemann, Hydrogen embrittlement and stress corrosion cracking in sour gas environments, *Current Solutions to Hydrogen Embrittlement Problems in Steels*, ASM, Metals Park, OH, 1982, pp. 299–307.
11. Troiano, A. R., The role of hydrogen and other interstitials in the mechanical behavior of metals: Campbell Memorial Lecture, *Trans. ASM*, 52:54 (1960).
12. Interrante, C. L., and L. Raymond, Hydrogen damage, In *Corrosion Tests and Standards: Application and Interpretation*, ASTM Manual Series MNL 20, ASTM, West Consohocken, PA, 1996, p. 227.
13. Raymond, L. (ed.), Hydrogen embrittlement of fasteners: Problems and solutions, In *First International Technology Transfer Conference*, Denver, CO.

40 Vibration- and Shock-Induced Loosening

Daniel P. Hess

40.1 INTRODUCTION

Threaded fasteners are used when a load-bearing connection that can be disassembled without destructive methods is required in a design. As a result, they have found widespread use in many machines and structures. However, assemblies that use fasteners are often subjected to dynamic environments. These environments include vibration, shock, and/or thermal cycling. Such conditions can lead to fastener loosening, which can result in increased maintenance and/or failure. This chapter focuses on shock- and/or vibration-induced fastener loosening.

Section 40.2 begins with basic comments on the bolted joint and on how friction, ideally, maintains preload. It is pointed out that the application of static and/or dynamic loads to a bolted assembly can result in loss of preload. The analogy of a block on an incline is used to show how this is possible.

Section 40.3 presents a survey of previous and ongoing theoretical analyses and experimental work focused on the problem of dynamic loosening of threaded fasteners. This work spans more than five decades. This review will provide the reader with a better understanding of various mechanisms of self-loosening. In addition, design guidelines for minimizing fastener loosening are presented.

Over the years, numerous techniques and devices have been developed that resist self-loosening. Many of these locking devices have well-established standards, while others are proprietary or fairly new. An overview of many of the existing anti-loosening devices is included in Section 40.4. This will provide the reader with a broad view of various locking features available today as well as assembly design considerations that can inhibit fastener loosening.

Descriptions of various vibration and shock tests that have been used to assess and quantify self-loosening properties are included in Section 40.5. Both standardized and nonstandardized tests are discussed. Such tests are primarily useful for comparing fastener locking capabilities under test-specific dynamic environments, which can be quite different from actual service conditions.

40.2 BOLTED JOINT AND BLOCK/INCLINE ANALOGY

As a bolted joint is assembled to a desired preload or clamping force, the bolt stretches, the joining components compress, and the potential energy in the system increases. If the friction at the thread interface and nut and bolt head interfaces can hold this energy, the preload in the joint is maintained. However, external loads, both static and dynamic, can alter the forces and moments in the system such that the holding friction forces are overcome and loosening occurs.

The simple analogy of a block on an inclined plane is often used to illustrate these concepts. Figure 40.1a shows a block loaded by gravity on an incline. Figure 40.1b shows the forces acting on the block: block weight W, normal contact force N, and friction force f. The weight W can be resolved into two components: $W \cos \beta$ perpendicular to the contact plane and $W \sin \beta$ parallel to that plane. The force $W \sin \beta$ acts to move the block down the plane but is resisted by the friction force f. The friction force has an upper limit that is equal to the normal force $N = W \cos \beta$ multiplied by the coefficient of friction μ. As a result, the block will not slide down the plane when the condition

DOI: 10.1201/9781003501183-46

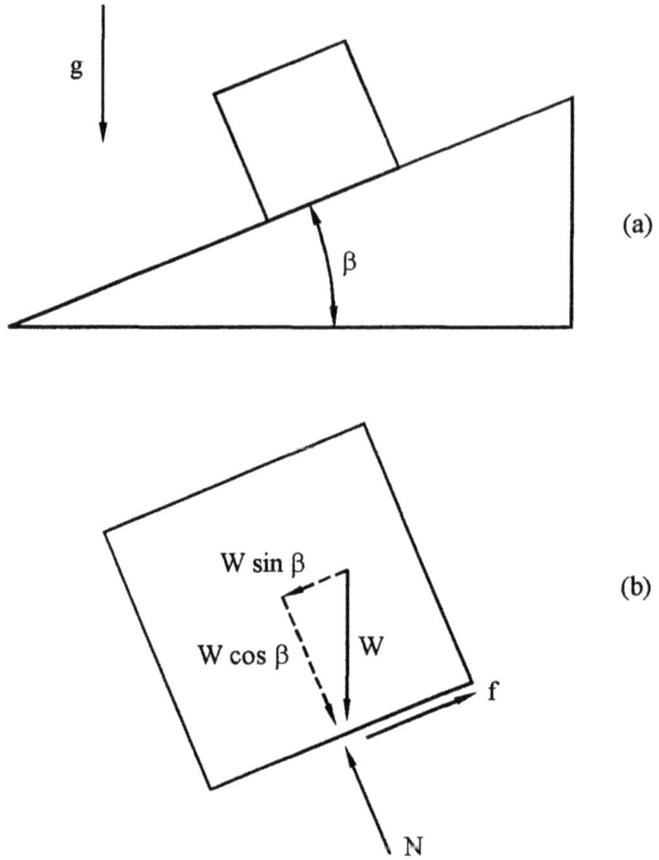

FIGURE 40.1 Simple analogy for bolted joint. (a) Block on incline loaded by gravity; (b) forces acting on block.

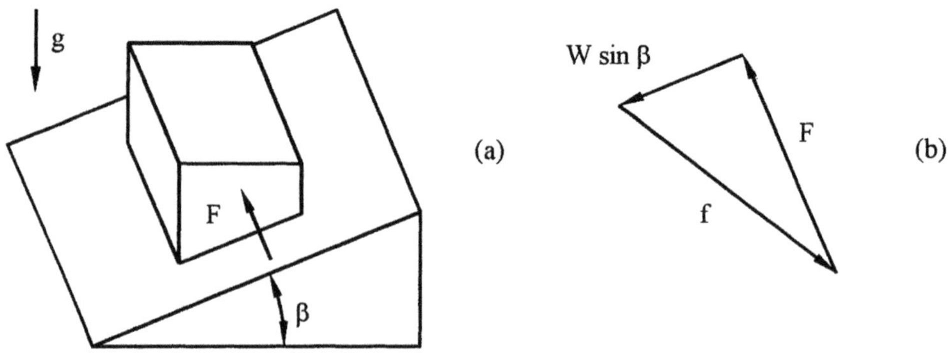

FIGURE 40.2 (a) Block on incline with external lateral force; (b) force balance in the plane of contact.

$$\mu W \cos \beta > W \sin \beta \tag{40.1}$$

holds true.

To illustrate the effect of an external static force on the system, consider a lateral force being applied to the block as shown in Figure 40.2a. In general, the friction force will resist all forces acting on the bock in the plane of contact. In this case, the friction force resists both force $W \sin \beta$ and lateral force F. The static force balance in the plane of contact is shown in Figure 40.2b. Note that the direction of the friction force opposes the direction of motion, which is now a combined down and lateral direction. The condition for no slip in this case becomes

$$\mu W \cos\beta > \left[F^2 + (W \sin\beta)^2\right]^{1/2} \tag{40.2}$$

As the lateral force F is increased, this condition will eventually not be met, and slip will occur. Both of these conditions are based on static force equilibrium balances. In other words, no dynamic effects, such as the inertia of the block and dynamic forces, are taken into account. This is an important point because many of the early theories on vibration-induced loosening are based on static torque balances and do not consider the dynamic effects.

The application of dynamic forces to either the block or the slider can result in further interesting behavior as exemplified by vibratory conveyors and feeders. These devices are used in manufacturing to supply parts or materials at a specified rate. They consist of a vibrating incline that transports parts up or down the incline to a desired destination. The most complete study on this subject, which includes dynamic analyses and experiments, is given by Gaberson (1972a,b). This and other work have resulted in a wealth of information on the interaction of vibration and friction of block on inclined plane-type systems. Similar phenomena occur in threaded component assemblies. However, due to the complex geometry of such systems, comprehensive dynamic modeling and analyses are quite involved. Nevertheless, as discussed in later sections, advances are being made in this area.

40.3 SURVEY OF WORK ON DYNAMIC LOOSENING

The following pages chronologically review the research and progress made toward understanding vibration- and shock-induced loosening. Work included ranges from as early as 1945 to current and ongoing funded research planned into the next century.

40.3.1 Goodier and Sweeney (1945)

Goodier and Sweeney (1945) of Cornell University were probably the first to theoretically and experimentally study the loosening of a bolt and nut under axial dynamic loading. Their experiments were performed with a set of ¾–10 nuts and bolts that were repeatedly loaded and unloaded in tensile test machines. As shown in Figure 40.3, microscopes attached to the bolt focused on pointers fixed to the nut were used to measure small nut rotation on the bolt (or bolt turn within the nut). Data showing loosening by as much as 0.0055 revolution in 500 cycles of load fluctuation were presented. This corresponds to about 73% loss in bolt tension for the tests. Data for individual loading–unloading cycles show loosening rotation during loading and tightening rotation during unloading, with a small net loosening rotation of the nut. This type of relative motion during loading and unloading is referred to as frictional ratcheting.

The theory of loosening proposed by Goodier and Sweeney is based on the idea that an increase in load causes a bolt thread to move radially inward and a nut thread radially outward. The direction of the friction forces will be the same as the direction of relative motion, which will be radial if there is no circumferential motion. However, the nut cannot remain in equilibrium if the friction is radial because circumferential friction is required to counteract the loosening torque of the normal component of thread pressure. Therefore, a load increase on a bolt causes loosening rotation of the nut, which causes the direction of both the thread motion and friction to be no longer exactly radial. This theory, which is based on static equilibrium conditions, predicts a net loosening twisting per cycle. Goodier and Sweeney explained that the apparent tightening observed in the experiments during unloading is due to elastic bolt twist.

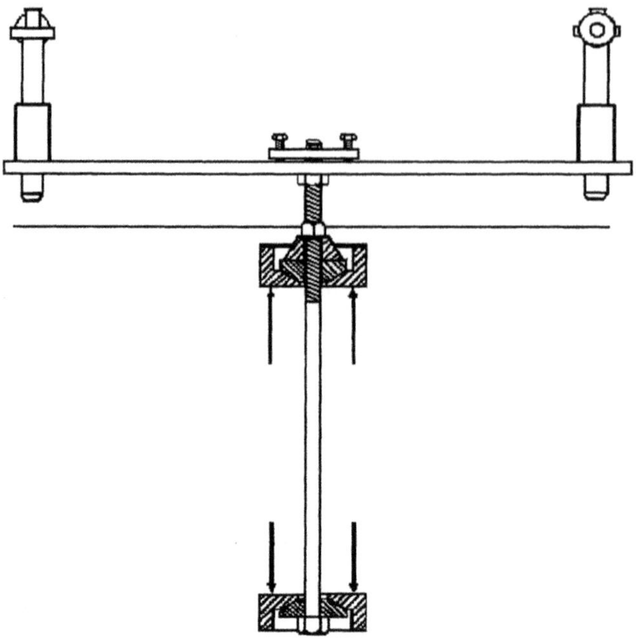

FIGURE 40.3 Test apparatus of Goodier and Sweeney (1945).

Based on force and moment equilibrium conditions, Goodier and Sweeney derived expressions for the torque required to loosen a bolt and nut during loading and unloading portions of a cycle. These expressions do not assume purely circumferential friction as do most nut and bolt torque equations. During loading when the nut is expanding the external torque just sufficient to loosen is

$$T_{off} = Wr\frac{\mu(\sin\gamma\cos\alpha - \cos\gamma\sin\alpha\sin\beta) - \sin\alpha\cos\beta}{\mu(\sin\gamma\sin\alpha + \cos\gamma\cos\alpha\sin\beta) + \cos\alpha\cos\beta} + p\mu\sin\gamma \qquad (40.3)$$

where

W is the load
r is the pitch radius
β is the half-angle of the thread
α is the helix angle
μ is the thread coefficient of friction
p accounts for the nut radius and friction, which are different than r and μ
γ is the angle of the friction force from the radial direction
Note that if the friction is assumed to be entirely circumferential, then γ is 90°.

The effect of radial nut and bolt motion on this loosening torque is most easily assessed from a plot of T_{off} versus angle 7. Such a plot is shown in Figure 40.4 with T_{off} normalized by Wr. The plot shows that T_{off} decreases as y decreases from 90°. Clearly, the effect of loading and corresponding radial motion is comparable to the effect of a lateral force acting on the block in the block-and-incline analogy discussed earlier. Note that T_{off} becomes less than zero for $\gamma < 8°$. This means that loosening will occur without any external torque if $\gamma < 8°$. Goodier and Sweeney estimated the angle γ to be about 6° for the nut, bolt, and loading conditions used in their experiments. Since this is within the range of negative bolt torque for loading, they conclude that the slip of the nut is well accounted for by this theory of loosening.

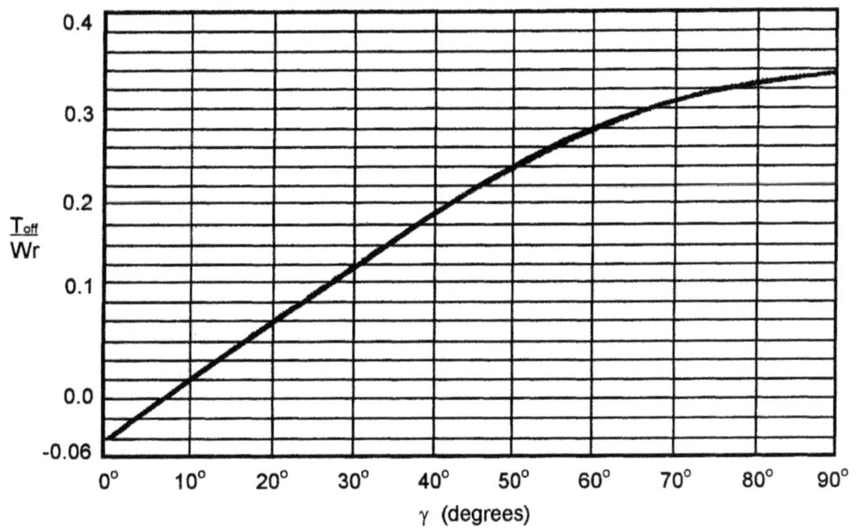

FIGURE 40.4 Theoretical nut torque curve. (Adapted from Goodier and Sweeney, 1945.)

40.3.2 Sauer et al. (1950)

The alternating axial loading in the experiments of Goodier and Sweeney (1945) was obtained by slow repeated use of a standard tensile testing machine. Sauer et al. of Pennsylvania State College felt it would be worthwhile to perform an experimental study on fastener loosening that subjected a bolt and nut assembly not to quasi-static loading but to vibratory conditions more representative of actual practice.

Their apparatus (Sauer et al., 1950) used a Sonntag universal fatigue testing machine that could produce up to 1,000 lb of 30 Hz sinusoidal axial load superimposed on the static preload of a nut and bolt assembly fixture. The top end of the fixture was attached to the eccentric rotating counterweight of the Sonntag machine using flex plates such that only vertical alternating load was transmitted to the bolt assembly fixture. The bottom end of the fixture was fastened to springs. The extension of these springs provided a measure of the initial static bolt tension. An optical lever system consisting of point light sources, small mirrors, and curved scales was used to measure the loosening of the nut relative to the bolt.

Their experimental tests examined the effect of alternating cycles of axial loading, dynamic to static load ratio, contact surface condition, and misalignment on the loosening of threaded components. For a given set of conditions, the same test was performed for anywhere from four to 20 nut and bolt specimens. The data for a given test consisted of a plot of loosening rotation in degrees versus number of cycles of dynamic loading. Since the data showed considerable scatter, it was necessary to average loosening curves to assess the effect of different variables on loosening.

The rate of loosening was generally found to be initially large but usually fell off rapidly after 2,000–3,000 cycles. It was found that loosening increases as the dynamic to static load ratio increases. Representative average loosening curves for different levels of static bolt tension ranging from 250 to 1,000 lb with the magnitude of the dynamic load fixed at 250 lb are shown in Figure 40.5. These data clearly show that for a given dynamic loading, the amount of loosening will decrease with increase of bolt tension, which was also shown to be proportional to tightening torque.

Tests performed with "once-used" or carefully cleaned and polished surfaces showed considerable reductions in loosening and scatter of data. Further tests with spherical seats to improve fastener alignment also indicated some reduction in loosening.

Sauer et al. (1950) recommended the following steps to overcome vibration loosening:

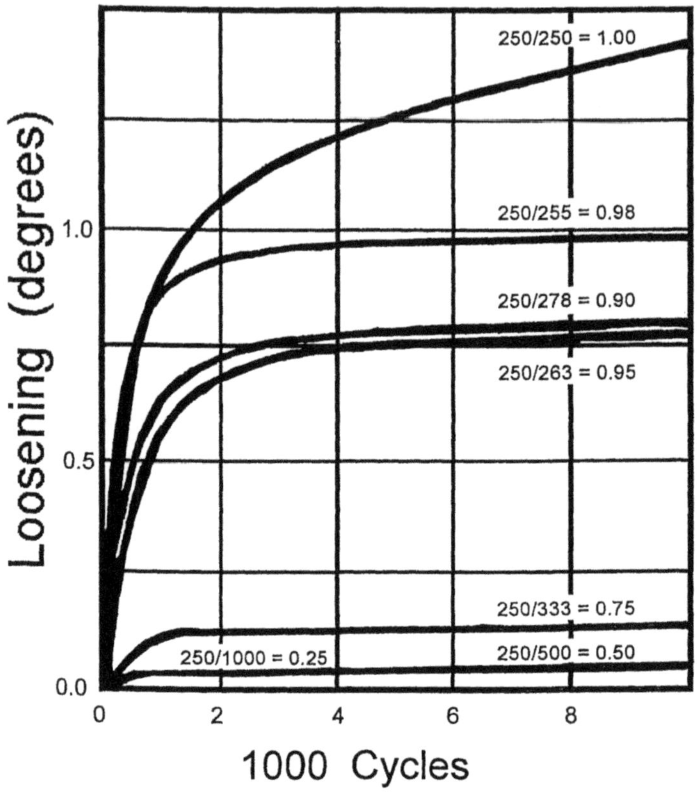

FIGURE 40.5 Loosening angle versus loading cycles data for different static bolt tension. (Adapted from Sauer et al., 1950.)

1. Properly specify the working loads so that the dynamic load will not approach or exceed the static tensile load originally developed in the tightening operation. Their results indicate that it is desirable to keep the dynamic/static load ratio in the range of 0.7 or less.
2. Adequately control the method of manufacture, fabrication, assembly, and maintenance so as to have all contact surfaces at the threads, base of nut, and head of bolt in the best possible condition of cleanliness and smoothness. Smooth, clean surfaces provide higher frictional resistance to loosening.
3. Improve alignment, as it results in better mating of the contacting parts in somewhat the same manner as does removal of dirt, grease, or surface irregularities.

40.3.3 Clark and Cook (1966)

Clark and Cook of the University of Michigan investigated the effect of fluctuating torque on loosening of a tightly seated bolt. They pointed out (Clark and Cook, 1966) that while it is true that in many machine situations, fluctuating tension remains the primary component of load on the bolt, there are also many situations in which two mating parts held by a bolt rotate a minute amount with respect to one another. Under such conditions, the bolt may be observed to loosen and eventually come out. This can be physically demonstrated with a bolted joint as shown in Figure 40.6. Here, one may see that two primary elements are needed to cause loosening of a bolt by oscillating torque or rotation. The first of these is the stationary portion, marked A, which is rigidly attached to some massive foundation. To this base is attached an ear E having a body hole so that the bolt B may be put through the hole and screwed tightly into part C. When a fluctuating torque is imposed on part C, the bolt can loosen.

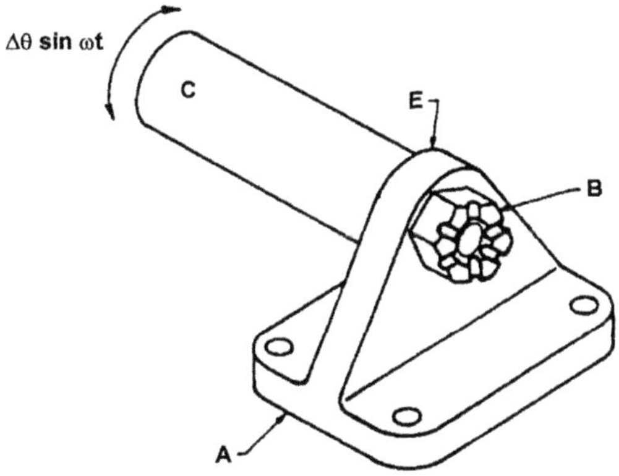

FIGURE 40.6 Bolted joint used for testing effect of oscillatory rotational motion on loosening. (Adapted from Clark and Cook, 1966.)

The experimental apparatus used consisted of a 100 lb capacity Baldwin-Sonntag fatigue-testing machine fitted with its standard torsion fixtures. This machine applied a 30 Hz oscillatory torque to part C of the bolt assembly shown in Figure 40.6. Part C was fitted with strain gages so that the axial load in the joint could be measured. The original counter supplied with the Baldwin-Sonntag fatigue-testing machine was replaced with a higher speed counter more appropriate for the bolt loosening tests.

Care was exercised in this work to obtain consistent surface and friction conditions by using a consistent greasing procedure prior to each test. As a result, little scatter was found in data from a single sample, although some scatter was usually found in data obtained from nominally identical bolts.

The test results show that bolt loosening is a function of both oscillatory torque amplitude and bolt initial shank stress. A representative plot of cycles to loosening versus oscillatory torque amplitude curves for a ½–13 standard cap screw at different initial preloads is shown in Figure 40.7. Examination of these bolt loosening curves shows that in some respects they are similar to fatigue curves. As the amplitude of the applied oscillatory torque is reduced, a larger number of cycles is required for bolt loosening, until eventually the bolt will apparently withstand a certain level of fluctuating torque indefinitely. This finding led Clark and Cook (1966) to the conclusion that accidental overloads contribute more to bolt loosening than do continuous repetitive oscillatory torques of small amplitude.

They further point out that loosening curves such as those shown in Figure 40.7 may be modeled by the mathematical expression

$$(T-\sigma)(N-1)=C \tag{40.4}$$

where

T is the amplitude of the oscillatory torque
σ is initial stress in the bolt
N is the number of cycles to loosen
C is a constant dependent on bolt design and material.

40.3.4 ESNA (1966, 1967)

Researchers from Elastic Stop Nut Corporation of America (ESNA) performed extensive dynamic tests of bolted assemblies subjected to vibration and shock, which are documented in reports from ESNA (Baubles et al., 1966; Faroni, 1967). This testing resulted in the first theory on shock-induced

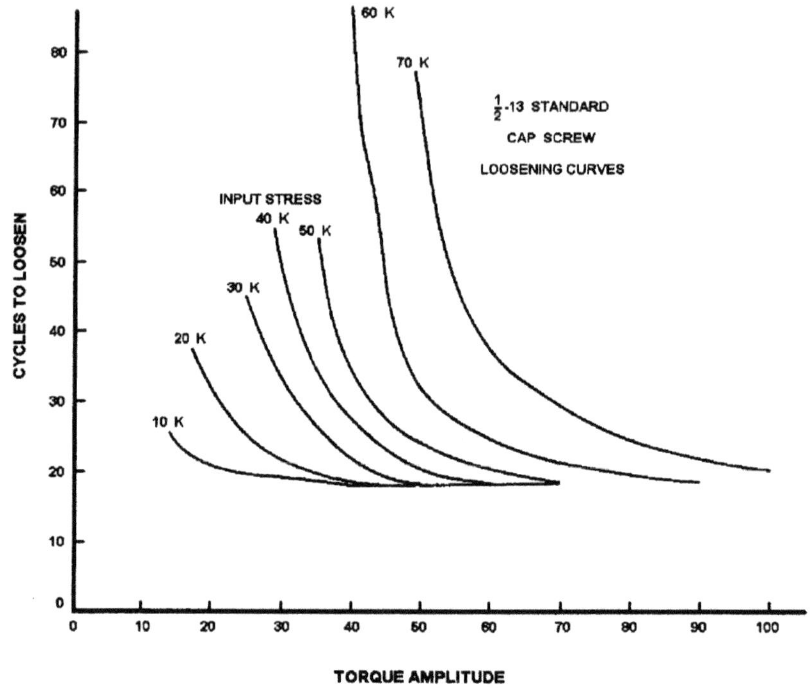

FIGURE 40.7 Oscillatory torque amplitude versus cycles to loosen data for different initial bolt stress. (Adapted from Clark and Cook, 1966.)

loosening, which is based on the idea that shocks or impacts to a bolted assembly will cause the components to resonate and loosen. This can be demonstrated using the block on an incline analogy as shown in Figure 40.8. If static fiction exceeds the component of weight that tends to cause sliding, the block remains at rest. If the plane is vibrated or if mechanical shocks are applied to it, the effective coefficient of friction is reduced, and the weight can slide down the plane.

The apparatus used in the testing consisted of spool-like arbors reciprocating within a slotted fixture as shown in Figure 40.9. This fixture is described in MIL-STD-1312-7A (U.S. Department of Defense, 1984). The arbors were held together by test nuts and bolts. The slotted fixture was vibrated at 30 Hz with a modified Sonntag universal fatigue-testing machine and at 60 Hz with an electromagnetic shaker. As the fixture vibrates, it subjects each spool to two impacts per cycle. A fixture was also developed to vibrate the arbors vertically, causing the impacts to be directed parallel to the bolt axis rather than perpendicular as with the fixture in Figure 40.9.

Test data show that the number of impacts and the severity of impacts are the variables that promote a loosening failure. Frequency of impact was found to have no effect except that a certain number of blows can be struck in less time if the frequency is raised. Direction of impact, perpendicular or parallel to the bolt axis, was found to have no significant effect by itself. Increasing bolt tension and increasing bolt length were found to tend to prevent loosening. Adding damping material, such as thread lubricant, rubber, or plastic material, to the thread interface was also found to tend to prevent loss of fastener tension.

This theory of shock-induced loosening assumes that bolts resonate. However, calculations of resonant frequencies for typical fasteners are on the order of 50 kHz or higher. Since mechanical vibrations of typical machines are well below this range, ESNA researchers believe that low-frequency impacts excite the high resonant frequencies of fasteners. Each impact results in a transient bolt vibration as shown in Figure 40.10. The energy associated with bolt resonance results in wear at critical bearing surfaces. The bolt vibration tends to hammer down and wear away minute metal

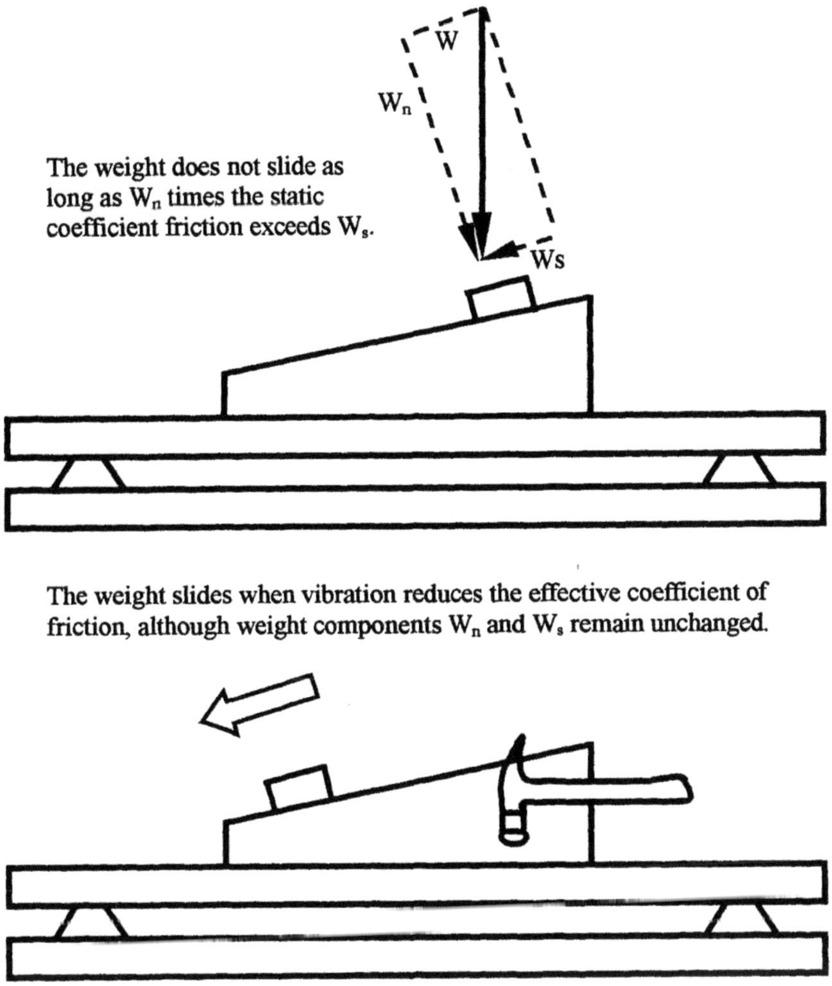

FIGURE 40.8 Block on incline analogy of loosening due to an applied shock or impact. (Adapted from Faroni, 1967.)

projections on all bearing surfaces. This results in a slow decrease in bolt tension without nut turn. Eventually, when bolt tension has dropped to a critical level in relation to vibration intensity, the nut begins to turn with rapid loss of preload. Figure 40.11 shows the described sequence of loosening mechanisms for different quality fasteners and preloads. Figure 40.11a includes the relaxation of bolt tension that occurs under static conditions.

This theory supports the experimental observation that adding damping to the bolt thread interface tends to reduce loosening, since adding such damping will reduce the duration and energy of the bolt resonant transient vibration.

40.3.5 Gambrell (1968)

Gambrell (1968) of the University of Alabama performed a series of experiments to examine how loosening is influenced by bolt thread series (coarse or fine), initial preload, lubrication, dynamic/static load ratio (DSR), frequency of loading, and number of cycles of load application. The apparatus is diagrammed in Figure 40.12. The test joint consisted of steel components with machined and ground bearing surfaces. Axial cyclic tension was applied to the test joint by a variable-speed

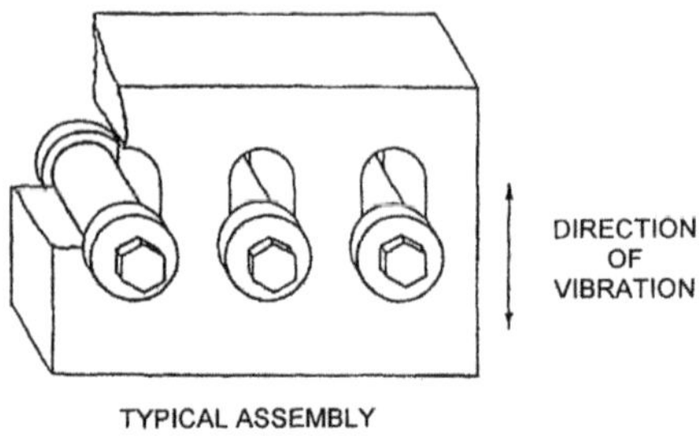

FIGURE 40.9 Test fixture used by Baubles et al. (1966) and specified in MIL-STD-1312–7A (U.S. Department of Defense, 1984.)

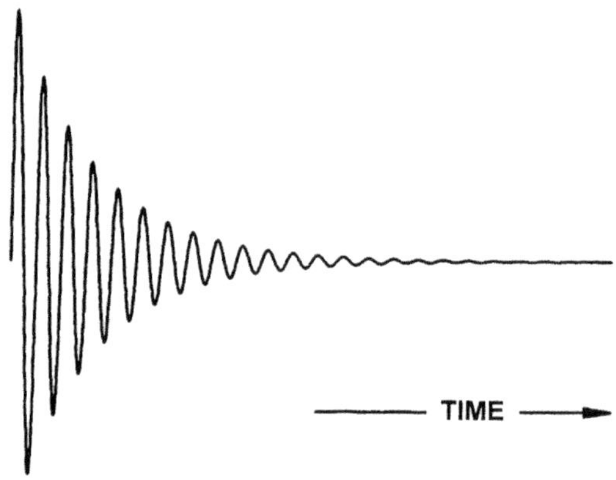

FIGURE 40.10 Transient bolt vibration.

electric motor driving a cam and follower that activates a load lever. The eyelid load was measured with a strain gage load cell. Both $^5/_{16}$–18 coarse thread and $^5/_{16}$–24 fine thread bolts and nuts were used in the test joint. The fasteners were preloaded to a yield stress of 55,000 psi using the elongation method, i.e., using a ball-end micrometer together with preload versus bolt elongation calibration curves.

Loosening by plastic deformation was prevented by limiting all loads to a value below the proof loads of the fasteners. Loosening by smoothing of joint bearing surfaces during cyclic loading was eliminated by grinding the surfaces of the test joint. Thorough cleaning eliminated any effects of grit on loosening. Loosening by isolated thread yielding could not be eliminated and therefore was reflected in the results. Impact loading was prevented by maintaining a minimum load of 200 lb on the test joint as the cam rotated.

Tests were performed for both the fine and coarse thread fasteners for DSR values ranging from 0.9 to 1.9 and for loading frequencies ranging from 3 to 20 Hz. Both dry and lubricated (SAE-30 motor oil) contacts were examined. All test results were based on a statistical analysis of data, and all statements of significance are at the 0.05 confidence level. Data showing the effect of DSR on loosening are given in Figure 40.13. The results of this work are summarized as follows:

1. For DSR < 1, no difference occurs in the loosening of coarse and fine thread bolts.
2. For DSR > 1, fine threads loosen less than coarse threads.
3. An increase in DSR increases the loosening of both thread types.
4. Loosening of fine threads cannot be attributed to lubrication. However, lubrication greatly increases the loosening tendency of coarse threads.

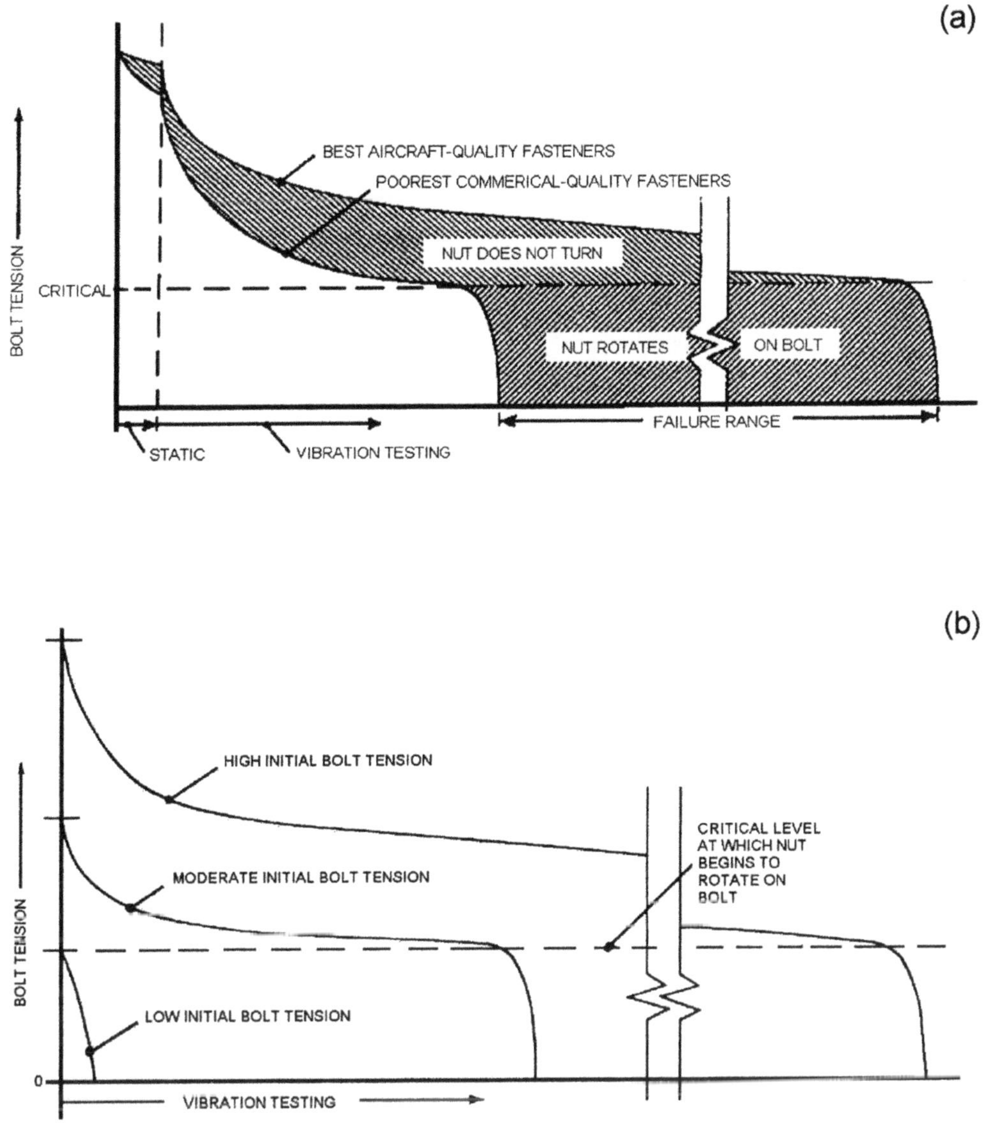

FIGURE 40.11 Sequence of loosening mechanisms (a) for different quality fasteners and (b) different preloads. (Adapted from Faroni, 1967.)

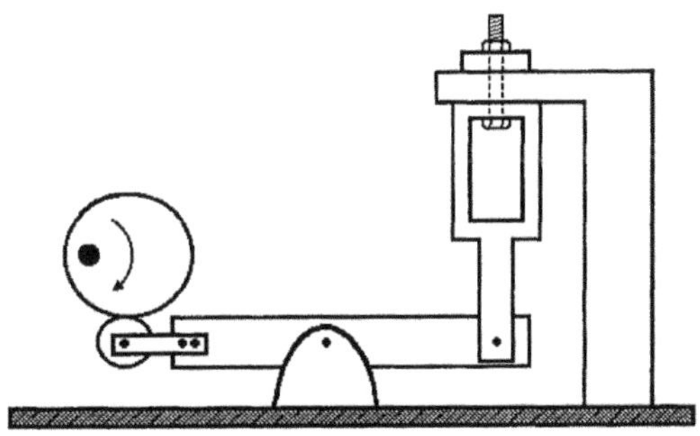

FIGURE 40.12 Test apparatus of Gambrell (1968).

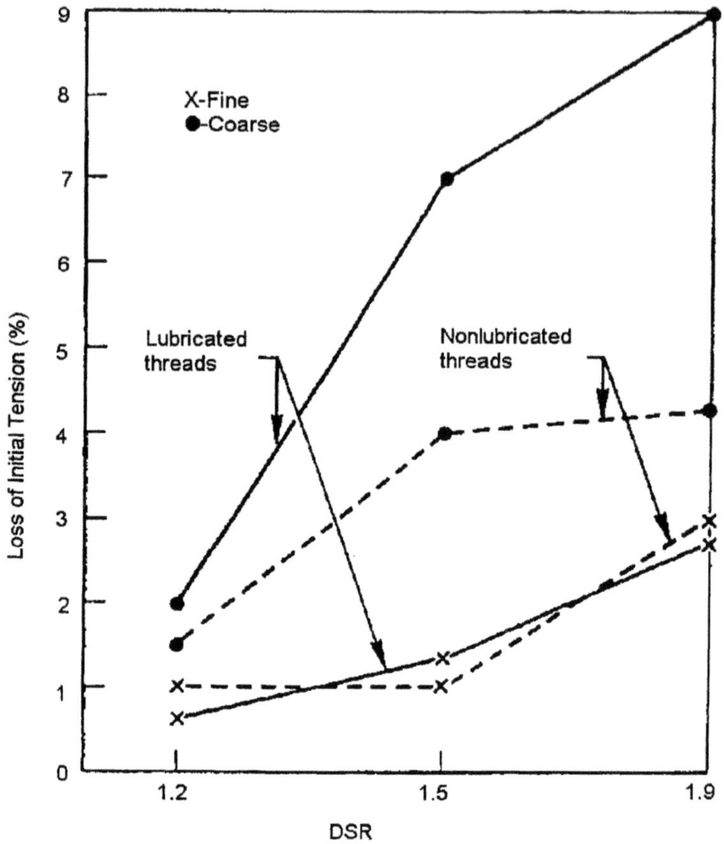

FIGURE 40.13 Loss of bolt tension versus DSR data for different thread series and lubrication conditions. (Adapted from Gambrell, 1968.)

5. Severe loosening, when present, occurs during the first 3,000 cycles. As the number of cycles increases from 3,000 to 5,000, the rate of loosening steadily decreases.
6. Within the limited range of frequencies (3–20 Hz) tested, frequency has no effect on loosening.
7. Where unpredictable loading exceeds design specifications of a joint, the choice of fine thread ensures that the joint remains tight to the greatest extent for the greatest number of cyclic load applications.

40.3.6 Junker (1969)

Junker (1969) of SPS (Standard Pressed Steel Company) was the first to study the influence of transverse vibration on the self-loosening of fasteners. He found that preloaded fasteners loosen as a result of rotation as soon as relative motion occurs between mating threads and/or bearing surfaces of a joint. His theory for loosening is that if relative motion occurs between threaded surfaces and/or other contact surfaces of the clamped and clamping parts because of an external force, the direction of which is either transverse or radial, the bolted connection loosens because the combined external force and circumferential component of the normal thread force overcome the friction forces holding the joint. He found that transverse dynamic loads generate a more severe condition for self-loosening than axial loads, due to the fact that the relative radial thread motion in axial loaded joints is substantially smaller than the relative thread and contact surface motion in transverse loaded joints. Transverse forces cause rocking motions of the screw and nut.

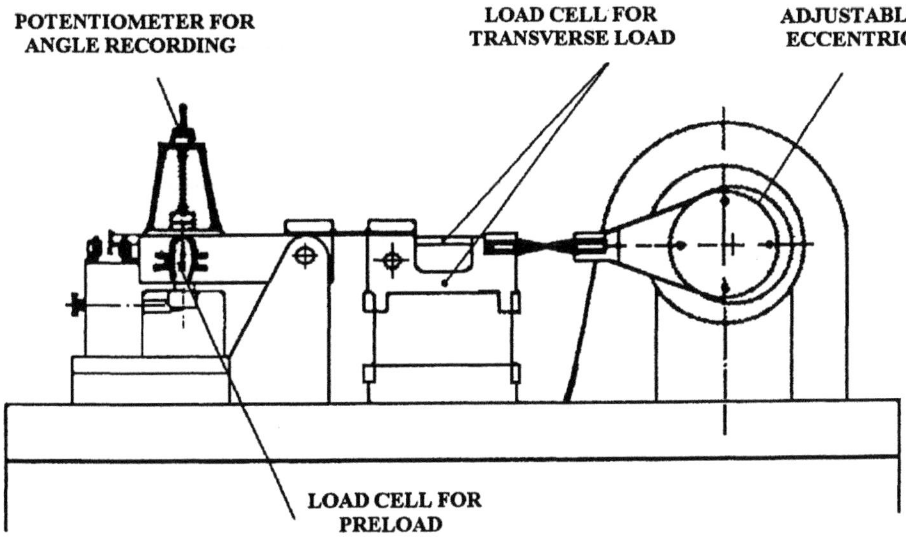

FIGURE 40.14 Test apparatus of Junker (1969).

To avoid fastener loosening, Junker recommended that joints be designed such that no relative movement occurs between the contact surfaces or the thread flanks. This requires that a sufficient residual clamping force remain after scatter of preload and embedding of contacting components are taken into consideration. The residual clamp load should never fall short of the value equal to the transverse load divided by the friction coefficient. Unfortunately, there are cases when these requirements can be met only if uneconomically overdimensioned fasteners are used. Hence, relative movements cannot always be excluded using this design guideline. In such cases, it was recommended that fasteners with locking elements be used.

Since no dynamic test method existed for quantitatively measuring the locking ability of locking elements, Junker (1969) developed a well-instrumented testing machine that subjects fasteners to transverse dynamic loading. The Junker testing machine is illustrated in Figure 40.14. Transverse sliding motion between two parts fastened with a test screw was generated by an adjustable eccentric. The resulting transverse force depends on the eccentric adjustment and is independent of frequency and starting speed of the machine. The eccentric was driven by a 5 hp AC motor with a belt drive allowing loading frequencies up to 75 Hz. The applied transverse force and joint preload were measured with load cells, the transverse displacement was measured with a linear variable differential transformer (LVDT), and the angular displacement of the test bolt was measured with a potentiometer. Data were obtained for $^3/_8$ in. test screws preloaded to 75% of the proof load. Fasteners with and without locking features were tested.

Junker explained various possible relations found in tests between transverse force and displacement with the schematic in Figure 40.15. The dashed lines correspond to a screw with narrow thread tolerance and no head slip. The screw is bent to the left by the top plate at A_1. When the plate moves to the right, the screw bends to the right until thread slip occurs at B_1. At D_1 the thread allowance is used up, so the screw bends until E_1. The same process is repeated when the top plate moves back to the left. The dash-dot line corresponds to a screw with wide thread tolerance and no head slip. For this case, thread slip starts at B_2 and ends at E_2. The solid lines correspond to a screw with head slip. The screw is bent to the left at A_3. When the plate moves to the right, the screw bends to the right until thread slip occurs at B_3. Before thread slip can proceed to D_1' for narrow thread tolerance or to D_2' for wide thread tolerance, head slip starts at C_3 until E_3. Distances W_1 and W_2 represent the part where both thread and head slip occur for narrow and wide thread tolerances, respectively. Junker states that fastener rotation can take place only on these distances because only there is the screw

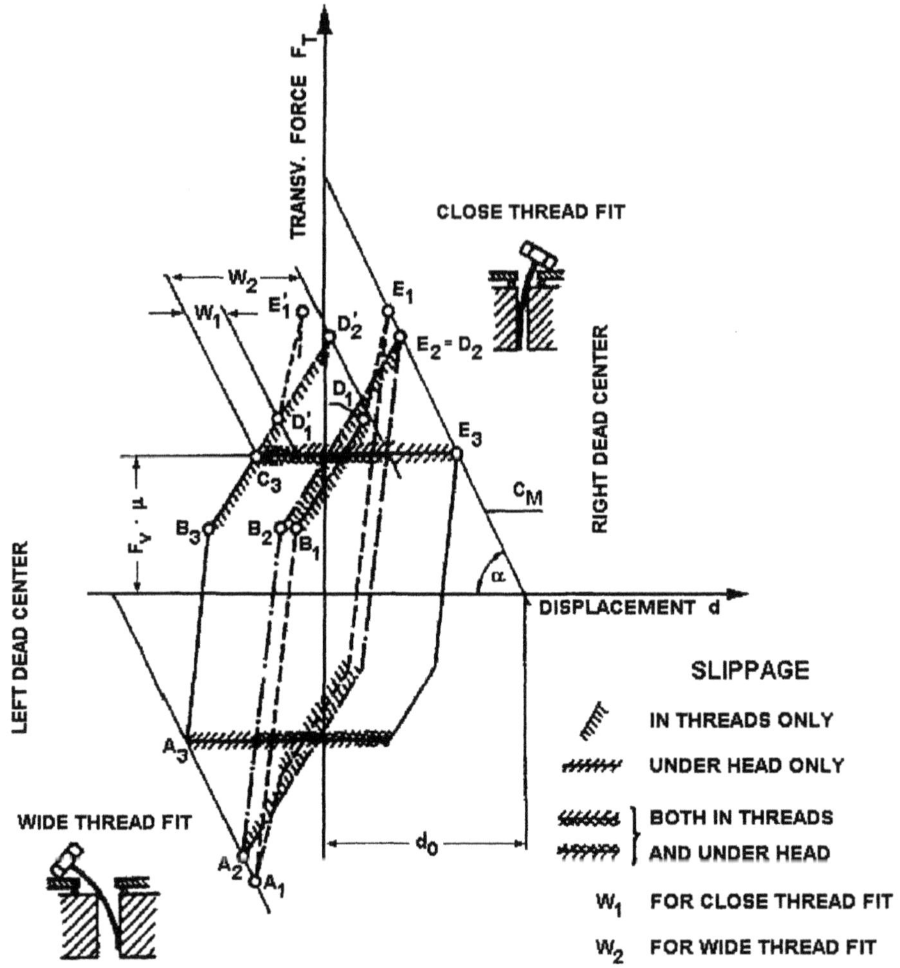

FIGURE 40.15 Schematic description of the self-loosening process under transverse vibration. (Adapted from Junker, 1969.)

free of friction. Comparing the magnitudes of these distances, one can see that screws with close thread fit should have better resistance to loosening.

Figure 40.16 shows five cycles of data at 100%, 80%, 60%, 40%, and 20% original preload for a nonlocking screw (specimen A1) and a screw with a serrated head (specimen C2). Head slip occurs with the nonlocking screw in the fully preloaded condition but does not occur until the preload is decreased to 40% for the screw with a serrated head.

The transverse force versus displacement data and the number of cycles to complete loosening were found to be essentially the same for loading frequencies of 0.17 and 50 Hz. This suggests that loosening is independent of frequency. However, Junker pointed out that if the forces causing the relative motion within a joint are inertial forces, the motions are a function of the square of the frequency, and in this case, the frequency indirectly influences the loosening process.

Junker proposed using the vibration product, i.e., the product of transverse force times displacement, as a criterion for the locking ability of a locking element. In an effort to find a test method for quantitatively assessing locking elements, data from a series of tests were presented in vibration product and loosening angle versus preload plots as well as vibration product and preload versus number of loading cycles plots. Finally, a method for large-scale testing and inspection testing was

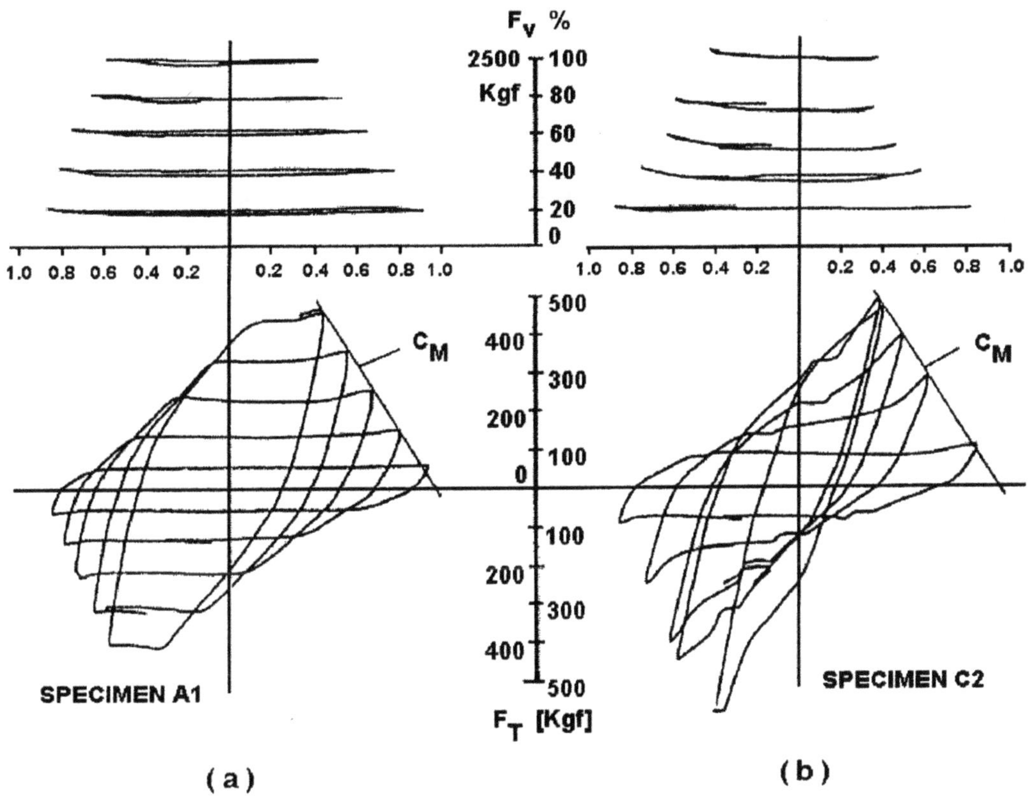

FIGURE 40.16 Preload and transverse force versus displacement data for (a) a nonlocking screw, and (b) a free-spinning locking screw. (Adapted from Junker, 1969.)

proposed that produces preload versus number of cycles for various maximum machine vibration energy levels. The maximum vibration energy was varied by adjusting the eccentric.

40.3.7 Koga (1970, 1973)

Koga (1970) of the Institute of Vocational Training in Kodaira, Japan, studied the loosening of threaded fasteners subjected to repeated impacts. He performed experiments and developed a theory of loosening based on the theory of propagation of stress waves.

The experimental apparatus is diagrammed in Figure 40.17. The bolted assembly consisted of parts *1–5*, where *1* and *2* are the test bolt and nut. The nominal size and pitch of thread used were 12×1.75 mm. To eliminate the effect of friction on the nut base, a ball thrust bearing *3* was inserted between the nut and a compression cylinder *4*. The threads were lubricated with machine oil for all tests. Strain gages *6* were attached to the compression cylinder and connected to a strain meter with wires 9. Four sash cords *10* were used to lift and drop the bolt assembly along a guide cylinder *8*, and an impact piece 5 of the assembly struck a fixed stopper 7. The resulting impact stress caused on the bottom of the impact piece propagated throughout the assembly according to the theory of propagation of stress waves within discontinuous sections.

Koga's theory of fastener loosening by impact can be summarized as follows:

1. Compressive and impact stresses are produced on the pressure flanks of an external thread and an internal thread.

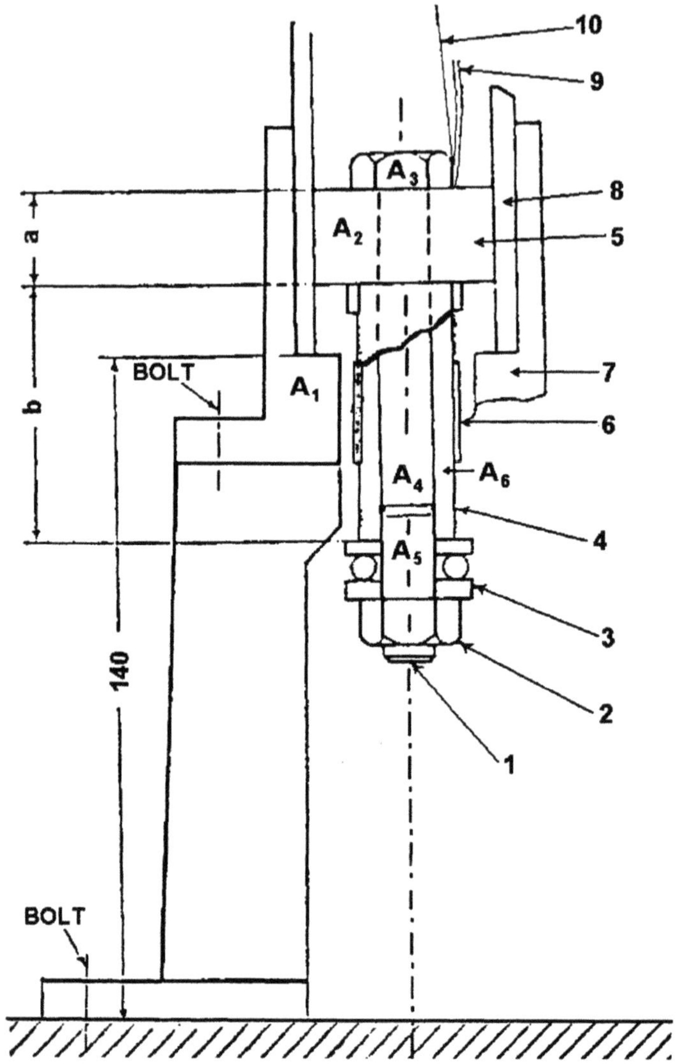

FIGURE 40.17 Test apparatus of Koga (1970).

2. The stress waves from the external pressure flank reflect on its clearance flank and the screw end and change into tensile waves, which reach the pressure flank.
3. The tensile waves decrease the compressive force on the pressure flank.
4. This causes the shearing stress at the thread interface, which is generated by circumferential friction, to decrease.
5. The decrease in shearing stress results in slip and what Koga called "initial loosening."

When the impact force caused on the thread surface is larger than the thread pressure, the nut jumps vertically on the surface in reaction to the impact and again is turned axially by the compression member. As a result, the nut unscrews. Koga referred to this as "reaction loosening."

Koga qualitatively showed that changing the angle of thread and the details of the screw end can alter the amount of tensile waves propagated. Experiments were performed with specially manufactured fasteners to test his hypothesis. In general, he found that it was harder to loosen fasteners with larger thread angle and more length between the nut and the screw end.

In a later report, Koga (1973) used three-dimensional analytical geometry to quantitatively show the effect of thread angle on loosening. Numerical calculations for 10 metric screw threads were performed for thread angles from 55° to 65°. The best resistance to loosening was found when the pitch-to-diameter ratio was 0.15, 0.125, or 0.1 and the thread angles were 63°, 62.5°, or 62°, respectively. When the thread angle was 60° or less, a comparatively strong loosening action was obtained.

40.3.8 Finkelston (1972)

Suggestions provided by Finkelston (1972) of SPS for increasing vibration resistance to loosening are given as follows:

1. Increase grip friction in the joint by increasing the preload or the number of bolts in the joint.
2. Eliminate relative motion in the joint by designing rigid members with little or no clearance.
3. Use fasteners with high resistance to loosening.

Finkelston pointed out that the first two may be uneconomical in designs optimized for weight and size.

A series of tests performed on a Junker machine (see Figure 40.14) were also reported. The factors investigated, in order of decreasing importance to fastener vibration resistance, are

1. Amplitude of dynamic motion (generated by the dynamic load). As vibration amplitude is decreased, vibration life increases. However, at a certain amplitude, life becomes infinite. This amplitude is analogous to the fatigue endurance limit.
2. Effectiveness of fastener locking method or feature. Test results in Figure 40.18c show that prevailing torque not only reduces the rate of loosening but also causes the loosening process to stop when the effective torque counteracts the loosening torque.
3. Thread pitch. Test results in Figure 40.18b show that fine thread nuts endured twice as many vibration cycles under severe loading conditions as coarse thread nuts with other conditions being equal.
4. Initial preload. Increasing preload in a fastener increases the friction forces in the joint and thereby increases its vibration resistance as shown by the test results in Figure 40.18a.
5. Bearing surface conditions. The hardness, roughness, and material properties of the mating thread surfaces and contacting surfaces of the clamped members also influences preload loss in joints under dynamic loading. Some degree of brinelling or embedding of the clamped surfaces takes place statically, and dynamic loading can cause additional plastic deformation of the joint surface. Test results showed that embedding can result in preload loss of as much as 20% in soft steel (Rb 90) and 5% for relatively hard steels (Rc 40). This preload loss generally occurred within the first 10 cycles of dynamic loading, until a stable load-carrying surface was developed.

40.3.9 Walker (1973)

Walker (1973) of SPS was the first to use fractional factorial experiment design concepts to investigate the influence of several variables or factors on the vibration life of fasteners. This approach allows one to efficiently investigate not only the effect of single variables but also the effect of combinations or interactions of variables. The factors evaluated were preload, prevailing torque, thread pitch, surface coating, and thread pitch diameter clearance.

The tests were performed with two different machines: a Junker machine (see Figure 40.14) and the apparatus described in MIL-STD-1312-7A (see Figure 40.9). The results, described below, from tests on both machines were in good agreement. Fasteners of ¼ and ⅜ in. diameter were tested.

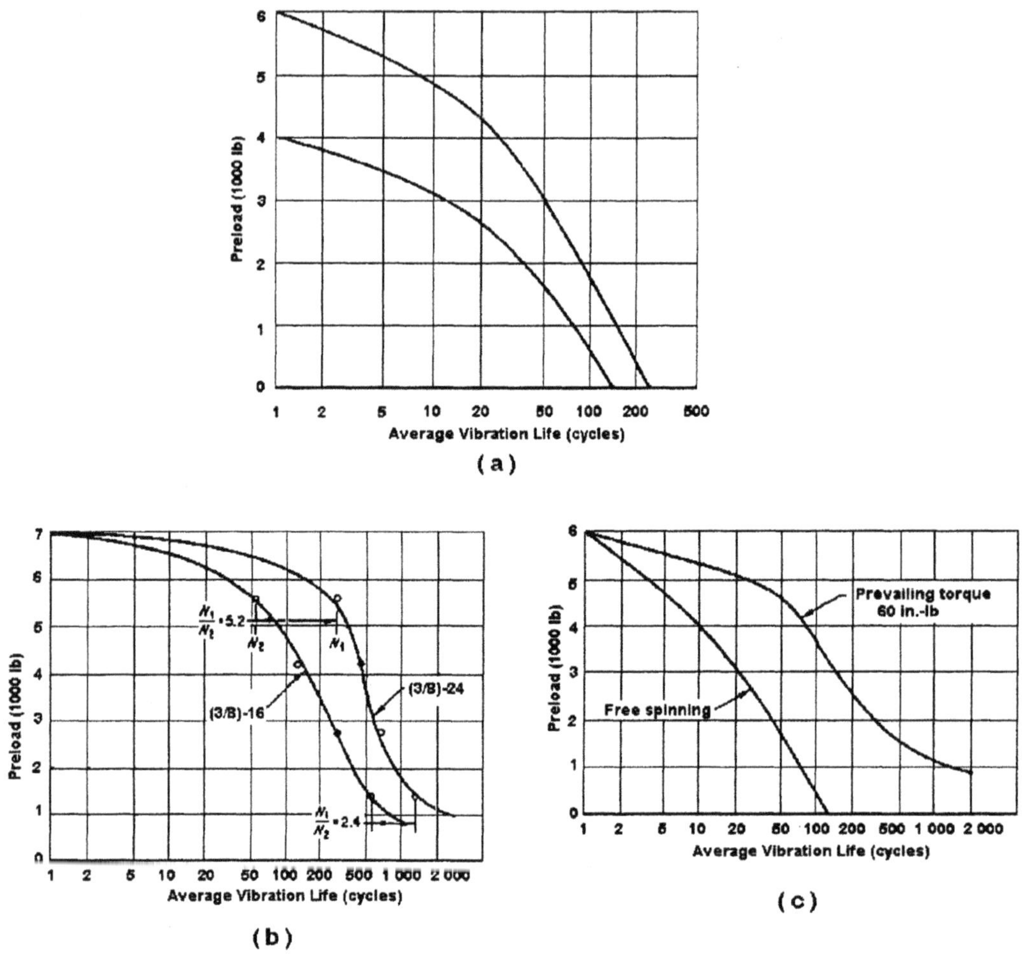

FIGURE 40.18 Preload versus average vibration life data for different (a) initial preload, (b) thread pitch, and (c) prevailing torque level. (Adapted from Finkelston, 1972.)

Thread pitch and prevailing torque and the interaction of the two were found to be the most significant factors. Thread pitch separately had the biggest influence. Fine pitch threads had better vibration resistance than coarse pitch threads. The second ranked factor was prevailing torque. Nuts with prevailing torque had better vibration resistance than nuts without prevailing torque. The third-ranked item was the interaction of thread pitch and prevailing torque. This interaction produced an additional improvement in vibration resistance beyond that found from the individual factors of thread pitch and prevailing torque.

The interaction of fine thread pitch and surface coating was found to exert a negative influence on the vibration life of fasteners. This result was unexpected since fine thread pitch showed a significant positive effect while the magnitude of the negative effect for positive surface coating (low torque coefficient) was not nearly so significant.

The interactions of thread pitch diameter clearance with preload, prevailing torque, and surface coating were next in the order of importance for influencing vibration resistance. An increase in vibration life was found by reducing thread diameter clearances when high preload, high prevailing torque, and low torque coefficients were incorporated in the fastener system.

The last two factors that seemed to have a substantial influence were preload and surface coating. A high preload value produced longer vibration life, while a positive surface coating (low torque coefficient) resulted in lower vibration life.

40.3.10 Sakai (1978, 1979)

In an effort to understand the mechanisms of bolt loosening under different types of loading, Sakai (1978a, 1978b, 1979) of Toyota Motor Company performed a series of theoretical analyses and experiments on bolted assemblies. His first report (Sakai, 1978a) focused on bolts subjected to transverse dynamic loading. Static force and moment balances for both straight slip and arc slip were examined (see Figure 40.19). The necessary condition for a bolt to loosen under a straight slip of clamped parts was derived as

$$\frac{\mu_s}{\cos\alpha} + \frac{d_w \mu_w}{d_e} < 1.5 \tan\beta \tag{40.5}$$

where μ_s, μ_w, d_w, d_e, α, and β are the thread friction coefficient, bearing surface friction coefficient, effective bearing surface diameter, effective thread diameter, half thread angle, and lead angle, respectively. The necessary condition for a bolt to loosen under arc slip was given as

$$\frac{\mu_s}{\cos\alpha} + \frac{d_w \mu_w}{d_e} \mp \frac{2R_o \mu_{ws}}{d_e A_w}\left(A_w - \int_{A_w} \cos\theta \, dA_w \right) < 1.5 \tan\beta \tag{40.6}$$

and the condition for a bolt to tighten under arc slip was given as

$$\frac{\mu_s}{\cos\alpha} + \frac{d_w \mu_w}{d_e} - \frac{2R_o \mu_{ws}}{d_e A_w}\left(A_w - \int_{A_w} \cos\theta \, dA_w \right) < -1.5 \tan\beta \tag{40.7}$$

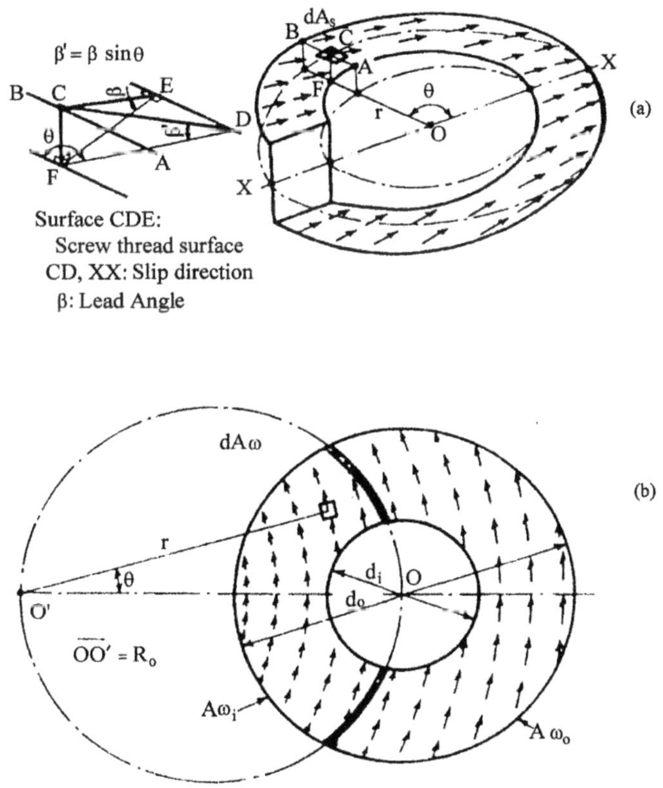

FIGURE 40.19 Transverse slip. (a) Straight slip on screw thread surface; (b) arc slip on bearing surface. (Adapted from Sakai, 1978a.)

where R_o and A_w are the radius of the arc slip and the area of the bearing surface. Angle θ is defined in Figure 40.19b. From these expressions, it was calculated that the bolt will not rotate loose due to the slipping of clamped parts under transverse load unless the friction coefficient is smaller than 0.03. Without slipping of the clamped parts, the friction coefficient must be less than 0.017 for loosening to occur.

Experiments were performed on a Junker transverse testing machine for straight transverse slip. Measurements from tests with this machine were used to estimate the thread and bearing friction coefficients. The friction coefficients were found to range from 0.005 to 0.02 when slip occurred at the bearing surface and from 0.10 to 0.14 when no slip occurred at the bearing surface. These results suggest that slip at the bearing surface is necessary for loosening and that some minimum slippage or "critical slippage" between the clamped parts is required for relative slip to occur at the bearing surface.

Sakai derived expressions for this critical slippage that were found to be in good agreement with measured test results as shown in Figure 40.20. These data show that no loosening occurs when there is no slip at the bearing surface. However, when slip occurs, the bolt loosens progressively.

The apparatus shown in Figure 40.21 was used to experimentally examine arc slip. The radius of the arc was varied from 25 to 200 mm. The slip was generated by an electrohydraulic vibration machine that was displacement-controlled. The relative rotating angle between the bolt and female thread was measured with a potentiometer as shown. Measurements showed that the rate of loosening decreased with the radius of arc slip. Theoretical calculations for the rate of loosening angle were shown to predict the same result.

In his second report, Sakai (1978b) examined the loosening mechanisms of bolts that were tightened at the center of rotation, such as a pulley or a gear bolted to a rotating shaft, and subjected to alternating torque or relative rotation. A theoretical analysis and experimental tests were performed. The apparatus shown in Figure 40.21 was used with the radius of arc set to zero.

It was shown that for a bolt to rotate loose, it is necessary that

$$\frac{d_2\mu_s}{d_w \cos\alpha} - \frac{p}{\pi d_w} < \mu_w < \frac{d_2\mu_s}{d_w \cos\alpha} + \frac{p}{\pi d_w} \tag{40.8}$$

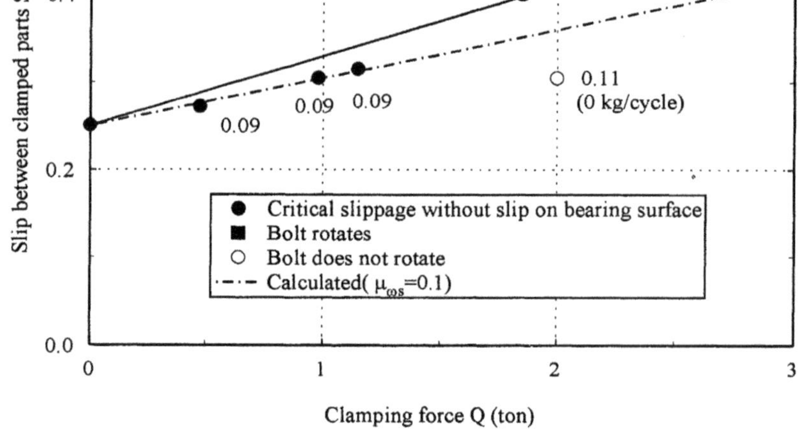

FIGURE 40.20 Critical slip between clamped parts without slip on bearing surface versus clamping force data. (Adapted from Sakai, 1978a.)

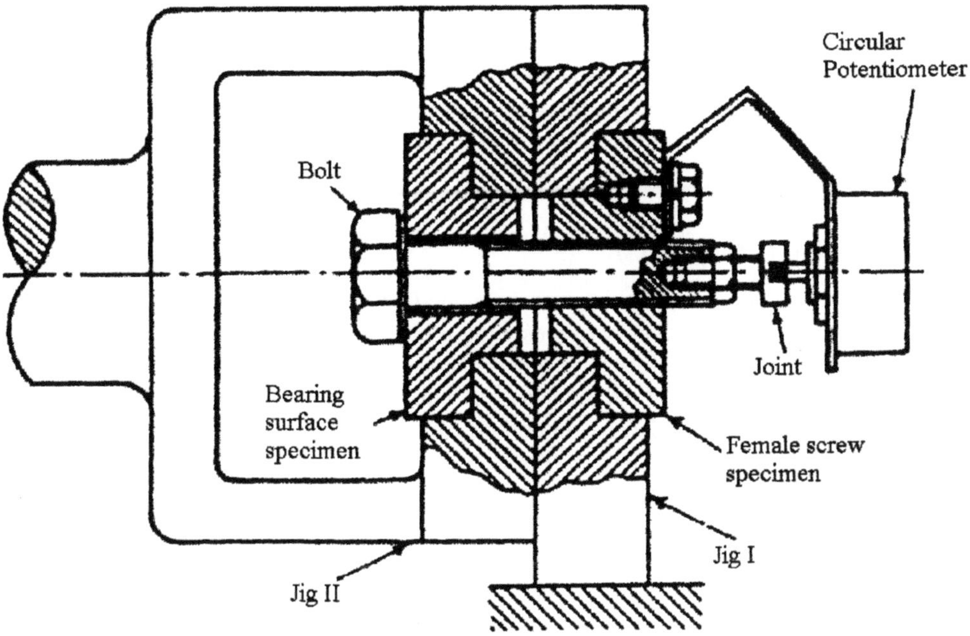

FIGURE 40.21 Test apparatus to study arc slip and component rotation. (Adapted from Sakai, 1978a, 1978b.)

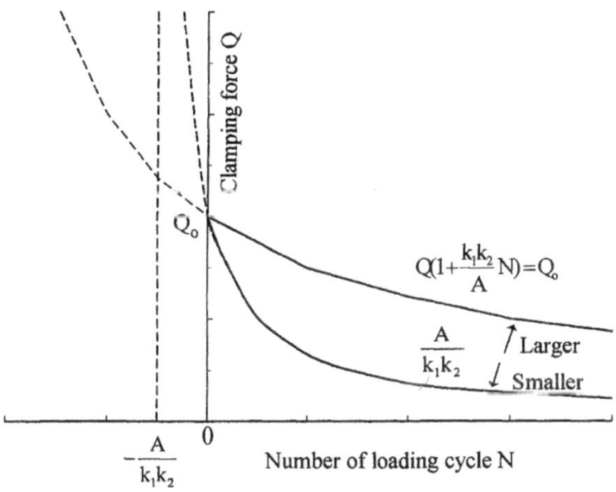

FIGURE 40.22 Relation between clamping force and number of loading cycles when fretting wear occurs on clamped parts. (Adapted from Sakai, 1978b.)

where d_2, d_w, μ_s, μ_w, and p are the effective diameter of the thread, equivalent diameter of the bearing surface, thread friction coefficient, bearing surface friction coefficient, and thread pitch, respectively. An expression was derived for the "critical angle," which is the minimum relative rotation angle between the clamped parts that is necessary to make the bolt rotate loose.

Some of the experimental results showed a significant decrease in clamping force without bolt rotation. This was thought to be due to fretting wear because fine metal particles were observed on the contact surfaces. Sakai derived an expression, shown in Figure 40.22, relating the change in clamping force to number of loading cycles due to fretting. In Figure 40.22, A is the area of the contact surface, k_1 is the coefficient of fretting, and k_2 is the effective spring constant of the bolt and clamped parts.

In a third report, Sakai (1979) investigated the loosening of bolts tightened over their yield point and subjected to tensile load. Theoretical analysis and static and dynamic experiments were reported. It was concluded that tightening bolts over their yield point has no disadvantage from the point of view of loosening. It was also found that the fatigue life of bolts tightened over their yield point was longer than that of bolts tightened in the elastic range. The assumed reason for this was that the larger clamping force, the smaller the load on the bolt, which corresponds to a longer fatigue life.

40.3.11 Haviland (1980, 1983)

Haviland (1980, 1983) of Loctite Corporation provided an explanation for loosening of bolts and nuts in a simplified manner. His theory of loosening is based on the movement of the clamped components and the resulting side-sliding on screw threads.

Haviland pointed out that all standard bolts and nuts are made with a clearance between them for assembly. For example, Class 2A has lateral clearance on the order of 0.0013–0.0114 in., which means that the nut can be moved sideways this amount. The block on an incline analogy with transverse loading or "side-sliding" was used to explain loosening as discussed earlier. The cross section of a bolt and nut shown in Figure 40.23 was used to describe the ratcheting action of side-sliding in threads. As the nut is moved into the page, the right side tends to loosen and the left side tends to tighten. Since the left side will move with greater difficulty, it acts as a pivot point about which the nut rotates loose. When the nut is pulled from the page, the right side becomes the pivot point about which the nut rotates loose. The net effect each time the nut is cycled sideways is a ratcheting loosening motion.

Experiments were performed with a simple composite cantilever beam (Figure 40.24) composed of two steel blades bolted together. Deflecting the end of the beam causes the two blades to slide as shown, which induces side-sliding on the threads of the bolts and nuts. The resulting data in Figure 40.25 show that the larger the side-sliding motion, the smaller the number of cycles needed to completely loosen the bolts and nuts. Also, side-sliding is less likely with

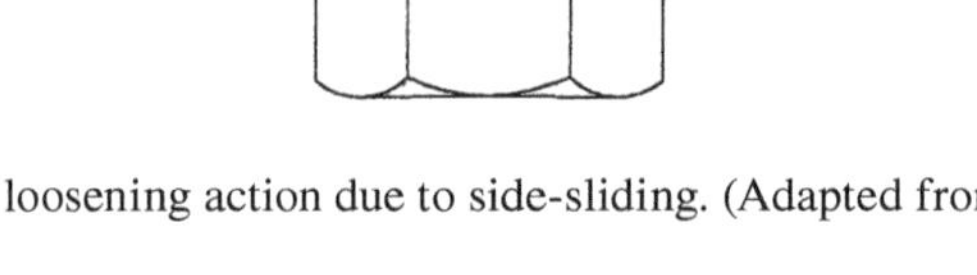

FIGURE 40.23 Ratcheting loosening action due to side-sliding. (Adapted from Haviland, 1980.)

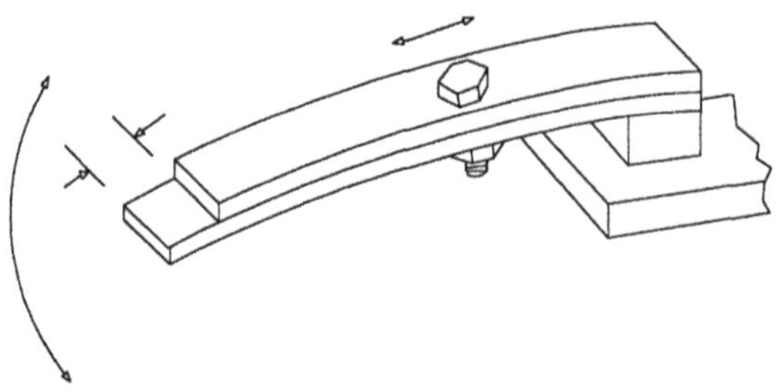

FIGURE 40.24 Test apparatus or "direct loaded side-slide demonstrator" of Haviland (1980).

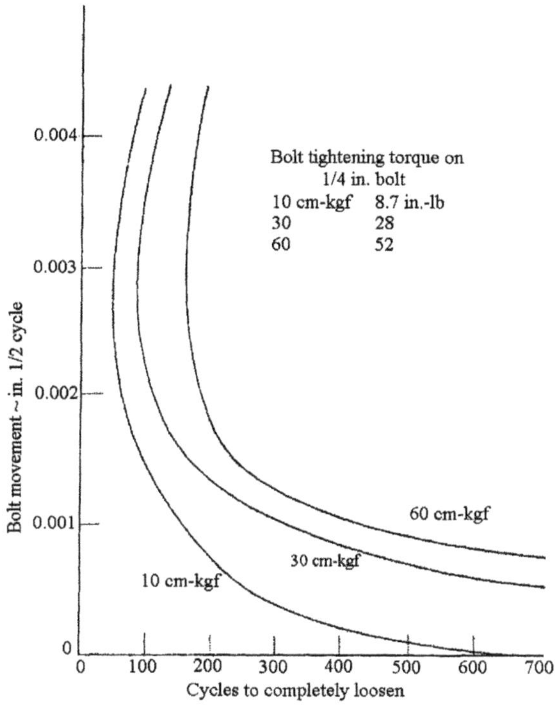

FIGURE 40.25 Bolt movement versus cycles to loosen data. (Adapted from Haviland, 1983.)

higher clamping forces, but if side movement does occur, loosening will occur regardless of the magnitude of the clamping force.

Side-sliding was shown to be a common phenomenon in bolted assemblies. It can be caused by assembly bending (see Figure 40.26a) or internal pressures (see Figure 40.26b). Side-sliding can also be caused by differential thermal expansion (see Figure 40.26c). For example, the differential expansion on the head of a six-cylinder in-line engine has been measured as high as 0.060 in. under extreme temperature cycling. In addition, side-sliding can be produced by shocks or impacts (see Figure 40.26d) or directly by axial forces on a nut, which cause it to dilate as shown in Figure 40.26e.

40.3.12 ASME/ANSI B-18 Subcommittee (1983)

In 1983, a subcommittee was formed at the request of the ASME Fastener Standards Committee to test various theories on vibration loosening of bolted joints (Bickford, 1983, 1984). This was in response to the facts that vibration loosening was considered the most common mode of bolted joint failure and engineering design problems involving vibration loosening were not very well understood. The name of this new subcommittee was the Joint ASME/ANSI B-18 Subcommittee on Loosening Mechanisms of Bolted Joints Under Vibration.

This subcommittee developed a proposal for a comprehensive research program aimed at better understanding vibration-induced fastener loosening. This proposal indicated that the U.S. Air Force had identified 76 variables that can affect the friction between nuts and bolts. Planned research by the subcommittee on these variables included analysis of finite element and probabilistic models as well as exploratory and milestone tests. Planned test variables included thread pitch, clearances, fit, length-to-diameter ratio, direction of vibration, frequency of vibration, and joint damping properties. The block diagram in Figure 40.27 illustrates the test and evaluation procedure planned by the subcommittee. Unfortunately, extended funding for this comprehensive research program was not obtained, and the subcommittee eventually disbanded.

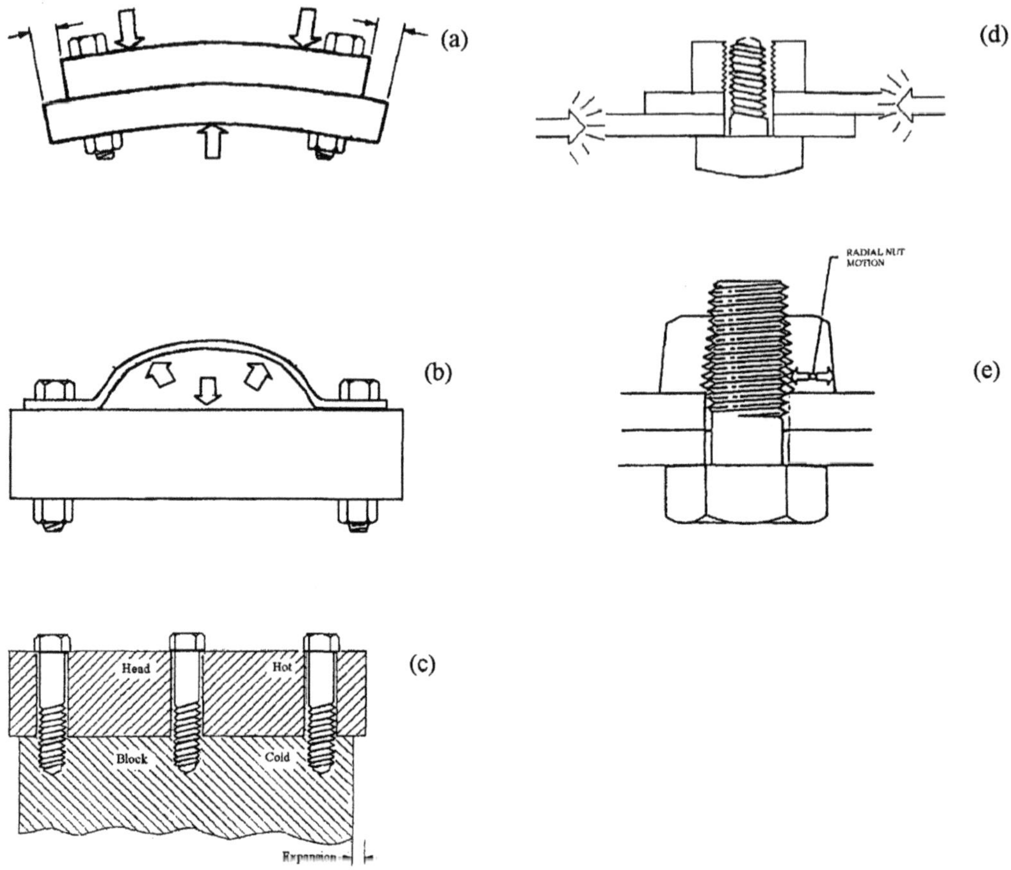

FIGURE 40.26 Examples of side-sliding bolted assemblies. (a) Bending; (b) internal pressures; (c) differential thermal expansion; (d) shock or impact; (e) nut dilation from axial load. (Adapted from Haviland, 1980, 1983.)

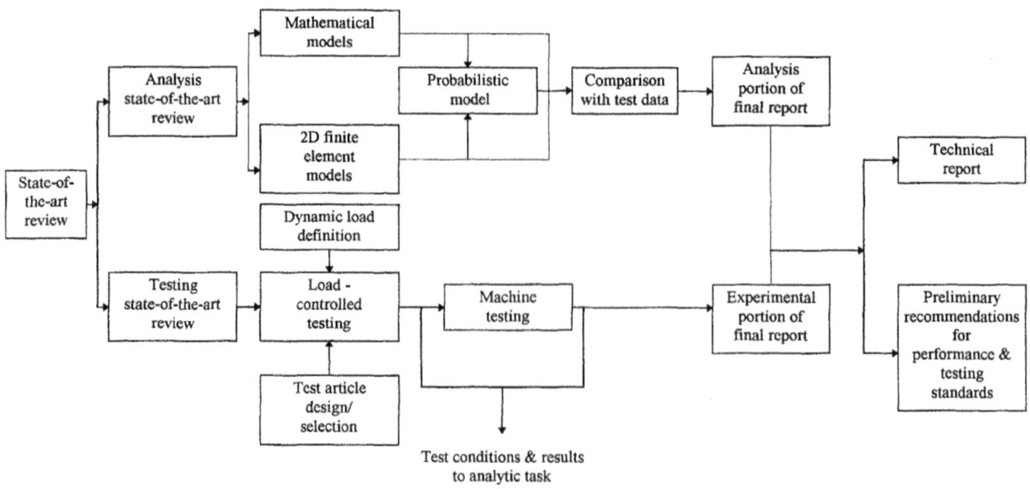

FIGURE 40.27 The ASME/ANSI B-18 Subcommittee's planned test and evaluation procedure to study vibration loads in bolted joints. (Adapted from Bickford, 1984.)

40.3.13 Kerley (1982, 1987)

Kerley (1982) of NASA GSFC (Goddard Space Flight Center) provided a report on time-related failures of bolts that includes failure by loosening. This included results from tests that were performed to verify the Haviland theory. Shocks and harmonic and random excitation were applied to a test apparatus for a sounding rocket. The vibration of the panels, struts, and bolts was recorded. These tests showed that low-frequency excitation causes bolts to loosen, whereas high-frequency excitation does not. However, even though low frequencies cause bolts to break away and back off, one should be aware that high frequencies can cause subharmonics that may be sufficient to cause bolts to back off.

A later publication from Kerley (1987) reported on his application of the retroduction method to analyzing and testing the loosening of threaded fasteners during dynamic loading. This work was performed in support of the ANSI/ASME Subcommittee on Loosening Mechanisms of Bolted Joints Under Vibration. Due to the extent of the bolt loosening problem, retroduction methods were used as a structured way to organize an attack on the problem. Retroduction is a process of dialectic questioning that is based on "given" items (items that are desired) and "known" items (principles of research that have been tried, used, and proved in other forms of research) as well as the "final cause" (e.g., why bolts and nuts vibrate apart). Once dialectic questions are posed, analysis and synthesis are used to abstract these questions to solve the given items and the final cause.

Figure 40.28 lists the given items, the known items, and the dialectic questions. The given items were specified by the ANSI/ASME Subcommittee on Loosening. Also, there were a total of 21

DIALECTIC QUESTIONS

1. What motions of the structures holding the bolts can cause the bolts to back off?
2. What are the forms of the structures used to hold the bolts?
3. What is the effect of lubrication on the threads?
5. While studying the many loads that could be applied to the bolts from the structure, is it possible to break down the loads into several different types of testing and evaluation?
6. What available vibration testing machines would be most desirable for testing bolts and nuts? What static testing machines?
7. What methods are available for measuring the exact moment that the nut begins to back off the bolt?
9. What types of vibration loads can be analyzed? What kind of tests can be performed to correlate with analytical techniques?
14. How important is preload, and how is it measured?
15. How important is a series of torquing, loading, unloading, torquing, loading, and unloading?
16. How many tests are necessary to establish a reliable trend or to bring the testing to a point of mathematical analysis?
18. What is the effect of the geometry of the threads?

GIVEN

1. Define loosening.
2. Define the loading of primary interest as transverse (clamped part loading or movement tending to "shear" the bolt, transverse slip between the clamped parts).
3. Define the bolt size range or interest (diameter).
4. Define the generic point of primary interest (bolt and nut, two clamped parts, rigid relative to the bolt axial stiffness).
5. Define the dynamics of the intended loading.
6. Work to sharpen the statement of objectives.

KNOWN

1. Determine what happens to a normal bolt before studying exotics and special types of locking devices ....
2. Start with low-vibration loads and build up until the bolt backs off. In this way, a time threshold of backing ....
3. Establish standard input wave patterns for testing, from past experience. In order to compare many different ....
4. Establish early in the testing program a series of structures that can be used to test the bolts ....
5. Force depends on acceleration and only partially on displacement (F=ma). The time function in the force input is important. It can vary all the way from a ....
6. Throughout the testing, constantly look for ways that the testing could be used as a proof of the dialectic ....
7. It is better to *know one part of the program well and with assurance* than to give a hopeful all-encompassing solution ....
8. Attempt to work with existing testing machines and data ....
9. Search the research field of past and existing programs engaged in vibration work. This can lead to help ....

FIGURE 40.28 Retroduction applied to the problem of vibration-induced loosening of bolts. (Adapted from Kerley, 1987.)

dialectic questions; however, only the 11 listed in Figure 40.28 were essential at first to establish fundamental principles. Figure 40.29 illustrates the path of reasoning that led to testing.

Question 3 was answered by considering that previous tests showed consistent results only when threads were lubricated. Question 18 was answered by considering that the vast majority of all bolts in the United States have the 60° thread and that the small number of bolts with a different thread geometry can be tested later. Question 1 was answered by given item 2. To answer question 2, Figure 40.29 shows various general assembly models of a typical bolt in pure shear. In response to question 6, the inertial loaded cantilevered beam fixture in Figure 40.30 was used in the testing because it can produce a fundamental waveform on the bolt without feedback or secondary motions caused by other members of the structure. Question 5 was answered by considering that more than 90% of the loads that cause horizontal shear on bolts are sine and random; also, inertial loading was chosen instead of direct loading because it is more common in real structures. In response to question 7, loosening was detected by the eye and by a change in natural frequency. In response to question 14, preload was taken as a function of torque. In response to questions 15 and 16, the tests were found to provide consistent results. Question 9 is addressed later.

The cantilever beam fixture consisted of two $^1/_{16} \times 1 \times 15$ in. steel beams held together with a ¼–20 UNC Class 3 bolt and Class 2 nut. Accelerometers were used to measure the input motion and

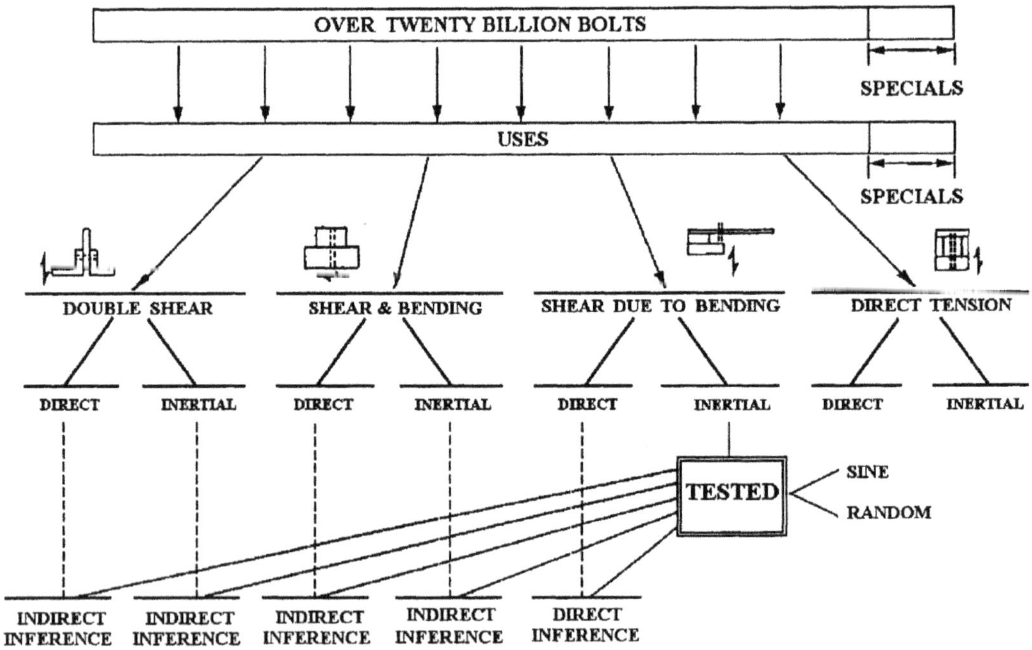

FIGURE 40.29 The path of reasoning leading to bolt vibration tests. (Adapted from Kerley, 1987.)

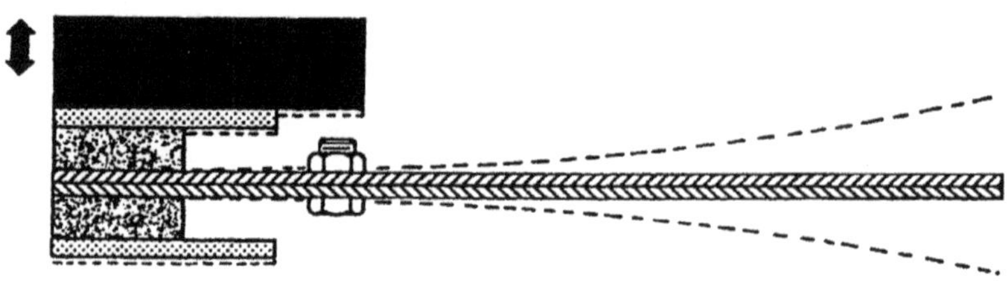

FIGURE 40.30 Inertial loaded cantilever beam apparatus of Kerley (1987).

the motion near the bolt. The sine load tests were sine sweep (20–2,000 Hz), single-frequency sine dwell (24.7, 25.9, 107, 941, and 947 Hz), and two-frequency sine dwell with amplitudes ranging from 0.5 to 50 g. The random load tests used broadband random (20–2,000 Hz), low frequency random (20–400 Hz), high frequency random (1,000–2,000 Hz), and combined low and high frequency random with amplitudes ranging from 1 to 20 g_{rms}.

In general, the tests showed that the nut backed off the bolt with low-frequency vibration when the amplitude of bending was high. The bolt completely loosened in 24 s when subjected to a sine dwell of 2 g at the fundamental frequency of the beam, 24.7 Hz. Driving the beam at its fundamental frequency, the lowest resonant frequency provides the largest beam deflection and shear load on the bolt. A slight shift in excitation frequency to 25.9 Hz caused it to drop out of resonance, and the bolt did not loosen. Loosening also was found for sine tests at the second harmonic of 107 Hz; at 50 g, loosening occurred in 10 s; however, no loosening occurred at 10 g. This test demonstrated that even with a second harmonic, which normally vibrates at a much lower amplitude, the bolt will loosen if the amplitude is allowed to get high enough. This supports the theory of some engineers that it is not the waveform or frequency, but the amplitude, that causes bolts to loosen in vibration. No loosening was observed for sine tests at higher frequencies. Two frequency sine tests of 24.7 and 944 Hz at 2 g caused loosening in 1 min and 27 s, respectively. The presence of a high frequency on top of a low frequency tended to dephase the low frequency, change its waveform and overall displacement amplitude, and increase the time of loosening.

Loosening occurred in 2.75 min for a broadband random test at 20 g_{rms} and in 2 min for a low-frequency random test at 10 g_{rms} but did not occur for high-frequency random tests at 20 g_{rms}. For a combined low and high-frequency random test at 15 g_{rms}, the bolt loosened in 2 min 21 s.

Figure 40.28 reveals that this testing provided only a partial solution to the problem of why bolts and nuts separate during vibration. The next step is to return to question 9 and apply analytical techniques. For example, one can synthesize data from the inertial loaded beam to analyze a direct loaded beam. From acceleration measurements taken along the beam during vibration, one can calculate the shear force mathematically as shown in Figure 40.31b. Then, assuming that the shearing force is causing the bolts to back off, one can calculate the static load in Figure 40.31a needed to give the same shearing force at the bolt in Figure 40.31b. This analysis can be verified with additional testing. This approach can be extended to analyze more complex structures. Figure 40.32 illustrates this process for a typical bent, which is the next structural form above a cantilever. A single load is applied at the center as shown in Figure 40.32a. The shearing load on the bolts can be calculated as shown in Figures 40.32b and 40.32c. Thus, since a direct mathematical relationship exists between the bent and the cantilever, the tests on the cantilever apply to the bent. Although the bent can now be tested to verify such calculations, this demonstrates that all structures do not have to be tested to get universal results. The method of retroduction can be applied to even more complex structures.

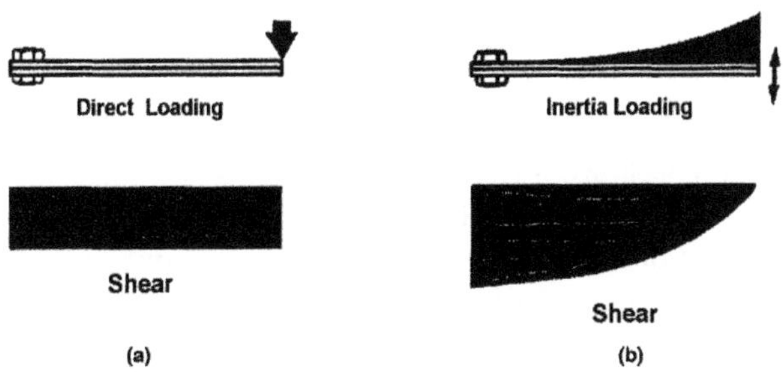

FIGURE 40.31 Synthesis of inertial loaded beam to analyze direct loaded beam. (a) Shear from direct loading; (b) shear from inertial loading. (Adapted from Kerley, 1987.)

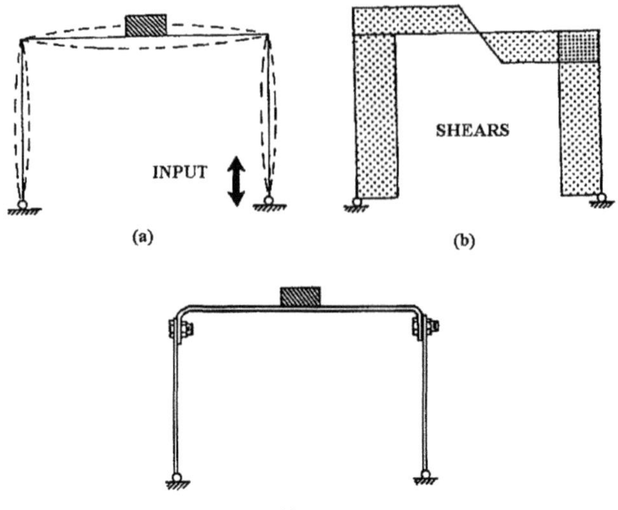

FIGURE 40.32 Analysis of a bent. (a) Input and motion; (b) shear curves; (c) assessment of shear on bolts. (Adapted from Kerley, 1987.)

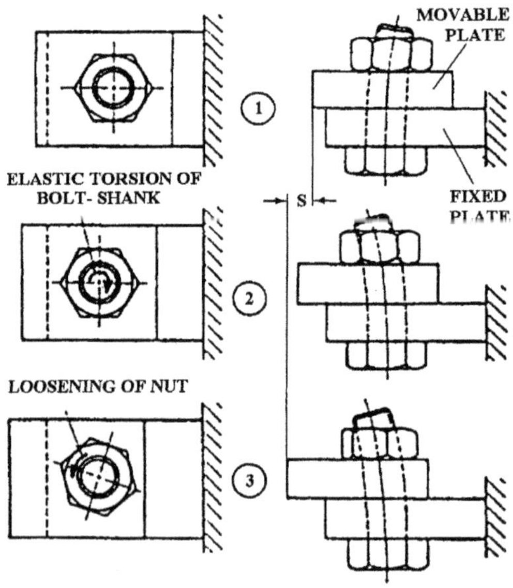

FIGURE 40.33 Mechanism of self-loosening of bolt-nut joint. (Adapted from Yamamoto and Kasei, 1984.)

40.3.14 Yamamoto and Kasei (1984)

Yamamoto and Kasei (1984) of Shinshu University proposed a new mechanism of self-loosening of bolt and nut joints subjected to transverse vibration. The mechanism attributes self-loosening rotation of a nut to the accumulation and release of potential energy due to torsional bolt deformation. A mathematical analysis was provided with supporting data from experiments using a somewhat different transverse testing apparatus than the Junker machine.

This mechanism of self-loosening is explained with reference to Figure 40.33, which shows the behavior of a bolted joint during one half-cycle of vibration. Two plates are clamped with a nut and

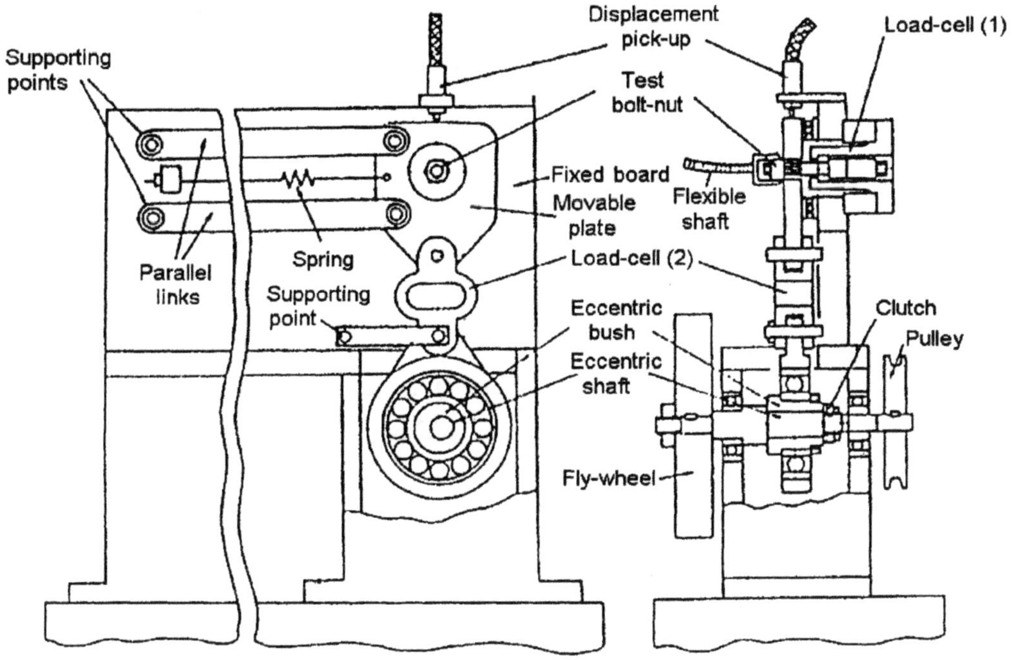

FIGURE 40.34 Test apparatus of Yamamoto and Kasei (1984).

bolt. One plate is fixed, and the other is subjected to transverse vibration. The bolt head is fixed against the fixed plate. Configurations 1 and 3 in Figure 40.33 illustrate the extreme right and left positions of the transverse motion, respectively. In the transition from configuration 1 to configuration 2, the bolt threads are forced to slide on the nut threads by the increase of the bolt shank deflection, while the nut sticks to the moving plate. The bolt threads slide on the nut threads and produce an elastic torsional deflection in the bolt shank as shown by an arrow in Figure 40.33. In the transition from configuration 2 to 3, the bearing surface of the nut is forced to slide on the moving plate, while the bolt threads stick to the nut threads. The torsional deflection releases and produces a loosening rotation of the nut as shown by an arrow in Figure 40.33. This mechanism also occurs in the other half of the vibration cycle such that loosening occurs twice in one cycle.

The experimental apparatus built to study this mechanism of loosening is shown in Figure 40.34. A test bolt and nut clamped a fixed board to a moving plate with ball bearings sandwiched in between. The moving plate was suspended by two linkages and was driven by an adjustable dual eccentric. The clamping force and bolt torque were measured with strain gages attached to a cylindrical load cell. The bolt head was clamped to this load cell. The transverse force was measured with a loop-shaped strain gage load cell. The displacement of the moving plate was measured by an inductive displacement sensor. The angle of rotation of the test nut was measured with a potentiometer through a flexible shaft. The test bolt and nut were lubricated with spindle oil. Tests were performed with M10 thread bolts and nuts with 5H/5h tolerance class.

A theoretical analysis of this mechanism was developed by Yamamoto and Kasei. Analytical expressions relating clamping force F, loosening angle ϑ, torque T, and transverse force W with transverse displacement v were derived.

Changes in clamping force ΔF, loosening angle θ, torque T, and transverse force W versus transverse displacement v data are shown in Figure 40.35. Points a, c, and d in the computed W versus v diagram correspond to configurations 3, 2, and 1 in Figure 40.33. Point b corresponds to the configuration where $v=0$. The bolt torque and loosening angle versus transverse displacement data strongly support this mechanism for loosening.

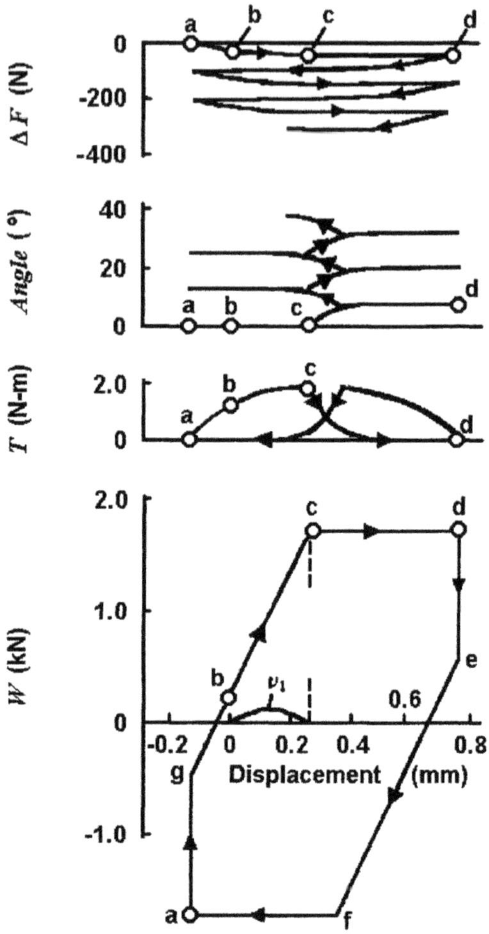

FIGURE 40.35 Change in clamping force, loosening angle, torque, and transverse force versus transverse displacement data. (Adapted from Yamamoto and Kasei, 1984.)

40.3.15 Vinogradov and Huang (1989)

Vinogradov and Huang (1989) of the University of Calgary appear to have been the first to develop a dynamic model of a bolted joint to study loosening. They performed numerical simulations to investigate the effect of frequency of dynamic excitation on self-loosening phenomena. A bolt was modeled as a rigid body, and a nut was modeled as a discretized five-mass system connected with springs and dashpots as shown in Figure 40.36. The discretization of the nut involved dividing it into a number of rings, each of which was loaded by a force that is a fraction of the preload as shown in Figure 40.37. The displacement of each ring or lumped mass is denoted by a coordinate u_i. The bolt was subjected to a kinematic harmonic excitation at an angle ϑ. A Coulomb friction model was used to model the thread friction.

Numerical simulations of this model were performed for standard ⅜–16 UNC and ⅜–24 UNF fasteners preloaded to 1,000 N and subjected to a kinematic excitation with various frequencies and amplitudes. The friction coefficient was taken as 0.15, and the *nth* damping constant was taken as $0.0002\omega_n$, where ω_n is the *n*th natural frequency.

The macroslip or loosening of the nut was assessed from the displacement of the discrete masses. The effect of excitation frequency on self-loosening is shown by the data in Figure 40.38, in which the ratio of inertial force to friction force is plotted versus excitation frequency. The shaded region

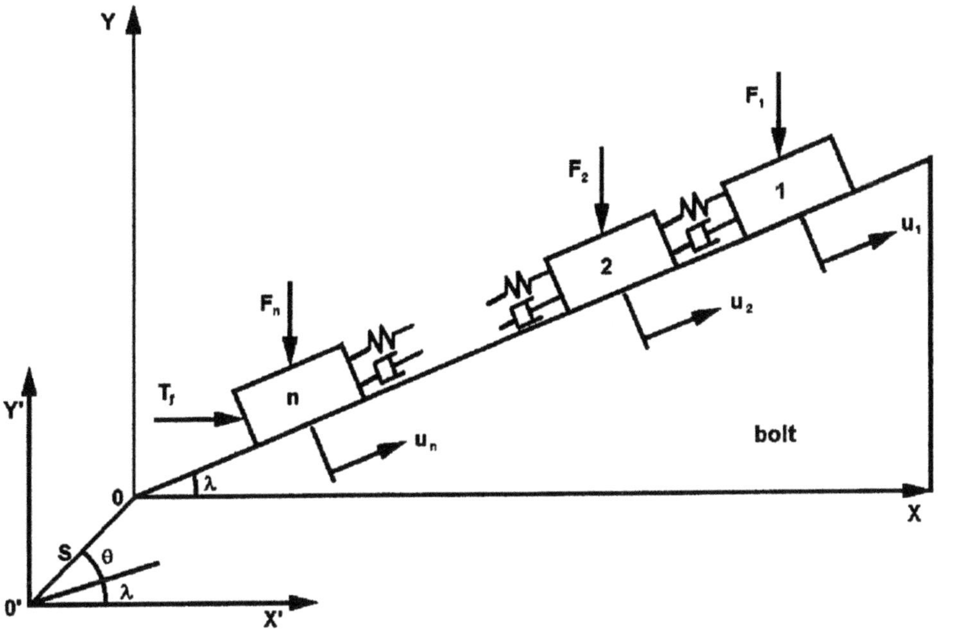

FIGURE 40.36 Dynamic nut and bolt model. (Adapted from Vinogradov and Huang, 1989.)

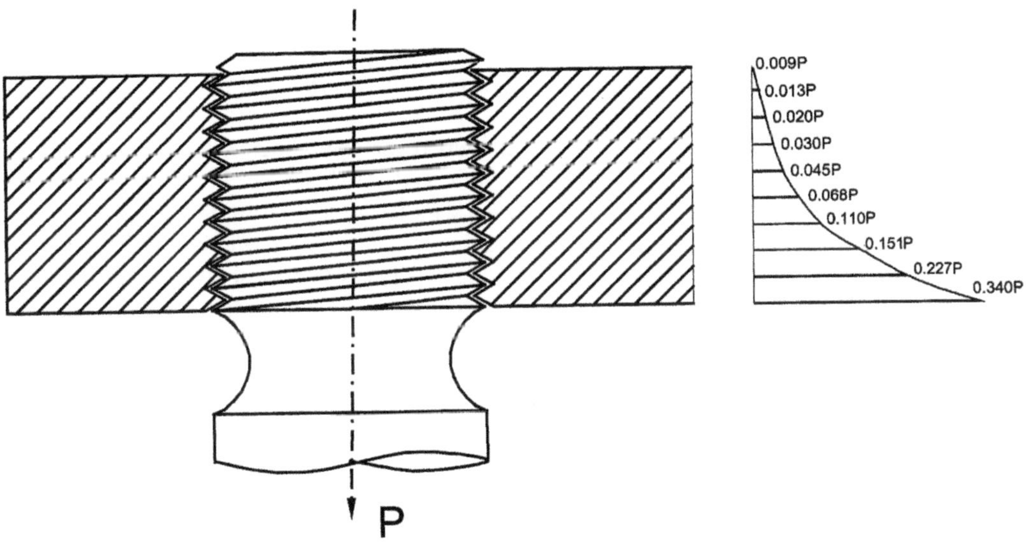

FIGURE 40.37 Preload distribution in threads used in dynamic model. (Adapted from Vinogradov and Huang, 1989.)

corresponds to self-loosening. The data show that at some frequencies the joint is more susceptible to self-loosening than at other frequencies. However, the frequency range where this occurs is much higher than the operating frequencies encountered in most machinery. This suggests that the mechanism of self-loosening due to inertial forces caused by the dynamics of an elastic nut or bolt may not be important in most applications. However, the assumption of a static friction model may be unrealistic, and the implementation of a dynamic friction model may result in lowering the

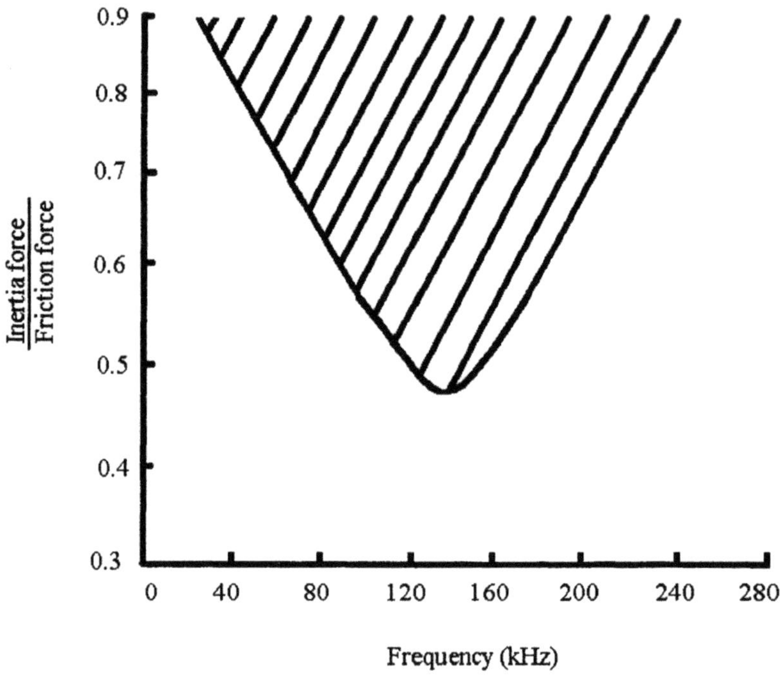

FIGURE 40.38 Ratio of inertial force to friction force versus excitation frequency data. (Adapted from Vinogradov and Huang, 1989.)

boundary of frequencies causing self-loosening. Also, the Vinogradov and Huang model includes only the inertial effects of a nut and does not include the inertial effects of a bolt or, perhaps more significantly, the clamped components in a joint.

Excitation amplitude versus macroslip plots and preload versus number of excitation cycles were also used to assess the rate of loosening, thread pitch, and thread angle. It was found that once macroslip occurs, the rate of self-loosening increases with the reduction in preload. For some frequencies, inertial forces increased the rate of loosening. In addition, it was found that transverse vibration was more dangerous than axial, fine thread fasteners resisted loosening better than coarse thread fasteners, and fasteners with larger thread angle resisted loosening better. It was also noted that the inertial mechanism of self-loosening with this joint model is triggered by the nonuniform distribution of preload along the thread. This suggests that fasteners with a more even distribution of preload would better resist vibration loosening.

40.3.16 Daadbin and Chow (1992)

Daadbin and Chow (1992) of the University of Newcastle upon Tyne developed a dynamic model to study fastener loosening caused by impact loads. Numerical simulations of the dynamic model were performed to assess the effect of the duration of impact, initial preload, lead angle, and thread friction.

The dynamic model is shown in Figure 40.39. The mass m and the plane represent the nut and the bolt head. The lead angle is represented by β. The thread interface is modeled with a spring of stiffness K and a dashpot with a damping constant of C. The spring is initially compressed from the application of preload P. After the application of the impact F, the reaction from the spring during rebound will push the mass upward. If the displacement of the mass m in the y direction exceeds the initial compression of the spring, it will separate and will be in free flight until it returns and falls on the spring and dashpot.

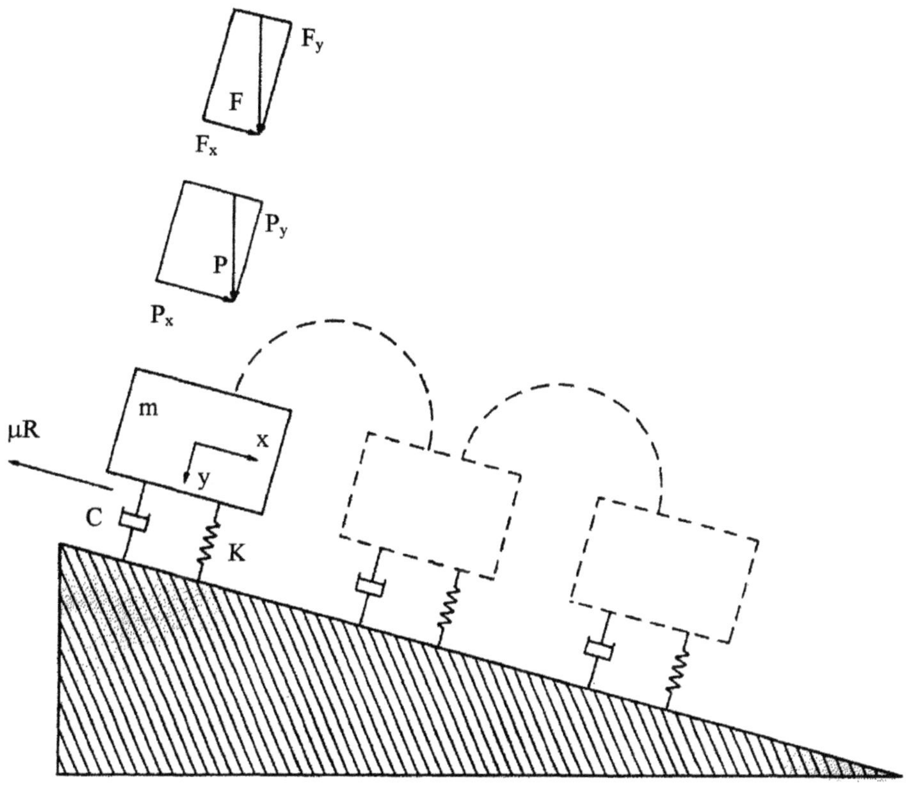

FIGURE 40.39 Dynamic nut and bolt model. (Adapted from Daadbin and Chow, 1992.)

The preload is a function of displacement in the x direction and is defined as $P = P_0 - \alpha x$, where P_0 is the initial preload and α is a coefficient that determines the rate of loosening with respect to x. The impact force F is defined as a half sine wave. When the impact force is applied to the mass, the mass will vibrate on the incline and possibly separate if the impact is strong enough to overcome the preload. The equations of motion of mass m in the x and y directions during contact are

$$m\ddot{y} = (F + P + mg)\cos\beta - C\dot{y} - Ky \tag{40.9}$$

$$m\ddot{x} = (F + P + mg)\sin\beta - \mu R \tag{40.10}$$

where μ is the friction coefficient at the contact, and the normal contact force is $R = \left[(C\dot{y})^2 + (Ky)^2\right]^{1/2}$. When the mass separates from the incline, the equations of motion become

$$m\ddot{y} = (F + P + mg)\cos\beta \tag{40.11}$$

$$m\ddot{x} = (F + P + mg)\sin\beta \tag{40.12}$$

These equations were solved numerically using a Runge-Kutta method.

The numerical results showed that the initial preload, lead angle, duration of impact force, and friction were the most important factors for loosening. The effect of impact force duration is shown in Figure 40.40a. Note, for example, that with an initial preload of 34 N, an 8% loss of preload

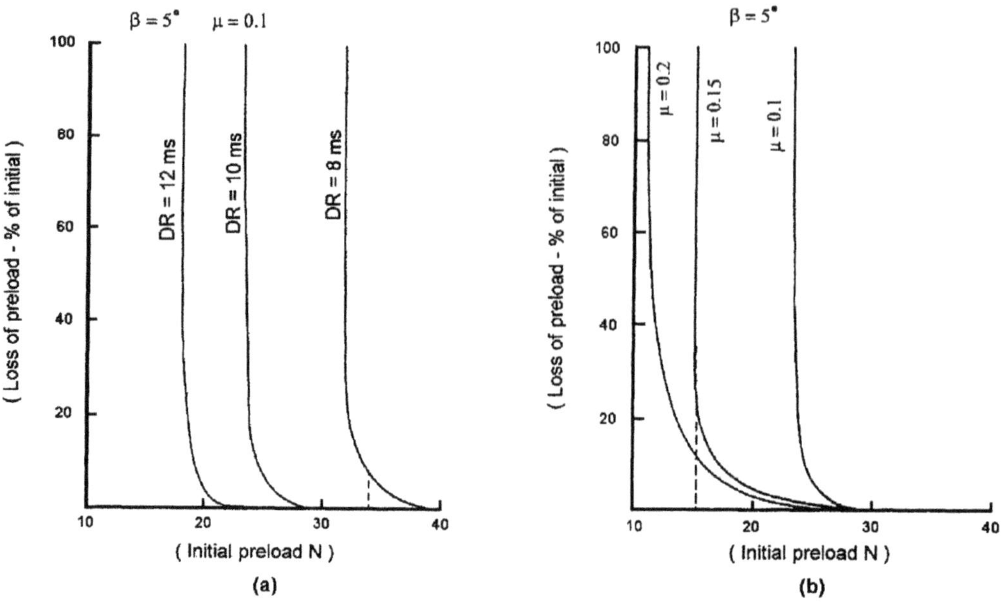

FIGURE 40.40 Loss of preload versus initial preload data for different (a) impact durations and (b) coefficients of friction. (Adapted from Daadbin and Chow, 1992.)

occurs for an impact of 8 ms duration compared with no loss of preload for impacts of 10 or 12 ms duration. These data show that the duration of impact is inversely proportional to the rate of loosening. An impact of shorter duration will cause higher excitation frequency, which creates larger inertial forces that accelerate loosening.

Figure 40.40b shows the effect of friction. For example, if the initial preload is 15.2 N, the loss of preload is 12%, 30%, and 100% for friction coefficients of 0.20, 0.15, and 0.10, respectively. As one would expect, the higher the friction, the better the resistance to loosening. The friction coefficient is influenced by roughness, material, coatings, and treatments. The friction force increases with preload.

It was also found that larger lead angles are less shock-resistant than smaller lead angles. This is supported by the fact that the theoretical internal loosening torque is directly proportional to the lead angle. In summary, for a fastener undergoing impact loading, the finer the thread, the longer the impact duration, and the larger the coefficient of friction, the smaller the loss of preload.

40.3.17 Ramey and Jenkins (1995)

Ramey and Jenkins (1995) of Auburn University employed the Taguchi method for the design of experiments and fractional factorial methodology to analyze the loosening of bolted assemblies due to vibration. Although there are probably 80–100 parameters that influence bolt loosening, 11 bolt parameters that were felt to be dominant were identified. These include bolt size (diameter), lubrication on bolt, hole tolerance, initial preload, locking device, grip length, thread pitch, lubrication between mating surfaces, class of fit, joint configuration, and mass of configuration. In addition to these design parameters, there are several load or noise parameters including vibration direction, vibration level, vibration frequency, and duration of vibration. Tests were performed with assemblies subjected to axial and transverse vibrations. Random vibrations were used with two different *g* levels for a duration of 2 min.

To keep the testing program within reasonable time and financial limitations, only two values of each test parameter were used. The design and load/noise parameter values used in the experiments are summarized in Table 40.1.

TABLE 40.1
Test Parameters and Values Used by Ramey and Jenkins (1995)

Test Parameter	Load/Noise Parameter	Parameter Value No. 1 (Lower Values)	Parameter Value No. 2 (upper Values)
Bolt size		¼ in.	¾ in.
Lubrication on threads		None	Tri-Flow
Hole tolerance		Oversized fit	Tight fit
Bolt preload		40% P_y	80% P_y
Locking device		Plain nut	Self-locking nut
Grip length[a]		½, 1 in.	1, 2 in.
Thread pitch[b]		20, 10	28, 16
Lubrication on mating parts		None	Tri-Flow
Class of fit		2	3
Joint configuration		Eccentric	Concentric
Mass of configuration		Mass of specimen	Mass of specimen + additional mass
	Vibration direction	Axial	Transverse
	Vibration amplitude	27 g_{rms}	40 g_{rms}

[a] ½–1 in. for ¼ in. bolts and 1–2 in. for ¾ in. bolts.
[b] 20–28 for ¼ in. bolts and 10–16 for ¾ in. bolts.

The principal contribution of Taguchi methods to this work is the concept of the orthogonal array. The orthogonal array allows the experiment to render the maximum amount of data with a minimum amount of testing. Using these concepts, the test matrix shown in Table 40.2 was developed.

The test specimens used were one-piece and two-piece cantilevers as shown in Figure 40.41. The two piece cantilever configuration was designed to introduce axial load and a prying action on the bolt when vibrated in the bolt's axial direction; shear and torsion are induced when the system is vibrated in the bolt's transverse direction. The one-piece cantilever configuration introduces axial load in the bolt when vibrated in the axial direction, and shear when vibrated in the transverse direction. Additional masses were used to achieve the desired mass of configuration and grip length desired when necessary. All test specimens were made of 6061-T6 aluminum, and all additional masses were made of A36 steel. The bolts and nuts were all commercial grade fasteners. The test specimens were mounted on a shaker table that provided the vibration. The motion of the test specimens was measured with accelerometers as shown in Figure 40.41. The bolt load before and after applications of vibration was determined from bolt length measurements with precision micrometers and an ultrasonic instrument. Nut on-torque and off-torque were also measured and used to assess bolt load.

A test response parameter or p value was defined to provide a dimensionless measure of loosening. For example, when bolt length/load was used as a measure of loosening, the p value was defined as

$$p_{\text{load}} = \frac{\text{load}_{\text{initial}} - \text{load}_{\text{residual}}}{\text{load}_{\text{initial}}} \quad (40.13)$$

where $\text{load}_{\text{initial}}$ and $\text{load}_{\text{residual}}$ are the bolt loads before and after the application of vibration. The higher the p value, the more significant the bolt loosening; a p value of 1 indicates complete loosening, and a p value of 0 indicates no loosening.

The main conclusions and recommendations drawn from analysis of this experimental investigation are as follows:

TABLE 40.2
Test Matrix

Test Setup	Fastener Size (Diameter)	Lubrication on Threads	Hole Tolerance (A)	Initial Preload (% Yield)	Locking Device (B)	Grip Length (In.)	Pitch (threads/In.)	Lubrication on Mating Matls.	Class of Fit	Joint Config. (C)	Mass of Config. (D)	Random Vibration: Axial Direction E	Random Vibration: Axial Direction F	Random Vibration: Transverse Direction E	Random Vibration: Transverse Direction F
1	¼″	None	OF	0.4	PN	0.5	20	None	2	A	X				
2	¼″	None	OF	0.4	PN	1.0	28	Tri-Flow	3	B	Y				
3	¼″	None	TF	0.8	SL	0.5	20	None	3	B	Y				
4	¼″	Tri-Flow	OF	0.4	SL	0.5	28	Tri-Flow	2	A	Y				
5	¼″	Tri-Flow	TF	0.8	SL	1.0	20	Tri-Flow	2	B	X				
6	¼″	Tri-Flow	TF	0.8	PN	1.0	28	None	3	A	X				
7	¾″	None	TF	0.8	PN	1.0	16	Tri-Flow	2	B	X				
8	¾″	None	TF	0.8	SL	2.0	16	None	2	A	Y				
9	¾″	None	OF	0.4	SL	2.0	10	Tri-Flow	3	A	X				
10	¾″	Tri-Flow	TF	0.8	PN	1.0	10	Tri-Flow	3	A	Y				
11	¾″	Tri-Flow	OF	0.4	PN	2.0	10	None	2	B	Y				
12	¾″	Tri-Flow	OF	0.4	SL	1.0	16	None	3	B	X				

Source: Adapted from Ramey and Jenkins (1995).

A: OF, Oversize Fit; B, Concentric Joint Configuration; B: PN, Plain Nut; C: A, Eccentric Joint Configuration; D: X, Small Mass; E, Low g-Level; F, High g-Level; SL, Self Locking Nut; TF, Tight Fit; Y, Large Mass.

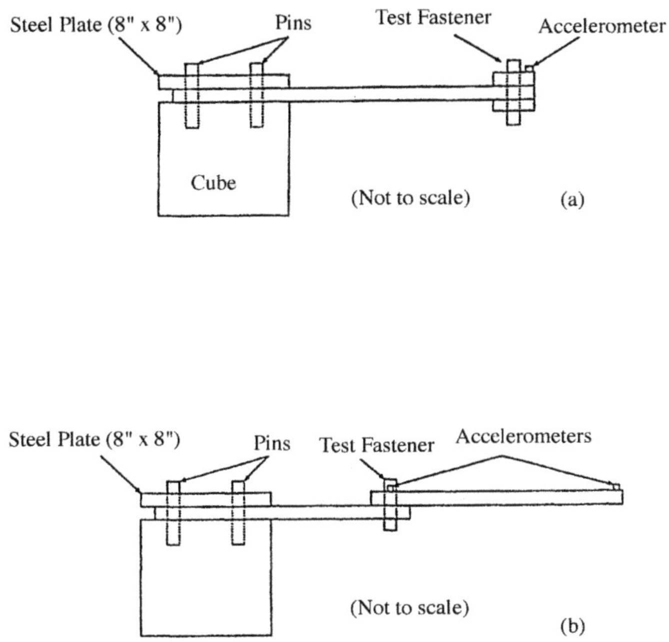

FIGURE 40.41 Test fixtures used by Ramey and Jenkins, (a) Typical one-piece cantilever; (b) typical two-piece cantilever. (Adapted from Ramey and Jenkins, 1995.)

1. The average p value for all tests in the test matrix was 0.003; hence, very little loosening occurred.
2. Numerous negative p values were obtained. This may be due to experimental error or perhaps tightening.
3. Microwelding in the two-piece test specimens mitigated relative movement of the test specimen pieces at the joint and thus mitigated bolt loosening.
4. The test data indicated that transverse loading on the test specimens was more inducive to bolt loosening than axial loading.
5. The test data indicated that a prevailing torque nut was superior to the plain nut at mitigating bolt loosening.
6. Much higher than normal bolt preloads, lighter than normal lubrication, and significant microwelding all contributed to reduce bolt loosening in executing the test matrix.
7. Additional testing results indicated that (1) transverse loadings are much more detrimental to bolt loosening than axial loads, (2) severity of vibration loadings has a major impact on bolt loosening, (3) larger bolts in a given vibration environment are more resistant to loosening than smaller bolts, and (4) more testing is needed to determine the effects of bolt lubrication and bolt preload on bolt loosening.
8. Additional small-scale preliminary testing using standard off-the-shelf bolts and nuts should be conducted to more fully identify the parameters having significant impact on bolt loosening due to vibration. This should include seeking alternative test configurations and specimens and a robust and sensitive bolt load monitoring device.
9. Future testing should probably use steel specimens to minimize microwelding problems.

40.3.18 Hess and Coworkers (1996)

Hess and coworkers of the University of South Florida performed experiments and dynamic analyses of threaded components subjected to vibration. This work began with a fundamental investigation

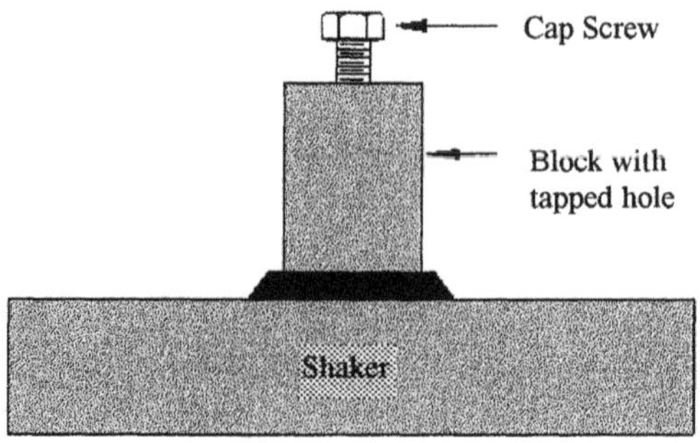

FIGURE 40.42 Test apparatus of Hess and Davis (1996).

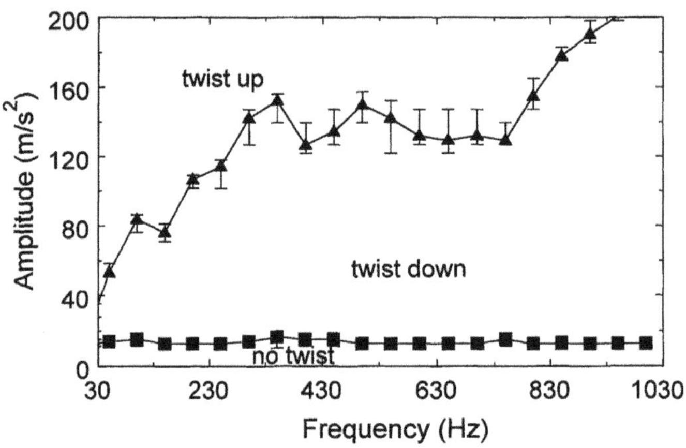

FIGURE 40.43 Vibration amplitude versus frequency data. (Adapted from Hess and Davis, 1996.)

of threaded components loaded only by gravity, i.e., without preload. This effort served as a first step and basis toward modeling and understanding the dynamics of more realistic bolted assemblies with preload.

Hess and Davis (1996) presented results from a series of experiments with threaded fasteners loaded by gravity and subjected to axial harmonic vibration. The apparatus consisted of a cap screw loosely twisted into a tapped hole in a base that was attached to a shaker as shown in Figure 40.42. The motions of the cap screw and base were measured with miniature accelerometers (not shown). Varying the amplitude and frequency of vibration, it was found that significant relative twisting motion can occur both with and against gravity and that the vibration can be tuned so that either action occurs. Figure 40.43 is a representative input vibration amplitude versus frequency diagram that shows regions of no twist, twist down with gravity, and twist up against gravity for a standard ¼–28 UNF ½ in. long 316 stainless steel cap screw. A no-twist behavior occurs at low input amplitudes for all frequencies. Increasing the amplitude for a given frequency results in a transition from no twist to twist down to twist up; increasing the frequency for a given moderate to high amplitude results in a transition from twist up to twist down. The effects of various parameters including material, bolt mass, and thread fit were examined. It was found that the direction of twist depends on these parameters as well as on the frequency and amplitude of vibration.

The physical explanation of these phenomena involves the nonlinear dynamic interaction of the vibration and friction and the resulting patterns of momentary sliding, sticking, and separation between the threaded components. Simultaneous measurements of the screw and base vertical accelerations have provided significant insight into the forms of motion associated with the different twisting behaviors. Figure 40.44 illustrates representative acceleration measurements.

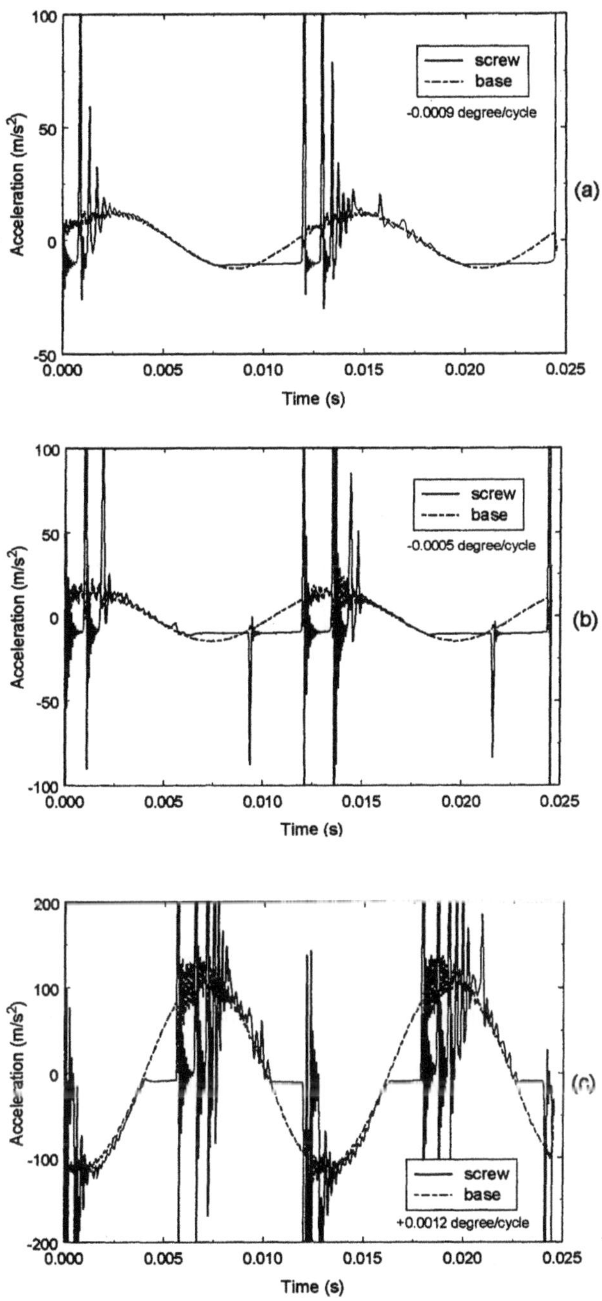

FIGURE 40.44 Acceleration measurements for (—) screw; (---) base. (a) Twist down without top impact; (b) twist down with top impact; (c) twist up. (Adapted from Hess and Davis, 1996.)

The acceleration of the screw is overlaid on the acceleration of the base, which is essentially harmonic. The acceleration traces in Figure 40.44a were measured during twist down. When the acceleration of the base reaches $-1g$, the screw and base separate, and the screw continues in flight (constant acceleration) until it impacts the base. Since the screw impact is positive and the base impact is negative, the bottom portion of the threads of the screw contact the bottom portion of the threads of the base. This was referred to as a bottom impact. The first bottom impact is followed by at least one bounce (i.e., another flight and bottom impact). Each impact results in transient motions.

After the last bottom impact and the resulting transients, the acceleration of the base and screw coincide until they reach –1g, at which point the cycle of motion repeats.

Another acceleration trace that was frequently observed during twist down is shown in Figure 40.44b. In this case, a negative screw impact (and a positive base impact) occurs while the screw is in flight prior to a bottom impact. This type of impact can result only if the top portions of the threads come in contact with each other. This was called a top impact. Following the top impact the screw motion is as follows: flight, first bottom impact, at least one bounce, transients, and then the same acceleration as base until –1g, when the whole cycle of motion repeats.

Acceleration data taken during twist up are given in Figure 40.44c. In this case, the screw and base separate when their accelerations decrease to —1g, and a top impact occurs while the base acceleration is still less than the screw acceleration. At least one bounce follows the top impact, and after transients the screw and base accelerations coincide. Here the top portions of the threads of each component are in contact. This continues until the accelerations increase to –1g, at which point the screw goes into flight until a bottom impact occurs, followed by a couple of consecutive bounces. After transients, the accelerations coincide with the bottom portions of the threads of each component in contact until the accelerations decrease to –1g, and the cycle repeats.

Hess (1996a) developed a kinematic model (see Figure 40.45) of this threaded component system. The model can be envisioned by unwrapping the threads of the tapped hole in the base and viewing them as upper and lower inclined constraints and viewing the screw as a lumped mass between the inclines. The upward or downward sliding displacement, x, of the mass is related to the upward or downward twisting angle, ϑ, of the cap screw in the tapped hole as $\vartheta = x \cos\beta / r_p$, where r_p is the pitch radius of the screw, β is the lead angle, d is the normal thread clearance, and s is the vibration input.

Kinematic analysis of this model showed that when the acceleration amplitude of the base is less than 1g, no twisting occurs. However, for amplitudes greater than 1g, both separation and twisting of the components occur for only portions of each cycle of vibration. Specific zones in a cycle were identified during which twisting down with gravity or twisting up against gravity can occur; these were defined as the slide L definite zone and the slide G definite zone, respectively. It was shown that as long as the dynamic coefficient of friction is less than the tangent of the thread lead angle, downward twisting occurs just as the acceleration amplitude of the base is increased above 1g. This results because the components are in contact during the slide L definite zone.

It was found that sustained contact during the G definite zone does not occur until the vibration level increases such that the instantaneous acceleration is more negative than –1g when the top portions of the threads of the screw and base impact. When this happens, an upward twisting action against the load occurs. Since both types of slide can occur within a cycle of vibration, the amount of twist up in a cycle must exceed the amount of twist down in a cycle for a net twist up to occur.

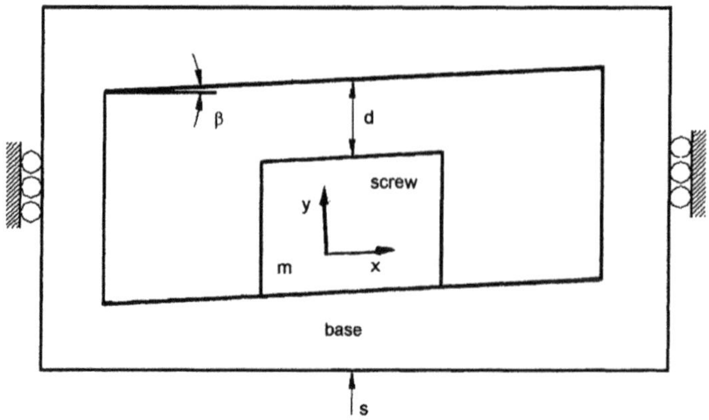

FIGURE 40.45 Kinematic model. (Adapted from Hess, 1996a.)

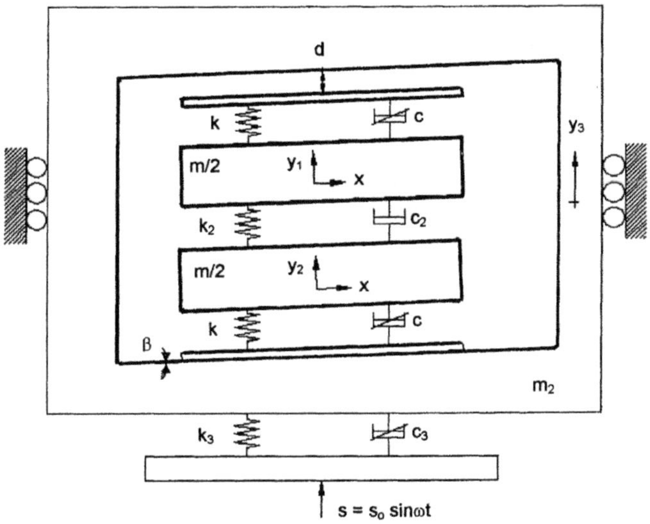

FIGURE 40.46 Dynamic model. (Adapted from Hess and Sudhirkashyap, 1996a.)

Since extending this kinematic analysis to more general conditions would be a formidable task, Hess and Sudhirkashyap (1996a) developed a dynamic model for threaded fasteners loaded by gravity and subjected to axial vibration that could be more readily extended to bolted assemblies with preload. A sketch of the dynamic model is shown in Figure 40.46. As with the kinematic model, the base constrains the bolt. The dynamics of the individual components are modeled with one dominant elastic mode for each component. The dynamic model accommodates one longitudinal elastic mode for the cap screw and one vertical mode of vibration for the base. The thread interfaces are modeled with an elastic spring in parallel with a nonlinear damping element to avoid unrealistic discontinuous forces at the onset of contact and separation (Hunt and Crossley, 1975). The motion of the screw is defined by coordinates x, y_1, and y_2. When the screw contacts either surface of the base, the normal contact force will be the sum of the thread interface spring and damping forces. The friction force is defined as the product of the coefficient of friction and the normal contact force.

Because of various possible contact and no contact configurations of the system, the governing equations of motion depend on the relative position of the components. The model parameters were estimated, and the nonlinear constrained equations of motion were solved numerically. The relative accelerations, velocities, and displacements were used to determine the sequence of motion forms experienced in a cycle of vibration. The possible motion forms include stick, flight, twist down, twist up, bottom impact, and top impact.

Figures 40.47 and 40.48 show computed vertical accelerations and twists of the screw and base for the three different vibration amplitudes used in the measurements in Figure 40.44. The simulations capture the essential behavior observed in the experiments. Further, the effect of thread fit, screw mass, thread friction, and vibration level and frequency were found to be in agreement with the experimental observations.

Hess and Sudhirkashyap (1996b) extended this approach to examine the dynamics of a moderately preloaded single-bolt assembly subjected to axial vibration. The apparatus used in this study is shown in Figure 40.49. It consists of a ¼–28 UNF 1.25 in. long bolt and a tapped base with a clamped mass and a helical spring sandwiched in between. The mass of the clamped component ranges from 0.043 to 0.107 kg. The preload increases as the spring is compressed. The contact stiffness between the base and clamped mass is varied by using springs of different stiffness, ranging from 693 to 75,800 N/m. The system is harmonically excited in the axial direction with a shaker over a frequency range of 50–1,000 Hz and up to 100g acceleration. The direction and amount of

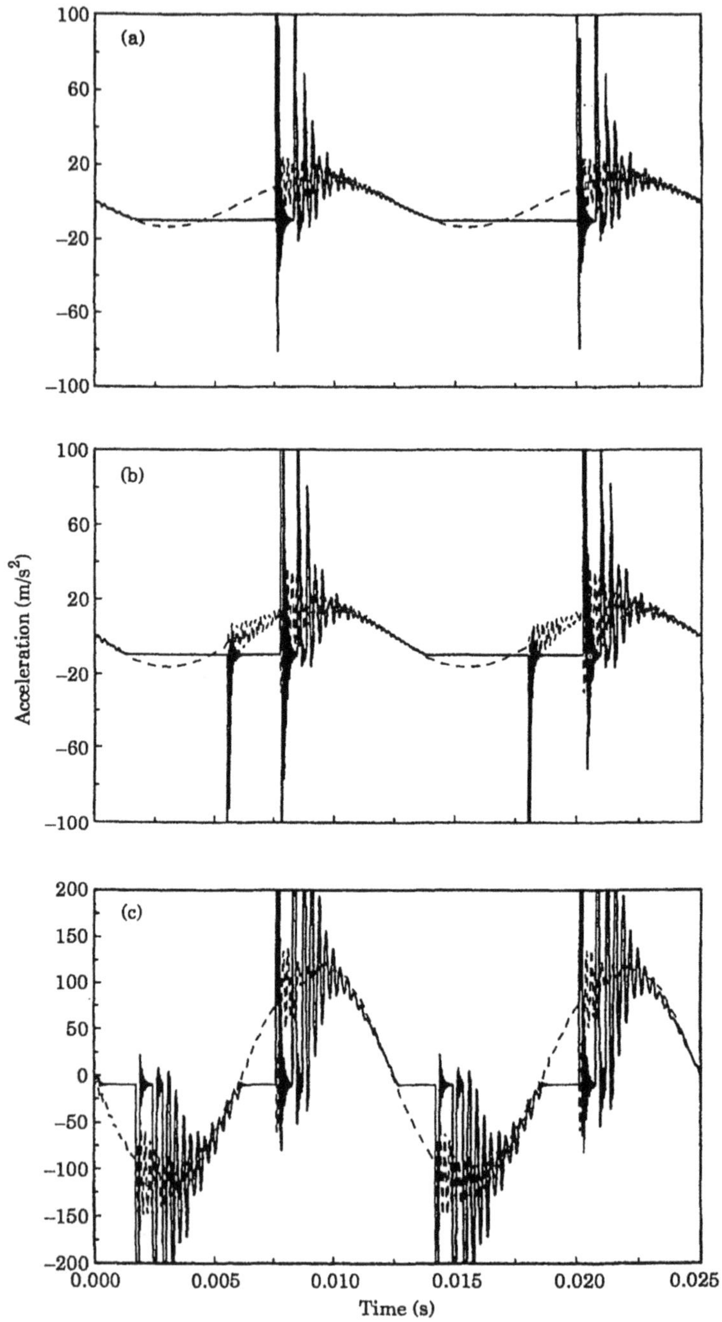

FIGURE 40.47 Simulation of accelerations for (—) screw; (---) base. (a) Twist down without top impact; (b) twist down with top impact; (c) twist up. (Adapted from Hess and Sudhirkashyap, 1996a.)

bolt twist are discernible to the human eye. The axial motions of the bolt, base, and clamped component are measured with miniature accelerometers (not shown). The components are run in before testing and are cleaned with a solvent in an ultrasonic cleaner prior to each test. The twist action observed within the first few seconds of each test is recorded.

Tests were performed for different clamped masses, interface stiffness, preload, thread fit, and vibration level and frequency. These tests revealed that a single-bolt assembly with moderate

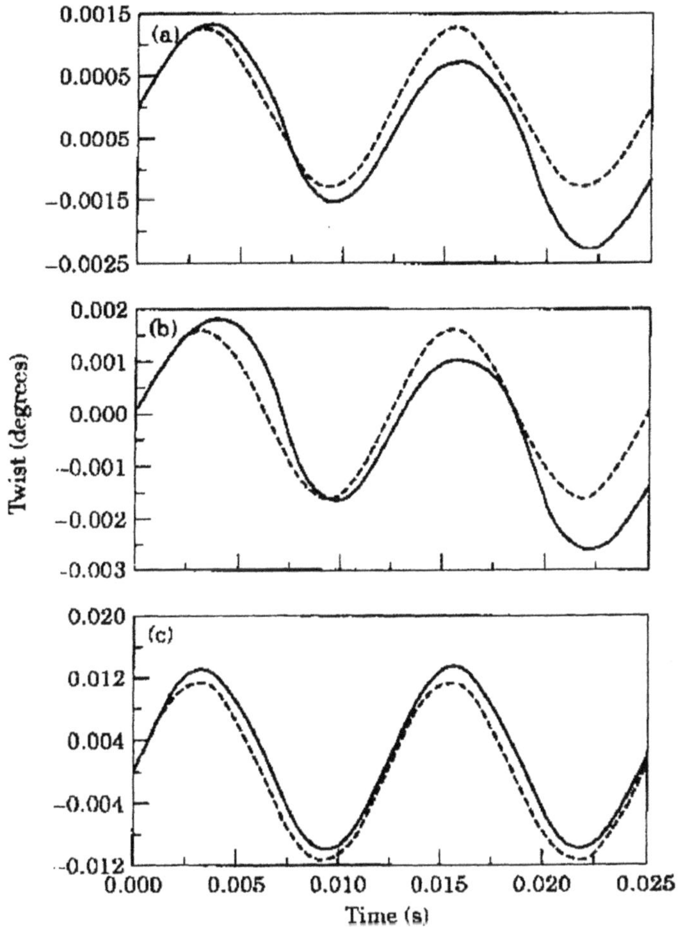

FIGURE 40.48 Simulation of twists for (—) screw; (---) base. (a) Twist down without top impact; (b) twist down with top impact; (c) twist up. (Adapted from Hess and Sudhirkashyap, 1996a.)

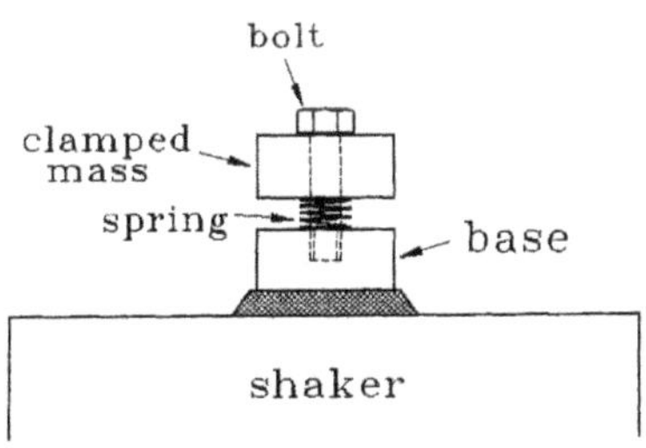

FIGURE 40.49 Test apparatus of Hess and Sudhirkashyap (1996b).

preload can, on average, loosen or tighten in the presence of vibration and that the applied vibration, as well as other system parameters, can be tuned so that either action occurs. A representative input amplitude–frequency diagram is shown in Figure 40.50. These data show that no twist occurs for any of the frequencies at small input vibration amplitudes. As the amplitude increases, a loosening action occurs over all frequencies. As the amplitude increases further, a tightening action develops at higher frequencies. The effect of increasing preload was found to be a shift of these three regions

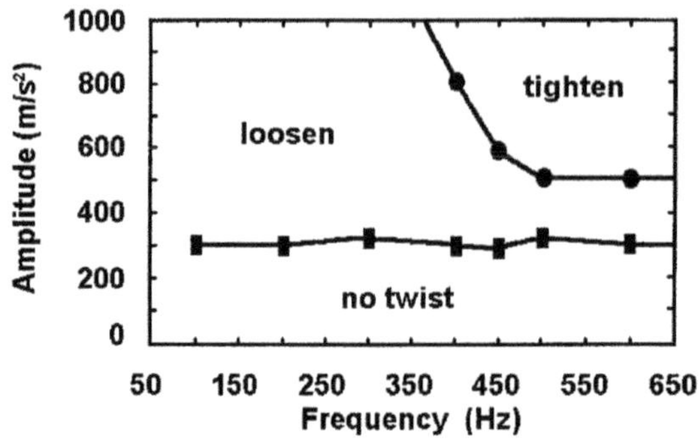

FIGURE 40.50 Vibration amplitude versus frequency data. (Adapted from Hess and Sudhirkashyap, 1996b.)

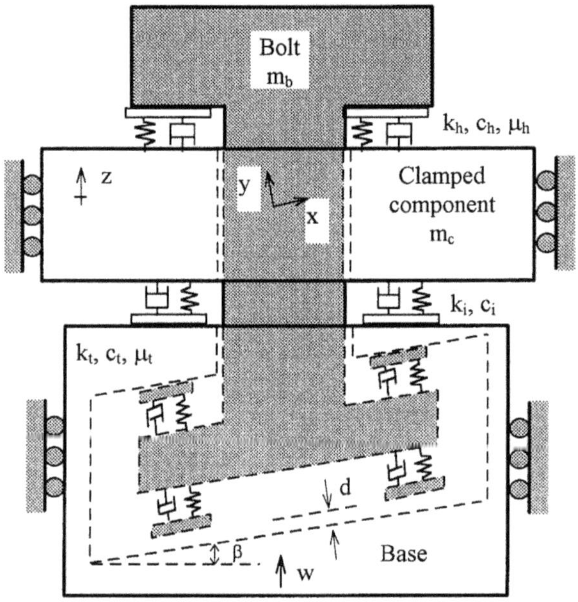

FIGURE 40.51 Dynamic model of single-bolt assembly. (Adapted from Hess and Sudhirkashyap, 1996b.)

to higher input amplitudes. The effect of increasing the spring stiffness was found to be a shift of the tightening region to higher frequencies. The effect of increasing clamped mass was to shift the tightening region to lower frequencies and somewhat lower amplitudes as well as to shift the loosening region to lower amplitudes. It was also found that as vertical thread clearance is decreased, tightening occurs at lower input amplitudes and the loosening region decreases.

The dynamic model for this system can be visualized by unwrapping the threads of the tapped hole in the base and viewing them as upper and lower inclined constraints on the bolt threads, modeling the bolt as a lumped mass m_b constrained between the upper base threads and the clamped component, and modeling the clamped component as a lumped mass m_c constrained between the bolt head and the base. A sketch of this model is shown in Figure 40.51. In the model, the bolt is considered rigid with elastic interfaces. This model is sufficient for bolts with moderate preload such as those in electronic equipment and joints with gaskets. One of our current goals is to incorporate into the model both the axial and torsional elastic bolt characteristics found in more heavily preloaded bolts.

When in contact, the interfaces are modeled with elastic stiffness k_h, k_i, and k_t in parallel with dampers c_h, c_i, and c_t. Nonlinear damping elements are used to avoid unrealistic discontinuous forces at the onset of contact and separation (Hunt and Crossley, 1975). The friction forces at the thread interface and the bolt head interface are modeled with a friction model described below.

The system input and normal motion of the base is $w=w_0 \sin \omega t$. Coordinate z describes the normal motion of the clamped component. The coordinates x and y describe the motion of the bolt. Coordinate x represents the circular displacement of the bolt at the pitch diameter. This displacement is proportional to the twisting angle $\vartheta=(x \cos \beta)/r_p$, where r_p is the pitch radius of the bolt. The lead angle of the bolt thread is β, and the thread clearance without preload is d_0.

The equations of motion for this model depend on the relative positions of the bolt, the clamped component, and the base. There exist six possible forms for the governing equations of the bolt and four possible forms for the clamped component. In one configuration, the bolt is in tension, the clamped component is in compression, and all interfaces are in contact (see Figure 40.51). This condition is expressed mathematically as $(y-w \cos \beta) \geq -\delta_t$, $(y \cos \beta + x \sin \beta - z) \leq \delta_h$, $(z-w) \leq \delta_i$, where δ_t, δ_h, and δ_i are the static deflections at the thread interface, bolt head, and clamped component interface, and clamped component and base interface, respectively. The governing equations of motion for this configuration are

$$m_b \ddot{y} = -N_u + N_h \cos \beta - F_h \sin \beta - m_b g \cos \beta \tag{40.14}$$

$$m_b \ddot{x} = F_u + N_h \sin \beta + F_h \cos \beta - m_b g \sin \beta \tag{40.15}$$

$$m_c \ddot{z} = -N_u + N_i - m_c g \tag{40.16}$$

where the normal contact forces are

$$N_h = \left(z + \delta_h - y\cos\beta - x\sin\beta\right)\left[k_h + c_h\left(\dot{z} - \dot{y}\cos\beta - \dot{x}\sin\beta\right)\right] \tag{40.17}$$

$$N_i = \left(w + \delta_i - z\right)\left[k_i + c_i\left(\dot{w} - \dot{z}\right)\right] \tag{40.18}$$

$$N_{tt} - \left(y + \delta_t - w\cos\beta\right)\left[k_t + c_t\left(\dot{y} - \dot{w}\cos\beta\right)\right] \tag{10.19}$$

and the static deflections are

$$\delta_h = \frac{k_i \delta_i - m_c g}{k_h} \tag{40.20}$$

$$\delta_t = \left(\frac{k_h \delta_h - m_b g}{k_t}\right) \cos\beta \tag{40.21}$$

$$\delta_i = \frac{P}{k_i} - \left(\frac{\left(x - w\ \sin\beta\right)\sin\beta}{1 + k_i/k_h + k_i/k_t}\right) \tag{40.22}$$

where P is the initial preload in the joint. The clamping force is defined by $F_c = k_i \delta_i$. The thread interface friction force, F_{tt}, and the bolt head interface friction force, F_h, both oppose the circular sliding or twist of the bolt. The nonlinear contact forces couple the axial motions of the system components with the twisting motion of the bolt.

The friction forces in the assembly are modeled with a friction model introduced by Karnopp (1985). The standard Coulomb friction model is discontinuous and has numerical limitations when the relative sliding velocities at the bolt thread and head interfaces approach zero. A number of friction models exist that remove these limitations (for an excellent review, see Armstrong et al., 1994); however, many of these models do not accommodate true stick. A friction model that includes stick and is both physically realistic and numerically manageable was proposed by Karnopp (1985).

This friction model defines a small velocity region between $-v_{min}$ and v_{min} surrounding zero sliding velocity. Inside this region, velocity is considered to be zero and the friction is determined by the forces acting on the system or by the friction breakaway value, whichever is less. Outside this region, the friction depends on the coefficient of friction and the normal contact forces.

For the single-bolt assembly configuration in which the bolt is in tension, the clamped component is in compression and all interfaces are in contact (see Figure 40.51), there are friction forces at the bolt head and thread interfaces. To correctly determine the friction forces at these two interfaces, four cases must be considered: (1) Both threads and head stick, (2) only the threads stick, (3) only the bolt head sticks, and (4) both the threads and head slip. For mathematical expressions of these friction forces, refer to Hess et al. (1996).

The model parameters have been determined from a combination of direct measurement and estimation from acceleration measurements and bolt twist observations and are given by Hess et al. (1996). The nonlinear constrained equations of motion of the dynamic bolt assembly model were solved numerically. These simulation results were found to agree qualitatively with the experimental results. Extended time traces of clamping force for various conditions were also computed (Hess et al., 1996). Figure 40.52 shows time traces for a range of moderate preloads. These data show that, at least for axial vibration, a steady clamping force is reached with time, and loosening does not lead to complete loss of preload.

This work shows that substantial axial vibration levels are necessary to achieve an increase in clamping force. However, recent experiments show that lower levels of combined axial, transverse, and angular vibration, as found in structures with bending, longitudinal, and torsional modes of vibration, are necessary to achieve significant tightening (Peters, 1996). For example, a vibration acceleration level of 25g is capable of increasing the clamping force in a single-bolt joint on a cantilevered beam from 80 to 120N.

This previous work was focused on understanding the nonlinear dynamics of threaded fasteners in single-bolt assemblies with moderate preload and simple component geometry and loading. This work is currently being extended to accommodate higher preload and combined axial and transverse dynamic loading. In addition, funding is already in place to extend this work to assemblies with multiple fasteners and more complex assembly component dynamics and geometry (Hess, 1996b). The goal is to develop general design guidelines and criteria (e.g., fastener positioning with respect to mode shapes, component geometry, etc.) that minimize maintenance and failure due to fastener loosening and that are applicable to a broad range of systems. This approach is novel in that it incorporates the effects of the dynamics of fasteners and other assembly components in the design of assemblies, including fastener selection and placement.

40.3.19 Johnson (1996)

Johnson (1996) of Johnson Gage Company has reported test results indicating that failure to maintain dimensional size conformance in standard fasteners can lead to degraded vibration resistance. He states that dimensional size conformance is mandatory for vibration resistance. Holmes (1988) also alludes to this problem in his discussion of the battle between assemblers who want closer tolerances and fastener manufacturers who want to protect their livelihood by prolonging the life of their dies.

Johnson and Hess are currently planning collaborative testing using an inertial loaded cantilever beam type of apparatus to assess dimensional size conformance. The test specimens for these tests consist of master bolts and nuts with precisely controlled dimensional size.

40.4 FASTENER LOCKING TECHNIQUES

Many bolted assemblies that are subjected to vibration and shock use fasteners with locking features to minimize loosening. Although there are a significant number of fastener locking techniques

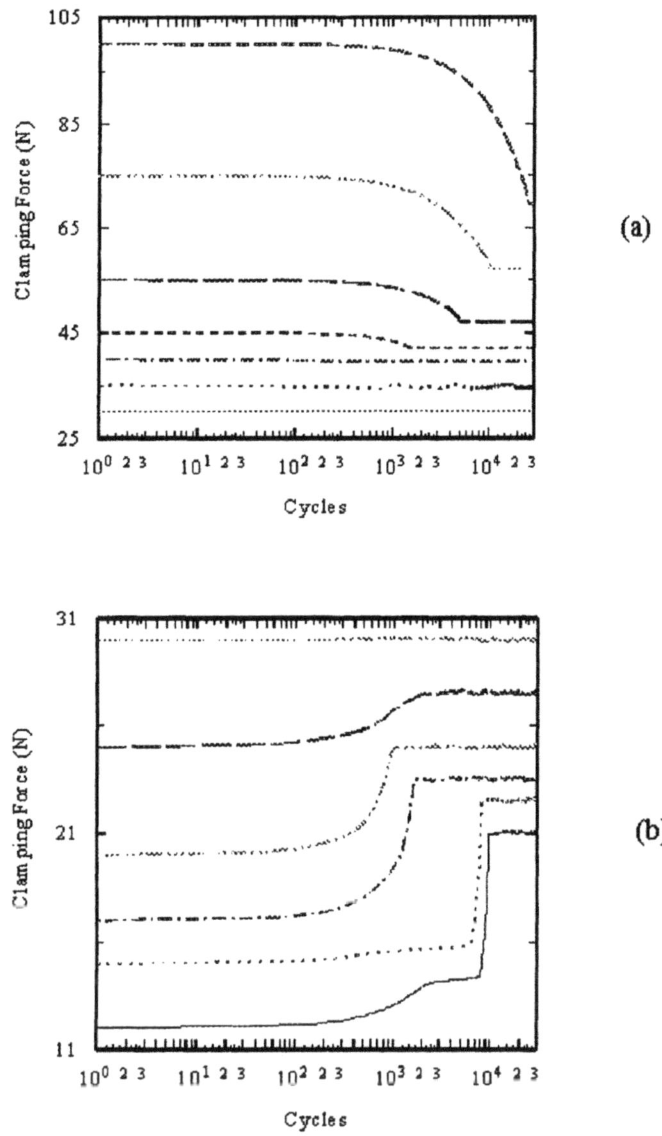

FIGURE 40.52 Clamping force versus vibration cycles for moderate preloads. (a) Loosening; (b) tightening. (Adapted from Hess et al., 1996.)

and locking fastener designs, these can be divided into four general groups: (1) free running preload-independent locking fasteners, (2) free running preload-dependent locking fasteners, (3) prevailing torque locking fasteners, and (4) chemical locking.

Free running locking fasteners assemble similarly to standard fasteners without any locking features until seated. At this point, the resistance to loosening is obtained by an additional accessory device, which may or may not be preload-dependent. Prevailing torque locking fasteners require additional torque to assemble and disassemble because of distorted threads or various other features. Prevailing torque is defined as the torque required to turn such fasteners with zero preload. Chemical locking is accomplished by causing a chemical reaction, with adhesives or corrosion, that increases thread friction. The next four subsections describe locking fasteners and techniques in the four groups.

FIGURE 40.53 Using lock wire to secure bolts. (Adapted from Bickford, 1995.)

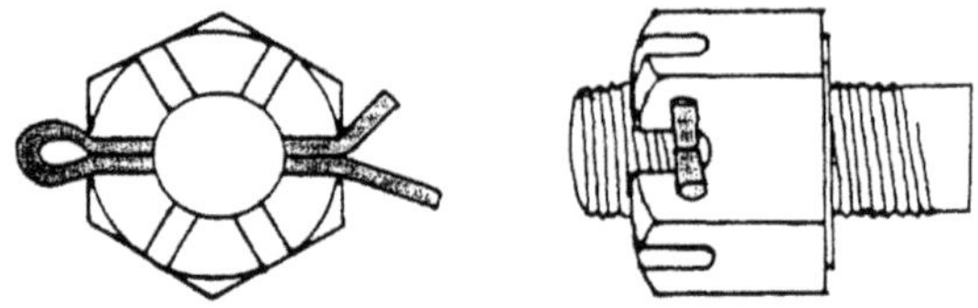

FIGURE 40.54 Using cotter pins to secure slotted nuts. (Adapted from Bickford, 1995.)

In addition to fastener locking designs and techniques, there are a number of assembly design considerations that can lead to improved resistance to vibration-induced loosening. These assembly design considerations are discussed in Section 40.4.5.

40.4.1 Free Running Preload-Independent Locking Fasteners

40.4.1.1 Jam Nuts

Jam nuts are used in addition to a bolt and nut. The jam nut is first tightened down, then the standard thicker nut is tightened against it.

40.4.1.2 Lock Wire

The use of a lock wire consists of feeding wire through a hole in a bolt head and tying neighboring bolts together such that a loosening action of any bolt would tend to tighten its neighbor (see Figure 40.53). A disadvantage of lock wire is that its installation and removal present a maintenance burden.

40.4.1.3 Cotter Pins

Cotter pins are usually used together with a slotted or castellated nut as shown in Figure 40.54. Once the nut is tightened down and aligned with a hole drilled through the bolt shank, the pin is inserted and set.

40.4.1.4 Welding

Nuts and bolts can be tack welded together or to joint surfaces. Also, Bickford (1995) points out that to preserve the threaded components, a plate with a hexagonal hole can be placed over the nut, and this plate, or keeper as it is called, is then welded to the joint surface.

40.4.1.5 Stage 8 Locking Safety Bolts

The Stage 8 fasteners (Bennett, 1996) use a locking arm, a retainer clip, and a groove cut in the head of the bolt or nut. A locking arm is placed over the bolt head or nut, and the retainer clip is installed in the groove to hold the locking arm in position to prevent fastener loosening (see Figure 40.55). Installing a Stage 8 locking safety bolt is a three-step process:

1. Torque fastener to specifications.
2. Position mating or locking arm on the fastener such that it abuts a solid immovable part of the assembly.
3. Install retainer clip in groove to hold locking arm in position.

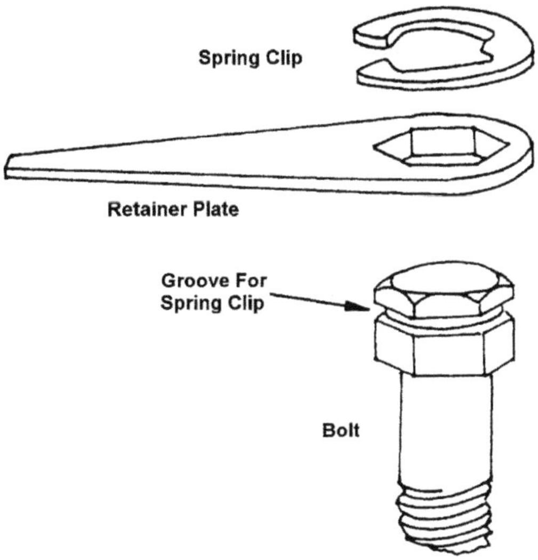

FIGURE 40.55 Stage 8 locking safety bolts. (Adapted from Bennett, 1996.)

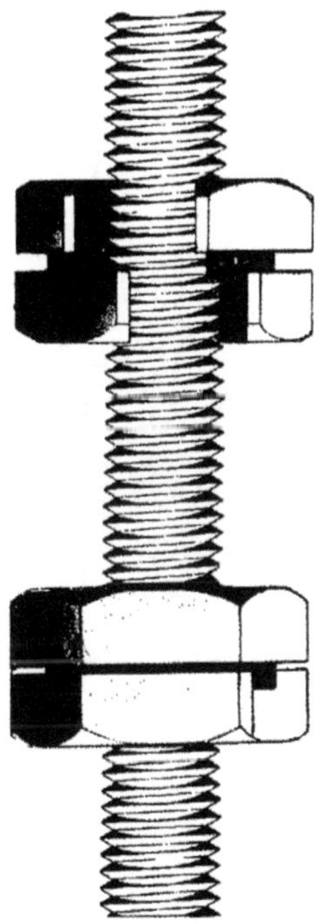

FIGURE 40.56 Slip-on lock nuts. (Adapted from Aldridge, 1996.)

In addition, the manufacturer recommends lubricating the fastener threads to provide better control of preload using a torque wrench. These fasteners are available in a wide range of sizes with a variety of locking arm shapes.

40.4.1.6 Slip-On Lock Nuts

The Slip-On lock nut (Aldridge, 1996) consists of an assembly of two permanently joined, independently rotatable slotted nuts, designed for direct slip-on installation and jam-type locking action (see Figure 40.56). The nuts are hexagonal with one of the sides removed in each nut to form the slot.

Users install the nut assembly by rotating both nuts to the open position. In the open position, with the open slots of the two halves aligned, users slip the assembly on or off a threaded rod or bolt, thereby eliminating rundown or complete removal of the nut. The two halves are turned to close the slot, and the nut assembly is tightened. Additional turning of the halves in the closing direction provides a jam-type locking action.

40.4.1.7 Boydbolt Fastening System

The Boydbolt Fastening System is a locking fastener designed by Rexnord Specialty Fasteners of Torrance, California and is shown in Figure 40.57. The bolt contains a spring-loaded plunger that presses two steel balls into groves in the nut thereby locking the fastener. During installation and removal, the plunger is displaced (see bolt head in Figure 40.57) such that the pressure on the balls is released via grooves in the plunger.

40.4.1.8 TrueLock Bolts

The TrueLock bolt was designed by TrueLock Company of Herndon, Virginia and is shown in Figure 40.58. Both the lock crown and the nut contain a series of grooves. These are designed such

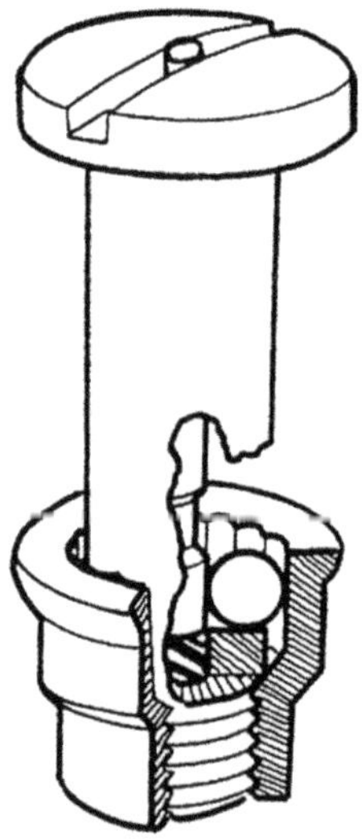

FIGURE 40.57 Boydbolt fastening system. (Adapted from Bickford, 1995.)

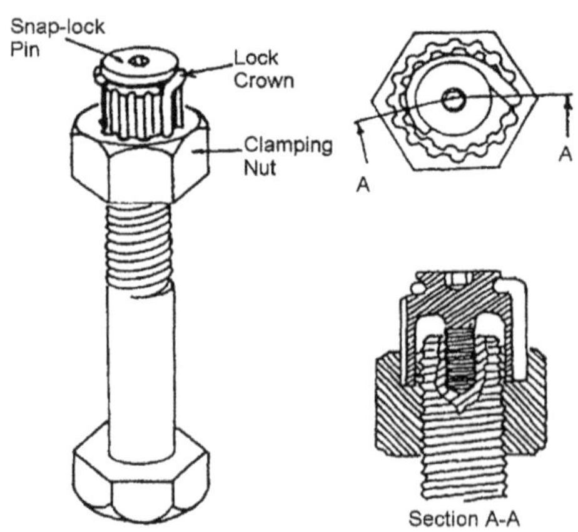

FIGURE 40.58 TrueLock bolts. (Adapted from Bickford, 1995.)

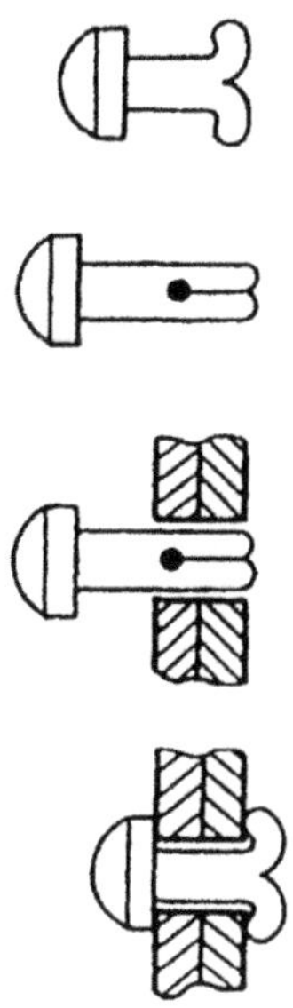

FIGURE 40.59 Shape memory alloy used for a blind fastener. (A) Original shape; (B) straightened shape after cooling; (C) installed in joint; (D) recovered shape after heating. (Adapted from Funakubo, 1987.)

that only one pair of grooves can be aligned at one time. Once the nut is tightened down, the lock crown is screwed on the end of the bolt with a fine pitch threaded pin and is tightened until one pair of grooves is aligned. A snap lock pin is then secured in the hole formed by the aligned grooves and is snapped into a retaining ring on the lock crown.

40.4.1.9 Shape Memory Alloys

Shape memory alloys (SMAs) are a family of metals whose shape can be significantly changed with temperature. This property is unrelated to thermal expansion and is far more dramatic (Borden, 1991). SMAs have been used for blind fasteners as shown in Figure 40.59 (Funakubo, 1987). A material with a critical temperature below room temperature is used. The fastener shape shown in Figure 40.59a with the ends spread apart is memorized. Before it is installed, the fastener is cooled with liquid nitrogen or dry ice, which straightens the ends as shown in Figure 40.59b. The fastener is then installed in the hole (Figure 40.59c). As the fastener returns to room temperature, the original shape is recovered to lock the fastened components (Figure 40.59d).

Intrinsic Devices Inc. of San Francisco, California, specializes in SMA devices for fastening (Borden, 1996). They make a family of SMA rings under the name UniLok. UniLok rings can be used to lock nuts in position on bolts as shown in Figure 40.60. An expanded UniLok ring is placed on a slotted cylindrical projection on the top of a nut. When the ring is heated to room temperature, it shrinks in diameter, compressing the nut fingers against the male thread, thereby locking the fastener. The locking ring can be removed in two ways: The ring can be cut off, or it can be restored to its expanded shape by cooling with liquid nitrogen.

40.4.2 Free-Running Preload-Dependent Locking Fasteners

40.4.2.1 Lock Washers

Lock washer features include spring action and/or toothed edges or serrated faces (Figure 40.61). Spring action lock washers include helical spring, Belleville (or cone), and wave washers. These washers are installed under a screw head or nut and are compressed when the fastener is tightened. Spring washers remain elastic and provide a reaction force even in the presence of fastener or

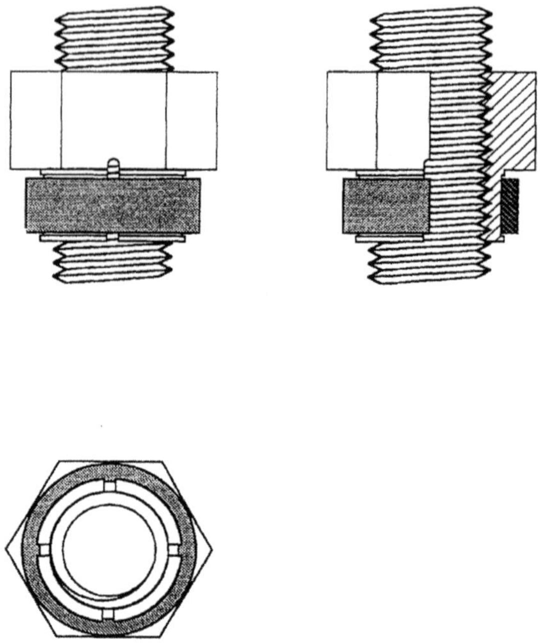

FIGURE 40.60 UniLok shape memory alloy lock nut ring. (Adapted from Borden, 1996.)

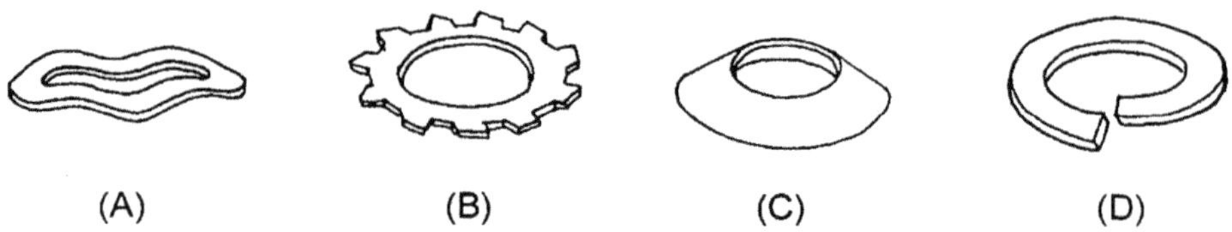

FIGURE 40.61 Lock washers. (A) Wave; (B) toothed edge; (C) Belleville; (D) helical spring.

assembly deflection. As long as such deflections remain less than the spring washer deflection, these washers can help to maintain some level of fastener tension. The main shortcoming of spring washers is that they may not be effective until after a significant amount of preload is lost.

Toothed and serrated washers are also installed under a screw head or nut. When the fastener is tightened, the washer digs into both the fastener and the clamped component. This increases the friction in the assembly and can prevent relative motion and loosening. However, washers of this type can easily damage the bearing surfaces of the joint.

40.4.2.2 Sems

Sems are screws with preassembled lock washers (Figure 40.62). These fasteners function just like lock washers but are more convenient to use. ANSI/ASME B18.13 (IFI, 1988) provides dimensional standards for sems. Nuts are also available with preassembled lock washers; however, there are no national standards for such fasteners.

40.4.2.3 Toothed or Serrated Lock Nuts and Screws

There are a variety of nuts and screws with toothed edges and/or serrated faces designed into the bearing surfaces of the fasteners (see Figure 40.63). When tightened, these fasteners dig into the clamped components. Although they can prevent loosening, they can also damage the surfaces of the joint.

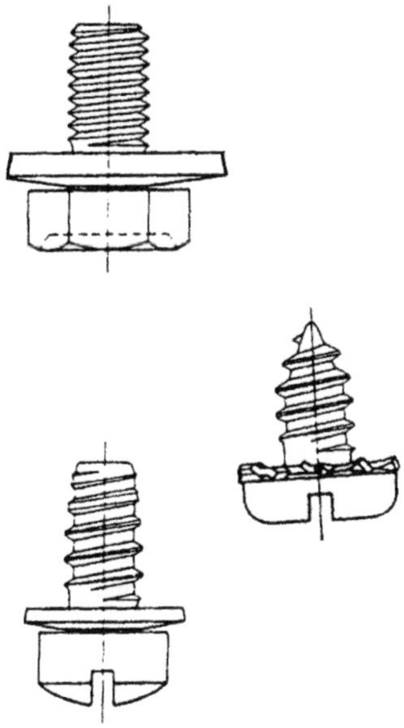

FIGURE 40.62 Sems.

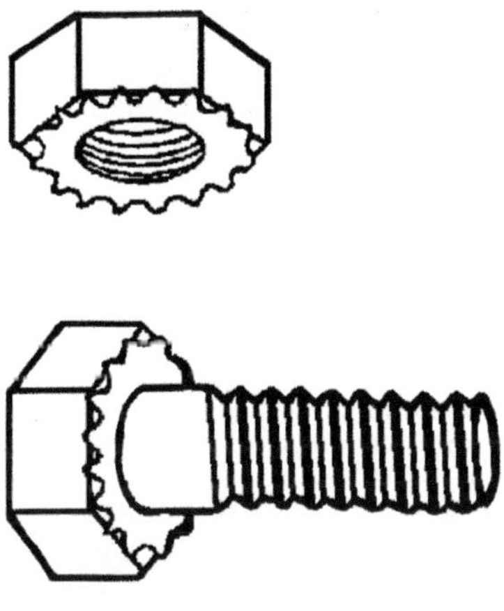

FIGURE 40.63 Toothed and serrated lock nut and screws.

40.4.2.4 Spring Action Lock Nuts and Screws

A variety of fasteners use an elastic spring action to achieve locking. These include dished or flanged bolt heads (Figure 40.64a) as well as recessed and flanged nuts (see Figure 40.64b).

40.4.2.5 Spiralock

Spiralock is a proprietary internal thread form that incorporates a 30° wedge ramp as shown in Figure 40.65 (Palm, 1996). Spiralock threaded holes and nuts accept standard male fasteners. When clamp load is applied during assembly, the Spiralock thread form tightly constrains the standard male fastener by drawing the crests of the male thread against the 30° wedge ramp. This can significantly reduce the transverse thread movement obtained with standard 60° thread forms that result from manufacturing tolerances (Holmes, 1988). This wedge-locking effect of Spiralock thread form also distributes the bolt load more evenly than the standard thread form as shown on the right in Figure 40.65. However, the Spiralock thread requires approximately 20% higher assembly torque than a standard thread to produce the same axial preload.

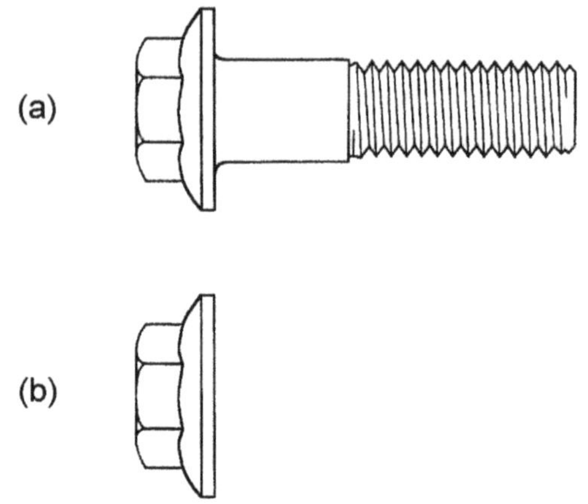

FIGURE 40.64 Spring action lock nuts and screws. (a) Flanged bolt head; (b) flanged nut.

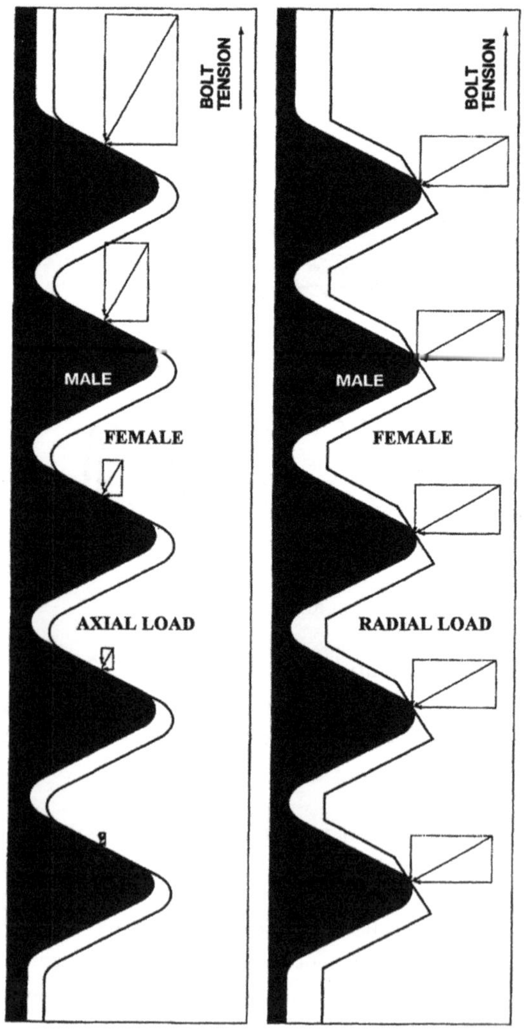

FIGURE 40.65 Spiralock thread form (right) compared with standard thread form (left). (Adapted from Palm, 1996.)

In a report prepared for Chrysler Corporation, Nayak (1978) of Massachusetts Institute of Technology presented calculations showing that

1. Bolt load is more evenly distributed over the engaged threads with the Spiralock thread form than with the standard thread form.
2. The maximum stresses at the root of the bolt thread are of the same order of magnitude in both cases.
3. The moment or torque required for relative rotation or loosening is significantly higher for the Spiralock form.

A report by Kerley (1984) of NASA GSFC presented results from tests with Spiralock nuts under vibration and static load conditions. The Spiralock nuts showed better resistance to loosening than standard nuts when subjected to both high amplitude sine and random vibration. As with standard thread nuts, static tests using Spiralock nuts showed that friction factors were more consistent when lubricant was used. More recently, Harrel (1992), an engineering consultant for Lawrence Livermore National Laboratory, reported on a finite element analysis comparing the Spiralock and UNF thread forms. The analysis showed that the Spiralock thread form distributes load more evenly than the standard UNF thread form. However, the calculated pullout strength with the Spiralock thread was found to be noticeably smaller than with the standard UNF thread.

40.4.2.6 Cam-Type Locking Devices

Disc-Lock (White, 1996) and Nord-Lock (Hosko, 1996) produce locking washers that use inclined cams to oppose loosening motion. The washers have inclined cams on one side and a series of ridges on the other. Two washers are installed as shown in Figure 40.66 such that the ridges face the nut and joint bearing surfaces and the inclined cams mate. The washer incline cam angle is larger than the fastener pitch angle, and the washer ridges dig into the nut and joint bearing surfaces. When the nut attempts to loosen, the pair of washers expand more than the nut and bolt. Tests with a Junker-type machine show that these washers can not only prevent vibration- and shock-induced loosening but can actually cause tightening, at least for moderate preloads (see Figure 40.67). Disc-Lock also produces preassembled cap screws and nuts with this type of cam action built in as shown in Figure 40.68.

40.4.3 Prevailing Torque Locking Fasteners

Prevailing torque locking fasteners provide some form of interference between threads. These fasteners can be grouped as (1) all-metal fasteners with modified or distorted threads and (2) fasteners

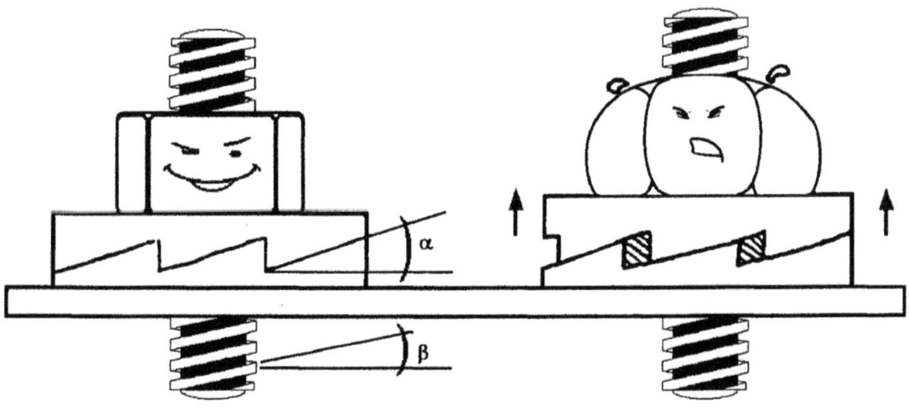

FIGURE 40.66 Cam-type locking washers. (Adapted from Hosko, 1996.)

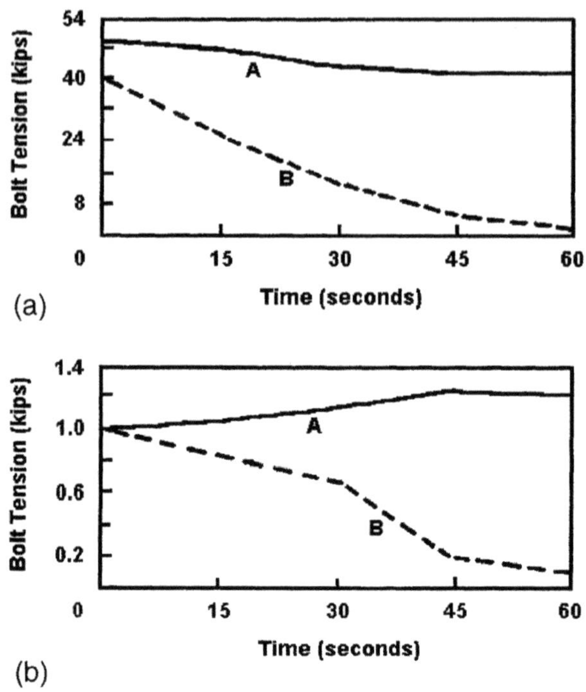

FIGURE 40.67 Test results from a Junker type test machine with cam-type locking washers (line A). (a) cam-type washer reduced the rate of vibration-induced loosening; (b) cam-type washer increased bolt tension during loosening rotation. (Adapted from White, 1996.)

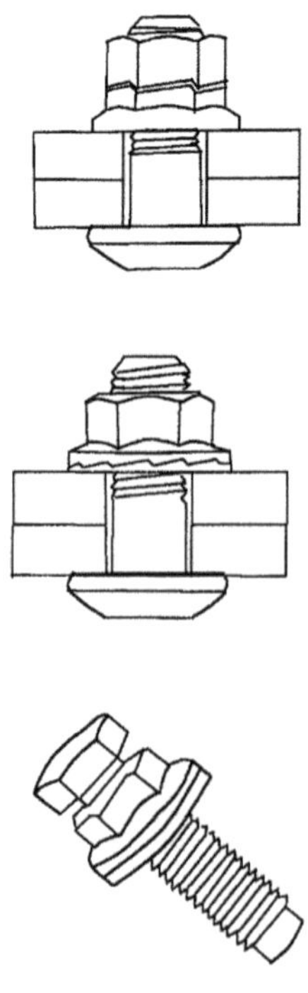

FIGURE 40.68 Disc-Lock preassembled cap screws and nuts.

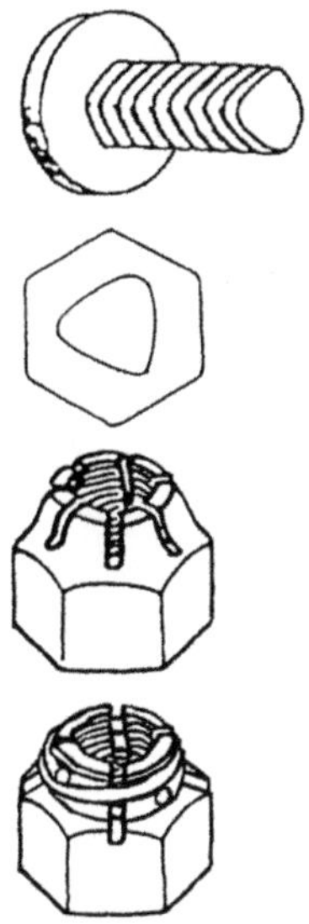

FIGURE 40.69 All-metal prevailing torque fasteners. (a) Bolt with distorted threads; (b) nut distorted by squeezing; (c) nut slotted and squeezed; (d) nut with helical spring.

with an added ring, patch, strip, or coating of nonmetallic material. The all-metal type can be used over a broader temperature range.

Today, there are a large number of different proprietary prevailing torque locking fasteners available. Fortunately, the Industrial Fasteners Institute (IFI, 1988) has established standards for prevailing torque fasteners. IFI Standard 100/107 specifies the dimensional envelopes, strength grades, and prevailing torque values for prevailing torque nuts. IFI Standard 124 provides specifications for prevailing torque lock screws.

40.4.3.1 All-Metal Prevailing Torque Fasteners

Figure 40.69 illustrates a few different types of all-metal prevailing torque bolts and nuts. The threads of the bolts generally are rolled to have offset threads or are distorted by squeezing (Figure 40.69a). The threads of all-metal nuts are generally distorted by squeezing the nut hole into an oval or triangular shape (Figure 40.69b) or by slotting and squeezing (Figure 40.69c). Other nut designs include an accessory device, such as a helical spring on a slotted nut as shown in Figure 40.69d (Perardi, 1996), to provide thread interference.

40.4.3.2 Prevailing Torque Fasteners with Nonmetallic Material

Figure 40.70 illustrates several different types of bolts and nuts with an added nonmetallic material. These include nuts with a nylon ring to create thread interference and to increase damping (Figure 40.70a) as well as bolts and nuts with nonmetallic patches (Figure 40.70b) and strips (Figure 40.70c) inserted. Research supporting the nylon ring design by the Elastic Stop Nut Corporation of America (Baubles et al., 1966; Faroni, 1967) has already been presented. Long-Lok Fastener Corporation (Ammon, 1996) distributes a product handbook that provides a comparison of fasteners with an array of different nonmetallic inserts and coatings.

40.4.4 Chemical Locking

40.4.4.1 Adhesives

The most popular adhesives used for threadlocking are anaerobic adhesives. These can be applied to standard nuts and bolts by the user at the time of installation, or one can obtain fasteners with

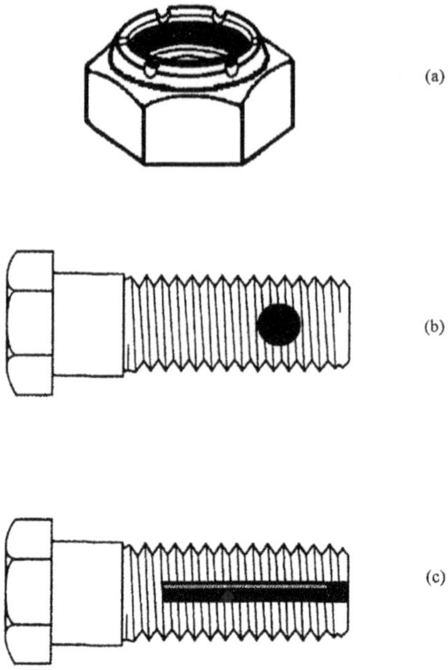

FIGURE 40.70 Prevailing torque fasteners with nonmetallic material. (a) Nylon ring; (b) patch; (c) strip.

an adhesive preapplied. The hardener in preapplied threadlocking adhesives is encapsulated and remains inert until during assembly, when it is released due to pressure. Threadlocking adhesives have been considered to provide a more permanent bond. However, low and medium strength adhesives are available so that fasteners can be removed using hand tools without damaging the threads.

The Industrial Fasteners Institute has established IFI Standard 125 for lock screws with chemical additives (IFI, 1988). This standard provides specifications for breakaway torque and prevailing torque for preapplied and at-time-of-installation chemical coatings.

A publication by Haviland (1983) of Loctite Corporation includes a useful guide for selecting threadlocking adhesives.

40.4.4.2 Corrosion

Rust and other forms of corrosion can also provide locking in fasteners. Bickford (1995) points out that the United States Marines reportedly drive their tanks through saltwater to corrode, and thereby effectively lock, the fasteners on the tread assembly.

40.4.5 Assembly Design Considerations

Vibration- or shock-induced loosening of fasteners can also be reduced or eliminated through careful design of the entire assembly. Too often, fastener placement and orientation in assemblies are not given careful thought at the design stage.

For example, remembering that axial vibration is far less of a problem than transverse vibration, Bickford (1995) points out that one way to minimize slip is to orient the bolts and joints so that bolt axes are parallel to the expected direction of vibration. For situations where this orientation cannot be achieved or loading is multidirectional, a designer can shape the joint such that relative slip is minimized as shown, for example, in Figure 40.71.

Shear force calculations could also be used to optimize fastener placement and orientation such that fasteners are placed where shear forces acting on the fasteners are minimum. For example,

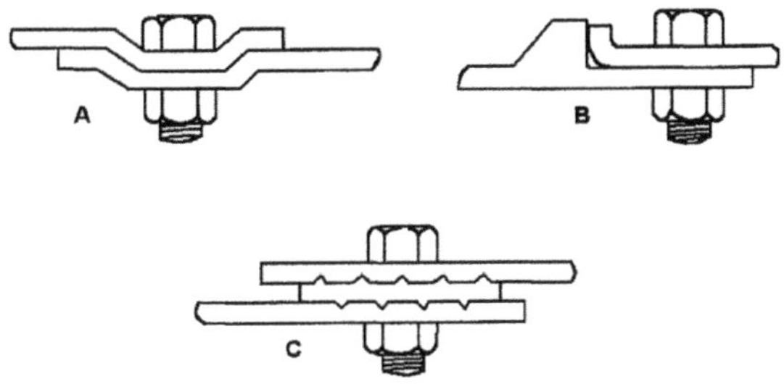

FIGURE 40.71 Sample joint designs A, B, C that are conducive to fastener vibration resistance. (Adapted from Bickford, 1995.)

the shear force calculations for a bent in Figure 40.32 suggest that if the bent were fastened at the middle of the bent, instead of the ends, the shear force on the fastener would be close to zero.

The placement and orientation of fasteners with respect to the vibratory mode shapes of a structure are also important because the vibration level and associated shear forces vary with position as defined by a particular mode shape. This is an area of ongoing research (Hess, 1996b), the goals of which are to develop general design guidelines and criteria (e.g., fastener positioning with respect to mode shapes, component geometry, etc.) that minimize maintenance and failure due to fastener loosening.

40.5 STANDARD DYNAMIC TESTS FOR ASSESSING SELF-LOOSENING

According to a recent publication from Division VI Aerospace Fasteners of the Industrial Fasteners Institute (IFI, 1995), the MIL-STD-1312-7A and a vibration endurance test specified in NAS 1675 are the only two dynamic tests for assessing loosening of dynamically loaded bolted joints for which standards have been developed. The U.S. Department of Defense Military Standard MIL-STD-1312-7A (U.S. Department of Defense, 1984), also known as the ALMA test, is well documented and briefly described below. The Aerospace Industries Association of American National Aerospace Standard NAS 1675 (AIAA, 1989), commonly referred to as the Boomer-Banger test, is also described. It is well appreciated (IFI, 1995) that such fastener tests are not necessarily representative of actual usage conditions and that vibration test requirements generally do not even approximate the conditions that result in vibration-induced failures in an actual assembly or structure.

The IFI *Test* Application Handbook (IFI, 1995) points out that in addition to fatigue failure, self-loosening is the most frequent cause of failure of dynamically loaded bolted joints. The fastener industry has been searching for the ultimate, definitive, and repeatable vibration test method for 25 years. However, until a new method is devised, MIL-STD-1312-7A and the vibration endurance test specified in NAS 1675 will continue to be the primary methods used.

For completeness, it is important to note that SPS has manufactured about 24 Junker transverse testing machines (Coladonato, 1996). Most automotive companies in the United States and some abroad have one or two of these machines. Although no standards have been developed for them, user guidelines and general test procedures are provided in the SPS *Vibration Test Machine* Users' Manual (Carpenter, 1974). Unfortunately, SPS no longer manufactures the machine; however, they still do provide testing on two in-house machines as part of their contract research services.

The composite cantilevered beam originally used by Haviland (1980, 1983) with direct loading and later used by Kerley (1982, 1987) with inertial loading is probably the best prospect for a realistic vibration test standard because this type of structure is the most common for causing shear loading on a bolt as found in joints and panels of most major buildings, aircraft, cars, homes, and household utilities. Further, inertial loading of the beam with a standard laboratory shaker is not as severe as the shock and impact loads in the existing standard tests and should provide more realistic and repeatable results.

40.5.1 MIL-STD-1312-7A

The Military Standard MIL-STD-1312-7A (U.S. Department of Defense, 1984) provides a procedure for conducting accelerated vibration tests on fasteners and fastener systems under preload. The test can be used to assess the life of fasteners and to compare the performance of fasteners subjected to dynamic loading. Failure is normally indicated as a minimum amount of loosening rotation or by the presence of cracking.

The test requires a machine, such as an electromagnetic shaker, that is capable of vibrating the test fixture shown in Figure 40.9 harmonically at a frequency of 30 Hz with a peak-to-peak amplitude of 0.450±0.015 in. The test fasteners clamp spool-like arbors that are constrained within a slotted fixture (Figure 40.9). As the fixture vibrates, it subjects each spool to a reciprocating motion with two impacts per cycle. The direction of impact is perpendicular to the fastener axis. Specifications for fixture details are provided for fasteners ranging in size from 0.190 to 0.625 in. in diameter.

Because of the fastener length requirements, the test cannot be used for testing blind fasteners or other installation formed fastening systems (IFI, 1995).

40.5.2 Vibration Endurance Test Specified in NAS 1675

The NAS 1675 specification (AIAA, 1989) establishes requirements for self-locking blind fasteners that are capable of being installed when only one side of an assembly is accessible. The vibration endurance test specified in this standard uses a 6 in. diameter 2024-T4 aluminum test plate in which the test fasteners are installed as shown in Figure 40.72b. Details on the number of fasteners used as

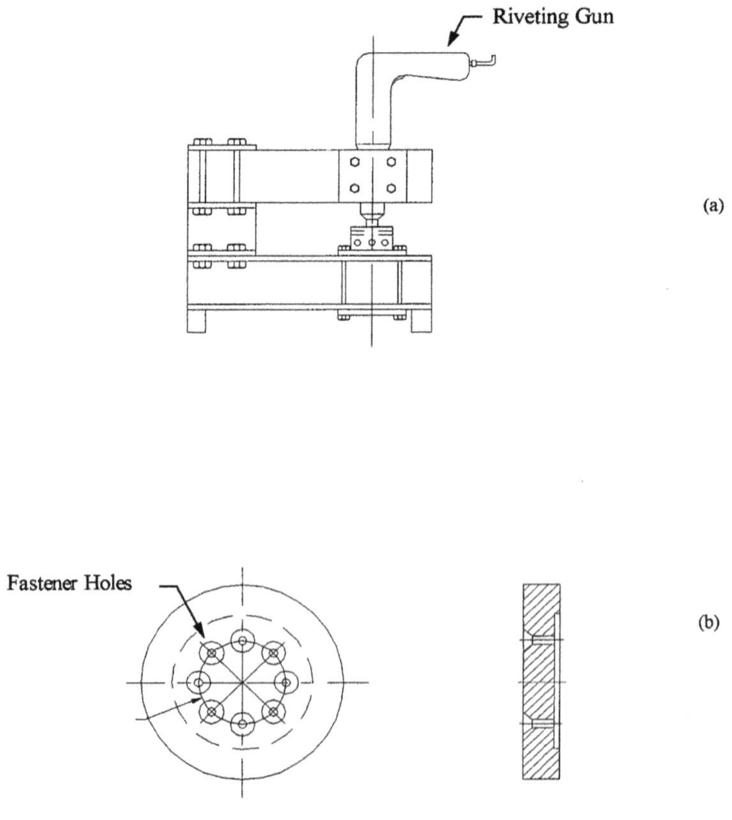

FIGURE 40.72 Vibration endurance (a) apparatus and (b) test plate. (Adapted from AIAA, 1989).

well as the fastener grip range are given in the standard. Lines are scribed on the fasteners and the test plate to assess relative motion or loosening.

The complete test apparatus is shown in Figure 40.72a. A Jiffy 500 riveting gun with a 1¼ in. diameter hammer is mounted above the test plate with 3/32 in. clearance. The gun impacts the plate to provide a severe vibration environment. The total duration of the test is 20 min with ten interruptions for fastener examination. Fasteners fail this vibration endurance test if more than one complete loosening rotation or fracture occurs within the 20 min test duration.

Vibration life has been found to vary significantly with the wood used in the apparatus. Fir has been found to be the preferred wood (IFI, 1995).

ACKNOWLEDGMENT

The author gratefully acknowledges the support of the National Science Foundation under Grants Nos. MSS-9210331 and CMS-9501824. The author also acknowledges the efforts of Mr. Yubo Dong, Mr. Niranjan Pai and Mr. Scott A. Tetzlaff in preparing the figures.

REFERENCES

AIAA. (1989). *NASA 1675, Fastener—Blind, Internally Threaded, External Sleeve, Self-Locking*, National Aerospace Standard, Aerospace Industries Association of America, Washington, DC.

Aldridge, J. (1996). *Slip-On Lock Nut Company*, Battle Creek, MI, Private correspondence.

Ammon, R. (1996). *Long-Lok Fastener Corporation*, Cincinnati, OH, Private correspondence.

Armstrong, B., P. E. DuPont, and C. Canudas de Wit. (1994). A survey of models, analysis tools and compensation methods for the control of machines with friction, *Automatica 30:*1083–1138.

Baubles, R. C., G. J. McCormick, and C. C. Faroni. (1966). Loosening of Fasteners by Vibration, Report No. ER272-2177, Elastic Stop Nut Corporation of America (ESNA), Union, NJ.

Bennett, B. (1996). *Stage 8 Locking Safety Bolts*, San Rafael, CA, Private correspondence.

Bickford, J. H. (1983). Junker's Theory of Vibration Loosening and the ASME/ANSI B-18 Subcommittee's Proposed Research Program to Study Loosening Mechanisms of Bolted Joints Under Vibration Loads, Raymond Engineering Inc., Middletown, CT.

Bickford, J. H. (1984). Fastener groups seek answers to bolted-joint problems, *Machine Design*, July 26, pp. 77–80.

Bickford, J. H. (1995). *An Introduction to the Design and Behavior of Bolted Joints*, 3rd ed., Marcel Dekker, New York, pp. 527–564.

Borden, T. F. (1991). Shape-memory alloys: Forming a tight fit, *Mech. Eng. 113*(10):66–72.

Borden, T. F. (1996). *Intrinsic Devices Inc.*, San Francisco, CA, Private correspondence.

Carpenter, M. R. (1974). SPS 30,000 lb. Vibration Test Machine Users' Manual, Report No. 5534, SPS Laboratories, Jenkintown, PA.

Clark, S. K., and J. J. Cook. (1966). Vibratory loosening of bolts, Society of Automotive Engineering, Paper No. 660432, pp. 1–10.

Coladonato, M. A. (1996). *SPS Laboratories*, Jenkintown, PA, Private correspondence.

Daadbin, A., and Y. M. Chow. (1992). A theoretical model to study thread loosening, *Mech. Mach. Theory* 27:69–74.

Faroni, C. C. (1967). Maintaining Tightness of Threaded Fasteners, Amerace Corporation, ESNA Division, Union, NJ.

Finkelston, R. J. (1972). How much shake can bolted joints take, *Mach. Des. 44:* 122 125.

Funakubo, H. (1987). *Shape Memory Alloys*, Gordon and Breach, New York, pp. 204–207.

Gaberson, H. A. (1972a). Particle motion on oscillating conveyors: Part I. The equations of motion and the rules for predicting motion form transitions, *ASME J. Eng. Ind. 94:*50–56.

Gaberson, H. A. (1972b). Particle motion on oscillating conveyors: Part II. Practical solutions to the equations of motion and the extension of the theory to beds of granular material, *ASME J. Eng. Ind. 94*:51–63.

Gambrell, S. C. (1968). Why bolts loosen, *Mach. Des. 40*:163–161.

Goodier, J. N., and R. J. Sweeney. (1945). Loosening by vibration of threaded fastenings, *Mech. Eng. 67:* 798–802.

Harrel, A. (1992). Mechanical Engineering Consultant, Bethel Island, CA, Spiralock Thread Form Finite Element Analysis Report, Prepared for Lawrence Livermore National Laboratory.
Haviland, G. S. (1980). *A Logical Approach to Secure Bolting and Locking*, Loctite Corporation, Newington, CT.
Haviland, G. S. (1983). Designing with threaded fasteners, *Mech. Eng. 105:*11–31.
Hess, D. P., and K. D. Davis. (1996). Threaded components under axial harmonic vibration: Part I. Experiments, *ASME J. Vib. Acoust. 118*:411–422.
Hess, D. P. (1996a). Threaded components under axial harmonic vibration: Part II. Kinematic analysis, *ASMEJ. Vib. Acoust. 118:*423–429.
Hess, D. P. (1996b). Research on vibration-induced fastener loosening, *Fastener Technol. Int. XIX*:62.
Hess, D. P., and S. V. Sudhirkashyap. (1996a). Dynamic analysis of threaded fasteners subjected to axial vibration. *Sound Vib. 193*:1079–1090.
Hess, D. P., and S. V. Sudhirkashyap. (1996b). Dynamic loosening and tightening of a single-bolt assembly, *ASME J. Vib. Acoust., 119*:311–316.
Hess, D. P., S. Basava, and I. A. Rasquinha. (1996). Variation of clamping force in a single-bolt assembly subjected to axial vibration, *Proceedings of the 1996 ASME Int. Mech. Eng. Congress*, ASME Volume DE-Vol. 90, Atlanta, GA, pp. 97–102.
Holmes, H. (1988). A spiral lock for threaded fasteners, *Mech. Eng. 110*(10):36–39.
Hosko, M. J. (1996). *Nord-Lock*, Lake Michigan Wire Technologies Inc., Muskegon, MI, Private correspondence.
Hunt, K. H., and F. R. Crossley. (1975). Coefficient of restitution interpreted as damping in vibro-impact, *J. Appl. Mech. 42*:440–445.
IFI. (1988). *Fastener Standards*, 6th ed., Industrial Fasteners Institute, Cleveland, OH, Section F, pp. F1–F24.
IFI. (1995). Test Application Handbook Relating to Mechanical Fasteners, Developed by Division VI Aerospace Fasteners, Industrial Fasteners Institute, Cleveland, OH, pp. 12–13.
Johnson, S. P. (1996). *The Johnson Gage Company*, Bloomfield, CT, Private correspondence.
Junker, G. H. (1969). New criteria for self-loosening of fasteners under vibration, *SAE Trans. 78*:314–335.
Karnopp, D. (1985). Computer simulation of stick-slip friction in mechanical dynamic systems, *ASME J. Dynamic Syst., Meas. and Control 107:*100–103.
Kerley, J. J. (1982). The use and misuse of six billion bolts per year, *Presented at the 35th Meeting of Mechanical Failures Prevention Group*, National Bureau of Standards, NASA, Goddard Space Flight Center, Engineering Services Division, Greenbelt, MD.
Kerley, J. J. (1984). *Report on Dynamic and Static Testing of Kaynar Spiralock Microdot Nuts*, NASA, Goddard Space Flight Center, Engineering Services Division, Greenbelt, MD.
Kerley, J. J. (1987). An Application of Retroduction to Analyzing and Testing the Backing Off of Nuts and Bolts during Dynamic Loading, NASA Tech. Memo. 4001, Goddard Space Flight Center, Greenbelt, MD.
Koga, K. (1970). Loosening by repeated impact of threaded fastenings, *Bull. JSME 13:*140–149.
Koga, K. (1973). The effect of thread angle on loosening by impact, *Bull. JSME 16:*1010–1019.
Nayak, P. N. (1978). Vibration Resistance and Stress Distribution in Threaded Fasteners, Report prepared for Chrysler Corporation; Massachusetts Institute of Technology, Cambridge, MA.
Palm, E. P. (1996). *Spiralock Corporation, a Detroit Tool Industries Company*, Madison Heights, MI, Private correspondence.
Perardi, R. C. (1996). Vargal Lock Nut, Gaily, Forno Canavese, Italy, Private correspondence.
Peters, J. C. (1996). Experiments investigating the effect of structural dynamics on the dynamics of threaded fasteners, M.S. Thesis, Univ. South Florida, Tampa, FL.
Ramey, G. E., and R. C. Jenkins. (1995). Experimental Analysis of Thread Movement in Bolted Connections Due to Vibrations, Final Report for NASA Research Project NAS8-39131, Marshall Space Flight Center, AL.
Sakai, T. (1978a). Investigations of bolt loosening mechanisms, 1st Report: Bolts of transverely loaded joints, *Bull. JSME 21:*1385–1390.
Sakai, T. (1978b). Investigations of bolt loosening mechanisms, 2nd Report: Center bolts of twisted joints, *Bull. JSME 21:*1391–1394.
Sakai, T. (1979). Investigations of bolt loosening mechanisms, 3rd Report: Bolts tightened over their yield point, *Bull. JSME 22:*412–419.
Sauer, J. A., D. C. Lemmon, and E. K. Lynn. (1950). Bolts: How to prevent their loosening, *Mach. Des. 22:*133–139.
U.S. Department of Defense. (1984). *Military Standard, MIL-STD-1312-7A, Fastener Test Methods—Method 7: Vibration*, U.S. Department of Defense, Washington, DC.

Vinogradov, O., and X. Huang. (1989). On a high frequency mechanism of self-loosening of fasteners, *Proc. 12th ASME Conf. Mechanical Vibration and Noise*, Montreal, Quebec, pp. 131–137.

Walker, R. A. (1973). *The Factors Which Influence the Vibration Resistance of Fasteners*, Standard Pressed Steel Co. (SPS), Jenkintown, PA.

White, G. (1996). Disc-Lock International, Culver City, CA, Private correspondence.

Yamamoto, A., and S. Kasei. (1984). A solution for self-loosening mechanism of threaded fasteners under transverse vibration, *Bull. Jpn. Soc. Precision Eng. 18:*261–266.

41 Vibration Loosening Analysis and Simulation of Preloaded Threaded Fasteners

Massimiliano De Agostinis and Stefano Fini

41.1 INTRODUCTION

Loosening is widely regarded as one of the main failure modes of threaded fasteners, as confirmed by a survey of 100 automobile dealers in the Detroit Metropolitan area, which revealed how an average of 23% of the service issues were (still in 1988) related to loosening of fasteners [1]. Indeed, some automakers reported peaks of service issues related to loose fasteners close to 40%, the greatest areas of concern being rocker arms, the oil pan, engine mounts, and the body in general. Over the years, the literature consistently reported several different failure scenarios of threaded connections whose root cause can be traced to loosening, e.g. [2] where incorrect preload triggered vibrational loosening of screws securing an aircraft engine support, which in turn led to fatigue failure of the screws and detachment of the engine during flight; [3] dealing with the derailment of an express passenger train which occurred on 23 February 2007 near Grayrigg in Cumbria (UK), on the West Coast Main Line: here the bolts holding the third permanent way stretcher bar to the right-hand switch rail became loose and subsequently completely undone. Therefore, rapid deterioration of the condition of the remaining stretcher bars and their fasteners occurred, which in turn led to the left-hand switch rail becoming totally unrestrained, eventually causing the wheels to climb over the rails and the vehicles to derail. In 2010, an accident resulting from a collapse at an amusement park in eastern Shenzhen caused six fatalities and injured ten people due to a loosening issue which triggered the fatigue failure of a bolt [4]. In Ref. [5], the authors documented a failure happened on the transmission of a wheel-loader which was traced to the self-loosening of a ring nut preloading a couple of tapered bearings. Because of the loss of preload, the bearings stopped providing the required radial support to the pinion shaft which, in turn, failed by fatigue. Finally, Ref. [6] as of 2004, the Italian National Institute of Statistics reported how nearly 5% of all documented road accidents that took place in Italy over the 2003–2004 period and due to mechanical failure of vehicles were triggered by wheel detachments, often caused by loosening phenomena.

These, among other evidence, remind how bolt loosening should be among the mechanical designer's priority, one to be addressed since the early stages of the design process. Bolt loosening has now been investigated since about 80 years ago, when the pioneering works by Goodier et al. [7] and Sauer et al. [8] firstly documented the phenomenon.

Over the years, some literature reviews dealing with the self-loosening phenomenon have been published [4,9]: readers interested in the historical development of scientific research on the topic are referred to these works.

This chapter aims at giving a basic understanding of the phenomenon of bolt self-loosening, starting with a definition of self-loosening and, more specifically, vibrational loosening; then, illustrating the basic analytical tools that have been used to study self-loosening and, subsequently giving a glance on the most recent numerical techniques which have been adopted by scholars. Based on the literature, the authors finally illustrate some dos and don'ts useful for the design engineer to avoid self-loosening.

DOI: 10.1201/9781003501183-47

41.2 TYPES OF BOLT LOOSENING AND THEIR CAUSES

Bolts may become loose due to different causes, generally split into two categories: non-rotational and rotational loosening. Non-rotational loosening [4] is generally associated with the relaxation of the initial axial preload, which might be related to: (1) embedding of the contact surfaces, (2) local plastic phenomena, and (3) fretting wear, among others. This phenomenon takes place right after the completion of the tightening process, even in the absence of external loads and usually leads to the loss of some percentage points of the initial preload force: technical standards account for it in different ways (e.g., see VDI 2230 [10]). Non-rotational loosening is out of the scope of the present chapter.

Rotational loosening happens when the screw or nut turns backward under the action of external cyclic loads, inducing a certain amount of preload loss per cycle. It is generally regarded as a more serious issue, since it can induce a continuous preload loss, which may eventually trigger other failure modes, such as fatigue. Since this chapter focuses on rotational loosening (or vibration loosening), we first need to clarify which loading scenarios may trigger it.

Axial vibration (Figure 41.1a) has been among the first causes of vibration loosening to be investigated [7]: nonetheless, whether this loading scenario can induce vibration loosening has since been controversial. The underlying principle put forth by the earliest research is that axial loads, superimposed on the preloaded joint, entail radial shrinkage of the screw and expansion of the nut, owing to Poisson's effect. Due to this, friction on the thread may be overcome, thus leading to a certain loss of preload through load cycles. Recently, the effectiveness of this loading mode toward generating a significant rotational loosening has been questioned [11] and the associated preload losses are mostly attributed to local plastic deformation and fretting wear mechanisms.

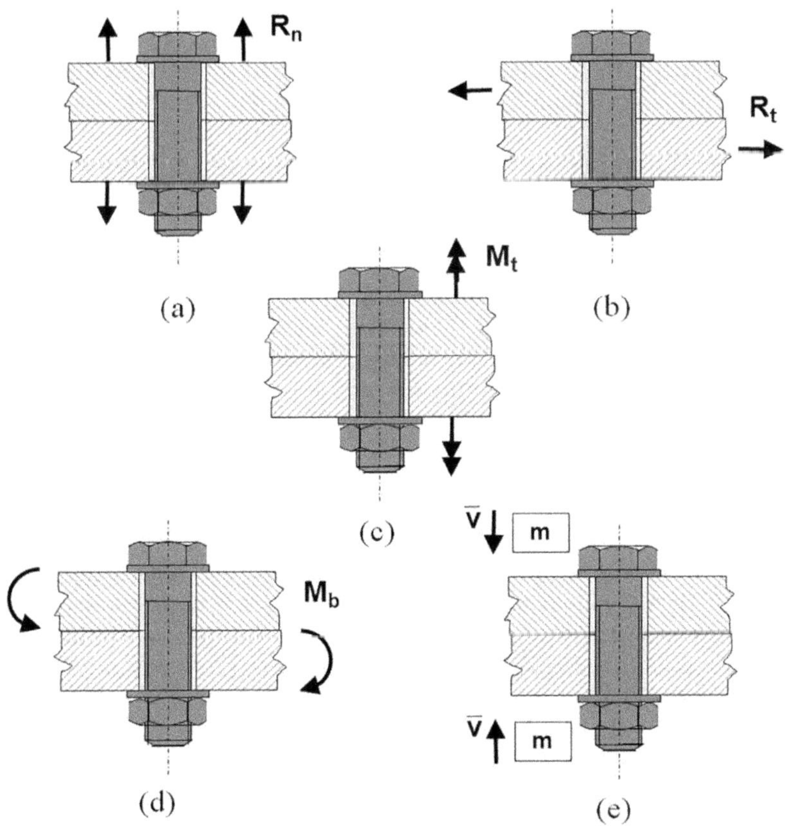

FIGURE 41.1 Several types of cyclic external loading scenarios potentially leading to vibration loosening. (a) axial tensile; (b) transverse load; (c) torque; (d) in-plane bending; (e) axial compressive impact.

Transverse vibration (Figure 41.1b), generated by loading the joint by fluctuating forces, normal to the bolt axis, can induce severe preload losses. Due to that, considerable amount of research has been devoted to the understanding of its basic mechanisms. In the literature, two distinct theories have been put forth to justify the onset and evolution of rotational loosening induced by transverse loads: (1) the complete slippage theory [12–16] and (2) the partial slippage theory [17,18].

With slight changes, the two theories cited above may also be used to study torsional vibration (Figure 41.1c) of bolted joints [19]. Rotational loosening induced by bending vibration (Figure 41.1d) has also been studied by a few researchers, relying on experimental approaches comprising either joints placed on a cantilever beam [20] or flanged joints [21].

Finally, Figure 41.1e refers to impact loading which may, in turn, generate rotational loosening of the joint. The literature showed that impacts can induce stress waves that translate into shear stresses at the thread interface [22], eventually inducing effects very similar to those given by axial, transverse, or torsional vibration.

In light of its practical importance, the transverse vibration loading mode will be treated in detail in this chapter. Since the complete slippage theory is easier to treat analytically, yet able to give a basic understanding of the vibration loosening phenomenon, it is the object of the following paragraph. The reader is therefore referred to the relevant literature for insights on the more recent (and complex) partial slippage theory.

41.3 ANALYTICAL APPROACH: THE COMPLETE SLIPPAGE THEORY

In his pioneering work of 1969, Junker put forth the basic principle upon which the whole theory of complete slippage relies: if the joint undergoes transverse slip—e.g. driven by the external forces R_t in Figure 41.1b—then friction under the nut face and/or in the thread contact has been overcome, thus leaving the joint basically free from friction in the tangential direction, i.e., free to untighten, as an effect of the residual shank torque [23] and additional torque generated by the transverse sliding. Applications qualitatively including a similar effect are that of a cork, which is easier to remove from a bottle when it is rotated firstly, and then pulled; that of a car braking on a bend—when if the frictional force is totally consumed for deceleration, then the car slides along the tangent to the bend. Junker did not give analytical formulae in support of such theory, which were then derived—similarly in many respects—by Sakai [13] and Yamamoto and Kasei [14], among others.

First, the forces (torques) promoting or resisting loosening shall be identified. Based on the well-known torque-tension equation [10] (here taken in the form of Eq. (41.1), also known as Motosh's equation), the first term on the right hand of the equation is relevant to the torque generating the axial preload of the bolt: this contribution works in the loosening direction as an effect of the lead angle, whereas the other two (frictional) counteract loosening.

$$T = F_i \cdot \left(\frac{p}{2\pi} + \frac{\mu_{\text{th}} \cdot d_2}{2\cos\alpha} + \frac{\mu_b \cdot d_b}{2} \right) \tag{41.1}$$

As noted by Junker, if transverse slip occurs, the nut face and the threads slide upon the respective mating surfaces, (nearly) nullifying the frictional resistance of these interfaces along any other direction. Moreover, a further loosening torque arises due to the different resistance offered between the ascending and descending sides of the thread upon application of a transverse displacement (see Figure 41.2), as illustrated by Sakai [13].

Referring to Figure 41.2, limited to one full turn of a square thread (α=0) for the sake of clarity, one must note that the lead angle along the x-axis (being x the direction along which the transverse displacement is imposed) is given by:

$$\tan\beta' = \tan\beta \cdot \sin\theta \tag{41.2}$$

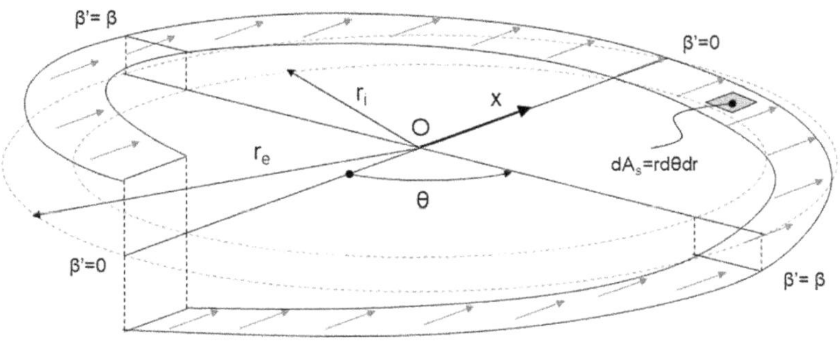

FIGURE 41.2 Linear slip occurring on a square thread, as explained by Sakai [13].

whereas the area of the contact surface (or slip area) is:

$$A_S = \pi \frac{\left(r_e^2 - r_i^2\right)}{\cos\beta} \tag{41.3}$$

By isolating a small element of area $dA_S = r \cdot d\theta \cdot dr$ and integrating, the following torque can be calculated:

$$\begin{aligned} T_S = & \int_{r_i}^{r_e} \int_0^{r_i} \frac{F_i}{A_S} \cdot dA_S \cdot \left(\mu_{\text{th}} \cos\beta + \sin\beta'\right) \cdot \cos\beta' \cdot r \cdot \sin\theta \\ & - \int_{r_i}^{r_e} \int_0^{\pi} \frac{F_i}{A_S} \cdot dA_S \cdot \left(\mu_{\text{th}} \cos\beta - \sin\beta'\right) \cdot \cos\beta' \cdot r \cdot \sin\theta \end{aligned} \tag{41.4}$$

In which the frictional components are computed by means of the lead angle β (they depend on the component of force that is normal to the helix), whereas the components related to the slope of the wedge must be computed by using β' (note that for $\theta = 0 + k\pi$ the local slope in the x direction is $\beta' = 0$, whereas it equals β for $\theta = \pi/2 + k\pi$). As these latter components combine with the same sign on the ascending (first integral) and descending (second integral) sides of the helix, an overall torque acting in the direction of loosening is generated.

Since $\beta \approx 3°$ for conventional sizes and coarse thread (e.g. M10), it follows that $\beta' = \beta \sin\theta$ and $A_S = \pi\left(r_e^2 - r_i^2\right)$; therefore, the above may be simplified as follows:

$$T_S = \frac{2F_i\beta}{A_S} \cdot \int_{r_i}^{r_e} \int_0^{\pi} r^2 \cdot \sin^2\theta \cdot d\theta \, dr \tag{41.5}$$

Integration of Eq. (41.5) gives:

$$T_S = \frac{F_i\beta}{3} \cdot \frac{\left(r_e^2 + r_e r_i + r_i^2\right)}{\left(r_e + r_i\right)} \tag{41.6}$$

By replacing $r_e + r_i = d_2$ (pitch diameter), as proposed by Eccles [9] and expressing β as a function thereof ($\beta \cong p/\pi d_2$), Eq. (41.6) can be recast as follows:

$$T_S \cong \frac{F_i p}{4\pi} \tag{41.7}$$

which can eventually be superimposed to the pitch torque in Eq. (41.1) to give the total amount of loosening torque in the case of transverse sliding:

$$T_{\text{loosen}} \cong \frac{F_i p}{4\pi} + \frac{F_i p}{2\pi} = \frac{3}{4}\frac{F_i p}{\pi} \tag{41.8}$$

Under the assumption of full sliding at both the thread and the nut face, the friction coefficients vanish $\mu_{\text{th}} = \mu_b \to 0$, therefore the only component that can counteract the loosening torque can be given by the prevailing torque provided by a locknut (if any). Even if Eq. (41.8) has been derived based on a square thread, it must be noted that, as the triangular thread consists of surfaces that would be set apart by $\alpha = 30°$ each from the square thread helix, symmetrically with respect to the bolt axis, the effect thereof would be balanced on the whole thread surface, so that Eq. (41.8) still holds true for triangular threads.

By looking at Eq. (41.8), it can be immediately observed that the loosening torque grows proportional to the thread pitch: this outcome is in line with the experimental results that will be analyzed in the following.

By following an approach like that of Sakai, Yamamoto and Kasei [14] derived an equation for calculating the transverse force required to cause slip on the nut face, allowing for bolt bending. In their work, they illustrated how rotational loosening is a consequence of the elastic torsion generated in the screw shank when a transverse sliding is imposed to the joint: Figure 41.3 represents the basic scheme they adopted.

Let the screw head be fixed to the bottom, fixed plate (red), while the top moving plate (blue) travels horizontally by a total stroke s. Let stages 1 and 3 of Figure 41.3 be the dead points of the stroke, while stage 2 is the critical stage, i.e. the point, on the right half of the stroke, where the transverse force overcomes the frictional shear resistance of the nut-plate contact. From stage 1 to stage 2, the screw-nut contact on the thread region is forced to slide, due to the bending of the screw, thus the screw shank accumulates (loosening) torque owing to a mechanism similar, in principle, to that described by Sakai [13]. During this phase the nut does not rotate, being restrained by the frictional strength at the interface with the moving plate—which still holds. Next, when stage 2 is passed and through stage 3, the nut face frictional restraint is overcome, thus the nut is allowed to rotate counterclockwise, entailing a loss of axial preload. The same happens twice per cycle, giving

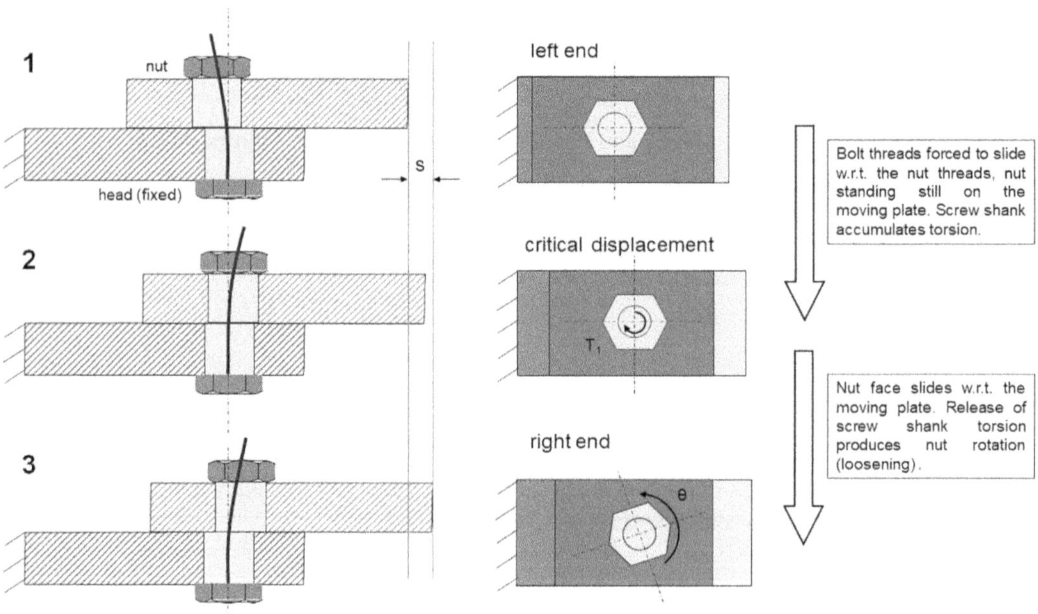

FIGURE 41.3 Scheme of the loosening process of a single-bolt joint subjected to transverse displacement. (Adapted from Yamamoto and Kasei [14].

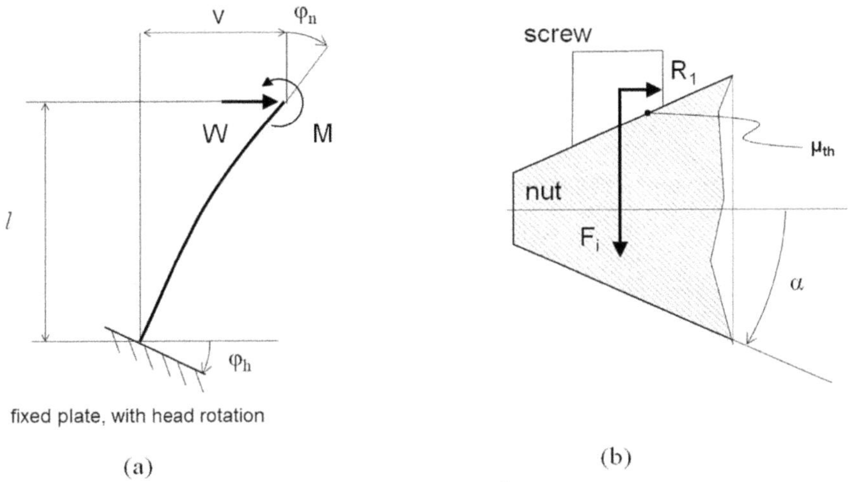

FIGURE 41.4 (a) Deformed shape of the bolt due to transverse displacement, (b) forces on the thread contact due to forced transverse sliding [14].

rise to an overall loss of preload we call ΔF. We now seek a relationship between such preload loss and the imposed transverse displacement v.

The drawing in Figure 41.4a depicts the screw as a cantilever beam subjected to bending: we assume that such beam be loaded—at its free end—by both a transverse shear force W and a concentrated torque M. Let the free length be $l = L + m/2$, where L is the plate thickness clamped by the bolt and m is the height of the nut. The screw head cannot be considered rigid; thus, its compliance (inducing rotation φ_h in Figure 41.4a) is accounted for through compliance k [rad/Nmm]. Displacements at the free end can be computed as follows, where I [mm^4] is the bending moment of inertia (in Ref. [14], $I = \pi d_2^4 / 64$) and E [MPa] is the Young's modulus of the screw:

$$v = \frac{\mathrm{Wl}^3}{3\mathrm{EI}} + \frac{\mathrm{Ml}^2}{2\mathrm{EI}} + \mathrm{kl} \cdot (\mathrm{Wl} - M) \tag{41.9}$$

$$\varphi_h = \frac{\mathrm{Wl}^2}{2\mathrm{EI}} + \frac{\mathrm{Ml}}{\mathrm{EI}} + k \cdot (\mathrm{Wl} - M) \tag{41.10}$$

Now, looking at Figure 41.4b, which represents the forces arising when the screw thread is forced to climb up the thread of the nut (in the transverse direction), by enforcing the equilibrium condition, one can retrieve R_1 as follows:

$$R_1 = F_i \cdot \frac{\mu_{\mathrm{th}}}{\cos^2 \alpha} \cdot (1 + \mu_{\mathrm{th}} \tan \alpha) \tag{41.11}$$

At sites where the screw thread is forced to slide down the nut thread, relying on a similar scheme, the following force can be calculated:

$$R_2 = F_i \cdot \frac{\mu_{\mathrm{th}}}{\cos^2 \alpha} \cdot (1 - \mu_{\mathrm{th}} \tan \alpha) \tag{41.12}$$

Based on the knowledge of these two forces, and on the scheme of Figure 41.5, the magnitude of moment M acting at the free end (nut side) of the beam in Figure 41.4a can be calculated, according to Eq. (41.13). For the sake of representation, the axial preload force F_i is split into four concentrated forces in Figure 41.5.

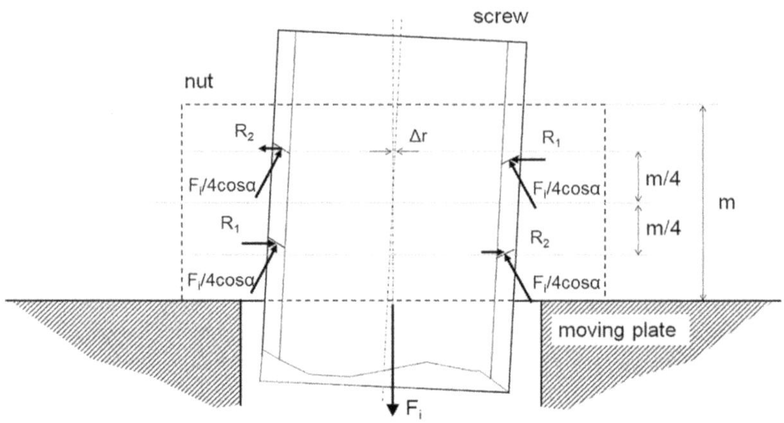

FIGURE 41.5 Simplified scheme of the forces acting on the thread contact between nut and screw due to forced transverse sliding. (Adapted from Yamamoto and Kasei [14].)

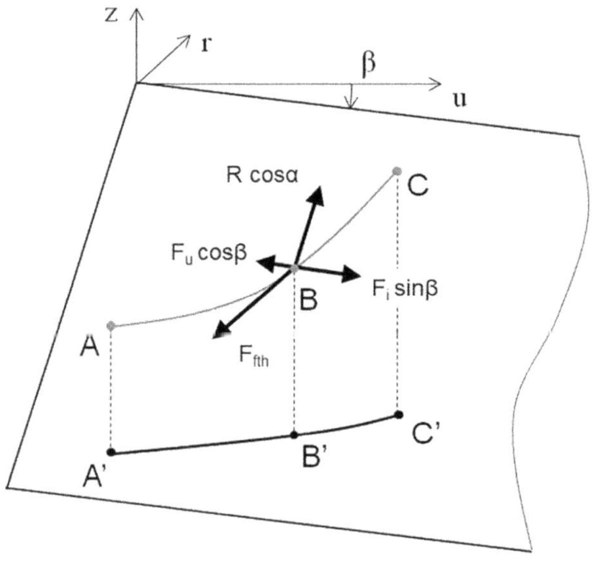

FIGURE 41.6 Sliding trajectory of an idealized contact point at the screw-nut thread interface. (Adapted from Yamamoto and Kasei [14].)

$$M = F_i \cdot \frac{m}{4} \cdot \frac{\mu_{\text{th}}}{\cos^2 \alpha} \tag{41.13}$$

In Figure 41.5, the radial displacement at the point of application of the forces is $\Delta r = \varphi_h \cdot m/4$. Let's now look at Figures 41.6 and 41.7, to derive a relationship between the applied forces and the displacements of both the screw and the nut. Figure 41.6 shows the trajectory of an idealized point of contact between the screw and the nut thread, on the side where the screw thread is forced by force R to climb up the thread of the nut from point A to point C. The forces here represented are those applied to the screw. Force F_u projected along the thread helix ($F_u \cos\beta$) is generated by the torsional springback of the screw shank, whereas the remaining is the component of the axial preload along the thread helix ($F_i \sin\beta$) and the frictional force F_{fth} that counteracts the climbing motion. For the sake of simplicity, the equilibrium condition may be cast on a plane normal to the screw axis (plane r-u in Figure 41.6), where, based on the equilibrium conditions, Eq. (41.14) can be cast (see A′B′C′ trajectory in Figure 41.6).

$$\frac{dr}{du} = \frac{R\cos^2\alpha}{F_i \sin\beta\cos\beta - F_u\cos^2\beta} \tag{41.14}$$

Based on Eqs. (41.11) and (41.12), since product $\mu_{\mathrm{th}}\tan\alpha$ is small, the following may be assumed (Eq. 41.15):

$$R = \frac{R_1 + R_2}{2} \cong F_i \cdot \frac{\mu_{\mathrm{th}}}{\cos^2\alpha} \tag{41.15}$$

Under the assumption of concentrating the thread forces on four points (see Figure 41.5) and since β is small, by integration of Eq. (41.14), one can obtain:

$$u = F_i \cdot \frac{\sin\beta}{K_u} \cdot \left(1 - e^{-r\cdot\frac{K_u}{F_i\cdot\mu_{\mathrm{th}}}}\right) \tag{41.16}$$

where K_u [N/mm] is the force per unit circumferential displacement measured at the pitch diameter d_2.

By extending the same analysis to the nut face contact (Figure 41.7), one can cast Eq. (41.17), where U is the circumferential displacement taken on the nut face contact diameter d_b

$$U = U_0 \cdot e^{-2\frac{d_2}{d_b}\cdot\frac{K_u}{W_1}\cdot(v - v_0)} \tag{41.17}$$

where W_1 is the transverse force at the critical stage (stage 2 in Figure 41.3), U_0 and v_0 are the initial positions of the contact point (point D in Figure 41.7). It must be noted that the torsional stiffness

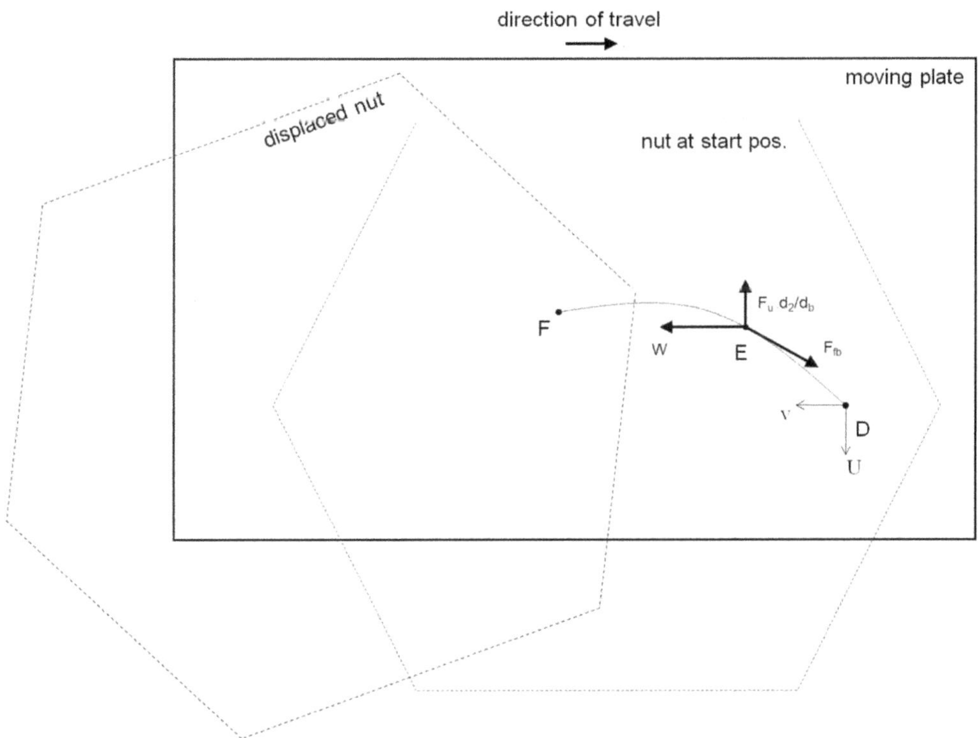

FIGURE 41.7 Sliding trajectory of an idealized contact point at the interface between the nut and the moving plate. Top view from the nut side.

of the shank K_u is expressed on the pitch diameter of the thread, while the equilibrium condition by which Eq. (41.17) is derived, is enforced on the bearing diameter d_b. Therefore, $F_u = K_u \cdot \frac{d_2}{d_b} \cdot U$ must be assumed.

Once the quantities expressed by Eqs. (41.16) and (41.17) are known, it is possible to back-calculate the transverse force W, the torque T, and the loosening angle θ. Based on the latter, the loss of preload over a cycle ΔF can eventually be computed.

$$W = \frac{1}{\frac{l^3}{3\mathrm{EI}} + \mathrm{kl}^2} \cdot v + \frac{\frac{l}{2\mathrm{EI}} + k}{\frac{l^2}{3\mathrm{EI}} + \mathrm{kl}} \cdot F_i \cdot \frac{m}{4} \cdot \frac{\mu_{\mathrm{th}}}{\cos^2 \alpha} \tag{41.18}$$

The critical value of $W_1 = F_i\, \mu_b$ is attained at stage 2 and holds through stage 3. As for the torque, the maximum torque in the shank, attained at stage 2, can be calculated according to Eq. (41.19):

$$T_1 = K_u \cdot \frac{d_2}{2} \cdot u_1 = F_i \cdot \frac{d_2}{2} \cdot \sin\beta \cdot \left(1 - e^{-2\Delta r_1 \cdot \frac{K_u}{F_i \cdot \mu_{\mathrm{th}}}}\right) \tag{41.19}$$

where u_1 is given by Eq. (41.20) below:

$$u_1 = F_i \cdot \frac{\sin\beta}{K_u} \cdot \left(1 - e^{-2\Delta r_1 \cdot \frac{K_u}{F_i \cdot \mu_{\mathrm{th}}}}\right) \tag{41.20}$$

Due to the sliding taking place at the nut face, from stage 2 to stage 3, the torque in the shank drops down to:

$$T_2 = K_u \cdot \frac{d_2^2}{2d_b} \cdot U = K_u \cdot \frac{d_2^2}{2d_b} \cdot U_0 \cdot e^{-\left(\frac{d_2}{d_b}\right)^2 \cdot \frac{K_u}{W_1} \cdot (v - v_0)} = F_i \cdot \frac{d_2}{2} \cdot \sin\beta \cdot \left(1 - e^{-2\Delta r_1 \cdot \frac{K_u}{F_i \cdot \mu_{\mathrm{th}}}}\right) \cdot e^{-\left(\frac{d_2}{d_b}\right)^2 \cdot \frac{K_u}{W_1} \cdot (v - v_0)} \tag{41.21}$$

where $U_0 = u_1 \cdot \frac{d_b}{d_2}$ is assumed as the tangential displacement attained on the pitch diameter at the end of the first stage u_1, rescaled at the bearing diameter d_b. Therefore, assuming the initial position in the transverse direction (corresponding to point D in Fig. 41.7) as $v_0 = 0$, the loosening angle θ can finally be calculated based on the drop of torque $\Delta T = T_1 - T_2$. According to Eq. (41.22), where v is the distance traveled in the transverse direction by the moving plate, we get:

$$\theta = \frac{\Delta T}{K_u} \cdot \frac{2}{\pi \cdot d_2^2} = F_i \cdot \frac{\sin\beta}{\pi\, d_2 K_u} \cdot \left(1 - e^{-2\Delta r_1 \cdot \frac{K_u}{F_i \cdot \mu_{\mathrm{th}}}}\right) \cdot \left(1 - e^{-\left(\frac{d_2}{d_b}\right)^2 \cdot \frac{K_u}{W_1} \cdot v}\right) \tag{41.22}$$

If the complete release of the (additional) shank torsion is achieved during stages 2–3, the total loosening angle can be calculated as follows:

$$2\theta_1 = 2\frac{u_1}{\pi d_2} \tag{41.23}$$

The total preload drop per cycle is finally given by:

$$\Delta F = 2 \cdot F_i \cdot \frac{\sin\beta}{K_u} \cdot \left(1 - e^{-2\Delta r_1 \cdot \frac{K_u}{F_i \cdot \mu_{\mathrm{th}}}}\right) \cdot K_a \cdot \tan\beta \tag{41.24}$$

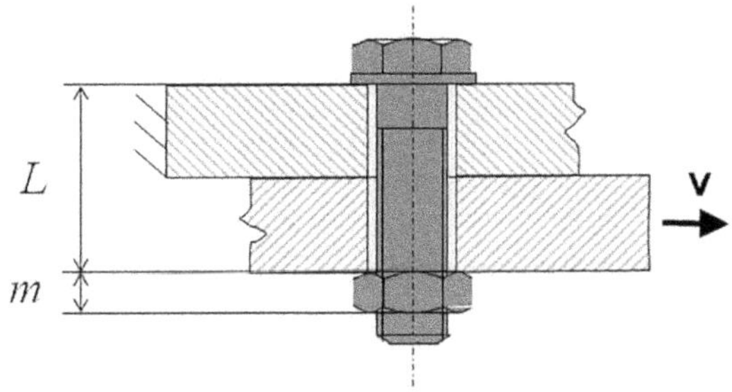

FIGURE 41.8 Reference dimensions of the threaded joint in example.

where K_a is the axial stiffness of the bolt shank, in N/mm.

The following numerical example illustrates the use of Eq. (41.24) for calculating the vibration-induced drop in bolt tension at the first cycle for the shown threaded joint system (Figure 41.8).

The M10 steel screw (ISO coarse metric thread with $\alpha=30°$) was initially tightened to a preload of $F_i=25$ kN in a steel joint with a grip length $L=20$ mm, the nut height $m=8$ mm, and is subsequently subjected to a transverse displacement $v=0.9$ mm (see Figure 41.8), with friction coefficients $\mu_{th}=\mu_b=0.15$. Assuming a head compliance of $k=1.93\times10^{-7}$ [Nmm^{-1}] from Ref. [14], a total loss of preload at the first cycle of $\Delta F=751$ N can be computed. Despite its relative simplicity, this approach allows to clarify the mechanisms underlying the rotational loosening phenomenon. Still dealing with the complete slippage theory, more advanced approaches [15,16], e.g. including nonlinearities [24] have been put forth in the last two decades. Moreover, it must be noted that the theory reported above assumes that loosening starts with the complete transverse slip at the bolt head surface. In contrast to this, Pai and Hess [17] were the first authors who identified that a fastener could undergo loosening under dynamic transverse loading only due to the accumulation of localized slip at the head and thread contact surfaces, thus introducing the term localized slip. To give a clearer overview of the effect of some design variables on the vibration-loosening response of a bolted joint, an excerpt of the data presented in Ref. [25] is synthesized and reported in Figure 41.9a–d. It can be seen how finer thread, tighter-fit hole, higher initial preload and more compliant screws all lead to better anti-loosening behavior.

The data in Figure 41.10 summarizes some results reported by Nassar and Housari in [26], regarding the effect of thread fit level and through hole clearance on the self-loosening rate of hex-head, ½ -13, grade 5 bolts. The through hole clearance is expressed as a percentage of the bolt nominal diameter. It can be seen how tighter thread fit and tighter through holes lead to an increased number of cycles to complete loosening.

The data reported in Figure 41.11 shows the relationship between the friction coefficients of the joint and the loosening rate, according to an analytical model developed by Housari and Nassar [27]. This model was validated by experimenting on two sets of M12, 8.8 grade bolts tested on a Junker apparatus with different states of lubrication. The data clearly shows the relationship between different combinations of friction coefficients and the self-loosening behavior of the joint, suggesting that the proper combination of friction coefficients must always be chosen based not only on assembly requirements but also on self-loosening behavior.

41.4 SIMULATION: THE NUMERICAL APPROACH

This chapter aims at giving a basic understanding of the strategies that have been adopted to simulate vibrational loosening by means of the finite element method, the relevant settings and the possible outcomes. Since the first decade of the 2000s, FE models based on a simplified approach have been proposed first [28], shortly followed by more refined complete models [29]. In their landmark paper of 2001, Pai and Hess presented a study, based on a finite element method, to simulate

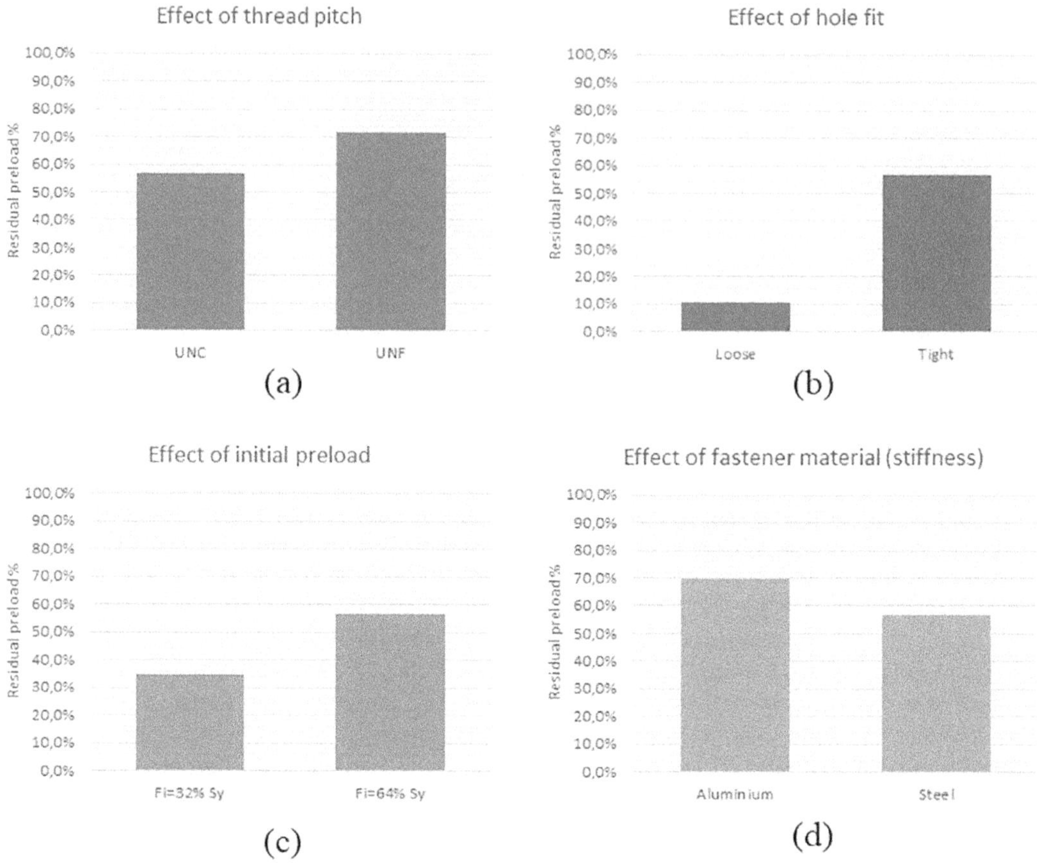

FIGURE 41.9 Residual preload force of a ¼″ fastener after testing on a Junker apparatus for 1750 cycles at 15 Hz and 1.5 mm amplitude in dry conditions: (a) effect of the thread pitch UNC vs. UNF; (b) effect of the through hole diameter 6.53 vs. 6.75 mm; (c) effect of initial preload level F_i=32% S_Y vs. F_i=64% S_Y; (d) effect of the fastener material (low/high stiffness) Aluminium vs. Steel.

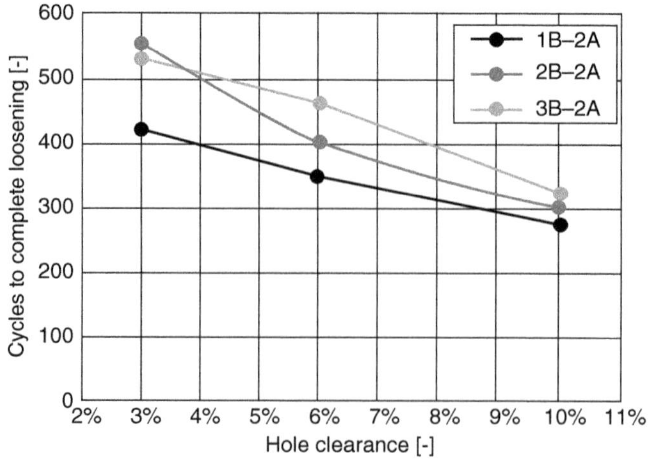

FIGURE 41.10 Number of cycles to complete loosening for different thread fit levels (1B looser—3B tighter) and through hole clearance, from Ref. [26].

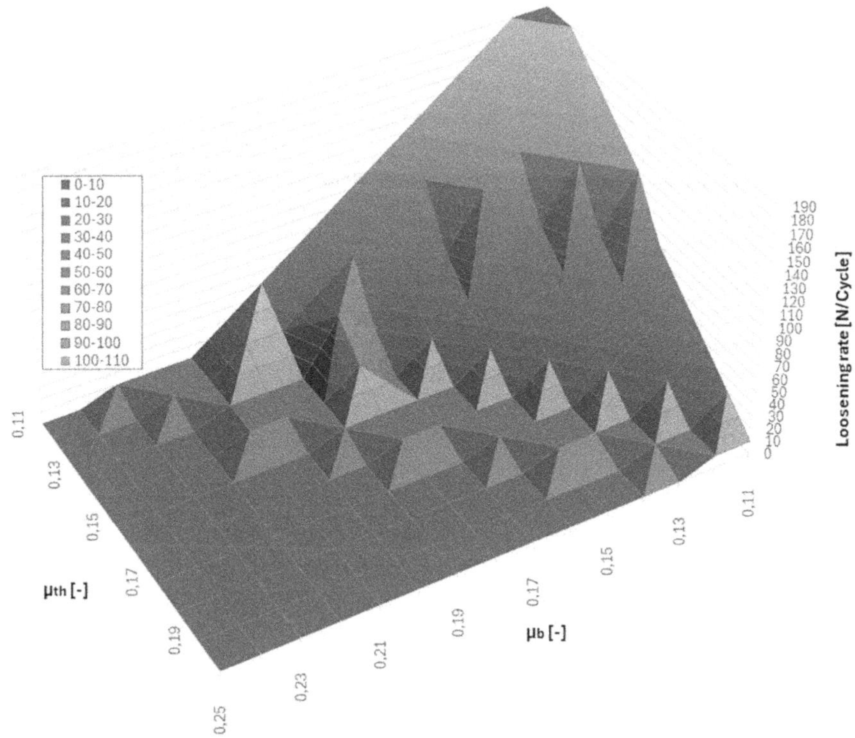

FIGURE 41.11 Effect of bearing and thread friction coefficient on the loosening rate, from Ref. [27].

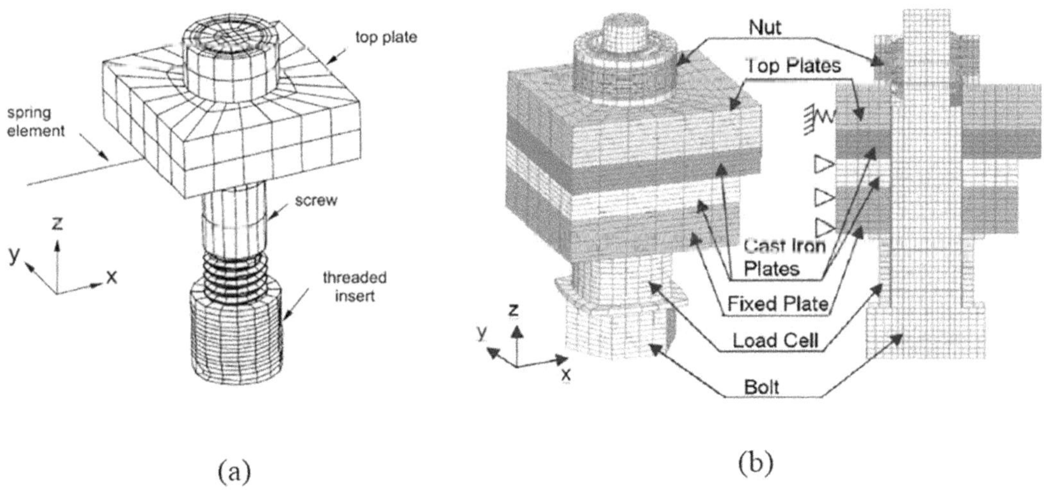

FIGURE 41.12 (a) Simplified FE model by Pai and Hess (2002) [28], (b) complete FE model by Zhang et al. (2007) [29].

rotational loosening under conditions of complete or localized slip. This model aimed at simulating a single bolt joint under the conditions of Junker's test [12] relying on a simplified geometry, i.e. one that only consisted of the screw, the nut and the moving plate (Figure 41.12a). About 12 external threads and 5 ½ internal threads were included in the model. Meshing was carried out mainly by 8-node and 20-node hexahedral elements, with only a few 10-node tetrahedral elements (for the inner portion of the screw shank), giving a total of 6×10^3 nodes: the model is comparatively coarse, but being the loosening results driven by displacements rather than stresses, such mesh density was deemed accurate by the authors of [28]. It must be noted that the screw is here modeled as a single body, i.e., without separating the helix by the core. The moving plate lateral faces are restrained so that they remain plane (i.e., infinite width/length of the plate), whereas its bottom face is constrained by frictionless support, to let it translate free in the transverse direction (x-axis in Figure 41.12a). Initial preload of the screw was applied by means of calibrated interference between the screw underhead surface and the top surface of the plate (note: this method is now obsolete due to the implementation of specific tools for preload generation in almost all commercial FE packages e.g. *bolt-pretension* in Ansys®).

Frictional contacts are defined between the screw head and the moving plate, the hole and the screw shank, the screw and nut threads. The transverse motion is applied to the moving plate by means of a spring whose other end is assigned a displacement, the aim of this spring is to replicate the stiffness of the actual loading chain. Loosening was modeled as a quasi-static process based on previous results that show loosening in the transverse test apparatus to be independent of the loading frequency [12]. All the materials are modeled as linear—elastic: nonetheless, owing to the contact nonlinearities, the solution must rely on the Newton-Raphson method. Each load-cycle is divided into a minimum of 32 load increments to enhance convergence.

Such a simplified model allows to capture the essential features of the loosening process, for example clarifying the role that the thread and underhead friction coefficients play in the definition of loosening scenarios (e.g. complete head slip with localized thread slip, complete slip everywhere, and so on). Other parametric studies can be carried out based on a similar model, e.g. evaluating the influence of several geometrical features on the loosening response: screw length, flank angle, pitch, and others. For instance, the model allowed Pai and Hess to conclude that longer fasteners are more likely to undergo loosening than shorter ones for given boundary conditions, owing to the higher bending moment that develops at the frictional constraints (thread and underhead), which is in line with the findings of the analytical models reported in the previous paragraph. Since a simplified model like that of [28] is just intended for replicating the Junker's test condition, it does not allow to simulate real applications in which friction between the plates is non-zero and the bottom (or fixed) plate is flexible. Moreover, the relatively coarse mesh does not allow to determine the stress state with sufficient accuracy, if the elastic-plastic behavior of the parts is considered.

The model represented in Figure 41.12b was introduced by Zhang et al. in 2007 [29], to replicate a slightly different experiment, involving a M12 screw with a load cell for measuring the axial preload of the screw and the presence of friction between the plates. A major difference with Pai's model is that the thread helix is modeled as a separate body and fixed by a (bonded) contact to the core of the screw. The materials' behavior is linear elastic. The contacts between all the plates, as well as those in the thread and underhead regions are modeled as frictional (the authors used an *infinitesimal-sliding* approach in the Abaqus® nomenclature). The elements are 8-node hexahedral, with a mesh size of about 0.5 mm at the finest, for a total node count of approximately 2.2×10^4. The axial preload was applied to the screw by means of a fictitious thermal expansion of the load-cell sleeve. To simulate the transverse load, one end of the lower plate was fixed whereas the opposite end of the upper plate was assigned a displacement. The load-cycle was divided into four stages: (1) zero position to dead end on the one side, (2) return to zero, (3) dead end on the opposite side and (4) return. To avoid convergence issues that may be induced by the sudden frictional sliding of any interface, a low stiffness spring (100 N/mm) was introduced in the model. Each load step was divided into four to ten increments of load, to enhance convergence. Besides the output provided by the simplified model, this slightly more sophisticated—so-called complete—model allows to evaluate the effect of frictional parameters at the plate-plate interfaces, as well as to more accurately account for the stiffness of the whole system. Similar approaches have been used later by other researchers, e.g. Dinger and Friedrich [18].

41.5 FINAL REMARKS

Right from the start, scientists investigated the effect of different parameters of the joint towards the definition of the vibration loosening phenomenon, trying to find ways to minimize the loss of preload. In his pioneering work [12], Junker himself studied the effect of several variables, such as the type of loading (highlighting that transverse vibration is the most harmful), the vibration frequency (preload loss does not depend onto it), the thread pitch (the finer the better) and fit (the tighter the better). As shown in the previous paragraphs, based on the work of several authors who experimentally analyzed the effect of different parameters (and their interactions) on the loosening performance of threaded joints, one may draw the following conclusions:

- Initial preload should be as high as possible, see Figure 41.9c, also in line with the provisions of the standards (e.g. Ref. [10]).
- Low modulus fastener materials (e.g. Titanium) can offer significant advantages, if they are compatible with the application, see Figure 41.9d. Note that low-modulus fasteners can also improve the fatigue behavior of the joint, in light of the smaller fraction of variable loads borne by the fastener, owing to its smaller stiffness in comparison to that of the plates.
- Fine threads (smaller pitch) shall be preferred, which is in line with the analytical findings illustrated above—see Figure 41.9a.
- A smaller residual torque [23] in the shank can provide a better performance (although the friction coefficients shall not be made too small to obtain this, see Ref. [10] and Figure 41.11).
- A tighter fit hole shall be preferred in joints undergoing transverse vibration (see Figure 41.9b), because it helps minimizing slip, as well as a tighter thread fit (see Figure 41.10).

BIBLIOGRAPHY

[1] Holmes, H. Seeking the perfect locking method for threaded fasteners. *Automation* 35 (3), (1988), 28–30.
[2] South African Civil Aviation Authority, Aircraft Accident Report and Executive Summary Reference CA18/2/3/8383, Nov 7th, 2007.
[3] Department for Transport UK. Rail Accident Investigation Branch Reports. Report 20/2008, v5 July 2011. https://www.gov.uk/raib-reports, 2020.
[4] Gong, H., Ding, X., Liu, J., Feng, H. Review of research on loosening of threaded fasteners. *Friction* 10(3): 335–359 (2022).
[5] Croccolo, D., De Agostinis, M., Fini, S., Olmi, G. A numerical and experimental approach to the design and failure analysis of a pinion shaft for wheel loaders. *Proc Inst Mech Eng Pt C J Mechan Eng Sci* 232(8) 1493–1504 (2018).
[6] Sistema Statistico Nazionale - Istituto Nazionale di Statistica, Automobile Club d'Italia, Statistica degli incidenti stradali, anni 2003–2004, (in Italian).
[7] Goodier, J. N., Sweeney, R. J. Loosening by vibration of threaded fastenings. *Mech. Eng* 67, 798–802 (1945).
[8] Sauer, J. A., Lemmon, D. C., Lynn, E. K. Bolts: How to prevent their loosening. *Mach Des* 22: 133–139 (1950).
[9] Eccles, M. Tribological aspects of the self-loosening of threaded fasteners. Ph.D. Thesis. Lancashire (England): University of Central Lancashire, (2010).
[10] Association of German Engineers, VDI 2230 Part 1. Systematic Calculation of High Duty Bolted Joints. 2003.
[11] Liu, J. H., Ouyang, H. J., Feng, Z. Q., Cai, Z. B., Liu, X. T., Zhu, M. H. Study on self-loosening of bolted joints excited by dynamic axial load. *Tribol Int* 115(5): 432–451 (2017).

[12] Junker, G. H. New Criteria for Self-Loosening of Fasteners under Vibration. SAE Technical Paper 690055, 1969, Vol. 78, pp. 314–335.
[13] Sakai, T. Investigations of bolt loosening mechanisms: 1st report, on the bolts of transversely loaded joints. *Bull JSME* 21(159): 1385–1390 (1978).
[14] Yamamoto, A., Kasei, S. A. solution for self-loosening mechanism of threaded fasteners under transverse vibration. *Bull Jpn Soc Precis Eng* 18(3): 261–266 (1984).
[15] Nassar, S. A., Housari, B. A. Effect of thread pitch and initial tension on the self-loosening of threaded fasteners. *J Press Vess Technol* 128(4): 590–598 (2006).
[16] Nassar, S. A., Yang, X. J. A mathematical model for vibration induced loosening of preloaded threaded fasteners. *J Vib Acoust* 131(2): 021009 (2009).
[17] Pai, N. G., Hess, D. P. Experimental study of loosening of threaded fasteners due to dynamic shear loads. *J Sound Vib* 253(3): 585–602 (2002).
[18] Dinger, G., Friedrich, C. Avoiding self-loosening failure of bolted joints with numerical assessment of local contact state. *Eng Fail Anal* 18(8): 2188–2200 (2011).
[19] Sakai, T. Investigations of bolt loosening mechanisms: 2nd report, on the center bolts of twisted joints. *Bull JSME* 21(159): 1391–1394 (1978).
[20] Haviland, G. S. Designing with threaded fasteners. *Mech Eng* 105(10): 17–31 (1983).
[21] Ishimura, M., Sawa, T., Karami, A., Nagao, T. Bolt-nut loosening in bolted flange connections under repeated bending moments. In *ASME 2010 Pressure Vessels and Piping Division -PVP Conference*, Bellevue, Washington, DC, 2010, pp. 405–413.
[22] Koga, K. Loosening by repeated impact of threaded fastenings. *Bull JSME* 13(55): 140–149 (1970).
[23] Croccolo, D., De Agostinis, M., Fini, S., Olmi, G., Robusto, F., Cavalli, O., Vincenzi, N. *Experimental Measurement of the Shank Torque as a Function of the Stiffness and Frictional Characteristics of the Bolted Joint*, American Society of Mechanical Engineers, Pressure Vessels and Piping Division (Publication) PVP, 2018, p. 2.
[24] Yang, X., Nassar, S. A., Wu, Z. Criterion for preventing self-loosening of preloaded cap screws under transverse cyclic excitation. *J Vib Acoust* 133: 041013 (2011).
[25] Sanclemente, J. A., Hess, D. P. Parametric study of threaded fastener loosening due to cyclic transverse loads. *Eng Fail Anal* 14 (2007) 239–249.
[26] Nassar, S. A., Housari, B. A. Study of the effect of hole clearance and thread fit on the self-loosening of threaded fasteners. *ASME J Mech Des* 129: 586–594 (2007).
[27] Housari, B. A., Nassar, S. A. Effect of thread and bearing friction coefficients on the vibration-induced loosening of threaded fasteners. *ASME J Vib Acoust* 129: 484–494 (2007).
[28] Pai, N. G., Hess, D. P. Three-dimensional finite element analysis of threaded fastener loosening due to dynamic shear load. *Eng Fail Anal* 9: 383–402 (2002).
[29] Zhang, M., Jiang, Y. Y., Lee, C. H. Finite element modeling of self-loosening of bolted joints. *J Mech Des* 129(2): 218–226 (2007).

Part VII

Testing and Inspection of Bolts and Joints

42 Field Testing of Structural Bolts and Installation Tools Using a Bolt Tension Calibrator

John W. Wilhelm, Jr.

42.1 INTRODUCTION

The bolt tension calibrator was developed to provide a practical and affordable tool to allow steel erectors to meet bolting specifications by testing high strength structural bolts in the field. The original bolting and installation specifications permitted the replacement of steel rivets with high strength bolts. Developed in the early 1950s by the Research Council on Structural Connections (RCSC) and endorsed by the American Institute of Steel Construction (AISC), these specifications have been modified several times over the years as research and new bolting products dictated. Through all of these changes, the bolt tension calibrator has continued to provide meaningful test data because it measures bolt tension. Although in most cases torque is used to install bolts and nuts, bolt tension, not torque, holds the joint together. An understanding of the relationship of torque to tension is a key element in successful bolting operations (these testers can also be used to test fasteners and a variety of installation tools in fields outside the construction industry).

42.2 WHAT IS A BOLT TENSION CALIBRATOR?

The bolt tension calibrator consists of three basic components: housing, piston, and hydraulic gage. The housing and piston are assembled together to create a hydraulic reservoir. A pressure gage is attached to the load cell so that when the cell is compressed, the gage will register hydraulic pressure. By calculating the internal area of the load cell under pressure, a dial for the gage can be made to indicate compression in pounds force (Force in pounds = pressure in pounds per square inch × area in square inches).

The load cell has a hole through the center so that with the use of fixtures on either side, a bolt and nut can be inserted. Tightening the nut and bolt will compress the load cell, causing the gage to indicate pressure within the cell and therefore axial tension on the bolt. A set of fixtures, or adapters, consists of a plate that attaches to the face of the unit and a bushing that fits into the back. Normally, bolts are inserted from the rear so that the hexagon head fits into the slot of the bushing. The bushing is pinned to the calibrator so the bolt cannot rotate as it is tightened. The bolt protrudes through the front plate, and the nut is threaded onto the bolt, so tightening the bolt compresses the load cell. Adapter sets are available for a variety of sizes and head configurations. A variety of testers are available to test structural bolts from ½ to 1½ in. in diameter. The calibrators are portable and can be clamped easily to structural members (see Figure 42.1).

Calibration of the tester can be accomplished in a compression press. Application of a known compressive force will give a direct reading on the gage in pounds of force that can be compared to the applied load. The units are calibrated to ±1% of gage reading. Gage adjustments usually require trained personnel.

DOI: 10.1201/9781003501183-49

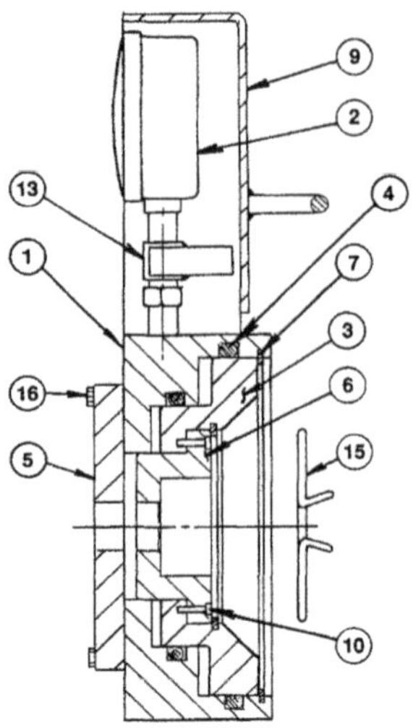

FIGURE 42.1 Schematic view of a bolt tension calibrator.

42.3 THE NEED FOR FIELD TESTING

The importance of testing fasteners and installation methods at the job site should not be underestimated. Critical components are collected from a wide variety of suppliers. While improved manufacturing and testing of these components provide assurance that appropriate specifications are met, on-site testing can provide the final step to successful installation of high strength bolts. Even when not required by local regulations, preinstallation testing provides proof that the correct components have been used. Testing and documentation of results will usually satisfy the needs of the most demanding inspectors. The bolt tension calibrator also provides a means of solving problems that arise during the bolting operation.

With so many tasks to be coordinated in any construction project, it was not unusual that a Federal Highway Administration (FHWA) report on bridge bolting listed 20 separate problems they encountered in their survey of bolting practice. The list ranged from such basic problems as thread stripping, bolt failure before preload, and improper use of lubricants to procedural problems such as unclear specifications. A bolt tension calibrator cannot correct a poorly planned project, but it can provide a practical method of answering bolting questions. Tests prior to installation can save time and money and reduce the need for costly repairs [1, pp. 1–15–1–20].

The use of galvanized nuts and bolts provides examples of several of these problems. To accommodate the thick zinc coating, nuts that are tapped oversize are often used. However, if the nuts are too large, threads may strip before proper bolt tension is obtained. In addition, lubrication is usually required to prevent galling of the threads. Insufficient lubrication may substantially increase the torque needed to tighten the bolt to required tension. This occurs so often that one of the first questions we ask clients experiencing installation problems is whether the fasteners are galvanized. The ability to test such variables at the site can help determine the cause of the problem by reducing guesswork. Figure 42.2 illustrates this problem.

The current AISC specs indicate the need to check bolts in slip-critical joints "to confirm: (1) the suitability to satisfy the requirements of (minimum tension) of the complete fastener assembly, including lubrication if required.... (2) calibration of wrenches, if applicable, and (3) the understanding and proper use by the bolting crew of the method to be performed" [2, p. 11]. Work by the FHWA has led to the introduction of job site rotational capacity tests as well. To quote Ref. [1] (p. 6–2): "Specifications require the R-C test to also be conducted on the project site or at the point where installation takes place."

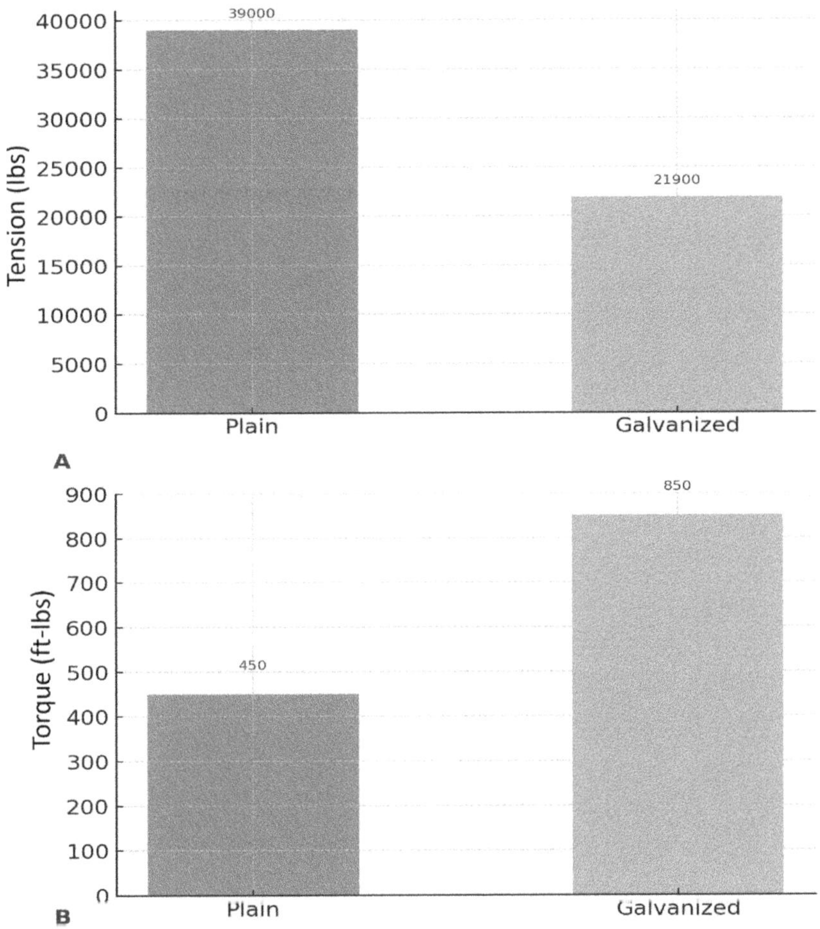

FIGURE 42.2 (a) Bolt tension in ⅞ in. A325 bolt with applied torque of 470 ft-lb. Plain finish=39,000 lb. Galvanized finish=21,900 lb. (b) Torque required to obtain 39,000 lb tension in ⅞ in. A325 bolt. Plain finish=450 ft-lb. Galvanized finish=850 ft-lb.

42.4 TORQUE AND TENSION

When torque, a rotational force usually measured in foot-pounds or newton-meters, is applied to a nut that is gripping plies of material, the bolt is stretched and put in tension. This tension, or preload, holds the assembled joint together during service. Therefore, torque and tension can be related. In fact, the original bolting specs developed in 1951 used torque to tension tables.

The torque–tension relationship depends on the surface conditions of the threads and washers and the fit of the bolt and nut. Unless batches of parts are manufactured to the same conditions, variations in the "friction factor" will cause wide variations in installed bolt tension. According to an AISC specification [2, p. 26], "Variation in the torque to tension relationship of as high as plus or minus 40% must be anticipated unless the relationship is established individually for each bolt lot, diameter and fastener condition." As a result, torque to tension tables were eliminated from their specs in 1954. In 1991, FHWA stated [1, p. 2–19]: "High-strength bolts used in bridge construction are required to be installed to initial pretension equal to or greater than 0.7 times their tensile strength, not a particular torque value. Torque may be used only if the requirements for calibrated wrench installation are followed."

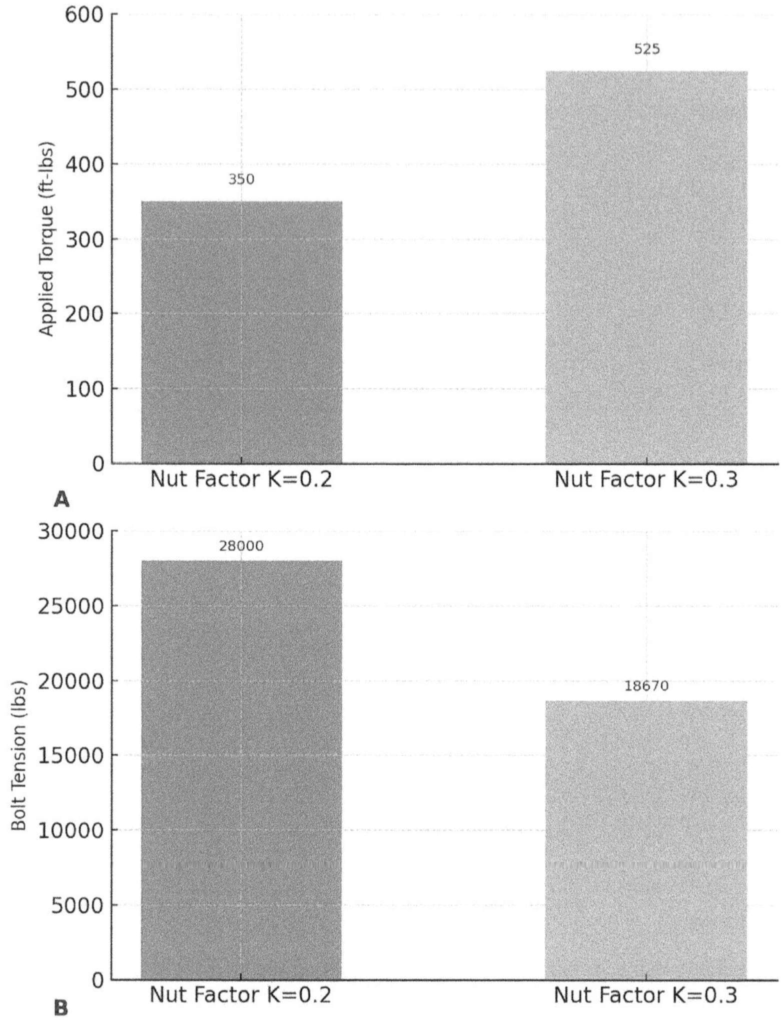

FIGURE 42.3 (a) Torque to achieve 28,000 lb tension in ¾ in. A325 bolt. With nut factor $K=0.2$, torque=350 ft-lb. With nut factor $K=0.3$, torque=525 ft-lb. (b) Bolt tension in ¾ in. A325 bolt with application of 350 ft-lb. With nut factor $K=0.2$, tension=28,000 lb. With nut factor $K=0.3$, tension=18,670 lb.

While estimates of the dimensionless friction factor are available for various joint conditions, this factor can be determined only by testing the actual components. Charts and various rules of thumb should be used only as guides. It is generally accepted that from 70% to 90% of the energy used to tighten a bolt is used to overcome friction in the joint. As a result, small changes in the friction will result in large changes in bolt tension, as shown in Figure 42.3. These changes can be large enough to cause bolts to be tightened below specifications or past safe load limits despite the application of the same torque.

Lubrication not only plays an important role in tightening the bolt, but it also has a direct effect on the bolt's axial ductility. More ductile bolts can be stretched further beyond yield or proof loads before failure tension without failing. Bolts are subjected to both tensile and torsional stresses during the tightening process. Torsion stresses reduce the axial load capacity of the bolt compared to its capacity under direct tension. Reducing torque during installation will reduce the chance for bolt failure. For this reason, "bolts that are well lubricated show more ductility than bolts that are dry or rusty" [3, p. 64].

FIGURE 42.4 Bolt tension calibrator.

42.5 TESTING HIGH STRENGTH BOLTING METHODS

A variety of bolting methods are available, each having its own advantages. The success of each method depends on understanding and following the procedure exactly. Failure to follow the proper routine can result in the installation of fasteners that may fail in service. Regardless of the method used, the bolt tension calibrator provides a means of testing the fasteners, the tools, and the procedure itself before the work is started (see Figure 42.4).

Normally, a sample of three to five sets of bolts, nuts, and washers of each size and grade should be tested. When inserted in the calibrator, the end of the bolt should be at least flush with the face of the nut, after the nut has been threaded on the bolt. For bolts that are so long that the impact socket cannot reach the nut, add washers, shims, or spacers. The washer from the test set should be placed over the bolt last so that it will bear on the nut. For bolts that are too short, the substitution of a bolt of sufficient length may be acceptable. The calibrator can be attached to a convenient column or beam and the tests performed where the bolting will be done. Testing for any of the following procedures confirms not only that the bolts and nuts are capable of obtaining the correct tension but also that the installation method is working properly. You must consult your own appropriate specifications for exact testing details for each method.

42.5.1 Turn-of-Nut

The turn-of-nut method calls for rotation of the nut a specified turn past an initial snug condition. Theoretically, rotation of the nut will stretch the bolt to develop the correct minimum bolt tension. The snug condition has been generally defined as the point when the joint plates are in positive contact. This can be obtained by the first few blows of an impact wrench or by the full effort by an operator pulling on a spud wrench. The degree of rotation varies from 120° to 240° depending on the length of the bolt. The calibrator can be used to verify the snug condition and to confirm that the required rotation of the nut will generate the correct minimum bolt tension.

Although turn-of-the-nut does not require the use of a hard washer on the job, hard washers are recommended when testing on the calibrator to prevent excessive galling of the front plate. Tighten the bolt using the appropriate snug definition, then mark the bolt and nut to provide a reference point. Tighten the bolt and nut the prescribed rotation, and verify that the tension as shown on the gage is at least 5% greater than the minimum requirement.

The proper implementation of the turn-of-nut method depends greatly on the snug condition. For example, if the joint is not pulled together, the initial rotation will simply draw the material together rather than tightening bolt. At the other extreme, bolts that are highly lubricated (such as hot-dip galvanized bolts) reduce the bolt and nut friction so much that snugging the bolt can tighten it almost to specification. Subsequent rotation will overtighten the bolt.

It should be noted that the calibrator is not as rigid as steel plates, and therefore there will be some difference in the rotation/tension ratio. This is due to some flexing of the components and compression of the hydraulic seals.

42.5.2 Calibrated Wrench Method

Calibrated impact wrenches are tools that have been tested in a bolt tension calibrator using a sample of three bolts, nuts, and washers from each lot. Hardened washers must be used, and the bolts and nuts are inserted into the tester as described above. The impact wrench is then operated in exactly the same manner as it would be on the actual joint under construction, paying particular attention to air pressure and air flow. The tool must shut off or reach its capacity when the calibrator registers at least 5% more than the required minimum bolt tension.

The success of this method depends on controlling the tightening variables such as surface condition of the bolt and nut, thread condition, lubrication, air supply to the tool, efficiency of the tool and the operator. In other words, the torque–tension relationship must be controlled. These tests must be performed daily in order to control these factors.

42.5.3 Alternative Design Bolts

Some bolts, usually referred to as tension control or twist-off bolts, use a special feature to determine if the bolt has been properly installed. The end of the bolt will twist off at correct tension for that size and grade. These bolts and their special tightening tools can be tested in the bolt tension calibrator using the same process. Sample fasteners are inserted into the tester and tightened to make sure that the end of the bolt shears off at the proper bolt tension. Typically, these bolts have a button head instead of the usual hexagon head, so a non-slotted back bushing should be used in the calibrator. The bolt does not turn during the tightening process because of the friction under the bolt head. (For this reason, it may be necessary to remove any excess oil under the head.) Just as in the calibrated wrench method, the entire assembly method can be checked.

Again, understanding the torque–tension principles and the use of a bolt tension calibrator can facilitate installation. These bolts are manufactured to control the torque/tension ratio very closely. They are usually sold as a set of a bolt, nut, and washer. Consistent lubrication is critical to their proper use. Many bolt manufacturers caution that changes in lubrication will directly affect the resulting bolt tensions.

42.5.4 Direct Tension Indicators (DTIs)

Another bolting development is the direct tension indicating washer or DTI. These washers have bumps or protrusions on one side that deform uniformly under load from the bolt. Typically, the gap between the fastener and the washer is measured with a feeler gage to determine if the correct bolt tension has been obtained.

Again, special adapters may be needed in the calibrator to accommodate the use of the feeler gage under the bolt head. The bolt, nut, and washer should be assembled into the calibrator according to the manufacturer's procedures and tightened. Depending on the application, the bolts can be installed from the front or rear of the tester. The test should be set up exactly as the DTI will be used in the structure. In most cases, the washers are placed under the stationary bolt head, and the nut is turned. If the washer is placed under the wrong element, or the protrusions bear on the plate, inaccurate readings may occur. In any event, two wrenches will be needed, one to tighten the nut and the other to keep the head from turning. Tighten the bolt to a snug condition and check the gaps with the appropriate feeler gage. Then complete tightening the assembly to the required minimum tension plus 5%. Using the appropriate feeler gage and following the manufacturer's procedures, verify that the washers have deformed correctly.

42.6 INSPECTION TORQUE

The best inspection method tests bolts and installation tools prior to the actual work and monitors the job once it is under way. Nevertheless, situations arise where inspection after assembly is required. An arbitration inspection method is outlined in the Research Council's Specification for Structural Joints [2, p. 18].

Using a bolt tension calibrator and a manual torque wrench, an inspection torque can be determined easily. Tighten three representative bolts to 15% of their required tension to establish an initial condition. The surface under the part that turns should simulate the surface conditions of the job as closely as possible. Finish tightening the bolt by any means to the required bolt tension. Nut rotation during the final tightening cannot exceed 1½ times the rotation allowed. Then apply enough torque with the job inspection wrench to turn the nut at least 5° in the tightening direction. The average torque for three tests is the job inspection torque.

At least 10% of the bolts in question, but not less than two, must be inspected. If any of the nuts or bolt heads turns from the application of the job torque, then all the bolts must be retightened.

42.7 ROTATIONAL CAPACITY TESTING

Due in large part to a study by the Federal Highway Administration, rotational capacity testing has been added to the list of on-site tests. This test is designed to check the residual ductility, effects of lubrication, and resistance to thread stripping of the fastener. These characteristics help ensure that the bolts maintain the proper tension during service.

Following the same procedures as above, sample bolts and nuts are installed into the calibrator and first tightened to a snug condition. Snug is defined by a specific tension in pounds (kips) depending on the bolt diameter. Lines can be marked on the front plate so that rotation of the nut can be determined. Tighten the bolt with the manual torque wrench to the minimum required tension, and record the torque with the nut in motion. Continue tightening to a prescribed rotation from the initial snug condition. If either the bolt fails or the nut strips, then the assembly shall have failed the test.

The final bolt tension recorded at this point shall be at least 15% greater than the minimum specification. If the bolt does not achieve this tension, then it will have failed the test. The bolt and nut are then removed and inspected for signs of failure. Any components with indications of failure will also fail the test. Last, a comparison to a computed torque value is made, and those bolts requiring more torque than calculated also fail the test. The torque computation uses the formula given in the next section and prevents the use of bolts that have a friction factor K greater than 0.25.

42.8 FIELD APPLICATION INFORMATION

It would appear that many field personnel still do not fully understand the relationship between torque and tension. A common misconception deals with the perception that a bolt of a given diameter has a unique installed torque. Although the torque charts as originally proposed by the RCSC have not been allowed for years, many projects use torque specifications for installed bolts. The purpose of this chapter is to assist in problem solving when any specification is not being met, not to debate the merits of various specifications.

An accepted formula to calculate torque is

$$T = Kdp/12$$

While we recommend that testing should determine the relationship of tension to torque, this formula can be a valuable tool in analyzing bolting problems. In this formula, T=torque in foot-pounds, K=nut factor, d=bolt diameter in inches, and p=bolt tension in pounds [3, p. 60]. The friction factor K is a dimensionless number that accounts for such variables as plating, lubrication, surface condition, and thread condition. The formula can be instructive in explaining why tension may vary even though applied torque remains constant, but it should not be used to set specifications for critical applications.

Calculations are straightforward. Using the data in Figure 42.3, a solution would be as follows:

$$T = (0.2 \times 0.750 \times 28{,}000)/12; \qquad T = 350 \text{ ft-lb}$$

We can also solve for other variables, such as K.

$$K = 12T/dp; \qquad K = (12 \times 350)/(0.75 \times 28{,}000); \qquad K = 0.2$$

An accepted value for K is 0.20. (This formula is sometimes used without the factor of 12. In this example, K would then be 0.0167.) When questions arise regarding torque we will usually calculate the nut factor to determine if it is within the industry norm. Significant differences in friction will lead to investigating the variables listed above, such as lubrication and thread condition.

For example, we recently talked to a client installing 1¼ in. diameter A490 bolts in a communications tower. Tightening these bolts in our calibrator with a hydraulic torque wrench, he was able to obtain only 31,000 lb tension, while the wrench indicated it was generating torque in excess of 1,500 ft-lb. Using the torque conversion formula, the nut factor K was over 0.50. We informed the contractor that a factor of 0.20 was usually accepted as normal and that the Federal Highway specs refer to a maximum factor of 0.25. We suggested that he investigate bolt and nut conditions to determine why friction was about twice the norm.

Correct selection and use of torque wrenches is also an important ingredient of the testing process. Direct-reading wrenches are preferred over the preset or click-type wrench because they are more accurate for this type of work. Readings taken while the nut is in motion may not be the same as readings taken when the wrench has stopped. Regardless of the method used, consistent use is very important in obtaining good torque readings. Refer to the manufacturer's instructions for the correct method of using a torque wrench.

One key to successful problem solving is listing the potential problem areas and testing one variable at a time under controlled conditions. More often than not, bolting problems involving torque and tension are caused by a simple factor that is overlooked. A thorough understanding of the elements that affect torque and tension will quickly point out the direction of analysis. Factors to be checked include air pressure and air flow to the impact wrench, lubrication of the components, thread condition, and accuracy of the test equipment. The use of a bolt tension calibrator at the job site will provide the testing capability necessary for the successful installation of high strength bolts in structural connections.

REFERENCES

1. Federal Highway Administration, *High-Strength Bolts for Bridges*, Report No. FHWA-SA-91-031, May 1991.
2. AISC, *Specification for Structural Joints Using ASTM A325 or A490 Bolts*, American Institute of Steel Construction, Inc., Chicago, IL, November 1985.
3. SSTC, *Structural Bolting Handbook*, Steel Structures Technology Center, Inc., Novi, MI, 1995.

43 Torque–Tension Audits

Ralph S. Shoberg

43.1 INTRODUCTION

The art of auditing torque on tightened fasteners has changed significantly during the past three decades. Prior to 1968, the primary method used to audit fasteners attempted to observe the "breakaway torque" as a previously tightened fastener was advanced in the tightening direction. Typically, the torque reading was obtained with a bending beam or dial indicator type of torque wrench. Although this technique is still used today for noncritical applications, the readings obtained are known to have little or no correlation with the actual tension developed by the installation torque.

Other than the fact that, in general, higher values of breakaway torque indicate that higher installation torque may have been applied, the breakaway torque measurement cannot be relied upon to verify fastener tension.

Following the introduction of the first rotary socket wrench torque transducers and the first battery-powered portable peak meters in 1968, there has been continuous development of fastener torque measurement technology. The comparison of dynamic-applied torque to hand torque audits was possible only after the first rotary socket wrench torque sensors were introduced with suitable peak torque reading instruments.[1] The techniques for measurement and interpretation of torque signatures have become highly refined, and capabilities have been developed both to interpret tool torque control and to make other more meaningful measurements that can be correlated with actual fastener tension.

If the friction coefficients and other variables associated with a given bolted joint are uniform and repeatable, it is possible to demonstrate that fastener assemblies tightened to a specified torque will achieve clamp loads with sufficiently low scatter for reliable performance of the assembly. Tightening to a specified torque is still the most widely used assembly process.

Torque-turn tightening strategies add a significant degree of precision to the capability to achieve a specified initial preload. They are clearly needed for applications where personal injury or loss of life or other costly consequences are likely if loosening or fatigue failure should occur.

The use of torque as the sole means of achieving fastener initial preload always indicates that the person responsible for the specification has accepted the scatter in preload results that will naturally occur. It should also be recognized that torque tightening strategies are based on statistical variations of a known degree of scatter, and there is no provision to allow for problems introduced when parts with characteristics outside of the normal population show up on the assembly line.

Torque-only tightening strategies assume that the materials have known torque–tension characteristics with a sufficiently small six sigma (6σ) scatter that the desired preload will be achieved if the proper, specified torque is applied. Hand torque audits are used to verify the installation torque. Torque-angle signature analysis methods can now be used to audit the installation torque, and with some additional study, they can be used to audit actual tension on tightened fasteners.

Dynamic torque measurements and tool certification audits following guidelines such as ISO 5393 are used to verify the capability of tools to consistently apply the specified torque over a broad range of joint friction and spring rate conditions.

Fastener engineers and mechanical designers have much to learn from the art of "torque-turn signature analysis." Experienced practitioners of the art of fastener signature analysis are capable of stripping away much of the mystery and uncertainty that is often associated with the strength and reliability of threaded fasteners and bolted joints.

 DOI: 10.1201/9781003501183-50

43.2 HAND TORQUE AUDIT—TOOL TORQUE CAPABILITY

Hand torque audits are used to verify the amount of torque that has been applied to a fastener. Torque audits are not capable of predicting fastener tension accurately. Measurement of applied torque alone does not provide sufficient information to permit estimation of clamp load achieved. It is a fundamental rule of analysis that if more than one variable can significantly affect a result, then more than one measurement must be made to independently verify the desired result.

43.3 TOOL TORQUE CAPABILITY—ISO 5393

The primary focus of ISO 5393 is to provide a standard method for evaluating the capability of assembly tools to tighten fasteners to a specified torque. The performance of a tool is first checked on "soft" joints where the angle of turn from 10% of the maximum torque to maximum torque is greater than 360°. Next the ability of the tool to deliver controlled torque on "hard" joints where the tightening angle is 30° is compared to the soft joint values to establish a tool torque control capability index.

A complete laboratory for testing of tool torque capability has been established by Ford Motor Company at Lawrence Technological University in Southfield, Michigan.

43.4 RELEASE ANGLE—TENSION AUDIT

The tension or clamping force developed, and not the torque applied, determines whether the fastener has been properly tightened. Since variations in friction coefficients can produce as much as 3:1 or greater variation in clamping force, it is necessary that some method other than monitoring applied torque be used to confirm the performance of tools in achieving fastener "tightness." This is particularly important in the special case of pulse or impact tools where the fastener is tightened with a series of torque impulses. The short duration pulses transfer energy into the bolted joint in a discontinuous process that can be greatly influenced by both the joint spring rate and the frictional characteristics of the fastener thread and underhead bearing surface.

Since it is possible to demonstrate "breakaway torque" correlation between "dynamic torque" applied by hand or electric/pneumatic tools with a continuous energy transfer process, many individuals have attempted to use the same process to audit fasteners tightened with pulse and impact tools. Due to the physics of the energy transfer process and the variable frictional characteristics and joint energy input requirements to which tools must respond (i.e., ISO 5393), it should be clear that such efforts are doomed to failure.

Those who have tried to accomplish this task have experienced great frustration. Not understanding the nature of the problem, some have attempted joint-specific "fixes." Reducing the frequency response (filtering) of the dynamic measuring instruments to force the dynamic readings to be closer to the static breakaway readings is the only way to obtain such correlation. For a given pulse or impact tool the method immediately fails as soon as the tool is applied to a fastener where the joint rate vs. friction characteristics differ from the initial setup conditions.

If the peak torque pulses are read with very high-frequency response instruments, the peak torque readings can be used to verify the setup of the pulse tool. For such tests to be meaningful, the joint spring rate and friction coefficients must be controlled or standardized. At the present time no standard, ISO or otherwise, exists as a guide to define the test joints for qualification of pulse or impact type tools. Thus the individual user must develop his own standards to meet his specific tool testing requirements.

In the final analysis, the actual tension and preload developed by the tightening process is the result that must be verified.

Examining the torque-angle signature of a fastener basically means looking at tightening and loosening curves or plots of torque versus angle as the fastener is first installed and then loosened. These curves are studied initially in the elastic tightening region where the fastener has not gone beyond yield, as shown in Figure 43.1.

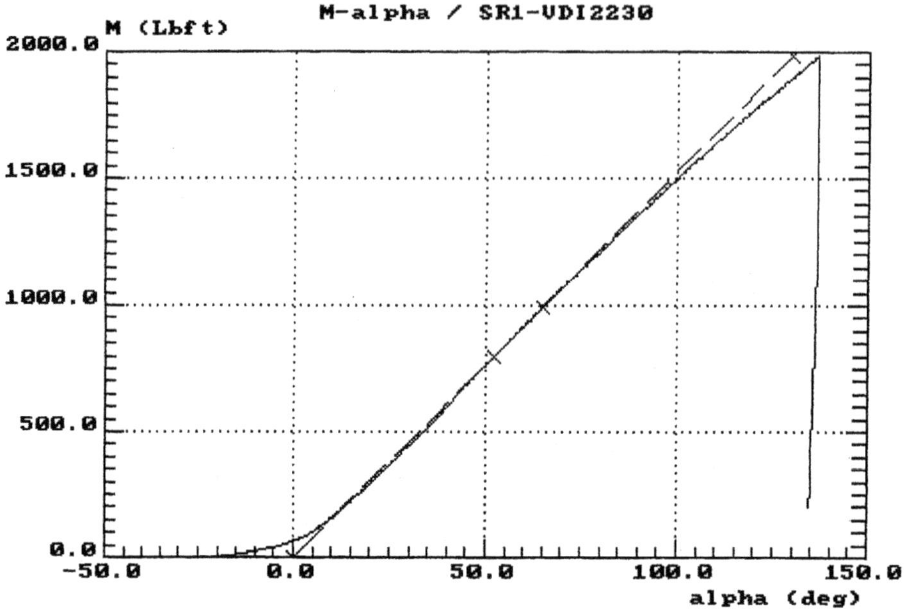

FIGURE 43.1 Torque-angle signature with embedment.

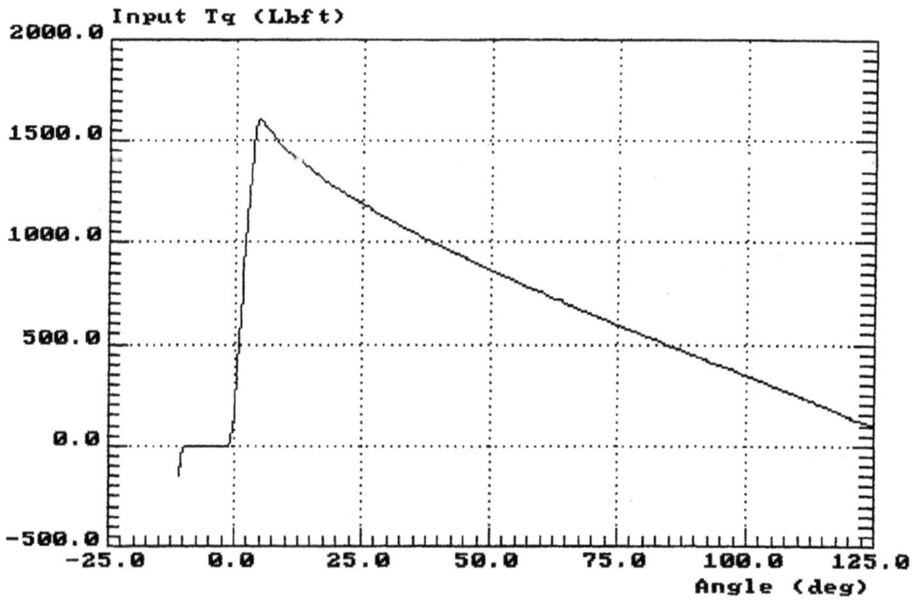

FIGURE 43.2 Loosening torque-angle sequence.

When a fastener is loosened, a torque-angle loosening signature, as shown in Figure 43.2, can be recorded. The release signature shows the "release" of the fastener stretch and also the release of the compression in the clamped parts. Analysis of this signature provides a direct method for verification of preload or tightness. A line tangent to the elastic release portion of the curve projected to zero torque locates the elastic origin. The release angle, measured from the point where loosening starts to the projected elastic origin, is a direct measure of the tension released from the bolted joint.

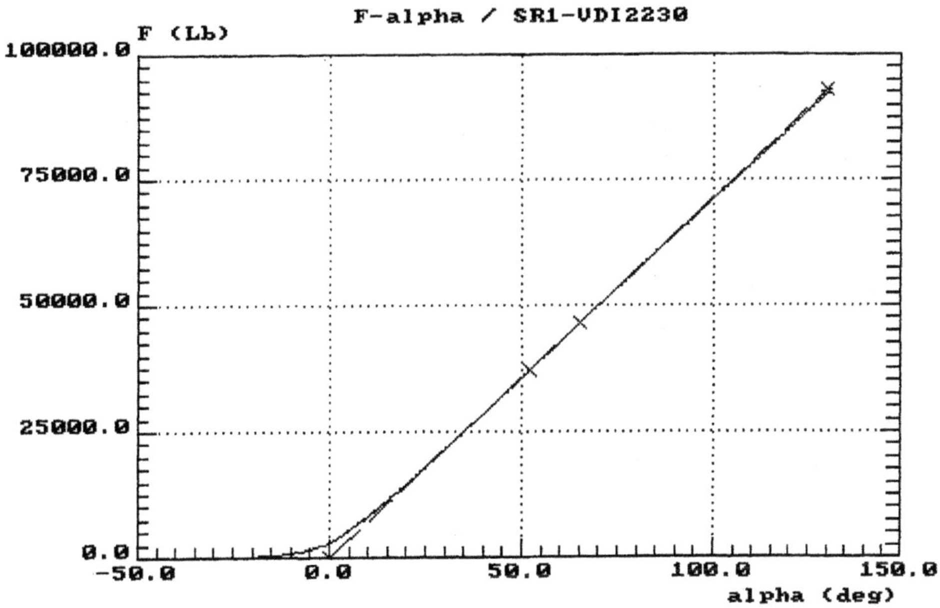

FIGURE 43.3 Clamping force vs. angle of turn from elastic origin.

The tangent line must be drawn on the straight-line portion of the curve after the initial peak release torque due to static friction or thread-locking adhesive has been broken free. The starting point is the angle where initial loosening motion begins. The total release angle is measured from the initial loosening point to the projected elastic origin. (*Note:* If a significant prevailing torque is present after loosening the fastener, the elastic origin must be located at the prevailing level, not zero torque.)

The torque-angle signature shown in Figure 43.1 has been plotted as an M-alpha (i.e., torque-angle) diagram with the tangent line, locating the elastic origin, drawn at 50% of the maximum torque to set the elastic tightening slope below the onset of embedment of the nut. The bolt is an M30×3.5 of strength class 11.9.

The clamping force signature, plotted on the F-alpha (i.e., preload-angle) diagram confirms that the clamping force increases linearly with the angle of turn from the projected elastic origin. In the example shown in Figure 43.3, the elastic tightening angle is approximately 125°.

The loosening torque-angle signature (refer to Figure 43.2) has a projected release angle of approximately 125°. The F-alpha diagram (refer to Figure 43.3) confirms the fact that even after embedment occurs the clamping force increases directly in proportion to the angle of turn from the elastic origin.

Experiments with strain gage bolts or force washers, where the clamping force is measured along with the torque and angle during tightening, will verify that this theory is correct for any given fastener.

To apply torque-angle signature analysis, a torque-angle transient recorder is used for curve measurement and plotting. The transient recorder can provide curves on-screen for analysis as well as print them out for detailed study. Tightening, audit, and release angle signatures for a given fastener can be simultaneously displayed and printed.

Figure 43.4 shows a release angle study performed on an automotive wheel nut. A tool with a torque and angle sensor connected to the transient recorder is used to loosen the nut, record the torque and angle values, and plot the data. The resulting printed curve shows an extremely high release torque. The high initial breakaway loosening peak torque region is disregarded, as this is simply an indication of the static torque required to start loosening motion. (The high value of release torque is

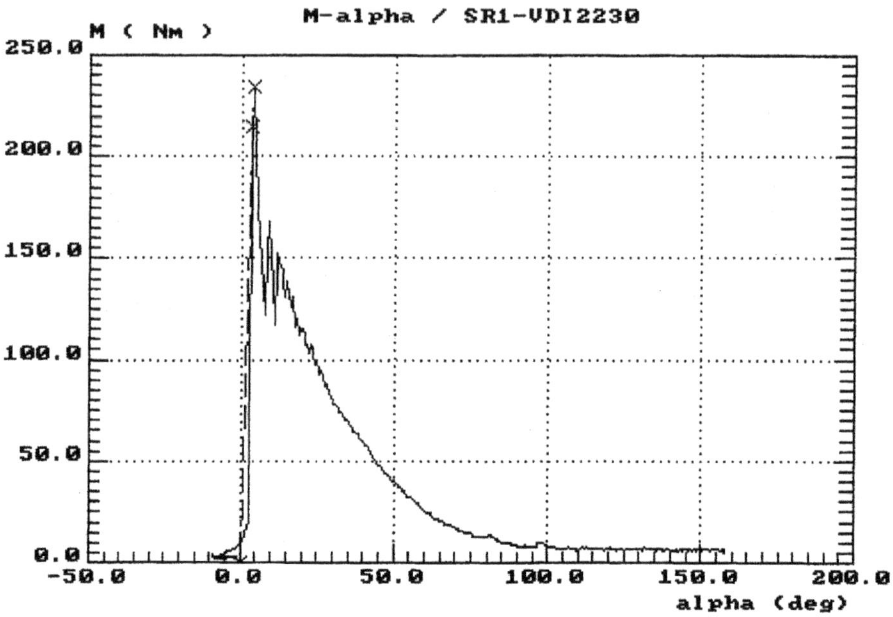

FIGURE 43.4 Wheel nut loosening sequence.

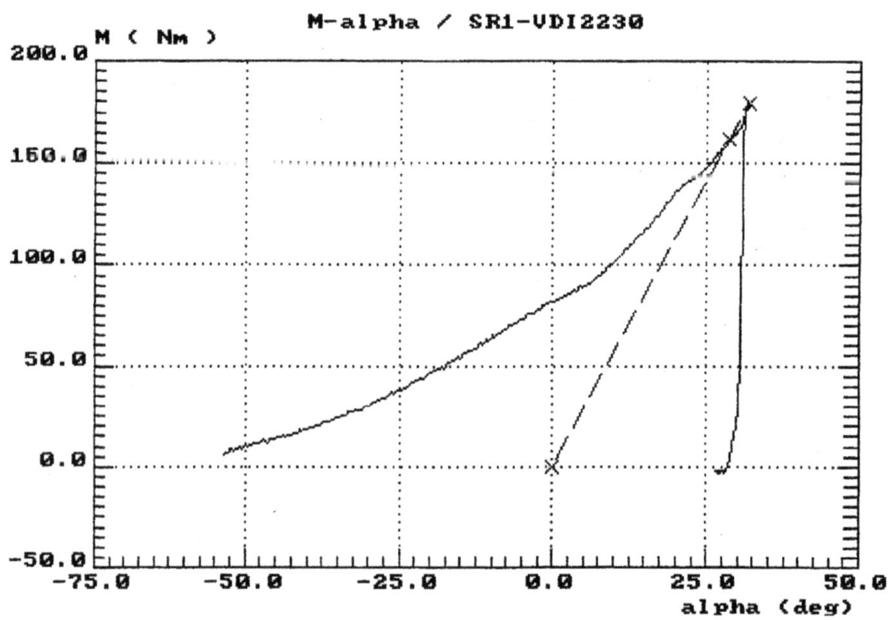

FIGURE 43.5 Wheel nut installation signature.

significant from the point of view that it illustrates the high thread friction due to thread pitch distortion on the wheel nut, a factor that helps prevent vibratory loosening on typical wheel nuts.)

The elastic release angle for the wheel nut shown in Figure 43.4 is approximately 40°. The nut had been tightened to a peak torque of 206 N·m (152 lb-ft), which is 75 N·m (55 lb-ft) greater than the vehicle manufacturer's specification. The wheel nut was originally tightened to a torque of 160 N·m (118 lb-ft), which did not appcar to get past the tightening alignment zone as illustrated by the signature shown in Figure 43.5.

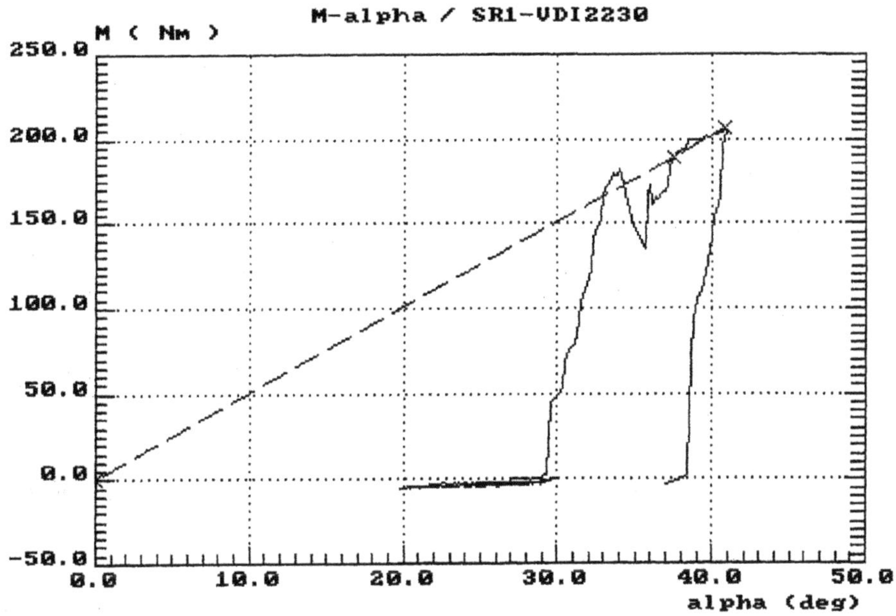

FIGURE 43.6 Wheel nut torque audit signature.

The installation was followed by a hand torque breakaway audit where the nut was advanced about 7°–8° in the tightening direction as shown in Figure 43.6. The *M*-alpha diagram for the audit shows that the final torque was about 206 N·m (152 lb-ft), with a projected elastic tightening angle of 40°.

Applying the release angle method, a line is projected tangent to the elastic release portion of the curve to zero torque. This release angle, measured from the release torque point to the point where the tangent line crosses the zero torque or prevailing torque level, is directly proportional to the tension or clamping force released. Comparing Figure 43.6 with Figure 43.4, a significant correlation is seen to exist between the release angle determined by loosening the fastener and the *M*-alpha diagram as applied to the torque-angle signature for the breakaway audit. The loosening signature in Figure 43.4 was recorded after the audit plotted in Figure 43.6.

43.5 LOOSENING—EMBEDMENT OR LOSS OF PRELOAD

The release angle method has been successfully used to study fastener loosening problems. The basic procedure involves recording and analysis of the torque-angle signatures for tightening and then loosening the fasteners that are to be tested.

First, the torque-angle tightening curve is plotted, the elastic origin is located, and the amount of angle of turn from the elastic origin is determined. After the assembly has been allowed to relax—for example, to sit overnight or run on a dynamic field test—the fastener is loosened and the loosening curve is analyzed. The release angle is determined and compared to the tightening angle, and if they are not equal, it is evaluated to see how much tension was lost by relaxation or loosening.

In one release angle study, a fastener had a tightening angle of 120°. After the part had sat overnight, the release angle was 20°. The manufacturer was already aware there was a major problem because the parts were literally falling apart somewhere between the assembly factor and the auto plant where they were delivered for final assembly in vehicles. The signature analysis study showed that creep or relaxation in the threads was causing an approximately 80% loss in clamping force over a 12 h period. The release angle method provided a quantitative answer as the amount of clamping force lost and clearly showed that the parts needed to be redesigned.

The release angle method is particularly valuable for studying short grip length fasteners holding composite or plastic parts. These parts are generally too small to allow for the use of strain gages or ultrasonic stretch measurements to confirm fastener preload.

For these applications, a torque-angle signature curve for tightening is recorded, then the parts are put in an environmental chamber and load/temperature cycled. Following the test load cycle the release angle signature is recorded. Analysis of the release angle signature in comparison to the tightening signature is used to directly estimate the percentage of initial clamp load lost due to embedment or creep of the plastic part in response to applied loads or temperature cycles. By changing the geometric shape and washer size, the effects can be quantitatively measured and compared.

43.6 MEASUREMENTS VERIFY FASTENER TORQUE AND TENSION

Torque-angle signature analysis is particularly useful for studying all critical fastener assemblies where it is important in terms of safety or reliability that proper preload be obtained initially and maintained throughout the operating life of an assembly.

Threaded fasteners are an important element in the product assembly process for many industries, including automobile, truck, aircraft, aerospace, machine tool, and construction. In the hands of a skilled operator, signature analysis methods can accurately estimate preload or tension within 5% or better.

In addition to analyzing fastener problems such as loosening and embedment, torque-angle signature analysis can also be used to evaluate the performance of tightening tools in applying the desired clamping force on fasteners. This technique is particularly applicable for evaluating pulse tools and impact tools.

Due primarily to ergonomic considerations, pulse tools have recently been extensively evaluated for use in high volume assembly operations. Unfortunately, the limitations of these tools related to their energy transfer characteristics are not generally well understood. The tightening results in terms of torque-tension relationships or achieved preload are very much dependent on the joint friction and spring rates.

Pulse and impact tools are particularly sensitive to joint rate and friction variations. Since friction coefficients are a function of velocity as well as surface pressure, tightening results with pulse and high rpm tools must be carefully evaluated to ensure suitable tightening process capability. Pulse and impact tools move fasteners at high speeds with a great deal of stick-slip, chatter, and unique frictional characteristics that are not seen with steady, continuous tightening processes. These factors can lead to a deceptively high torque reading but with minimal clamping force created. By checking the assembled joint with a release angle study, the user can ensure that an adequate angle of turn, and thus proper clamping force, is being achieved.

Clearly, this method of audit provides a direct measure of the capability of a given tool to develop tension in the tightened fastener. The results of release angle audits, being directly related to the tension achieved, are significantly more meaningful than the torque magnitudes obtained from breakaway torque audits. An improved version of the breakaway torque audit, which uses the torque-angle signature of the audit, can be used to directly estimate fastener tension. This analysis process correlates precisely with the release angle signature method. The only limitation is that the breakaway audit must be conducted in the elastic tightening region for the bolted joint where bolt yield or thread strip is not present.

The following series of tightening, breakaway torque audits, and release signatures illustrate the basic concepts of torque and tension audit using torque-angle signatures. Understanding of the engineering mechanics of threaded fasteners is greatly enhanced through the use of the concept of the "elastic origin" and the application of M-alpha and F-alpha diagrams to the audit process.

In the example illustrated in Figure 43.7, the M12 × 1.75 fastener was tightened to 60 lb-ft (81 N·m). The signature has been recorded with a "record threshold" of 20 lb-ft (27 N·m). The plot shows both torque and tension versus angle of turn, with zero angle located at the threshold.

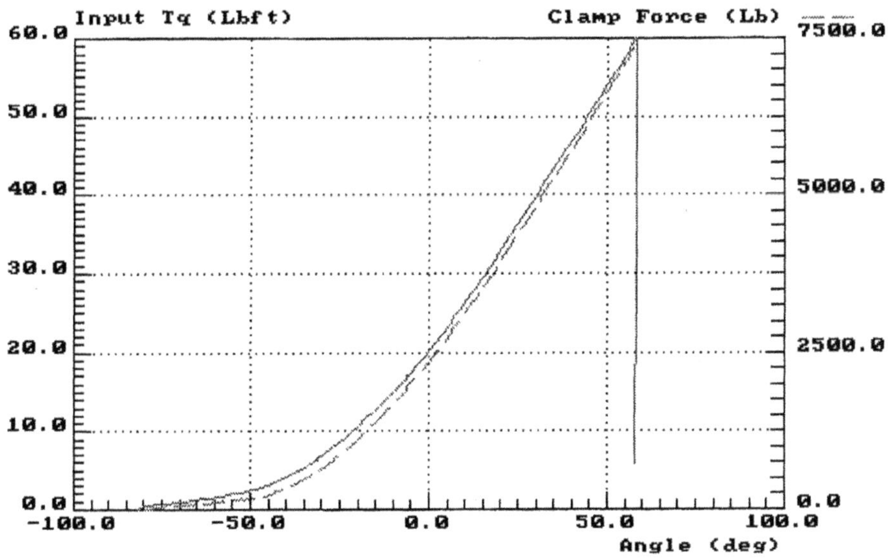

FIGURE 43.7 Torque and clamp force vs. angle.

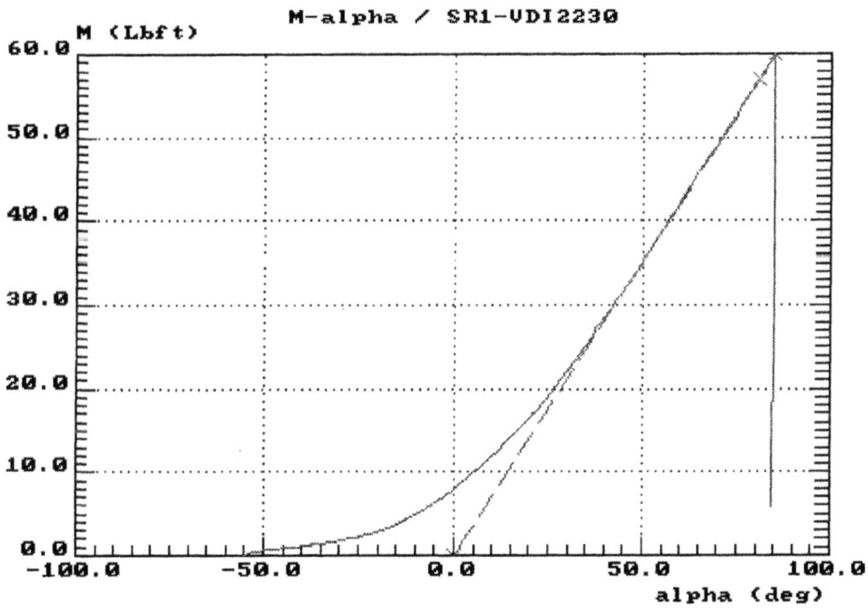

FIGURE 43.8 M-alpha diagram SR1-VDI2230.

The signature analysis software can automatically locate the "elastic origin" on the *M*-alpha diagram by projecting a tangent line from the final point on the torque-angle curve to the zero torque or, if present, the prevailing torque level. The *M*-alpha diagram for the installation tightening signature, shown in Figure 43.8, illustrates that the torque resulted in a projected elastic tightening angle of approximately 85°, resulting in a clamping force of about 7,500 lb (33,360 N).

The corresponding *F*-alpha curve, shown in Figure 43.9, confirms the relationship between torque and angle with the concept of the elastic origin. Note that the 85° elastic tightening angle for the bolt results in approximately 7,500 lb (33,360 N) of clamping force.

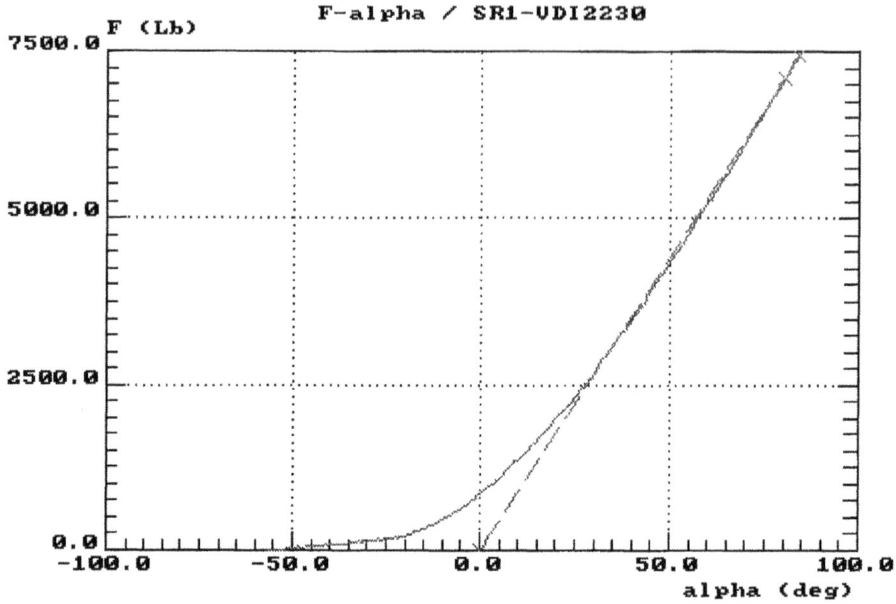

FIGURE 43.9 *F*-alpha diagram SR1-VDI2230.

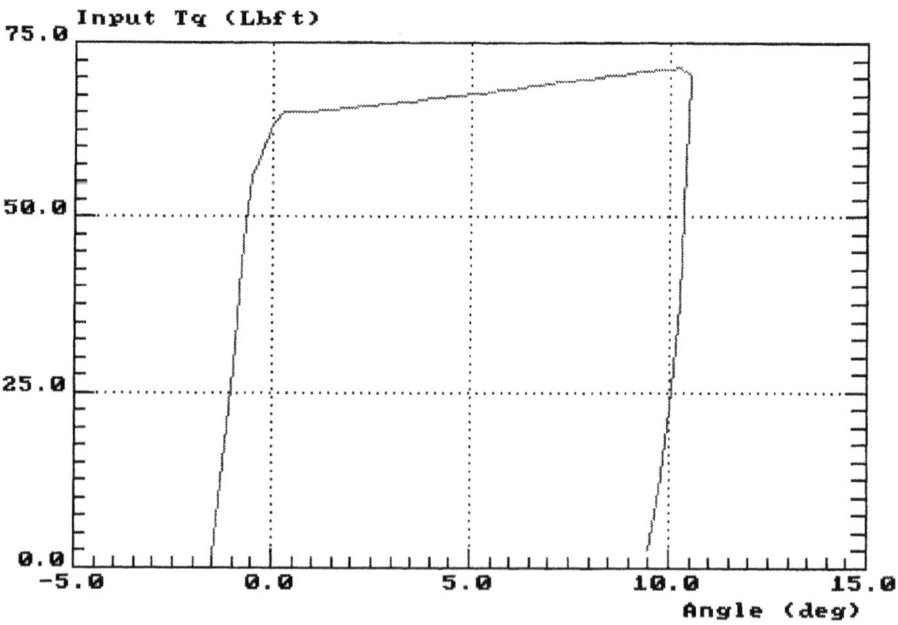

FIGURE 43.10 Breakaway torque audit.

The next signature is the "breakaway" torque audit, shown in Figure 43.10, on the bolted joint tightened for the example shown in Figure 43.9. Breakaway hand torque audits are often used to attempt to correlate dynamic installation torque with the measured breakaway point. In this example, the fastener was torqued in the tightening direction until an additional angle of turn of about 12°–13° was attained. Note that the head of the fastener started to move at about 55 lb-ft (74.5 N·m). Actual breakaway and continuation of the tightening process occurred at about 65 lb-ft (88 N·m) of applied torque. These observations confirm the installation torque of 60 lb-ft (81 N·m) as is normally done for a breakaway torque audit.

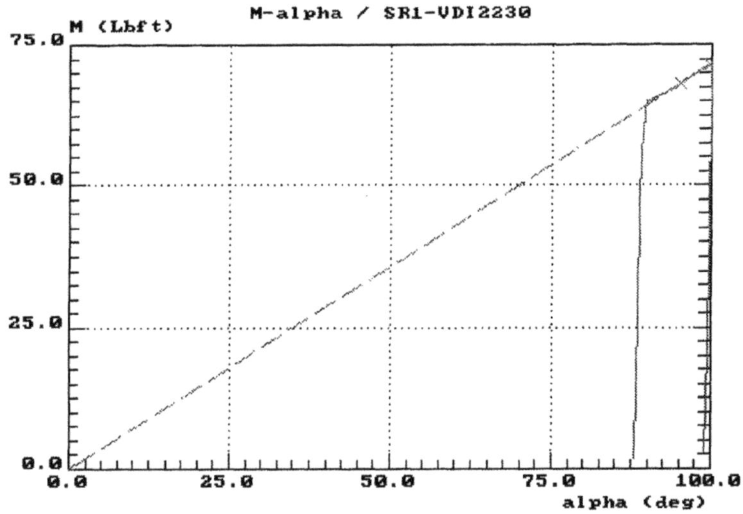

FIGURE 43.11 Breakaway torque audit *M*-alpha SR1-VDI2230.

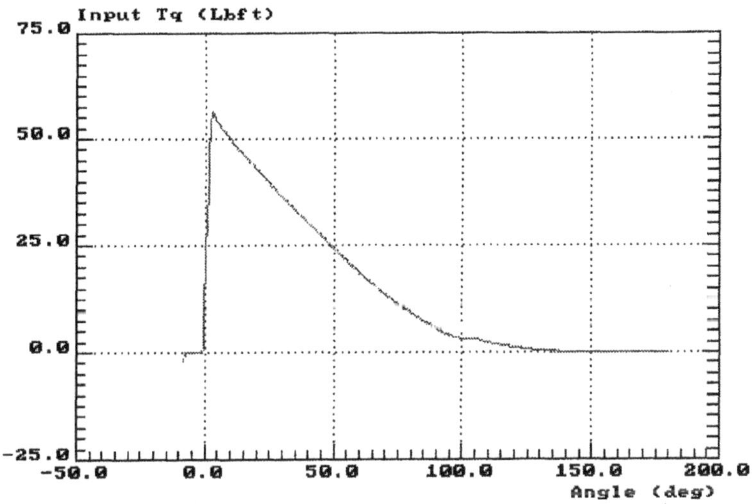

FIGURE 43.12 Release signature.

The signature analysis diagram shown in Figure 43.11 is one of the most significant analysis tools developed in the past 10 years. This diagram shows how it is possible to both audit installation torque and correlate the signature of the audit curve directly with fastener tension. The projection of the tangent to the torque-angle signature curve, locating the elastic origin, is the key to significant improvement of the hand torque audit process.

The *M*-alpha diagram for the audit signature clearly shows the torque breakaway point related to the installation torque and also shows the 85° initial tightening angle, which correlates with a preload of 7,500 lb (33,360 N) of champing force. Note that the breakaway audit increased the tightening angle to approximately 95°–100° projected from the elastic origin, with an expected proportionate increase in preload.

If the release torque-angle signature is recorded when a fastener is loosened, as shown in Figure 43.12, the elastic tightening angle can be estimated, and thus the approximately fastener tension released can be directly estimated if the *F*-alpha slope for the joint has been established.

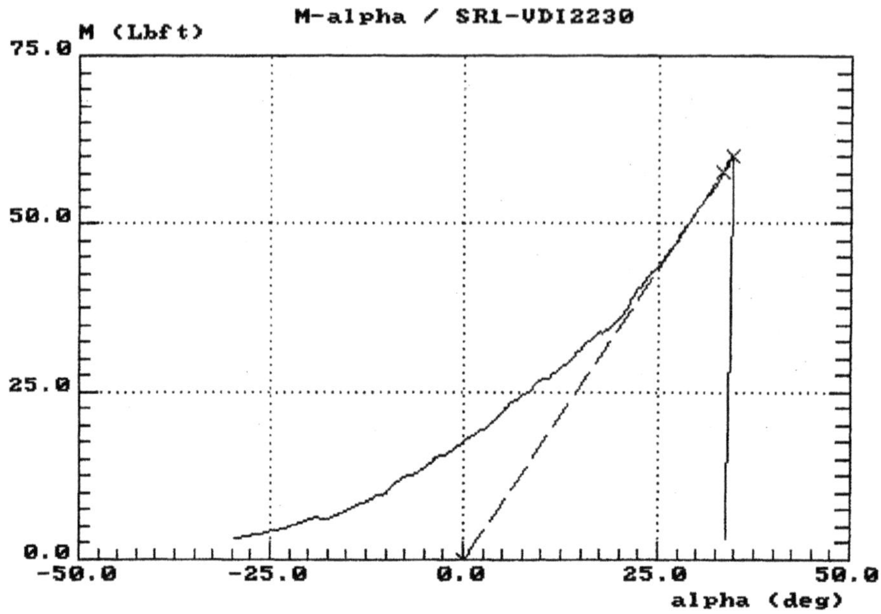

FIGURE 43.13 *M*-alpha diagram for a tightened dry fastener.

M-alpha plots and release angle plots can be used to directly estimate bolt tension, or preload, which is the ultimate goal of the fastener tightening process. In the example shown in Figure 43.12, the release angle of approximately 95° confirms the tightening angle measured on the *M*-alpha diagram for the hand torque audit.

To continue the examples of torque-angle signature analysis to audit installation torque and preload, the original fastener used in the initial example above was stripped of all thread and underhead lubricants to create higher friction coefficients in the thread and underhead regions. The *M*-alpha diagram for tightening to 60 lb-ft (81 N·m), as shown in Figure 43.13, indicates a tightening angle of only 25° projected from the elastic origin. Compared to the lubricated fastener, where the tightening angle was 85°, the predicted preload of 2,200 lb (9,786 N) was confirmed by the clamping force measurement.

The breakaway audit of the "dry" tightened fastener, shown in Figure 43.14, confirms that the installation torque was approximately 60 lb-ft (81 N·m) and also reveals the expected very low angle of turn from the elastic origin.

The torque-angle signature method of analysis applied to tightening and loosening curves is plain, simple, and straightforward. It is a basic engineering analysis technique using fundamental stress, deflection, and material strength properties to model and measure the bolted joint tightening process. The torque-angle signatures can be analyzed to determine installation torque, thread strip, underhead embedment, bolt yield, and, most importantly, fastener tension. The M-alpha plots with reference to the elastic origin are directly linked to the SRI Bolted Joint Design software [1].

The technique can be applied to fasteners of all sizes and all grip lengths. While there may be 75–100 factors that can alter the tightness of a given bolted joint, the torque-angle signature analysis method provides a practical method for direct verification of clamping force to ensure a quality fastener assembly.

In addition to direct estimation of tension achieved, it is possible for the signature analysis method to detect bolt yield, thread strip, and embedment problems. A comparison of the release

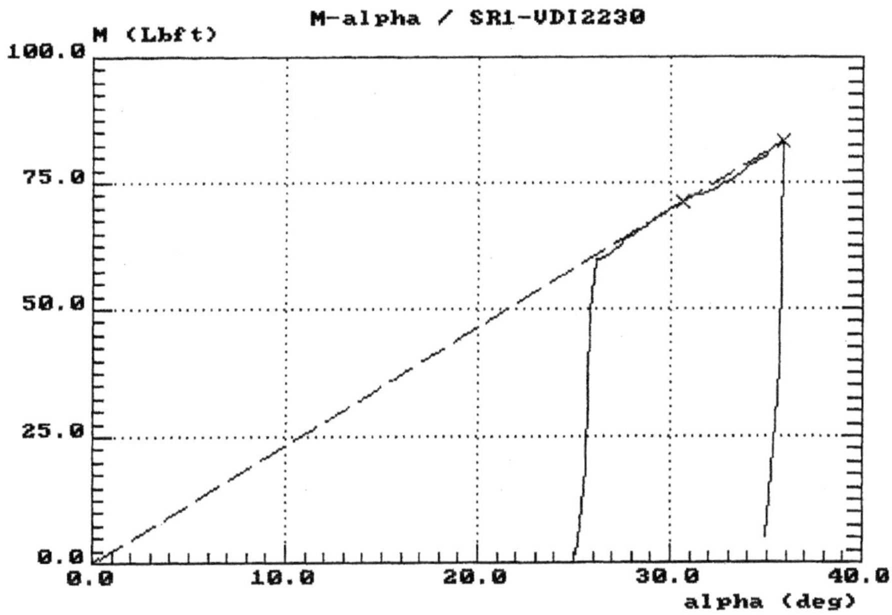

FIGURE 43.14 Breakaway *M*-alpha diagram for a dry fastener.

angle signature and the installation torque angle can be used to evaluate the clamp load retained after a dynamic test.

Material creep and embedment phenomena, which lead to loss of preload, are readily analyzed and quantitatively evaluated through the use of the release angle analysis methods. The methods are particularly powerful in evaluating and comparing the tightening capability of various manufacturers' pulse tools for application on a given assembly or bolted joint.

GLOSSARY OF IMPORTANT TERMS

D_{km}	Effective diameter for the friction in the bolt head or nut bearing area (in., mm)
F	Force, general (lb, N)
F_M	Initial clamp load (assembly preload) (lb, N)
F_V	Preload general (lb, N)
K	Nut factor for basic equation $T=KdF$
M_A	Tightening torque for the assembly until preloading of bolt reaches F_M (lb-ft, N·m)
P	Pitch of the bolt thread (in., mm)
T	Torque, general (lb-ft, N·m)
d	Bolt diameter = outside diameter of thread (nominal diameter) (in., mm)
d_h	Bore diameter of the clamped parts; inner diameter of the substitution cylinder (in., mm)
d_w	Outer diameter of the plane head bearing surface (at the inlet of the transition radius of the head) (in., mm)
d_2	Pitch diameter of bolt thread (in., mm)
μ	Coefficient of friction, general
μ_G	Coefficient of friction in thread
μ_{ges}	Average coefficient of friction for thread and bolt head bearing surface
μ_K	Coefficient of friction for bolt head bearing surface

NOTE

1 Historical note: The first commercially available socket wrench torque transducers and battery-powered peak meters were designed and developed by Macit Gurol and Ralph Shoberg, cofounders of GSE, Inc.

REFERENCE

1. SRI Bolted Joint Design Software, Version 6.1+, RS Technologies, Farmington Hills, MI, w/Hexagon Industriesoftware GmbH, Kirchheim/Teck, Germany.

44 Statistical Design and Analysis of Multivariable Experiments

Bernard Clément

44.1 EXPERIMENTATION STRATEGY

44.1.1 Introduction

Engineers and scientists use experiments as a fundamental strategy to increase their knowledge. Every day, thousands of experiments are run in practically every field of human activities and scientific research, in the life sciences, in the social sciences, and in the physical and engineering sciences.

The art and practice of statistically designed experiments (SDE or DOE) are, surprisingly enough, not well known among engineers and scientists. The basic concepts for designing efficient and effective experiments in a multivariable (factors) context are a highly valuable tool for anyone who is planning a data collection or testing program.

SDE embodies a wide range of efficient experimental protocols based on testing multiple factors (or input variables) in a planned sequence of runs called an experiment. When you test many factors at the same time, you can get more out of your research project resources. SDE strategies are flexible and can be used in the design of products and processes, in laboratory studies, and in production.

The results of SDE have led manufacturing companies to alter the way they operate existing equipment or to modify existing procedures, increasing the capabilities of their production lines with less rework or scrap. SDE is used in all manufacturing industries, in designing prototypes, and in research laboratories. From an applied point of view, SDE can be used in studies such as process design, measurement processes, manufacturing processes, and mathematical process modeling.

The subject of SDE is a vast one that cannot be realistically covered in one chapter. The reader will find additional information in the references given at the end of this chapter. My aim is to focus on some of the most useful statistical designs for practical applications: fractional factorial designs and response surface designs.

44.1.2 Aim, Role, and Benefits

The basic aim of any experiment is to increase the knowledge level about processes and products. The knowledge can be obtained through a historical database or with a planned experiment. The former strategy is long-term and is fraught with difficulties such as undocumented changes in the data collection process, possible high multicollinearity among the input variables, and lack of desirable changes in the levels of each variable. Planned experimentation is a short-term project and does not suffer any of these difficulties.

The role of SDE is to answer some fundamental questions:

1. What are the critical input variables affecting a response variable?
2. What is the mathematical relationship between the critical input variables and the response variable?
3. How do you control the response variable at a desirable level (nominal, maximum, minimum) by controlling the levels of the input variables?

DOI: 10.1201/9781003501183-51

From both a statistical and a practical point of view, the benefits of SDE are well documented, and many case studies have been published [1,2]. SDE basically makes it possible to

1. Minimize the cost of experimentation by minimizing the number of runs.
2. Ease the interpretation and communication of results in a clear and concise manner.
3. Decrease the time for design of processes and products.
4. Improve processes, resulting in high-quality products even if the process conditions are less than optimal or components are less than perfect.

44.1.3 Test Strategy Guideline

Although knowledgeable about their area of expertise, engineers and scientists do not always use the best test strategies in designing experiments. The approach diagrammed in Figure 44.1 is recommended because it represents the collective wisdom and practice from the best references and researchers in the field of statistical experimental designs. It is a "top down" approach, starting with many factors (low knowledge) and using two main steps: (1) a screening phase to identify the critical factors, followed by (2) a modeling phase. At each stage, specialized statistical designs are used.

A good testing strategy

1. Allows many factors to be manipulated (controlled) during the experiment.
2. Varies several factors from run to run.
3. Varies each factor with a small number of levels:
   a. Two levels during the screening phase.
   b. Three to five levels for the response surface modeling phase.

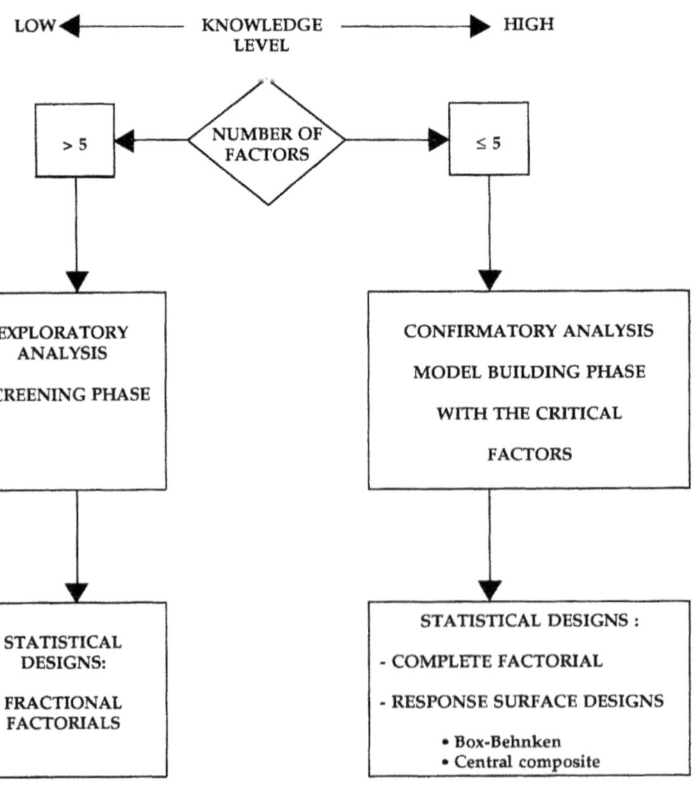

FIGURE 44.1 Test strategy guideline.

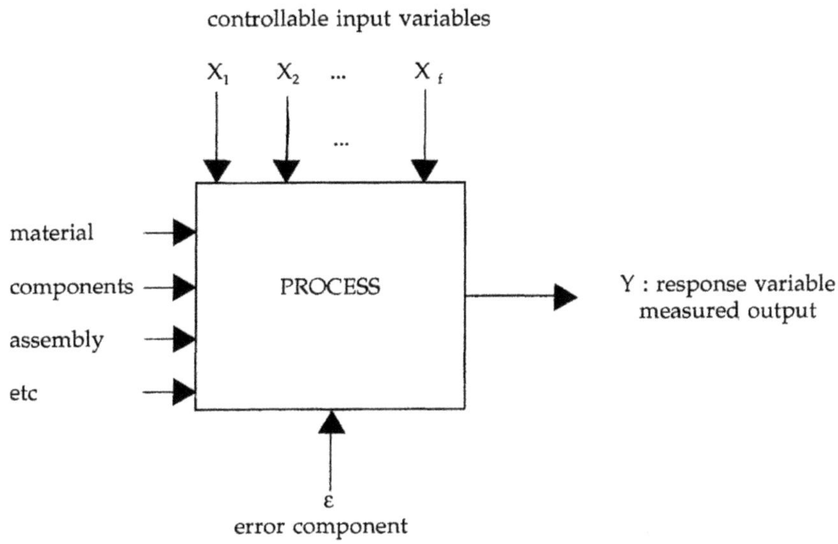

FIGURE 44.2 A schematic view of a process.

Some statistical designs with these properties are presented in this chapter. But these requirements are in sharp contradiction with often seen testing strategies such as that of varying one factor at a time (OFAT). The OFAT strategy has many undesirable characteristics and should be strongly discouraged. For example, it is impossible to estimate the effect of each factor independently of the others; it is not possible to detect interaction effects; and there is a loss in the ability or sensitivity in detecting the effect of each factor.

44.1.4 The Process of Experimentation

One of the best ways to introduce the concept of experimentation is to think of it as a process for studying a process (design, manufacturing, measurement, etc.) or a phenomenon. Loosely defined, a *process* is any system that is repeatable with controlled (or manipulated) *input variables* (or *factors*) and with a measured output known as a *response variable.* The output is always subject to some degree of variation for any given set of inputs. This situation is a combination of measurement error in the response and in the controlled variables plus all the other unsuspected or nonmeasurable factors. This is known as the *error component* of the process. Any process under study can be represented as in Figure 44.2.

The process of experimentation is put in motion by considering a sequence of steps, activities, and decisions that can be represented by a flow diagram as shown in Figure 44.3.

Planning and running an experiment is a project to be managed with dedication and is the result of team effort. There are many decisions to be made, especially during the planning steps. Precise information is given as this chapter progresses. Table 44.1 presents a summary check list and some managerial and statistical guidelines for experimentation. These are general issues and are independent of subject matter application.

44.2 SOME STATISTICAL DESIGN CONCEPTS

44.2.1 Basic Terminology

It is necessary to define some basic terms in order to classify statistical experimental designs. The basic terms of statistical experimental designs are presented in Table 44.2. Some of the most useful designs for applications in the physical and engineering sciences are described in Table 44.3.

PHASE	STEP	COMMENTS
	1 Define the PROBLEM/PROCESS and the objectives for the experiment	• specific, measurable and practical objectives related to the response variable • questions to be answered
	2 Choose the RESPONSE variables to be measured	• the measurement process must give precise and accurate results • there can be several response variables
PLANNING	3 Choose the INPUT VARIABLES and their settings	• see Table 1 for more information on this step
	4 Define and compare several alternative STATISTICAL DESIGNS	• see Table 1 for more information on this step
	5 PREPARE for the experiment	• project management: material, people, schedule etc.
RUNNING	6 CARRY out the experiment	• make sure to carry out the experiment as planned
ANALYSIS	7 ANALYZE the results	• see Table 1 for more information
TRANSFER	8 ACT on the results	• product and process information gained in relation with the objectives

FIGURE 44.3 A process view of experimentation.

TABLE 44.1
Guidelines and Recommendations for Running an Experiment

Step 1: Every experiment will involve resources: time, material, and people. It is a good idea to quantify the return on investment that the project will generate by examining several alternatives and plausible scenarios. Objectives are usually formulated in terms of a response variable such as reduced scrap, improved first time yield, increased purity, or reduced cycle time. In a screening experiment, the aim is to identify the critical input factors controlling the response variable. The objectives must be prioritized if the resources are insufficient.

Step 2: The response on quality characteristics must be specified: strength, dimensions, yields, etc. Several quality characteristics can be measured in accordance with the same experimental design if the set of factors is the same for all responses. Each response will be analyzed separately. What measurement process will be used to obtain the response data? What are the stability, accuracy, and precision of the measurement process? If no information is available, a measurement process evaluation must be undertaken. This is a separate experimental project known as a Repeatability and Reproducibility (R & R) study. This subject is not treated in this chapter, but the reader will find valuable information on the topic in the list of references at the end of the chapter.

(*Continued*)

TABLE 44.1 (*Continued*)
Guidelines and Recommendations for Running an Experiment

Step 3: The input variables to be controlled by the experimenter are called factors of the experiment. They are best chosen in a team brainstorming session involving design engineers, process engineers, quality engineers, statisticians, technicians, and operators involved in the experimentation project. An Ishikawa or fishbone diagram (Figure 44.9) is recommended for displaying the factors. It is best to list all the variables even if the team should select a subset of, say, no more than 15 factors for the experiment. Although statistical designs can be defined for any number of input variables, there are practical and managerial considerations for not undertaking an experiment with several dozen input variables. The number of runs will increase, but, more importantly, the management aspect will become cumbersome.

Step 4: The factors can be qualitative (e.g., type of material) or quantitative (e.g., temperature). In a screening experiment, choose two well-separated levels for each factor. For modeling, there will be three or five levels depending on the design.

Step 5: The choice of statistical design will depend on a number of considerations: number of input variables, number of levels, (experimental space), size of effects to be detected, size of the error component, etc. According to Figure 44.1, if you are dealing with a screening experiment to identify the critical factors, a fractional factorial design with every factor set at two levels is recommended. On the other hand, if you have five or fewer input factors, statistical designs such as complete factorial, Box-Behnken (three levels), or central composite (five levels) are recommended. These designs are presented in Section 44.2.

Step 6: An experiment requires project management skills. It is a good idea to identify a project leader and select the team members at the beginning of the project. Having chosen the statistical design, the success and completion of the experiment will involve scheduling, data collection, statistical analysis of the results, and report writing.

Step 7: To ensure that the experiment is run as designed, it is best to create a table showing the experimental plan: the factors to be tested, the levels of each factor, and the treatment combinations of each run. Each run should be monitored closely to make sure the design is carried out as planned.

Step 8: The statistical analysis of the response can be done using a combination of the following methods: estimation of main effects and interaction effects, analysis of variance table, regression analysis, normal probability plot of effects, a response plot, and residual analysis. In the end, the few vital or critical factors affecting the response variable of the process will be clearly identified. The statistical analysis is presented in Section 44.3.

Step 9: What does the analysis tell you about your product or process? How can you change your product or process to incorporate what you have learned? Additional experimentation can be called for if the results of the analysis generate new objectives and questions.

TABLE 44.2
Basic Terms of Design of Experiment

Experimental unit: Smallest piece of materials, objects, or units to which treatments can be applied; they may be living or not, natural or fabricated; if the units are not homogeneous, a blocking or stratification scheme must be employed and taken into account in the design.

Factors or primary variables: Controllable input variables whose effects on the response variable are to be evaluated directly; they created the need for the experiment in the first place.

Background variables: Variables that are not of primary interest and cannot and perhaps should not be held constant. For example, if an experiment is run over a long time or with different equipment or with different technicians; background variables may be also related to the heterogeneity of the experimental units.

Experimental error: The total effect of all uncontrollable variables known and unknown.

Treatment or treatment combination: A combination of the primary variables where each variable is set at some specific level or value.

Block/blocking: Relatively homogeneous set of conditions or experimental units for which different treatments are compared.

Experimental design: Complete specification of experimental test runs, including blocking, randomization, replication, and the assignment of factor-level combinations to experimental units. There are two structures involved: the assignment method and the treatment structure.

(*Continued*)

TABLE 44.2 (*Continued*)
Basic Terms of Design of Experiment

Randomization: Assigning treatments, running tests, and taking measurements are done randomly; the aim of randomization is to balance out uncontrolled variables.
Repeatability: Taking replicate measurements on the same experimental unit.
Replication: Taking readings on samples prepared independently of one another with the same treatment on different experimental units, during different runs; it is the main tool to obtain an estimate of the experimental error.
Local control: The application of randomization, replication, and blocking to reduce the experimental error.
Response: Outcome or measured result of a run or test.
Test: Combination of factor levels that yield an observed value of the response.
Level: Specific value, numerical or qualitative, of a factor.
Effect: Change in the average response between two experimental conditions.
Main effect: Change in the average response between two or more levels of a factor.
Interaction effect: Joint factor effect in which the effect of each factor depends on the levels of the other factors.

TABLE 44.3
Class of Statistical Designs

Assignment method
Completely randomized: The treatments are allocated at random and without any restriction to the experimental units.
Randomized blocks: The treatments are allocated at random within each block of homogeneous experimental units or set of conditions.
Balanced incomplete blocks: A special case of blocking where the block size is smaller than the number of treatments but each treatment is replicated an equal number of times across all blocks; moreover, each pair of treatments appears a fixed number of times within each block.
Latin squares, Greco-Latin squares, hyper-Latin squares: Blocking designs to remove the effect of two, three, or more blocking variables.
Split plot: The experimental units can be considered of two different sizes when introducing the variables. For example, the first variable might be alloy composition (whole plot); then each alloy could be split into smaller parts (subplots or split plots) to which different coatings (second variable) can be applied.
Repeated measures: The same experimental unit is measured at different points in time or space; the time factor is called a within factor in contrast to the other factors, which are called between factors.

Treatment structure
Complete factorial or fully crossed: All treatment combinations of the primary variables are run; the special class where all variables are set at two levels is especially important in practical applications.
Fractional factorial: A selected subset of treatment combinations from a complete factorial. Rules for selecting good fractional factorials according to specific criteria are given in the remainder of this chapter in cases where all input variables are set at two levels, denoted as 2^{k-p} in the literature.
Nested: When the levels of a second factor are specific to the levels of a first factor, the second factor is said to be nested within the first factor.
Response surface: Special designs for quantitative input variables used in fitting a second-order polynomial. The most important designs of this class are the central composite designs and the Box-Behnken designs.
Mixture designs: Designs in which the input variables are the percentage of the ingredients that make up a material and must add up to 100%.

44.2.2 Classification of Statistical Designs

Statistical designs are usually classified with reference to the assignment method or the treatment structure or a combination of the two. Some of the major design categories are listed in Table 44.3.

In this chapter, we restrict our attention to complete and fractional factorial designs and response surface designs. Moreover, we assume that the assignment method is completely randomized with

TABLE 44.4
Number of Treatment Combinations in a Complete Factorial

Number of factors (f)	5	5	10	10	15	15
Number of levels of each factor (L)	2	3	2	3	2	3
Number of runs (E) ($E=L^f$)	32	243	1024	59,049	32,768	14,348,907
Design notation	2^5	3^5	2^{10}	3^{10}	2^{15}	3^{15}

no restriction. Response surface designs are presented with one blocking variable. The analysis of data with more complex assignment methods is not treated. The reader can find additional information in Refs. [3–5]. Each of the designs is appropriate for certain types of applications. Depending on the design type, the information sought can be the estimation of main and interaction effects, the estimation of variance components, or the mapping of a response surface.

44.2.3 Complete and Fractional Designs

Fractional and complete factorial designs are some of the most useful designs in engineering and in the physical sciences. A complete factorial is easily constructed, but the number of treatment combinations grows exponentially with the number of factors as shown in Table 44.4.

So, except for the case where the number of factors is five or less and the number of levels for each factor is two, complete factorial experiments are unthinkable. Even when applicable, complete factorial experiments can be replaced with suitable fractional experiments. For example, five factors set at two levels can be very suitably studied, without losing much information, in 16 well-chosen runs instead of 32 runs.

A large number of factors can be studied efficiently only with a fractional factorial design. The ways in which to choose the right fractional factorial design are tackled in this chapter. As an example, it is shown that ten factors at two levels can be studied with only 32 runs instead of the 1024 runs required by a complete factorial. This is a considerable saving of time and resources. This type of screening experiment in only 32 runs will, after the analysis, give clear identification of and information about the critical variables. The main question to be answered is: How does one construct a "good" fractional factorial design? To answer this question, we need to discuss complete factorial designs where each factor is at two levels, that is, 2^k *factorial experiments*.

44.2.4 2^K Factorial Designs

We consider the case where all k factors are set at two levels and are varied simultaneously in a complete fully crossed pattern. These designs permit the estimation of individual (main) effects and interaction effects of all k factors. We illustrate the basic notations and definitions for the case of four factors. The method is easily adapted to any number of factors.

The factors are identified with a letter—*A*, *B*, C, *D*. The two levels of each factor are denoted by + and − signs. There are 16 treatment combinations (±, ±, ±, ±) defining the 2^4 factorial design shown in Table 44.5. The + and − signs represent +1 and −1, the coded form of the factors used in the statistical analysis.

44.2.4.1 Some Properties of the Design Matrix

The design is *balanced* because the numbers of + and − signs in each column are equal. Each column can be considered as a vector of +1 and −1 elements, and the design is *orthogonal* because the scalar product of every pair of columns is zero.

TABLE 44.5
2^4 Factorial Experiment

Run	Design Matrix Factors *A*	*B*	*C*	*D*	Response Variable *Y*
1	–	–	–	–	Y_1
2	–	–	–	+	Y_2
3	–	–	+	–	Y_3
4	–	–	+	+	Y_4
5	–	+	–	–	Y_5
6	–	+	–	+	Y_6
7	–	+	+	–	Y_7
8	–	+	+	+	Y_8
9	+	–	–	–	Y_9
10	+	–	–	+	Y_{10}
11	+	–	+	–	Y_{11}
12	+	–	+	+	Y_{12}
13	+	+	–	–	Y_{13}
14	+	+	–	+	Y_{14}
15	+	+	+	–	Y_{15}
16	+	+	+	+	Y_{16}

TABLE 44.6
Table of Signs for Calculating Effects in a 2^4 Factorial

	Design and Main Effect				Two-Factor Interaction						Three-Factor Interaction				Four-Factor Int.	
	1	2	3	4	5	6	7	8	9	10	11	12	13	13	15	
Column Run	*A*	*B*	*C*	*D*	*AB*	*AC*	*AD*	*BC*	*BD*	*CD*	*ABC*	*ABD*	*ACD*	*BCD*	*ABCD*	Response *Y*
1	–	–	–	–	+	+	+	+	+	+	–	–	–	–	+	Y_1
2	–	–	–	+	+	+	–	+	–	–	–	+	+	+	–	Y_2
3	–	–	+	–	+	–	+	–	+	–	+	–	+	+	–	Y_3
4	–	–	+	+	+	–	–	–	–	+	+	+	–	–	+	Y_4
5	–	+	–	–	–	+	+	–	–	+	+	+	–	+	–	Y_5
6	–	+	–	+	–	+	–	–	+	–	+	–	+	–	+	Y_6
7	–	+	+	–	–	–	+	+	–	–	–	+	+	–	+	Y_7
8	–	+	+	+	–	–	–	+	+	+	–	–	–	+	–	Y_8
9	+	–	–	–	–	–	–	+	+	+	+	+	+	–	–	Y_9
10	+	–	–	+	–	–	+	+	–	–	+	–	–	+	+	Y_{10}
11	+	–	+	–	–	+	–	–	+	–	–	+	–	+	+	Y_{11}
12	+	–	+	+	–	+	+	–	–	+	–	–	+	–	–	Y_{12}
13	+	+	–	–	+	–	–	–	–	+	–	–	+	+	+	Y_{13}
14	+	+	–	+	+	–	+	–	+	–	–	+	–	–	–	Y_{14}
15	+	+	+	–	+	+	–	+	–	–	+	–	–	–	–	Y_{15}
16	+	+	+	+	+	+	+	+	+	+	+	+	+	+	+	Y_{16}

Note: Each ± sign should be interpreted as ±1 when performing the computations.

The matrix can be expanded with additional columns by multiplying the components termwise. The result is Table 44.6, a matrix of 15 orthogonal columns used to compute the main effects and the interaction effects. The expanded matrix defines a model for the orthogonal decomposition of the response as a function of 15 effects in addition to the grand average.

44.2.4.2 Main and Interaction Effects

In addition to the grand average, there are four main effects, six two-factor interactions, four three-factor interactions, and a single four-factor interaction, giving a total of 16 effects. All these effects are orthogonal to one another. The effects are easily obtained with the table of signs shown in Table 44.6. The signs for the interaction effects or higher order effects are obtained by multiplying the appropriate columns from the design matrix.

44.2.5 Designing a 2^{k-p} Fractional Factorial Design

When there are many factors, a complete factorial experiment involves too many runs, even at two levels. It is necessary to have a statistical design requiring fewer runs than the complete factorial. The solution to this problem is to construct a carefully prescribed subset of all possible combinations. These designs are denoted 2^{k-p} and involve k factors and a well-chosen fraction $1/2^P$ of all 2^k combinations.

There will be 2^{k-p} runs, and you want $k-p$ small enough—say 3, 4, or 5—that the experiment can be run with 8, 16, or 32 treatment combinations. As a general rule, two or three replicates of each treatment combination are needed to obtain an estimate of the experimental error. If resources are insufficient, you can replicate only some treatments, but the experimental error should be estimated with at least five degrees of freedom. Also, you can use high-order interactions to estimate the experimental error, but this last option is more easily available in a complete factorial design than in a highly fractional design.

44.2.5.1 Principle for Fractioning and Design Resolution

The basic idea is to start with a complete factorial with only a few factors—say three, four, or five. These designs are the 2^3, 2^4, and 2^5 experiments involving 8, 16, or 32 runs. Then we add additional factors by confounding them with high-order interactions. The degree of confounding must be kept at a reasonable level for a clear interpretation of the effects.

This degree of confounding is defined by the design resolution. When some or all main effects are confounded with two-factor interactions, the design is said to be of resolution III. When all main effects are clear of two-factor interactions, the design is of resolution IV. When all main effects and two-factor interactions are not confounded, the design is of resolution V.

Obtaining a design of resolution IV is generally sufficient for most experimental projects. Designs of resolution V are mostly used in modeling processes with response surface designs after an initial screening experiment of resolution IV has been conducted. Designs of resolution IV can be obtained with two reflected designs of resolution III. A *reflected* (foldover) design of an initial design is obtained by interchanging the + and — signs of the initial design. We shall exploit this technique in the case study.

44.2.5.2 One-Half Fraction: 2^{k-1}

To design a one-half fraction of a 2^k design, we first write the full factorial involving $k-1$ factors. For example, if we have four factors, A, B, C, and D, in a complete factorial of 16 runs, a fifth factor E is added by confounding its levels with the $ABCD$ interaction. $ABCD$ is called the generator of E. More generally, the highest order interaction is always used to define the levels of one additional factor.

44.2.5.3 One-Quarter Fraction: 2^{k-2}

For a one-quarter fraction, for example, in the 16-run design with four factors, *A, B, C, D*, a fifth factor *E* and a sixth factor *F* are generated with

$$E = ABC,\ F = BCD$$

This choice is best to keep the design resolution at a maximum. Multiplying these generators together, we have

$$EF = (ABC)\,(BCD) = AB^2C^2D = AD$$

since $B^2 = C^2 = I$, the identity column. We can also write $ADEF = I$. This is called the *defining relation* of the fractional design. The resulting design is of resolution IV because the product of four letters is the identity vector.

44.2.5.4 Tables of Fractional Designs

Although the underlying principles for designing a fractional design are rather straightforward, it is best for the researcher to refer to published tables to avoid mistakes and reduce the time required in designing a two-level fractional experiment. One such set of tables has been devised by Donald J. Wheeler and published by SPC Press [6,7]. The eight-run designs (Table 44.7) and the 16-run designs (Table 44.9) for studying 3–15 factors are reproduced here with the kind permission of the editor. This set of tables contains many more designs like the 32-run design as well as hybrid designs with factors varying at different numbers of levels.

TABLE 44.7
Eight-Run Designs: Basic and Reflected

Column	1	2	3	4	5	6	7
Factor	*A*	*B*	*C*	*D*	*E*	*F*	*G*
				Basic 8-Run Design			
Run							
1	−	−	−	−	−	−	−
2	−	−	−	+	+	+	+
3	−	+	+	+	+	−	−
4	−	+	+	−	−	+	+
5	+	+	−	−	+	+	−
6	+	+	−	+	−	−	+
7	+	−	+	+	−	+	−
8	+	−	+	−	+	−	+
				Reflected 8-Run Design			
Run							
1	+	+	+	+	+	+	+
2	+	+	+	−	−	−	−
3	+	−	−	−	−	+	+
4	+	−	−	+	+	−	−
5	−	−	+	+	−	−	+
6	−	−	+	−	+	+	−
7	−	+	−	−	+	−	+
8	−	+	−	+	−	+	−

Source: Ref. [7].

44.2.6 The Eight-Run Design with All Factors at Two Levels

The eight-run design is defined by a matrix of eight runs and seven columns. Each row corresponds to a run, and each column defines a specific contrast or comparison. The matrix contains all the information for designing the experiment and for the ensuing statistical analysis. The − and + signs correspond to two well-chosen levels for each factor. One noted property of the matrix is that the number of sign changes is 1 for *A*, 2 for *B*, etc. This can be exploited in assigning factors to the labels *A*, *B*, etc. If the experiment is not randomized, the experimenter would assign label *A* to the factor that is most difficult to change, while label *G* would be assigned to the factor easiest to change.

44.2.6.1 Using Tables 44.7 and 44.8

The matrix can accommodate two to seven factors. The assignment of factors to the columns is done by consulting Table 44.8. For example, if we have three factors, they will be labeled *A*, *B*, and *D*, and columns 1, 2, and 4 will be used to define the test combination structure. The design is a fully crossed factorial. The other columns (3, 5, 6, 7) will contain interaction effects, namely −*AB*, −*AD*, −*BD*, and *ABD*, respectively. With two factors, *A* and *B*, the design is a twice replicated 2^2 complete factorial.

If we have more than three factors, the column assignment is obtained from Table 44.8. For example, with five factors, columns 1, 2, 3, 4, and 5 are used for the test structure, and columns 1–7 are used for the statistical analysis and the computation of effects.

TABLE 44.8
Factor Assignment and Contrast Interpretation for the Eight-Run Basic Design

Design	Column	1	2	3	4	5	6	7
2^{7-4}, Resolution III	Seven factors	*A*	*B*	*C*	*D*	*E*	*F*	*G*
	Main effect	*A*	*B*	*C*	*D*	*E*	*F*	*G*
	Two-factor interactions	−*BC*	−*AC*	−*AB*	−*AE*	−*AD*	−*AG*	−*AF*
		−*DE*	−*DF*	−*DG*	−*BF*	−*BG*	−*BD*	−*BE*
		−*FG*	−*EG*	−*EF*	−*CG*	−*CF*	−*CE*	−*CD*
2^{6-3}, Resolution III	Six factors	*A*	*B*	*C*	*D*	*E*	*F*	
	Main effect	*A*	*B*	*C*	*D*	*E*	*F*	
	Two-factor interactions	−*BC*	−*AC*	−*AB*	−*AE*	−*AD*	−*BD*	−*AF*
		−*DE*	−*DF*	−*EF*	−*BF*	−*CF*	−*CE*	−*BE*
								−*CD*
2^{5-2}, Resolution III	Five factors	*A*	*B*	*C*	*D*	*E*	—	—
	Main effect	*A*	*B*	*C*	*D*	*E*	—	—
	Two-factor interactions	−*BC*	−*AC*	−*AB*	−*AE*	−*AD*	−*BD*	−*BE*
		−*DE*					−*CE*	−*CD*
2^{4-1}, Resolution IV	Four factors	*A*	*B*	—	*D*	—	—	*G*
	Main effect	*A*	*B*	—	*D*	—	—	*G*
	Interactions	*BDG*	*ADG*	−*AB*	*ABG*	−*AD*	−*AG*	*ABD*
				−*DG*		−*BG*	−*BD*	
2^3, Complete factorial	Three factors	*A*	*B*	—	*D*	—	—	—
	Main effect	*A*	*B*	—	*D*	—	—	—
	Interactions			−*AB*		−*AD*	−*BD*	*ABD*
2^2, Complete factorial twice replicated	Two factors	*A*	*B*	—	—	—	—	—
	Main effect	*A*	*B*	—	—	—	—	—
	Interaction			−*AB*				

Source: Ref. [7].
Note: With the reflected design, all two-factor interactions have their signs changed from − to +.

Obviously, with five, six, or seven factors the design resolution is III, and the main effects are confounded with two-factor interaction effects. This is not desirable, and we recommend the following strategy. The experimenter should run a basic design and a reflected design for a total of 16 runs. Then, with some simple computations to be illustrated in the case study, the main effects will be separated from two-factor interactions to reach a resolution IV design.

44.2.7 The 16-Run Design with All Factors at Two Levels

The 16-run basic design is defined by a 16×15 matrix of+and−signs in Table 44.9, which is a rearrangement of Table 44.6 that keeps the same properties. The reflected design, not shown, would be defined by a complete reversal of the signs in Table 44.9.

TABLE 44.9
16-Run Basic Design

	1	2	3	4	5	6	7	8	9	10	11	12	13	14	15
Column Factor	*A*	*B*	*C*	*D*	*E*	*F*	*G*	*H*	*I*	*J*	*K*	*L*	*M*	*N*	*O*
Run															
1	−	−	−	−	−	−	−	−	−	−	−	−	−	−	−
2	−	−	−	−	−	−	−	+	+	+	+	+	+	+	+
3	−	−	−	+	+	+	+	+	+	+	+	−	−	−	−
4	−	−	−	+	+	+	+	−	−	−	−	+	+	+	+
5	−	+	+	+	+	−	−	−	−	+	+	+	+	−	−
6	−	+	+	+	+	−	−	+	+	−	−	−	−	+	+
7	−	+	+	−	−	+	+	+	+	−	−	+	+	−	−
8	−	+	+	−	−	+	+	−	−	+	+	−	−	+	+
9	+	+	−	−	+	+	−	−	+	[illegible]	−	−	+	+	−
10	+	+	−	−	+	+	−	+	−	−	+	+	−	−	+
11	+	+	−	+	−	−	+	+	−	−	+	−	+	+	−
12	+	+	−	+	−	−	+	−	+	+	−	+	−	−	+
13	+	−	+	+	−	+	−	−	+	−	+	+	−	+	−
14	+	−	+	+	−	+	−	+	−	+	−	−	+	−	+
15	+	−	+	−	+	−	+	+	−	+	−	+	−	+	−
16	+	−	+	−	+	−	+	−	+	−	+	−	+	−	+

Source: Ref. [7].

TABLE 44.10
Factor Assignment and Contrast Interpretation for the 16-Run Basic Design[a]

15 Factors; Design 2^{15-11}; Resolution III

Column	1	2	3	4	5	6	7	8	9	10	11	12	13	14	15
main effect	*A*	*B*	*C*	*D*	*E*	*F*	*G*	*H*	*I*	*J*	*K*	*L*	*M*	*N*	*O*
interaction	−*BC*	−*AC*	−*AB*	−*AE*	−*AD*	−*AG*	−*AF*	−*AI*	−*AH*	−*AK*	−*AJ*	−*AM*	−*AL*	−*AO*	−*AN*
	−*DE*	−*DF*	−*DG*	−*BF*	−*BG*	−*BD*	−*BE*	−*BJ*	−*BK*	−*BH*	−*BI*	−*BN*	−*BO*	−*BL*	−*BM*
	−*FG*	−*EG*	−*EF*	−*CG*	−*CF*	−*CE*	−*CD*	−*CK*	−*CJ*	−*CI*	−*CH*	−*CO*	−*CN*	−*CM*	−*CL*
	−*HI*	−*HJ*	−*HK*	−*HL*	−*HM*	−*HN*	−*HO*	−*DL*	−*DM*	−*DN*	−*DO*	−*DH*	−*DI*	−*DJ*	−*DK*

(*Continued*)

TABLE 44.10 (*Continued*)
Factor Assignment and Contrast Interpretation for the 16-Run Basic Design[a]

	−JK	−IK	−IJ	−IM	−IL	−IO	−IN	−EM	−EL	−EO	−EN	−El	−EH	−EK	−EJ
	−LM	−LN	−LO	−JN	−JO	−JL	−JM	−FN	−FO	−FL	−FM	−FJ	−FK	−FH	−FI
	−NO	−MO	−MN	−KO	−KN	−KM	−KL	−GO	−GN	−GM	−GL	−GK	−GJ	−Gl	−GH

14 Factors; Design 2^{14-10}; Resolution III

Column	1	2	3	4	5	6	7	8	9	10	11	12	13	14	15
main effect	A	B	C	D	E	F	G	H	/	J	K	L	M	N	–
interaction	−BC	−AC	−AB	−AE	−AD	−AG	−AF	−Al	−AH	−AK	−AJ	−AM	−AL	−BL	−AN
	−DE	−DF	−DG	−BF	−BG	−BD	−BE	−BJ	−BK	−BH	−Bl	−BN	−CN	−CM	−BM
	−FG	−EG	−EF	−CG	−CF	−CE	−CD	−CK	−CJ	−Cl	−CH	−DH	−Dl	−DJ	VCL
	−HI	−HJ	−HK	−HL	−HM	−HN	−IN	−DL	−DM	−DN	−EN	−El	−EH	−EK	−DK
	−JK	−IK	−IJ	−IM	−IL	−JL	−JM	−EM	−EL	−FL	−FM	−FJ	−FK	−FH	−EJ
	−LM	−LN	−MN	−JN	−KN	−KM	−KL	−FN	−GN	−GM	−GL	−GK	−GJ	−Gl	−FI
															GH

13 Factors; Design 2^{13-9}; Resolution III

Column	1	2	3	4	5	6	1	8	9	10	11	12	13	14	15
main effect	A	B	C	D	E	F	G	H	/	J	K	–	M	N	–
interaction	−BC	−AC	−AB	−AE	−AD	−AG	−AF	−Al	−AH	−AK	−AJ	−AM	−CN	−CM	−AN
	−DE	−DF	−DG	−BF	−BG	−BD	−BE	−BJ	−BK	−BH	−Bl	−BN	−Dl	−DJ	−BM
	−FG	−EG	−EF	−CG	−CF	−CE	−CD	−CK	−CJ	−Cl	−CH	−DH	−EH	−EK	−DK
	−HI	−HJ	−HK	−IM	−HM	−HN	−IN	−EM	−DM	−DN	−EN	−El	−FK	−FH	−EJ
	−JK	−JK	−IJ	−JN	−KN	−KM	−JM	−FN	−GN	−GM	−FM	−FJ	− GJ	−Gl	−FI
			−MN									−GK			−GH

12 Factors; Design 2^{12-8}; Resolution III

Column	1	2	3	4	5	6	7	8	9	10	11	12	13	14	15
main effect	A	B	C	D	E	F	G	H	I	—	K	—	M	N	—
interaction	−BC	AC	−AB	−AE	−AD	−AG	−AF	−Al	−AH	−AK	−Bl	−AM	CN	−CM	−AN
	−DE	−DF	−DG	−BF	−BG	−BD	−BE	−CK	−BK	−BH	−CH	−BN	−Dl	−EK	−BM
	−FG	−EG	−EF	−CG	−CF	−CE	−CD	−EM	−DM	−Cl	−EN	−DH	−EH	−FH	−DK
	−HI	−IK	−HK	−IM	−HM	−HN	−IN	−FN	−GN	−DN	−FM	−El	−FK	−Gl	−FI
			−MN		−KN	−KM				−GM		−GK			−GH

11 Factors; Design 2^{11-7}; Resolution III

Column	1	2	3	4	5	6	7	8	9	10	11	12	13	14	15
main effect	A	B	C	D	E	F	G	H	—	—	K	—	M	N	—
interaction	−BC	−AC	−AB	—AE	−AD	−AG	—AF	−CK	−AH	−AK	−CH	−AM	−CN	−CM	−AN
	−DE	−DF	−DG	−BF	−BG	−BD	−BE	−EM	−BK	−BH	−EN	−BN	−EH	−EK	−BM
	−FG	−EG	−EF	−CG	−CF	−CE	−CD	−FN	−DM	−DN	−FM	−DH	−FK	−FH	−DK
			−HK		−HM	−HN			−GN	−GM		−GK			−GH
			−MN		−KN	−KM									

10 Factors; Design 2^{10-6}; Resolution III

Column	1	2	3	4	5	6	7	8	9	10	11	12	13	14	15
main effect	A	B	C	D	E	—	G	H	—	—	K	—	M	N	—
interaction	−BC	−AC	−AB	−AE	−AD	−AG	−BE	−CK	−AH	−AK	−CH	−AM	−CN	−CM	−AN
	−DE	−EG	−DG	−CG	−BG	−BD	−CD	−EM	−BK	−BH	−EN	−BN	−EH	−EK	−BM
			−HK		−HM	−CE			−DM	−DN		−DH			−DK
			−MN		−KN	−HN			−GN	−GM		−GK			−GH
						−KM									

(*Continued*)

TABLE 44.10 (*Continued*)
Factor Assignment and Contrast Interpretation for the 16-Run Basic Design[a]

9 Factors; Design 2^{9-5}; Resolution III

Column	1	2	3	4	5	6	7	8	9	10	11	12	13	14	15
main effect	*A*	*B*	C	*D*	—	—	*G*	*H*	—	—	/C	—	*M*	*N*	—
interaction	−*BC*	−*AC*	−*AB*	−*CG*	−*AD*	—*AG*	−*CD*	−*CK*	−*AH*	−*AK*	−*CH*	−*AM*	−*CN*	−*CM*	−*AN*
			−*DG*		−*BG*	−*BD*			−*BK*	−*BH*		−*BN*			−*BM*
			−*HK*		−*HM*	−*HN*			−*DM*	−*DN*		−*DH*			−*DK*
			−*MN*		−*KN*	−*KM*			−*GN*	−*GM*		−*GK*			−*GH*

8 Factors; Design 2^{8-4}; Resolution IV

Column	1	2	3	4	5	6	7	8	9	10	11	12	13	14	15
main effect	*A*	*B*	—	D	—	—	*G*	*H*	—	—	/C	—	*M*	/V	—
interaction			−*AB*		−AD	—*AG*			−*AH*	−*AK*		−*AM*			−*AN*
			−*DG*		−*BG*	−*BD*			−*BK*	−*BH*		−*BN*			−*BM*
			−*HK*		−*HM*	−*HN*			−*DM*	−*DN*		−*DH*			−*DK*
			−*MN*		−*KN*	−*KM*			−*GN*	−*GM*		−*GK*			−*GH*

7 Factors; Design 2^{7-3}; Resolution IV

Column	1	2	3	4	5	6	7	8	9	10	11	12	13	14	15
main effect	*A*	*B*	—	D	—	—	G	*H*	—	—	—	—	/W	*N*	—
interaction			−*AB*		−*AD*	—*AG*			−*AH*	−*BH*		−*AM*			−*AN*
			−*DG*		−*BG*	−*BD*			−*DM*	−*DN*		−*BN*			−*BM*
			−*MN*		−*HM*	−*HN*			−*GN*	−*GM*		−*DH*			−*GH*

6 Factors; Design 2^{6-2}; Resolution IV

Column	1	2	3	4	5	6	7	8	9	10	11	12	13	14	15
main effect	*A*	*B*	—	*D*	—	—	G	*H*	—	—	—	—	—	/V	—
interaction			−*AB*		−*AD*	−*AG*			−*AH*	−*BH*		−*BN*			−*AN*
			−*DG*		−*BG*	−*BD*			−*GN*	−*DN*		−*DH*			−*GH*
						−*HN*									

5 Factors; Design 2^{5-1}; Resolution V

Column	1	2	3	4	5	6	7	8	9	10	11	12	13	14	15
main effect	4	*B*	—	D	—	—	—	*H*	—	—	—	—	—	—	0
interaction			−*AB*		−*AD*	−*BD*	−*HO*		−*AH*	−*BH*	−*DO*	−*DH*	−ßO	—*AO*	
			+DHO		+*BHO*	+*AHO*	+*ABD*		+*BDO*	+*ADO*	+*ABH*	+*ABO*	+*ADH*	+*BDH*	

4 Factors; Design 2^4; Complete Factorial

Column	1	2	3	4	5	6	7	8	9	10	11	12	13	14	15
main effect	*A*	*B*	—	D	—	—	—	*H*	—	—	—	—	—	—	—
interaction			−*AB*		−*AD*	−*BD*			−*AH*	−*BH*		−*DH*			
							+*ABD*				+*ABH*		+*ADH*	+*BDH*	−*ABDH*

3 Factors; Design 2^3 Twice Replicated; Complete Factorial

Column	1	2	3	4	5	6	7	8	9	10	11	12	13	14	15
main effect	*A*	*B*	—	*D*	—	—	—	—	—	—	—	—	—	—	—
interaction			−*AB*		−*AD*	−*BD*	*ABD*								

2 Factors; Design 2^2 Replicated 4 Times; Complete Factorial

Column	1	2	3	4	5	6	7	8	9	10	11	12	13	14	15
main effect	*A*	*B*	—	—	—	—	—	—	—	—	—	—	—	—	—
interaction			−*AB*												

Source: Ref. [7].

[a] In the reflected design, all two-factor interactions have their signs changed from − to +.

Table 44.10 is used for any number of factors from four to 15. It gives the assignment of factors to specific columns where the main effects are located. All 15 columns are used for estimating the effects. If the number of factors is eight or less, the design is resolution IV and the main effects are not confounded with interaction effects. With 9–15 factors, the design is resolution III. To obtain a resolution IV with 9–15 factors, you should run a pair of reflected 16-run designs and combine their results to separate the main effects from the effect strings of two-factor interactions. An example of this strategy is given in the case study, Section 44.4.

44.2.8 Choice of Design

The choice of a statistical experimental design is a crucial decision for any experimenter. It is a major factor in whether or not the objectives of the experiment will be accomplished.

The recommendations given in this section cannot realistically cover all possible situations that an experimenter is likely to encounter. The aim is to focus on some experimental designs that are more appropriate in the physical and engineering sciences than in other fields such as the life sciences or social sciences. The experimental designs recommended here are fractional factorial designs to screen the factors and response surface designs with continuous factors. The corresponding matrices have been presented previously in this chapter for the 2^{k-p} designs. The response surface designs are presented in Section 44.2.9.

Our recommendation is to use the test strategy of Figure 44.1. We assume that the experimenter is starting with a low level of knowledge and many factors. Our approach consists of a screening phase followed by a modeling phase. This sequential approach makes good use of the available resources. The second phase is optional and depends on the objectives of the experiment.

In both phases, optimization problems for the response variable can be formulated for the mean value or its variance. In the screening phase, optimal values are found from a finite list of treatment combinations at one or more vertices of the experimental space of critical factors. This is so because the assumed statistical model is first-order. In the modeling phase, optimal values are obtained as stationary points within the continuous experimental space because the model is second-order.

In Table 44.11, explicit recommendations are given for choosing a fractional design. The number of treatment combinations excludes repeat tests needed to estimate the experimental error. As a general rule, two or three replicates of all or some treatment combinations are recommended. This permits statistical testing of hypotheses for judging the significance of the effects.

TABLE 44.11
Choice of a Minimum Resolution IV Design with Factors at Two Levels Not Including Repeat Tests

Number of Factors	Number of Runs
2	4
3	8
4	8
5–7	16, or 2 reflected designs of 8[a]
8	16
9–15	32, or 2 reflected designs of 16[a]
16	32
17–31	64, or 2 reflected designs of 32[a]
32 or more	Reduce the number of factors

[a] The two designs are composed of a basic design and a reflected design; each is the reflection of the other.

44.2.9 Response Surface Designs

The main objective of response surface designs is to provide empirical maps showing how quantitative factors, under the experimenter's control, influence the response and locate optimal values. These designs are used when the critical factors have been identified. This is usually accomplished with a 2^{k-p} design. The number of critical quantitative factors is generally between two and five.

The first class of response surface designs is the central composite, where each factor takes five coded levels: $-x$, -1, 0, 1, x; the specific value of x depends on the number of input variables and other criteria. The design is a subset of a 5^k complete factorial. The second group of response surface designs is composed of the Box–Behnken designs where each factor is at three levels: −1, 0, 1.

A second-order polynomial is fitted to the response and is used to find the points in the factor space at which the response takes optimum (min, max) values. The polynomial is fitted by a regression approach, and a statistical software program is required for the analysis.

Tables 44.12 and 44.13 list some useful response surface designs. The reader will find additional response surface designs in Refs. [8] and [9]. The designs are in coded form and allow the possibility of blocking to reflect secondary background variables. The block variable is assumed not to interact with the main variables.

TABLE 44.12
Box–Behnken Designs for Three, Four, or Five Variables

Number of Variables	Design					Number of Points or Replicates	Blocking
	x_1	x_2	x_3	x_4	x_5		
3	±1	±1	0			4	Not possible
	±1	0	±1			4	
	0	±1	−1			4	
	0	0	0			3	
V4	±1	±1	0	0		4	Block 1
	0	0	±1	±1		4	
	0	0	0	0		1	
	±1	0	0	±1		4	Block 2
	0	±1	±1	0		4	
	0	0	0	0		1	
	±1	0	±1	0		4	Block 3
	0	±1	0	±1		4	
	0	0	0	0		1	
V5	±1	±1	0	0	0	4	Block 1
	0	0	±1	±1	0	4	
	0	±1	0	0	±1	4	
	±1	0	±1	0	0	4	
	0	0	0	±1	±1	4	
	0	0	0	0	0	3	
	0	±1	±1	0	0	4	Block 2
	±1	0	0	±1	0	4	
	0	0	±1	0	±1	4	
	±1	0	0	0	±1	4	
	0	±1	0	±1	0	4	
	0	0	0	0	0	3	

TABLE 44.13
Central Composite Designs for Two, Three, Four, or Five Variables

Two Variables in Two Blocks of Six Runs

Block	Run	x_1	x_2	Block	Run	x_1	x_2
1	1	−1	−1	2	7	−1.41	0
1	2	−1	1	2	8	1.41	0
1	3	1	−1	2	9	0	−1.41
1	4	1	1	2	10	0	1.41
1	5	0	0	2	11	0	0
1	6	0	0	2	12	0	0

Three Variables in Three Blocks of Six Runs

Block	Run	x_1	x_2	x_3
1	1	−1	−1	1
V1	2	1	−1	−1
1	3	−1	1	−1
1	4	1	1	1
1	5	0	0	0
1	6	0	0	0
2	7	−1	−1	−1
2	8	1	−1	1
2	9	−1	1	1
2	10	1	1	−1
2	11	0	0	0
2	12	0	0	0
3	13	−1.63	0	0
3	14	1.63	0	0
3	15	0	1.63	0
3	16	0	−1.63	0
3	17	0	0	0
3	18	0	0	0

Four Variables in Three Blocks of Ten Runs

Block	Run	x_1	x_2	x_3	x_4
1	1	−1	1	−1	−1
1	2	−1	−1	1	1
1	3	−1	1	−1	1
1	4	−1	1	1	−1
1	5	1	−1	−1	1
1	6	1	−1	1	−1
1	7	1	1	−1	−1
1	8	1	1	1	1
1	9	0	0	0	0
1	10	0	0	0	0

Four Variables in Three Blocks of Ten Runs

Block	Run	x_1	x_2	x_3	x_4
2	11	−1	−1	−1	1
2	12	−1	−1	1	−1
2	13	−1	1	−1	−1
2	14	−1	1	1	1
2	15	1	−1	−1	−1

(Continued)

TABLE 44.13 (*Continued*)
Central Composite Designs for Two, Three, Four, or Five Variables

Block	Run	x_1	x_2	x_3	x_4
2	16	1	−1	1	1
2	17	1	1	−1	1
2	18	1	1	1	−1
2	19	0	0	0	0
2	20	0	0	0	0
3	21	−2	0	0	0
3	22	2	0	0	0
3	23	0	−2	0	0
3	24	0	2	0	0
3	25	0	0	−2	0
3	26	0	0	2	0
3	27	0	0	0	−2
3	28	0	0	0	2
3	29	0	0	0	0
3	30	0	0	0	0

Five Variables in Three Blocks

Block	Run	x_1	x_2	x_3	x_4	x_5
1	1	−1	−1	−1	−1	1
1	2	−1	−1	−1	1	−1
1	3	−1	−1	1	−1	−1
1	4	−1	−1	1	1	1
1	5	−1	1	−1	−1	−1
1	6	−1	1	−1	1	1
1	7	−1	1	1	−1	1
1	8	−1	1	1	1	−1
1	9	1	−1	−1	−1	−1
1	10	1	−1	−1	1	1
1	11	11	−1	1	−1	1
1	12	1	−1	1	1	−1
1	13	1	1	−1	−1	1
1	14	1	1	−1	1	−1
1	15	1	1	1	−1	−1
1	16	1	1	1	1	1
1	17	0	0	0	0	0
1	18	0	0	0	0	0
1	19	0	0	0	0	0

Five Variables in Three Blocks

Block	Run	x_1	x_2	x_3	x_4	x_5
2	20	−1	−1	−1	−1	−1
2	21	−1	−1	−1	1	1
2	22	−1	−1	1	−1	1
2	23	−1	−1	1	1	−1
2	24	−1	1	−1	−1	1
2	25	−1	1	−1	1	−1
2	26	−1	1	1	−1	−1
2	27	−1	1	1	1	1

(*Continued*)

TABLE 44.13 (*Continued*)
Central Composite Designs for Two, Three, Four, or Five Variables

Block	Run	x_1	x_2	x_3	x_4	x_5
2	28	1	−1	−1	−1	1
2	29	1	−1	−1	1	−1
2	30	1	−1	1	−1	−1
2	31	1	−1	1	1	1
2	32	1	1	−1	−1	−1
2	33	1	1	−1	1	1
2	34	1	1	1	−1	1
2	35	1	1	1	1	−1
2	36	0	0	0	0	0
2	37	0	0	0	0	0
2	38	0	0	0	0	0
3	39	−2.15	0	0	0	0
3	40	2.15	0	0	0	0
3	41	0	−2.15	0	0	0
3	42	0	2.15	0	0	0
3	43	0	0	−2.15	0	0
3	44	0	0	2.15	0	0
3	45	0	0	0	−2.15	0
3	46	0	0	0	2.15	0
3	47	0	0	0	0	−2.15
3	48	0	0	0	0	2.15
3	49	0	0	0	0	0

44.3 STATISTICAL ANALYSIS

44.3.1 Statistical Models

A *statistical model* is an equation together with additional assumptions that link a response (or dependent) variable with a set of explanatory (or independent) variables. Statistical models can be classified in many ways depending on the link function and the nature of the explanatory variables. A model is said to be linear if the link function is linear in the unknown coefficients of the model. This class covers the vast majority of applicable statistical models. Models can also be classified with respect to the type of explanatory variables. A *regression model* is one in which all the variables are quantitative. If all the variables are qualitative, the model is called an *analysis of variance (ANOVA) model.* An analysis of variance model can be transformed into a regression model by introducing dichotomous 0–1 variables with each value (level) in all qualitative variables. The model can also be classified as fixed effect or random if the explanatory variables are fixed or random. We shall deal only with fixed explanatory variables.

The general form of all statistical models is

$$y = f\left(x_A, x_B, \ldots; \beta_0, \beta_1, \ldots\right) + \varepsilon \tag{44.1}$$

where f is the link function of the coded variables x_A, x_B, … of the controllable factors. The function has a known form but unknown coefficients β_0, β_1, …. It represents the mean value of the response y over varying conditions of the factors. The other component ε is called the experimental error and is assumed to have a zero mean value and a constant variance σ^2. The error component represents the effect of all the other factors that are not controllable, including the measurement error. The error is characterized by the variance parameter σ^2.

In the remainder of Section 44.3, we restrict the statistical analysis to two-level fractional designs. The computations are much simpler because of the design structure and the associated statistical models. The main steps of the analysis are illustrated with the 16-run design defined in Tables 44.9 and 44.10. The basic ideas and computations are directly applicable to all 2^{k-p} designs. The statistical model for the 16-run design given in Table 44.9 can be written as

$$y = \beta_0 + \beta_1 x_1 + \beta_2 x_2 + \ldots + \beta_{15} x_{15} + \varepsilon \tag{44.2}$$

where y is the response variable; x_1, x_2, …, x_{15} are the coded variables (±1) defined by the columns of Table 44.9; β_1, …, β_{15} are the coefficients and are called main or interaction halfeffects; ε is the error component.

Equation (44.2) can be exploited with 4–15 factors, and Table 44.10 is needed to give the specific meaning of the coefficients, β_j and the variables x_j. In Table 44.14, we show three particular cases of the model with four, eight, and ten factors. The column variables are written directly with the labels of the factors instead of the coded variables.

The steps in the analysis of all statistical models are

1. Specification of the model
2. Estimation of the parameters
3. Computation of the analysis of variance (ANOVA) table
4. Testing of hypotheses and computation of confidence intervals
5. Model checking
6. Displaying the results.

A regression analysis approach can be used with all statistical models. The subject of regression analysis has many theoretical and practical aspects. One cannot hope to cover this important

TABLE 44.14
Interpretation of the Model with Four, Eight, and Ten Factors

Term in the Model	Four Factors (A, B, D, G)	Eight Factors (A B, D, G, H, K, M, N)	Ten Factors (A, B, C, D, E, G, H, K, M, N)
x_1	A	A	A − (BC+DE)
x_2	B	B	B −− (AC+EG)
x_3	−AB	− (AB+DG+HK+MN)	C − (AB+DG+HK+MN)
x_4	D	D	D − (AE+CG)
x_5	−AD	− (AD+BG+HM+KN)	E − (AD+BG+HM+KN)
x_6	−BD	− (AG+BD+HN+KM)	− (AG+BD+CE+HN+KM)
x_7	ABD	G	G−(BE+CD)
x_8	H	H	H − (CK+EM)
x_9	−AH	− (AH+BK+DM+GN)	− (AH+BK+DM+GN)
x_{10}	−BH	− (AK+BH+DN+GM)	− (AK + BH+DN+GM)
x_{11}	ABH	K	K − (CH+EN)
x_{12}	−DH	− (AM+BN+DH+GK)	− (AM+BN+DH + GK)
x_{13}	ADH	M	M − (CN + EH)
x_{14}	BDH	N	N − (CM+EK)
x_{15}	−ABDH	− (AN+BM + DK+GH)	− (AN+BM+DK+GH)

statistical method in any great detail within this chapter. The reader can find additional information in some of the references [e.g., 10,11]. Moreover, a good commercial statistical software package is essential to perform the analysis, even with a small set of observations.

Regression analysis can be used with data from observational studies or with data from designed experiments. It is the method of choice in a case where observations involve multiple variables at least one of which is identified as the response variable. A closely related method is the ANOVA decomposition of the response in terms of the different components of the model and the experimental error.

44.3.2 Estimation

44.3.2.1 Effects Estimates and Estimated Model

Let x_{ij} denote the i row (trial), $i=1, 2, \ldots, 16$ of column variable x_j, $j=1, 2, \ldots, 15$, in Table 44.9. These quantities have the following properties:

$$x_{ij} = \pm 1 \quad \sum_{i=1}^{16} x_{ij} = 0, \quad \sum_{i=1}^{16} x_{ij}^2 = 16$$

$$\sum_{i=1}^{16} x_{ij} x_{ij'} = 0, \qquad j, j' = 1, 2, \ldots, 16; j \neq j' \tag{44.3}$$

Assume that we have an equal number n of repeat measurements $y_{i1}, y_{i2}, \ldots, y_{in}$ of the response at each trial, $i=1, \ldots, 16$. Let

$$\overline{y}_i = \frac{1}{n} \sum_{\alpha=1}^{n} y_{i\alpha} = \text{mean value at the } i\text{th trial} \tag{44.4a}$$

$$y_i = \frac{1}{16} \sum_{i=1}^{16} \overline{y}_i = \text{overall mean} \tag{44.4b}$$

$$s_i^2 = \frac{1}{n-1} \sum_{\alpha=1}^{n} \left(y_{i\alpha} - \overline{y}_i \right)^2 = \text{variance at the } i\text{th trial} \tag{44.4c}$$

Because the model matrix (Table 44.9) is orthogonal, the least squares estimates $\hat{\beta}_j$ of the model parameters (half-effects) are easily expressed as

$$\hat{\beta}_j = \frac{1}{16} \sum_{i=1}^{16} x_{ij} \overline{y}_i, \qquad j = 1, 2, \ldots, 15 \tag{44.5a}$$

$$\hat{\beta}_0 = \overline{\overline{y}} \tag{44.5b}$$

The estimated model $\hat{y}$ is

$$\hat{y} = \hat{\beta}_0 + \hat{\beta}_1 x_1 + \hat{\beta}_2 x_2 + \ldots + \hat{\beta}_{15} x_{15} \tag{44.6a}$$

and the predicted value of $\hat{y}_i$ at the ith trial becomes

$$\hat{y}_i = \hat{\beta}_0 + \hat{p}_1 x_{i1} + \hat{\beta}_2 x_{i2} + \ldots + \hat{\beta}_{15} x_{i,15} \tag{44.6b}$$

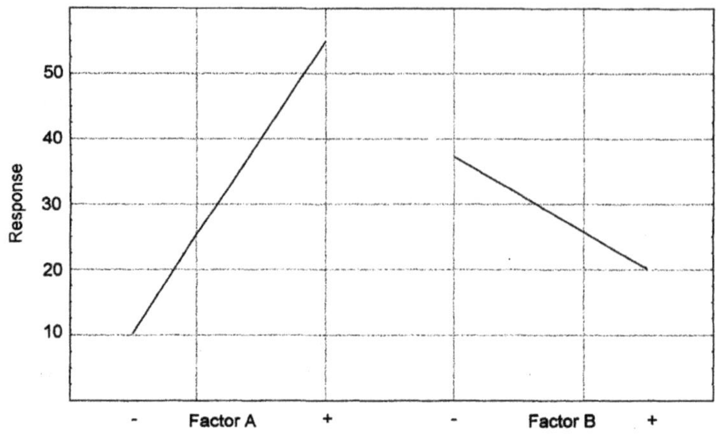

FIGURE 44.4 Main effect plot.

44.3.2.2 Error Estimate

The experimental error is estimated with $\hat{\sigma}^2$:

$$\hat{\sigma}^2 = \frac{1}{16}\sum_{i=1}^{16} s_i^2 = \frac{1}{16(n-1)}\sum_{i=1}^{16}\sum_{\alpha=1}^{n}(y_{i\alpha} - \bar{y}_i)^2 \tag{44.7}$$

This estimate is based on replicates and the assumption of constant variance over the experimental space. If no replicates are available, the error can be estimated only by making some special assumptions on the model. For example, high-order interactions are often used to obtain an error estimate. These estimates are computed with the residuals $y_i - \hat{y}_i$ and their sum of squares.

44.3.2.3 Interpretation of Main Effects

The main effect is the change in the response that occurs when the factor goes from − to +. The main effect is equivalent to twice the slope as the coded variable changes from −1 to 1. (See Figure 44.4.)

The corresponding half-effect is the beta coefficient in the model, and it measures the change of the response as the factor-coded variable goes from 0 to 1.

44.3.2.4 Interpretation of Two-Factor Interaction Effect

A two-factor interaction effect, say *AB*, measures the change in the response when both factors are varied. The + sign in the *AB* column applies when both factors are (−, −) or (+, +). The − sign is associated with (+, −) or (−, +). The interaction measures a differential effect of one factor when a second factor is varied. The best way to understand an interaction effect is with a graph of the response as shown in Figure 44.5.

Numerically, the interaction is a measure of nonparallelism of the two lines. In other words, it is the difference between the mean value of the response at points 1 and 2 and the mean value of the response at points 3 and 4. If the interaction is zero, then the lines are parallel. A strong interaction effect is present when one line has a positive slope and the other has a negative slope.

44.3.2.5 Interaction of Higher Interaction Effects

Higher interaction effects are much more difficult to interpret. The best way to understand them is with a graph. For example, a third-order interaction effect *ABC* can be represented as shown in Figure 44.6 by showing the value of the response for the eight combinations (±, ±, ±) of factors *A*, *B*, and C.

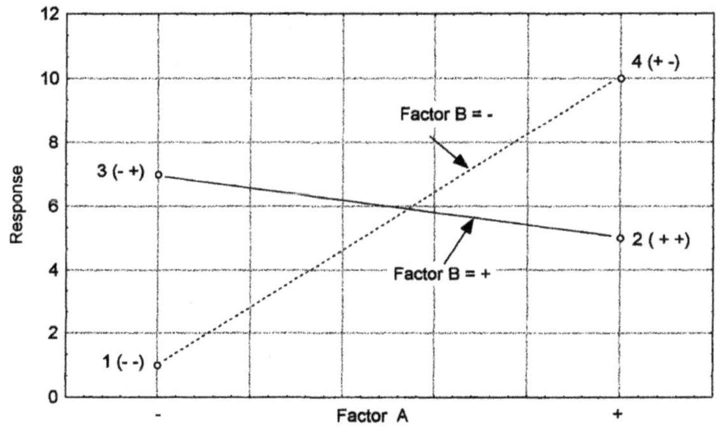

FIGURE 44.5 *AB* interaction plot.

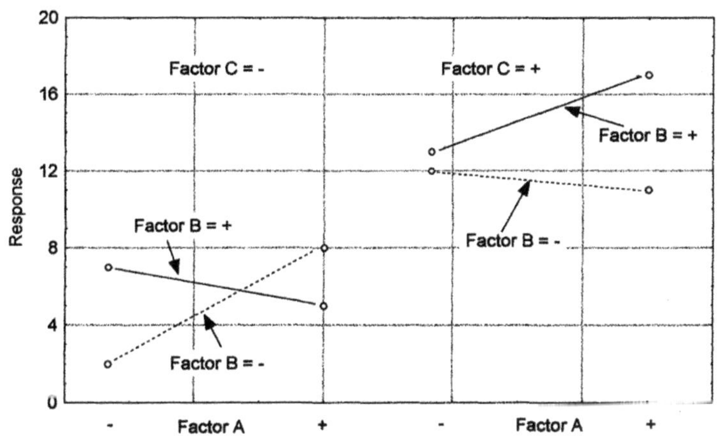

FIGURE 44.6 *ABC* interaction plot.

44.3.3 Analysis of Variance

44.3.3.1 Basic Decomposition

The idea is to decompose the total sum-of-squares variation of the response into various component sums of squares of the different terms of the model. We again restrict ourselves to the case of the 16-run design (Tables 44.9 and 44.10).

The analysis of variance decomposition equation is

$$\sum_{i=1}^{16}\sum_{\alpha=1}^{n}\left(y_{i\alpha}-\bar{\bar{y}}\right)^2 = n\sum_{i=1}^{16}\left(\hat{y}_i-\bar{\bar{y}}\right)^2+\sum_{i=1}^{16}\sum_{\alpha=1}^{n}\left(y_{i\alpha}-\bar{y}_i\right)^2 \tag{44.8a}$$

$$SS_{TOT} = SS_{MOD} + SS_{REP} \tag{44.8b}$$

where

$y_{i\alpha}$ is the observed value of y, $i=1, 2, \ldots, 16$; $\alpha=1, 2, \ldots, n$
$\bar{y}_i$ is the mean value of y at the ith trial [Eq. (44.4a)]

$\bar{\bar{y}}$ is the overall mean of y of all observations [Eq. (44.4b)]
$\hat{y}_i$ is the predicted value of y at the ith trial [Eq. (44.6b)]

and

$$SS_{TOT} = \sum_{i=1}^{16}\sum_{\alpha=1}^{n}\left(y_{i\alpha} - \bar{\bar{y}}\right)^2 \tag{44.9a}$$

is the *total sum-of-squares variation* based on $16n-1$ degrees of freedom.

$$SS_{MOD} = n\sum_{i=1}^{16}\left(\hat{y}_i - \bar{\bar{y}}\right)^2 \tag{44.9b}$$

is the *model* (or *explained) variation* based on 15 degrees of freedom.

$$SS_{REP} = \sum_{i=1}^{16}\sum_{\alpha=1}^{n}\left(y_{i\alpha} - \bar{y}_i\right)^2 = 16(n-1)\hat{\sigma}^2 \tag{44.9c}$$

is the *within variation* of replication based on $16(n-1)$ degrees of freedom.
If $n=1$ (no repeats), then $\hat{y}_i = y_i$ and Eq. (44.8) is reduced to

$$\sum_{i=1}^{16}\left(y_i - \bar{y}\right)^2 = \sum_{i=1}^{16}\left(\hat{y}_i - \bar{y}\right)^2 \tag{44.10}$$

and the estimate of experimental error is not available. The only way to estimate the error is to use a modified model containing a reduced number of effects (parameters).

3.3.2 Decomposition of SS_{MOD}

The model sum of squares SS_{MOD} can be further decomposed using the orthogonal property of the model:

$$SS_{MOD} = \sum_{j=1}^{15} SS(x_j) \tag{44.11}$$

where

$$SS(x_j) = SS\left(\hat{\beta}_j\right) = 16n\left(\hat{\beta}_j\right)^2, \; j = 1, 2, \ldots, 15 \tag{44.12}$$

is the explained variation of the jth column effect.

44.3.3.3 Analysis of Variance Table

Combining Eqs. (44.8), (44.11), and (44.12), we have

$$\sum_{i=1}^{16}\sum_{\alpha=1}^{n}\left(y_{i\alpha} - \bar{\bar{y}}\right)^2 = 16n\sum_{j=1}^{15}\hat{\beta}_j^2 + 16(n-1)\hat{\sigma}^2 \tag{44.13}$$

and all the information can be summarized into an analysis of variance table (Table 44.15).
The F ratio is used for testing the statistical significance of effects.

TABLE 44.15
Analysis of Variance

Source of Variation	Sum of Squares	Degrees of Freedom	Mean Square	*F* Ratio
Model	SS_{MOD}	15	$SS_{MOD}/15$	$SS_{MOD}/15\hat{\sigma}$
x_1	SS (x_1)	1	SS (x_1)	$SS(X_1)\hat{\sigma}^2$
x_2	SS (x_2)	1	SS (x_2)	$SS(X_2)\hat{\sigma}^2$
:	:	:	:	:
x_{15}	SS (x_{15})	1	SS (x_{15})	$SS(X_{15})/\hat{\sigma}^2$
Error	SS_{REP}	$16\,(n-1)$	$\hat{\sigma}^2$	—
Total	SS_{TOT}	$16\,n-1$	—	—

44.3.4 Statistical Tests

44.3.4.1 Overall Test

The overall test of the statistical significance of the model is to reject the null hypothesis

$$H_0 : \beta_1 = \beta_2 = \ldots = \beta_{15} = 0 \text{ (no effect)}$$

at the α significance level ($0<\alpha<1$) if

$$\frac{SS_{MOD}\,/\,15}{\hat{\sigma}^2} > F_{15,16(n-1),\alpha} \tag{44.14}$$

where $F_{15,16n-1,\alpha}$ is the $(1-\alpha)$th quantile of the F distribution with 15 degrees of freedom in the numerator and $16(n-1)$ degrees of freedom in the denominator. The α level is generally chosen in the range $0<\alpha<0.10$.

44.3.4.2 Test of Individual Effect

The test of the statistical significance of an individual effect β_j is to reject the hypothesis

$$H_0 : \beta_j = 0$$

at the α level if

$$\frac{SS\left(\hat{\beta}_j\right)}{\hat{\sigma}^2} > F_{1,16(n-1),\alpha} \tag{44.15}$$

when $F_{1,16(n-1)\alpha}$ is the $(1-\alpha)$th quantile of the F distribution with one degree of freedom in the numerator and $16(n-1)$ degrees of freedom in the denominator.

44.3.4.3 Judging the Effects When $n=1$

When $n=1$, the estimate of the experimental error $\hat{\sigma}^2$ is not available, because the within sum of squares is not defined [Eq. (44.9c)]. To judge the effects, a useful method proposed by C. Daniel is to plot the estimated effects on a half-normal probability graph. The technique, preferably carried out with software, is based on the following computations.

Let $\hat{\beta}_1, \hat{\beta}_2, \ldots, \hat{\beta}_{15}$ be the estimated half-effects of the fitted model and $\left|\hat{\beta}\right|_{(j)}$ be the absolute values ranked in increasing order:

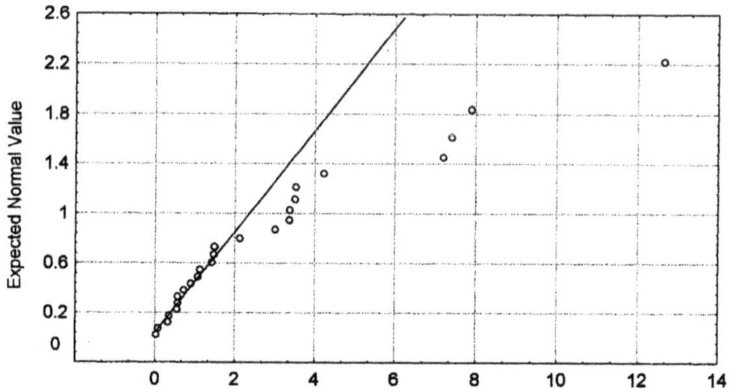

FIGURE 44.7 Half-normal probability plot.

$$\left|\hat{\beta}\right|_{(1)} \leq \left|\hat{\beta}\right|_{(2)} \leq \ldots \leq \left|\hat{\beta}\right|_{(15)} \tag{44.16a}$$

Let

$$q_j = \frac{j-0.5}{30} + 0.5, \quad j = 1, 2, \ldots, 15 \tag{44.16b}$$

$$z_j = 4.91\left[q_j^{0.14} - (1-q)^{0.14}\right] \tag{44.16c}$$

The z_j values are the q_jth quantile of a half-normal distribution (absolute value of a normal distribution). Equation (44.16c) is a good approximation of the inverse of the cumulative function of a normal distribution and does not require a table look-up. The half-normal plot is the graph of $(\left|\hat{\beta}\right|_{(j)}, z_j)$ in Figure 44.7.

A line through the origin is drawn to encompass "small" effects. The remaining outlier points are considered "important." The underlying idea of this procedure is the assumption that if the model contains no real effects, the estimates' behavior can be considered as coming from a normal distribution and all the estimates would fall near a line.

44.3.4.4 Explanatory Indicators

As an aid to interpreting the model and estimated coefficients, we can compute several quantities,

$$R^2 = \frac{SS_{MOD}}{SS_{TOT}} \tag{44.17}$$

measuring the overall fraction of the variation explained by all the combined effects and

$$R^2(x_j) = \frac{SS(\hat{\beta}_j)}{SS_{MOD}}, \quad i = 1, 2, \ldots, 15 \tag{44.18}$$

measuring the explanatory power of the jth effect in the model.

A very useful plot closely related to the half-normal probability plot is a graph of the $R^2(x_j)$ values in decreasing order called a scree plot (or Pareto plot) (Figure 44.8).

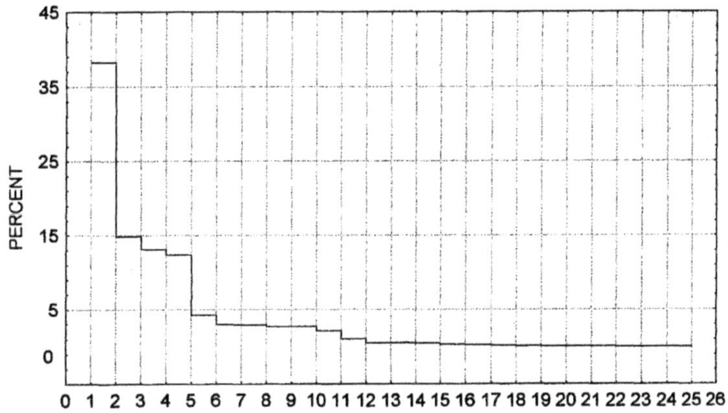

FIGURE 44.8 Pareto plot of effects.

44.3.4.5 Interpretation Principles

The researcher can use three interpretation principles in conjunction with the statistical analysis:

1. The principle of *effect parsity* says that out of many effects, only a handful will emerge as important. This is a form of the Pareto principle ("vital few, trivial many") that is applicable in experimental design.
2. The principle of *effect heredity* says that an important interaction effect will involve at least one factor that has a strong main effect.
3. The principle of *effect hierarchy* says that high-order interaction effects are much less likely than main effects and two-factor interactions.

Beyond these principles is the distinction between statistical significance and practical significance. A statistically significant effect is not important in practice if its influence on the response variable is relatively small [Eq. (44.18)] compared with the other statistically significant effects.

44.3.4.6 Model Simplification

All the previous statistical analysis and graphing procedures should help the researcher identify the critical factors from the initial set. In general, a relatively small number of important factors will emerge. A factor will be judged important if it has a large main effect or is involved in an important interaction effect. These factors are the basis for interpreting the response variable.

Consequently, a simplified model can be written down without any additional computations. As an example, suppose a 16-run experiment is performed with eight factors *A, B, D, G, H, K, M, N* (Table 44.11) and the resulting analysis shows $\hat{\beta}_1, \hat{\beta}_2, \hat{\beta}_3, \hat{\beta}_{15}$ to be practically and statistically dominant. The interpretations of these half-effects are (Table 44.11) as follows:

$\hat{\beta}_1$: main effect of A
$\hat{\beta}_2$: main effect of B
$\hat{\beta}_3$: interaction effect – $(AB+DG+HK+MN)$
$\hat{\beta}_{15}$: interaction effect – $(AN+BM+DK+GH)$.

Since *D, G, H, K, M,* and *N* have no significant main effects, we can drop their two-factor interactions *DG, HK, MN, DK, GH* in $\hat{\beta}_3$ and $\hat{\beta}_{15}$. The other two-factor interactions, *AB, AN,* and *BM*, are

TABLE 44.16
Computations of Effects of 10 Factors in Two Reflected Designs of 16 Runs

Column	Basic Design Effect (1)	Reflected Design Effect (2)	Main Effect, $\frac{(1)+(2)}{2}$	Interaction Effect, $\frac{(2)-(1)}{2}$
1	$A-(BC+DE)$	$A+(BC+DE)$	A	$BC+DE$
2	$B-(AC+EG)$	$B+(AC+EG)$	B	$AC+EG$
3	$C-(AB+DG+HK+MN)$	$C+(AB+DG+HK+MN)$	C	$AB+DG+HK+MN$
4	$D-(AE+CG)$	$D+(AE+CG)$	D	$AE+CG$
5	$E-(AD+BG+HM+KN)$	$E+(AD+BG+HM+KN)$	E	$AD+BG+HM+KN$
6	$-(AG+BD+CE+HN+KM)$	$+(AG+BD+CE+HN+KM)$	—	$AG+BD+CE+HN+KM$
7	$G-(BE+CD)$	$G+(BE+CD)$	G	$BE+CD$
8	$H-(CK+EM)$	$H-(CK+EM)$	H	$CK+EM$
9	$-(AH+BK+DM+GN)$	$(AH+BK+DM+GN)$	—	$AH+BK+DM+GN$
10	$-(AK+BH+DN+GM)$	$(AK+BH+DN+GM)$	—	$AK+BH+DN+GM$
11	$K-(CH+EN)$	$K+(CH+EN)$	K	$CH+EN$
12	$-(AM+BN+DH+GK)$	$(AM+BN+DH+GK)$	—	$AM+BN+DH+GK$
13	$M-(CN+EH)$	$M+(CN+EH)$	M	$CN+EH$
14	$N-(CM+EK)$	$N+(CM+EK)$	N	$CM+EK$
15	$-(AN+BM+DK+GH)$	$(AN+BM+DK+GH)$	—	$AN+BM+DK+GHH$

kept because A and B have main effects. The critical factors are therefore A, B, N, and M, since N and M are involved in interaction effects. We would ideally keep their main effects $(\hat{\beta}_{13}, \hat{\beta}_{14})$ in the reduced model:

$$y = \hat{\beta}_0 + \hat{\beta}_1 x_A + \hat{\beta}_2 x_B + \hat{\beta}_3(-x_A x_B) + \hat{\beta}_{13} x_M + \hat{\beta}_{14} x_N + \hat{\beta}_{15}(-x_A x_N - x_B x_M) \tag{44.19}$$

where x_A, x_B, x_M, and x_N are the coded variables (±1) of the corresponding factors. Dropping the coefficients $\hat{\beta}_{13}$ and $\hat{\beta}_{14}$ would probably not reduce the explanatory power of model (44.19).

The final interpretation of the response variable would be made with the 16 treatment combinations (±, ±, ±, ±) of factors *A, B, M, and N.* Since the experiment was a 16-run one, it can be considered a complete factorial on these factors, with the other factors *D, G, H, K* serving the useful role of replications because their effects are of the same order of magnitude as the experimental error.

44.3.4.7 Combining Basic and Reflected Designs

When the initial basic design of 16 runs is used with 9–15 factors, we have a resolution III design. Our recommendation (Table 44.11) is to follow up with an experiment based on the reflected 16-trial design. The combined experiments will result in a global resolution of level IV.

Simple computations will permit us to separate the main effects and the chains of two-factor interactions illustrated in Table 44.16 with a ten-factor experiment. (See Table 44.11 for column interpretation.) The values in columns 1 and 2 are 2 $\hat{\beta}_j$, where $\hat{\beta}_j$ is defined by Eq. (44.5a).

44.3.5 Plots

A variety of plots are essential for communicating the results and for checking the model assumptions. Main effect and interaction effect plots were presented in earlier figures, and further examples are presented in the case study, Section 44.4.

44.3.5.1 Residual Plots

The residuals,

$$r_{i\alpha} = y_{i\alpha} - \hat{y}_i,\ i = 1, \ldots, 16; \alpha = 1,\ 2,\ \ldots,\ n \qquad (44.20)$$

should plot around a straight line on a normal probability scale. As an additional check, the scaled residuals, $r_{i\alpha} / \hat{\sigma}$, should have values between −2 and +2 for 95% of all observations. This simple graphical test is a check on the assumption that the experimental error is normally distributed with a constant variance. This should be true whether the predicted values are computed with the complete model or with a reduced model. See Figure 44.14 in Section 44.4 for an example.

44.3.5.2 Response Plot

The response plot is made with the different treatment combinations of the active factors and the corresponding reduced model. The different treatment combinations are put on the x-axis at equally distanced points, and the estimated response is put on the y-axis. The observed values of the response can be optionally superimposed on the plot. The different treatment combinations can be arranged in such a way that the response variable is increasing from one combination to the next. An example of a response plot is given in Figure 44.13 in Section 44.4.

44.3.5.3 Tukey Comparisons

Although distinct, the different treatment combinations of the active factors are not all statistically different. A test of the statistical significance between any two distinct combinations is obtained with the Tukey multiple differences post hoc comparison test. A minimum significant difference, denoted HSD, is defined by

$$\text{HSD} = g_{k,v,\alpha} \frac{\hat{\sigma}}{\sqrt{m}} \qquad (44.21)$$

where

k = number of group treatment combinations
v = degrees of freedom in estimating σ
α = significance level
m = number of observations in the group
$\hat{\sigma}$ = estimation of σ based on the within-group variation
$g_{k,v,\alpha}$ = Tukey tabulated constants [6].

Two distinct different group combinations l and l' are statistically significant if

$$\left|\hat{y}_l - \hat{y}_{l'}\right| > \text{HSD} \qquad (44.22)$$

where $\hat{y}_l$, $\hat{y}_{l'}$ are the predicted response values. The researcher can classify the treatment combinations in heterogeneous groups and establish trade-offs within and between these groups. The response plot can be enhanced by adding the HSD value to the graph.

44.4 A CASE STUDY IN BOLTING

44.4.1 The Problem

The problem of preload in bolting technology is a complex phenomenon. In a document prepared for the Bolting Technology Council, Bickford [12] lists more than 50 parameters (factors) that affect

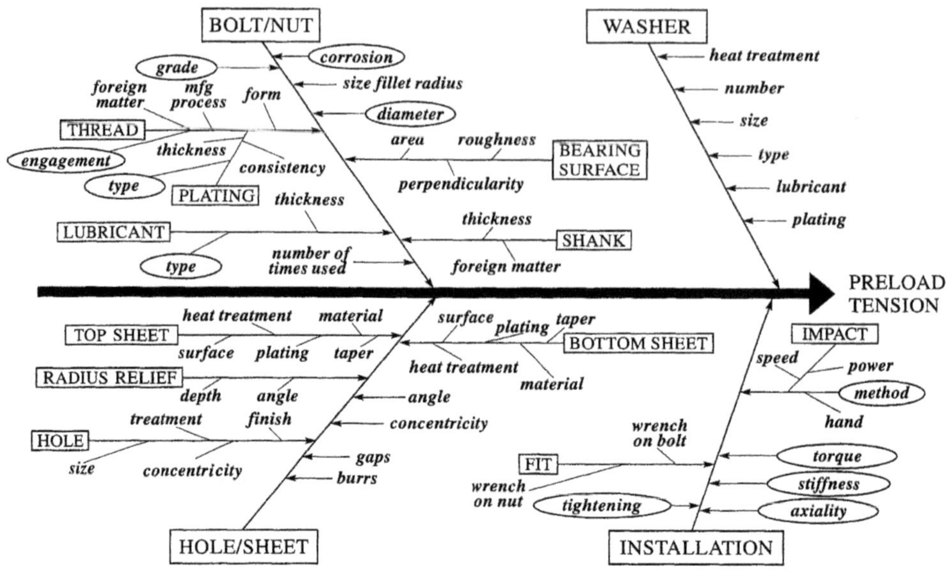

FIGURE 44.9 Ishikawa diagram.

the torque–tension (preload) relationship of a joint fastened together with threaded products. This information is shown in the Ishikawa diagram (Figure 44.9). The list includes such things as the bolt and joint materials involved; the hardness and surface roughness of each part; the type and amount of lubricant used; the grade, diameter, and corrosion of the bolt–nut combination; the installation method; etc. The assembly and behavior of a single-bolt joint is very often unpredictable, and designers have probably designed bolted joints much sturdier than would be required if we had a better understanding of bolted joints.

In 1988 and 1990, the Bolting Technology Council [12,13] sponsored two research projects [14,15]. In both projects, a designed experiment approach was used to increase the understanding of factors that can have a significant effect on the amount of preload achieved in a single bolt when we tighten it to a given torque.

In the first experiment, 11 factors (circled in Figure 44.9) were selected. Two (one basic, one reflected) 16-run fractional designs were used to gain an understanding of the problem.

In the second project, an essentially identical experiment was undertaken to confirm the results of the first project. Several minor changes were made, but it was decided to test two diameters (⅜ and ¾ in.) separately in two designed experiments with ten factors. This latter project is summarized in the rest of this chapter for the ⅜ in. bolt. Our aim is to illustrate an application of the methodology of designed experiments in bolting technology and to learn some new results about bolting.

44.4.2 The Experiment

The main features of the factors, levels, responses, materials, and test procedures are first briefly described.

44.4.2.1 Factors and Levels

Using expert judgment and taking into account the resources available, ten factors were selected as most promising, to study the torque–tension relationship (Figure 44.9). The factors are listed in Table 44.17 with their respective low (–) and high (+) levels. The labels are those suggested in Table 44.10.

TABLE 44.17
Factors and Levels

Label	Factor Name	Low Level (−)	High Level (+)	Note
A	Corrosion	As is	3.5% salt solution	1
B	Lubricant	As is	MoS_2	2
C	Axiality	5°	0°	3
D	Plating	As is	Zinc-coated	4
E	Torque	50% of table	75% of table	5
G	Tightenings	1	4	6
H	Grade of bolt	5	8	7
K	Stiffness of joint	Solid	Spring	8
M	Thread engagement	Full	Half	9
N	Installation method	Torque wrench	Power dyne	10

Notes:

1. (−) The fasteners were used in their "as-received" conditions.
   (+) The bolts were dipped in a solution containing 3.5% by weight of table salt in water; immersion time was 3 h for unplated bolts and 100 h for plated ones.
2. (−) Unlubricated or "as is."
   (+) A thin coating of molybdenum disulfide (MoS_2) was applied to the threaded end of the bolts. All the fasteners were cleaned and degreased prior to testing.
3. (−) A 5° misalignment using a wedge block.
   (+) No misalignment.
4. (−) Plain bolts, nuts, and washers.
   (+) Plated bolts, nuts, and washers.
5. Recommended seating torques (Industrial Products Division, SPS technologies):

	Low (−) 50%	High (+) 75%
For ⅜″–16 Grade 5, $T = 38$ ft-lb	19	28.5
For ⅜″–16 Grade 8, $T = 50$ ft-lb	25	37.5

6. (−) Full torque applied once.
   (+) Full torque applied four times with an interval of about 3 s in between.
7. (−) Grade 5 or A325.
   (+) Grade 8.
8. (−) With solid blocks.
   (+) With Belleville spring washers.
9. (−) Full thread engagement.
   (+) Half thread engagement in the nut.
10. (−) Click-type torque wrench.
    (+) PS602 Power Dyne.

44.4.2.2 Response Variable *Y*

The response variable is the preload measured in pounds but transformed into kilograms per square inch in all equations, tables, and graphs.

44.4.2.3 Materials and Method

A stack of steel blocks, solid washers, and Belleville springs, together with a load transducer, make up a fixture. The different components can be readily interchanged for each run. The head of the bolt fits in an appropriate slot at the base of the fixture, which is itself fastened to a heavy steel mount. This allows the operator to work unassisted during the torquing operation. All of the tests were performed by the same operator. Details of the experimental rig are given in Ref. [14].

44.4.2.4 Statistical Design

A complete factorial experiment with ten factors varying at two levels would require 1024 runs. Following the recommendations given in Table 44.11, a resolution IV statistical design can be achieved with a 32-run design or with two series (blocks) of 16 runs, defined respectively by a basic design and a reflected (foldover) design. The second alternative was chosen for managerial reasons and because a statistical analysis can be performed after each design. The two analyses are combined as in Table 44.16 to get the global result.

Three replicates of each run were made to get an estimate of the experimental error. Also, the replicates made it possible to investigate dispersion effects in addition to location effects.

The basic design was defined in Table 44.10 for the ten-factor case. The test combinations were defined with columns: 1 (*A*), 2 (*B*), 3 (*C*), 4 (*D*), 5 (*E*), 7 (*G*), 8 (*H*), 11 (*K*), 13 (*M*), 14 (*N*). Table 44.9 was then used to find the level (– or +) of each factor. The reflected design is obtained by a complete reversal of+and–signs in the basic design. All tests with the same level of torque were run in one batch because the adjustment of the torque was time-consuming. Therefore, the tests were not completely randomized.

44.4.2.5 Experimental Data

The resulting data are listed in Table 44.18 for 96 observations (16×2×3). The replicated values of the preload (in ksi) are denoted y_i, y_2, y_3; $\overline{y}$ and s are respectively the mean value of the response and the standard deviation of preload at each run.

44.4.3 Statistical Analysis

The models and analysis of variance are given in Tables 44.19 and 44.20. In Table 44.21, the main half-effects and the chains of two-factor interaction half effects are obtained by combining the two designs. (See Table 44.16.) The half-effects are ranked in decreasing order of absolute value and are presented in the Pareto diagram in Figure 44.10. The main effect plot (Figure 44.11) shows the dominant effect of torque as was expected. But the ranking of the other factors was unknown before the experiment. The last four factors—plating, tightening, stiffness of bolt, and corrosion—do not have any practical influence on the preload. In Figure 44.12, several interaction plots are presented. Note that thread engagement interacts with torque, grade of bolt, and installation method. Also, there is an interaction between lubricant and grade of bolt.

44.4.3.1 Fitted Models

Each estimated effect in Table 44.21 has a standard error of 0.68, so any effect greater than 2×0.68=1.36 is statistically significant at the 0.05 level. However, some of the effects have a much greater impact than others on the preload response. There are several statistical models that could be proposed, but the objective is to use a model that is based on the most important significant effects but simple enough for easy understanding. We can rely on the interpretation principles that were previously put forward for choosing a model. A closer look at Table 44.21 will help in that decision-making process.

Twenty-five effects were estimated: ten main effects and 15 chains of two-factor interactions. Two models are proposed with an increasing number of retained effects. Model 1 is based on the four most important effects, which happen to be main effects: torque (*E*), lubricant (*B*), grade (*H*), and installation method (*N*). The second model retains two additional main effects, thread engagement (*M*)

TABLE 44.18
Data for ⅜ in. Bolts

Run	Design	y_1	y_2	y_3	$\bar{y}$	s
1	Basic	31.35	36.13	33.03	33.51	2.42
2	Basic	24.26	30.71	25.29	26.75	3.47
3	Basic	58.97	44.90	58.84	54.24	8.08
4	Basic	50.45	52.26	55.23	52.65	2.41
5	Basic	74.84	63.48	68.39	68.90	5.69
6	Basic	67.74	70.84	75.10	71.23	3.69
7	Basic	53.81	68.13	62.58	61.51	7.22
8	Basic	29.03	27.61	23.61	26.75	2.81
9	Basic	48.39	49.42	54.45	50.75	3.24
10	Basic	66.71	74.97	76.90	72.86	5.41
11	Basic	44.52	48.52	49.81	47.61	2.76
12	Basic	64.00	55.74	54.71	58.15	5.09
13	Basic	19.61	21.42	21.03	20.69	0.95
14	Basic	49.68	41.16	45.16	45.33	4.26
15	Basic	32.90	49.81	47.48	43.40	9.16
16	Basic	61.94	55.48	64.52	60.65	4.65
1	Reflec.	82.84	87.74	80.52	83.70	3.69
2	Reflec.	49.81	58.19	54.45	54.15	4.20
3	Reflec.	23.35	23.23	18.19	21.59	2.94
4	Reflec.	48.65	56.90	57.03	54.19	4.81
5	Reflec.	28.52	28.65	29.55	28.90	0.56
6	Reflec.	32.77	34.97	36.13	34.62	1.70
7	Reflec.	51.61	50.71	57.16	53.16	3.49
8	Reflec.	92.00	93.68	72.26	85.98	11.91
9	Reflec.	33.16	44.39	40.77	39.44	5.73
10	Reflec.	14.84	16.77	11.61	14.41	2.61
11	Reflec.	39.87	39.48	35.48	38.28	2.43
12	Reflec.	46.97	52.90	45.81	48.56	3.81
13	Reflec.	114.58	119.74	111.10	115.14	4.35
14	Reflec.	53.16	55.48	57.16	55.27	2.01
15	Reflec.	45.68	41.94	46.45	44.69	2.42
16	Reflec.	42.97	45.16	50.45	46.19	3.85

TABLE 44.19
Model Estimates and Analysis of Variance for the Basic Design

Source of Variation	Beta Coeff.[a]	Sum of Squares	Degrees of Freedom	Mean Square	F Ratio	Sign.	R^2
Model	–	11,849.115	15	789.941	32.147	—[b]	0.938
x_0	49.685	—	—	—	—	—	%[c]
x_1	0.244	2.867	1	2.867	0.117		0.02
x_2	7.536	2,715.711	1	2,725.711	110.895	—[b]	23.00
x_3	–0.120	0.694	1	0.694	0.028		0.00
x_4	2.665	340.854	1	340.854	13.868	—[b]	2.88

(Continued)

TABLE 44.19 (*Continued*)
Model Estimates and Analysis of Variance for the Basic Design

Source of Variation	Beta Coeff.[a]	Sum of Squares	Degrees of Freedom	Mean Square	*F* Ratio	Sign.	R^2
x_5	9.648	4,467.757	1	4,467.759	181.769	—[b]	37.70
x_6	−1.589	121.126	1	121.126	4.928	—[b]	1.02
x_7	0.933	41.757	1	41.757	1.699		0.00
x_8	3.181	485.713	1	485.713	19.761	—[b]	4.01
x_9	0.809	31.412	1	31.412	1.278		0.00
x_{10}	−2.901	403.854	1	403.854	16.431	—[b]	3.41
x_{11}	2.379	271.748	1	271.748	11.056	—[b]	2.29
x_{12}	0.928	41.311	1	41.311	1.681		0.00
x_{13}	2.084	208.458	1	208.458	8.481	—[b]	1.76
x_{14}	7.206	2,491.786	1	2,492.786	101.419	—[b]	21.04
x_{15}	2.110	213.659	1	213.659	8.692	—[b]	1.80
Error —		786.536	32	24.529	—		0.062
Total —		12,636.452	47	—	—	—	—

[a] Each beta coefficient is estimated with a standard error of 0.715.
[b] Significant at level $\alpha = 0.05$.
[c] Percent of SS_{MOD}.

TABLE 44.20
Model Estimates and Analysis of Variance for the Reflected Design

Source of Variation	Beta Coeff.[a]	Sum of Squares	Degree of Freedom	Mean Square	*F* Ratio	Sign.	R^2
Model	—	29,321.673	15	1,954.778	96.548	—[b]	0.978
x_0	51.142	—	—	—	—	—	%[c]
x_1	0.896	38.503	1	38.503	1.90	—	1.31
x_2	8.223	3,245.750	1	3,245.750	160.38	—[b]	11.67
x_3	−6.853	2,254.061	1	2,254.061	111.38	—[b]	7.69
x_4	1.573	118.724	1	118.724	5.87	—[b]	0.40
x_5	15.637	11,736.577	1	11,736.577	579.93	—[b]	40.03
x_6	−3.755	676.726	1	676.726	33.44	—[b]	2.31
x_7	2.062	204.064	1	204.063	10.08	—[b]	0.01
x_8	11.622	6,483.263	1	6,483.263	320.35	—[b]	22.11
x_9	−0.465	10.369	1	10.369	0.51		0.00
x_{10}	3.819	700.206	1	700.206	34.60	—[b]	2.39
x_{11}	0.562	15.154	1	15.154	0.75		0.00
x_{12}	−0.509	12.414	1	12.414	0.61		0.00
x_{13}	4.944	1,173.251	1	1,173.251	59.97	—[b]	4.00
x_{14}	−7.169	2,467.197	1	2,467.197	121.91	—[b]	8.41
x_{15}	1.971	186.402	1	186.402	9.21	—[b]	0.64
Error		647.616	32	20.238	—	—	0.022
Total		29,970.27	47	—	—	—	—

[a] Each beta coefficient is estimated with a standard error of 0.649.
[b] Significant at level $\alpha = 0.05$.
[c] Percent of SS_{MOD}.

TABLE 44.21
Half–Effects Estimates Ranked in Absolute Value

Rank	Main/Interaction	Half-Effect	Interaction Guess
1	*E*: Torque	12.642	
2	*B*: Lubricant	7.879	
3	*H*: Grade	7.401	
4	*N*: Installation method	7.188	
5	$CK+EM$	4.220	EM
6	*M*: thread engagement	3.514	
7	$AB+DG-HK+MN$	3.486	MN
8	$-AK+BH-DN+GM$	3.366	BH
9	*C*: Axiality	3.360	
10	$AD+BG+HM+KN$	2.995	HM
11	*D*: plating	2.672	
12	G: tightening	1.497	
13	$-CH-EN$	1.471	
14	$CN+EH$	1.430	
15	$AH-BK+DM-GN$	−1.135	
16	$AG+BD-CE-HN-KM$	−1.085	
17	*K*: Stiffness	0.910	
18	$AM-BN+DH-GK$	0.720	
19	*A*: corrosion	0.570	
20	$BE-CD$	0.565	
21	$AC-CG$	−0.545	
22	$-AC+EG$	0.345	
23	$-BC+DE$	0.325	
24	$-AN+BM-DK-GN$	−0.070	
25	$-CM-EK$	0.020	

Note: The levels of factors *C*, *K*, *N* were interchanged after the global analysis so that their main half-effects would be positive. Consequently, the two-factor interactions of *C*, *K*, *N* with all the others undergo a sign inversion. The interactions *CK*, *CN*, *KN* remain intact.

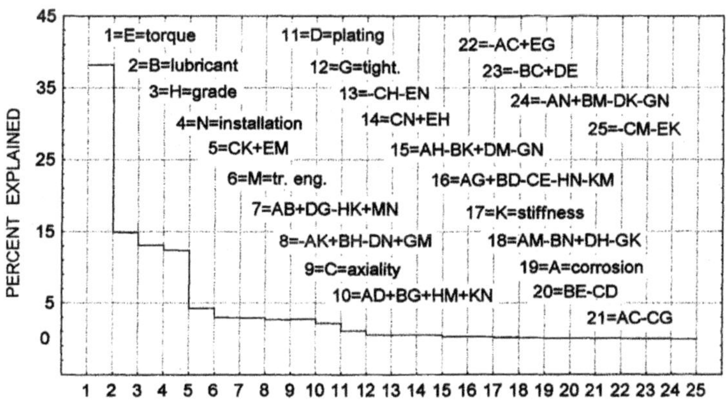

FIGURE 44.10 Pareto plot of effects.

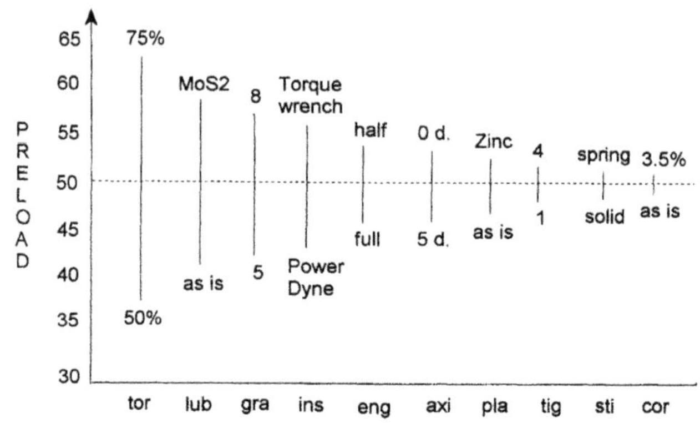

FIGURE 44.11 Main effects.

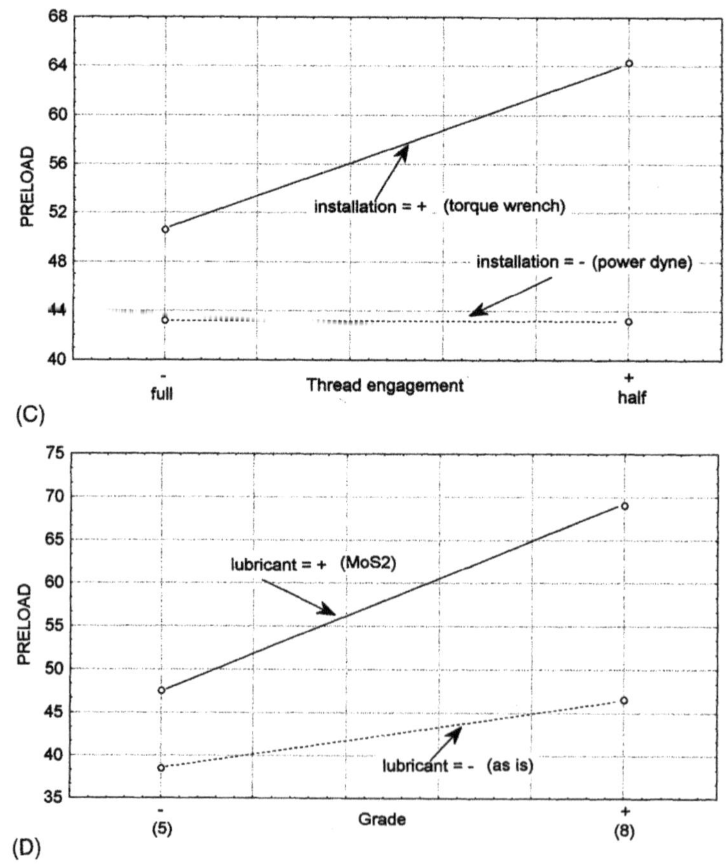

FIGURE 44.12 Interaction plots.

and axiality (C), plus the intermediate two-factor chain interaction effects. In these chains, the most likely interactions are those coming from the retained main effects: *EM, MN, BH*, and *HM*. The factors are coded with the first three letters of their names.

Model 1:

$$\text{Preload} = 50.41 + 12.64 \times \text{tor} + 7.88 \times \text{lub} + 7.40 \times \text{gra} + 7.19 \times \text{ins}$$
$$R^2 = 0.739;\ \text{adjusted } R^2 = 0.727 \tag{44.23}$$

Model 2:

$$\begin{aligned} \text{Perload} &= 50.41 + 12.64 \times \text{tor} + 7.88 \times \text{lub} + 7.40 \times \text{gra} + 7.19 \times \text{ins} \\ &\quad + 3.51 \times \text{eng} + 3.365 \times \text{axi} + 4.22 \times \text{tor} \times \text{eng} \\ &\quad + 3.49 \times \text{ins} \times \text{gra} + 3.36 \times \text{lub} \times \text{gra} + 2.995 \times \text{gra} \times \text{eng} \\ R^2 &= 0.905;\ \text{adjusted } R^2 = 0.894 \end{aligned} \tag{44.24}$$

44.4.3.2 Predicted Values

If we retain model 2, which is composed of six main effects and four two-factor interactions, we can predict the preload variable for each of the 64 combinations that they define. The predicted values from Eq. (44.24) are given in Table 44.22 and in the response plot in Figure 44.13. Note that 32 of the 64 treatment combinations were not run, so the observed preload is not available for those combinations. The normal probability plot of the residuals (Figure 44.14) shows no obvious deviation.

TABLE 44.22
Observed and Predicted Response

	Group	tor	Lub	gra	ins	eng	axi	Predicted	Observed
1	1	−1	−1	−1	−1	−1	−1	22.49	20.69
2	2	−1	−1	−1	−1	−1	1	29.23	—
3	3	−1	−1	−1	−1	1	−1	15.08	14.41
4	4	−1	−1	−1	−1	1	1	21.82	—
5	5	−1	−1	−1	1	−1	−1	29.89	—
6	6	−1	−1	−1	1	−1	1	36.62	33.51
7	7	−1	−1	−1	1	1	−1	22.49	—
8	8	−1	−1	−1	1	1	1	29.22	34.62
9	9	−1	−1	1	−1	−1	−1	17.59	—
10	10	−1	−1	1	−1	−1	1	24.32	28.90
11	11	−1	−1	1	−1	1	−1	22.17	—
12	12	−1	−1	1	−1	1	1	28.90	26.75
13	13	−1	−1	1	1	−1	−1	38.96	39.44
14	14	−1	−1	1	1	−1	1	45.69	—
15	15	−1	−1	1	1	1	−1	43.52	45.33
16	16	−1	−1	1	1	1	1	50.26	—
17	17	−1	1	−1	−1	−1	−1	31.53	26.75
18	18	−1	1	−1	−1	−1	1	38.26	—
19	19	−1	1	−1	−1	1	−1	24.12	21.59
20	20	−1	1	−1	−1	1	1	30.85	—

(Continued)

TABLE 44.22 (*Continued*)
Observed and Predicted Response

	Group	tor	Lub	gra	ins	eng	axi	Predicted	Observed
21	21	−1	1	−1	1	−1	−1	38.92	—
22	22	−1	1	−1	1	−1	1	45.65	58.15
23	23	−1	1	−1	1	1	−1	31.51	—
24	24	−1	1	−1	1	1	1	38.25	44.69
25	25	−1	1	1	−1	−1	−1	40.09	—
26	26	−1	1	1	−1	−1	1	46.82	46.19
27	27	−1	1	1	−1	1	−1	44.65	—
28	28	−1	1	1	−1	1	1	51.39	47.61
29	29	−1	1	1	1	−1	−1	61.45	54.19
30	30	−1	1	1	1	−1	1	68.17	—
31	31	−1	1	1	1	1	−1	66.02	61.51
32	32	−1	1	1	1	1	1	72.74	—
33	33	1	−1	−1	−1	−1	−1	39.34	—
34	34	1	−1	−1	−1	−1	1	46.07	53.16
35	35	1	−1	−1	−1	1	−1	48.81	—
36	36	1	−1	−1	−1	1	1	55.54	52.64
37	37	1	−1	−1	1	−1	−1	46.73	38.28
38	38	1	−1	−1	1	−1	1	53.47	—
39	39	1	−1	−1	1	1	−1	56.20	60.65
40	40	1	−1	−1	1	1	1	62.94	—
41	41	1	−1	1	−1	−1	−1	34.44	43.40
42	42	1	−1	1	−1	−1		41.17	—
43	43	1	−1	1	−1	1	−1	55.88	48.56
44	44	1	−1	1	−1	1	1	62.61	—
45	45	1	−1	1	1	−1	−1	55.79	—
46	46	1	−1	1	1	−1	1	62.52	54.24
47	47	1	−1	1	1	1	−1	77.25	—
48	48	1	−1	1	1	1	1	83.97	85.98
49	49	1	1	−1	−1	−1	−1	48.36	—
50	50	1	1	−1	−1	−1	1	55.10	55.27
51	51	1	1	−1	−1	1	−1	57.83	—
52	52	1	1	−1	−1	1	1	64.57	50.75
53	53	1	1	−1	1	−1	−1	55.76	54.15
54	54	1	1	−1	1	−1	1	62.49	—
55	55	1	1	−1	1	1	−1	65.24	68.90
56	56	1	1	−1	1	1	1	71.97	—
57	57	1	1	1	−1	−1	−1	56.93	71.23
58	58	1	1	1	−1	−1	1	63.65	—
59	59	1	1	1	−1	1	−1	78.38	83.70
60	60	1	1	1	−1	1	1	85.11	—
61	61	1	1	1	1	−1	−1	78.28	—
62	62	1	1	1	1	−1	1	85.01	72.86
63	63	1	1	1	1	1	−1	99.74	—
64	64	1	1	1	1	1	1	106.47	115.14

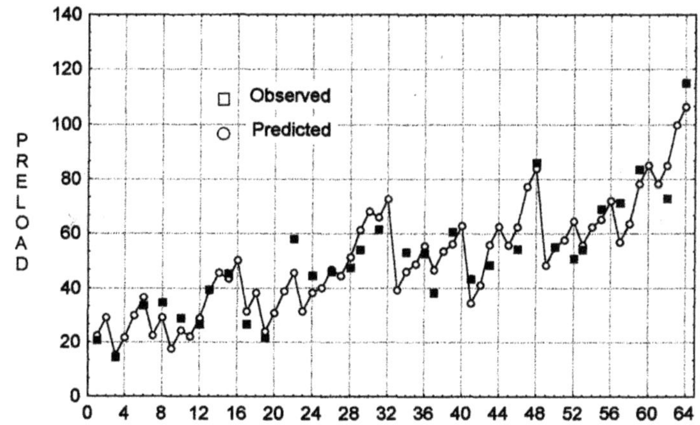

FIGURE 44.13 Response plot.

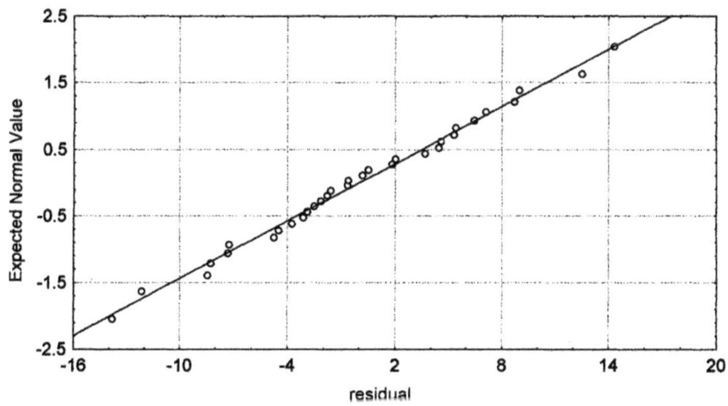

FIGURE 44.14 Normal probability plot of residuals.

44.4.3.3 Other Results

44.4.3.3.1 Dispersion Effects

The dispersion effects can be analyzed by considering the standard deviation of preload at each run as a second response variable. The critical factors, in order of importance, are grade of bolt, installation method, and torque.

44.4.3.3.2 3/8 in. Bolt

The ranked factors influencing the location effects are torque, grade of bolt, lubricant, axiality, and installation method. The thread engagement was not an important factor when compared to the ⅜ in. bolt. No factor showed any impact on the dispersion effects.

REFERENCES

1. Symposium on Taguchi Methods, 1987, Proceedings of the *Fifth Symposium on Taguchi Methods,* American Supplier Institute, Inc., Dearborn, MI.
2. Schmidt, S. R., and R. G. Launsby, *Understanding Industrial Designed Experiments*, 3rd ed., Air Academy Press, Colorado Springs, CO, 1995.

3. Hicks, C. R., *Fundamental Concepts in the Design of Experiments*, 4th ed., Holt, Rinehart, and Winston, New York, 1993.
4. Box, G. E. P., W. G. Hunter, and J. S. Hunter, *Statistics for Experimenters*, Wiley, New York, 1978.
5. Montgomery, D. C, *Design and Analysis of Experiments*, 3rd ed., Wiley, New York, 1991.
6. Wheeler, D. J., *Understanding Industrial Experimentation*, 2nd ed., SPC Press, Knoxville, TN, 1990.
7. Wheeler, D. J., *Tables of Screening Designs*, 2nd ed., SPC Press, Knoxville, TN, 1989.
8. Box, G. E. P., and N. R. Draper, *Empirical Model-Building and Response Surfaces*, Wiley, New York, 1987.
9. Myers, R. H., and D. C. Montgomery, *Response Surface Methodology: Process and Product Optimization Using Designed Experiments*, Wiley, New York, 1995.
10. Draper, N. R., and H. Smith, *Applied Regression Analysis*, 2nd ed., Wiley, New York, 1981.
11. Neter, J., N. Wasserman, and M. H. Kutner, *Applied Statistical Models*, 3rd ed., Irwin, Boston, MA, 1990.
12. Bickford, J. H., *Request for Proposal*, Bolting Technology Council, New York, 1987.
13. Bickford, J. H., *Bolting Technology Council's Preload Project*, Fastener Technology International, August 1993, pp. 60–61, New York..
14. Clément, B., and A. Bazergui, A Study of the Preload Relationship in Bolting Technology: Experimental Design and Analysis, Report submitted to the Bolting Technology Council, 1990, New York.
15. Bazergui, A., and B. Clément, Experimental Design and Analysis of Preload Response with a Set of 10 Controllable Parameters in Bolting Technology Council. Report submitted to the Bolting Technology Council, 1992, New York.

45 Case Study
Rupture of a Wind Turbine Composite Blade-to-Hub Bolted Joint

Jereme Kent, Chelsea Blub, Sherif Gindy, and Sayed A. Nassar

45.1 INTRODUCTION

This chapter outlines a case study of the sudden rupture of a flanged composite blade-to-steel joint during normal operation. The severed 43.5-m blade has a mass of 6,000 kg (i.e. static weight=58.8 kN); the blade fell to the ground next to the base of a wind turbine tower without causing any fatality, injury, or damage to other structures. The entire wind farm was completely shut down as a safety precaution until the root cause of joint failure was identified, rectified, and test validated. At the time of writing, this chapter manuscript for inclusion in this Handbook of Bolts and Bolted Joints, second edition, the affected wind farm is still shut down. Manufacturers' specifications for blade flange stud preloading are to ensure sufficiently high level of uniform initial clamp load distribution in all 54 studs on a bolt circle of about 2 m in diameter; the selection of initial preload level should take into account the elastic interaction effect among the 54 studs during bolt-up, subsequent in-service loads, and the environmental factors during normal operation of the turbine. The turbine had been constructed 3 years earlier and had been operating normally up until that moment when the blade fell to the ground. It was operating at rated power and wind speeds were around 8–10 m/s, well within the operating range of the turbine.

Preventing flange separation under service loads and/or environmental factors would protect flange studs from being exposed to high amplitudes of dynamic loads that could lead to premature fatigue failure of flange studs. The opposite is true; clamp load loss would lead to joint separation, and stud exposure to large dynamic load amplitudes.

The safety and reliability of complex threaded joints are determined by many dozens of system variables and parameters [1-9]. Parameters of a threaded system are typically divided into six groups of variables that respectively belong to: (1) the joint, (2) the threaded fastener, (3) the assembly tool, (4) the preloading process, (5) in-service loads, and (6) the environmental factors. Reliable design and safe operation of a critical threaded system necessitate simultaneous consideration of their effect on the initial level of fastener preload as well as the effect of post-assembly service loads and environmental factors on the residual clamp load value (stud tension) in each of the 54 studs in turbine blade joint.

Preliminary SEM testing of ruptured flange studs shows signs of fatigue failure in those studs located at, or near, location(s) with uneven flange contact surfaces, and/or, having localized flange gaps, which suggests possible localized joint separation at those locations under service loads. A prototype strain gage-based health monitoring RF system is developed and implemented on one stud to be mounted at such location, in order to wirelessly transmit real-time data to a display receiver in the control room during normal operation of the wind turbine. The health monitoring system could be moved around to various suspect locations on the flange in order to collect and analyze the overall 54-stud joint signature in terms of clamp load variations and moment levels acting on each stud location during normal wind turbine operation. Collected data from the novel Radio Frequency health monitoring system is used for reliable design, analysis, and test validation of the threaded fastener systems of composite wind turbine blade joints that ensure safe and reliable wind turbine operation.

DOI: 10.1201/9781003501183-52

45.2 INITIAL DATA COLLECTION

Immediately following the failure, the affected turbine owner shut down all other units within their fleet matching the damaged unit's make and model. The owner dispatched a team to complete a field survey of the blade debris as well as a drone survey, which showed all debris was contained within a 225-foot radius of the centerline of the turbine foundation. All debris of significance was labeled and collected as evidence to be reviewed.

During this investigation, video surveillance was utilized to verify how the blade failed during normal operation. Figure 45.1 shows a conceptual mapping of the failure location.

The video footage indicates that there was an initial failure of the studs as the blade began a new revolution. Generally, at this point in the revolution, the blade's leading edge is in compression and the trailing edge is in tension. It was observed that the blade ultimately failed at the trailing edge and then hinged and ultimately failed at the leading edge as the blade continued the downward revolution.

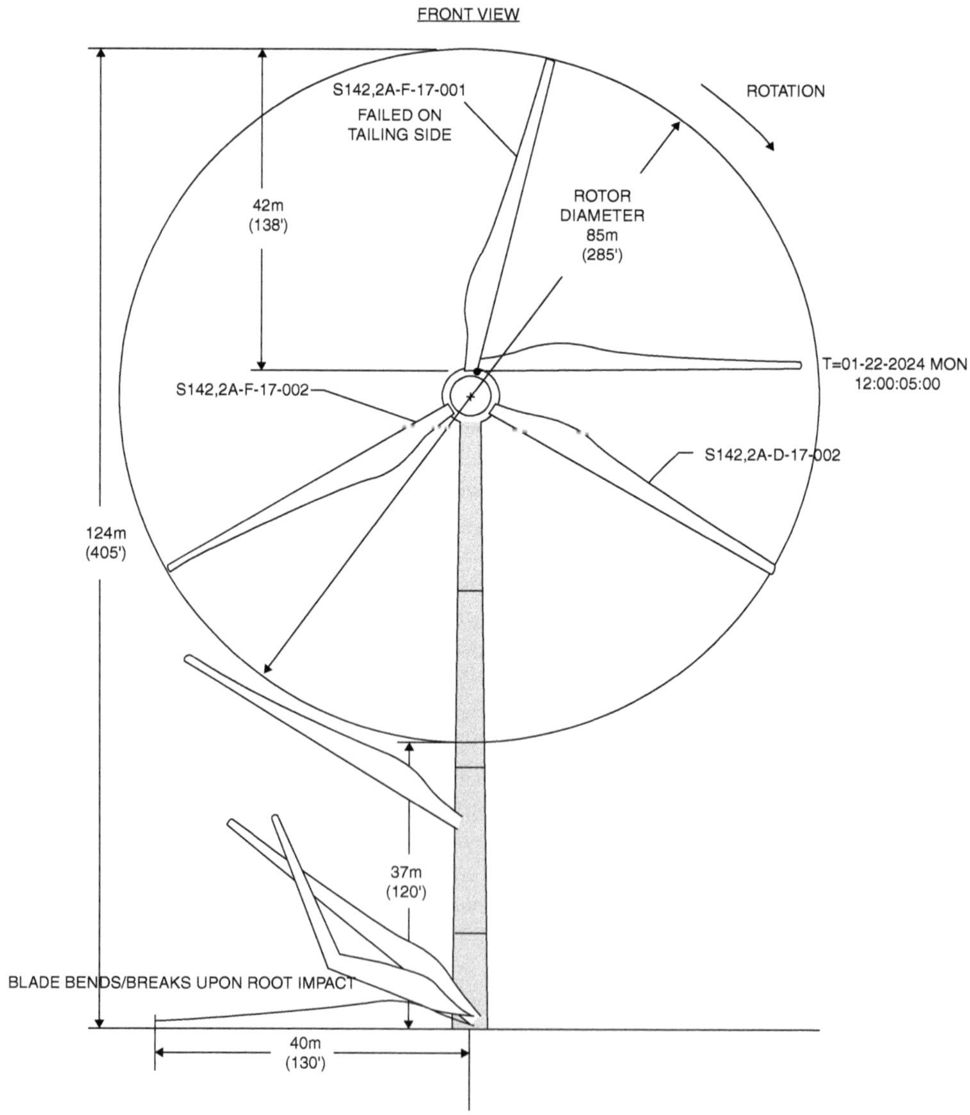

FIGURE 45.1 Mapping of failed blade joint.

At the time of failure, SCADA showed that the turbine was operating at its rated output (1,500 kW) and wind speeds were not excessive for the unit (8–10 m/s). Local weather station data sets also indicated that the weather patterns locally were within operating parameters for the unit with no abnormal conditions.

To initiate the ground investigation, the owner hired a third party to complete a thorough review of the blade remains and root. The blade remains indicated that there was a bolting failure, showing signs of fatigue and ultimate failures. Studs and relevant materials were saved for future laboratory analysis.

In tandem, the owner completed an engineering inspection following procedures established by third-party structural and foundation engineers. Upon this successful ground inspection, the turbine supplier and the owner worked in tandem to gain access to the blade-to-hub interface.

Upon entry to the hub, studs were collected and visually inspected. The collected sample indicated that many studs suffered ultimate failure, but localized areas also showed signs of fatigue. With the two blades still attached, there was a lack of the OEM specified lubricant identified between the nut and stud. Additionally, it was noted that there was visible lubricant egress in clamped connections, indicating that there may not be good joint contact throughout the blade-to-hub interface.

These findings provided cause to investigate all turbines within the comparable fleet for causal similarity, including lubricant identification, maintenance records, feeler gauge analysis at jointed interfaces, and stud sampling in high fatigue zones. This fleet analysis and the failed unit analysis were completed in parallel and showed consistency with the potential root cause of failure, outlined in future sections.

45.3 DESCRIPTION OF THE FLANGED BOLTED SYSTEM

The blade attachment scheme is a complicated composite interface that experiences relatively low cyclic loading but has high complex fatigue cycle counts due to the multiple variations in loading during operations. The flanged system experiences rotor rotation, blade pitch, blade vibration, and load variations due to changes in wind speeds that occur naturally and through changes in wind speed throughout the rotation cycle. The flanged system is also assembled both in part in the factory and in part in the field by the installing contractor in field conditions. Figure 45.1 shows a cross section of the composite-to-composite flanged blade joint (Figure 45.2).

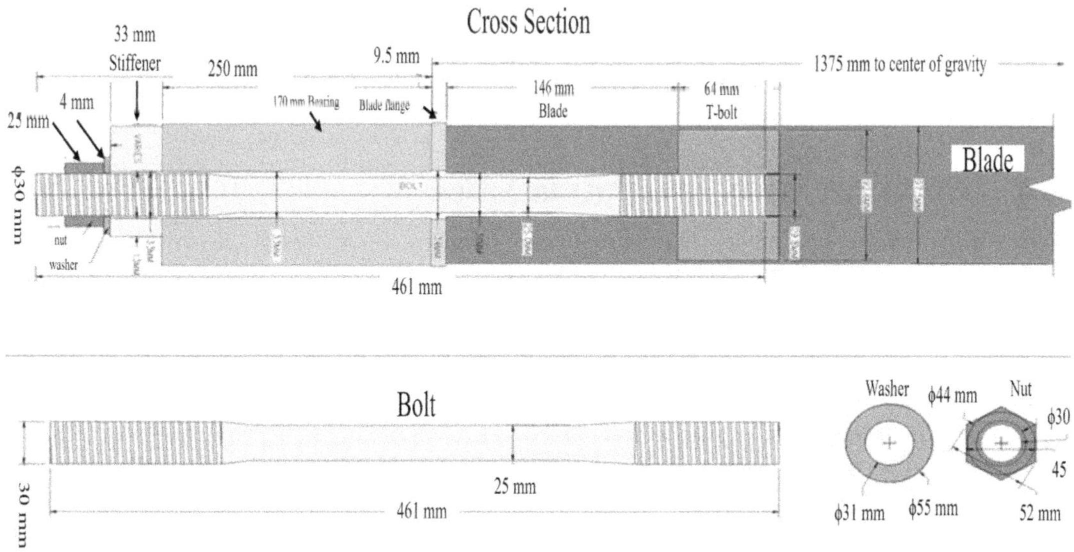

FIGURE 45.2 Cross section of the composite-to-composite flanged blade joint.

The flanged, working from blade to hub, consists of the following components:

1. *Fiberglass blade*: A hollow, two-shell, 6,000 kg blade made up of fiberglass and balsa wood. The root end is a solid fiberglass ring with an external radius of approximately 1,890 mm and a flange (ring) thickness of approximately 100 mm. The end of the blade is typically saw cut to ensure a flat and square interface.
2. *Steel T-bolt*: A steel cylinder that has been drilled and tapped to accommodate the blade stud. The T-Bolt is embedded in the fiberglass blade. There are 54 bolts embedded into the fiberglass blade. They are tapped for a M30 stud.
3. *Steel blade flange*: A steel ring that is approximately 9.5 mm thick. The ring contains 54 holes to accommodate the blade studs. The ring is typically held in place during shipping with screws embedded into the blade. The ring is typically a two-piece semi-circle design made out of uniform thickness plate.
4. *Cast steel bearing ring*: The outer ring of the blade pitch bearing is a 170 mm thick casting that has 54 cast holes. It is typically machined flat on the blade side.
5. *Steel stiffener plate*: An approximately 33 mm thick plate that matches the bolt pattern and is intended to help reduce stress concentrations on the blade bearing ring. The stiffener was added to the original design by the manufacturer based on field observations. The Stiffener Plate is made from a uniform thickness plate. The majority of the holes in the Stiffener Plate are machined or drilled. Some of the notches in the plate used to interface with sensors and locking pins may be stamped. The stiffener plate is made up of two semi-circles.
6. *Steel stud and washer assembly*: A grade 10.9 M30×461 mm blade stud with matching high-strength nut and washer. The blade stud inserts through the stiffener plate, the bearing, the blade flange, the blade, and screws into the T-Bolt embedded in the Blade. The stud is typically hand installed in the blade T-Bolt prior to installation of the whole blade into the hub assembly during construction. The stud has a tapered shank with a 25 mm cross section. The target preload for the stud is 300 kN which is achieved through torque. The stud is field lubricated and 1,070 NM of torque is applied through a rolling triangle across all studs in a 50%, 75%, and 100% torque sequence.

Overall, the flanged assembly consists of 6 different load-transferring interfaces across multiple materials. The 54 blade studs are designed to compress all of the interfaces together to enable the composite flange to have uniform contact and load transfer of the stud preload. The flange is designed to be retorqued after a 500 h operation break-in period and then periodically inspected and torque tested over its operating life.

45.4 FAILURE LOCATION AND FAILURE MODE

The company initially brought in a blade structural expert to assess the condition of the blade immediately after failure. The blade itself showed no signs of structural failure. Security video viewed by the company confirmed that the blade was intact when it separated from the turbine. The supporting pieces of information led the company to conclude that the blade failed at the bolted connection to the hob.

The studs that were collected told a compelling story. A number of 54 studs had experienced significant fatigue and had nearly completely failed as a result of fatigue. Various ranges of ultimate failure in the cross sections of the blade studs show that a material number of the studs were ripe for a cascading "zipper" failure. The remaining studs experienced an ultimate failure condition that would be expected after a large enough number of the studs failed and transferred the load to the remaining studs as the blade came over the top of the turbine and cycled loads from compression to tension (and vice versa) across the bolted circle (Figures 45.3–45.7).

FIGURE 45.3 Stud Fatigue Failure at first fully engaged thread.

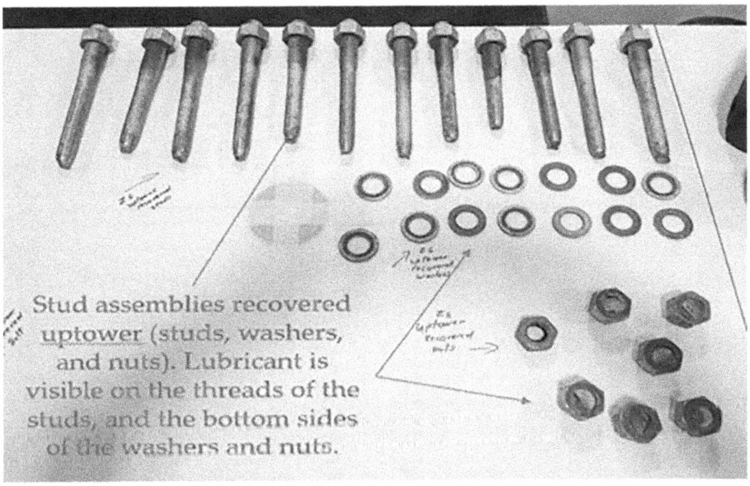

FIGURE 45.4 Detailed view of failed stud from hole 45.

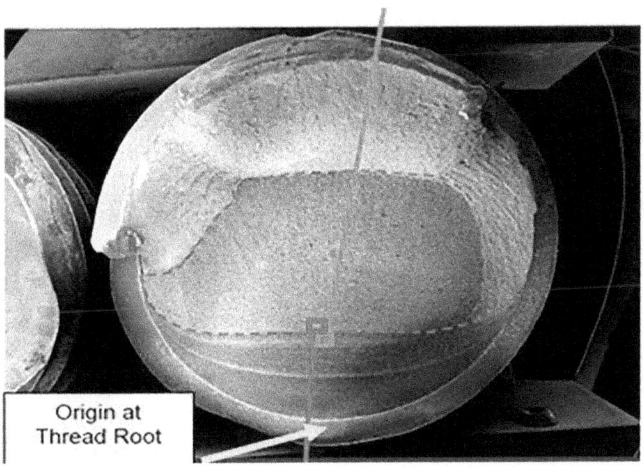

FIGURE 45.5 Side view of failed stud with visible cracks.

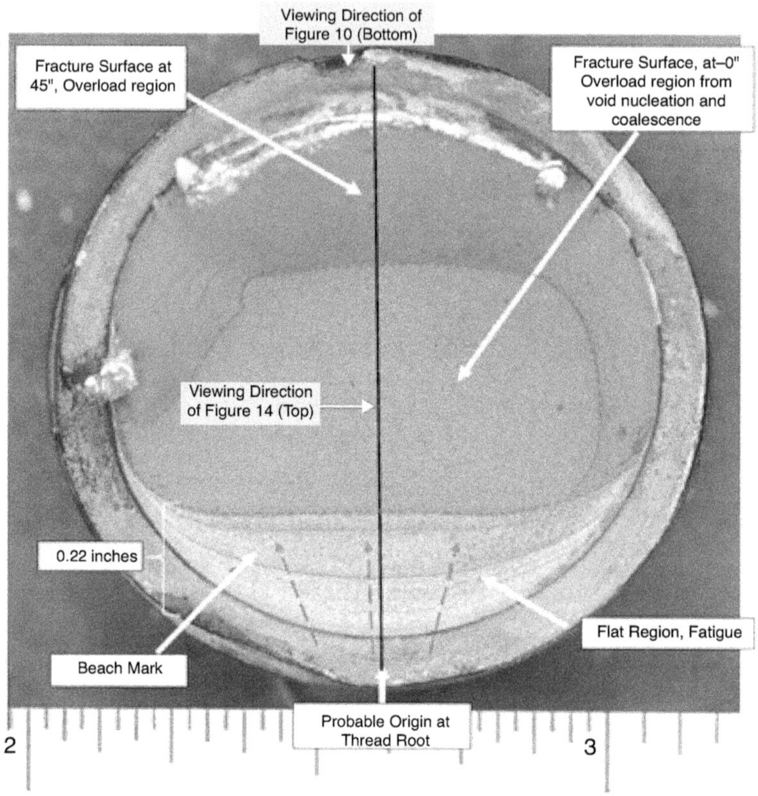

FIGURE 45.6 Cross section of a failed stud showing fatigue cracks.

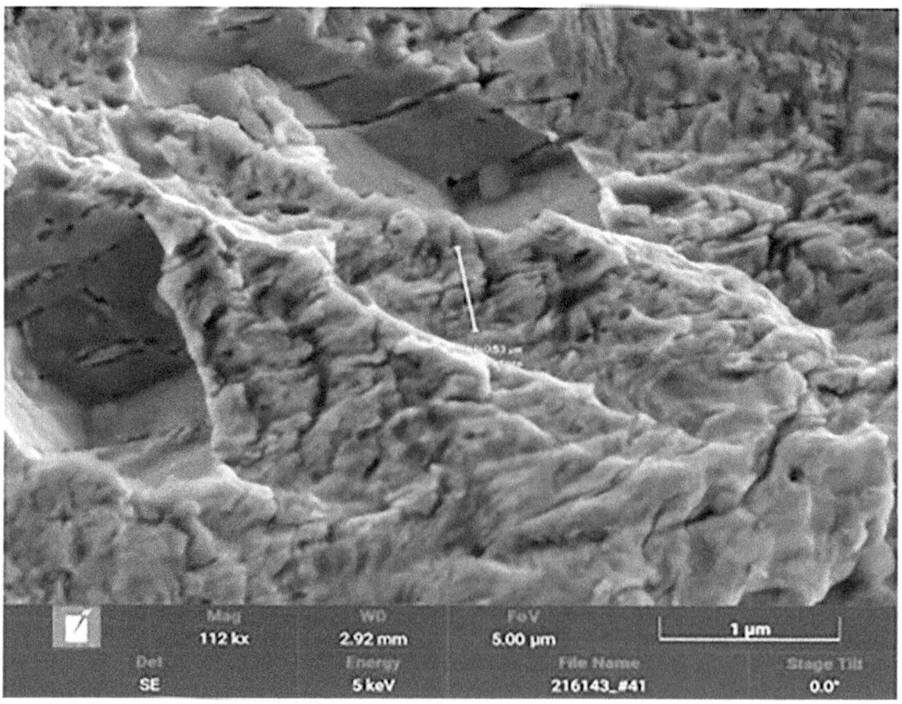

FIGURE 45.7 Metallurgical analysis of failed stud cross section.

45.5 LABORATORY TESTING

Upon completion of the initial site analysis, a third-party laboratory was contracted to complete a metallurgical analysis of four studs relating to the failure to determine any immediate metallurgical cause of the failure of potential contributing factors. The same lab was also contracted to complete a fatigue striation spacing estimate of a fatigue-failed stud to estimate cycles leading to the ultimate failure.

Throughout this section, the stud location relative to the failure zone is significant. For reference, the studs identified in the lab reports as holes 42–50 are in the high fatigue zone.

45.5.1 Metallurgical Analysis

The studs included in the metallurgical analyses include two studs from the failed turbine blade, one stud from a non-failed blade at the affected turbine, and one stud from the original project inventory to be used as a baseline (Table 45.1).

The first stud tested was from hole 45. This stud is within the identified fatigue zone of the failed blade. This stud was found to have failed by cracking that initiated at a stress concentration located at the first fully engaged thread on the hub side of the interface. Ultimately this stud failed by ductile overload due to the reduced available cross section due to fatigue cracking. There was presence of molybdenum disulfide lubricant on the stud and nut adjacent to its fracture surface. Figure 45.8 shows the failed stud's cross section.

The second stud tested was found on the ground and the corresponding hole location is unknown. The failure mechanism was identical to that of the stud from hole 45. Figure 45.9 shows the failed stud's cross section, while Figure 45.10 shows the side view.

TABLE 45.1
Results of Dimensional Measurements Performed on Sampled Studs

Sample	Bolt Description	Threads per Inch[a]	Major Diameter at Hub Side[b] (inches)		Major Diameter at Blade Side[b] (inches)		Specification from Drawing for Sinoma Studs[c] (inches)	Overall Length[d] (inches)	Specification from Drawing for Sinoma Studs[c] (inches)
			1	2 (90° from 1)	1	2 (90° from 1)			
216115	Blade 2 Hole 46	3.5	1.175	1.171	1.173	1.169	1.18 (30 mm)	18.12	18.15(461 mm)
216101/ 216143	Blade 3 Hole 45	3.5	1.174	1.172	1.176	1.174		16.88 (+1.63)	
216105	Blade 3 Bolt found on ground	3.5	1.171	1.166	1.150[e]	1.146		16.92	
216124	Bolt#102	3.5	1.176	1.176	1.175	1.174		18.12	

[a] Thread per inches measured using a pitch gage on both the blade and hub ends.
[b] Major diameters measured using a calibrated caliper.
[c] Engineering drawing of sinoma Studs provided by one energy see Appendix B.
[d] Overall length measured using a flexible tape measure. In cases where fractures occurred the individual length of both pieces are provided.
[e] Note these measurements were performed on a position of the bold that was starting to taper as the blade end of this bold was embedded in an undercovered/unmatched barrel nut.

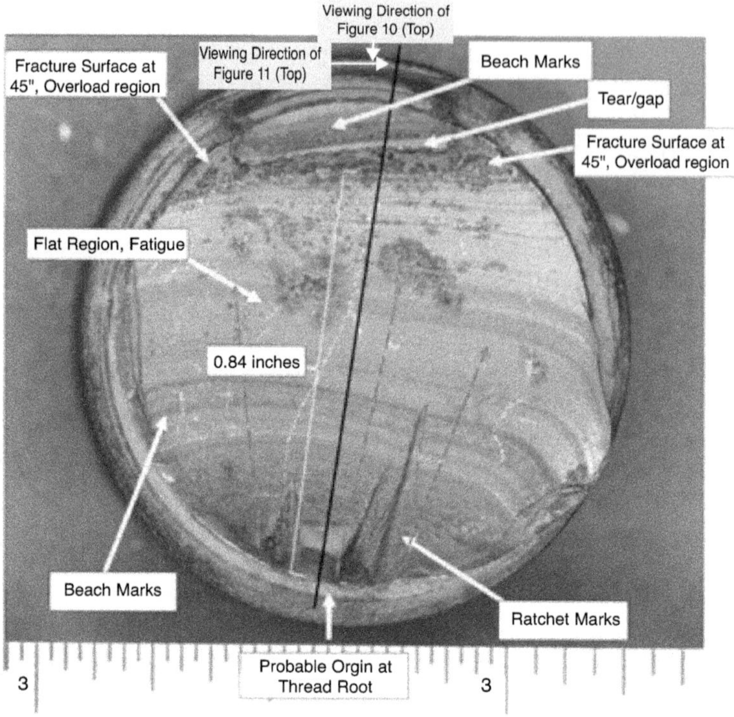

FIGURE 45.8 Cross section of the first failed stud showing ductile overload.

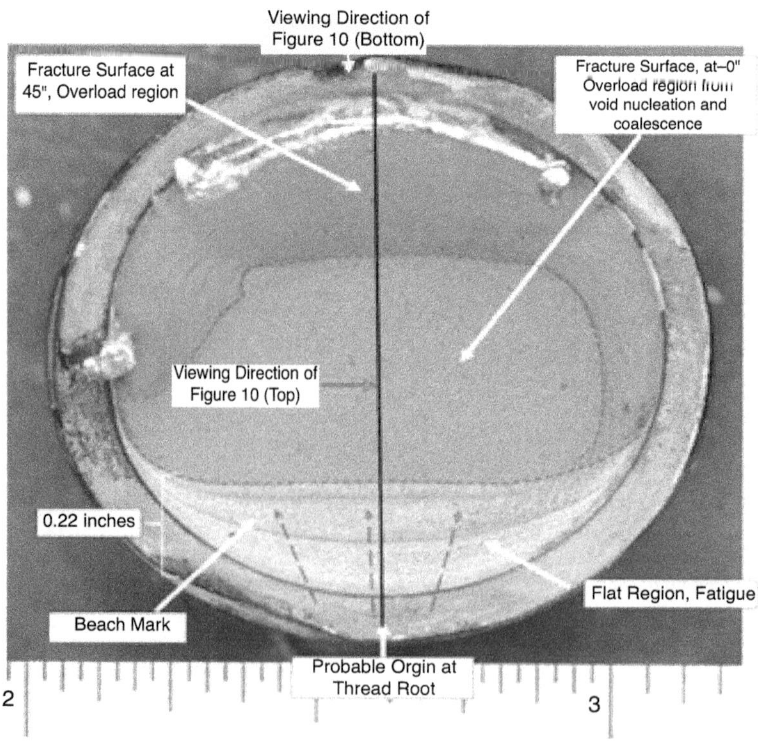

FIGURE 45.9 Cross section of the second failed stud with identical failure mechanism.

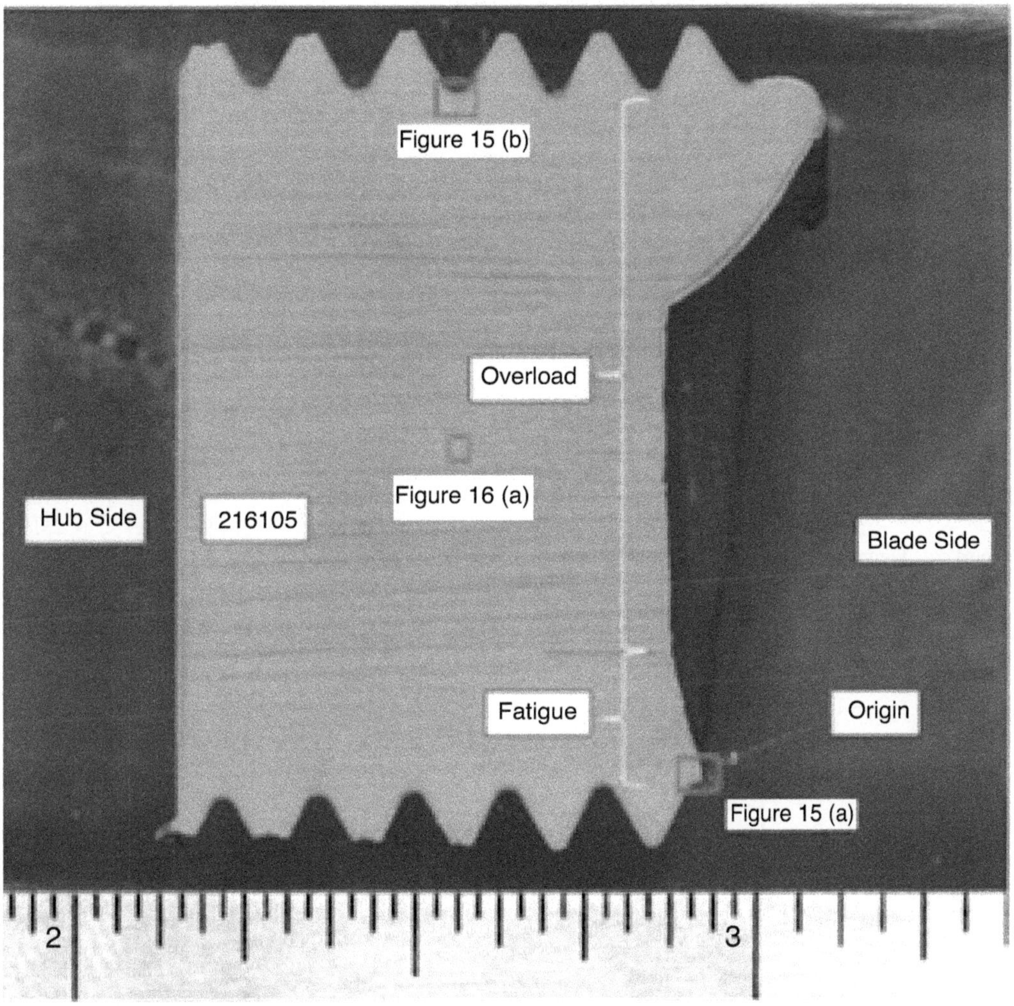

FIGURE 45.10 Side view of the second failed stud.

The non-failed blade stud was from blade 2, hole 46. This stud did not show evidence of cracking. The stud from the project inventory showed no deviations from the manufacturer's specification.

In conclusion, none of the operable studs showed signs of corrosion or secondary cracking within the threads. The studs were found to be in compliance with material Class 10.9 per ISO standard 898-1 2013 and OEM specifications. There was no evidence of the stud collection showing metallurgical deficiencies.

45.5.2 Fatigue Striation Spacing Estimate

Utilizing the failed blade's stud 45, which was also used in the metallurgical analyses, the lab performed fractography using scanning electron microscopy (SEM). During this inspection, the fracture surface of stud 45 was analyzed and estimates of striation spacing to cycles were estimated. The average spacing for all estimates was 4.80×10^{-6} inches, which corresponds to 2.30×10^{5} cycles to traverse 1.0 inches. Table 45.2 shows all collected SEM data.

For historical comparison, the failed turbine was determined to have operated for 3.5 years with a total of more than 21 million full revolutions (from turbine-generated data sets.)

TABLE 45.2
Summary of Striation Spacing Estimates Made From SEM Imagery of the Fracture Surface of the Bolt

Location#	Figure	Line	Striation Number	Measurement Distance (μm)	Measurement Distance (in)	Striation Spacing (in/striation)	Number of Striations in an Inch
Location 1	Figure 45.2	L1	16	2.04	8.03E 05	5.02E 06	199,216
		L2	33	4.03	1.59E 04	4.81E 06	207,990
		L3	9	0.62	2.44E 05	2.71E 06	368,709
Location 2	Figure 45.3	L1	11	1.52	5.98E 05	5.44E 06	183,816
		L2	18	2.64	1.04E 04	5.77E 06	173,182
		L3	8	1.19	4.69E 05	5.86E 06	170,756
		L4	26	4.56	1.80E.04	6.90E 06	144,824
		L5	7	0.84	3.31E 05	4.72E 06	211,667
Location 3	Figure 45.4	L1	33	4.93	1.94E 04	5.88E 06	170,020
		L2	20	2.02	7.95E 05	3.98E 06	251,485
Location 4	Figure 45.5	L1	23	1.94	7.64E 05	3.32E 06	301,134
Location 5	Figure 45.6	L1	9	0.53	2.09E 05	2.32E 06	431,321
Location 6	Figure 45.7	L1	16	1.80	7.09E 05	4.43E 06	225,778
		L2	8	0.71	2.80E 05	3.49E 06	286,197
		L3	15	2.24	8.82E 05	5.88E 06	170,089
Location 7	Figure 45.8	L1	6	0.86	3.39E 05	5.64E 06	177,209
		L2	6	0.92	3.62E 05	6.04E 06	165,652
Location 8	Figure 45.9	L1	7	0.81	3.19E 05	4.56E 06	219,506
		L2	7	0.79	3.11E 05	4.44E 06	225,063
		L3	6	0.53	2.09E 06	3.48E 06	287,547
Location 9	Figure 45.10	L1	6	1.43	5.63E 05	9.38E 06	106,573
		L2	11	1.71	6.73E 05	6.12E 06	163,392
Location 10	Figure 45.11	L1	13	1.06	4.17E 05	3.21E 06	311,509
		L2	11	1.07	4.21E 05	3.83E 06	261,121
		L3	10	0.75	2.95E 05	2.95E 06	338,666
		L4	12	1.38	5.43E 05	4.53E 06	220,869
Average	–	–	–	–	–	4.80E 06	229,742
Minimum	–	–	–	–	–	2.32E 06	106,573
Maximum	–	–	–	–	–	9.38E 06	431,321

45.6 IN-SITU INVESTIGATIONS

To fully understand the failure, a fleetwide analysis was completed to determine if any identified observations were consistent at other turbines. This investigation included the following procedures at 24 wind turbines (72 blades):

- Torque check
- Gap inspection

During the torque check, all turbines were retorqued at the specified torque values. Any movement was noted and saved for the record. There were indications that proper torque checks had not been applied to the fleet, including missing witness marks or significant movement of studs.

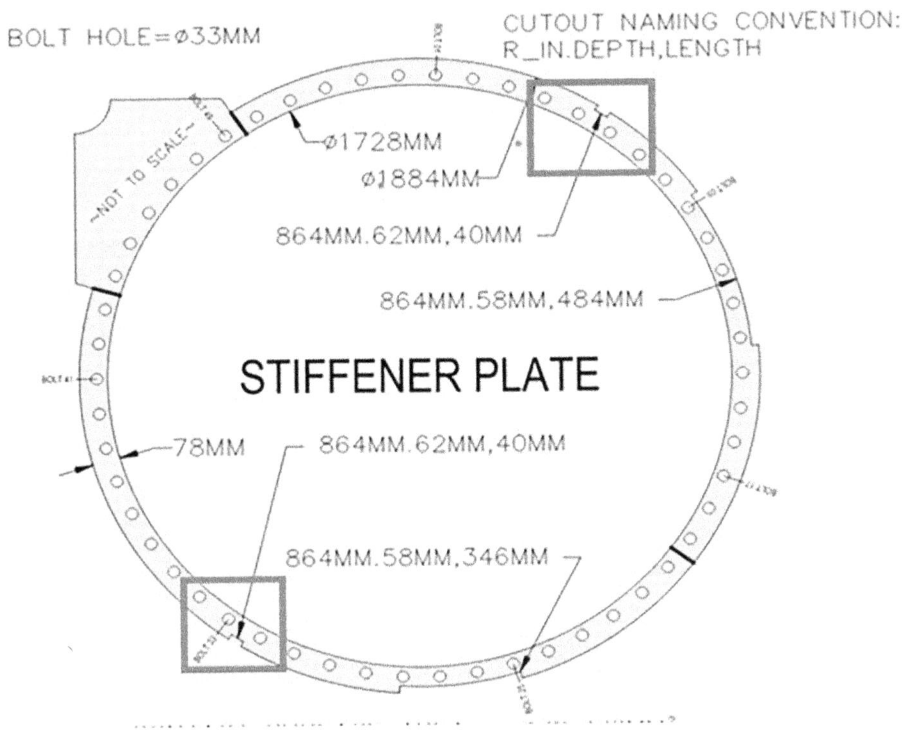

FIGURE 45.11 Correlation between gap locations and pitch block placement in the flange.

During the gap inspection, a feeler gauge of 0.25 mm thickness was inserted circumferentially at the following interfaces:

- *Inside blade*: Blade/Plate
- *Inside blade*: Plate/Bearing
- *Outside blade*: Blade/Plate
- *Outside blade*: Plate/Bearing
- *Outside blade*: Bearing/Stiffener.

All depths of penetration were collected to build gap profiles where applicable.

Figure 45.11 shows that there is a strong correlation between gap locations to be at the same two locations where the two respective pitch blocks are placed in the flange (identified with red boxes in Figure 45.11). Overall, the wind turbine farm had 22 blades that had observed gaps in their respective clamped flange connection with the hub; of those, there were 20 blades that showed gaps between the stiffener plate and bearing. More than 50% of the observed gaps correlated to the area of associated bolt fatigue failure in the failed blade.

45.7 ROOT CAUSE ANALYSIS

To go from data collection and analysis to root cause the Company, with the help of external technical advisors, constructed a Why-Why simplified schematic shown in Figure 45.12 diagram. The diagram starts with "Blade impacts ground" and works back through the range of "Why's". A simplified version of the Why-Why diagram in Figure 45.12.

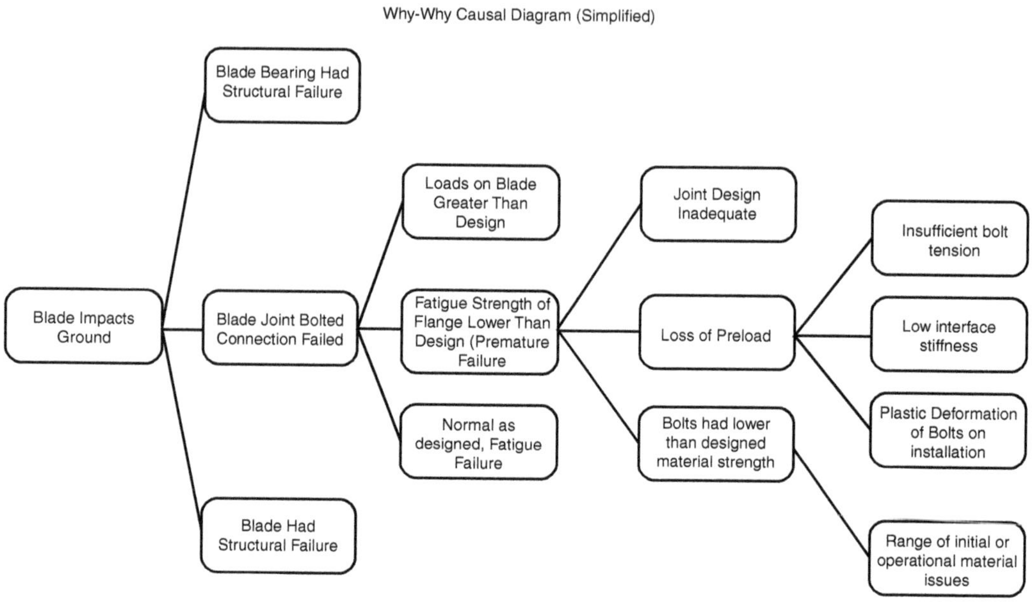

FIGURE 45.12 A simplified version of the Why-Why schematic for root cause analysis.

The blade and bearing structural failures were quickly ruled out based on the evidence collected. That focused the investigation on the bolted composite flange. The turbine was operating under normal conditions and was new and not near the as-designed end of life, which focused the investigation on why the flange did not have the as-designed fatigue strength.

The extensive laboratory material performed on the studs ruled out the stud materials or in-situ condition as a cause. Given the extensive operational history of this turbine design the Joint Design was unlikely to be inadequate. They focused the investigation on the loss of preload in the bolted flange.

The torque records for initial installation were consistent with prudent installation practices. The investigation noted that there was a change in supplier of the molybdenum disulfide lubricant used on the studs from the original manual. The lubricant was confirmed to be on the studs through laboratory investigation. Preliminary testing of nut-factors showed similar nut-factors across lubricants but the anomaly is notable as a contributing cause that cannot be completely ruled out.

Torque records for the ongoing service of the flange were lacking. This lack of information has to be considered. Failure to provide ongoing proper service could result in a loss of preload. Since there is no forensic way to know after the fact if the studs were properly serviced the suboptimal service has to also be noted as a potential contributing factor. In an extreme scenario, this could also be a primary cause. The condition of the other two blades on the same turbine suggests that this is likely a contributing factor and not a primary cause.

There is no evidence the studs were ever over-torqued to the point of plastic deformation. Bench testing shows that torque values would have had to have been significantly higher to cause plastic deformation.

By the process of elimination, flange stiffness became the primary focus of the investigation. The extensive investigation of every blade flange assembly in the fleet showed a consistent pattern of measurable gaps between the stiffener plate and the bearing. These gaps were, in some cases, significant and spanned multiple studs and penetrated across the entire flange face. All of these flanges were confirmed to be properly torqued as part of the investigation. Gaps this significant in nature are clearly not intended in proper flange design. It is notable that the stiffener plate was not part of the original turbine design and was part of a retrofit.

Ultimately these gaps were observed with very high frequency in similar locations on the other turbines in the Owner's fleet that had the same version of the retrofit. When all of the gap observations were aligned and superimposed a pattern became apparent. Gaps this significant in scale transfer almost all of the fatigue loading from the flange to the studs alone and will result in premature failure of the studs due to higher than intended fatigue loads. The cluster of gaps was centered on a "notch" in the stiffener plate. This notch, unlike the other notches in the stiffener plate, did not have rounded corners. It appears to have been stamped during production. Regardless of the exact manufacturing process, it is clear that the process used to create this notch also deformed the flatness of the stiffener plate. The thickness of the stiffener plate, ironically, made the plate very stiff. Calculations showed that if there was a 0.25 mm bow in the stiffener plate it would take a span of eight studs to correct the gap. Thus, for smaller local deformations, it would be impossible with normal stud preload to close the gap.

The presence of a significant air gap in the flanged assembly as a result of a lack of flatness of the retrofitted stiffener plate was determined to be the primary cause of significant loss of localized clamping around respective nearby stud(s), which eventually led to stud fatigue failure due to exposure to high in-service force and moment amplitudes that the stud was not designed to withstand. Given the significant scale of the measured gaps, the 10:1 lifetime reduction is both reasonable and likely. For those studs, they effectively acted as though there was no flange at all and the studs carried 100% of the fatigue loading themselves.

45.8 REAL-TIME RF-BASED HEALTH MONITORING OF TURBINE BLADE-TO HUB JOINTS

A novel wireless technology is procured from SensorData Technologies to strain gage the studs to continuously measure and transmit and display data on clamp load (stud tension F_z) and external bending moments M_x and M_y to a receiver control room on the ground level.

45.8.1 The Technology

Measurement technology of choice was the Strain-Gauge Technology, as it is the most accurate, repeatable, and less invasive. Measured signals are to be transmitted wirelessly to a ground receiver. Three Wheatstone bridges, connecting strain gauges are placed on the stud. Each bridge is connected to a module placed on the hub close to the stud. The module includes a battery to power the Wheatstone bridge and an onboard wireless transmitter. Measurements are wirelessly sent to a ground station within 500 ft from the wind turbine tower.

45.8.2 Instrument (Sensor)

SensorData has instrumented (strain gauged) an identical but new (fresh) stud that was supplied by the Wind Turbine owner (One Energy). Some minor alterations have been implemented to lay the strain gauges and rout out bridge wires.

45.8.3 In-Situ Measurements

Measurement is to be done by One Energy on an actual wind turbine, not on a simulated test setup to obtain true representative data. Dynamic inertia of measurement system components is to be considered carefully in the layout and securing method.

45.8.3.1 Parameters to Measure

Axial force in stud and two perpendicular bending moments M_x and M_y : each is measured by a separate Wheatstone bridge.

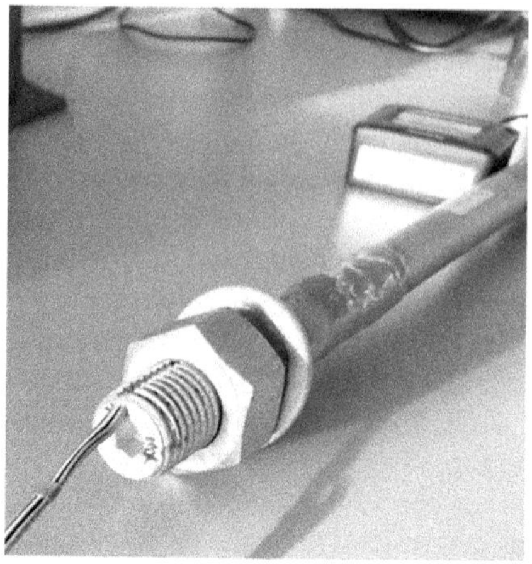

FIGURE 45.13 Bridge wires exit from stud.

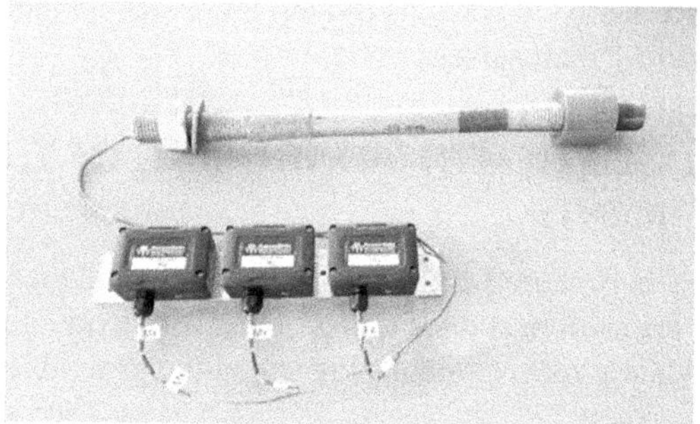

FIGURE 45.14 Stud with three onboard F_z, M_x, M_y bridges and wireless modules (for measuring stud tension F_z and two bending moments M_x and M_y).

FIGURE 45.15 Calibration of stud tension bridge (F_z).

45.8.3.2 Technology

Measurement technology of choice was the Strain-Gauge Technology, as it is the most accurate, repeatable and less invasive. Measured signals are to be transmitted wirelessly to a ground receiver. Three Wheatstone bridges, connecting strain gauges are placed on the stud. Each bridge is connected to a module placed on the hub close to the stud. The module includes a battery to power the Wheatstone bridge and an onboard wireless transmitter. Measurements are wirelessly sent to an earth station ~500 feet away (Figures 45.13–45.15).

45.9 OBSERVATIONS AND WIND TURBINE INDUSTRY INFORMATION GAPS

The editor of this *Handbook of Bolts and Bolted Joints* notes in the book's preface that bolted joint failures are often related to the lack of wholistic system's approach to the design, analysis, and validation of threaded joint systems that include (1) the joint, (2) the threaded fastener, (3) the assembly tool, (4) control process of bolt tightening, (5) post assembly loads, and (6) the environmental parameters. Unintended consequences often lead to the system failure. Pending validation using the Health monitor in Section 45.9 of this chapter, this case study appears to be an example of that: a stand-alone flange retrofitting created gaps that eventually caused loss of stud preload which created localized joint separation that dramatically increased the amplitude of dynamic forces and bending moments on bolt during turbine operation, that reduces the bolt fatigue life. In other words, a retrofit was implemented without ensuring flatness tolerance, which increased the susceptibility of the system to localized joint separation under in-service loads that eventually led to catastrophic rupture of the blade joint.

Through the process of the root-cause investigation, the Company and its advisors noted a number of industry information gaps. Those gaps are discussed briefly in this section.

1. There is significant laboratory testing on nut factor variations and over time for smaller fasteners but there is a lack of available published information for larger-diameter fasteners (30 mm+)
2. There is little published information on how stud nut factors change based on the original lubricant over long-term periods when bolts are "retorqued" during their life.
3. There is not a standard protocol to reliably re-determine the in-situ condition of fasteners unless you have access to original data from testing those specific fasteners. Natural variations in materials make it nearly impossible to compare tension in neighboring studs using non-destructive test equipment.
4. There is a lack of published information about the accuracy of the industry method of using 110% of the design torque to loosen the stud in order to verify that it was not over-torqued. The method seems to be used largely as a - the best we have - tool but there is no published research to support that.
5. There is, as is widely known, generally no way to determine how much fatigue life of a stud has been used based on laboratory testing until the stud has begun physically failing. This makes it nearly impossible to pull sample studs to assess how much fatigue life has been used on a stud as a preventative tactic and often forces a "replace them all" approach.
6. There is no easily deployable NDT method that can be used to confirm flange contact across multiple surfaces in a composite flange. The reflective nature of each layer makes it nearly impossible to determine if the composite flange has good internal surface contact in situ. The Company is still exploring X-ray investigations but the scale of that equipment, even if it produces valuable information, makes it impractical as a in-field, up tower, inspection tool currently.

7. Striation counting of beaching marks as a time estimate seems to be the only tool available to understand failure timing but also seems to be a very difficult-to-use tool that may or may not provide informative results.
8. In situ sensors and condition monitoring are needed to be able to understand the flange behavior and the loading on the studs. There is not a generally accepted common solution to this in the industry currently. One such solution, which was developed by SensotData Technologies, is currently in the process of being deployed by the Company as a tool to characterize other flanges and to confirm that retrofits on the gapped areas have the desired effect on the blade studs.

REFERENCES

[1] Juvinall, R., and Marshek, M., 2000, *Fundamentals of Machine and Component Design*, 3rd edition, John Wiley & Sons, New York.

[2] Lee, Y., Pan, J., Hathaway, R., and Barkey, M., 2005, *Fatigue Testing and Analysis (Theory and Practice*, pp. 103–17), Elsevier, Philadelphia, PA.

[3] Gerber, W. Z., 1874, "Calculation of the Allowable Stresses in Iron Structures," *Z. Bayer Archit Ing Ver*, 6 (6), pp. 101–110.

[4] Goodman, J., 1899, *Mechanics Applied to Engineering*, Longmans Green, London, pp. 438–442.

[5] Miner, M. A., 1945, "Cumulative Damage in Fatigue," *Journal of Applied Mechanics*, 67, pp. A159–A164.

[6] Bennett, J. A., 1946, "A Study of the Damaging Effect of Fatigue Stressing on X4130 Steel," *Proceedings of American Society for Testing and Materials*, 46, pp. 693–714.

[7] ASTM E739-91, 1998, "Standard Practice for Statistical Analysis of Linear or Linearized Stress-Life (S - N) and Strain-Life (ε - N) Fatigue Data", American Society for Testing and Materials. www.astm.org.

[8] Nassar, S.A. and Ganganala, M., 2011, "Effect of Separating Load Eccentricity on the Clamp Load Loss in a Bolted Joint Using a Strain Hardening Model", ASME *Journal of Pressure Vessel Technology*, 133, p. 021206.

[9] Mansour, R.A., "Fatigue of Preloaded Threaded Fasteners", 2012, MS Thesis, Department of Mechanical Engineering, Oakland University, Rochester, MI.

Index

For Product Safety Concerns and Information please contact our EU
representative GPSR@taylorandfrancis.com
Taylor & Francis Verlag GmbH, Kaufingerstraße 24, 80331 München, Germany

www.ingramcontent.com/pod-product-compliance
Ingram Content Group UK Ltd.
Pitfield, Milton Keynes, MK11 3LW, UK
UKHW052055230926
475702UK00031B/2868

9781032817521